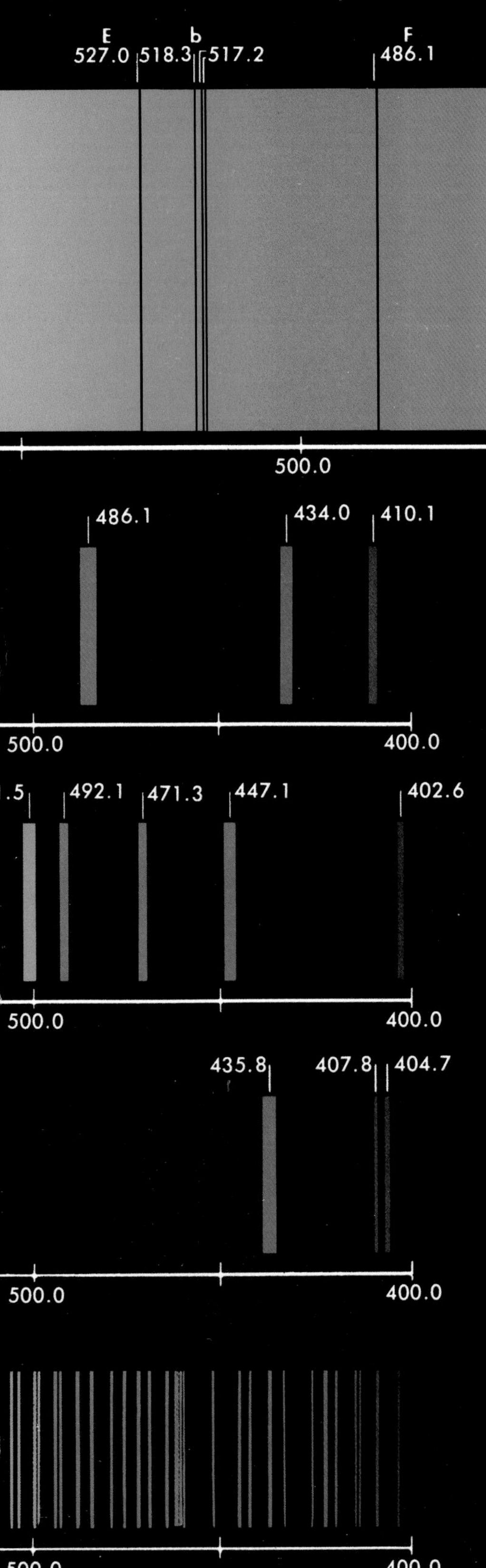

Such diverse and fundamental information on the nature of matter as the composition of distant stars and the structure of atoms and molecules has been obtained by analysis of the light emitted from substances heated to incandescence.

In the SPECTROSCOPE, such light, passed through a slit and a prism, is broken up into its component wavelengths, which are observed as colored lines (i.e., light of different energies) characteristic of the differences between the various electron energy levels of the atoms. This EMISSION SPECTRUM is CONTINUOUS when the images of the wavelengths are uninterruptedly overlapping; it is a LINE SPECTRUM when only certain specific wavelengths are emitted, as shown here for the elements hydrogen, helium, mercury, and uranium.

On the solar spectrum across the top of this plate appears a series of dark lines—FRAUNHOFER LINES—forming an ABSORPTION SPECTRUM. Some of the light from the intensely hot interior of the sun is absorbed by the cooler gases of its outer layers as the light energies raise the atoms in the cooler layers to higher energy states; bright lines are not, therefore, seen for these changes.

The spectra are calibrated in nanometers (1 nm = 10^{-9} m); the letters are arbitrary designations introduced by Fraunhofer for lines important in spectroscopy.

Definitions, Conversion Factors, Equivalents, and Physical Constants

Length (Distance)

1 inch = 2.540×10^{-2} m = 2.540 cm
1 microinch (μin.) = 10^{-6} in.
1 foot = 0.3048 m
= 30.48 cm
= 304.8 mm
1 yard = 0.9144 m
= 91.44 cm
= 3 ft
1 nautical mile = 6076 ft
= 1.151 statute mi
1 statute mile = 5280 ft
= 1760 yd
= 1609 m = 1.609 km
1 light-year = 9.461×10^{15} m
= 5.879×10^{12} statute mi

1 meter = 100 cm = 1000 mm
= 10^{-3} km
= 1.094 yd
= 3.281 ft
= 39.37 in.
1 kilometer = 1000 m
= 0.6214 mi
= 3281 ft = 1094 yd
1 centimeter = 0.01 m
1 millimeter = 10^{-3} m
1 micrometer (μm) = 10^{-6} m
1 nanometer (nm) = 10^{-9} m
1 angstrom (Å) = 10^{-10} m
= 10^{-8} cm

Area

1 in^2 = 6.452 cm^2
1 ft^2 = 929.0 cm^2
1 m^2 = 10.76 ft^2

1 acre = 43,560 ft^2
= 0.4047 hectare
1 hectare = 10,000 m^2
= 2.471 acres
1 square mile = 640 acres

Volume

1 cubic meter = 10^6 cm^3
= 10^3 liters
= 35.31 ft^3
= 264.2 gal
1 cubic foot = 1728 in.^{3}
= 7.481 gal
= 0.02832 m^3
= 28.32 liters
1 cubic inch = 16.39 cm^3

1 liter = 1000 cm^3
= 10^{-3} m^3
= 61.02 in.^{3}
= 0.2642 gal
= 1.057 qt
1 quart = 946.5 cm^3
1 U.S. gallon = 231.0 in.^{3}
= 0.1337 ft^3
= 3.785 liters
= volume of 8.337 lb of water at 60° F

Angle

1 degree (°) = $^1/_{360}$ the angular measure around a point (revolution)
= 60 minutes of arc
= 3600 seconds of arc
1 minute of arc = 60 seconds of arc

1 radian = 57.30°
= 0.1592 rev
1 revolution = 360°
= 2π rad

Speed

1 m/s = 3.60 km/h
= 3.281 ft/s
1 km/h = 0.6214 mi/h
1 ft/s = 0.3048 m/s
= 30.48 cm/s

1 mi/h = 1.467 ft/s
= 1.609 km/h
30 mi/h = 44 ft/s
1 knot = 1 nautical mi/h
= 1.151 statute mi/h
= 1.690 ft/s

(continued on endpapers at back of book)

PHYSICS PRINCIPLES AND APPLICATIONS

FIFTH EDITION

Norman C. Harris
Professor Emeritus of Higher Education
The University of Michigan

Edwin M. Hemmerling
Chairman Emeritus, Division of Science,
Mathematics, Engineering, and Technology
Bakersfield College

A. James Mallmann
Professor of Physics
Milwaukee School of Engineering

GREGG DIVISION
McGRAW-HILL PUBLISHING COMPANY

New York • Atlanta • Dallas • St. Louis • San Francisco
Auckland • Bogotá • Caracas • Hamburg • Lisbon • London
Madrid • Mexico • Milan • Montreal • New Delhi • Paris
San Juan • São Paulo • Singapore • Sydney • Tokyo • Toronto

Sponsoring Editor: Stephen M. Zollo
Editing Supervisor: Paul Farrell
Design and Art Supervisor: Caryl Valerie Spinka
Production Supervisor: Albert H. Rihner

Production Services: York Production Services
Cover Design: Larry Didona
Cover Photograph: Ulof Björg Christianson

Library of Congress Cataloging-in-Publication Data

Harris, Norman C.
Physics: principles and applications.

Rev, ed. of: Introductory applied physics. 4th ed. © 1980.
1. Physics. I. Hemmerling, Edwin M. II. Mallmann, A. James. III. Harris, Norman C. Introductory applied physics. IV. Title.
QC21.2.H37 1989 530 89-12826
ISBN 0-07-026851-7

Physics: Principles and Applications, Fifth Edition

2 3 4 5 6 7 8 9 0 RMTRMT 8 9 6 5 4 3 2

ISBN 0-07-026851-7

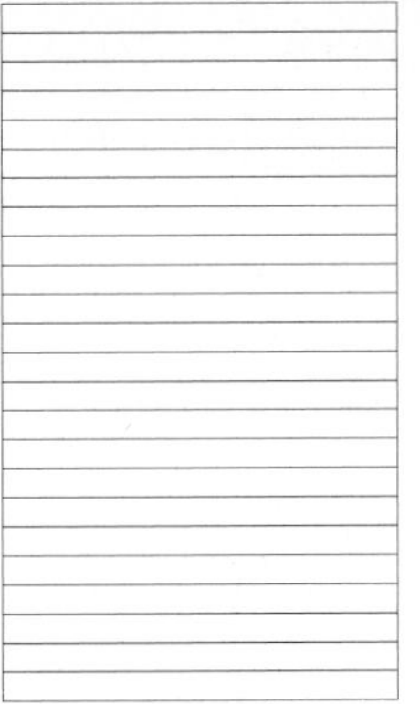

CONTENTS

PART 6 LIGHT AND OPTICS 520

PART 7 ELECTRICITY AND MAGNETISM 597

PART 8 MODERN PHYSICS 815

APPENDIXES

PREFACE

Physics: Principles and Applications is a basic text for students enrolled in programs of collegiate technical education. The book is *introductory* in the sense that it is intended for first-year physics courses in colleges and technical institutes. It provides essential rigor and coverage of all major divisions of physics, giving proper emphasis to basic scientific principles. But beyond that, and distinguishing *Physics: Principles and Applications* from most other college physics texts, there is a strong and continuing emphasis on the *applications of physics*—the essential relationships between physics and industry, technology, agriculture, commerce, and medicine.

This fifth edition of *Physics: Principles and Applications* embodies a complete revision and updating of the entire book. The unifying theme of *energy and matter* has been retained, but every chapter has been re-written, with the addition of new content, new approaches to learning, scores of new illustrations, and a twofold increase in the number of problems. All of the major divisions of physics are included—mechanics, measurement, properties of matter, heat and thermodynamics, wave motion and sound, light and optics, electricity and magnetism, electronics, and high-energy (particle) physics. The organization of the text into eight major parts will facilitate instructor selection of topics for courses of shorter duration than a full academic year.

The attention of instructors is especially directed to the extended treatment given to the following applications of physics: analysis of basic machines and their relationship to complex machines; heat engines, refrigeration, and air conditioning; satellite mechanics; acoustics of buildings; industrial electronics, including automated processes and robotics; and nuclear physics.

The mathematics requirement of the text has purposely been limited to algebra, plane geometry, and simple trigonometry. A comprehensive review of these mathematics skills is provided in Chapter 1. The use of a pocket electronic calculator by all students is encouraged and expected. Students whose mathematical skills are already at collegiate level could perhaps omit the mathematics review section, but the authors recommend it for the entire class.

Illustrative Problems. Each chapter contains a large number of completely solved *Illustrative Problems*. These are step-by-step solutions and they will guide student learning. They show students how to ''set up'' a problem, how to use diagrams in solving problems, how to work with fundamental and derived units, and how to use dimensional analysis in problem solving.

Units of Measurement. This edition maintains a balance between the SI-metric system with its meter-kilogram-second (mks) units, and the English system with its foot-pound-second (fps) units. The English-engineering system (foot-slug-second units) is also introduced and used with some frequency. The fact that most physicists, including the authors of this book, favor the SI-metric system does not alter the fact that metrication is proceeding very slowly in the United States—in fact, not at all in some industries. Technicians and engineers grounded in basic physics in which only the SI-metric system is used will be at a significant disadvantage in most technological fields (in the United States) for years to come. Even though dealing with two systems of measurement requires more instructor and student effort, preparation for careers in engineering technology (and other professional fields) in the U.S. requires thorough familiarity with both systems. Pretending that the English system is no longer important is just that—a pretense.

Added Features in this Edition. A new page format for this edition has allowed a significant increase in the number of diagrams and line drawings. The wide margins also provide space for special notes—some for motivational suggestions, emphasis, or review; some that highlight applications; and some emphasizing the history of physics and technology.

New Coverage. Every chapter has been extensively rewritten, but special mention is made of the following:

1. Increased content of the sections on vibratory motion and waves.
2. The extended treatment of elementary principles of rocket propulsion, aerospace travel, and satellite mechanics.
3. Completely new coverage of solid-state electronics and applications to telecommunications, computers, industrial automation (robotics), and medical research and diagnosis.
4. A new section on fiber optics, with applications.
5. The latest developments in superconductivity.
6. Up-dated chapters on atomic and nuclear physics.

New Problems. This edition contains some five hundred new end-of-chapter problems, about evenly divided between basic physics principles and applications. As an aid in making assignments, problems have been divided into three groups. *Group One* consists of relatively easy, or ''warm-up'' problems. Most students should be able to solve them from a one-time study of the chapter. *Group Two* problems are at the ''standard expectation'' level—problems that most students should be able to solve after in-depth text study and in-class lectures, demonstrations, and discussions. *Group Three* consists of ''challenge problems'' to test the ingenuity of the most capable students of physics.

Answers to all odd-numbered problems are provided in Appendix IX at the end of the book.

To The Student. Although this book's primary function is to serve as your classroom text, you will find that much of its content will later relate specifically to situations you will face on the job in research and development, or in engineering technology, agriculture, business, and allied health fields. The extensive coverage of both the principles and the applications of physics makes this book a valuable reference for the analysis of day-to-day problems you will meet later in your career.

Instructor's Manual. An *Instructor's Manual* is available to faculty members using the book as a class text. It contains, for each of the book's thirty-four chapters, these features:

1. A suggested list of instructor demonstrations and student activities.
2. Items for in-class quizzes—short-answer items and selected ''easy'' and ''medium-level'' problems.
3. And, to save untold hours of instructor time—complete solutions to all end-of-chapter problems in the book.

Acknowledgements. The authors sincerely appreciate the many helpful suggestions provided by professors who have used the fourth and prior editions of this book (entitled *Introductory Applied Physics*) since its first edition in 1955. We look forward to receiving suggestions for further improvement, as instructors put this newly titled edition to classroom use. We especially wish to acknowledge, for their valuable assistance on this edition, the following persons:

Robert Greenler, Professor of Physics, University of Wisconsin, Milwaukee
Jeff Hock, Associate Professor of Physics. Marquette University
Darrell Seeley, Associate Professor of Physics, Milwaukee School of Engineering
Alan Drew, Gas and Electric Operations Manager, Pacific Gas and Electric Company, California

The authors also wish to express their appreciation to those who reviewed early versions of this book in manuscript form: Alfred Amatangelo of the Central Maine Vocational Institute, Auburn, Maine; Gordon A. Yount of Catawba Valley Technical College, Hickory, North Carolina; and Dr. Sigmund P. Harris, Professor Emeritus, Los Angeles Pierce College, Los Angeles, California.

If there is excellence in these pages, all the above persons share in it. If there are inadequacies, errors, or inconsistencies, only the authors are to blame.

The valuable assistance of many professionals in engineering, industrial, commercial, and health fields is also gratefully acknowledged. More than one-hundred corporations, research and development laboratories, manufacturers, public utilities, agri-business firms, hospitals and clinics, and U.S. government agencies responded to requests for technical information, photographs, and diagrams.

NORMAN C. HARRIS
EDWIN M. HEMMERLING
A. JAMES MALLMANN

INTRODUCTION, MEASUREMENTS, AND GRAPHIC METHODS

An automated microprobe system being used to test an aircraft engine seal. It can detect variations of only a few millionths of an inch. (United Technologies—Pratt and Whitney)

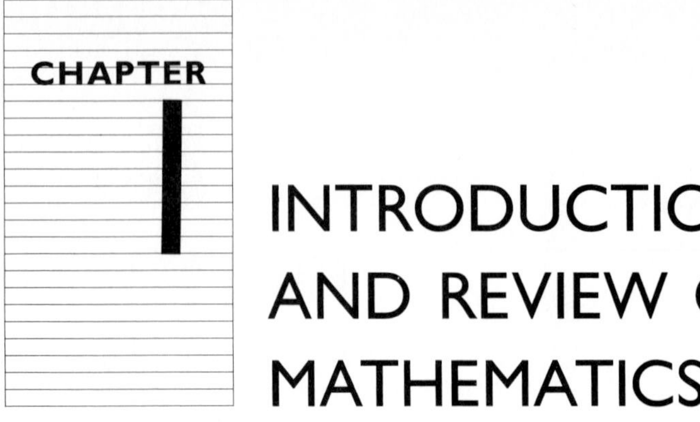

INTRODUCTION AND REVIEW OF MATHEMATICS

Physics in its pure form is a *science*. ''Science'' comes from a Latin root word meaning ''having knowledge,'' or ''knowing.'' Science is a way of looking at and interpreting nature and nature's laws and forces. *Applied physics* addresses itself partly to the knowing and to the looking at nature, but partly also to the ways and techniques by which scientific knowledge can be put to work to achieve goals desired by individuals or by society. The application of physics to the problems and needs of people is known as *technology*.

1.1 ■ Physics Is the Science of Matter and Energy

Matter is the stuff of which everything you see and touch is made. *Energy* represents the *ability* to do work, but the expenditure of energy does not always mean that work is done, the potential for conversion to work being dependent on how the energy is organized. Advanced technological societies such as ours depend on high-quality *materials* of all kinds—metals, plastics, fuels, fibers, synthetics, wood products, stone, chemicals, medicines, and food—and they use vast amounts of energy in finding, mining, refining, growing, shaping, and fabricating these materials. Physics is the study of both matter and energy, and of the factors which make industry and advanced technology possible.

1.2 ■ Matter Exists in Three Forms

For convenience in study, matter is ordinarily classified as existing in one of three *states*—as a *solid*, a *liquid*, or a *gas*.*

Solids and their properties are of great importance industrially. Most metals are solids at normal temperatures. Many new metallic alloys have been developed in recent years for use in devices and machines for space exploration. Certain alloys of magnesium, titanium, and nickel are examples of new ''exotic'' metals.

Liquids, and their properties such as pressure, surface tension, viscosity, and specific gravity, will be studied. The applications of hydrostatics and hydraulics to many industries will be explained. Two-thirds of the earth's surface is covered by the oceans, and we are only beginning to realize their importance as a natural resource for food, minerals, chemicals, and perhaps eventually, energy.

Gases are highly important in today's industrial economy. Their use in welding and as anesthetics is well known. Gases expanding as a result of heat serve as the working substance in heat engines—steam and internal combustion engines, jet en-

* In ''modern physics'' some scientists suggest a fourth state—the *plasma state*—involving gaseous mixtures of electrified atomic particles at temperatures reaching millions of degrees.

gines, and rocket motors. The properties of several gases make them suitable for use as refrigerant vapors in refrigerating and air-conditioning systems. Aside from industrial uses, the very atmosphere we breathe is a mixture of gases which exerts a considerable pressure at the earth's surface. Current problems of air pollution around the world are now serious enough to cause scientists, engineers, medical doctors, and the general public to become increasingly concerned about the thin envelope of air which supports life on this planet.

Solids, liquids, and gases as forms of matter will be discussed in detail in later chapters.

1.3 ■ Different Forms of Energy

Energy shows up in many forms. *Potential energy* is energy due to position, like that possessed by water at the top of a dam or by a pile-driver hammer at the top of its travel; or energy due to an unusual arrangement of matter, as in a coiled spring or in compressed air. *Kinetic energy* is energy of motion, for example that of a rushing stream, a pile-driver hammer as it falls, a speeding automobile, or a rocket exhaust.

Perhaps the most common form of *potential energy* is that due to the force of gravity.

Kinetic energy is energy of motion.

Before an object or substance can possess potential energy, *work* must be done on it. For example, in winding up a spring or in compressing air, the coiled spring and the compressed air possess potential energy. By the use of a suitable engine, this potential energy can, with some losses, be converted back into work. For the purposes of physics the definition of an *engine* is *a device to convert energy into work.*

Some examples of forms of energy in common industrial use are: *electric energy, heat energy, chemical energy, hydraulic energy,* and *nuclear energy.*

Energy Sources Petroleum, natural gas, coal, and oil shale are known as *fossil fuels*. With fossil fuels, except coal, getting dangerously scarce and very costly, great interest and much research and exploratory activity are focused on developing energy from sources which are "renewable" or inexhaustible. Examples of these sources are: *solar energy, wind energy, geothermal energy* (heat from the interior of the earth), and *biomass energy* (energy from plants, wood, garbage, etc.). Nuclear energy is undergoing intensive development also.

Other sources of renewable energy are hydropower and tidal and ocean-thermal power.

The study of energy involves the concepts of *mass, force,* and *motion,* and leads to an understanding of *heat, work,* and *power*. These terms, and many more, will all be defined and discussed in detail in later chapters.

Heat Energy and Mechanical Work The study of heat and temperature is a vital part of technical physics. Heat engines run most of the nation's transportation systems and generate a major share of the nation's electricity; and heat energy warms our homes and factories. Reversed heat engines perform refrigeration and air-conditioning jobs, and heat is a basic factor in the manufacture of steel and in most other metallurgical processes. Many of the machines of industry and nearly all of those in agriculture are powered by heat engines, and the electricity that powers the rest of them is, in large part, generated in steam plants which convert heat energy to electric energy. The exact relationship between heat and mechanical work will be explained in detail in a later chapter.

A unique property of heat engines (most of them powered by fossil fuels) is that they are mobile—they *and their fuel* can be carried by the vehicle they energize.

Energy and Wave Motion Wave motion is an interesting study, and we shall encounter it as we discuss the fields of *sound* and *acoustics* and again in *light* and *optics,* and yet again in electricity, electronics, and "modern physics." The communications and entertainment industries, including telephony, sound movies, phonograph and tape recordings, radio, and television, require a knowledge of how people hear and how hearing in rooms can be improved by the application of acoustical materials.

Modern communications systems are utilizing the optical properties of certain fibers to carry message traffic by light waves instead of by electrons. The field of *fiber optics* is already a growth industry.

The motion-picture and television industries, camera and video recorder manufacturing, optical-instrument manufacturing, and the preparation and fitting of eye-

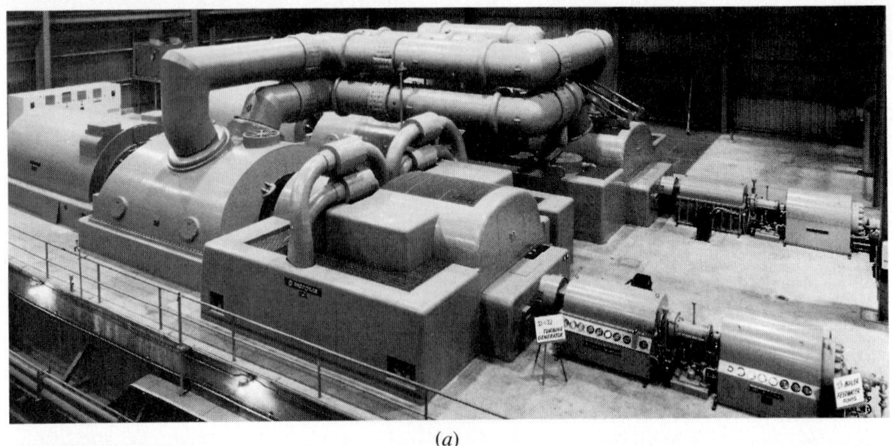

(*a*)

(*b*)

Fig. 1.1 Energy is the foundation of industry. (*a*) Steam turbine–driven electric generators produce commercial electric power at the Morro Bay (California) Power Plant. (Pacific Gas and Electric Company) (*b*) Heat energy is essential in all of heavy industry. Here molten iron at 2600°F is being charged into a basic oxygen furnace, where it will be refined into steel. (Bethlehem Steel Corporation)

glasses for the millions who have subnormal vision all depend on the basic principles of light and optics.

Electric Energy—the Foundation of Technology What is electricity? Despite our near total dependence on electricity, it is little understood by the majority of people. Electricity is *energy* in the form of *electron flow,* or the flow of electric charge. It will be the purpose of several chapters of this book to explain in detail the basic principles and common industrial practices in the production and use of electric energy. We shall develop electrical theory at some length, then devote considerable space to lessons in electric circuits, motor and generator principles, construction and use of electrical instruments, and methods of generating and distributing electric energy. Alternating-current circuits will be explained, and the principles of commercial power distribution will be outlined. Single-phase and three-phase circuits will be discussed, as will motors operating in these circuits. Electrical communications and electronics will be treated and many of their industrial applications explored. Transistors, computers, oscilloscopes, and electronics in medicine will be covered in considerable detail.

WHY STUDY PHYSICS?

Technology as a Career One of the consequences of an accelerating rate of technological change is the need in advanced industrial nations for hundreds of thousands of highly trained *technicians* and *technologists*. Persons in these occupational classifications usually prepare for their careers in college programs of two or four years' duration in community colleges, technical colleges, and four-year engineering schools. Technicians and technologists are involved in manufacturing and construction; in science and engineering research and development; in aerospace development and operations; in defense industries; in energy resources development; in the operation and repair of complex machinery and electric/electronic equipment; in business and finance; in medicine and allied health fields; in agriculture; and in public and human services. For all of these careers, the proper starting point of the technical training program is an introductory but thorough course in physics.

Scientists conceptualize. Engineers design and specify. Technicians work with both scientists and engineers to transform concepts and designs into practical reality.

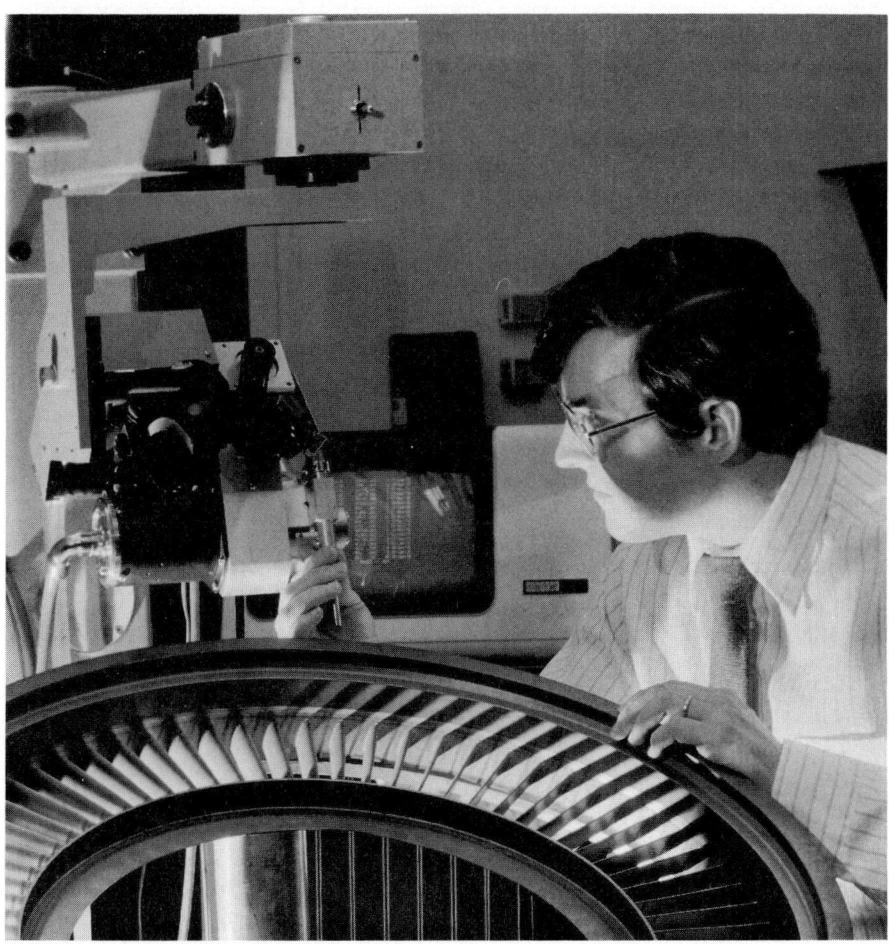

Fig. 1.2 A compressor stator assembly for a turbo-fan jet engine is prepared for a measurement of residual stress by x-ray diffraction. Technicians and technologists are indispensable to high-tech industry.

In this text you will learn a great deal about the science which is basic to all technology and about the applications of physics to the career field you may be considering.

MATHEMATICS REVIEW

Ability in computation often determines the rate at which technicians advance to positions of greater responsibility. The purpose of the following sections is to review some elementary principles of decimal notation and to illustrate some concepts of algebra, geometry, and trigonometry essential to technical calculations. Construction and interpretation of graphs will be explained also, since they are used frequently in technical work.

1.4 ■ Decimal (and Scientific) Notation and Exponents

Consider the number 10,385.6491. Spreading it out across the page and numbering the digits both ways from the decimal point affords a quick review of decimal terminology.

ten-thousands	thousands	hundreds	tens	units		tenths	hundredths	thousandths	ten-thousandths
1	0	3	8	5	•	6	4	9	1

A convenient way of handling decimals and calculations with them is to deal with exponents. An *exponent* is a small number printed to the right and above another number. This small superscript tells how many times that other number is to be taken as a factor. Thus, in the expression $5 \times 5 \times 5 \times 5$, five is taken as a factor four times. We would express this as 5^4. Here are two more examples:

$$4^2 = 4 \times 4 = 16$$
$$2^5 = 2 \times 2 \times 2 \times 2 \times 2 = 32$$

A review of the laws of exponents is provided in Sec. 1.12, p. 13.

An *exponent indicates the power to which a given number is to be raised*. We read 10^4 as "ten to the fourth power."

Table 1.1 lists some powers of 10 and their decimal values.

Scientific Notation Exponential notation frequently simplifies numerical calculations, especially if very large or very small numbers are involved. The method is also known as *scientific notation,* since it is the method used by scientists for numerical calculations. When numbers are written in scientific notation, they are often written as numbers between 1 and 10 times 10 to a power.

Table 1.1 Some Powers of 10 and Their Decimal Values

$10^6 = 1{,}000{,}000$	$10^{-1} = 0.1$,	read as	"one-tenth"
$10^5 = 100{,}000$	$10^{-2} = 0.01$,	read as	"one-hundredth"
$10^4 = 10{,}000$	$10^{-3} = 0.001$,	read as	"one-thousandth"
$10^3 = 1000$	$10^{-4} = 0.0001$,	read as	"one ten-thousandth"
$10^2 = 100$	$10^{-5} = 0.00001$,	read as	"one-hundred-thousandth"
$10^1 = 10$	$10^{-6} = 0.000001$,	read as	"one-millionth"
$10^0 = 1$			

Analysis of the above table reveals that increasing the power of 10 by 1 unit is equivalent to moving the decimal point one place to the right; and conversely, decreasing the power of 10 by 1 unit is equivalent to moving the decimal point one place to the left. Consequently, the number 18,750,000 can be expressed by any of the following:

$$18{,}750 \times 10^3$$
$$18.75 \times 10^6$$
$$1.875 \times 10^7$$

In like manner the decimal fraction 0.0000785 can be correctly expressed in scientific notation as

$$7.85 \times 10^{-5} \qquad \text{or as} \qquad 78.5 \times 10^{-6}$$

You have probably observed that, in these examples the value of the exponent or power of 10 is equal to the number of places the decimal point is moved—to the right for increasing values of positive exponents, and to the left for increasing values of negative exponents. Here are some examples:

$3.48 \times 10^0 = 3.48$ decimal point not moved at all
$3.48 \times 10^1 = 34.8$ decimal point moved right 1 place
$3.48 \times 10^2 = 348.$ decimal point moved right 2 places
$3.48 \times 10^3 = 3480.$ decimal point moved right 3 places
$3.48 \times 10^{-1} = 0.348$ decimal point moved left 1 place
$3.48 \times 10^{-2} = 0.0348$ decimal point moved left 2 places
$3.48 \times 10^{-3} = 0.00348$ decimal point moved left 3 places

Setting up problems for a computer or pocket calculator by using scientific notation may greatly simplify the calculations involved in solving them.

Illustrative Problem 1.1 Compute the value of

$$\frac{648.5 \times 0.000397}{0.0016}$$

Solution Express each factor as follows, using powers of 10:

$$\frac{6.485 \times 10^2 \times 3.97 \times 10^{-4}}{1.6 \times 10^{-3}}$$

Divide 10^{-3} into 10^{-4}, leaving 10^{-1}, which, when multiplied by the 10^2 in the numerator, gives 10^1. We now have

$$\frac{6.485 \times 3.97 \times 10^1}{1.6} = 160.9 \qquad \textit{answer}$$

1.5 ■ Using Scientific Notation with Very Large and Very Small Numbers

Large numbers and small numbers are frequently encountered in scientific and technical work. For example, the speed of light in a vacuum is approximately 983,650,000 ft/sec (ft/s) or 299,800,000 meters/sec (m/s). The rest mass of a proton is approximately 0.000,000,000,000,000,000,000,000,001,673 kilograms (kg). Writing such large and small numbers in ordinary notation would be very cumbersome and inconvenient, and at a glance, would give little indication of the actual value of the number. Furthermore, if such large and small numbers enter into mathematical operations involving multiplication and division, the ensuing calculations would fill the page with zeros.

Fig. 1.3 A typical pocket electronic calculator for use by students of physics, technology, and engineering. (Texas Instruments)

Such numbers are readily handled using scientific notation. The speed of light becomes 9.8365×10^8 ft/s, or 2.998×10^8 m/s. The rest mass of the proton becomes 1.673×10^{-27} kg.

Scientific Notation and Pocket Calculators A pocket electronic calculator intended for scientific and technical use will handle extremely large and small numbers with great ease. Figure 1.3 illustrates a typical model of such a calculator. Suppose you want to multiply 4,565,800,000 by 3500. If your calculator has a 10-digit display you can enter the 4,565,800,000. Now press the × (times) key and then enter 3500. Press the = key and the answer 1.59803 13 will appear. This is understood to mean 1.59803×10^{13}. To perform calculations by *entering* scientific notation, use the EE key as in the following example:

Multiply $(25.8 \times 10^7) \times (3.25 \times 10^4)$
Enter 25.8, then press EE and the 7 key
Now press the × key
Enter 3.25, then press EE and the 4 key
Press the = key for the answer: 8.385 12 or 8.385×10^{12}

Another example illustrates a calculation with very small numbers. Evaluate $6.08 \times 10^{-5}/14.2 \times 10^4$

Enter 6.08, then press EE, the +/ − key, and the 5 key. The +/ − key is for entering a negative exponent.
The display should now read 6.08 −05
Now press the ÷ key
Enter 14.2, then press EE, and the 4 key
Press the = key for the answer 4.2817×10^{-10}

Note: Different calculators may require different operations.

ALGEBRA REVIEW

It is assumed that students of applied physics have completed at least a basic course in algebra. Those whose ability in mathematics is limited should, of course, be enrolled in a mathematics course concurrently and should be acquiring the mathematical skills needed for the study of physics. The brief review presented here merely relates some elementary concepts of algebra to the solution of simple physics problems. Since technicians often work with formulas, our discussion begins with several such examples.

1.6 ■ Algebra Applied to Formulas

A formula is a mathematical statement of equality in which letter symbols are combined with numbers to express a physical relation in a more convenient form for problem solving than is possible by the use of words. As an example, take the verbal statement, "Distance traveled is equal to the average velocity multiplied by the time." By letting s stand for distance, $\bar{v}$ for average velocity, and t for time, we can write the simple formula

$$s = \bar{v}t \tag{1.1}$$

Another rather familiar example may be drawn from a problem in finance. A certain sum of money, called the *principal,* draws *simple interest* at a certain rate for a stipulated *time*. The interest which accrues in a given time is equal to the product of the principal, the rate, and the time. The principal and the interest are usually expressed in dollars, the time in years, and the rate in percent per year. As a formula, letting i stand for the accrued interest, P for the principal, r for the rate, and t for the time, we find that the interest accrued is

$$i = Prt \tag{1.2}$$

As a final example, a formula from the field of electricity will be chosen. It is known that the amount of current in a wire, measured in units called *amperes,* is equal to the electrical pressure, measured in units called *volts,* divided by the resistance of the wire, measured in units called *ohms.* As a word equation, then

$$\text{Amperes} = \frac{\text{volts}}{\text{ohms}}$$

Let I stand for current, V for voltage, and R for resistance. The formula (known as *Ohm's law*) then becomes

$$I = \frac{V}{R} \tag{1.3}$$

Solving Problems with Formulas In formulas, the various letters and symbols can be thought of as *variables,* the value of any one depending on or *varying with* the values assigned to the others. For example, in Eq. (1.1) above, the value of s depends on the values assigned to $\bar{v}$ and t. This formula as written is said to be solved *explicitly* for s. Should s and t be given and the value of $\bar{v}$ desired, the formula must first be solved explicitly for $\bar{v}$. Then the known values can be substituted for s and t and the numerical value of $\bar{v}$ can be determined. Solving Eq. (1.1) for $\bar{v}$ requires the use of one of the basic axioms of algebra, which states that *when equals are divided by equals, the results are equal.* The process involves dividing both sides of the equation (or formula) by t:

Formulas are equations, but not all equations are formulas. A formula is an equation that expresses a mathematical or scientific fact or rule or law, and it will yield a predictable answer when values are substituted in it and the computation is carried out.

$$\frac{s}{t} = \frac{\bar{v}\cancel{t}}{\cancel{t}}$$

The t's on the right side cancel out, leaving

$$\frac{s}{t} = \bar{v} \qquad \text{or} \qquad \bar{v} = \frac{s}{t}$$

If it were desired to solve Eq. (1.3) explicitly for R, we would use another axiom, namely, that *when equals are multiplied by equals, the results are equal.* First multiply both sides by R, obtaining

$$RI = \frac{V\cancel{R}}{\cancel{R}}$$

Then divide both sides by I,

$$\frac{R\cancel{I}}{\cancel{I}} = \frac{V}{I}$$

and obtain the formula solved explicitly for R, as

$$R = \frac{V}{I}$$

Illustrative Problem 1.2 If a waffle iron has an effective electrical resistance of 8.6 ohms (Ω) while operating on a household voltage of 120 volts (V), what current, in amperes (A), does it draw?

Solution Note that Eq. (1.3) is already in a form to calculate the current, I. Substitute the given values for V and R, obtaining

$$I = \frac{V}{R} = \frac{120\ \text{V}}{8.6\ \Omega} = 14\ \text{A} \qquad \textit{answer}$$

NOTE: *Answers to all illustrative problems in this book have been obtained by a pocket electronic calculator. This is true also of the answers to the odd-numbered problems at the back of the book.*

See Sec. 2.13, p. 39 for an explanation of how many significant digits should be retained in answers to problems.

1.7 ■ Algebraic Symbols

Algebra replaces the numerical units of arithmetic with letters which can represent *any* arithmetic numbers. Letters of the alphabet are used as symbols to represent sets of numbers. The letters are called *variables*. In some cases, where physics relationships are involved, a letter will also be assigned to represent a *constant* value—a value which remains unchanged over time—for example, g, the symbol for the acceleration of gravity.

Algebra is a *language;* and like any other language it has rules and symbols that must be learned and a "vocabulary" that must be acquired before one can use it effectively.

The four fundamental processes, indicated by the symbols + (plus), − (minus), × (times), and ÷ (divided by), have the same meaning and application in algebra that they have in arithmetic. If x and y represent two numbers, we would write their sum as

$$x + y$$

If y is to be subtracted from x, the difference is written

$$x - y$$

To indicate the product of numbers represented by x and y we write

$$(x)\,(y) \text{ or simply } xy$$
$$\text{or sometimes} \quad x \cdot y$$

Sometimes the "times sign" (×) is used as an indicator for multiplication. The quotient of x divided by y is generally written as

$$\frac{x}{y} \quad \text{or as } x/y \quad \text{not as} \quad x \div y$$

1.8 ■ Equations

An *algebraic expression* is a combination of variables and constants, together with certain indicated operational symbols such as addition, subtraction, multiplication, or division. Some algebraic expressions also include operations like taking roots, raising to powers, and trigonometric relationships.

An *algebraic equation* is a statement that two algebraic expressions are equal or that they are equivalent to the same number. We will use two kinds of equations in this book, illustrated by the two following types:

Type 1 examples: $3x + 4x = 7x$; $s = \bar{v}t$
Type 2 examples: $3x + 4x = 35$; $x^2 + 3x = 4$

Equations of the first type are called *identities*. Equations of the second type are called *conditional equations*.

The statement

$$3x + 4x = 7x$$

is true for *any* value that might be assigned to x and is an *identity*.

The equation relating distance, velocity, and time units is true for all sets of consistent units, and is also an identity. Thus, if we know $\bar{v} = 60$ miles per hour (mi/h) and $t = 3$ hours (h), we use

$$s = \bar{v}t$$

to determine that

$$s = \left(60 \frac{\text{mi}}{\text{h}}\right)(3\text{ h}) = 180\text{ mi} \qquad \textit{answer}$$

Equations like $s = \bar{v}t$, relating physical quantities in predictable relationships, are called *formulas* (Sec. 1.6).

In contrast to these identities, the conditional equation

$$3x + 4x = 35$$

is true if, and only if, the value of x is 5. The number 5 is called the *root* of the equation, and the expression is a true equation *only on the condition* that x is assigned a value of 5.

In like manner, the equation

$$x^2 + 3x = 4$$

is true if, and only if, $x = 1$ or -4.

1.9 ■ Evaluation of Algebraic Expressions

A very important process that is often encountered in physics is the *evaluation of algebraic expressions*. In this process the value of an algebraic expression is determined when the numerical values of the literal factors are known. Whenever we evaluate an algebraic expression, we substitute numbers for letters. Hence, the process is called *substitution*.

Illustrative Problem 1.3 If $a = 1$, $b = 2$, $c = 3$, find the value of $2a - 4b + 3c$.

Solution

$$\begin{aligned} 2a - 4b + 3c &= \\ 2(1) - 4(2) + 3(3) &= \\ 2 - 8 + 9 &= 3 \qquad \textit{answer} \end{aligned}$$

Illustrative Problem 1.4 If $x = 2$, $y = -1$, $z = 3$, find the value of $8xy - 5yz - 6xz$.

Solution

$$\begin{aligned} 8xy - 5yz - 6xz &= \\ 8(2)(-1) - 5(-1)(3) - 6(2)(3) &= \\ -16 + 15 - 36 &= -37 \qquad \textit{answer} \end{aligned}$$

Illustrative Problem 1.5 If $u = -3$, $v = 2$, $w = -1$, find the value of $u(v - w) - v(u + w)$.

Solution

$$\begin{aligned} u(v - w) - v(u + w) &= \\ -3(2 + 1) - 2(-3 - 1) &= \\ -3(3) - 2(-4) &= \\ -9 + 8 &= -1 \qquad \textit{answer} \end{aligned}$$

1.10 ■ Solving Equations

The value or values for a letter symbol in an equation which make it true, or "satisfy" it, are called the *solution* to the equation. The process of finding these values is called *solving the equation*.

Engineers and technicians spend much of their working time in solving equations. Often, the equation to be solved must first be formulated, either from mathematical reasoning or from on-the-site or laboratory observations and research.

In solving equations use is often made of a number of "self-evident" truths or agreed-upon rules called *axioms*. One of these is:

> ***Any operation performed on one side of the equals sign of an equation must also be performed on the other side.***

Adding the same quantity to both sides of an equation or subtracting the same quantity from both sides does not alter the equation, and it will still have the same roots. In like manner, multiplying or dividing both sides of an equation by the same quantity will not change the validity or the roots of the equation. Operations such as raising to the same power or taking the same root of both sides of an equation also will not change the truth of the equation.

Illustrative Problem 1.6 Solve the equation $2x - 9 = 21$, for x.

Solution Add 9 to both sides of the equation:

$$2x - 9 + 9 = 21 + 9$$
$$2x = 30$$

Now divide both sides by 2:

$$\frac{2x}{2} = \frac{30}{2} \quad \text{and} \quad x = 15 \qquad \textit{answer}$$

To verify that 15 is truly the root of the equation, substitute it back in the original equation. Thus

$$2x - 9 = 21$$
$$2(15) - 9 = 21$$
$$30 - 9 = 21$$
$$21 = 21$$

Illustrative Problem 1.7 Solve the equation

$$\frac{5x + 10}{2} = 3(31 - x)$$

Solution First, multiply both sides by 2:

$$2\left(\frac{5x + 10}{2}\right) = 2[3(31 - x)]$$

$$5x + 10 = 6(31 - x)$$

Now remove the parentheses by multiplying each term inside them by 6.

$$5x + 10 = 186 - 6x$$

Next add $6x - 10$ to both sides. The result is

$$5x + 10 + 6x - 10 = 186 - 6x + 6x - 10$$

Collect like terms:

$$11x = 176$$

Finally, divide both sides by 11.

$$\frac{11x}{11} = \frac{176}{11}$$

$$x = 16 \qquad \textit{answer}$$

1.11 ■ Transposing

It will be noted in the last two illustrative problems that when the proper term is added to or subtracted from both sides of an equation, terms disappear from one side of the equation and reappear on the other side with signs changed. This property leads to a short cut in solving some equations. The process is called *transposing* and it can be expressed by the following rule:

Transposing is a short cut in the solution of equations. However, one must not lose sight of the fact that what is actually being done in transposing is the application of the axiom stated in Sec. 1.10.

> ***If any term in an equation is moved from one side to the other with its sign changed, an equivalent equation will result.***

The key word here is "term." A term is a number or symbol separated from another by a + or a − sign. For example, in the expression $7x + 3$, $7x$ and 3 are *terms* but 7 alone is not a term and neither is x alone a term.

Illustrative Problem 1.8 Solve the equation

$$7x - 6 = 3x + 14$$

Solution Transpose both the −6 and the $3x$ terms. (Remember to change signs.)

$$\begin{aligned} 7x - 6 &= 3x + 14 \\ 7x - 3x &= 14 + 6 \\ 4x &= 20 \\ x &= 5 \qquad \textit{answer} \end{aligned}$$

1.12 ■ Laws of Exponents

The solutions to many problems in science and technology involve multiplying a number by itself many times. A convenient bit of symbolism is used to indicate this process. The symbol or indicator is the *exponent,* discussed in Sec. 1.4 as it applies to the *base* 10. However, any number can serve as a *base* and the superscript or exponent (if *positive*) indicates *how many times the base is taken as a factor* in repeated multiplication. (For *negative* exponents, see Law 6, below.)

A positive exponent is a superscript indicating how many times the "base" is taken as a factor.

The numbers 2^2, 2^3, 2^4, . . . are called powers of 2. Thus 2^5 is read "two to the fifth power" or "the fifth power of 2," and it expresses $2 \times 2 \times 2 \times 2 \times 2$, or 2 taken as a factor five times.

The following laws of exponents dealing with the product and quotient of two powers will be found useful in solving problems, both those in this book and the ones technicians will encounter on the job.

Some Laws of Exponents

1. The exponent of the *product* of *powers with the same base* is the *algebraic sum* of the exponents of the factors. In other words, the exponents are added.

$$(x^m) \cdot (x^n) = x^{m+n}$$

Some examples will illustrate:

$$(x^4) \cdot (x^7) = x^{4+7} = x^{11} \qquad 2x^3(4x^6) = 8x^9 \qquad (5x^5) \cdot (-6x^{-2}) = -30x^3$$

2. The exponent of the *quotient* of two powers with the same base is the *difference* of the exponents of the powers. In general,

$$\frac{x^m}{x^n} = x^{m-n} \qquad \text{(if } x \text{ does not equal zero)}$$

Some examples are:

$$\frac{x^8}{x^2} = x^6 \qquad \frac{10x^2}{2x^6} = 5x^{-4}$$

In working with exponents, careful attention must be given to the *location* of the exponent. Is it inside or outside a parenthesis; in the numerator or in the denominator of a fraction?

3. The exponent of a power-of-a-power is the *product* of the exponents of the powers. In general,

$$(x^m)^n = x^{mn} \qquad \text{and} \qquad (x^m y^n)^k = x^{mk} y^{nk}$$

Examples:

$$(x^2)^3 = x^6 \qquad (5x^2y^3)^4 = 625x^8y^{12}$$

4. *Any number* (except zero) raised to the *zero power* is equal to 1. This follows from

$$\frac{x^n}{x^n} = x^{n-n} = x^0$$

Examples:

$$\frac{12x^3}{x^3} = 12x^{3-3} = 12x^0 = 12$$

But $(5x)^0 = 1$ (by *definition*)

Any number (except zero) to the zero *th* power = 1.

5. The roots of an equation will not change if both sides are raised to the same power.

Illustrative Problem 1.9 Solve the equation (formula)

$$T = 2\pi\sqrt{m/k}$$

for m.

Solution In order to eliminate the square root sign, square both sides of the equation, with the result

$$T^2 = 4\pi^2\left(\frac{m}{k}\right)$$

Then multiply both sides by k, obtaining

$$kT^2 = 4\pi^2 m \qquad \text{or} \qquad 4\pi^2 m = kT^2$$

Finally, divide both sides by $4\pi^2$ to get

$$m = \frac{kT^2}{4\pi^2} \qquad \textit{answer}$$

6. Negative exponents are indicative of *reciprocals*. Any quantity with a negative exponent is equivalent to the reciprocal of that quantity with the corresponding positive exponent. Examples are:

$$x^{-1} = \frac{1}{x}, \quad x^{-3} = \frac{1}{x^3}, \quad x^5 \cdot x^{-2} = \frac{x^5}{x^2} = x^3,$$

$$\frac{x^4}{x^{-2}} = x^4x^2 = x^6, \quad x^{-n} = \frac{1}{x^n}, \quad \text{and} \quad \frac{1}{x^{-n}} = \frac{1}{1/x^n} = x^n$$

7. Fractional exponents describe *roots* of numbers. See the next section on fractional exponents and radicals.

1.13 ■ Fractional Exponents and Radicals

Exponents need not be whole numbers—they can be fractional numbers as well. Fractional exponents can be defined in a manner which will satisfy the laws of exponents. For example,

$$a^{1/2} \cdot a^{1/2} = a^{1/2+1/2} = a^1 = a$$

and

$$a^{1/3} \cdot a^{1/3} \cdot a^{1/3} = a^{1/3+1/3+1/3} = a^1 = a$$

From which we note that $a^{1/2}$ is one of *two* equal factors of a, and $a^{1/3}$ is one of *three* equal factors of a. To avoid the awkward expression "equal factor of," we use the terminology "root of," as in

$a^{1/2} = \sqrt{a}$ read, "the square root of a"
$a^{1/3} = \sqrt[3]{a}$ read, "the cube root of a"
$a^{1/q} = \sqrt[q]{a}$ read, "the qth root of a"

In general, $a^{1/q}$ is defined as the "q^{th} root of a".

The operating symbol $\sqrt{\ }$ is called a *radical,* with the number placed in its "vee" being the desired root. The absence of a number in the vee indicates that the square root is desired.

So far we have considered only those fractional exponents with *one as a numerator*. But consider the situation when $a^{1/q}$ is itself taken as a factor p times. The result is

$$a^{1/q} \cdot a^{1/q} \cdot a^{1/q} \ldots (p \text{ times}) = a^{1/q+1/q+1/q \cdots} = a^{p(1/q)} = a^{p/q}$$

This can be written

$$a^{p/q} = (\sqrt[q]{a})^p = \sqrt[q]{a^p}$$

Illustrative Problem 1.10 Reduce each of the following expressions to its simplest numerical value.

(*a*) $(36)^{1/2}$ (*c*) $(64)^{2/3}$ (*e*) $\left(\frac{4}{9}\right)^{-3/2}$

(*b*) $(-8)^{1/3}$ (*d*) $(81)^{-1/2}$

Solutions

(*a*) $36^{1/2} = \sqrt{36} = 6$
(*b*) $(-8)^{1/3} = \sqrt[3]{-8} = -2$
(*c*) $(64)^{2/3} = (\sqrt[3]{64})^2 = (4)^2 = 16$

(*d*) $(81)^{-1/2} = \dfrac{1}{81^{1/2}} = \dfrac{1}{\sqrt{81}} = \dfrac{1}{9}$

(*e*) $\left(\dfrac{4}{9}\right)^{-3/2} = \dfrac{1}{(4/9)^{3/2}} = \dfrac{1}{(\sqrt{4/9})^3} = \dfrac{1}{(2/3)^3} = \dfrac{1}{(8/27)} = \dfrac{27}{8}$

1.14 ■ Raising to Powers and Extracting Roots

The process of raising to a power or of extracting a root of a number used to be tedious until the advent of the modern hand-held calculator. Until then, such powers and roots were obtained by using tables of logarithms and antilogarithms, or by actually multiplying the base by itself the indicated number of times (raising to a power), or by finding square roots and cube roots by long and complex arithmetic processes. The very tediousness of these operations resulted in frequent errors.

For a brief review of logarithms, see Sec. 1.15 following.

Skill with a pocket electronic calculator is almost a necessity for success in college physics, *and* for success on the job as a technologist or technician.

These calculations are no longer necessary if you have a pocket calculator with a y^x key and an INV key. Raising to powers and finding roots are as simple as multiplying and dividing.

Illustrative Problem 1.11 Find the value of $(57.834)^5$. (Note that the fifth digit, though supplied, may be *doubtful* if the given number is the result of a measurement.)

Solution

1. Enter the base, 57.834, on the calculator
2. Press the y^x key
3. Enter the power, 5
4. Press the = key to get $(57.834)^5 = 6.470177 \times 10^8$

The calculator retains many digits, but since the base is accurate to only 5 digits, only five digits of the answer should be retained. The recorded answer should be 6.4702×10^8. *answer*

NOTE: *Difficulty will be encountered if you try letting* y *be negative in a hand-held calculator operation. An "Error" will be indicated in the answer display. In that event (* y *being negative), find the required value by entering a positive value for* y, *and then adjust for the negative value by using the laws of algebraic signs.*

The process of taking roots of numbers comes up quite frequently in physics. The order of events on the hand-held calculator is as follows: (1) Enter the base y. (2) Press the INV key. (3) Press the y^x key. (4) Enter the desired root, x. (5) Press the = key for the answer.

Illustrative Problem 1.12 Find the value of $\sqrt[6]{252.38}$

Solution

1. Enter the base, 252.38
2. Press the universal root key $\sqrt[x]{y}$
3. Enter the desired root, in this case, $x = 6$
4. Press the = key to get $\sqrt[6]{252.38} = 2.5138$ *answer*

1.15 ■ Logarithms

The concept of logarithms is directly related to that of an exponent. A logarithm *is* an exponent. Since it is an exponent, it must be an exponent of *some number*. That number is called the *base*.

If N denotes any positive number, and b denotes a base for an exponent, we can write $N = b^x$ for any value of b greater than 1. By definition, x is the *logarithm* of N to the base b; and N is the *antilogarithm* of x; or N is the number whose logarithm is x.

Common logarithms use the number 10 as a base. It follows, then, from the defining equation $N = b^x$, that since $100 = 10^2$, the logarithm (to base 10) of 100 is 2. We write this as log 100 = 2. (For *common logarithms* the word *log* assumes base 10.) It follows that: log 10 = 1; log 100 = 2; log 1000 = 3; log 10,000 = 4, etc. Logarithms are not limited to whole numbers or they would be quite useless for computational purposes. For example,

$$\begin{array}{lll} \text{Log } 200 = 2.301 & \text{since } 10^{2.301} = 200 & \text{(check these on your calculator)} \\ \text{Log } 700 = 2.845 & \text{since } 10^{2.845} = 700 & \\ \text{Log } 65 = 1.813 & \text{since } 10^{1.813} = 65 & \\ \text{Log } 8450 = 3.927 & \text{since } 10^{3.927} = 8450 & \end{array}$$

Calculators and computers have rendered logarithms obsolete for *computational* purposes. However, logarithms enter into various relationships in science and graphics. Our use of logarithms in this book will occur in the chapter on sound and acoustics. Any pocket calculator designed for scientific or technical computations will provide the logarithm of any number at the flick of a few keys.

Finding logarithms Find the log of 4865.8

1. Enter 4865.8 on the calculator keyboard
2. Press the log key
3. Read 3.6872

Find the log of 125,500

1. Enter 125,500
2. Press the log key
3. Read 5.099

Finding antilogs When the log (x) itself is known and the antilog (N) is desired, recall that $N = b^x$, and the base b is 10. To find the antilog of 2.4772, for example,

1. Enter 2.4772
2. Press INV key
3. Press the log key
4. Read 300.05

1.16 ■ The Quadratic Equation in One Unknown

The *degree* of an equation in a certain variable letter is the highest power of that letter that occurs in the equation. Thus, $2x^4 - 7x^2y^2 + 3xy^3 - 82 = 0$ is an equation of the fourth degree in x and the third degree in y. A rational equation of the second degree in one variable is called a *quadratic equation*. Every quadratic equation in one variable x can be arranged in the standard form

$$ax^2 + bx + c = 0 \qquad a \neq 0$$

where a, b, and c are real constants. Such an equation can be solved by using the *quadratic formula*

$$x = \frac{-b \pm \sqrt{b^2 - 4ac}}{2a} \tag{1.4}$$

This formula should be memorized, for it will be used in solving some problems in this text.

Illustrative Problem 1.13 Solve the equation $6x^2 - 20 = 7x$.

Solution Transpose the $7x$ to the left of the equality to get the equation in standard form.

$$6x^2 - 20x = 7x$$
$$6x^2 - 7x - 20 = 0$$

Here $a = 6$, $b = -7$, $c = -20$. Then

$$x = \frac{-b \pm \sqrt{b^2 - 4ac}}{2a}$$
$$= \frac{-(-7) \pm \sqrt{(-7)^2 - 4(6)(-20)}}{2(6)}$$

$$= \frac{7 \pm \sqrt{49 + 480}}{12}$$

$$= \frac{7 \pm \sqrt{529}}{12} = \frac{7 \pm 23}{12} = +2.5 \text{ or } -\frac{4}{3}$$

Both of these values (roots) "satisfy" the equation.

GEOMETRY REVIEW

Certain plane and solid geometrical figures are often referred to in the study of physics. The areas and perimeters of plane figures enter into many physics problems and so do volumes and surface areas of solid figures.

1.17 ■ Plane Geometrical Figures

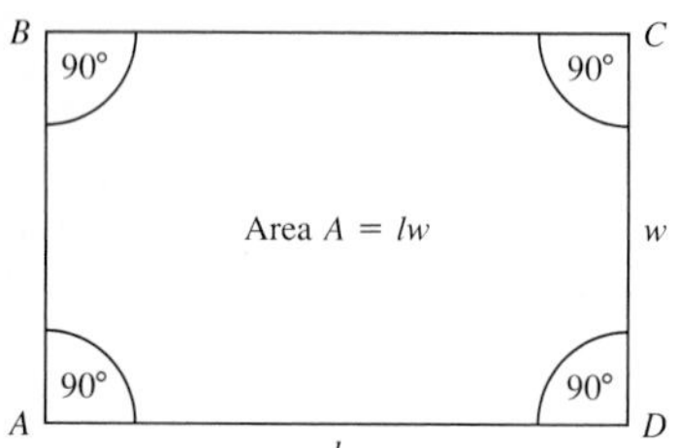

Fig. 1.4 The general rectangle, length l, width w.

The Rectangle Figure 1.4 shows a general rectangle. Note that all four angles are right angles (90°). Opposite sides of rectangles are equal in length, and consequently only two sides need to be designated. The length is lettered l, and the width, w, as shown. The area of a rectangle is simply the product of length and width. As a formula

$$A_{\text{(rectangle)}} = lw \tag{1.5}$$

Areas of all geometrical figures are expressed in square units, as square feet (ft^2), square inches (in.^2), square meters (m^2), and so on.

A *square* is a rectangle with all four sides equal, that is $l = w$. The area of a square is (see Fig. 1.5)

$$A_{\text{(square)}} = l \cdot l = l^2 \tag{1.6}$$

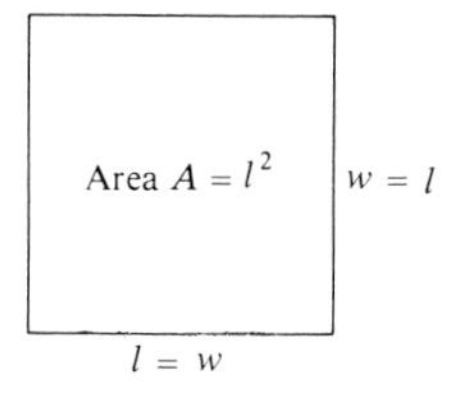

Fig. 1.5 The square. A rectangle with all four sides equal. $l = w$.

The General Triangle Figure 1.6 shows the typical general triangle and its nomenclature. Note that no two sides are equal and that no two angles are equal. The angles are designated by capital letters and the sides opposite them by corresponding lowercase letters. The side AC, on which the triangle appears to rest, is called the *base* and is lettered b. The perpendicular distance from the vertex B to side AC is called the *altitude*, lettered h. The area of a triangle, regardless of its shape, is equal to one-half the product of the base and the altitude, or

$$A = \frac{bh}{2} \tag{1.7}$$

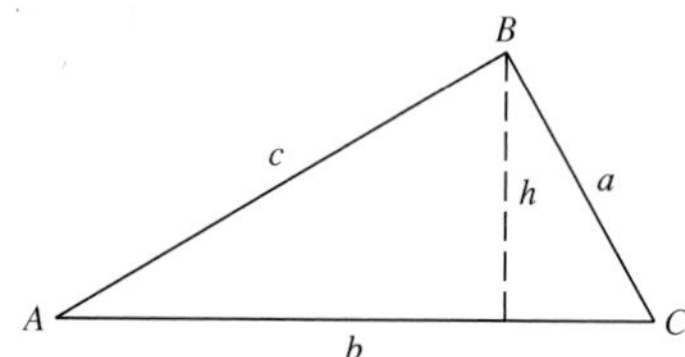

Fig. 1.6 The general triangle—no equal sides and no equal angles.

The Right Triangle Figure 1.7 shows a right triangle. Angle C is a 90° angle or *right angle*. The side c opposite the right angle is called the *hypotenuse*. Sides a and b are called *legs*.

If one leg b is taken as the base, the other leg a becomes the altitude h, since it is perpendicular to b by definition. The area of a right triangle is also

$$A = \frac{bh}{2} \tag{1.8}$$

The method illustrated here of designating the sides and angles of triangles should be carefully noted, since it is a standard labeling system.

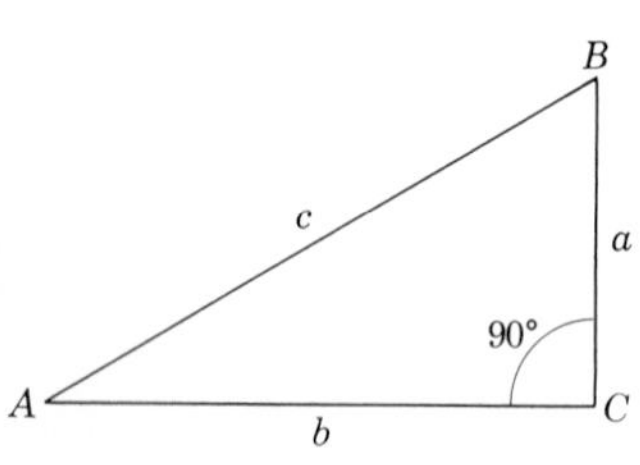

Fig. 1.7 The right triangle. Angle C is 90 degrees (90°).

Finding the Hypotenuse of a Right Triangle; Pythagorean Theorem The hypotenuse c of a right triangle can be found if the legs a and b are known. Pythagoras, a Greek mathematician and philosopher of the sixth century B.C., is credited with the discovery that *in right triangles, the square of the hypotenuse is equal to the sum of the squares of the other two sides*. As a formula

$$c^2 = a^2 + b^2 \tag{1.9}$$

or

$$c = \sqrt{a^2 + b^2} \tag{1.9'}$$

Geometrically, the relationship is shown in Fig. 1.8 where the area of the larger square, constructed on the hypotenuse as a side, is equal to the sum of the two smaller areas, each constructed on one of the legs as a side.

The theorem of Pythagoras is one of the earliest "laws" of mathematics to be found in the records from ancient Greece.

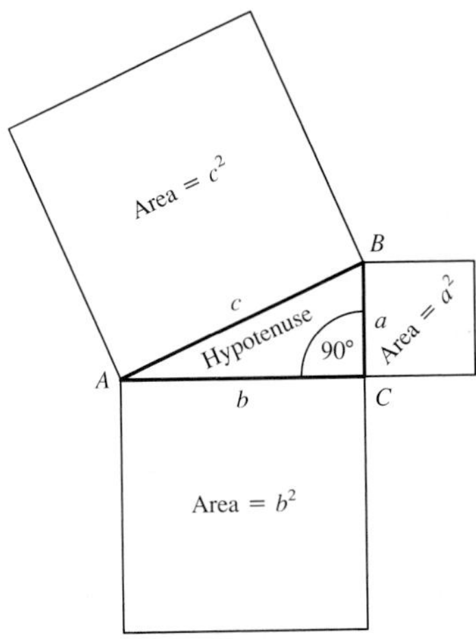

Fig. 1.8 Geometric representation of the theorem of Pythagoras. The square of the hypotenuse is equal to the sum of the squares of the other two sides: $c^2 = a^2 + b^2$.

The Circle The simple geometry of the circle can be studied with the aid of Fig. 1.9. A circle is defined as a closed curve in a plane, all points of which are equidistant from a point called the center O. The distance from the center O to the circle is the radius r. The *diameter* d is a line segment through the center of the circle with end points on the circle. The diameter has a length twice that of the *radius* r. The distance c around a circle is called the *circumference*. Regardless of the size of a circle, the ratio of the circumference to the diameter is always a constant. This constant value was given the name of a letter of the Greek alphabet, π, pronounced "pi." As a formula,

$$\frac{c}{d} = \pi$$

or

$$c = \pi d = 2\pi r \tag{1.10}$$

The value of π cannot be expressed as an exact decimal fraction. Worked out to four decimal places, its value is 3.1416. For most computations in applied physics we may use the value 3.14 without introducing a significant error in the results.

Area of a Circle It is often necessary to find areas of circular regions. The area of a circle is given by

$$A = \pi r^2 = \frac{\pi d^2}{4} \tag{1.11}$$

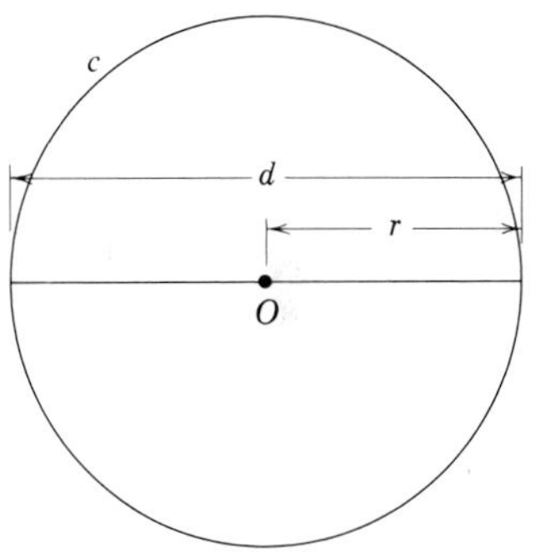

Fig. 1.9 The circle. The diameter is twice the radius: $d = 2r$. The circumference $c = \pi d = 2\pi r$. The area is $A = \pi r^2 = \pi d^2/4$.

1.18 ■ Solid Geometrical Figures

Of the many possible geometrical solids, only three will be briefly studied here. These three, which occur with some frequency in the application of physics to industrial and technical problems, are the *sphere,* the *cylinder,* and the *cone.*

The Sphere A sphere (Fig. 1.10) is a solid figure every point of whose surface is equidistant from a fixed point within called the *center*. The distance from the center to the surface is called the *radius*. The volume of a sphere is computed from the formula

$$V = \frac{4}{3}\pi r^3 = \frac{\pi d^3}{6} \tag{1.12}$$

and the surface area from

$$S = 4\pi r^2 \tag{1.13}$$

Of all solid figures, the sphere encloses the largest volume per unit of surface area.

Right Circular Cylinder Figure 1.11 illustrates a right circular cylinder. A cross section of such a cylinder is a circle of radius r. The length of the cylinder is h. Its volume is given by the formula

$$V = \pi r^2 h \tag{1.14}$$

The *lateral surface* area can be calculated from

$$S = 2\pi r h \tag{1.15}$$

and the total surface area, including top and bottom, from

$$A_T = 2\pi r(h + r) \tag{1.16}$$

In industry, storage tanks are often in the form of right circular cylinders. Tanks for storage of liquefied gases (refrigerants, rocket fuels and oxidizers, liquid air, etc.) must be able to withstand high pressures, and they are often spherically shaped, or cylindrical with hemispheres for both ends.

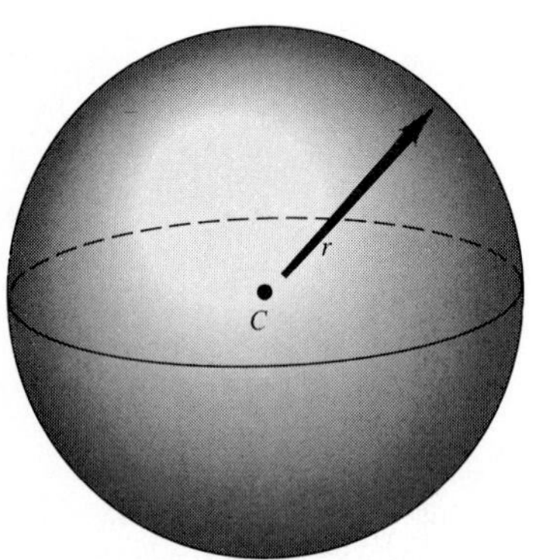

Fig. 1.10 The sphere. Every point on a spherical surface is equidistant from the center C. This distance is the radius r.

Cross section is a circle

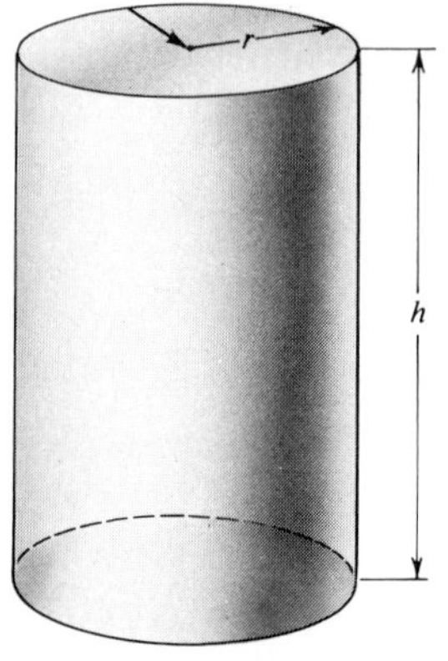

Fig. 1.11 A right circular cylinder. This is the geometrical form of countless tanks, tank trucks, and tank railroad cars.

Right Circular Cone A right circular cone is a solid figure generated by a right triangle as it rotates about one of its legs as an axis. In Fig. 1.12*a* right triangle ABC has been rotated around AC as an axis to form the cone. A line from the *vertex* A perpendicular to the base passes through the center C of the base circle. This line (h in the diagram) is called the *altitude* of the cone. The distance AB, labeled s, is called the *slant height* of the cone. It is, of course, the hypotenuse of the right triangle ABC.

The volume of a cone is equal to one-third the area of the base times the altitude. As a formula,

$$V = \frac{\pi r^2 h}{3} \tag{1.17}$$

The lateral surface area is given by

$$S = \pi r s \tag{1.18}$$

and the total surface area by

$$A_T = \pi r(s + r) \tag{1.19}$$

The Conic Sections A right circular cone can be cut by a plane at right angles (perpendicular to) the axis of the cone. The resulting cross section of the cone has the form of a *circle,* as illustrated in Fig. 1.12*a*. Cones can, of course, be cut by planes at almost any angle, resulting in a variety of cross sections, all of which are classified into three types—the *ellipse,* the *parabola,* and the *hyperbola*. Figure 1.12*b* for example, illustrates an ellipse. Because of these relationships to the cone, the circle, the ellipse, the parabola, and the hyperbola are known as *the conic sections*.

The circle is important in nearly all fields of physics. The parabola will be needed as projectiles, ballistic missiles, and telescopes are studied. The ellipse and the hyperbola are involved in aerospace and satellite mechanics, and in astronomy.

TRIGONOMETRY REVIEW

Certain problems of the right triangle are more readily solved by trigonometry than by any other method. The word *trigonometry* literally means "measuring three-sided figures."

1.19 ■ Basic Trigonometric Relations

The right triangle of Fig. 1.13 has vertices A, B, and C; angles α (alpha), β (beta), and γ (gamma); and sides a, b, and c. Angles are measured in degrees (°). This notation is conventional in trigonometry and should be memorized. The *right angle* (i.e., the 90° angle) is always at C, or in other words, $\gamma = 90°$.

There are six possible ratios that can be formed from the three sides of any right triangle. The fundamental principle of trigonometry lies in expressing angle relationships in terms of these ratios of the sides. The names and definitions of these ratios (functions) follow, and *they should be memorized,* since they will be used to describe a variety of electrical and wave phenomena, as well as several kinds of periodic motion in the study of mechanical systems.

$$\text{Sine } \alpha = \frac{\text{opposite side}}{\text{hypotenuse}} \qquad \text{Secant } \alpha = \frac{\text{hypotenuse}}{\text{adjacent side}}$$

$$\text{Cosine } \alpha = \frac{\text{adjacent side}}{\text{hypotenuse}} \qquad \text{Cosecant } \alpha = \frac{\text{hypotenuse}}{\text{opposite side}}$$

$$\text{Tangent } \alpha = \frac{\text{opposite side}}{\text{adjacent side}} \qquad \text{Cotangent } \alpha = \frac{\text{adjacent side}}{\text{opposite side}}$$

Using the approved abbreviations and substituting the corresponding letters for the sides, the *six trigonometric functions* of angle α become

$$\sin \alpha = \frac{a}{c} \qquad \csc \alpha = \frac{c}{a}$$

$$\cos \alpha = \frac{b}{c} \qquad \sec \alpha = \frac{c}{b}$$

$$\tan \alpha = \frac{a}{b} \qquad \cot \alpha = \frac{b}{a}$$

A similar set of trigonometric functions could be written for angle β.

Since the lengths of sides can be measured, the ratios representing the trigonometric functions of an angle can be evaluated as decimal numbers and used in numerical computations. Mathematicians have computed the values of the six trigonometric functions for all angles from 0 to 90°, so that reference may be made to these trigonometric tables for the solution of numerical problems. Appendix IV contains such a table in steps of whole degrees, accurate to three decimal places, for *sines, cosines,* and *tangents*.

Values of the trigonometric functions for any angle can also be found very easily by the use of a pocket calculator, as the following example shows.

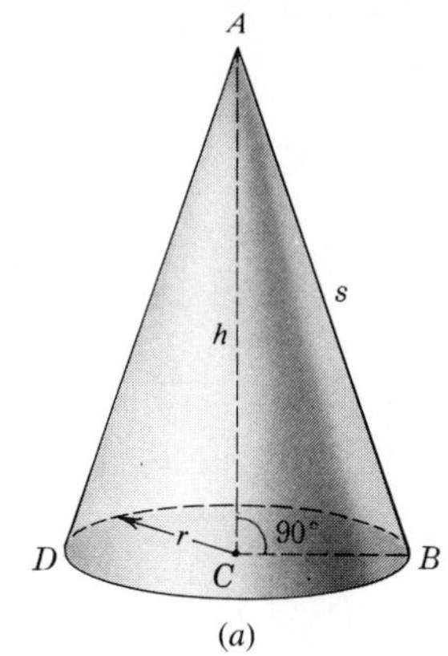

(*a*)

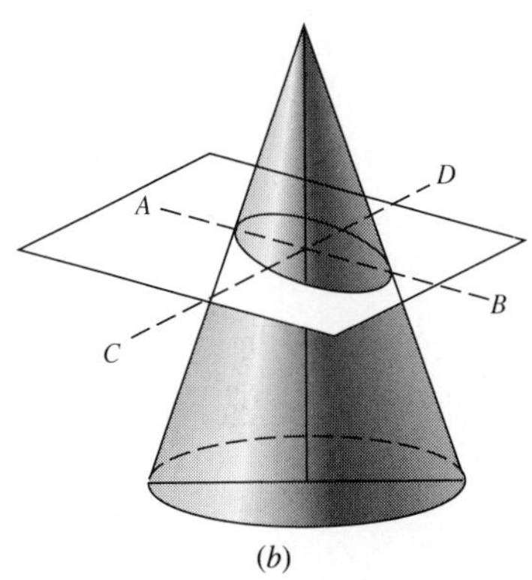

(*b*)

Fig. 1.12 (*a*) A right circular cone. Thrust bearings are shaped like a cross-section segment cut out of a cone. (*b*) An ellipse is a conic section. Here the plane determined by the intersecting straight lines *AB* and *CD* cuts through the cone, and the section is an ellipse.

Illustrative Problem 1.14 Find the values of the sine, cosine, and tangent of an angle of 42°.

Solution Enter the magnitude of the angle, 42°. Then press the *sin* key, to get

$$\sin 42° = 0.6691$$

In like manner, we obtain

$$\cos 42° = 0.7431$$
$$\tan 42° = 0.9004 \qquad \textit{answer}$$

Inverse Trigonometric Functions Just as we can write

$$\sin \alpha = 0.185, \text{ for example,}$$

we can also write

$$\sin^{-1} 0.185 = \alpha,$$

which says that "α is the angle whose sine is 0.185." Electronic calculators have keys marked "INV," to deal with "inverse" functions.

Illustrative Problem 1.15 Find the magnitudes of the angles with the following inverse functions: $\sin^{-1} 0.538$; $\cos^{-1} 0.538$; $\tan^{-1} 0.538$.

Solution

1. Enter 0.538 in the calculator
2. Press the INV key
3. Press the *sin* key to get

$$\sin^{-1} 0.538 = 32.55°$$

In like manner,

$$\cos^{-1} 0.538 = 57.45°$$
$$\tan^{-1} 0.538 = 28.28° \qquad \textit{answer}$$

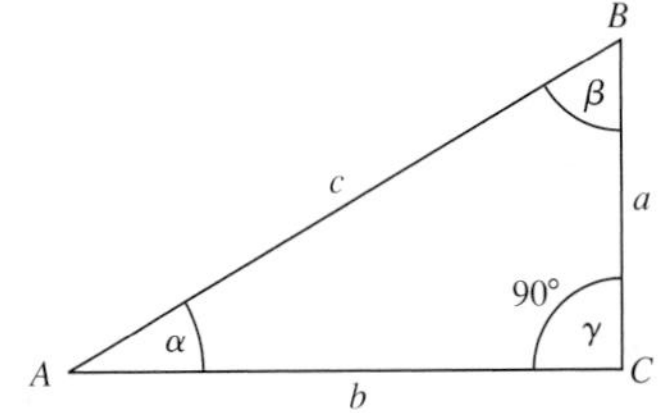

Fig. 1.13 A right triangle, labeled with the conventional symbols of trigonometry.

1.20 ■ Algebra and Trigonometry

The methods of algebra are used to solve expressions involving trigonometry. Figure 1.14 shows a right triangle with the two sides *AC* and *BC* known. Find the third side and the angles α and β.

Find angle α.

$$\tan \alpha = \frac{\text{opposite side}}{\text{adjacent side}} = \frac{18}{35} = 0.514$$

Trigonometry allows the determination of lengths and the sizes of angles that are in such locations that they cannot actually be measured. Surveying, navigation, and astronomy depend heavily on trigonometry, and its principles are used in building construction as well.

Thus α *is the angle whose tangent is* 0.514. It is common practice to express such a relation by either of the following:

$$\alpha = \tan^{-1} 0.514$$

or

$$\alpha = \arctan 0.514$$

both of which mean "α is the angle whose tangent is 0.514." From the table of Appendix III or from a pocket calculator, to the nearest whole degree,

$$\alpha = \tan^{-1} 0.514 = 27° \qquad \textit{answer}$$

Find angle β. The sum of the angles of any triangle is 180°. Since $\gamma = 90°$ and $\alpha = 27°$,

$$\beta = 180 - (90 + 27)$$
$$= 180 - 117 = 63° \qquad \textit{answer}$$

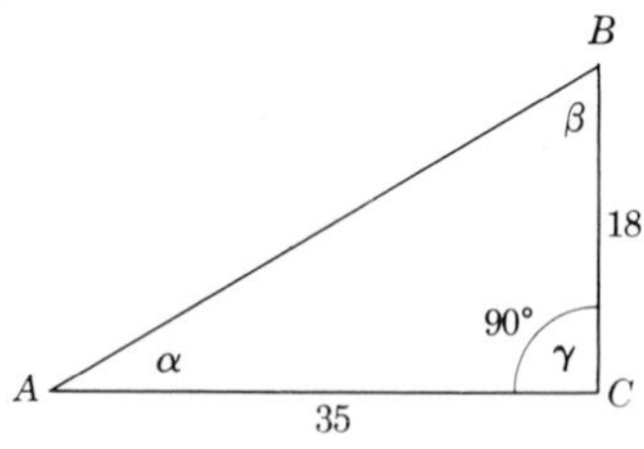

Fig. 1.14 A right triangle with two sides known. The hypotenuse and angles α and β are to be found.

Find hypotenuse AB. This could be done using the pythagorean theorem, that is,

$$AB = \sqrt{\overline{AC}^2 + \overline{BC}^2}$$

but it is more easily done using the cosine function, as follows:

$$\cos \alpha = \frac{\text{adjacent side}}{\text{hypotenuse}} = \frac{AC}{AB}$$

Substituting known values,

$$\cos 27° = \frac{35}{AB}$$

This is an algebraic equation and is solved by regular methods of algebra.

$$AB = \frac{35}{\cos 27°} = \frac{35}{0.891} = 39.3 \qquad \textit{answer}$$

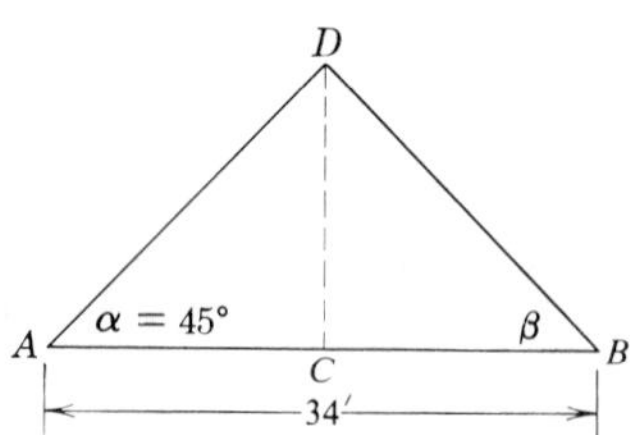

Fig. 1.15 Diagram of house rafter-length problem—Ilustrative Problem 1.16.

Illustrative Problem 1.16 A house is being constructed, and the plans call for a simple gable roof to be pitched at 45°. If the width of the house is 34 ft, what length are the rafters, assuming no overhang?

Solution The diagram of Fig. 1.15 shows the facts of the problem. $AB = 34$ ft, and angle α = angle $\beta = 45°$. Draw DC perpendicular to AB. The midpoint of AB is C, and therefore $AC = 17$ ft. In right triangle ACD, the hypotenuse AD represents a rafter.

$$\cos 45° = \frac{AC}{AD} = \frac{17\text{ ft}}{AD}$$

$$AD = \frac{17\text{ ft}}{\cos 45°}$$

$$= \frac{17\text{ ft}}{0.707} = 24\text{ ft} \qquad \textit{answer}$$

Trigonometry is by no means limited to the solution of right triangles, but for the time being we shall not need any more advanced analysis than this. Ordinarily only the first three functions defined above (sine, cosine, tangent) will be involved in technical physics problems.

THE USE OF GRAPHS

Just as a picture is often more meaningful than word explanations, so graphs are an effective aid to the interpretation of data and of physics principles and practices.

1.21 ■ Construction and Interpretation of Graphs

A graph is a type of picture or diagram that shows the relationship existing between two or more quantities. Newspapers and magazines frequently use graphs to interpret economic, business, health, and cost-of-living data. As a simple form of graph, consider Fig. 1.16, which plots electric energy production from a steam plant against time for the 12 months of the year. This is a "line graph."

A short study of the graph should give a clear picture of the production record of this plant for the given year. Peak months and slumps are clearly shown. Many hundreds of words would be necessary to present this production record as clearly and as simply as the graph does.

As a second example, look at Fig. 1.17, which plots the maximum horsepower output of an auto engine against the speed in revolutions per minute (rpm) at which it is tested. This graph has been "smoothed" so that a flowing curve results. It portrays results which should be typical of all engines of the same model when run under identical conditions. Note that this engine "peaks" at about 6000 rpm and that power output drops off sharply thereafter. Such a graph would be constructed by the test engineer from data obtained during actual runs on a test instrument called a *dynamometer*. What is this engine's maximum power output at 3000 rpm?

The two graphs just considered are both *line graphs,* a type which is well suited for presenting technical data. Other forms of graphs are *bar graphs, circle graphs,* and *pictographs,* which are especially suited to the portrayal and interpretation of business data and economic statistics, as well as medical and health data.

When you construct graphs or charts, think *carefully* about the title and the labeling of each axis. Make sure that the title tells exactly what the graph is intended to portray, and that the variables plotted and units used are accurately stated on each axis.

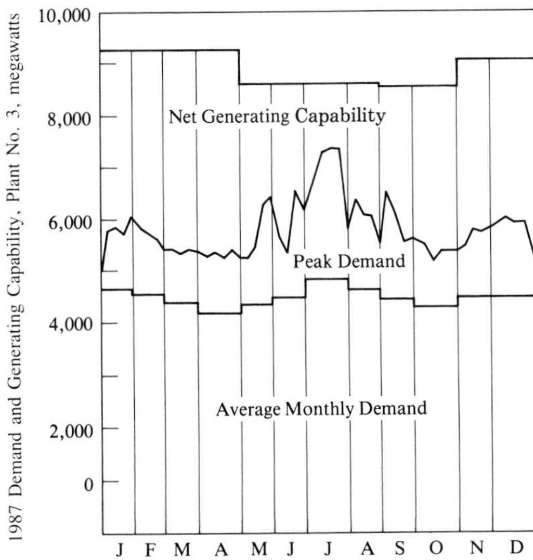

Fig. 1.16 Graphic representation of electric power plant data. Net generating capability, peak demand, and average monthly demand for electric power are shown in a manner that allows ready comparison.

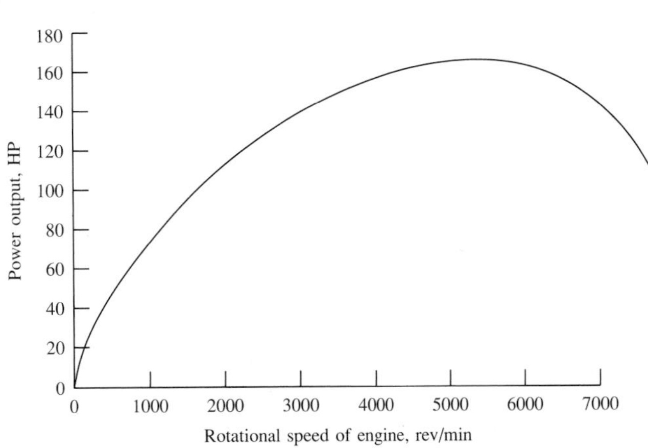

Fig. 1.17 Power-output versus engine-speed curve (smoothed) for an auto engine rated at 165 hp.

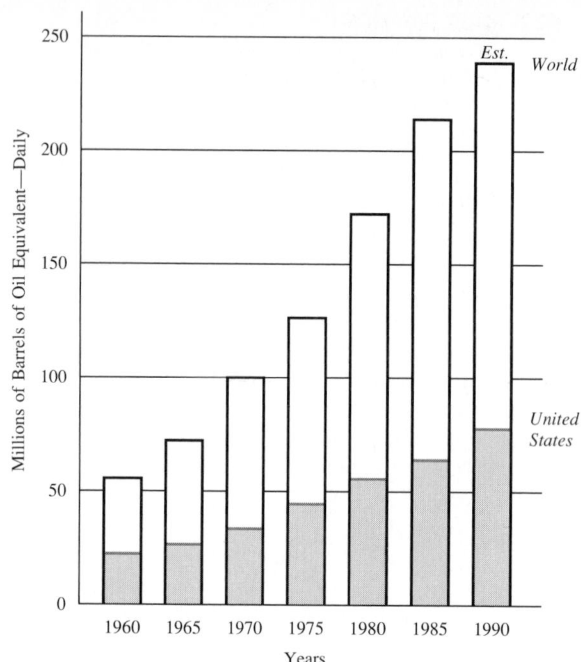

Fig. 1.18 World and United States daily energy demand 1960 to 1990 (est.). The energy amounts are expressed in terms of millions of barrels of oil equivalent (MBOE) daily. (Exxon Corp.)

Figure 1.18 portrays, by means of a bar graph, the world and United States energy demand from 1970 to 1990, in millions of barrels of oil-equivalent (MBOE) daily.

In general, graphs of physical data are constructed by plotting the two quantities against each other along two axes—one horizontal, the other vertical. *The quantity whose variation is of the most immediate interest is usually plotted along the vertical axis,* since its ''ups and downs'' are thus made more evident. Every graph should have a clear, well-phrased title, and each axis must be carefully labeled so there will be no doubt about the units or the quantities being plotted.

We shall have occasion to use graphs or ''curves'' many times in the presentation of physical principles and data throughout this book. Laboratory work frequently calls for the preparation or interpretation of graphs.

QUESTIONS AND EXERCISES

1. Identify at least six substances that normally are (*a*) solids, (*b*) liquids, (*c*) gases.

2. In your chosen career, or in the one you are thinking seriously about (tell what it is), in what ways do you think you will be involved with (*a*) electrical energy? (*b*) heat energy? (*c*) chemical energy? (*d*) solar energy? (*e*) mechanical energy and machines? (*f*) nuclear energy? (*g*) other applications of physics? (List them.)

3. Why are mathematical formulas a more convenient method of solving problems than the use of a statement in words?

PROBLEMS

Group One

1. Express the following in decimal form: (*a*) six ten-thousandths, (*b*) three and one-half thousandths, (*c*) five-millionths, (*d*) seventy-five thousandths.

2. A cylindrical shaft has a diameter of $3\frac{7}{16}$ in. Express this in decimal form accurate to three places.

3. Express the following in scientific (or exponential) notation, as numbers between 1 and 10 times 10 to a power. (*Example:* $6785 = 6.785 \times 10^3$.) (*a*) 22,520, (*b*) 25,850,000, (*c*) 629, (*d*) 0.00629, (*e*) 0.0000785, (*f*) 0.084

4. Solve the following algebraic equations (formulas) for the factors indicated:
(*a*) $V = IR$ for I (*c*) $s = \frac{1}{2} gt^2$ for t

(*b*) $\dfrac{F}{W} = \dfrac{L}{h}$ for W (*d*) $F = \dfrac{mv^2}{r}$ for m

5. From Eq. (1.1), if a rifle bullet impacts the target 400 yards (yd) away 0.55 seconds (s) after the gun was fired, what was its average *velocity* in feet per second (ft/s)?

6. A sum of $8450 was placed in a savings bank for 9 months at simple interest. The amount of interest earned was $367.50. Find the *annual interest rate*.

7. Find the area of a triangular plot of ground whose frontage along the street is 185 ft and whose depth (altitude of the triangle) is 140 ft.

8. A pulley has a circumference of 48 in. What is its radius to the nearest 0.01 in.?

9. A round concrete pier has a diameter of 2.5 ft. What is its circumference?

10. A sewage-treatment plant has a water aerator system with circular ponds 80 ft in diameter. What is the surface area of the water exposed to the air in each pond?

Group Two

11. Simplify the following expressions:
(*a*) $(3x)(5x)$ (*d*) $(2a^2b)^3$

(*b*) $(xy)(xyz)(-xz)$ (*e*) $\dfrac{-12x^2y^3}{4xy}$

(*c*) $(xy^3)^2$ (*f*) $\dfrac{x^3x^9}{x^2}$

12. Reduce each of the following expressions to its simplest equivalent:
(*a*) $(0.16)^{1/2}$ (*d*) 7^{-2}

(*b*) $\left(\dfrac{8}{125}\right)^{1/3}$ (*e*) $(-8)^{-2/3}$

(*c*) $(32)^{4/5}$ (*f*) $\left(\dfrac{9}{25}\right)^{-3/2}$

13. Use a hand-held calculator to find the values of:
(*a*) $(3.1416)^3$ (*c*) $\sqrt[3]{5813.82}$
(*b*) $(2.998 \times 10^8)^2$ (*d*) $\sqrt[4]{0.0019347}$

14. Find the value of (*a*) the angle whose tangent is 0.27103; (*b*) $\cos^{-1} 0.8291$; (*c*) $\sin^{-1} 0.9654$.

15. Many mechanical pencils use cylindrical lead that has a diameter of 0.02 in. What length of such a piece of lead would have a volume of 1 cubic inch (in.^{3})?

16. In a right triangle, what value (degrees) for the smallest angle will cause the shortest side to be half as long as the hypotenuse?

17. What is the diameter of a circular plot of ground whose area is exactly one acre (43,560 ft^2)?

18. How high must a 1-ft-diameter right circular cylinder be if its interior volume is to equal 1 gallon (gal)? (1 gal = 231 in.^{3}.)

19. A town needs an elevated spherical water tank that will hold 1 million gallons of water. Find the required diameter of the tank in feet.

20. A cylindrical railroad tank car for shipping chemicals is 6 ft in diameter. If it holds 10,000 gal of fluid cargo, what is its length?

21. Which has a greater volume of solid material—a hollow spherical shell with an outer diameter of 20 in. and an inner diameter of 19 in., or a completely solid sphere with a diameter of 10 in.?

22. Compute the numerical value of the following, expressing the result as a number between 1 and 10 times 10 to a power:

(*a*) $\dfrac{5.421 \times 728 \times 0.0036}{0.092}$

(*b*) $\dfrac{17.85 \times 440 \times 0.685}{27.6 \times 0.321}$

23. Find the hypotenuse of a right triangle whose sides are 8 ft and 5 ft. Use the theorem of Pythagoras.

24. The area of a square is 625 in.^{2}. What is the area of the largest circle that can be inscribed in the square?

25. Find the volume and the lateral surface area of a right circular cone whose height is 18 in. and whose base has a radius of 6 in.

26. The hypotenuse of a right triangle is 36 in., and its angle with the base is 42°. Side *a* is 24.1 in. How long is the base? Solve first, using trigonometric functions (cosine or tangent function), and then using Pythagoras' theorem. Note the relative convenience of the trigonometric method.

27. A conveyor track with rollers is 35 ft long. One end is on a dock, and the other end is secured to the deck of a ship tied up at the dock. If the ship's deck is 16 ft above the dock, find (*a*) the angle which the conveyor track makes with the dock and (*b*) the horizontal distance from the base of the conveyor to the ship's side.

28. A communications antenna tower is 350 ft high. It is to be steadied by guy wires 400 ft long. What angle will these cables make with the ground?

29. From a point 200 ft from the base of a tree, the angle of elevation of the treetop is observed to be 30°. How tall is the tree?

30. A mountain road climbs steeply a 30 percent grade, which means it rises 30 ft for every 100 ft of horizontal distance. Find the angle of the incline in degrees from the horizontal.

Group Three

31. A silo on a farm has a lower cylindrical section that is 24 ft high, with an interior diameter of 12 ft. On top of this cylindrical section is a hemispherical section that has an inner radius of 6 ft. (*a*) What is the total interior volume of the silo? (*b*) How many 8 ft × 10 ft × 4 ft truck loads of silage would fill the silo completely?

32. An automobile race track is to be designed with the following specifications: Two parallel, straight segments are to be connected by semicircular segments at either end. The radius of each semicircular segment is to be $\frac{1}{4}$ mi, and the total length of the track is to be 4 mi. How long should the straight segments be?

33. A new 600-ft tall antenna tower is to be erected on a *square* plot of land for a television station. Guy wires from the top of the tower to the ground are to make an angle of 60° with the ground. As a technician for the project you must determine the *minimum* lengths of the sides of the square plot. Four guy wires, equally spaced at the top, must be anchored in the ground so that their concrete bases are on the property lines of the square plot. What are the lengths of the sides of the square?

34. A circular building is to have a roof in the shape of a cone. The diameter of the building is 80 ft, and the roof overhang is negligible. The apex of the conical roof is 20 ft above the top of the circular building wall. Find the area of the roof.

35. Find the number of square feet of steel plate required to make a cylindrical tank 30 ft high and 20 ft in diameter. Include bottom and top of tank.

36. How many square feet of steel plate are required to construct a spherical refinery tank to hold 40,000 gal? (7.48 gal = 1 ft^3.)

37. Prepare a speed curve for a ship from the data given below. Use graph paper. Plot ship's speed in knots on the vertical axis and revolutions per minute (rpm) of the screw on the horizontal axis. "Smooth" the curve and write an interpretation including everything the graph tells you about the ship's performance.

Speed of Screw, rpm	*Ship's Speed, knots*
0	0
50	10
100	17
150	22
200	26
250	27.5
300	28.5
350	29.3
400	30
500	31

38. A loading chute delivers filled cartons from a warehouse to a freight-car door 25 ft below and 18 ft out from the warehouse wall. How long is the chute, and what angle (degrees) does it make with the warehouse wall?

39. A bicycle wheel has a diameter of 26 in. measured to the outer rim of the inflated tire. If there is no slippage between the tire and the road, how many revolutions of the wheel will occur as the cyclist rides a distance of one mile (5280 ft)?

40. The following data were obtained on road tests of fuel consumption at different speeds in a new model automobile. Plot a "smoothed" line graph of the data. Choose carefully the variable you plot on the vertical axis. From 40 to 100 mi/h what relationship seems to exist between fuel consumption and speed? Title the graph properly and label the axes carefully.

Speed, mi/h	*Fuel Consumption, gal/mi*
10	0.046
20	0.043
30	0.039
40	0.035
50	0.032
60	0.031
70	0.033
80	0.039
90	0.055
100	0.072

CHAPTER 2 PRECISION MEASUREMENT

The necessity for measurement is universally recognized, whether the physicist or technician is dealing with *macrophysics* (great distances, enormous quantities of matter, huge sources of energy) or with *microphysics* (the realm of molecules, electrons, atomic nuclei, crystals, or fog particles). In all of physics research, and in most applications of physics to industry and engineering, the necessity to measure with precision and accuracy is always present.

All measurements, no matter how carefully they are made, have a certain level of error involved. In dealing with these measurement errors physicists use the terms *precision* and *accuracy,* terms which do not mean the same thing. *Precision* has to do with the degree of refinement with which a measurement can be taken—the closeness with which several repeated measurements of the same quantity, under the same conditions, agree with each other. Precision also refers to the capability of the measuring instrument to provide closely spaced graduations or marks along its reading scale. For example, a micrometer with a scale that can be read to 0.0001 (one ten-thousandth) in. is a far more precise measuring instrument than a steel rule with markings 0.01 (one hundredth) in. apart.

Precision distinguished from *accuracy.* The two terms are related, but they do not mean the same thing. High accuracy implies small *systematic errors.* High precision implies small *random errors.*

Accuracy, on the other hand, refers to how close a measurement is to some accepted standard or "true value." A measurement is one of *precision* when the random errors are small. But relative to the "true value" it may not be very *accurate.* A precise instrument can give remarkably consistent readings (that is, with very small *random errors*), but this does not always imply great accuracy, since the instrument may exhibit *systematic errors* due to wear or poor construction, or the persons doing the measuring may be at fault (human error). Even so, it should be emphasized that when scientists are engaged in establishing an accepted or "true" value of a measurement, they will always use instruments capable of the greatest possible precision.

The method in most common use for expressing the results of a series of measurements of a given quantity consists simply in "averaging the results." A series of measurements is taken under conditions that vary as little as possible, and then the average or arithmetic mean of the readings is determined. The assumption here is that the *mean* or *average* value will better represent the true value than will any one of the several measurements taken alone, an assumption that is not necessarily true. However, how could the single, "most accurate" of the several readings be selected? An average result from several measurements tells us little or nothing about the precision with which the measurements were made.

Average, or *mean* of several measurements.

THE IMPORTANCE OF PRECISION MEASUREMENT

In industry, precision measurements accurate to 1/100,000 in. are necessary in some cases; 1/10,000 in. is routine in the aviation and space industry; and accuracy to 1/1000 in. is common practice in the automotive and heavy machinery field.

Precision measurement is essential to mass production and to all modern "high-tech" industries.

Mass production in industry requires complete interchangeability of parts. For example, in the automobile, appliance, and aerospace industries, component parts may be manufactured by dozens of subcontractors, and the finished product may be assembled in a plant thousands of miles away from the parts plants. Weeks or months later, these parts will be put together by assembly-line workers on a tight time schedule where every part must "fit." The parts must be manufactured to the close *tolerances* specified by the design engineer and stipulated on drawings and in specifications prepared by technical draftsmen. For example, a valve stem might have a design diameter of 0.3125 in., with a tolerance of plus or minus ($\pm$) 0.0003 in. The *allowable tolerance,* then, is plus or minus 3 ten-thousandths of an inch. All stems measuring between 0.3122 in. and 0.3128 in. are acceptable. Those with diameters beyond these limits are rejected.

Tolerance specifications state how much larger than or smaller than the ideal value a part can be and still function satisfactorily.

Accurate measurements are also important to business and finance, agriculture, medicine and health, and to many everyday activities such as travel, sports and recreation, cooking and baking, and communications and entertainment.

2.1 ■ Fundamental Units of Measurement

The *fundamental units* involved in mechanics are *length* (L), *mass* (M), and *time* (T).

All measurements are actually relative in the sense that they are *comparisons with some standard unit of measurement.* Although we use many different units of measurement, there are only three *fundamental units* needed for the study of mechanics—*length, mass,* and *time*. Other well-known measures are called *derived units* to distinguish them from fundamental units. For example, the familiar unit of *area,* the square foot, is a product of a length times a length. Another derived unit is the unit of *velocity* or speed, e.g., the foot per second, which is obtained by dividing a length unit by a time unit. Another is the unit of *density,* e.g., the kilogram per cubic meter, derived by dividing a mass unit by the cube of a length unit. In this manner all derived units *in mechanics* can be expressed in terms of length L, mass M, and time T. Three other fundamental units will be introduced later, in chapters on heat, electricity, and light.

Derived units are made up of various combinations of the fundamental units, and they express quantities or measurements essential to science, engineering, or commerce.

The examples of derived units of measurement just given are mostly those from the system familiar to most of us in the United States—the *English system* of measurement. Most other nations of the world use a different system called the *metric system*. The United States also uses the metric system in science. In other fields the United States is gradually adopting some features of the metric system, but change is coming very slowly. The Secretary of Commerce recommended in 1971 that the changeover be completed by 1981. In 1975 the Metric Conversion Act was passed by the Congress and signed by the President, calling for voluntary conversion to the metric system, but not setting any specific date for the changeover. Adopting the metric system (called *metrication*) in the United States is a somewhat controversial issue because first, staggering costs are involved (estimates range into billions of dollars), and second, because people are reluctant to change from a system they are comfortable with to a new system which is foreign to them. The suggested deadline of 1981 was not met, and it is impossible to predict at present how long it will be until the United States becomes a "metric nation." Industry and commerce in the United States are still heavily committed to the English system, and many fields of engineering and technology (building construction, transportation, steam and hydraulic energy, refrigeration and air conditioning, to name only a few) still operate for the most part within the traditional (English) system. Consequently, considerable emphasis on English-system units will be retained in this book.

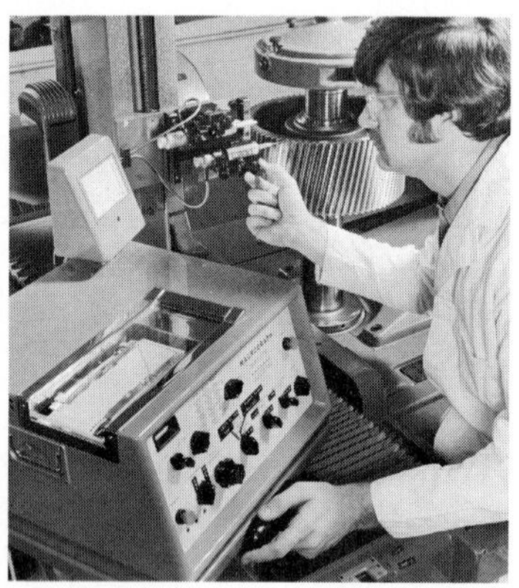

Fig. 2.1 Precision measurement in industry. A technician checks a helical gear with an electronically actuated "macrograph." Irregularities in the gear surfaces of only a few hundred-thousandths of an inch can be detected. (The Falk Corp.)

However, in *science,* the metric system has been the universally used system for more than a century. Engineering and technology make greater use of the metric system each year, and instruction in the metric system is now a part of the curriculum in public schools. This book will therefore give nearly equal emphasis to both systems of units, in discussions of both theory and practical applications. See Appendix I for a list of metric-system units and symbols.

SYSTEMS AND UNITS OF MEASUREMENT

2.2 ■ Measurement of Length

The Metric System The unit of length in the metric system is the *meter* (spelled *metre* in Great Britain and many other English-speaking countries). The standard meter was, until 1960, defined as the distance between two fine lines ruled on a platinum-iridium bar stored at the International Bureau of Weights and Measures in Sèvres, France. However, in October 1960, the International Committee on Weights and Measures held a General Conference in Paris and out of that meeting came an international agreement that the world's new standard of length would be based on the wavelength of the orange-red line of krypton-86.

More recently, at a 1983 conference, the meter was again redefined by the same international group—this time in terms of the velocity of light. The present definition of the *standard meter* is: *One meter is the distance that light travels* (in a vacuum) *during a time period of* 1/299,792,458 *second*. This is equivalent to defining the speed of light to be exactly 299,792,458 m/s.

The length of the standard meter is based on the speed of light. In a vacuum, light travels at 299,792,458 m/s.

The General Conference of 1960 gave the name *International System of Units* to the metric system. The *meter* (length), the *kilogram* (mass), and the *second* (time), along with the three other basic units which we shall define later in the book, make up the new International System. It is now commonly known as *SI-metric,* SI being the initials of the first two words of the French, *Système International d'Unités*. In mechanics, where the fundamental units are three only (length, mass, and time), the metric system is often referred to as the meter-kilogram-second (mks) system.

The *complete* metric system for all of science is called the SI-metric system, for *Système International d'Unités.*

The metric system is completely decimalized. Subdivisions of the meter include the *decimeter* (dm, 0.1 meter), the *centimeter* (cm, 0.01 m), the *millimeter* (mm, 0.001 m), the *micrometer* (μ, 0.000,001 m), and the *nanometer* (*nm,* 0.000,000,001 m). For greater lengths the *kilometer* (km, 1000 m) is used.

Ease of decimalization is one advantage of the metric system.

If you have never seen a meter stick or a centimeter ruler, you can visualize the meter as being about $3\frac{1}{2}$ in. longer than a yard (Fig. 2.2*a*). The inch is slightly longer than 2.5 cm (see Fig. 2.2*b*). Table 2.1 presents the more commonly used metric and English units, and Table 2.2 gives metric-to-English and English-to-metric equivalents.

The English System The standard unit of length in the English system is the *yard* (yd). The legal definition of the yard was first established in 1878, and it defined the yard as the distance between two ruled lines on a bronze bar at the British Exchequer when the metal temperature is 62°F. A later definition (1963) defines the yard as 0.9144 meter. In the United States, although the yard is also a standard unit of measure, the *foot* ($\frac{1}{3}$ yd) is a more commonly used unit. Other length units are the *inch* ($\frac{1}{12}$ ft) and the *statute mile* (5280 ft). In navigation, the *fathom* (6 ft) and the *nautical mile* (6076 ft) are used. The nautical mile is approximately equal to the distance along the surface of the earth at the equator or along any great circle, subtended by one minute of arc. However, since the earth is not a perfect sphere, one minute of arc does not always subtend the same distance on a great circle. Consequently, the nautical mile is now defined as 1852 meters.

Units of the English system for use in the United States.

This book gives equal emphasis to the English and the SI-metric systems of measurement.

The inch is subdivided into smaller units in two different ways. Common practice, where precision is not required, makes use of the $\frac{1}{2}$, $\frac{1}{4}$, $\frac{1}{8}$, $\frac{1}{16}$, $\frac{1}{32}$, $\frac{1}{64}$ system of subdividing. Precision measuring instruments are, however, calibrated to read in *thousandths of an inch* and *ten-thousandths of an inch*. Some test-laboratory instruments can even measure accurately to one-millionth (0.000001) of an inch. This unit is called the *microinch* (abbreviated μin.). Standards of measurement in this country are legally fixed by the Congress, and the National Bureau of Standards is charged with the responsibility of maintaining accurate originals of all units of measurement.

2.3 ■ Measurement of Mass and Weight

Mass is defined at this point as the *quantity of matter contained in a body* and is to be distinguished from *weight,* which is a *force* measured by *the pull of gravity on a body*. A more precise distinction between mass and weight will be made in Chap. 5.

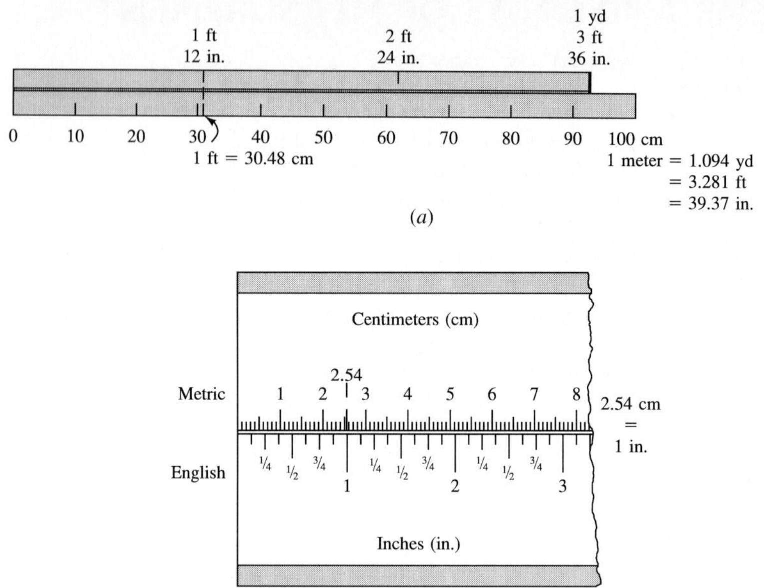

Fig. 2.2 Comparison of metric and English units of length measurement. (*a*) The meter (m) compared to the yard (yd). (*b*) The centimeter (cm) compared to the inch (in.). 1 in. = 2.54 cm.

(*b*)

Table 2.1 Units of measurement in the SI-metric and English systems

Units of Length		*Units of Mass*		*Units of Time*	
		SI-Metric System			
1 meter (m)	= the standard	1 kilogram (kg)	= the standard	1 second (s)	= the standard
1 decimeter	= 0.1 m	1 gram (g)	= 0.001 kg	1 minute (min)	= 60 s
1 centimeter (cm)	= 0.01 m	1 centigram (cg)	= 0.01 g	1 hour (h)	= 3600 s
1 millimeter (mm)	= 0.001 m	1 milligram (mg)	= 0.001 g	1 day	= 24 h
1 micrometer* (μm)	= 0.000,001 m	1 microgram (μg)	= 0.000,001 g		= 86,400 s
1 nanometer	= 0.000,000,001 nm			1 millisecond	= 0.001 s
1 kilometer (km)	= 1000 m			1 microsecond	= 0.000,001 s
1 megameter (Mm)	= 10^6 m			1 nanosecond	= 0.000,000,001 s
1 gigameter (Gm)	= 10^9 m				
		English System			
1 yard (yd)	= the standard	1 pound (lb) (still used as a mass in U.S. commerce and industry but not in science—actually a unit of weight)			
1 foot (ft)	= $\frac{1}{3}$ yd				
1 inch (in.)	= $\frac{1}{12}$ ft				
1 microinch (μin.)	= 0.000,001 in.	1 ounce (oz)	= $\frac{1}{16}$ lb	Same as in SI-Metric System	
1 statute mile	= 1760 yd	1 grain (gr)	= $\frac{1}{7000}$ lb		
	= 5280 ft	1 ton	= 2000 lb		
1 nautical mile	= 6076 ft (1852 m)	1 slug	= the engineering standard of mass—to be defined later		
	= 1 min of arc at earth's equator (approx.)				
1 fathom	= 6 ft				

* Formerly *micron* in the United States.

In the metric system, the standard unit of *mass* is the *kilogram* (kg), now defined as the mass of a standard cylinder of platinum-iridium in the International Bureau of Weights and Measures at Sèvres, France. One kg of water occupies 1,000.027 cm^3. The volume, 1000 cubic centimeters (cm^3), is called a *liter*. Subdivisions of the kilogram are the *gram* (g, $\frac{1}{1000}$ kg), the *centigram* (cg, $\frac{1}{100}$ g), the *milligram* (mg, $\frac{1}{1000}$ g), and the *microgram* (μg, $\frac{1}{1,000,000}$ g). The international standard kilogram is illustrated in Fig. 2.3. The basic unit of *weight* in the metric system is a force unit, the *newton,* to be defined in Chap. 5.

Weight measurements in the English system are stated in *pounds* as the basic unit. Subdivisions of the pound are the *(avoirdupois) ounce* ($\frac{1}{16}$ lb) and the *grain* (gr, $\frac{1}{7000}$ lb). Larger units are the *hundredweight* (cwt) and the *ton* (2000 lb). *Masses* in the English system are still expressed in pounds in some fields of U.S. industry and commerce. The *engineering unit* of mass, however, is the *slug,* to be defined later in Chap. 5. We will use the pound as a mass unit occasionally in mechanics and heat, since U.S. industry does. See Chap. 5 for further discussions of English-system units of mass and weight.

Fig. 2.3 The U.S. standard kilogram. It is an accurate copy of the international standard kilogram at the International Bureau of Weights and Measures at Sèvres, France. (National Bureau of Standards)

Table 2.2 Metric and English equivalents of length, mass, and volume

Metric-to-English		*English-to-Metric*	
Length			
1 meter (m)	= 39.37 in.	1 yard (yd)	= 0.9144 m (by definition)
	= 3.28 ft	1 foot (ft)	= 30.48 cm = 0.3048 m
1 centimeter (cm)	= 0.394 in.	1 inch (in.)	= 2.54 cm = 0.0254 m
1 millimeter (mm)	= 0.0394 in.		= 25.4 mm
1 micrometer (μm)	= 0.0000394 in.	1 microinch (μin.)	= 0.0254 micrometer (μm)
1 kilometer (km)	= 0.6214 mi	1 statute mile (mi)	= 1.609 km
	= 3281 ft	1 nautical mile	= 1.852 km
Mass			
1 kilogram (kg)	= 2.205 lb*	1 pound (lb)*	= 454 g
1 gram (g)	= 0.0353 oz*		= 0.454 kg
1 milligram (mg)	= 0.0154 gr*	1 ounce (oz)	= 28.35 g
1 kilogram	= 0.0685 slugs	1 grain (gr)	= 0.065 g
			= 65 mg
		1 ton	= 907 kg
		1 slug (sl)	= 14.6 kg

* In science and engineering, the pound, the ounce, and the grain are *not used as units of mass* and are regarded as *weights* or *forces*. However, in U.S. industry, commerce, and agriculture they are still often used as mass units as well as weight units.

Volume			
1 cu meter (m^3)	= 1.31 yd^3	1 cu yd (yd^3)	= 0.765 m^3
	= 35.4 ft^3		= 27 ft^3
1000 cm^3	= 1 liter	1 cu ft (ft^3)	= 28.3 liter
	= 1.057 qt		= 7.48 gallon (gal)
1 cu. cm.	= 0.001 liter	1 quart (qt)	= 0.946 liter
	= 1 millimeter (ml)	1 gallon (gal)	= 3.785 liter
1 liter	= 61.01 in.^{3}	1 cu in. (in.^{3})	= 16.39 cm^3 (ml)
	= 1.057 qt		

Miscellaneous	
1 acre	= 43,560 ft^2 = 4047 m^2 = 0.405 hectare
1 hectare	= 2.47 acres
1 square mile	= 640 acres = 2.59 square kilometers (km^2)
1 bushel (dry)	= 2219 in.^{3} = 0.0352 m^3 = 35.2 liter
1 gallon	= 231 in.^{3}

2.4 ■ Measurement of Time

Precise measurement of time. The *second* is the basic time unit for both the metric and the English systems of measurement.

Both the metric and English systems of measurement employ the same time unit, the *second,* which was once defined as 1/31,556,925.9747 of the tropical year 1900. Because of irregularities in the earth's orbital movements, all years are not of exactly the same duration, and slight variations in the earth's rotation cause the length of the day to vary slightly. However, for most purposes, the second can be defined as

$$\frac{1}{86{,}400} = \left(\frac{1}{24\text{ h/day}} \times \frac{1}{60\text{ min/h}} \times \frac{1}{60\text{ s/min}}\right)$$

of the mean solar day.

A much more recent time standard is based on the spinning motion of electrons in atoms. In 1967 the International Committee on Weights and Measures adopted a new definition of the second, making 1 s equal to the duration of 9,192,631,770 periods of vibration of the outermost electron of the atom of cesium-133 (see Fig. 2.4).

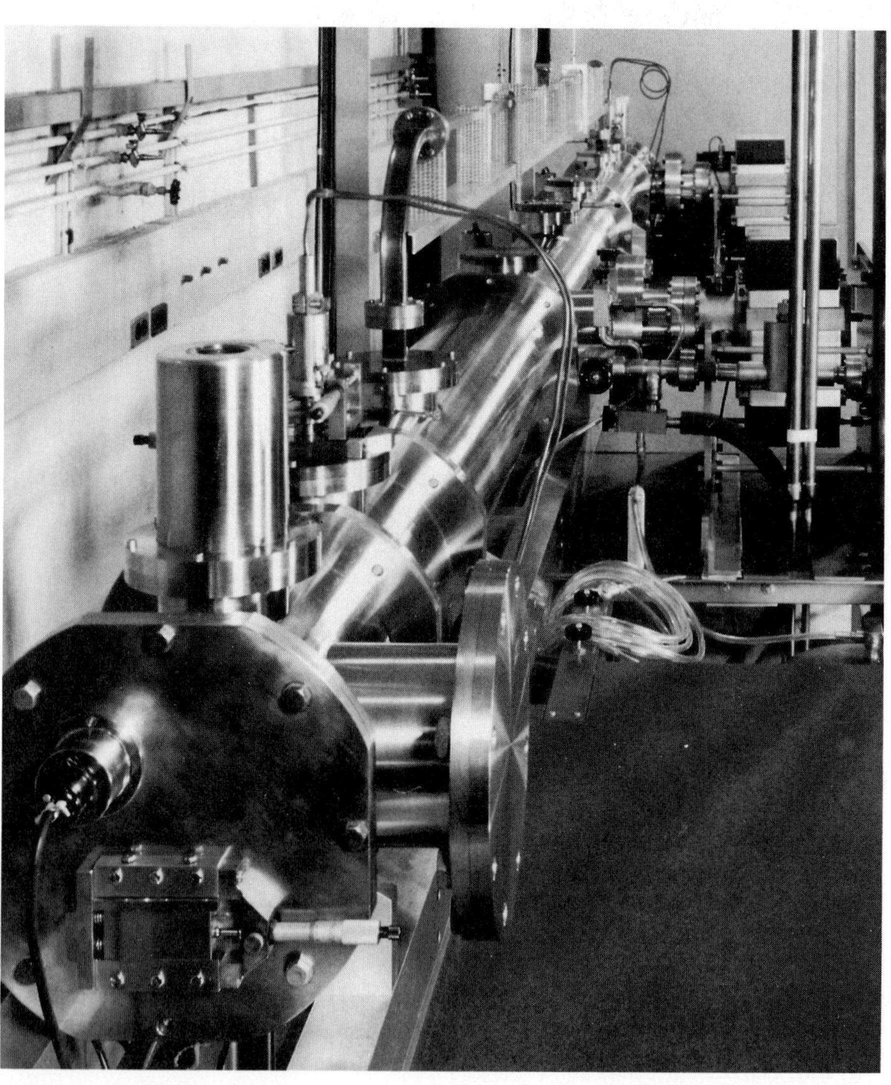

Fig. 2.4 An atomic clock in the Boulder, Colorado, laboratory of the National Bureau of Standards. It is actuated by the electromagnetic oscillations in the atom of cesium-133. Its margin for error is estimated to be no more than one second in 30,000 years.

Since scientists and technicians are nearly always interested in the time rate of change of physical phenomena, we shall have frequent need to use time units, principally the *second* (and the *microsecond*, μs, 0.000,001 s), and the *hour*.

Since the standard units of the metric system are the meter, the kilogram, and the second, it is often referred to as the *meter-kilogram-second* (*mks*) system. Recently, the term *SI-metric* system is often used as mentioned earlier in Sec. 2.2. References may sometimes be seen to the *centimeter-gram-second* (*cgs*) system, a designation used a few decades ago when the centimeter and the gram were looked upon as being basic units for scientific purposes. The cgs system is still used in some laboratory research, particularly in chemistry.

In like manner we sometimes refer to the English system as the *foot-pound-second* (*fps*) system. In some branches of engineering, a different system of English units, the *foot-slug-second* system, is used, which we shall introduce later.

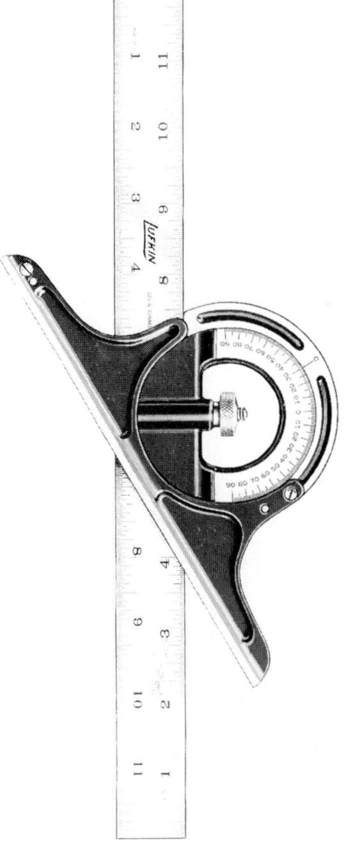

Fig. 2.5 An ordinary steel rule for English-system use. This model is fitted with a bevel protractor for measuring angles. (Lufkin Rule Co.)

SOME METHODS OF MEASUREMENT

2.5 ■ Measuring Instruments—Industrial and Scientific

The basic length-measuring instrument of the mechanic or machinist is the steel rule. Figure 2.5 shows a machinist's steel rule combined with a bevel protractor, graduated in thirty-seconds and sixty-fourths of an inch. Rules are frequently combined with straightedges to make squares (if the angle is 90°) or to measure any given angle. Some rules are marked off in decimal fractions, that is, in tenths and hundredths of an inch.

Rules of similar design are graduated (i.e., marked off) in the metric system and these will be seen more and more frequently as metrication is adopted in U.S. manufacturing industries.

Surveyors use long steel tapes marked off in meters in the SI-metric system and in feet in the English system. The English-system surveyor's tape is graduated in tenths and hundredths of feet, not in inches.

The best accuracy possible with an ordinary steel rule is of the order of 0.01 in., and this is far too crude for most machine work. For accurate work, the machinist makes use of the *micrometer caliper*. (Note that this is the same spelling as the word for 0.000,001 m given in Table 2.1. The pronunciation is different, however. The word used for a caliper is pronounced *micrómeter*, and the unit of length is pronounced *mícrometer*.) The word *micrometer* literally means *small* (*or fine*) *measurement*.

The micrometer caliper—a "workhorse" instrument in industry.

Figure 2.6 shows a standard micrometer caliper capable of reading to 0.001 in. The entire instrument is rugged in construction to ensure uniformly precise measure-

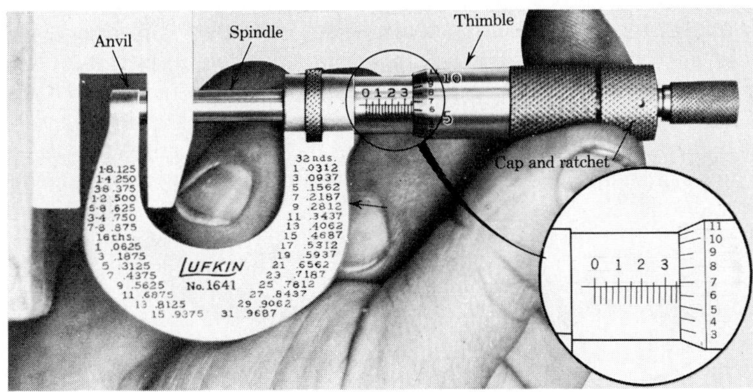

Fig. 2.6 A standard micrometer caliper for English-system measurements. Note the names of the parts. Machinists call this instrument a "mike." The reading here is 0.357 in. (Lufkin Rule Co.)

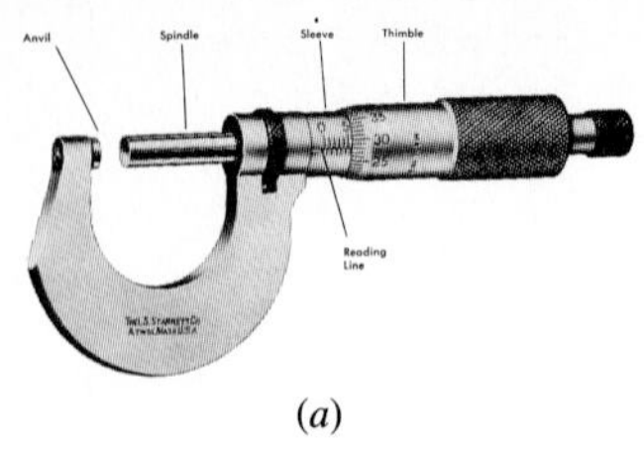

(*a*)

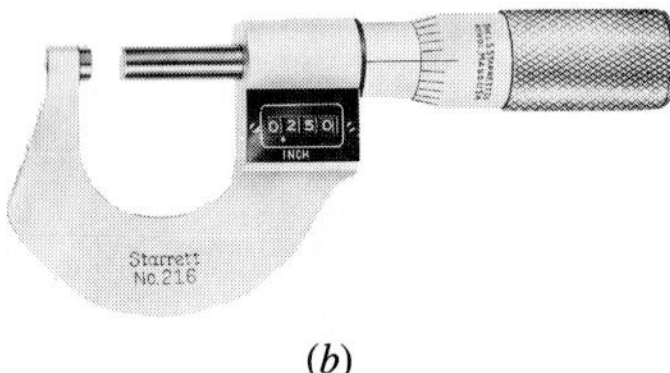

(*b*)

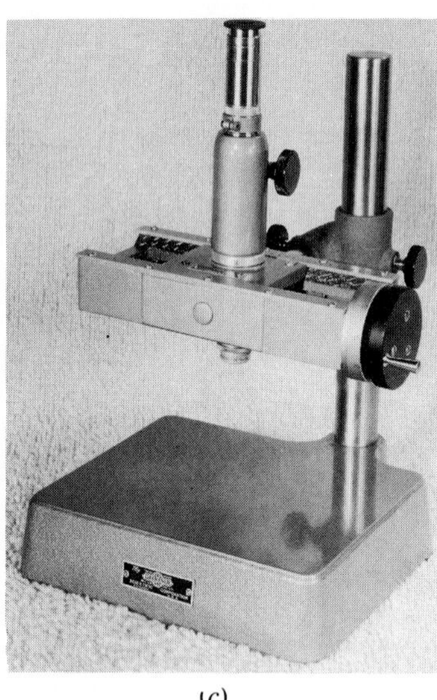

(*c*)

Fig. 2.7 (*a*) A micrometer for metric system readings, capable of reading to 0.01 mm. The illustration shows a reading of 5.78 mm. (The L.S. Starrett Co.) (*b*) English-system micrometer with direct dial reading feature, accurate to the nearest thousandth of an inch. (The L.S. Starrett Co.) (*c*) The micrometer principle employed on a Gaertner micrometer slide comparator, or traveling microscope. (Welch Scientific Co.)

ments. The work to be measured is inserted between the *anvil* and the *spindle*. The *thimble* is then turned clockwise; this turn advances the *spindle* toward the *anvil*. The spindle is threaded with 40 threads to the inch. The *pitch* of the screw is then $\frac{1}{40}$ in. = 0.025 in. Consequently the main scale (on the *hub*) is graduated in divisions of 0.025 in., and one full turn of the thimble advances it and the spindle one main scale division. The thimble scale has 25 equal divisions, so that measurements to the nearest 0.001 in. can be read directly. The ratchet is to be used as the measuring faces close on the work to avoid the possibility of damage to the instrument. The reading, as shown in the figure, is 0.357 in. to the nearest thousandth. To get this value, read on the main scale 0.3 in. plus 0.05 in. = 0.35 in. Add to this the 0.007 from the thimble scale to obtain the reading of 0.357 in. Some micrometer calipers have scales and screw pitch so designed that direct readings to 0.0001 in. can be taken.

Micrometer calipers are also commonly used for metric system measurements. Figs. 2.7*a* and 2.7*b* show two of these instruments. Micrometer calipers have many uses in the physics laboratory, as well as in the machine-tool industry.

The micrometer principle is used in many other measuring instruments, for example the depth "mike" for use in measuring the depth of cuts or holes in machined parts; and on the *micrometer comparator* or traveling microscope (Fig. 2.7*c*).

The *vernier caliper* is another instrument capable of considerable accuracy. Figure 2.8*a* shows a student-type vernier caliper for metric-system measurements. The *main scale* is marked off in centimeters, each of which is divided into 10 parts, or millimeters. The *vernier scale,* which is movable, contains 10 divisions also, but each one is nine-tenths as long as the smallest main-scale division. Therefore the 10 divisions of the vernier scale have the same length as the 9 divisions of the main scale (see Fig. 2.8*b*). Each vernier division is then $\frac{1}{10}$ mm shorter than a main-scale division. This difference is called the *least count* of the instrument.

In taking a measurement, the jaws are brought against the work. The distance between the zeros of the main and vernier scales is equal to the distance between the jaws for any setting. The main scale is therefore read at the zero of the vernier scale. Referring to Fig. 2.8*c*, read from the main scale 0.6; count over on the vernier scale (two divisions) to find that point where the two scales coincide. The reading is 0.62 cm. Similarly, Fig. 2.8*d* reads 0.37 cm. For both micrometer and vernier calipers, the zero should always be carefully checked before beginning to take measurements. An English system vernier caliper is shown in Fig. 2.9*a*, and a scale-reading exercise is given in Fig. 2.9*b*. Can you verify the reading as 0.883 in.?

Many measurements are of such a nature that they cannot conveniently be determined by either micrometer calipers or vernier calipers. For example, depths of certain machine cuts and grooves would be difficult if not impossible to measure with such instruments. Furthermore, where continual checking of a dimension is necessary, as on every tenth part produced in a production run, a speedier method is required, but no sacrifice in accuracy can be permitted. The *dial indicator,* shown in Fig. 2.10, is a specialized measuring device which can be assembled in a variety of ways to measure accurately and speedily dimensions difficult to determine otherwise. Measurements on a dial indicator are taken with respect to some fixed point or reference plane such as a lathe bed or a surface plate, and this arbitrary *zero* can be reset by the operator at any time. A variation of the dial indicator is the *comparator gauge,* which automatically registers how much a certain part is over or under the required dimension.

2.6 ■ Calibrating Measuring Instruments

Dial indicators, micrometer and vernier calipers, and comparators can and do get out of adjustment. Accurate technical work demands frequent comparison of these instruments against standards whose accuracy is unquestioned. This process is called *calibration*. Modern, well-equipped shops make use of a *surface plate* and *gauge blocks* for periodic calibration of all shop instruments. The surface plate is a heavy block of high-grade steel or of stone with a smooth surface ground as flat as possible.

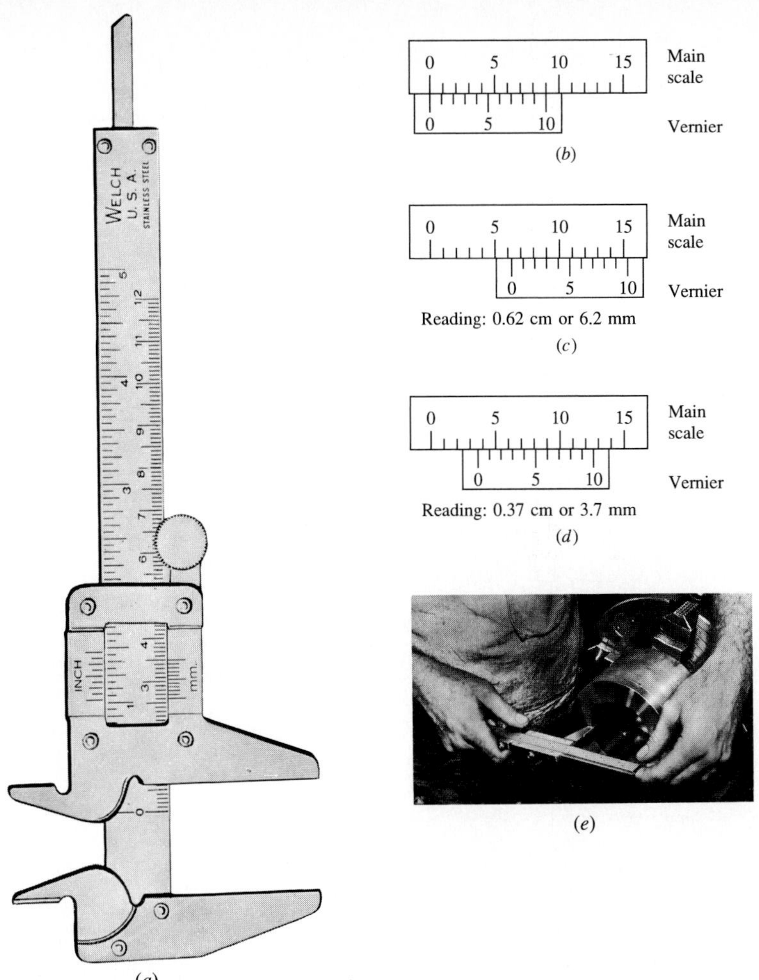

Fig. 2.8 Vernier calipers. (*a*) Student-type metric vernier caliper. (Welch Scientific Co.) (*b*) Vernier scale. Note that 10 divisions on the vernier scale span exactly 9 divisions of the main scale. (*c*) and (*d*) Exercise in reading vernier scales. The readings are: (*c*) 6.2 mm; (*d*) 3.7 mm; (*e*) Machinist's vernier caliper being used to check inside diameter of work on a lathe. (The L.S. Starrett Co.)

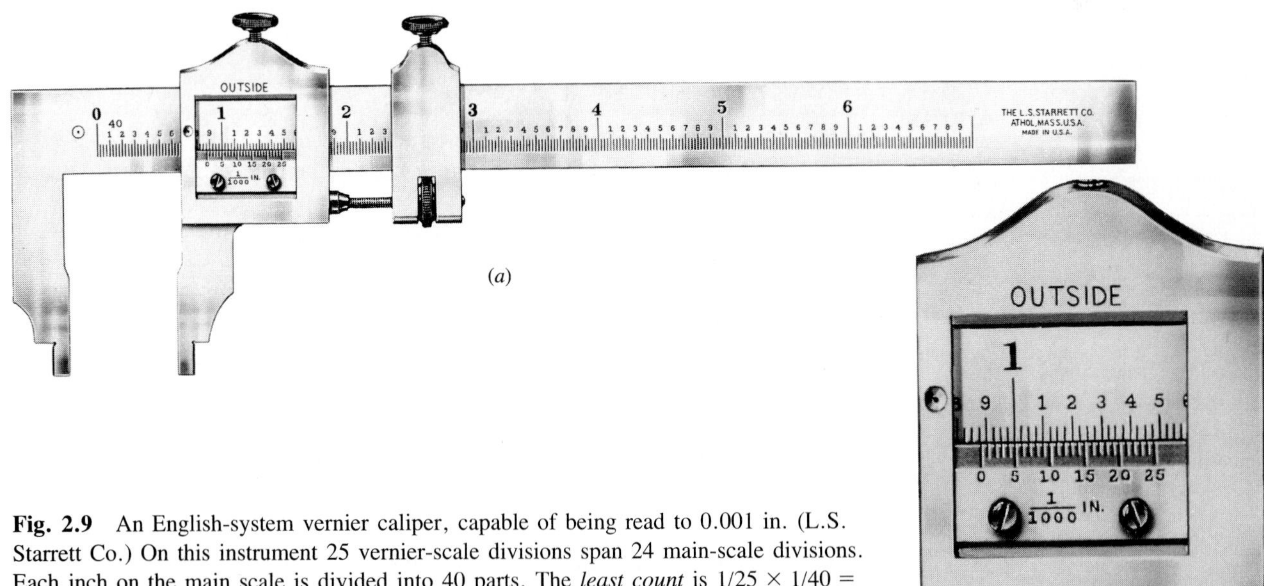

Fig. 2.9 An English-system vernier caliper, capable of being read to 0.001 in. (L.S. Starrett Co.) On this instrument 25 vernier-scale divisions span 24 main-scale divisions. Each inch on the main scale is divided into 40 parts. The *least count* is $1/25 \times 1/40 = 1/1000$. (*b*) Scale reading exercise for an English-system vernier caliper. The reading here is 0.833 in.

One typical full set of gauge blocks consists of 81 blocks, with which it is possible to obtain more than 100,000 practical combinations of dimensions. Figure 2.10 shows a complete set for a limited range of measurements, and also one application of the blocks.

Gauge blocks are made of case-hardened steel, carefully machined, finely ground and lapped, and with a surface finish so supersmooth that when "squeegeed" together they cling as if they were magnets. (This molecular attraction is called *cohesion* and is explained in Chap. 10.) Gauge blocks are manufactured to specified tolerances. Accuracy to 0.000,001 in. is not uncommon. They are calibrated by using the wavelength of a standard light source. They must be used at a fixed standard temperature to avoid errors due to expansion. Gauge blocks can be assembled in various jigs and with many accessories to perform a variety of precision measurement tasks.

Measuring instruments of great precision—gauge blocks, optical flats, and lasers.

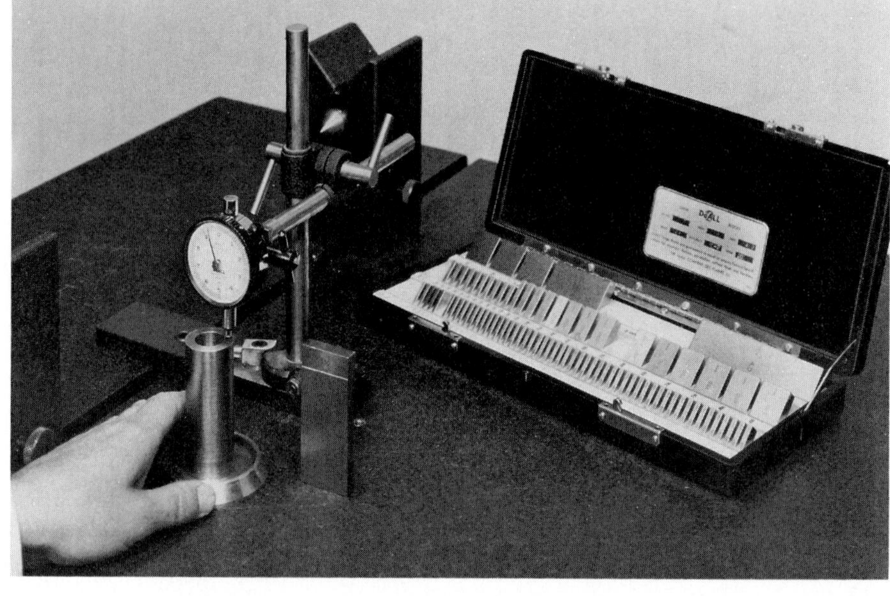

Fig. 2.10 A dial indicator in use with a set of gauge blocks. A steel bushing, selected at random from a production run, is being checked by the dial indicator, which in turn, is calibrated by the gauge blocks. (DoALL Company)

Electronic and electromagnetic-wave distance-measurement methods—radar and sonar.

Optical and Laser Methods The most accurate length measurements use principles involving the wavelength of light. For example, the orange-red line of krypton-86 has a wavelength of only a little more than one-half a millionth of a meter (more precisely 0.6058 millionths of a meter) in vacuum. Since the krypton-86 radiation can be produced in any properly equipped laboratory, its use for standardizing measurements is very important (Fig. 2.11). Well-adjusted helium-neon gas *lasers* are also showing promise of great precision, with convenience unmatched by the krypton-86 lamp. Laser beams are used for precision alignment of machine parts and alignment of the elements of construction projects. (See Chaps. 22 and 34.)

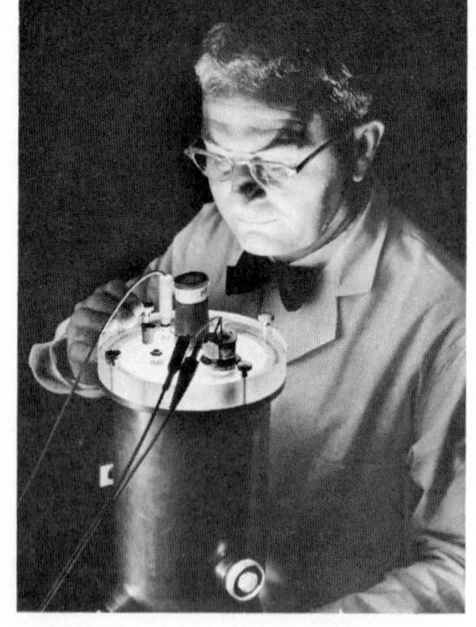

Fig. 2.11 The krypton-86 lamp at the National Bureau of Standards. The wavelength of the orange-red light emitted by Kr^{86} is precisely 0.6058 millionths of a meter or 0.0000238 in. Krypton lamps are often used to calibrate other instruments for precise measurement. (National Bureau of Standards)

Optical flats are round disks of highly polished glass with surfaces made as flat as is humanly possible by the most advanced scientific and technical methods. They are used to standardize or *calibrate* gauge blocks and other precision measuring instruments, using the principles of light-wave interference.

2.7 ■ Measuring Longer Distances

Before about 1930 only two reasonably accurate methods of measuring long distances were available—actual measurements by surveyors' tapes and associated mathematical calculations; and the use of astronomical observations and calculations (navigation). Developments in microwave technology, radio ranging (*radar*), and sound ranging (*sonar*) in recent decades have made accurate measurement of long distances commonplace. The theory and technique of radar and sonar measurements will be explained in later chapters. Also, later sections will discuss the measurement of fluid flow, temperature, mechanical energy, heat energy, electron flow, magnetism, sound intensity, light intensity, and atomic and nuclear radiation.

2.8 ■ Measurement of Mass

Masses are measured by *balances* of various designs. The *beam balance* of Fig. 2.12*a* uses the principle of the lever to compare an unknown mass (left pan) with a known mass in the pan at the right end of the beam. The "known mass" is made up of selected masses from a laboratory set. Actually, even though *masses* are being compared, the *force of gravity* is involved in the process. The assumption is that since the known mass and the unknown mass are both in essentially the same location on an equal-arm balance with respect to the earth and its gravitational field, the action of gravity on both is the same and therefore when the forces of gravity on both are equal (that is, when the instrument is "balanced"), the masses are equal. Modern laboratory balances are usually of the digital read-out type, as illustrated in Fig. 2.12*b*.

A different kind of balance, called an *inertial balance,* is not dependent on gravitational forces, and it measures mass by utilizing one of the essential properties of mass, namely *inertia*. The relationships among mass, inertia, and force will be the subject of a later chapter. It is not necessary that you have a complete understanding of mass, force, and inertia at this point.

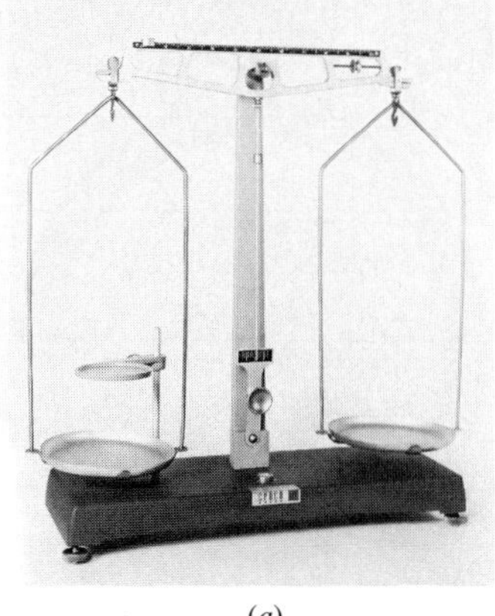

(*a*)

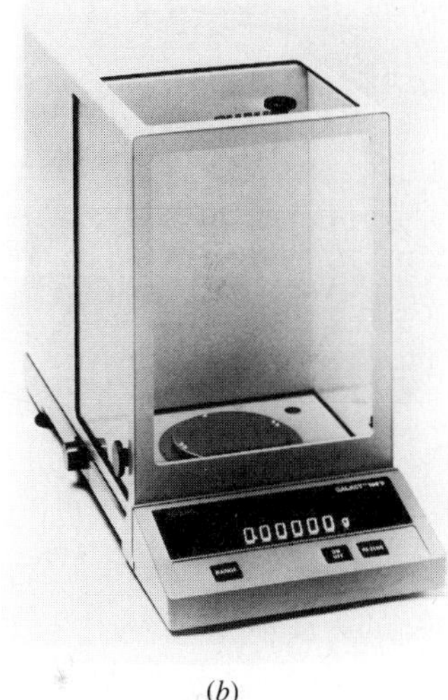

(*b*)

Fig. 2.12 Instruments for the determination of mass. (*a*) A standard laboratory balance (now replaced, in many laboratories, by modern digital read-out balances). (*b*) An electronic, digital read-out, analytical balance capable of readings to the nearest 0.01 mg. (Ohaus Scale Corporation)

2.9 ■ Time Measurement

The standard instruments for measuring time are clocks and watches. *Stop watches, stop clocks,* and *electronic timers* are standard physics laboratory items (see Fig. 2.13). Timing devices utilizing the natural frequency of vibration of crystals such as quartz are commonly used in science and industry and in many watches for personal use. Extremely accurate "clocks" are based on the electromagnetic vibrations of certain atoms (see Fig. 2.4). Time periods of longer duration such as the *mean solar day* and the *year* are determined by the earth's rotation on its axis and its revolution around the sun, as measured by astronomical methods at research laboratories like the United States Naval Observatory and the Royal Greenwich Observatory at Greenwich, England.

2.10 ■ Conversion of Units

Often the units in which a measurement is expressed or a price is quoted are not the ones we want to deal with. Engineering practice, market traditions, or just personal preference may dictate changing to a different set of units, either in the same measurement system, or from English units to SI-metric units, or vice versa. For example, a U.S. airline crew, metering jet fuel being received in Rome might wish to convert the *liter* (L) quantity into *gallons* (gal). Or, grain sacked in the United States in bags weighing 100 lb (CWT) may be measured in *kilos* (the weight of one kilogram of mass) on delivery in a metric country. Many problems in this book will be stated in such a manner that unit conversions will be necessary. Accomplishing these conversions quickly and correctly is not difficult after a little practice. Remember these guidelines:

Converting units from one measurement system to the other; or to more convenient units in the same system.

1. Units are to be handled according to the rules of algebra. Multiply, divide, and cancel them just as you would x's, y's, or any other algebraic quantity.

2. Consult tables to find equivalencies between pairs of units, for example, 1 km = 0.621 mi; 1 gal = 3.785 liter (L); 1 lb = 0.454 kilo (kilogram weight). (Not used in science.)

3. Note that each of these equivalencies can be rearranged to equal unity; or 1; 0.621 mi/km = 1; 3.785 L/gal = 1; 0.454 kilo/lb = 1.

Now, multiplying or dividing any quantity by 1 does not affect its numerical value, but does afford a method of changing the units. Study the following examples.

(*a*)

(*b*)

(*c*)

Fig. 2.13 Instruments for the laboratory determination of time duration. (*a*) Ordinary stopwatch with one-fifth second intervals. (*b*) Electric stop-clock for laboratory use (one-second intervals). (*c*) Digital timer, accurate to 0.01 second. (Daedalon Corporation)

Illustrative Problem 2.1 An importer is billed \$1.40 per kilo for a shipment of Colombian coffee. Express this billing at a cost in dollars per pound.

Solution Note (as above), 0.454 kilo/lb = 1. Multiplying by the equivalency gives

$$\text{Price per lb} = \frac{\$1.40}{\cancel{\text{kilo}}} \times 0.454\,\frac{\cancel{\text{kilo}}}{\text{lb}}$$

$$= \$00.636 \text{ per lb} \qquad \textit{answer}$$

Illustrative Problem 2.2 A tourist in Germany is buying gasoline. The pump meter reads 52 liters (L) when the tank is full. How many gallons is this?

Solution Note, as above, 3.785 L/gal = 1
Change the units to gallons by dividing by the equivalency,

$$\frac{52\,\cancel{\text{L}}}{3.785\,\cancel{\text{L}}/\text{gal}} = 13.74 \text{ gal} \qquad \textit{answer}$$

Illustrative Problem 2.3 Finally, a slightly more complicated situation. Express a speed of 60 km/h in ft/s. Note that this involves a change from metric to English units, and also involves changes in time units.

Solution Use the necessary equivalencies, as before: 1 km = 3281 ft; 1 h = 3600 s. Then,

$$\left(60\frac{\cancel{\text{km}}}{\cancel{\text{h}}}\right)\left(3281\frac{\text{ft}}{\cancel{\text{km}}}\right)\left(\frac{1\,\cancel{\text{h}}}{3600\text{ s}}\right) = 54.7 \text{ ft/s} \qquad \textit{answer}$$

COMPUTATIONS BASED ON MEASUREMENTS

2.11 ■ Exact and Approximate Numbers

Some numbers, such as *counting numbers,* are *exact,* like the number of wheels on an automobile (four) or the amount of money in your bank account on a given day (\$865.22). But many numbers are the result of measurements, and these are *approximate numbers,* since no measurement is ever absolutely accurate. Most of the numbers we deal with in physics and technology are approximate numbers, since they result from measurements.

Certain other numbers like π (3.1416) and $\sqrt{3}$ (1.732), though not the direct results of measurements, are also approximate numbers, since the results never "come out even" no matter how far the calculation of their values is carried out.

When dealing with measured data one must remember that they are approximations and treat them accordingly. Two approaches to calculations made from approximate numbers are suggested here.

2.12 ■ The Decimal-Place Approach

In the normal processes of multiplication and division the products and quotients obtained frequently have an apparent accuracy indicated by several places of decimals, which is not justified by the data available from measurements.

Suppose it were required to compute the area of a circle whose diameter could be measured accurately to only two decimal places as 2.45 in. The area

How many decimal places should be retained in calculations from measured data?

$$A = \frac{\pi D^2}{4} = \frac{3.1416 \times (2.45 \text{ in.})^2}{4}$$

$$= 4.714363 \text{ in.}^2 \quad \text{to six decimal places}$$

Six-place accuracy for the computed result is of course misleading, since accuracy to only two decimal places was inherent in the measured data. The answer obtained for the area should be rounded off to 4.71 in.^{2}.

As a general rule, *do not retain decimal places which imply greater accuracy than the original measured data justify*. In most industrial calculations, π may be rounded off to 3.14.

2.13 ■ Significant Digits

The *significant digit* approach to calculations from measured data.

Engineers and scientists treat the problem of accuracy in computations from the viewpoint of *significant figures* or *significant digits*. The number expressing the magnitude of a measurement should be recorded with *only as many digits as properly indicate the accuracy of the measurement*. Only *those* digits are *significant digits*. When computations are subsequently made using numbers recorded from such measurements, the significant digits expressed in the final result of the computation should be limited to the number in the *least accurate* of the original data. For example, if the outside diameter of a steel shaft is measured with a micrometer caliper, and 2.42 cm can be read directly off the main and thimble scales, with a third *estimated* digit of 7 from the thimble scale, the reading would be 2.427 cm. Thus, we have three certain digits and one *doubtful* (but reasonable) digit. Only one doubtful digit is retained in recording the results of a measurement.

In computations with measured data or numbers known to be approximate, the following guidelines should be followed:

Rules for dealing with significant digits.

1. In adding and subtracting, do not include columns beyond the first one that contains a doubtful digit.

2. In multiplying and dividing, retain no greater number of significant digits in the final result than were in the *least* accurate of the approximate numbers involved in the calculation. One doubtful digit can be retained in the answer, if desired.

3. Nonzero digits should always be regarded as being significant.

4. Zeros between nonzero digits (as 1.406) are significant.

5. Zeros before nonzero digits serve as placeholders, and are not significant digits. (For example, 0.00365 is expressed to three significant digits, not five. The two zeros are placeholders.)

6. Zeros at the end of a *whole number* are usually not significant. (*Example:* A 75,000-horsepower (hp) rating on a marine engine could easily be in error by plus or minus 500 hp. The stated rating is probably accurate to only two significant figures.)

7. Sometimes, however, zeros at the end of a whole number are significant. If we wish to indicate that the ending zeros in whole numbers are significant, we can write the number using scientific notation (see Sec. 1.4). For example, if we want to indicate that 1 light year = 5,880,000,000,000 statute miles, accurate to four significant figures, we can write 1 light year = 5.880×10^{12} statute miles. Now, since the 0 is to the right of the decimal, it becomes significant.

8. Zeros at the end of decimal fractions are intended to be significant. (*Example:* A recorded measurement of 2.450 cm means that the zero is significant. If it were not significant the result should have been recorded as 2.45.)

Some additional examples:

58,000	Probably correct to only two significant digits
0.0058	Correct to two significant digits
3.7825	Correct to five significant digits
4060	Correct to three significant digits
0.4060	Correct to four significant digits
0.004060	Also correct to four significant digits. (The two zeros immediately to the right of the decimal point serve as placeholders, not as significant digits.)

Illustrative Problem 2.4 A jet plane takes off and maintains a 25° angle of climb at an average speed of 415 mi/h. How far from the takeoff point is it (horizontal distance as measured along the ground) after 3.00 min?

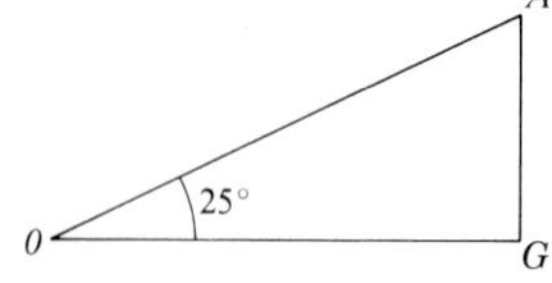

Fig. 2.14 Diagram of aircraft angle of climb for Illustrative Problem 2.4.

Solution Diagram the problem as shown in Fig. 2.14. Point O is at takeoff. OA is the distance traveled through the air, and OG is the horizontal distance traveled. Note that all the given data are *measured data* (approximate numbers).

First, determine the air-line distance traveled.

$$OA = 415\frac{\text{mi}}{\text{h}} \times \frac{3.00\ \text{min}}{60\ \text{min/h}} = 20.75\ \text{mi}$$

retaining four digits *temporarily*.

Now use the cosine function and write

$OG = OA \cos 25$	(Obtain cos 25 as 0.906 from
$= 20.75 \times 0.906$	table in Appendix IV or from a calculator)
$= 18.7995$ mi	if all digits are retained

The data given, however, justify only three significant figures (60 min/h is an exact number, not a measurement). The value of cosine 25° (0.906) is accurate to three significant figures, but the 25° itself is a measurement, so we cannot place any real confidence in the third digit (6) of its cosine. It (the 6) is a very *doubtful* digit.

Therefore the answer should be written to only three significant figures, and is

18.8 mi *answer*

Handling *doubtful digits.*

In this case, due to the uncertainty of measurement of the 25° angle of climb, the third digit is "doubtful."

It should be noted that although the decimal-place method and the significant-figure method are concerned with the same idea, they differ in their approach. The idea behind both is to prevent false or misleading degrees of accuracy from appearing in the results of computations, when the measured data which make up the problem have only limited accuracy. You should be familiar with both methods and use whichever your instructor prefers. In general, we will retain only *one doubtful digit* in the answers to problems throughout the book.

2.14 ■ Computing Devices

It is assumed that students will have access to either a pocket electronic computer or a desk calculator. Slide rules, though not in common use today, could certainly be used, if desired. Longhand calculations are not recommended on two counts: (1) the opportunities for error are too great, and (2) the method is so burdensome and time-consuming that the principles of physics being studied tend to fade into the background.

Electronic calculators often yield answers containing up to 10 digits. That many digits are rarely justified in engineering and technical work. Since most input data are measured and are therefore approximate values, the seemingly very precise answers yielded by such calculators are really a gross overstatement. You should use the guidelines given above to determine how many significant digits to retain from electronic calculator operations.

Electronic pocket calculators yield results with many digits. Discard all digits except those justified by the original data of the problem.

QUESTIONS AND EXERCISES

1. How does a beam balance measure *mass* when it is really the pull of gravity that actuates the beam or lever?

2. Describe how you could measure the height (with somewhat low precision) of a 10-story building without yourself or anyone assisting you leaving ground level. What simple tools and/or instruments would you need? Illustrate your method with a sketch.

3. What is meant by *fundamental units? derived units?* Give examples of three derived units and show how they are derived from fundamental units.

4. Explain in detail how gauge blocks, comparators, dial indicators, and micrometers might be used in ''tooling up'' for the manufacture and inspection of a new line of electric motors for household appliances. Remember that subcontracting, interchangeability of parts, and assembly-line methods will all be involved.

5. Despite the rather slow pace toward metrication in the United States, some industries and commercial and professional fields are changing over quite rapidly. Find out which ones are changing. Also identify some fields where little if any movement toward metrication is apparent. List some reasons for the resistance to change in these latter fields.

6. What is the justification for taking several ''readings'' of a quantity that is to be measured, rather than just one? Why is the arithmetic average or *mean* of the several readings considered to be a better result than any one of the readings alone? Would the average always be more *accurate* than any one of the several readings?

7. Based on some library reading, explain carefully the meaning of the terms *random errors* and *systematic errors,* as applied to measurements in science and technology. How is each related to *accuracy* and to *precision* in measurement?

8. If your laboratory has the equipment, examine carefully a set of gauge blocks, a dial indicator, a micrometer caliper, and a vernier caliper. Using the gauge blocks, calibrate the dial indicator. Also check the accuracy of the micrometer and vernier caliper readings on the 1-in. gauge block. Wipe the gauge blocks clean with a soft cloth before returning them to the case.

9. Explain the principle of the *vernier scale* of a vernier caliper. What is meant by the *least count* of such an instrument?

10. In a competition to perform the largest number of mathematical operations, who wins—a personal computer on line for one hour performing one operation per microsecond; or a student working for 100 years performing one operation per second?

11. Is an auto traveling at a speed of 1 in./ms breaking the legal speed limit of 55 mi/h?

12. The diameter of a fuel injector piston is certified by the manufacturer to be 0.6158 in. +0.0002 in., −0.0004 in. One of these is used in the physics laboratory as an assignment in precise measurement with a micrometer caliper. Group One took five separate measurements and recorded them as follows: 0.6151 in., 0.6152 in., 0.6149 in., 0.6153 in. and 0.6150 in. Group Two, with a different micrometer caliper, recorded these readings: 0.6164 in., 0.6154 in., 0.6157 in., 0.6163 in. and 0.6152 in.

If we assume the manufacturer's certification to be correct, both student groups made errors in the measurement. What kind of error seems to be associated with Group One? With Group Two? If you needed technicians to work in your measurements laboratory, would you employ Group One or Group Two?

PROBLEMS

NOTE: *Use tables on end papers for all data needed.*

Group One

1. The legal speed limit on some U.S. highways is 55 mi/h. What would be the km/h reading of a European car traveling at this speed?

2. Crates of tomatoes from Mexico are stamped 12 kilos. How many pounds is this?

3. If your car's gasoline tank holds 18 gal, how many liters would it take to fill it?

4. The shot used in college track meets weighs 16 lb. What is its mass in kilograms?

5. Upon inspecting the lengths of a production run of rollers for roller bearings, an inspector finds that they range from 1.995 to 2.003 cm. The specification on the blueprint for these rollers reads: 2, +0.001, −0.006. Can he OK the entire lot? Why?

6. A chemical plant ships acids in carboys (large glass bottles in wooden crates) containing 50 L. How many U.S. gallons is this?

7. Which is the greater distance—125 mi or 200 km? Which is the shorter distance—0.001 in. or 0.01 mm?

8. A bushel basket of potatoes is weighed and found to contain 56 lb. How many kilos is this?

Group Two

9. If your motorcycle can travel 55 mi on a gallon of gasoline, how many kilometers should it travel on a liter of the same fuel?

10. One ounce is the equivalent of how many milligrams?

11. Until quite recently prescription druggists expressed the weight of tablets and capsules in *grains* in the English system of measurement. Express the weight of a 5-grain aspirin tablet in milligrams of mass.

12. Convert the following: (*a*) 20 mm to inches, (*b*) 45 cm to inches, (*c*) 0.785 in. to centimeters. (*d*) 5.15 μin. to micrometers, (*e*) 60 nautical miles to kilometers, (*f*) 160 acres to hectares.

13. Express the following in decimal form: (*a*) $\frac{1}{1000}$ in., (*b*) $\frac{1}{10{,}000}$ in., (*c*) 1 μin., (*d*) 1 micrometer (μm), (*e*) $\frac{5}{8}$ in., (*f*) $\frac{22}{7}$.

14. A specification for a part manufactured in the United States reads, ''allowable tolerance: +0.007; −0.003 in.'' Express these figures in micrometers.

15. A cylindrical shaft has the following measurements: Diameter, 0.964 in., length, 3.405 in. Calculate its volume in cubic inches, retaining only that number of significant digits justified by the measurements given (see Eq. 1.14).

16. An airplane is said to fly at Mach 1 when it attains the speed of sound. At or near sea level the speed of sound is about 330 m/s. Express Mach 1 at sea level in miles per hour.

17. The weight for an official table tennis ball must be in the range from 37 to 39 grains. What is the acceptable range in milligrams of mass?

18. The length of 1 minute of arc on a great circle at the earth's surface is one nautical mile—approximately 6076 feet. What is the length of 1 second of arc along a great circle, in kilometers?

19. A mechanic working on a car finds that a $\frac{3}{4}$-in. wrench is needed. At the time, only a set of metric wrenches is available, in sizes 10, 11, 12, . . . , 18, 19, and 20 mm. Which of the metric wrenches would be best to use?

20. The cross section of a standard ''two-by-four'' piece of framing lumber is actually $1\frac{5}{8}$ in. by $3\frac{1}{2}$ in. Express its cross-section dimensions in millimeters.

21. A popular six-cylinder, transverse-mounted auto engine has a volumetric piston displacement of 3.8 L. What is the displacement in cubic inches?

22. A swimming pool measures 18 ft × 32 ft and slopes steadily from a depth of 3 ft at the shallow end to a depth of 8 ft at the deep end. What quantity of water is required to fill it (*a*) in gallons, (*b*) in liters?

23. The wavelength of the light in a certain region of the yellow spectrum band is 5.7×10^{-7} m. Express this length in (*a*) micrometers, (*b*) nanometers, (*c*) microinches.

24. A geosynchronous or ''hovering'' satellite is stationed 22,500 mi above a point on the earth's equator. How many kilometers is this? In one day (that is, for one revolution of the earth on its axis) what is the total distance traversed by this satellite in its (assumed) circular path? (See tables in end papers for the radius of the earth.)

25. A bicycle wheel has a diameter of 30 in. to the outer edges of the inflated tire. How many revolutions will it turn as the rider travels one kilometer?

26. A surface plate is measured with the following results: Length: 65.23 cm, width: 48.74 cm. Calculate its area in square centimeters and indicate the number of significant digits you would retain in the calculated result.

27. The approximate value of the speed of light in a vacuum is 186,000 mi/s in English-system units. Express this value in centimeters per microsecond.

28. The cylinders in a European auto engine are 8.6 cm in diameter and the piston moves up and down a distance of 9 cm. What is the total piston displacement of this 6-cylinder engine (*a*) in liters, and (*b*) in cubic inches?

29. A major league pitcher's throw has been ''clocked'' at 95 mi/h. The distance from where he releases the ball to home plate is 56 ft. How many seconds does it take for the ball to reach the plate?

30. A ship's position is described as being on a certain meridian at 36° 22′ 36″ north latitude. Express this latitude in degrees and decimals of a degree to five significant digits. [NOTE: 1° (degree) of arc = 60′ (minutes) of arc; and 1′ = 60″ (seconds).]

Group Three

31. A small car will go 36 mi on a gallon of gasoline. The car's owner has a cylindrical gasoline storage tank 2 m high and 1 m in diameter. How many miles can the car be driven on the amount of gasoline the storage tank will hold?

32. The instructions on the label for a gallon of exterior house paint claim that 1 gal of the paint will cover 450 ft^2 of surface. What average thickness of the coat of paint is the manufacturer assuming? (Paint does not "soak in" to the surface.)

33. A racing car can travel 6 mi/gal of fuel at racing speeds on the track. The driver would like to compete in 200-mi races with only one stop for refueling. What is the minimum capacity gasoline tank required? If the tank is cylindrical and 42 in. long, what must its diameter be?

34. How many liters will a spherical tank contain if its diameter is 5.3 m? (See Eq. 1.12.)

35. Calculate the total volumetric piston displacement of an eight-cylinder auto engine if the cylinder diameter (bore) is $3\frac{7}{16}$ in. and the length of the piston stroke is $4\frac{1}{4}$ in. Assume each cylinder is a right circular cylinder and use Eq. (1.14).

36. What is the *great-circle* distance in nautical miles between the equator and a point at 38° 42′ N latitude?

37. The velocity of light is 2.998×10^8 m/s. How long (in seconds) does it take light to travel from the moon to the earth, a distance of 384,500 km?

38. A metric system micrometer has two threads per millimeter along the spindle. There are 50 divisions around the thimble scale. What is the reading from the scales as shown in Fig. 2.15?

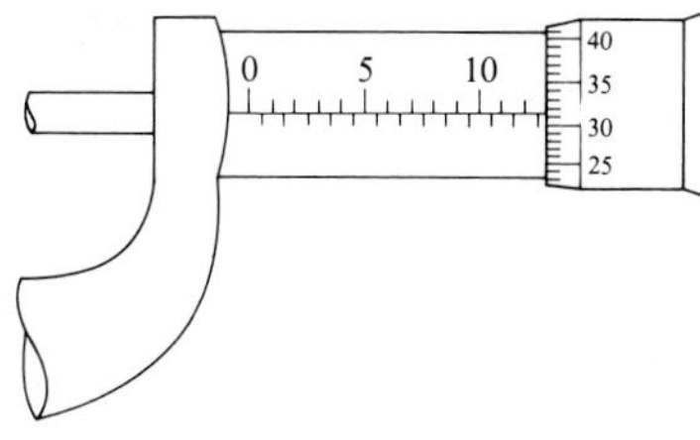

Fig. 2.15 Micrometer scale reading for Problem 38.

39. An English-system micrometer caliper has 50 threads to the inch along the spindle. There are 200 divisions around the thimble scale. To what accuracy can the instrument be read? (See Fig. 2.6.)

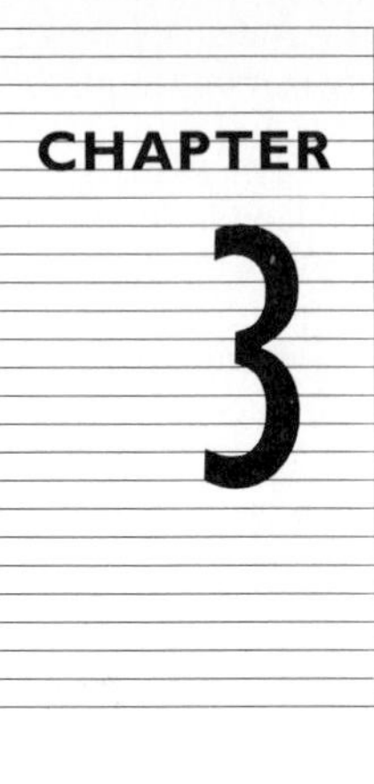

VECTORS AND GRAPHIC METHODS—STATICS

The old Chinese proverb that ''a picture is worth a thousand words'' is often true in the analysis and solution of physics and engineering problems. In some instances methods based on algebra and trigonometry become so involved in mathematical processes that the basic physics and engineering principles are pushed into the background. Many problems related to distance, time, direction, force, velocity, acceleration, and momentum are more readily solved by graphic methods than by mathematical analysis. And, in the graphical process the basic principles are more often clarified than is the case with complex mathematical computations. Both *fundamental units* and *derived units* lend themselves to graphical treatment.

In particular, problems dealing with travel on land and sea and in the air; with velocity and acceleration; and with forces acting on, and stresses within structures can be more easily understood and solved by using graphic methods than by using analytic methods. In such problems much use is made of arrows drawn to a definite length and in a stipulated direction.

VECTORS

3.1 ■ Vector and Scalar Quantities

Vector, defined—a quantity possessing both magnitude and direction.

Arrows used in diagrams and graphical solutions are handy devices for representing clearly the relationships of a physical problem. If the arrow is carefully drawn, its direction can indicate the direction of a motion or the line of application of a force. The *arrowhead* may also indicate the endpoint of the travel or the *point of application* of the force. And finally, the arrow itself may be drawn to an exact scale whose length will represent the magnitude, or size, of the distance, velocity, or force which it represents. Such arrows, drawn to careful scale, are called *vectors*.

A *vector quantity* is a quantity which possesses the attributes of both magnitude and direction. Examples of vector quantities are force, displacement, velocity, acceleration, momentum, and the magnetic field.

Scalar, defined—a quantity possessing only magnitude.

A *scalar quantity* possesses magnitude only and has no attribute of direction. Examples of scalar quantities are such things as the volume of the city water tank, auto production for the year, your annual earnings, the number of cattle on your farm, and the number of beds in the local hospital. In physics, mass, temperature, time, pressure, energy, and density are examples of scalars.

Our treatment here will be only an introduction to the topic, and of all the possible uses of vectors, we will deal in this chapter only with their application to the solution of problems involving changes of position (displacements), forces, and velocities. Some other applications will be presented in later chapters. In any one vector diagram all vectors must be of the same kind, that is, *force* vectors, or *displacement* vectors, or *velocity* vectors.

3.2 ■ Vectors in Displacement Problems—Vector Notation

If, when you and a friend part on a street corner, he tells you that he is going to walk 12 blocks, you have no idea where he will be at the end of his walk. After all, he could walk any combination of streets and directions he chooses. He might even arrive back at the starting point. His "12 blocks" is a *distance,* not a displacement. Distance alone is not a vector quantity, since no direction is specified.

But if he says he will walk 7 blocks *north* and then 5 blocks *east,* then you will know exactly where he will be at the end of his 12-block walk. This time he has provided enough information so that you can determine the net *displacement* he will experience from his initial location where you part on the street corner, to his final location at the end of his 12-block walk.

Displacement is a vector quantity whose magnitude may or may not equal the distance moved. Displacement is a net change of location.

An object that moves from some initial location to some final location experiences a displacement that is a vector quantity.

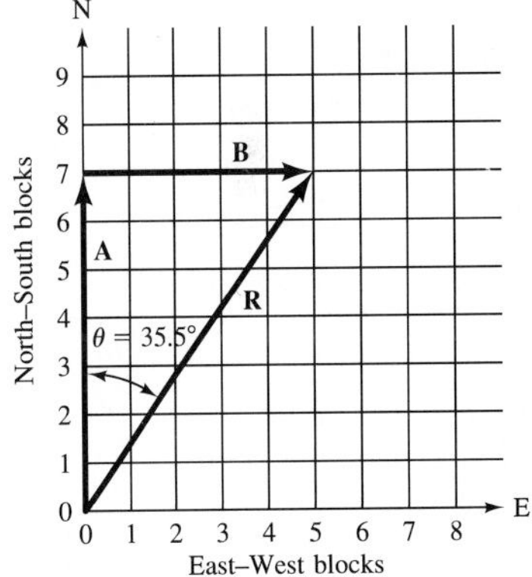

Fig. 3.1 Vector diagram for a 12-block walk. $A + B$ is the *distance* walked, but the vector **R** represents the *displacement*.

The *magnitude* of the displacement vector is the straight-line distance from the initial location to the final location. The *direction* of the displacement vector is given by an arrow that points from the initial location to the final location.

It is useful to display the information about the 12-block walk in a diagram. As suggested in the discussion above, it is convenient to use an arrow to represent a vector in a diagram. We will also use boldface letters such as **V** or **R** to represent vectors. The same symbols, but not in boldface, such as V or R, will be used to represent the *magnitudes* of the vector quantities.

A vector diagram for the 12-block walk is shown in Fig. 3.1. For the first 7 blocks of the walk, the person experiences a displacement given by the vector **A** in the diagram. For the second 5 blocks the displacement is given by the vector **B**. Note that for each of these two segments of the trip, the displacement vector points from the initial location to the final location *for that segment*. Also, for each segment of the trip, the length or magnitude of the displacement vector represents the distance walked, since the person walked in a straight line for each segment. But, for the entire 12-block walk the displacement is given by the vector **R** (called the *resultant*), whose magnitude and direction can be obtained graphically as follows.

A scale drawing of the given factors in the problem is shown in Fig. 3.1. The values for the magnitude of **R** and for the direction angle θ can be obtained by taking measurements with a scale and a protractor. The larger the diagram and the more carefully it is drawn, the more accurate the values for R and θ will be. Scaling off the value of R should give you 8.6 blocks; and measuring the angle θ should result in 35.5° (east of north). Vectors are shown in boldface type, scalars (magnitudes) in ordinary type.

For this example, the total distance walked is greater than the magnitude of the displacement vector.

These results can also be obtained mathematically from use of the Pythagorean theorem, since the vectors **A, B,** and **R** form a right triangle. Solving for the magnitude of the displacement:

$$\begin{aligned} R^2 &= A^2 + B^2 \\ &= (7 \text{ blocks})^2 + (5 \text{ blocks})^2 \\ &= 74 \text{ blocks}^2 \\ R &= \sqrt{74 \text{ blocks}^2} = 8.6 \text{ blocks} \qquad \textit{answer} \end{aligned}$$

The direction of the displacement vector **R** can be described by specifying the angle that it makes with the vector A.

Remember that $\tan^{-1}$ means "the angle whose tangent is". Use the $\tan^{-1}$ or the Inv/Tan key on your calculator.

Since

$$\tan\theta = \frac{B}{A} = \frac{5}{7} = 0.7143$$

$$\theta = \tan^{-1}(0.7143) = 35.5° \qquad \textit{answer}$$

3.3 ■ Vector Arithmetic

You learned how to do arithmetic with scalar quantities in elementary school. For many of the topics in this book you will need to be able to do *arithmetic with vectors*. Vector arithmetic is different from scalar arithmetic in a number of ways. For example, in scalar arithmetic, addition and subtraction are easier than multiplication and division. However, the procedure for multiplying a vector by a scalar, as explained below, is a simpler mathematical operation than either vector addition or subtraction. An ability to do the vector arithmetic discussed in the sections below will be essential for understanding some of the ideas and for solving some of the problems in later chapters.

3.4 ■ Addition of Vectors by Graphical Methods

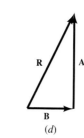

Fig. 3.2 Vector addition by graphical methods.

The addition of vectors is a more complex process than the addition of scalars because vector addition must take into account both the magnitudes and the directions of the vectors being added. The graphical methods for vector addition presented here will use the arrow notation introduced in the discussion of displacement vectors.

Consider the problem of adding the vectors **A** and **B** in Fig. 3.2*a*. Since a vector is completely specified by its magnitude and direction, it is acceptable to redraw a vector at different locations. For example, the vectors **A** and **B** in Fig. 3.2*b* are the same two vectors as in Fig. 3.2*a*. This property of vectors can be used to describe a procedure for adding the vectors **A** and **B**.

Draw the vector **B** so that the tail of **B** touches the head of **A**. The vector sum of **A** and **B** is obtained by drawing a vector arrow whose tail touches the tail of vector **A** and whose head touches the head of vector **B** as shown in Fig. 3.2*c*. Call this third vector **R**. We see that, *vectorially*

$$\mathbf{A} + \mathbf{B} = \mathbf{R}$$

You should note that the displacement problem of Fig. 3.1 was solved by a vector diagram of this type.

Now draw the vector **A** so that the tail of **A** touches the head of **B**. The sum of **B** and **A** is obtained by drawing a vector arrow whose tail touches the tail of **B** and whose head touches the head of **A** as shown in Fig. 3.2*d*. This is also the vector **R**. From this vector diagram we read

$$\mathbf{B} + \mathbf{A} = \mathbf{R}$$

There is evidence here of a general property of vector addition:

$$\mathbf{A} + \mathbf{B} = \mathbf{B} + \mathbf{A}$$

That is, vectors can be added in any order and the sum is the same.

This graphical method for vector addition is also called the *vector polygon method.*

The Vector Polygon Method of Vector Addition. Figure 3.3 shows how three or more vectors can be added by the vector polygon method to obtain the resultant **R**. The vectors **S, T, U,** and **V** (Fig. 3.3*a*) are added successively tail to head, and then the resultant vector **R** is drawn from the tail of the first (**S**) vector to the head of the final (**V**) vector (Fig. 3.3*b*).

The Parallelogram Method. A third method of adding vectors is to use a parallelogram for the vector diagram. Figure 3.4*a* shows two vectors **P** and **Q**. To add them place both tails together at O as in Fig. 3.4*b* and then draw a line parallel to **Q** from the head end of **P** and a parallel to **P** from the head end of **Q**. These parallels will meet at X, forming a parallelogram. From the origin at O draw the resultant vector $\mathbf{OX}(= \mathbf{P} + \mathbf{Q})$. The magnitude of **OX** can be scaled off the diagram and the angle XOQ can be measured as before.

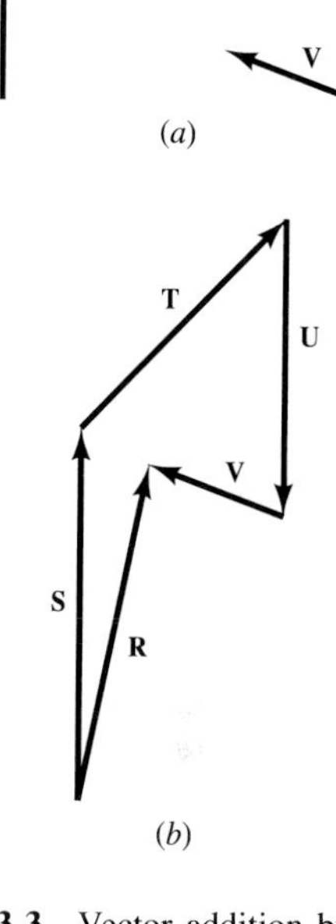

Fig. 3.3 Vector addition by the vector polygon method.

Illustrative Problem 3.1 A ship departs from port A (see Fig. 3.5) and steams due north for 650 nautical miles to point B, then turns northwest (45° to the left of due north) and steams 400 miles to C. Here it changes course again and steams southwest for 550 miles to point D, then steams 300 miles due east to point M, where it receives a radio message to return to home port at A. How far away and in what direction is home port?

Solution First sketch the problem, labeling all displacements (magnitude and direction) as shown in the diagram. Then plot the vector diagram on graph paper very carefully, choosing a scale which will nearly fill the sheet. The vector **R** represents the resultant displacement *from home port*. Careful measurement of angles and distances, and conversion to your scale should find that **R** represents about 660 nautical miles. Measure angle SMA with a protractor to determine the *direction of home port* from M. It is about 34°, so the course to home port is 34° east of south. Later, a method of referring all directions in air and sea navigation to true north will be explained.

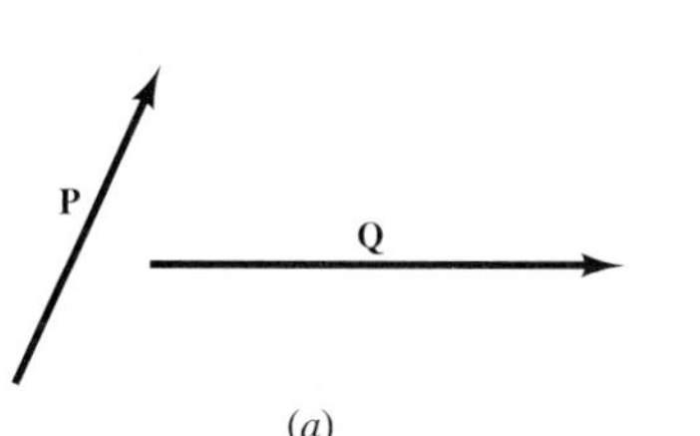

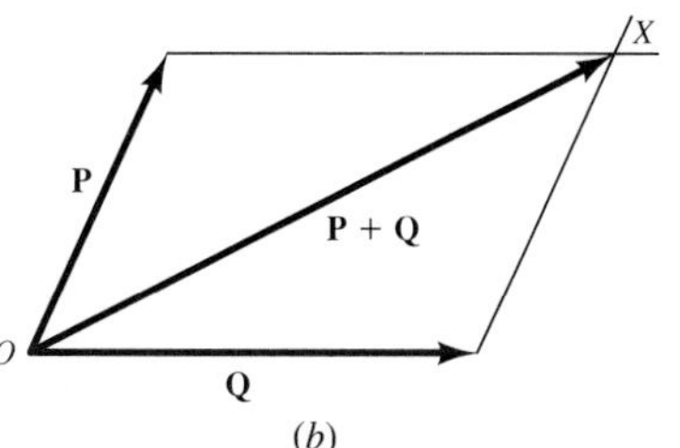

Fig. 3.4 The parallelogram method of adding vectors.

VECTORS APPLIED TO FORCE PROBLEMS

3.5 ■ The Parallelogram of Forces

Two forces AB and AD act at a common point A as shown in Fig. 3.6. Forces that act at a common point are said to be *concurrent*. The vector **AB** represents a force of 50 lb and is plotted 5 units long, each unit representing 10 lb. Vector **AD** represents a force of 95 lb and is therefore 9.5 units long. **AB** acts at 35° with the vertical and **AD** at 110° with the vertical. What are the magnitude and direction of the *net effect* (*resultant*) of these two forces? The resultant is found as follows. From B draw BE parallel to AD, and from D draw DF parallel to AB. (Use a set of parallel rules.) These lines intersect at C, and with the two original vectors, form the parallelogram $ABCD$. The diagonal AC of this parallelogram represents, to scale, the resultant **R** of the two given forces. Measuring **AC** and the angle YAC gives **R** as a force of 118 lb acting at an angle of 87° with the vertical.

3.6 ■ Equilibrium of Concurrent Forces

If, instead of finding the net effect, or resultant, of the two forces AB and AD, it is desired to determine a single force which will balance or cancel them out, we may reason as follows. If vector **AC** (Fig. 3.6) is the *net effect* of the two forces, then a single force equal in magnitude and *opposite in direction* to AC will cancel the net effect of the two forces. This canceling force is of course just **AC** turned around, or **CA.** Such a force is called an *equilibrant*. Its magnitude is the same as that of vector **AC,** but its direction is 267°, 180° farther around clockwise from the vertical (000°).

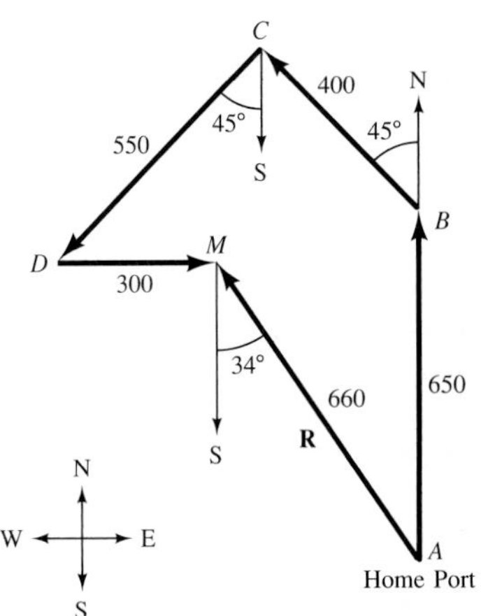

Fig. 3.5 Vector diagram for Illustrative Problem 3.1, a navigation problem at sea.

Concurrent force systems present in building structures must be in equilibrium.

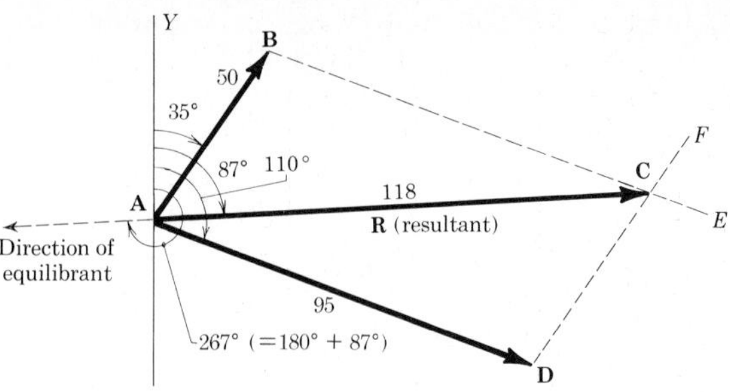

Fig. 3.6 Vectors applied to the solution of a force problem—the parallelogram of forces.

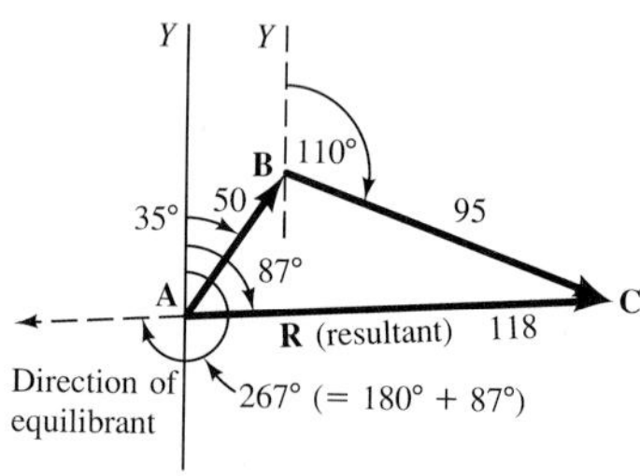

Fig. 3.7 Force systems in *equilibrium—resultant* and *equilibrant.* Determining resultant and equilibrant by the vector triangle method.

An *equilibrant* is a single force of such magnitude and acting in such a direction as to cancel out the net effect of a force system, or in technical terms, to put the force system into *equilibrium*. When a force system is in equilibrium, there is no net *unbalanced* force and consequently, no acceleration of the body on which the force system acts will occur. Conversely, a body subjected to the action of a force system which is not in equilibrium will undergo an acceleration proportional to the magnitude of the resultant of the forces and in the same direction as the resultant. *The equilibrant is always equal in magnitude to, but opposite in direction from, the resultant.*

3.7 ■ The Vector Triangle

Another, and sometimes more convenient, method of solving problems of concurrent forces is by means of the *vector triangle*. Take the same two forces as were used in Fig. 3.6. Draw **AB** as before representing the 50-lb force at 35° with the vertical (see Fig. 3.7). From the arrowhead end of this first vector, draw the vector **BC** representing the 95-lb force at 110° from the vertical. Now note carefully that if the triangle is *closed* by drawing the vector **AC**, we get **R**, the *resultant,* which, when measured, gives 118 lb at an angle of 87° with the vertical, the same answer obtained from the parallelogram method. If, on the other hand, we choose to close the triangle by drawing the vector **CA**, we get the *equilibrant,* which, as before, is 118 lb at 267° from the vertical.

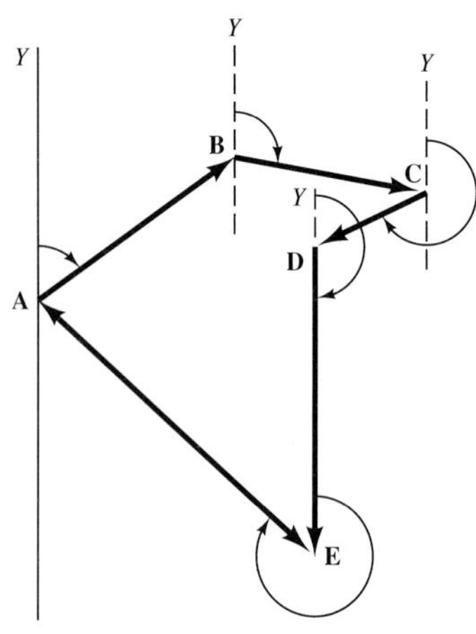

Fig. 3.8 The polygon of forces. When a system of coplanar concurrent forces is in equilibrium, the vector polygon will close.

3.8 ■ The Vector Polygon

Sometimes more than two concurrent forces act at a point and it is desired to find the resultant or the equilibrant. The method is merely an extension of the vector triangle method. Figure 3.8 shows four concurrent forces represented by the plotted vectors **AB, BC, CD,** and **DE**. The resultant is **AE**, *drawn from the starting point to the arrowhead end of the last vector*. The *equilibrant* is obtained by drawing **EA** *from the end of the last vector and "going home" to the starting point.*

René Descartes (1596–1650) made many contributions to the development of physics and mathematics. His concept of representing mathematical relationships on a grid with *x* (horizontal) and *y* (vertical) axes is one of the greatest single advances in the history of mathematics. Ever since Descartes' time, mathematicians, scientists, and economists have plotted "curves" using *Cartesian coordinates.*

3.9 ■ Resolution of Vectors into Rectangular Components

The solution of certain problems in mechanics is made easier by determining the so-called *rectangular components* of a vector. A *component* of a vector is another vector which indicates the net effect of the original vector in a specified direction. For example, in Fig. 3.9, the force vector **F**, which makes an angle θ (Greek *theta*) with the horizontal (x) axis, has components along the x and y axes, $\mathbf{F}_x$ and $\mathbf{F}_y$, respectively. $\mathbf{F}_x$, the component along the x axis, is obtained graphically by merely dropping the perpendicular *PA* from the end of the original vector to the x axis. $\mathbf{F}_y$ is obtained in a similar manner by constructing *PB* perpendicular to the y axis.

Analytically, the values of $\mathbf{F}_x$ and $\mathbf{F}_y$ are readily obtained from the simple trigonometry of the right triangle, as

$$\mathbf{F}_x = F \cos \theta \qquad \text{and} \qquad \mathbf{F}_y = F \sin \theta \tag{3.1}$$

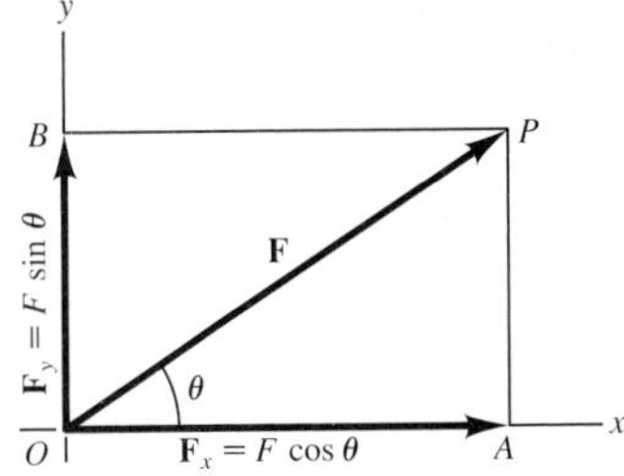

Fig. 3.9 Resolution of a force vector into rectangular (vertical and horizontal) components.

Many force problems are readily solved by resolving a known force into its components in specified directions. Two examples will be given, one rather simple, the other somewhat more complex.

The Lawnmower Problem In Fig. 3.10 a man is pushing on the lawnmower handle with a force of 75 lb when the handle is inclined at an angle of 40° with the horizontal. The pushing force can be resolved into its horizontal and vertical components either graphically, as shown in the diagram by the **H** and **V** vectors, or analytically, using Eq. (3.1).

$$\mathbf{H} = 75 \cos 40° = 57.5 \text{ lb}$$
$$\mathbf{V} = 75 \sin 40° = 48.2 \text{ lb}$$

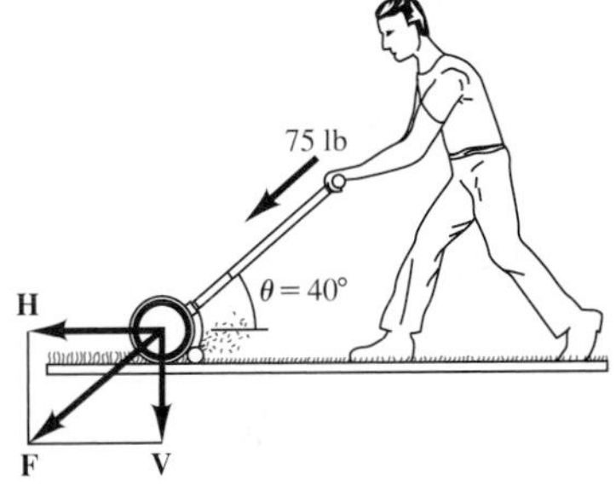

Fig. 3.10 The "lawnmower problem." Force acting at an angle to the desired direction of motion. The force vector **F** is resolved into **H** and **V** components. Only the **H** component pushes the mower.

The Sailboat Problem People from ancient times to the present day have wondered how it is possible for a sailboat to sail more or less against, or into, the wind. This is basically a problem in resolution of forces. Figure 3.11 shows the vector relationships of the problem. The wind is from the west, and the boat is headed in a generally northwest direction. With the sail set at the proper angle, the wind strikes it and is deflected at the same angle, as shown by the arrows labeled *W-W*. The resultant force on the sail, indicated by **F**, is normal (perpendicular) to the sail's surface. The *useful component* of **F** is that component in the desired direction of travel, represented by vector **P**. The other component, athwart the boat, shown by vector **K**, is the force that causes the boat to heel over. This force is counteracted by the force against a centerboard, or keel, by the water beneath the hull, by the rudder, or by the boat's occupants themselves, who lean far out over the windward side of a light sailboat during a heavy blow. With careful adjustment of sails and rudder it is possible for small sailboats to make good progress upwind when sailing only 20° "by the wind," or 20° from the "wind's eye."

In summary, to find the component of any given force in any desired direction, merely drop a perpendicular from the end of the given force vector to the desired line of action. If a 50-lb force is acting at an angle of 70° with the horizontal (Fig. 3.12), and its component along a line making an angle of 20° with the horizontal is desired, the solution is obtained graphically as 32 lb. The magnitude of force vector **F′** could also be obtained analytically from

$$\begin{aligned} F' &= F \cos (70° - 20°) = 50 \text{ lb} \times \cos 50° \\ &= 50 \text{ lb} \times 0.643 \\ &= 32 \text{ lb} \end{aligned}$$

Every vector can be expressed in terms of its components along a chosen set of axes.

STATICS

3.10 ■ Applications of Vectors to Conditions of Static Equilibrium

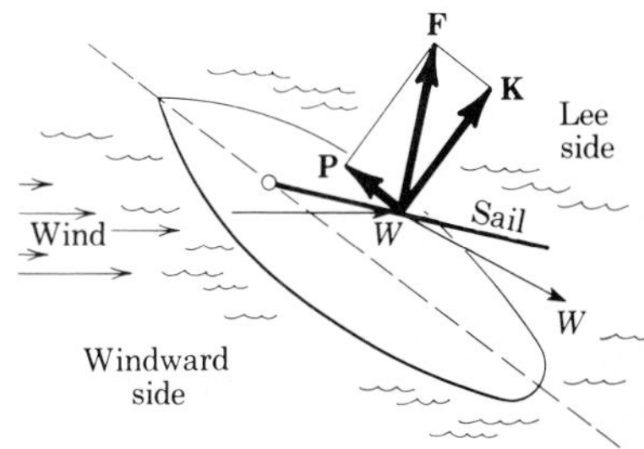

Fig. 3.11 The "sailboat problem." Vector diagram of forces acting on the mainsail of a yacht when sailing "close to the wind."

Buildings and other land-based structures are often designed to support great loads or withstand tremendous forces without undue vibration or bending or sway beyond established safe maximum values. The piers, girders, and beams of a multistory building, for example, are subject to tremendous stresses from the weight of the building and from the weight of the machinery, materials, and people it contains. Another example is provided by the piers, trusses, and cables of a bridge (see Fig. 3.13) which must support the weight of the bridge itself together with the traffic crossing it. Since forces *tend to cause motion,* the problem of the engineer and builder is to distribute the forces and stresses in the members of a structure in such a way that they cancel one another out, leaving each and every member, i.e., truss,

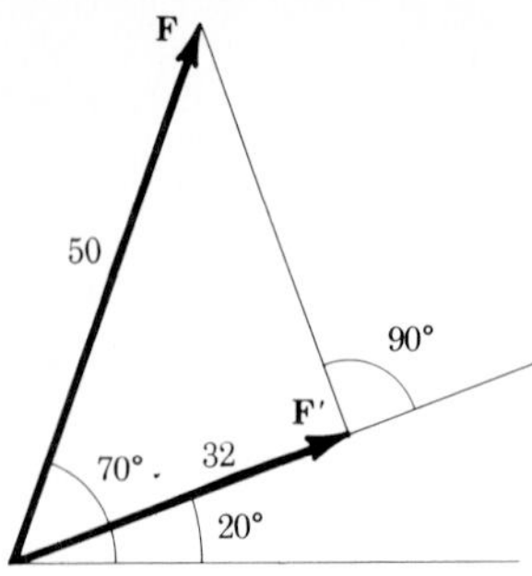

Fig. 3.12 Obtaining the component of a force along a specified direction.

girder, beam, cable, joist, in *equilibrium* under the action of whatever forces may be acting on it—in other words, a condition of *no movement*.

The section of physics that deals specifically with equilibrium conditions and the problems of forces in structures is called *statics*. A complete treatment of statics is found in textbooks of mechanical engineering; we shall discuss here only a few elementary examples that illustrate the principles. Some use of the principle of the lever will be required, and vector methods will be used.

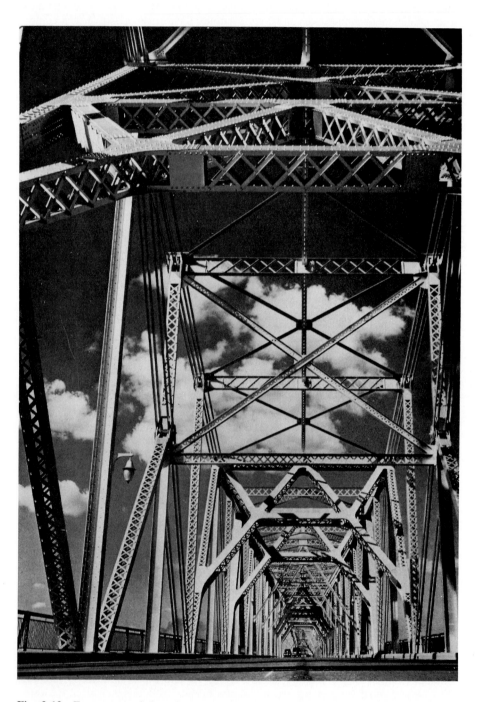

Fig. 3.13 Every truss, girder, pier, and cable of a bridge must be in equilibrium. The resultant of all forces that meet at any common point must be zero. (Reynolds Metals Co.)

Before considering specific equilibrium problems we need to discuss the idea of the center of gravity of an object and the three different kinds of equilibrium for an object.

All the particles that make up any object are attracted to the earth by gravitational forces. The sum of the very large number of such forces is called the *weight* of the object. For the types of problems we are about to consider, it will be convenient to treat the weight of an object as a single force acting at one point. This point is called the *center of gravity* of the object. The weight of an object acts like a force applied at the center of gravity.

For simple, symmetrical objects, the center of gravity is at the geometric center of the object. Sometimes the center of gravity is located at a point inside the object where there is solid mass, or it may be in an inner shell where there is no mass. For example, the center of gravity of a spherical shell like a basketball or tennis ball is located at the center of the ball where there is no solid mass. For some irregularly shaped bodies, the center of gravity is located at a point entirely outside the object, in empty space. As an example of this, when a person bends over 90° at the waist, the center of gravity would be at a point about waist high, but out in front of the hip line, several inches in space.

The location of the center of gravity for irregularly shaped bodies can be determined experimentally as follows: Suspend the object on a pivot (for example, A in Fig. 3.14*a*), and it will orient itself so that its center of gravity (c.g.) lies somewhere on a vertical line that passes through the pivot. If the object is now suspended on pivot B, at some different location on the object (Fig. 3.14*b*), the object will reorient itself so that the center of gravity is again on a vertical line that passes through the pivot. The point of intersection of any two such lines determines the location of the center of gravity of the irregularly shaped object.

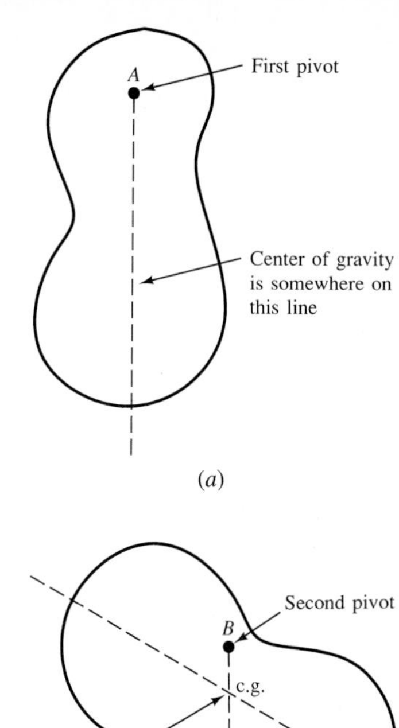

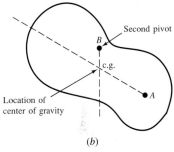

Fig. 3.14 Determining the center of gravity of an irregularly shaped body. An object free to pivot around a point by which it is supported will come to rest in a position with the center of gravity on a vertical line that passes through the support pivot.

The proper balancing of automobile tires is a practical application of the center-of-gravity idea. The mechanic attaches small weights to the rim of the wheel until the location of the *center* of gravity of the wheel with tire mounted on it coincides with the center of the axle on which the wheel will rotate. A balanced wheel and tire will rotate without experiencing undesirable vibrations. Out-of-balance wheels may cause ''shimmy'' of the entire vehicle and rapid wearing out of the tire treads.

In addition to the idea of the center of gravity of an object, there is also a point called the *center of mass* of the object. For the discussions of this chapter, the center of mass and the center of gravity are assumed to be at the same point.

Equilibrium. An object is said to be in equilibrium if the vector sum of the forces acting on it is zero. But not all objects experiencing a resultant force of zero are in equivalent equilibrium situations. The three possible (different) types of equilibrium are called *stable equilibrium, unstable equilibrium,* and *neutral equilibrium.*

Three different kinds of equilibrium are possible for an object at rest.

An object that returns to its original equilibrium location after being moved a small distance in any direction is said to be in stable equilibrium. A marble at rest at the bottom of a bowl, as shown in Fig. 3.15*a*, is an example of an object in *stable equilibrium*.

An object (initially in equilibrium) that continues to move away from its original equilibrium location when it is moved a small distance in any direction is said to be in unstable equilibrium. A marble at rest at the top, center spot on a solid sphere, as shown in Fig. 3.15*b*, is an example of an object in *unstable equilibrium*.

An object that remains at its new location when moved a small distance in any direction is said to be in neutral equilibrium. A marble at rest on a flat horizontal surface, as shown in Fig. 3.15*c*, is an example of an object in *neutral equilibrium*.

In each of these examples involving the marble we assume that the marble is moved in such a way that it stays in contact with the supporting surface.

The equilibrium status of an object can also be tested by considering what happens to the location of the center of gravity of the object when the object is moved a small distance in any direction. An object is in stable equilibrium if its *center of gravity is raised* (that is, if its potential energy is increased) when the object is moved. An object is in unstable equilibrium if its *center of gravity is lowered* when the object is moved. And finally, an object is in neutral equilibrium if the center of gravity is neither raised nor lowered when the object is moved.

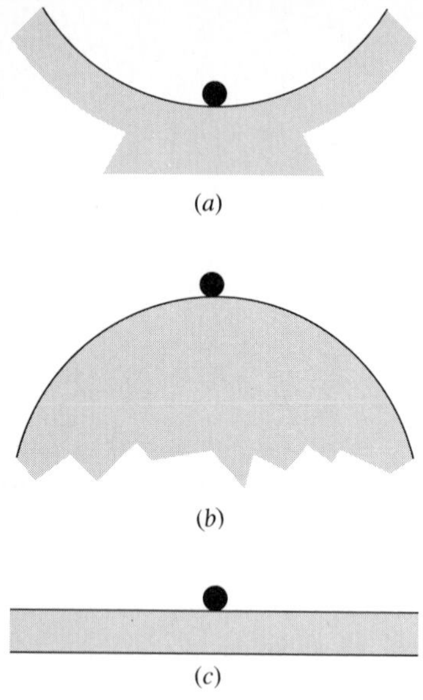

Fig. 3.15 Examples of (*a*) stable equilibrium; (*b*) unstable equilibrium; and (*c*) neutral equilibrium.

Static Equilibrium An object at rest that remains at rest even though it may be experiencing the action of two or more forces is said to be in *static equilibrium*. Any object that you see on earth that is at rest must be experiencing at least two forces, one of which will be its weight. *An object that is experiencing only one force cannot be at rest,* but it is possible for an object experiencing two or more forces to be in static equilibrium.

In the following introductory study of static equilibrium we will consider only objects experiencing *coplanar* forces—that is forces whose lines of action all lie in the same plane. Static equilibrium involving objects experiencing *non-coplanar* forces is an engineering subject well outside the scope of introductory physics.

Two sets of coplanar forces will be considered—*concurrent coplanar forces* and *nonconcurrent coplanar forces*. The lines of action of all the forces in a set of concurrent forces, by definition, meet at a point. A set of concurrent forces acting on an object cannot cause the object to rotate.

The lines of action of the forces in a set of nonconcurrent forces do not all meet at a point. Even if the vector sum of the forces in the nonconcurrent set were zero, those forces could cause the object to rotate. A special case of nonconcurrent coplanar forces is that in which all the forces acting on a body have lines of action that are parallel.

The vector sum of the forces acting on an object at rest must be zero.

Free Body Diagram In solving static equilibrium problems it is useful to *isolate* the body of interest and draw what is called a *free-body diagram* to show the forces that the object experiences due to its interactions with other objects. Be sure to show *all* such forces.

Isolating a body for vector treatment. Drawing a "free-body diagram."

For example, consider a ball hanging at the end of the cord in Fig. 3.16*a*. The ball experiences two forces, shown by the vectors **T** and **w**. One of the forces is the downward-directed weight **w** of the ball. The ball also experiences the upward-directed force labeled **T** as a result of its contact with the supporting cord.

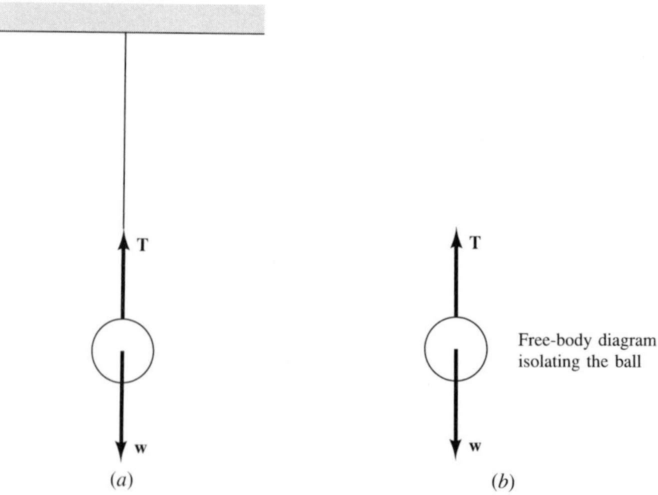

Fig. 3.16 A ball hanging from a cord is in equilibrium under the action of two oppositely directed, concurrent forces—the weight of the ball and the tension in the cord. In (*b*) the ball is "isolated" and the *free-body diagram* is drawn.

A free-body diagram isolates the object of immediate interest and utilizes vector arrows to indicate the forces it experiences because of its interactions with other objects. Figure 3.16*b* shows a free-body diagram for the ball. The forces **T** and **w** that act on the ball are a set of coplanar, concurrent forces. They are both in the plane of the page.

Figure 3.17*a* shows a scaffold which has a seat supported by two vertical ropes. The seat experiences two upward-directed forces $\mathbf{T}_1$ and $\mathbf{T}_2$ due to the supporting ropes, and it experiences a downward force **w** due to its weight. Figure 3.17*b* shows a free-body diagram for the scaffold. The forces $\mathbf{T}_1$, $\mathbf{T}_2$, and **w** are a set of coplanar, nonconcurrent (parallel) forces.

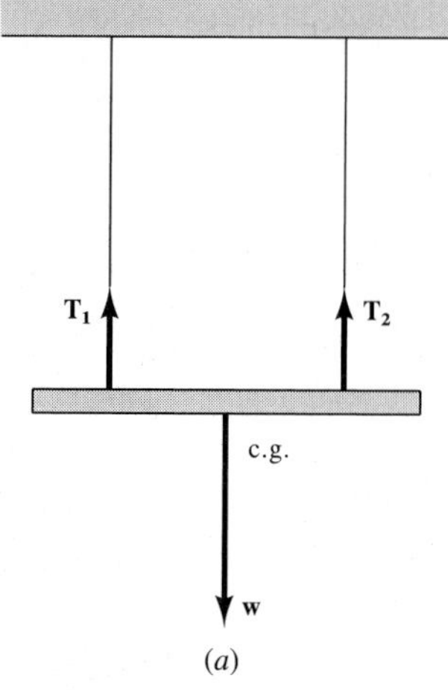

(*a*)

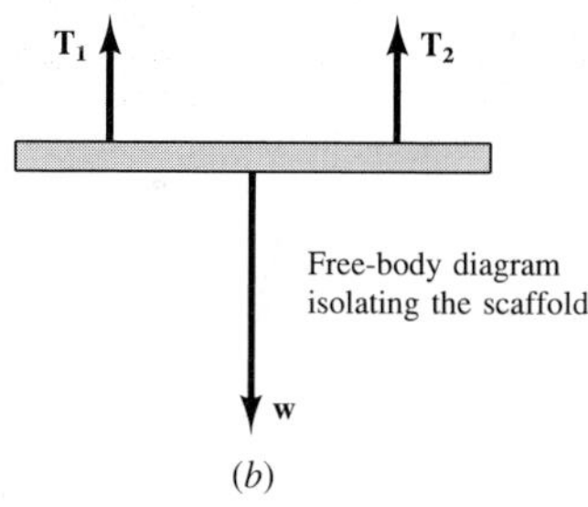

(*b*)

Fig. 3.17 Forces acting on a scaffold. (*a*) Sketch of the scaffold and supporting ropes. (*b*) The free-body diagram, isolating the scaffold.

3.11 ■ Equilibrium Due to Coplanar Concurrent Forces—Analytical Methods

An object at rest will remain at rest if the vector sum of the *concurrent forces* acting on it is zero. This condition for static equilibrium can be written as a mathematical statement:

$$\Sigma\ \mathbf{F} = 0$$

where the Greek letter Σ (sigma) represents the phrase "the sum of." The above equation is equivalent to the two following equations:

$$\Sigma\ F_x = 0 \qquad \Sigma\ F_y = 0$$

That is, for an object to be in static equilibrium, the sum of the x components of the forces acting on the object must be zero, and the sum of the y components of the forces acting on the object must be zero.

The following illustrative problems will show how vector addition, the idea of components of forces, free-body diagrams, and the conditions for static equilibrium are used to analyze and solve problems in static equilibrium.

Here are some guidelines to follow:

Guidelines for solving static equilibrium problems.

1. Read the problem statement carefully and identify the unknown quantity to be determined.
2. Draw a diagram that clearly shows the physical situation described in the problem statement. (Sometimes a diagram will be included with the problem statement.)
3. Draw and label vector arrows to show the forces acting on the objects in the diagram.
4. Isolate the object experiencing the unknown force, and draw a free-body diagram for it. Indicate in this diagram the directions that you choose to call the positive x and the positive y directions.
5. Find the x and y components of the forces in the free-body diagram of step 4.
6. Use the x and y components obtained in step 5 to write the equations for equilibrium: $\Sigma F_x = 0$, $\Sigma F_y = 0$.
7. Use the equations from step 6 to solve for the unknown.

Illustrative Problem 3.2 The pulley on the top portion of a crane used in the construction of a high-rise apartment building is supported by two cables that make an angle of 20° with the vertical as shown in Fig. 3.18. A single cable that weighs 3 lb is attached to the lower end of the pulley assembly and is used to lift a 200-lb I-beam. The 10-lb pulley assembly experiences a downward force **F** of magnitude 213 lb due to the influence of the lower cable and I-beam and due to its own weight. What must be the magnitude of the tension force in each of the upper cables when the pulley assembly is at rest (that is, in static equilibrium)?

Solution First draw a free-body diagram that shows the forces acting on the pulley assembly (see Fig. 3.19). Then write an equation for ΣF_y.

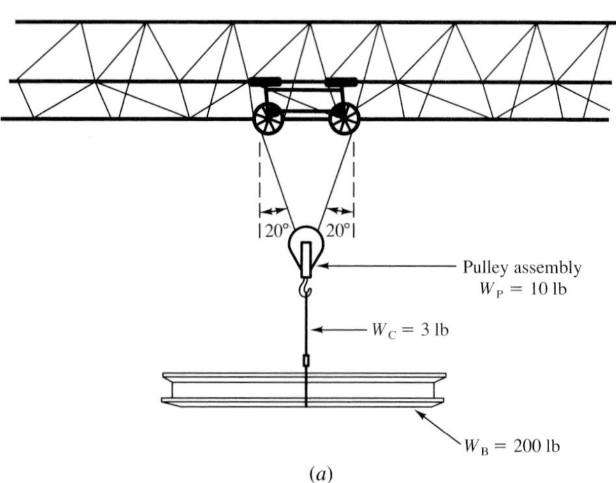

Fig. 3.18 (*a*) Diagram of crane boom and travelling hoist, with data for Illustrative Problem 3.2. (*b*) Huge crane boom lifting heavy truss on construction project near U.S. Capitol in Washington, D.C. (Bethlehem Steel Corp.)

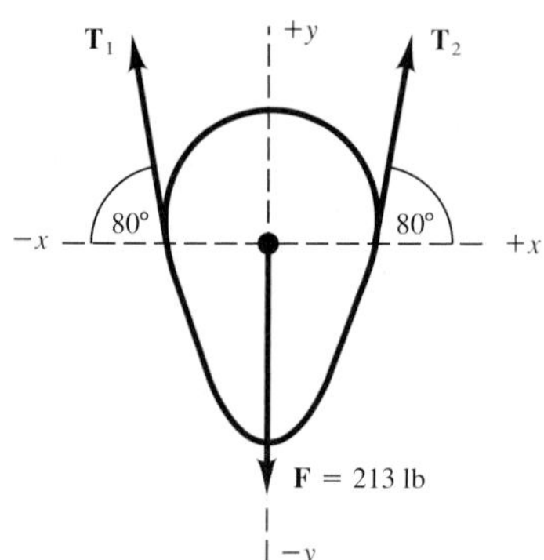

Fig. 3.19 Free-body diagram of the pulley in the hoisting crane of Illustrative Problem 3.2.

The pulley assembly is in static equilibrium, and therefore the vector sum of the forces acting on it must be equal to zero. That is, *as a vector equation*,

$$\mathbf{T}_1 + \mathbf{T}_2 + \mathbf{F} = 0$$

But if the vector sum of the forces must be zero, the algebraic sum of their x components and y components must also individually be zero. Consider first the x components:

$$(T_1)_x = -T_1 \cos 80$$
$$(T_2)_x = +T_2 \cos 80$$
$$F_x = 0 \quad \text{(since } F \text{ is vertical)}$$

Adding, $$(T_1)_x + (T_2)_x + F_x = 0$$

Substituting, $\quad -T_1 \cos 80 + T_2 \cos 80 = 0$

from which, $\quad T_1 = T_2$

Therefore, the magnitudes of the tensions in the two upper cables must be equal.

Now consider the y components:

$$(T_1)_y = +T_1 \sin 80$$
$$(T_2)_y = +T_2 \sin 80$$
$$F_y = -F = 213 \text{ lb}$$
$$(T_1)_y + (T_2)_y + F_y = 0$$
$$T_1 \sin 80 + T_2 \sin 80 - F = 0$$

But $T_1 = T_2$, so

$$\Sigma F_y = T_1 \sin 80 + T_1 \sin 80 - F = 0$$
$$2T_1 \sin 80 = F$$
$$T_1 = F/2 \sin 80$$
$$= 213 \text{ lb}/[2(0.985)]$$
$$= 108 \text{ lb}$$

Since $T_1 = T_2$, both T_1 and T_2 are equal to 108 lb. *answer*

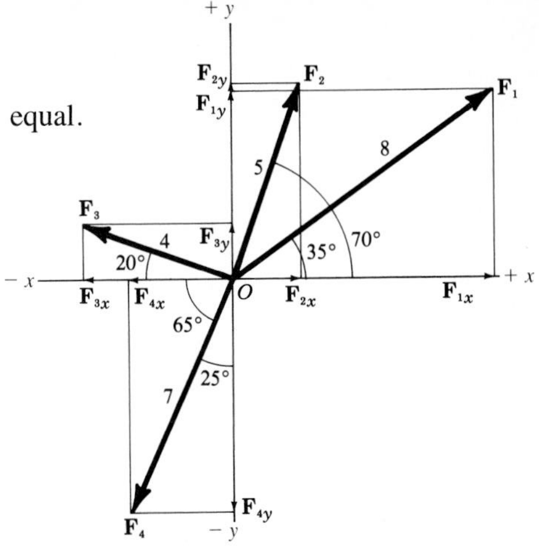

Fig. 3.20 Space diagram of a coplanar force system, concurrent at the origin of a set of xy axes.

The method of components for coplanar concurrent forces.

Three or More Coplanar Forces Concurrent at a Point When several forces act on a rigid body at a point, their resultant can be found *graphically* by the vector polygon method (Sec. 3.8). The *method of components,* however, offers the best *analytical* approach to the solution of such problems.

In Fig. 3.20 a force system of four forces all in the same plane (coplanar), and concurrent at point O, is shown. Forces are in pounds. To find the resultant of such a force system analytically, take the following steps in order:

1. Resolve each force into x and y components as shown.
2. Add (*algebraically*) all the x components to give the x component of the resultant (ΣF_x), and all the y components to give the y component of the resultant (ΣF_y).
3. ΣF_x and ΣF_y are then added *vectorially* to obtain the final resultant **R** (Fig. 3.21).

Components above the x axis or to the right of the y axis are considered positive; those to the left of the y axis or below the x axis are considered negative. The work for the exercise of Fig. 3.20 is summarized as follows:

x components		y components	
$\mathbf{F}_{1_x} = 8 \cos 35°$	$= +6.55$ lb	$\mathbf{F}_{1_y} = 8 \sin 35°$	$= +4.59$ lb
$\mathbf{F}_{2_x} = 5 \cos 70°$	$= +1.71$	$\mathbf{F}_{2_y} = 5 \sin 70°$	$= +4.70$
$\mathbf{F}_{3_x} = -4 \cos 20°$	$= -3.76$	$\mathbf{F}_{3_y} = 4 \sin 20°$	$= +1.37$
$\mathbf{F}_{4_x} = -7 \cos 65°$	$= -2.96$	$\mathbf{F}_{4_y} = -7 \sin 65°$	$= -6.33$
$\Sigma F_x =$	$+1.54$ lb	$\Sigma F_y =$	$+4.33$ lb

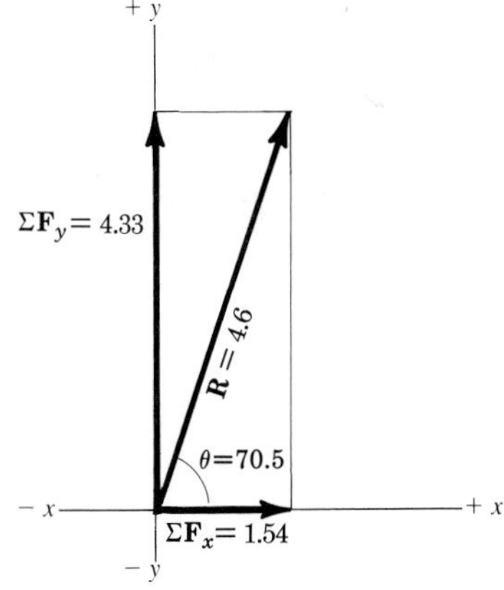

Fig. 3.21 The algebraic sum of x components (ΣF_x) and y components (ΣF_y) of the forces of Fig. 3.20, plotted in a vector diagram to find the resultant **R**.

These resultant components can now be combined vectorially as in Fig. 3.21. **R** is determined from the plot as 4.6 lb, and the angle θ is measured with a protractor and found to be 70.5°.

Analytically, R can also be determined from the theorem of Pythagoras as

$$R = \sqrt{(\Sigma F_x)^2 + (\Sigma F_y)^2}$$
$$= \sqrt{(1.54)^2 + (4.33)^2} = 4.60 \text{ lb} \qquad \textit{answer}$$

The angle θ can be calculated from the trigonometric relation

$$\tan \theta = \frac{4.33}{1.54} = 2.81$$

$$\theta = 70.4° \qquad \textit{answer}$$

In general,

$$\Sigma F_x = \Sigma \mathbf{F} \cos \theta$$
$$\Sigma F_y = \Sigma \mathbf{F} \sin \theta$$
$$\mathbf{R} = \sqrt{(\Sigma F_x)^2 + (\Sigma F_y)^2} \tag{3.2}$$
$$\theta = \tan^{-1} \frac{\Sigma F_y}{\Sigma F_x} \tag{3.3}$$

3.12 ■ Vectors Applied to Structural Problems—Graphical Methods with Coplanar Concurrent Forces

Many construction problems involve the equilibrium of pins, joints, or unions at which three or more coplanar forces act at a common point. A common example of this type of force system is the ordinary derrick or hoisting crane. Figure 3.22 shows a diagram of the typical arrangement in which the mast *MG* is steadied by guy wire *MT* while the load *w* is lifted by the cable passing over the pulley at *B*. The boom *AB* is hinged at *A* and supported by cable *MB*. In the following analysis, the weights of structural parts (boom, pulleys, cables) are *assumed* negligible, compared to the load *w*.

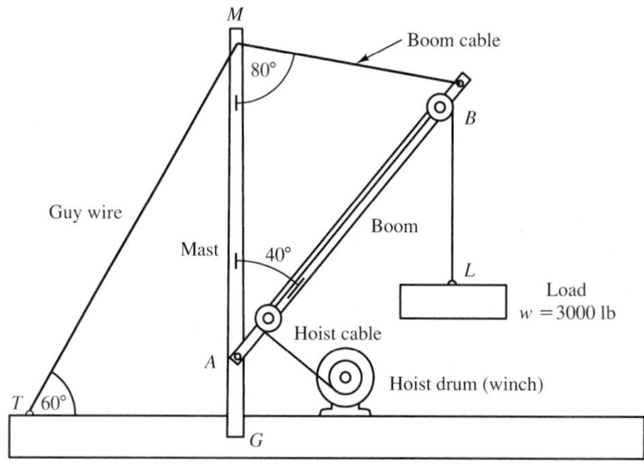

Fig. 3.22 Diagram of the elements of the "hoisting-crane problem." A study of the equilibrium of coplanar concurrent forces.

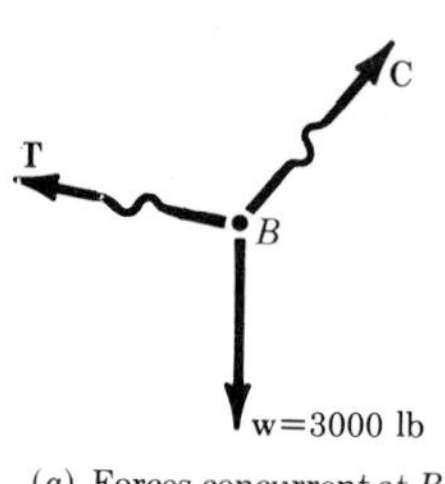

(*a*) Forces concurrent at *B*

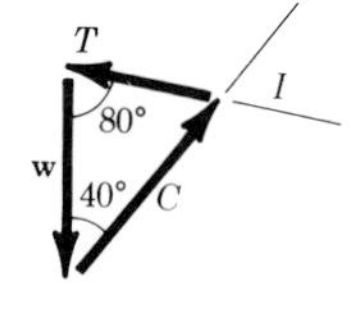

(*b*) Vector diagram for forces at *B*

Fig. 3.23 Vector solution of the "hoisting-crane problem" of Fig. 3.22. (*a*) Forces concurrent at *B*. (*b*) The vector triangle.

In solving such problems we *isolate* one point at a time. To *isolate* a point means to consider, *for the time being,* only those forces acting at that point and to omit temporarily all other conditions of the problem. Isolating point *B*, we see that three forces act on it, namely, the *tension* in *BL*, which is 3000 lb vertically downward, the *tension* in *BM*, and the reaction to *compression* in *BA*, whose directions are known but whose magnitudes are unknown. Of course there are forces acting in *MT* and *MG*, but these are not considered while *isolating* point *B*. Vector methods offer a ready solution in determining the magnitudes of the tension in *BM* and the compression in *BA*.

Refer to the diagrams of Fig. 3.23. The *space diagram* (Fig. 3.23*a*) shows the three forces acting on the point *B*. The force w is plotted to scale, but only the *directions* of *C* (compression) and *T* (tension) are known; so they are shown with a wavy section to indicate that their lengths are unknown. Constructing such a space diagram is an extremely important step in the analysis of equilibrium of concurrent forces. *Space diagrams* need not be drawn to accurate scale, since their purpose is to portray the general conditions of the problem rather than to provide measures for obtaining the final solution. They guide the drawing of the *vector diagram*.

To draw the *vector diagram* in Fig. 3.23*b* start with the vector **w**, since *both its direction and magnitude* are known. Draw it as shown vertically downward, just as the force acts on point *B*, to a scale representing 3000 lb. Now from the head of this vector draw a line in the direction of the force in the boom. This is the line *C* of the diagram, at 40° with the vertical and parallel to *BC* of the space diagram. Since the point *B* is in equilibrium, the three forces acting at *B* must cancel each other; i.e., they must have no net resultant. Therefore, when the vector representing the cable tension is added to the vector diagram, the vector triangle *must close*. The cable *MB* makes an angle of 80° with the vertical; so draw the line of vector **T** *back* toward the line of vector **C** at an angle of 80° with the vertical until **T** and **C** intersect at *I*. The vector triangle is now complete, and scaling values off the diagram gives *C*, the compression reaction force in the boom, as 3400 lb and *T*, the tension in the boom cable, as 2200 lb. All vector diagrams must be drawn to accurate scale, preferably on graph paper, and of a *size large enough* so scale values will be reasonably accurate. Do this yourself with this exercise and verify the results given here.

Now that the tension in *MB* is known, the solution for the tension in the guy wire *MT* and the compression in the mast *MG* can be worked out in a like manner by isolating the point *M*, drawing a space diagram for it, and solving the vector triangle for the forces acting there. (The exercise just indicated should be performed by the student at this time. For checking results, the answers are: *GM,* 4150 lb compression, and *MT,* 4400 lb tension.)

Illustrative Problem 3.3 If the maximum allowable compression in the boom *AB* of Fig. 3.24*a* is 5000 lb, what is the maximum load *w* which can be handled? At this load what is the tension in the horizontal cable *BC*? Neglect weight of boom.

Solution Isolate point *B* and draw the space diagram as shown in Fig. 3.24*b*. The compression in the boom, **C**, is known in both magnitude and direction, but **w** and **T** are known in direction only, so they are shown with wavy sections.

Now draw the vector diagram (Fig. 3.24*c*). First, draw vector **C** to scale for 5000 units at 60° with the vertical. Then draw the line of vector **w** vertically downward from the arrowhead end of **C**, length indefinite. Finally, close the vector triangle with the line of vector **T** drawn horizontally from the starting point of vector **C** until the line of **w** is cut. From the vector diagram as now completed, scale off the desired results:

w = load at maximum allowable boom compression = 2500 lb
T = tension in horizontal cable *BM* = 4330 lb *answer*

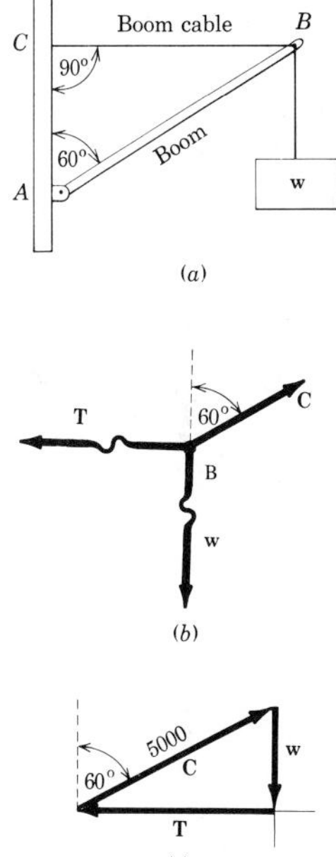

Fig. 3.24 Diagrams for Illustrative Problem 3.3. (*a*) Sketch of boom, boom cable, and load. (*b*) The space diagram. (*c*) The vector diagram.

Illustrative Problem 3.4 A weight of 4000 lb is being supported by two cables under the conditions illustrated in Fig. 3.25*a*. Find the tension in each cable.

Solution Three forces, tension in *CL*, tension in *BC*, and tension in *AC*, are all concurrent at *C*. The forces are coplanar, and since *C* is not being accelerated, the system is in equilibrium; i.e., the resultant of the three forces is zero.

Isolate point *C* and draw a *space diagram* as shown in Fig. 3.25*b*. Both the magnitude and the direction of **L** (load) are known, but only the *directions* of *A* and *B* are known, and they are shown with wavy segments to indicate that the magnitudes are unknown.

Now draw the *vector diagram* carefully (the entire problem should be set up on graph paper) to scale, as shown in Fig. 3.25*c*. (Make *large*, scale diagrams.) The vector **L** will be drawn 4000 units long to your chosen scale, vertically downward. **B** is drawn from the arrowhead end of **L** at an angle of 30° with the horizontal, and **A** is drawn *backward* from the "feathered" or tail end of the **L** vector, at an angle of

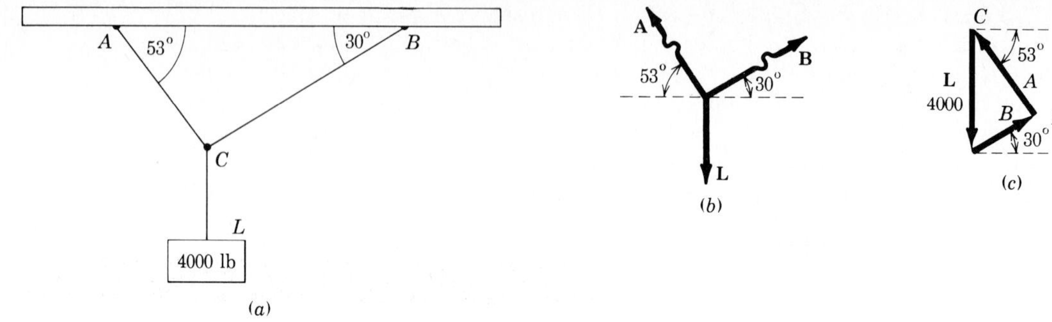

Fig. 3.25 Diagrams for Illustrative Problem 3.4. (*a*) Three coplanar, concurrent forces in equilibrium at *C*. (*b*) The space diagram. (*c*) The vector diagram.

53° below the horizontal, until it intersects the line of **B** at *I*. Scaling off the vectors **B** and **A** gives the following results.

$$\mathbf{A} = \text{tension in } AC = 3490 \text{ lb}$$
$$\mathbf{B} = \text{tension in } BC = 2420 \text{ lb} \qquad \textit{answer}$$

3.13 ■ Equilibrium of Coplanar Parallel Forces

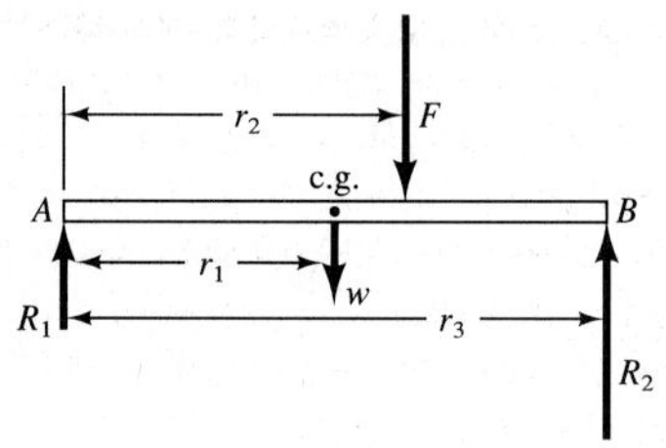

Fig. 3.26 Equilibrium of coplanar, parallel forces.

Up to this point we have been dealing with concurrent forces—several forces acting at a common point. What if the forces are parallel and act on a body at different points?

Consider the beam *AB* of Fig. 3.26, under the action of the parallel forces shown. The force w is the weight of the beam itself acting at a point which we have called the *center of gravity* (c.g.). (Sec. 3.10.) In Fig. 3.26, F is a force acting down on the beam, a force perhaps due to a heavy piece of machinery installed over the beam, and R_1 and R_2 are the reaction forces in the end piers which support the beam. All these forces act vertically, and since there are no forces acting in any other direction, if motion does occur it will have to be either vertical motion—the whole beam moving up or down—or a turning (rotation) of the beam about some point in its plane as a center.

Considering the first of these two possibilities, vertical motion either up or down can be ruled out if the sum of F and w equals the sum of R_1 and R_2. If this is true, all the vertical effects of the forces cancel out and therefore no acceleration of the beam up or down will take place.

Condition I for equilibrium of coplanar parallel forces.

Condition I for *equilibrium of parallel forces* may be stated: *The sum of the forces in one direction must be equal to the sum of the forces in the opposite direction.* Assigning one direction as positive (+) and the other negative (−), we obtain the following statement for Condition I:

The algebraic sum of the parallel forces acting on a body in equilibrium must be equal to zero.

Or, mathematically,

$$\Sigma F = 0 \tag{3.4}$$

Condition I for equilibrium might be satisfied, with no tendency for vertical motion of the center of gravity of the beam, but the distribution of the forces F, w, R_1, and R_2 might easily be such that there will be a net turning effect (torque) to rotate the beam. Such rotations must, of course, be prevented if equilibrium is to be attained.

Condition II for equilibrium (parallel forces) states *that there must be no net turning effect (torque) about any axis, either within the body or outside it.*

To carry this analysis a bit further, refer to Fig. 3.27, which shows a meter stick mounted on a support at P, in equilibrium under the action of five forces F_1, F_2, F_3, w (its weight, acting downward through P, its center of gravity) and R, the reaction at the support. Since the meter stick is not being accelerated vertically up or down, Condition I for equilibrium is satisfied; that is,

$$\Sigma F = R - F_1 - F_2 - F_3 - w = 0$$

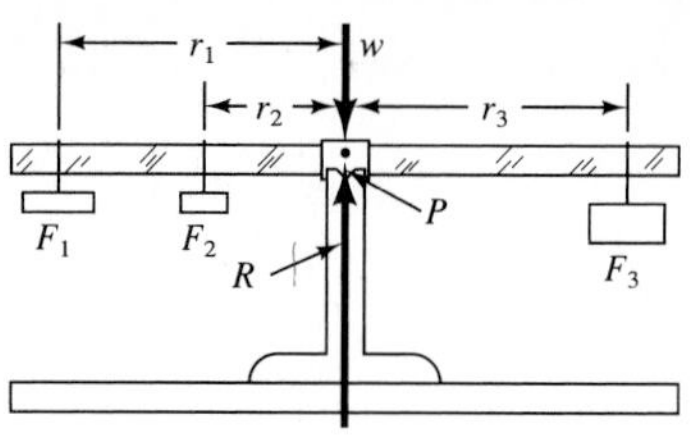

Fig. 3.27 Diagram for studying the conditions for equilibrium of coplanar, parallel forces.

What is necessary now to satisfy Condition II, i.e., to prevent rotation? Obviously the net turning effect *clockwise* around point P must be equal to the net turning effect *counterclockwise*. *Each force times the perpendicular distance to the pivot F* contributes a turning effect or *turning moment* (or *torque*). In general, the *moment of a force* is the product obtained by multiplying the force itself by the perpendicular distance from the line of action of the force to the chosen center of rotation. This is the principle of the common *lever,* which we will deal with in detail in Chap. 7. *Moments* are classified *clockwise* or *counterclockwise* according as the effect of the force would be to cause clockwise or counterclockwise rotation. In Fig. 3.27, we note that the moment F_3r_3 is a clockwise one and that F_1r_1 and F_2r_2 are counterclockwise moments about point P. R and w have no moments at all, since they pass through the pivot point and their moment (lever) arms around P are zero. Moments are vector quantities.

Parallel forces acting on an object can cause clockwise or counterclockwise rotation of the object.

Condition II for equilibrium of a body under the action of parallel forces can now be stated another way:

Condition II for equilibrium of coplanar parallel forces.

The sum of clockwise moments about any point in the plane of the forces must equal the sum of counterclockwise moments about the same point.

Letting the letter $\boldsymbol{M}$ stand for moments, we have

$$F_1r_1 = \boldsymbol{M}_1$$
$$F_2r_2 = \boldsymbol{M}_2$$
$$F_3r_3 = \boldsymbol{M}_3$$

Designating counterclockwise moments as positive (which is conventional) and clockwise moments as negative, we can write a shorter statement for Condition II for equilibrium:

The algebraic sum of moments about any point in the plane of the forces must be equal to zero.

Mathematically, for Fig. 3.27,

$$\boldsymbol{M}_1 + \boldsymbol{M}_2 - \boldsymbol{M}_3 = 0$$

or, in general,

$$\Sigma M = 0 \tag{3.5}$$

It is emphasized that there must be no net turning effect (moment) about *any point,* either within the body or anywhere else in the plane of the forces.

Returning now to the beam of Fig. 3.26, take moments about A (any other point would do as well, but A is convenient) and write the equation for the second condition of equilibrium:

$$R_2r_3 - Fr_2 - wr_1 = 0$$

Note that R_1 produces no moment when A is the moment center. The following problem will illustrate the foregoing principles.

Illustrative Problem 3.5 A steel I-beam (Fig. 3.28) weighs 450 lb and spans a distance of 26 ft. It is supported at each end by pillars. The beam supports a 2200-lb punch press 6 ft from the right end and a 4500-lb lathe 9 ft from the left end. Find the reaction forces in the supporting pillars.

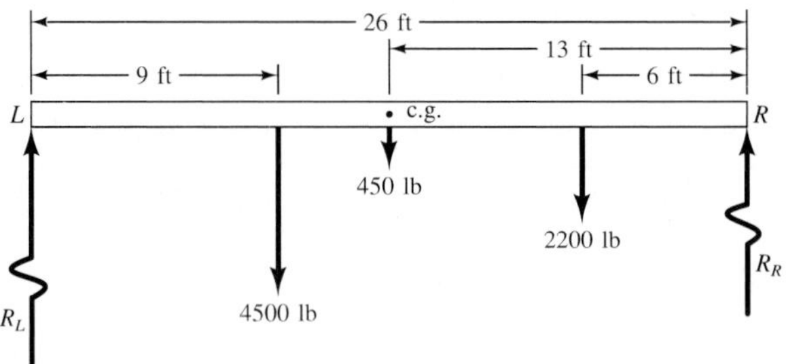

Fig. 3.28 Equilibrium of a loaded I-beam. The data supplied are for Illustrative Problem 3.5.

Solution First diagram the problem as shown, using vectors to show the forces involved. The lengths of R_L and R_R are unknown, and this is indicated in the diagram by the wavy lines.

Applying condition I, we obtain

$$\Sigma F = 0$$

$$R_L + R_R - 4500 - 450 - 2200 = 0$$

or

$$R_L + R_R = 7150 \text{ lb}$$

This equation for Condition I cannot be solved for either reaction, since both are unknown.

Applying Condition II, take moments about any point, say L, and solve for R_R. (L is chosen as the point about which moments are to be taken since then R_L has a zero moment, and the problem is thus simplified.)

$$\Sigma M_L = 0$$

$$26R_R - 2200 \times 20 - 450 \times 13 - 4500 \times 9 = 0$$

$$26R_R = 90{,}350$$

$$R_R = \frac{90{,}350}{26} = 3475 \text{ lb} \qquad \textit{answer}$$

which is the reaction in the right-hand pillar. Substituting this value in the equation for condition I above and solving for R_L gives

$$R_L = 7150 - 3475 = 3675 \text{ lb} \qquad \textit{answer}$$

the reaction in the left pillar.

VECTORS APPLIED TO VELOCITY PROBLEMS

Since velocity possesses the attributes of both magnitude and direction, it is a vector quantity. Therefore, velocities can be combined vectorially to yield a resultant velocity, and a given velocity can be resolved into components in specified directions.

3.14 ■ The Resultant of Two Velocities. Composition of Vectors

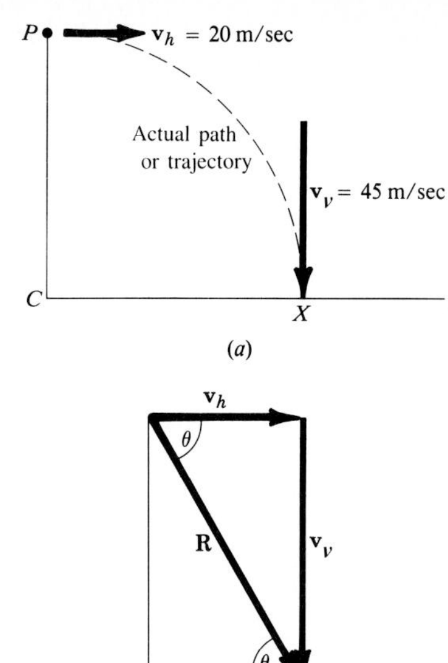

Fig. 3.29 Vectors applied to velocity problems. Horizontal velocity is unaffected by increasing vertical velocity. (*a*) Space diagram. (*b*) Vector diagram.

If you have ever thrown a ball or a rock off a high cliff you noticed that in addition to the horizontal velocity you gave it, the force of gravity immediately began to give it a vertical downward acceleration as well. In Fig. 3.29*a* the ball is thrown from cliff *PC* at a horizontal velocity $\mathbf{v}_h = 20$ m/s. Neglecting air resistance we can assume that the horizontal velocity is constant at this value. The ball begins immediately to pick up vertical velocity, $\mathbf{v}_v$, which is calculated to be 45 m/s on impact at *X*. What is the *resultant velocity* (magnitude and direction) just before impact?

The vector diagram for the problem is shown in Fig. 3.29*b*. This is the process of *vector addition* ($\mathbf{v}_h + \mathbf{v}_v = \mathbf{R}$) (Sec. 3.4). You should plot the vector diagram carefully on graph paper to check the following results:

$$\mathbf{R} = 49.2 \text{ m/s}$$
$$\theta = \tan^{-1} \tfrac{45}{20} = \tan^{-1} 2.25 = 66°$$

3.15 ■ Resolution of Velocity Vectors

Throwing baseballs off cliffs is not a frequent occurrence, but throwing them around the baseball diamond is. If an infielder fields a ground ball and throws to first base for the out, he must project the ball both toward first base (horizontal velocity) and slightly upward so that its drop due to gravity will be compensated. The diagram of Fig. 3.30 shows the factors involved (neglecting air resistance). The actual velocity of the ball leaving the third baseman's fingers is indicated by **v**. Note that it has a horizontal component, $\mathbf{v}_h$, and a vertical component, which at the instant of release, is labeled $\mathbf{v}_v$. The vertical velocity is affected by gravity, reducing to zero at the top of the ball's flight or *trajectory*. Directed downward after that, it increases to a maximum as the ball reaches the first baseman's mitt.

The composition and resolution of velocity vectors has important military applications in gunnery, aircraft operations, rocket flight, and ship movements. In later chapters we will discuss these in detail and develop *mathematical* methods of solving practical problems.

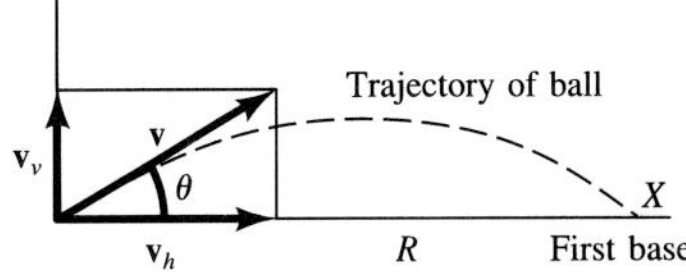

Fig. 3.30 Vector treatment of the velocity and trajectory of a thrown baseball. The initial velocity vector is resolved into horizontal and vertical components.

3.16 ■ Relative Velocity

A hungry passenger walking aft to the dining salon along the deck of a ship which is steaming on a westerly course at 15 knots (1 knot is 1 nautical mi/h) walks due east with a velocity of 5 ft/s *with respect to the ship*. What is the passenger's velocity with respect to the earth? Vectors may be used for a ready solution. In Fig. 3.31 let vector **SG** represent ship's velocity with respect to ground (earth), and vector **PS** represent passenger's velocity with respect to ship. Since **PS** is oppositely directed with respect to **SG**, the resultant or net velocity of the passenger with respect to ground is shown by the vector **PG**. Vectorially, this vector difference is written

Velocity is a vector quantity, since it possesses both magnitude and direction. Velocity problems can be solved by vector methods.

$$\mathbf{SG} - \mathbf{PS} = \mathbf{PG}$$

To obtain consistent units, change ship's velocity to ft/s. From Table 2.1, p. 30, note that a nautical mile is 6076 ft.

Ship's velocity

$$\mathbf{SG} = \frac{15(\text{naut. mi/h}) \times 6076(\text{ft/naut. mi})}{3600 \text{ s/h}}$$
$$= 25.3 \text{ ft/s west, with respect to the earth.}$$

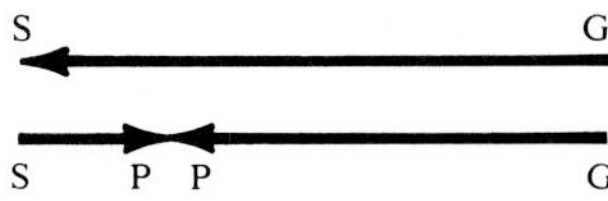

Fig. 3.31 Vectors illustrating a passenger's velocity with respect to that of ship as the passenger walks aft to the dining salon.

Consequently

$$\mathbf{PG} = 25.3 \text{ ft/s} - 5 \text{ ft/s}$$
$$= 20.3 \text{ ft/s, west, the passenger's velocity with respect to the ground.}$$

All motion is in fact relative, for although we commonly consider the earth to be stationary and refer all motions to it, actually the earth is in motion in a rather complex manner. It makes a complete rotation on its polar axis once every 24 h. Each $365\frac{1}{4}$ days it makes an elliptical journey around the sun, a distance of some 585 million miles. Other planets in our solar system have their own equally complex rotations and orbits around the sun. There is evidence also that the entire solar system is moving through the Milky Way galaxy toward the constellation Hercules at a prodigious speed. And finally, our galaxy seems to move through space *away* from other galaxies. Consequently the launchings of artificial satellites, interplanetary rockets, and other space vehicles are fraught with exceedingly complex problems in relative velocity.

Ships and aircraft move in a medium which is also moving.

Motion of a Body in a Moving Medium Ship captains charting their courses for various ports desire to execute a motion with respect to the earth. However, their ships move *through* water, a medium that because of tides, storms, or ocean currents, moves appreciably with respect to the land. Aircraft fly from point to point on earth but must make their way through the air, a medium that is continually moving with respect to the earth's surface. Relative-velocity problems like these are readily solved by the use of vector methods, as follows.

First sketch the vectors *separately,* freehand, *not assembled into a vector diagram* and not to exact scale, but with *approximate directions and relative lengths* indicated. Label each vector with a letter *at the arrowhead end* which stands for the *moving object or medium,* and a letter *at the tail* (feathered) end which indicates *the object or medium with respect to which it is moving*. Then, from these, construct the vector diagram accurately to scale, starting with a vector whose magnitude and direction are both known. Two examples illustrating the method are given. All directions in sea and air navigation are given in three digits, *measured clockwise from due north*—0° to 360°. Work both of these problems yourself on graph paper.

Illustrative Problem 3.6 A pilot flies on a compass heading of 045° at an indicated airspeed of 140 mi/h. During the flight a wind is blowing at 40 mi/h due south. What is the resultant velocity of the plane with respect to the ground?

Solution First sketch the vectors and label them as shown in Fig. 3.32*a*. Then draw the vector diagram Fig. 3.32*b*, accurately on graph paper, starting with the wind vector **AG**, and *putting like letters together,* that is, *A* to *A*. Close the vector triangle by drawing vector **PG**, which gives the plane's speed with respect to ground. The student should check the answer as 116 mi/h, actual course 059°.

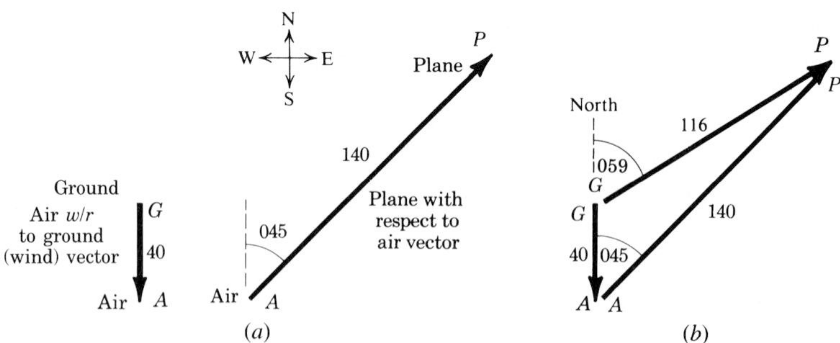

Fig. 3.32 Vector solution of a relative-velocity problem in aircraft piloting. (*a*) Preliminary sketch for identifying and labeling the vectors (not to scale). Note that the vectors are labeled with a letter at the arrowhead end which indicates the *moving* object or medium, and a letter at the "feathered" end indicating the object or medium *with respect to which* the motion takes place. (*b*) Vector diagram to solve for plane's velocity with respect to ground (vector **PG**). Data are for Illustrative Problem 3.6.

Illustrative Problem 3.7 A pilot whose private plane cruises at 350 mi/h (indicated airspeed) leaves San Francisco bound for St. Louis, 1700 miles due east. The pilot is advised that a 50-mi/h wind blowing southeast (i.e. 135°) will be encountered throughout the entire flight. (*a*) What compass heading should be used? (*b*) What is the plane's speed over the ground? (*c*) How long will it take to reach St. Louis?

Solution Sketch and label the vectors as in Fig. 3.33*a*. The wavy section in vector **PG** means that its *magnitude* is unknown. Showing vector **PA** dotted signifies that its *direction* is unknown.

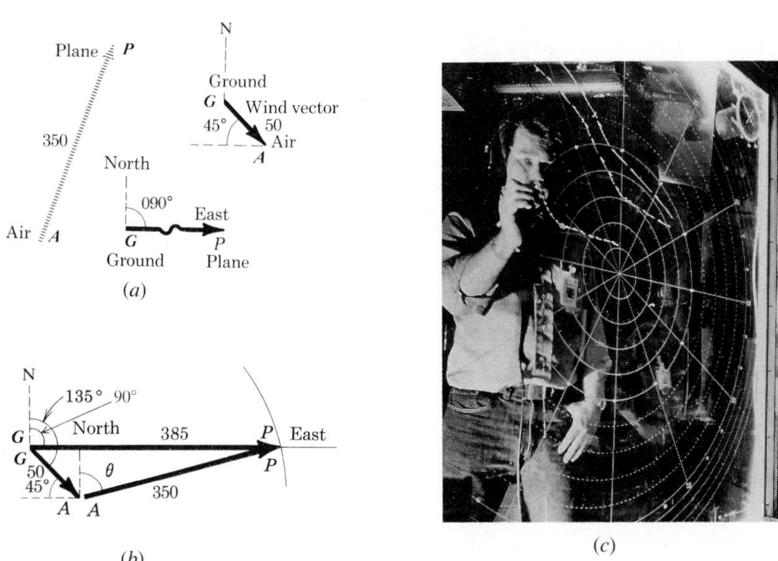

Fig. 3.33 Vector solution of the relative velocity problem of Illustrative Problem 3.7. (*a*) Preliminary sketch for identifying and labeling vectors. (*b*) Vector diagram to solve for the plane's velocity with respect to the ground (vector **PG**). Vector **PA** yields the *compass heading* necessary to make good the required *course,* which is due east. (*c*) Plotting vectors for a navigation problem aboard ship. (U.S. Navy photograph)

Now construct the *vector diagram* as shown in Fig. 3.33*b*. Start with the wind vector **AG**, since both its magnitude and direction are known. Then draw the *line of action* of vector **PG**, *putting G to G,* and extending the *P* end out *indefinitely*. Now put the *A* end of vector **PA** at the *A* end of vector **AG** and swing an arc of radius 350 units, the *magnitude* of vector **PA**. The point at which this arc cuts the *line* of vector **PG** locates the third vertex of the vector triangle.

From **PG** scale off the magnitude of the plane's velocity with respect to the ground as 385 mi/h. With a protractor, measure angle θ as 085°, the compass *heading* which the pilot must use to make good a *course* of 090° for the flight. The time required for the flight is:

$$t = \frac{s}{v} = \frac{1700 \text{ miles}}{385 \text{ mi/h}} = 4.42 \text{ h} \qquad \textit{answer}$$

Although *vector principles* are used in the solution of complex problems of relative motion—in gunnery, navigation (see Fig. 3.33c), and aerospace operations—in actual practice these problems are usually solved by modern computers, radar, and associated electro-optical equipment.

Vector *principles* are used in solving problems in relative velocity, but the "plots" needed by sea and aerospace navigation and by gunnery and missile operations are now determined by computers, using the inputs provided by modern high-technology electro-optical, radar, and sonar equipment.

QUESTIONS AND EXERCISES

1. Which of the following are vector quantities and which are scalar quantities?

The population of your city
Mass of the earth
Density of lead
Time for the earth to revolve around the sun
Force of a golf club hitting the ball
Weight of a bushel of corn
Capacity of your car's gas tank
Velocity of an airliner flying overhead
Weight of a sailboat's keel
A walk from your office to the bank
A push on your car's brake pedal

2. If a set of forces is known to be in equilibrium, what can be said of any one force with respect to the combined effect of all the others?

3. If three coplanar forces act at a point and the point is in equilibrium, why does the vector plot result in a closed triangle?

4. When a baseball or rock is thrown straight up, it is momentarily at rest at the very top of its flight: Is it in equilibrium at that point? Explain.

5. If a force in the positive "*x*-direction" of a pair of coordinate axes is plotted on a vector diagram with an equal force in the positive "*y*-direction," what is the direction of the resultant?

6. A balloon is 500 m above the ground and is stationary in still air. It would appear to be in equilibrium at the moment. Is it? If so, what forces are equally balanced?

7. In riding a department store escalator from one floor to another you decide to walk up the steps as they are moving up the incline. What effect does this have on your velocity? your displacement?

8. When a set of parallel vertical forces acts on a body, will the body always be in equilibrium if the sum of the upward forces equals the sum of the downward forces? Explain.

9. From various sources, technical and nontechnical, obtain several definitions of *center of gravity*. List them, giving the source of each, and discuss how the concept of center of gravity is related to equilibrium.

10. If a beam is in equilibrium under the action of three coplanar but nonparallel forces, show that the *lines of action* of these forces must intersect in a single point. Assuming a beam of finite weight, where will the center of gravity be with respect to the single-point intersection?

11. A steel ball is attached to the ends of two horizontal springs as shown in Fig. 3.34. Is the steel ball in stable, unstable, or neutral equilibrium?

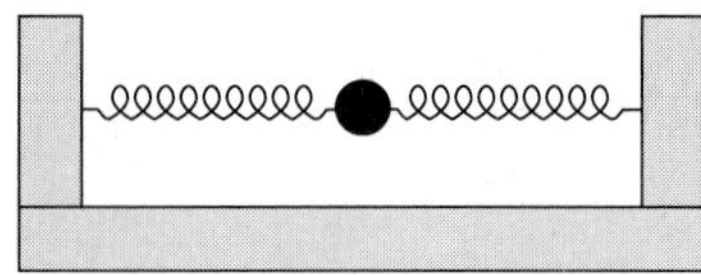

Fig. 3.34 Diagram for Question 3.11.

12. A ship several hundred miles off shore determines that the true bearing of the Golden Gate is 090° (due east). This (true) course is maintained until land is sighted, but the landfall is some 25 mi south of the Golden Gate sea buoy. Assume compass and steering were not at fault. What factors could have caused such an error? If those factors had been known, how could vectors have been used to plot a better course?

13. Explain in detail why a wheel or tire will *roll* downhill. Use vectors and analyze the forces acting on the wheel to explain the rolling action.

14. In effect, what are the reasons why a heavy keel is incorporated into sailboat hulls?

15. A skyscraper stands on a solid foundation, seemingly with only one force acting on it—its weight. But, if any object is subjected to an unbalanced force it will be accelerated. What other force is acting on the skyscraper? What magnitude and what direction does it have? (Neglect any wind forces.)

PROBLEMS

NOTE: *Draw careful diagrams for all problems.*

Group One

1. An auto is driven 20 km due east, then 8 km due north, then 15 km along a road running northwest, then 16 km due west. Use the vector polygon method and graph paper to find its displacement (distance and direction) from the starting point.

2. Use the vector-parallelogram method to find the resultant of these two concurrent forces: one of 125 lb acting vertically upward, and one of 65 lb acting at an angle 20° below the horizontal to the right.

3. An airplane takes off and flies for 3 h on course 045° at 350 mi/h. The pilot then changes course to 110° and flies at 450 mi/h for 2.5 h. How far is he from the starting point? (Plot all courses clockwise from due north, which is 000°.)

4. The pilot of a small business aircraft leaves from the home airport on a trip that requires landings at three cities. The first "leg" is straight east for 30 mi to the first city. Next, the flight is straight northeast for 85 mi to the second city. To get to the third city the flight is straight west for 60 mi. (*a*) In what direction, and how far will the return flight be from the third city to the home airport? (*b*) A week later stops are required only at the second and third cities. At what angle north of straight east should the heading be to travel in a straight line to the second city?

5. A 70-m displacement is at an angle of 35° with the positive x-axis. Find its x and y components graphically and then trigonometrically.

6. A velocity vector has a magnitude of 200 m/s in a compass direction of 055°. Find its north and east components.

7. Find the horizontal and vertical components of a 500-lb force which acts upward and to the right, making an angle of 50° with the horizontal. Plot it accurately to scale, and then check your answers trigonometrically.

8. A 5-lb book is at rest on the horizontal surface of a desk. A 4-lb book is on top of the 5-lb book. The 5-lb book experiences one force due to its own weight, a second force due to its contact with the desk, and a third force due to the 4-lb book pressing down on its top surface. The 4-lb book experiences one force due to its own weight, and a second force due to its contact with the 5-lb book below it. (*a*) Draw a free-body diagram for each of the books. (*b*) Use the first condition for equilibrium to determine the magnitudes and directions of the forces acting on the books.

9. A 100-lb crate is at rest on a ramp that makes an angle of 30° with the horizontal. (*a*) What are the magnitude and direction of the force experienced by the crate due to its contact with the ramp? (*b*) If the angle that the ramp makes with the horizontal is reduced from 30° to 20°, what will be the magnitude and direction of the force experienced by the crate due to its contact with the ramp?

10. A 500-lb box is held stationary on a moving van's loading ramp with a force of magnitude F that is directed parallel to the plane of the ramp as shown in Fig. 3.35.

Fig. 3.35 Diagram for Problem 3.10.

The cylindrical rollers on the ramp render friction forces negligible. What must be the magnitude of the force F?

Group Two

11. Three members of a bridge truss meet at the point A, as shown in the diagram of Fig. 3.36. AB and AC are under tensional stresses of 2200 lb each. What must be the stress in AD for equilibrium? (Solve by the vector-triangle method.)

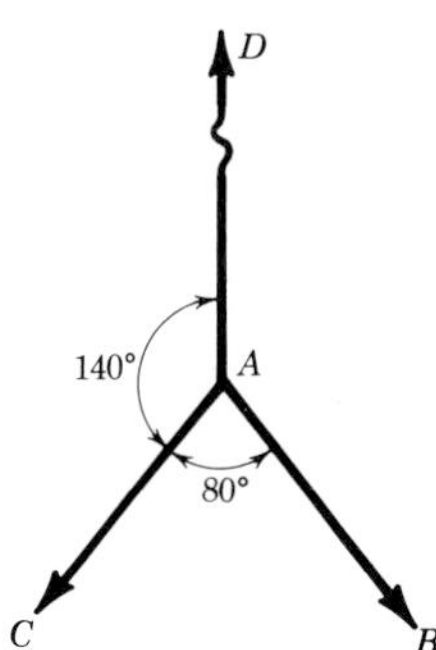

Fig. 3.36 Diagram for Problem 3.11. Concurrent forces in a bridge truss.

12. What horizontal force F must be applied to pull the supporting rope AC aside until the angle at the ceiling is 60° as shown in the diagram of Fig. 3.37?

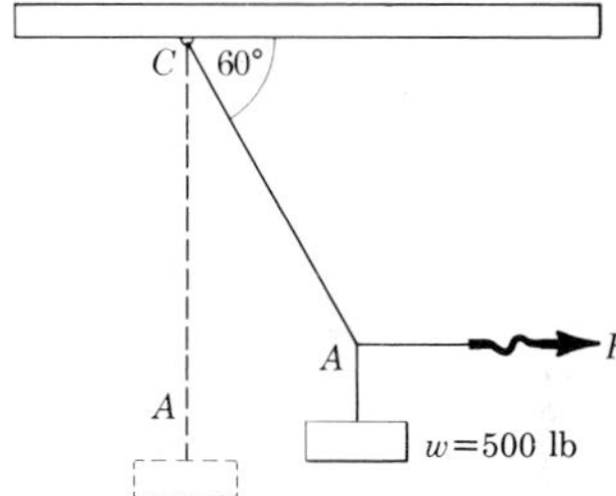

Fig. 3.37 Diagram for Problem 3.12.

13. Find the resultant of the force system shown in the diagram of Fig. 3.38 by the method of components. (Review Sec. 3.11.)

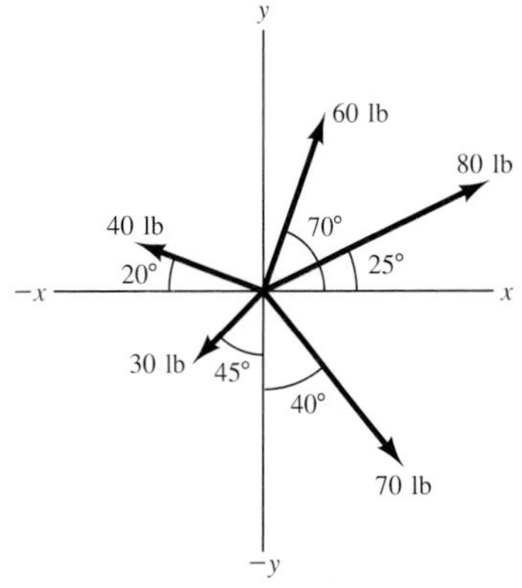

Fig. 3.38 Diagram for Problem 3.13.

14. The following four forces as diagrammed in Fig. 3.39 are in the same plane and are concurrent: 500 lb at 035°, 350 lb at 130°, 450 lb at 220°, and 285 lb at 260°. What are the direction and magnitude of the single force which will put this system in equilibrium? (Use the vector-polygon method.)

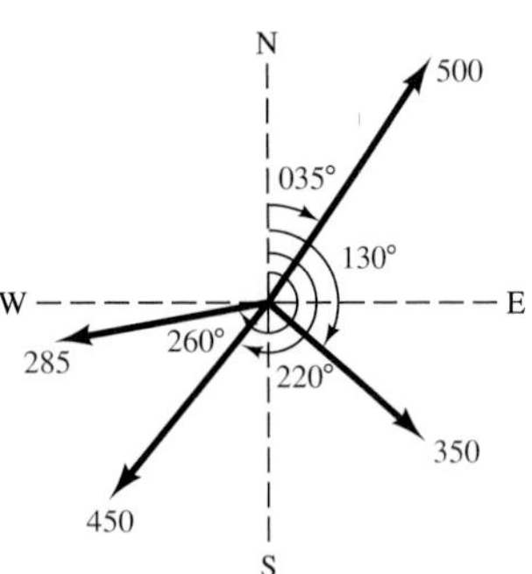

Fig. 3.39 Equilibrium of four coplanar, concurrent forces. Diagram for Problem 3.14.

15. A vertical cable attached to a crane (not shown) has a hook at its lower end that is used to lift a 1200-lb crate as shown in Fig. 3.40. The hook is located at the midpoint C of a 5-ft-long rope of negligible weight that is attached to the crate at points A and B. (*a*) What is the magnitude of the tension in the vertical cable D? (*b*) What are the tensions in segments AC and BC of the rope attached to the crate?

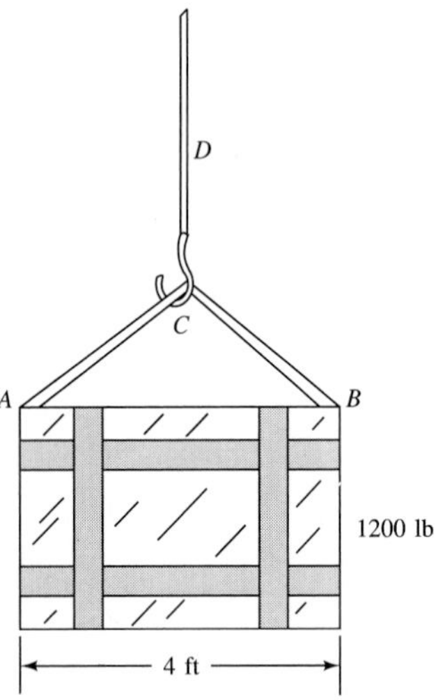

Fig. 3.40 Diagram for Problem 3.15.

16. A block A is held on an inclined plane, as shown in Fig. 3.41 by the cord P. If the weight of the block is 5 lb, find the normal force N with which the plane reacts against the block, and the tension in cord P. Neglect friction.

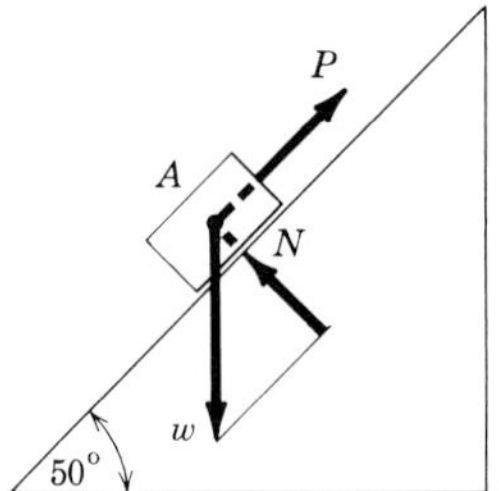

Fig. 3.41 Forces on a block on an inclined plane (Problem 3.16).

17. Find the compression in the boom AB and the tensions in the guy wire DC and boom cable BC of the crane shown in the diagram of Fig. 3.42. Neglect the weight of the boom.

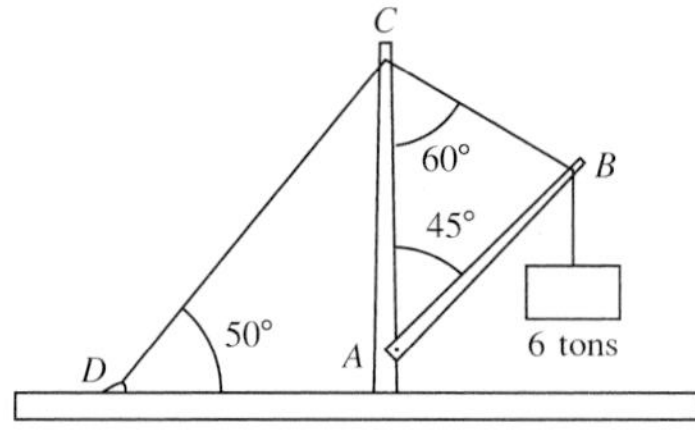

Fig. 3.42 Diagram for the hoisting crane problem of Problem 3.17.

18. Given the data shown in the diagram of Fig. 3.43, find the tensions in ropes AB and CD and also the value of angle θ when $\alpha = 45°$.

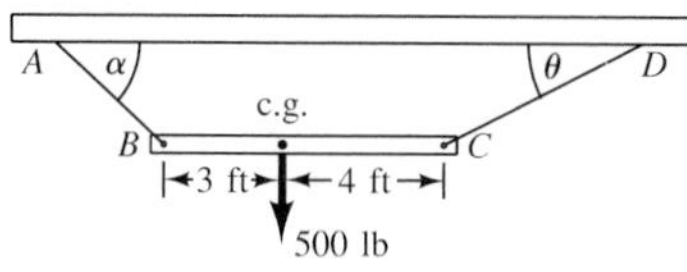

Fig. 3.43 Diagram for Problem 3.18.

19. A 12-ft-long beam that weighs 30-lb is supported from below at a point 1 ft from its left end and at a point 4 ft from its right end as shown in Fig. 3.44. Use the first and second conditions for equilibrium to determine the magnitude of the upward-directed forces that the board experiences at each of the two support points.

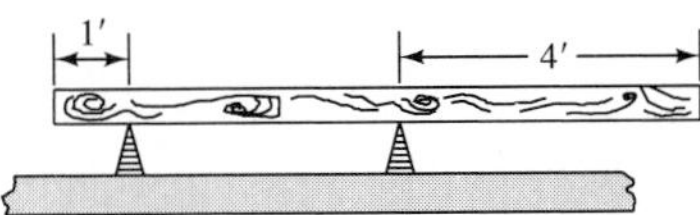

Fig. 3.44 Diagram for Problem 3.19.

20. A small bridge, supported by piers at each end, is being crossed by a 25-ton truck and semitrailer. If the bridge is 150 ft long and the center of mass of the truck and trailer is 50 ft from the left end, calculate the amount of the load carried by each pier.

21. The system of parallel forces shown in the diagram of Fig. 3.45 is in equilibrium. Find the value of the force F and its distance x from the 600-lb force. The markings on the beam are in feet. Neglect the weight of the beam.

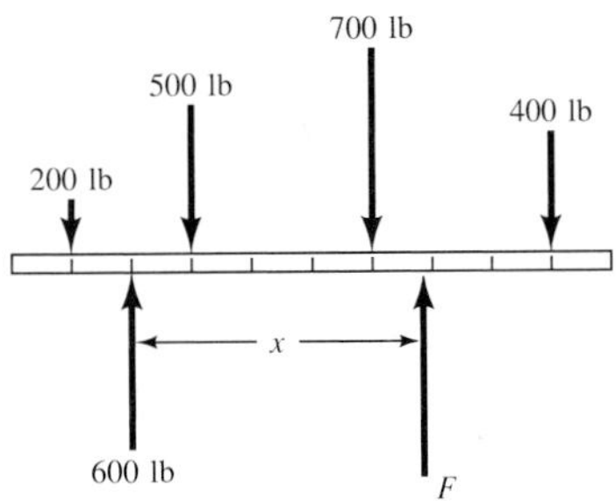

Fig. 3.45 Equilibrium of parallel forces. Diagram for Problem 3.21.

22. A river current flows south at 4 mi/h in a channel which is 0.5 mi wide. A swimmer who can swim at a steady rate of 1.5 mi/h in still water dives in and *heads* straight across the channel. (*a*) How far downstream is she when she reaches the far shore? (*b*) How far is she from her starting point? (*c*) How many minutes have elapsed since she left the starting point?

23. A ball is thrown from the passenger (right-hand) side of a moving auto with a velocity of 20 m/s with respect to the auto, in a direction perpendicular to the auto's instantaneous velocity, which is 80 km/h due east. Find the velocity (speed and direction) of the ball with respect to the ground. Neglect gravitational factors.

24. A ferryboat crosses a north-south channel 500 yd wide in which a strong current runs due north at 5 knots. The ferry is capable of 10 knots (nautical miles per hour) in still water. What must be its compass heading in order to reach a dock directly across (due east) from its departure point?

25. A Navy destroyer is headed due west at 30 knots. A 40-knot wind is blowing from the southeast. (*a*) Find the velocity of the wind with respect to the ship. (*b*) What angle does the flag make with the ship's course as it streams from the flagstaff in the relative wind?

Group Three

26. A pilot leaves airport A to fly to airport B, a distance of 450 mi due south. He sets the plane's nose due south and flies at an indicated airspeed of 225 mph. Unknown to him, a 40-mi/h wind from the northwest was blowing throughout the flight. At the expiration of 2 h he expects to see airport B, but despite excellent visibility it is nowhere in sight. How far and in what direction is he from airport B?

27. A jet transport plane will fly at 550 mph indicated airspeed, from airport A to airport B, 2400 mi distant. The true bearing (direction) of B from A is 072°. A 70-mi/h wind blowing due south is to be encountered throughout the entire flight. Find (*a*) the compass heading which must be maintained to stay on course to airport B and (*b*) the time required for the flight.

28. Two window washers stand on a 40-lb, 16-ft-long board to wash windows on a building. Two vertical cables, L and R, each attached to the board 1 ft from its ends, are used to support the board from above. One worker, who weighs 160-lb, stands at the midpoint of the board, and a second 175-lb worker stands 4 ft from the right-hand end of the board as shown in Fig. 3.46. Use the first and second conditions for equilibrium to determine

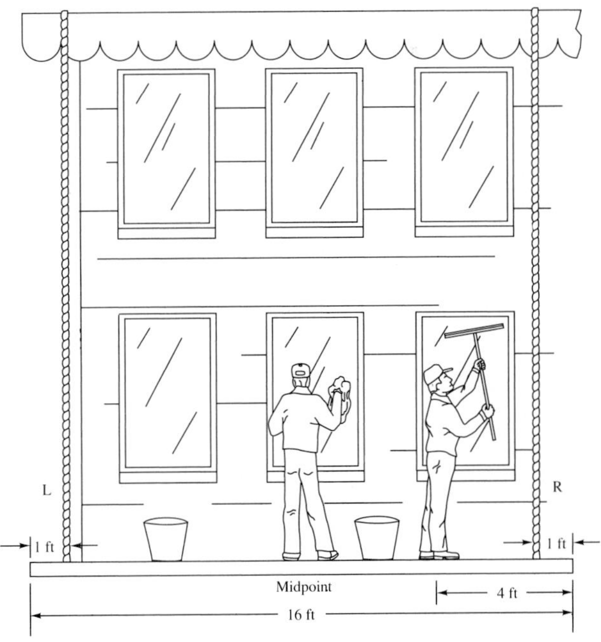

Fig. 3.46 Window washers want their scaffold to be in equilibrium. Sketch for Problem 3.28.

the magnitudes of the tensions in each of the two vertical cables.

29. The sketch of Fig. 3.47 shows the cross section of a roof which weighs 800 lb/ft of length along the dimension AC. T is a heavy water-cooling tower (weight 5 tons) located as shown. Calculate the load borne by each of the supporting walls A and B.

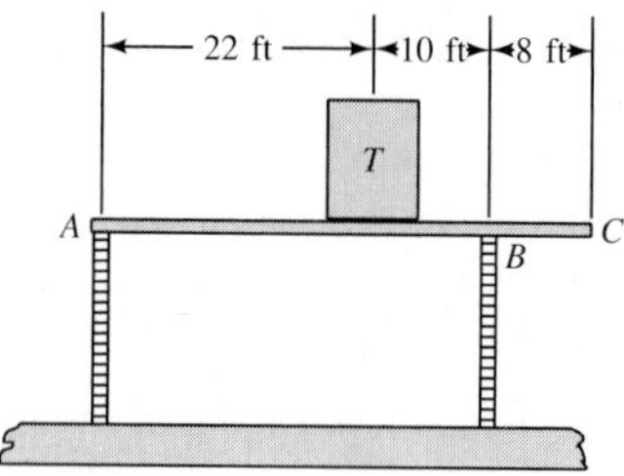

Fig. 3.47 Roof with heavy cooling tower. Diagram for Problem 3.29.

30. The lower end of a 16-lb ladder experiences a force F_f due to its contact with the floor. The force F_f has a component F_h parallel to the floor and a component F_v perpendicular to the floor as shown in Fig. 3.48. A roller at the upper end of the ladder prevents any noticeable frictional interaction between the ladder and the wall. Because of this the ladder experiences a force F_w exerted by the wall that is directed perpendicular to the wall. Use the first and second conditions for equilibrium to determine the magnitudes of F_h, F_v, and F_w.

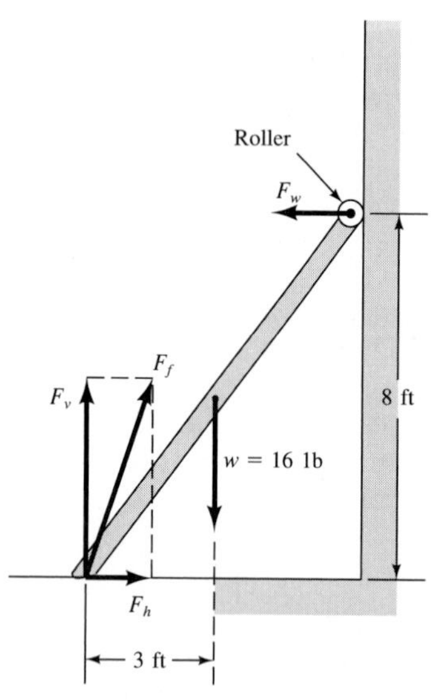

Fig. 3.48 The leaning ladder problem. Diagram for Problem 3.30.

31. A 12-ft-long board that weighs 30-lb is supported from below at a point 1 ft from its left end and at a point 4 ft from its right end as shown in Fig. 3.49. If a 120-lb person stands near enough to the right end of the board, the left-hand support will no longer need to contribute an upward force because the board will start to rotate about the right-hand support point. Use the first and second conditions for equilibrium to determine where the person must stand to cause the board to just begin rotating.

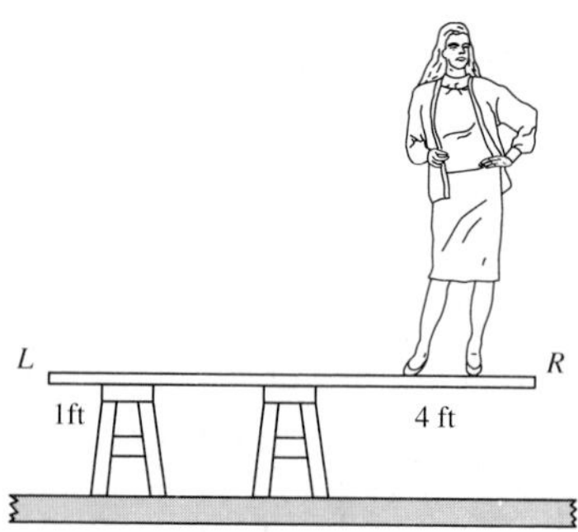

Fig. 3.49 Diagram for Problem 3.31.

32. A hardware store has a large sign hanging high over the sidewalk. The arrangement is shown in Fig. 3.50. The sign weighs 250 lb, and its center of gravity is at the geometrical center. Find (*a*) the tension in the supporting cable; and (*b*) the horizontal and vertical components of the force acting on the sign at the rigid wall hanger (point P). (Assume no wind forces.)

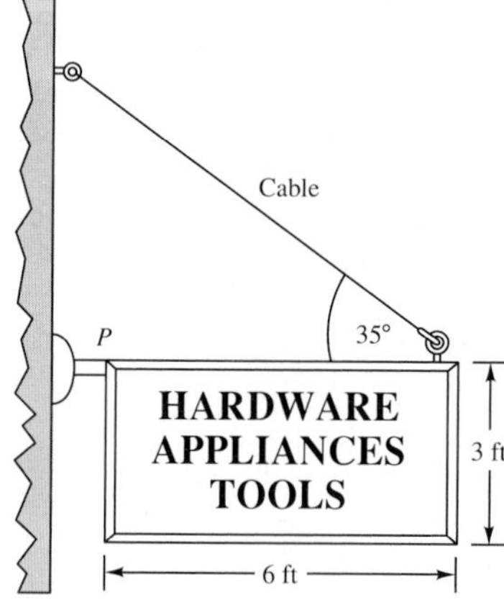

Fig. 3.50 The hanging sign. Sketch for Problem 3.32.

PART 2 MECHANICS

Conceptual design for a U.S. space station for the 1990s. One of the several submitted to the National Aeronautics and Space Administration. (Martin Marietta Corporation)

CHAPTER 4 MOTION

Physics has been defined as the science of *matter* and *energy*, but two other concepts are almost equally important in trying to understand order and disorder in the universe and in the world around us. *Force* and *motion* are the essential inputs to *mechanical energy*, the form of energy that will be treated in this major section of the book. Later sections will deal with heat energy, light and sound, electrical energy and electronics, and atomic and nuclear energy.

Kinematics is the study of motion without reference to the cause of the motion.

Motion will be dealt with at length in this chapter. The study of the motion of a particle or a system of particles without reference to the forces that act on the system is called *kinematics*. *Force* will be mentioned only briefly here. An in-depth treatment of force and its relationship to motion will be the subject of Chap. 5.

Motion means a *condition of not being at rest with respect to the surroundings*. A body in motion is constantly changing its position or location within a given space or with respect to a given *frame of reference*.

A force may start something moving or stop something which is already moving. Or, it may merely speed up (*accelerate*) or slow down (*decelerate*) a moving object. Or, it may change the *direction* of the motion. The essential thing to remember about an *unbalanced force* is that it invariably *causes a change in motion*. In other words it causes a change in the *velocity* of the body on which it acts.

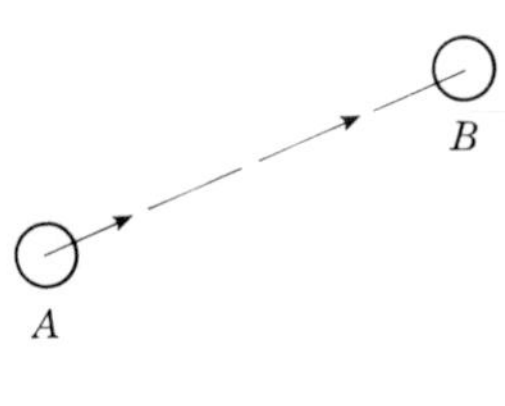

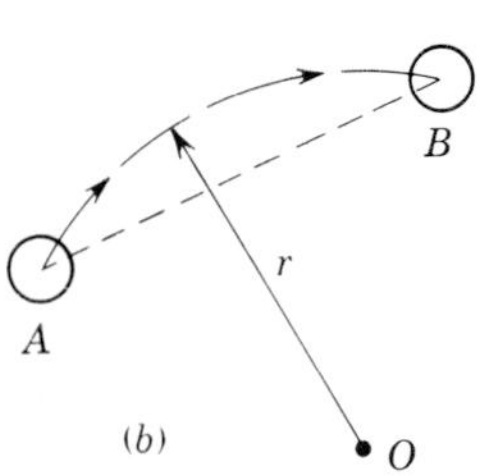

Fig. 4.1 Rectilinear (straight-line) motion and curvilinear motion. The sphere moves from point *A* to point *B* in both sketches. The *displacement* is *AB* in both cases, but the path *distance* is different. The curvilinear motion in (*b*) is on an arc of a circle with radius *r* and center *O*.

MOTION

4.1 ■ The Concept of Motion

Motion is defined as *a continuous change in position*.

A change in position is called a *displacement*. Straight-line motion is said to be *rectilinear*, and motion along a curved path is called *curvilinear* motion (Fig. 4.1).

As an airplane flies in level flight, as an automobile moves along a straight, level road, or as a piston moves back and forth in the cylinder of an engine, these bodies move along straight lines. Motion of this kind is said to be pure *translation*, a linear motion. Motion which, like that of a turbine rotor or a centrifuge, involves the spinning of the object about a stationary axis, is called pure *rotation*. Complex motions may involve both translation and rotation, of course, but for the present we shall stay with simple forms of motion, considering translation in this chapter and rotation in a later chapter.

You will have noted from studying Fig. 4.1 that *distance* and *displacement* are not synonymous terms. (See also Sec. 3.2.) In Fig. 4.1*a* distance traversed is equal to the displacement *AB*, but in Fig. 4.1*b*, the net displacement is much less than the distance traversed. The term *displacement* refers to the net change of position of a particle or object as a result of motion.

Motion is often thought of as the opposite of *rest*. An object at rest is not in motion, and vice versa. Early philosophers considered *rest* to be the normal state of things, with motion as a special condition. With the advent of systematic, experimental study of mechanics by Galileo Galilei and Sir Isaac Newton, however, it became apparent that just the opposite is true—motion is the normal state of things, and rest is a special condition.

In considering motion we are nearly always interested in the displacement that occurs or the distance traveled, or both. The time required for the motion is also an important consideration. And these two factors, when set up as a ratio (distance/time) give us the *rate* of the motion. The rate of motion is described by two terms—*speed* and *velocity*. Each of these concepts, in turn, has two possible interpretations. There is *average speed* and *instantaneous speed;* and there is *average velocity* and *instantaneous velocity*.

Motion is the normal state of things in nature. *Rest* is that special case of motion in which velocity is zero.

4.2 ■ Speed and the Units in Which it Is Measured

Speed is defined as the distance traveled divided by the time taken to execute the motion. When the motion is such that equal distances are covered in each succeeding unit of time, the speed is said to be *uniform* or *constant*. Rectilinear motion at uniform speed is the simplest kind of motion. *Speed* is defined as *the rate of motion,* and in many types of problems *speed* and *velocity* can be used interchangeably. The term *velocity* is used when both *rate* and *direction* of motion are specified, and the term *speed* is used when direction is not known or is not important in the problem. *Speed* is the magnitude factor of velocity. Velocity is a *vector* quantity; speed is a *scalar* quantity.

Both speed and velocity may be expressed in miles per hour (mi/h), feet per second (ft/s), meters per second (m/s), kilometers per hour (km/h), centimeters per second (cm/s), etc. A fast sprinter can move at 30 ft/s, an express train at 130 mi/h, an antiaircraft projectile at 950 m/s, and a high-performance jet aircraft at over 1000 mi/h.

Aircraft and ship speeds are expressed in *knots*. *One knot is one nautical mile per hour*. A nautical mile is 6076 ft.

Speed and *velocity* distinguished from one another. Speed is the *magnitude factor* of velocity. Speed is a *scalar*, and velocity is a *vector*. Both speed and velocity express the rate of motion, but velocity also stipulates the *direction* of the motion.

Speed and velocity are expressed in the same units.

Constant or Uniform Speed In any motion problem, three variables—*distance, speed,* and *time*—are involved. The defining equation for rectilinear motion at *uniform* speed is

$$\bar{v} = \frac{s}{t} \tag{4.1}$$

where $\bar{v}$ = uniform speed
s = distance traveled in time t
t = time required

The bar over the v signifies *uniform speed*.

Uniform speed and average speed.

Average Speed A body traversing a finite distance may have a variable rather than a uniform speed. By suitable analysis, however, an *average* speed for the entire motion can often be determined. If average speed is known for a motion problem, Eq. (4.1) can also be used, with $\bar{v}$ as the average speed. This speed formula, like any algebraic equation, can be solved for any one of the variables in terms of the others. The other two forms of the equation are

$$t = \frac{s}{\bar{v}} \tag{4.1$'$}$$

$$s = \bar{v}t \tag{4.1$''$}$$

Illustrative Problem 4.1 A commercial airliner flew from Kennedy International Airport, New York City, to Los Angeles International Airport in an elapsed time of 5 h 12 min. The airline distance is 2885 mi. What was the average speed in miles per hour?

Solution Since distance and time are known, and average speed is desired, we make use of Eq. (4.1), $\bar{v} = s/t$. Substituting known values,

$$\bar{v} = \frac{2885 \text{ mi}}{5.20 \text{ h}} = 555 \text{ mi/h} \qquad \textit{answer}$$

Illustrative Problem 4.2 A satellite in earth orbit has a speed of 19,000 mi/h, assumed constant. If its nearly circular orbit represents a total distance of 28,500 mi, how long does it take for one trip around the earth?

Solution Since distance and uniform speed are known, and time is desired, the use of Eq. (4.1′) is indicated. Substituting,

$$t = \frac{28{,}500 \text{ mi}}{19{,}000 \text{ mi/h}} = 1.5 \text{ h, or 90 min.} \qquad \textit{answer}$$

4.3 ■ Average Velocity

Average velocity versus uniform or constant velocity.

We often use the words *speed* and *velocity* interchangeably in everyday conversations. In physics, however, there is a clear distinction in the meanings of the two words. One difference is that speed is a scalar quantity and velocity is a vector quantity, as pointed out above. A second difference is that speed is defined in terms of the *distance* traveled during a time interval, while velocity is defined in terms of the *displacement resulting from the motion* during the time interval. Average velocity is defined by the following equation.

$$\mathbf{v} = \frac{\mathbf{s}}{t} \tag{4.2}$$

where $\mathbf{v}$ = the average *velocity* of the object during the time interval t. (The direction of $\mathbf{v}$ is the same as the direction of $\mathbf{s}$.)

$\mathbf{s}$ = the *displacement* (a vector quantity) experienced by the particle or object during the time interval t.

t = the time required for the object to experience the displacement $\mathbf{s}$.

Note that $\mathbf{v}$ and $\mathbf{s}$ are portrayed in boldface type in Eq. (4.2) to indicate that they are *vectors*.

For many problems it is convenient to express the displacement $\mathbf{s}$ in terms of the difference between the locations of the object at the beginning and at the end of the motion. Figure 4.2 shows an example for the rectilinear (one-dimensional) motion now being considered. Note that the displacement vector $\mathbf{s}$ points from the location of the object at time zero to its location t s later. In this example the object moves a distance of 3 m, from $x = 1$ m at time zero to $x = 4$ m at the end of the time interval, t s later. The displacement vector $\mathbf{s}$ points in the positive (+) x direction.

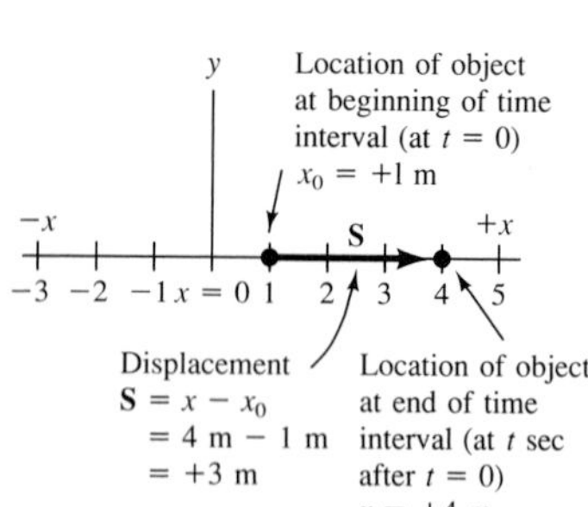

Fig. 4.2 Diagram showing displacement as a vector, plotted on an xy coordinate system.

We can now rewrite Eq. (4.2) so that it includes symbols that indicate the location of the object, as follows:

$$\mathbf{v} = \frac{x - x_0}{t} \tag{4.3}$$

where $\mathbf{v}$ and t have the same definitions as in Eq. (4.2)

x_0 = the x coordinate of the object at $t = 0$

x = the x coordinate of the object t s later at the completion of the motion

Equation (4.3) will give a positive value for **v** when the net motion is in the positive x direction and a negative value for **v** when the net motion is in the negative x direction.

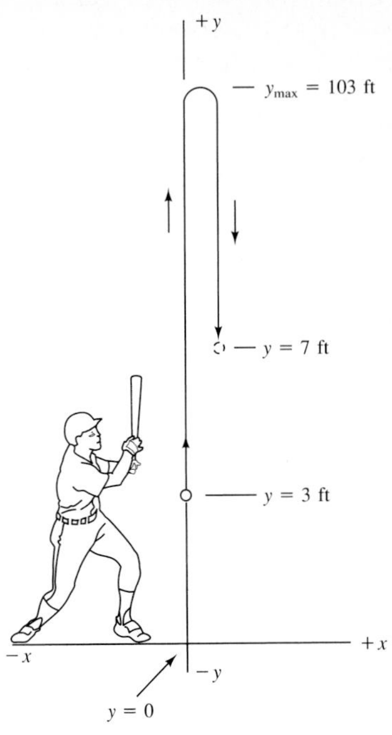

Fig. 4.3 A "pop-up" flyball is portrayed as rising vertically (positive y direction) and then falling straight downward (negative y direction). The average speed and the average velocity can both be determined (Illustrative Problem 4.3).

Illustrative Problem 4.3 A baseball player hits a pop-up. The ball is 3 ft above the ground when it leaves the bat and it travels rapidly straight upward. It reaches a height of 103 ft above the ground before it starts to fall back vertically downward to home plate where the catcher is waiting for it. The ball lands in the catcher's mitt 7 ft above the ground exactly 5.05 s after it left the bat. During the 5.05 s time interval, (a) what was the average *speed* of the ball? (b) The average *velocity* of the ball? (See Fig. 4.3.)

Solution

(***a***) The ball travels 100 ft upward and 96 ft coming back down. The total distance s traveled by the ball is 196 ft. The time required for the motion is 5.05 s. From Eq. (4.1),

$$\bar{v} = \frac{s}{t} \qquad \text{for average } \textit{speed}$$

$$= \frac{196 \text{ ft}}{5.05 \text{ s}} = 38.8 \text{ ft/s} \qquad \textit{answer}$$

(***b***) Let $t = 0$ just as the ball leaves the bat. Then $y_0 = 3$ ft. At the end of the motion, when $t = 5.05$ s, the y coordinate of the ball is 7 ft. The *displacement*, therefore, is only 4 ft. From Eq. (4.3), substituting the y direction of this motion for the x direction assumed in Eq. (4.3),

$$\mathbf{v} = \frac{y - y_0}{t} \qquad \text{(for average } \textit{velocity}\text{)}$$

$$= \frac{7 \text{ ft} - 3 \text{ ft}}{5.05 \text{ s}}$$

$$= +0.79 \text{ ft/s} \qquad \textit{answer}$$

The + sign tells us that the velocity is directed upward. Note that the average *speed* of the ball is approximately 49 times as great as the magnitude of the average *velocity* of the ball.

4.4 ■ Representing Average Speed or Velocity Graphically

Suppose a runner in an 800-m race intends to pace himself throughout the race at a uniform or constant speed and wants you to "clock" him. Markers are on the track at every 100 m, and as he passes each marker you note the elapsed time as recorded by an electronic timer. You might obtain data like those in Table 4.1.

Table 4.1 Time-distance measurements: 800-m race

Distance, m	0	100	200	300	400	500	600	700	800
Elapsed Time, s	0	14.5	29.2	43.0	58.5	76.5	91.5	107.0	120.0

From these data, the graph of Fig. 4.4 can be plotted, with the time scale on the horizontal axis and the distance scale on the vertical axis. Each plotted point (small circles) represents a position of the runner at the time indicated. Note that although some points fall slightly off the "straight line," the runner did achieve his goal of uniform speed quite well.

When any two quantities are plotted on a graph of this kind (horizontal and vertical axes intersecting at the zero point or origin of both axes), and the resulting plot is a

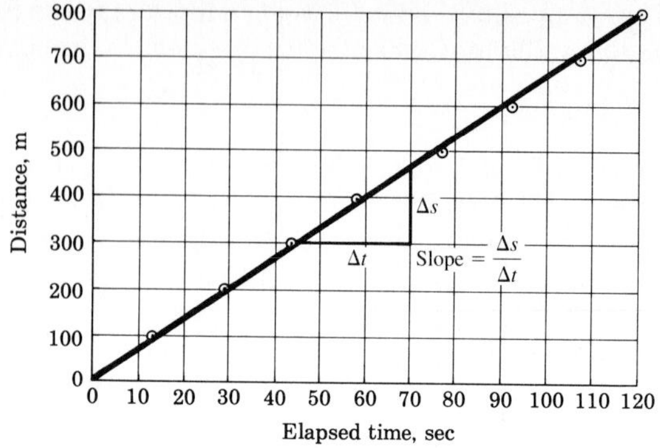

Fig. 4.4 "Smoothed" graphical plot of the data from Table 4.1, a distance vs. time plot of an 800-m run. For uniform or constant velocity the slope of the *s-t* graph determines the magnitude of the velocity.

straight line through the origin, the two quantities have a constant relationship to each other, or we say that they are *directly proportional* to each other. In this case, both from a visual inspection of the graph and from Eq. (4.1), the constant of proportionality is $\bar{v}$, the uniform speed. Taking the race as a whole, the average speed was

$$\bar{v} = \frac{s}{t} = \frac{800 \text{ m}}{120 \text{ s}} = 6.67 \text{ m/s}$$

Uniform or constant velocity can be determined from the slope of a distance-time graph.

The *slope* of the distance-time graph is determined by the ratio $\Delta s/\Delta t$, and it is therefore apparent that when an object is moving at uniform or constant velocity in a straight line its velocity can be calculated from the slope of its s-t graph.

4.5 ■ Instantaneous Velocity and Instantaneous Speed

In real situations speeds are not often uniform or constant. Runners slow down as they become tired; autos speed up and slow down with traffic, and brake to a complete stop at red lights and stop signs; and machine parts go through rapid cycles of speeding up, slowing to zero speed, and then reversing direction, as, for example, the pistons and valves in an internal combustion engine or a refrigeration compressor.

In Illustrative Problem 4.3 above, Eqs. (4.1) and (4.3) gave very different values for the average speed and the average velocity of the baseball during its 5.05-s trip. If we had computed the average speed and the average velocity for a shorter period of time, the difference between the two values would not have been so great. Indeed, as the length of the time interval t gets smaller and smaller, the difference between average speed and average velocity gets less and less. The reason for this is that the value for the distance traveled and the value for the magnitude of the displacement approach equality as the duration of the time interval t gets smaller and smaller.

Instantaneous speed is given by the reading from an automobile speedometer.

Speedometers and similar instruments can determine instantaneous speed and give either analog or digital readouts.

4.6 ■ Accelerated Motion

Motion in which velocity is changing is called *accelerated motion*.

Acceleration is ordinarily considered *positive* if the moving object is gaining speed, *negative* if speed is decreasing. Just as uniform velocity is the simplest form of motion, so uniform acceleration is the simplest type of accelerated motion.

Acceleration is defined as change in velocity per unit time or the time rate of change of velocity. It is a vector quantity.

If an object is moving in a straight line with an initial velocity v_1 at a given instant, and t s later its velocity has increased to v_2, the average acceleration is expressed by the formula

Acceleration defined. It is the time rate of change of velocity.

$$a = \frac{v_2 - v_1}{t} \tag{4.4}$$

Illustrative Problem 4.4 Sales literature for a new car claims that it will accelerate from 10 to 75 mi/h in 12 s. Find the average acceleration of which it is capable.

Acceleration is a vector quantity, possessing both magnitude and direction. It can be either *positive* (causing velocity to increase) or *negative* (causing velocity to decrease).

Solution The initial velocity v_1 is 10 mi/h. The final velocity v_2 is 75 mi/h. The time required t is 12 s. Using Eq. (4.4),

$$a = \frac{v_2 - v_1}{t}$$

and substituting,

$$a = \frac{75\text{ mi/h} - 10\text{ mi/h}}{12\text{ s}}$$

$$= \frac{65\text{ mi/h}}{12\text{ s}} = 5.4\,\frac{\text{mi/h}}{\text{s}} \qquad \textit{answer}$$

Note carefully the units in which acceleration is expressed. Since 65 mi/h equals 95.3 ft/s (see Notes below), the acceleration of the car referred to could be written as

$$a = \frac{95.3\text{ ft/s}}{12\text{ s}} = 7.9\,\frac{\text{ft/s}}{\text{s}}$$

NOTE: *A convenient ratio of miles per hour to feet per second is*

$$30\,\frac{\text{mi}}{\text{h}} = 30\,\frac{\text{mi}}{\text{h}}\left(\frac{5280\text{ ft}}{1\text{ mi}}\right)\left(\frac{1\text{ h}}{3600\text{ s}}\right) = 44\,\frac{\text{ft}}{\text{s}}$$

For example, what is the speed in mi/h of an airplane flying at 850 ft/s? With a pocket computer set the ratio 30 mi/h/44 ft/s and convert as follows:

$$\frac{30\text{ mi/h}}{44\text{ ft/s}} = \frac{v}{850\text{ ft/s}}$$

$$v = 580\text{ mi/h}$$

NOTE: *This important relationship (30 mi/h = 44 ft/s) should be memorized, since it will come into use many times.*

"Feet per second per second" is the most common form of expressing acceleration in the English system. It is usually written as ft/s^2.

The standard unit for acceleration in the metric system is the meter per second per second, written m/s^2. The unit cm/s^2 is sometimes used for smaller accelerations.

Illustrative Problem 4.5 The Bullet Train of the New Tokaido line in Japan can accelerate from 30 to 210 km/h in 32 s. Find the acceleration (assumed uniform) in meters per second per second (m/s^2).

Solution From Eq. (4.4)

$$a = \frac{v_2 - v_1}{t} = \frac{210{,}000\text{ m/h} - 30{,}000\text{ m/h}}{32\text{ s}}$$

Changing to, m/s^2, $\dfrac{5625 \text{ m/h}}{1 \text{ s}}$

$$= \frac{5625 \text{ m/h}}{1 \text{ s}} \times \frac{1 \text{ h}}{3600 \text{ s}}$$

$$= 1.56 \text{ m/s}^2 \qquad \textit{answer}$$

4.7 ■ Velocity and Accelerated Motion

Multiplying both sides of Eq. (4.4) by t results in

$$v_2 - v_1 = at$$

or

$$v_2 = v_1 + at$$

$$\text{Final velocity} = \text{initial velocity} + \text{change in velocity} \tag{4.5}$$

Now $v_2 - v_1$ equals the net gain in velocity in time t s. Let

$$v_2 - v_1 = v$$

Then

$$v = at \tag{4.6}$$

Change in velocity equals *acceleration* multiplied by *time*.

Acceleration can also be negative, as described by the following problem.

Illustrative Problem 4.6 The brakes of a certain car are capable of giving the car a uniform negative acceleration (*deceleration*) of 25 ft/s^2. (*a*) How long a time (seconds) does it take for this car to come to a stop from 50 mi/h? (*b*) How far does the car travel while being stopped?

Solution

(***a***) Using Eq. (4.6),

$$v = at$$

Solve for t, obtaining

$$t = \frac{v}{a}$$

It is now necessary to change the velocity as given in miles per hour, into feet per second. Remembering that 30 mi/h = 44 ft/s,

$$50 \text{ mi/h} = (50 \times \tfrac{44}{30}) \text{ ft/s} = 73.3 \text{ ft/s}$$

Now substituting the values for v and a in $t = v/a$,

$$t = \frac{73.3 \text{ ft/s}}{25 \text{ ft/s}^2} = 2.93 \text{ s} \qquad \textit{answer}$$

(***b***) The first step is to find the *average velocity* during the braking period. Assuming uniform deceleration, the average of the two velocities is merely their sum divided by 2:

$$\text{Average velocity} \qquad \bar{v} = \frac{v_2 + v_1}{2}$$

Applied to the present problem,

$$\bar{v} = \frac{73.3 \text{ ft/s} + 0 \text{ ft/s}}{2} = 36.7 \text{ ft/s}$$

Now, making use of Eq. (4.1″), $s = \bar{v}t$, and substituting values just determined,

$$s = 36.7 \text{ ft/s} \times 2.93 \text{ s} = 108 \text{ ft} \qquad \textit{answer}$$

4.8 ■ Distance and Velocity Related to Uniformly Accelerated Motion

Relation of Distance to Time There are many motions in which acceleration is not uniform. The mathematical treatment of these cases is beyond the scope of a basic text, and the discussions to follow concern themselves with uniform or constant acceleration.

If a body starts from rest, that is $v_1 = 0$, and is accelerated uniformly, the average speed is equal to one-half the final speed v_2:

$$\bar{v} = \text{average speed} = \tfrac{1}{2}v_2$$

From Eq. (4.6) we note that the final speed depends on the duration of the acceleration ($v = at$). Average speed, then, is $\frac{1}{2}v_2 = \frac{1}{2}at$. But *distance* equals *average speed* multiplied by *time,* and therefore the distance traveled in time t by a moving body *starting from rest,* whose uniform acceleration is a is

$$s = \tfrac{1}{2}at \times t$$

or

$$s = \tfrac{1}{2}at^2 \tag{4.7}$$

Relation of Velocity to Distance and Acceleration Suppose a rocket starts from rest and accelerates uniformly with a known acceleration. How can its velocity be determined at any specified distance from the starting point? Note that the problem involves a, v, and s.

From Eq. (4.6) we can write

$$t = \frac{v}{a}$$

and from Eq. (4.7), substituting for t,

$$s = \tfrac{1}{2}at^2 = \tfrac{1}{2}a \times \frac{v^2}{a^2} = \frac{v^2}{2a}$$

from which

$$v^2 = 2as \tag{4.8}$$

or

$$v = \sqrt{2as}$$

A rocket with a uniform acceleration of 7.0 m/s^2, for example, would, after traveling 1000 m, have a velocity

$$v = \sqrt{2 \times 7.0\frac{\text{m}}{\text{s}^2} \times 1000 \text{ m}} = 118 \text{ m/s}$$

We now have a group of formulas for uniform motion and uniformly accelerated motion *starting from rest*, which are collected here for ready reference:

Uniform motion:	$s = vt$	from (4.1″)
Accelerated motion:	$v = at$	(4.6)
	$s = \frac{1}{2}at^2$	(4.7)
	$v^2 = 2as$	(4.8)

Collected formulas for problems in uniform motion and accelerated motion, when the motion starts from rest.

4.9 ■ Generalized Equations of Accelerated Motion—Initial Velocity not Zero

It should be noted that the three formulas for accelerated motion listed above apply only when the accelerating object is assumed to *start from rest* (if increasing speed) or to *come to rest* (if decreasing speed). Accelerations are *positive* if speed is increasing, *negative* if speed is decreasing.

When an object has an initial velocity v_1 before the acceleration begins, we recall that

$$\text{Final velocity} = \text{initial velocity} + \text{change in velocity}$$

or

$$v_2 = v_1 + at \tag{4.5}$$

The distance traveled by a body which is moving initially with velocity v_1 and is then accelerated for a time t is the distance which would have been covered had the velocity remained unchanged, plus the distance resulting from the acceleration a acting for t s. As an equation, this becomes

$$s = v_1 t + \tfrac{1}{2}at^2 \tag{4.9}$$

Figure 4.5 illustrates graphically the relationships involved in both Eqs. (4.1) and (4.8).

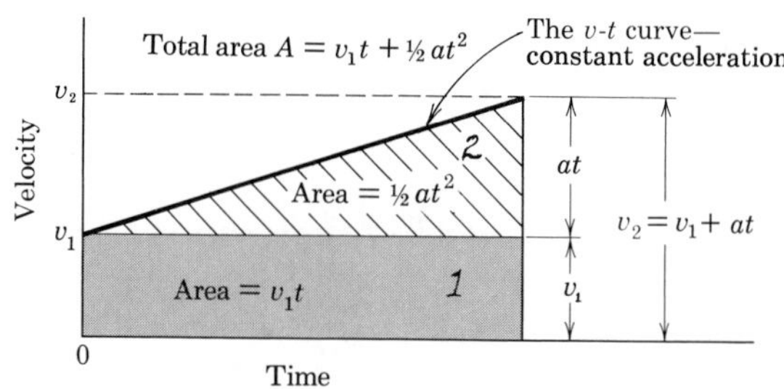

Fig. 4.5 Graph showing distance traversed by a body moving with an initial velocity v_1, which is then accelerated for a time t with an acceleration a. The shaded area (1) represents the distance covered due to the initial velocity, and the cross-hatched area (2) that due to the accelerated motion. The total area of the trapezoid represents the entire distance traveled in time t, or $s_{tot} = v_1 t + \frac{1}{2}at^2$.

An expression relating distance traveled to initial and final velocities may be derived as follows, the distance moved by such a body being the *average velocity* multipled by *time*:

$$s = \left(\frac{v_2 + v_1}{2}\right)t$$

from which

$$v_2 + v_1 = \frac{2s}{t}$$

But, from Eq. (4.5)

$$v_2 - v_1 = at$$

Multiplying these two equations,

$$(v_2 + v_1)(v_2 - v_1) = \frac{2s}{t} \times at$$

gives

$$v_2^2 - v_1^2 = 2as$$

or

$$v_2^2 = v_1^2 + 2as \tag{4.10}$$

These generalized equations for uniform and accelerated motion are collected here for ready reference:

Generalized equations for uniform motion and accelerated motion.

Uniform motion:

$$s = vt \qquad \text{from (4.1}'')$$

Distance = uniform (constant) velocity × time

Accelerated motion:

$$v_2 = v_1 + at \tag{4.5}$$

Final velocity = initial velocity plus gain in velocity due to acceleration

$$s = v_1 t + \tfrac{1}{2}at^2 \tag{4.9}$$

Total distance = distance due to initial velocity plus distance due to acceleration

$$v_2^2 = v_1^2 + 2as \tag{4.10}$$

or, $$s = \frac{v_2^2 - v_1^2}{2a} \tag{4.10'}$$

Distance in terms of initial and final velocities and uniform acceleration

The term *generalized*, associated with the above equations, means that these equations apply to the general case of initial and final velocities *not being zero*. In using these equations proper regard must be given to the *algebraic sign* (+ or −) of the acceleration.

Illustrative Problem 4.7 The velocity-time graph of Fig. 4.6 was plotted by observing the movements of an automobile in traffic. Study the graph carefully and see what it tells you.

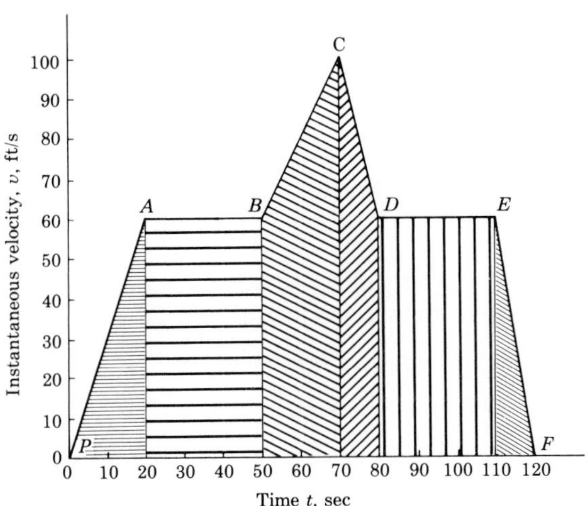

Fig. 4.6 Graph of instantaneous velocity vs. time for an automobile being driven in heavy traffic. Studying such a graph can tell you a great deal about the motions of the vehicle.

Solution First, you should note that the auto starts from rest ($v_1 = 0$) and accelerates uniformly (How do you know that the acceleration is uniform?) for 20 s, at which time its velocity has increased to 60 ft/s (about 41 mi/h). Then, at point *A* on the graph, it proceeds at constant velocity (zero acceleration) for 30 s at 60 ft/s, moving with the traffic. At point *B* it accelerates for 20 s (perhaps to pass another car), attaining a maximum speed of 90 ft/s (about 61.4 mi/h) at point *C*. It then slows down (negative, uniform acceleration) for 10 s to its former cruising speed of 60 ft/s (point *D*). It moves at this speed for 30 s and then, at point *E*, the auto undergoes uniform negative acceleration coming to a complete stop in 10 s.

The numerical values for the accelerations represented by the *PA*, *BC*, *CD*, and *EF* parts of the graph can be determined from the equation $v_2 - v_1 = at$. Solve this equation for a, obtaining

$$a = \frac{v_2 - v_1}{t}$$

The uniform acceleration for any one portion of the curve is merely the ratio of the *change in velocity* for that portion to the time required for that part of the motion.

For the motion represented by *PA*,

$$a_{PA} = \frac{60 \text{ ft/s}}{20 \text{ s}} = 3 \text{ ft/s}^2$$

In like manner, you should now calculate the accelerations a_{BC}, a_{CD}, and a_{EF}.

Finally, what can be learned from the graph about the distances traveled? We know that for *constant velocity*, distance traveled equals uniform velocity multiplied by time, or $s = vt$. Therefore, for the AB portion of the travel, the distance traversed is

Velocity-time curves provide useful information about bodies in motion.

$$s_{AB} = v_{AB}t_{AB} = 60 \text{ ft/s} \times (50 - 20) \text{ s}$$
$$= 1800 \text{ ft}$$

As was pointed out in Fig. 4.5 this is equivalent numerically to the shaded area under the AB portion of the graph.

For the PA part of the travel, where the car is uniformly accelerated, recall that $s = \frac{1}{2}at^2$, and that $a_{PA} = 3 \text{ ft/s}^2$. Therefore,

Distance traveled by a body in accelerated motion can be determined from the area under a velocity-time curve.

$$s_{PA} = \frac{3 \text{ ft/s}^2 \times (20 \text{ s})^2}{2}$$
$$= 600 \text{ ft}$$

A quick look at Fig. 4.6 shows that this is numerically equal to the shaded area under the PA part of the graph.

Now, as an exercise, calculate the distances traveled in the remaining portions of the travel and sum up everything for the total distance traveled. Do you find the result to be 6700 ft?

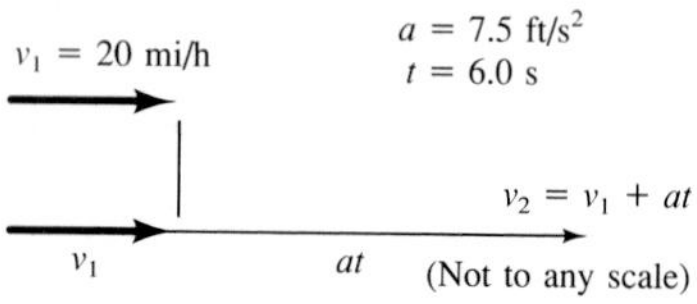

$v_1 = 20$ mi/h $\quad a = 7.5 \text{ ft/s}^2$ $\quad t = 6.0$ s

$v_2 = v_1 + at$

$v_1 \quad at \quad$ (Not to any scale)

Fig. 4.7 Sketch of vectors for Illustrative Problem 4.8 (not to any scale).

Illustrative Problem 4.8 Your speedometer reads 20 mi/h as you press the accelerator to the floor. If your car is capable of an acceleration of 7.5 ft/s^2, what should your speedometer read after exactly 6.0 s? (See sketch, Fig. 4.7.)

Solution v_1, a, and t are given, and v_2 is desired, so select Eq. (4.5). The value of a is positive (+), since the car is accelerating. Figure 4.7 shows the vectors involved.

$$v_2 = v_1 + at$$

Now since $20 \times \frac{44}{30} = 29.3$ (see Sec. 4.6)

$$20 \text{ mi/h} = 29.3 \text{ ft/s}$$

Substituting,

$$v_2 = 29.3 \text{ ft/s} + (7.5 \text{ ft/s}^2 \times 6.0 \text{ s})$$
$$= 74.3 \text{ ft/s}$$

or

$$v_2 = 74.3 \text{ ft/s} \times \frac{30 \text{ mi/h}}{44 \text{ ft/s}}$$
$$= 51 \text{ mi/h} \qquad \textit{answer}$$

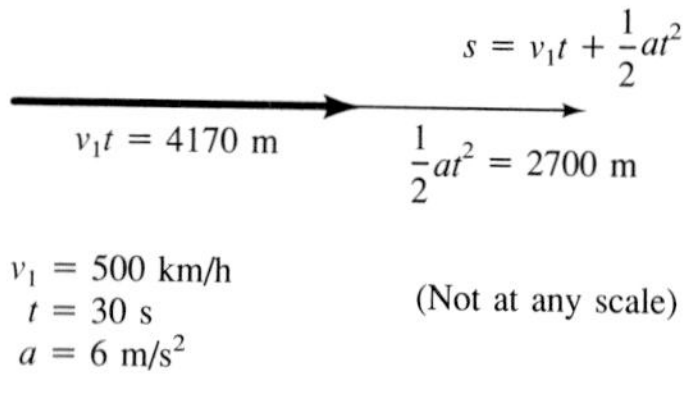

$s = v_1t + \frac{1}{2}at^2$

$v_1t = 4170$ m $\quad \frac{1}{2}at^2 = 2700$ m

$v_1 = 500$ km/h
$t = 30$ s
$a = 6 \text{ m/s}^2$

(Not at any scale)

Fig. 4.8 Sketch of vectors for Illustrative Problem 4.9 (not to any scale).

Illustrative Problem 4.9 A jet plane is flying at a steady 500 km/h when the pilot goes to full power. The additional thrust gives an acceleration of 6 m/s^2. If full power is left on for 30 s, how far (km) will the aircraft travel during the acceleration period? (See sketch, Fig. 4.8.)

Solution v_1, a, and t are given and s is desired. The use of Eq. (4.9) is therefore indicated. Consistent units must first be obtained.

Now,

$$500 \text{ km/h} = 139 \text{ m/s}$$

From Eq. (4.9),

$$s = v_1t + \tfrac{1}{2}at^2$$

Substituting gives

$$s = 139 \text{ m/s} \times 30 \text{ s}$$
$$+ \frac{6 \text{ m/s}^2 \times (30 \text{ s})^2}{2} \qquad (a \text{ is positive})$$

$$= 4170 \text{ m} + 2700 \text{ m}$$
$$= 6870 \text{ m}$$
$$= 6.87 \text{ km} \qquad \textit{answer}$$

4.10 ■ The Accleration of Gravity—Free Fall

The laws and equations of uniformly accelerated motion apply to freely falling bodies, and the force of gravity which causes the acceleration is a constant at any given locality and becomes a known quantity in the solution of all free-fall problems.

Galileo (1564–1642) is credited with the discovery that all bodies acted upon by the force of earth's gravitation *fall with the same acceleration* if air resistance is neglected (Fig. 4.9). It is true that the earth pulls with twice the force on a 100-kg body that it does on a 50-kg body, but the 100-kg body has twice as much mass which has to be set in motion. Its resistance to a change in motion (*inertia*) is twice as great, and therefore the accelerating force per unit mass is the same for both bodies.

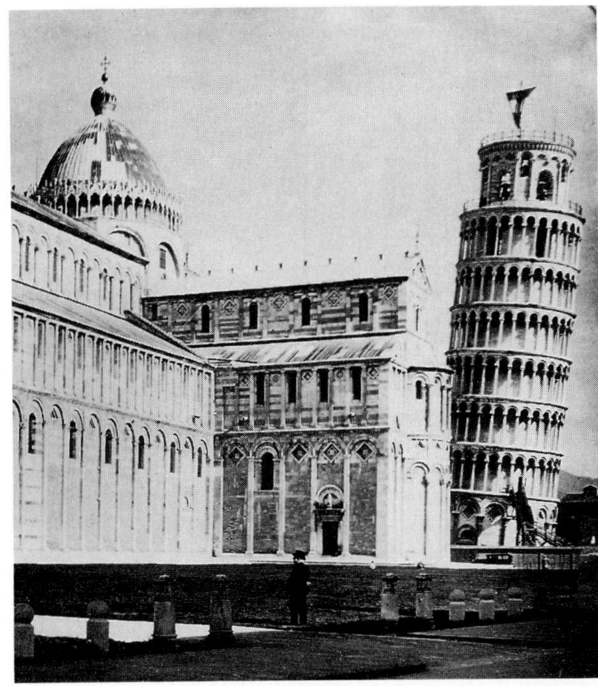

Fig. 4.9 The famous leaning tower in Pisa, Italy. Galileo Galilei may have conducted some of his experiments with falling bodies from this tower.

This tower, on the grounds of the University of Pisa, leaned sufficiently in Galileo's time that falling body experiments could have been conducted from it. There is no definite evidence that Galileo used it however. He *may have*, but proof is lacking. It is probable that most of his experiments involved measuring the time required for balls of various sizes and weights to roll down a smooth inclined plane. (Why would this have been a better experimental approach, in Galileo's time, than attempting to observe actual falling bodies?) Today, the Leaning Tower of Pisa leans so alarmingly that efforts are being made to shore up the collapsing foundation, lest it topple during this century.

The acceleration due to the force of earth's gravity is approximately 9.81 m/s^2 (981 cm/s^2) at sea level at the equator. This value is called g, the *acceleration of gravity*. In the English system the value of g is approximately 32.2 ft/s^2. Slight variations exist due to altitude and latitude, but they are too small to affect the practical problems of technical physics on the earth's surface. Air resistance definitely affects the rate of fall, and its effect must be allowed for in all real problems such as those encountered in predicting the flight of projectiles and rockets and the trajectory of bombs, as well as in the design of parachutes. Objects "falling" toward the earth from outer space encounter air resistance so great that the heat generated usually consumes them before they hit the earth's surface. The shooting stars (*meteors*) seen frequently in the August sky are actually "falling bodies" from outer space burning themselves up in the heat generated by atmospheric friction. Reentry of satellites and spacecraft into the earth's atmosphere is one of the very difficult problems which had to be solved before manned flights in space were possible. Heat shields capable of withstanding the very high temperatures produced by atmospheric friction are provided on all such "reentry vehicles." The shields are made of specially developed ceramic tiles.

The acceleration due to earth's gravity is $g = 9.81$ m/s^2, or 32.17 ft/s^2.

Table 4.2 shows the distances fallen and the velocities attained at the end of each second by a body in free fall from rest over a total time period of 5 s (air resistance neglected). Values are for the earth's surface.

Table 4.2 Data for a freely falling body at the earth's surface

Elapsed Time,	*Distance Fallen, s*		*Instantaneous Velocity*	
s	*m*	*ft*	*m/s*	*ft/s*
0	0	0	0	0
1 ($s = \frac{1}{2}gt^2$)	4.9	16	9.8 (approx)	32 (approx)
2	19.6	64	19.6	64
3	44.1	144	29.4	96
4	78.4	256	39.2	128
5	122.5	400	49.0	160

NOTE: $g = 9.8$ m/s^2 = 32 ft/s^2 approx. More exactly $g = 32.2$ ft/s^2, or 9.81 m/s^2 (air resistance neglected).

Collected Equations for Free Fall By merely replacing a by g and s by h in Eqs. (4.6), (4.7), and (4.8), we obtain a set of formulas for freely falling bodies for the special case where $v_1 = 0$, that is, when the fall starts from rest.

$$v = gt \tag{4.11}$$
$$h = \tfrac{1}{2}gt^2 \tag{4.12}$$
$$v^2 = 2gh \tag{4.13}$$

or

$$h = \frac{v^2}{2g} \tag{4.13'}$$

A set of equations for solving problems involving freely falling bodies.

A set of general formulas for cases where initial or final velocities *are not zero* is easily obtained from Eqs. (4.5), (4.9) and (4.10):

$$v_2 = v_1 \pm gt \tag{4.14}$$
$$h = v_1t \pm \tfrac{1}{2}gt^2 \tag{4.15}$$
$$v_2^2 = v_1^2 \pm 2gh \tag{4.16}$$

Algebraic signs are applied as follows: If initial velocity is *upward* the minus (−) sign is used, since the acceleration g is *downward*. If initial velocity is downward, the plus (+) sign is used.

Illustrative Problem 4.10 The hammer of a pile driver is released at the top of its boom and falls vertically a distance of 38 ft. What is its velocity as it hits the piling?

Solution The hammer starts from rest, so $v_1 = 0$; s is known, and $g = 32.2$ ft/s^2. From Eq. (4.13),

$$v^2 = 2gh$$

Substituting,

$$v^2 = 2 \times 32.2 \text{ ft/s}^2 \times 38 \text{ ft} = 2450 \text{ ft}^2/\text{s}^2$$
$$v = \sqrt{2450 \text{ ft}^2/\text{s}^2} = 49.5 \text{ ft/s} \qquad \textit{answer}$$

4.11 ■ Free Fall and Air Resistance—Terminal Velocity

Air is a fluid, and any object moving through air is subject to the frictional force of the airstream acting along the surfaces of the moving object. These frictional retarding forces, sometimes called *drag*, oppose the downward motion of a falling body.

Since air friction increases as the speed of a body moving through air increases, as a falling body is accelerated downward by the force of gravity a point is eventually reached at which the air resistance force (acting upward) becomes equal to the gravitational force (acting downward). There is then no *unbalanced* force on the body, and its *acceleration* toward the earth becomes zero. Its downward (falling) velocity becomes constant. This constant velocity reached by bodies falling in air is called *terminal velocity*.

Different bodies have different terminal velocities, depending on their shape, mass-to-surface-area ratio, roughness of surface, etc. Raindrops fall at about 7–10 m/s when there is no wind; a human being free falls at about 60–80 m/s.

It should be noted, therefore, that the equations developed above for free fall [Eqs. (4.11) through (4.16)] are theoretical or *idealized* formulas that yield correct results only if the fall were to occur in a vacuum. The errors are relatively small, however, if the mass-to-surface-area ratio of the falling object is large, as, for example, a lead ball.

Figure 4.10 depicts a simple laboratory experiment that confirms Galileo's finding that all bodies acted on by the earth's gravity fall with the same acceleration, *if they fall in a vacuum*. In contrast, Fig. 4.11 compares graphically the speed-time curves of two falling bodies subject to the acceleration of earth's gravity—one body falling in a vacuum, and the other falling in air.

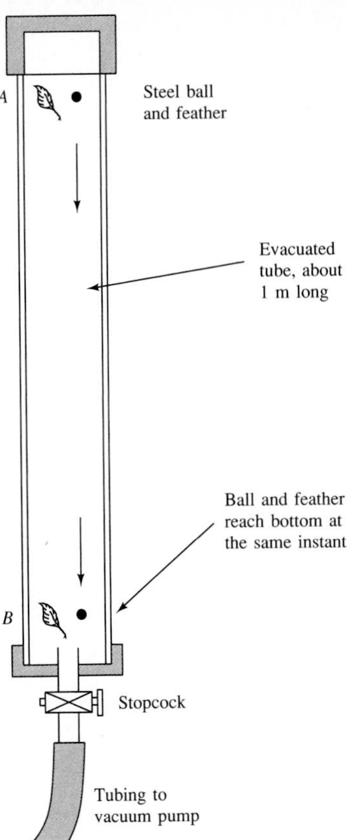

Fig. 4.10 Sketch of laboratory apparatus to show that, in a vacuum, all objects fall with the same acceleration under the influence of the earth's gravity. A steel ball and a feather are placed in a long glass tube, which is then sealed and connected to a vacuum pump. After the tube is evacuated, the tube is quickly inverted and the ball and feather can be seen falling together, reaching the bottom simultaneously.

4.12 ■ Vertical Motion Is Affected by the Force of Gravity

The vertical motion of objects thrown or projected upward or downward is subject to *acceleration* due to gravity—negative acceleration if projected upward, positive acceleration if projected downward. The *vertical component* of all motions in the earth's gravitational field is likewise affected. Consider the following, for example.

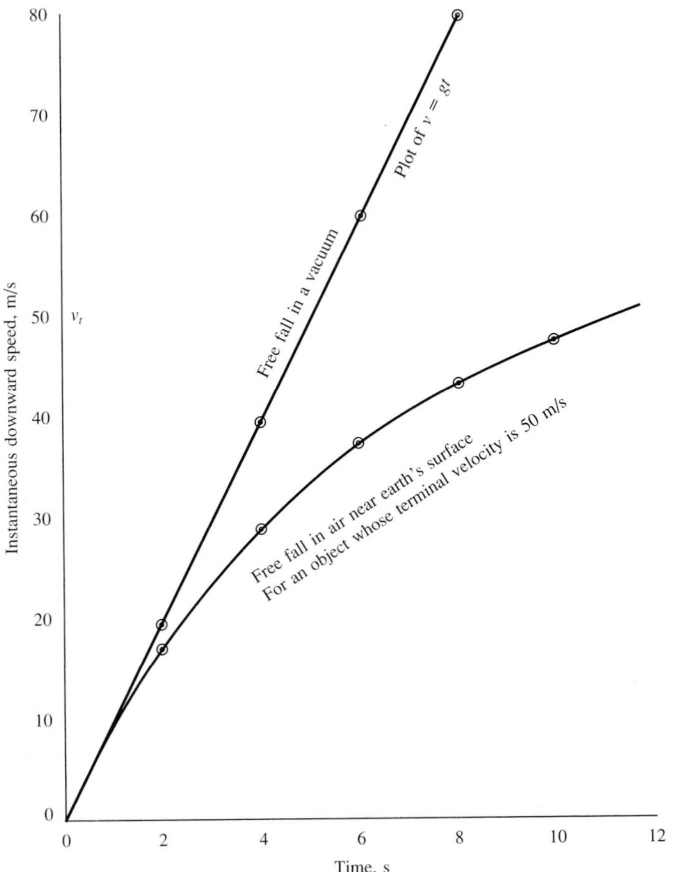

Fig. 4.11 Speed-time curves of two different falling bodies subject to earth's gravity—one body assumed to be falling in a vacuum, and the other falling in the earth's atmosphere with an assumed terminal velocity of 50 m/s.

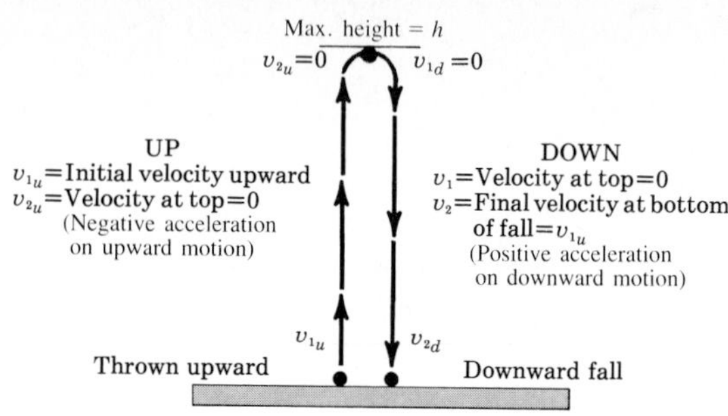

Fig. 4.12 A ball thrown vertically upward with initial velocity v_1 travels upward a distance h while it is being decelerated (negative acceleration) by the force of gravity. When its upward velocity is reduced to zero, it then becomes a freely falling body, starting from rest at its maximum height position. It falls vertically to earth, being positively accelerated by the force of gravity. Its velocity as it arrives back at the level from which it was thrown is equal and opposite to the initial upward velocity (see Illustrative Problem 4.11).

Illustrative Problem 4.11 A baseball is thrown straight up, and 5.4 s elapse before it hits the ground. (*a*) How high did it go? (*b*) With what velocity was it thrown? (See Fig. 4.12.)

Solution

(***a***) At the top of its flight its velocity, momentarily, will be zero, and we can consider its return to earth as if it were a falling body dropped from that height. Let h stand for the maximum height the ball attains. The flight upward is *decelerated* by gravity over the upward distance, and the flight downward is accelerated by the same force over the same distance. The *time of rise* is equal to the *time of fall*, and the velocity as it returns to the ground (neglecting air resistance) is equal to that with which it was thrown upward. Let T = total elapsed time, and t = time of fall. Note that

$$t = \frac{T}{2} = \frac{5.4 \text{ s}}{2} = 2.7 \text{ s}$$

Select Eq. (4.12) since it involves t, g, and h:

$$h = \tfrac{1}{2}gt^2$$

Substituting,

$$h = \frac{9.81 \text{ m/s}^2}{2} \times (2.7 \text{ s})^2 = 35.8 \text{ m} \qquad \textit{answer}$$

(***b***) To find the velocity with which it was thrown, which, it will be recalled, is equal to the velocity with which it hits the ground, select Eq. (4.11):

$$v = gt$$

where again $t = T/2$.

$$v = 9.81 \text{ m/s}^2 \times 2.7 \text{ s} = 26.5 \text{ m/s} \qquad \textit{answer}$$

These problems illustrate the use of the equations for freely falling bodies. Study carefully all the steps in their solution.

Illustrative Problem 4.12 A baseball is thrown straight up from the top of a tall building with an initial velocity of 30.0 m/s (see Fig. 4.13). As it comes back down it barely misses the railing and falls to the street 290 m below. (*a*) How far up does it go before starting to fall downward? (*b*) What is the total time of flight? (*c*) With what velocity does it hit the street? (Neglect air resistance.)

Solution

(***a***) The initial upward velocity is known. So is the value of g (9.81 m/s^2), and so is the final upward velocity, which is zero. The height h is desired. Equation (4.16) involves all these factors, so use it and solve for h. Note that the acceleration is negative in this case since it is oppositely directed to the motion. The motion, being upward, is taken as positive.

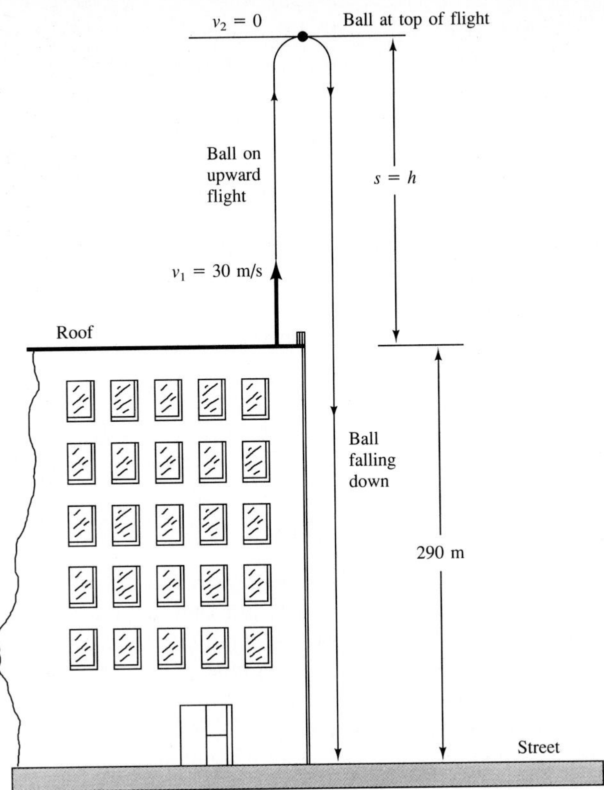

Fig. 4.13 Diagram of ball thrown upward from building roof and allowed to fall all the way to street level (Illustrative Problem 4.12).

$$v_2^2 = 0 = v_1^2 + 2gh \qquad \text{and} \qquad v_1^2 = -2gh$$

Solving for h, remembering that g is negative, gives

$$h = \frac{v_1^2}{-2(-g)}$$

Substituting values we get,

$$h = \frac{(30.0 \text{ m/s})^2}{(-2)(-9.81 \text{ m/s}^2)}$$

$$= 45.9 \text{ m} \qquad \textit{answer}$$

(***b***) Total time of flight, T, equals the time in upward flight, t_u, plus the time of fall from the maximum height attained, t_f. But the time in upward flight, t_u, is equal to the time for the ball to fall back down to the level where it was thrown, t_d. We now know h from part (*a*), so we can use Eq. (4.12) to find t_d.

$$h = \tfrac{1}{2}gt_d^2$$

Solving for t_d,

$$t_d = \sqrt{2h/g}$$

Substituting values,

$$t_d = \sqrt{\frac{2 \times 45.9 \text{ m}}{9.81 \text{ m/s}^2}} = 3.06 \text{ s}$$

the time of fall to reach the roof of the building on the way down. But $t_d = t_u = 3.06$ s, the time for upward travel.

Use the same equation to find t_f, recalling that the total distance of fall is now 290 m plus 45.9 m, or 335.9 m.

Substituting,

$$t_f = \sqrt{\frac{2 \times 335.9 \text{ m}}{9.81 \text{ m/s}^2}}$$

$$= 8.28 \text{ s}$$

Total time

$$T = 3.06 \text{ s} + 8.28 \text{ s} = 11.3 \text{ s} \qquad \textit{answer}$$

(*c*) Treat this part of the problem as a free fall from a height of 335.9 m, and use Eq. (4.13):

$$v_2^2 = 2gh \qquad \text{since} \qquad v_1 = 0$$

Substituting,

$$v_2^2 = 2 \times 9.81 \text{ m/s}^2 \times 335.9 \text{ m}$$
$$= 6590 \text{ m}^2\text{/s}^2$$
$$v_2 = \sqrt{6590 \text{ m}^2\text{/s}^2} = 81.2 \text{ m/s} \qquad \textit{answer}$$

4.13 ■ Missiles and Projectiles

Thus far we have been dealing with objects going straight up or falling straight down, under the influence of only one force, that of gravitational attraction. Now, the case of combined horizontal and vertical motion—motion in two dimensions but in a vertical plane—will be considered. Some extremely interesting and practical problems arise in motions of this sort.

To begin, several terms need to be defined. The word *missile* is one of these. It is derived from Latin roots meaning ''to send on a mission''. As interpreted today, a missile is any object that is shot, thrown, launched, or propelled toward a target. A missile may, or may not, incorporate its own propulsion and guidance systems.

All projectiles are missiles, but not all missiles are projectiles.

A *projectile* is an object (missile) that is thrown, shot, catapulted, or launched into the atmosphere or into space, and then left with no further propelling, guidance, or other forces acting on it except the force of gravity, the effects of the earth's rotation, and air resistance. In other words, a projectile is a missile dominated by gravitational and air-resistance forces. All thrown, batted, or driven balls are projectiles.

A ballistic missile is not a projectile in the early stages of its flight, but becomes a projectile later in the flight, when its propulsion system and guidance systems have exhausted their fuel.

A *ballistic missile* is a missile which, though it is usually propelled and guided by self-contained devices early in its flight, exhausts these and becomes a true projectile in later stages, subject only to gravitational and air-resistance forces.

''Smart bombs'' incorporate guidance systems of one kind or another; that is, they are ''guided missiles.'' Some types ''home in'' on the target using radio or radar signals; others are guided by laser beams. They are not projectiles, since their flight is influenced by other forces as well as by gravity. One type of missile can be guided to its target by means of signals sent back to its guidance system through a thin fiber-optic thread which it tows in flight.

Ball games of many kinds are played with projectiles.

Most of the balls with which modern games are played are projectiles. The third baseman, in ''rifling'' a throw to first to catch the runner, in effect is solving a problem in projectile motion in two dimensions. He must aim the ball upward at the correct angle to compensate for the acceleration of gravity so that with his throwing velocity and the distance to first base as factors, the ball will arrive there one foot off the ground in the absolute minimum of time. Examples from golf, basketball, and tennis could also be cited.

In the treatment to follow we shall assume that the distance traversed by a projectile is short enough for the earth's surface underneath to be considered a flat plane. Further, all such effects as projectile spin, air resistance, the earth's rotation, etc., will be neglected.

4.14 ■ Horizontally Launched Projectiles

One familiar case of projectile motion is that which occurs when an object such as a ball, bullet, or bomb is projected or released with an initial horizontal velocity. It immediately begins acquiring vertical velocity downward under the pull of gravity, but its horizontal velocity is not affected thereby if air resistance is assumed zero.

It is easy to set up and observe a simple experiment of this kind. Suppose two identical steel balls A and B are on the edge of a table (Fig. 4.14): B is projected horizontally, perhaps by a flip of the finger, while A is just rolled over the edge of the table (toward the reader) at the same instant. It will be observed that for any value of the horizontal velocity, v, fast or slow, the two balls hit the floor at the same instant. Listen for the impact on the floor. Both balls undergo *free fall* through the same vertical distance h, and the fact that ball B has a horizontal velocity and ball A does not, makes no difference in the vertical motion. *The vertical and horizontal motions are independent of each other. Each takes place as if the other were zero.*

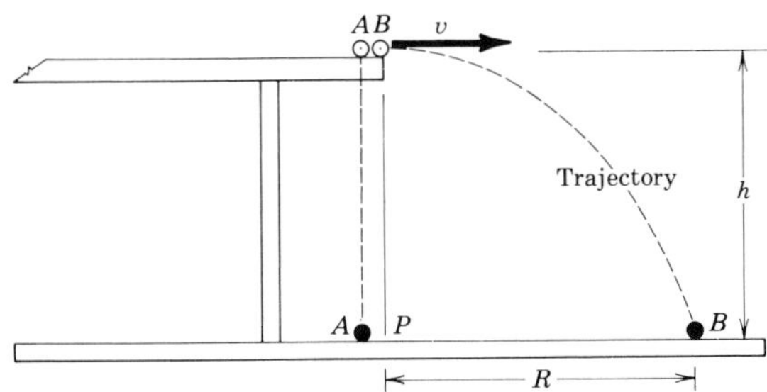

Fig. 4.14 Horizontal and vertical motions of a simple projectile (air resistance assumed zero).

The actual flight path of ball B is called the *trajectory*, and if air resistance is neglected, this curve is a *parabola*.

Both the vertical component and the horizontal component of projectile motion occur as if the other did not exist.

We have already developed and used equations which describe the motion of ball A. Let us now do so for ball B, noting that they will then apply to any projectile with an initial horizontal velocity v which has no forces acting on it but the force of gravity.

First, the *time of flight* t for ball B is equal to that for ball A, and is simply the time required for the ball to fall through a vertical height h. Using Eq. (4.12),

$$h = \tfrac{1}{2}gt^2$$

and the time of flight

$$t = \sqrt{2h/g} \tag{4.17}$$

The ball B has a *horizontal range* R, as follows:

$$R = vt = v\sqrt{2h/g} \tag{4.18}$$

A military application of the above principles is in aircraft bombing, when bombs are released at high altitude with the aircraft in horizontal flight. Figure 4.15 shows the basic elements of the problem. At release, (point P), the bombs have a horizontal velocity $\mathbf{v}$ equal to that of the aircraft. The straight-line distance between the aircraft and the target is called the line of sight (LOS). The angle between the LOS and the horizontal flight path of the aircraft is called the *angle of depression* θ. In order to strike the target, the bomb must be released when the angle of depression θ is given by

$$\tan\theta = \frac{h}{R} = \frac{h}{v\sqrt{2h/g}} \qquad \text{[from Eq. (4.17)]}$$

or, simplifying, when

$$\tan\theta = \frac{1}{v}\sqrt{\frac{gh}{2}} \tag{4.19}$$

It is again emphasized that wind conditions and air resistance are not considered in the simplified treatment here given. Modern bombsights make speedy computations of such problems and take into account many possible sources of error. At night, or under conditions of poor visibility, they compute the proper release point from radar information. Ordinary bombs are true projectiles.

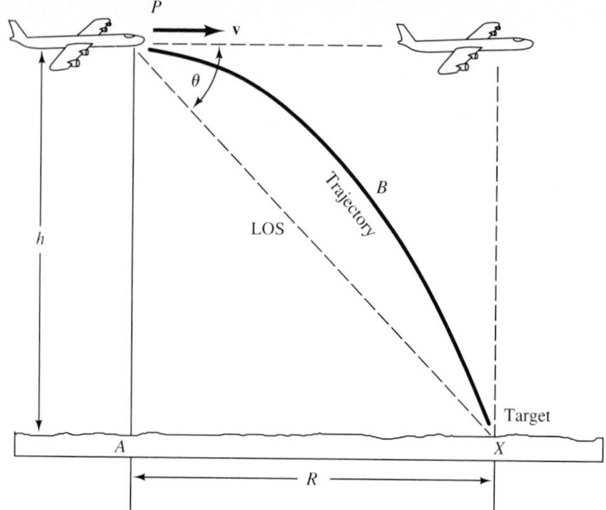

Fig. 4.15 High-altitude horizontal bombing is a military application of the laws of projectile motion. The horizontal and vertical motions of the bomb occur independently of one another. The equations derived from this diagram assume air resistance is not a factor.

4.15 ■ Launcher and Target on the Same Horizontal Plane—Projectile Launched at an Upward Angle

This description and the sketch of Fig. 4.16 describe the standard problem for golf, tennis, baseball, football passes and punts, and gunnery (neglecting air resistance).

In many games, such as baseball, golf, and tennis, the ball leaves a point near the ground at an angle and rises as it moves horizontally. It reaches a maximum altitude and then descends to a target on the same level—another player, the fairway, or the court. Figure 4.16 shows the basic elements of the problem. The ball at A has an initial velocity v at an angle θ with the horizontal. The horizontal component of its velocity is $v_x = v \cos \theta$, and the vertical component is $v_y = v \sin \theta$. The ball rises to a maximum height h at P. It comes back to the launch level at X. The horizontal distance traveled (range) is R. The curve APX is its trajectory and it would be a parabola if air resistance were absent. If the acceleration of gravity were zero, the ball would move out along the line AB indefinitely.

The horizontal component of the velocity remains constant (air resistance neglected), but the vertical component is always affected by the acceleration of gravity, and its value at any time t is expressed by

$$v_{y_t} = v \sin \theta - gt$$

where g is the acceleration of gravity and t is the number of seconds which have elapsed since the ball left point A. The height attained by the ball at any time t is, from Eq. (4.15) (see Fig. 4.17)

Height, $$y_t = vt \sin \theta - \tfrac{1}{2}gt^2 \qquad (4.20)$$

Now, when the ball arrives at the target X, $y = 0$ and $t =$ total time of flight, T. Putting $y_t = 0$ in Eq. (4.20) and solving for t, we obtain

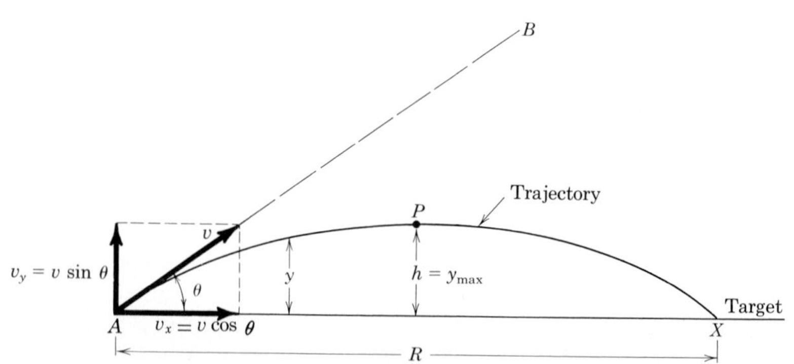

Fig. 4.16 Projectile flight when launcher and target are on the same horizontal plane. This is often the case in games and sports, as well as in military gunnery both on land and sea (air resistance neglected).

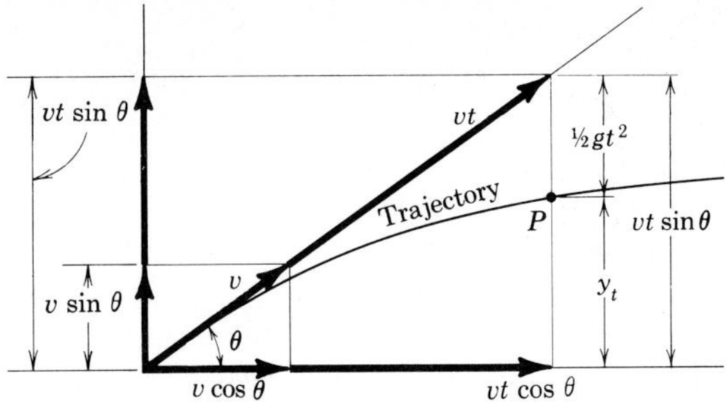

Fig. 4.17 Height y_t of projectile or ball at any time t after launch. With projectile or ball at P, after t s, its height is $y_t = vt \sin\theta - \frac{1}{2}gt^2$.

Time of flight $$t = T = \frac{2v \sin\theta}{g} \tag{4.21}$$

the total time of flight. For the range R,

$$R = v_xT = v \cos\theta \times \frac{2v \sin\theta}{g}$$

$$= \frac{2v^2 \sin\theta \cos\theta}{g}$$

or, since $2 \sin\theta \cos\theta = \sin 2\theta$,

Range, $$R = \frac{v^2 \sin 2\theta}{g} \tag{4.22}$$

which is the basic formula for *range* when *initial velocity* and *angle of elevation* are known. The maximum height to which the ball rises is given by

Maximum height $$h = \frac{v^2 \sin^2\theta}{2g} \tag{4.23}$$

Sporting and military applications of the above principles are numerous. Some illustrative problems follow.

Remember—none of these equations [Eqs. (4.21), (4.22), (4.23)] takes air resistance into account. In reality, range, height, and time of flight are all seriously affected by air resistance, the magnitude of the effect being dependent on the size, density, and conformation of the projectile. In military operations corrections for air resistance, wind, earth rotation, and projectile spin are made by fire-control computers. Why put "dimples" on a golf ball? Because their presence and arrangement affect the flight of the ball to a remarkable extent. A properly "dimpled" ball, when driven by "Iron Byron," the robot golfer, will fly 50 to 100 yd farther than a perfectly smooth ball of the same size and weight and interior construction.

Illustrative Problem 4.13 A golfer selects a club which will give the ball, when hit properly, an angle of elevation of 35°. His target is the flag stick 160 yd away on the green. With what velocity (ft/s) must the ball leave the club in order to land in the target area?

Projectile theory has many applications in sports and in military operations. Advanced mathematical methods, computer modeling, and experimental studies yield accurate corrections for air resistance.

Solution Note that Eq. (4.22) contains the elements of the problem. First, solve it for v:

$$Rg = v^2 \sin 2\theta$$

from which

$$v^2 = \frac{Rg}{\sin 2\theta} \quad \text{and} \quad v = \sqrt{\frac{Rg}{\sin 2\theta}}$$

Substituting values yields

$$v = \sqrt{\frac{160 \text{ yd} \times 3 \text{ ft/yd} \times 32.2 \text{ ft/s}^2}{\sin 70°}}$$

$$= \sqrt{\frac{15460 \text{ ft}^2/\text{s}^2}{0.94}} = \sqrt{16450 \frac{\text{ft}^2}{\text{s}^2}}$$

$$= 128 \text{ ft/s} \qquad \textit{answer}$$

Illustrative Problem 4.14 Artillery range to a target is 15,000 yd and the muzzle velocity of the gun being used is 2200 ft/s. (*a*) Find the required theoretical angle of elevation of the gun; and (*b*) the maximum range of this gun over level ground. Neglect air resistance.

Solution

(***a***) R, v, and g are known, and θ is desired. Select Eq. (4.22), and solve it for sin 2θ:

$$\sin 2\theta = \frac{Rg}{v^2}$$

Substituting,

$$\sin 2\theta = \frac{15{,}000 \text{ yd} \times 3 \text{ ft/yd} \times 32.2 \text{ ft/s}^2}{(2200 \text{ ft/s})^2}$$

$$= 0.299$$

From a pocket calculator,

$$2\theta = 17.4°$$

and

$$\theta = 8.7° \qquad \textit{answer}$$

(***b***) To find maximum theoretical range, make use of Eq. (4.22),

$$R = \frac{v^2 \sin 2\theta}{g}$$

For any given value of v, R will be a maximum when sin 2θ is a maximum, since g is a constant. Sin 2θ is a maximum when $2\theta = 90°$, making sin $2\theta = 1$. Consequently, maximum range will occur when $\theta = 45°$. Solving,

$$R_{max} = \frac{(2200 \text{ ft/s})^2 \times \sin 90°}{32.2 \text{ ft/s}^2}$$

$$= 150{,}300 \text{ ft} = 50{,}100 \text{ yd} \qquad \textit{answer}$$

This indicated maximum range is not realistic for such a gun, as air resistance would reduce the theoretical range a great deal. (See Figs. 4.18 and 4.19.)

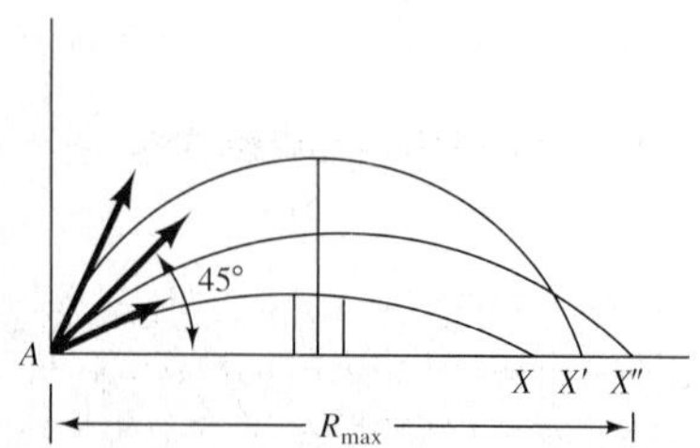

Fig. 4.18 The theoretical maximum range of a projectile is attained when it is fired (thrown) at an angle of 45° with the horizontal (air resistance and wind effects neglected).

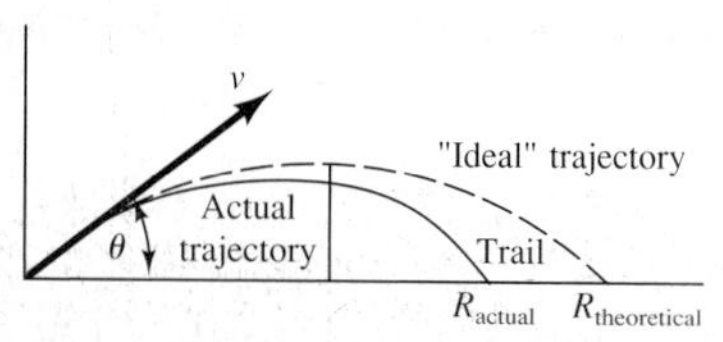

Fig. 4.19 Effect of air resistance on the flight of projectiles, balls, and ballistic missiles. The ideal parabolic trajectory (dashed line) is not even roughly approximated by most projectiles. Lightweight, high-velocity projectiles are affected appreciably more than heavy, slow-moving projectiles like the shot used in track meets. Wind conditions are also a major factor in projectile flight, as any gunner or golfer can testify.

The Mid-Air Collision Gun Some essential properties of projectile motion can be observed in the laboratory with a simple apparatus, as diagrammed in Fig. 4.20. G indicates the "muzzle" of a spring-loaded "gun" capable of shooting a steel ball B a distance of some 25–30 feet. M is an electromagnet that holds an identical ball B' until the exact instant that ball B is fired by the gun. (An electrical circuit—not shown—energizes the magnet, and the circuit is broken as ball B leaves the gun.) Gun G is aimed *directly* at ball B', as shown in the diagram. A series of shots is now fired at various "muzzle velocities." The magnitude of the muzzle velocity v is varied by adjusting the spring compression of the gun. For all values of the muzzle velocity it is observed that the projectile ball and the falling ball collide in mid-air! By referring to Figs. 4.20 and 4.17 and to Eq. (4.20) see if you can show why the mid-air collision is predictable.

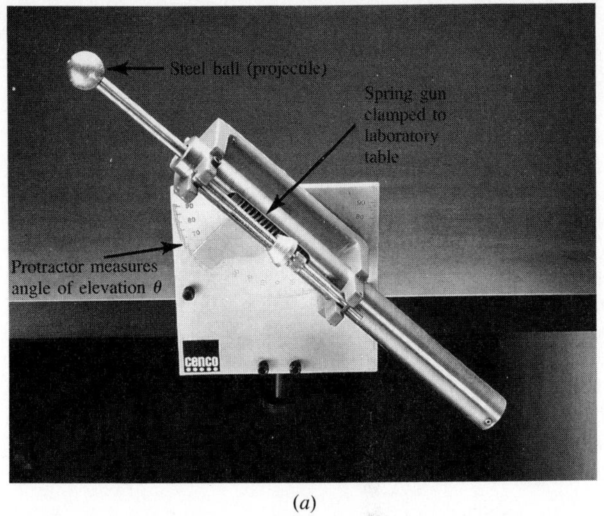

(a)

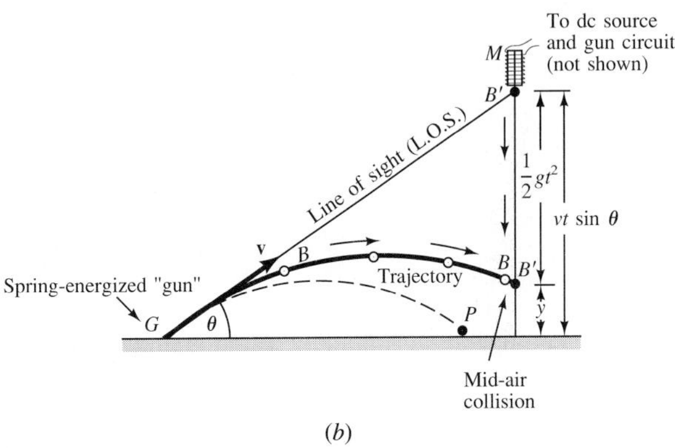

(b)

Fig. 4.20 (*a*) One model of a spring-loaded gun, suitable for in-laboratory projectile motion (ballistic) studies. (b) Sketch of such a gun in use, showing that a projectile is actually a falling body, and illustrating a mid-air collision. (Photo of spring gun courtesy of Morris and Lee, Inc., and Central Scientific Company.)

Query: What if the spring compression were sufficiently weak that ball *B* would impact on the floor at some point such as *P* (Fig. 4.20), far short of the point of impact of the falling ball *B'*? (*Hint:* Suppose point *P* is near the edge of a vertical cliff.)

QUESTIONS AND EXERCISES

1. Explain carefully the difference between *velocity* and *speed*. Give some examples of both concepts.

2. *Velocity* and *acceleration* are both vector quantities. Why? Explain the units in which each is measured. Define acceleration in terms of velocity.

3. How does instantaneous speed differ from average speed? From uniform or constant speed?

4. Using a diagram, explain the difference between *distance* and *displacement*.

5. Can an elevator experience downward acceleration while it is actually moving upward? Explain why or why not.

6. In the units of acceleration, why does a unit of time to the second power appear in the denominator, as in m/s^2? Explain fully.

7. Show by a simple sketch how the (uniform) speed of a moving object can be determined from its distance-time graph.

8. When dealing with *instantaneous* speed, why is it necessary to describe it in terms of $\Delta s/\Delta t$, instead of simply s/t?

9. Discuss completely, with sketches, why it is necessary to adjust the sights of a gun when shifting to a target at a different range.

10. During the stroke of a piston in an engine cylinder, at what points is the piston speed the greatest? At what points is the piston's acceleration the greatest?

11. Much of Galileo's work with "falling bodies" was not done by actually dropping objects and observing their fall under the influence of gravity. Do some library reading (in a book on the history of physics) and find out how he used the technique of rolling objects down inclined planes to deduce the laws of falling bodies. Why do you suppose he resorted to this technique?

12. The story is told of the "stupid" hunter and the "smart" monkey. A hunter spied a monkey hanging from a tree limb a long way off, and forgetting that his rifle was sighted in only for targets at point-blank (very close) range, he aimed directly at the monkey and fired. The monkey, accustomed to guns and hunters, watched for the flash at the gun barrel and immediately let go of the limb, becoming an object in free fall. Was he a "smart monkey"? Explain what happened.

PROBLEMS

NOTE: *Air resistance is to be neglected in setting up and solving all of these problems.*

Group One

1. A supersonic aircraft travels at 2900 ft/s. At this average speed how many hours would it take to fly across the continental United States, a distance of 3100 mi?

2. Raindrops with an average diameter of 1 mm have a terminal velocity in air of about 6 m/s. How long would it take for a drop falling with this velocity to fall 2 km from a cloud to the ground?

3. Which auto is moving faster—one at a speed of 60 mi/h or another at a speed of 96 ft/s?

4. The average speed of a nitrogen molecule in air at room temperature is approximately 500 m/s. How long would it take for such a molecule to travel the 300-ft length of a football field if it could avoid all collisions during the trip?

5. How much time is required for a 110-ft-long jet airplane with a speed of 550 mi/h to travel a distance equal to its own length?

6. Batters at the plate in baseball have to make split-second decisions about swinging on the ball. The distance from the point where the ball leaves the pitcher's hand to the center of home plate is about 58 ft, and a fastball pitcher can throw at 98 mi/h. Under these conditions, what total time does the batter have, first to study the pitch, then to decide to swing, and finally to bring the bat across the plate to meet the ball?

7. A rock is dropped off a sheer cliff into a lake. The splash is seen exactly 5.0 s after the rock is dropped. (*a*) How far is it down to the water? (*b*) What is the velocity of the rock as it hits the water?

8. The speedometer of an automobile registers 20 km/h at a given instant, and 7 s later it is passing 70 km/h. Find the acceleration of this car in meters per second per second, assuming it to be uniform.

9. A distance runner paces herself at a constant speed of 22.0 ft/s. What will be her time for the mile run?

10. A spaceship is moving with a steady speed (relative to the earth) of 4000 km/h. Its main rocket engine is given a "burn time" of 7 s, after which its speed is 6400 km/h. Assuming it to be uniform, find the acceleration during the "burn", in km/h · s.

Group Two

11. A catapult on an aircraft carrier deck starts a plane from rest and uniformly accelerates it to 185 mi/h at the instant of launching, 2.25 s later. Find the acceleration in feet per second per second.

12. A baseball pitcher can give the ball an acceleration of 600 ft/s^2 during delivery. The distance from the start of forward motion to the point where the ball leaves the pitcher's fingers is 8.5 ft. With what speed does the ball start toward the batter?

13. Logs from a greased chute are discharged horizontally at 40 ft/s at the edge of a vertical cliff overhanging a mountain lake. The vertical drop to the water is 125 ft. How far from the base of the cliff do the logs hit the water?

14. A motorist travelling at 90 km/h sees trouble ahead and slams on the brakes. If the negative acceleration is 8 m/s^2 and if 0.20 s is allowed for the driver's reaction time, how far ahead must the trouble have been noticed in order to stop in time? (Answer in meters.)

15. A girl drops a stone down a well that is 70 m deep. How many seconds after releasing the stone will she hear it splash into the water? Take the speed of sound in air as 344 m/s.

16. An auto can start from rest and accelerate at a rate half that of a freely falling body. How many seconds does it take this auto to attain a speed of 50 mi/h, from rest?

17. If an antiaircraft projectile is to reach a height of 3 km, what must be the vertical component of its velocity as it emerges from the gun barrel?

18. A baseball is thrown in to the catcher from left field, a distance of 250 ft, at an average horizontal-component velocity of 80 ft/s. How high does it rise in its flight?

19. At what muzzle velocity (ft/s) must a projectile be shot if it is to strike a target on the same horizontal plane which is 5000 yd away, if the angle of elevation of the gun is 25°?

20. An auto moving at 50 km/h is braked uniformly to a stop in a distance of 100 m. What is its negative acceleration?

21. A bullet is fired into a log from a rifle and it penetrates the wood a distance of 0.75 ft. The muzzle velocity of the bullet is known to be 2300 ft/s. Calculate the average deceleration of the bullet in the wood.

22. A test rocket is being fired vertically upward. The burn time of its engine is 40 s, during which time it provides a constant acceleration to the rocket of 140 m/s^2. After 40 s, the only accelerating force on the rocket is that due to earth's gravity, which is assumed constant at g. Calculate (*a*) the maximum speed reached by the rocket;

(*b*) the maximum altitude it attains; and (*c*) the total time elapsed until the rocket hits the earth.

23. Find the theoretical range of a 5-in. artillery piece, if the muzzle velocity is known to be 1400 ft/s and the angle of elevation is 32°. Assume a horizontal plane of action.

24. A woman driving a car sees a child dart out from the curb ahead of her. She slams on the brakes and the car rapidly decelerates from 50 mi/h to a dead stop in 10 s, just in time to avoid hitting the child. How far away was the child when the woman applied the brakes? Assume uniform deceleration.

Group Three

25. The quarterback on a football team throws downfield at an angle of 30° with the horizontal. His receiver gathers it in 50 yd downfield from the passer, at the same height above ground as the passer's throwing hand. (*a*) What was the ball's velocity as it left the passer? (*b*) How high above the passer's throwing hand did it rise?

26. In a "longest drive" contest the winner drove a golf ball a distance of 365 yd. It was observed to leave the tee at an angle of elevation of 15°. (*a*) What was its velocity as it left the tee, and (*b*) how high did it rise in flight? Assume level terrain.

27. A firefighter directing a stream of water from a fire hose nozzle finds that, to make the water stream hit a burning wall 100 ft away on the same horizontal plane as the nozzle, an angle of elevation of 25° is necessary. Find the speed of the water stream emerging from the nozzle.

28. A Navy cruiser's 8-in. guns are "on target" at 16,000 yd. If the muzzle velocity is 1500 ft/s, (*a*) what should the angle of elevation of the guns be? (*b*) Find the time of flight of the projectiles.

29. Companies with large fleets of trucks often install monitors in them to record the day's trips and to check on the driving habits of the drivers. The speed-time graph of Fig. 4.21 might have been plotted from an analysis of a short section of a monitor's recording tape. Interpret this 7-min segment in detail. If this were a heavy truck operating under city traffic conditions, what would you infer about the driver?

30. A high-altitude bomber approaches on a bombing run at 48,000 ft and 650 mi/h. (*a*) How far (horizontally) from the target is the correct release point for the bombs? (*b*) What is the correct angle of depression at release?

31. A pile-driver hammer is in free fall for 40 ft before it hits the top of the piling and drives it into the sand and mud a distance of 0.85 ft. What is the deceleration of the pile-driver hammer, assuming it is uniform?

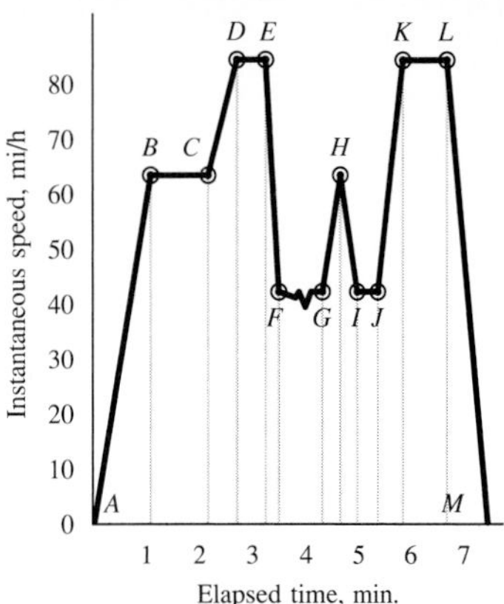

Fig. 4.21 Speed-time graph for a truck being driven in traffic (Problem 29).

32. You are sent in to get off a punt for your football team. You get off a good kick at an angle of elevation of 35° and an initial velocity of 70 ft/s. The opposing team's runback specialist is facing you 210 ft straight down the field. If the football also goes straight downfield, how fast must the runback specialist move (average speed) in order to catch the ball at shoulder height, assuming he starts running as the ball leaves your foot?

33. The radius of the earth's orbit (assumed circular) around the sun is 93,000,000 mi, and the earth completes this orbit every 365.25 days of 24 h each. Find the linear speed of the earth in its orbit in miles per second.

34. The frictional forces between rubber tires and dry pavement are such that the maximum possible negative acceleration that brakes can give an auto is about 21 ft/s^2. In a "panic stop," if brakes are applied when the speedometer reads 60 mi/h, how far will the auto go before coming to a complete stop? What is the total elapsed time of the panic stop?

35. A baseball has been thrown by an outfielder, aimed at home plate. At a certain point and instant in its trajectory its *velocity* has a magnitude of 85 ft/s and a direction which is at an angle of 34° above the horizontal. (*a*) Sketch and evaluate the vector that represents its *change in velocity* during the next second. (*b*) What is the *velocity* (magnitude and direction) at the end of one more second? (You may want to draw these vectors to scale on graph paper.)

36. A golf ball is projected vertically upward with an initial velocity of 40 m/s. (*a*) How high does it go? (*b*) How long does it take to reach maximum height? (*c*) Six seconds after being projected, where is it and what is its velocity?

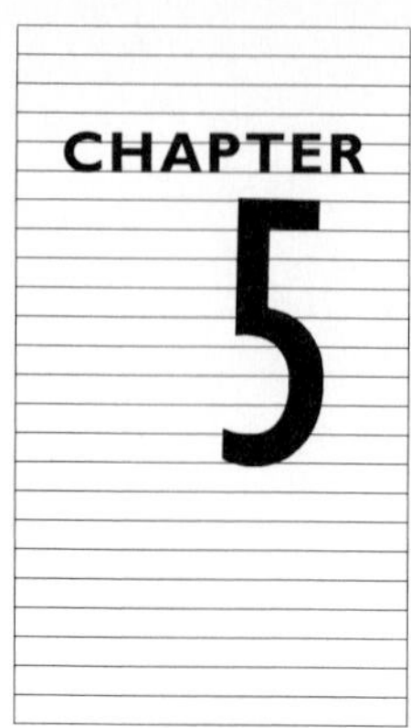

CHAPTER 5 FORCE AND MOTION—NEWTON'S LAWS

Applying and managing forces is a lifetime career for many engineers and technicians.

Forces often result in mechanical energy, and they do the work of the world by means of machines and engines.

Gravitational attraction is an *action-at-a-distance* force.

Scientists, engineers, and technicians are continually working with forces; often to utilize them, sometimes to magnify them, and sometimes to mitigate their effects. Winds, waves, lightning, tides, and ocean currents all exhibit forces that we would like to be able to control for the vast amounts of energy they can produce. Other well-known forces, such as those from expanding steam, hot combustion gases, electrostatics and electromagnetism, running water, and the earth's gravity are readily adapted to our needs, and presently they are the primary sources of energy for advanced technological societies.

Forces can be applied by means of the tension or compression of rigid bodies; by molecular action as in gas pressure; by impact, as of a speeding bullet or a strong wind or a stream of water; and by reaction as in hydro- and steam turbines. Some of the most universal forces however, for example gravitational force, magnetic force, and the forces within the atom are not the result of direct contact, but are "action-at-a-distance" forces. The *results* of these *action-at-a-distance* forces are, for the most part, understandable, but their cause and the mechanisms by which they operate still await a completely satisfactory explanation. The action-at-a-distance force that concerns us here is the force of gravitational attraction.

5.1 ■ The Force of Gravitation

Newton was born in the year that Galileo died, which seems altogether fitting, since Galileo's experimental and theoretical work provided a base from which the genius of Newton eventually brought order to the science of mechanics. His *Principia Mathematica*, published in 1686, is still considered to be one of the three or four most valuable contributions to science. In addition to mechanics, Newton was a recognized leader in many other fields—light and optics, astronomy, mathematics, and philosophy, to name a few.

One of the basic forces of nature is the force of gravity. We can use it, measure it, deplore it sometimes, and define it; but we cannot explain it or eliminate it. The *laws* of gravity are known but not the *cause* of gravity.

You will recall the extended discussion of the *acceleration of gravity* in Chap. 4. Here we are concerned with the *force* of gravity and its relationship to *mass* and *weight*.

Sir Isaac Newton (1642–1727), the British natural philosopher and mathematical genius, studied gravitation over many years and formulated many of the so-called laws of motion and gravitation. He found that gravitation is *universal*, i.e., it applies to all bodies throughout the known universe; and that it is *mutual* in the sense that as the earth attracts the moon, so does the moon attract the earth. These general laws of gravitation will be presented in a later chapter (see Chap. 8). For the present, we are concerned only with the earth's gravity and with the fact that the earth's gravitational force causes a predictable change in motion as it acts on bodies, unless there is a counteracting force of some kind.

5.2 ■ The Force of Gravity on Mass Is the Measure of Weight

In the discussion of measurement (page 29), *weight* was defined as *the pull of earth's gravity on the mass of a body*. If you *weigh* 175 lb, this merely means that you and the earth attract each other with a *force* of 175 lb. The units in which forces and weights are measured need careful study, however, since past practices, from many different countries, still affect current usage.

The earth's gravity pulls on you, and you exert an equal and opposite pull on the earth.

In countries using the metric system the approved *force* unit is the *newton* (defined later in this chapter). The older unit, "kilogram of force" is no longer used in scientific and technical work. Prior to the adoption of the entire SI-metric system, the kilogram was used both as a unit of mass and as a unit of force, or weight in many metric countries. In fact, everyday (commercial and nonscientific) use in some metric nations still finds the kilogram as a unit of weight, as in the purchase of "a kilo of potatoes." In this book, when we use the word "kilogram" or its abbreviation, "kg," we will mean *mass,* not *force*.

The newton (N) is the SI-metric unit of force and weight.

In the English (engineering) system the *pound* is the unit for both *weight* and *force* in everyday use as well as in technical work, but there is still lack of complete standardization in the United States when it comes to the unit of *mass*. Industry and commerce still use the pound as a mass unit in some applications, while engineering practice ordinarily uses a unit called the *slug,* which will be defined and explained later in this chapter.

The pound (lb) is the English-system unit of force and weight.

Despite the fact that the same words are sometimes used to describe forces and masses in the English system (and even to some extent in metric countries, for commercial transactions), it is strongly emphasized that *mass and weight do not mean the same thing*. *Mass* denotes the quantity of matter in a body, and the mass of a body is a constant no matter where it is measured—on the earth's surface, on the moon, or in a space ship on the way to Mars. *Weight* is merely the word we use to describe *the force of gravitational attraction between a body and the earth*. The moon's gravity is much less than that of the earth, since its mass is much smaller, and your 175-lb weight on earth would be only about one-sixth that, or 29 lb, on the moon. An astronaut whose weight is 180 lb on earth becomes weightless when his space vehicle travels beyond the earth's effective gravitational field, but his mass is the same as it was on earth.

An object is weightless in outer space, but its mass is everywhere the same.

Mass and weight are related on the surface of the earth by the constant g, the acceleration of earth's gravity.

Weights are commonly measured by *spring balances,* calibrated to read in pounds and ounces (English system); and in newtons in the metric system. Figure 5.1 shows a typical spring balance or pair of scales.

Mass is measured by the methods and apparatus described in Sec. 2.8, p. 37.

The force of gravity, though it is of extreme importance to us here on earth, is only one of the many types of forces that can act on bodies, and it is therefore necessary to introduce and explain some general laws of force and motion. The study of motion (Chap. 4) is characterized by the technical term *kinematics* (from the Greek *kinema,* meaning motion). The study of the relationship of *force and motion,* since energy and power result from this relationship, is given the technical term *dynamics* (from the Greek *dynamikos,* meaning powerful).

Fig. 5.1 A laboratory spring balance, graduated in newtons and pounds. Sometimes called "scales," spring balances measure gravitational force, or *weight*. They do not measure *mass*. (Ealing Corporation)

NEWTON'S LAWS OF MOTION

The laws of force and motion are the basis of classical mechanics, and they apply at any time—on earth or in space, today or yesterday, whether people are there to observe or not—in other words they are *universal laws* of force and motion.

5.3 ■ Force Is the Cause of Change in Motion

Kinematics is the study of motion without regard to its cause; and *dynamics* is the study of *force and motion.*

It is a matter of common experience that most ordinary objects are at rest with respect to the earth and that they tend to remain at rest unless acted upon by some force

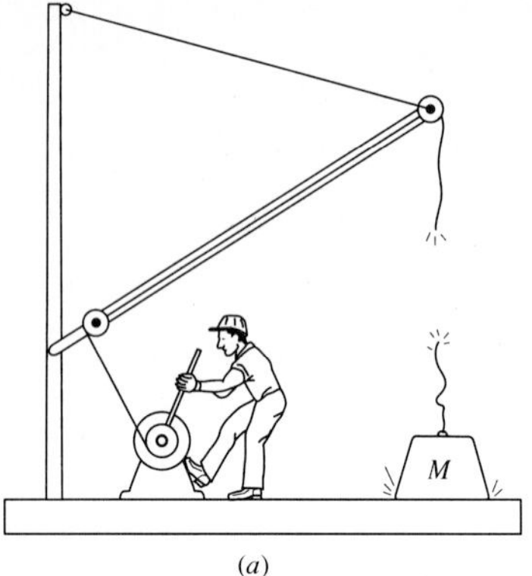

(a)

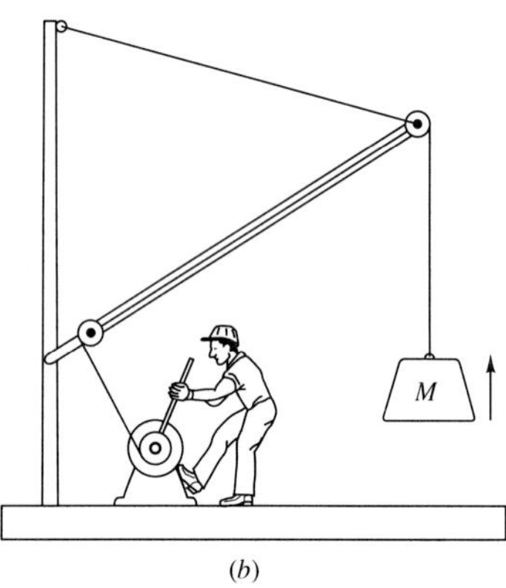

(b)

Fig. 5.2 Inertia is that property of mass that resists a change in motion. In (*a*) the hoist operator has attempted to set the mass *M* in motion too quickly, with the result that the cable snapped. In (*b*) the strain is taken more slowly and the mass is successfully hoisted.

external to themselves. Galileo first proposed the idea that forces cause or inhibit motion and that unless an external force acts on a body, it tends to remain at rest if it is at rest and tends to remain in motion if it is in motion. Newton (born in 1642, the year Galileo died) refined and sharpened these ideas about motion and rest, and formulated a basic law of motion which he called the *law of inertia*.

5.4 ■ Force Is Necessary to Overcome Inertia

The property of *resisting a change in motion* is known as *inertia*. In fact, *inertia* is the most outstanding property of *mass*.

Inertia is that property of mass that resists a change in motion.

An object which is moving in a straight line resists having the *direction* of its motion changed even if the speed is unchanged. Spinning wheels display *rotational inertia,* a fact which the gyroscope, with its many applications in compasses, gyro-stabilizers, and aerospace guidance systems illustrates very well.

The greater the resistance that a body offers to being set in motion from rest, or to having its rate or direction of motion changed, the greater is its mass. Force is required to overcome inertia.

5.5 ■ Newton's First Law of Motion—Inertia

Newton's first law may be stated as follows:

A body which is at rest will remain at rest and a body which is in motion will remain in motion at the same speed and in the same direction unless acted upon by an unbalanced force.

The property of inertia is possessed by all matter and is independent of any forces (such as gravity) which may be acting. A body has inertia even out in interstellar space completely away from the gravitational forces of the earth or any other planet. *Mass* is the quantitative measure of *inertia*.

Some common examples of inertia will be useful in clarifying the concept. A stalled auto on a level street requires a great effort to *start* it moving, but once moving, only a modest push will keep it rolling at constant velocity. If an auto at high speed is braked to a stop suddenly, passengers, boxes, and any loose objects all tend to continue "in motion at the same speed and in the same direction." Seat belts and other restraints can provide the force to nullify this inertia of motion of the passengers. Figures 5.2, 5.3, and 5.4 illustrate the effects of inertia. One of the basic disadvantages of a reciprocating engine, compared to a turbine type (rotary) engine, is the *inertia of motion* possessed by the pistons during each and every stroke. The forces required of the engine parts to stop the pistons' motion in one direction and then accelerate them in the reverse direction create strain, wear, vibration, noise, and inefficient operation of the reciprocating engine.

Mass resists a change in motion and this resistance is called *inertia.* Force is required to overcome inertia.

Mass versus Weight A further *distinction between mass and weight* is now possible, using the inertia concept. On earth the weight of a body is proportional to its mass, and at sea level, weight and mass are often expressed *numerically* by the same units in U.S. industry and commerce (the *pound* for example). But out in space, where a body has no measurable weight, it still has mass, and it still has inertia.

Objects possess inertia of rest and also inertia of motion.

Although a baseball would seem weightless in space, force would be required to throw it (overcoming its inertia of rest), and if caught, it would smack into the catcher's mitt and require force there to overcome its inertia of motion. A prankster (in space) might hand you one regular baseball and another made from solid lead around which the usual horsehide cover has been sewn. Holding them, you could not tell the difference since both would be *weightless* in space. Batting the regular one would give almost the same feel as batting a baseball on earth, since the same force would be required to overcome its inertia. Hitting the lead "baseball" would break the bat, perhaps sprain your wrists, and generally be quite a surprise. *Mass* and *inertia* are *universal* properties. Mass, wherever it is, resists a change in its motion, and this resistance to a change in motion Newton called *inertia*. Weight, on the other hand, is the *force of gravitational attraction* on mass, and in the absence of gravitational forces, objects are weightless.

Thus we see that the preliminary definition of mass as *the quantity of matter in a body* (see Sec. 2.3) is inadequate for scientific and technical purposes. In mechanics we define mass as follows:

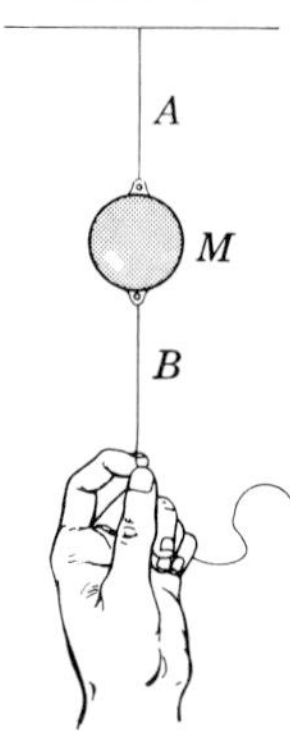

Fig. 5.3 A laboratory demonstration of inertia. M is a heavy ball suspended by string A. String B is of the same twine. If B is pulled slowly and steadily, A will eventually break; but if B is jerked suddenly, B will break. The ball M resists a change in motion, even though that potential motion is downward. The heavy ball has a great deal of inertia.

> ***Mass is the quantitative measure of an object's resistance to a change in motion.***

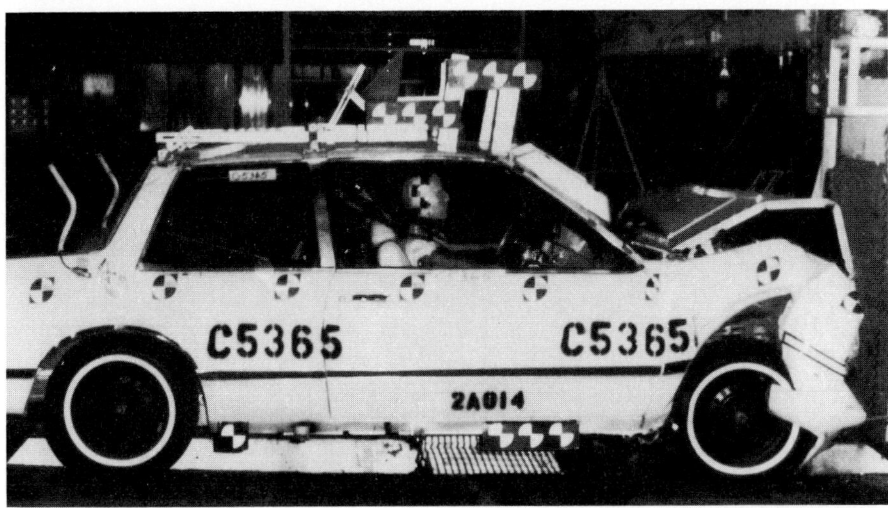

Fig. 5.4 Inertia of motion and auto accidents. The car shown here has been test-crashed against a solid barrier at 30 mi/h. Note that the passenger-restraint device has prevented the (dummy) driver's inertia of motion from crashing him against the windshield, even though the inertia of motion of the car itself has resulted in a badly damaged front end. (General Motors Corporation)

5.6 ■ Newton's Second Law—Acceleration

When an unbalanced force *does* act on a body, a change in motion (acceleration) is produced. The greater the force acting, the greater the acceleration; but the greater the mass of the body, the greater its resistance to a change in motion (inertia). The preliminary definition of force given on p. 70 can now be stated more precisely and in terms that provide a quantitative definition.

Mass and *inertia* are universal properties, whereas *weight* is apparent only when gravitational attraction is present.

Newton's second law states that:

> ***A body which is acted upon by an unbalanced force is given an acceleration in the direction of the force which is proportional to the force acting and inversely proportional to its mass.***

Acceleration is the result of the action of an unbalanced force on mass.

Mathematically, this law may be stated as follows:

$$a \propto \frac{F}{m}$$

where
a = acceleration produced
F = net force acting on the body
m = mass of the body

$\propto$ is read "is proportional to". By choosing units which are consistent with observed natural phenomena, we can write the defining equation for Newton's second law:

$$F = ma \tag{5.1}$$

$$\text{Force} = \text{mass} \times \text{acceleration}$$

This formula is valid only when F, m, and a are expressed in consistent units. The following discussion will define the units required.

5.7 ■ Units and Newton's Second Law

The metric unit of mass has already been defined as the *kilogram* (see Sec. 2.3). A kilogram of any matter has the same mass as the prototype platinum-iridium cylinder at the International Bureau of Weights and Measures at Sèvres, France (see Fig. 2.3).

The basic unit of mass in the English system used to be the *pound*. It is represented by a prototype block of platinum in the British Exchequer in London. In the United States, the pound is still used as a mass unit in some branches of industry, commerce, and engineering. However, in all of science and in most engineering applications, the pound is a unit of *force*, not of mass. The standard unit of mass in English system is now the *slug*, to be defined below.

Further, we have defined the unit of length as the *meter* in the metric system and as the *yard* in the English system, these units also being standardized by prototypes in France, Great Britain, and the United States. (It will be recalled that although the *yard* is the English-system *standard*, the *foot*, $\frac{1}{3}$ yd, is actually the unit in most common use.)

The concept of *acceleration* (change in velocity per unit time) is expressed in terms of length L and time T in metric units, as the meter per second per second (m/s^2) and in English units as the foot per second per second (ft/s^2). With *mass* and *acceleration* already possessing agreed-upon units in the metric system, all that remains to make Eq. (5.1) dimensionally correct is to define a *new unit of force*.

Newton's second law is the basis for defining units of force, weight, and acceleration.

Metric-System Force Unit, the Newton, and the SI-Metric System of Units The fundamental unit of mass in the metric system is the *kilogram;* of length, the *meter*; and of time, the *second*. We now define a new unit of force, called the *newton*, which makes the equation $F = ma$ dimensionally correct.

> ***One newton is that force which produces an acceleration of one meter per second per second when it acts alone on a mass of one kilogram.***

The newton (N) is the basic unit of force in the meter-kilogram-second (mks or SI-metric) system of units. It is an *absolute* or *universal* unit of force, in that it has no relationship to the earth's gravity or to the acceleration caused by gravity (see Fig. 5.5). Setting

$$F \text{ (newtons)} = m \text{ (kg)} \times a \text{ (m/s}^2\text{)}$$

it is apparent that, dimensionally,

The dimensions of the newton (N), the SI unit of force.

$$1 \text{ newton (N)} = 1 \frac{\text{kg} \cdot \text{m}}{\text{s}^2} \qquad \textit{(Memorize this!)}$$

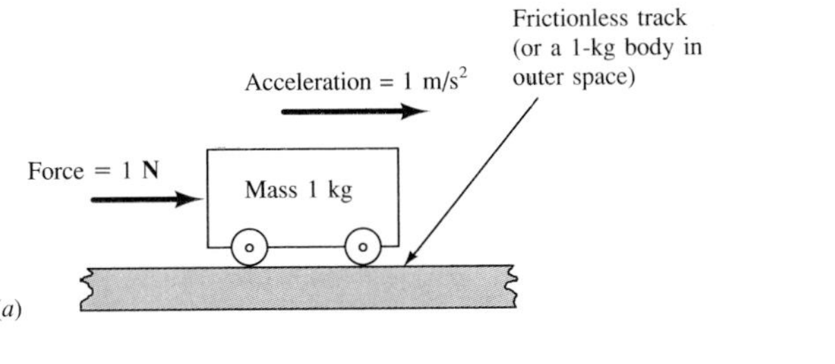

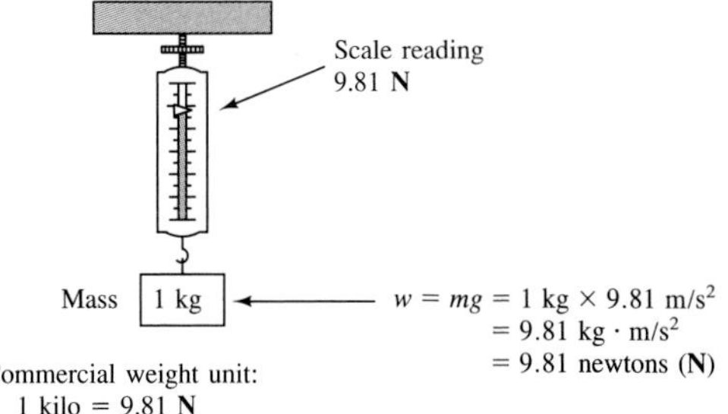

Fig. 5.5 Illustrations of Newton's second law—SI-metric units. (*a*) A force of 1 newton (N) causes an acceleration of 1 m/s^2, when applied to a mass of 1 kg. The units of the newton are therefore kg · m/s^2. (*b*) One kilogram of mass *weighs* 9.81 N at the earth's surface, where the value of the acceleration of gravity, $g = 9.81$ m/s^2. Weight, $w = mg$.

The *weight* of 1 kg of mass at sea level is equal to 9.81 *newtons of force,* since the acceleration of gravity at sea level is 9.81 m/s^2. The older unit of force in the metric system, the *kilogram of force* (kg-*f*), is still used as a measure of weight (the *kilo*) in some metric countries. You will recall that when a *mass* of one kilogram is acted upon by a *force* of one kilogram the acceleration produced is the acceleration of gravity, $g = 9.81$ m/s^2. One kilogram of force is equivalent to 9.81 newtons (N). In time, it is possible that the *kilo* will become obsolete, and metric countries will weigh all commodities in newtons.

You may occasionally come across references to the centimeter-gram-second (*cgs*) system of units, which used to be the standard. This system is not much used in science and engineering today, and we will make only occasional references to its units in this book.

English-Engineering System of Units: The Slug Engineers in English-system countries use another approach to the definition of consistent units for Newton's second law. Instead of choosing and defining a new *force unit* to fit $F = ma$, engineers decided instead to define a new *mass unit*. With the *pound* defined as the basic *force* unit and the foot per second per second as the basic unit of acceleration, a new mass unit, the *slug* (inertia is akin to being *sluggish*) is defined:

The slug is the English-system unit of mass.

One slug is that unit of mass which when acted on by a force of one pound will be accelerated at the rate of one foot per second per second.

Setting

$$F \text{ (lb)} = m \text{ (slugs)} \times a \text{ (ft/s}^2\text{)}$$

we find the dimensions of the slug (sl) to be

$$1 \text{ sl} = 1 \text{ lb} \cdot \text{s}^2/\text{ft} \qquad (\textit{Memorize this!})$$

and the dimensions of the pound,

$$1 \text{ lb} = 1 \frac{\text{sl} \cdot \text{ft}}{\text{s}^2}$$

Since 1 lb force accelerates 1 sl mass at 1 ft/s^2, and 1 lb force accelerates 1 lb mass at 32.2 ft/s^2 (a body in free fall acted on only by the force of gravity), it follows that 1 slug mass has a weight of about 32.2 lb (see Fig. 5.6).

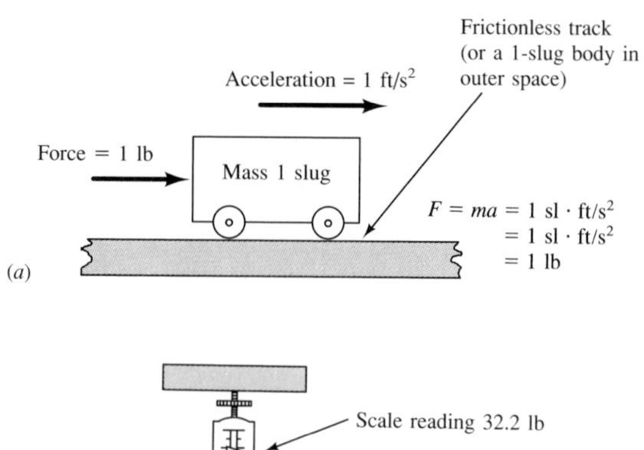

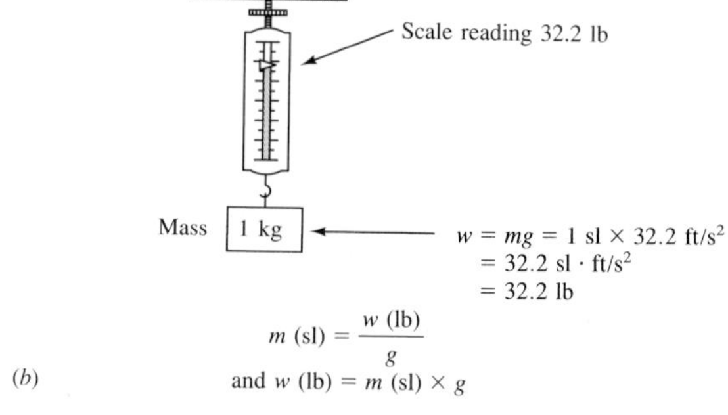

Fig. 5.6 Illustrations of Newton's second law—English-system units. (*a*) A force of 1 lb causes an acceleration of 1 ft/s^2, when applied to a mass of 1 slug. The units of the slug are therefore lb · s^2/ft. (*b*) One slug of mass *weighs* 32.2 lb at the earth's surface, where the value of the acceleration of gravity, g = 32.2 ft/s^2. Again, $w = mg$.

$$m(\text{sl}) = \frac{w(\text{lb})}{32.2 \text{ ft/s}^2}$$

The dimensions of the pound and the slug.

In the English-engineering system of units forces are measured in pounds, masses in slugs, and accelerations in feet per second per second.

This book will make frequent use of both the mks (SI-metric) and the foot-slug-second (engineering) systems. These units are called *absolute units.*

Absolute units of *force* (newtons and slugs) must always be used with $F = ma$. Also, even though the pound is sometimes used as a unit of mass in U.S. commerce and industry, the pound *cannot be used* as a mass unit in calculations involving $F = ma$. The pound is properly a unit of *force* or *weight.*

It must be clearly understood that the equation $F = ma$ is dimensionally correct only when *absolute units* of force (newtons) are used, or when the slug of mass is used with the (English) engineering system.

Gravitational Units In contrast to the *absolute units* discussed above is the use of a system of units referred to the surface of the earth where the *earth's gravity* is an ever-present force factor. In order to understand clearly the distinction between *absolute units,* i.e., those units based on mass/inertia and acceleration considerations, and *gravitational units,* i.e., those based on the force of gravity at the earth's surface, we must recall that *weight* is the *force* exerted by gravity on *mass*; and that *this force would give the mass an acceleration equal to* g (at the earth's surface), in the absence of any other forces.

Gravitational units are those based on *weight*—the pound (lb), or the kilogram-force (kg · *f* or *kilo*). The equation $w = mg$ relates weight and mass.

In general, absolute units are always used in scientific work and in engineering research. Gravitational units are used in commerce and industry in the United States and in some other countries.

The relationship of gravitational units to absolute units is governed by Newton's second law. It can be stated in words as follows:

Weight (force) equals mass times the acceleration of gravity.

As an equation

$$w = mg \tag{5.2}$$

and

$$m = \frac{w}{g} \tag{5.2'}$$

To use Newton's second law ($F = ma$) with gravitational units, it must be modified by substitution from Eq. (5.2′), giving

$$F = \frac{w}{g}\, a \tag{5.3}$$

The following problems illustrate the use of Newton's second law, and methods of applying it to both absolute and gravitational units.

Illustrative Problem 5.1 An elevator weighs 5000 lb fully loaded and is operated by a cable in which a tension of 7500 lb must not be exceeded. What is the greatest upward acceleration possible? (See Fig. 5.7.)

Solution Since the loaded elevator weighs 5000 lb (force), only 2500 lb force (7500 − 5000) of the allowable cable tension is available to accelerate the elevator upward. All units should first be changed to *absolute* units, in order that Newton's second law can be used. From Eq. (5.2′) the *mass* of the elevator is

$$m = \frac{w}{g} = \frac{5000 \text{ lb}}{32.2 \text{ ft/s}^2} = 155.3 \text{ sl}$$

When a force of 2500 lb is applied to this mass, the acceleration is

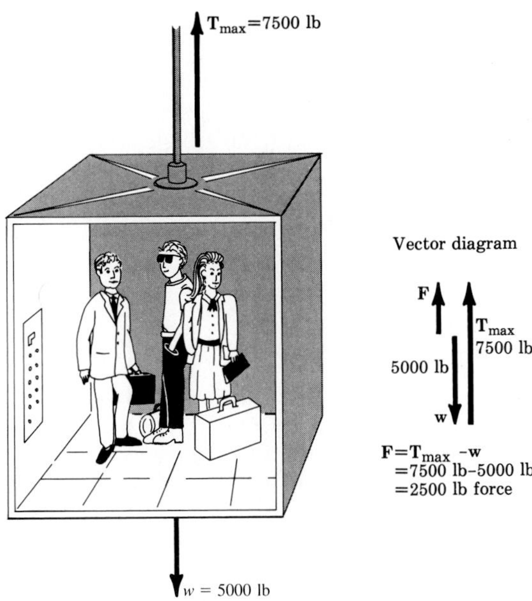

Fig. 5.7 Sketch of the forces acting on an elevator, together with a vector diagram (Illustrative Problem 5.1).

$$a = \frac{F}{m} = \frac{2500 \text{ lb}}{155.3 \text{ sl}} = 16.1 \text{ ft/s}^2 \qquad \textit{answer}$$

Illustrative Problem 5.2 A crate of melons from Mexico is labeled 20 ''kilos.'' (Although the kilogram is properly a mass unit, it is still used in commerce as a weight unit.) What would be the weight of this crate of melons in newtons?

Solution The crate of melons actually has a *mass* of 20 kg. Its weight in newtons would be given by Eq. (5.2),

$$\begin{aligned} w &= 20 \text{ kg} \times 9.81 \text{ m/s}^2 \\ &= 196 \text{ kg} \cdot \text{m/s}^2 \\ &= 196 \text{ N} \end{aligned}$$

Illustrative Problem 5.3 An auto has a mass of 1800 kg. What applied force in newtons would be required to give it an acceleration of 1.5 m/s², assuming zero road friction and air resistance?

Solution Applying Newton's second law, the unbalanced force is given by

$$\begin{aligned} F = ma &= 1800 \text{ kg} \times 1.5 \frac{\text{m}}{\text{s}^2} \\ &= 2700 \text{ N} \qquad \textit{answer} \end{aligned}$$

At this point, for convenience and review, and *for ready reference throughout the study of applied physics,* Table 5.1 is provided (see next page), bringing together the important definitions, units, constants, and conversion factors that are essential to an understanding of mechanics, for both metric countries and English-system countries.

Illustrative Problem 5.4 A space vehicle whose total mass (including rocket motors and fuel) is 62,000 sl (assumed constant in the flight) must be given a vertical upward acceleration sufficient for it to attain an altitude of 10 mi in 90 s after liftoff. What average thrust, in pounds-force, must the rocket engines develop?

Solution (See Fig. 5.8.) The weight (downward force) of the vehicle is $w = mg =$ 62,000 sl × 32.2 ft/s² = 1,996,000 lb. Consequently, the rocket engine must develop a thrust of 1,996,000 lb merely to lift the vehicle off the pad against the force of

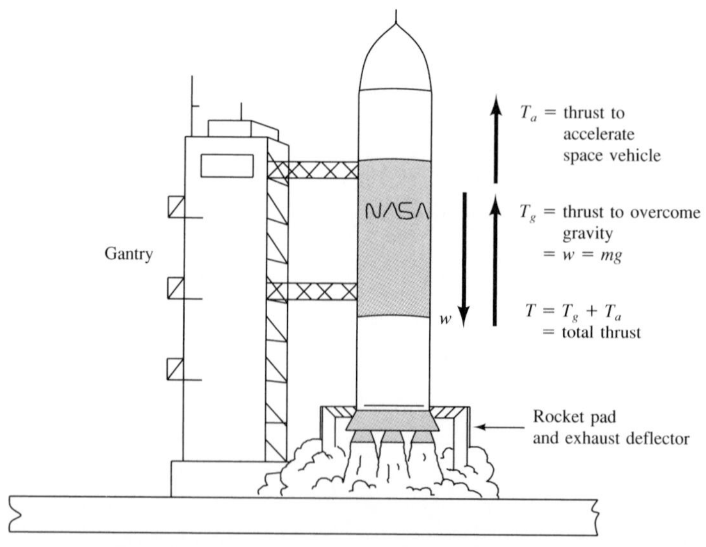

Fig. 5.8 Forces acting on a space rocket to cause lift-off and vertical acceleration. The thrust of the rocket motor must be sufficient to overcome the force of gravity *and* to provide acceleration to the mass of the vehicle (Illustrative Problem 5.4).

Table 5.1 Definitions, units, constants, and conversion factors for the study of mechanics

Definitions of Basic Terms

Mass	The quantity of matter in a body. Mass resists a change in motion. This resistance is called *inertia*. Mass is the measure of inertia.
Force	A push or a pull which tends to cause a change in motion. That which is required to overcome inertia.
Weight	The force of the earth's gravity on mass
Motion	Continuous change of position or location
Velocity	Rate of motion in a specified direction—a vector quantity
Speed	Rate of motion, no direction stated—a scalar quantity
Acceleration	Rate of change of velocity—a vector quantity
Inertia	That property of mass which resists any change in motion

Mass Units and Their Definitions

Kilogram (kg)	The mass of the prototype cylinder at the International Bureau of Weights and Measures, Sèvres, France
Slug (sl)	A mass which, when acted upon by a force of 1 lb, is accelerated at the rate of 1 ft/s^2. The English-engineering unit of mass. The slug is equivalent to 1 lb · s^2/ft. (Memorize this.) The slug weighs 32.2 lb at the earth's surface.
Pound (lb)	Not properly a mass unit but still used as such in some English-system engineering, commercial, and industrial situations. We will use it as a mass unit in some engineering applications. Properly a force unit.

Force Units and Their Definitions

Absolute Units (For these units, $F = ma$)

SI-metric (mks)	newton (N)	1 N of force gives 1 kg of mass an acceleration of 1 m/s^2
Engineering	pound (lb)	1 lb of force gives 1 *slug* of mass an acceleration of 1 ft/s^2

Gravitational Units (Weight) $\left(\text{For these units, } F = \frac{w}{g} a, \text{ since } m = \frac{w}{g}\right)$

Metric	kilogram-force ($kg \cdot f$)	1 $kg \cdot f$ gives 1 kg of mass an acceleration of 9.81 m/s^2
	gram-force (g-*f*)	1 g-*f* gives 1 g of mass an acceleration of 981 cm/s^2
English	pound (lb)	1 lb-force gives 1 lb of mass an acceleration of 32.2 ft/s^2

Conversion Factors and Constants

1 kilogram-force ($kg \cdot f$)	= 981 N	= 2.2 lb	g_{earth} = 9.81 m/s^2
1 kilogram (kg)	= 0.0685 sl		= 981 cm/s^2
1 newton (N)	= 0.225 lb (force)		= 32.2 ft/s^2
1 pound (lb) force	= 0.454 $kg \cdot f$	= 4.45 N	
1 slug (mass) weighs	= 32.2 lb	= 14.59 kg	

Dimensions

$$m(\text{kg}) = \frac{w(\text{N})}{9.81 \text{ m/s}^2} \qquad m(\text{sl}) = \frac{w(\text{lb})}{32.2 \text{ ft/s}^2}$$

$w(\text{N}) = m(\text{kg}) \times 9.81 \text{ m/s}^2$
$w(\text{lb}) = m(\text{sl}) \times 32.2 \text{ ft/s}^2$

gravity. Let this thrust be $T_g = 1{,}996{,}000$ lb. Let the additional thrust (for acceleration) be T_a, assumed constant during the 90-s interval. Consequently, *total thrust*

$$T = T_g + T_a$$

The acceleration required to attain the 10-mi altitude in 90 s may be obtained from Eq. (4.8), $s = \frac{1}{2}at^2$, since $v_1 = 0$. Transposing and substituting,

$$a = \frac{2s}{t^2} = \frac{2 \times 10 \text{ mi} \times 5280 \text{ ft/mi}}{(90 \text{ s})^2}$$

$$= 13.04 \text{ ft/s}^2$$

From $F = ma$ (Newton's second law),

$$T_a = 62{,}000 \text{ sl} \left(\frac{\text{lb} \cdot \text{s}^2}{\text{ft}}\text{/sl}\right) \times 13.04 \text{ ft/s}^2$$

$$= 808{,}500 \text{ lb}$$

$$\text{Total thrust} = T_g + T_a$$

$$= 1{,}996{,}000 \text{ lb} + 808{,}500 \text{ lb}$$

$$= 2{,}804{,}500 \text{ lb} \qquad \textit{answer}$$

Illustrative Problem 5.5 The sketch of Fig. 5.9 shows a small car of mass m_1 that can move along a smooth table without friction. It will roll across the table when hanging mass m_2 is released to the force of gravity. The pulley P is assumed massless and frictionless. The rolling car has a mass of 5 kg and m_2 is 1 kg. Find the acceleration of the system when m_2 is released, and the tension T in the cord as the system is accelerating.

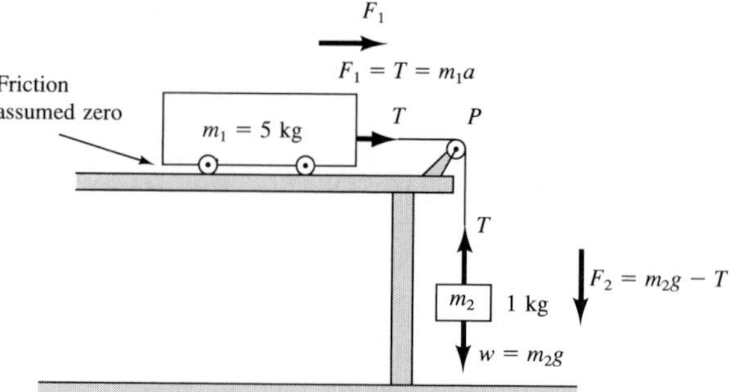

Fig. 5.9 The force of gravity on a hanging mass (that is, its weight) accelerates a system made up of the mass itself and a rolling car. Friction assumed negligible (Illustrative Problem 5.5).

Solution The sketch shows the forces acting in the system. Since the pulley is frictionless and massless, the tension pulling to the right on m_1 is the same as that pulling upward on m_2. Designate this tension T. The cord is assumed not to stretch, so the accelerations of the car and the mass m_2, though differently directed, are of the same magnitude. Let a be the acceleration of the system.

Analyzing the forces acting in the system:
At m_1 we can write: $F_1 = T = m_1a$, directed to the right. At m_2 the net force available to accelerate it downward is the difference between its weight m_2g and the upward tension T in the cord: $F_2 = m_2g - T = m_2a$, directed downward. From these two equations, eliminating T,

$$m_1a = m_2g - m_2a$$

and

$$a = \frac{m_2g}{m_1 + m_2}$$

Substituting, $$a = \frac{1 \text{ kg} \times 9.81 \text{ m/s}^2}{6 \text{ kg}} = 1.64 \text{ m/s}^2 \qquad \textit{answer}$$

To find the tension in the cord, write Newton's second law for the rolling car alone,

$$T = 5 \text{ kg} \times 1.64 \text{ m/s}^2$$
$$= 8.2 \text{ kg} \cdot \text{m/s}^2 = 8.2 \text{ N} \qquad \textit{answer}$$

An alternate solution for the acceleration of the system depends simply on the realization that the *total mass* of the system to be accelerated is $m_1 + m_2$, which is 5 kg + 1 kg = 6 kg. The *total force* available to accelerate the system is the weight of the hanging mass $w = m_2 g$. Using Newton's second law, solving for a, the acceleration of the *system*,

$$a = \frac{F_{\text{tot}}}{m_{\text{tot}}} = \frac{9.81 \text{ kg} \cdot \text{m/s}^2}{6 \text{ kg}}$$
$$= 1.64 \text{ m/s}^2 \text{ as before}$$

5.8 ■ Newton's Third Law—Action and Reaction

Although we have discussed the action of individual forces in the foregoing, it is a matter of fact that *forces never exist singly but always in pairs* (see Fig. 5.10*a*). As you walk, forces are applied by your leg muscles through your shoes to the sidewalk. But your motion is caused by the sidewalk's pushing back on your feet with forces equal and opposite to those applied by your feet to the sidewalk. These forces are due to the earth's *elasticity,* and the desirability of these *reactive* forces can easily be seen if you try to walk on a nonelastic surface such as deep mud or quicksand. If friction is absent, as on smooth ice, your shoe soles do not exert an appreciable horizontal component of force on the ice, and consequently there are no horizontal reactive forces transmitted to your shoe soles, and floundering is the result. An automobile seems to push itself along, but actually the road under the wheels is exerting forces on the wheels, just as the wheels are exerting forces on the road (see Fig. 5.10*b*). It is the reaction force of the road on the wheels that moves the car.

Forces always occur in pairs. An action force always elicits an equal and opposite reaction force.

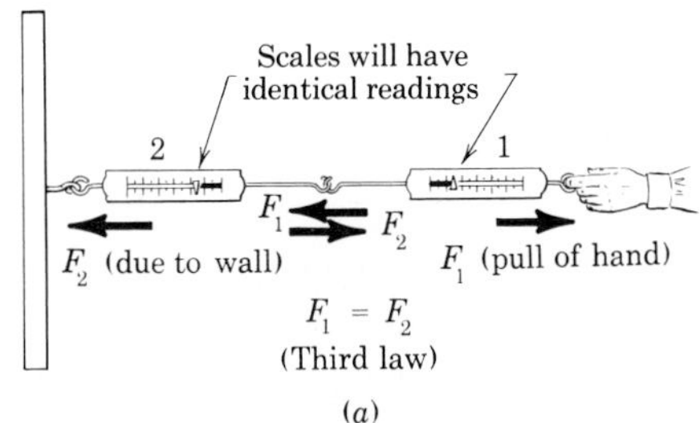

(*a*)

(*b*)

Fig. 5.10 Action and reaction. (*a*) Forces always exist in pairs, not singly. (*b*) The car's tires and wheels supply the action force, but without the reaction force provided by the road surface, the auto would not be accelerated forward. At a steady speed, the reaction force supplied by the roadway is just equal to the combined retarding forces of air resistance and road friction on the auto.

Most of us have seen a squirrel in a revolving cage, running but not getting anywhere; and we know that when exploding powder expels a projectile from the muzzle of a gun, the gun always recoils. These, too, are examples of action and reaction. Figure 5.11 shows another example.

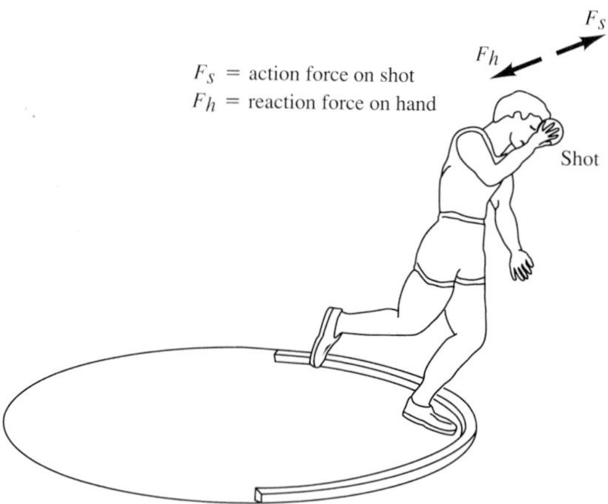

Fig. 5.11 A shot putter at the moment of greatest force applied to the shot. His hand and wrist are applying a force F_s to the shot, and the shot's inertia (resistance to having its state of motion changed) is applying an opposite reaction force F_h back on his hand. If it were not for this reaction force supplied by the shot, the athlete's forward motion would carry him over the foul board and out of the ring on every toss.

Fig. 5.12 Rocket propulsion is an example of action and reaction—Newton's third law. The missile is propelled by an action force equal to the reaction force of the mass of gases being accelerated to the rear. (Official U.S. Navy photo)

The earth's gravitational pull (force) on the moon holds the moon in orbit, but the moon, in turn, has a reacting force on the earth, the most easily observed result being the tides in the oceans.

An occurrence all too frequently observed is the collision of two automobiles. Even if they are both traveling in the same direction and one overtakes the other and smashes the rear deck of the car in front, this *action* (a force) is accompanied by a *reaction* force back against the overtaking vehicle, damaging its grille, headlights, and engine hood, and probably injuring its occupants.

The fact that forces occur in equal and opposite pairs was recognized by Newton and formulated into another law of motion. Newton's third law of motion states:

> ***When one body exerts a force on another body, the second body exerts an equal and opposite force on the first body.***

Action-reaction pairs of forces never act on the same body.

It should be noted that action-reaction pairs of forces never act on the *same* body, and therefore do not nullify each other.

Rockets and jet aircraft are propelled by the reaction force from hot gases expelled to the rear at high velocity. Simply put, rockets and jet planes are propelled by the force that is required to throw a part of their mass out of the rear end of the vehicle.

In addition to the examples cited above, many steam and hot-gas turbines and hydraulic turbines utilize the reaction principle in their design. Rotary lawn sprinklers are turned by the reactive force of the escaping water. Jet-propelled aircraft and the huge rockets used for putting earth satellites into orbit or for propelling space vehicles are pushed forward by reaction from the force of expanding gases accelerating to the rear (Fig. 5.12). There is still some misunderstanding of the principle of jet and rocket propulsion. One popular and mistaken view is that the push of the escaping gases against atmospheric pressure creates the forward thrust. This is not the case at all; in fact, rockets and rocket-type aircraft *which carry their own oxygen for combustion* will operate at higher speeds and greater efficiency in space, where there is no atmosphere, than they will within the earth's atmosphere. The actual mechanics of jet engines and rocket engines is explained in Chap. 17.

Two additional examples of Newton's third law—action and reaction—are often experienced. In the game of billiards the cue ball collides with another (stationary) ball. The *action force* exerted by the cue ball sets the stationary ball in motion, and the *reaction force* exerted by the stationary ball on the cue ball slows up and/or changes the direction of the cue ball (Fig. 5.13).

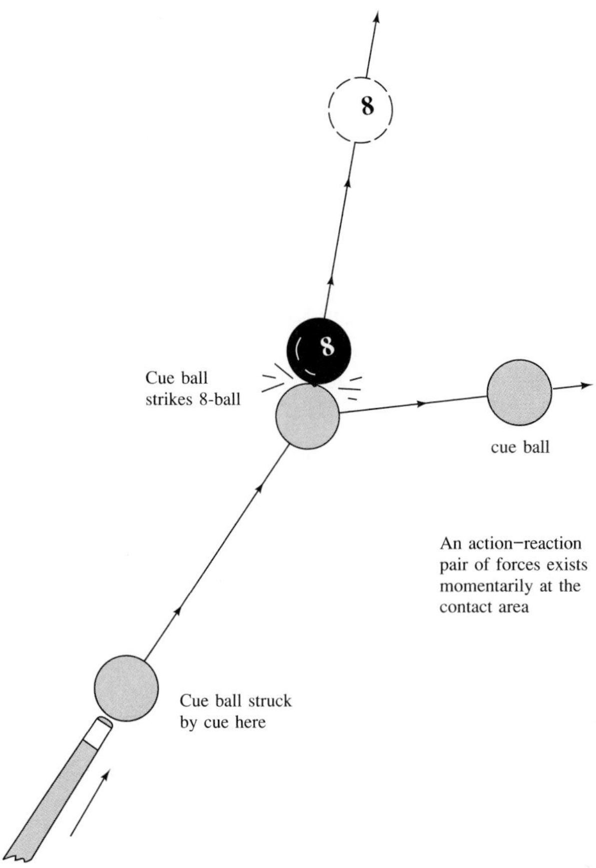

Fig. 5.13 Colliding billiard balls. At impact the cue ball exerts a force on the 8 ball, giving it an acceleration. But its inertia and elasticity provide an equal and opposite reaction force on the cue ball, with the approximate results depicted in the sketch.

A person steps easily *off a dock* into a small boat, because the stationary dock provides a reaction force for the action force of the person's leg muscles and feet. The horizontal component of the reaction force of the dock propels the person toward and into the boat. Disembarking *from the* (*unmoored*) *boat* is a different matter entirely. As the boater's muscles apply an *action force* (through the shoe soles) to the hull or gunwale of the boat, the boat is free to move and it is accelerated away from the dock. Since the small boat provides little *reaction force* to the boater's shoe soles, the boater gains an inadequate acceleration toward the dock and may well end up in the water (Fig. 5.14).

Remember:

***Action and reaction forces are equal and opposite.* They always occur in pairs, never singly. They act on *different* objects, not on the same object.**

5.9 ■ Dimensional Analysis

Some quantities have no dimensions and no units. Examples of such *dimensionless quantities* are: the values of trigonometric functions (sines, cosines, etc.), π, and

Fig. 5.14 Stepping off an unmoored boat to a dock is an invitation to a swim. The muscular effort of the person applied through the shoe sole to the deck of the boat is a force on the boat. The unmoored boat accelerates away from the dock, and although the reactive force (applied by the deck on the person's shoe sole) accelerates the person, the latter motion may not be sufficient to land the person on the dock.

certain other physical and mathematical constants. All ratios of like quantities are dimensionless since the units cancel as the value of the ratio is determined.

In problem solving, care with dimensions is fully as important as care with numerical calculations.

Throughout this book, emphasis will be placed on the importance of dimensions and units. Applied physics is a problem-solving science, and in the solution of any problem two parts of the final answer are of equal importance: the numerical part, and the *dimensions* and *units*. Needless to say, an answer is correct only if the numerical value *and the dimensions* and units are correct. The student will have noted that the illustrative problems distributed through the text have put as much emphasis on carrying out indicated operations with units as on the computing of the numerical answer.

Frequently it becomes necessary to change a given quantity from one unit of measurement to another. Two examples will be given here.

Illustrative Problem 5.6 Change 45 mi/h to its equivalent value in feet per second.

Solution At the outset certain equivalences must be established, either from memory or by consulting tables. In this case we need to know that 1 mi = 5280 ft, and that 1 h = 60 s. Now set up the conversion equation,

$$45\ \text{mi/h} = \frac{45\ \cancel{\text{mi}}}{1\ \cancel{\text{h}}} \times \frac{5280\ \text{ft}}{1\ \cancel{\text{mi}}} \times \frac{1\ \cancel{\text{h}}}{60\ \cancel{\text{min}}} \times \frac{1\ \cancel{\text{min}}}{60\ \text{s}} = 66\ \text{ft/s} \qquad \textit{answer}$$

Note that we have really multiplied 45 mi/h by (1) × (1) × (1). Each of the units on the right side of the equation cancels out except the desired *ft* and *s* units. Note the cancellation "slash marks".

Illustrative Problem 5.7 Convert 0.0753 tons/cubic foot to kilograms/cubic meter. Use the following equivalences: 1 ton = 2000 lb; 1 lb = 454 g; 1 kg = 1000 g; 1 ft = 30.5 cm; 1 m = 100 cm.

Solution

$$0.0753\ \text{ton/ft}^3 = \frac{0.0753\ \cancel{\text{ton}}}{1\ \cancel{\text{ft}^3}}\left(\frac{2000\ \cancel{\text{lb}}}{1\ \cancel{\text{ton}}}\right)\left(\frac{454\ \cancel{\text{g}}}{1\ \cancel{\text{lb}}}\right) \times \left(\frac{1\ \text{kg}}{1000\ \cancel{\text{g}}}\right)\left(\frac{1\ \cancel{\text{ft}}}{30.5\ \cancel{\text{cm}}}\right)\left(\frac{100\ \cancel{\text{cm}}}{1\ \text{m}}\right)^3 = 2410\ \text{kg/m}^3 \qquad \textit{answer}$$

Note how all units cancel out except those needed for a (correct) result.

Illustrative Problem 5.8 A body whose weight is 4000 lb is acted on by a horizontal force of 250 lb. Neglecting all frictional forces, what is its acceleration?

Solution Suppose it were forgotten that the acceleration formula, $F = ma$, requires that F and m be in *absolute* units. *Erroneously* (!) substituting values directly from the problem statement (as if weight were an absolute unit rather than a gravitational unit) gives

(Erroneously)
$$\underset{\text{(Force)}}{250\ \text{lb}} = \underset{\text{(Weight)}}{4000\ \text{lb}} \times a$$

Solving for a,

$$a = \frac{250\ \cancel{\text{lb}}}{4000\ \cancel{\text{lb}}}$$

$$= 6.25 \times 10^{-2} \qquad \text{a dimensionless number! (obviously wrong).}$$

Knowing that the dimensions of the answer for a (acceleration) *must be* feet per second per second, we conclude that a mistake has been made and proceed to reanalyze the problem. When the proper relationships are recalled (in the engineering system, mass must be in *slugs* when force is in pounds; or, recalling that $m = w/g$) the problem is correctly set up as follows:

(Correctly)
$$250\ \text{lb force} = \left(\frac{4000\ \text{lb}}{32.2\ \text{ft/s}^2}\right)a$$

$$a = \frac{250\ \text{lb} \times 32.2\ \text{ft/s}^2}{4000\ \text{lb}} = 2.01\ \text{ft/s}^2 \qquad \textit{answer}$$

Remember, the units of the slug are:

$$\frac{\text{lb} \cdot \text{s}^2}{\text{ft}}, \text{ or } \frac{\text{lb}}{\text{ft/s}^2}.$$

Consider another example, with SI-metric units.

Illustrative Problem 5.9 An elevator has a mass of 1000 kg. What force (tension) in the hoisting cable must be applied to accelerate it upward at 2 m/s^2? (See Fig. 5.7 for a sketch of a similar problem.) Answer in newtons.

Solution Since the answer is required in newtons we recall the units of the newton: 1 N = 1 kg · m/s^2. Consequently, we know that unless the final answer has these dimensions, it cannot be correct. Suppose it were forgotten that a load (weight) is a *force*, and that the mathematical expression for weight is $w = mg$. In that case we might *erroneously* set up the problem as follows:

$$\text{Tension in cable} = \text{tension to support load} + \text{force to accelerate}$$

(Erroneously)
$$T = 1000\ \text{kg} + 1000\ \text{kg} \times 2\ \text{m/s}^2$$

Attempting to add the quantities on the right side of the equation is futile since they *do not have the same dimensions*. This very fact tells us that we have erred in setting up the problem. The "tension to support load" (weight) must be expressed in *absolute units* (newtons), since it is a force, not a mass, and it has to be added to the force to accelerate ($F = ma$), which expression requires absolute units.

Quantities with different dimensions or units cannot be added or subtracted.

(Correctly)

$$\begin{aligned} T &= mg + ma \\ &= 1000 \text{ kg} \times 9.81 \text{ m/s}^2 + 1000 \text{ kg} \times 2 \text{ m/s}^2 \\ &= 9810 \text{ kg} \cdot \text{m/s}^2 + 2000 \text{ kg} \cdot \text{m/s}^2 \\ &= 11{,}810 \text{ kg} \cdot \text{m/s}^2 = 11{,}810 \text{ N} \qquad \textit{answer} \end{aligned}$$

An answer with the wrong dimensions or units is always wrong; but an answer with correct units is not necessarily right.

It should be emphasized that ending up with the correct units *does not assure that the numerical value of the answer is correct*; but, ending up with wrong or inconsistent units automatically indicates that the solution is *wrong*. Correct dimensions are a necessary condition for a correct answer, but they are not a sufficient condition.

QUESTIONS AND EXERCISES

1. When you think about the units of mass, inertia, weight, force, and acceleration, and their definitions, it might seem to you that they are all defined in terms of one another. Which one(s) is (are) "pegged" to an external standard, and what is that standard?

2. Astronauts would experience weightlessness if they were sufficiently far from the gravitational field of the earth or the moon. How is it that they can have mass but no weight? How would your life and actions be changed in a weightless environment?

3. From various fields of industry, engineering, agriculture, or medicine, list three examples of practical applications dependent on each of Newton's first, second, and third laws of motion.

4. It was mentioned early in the chapter that the laws of gravity are well known, but not the cause of gravity. The same statement was made about action-at-a-distance forces in general. Do some library reading about *fields* and *field theory*, and give a class report on how these concepts help explain action-at-a-distance forces.

5. A spring scale is illustrated in Fig. 5.1. In the laboratory, experiment with one for a while, weighing various objects. What is the central assumption on which the scale readings are based? What is the scale actually measuring? What would it measure if it were used in outer space?

6. Read a bit in a book on the history of physics and then refer to Newton's statement about "standing on the shoulders of giants". Name a half-dozen of the "giants" he probably had in mind.

7. Suppose you are riding a bicycle at a steady speed down a straight road. You come to a corner and without reducing speed you turn the corner. Your speed has not changed but your *motion* has. But force is required to cause a change in motion. Where does the force come from that produces your change in motion? (Since your *direction* changed, there had to be acceleration.)

8. If you had been in charge of engineering standards at the time, would you have "invented" the slug as a new mass unit in order to retain the pound as a force unit, or would you have made the pound the English-system *mass* unit and "invented" a new force unit? (Read about an English-system unit called the *poundal*—now obsolete.)

9. In shotputting, the athlete rarely oversteps the front edge of the ring on really good tosses, but often does on weak throws. Can you offer a rational explanation for this observation?

10. Figure 5.13 shows a collision between two billiard balls. The *action* force of the cue ball on the 8 ball is easily explained because the cue ball is moving at high velocity and has an impact force. But what about the *reaction* force from the 8 ball that changes the motion of the cue ball? What is its source? Explain in detail.

PROBLEMS

Group One

1. A truck is towing a 3500-lb automobile on a level road and accelerating it at 3.4 ft/s^2. What is the tension in the towline, in pounds? Neglect road friction.

2. A thrust of 50,000 N accelerates an airplane horizontally at 5 m/s^2. Find the mass of the airplane. Neglect air resistance forces.

3. A truck has a mass of 8000 kg. If friction forces are negligible, what force would the engine have to deliver to the drive wheels to give the truck an acceleration of 1.8 m/s^2 on a level road?

4. At the earth's surface what is the weight of a metal casting whose mass is 45 kg?

5. An air-conditioning unit is crated for shipping and the total weight is recorded as 1850 lb. What is its mass in slugs?

6. If your weight at the earth's surface is 155 lb, what would your *mass* be on the surface of the moon?

7. The acceleration of gravity on the surface of Mars is two-fifths that on earth. A space probe whose mass is known to be 4000 kg arrives and lands on Mars. How much will it weigh there?

8. A net force of 65 lb acts on an object whose mass is 200 sl. Find the resulting acceleration.

9. A mass of 2 kg is thrown upward with an acceleration of 15 m/s^2. How much force had to be supplied? What force would have to be supplied to throw it *downward* with the same acceleration?

10. A small towing tractor in a factory can exert a drawbar pull of 1800 N. When hooked to a trailer whose mass (loaded) is 500 kg, what acceleration will result? (Assume no friction.)

Group Two

11. An elevator is designed for an upward acceleration of 1.5 m/s^2. Fully loaded its mass is 6000 kg. What tension will exist in the cable as it is accelerated upward at the design rate? (HINT: Remember that there is tension in the cable just to support the weight of the elevator.)

12. A parachutist whose mass is 80 kg is in free fall at 90 m/s. His parachute opens and he slows to a velocity of 6 m/s in a vertical distance of 100 meters. What force in newtons did his parachute exert on him?

13. A 2000-kg auto is traveling at 80 km/h on a level road. What is the least distance in which it can be stopped if the average net braking force is 8000 N?

14. A space vehicle whose total mass is 3.5×10^6 kg is launched vertically from the earth's surface by a rocket engine that exerts a total thrust of 4.4×10^7 N. Find the initial upward acceleration.

15. A woman stands on a scale on the floor of an elevator. With the elevator at rest the scale reads 125 lb. What will the scale read (*a*) as the elevator accelerates upward at 2.5 ft/s^2? (*b*) as the elevator accelerates downward at 3.0 ft/s^2? (*c*) as the elevator moves upward with a constant speed of 10 ft/s?

16. The takeoff speed of a commercial jet plane whose total (loaded) mass is 80,000 kg is 300 km/h. A runway "run" of 35 s is required. (*a*) Find the average thrust provided by the engines. (*b*) What was the length of the "run," from standstill to the lift-off point? Assume uniform acceleration and zero frictional forces.

17. A 2800-lb auto moving at a speed of 50 ft/s slows uniformly to a stop in 7.0 s. (*a*) Find the (negative) acceleration of the auto. (*b*) What average force did the brakes provide? (*c*) How far did the car move after the brakes were applied?

18. An auto, moving initially with a speed of 30 mi/h, collides with a heavy truck and is decelerated to a complete stop in 0.3 s. (*a*) What was the average deceleration of the auto? (*b*) How many *g*'s (that is, what multiple of the acceleration of gravity) was the driver subjected to, assuming that she was held in place by a seat belt?

19. A rifle imparts a muzzle velocity of 2600 ft/s to a 150-grain bullet as the bullet is accelerated through the 26-in. barrel. (*a*) Find the average acceleration of the bullet while in the barrel, and (*b*) the average accelerating force. (Assume no friction between bullet and the bore of the rifle.)

20. A hoisting cable has a breaking strength of 20,000 lb. A hopper of wet concrete mix weighing 6000 lb is being hoisted. Find the maximum upward acceleration that can be given the loaded hopper without breaking the hoisting cable.

21. A 12.0 g bullet at a speed of 800 m/s strikes and goes through a 3-cm-thick wooden plank. After it emerges from the wood its speed is 100 m/s. Find (*a*) the deceleration (assumed uniform) of the bullet while in the plank; (*b*) the decelerating force on the bullet; and (*c*) the length of time it was in the plank.

22. A loaded elevator has a mass of 2500 kg while it is motionless at the eighteenth floor of a building. What is the tension in its support cable? It starts downward with a uniform acceleration of 2.5 m/s^2. What is the tension in the cable during this acceleration? Finally, after a descent of sixteen floors at a steady speed of 8 m/s, what is the tension in the cable as the elevator slows to a stop at the main lobby under a deceleration of 2.0 m/s^2.

23. A hoisting cable will withstand a tension of 19,800 N. If it is used to raise a load of 1500 kg from the ground level to a work station 25 m high, what is the least permissible time for the operation?

24. A .30-caliber rifle is fired into an oak tree. The bullet has a mass of 10.5 g and a muzzle velocity of 750 m/s. It penetrates the wood to a distance of 22 cm. Assume that the retarding force on the bullet is constant and find (*a*) the value of the decelerating force; and (*b*) the length of time for the bullet to come to a stop.

25. A constant horizontal force acts on a body resting on a smooth horizontal plane (no friction). The force is parallel to the plane and has a magnitude of 20 lb. The body is initially at rest, and after 10 s it has moved 100 yd. (*a*) Find the mass of the body in slugs. (*b*) What is its velocity? (*c*) If the original force is removed, what oppositely directed (and steady) force would be required to bring it to rest in 3 s?

Group Three

26. An electron leaves the cathode of an electron tube at (assumed) zero speed and is immediately accelerated in an electric field. It arrives at the anode 0.062 cm distant with a speed of 6.5×10^6 m/s. Assume the mass of the electron is its *rest mass*—9.11×10^{-31} kg. Find the magnitude of the force (newtons) acting on the electron during this acceleration.

27. A small car rolls across a smooth horizontal surface without friction. It is pulled along by a spring scale attached to it and the force is parallel to the surface on which the car rolls. Pulling the car with a constant acceleration of 0.25 m/s^2 causes the scale to read 3.1 N. Find the mass of the car.

28. A train is going up a grade that rises 3 ft for every 100 ft along the track. The observation-dome car is at the end of the train and its total loaded *weight* is 84,000 lb. The train is leaving a station and is accelerating at 1.5 ft/s^2. Neglect friction and find the drawbar pull on the coupling at the dome car. (Answer in pounds.)

29. Two blocks are being pulled across a smooth table (friction negligible) by a cord angled upward at 28° as shown in the diagram of Fig. 5.15. Block m_1 has a mass of 4.0 kg and m_2 a mass of 2.5 kg. The acceleration of the block system is measured as 1.5 m/s^2. Find the value of the force P, and the tension in the cord between the two blocks.

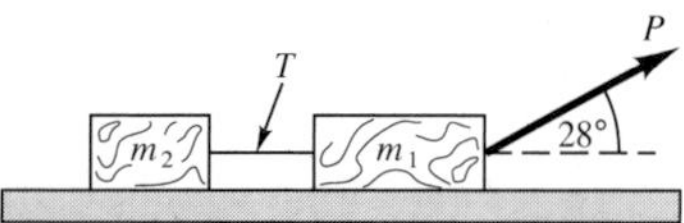

Fig. 5.15 Sketch for Problem 29.

30. If the system shown in Fig. 5.16 is released, what acceleration will result? (HINT: Remember that the entire system mass has to be accelerated, but the pull of gravity on the mass hanging vertically is the only force available to do the accelerating.) Neglect friction and assume a massless pulley. Be careful with units!

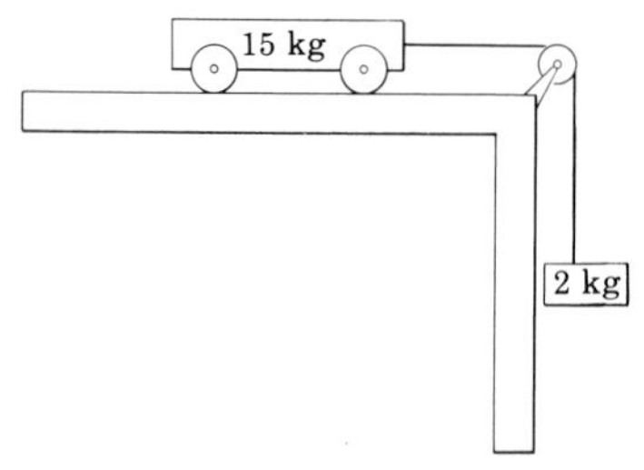

Fig. 5.16 Sketch for Problem 30.

31. A rotary oil-well drilling rig is handling 4200 ft of 6-in. pipe, which weighs 5 lb per foot of length. The pipe is being lowered into the hole at a uniform velocity of 10 ft/s. The brake on the cable drum is applied, stopping the entire assembly in a distance of 12 ft. Calculate the maximum force in pounds on the supporting hook.

32. A baseball bat is in contact with the ball for 0.02 s and changes the ball's velocity from 85 ft/s northward to 150 ft/s southward. The ball weighs 5.25 oz. What was the average force of the blow in pounds?

33. Figure 5.17 illustrates an apparatus called an *Atwood's machine*. If $m_1 = 2.5$ kg and $m_2 = 3.1$ kg, find (*a*) the acceleration of the mass-and-cord system when m_2 is released and (*b*) the tension in the cord while the system is running. Assume the wheel and the cord have zero mass and that system friction is zero.

34. If the acceleration of an Atwood's machine system can be accurately determined, the value of g, the acceleration of gravity, can be calculated. With an arrangement like that of Fig. 5.17 above, m_1 and m_2 were both taken as 10 kg. An additional mass of 100 g was placed on top of m_2. When the system was released the (m_2 + 100 g) mass descended a distance of 2.0 m in 6.42 s. Calculate g from this experiment.

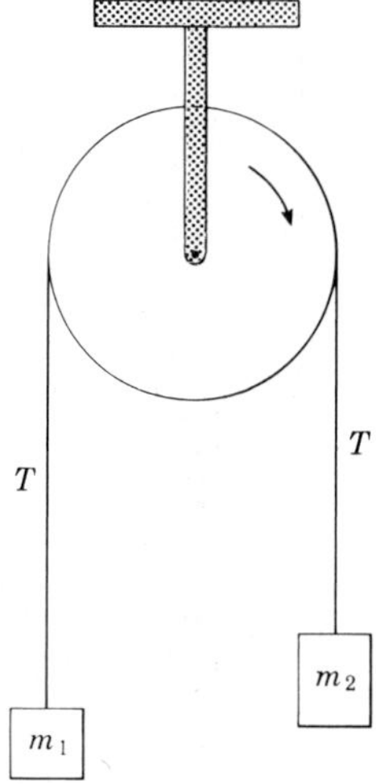

Fig. 5.17 Diagram of Atwood's machine (Problem 33).

35. A proton, mass 1.6725×10^{-27} kg, enters a force field where it is subjected to a steady unbalanced force of 3.6×10^{-22} N. Find the magnitude of its acceleration. How many g's is this?

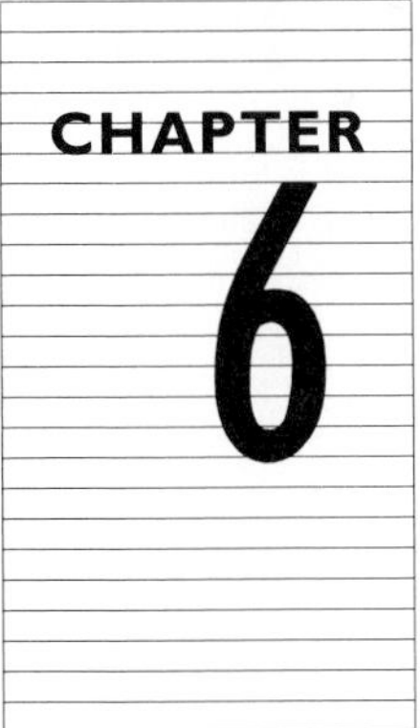

CHAPTER 6 WORK, ENERGY, MOMENTUM, AND POWER

In everyday language *energy* represents the capability to "get things done," and *work* is the process by which things get done.

Most industrial and commercial work in modern nations is done by machines. Dexterity and skill in the operation of machines is very important in some industrial and business processes. All of the machines and tools of industry, commerce, and agriculture require vast amounts of energy.

Machines do work and require vast amounts of input energy

Anyone who is contemplating a professional or technical career needs to know a great deal about energy, work, and power. This chapter will provide a basic foundation for understanding these concepts, and will illustrate a number of applications of energy principles.

Fig. 6.1 Painting automobiles with robots. Programmed by computer for each model of car coming down the line, these robots ensure constant paint thickness and complete coverage. They perform tirelessly in an environment that human workers find undesirable. (General Motors Corp.)

WORK

6.1 ■ Work Defined

Work performed by human muscle tends to be measured in terms of human fatigue, or in terms of the hours in a workday. With machines, on the other hand, a more exact and technical definition of work is necessary, one which can be standardized to mean the same net accomplishment throughout industry. Technically, *work* is defined in terms of *force applied* and the *distance through which it acts*.

Work, defined

$$\text{Work} = \text{force} \times \text{distance}$$

or

$$W = Fs \text{ (when } F \text{ and } s \text{ are parallel)} \qquad (6.1)$$

Work is a scalar quantity equal to the product of the *magnitudes* of F and s.

In a scientific sense a force must act through a measurable distance before work is accomplished. You might exert the maximum muscular force of which you are capable in an effort to push a truck, but if it does not move, no work is done, even though you soon become fatigued (Fig. 6.2).

Regardless of the force (effort) applied, no work is done unless there is a resulting displacement

The analysis above and Eq. (6.1) both assume that the resulting motion has the same direction as the applied force. This is by no means always the case. The more general situation is illustrated in Fig. 6.3*a*, where the direction of the force pulling the sled is different from that of the sled's motion. A simple vector diagram (Fig. 6.3*b*) provides the component of the force (see Sec.3.9) in the direction of the motion. *Useful work* is done *only by the component of the force in the direction of the displacement*. As a more general proposition then, work done is given by

The general definition of work, as an equation

$$W = F \cos \theta \cdot s = Fs \cos \theta \qquad (6.2)$$

where θ is the angle between the applied force vector and the resultant displacement vector. If the force were at right angles to the displacement, $F \cos \theta = 0$, and therefore no work would be done (Fig. 6.4).

6.2 ■ Units of Work

Metric Units

SI unit of work—the *joule* (J)

In the mks (SI-metric) system of units, forces are measured in *newtons* and distances in *meters*. The mks unit of work and of energy is therefore the *newton-meter*, abbre-

Fig. 6.2 Pushing on a truck stuck in the mud requires a great expenditure of force. You soon become fatigued, but no work is done because your force has not moved the truck through any distance.

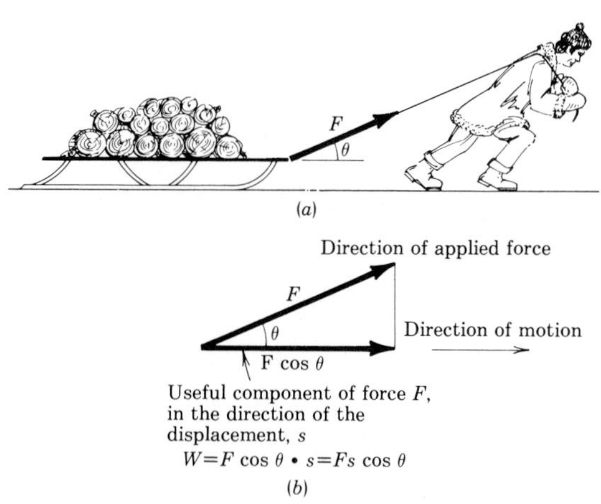

Fig. 6.3 Force and work. Useful work is done only by that component of the applied force that is in the direction of the motion. (*a*) Sled being pulled by a force applied at an angle θ with respect to the direction of the motion. (*b*) Vector diagram of the sled-pulling problem.

viated N · m. The newton-meter has been given the name *joule* (symbol J), in honor of James Prescott Joule, a British scientist who made many contributions to the branch of physics called *thermodynamics*.

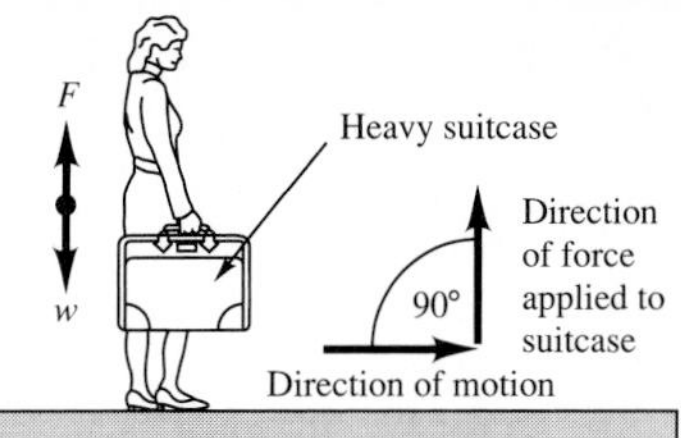

Fig. 6.4 Force and work. When the force applied is at right angles to the motion, $F \cos \theta = 0$, and no work is done. The force may be great, resulting in early fatigue, but work, in the sense of physics and engineering, has not been accomplished.

1 joule = 1 newton-m (N · m)

One joule is the work accomplished (or energy expended) when a force of one newton acts through a distance of one meter.

Illustrative Problem 6.1 A hoist lifts a mass of 1500 kg through a distance of 18 m at constant velocity. How many joules of work are done by the hoist? (See Fig. 6.5.)

Solution

$$W \text{ (J)} = F \text{ (N)} \times h \text{ (m)}$$

In this case, the force applied equals the *weight* of the 1500-kg mass. The height h is the distance traveled.

$$F = w = mg \text{ (Recall that weight equals mass} \times g\text{; Eq. (5.2).)}$$
$$= 1500 \text{ kg} \times 9.81 \text{ m/s}^2$$
$$= 14{,}700 \text{ N}$$

Therefore

$$W = 14{,}700 \text{ N} \times 18 \text{ m}$$
$$= 2.65 \times 10^5 \text{ J (or N} \cdot \text{m)} \qquad \textit{answer}$$

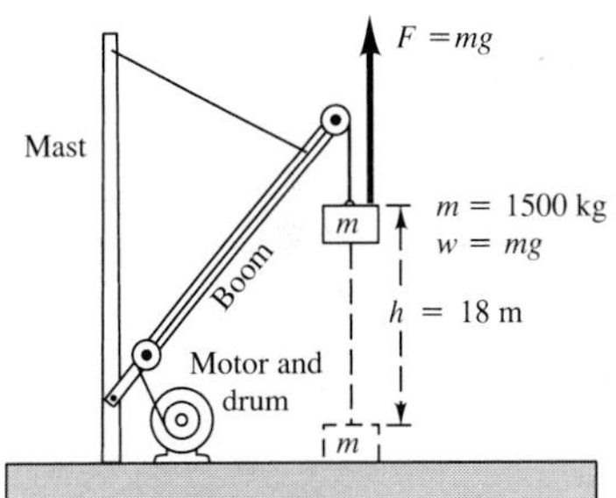

Fig. 6.5 Work done against gravity. In order to lift a given mass at constant velocity, a force equal to its weight must be applied, acting vertically. But $w = mg$, so the force applied to lift the object must be equal to mg.

English System Units Since forces are measured in pounds and distances in feet in the English system, the practical unit of work is the *foot-pound* (ft · lb). Suppose (Fig. 6.6) a horizontal force of 25 lb pushes a piano 10 ft on a level floor. The work accomplished is

$$W = Fs = 25 \text{ lb} \times 10 \text{ ft} = 250 \text{ ft} \cdot \text{lb}$$

Work Done in Lifting When objects are being lifted at constant velocity, it is easy to compute the work done, since the weight of the object is actually the force of gravity pulling on it, and this must be equalled by the force applied to lift it. If a 60-lb casting is lifted from the floor to the bed of a milling machine 4 ft high, the work done in lifting (Fig. 6.7) is 4 ft × 60 lb = 240 ft · lb. Note that, in the English system, *weights* are generally used, not masses. Weights are equivalent to forces for lifting, and the relation $w = mg$ is not used.

English-system unit of work—the *foot-pound* (ft · lb)

However, if the mass of an object to be lifted is given in engineering units (slugs), the *force* to lift it is given by Eq. (5.2).

$$\text{Force (weight)} = mg = \text{slugs} \times 32.2 \text{ ft/s}^2$$

The work done (ft · lb) in hoisting a mass of m slugs through a height of h feet is given by

$$W = mgh \tag{6.3}$$

Work done against the force of gravity

It should be noted that, in computing the work done in lifting objects against the force of gravity, the assumption is that they are *lifted at constant velocity*. If they were jerked upward rapidly, they would be *accelerating* and, in addition to the force required to lift them ($F_w = mg$), an additional amount of force ($F_a = ma$) would be necessary to provide the acceleration.

Note that W is used for *work* and w for *weight*.

Table 6.1 SI-Metric and English System Units of Work and Energy
(Energy units are the same as work units)

Measurement System	*Unit of Force*	*Unit of Distance*	*Unit of Work/Energy*
SI-metric (mks)	newton (N)	meter (m)	N · m (*joule*, J)
(metric—cgs)	dyne	centimeter	dyne-cm (erg)
			kilowatt-hour (kW · h)
English-engineering	pound (lb)	foot (ft)	foot-pound (ft · lb)
			horsepower-hour (hp · h)

Conversion Factors: 1 ft · lb = 1.356 J 1 J = 0.738 ft · lb

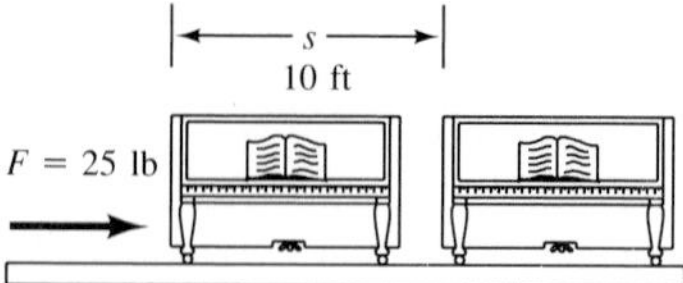

Fig. 6.6 Work in the English system of measurement. When a 25-lb force is applied to move a piano on a level floor and the piano moves 10 ft, the work done is $W = 10 \text{ ft} \times 25 \text{ lb} = 250 \text{ ft} \cdot \text{lb}$.

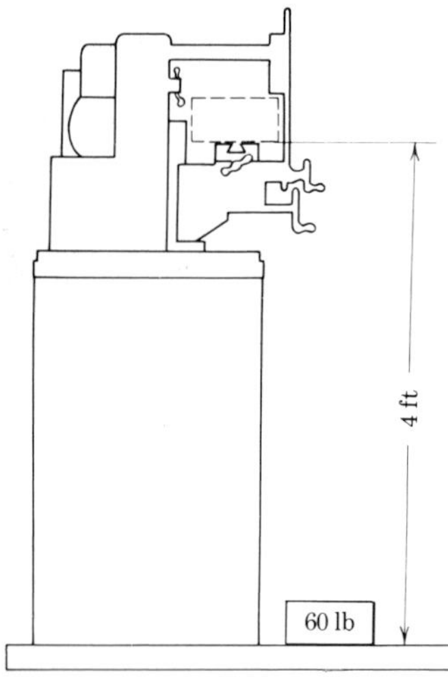

Fig. 6.7 Lifting requires that work be done against the force of gravity. The net work done in lifting is equal to the *weight* of the object lifted times the *vertical distance* through which it is lifted. In this case, $W = 4 \text{ ft} \times 60 \text{ lb} = 240 \text{ ft} \cdot \text{lb}$.

Illustrative Problem 6.2 An atmospheric cooling tower, mass 625 sl, is hoisted to the roof of a tall building 360 ft from ground level. How much work is done against the force of gravity? (Remember: $1 \text{ sl} = 1 \text{ lb} \cdot \text{s}^2/\text{ft}$.)

Solution From Eq. (6.3),

$$W = 625 \frac{\text{lb} \cdot \text{s}^2}{\text{ft}} \times 32.2 \text{ ft/s}^2 \times 360 \text{ ft}$$

$$= 7{,}245{,}000 \text{ ft} \cdot \text{lb} \qquad \textit{answer}$$

Summarizing, the units of work are based on the product of a force and a displacement (distance). In the SI-metric system the basic unit of work is the newton-meter (N · m), named the *joule*. (Another metric unit, obtained by multiplying the *dyne* of force by the *centimeter* of displacement—the dyne-centimeter, or *erg*—is a unit of the old centimeter-gram-second-(cgs)-system, but it is not used in the SI or mks system (one joule = 10^7 ergs).)

The basic unit of work in the English-engineering system is the foot-pound (ft·lb).

Two very important industrial units of work are merely given mention at this point—the *kilowatt hour* (kW · h), and the *horsepower-hour* (hp · h). They will be defined and used in later chapters.

Table 6.1 lists common units of work and energy and gives some conversion factors from SI-metric to English units.

ENERGY

Matter *under certain conditions* has the capacity or ability to do work even though none is being accomplished at the moment. A rushing stream, or water behind a dam, for example, can be put to work at a variety of jobs. A heavy weight suspended at a great height could do work if it were allowed to fall; a compressed spring possesses a latent capacity for doing useful work and so does compressed air. Fuels like gasoline and natural gas do work as the result of combustion. This *capacity to do work* is one attribute of energy. It should be emphasized however, that there are many forms of energy, and that much energy is not available to do useful work.

6.3 ■ Classifications of Energy

The running stream possesses energy of motion; the suspended weight and the impounded water, energy of position; the compressed spring, energy due to molecular arrangement; and fuels possess chemical energy.

Work and energy. Energy has the capability of doing work, but does not always result in work.

Energy of motion is called kinetic energy (KE), and energy due to position or molecular arrangement is called potential energy (PE).

Bodies also possess *rest energy*, merely as an attribute of *mass*. In a later chapter the concept of rest energy will be elaborated. Bodies at elevations above sea level possess gravitational potential energy, GPE.

Gravitational potential energy (GPE) is possessed by bodies at elevations above sea level or above a base datum plane of "zero energy."

Other forms of energy to be studied in detail later include: *heat energy, electrical energy, nuclear energy, radiant (light) energy,* and *sound energy*. *Chemical energy* is of tremendous importance also, but its study falls within the discipline of chemistry rather than that of physics.

A body possessing kinetic energy does work when it is stopped or slowed down. A hammer possesses a considerable amount of kinetic energy as it strikes a nail, and most of this kinetic energy is converted into the useful work of driving the nail. Other examples of kinetic energy are a moving auto, a projectile in flight, a rotating flywheel, or a plunging fullback.

A body or system that possesses potential energy has had work done on it at some previous time, and its present potential energy is a measure of the energy once expended to give it its present position or molecular arrangement. The potential energy which it possesses can often be converted back into useful work (not without losses, however) by means of a suitable machine.

6.4 ■ Measurement of Energy

Energy is expressed in the same units as work, namely, *joules* in the metric system, or *foot-pounds* in the English system. If an object possesses 10 ft · lb of energy, it is *theoretically capable* of doing 10 ft · lb of work. If work is expended on an object to lift it, potential energy is acquired as it is lifted. Then if it is allowed to fall, as it gains velocity the potential energy is converted into kinetic energy, which, in turn, can be converted into work as the falling body is stopped (see Fig. 6.8). In other words, potential energy (PE) and kinetic energy (KE) can be interchangeable.

Interchangeability of KE and PE.

Since *direction* is unspecified in energy relationships, energy (like work) is a *scalar quantity*.

Energy and work are scalar quantities

6.5 ■ Gravitational Potential Energy (GPE)—Work Done Against Gravity

Potential energy often consists of the work done on a body to raise it to an elevated position with respect to some chosen level of zero energy (usually the earth's surface). It may therefore be readily reasoned that if an object has a mass m, and if it is slowly raised to a height h, the work done on it is $W = Fh$. Since the force required

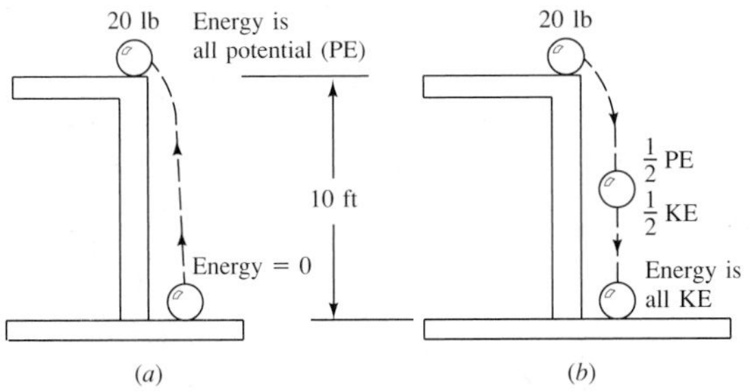

Fig. 6.8 Potential energy and kinetic energy compared. (*a*) Work done in lifting an object results in increased potential energy of the object. If the 20-lb (weight) ball is lifted to a 10-ft-high shelf, potential energy (with respect to the floor) is acquired. The amount of potential energy gain is exactly equal to the work done. PE = 20 lb × 10 ft = 200 ft · lb. (*b*) As the ball falls from the shelf, accelerated by the force of gravity, the potential energy is converted into kinetic energy. Midway in the fall the energy possessed by the ball is half potential, half kinetic. What becomes of the kinetic energy as the ball hits the floor? What has happened to the potential energy it had when it was on the shelf?

An object above the earth's surface possesses gravitational potential energy (GPE)

to lift an object of mass m is equal to its weight ($w = mg$), the gravitational potential energy acquired can be expressed as follows:

$$\text{GPE} = mgh \tag{6.4}$$

Potential energy calculated from Eq. (6.4) will be expressed in the same units in which work is expressed, i.e., joules (mks system) and foot-pounds (English system).

One caution is necessary. The above equation is dimensionally correct for the English system only when mass is given in *slugs*. Since masses are still expressed in *pounds* in some nonscientific and industrial fields, you should be careful with units. Where that is the case, merely multiply the given "mass" in pounds (actually *weight*) by the height, to obtain the GPE in ft · lb, since (Eq. 5.2) $w = mg$. In other words, $\text{GPE}_{(\text{ft}\cdot\text{lb})} = w_{(\text{lb})} \times h_{(\text{ft})}$. Also, for SI units,

$$\text{GPE}_{(\text{J})} = w_{(\text{N})} \times h_{(\text{m})}.$$

Gain in potential energy is not dependent on the *path taken* to reach a given point above zero reference level; only the increase in elevation matters.

Path Makes No Difference If frictional and other forces are assumed to be zero, the path used to raise a body from one level or elevation to another makes no difference in the work done, or in the GPE gained. In terms of Fig. 6.9*a*, this means that if all forces except gravitational attraction are neglected, the work done on a mass at A by a force moving it upward along *any path* such as $ABCP$ for example, is the same as that done if it were simply raised from A' to P. The net increase in potential energy from A to P is also the same, regardless of path. The following analysis should help you think this through.

From A to B there is a straight lift, and $W_{AB} = mgh_1$. From B to C no work at all is done if friction is zero. On the curved path CP at point X, consider any very small distance Δs*, along the curved path. The enlarged vector diagram of Fig. 6.9*b* shows the forces acting on the body as it is being moved through the path distance Δs, resulting in a gain in elevation Δh. The force F_u required to move the body up the slope without acceleration and in the absence of friction is exactly equal to F_d, the component of the force of gravity along the down slope. But $F_d = mg \sin \theta$. Therefore

$$F_u = mg \sin \theta$$

The net work done for the segment Δs is $\Delta W = F_u \, \Delta s$. But

$$\Delta s = \frac{\Delta h}{\sin \theta}$$

and therefore

$$\Delta W = \frac{mg \cancel{\sin \theta} \, \Delta h}{\cancel{\sin \theta}} = mg \, \Delta h$$

This is precisely the work which would have been done (and the gain in GPE which would result) if the object had been moved through the *vertical path* Δh instead of the curved path Δs.

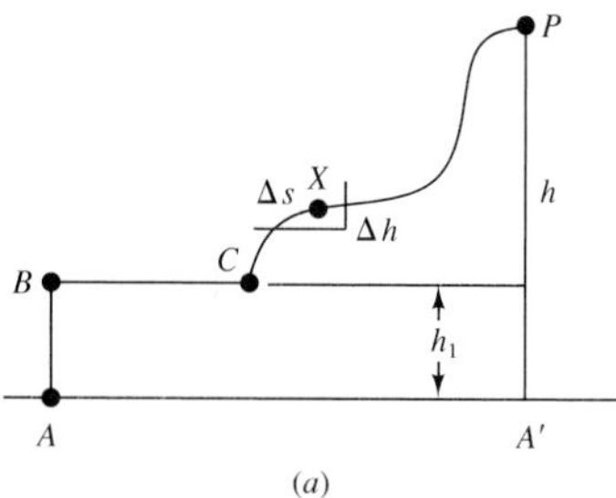

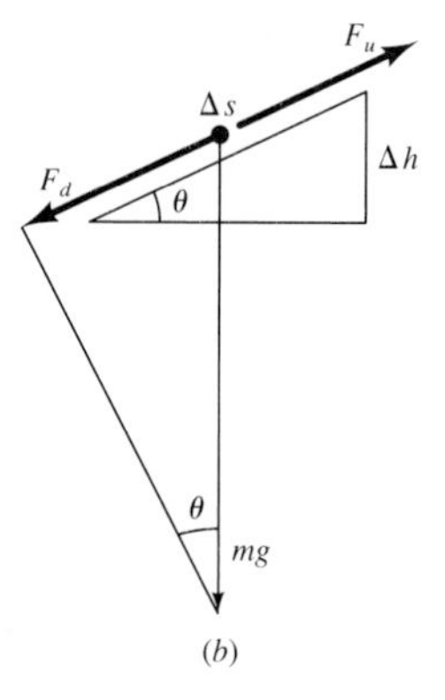

Fig. 6.9 Work done against gravity on a system and the resultant GPE gained by the system do not depend on the path of the motion (friction neglected). (*a*) The work done in moving an object along path *ABCXP* is equal to the work done in raising the same object vertically along path *A′P*. (*b*) Vector relations at point *X* along the curved path *CP*.

The work done against gravity as a mass m is lifted through a height h is W = mgh, and is independent of the path of the motion.

* The symbol Δ means "a very small amount of." Δs is read "delta *s*." It *does not* mean delta multiplied by *s*.

6.6 ■ Kinetic Energy

Kinetic energy is energy of motion

The *energy of motion* of a moving body is precisely the amount of work which would have to be done on it by the force that accelerates it from rest to its present velocity or, conversely, the amount of work which the body could do against a resisting force which is decelerating it to zero velocity.

An expression for calculating kinetic energy* can be obtained by recalling Newton's second law. The force F necessary to impart an acceleration a to a mass m is

$$F = ma \tag{5.1}$$

Multiplying both sides of this equation by s,

$$Fs = mas$$

But $$v^2 = 2as \tag{4.9}$$

and therefore $$as = v^2/2$$

Substituting $$Fs = \tfrac{1}{2}mv^2$$

But the product Fs represents the work done on a mass m to accelerate it from rest to velocity v and is therefore the energy of motion (kinetic energy) possessed by the mass m when it is moving at the velocity v. Consequently,

$$\text{KE} = \tfrac{1}{2}mv^2 \tag{6.5}$$

This equation can be used *directly with SI-metric units*. Just make sure that mass is in kilograms and velocity is in meters per second. The answer will be in newton-meters (joules), as this dimensional analysis shows:

KE units (metric):

$$\text{kg} \times \left(\frac{\text{m}}{\text{s}}\right)^2 = \frac{\text{kg} \cdot \text{m}}{\text{s}^2} \times \text{m} = \text{N} \cdot \text{m} = \text{J}$$

Units of kinetic energy

In the English (engineering) system however, you must be careful. The kinetic energy equation is derived, as noted above, from Newton's second law, and that law ($F = ma$) requires that all units be *absolute*, not gravitational units. It is necessary therefore, to use *slugs* for mass, not *pounds*. Dimensionally,

$$1 \text{ sl} = 1\frac{\text{lb} \cdot \text{s}^2}{\text{ft}} \qquad \text{(see Sec. 5.7)}$$

Using slugs for mass, kinetic energy will then be in foot-pounds, as this analysis shows:

KE units (English):

$$\frac{\text{lb} \cdot \text{s}^2}{\text{ft}} \times \left(\frac{\text{ft}}{\text{s}}\right)^2 = \text{lb} \cdot \text{ft or ft} \cdot \text{lb}$$

In industrial practice, and sometimes even in engineering practice, design and operational problems start off with data in pounds, not in slugs. When this is the case, and when kinetic energy is being calculated, recall that $m = w/g$ (Eq. 5.2) and use the following alternative kinetic energy formula (English system):

$$\text{KE} = \frac{1}{2} \cdot \frac{w}{g} v^2 = \frac{wv^2}{2g} \tag{6.5$'$}$$

If, in a metric country, weight in *newtons* is given (instead of mass in kilograms), the same form (Eq. 6.5′) should be used.

* The word *kinetic* is from a Greek word meaning *motion*.

Fig. 6.10 Schematic representation of a pile driver (Illustrative Problem 6.3).

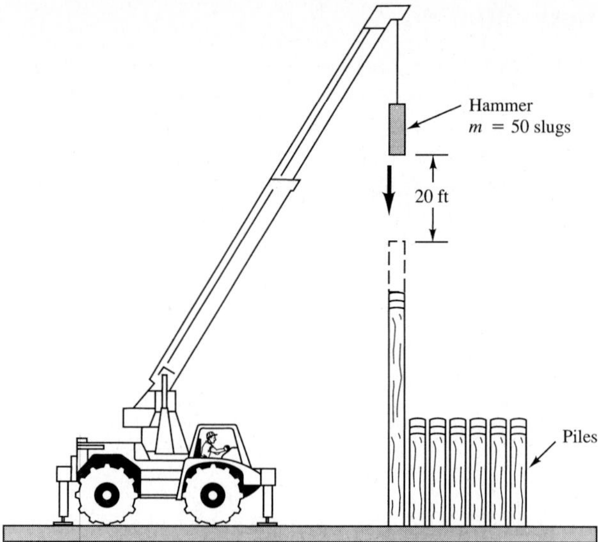

It should be apparent that in problems where both potential and kinetic energy are involved, both must be expressed in the same units, i.e. absolute or gravitational, before adding, subtracting, or otherwise combining them in any way.

Illustrative Problem 6.3 Suppose the hammer of a pile driver (Fig. 6.10) has a mass of 50 sl and that it falls 20 ft before it hits the pile. What kinetic energy does the hammer have just before it hits?

Solving problems using energy principles

Solution We will solve the problem by two different methods.
(*a*) *First method* (using energy principles): Remembering that the kinetic energy of the hammer at the bottom of its fall must be equal to the gravitational potential energy at the top, write Eq. (6.4), GPE = mgh. Remember that slugs have the units (lb · s^2)/ft, and write

$$\text{GPE} = 50\,\frac{\text{lb}\cdot\text{s}^2}{\text{ft}} \times 32.2\frac{\text{ft}}{\text{s}^2} \times 20\text{ ft}$$

$$= 32{,}200\text{ ft}\cdot\text{lb} \qquad \textit{answer}$$

(*b*) *Second method* (using acceleration and velocity principles, as developed in Chap. 4). Before substitution of numerical values in the kinetic energy formula, the velocity at the moment of impact must be found. Recall (Eq. 4.14) that the velocity of a freely falling body after falling a distance s is

$$v^2 = 2gs$$

Substitute given values,

$$v^2 = 2 \times 32.2\,\frac{\text{ft}}{\text{s}^2} \times 20\text{ ft} = 1290\left(\frac{\text{ft}}{\text{s}}\right)^2$$

Now, substituting this value of v^2 in Eq. (6.5),

$$\text{KE} = \frac{50\text{ sl}\left(\text{that is, }\dfrac{\text{lb}\cdot\text{s}^2}{\text{ft}}\right) \times 1290\,\dfrac{\text{ft}^2}{\text{s}^2}}{2}$$

$$= 32{,}200\text{ ft}\cdot\text{lb} \quad \text{(as before)} \qquad \textit{answer}$$

The comparative simplicity of the first method should be carefully noted. Solving problems in mechanics from *energy considerations,* when they are applicable, is usually preferable to applying acceleration principles to their solution. These considerations are often referred to as the *principle of work,* or the *work-energy theorem.* This principle constitutes a powerful tool in the solution of problems in mechanics.

The work-energy theorem—a powerful tool in problem solving

The Work-Energy Theorem (Principle)

The total mechanical energy of a body or system at any time t
= the total energy at some former time t_0
+ change in KE in the time interval
+ change in GPE in the time interval
+ work done on the body by forces other than gravity

In using this principle, increases in energy are treated as positive and decreases in energy are negative. In cases where the initial energy is assumed zero (for example, object at rest and located at a level of zero gravitational potential energy), the work-energy theorem can be stated as follows:

$$\textbf{KE change} + \textbf{GPE change} = \textbf{work done on body} \tag{6.6}$$

It must be remembered that ''work done on the body'' can be either positive or negative. Work done by an accelerating force (a propellant, for example) is positive, while work done by frictional forces applied to the body is negative.

Illustrative Problem 6.4 A rocket with an attached satellite to be placed in orbit has a total mass of 100,000 kg. It is launched vertically with the rocket motors providing a constant thrust of 1.6 million N. What is the rocket's velocity as it passes through an altitude of 5000 m? Neglect effects of air resistance and loss of mass due to fuel used.

Solution
(*a*) *By the work-energy theorem.* The total energy (KE + PE) at 5000 m equals the work done by the thrust force.

$$\tfrac{1}{2}mv^2 + mgh = Fs$$

Substituting,

$$\tfrac{1}{2}(100{,}000 \text{ kg}) \times v^2 + 100{,}000 \text{ kg} \times 9.81 \frac{\text{m}}{\text{s}^2} \times 5000 \text{ m} = 1.6 \times 10^6 \frac{\text{kg} \cdot \text{m}}{\text{s}^2} \times 5000 \text{ m}$$

Simplifying,

$$v^2 = \frac{2 \times 5000 \text{ m}\left(1.6 \times 10^6 \dfrac{\text{kg} \cdot \text{m}}{\text{s}^2} - 981{,}000 \dfrac{\text{kg} \cdot \text{m}}{\text{s}^2}\right)}{100{,}000 \text{ kg}}$$

$v = 249$ m/s *answer*

(*b*) *By Newton's second law (acceleration principles).* The *net upward force* is the thrust of the rocket motor minus the *weight* (not mass) of the rocket and its payload. Substituting in $F = ma$,

$$1{,}600{,}000 \frac{\text{kg} \cdot \text{m}}{\text{s}^2} - 100{,}000 \text{ kg} \times 9.81 \frac{\text{m}}{\text{s}^2} = 100{,}000 \text{ kg} \times a$$

$$a = \frac{619{,}000(\text{kg}\cdot\text{m})/\text{s}^2}{100{,}000\text{ kg}} = 6.19\text{ m/s}^2$$

Now, using

$$v^2 = 2as$$

and substituting,

$$v^2 = 2 \times 6.19\frac{\text{m}}{\text{s}^2} \times 5000\text{ m}$$

$$v^2 = 249\text{ m/s} \quad \text{(as before)} \qquad \textit{answer}$$

Velocity Attained by a Body in Free Fall Frequently, in mechanics, it is necessary to be able to calculate the velocity of a body that falls from rest through a vertical height h. The work-energy theorem provides a ready solution to this problem.

KE change + GPE change = work done on the body

1. Since the original KE of the body (at rest) = 0, the KE change after reaching velocity v is: $\frac{1}{2}mv^2$.
2. In falling through an elevation h, GPE change is: $-mgh$. (The sign is minus, since the body loses GPE.)
3. If air resistance (friction) is assumed zero, work done on the body is zero.

Therefore we write: $\frac{1}{2}mv^2 + (-mgh) = 0$

Solving for v:

Free-fall velocity, $v = \sqrt{2gh}$

$$v = \sqrt{2gh} \tag{6.7}$$

This expression is an important one in mechanics, and it should be memorized. Note that it assumes air resistance is zero. The following problem illustrates its use.

Illustrative Problem 6.5 A waterfall is 150 ft high. Assuming free fall and no air resistance, what is the velocity of the falling water stream as it hits the pool at the foot of the fall?

Solution Substituting $v = \sqrt{2gh}$ in Eq. (6.7),

$$v = \sqrt{2 \times 32.2\text{ ft/s}^2 \times 150\text{ ft}}$$
$$= 98.3\text{ ft/s} \qquad \textit{answer}$$

6.7 ■ Energy and Its Transformations

Energy converted from PE to KE and back again—the pendulum

So far the discussion has involved two forms of energy, potential and kinetic, and we have shown how one can be transformed into the other, how potential energy is the result of work done, and how work can be done by an object possessing kinetic energy. A swinging pendulum is a good example of potential energy being transformed into kinetic energy and back again to potential energy (see Fig. 6.11).

Illustrative Problem 6.6 A pendulum bob is pulled to the side, or displaced from its equilibrium position, until its center is 0.5 m higher than the "dead center," or equilibrium position. Find the velocity of the bob as it passes through dead center. (Neglect air resistance and friction; see Fig. 6.11.)

Solution The only energy possessed by the system as the bob is released is potential energy from the 0.5-m elevation of the bob above dead center. The kinetic energy as

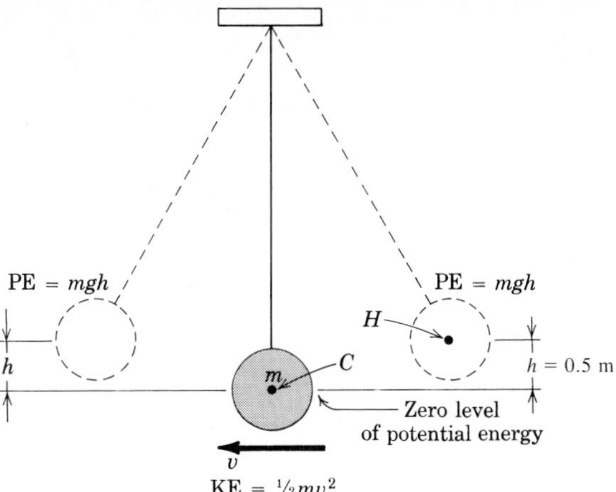

Fig. 6.11 A swinging pendulum illustrates the transformation from potential energy to kinetic energy and back again.

the bob passes the equilibrium point (no remaining potential energy) is equal to the potential energy as the bob was released,

$$KE_{(\text{point } C)} = PE_{(\text{point } H)}$$

In absolute units,

$$\frac{\not{m}v^2}{2} = \not{m}gh$$

Note that mass m cancels out—mass of bob makes no difference—giving

$$v = \sqrt{2gh}$$

Substituting,

$$\begin{aligned} v &= \sqrt{2 \times 9.81 \text{ m/s}^2 \times 0.5 \text{ m}} \\ &= 3.13 \text{ m/s} \qquad \textit{answer} \end{aligned}$$

The KE at "dead center" is converted to PE again as the bob swings left, reaching the height h (friction neglected).

Energy transformations in nature, and in industry and commerce

Many forms of energy can readily be converted into other forms. The sun's heat energy is the result of nuclear and chemical energy changes in its interior and on its surface. This radiant energy travels through space with the speed of light—2.998×10^8 m/s (186,300 mi/s). When this radiation is stopped by the earth, it is reconverted into heat, which evaporates water from the oceans and causes winds and storms of tremendous energy. Rain and snow are deposited at higher elevations where the mass of water results in great potential energy. On its way to the sea this potential energy becomes kinetic energy. At hydroelectric plants it can be converted into electric energy, which can be reconverted into heat, light, sound, and mechanical energy at the flip of a switch.

From the viewpoint of industry, today most of the energy we use comes from coal, oil, and natural gas—stored *chemical energy* which we convert into heat energy to fire boilers, run engines, make electricity, and power motorcars and trucks. Industrial use of nuclear energy increased steadily until recently, when concerns about nuclear reactor safety and disposal of radioactive wastes slowed the construction of new plants. Later chapters will discuss atomic and nuclear energy at some length. Alternative sources of energy—solar, wind, geothermal, and tidal—are currently under development in this country and abroad. These are often referred to as *renewable energy sources*.

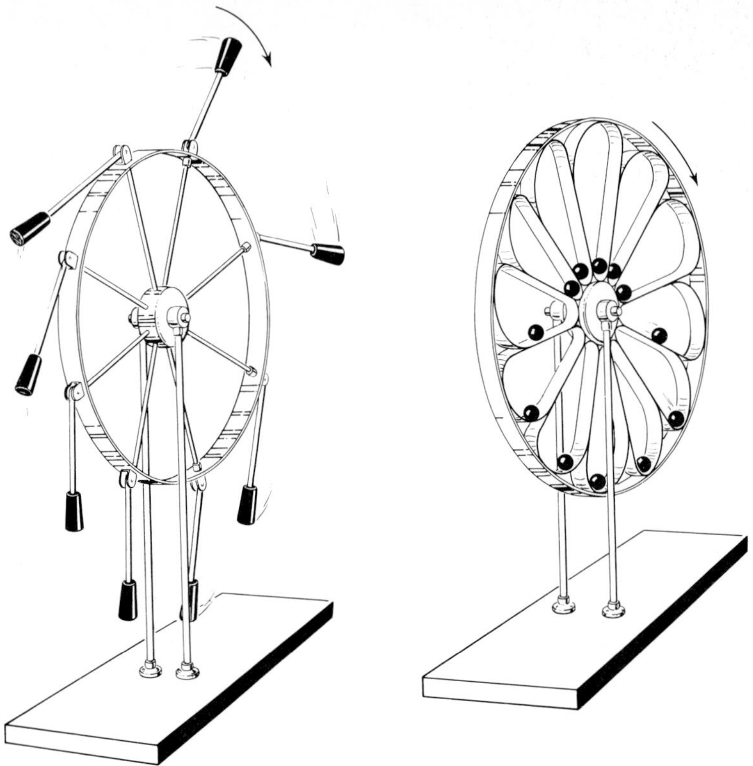

Fig. 6.12 Two schemes for a "perpetual motion" machine. The inventor's idea can be perceived by studying the construction. No such machine has ever been devised that will actually do any useful work. As a matter of fact, energy would have to be *supplied* to such machines to overcome friction.

6.8 ■ Law of Conservation of Energy

One statement of the law of conservation of energy

Although energy can be changed from one form to almost any other form by the application of a suitable machine, no machine or device exists which can actually *create* energy. "Perpetual motion" machines (Fig. 6.12) have challenged the inventive genius of many generations, but none has yet operated successfully *to do useful work*. The following statement seems to be an immutable law of nature: *Energy can be changed in form but cannot be created or destroyed*. This important law was first clearly stated by the German physicist Hermann von Helmholtz (1821–1894) and is referred to as the ***law of conservation of energy***.

The Law of Conservation of Energy:

Energy can be changed in form, but it cannot be created or destroyed.
Or, an alternative statement of the law:
The total amount of energy in an isolated system is constant.

What about "lost" energy?

It is true that in energy transformations, some energy always *seems to be lost*. Careful analysis, however, reveals the fact that when energy of one form disappears, other forms of energy appear in amounts equal to the apparent discrepancy. For example, if the electric energy input to a certain motor is carefully measured and compared with the mechanical energy output, a loss of from 4 to 6 percent seems to have occurred. Actually, however, this "lost energy" has gone into heat, and the law of conservation of energy for this particular energy transformation could be stated as follows:

Energy input (electric) = energy output (mechanical) + heat energy ("lost")

The heat energy is lost only in the sense that it is not useful for the purpose at hand.

MOMENTUM

A concept which is related to energy and also to Newton's laws of motion is the idea of *momentum,* which Newton called *quantity of motion*. The *linear momentum* of a body is defined as the product of its mass and its velocity. Momentum is designated by the letter **p.**

Quantity of motion—momentum. *Momentum* equals *mass* times *velocity.*

$$\mathbf{p} = mv \qquad (6.8)$$

$$\text{Momentum} = \text{mass} \times \text{velocity}$$

Momentum is a *vector quantity,* since it is a product of *velocity* (a vector quantity) and mass. Its symbol **p** is in boldface to indicate its vector nature.

Comparison of momentum with kinetic energy

By comparison with Eq. (6.5), it is seen that both kinetic energy and momentum are proportional to the mass of the moving body. Further examination of the two equations shows that *kinetic energy* is proportional to the *square of the velocity,* while *momentum* increases as the *first power of the velocity*. Kinetic energy is a measure of the work a moving body will do as it is being stopped.

Momentum is a measure of the impulse or impact of a moving body when it hits another body.

The quantity of motion (momentum) possessed by a moving body depends on both its mass and its velocity. A heavy object even when moving slowly, e.g., a bulldozer, will have great momentum; and a small object moving with high speed, e.g., a bullet, also possesses a great deal of momentum. Objects possessing momentum exert forces on other objects with which they come in contact (the second law of motion), and in turn the objects they collide with will exert a reaction force (third law) back on them due to inertia (first law).

Units of momentum—SI and English systems

The units of momentum are those of mass times velocity—kg · m/s in the SI system, and sl · ft/s in the English-engineering system.

Common experience tells us that if a force is applied to an object for a short time, a small change in its motion will occur; and if the same force is applied for a longer time, a greater change in motion will take place. The all-important principle of follow-through, stressed by coaches of golf, tennis, and baseball, is recognition of this fact. The longer the time the club, racquet, or bat can be kept in contact with, that is, applying force to, the ball, the greater will be the change in momentum and therefore the speed of the ball (Fig. 6.13).

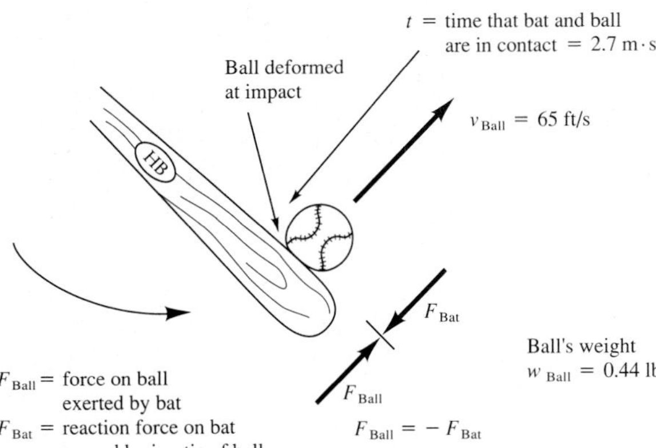

Fig. 6.13 Bat and baseball during the brief time of contact. The impulse imparted by the bat is $F_{bat}t$, where t is the time of contact. This impulse imparts a momentum to the ball, $\mathbf{p} = w_{ball} \times v_{ball}/g$. The ball is temporarily deformed at impact, but its elasticity restores its shape within a few milliseconds after contact ends. The momentum of the bat-ball system is conserved at impact.

6.9 ■ Impulse

Impulse—the effect of force acting for a finite time. Impulse produces a change in momentum.

The product of *force acting and the time which it acts* is called *impulse*. Comparing work with impulse, we have

$$\text{Impulse} = Ft \tag{6.9}$$

and

$$\text{Work} = Fs \tag{6.1}$$

To show that the momentum of a moving body is a measure of the impulse it surrenders when it hits another object, write Newton's second law,

$$F = ma \tag{5.1}$$

Multiplying both sides by t gives

$$Ft = mat$$

But *change in velocity*

$$\Delta v = v_2 - v_1 = at \tag{4.6}$$

and therefore

$$Ft = mv_2 - mv_1 \tag{6.10}$$

The impulse-momentum equation. Units of momentum and impulse.

or

$$Ft = m(v_2 - v_1) = m\,\Delta v = \Delta mv$$
$$\textit{Impulse} = \textit{change in momentum}$$

Equation (6.10) is called the *momentum equation*. You must remember that Newton's second law, $F = ma$, calls for absolute units of force, and since Eq. (6.10) is derived from it, the *momentum equation also calls for force units:* newtons or pounds. For example, factoring Eq. (6.10),

$$F \times t = m(v_2 - v_1)$$

we have

Newtons (force) × seconds = kilograms (mass) × meters per second

or,

Pounds (force) × seconds = slugs (mass) × feet per second

An alternative form of the momentum equation for use with *gravitational* (weight) units is

$$Ft = \frac{w(v_2 - v_1)}{g} \tag{6.11}$$

where F will be in pounds, w also in pounds, and g is again the acceleration of gravity, 32.2 ft/s^2. Dimensionally, Eq. (6.11) checks out (English system) as

$$F\ (\text{lb}) \times t\ (\text{s}) = \frac{w\ (\text{lb}) \times (v_2 - v_1)(\text{ft/s})}{g\ (\text{ft/s}^2)}$$

or

$$\text{lb} \cdot \text{s} = \text{lb} \cdot \text{s}$$

Equation (6.11) should also be used if w and F are both in newtons. In that case, dimensionally,

$$\text{N} \cdot \text{s} = \frac{\text{N(m/s)}}{\text{m/s}^2} \qquad \text{or} \qquad \text{N} \cdot \text{s} = \text{N} \cdot \text{s}$$

In the SI-metric system, recall that a newton is dimensionally equivalent to 1 kg · m/s^2. In metric units, Eq. (6.10) may be analyzed dimensionally as follows:

$$F\left(\text{kg}\,\frac{\text{m}}{\text{s}^2}\right) \times t\ (\text{s}) = m\ (\text{kg}) \times v\left(\frac{\text{m}}{\text{s}}\right)$$

or

$$\frac{\text{kg}\cdot\text{m}}{\text{s}} = \frac{\text{kg}\cdot\text{m}}{\text{s}}$$

Illustrative Problem 6.7 The standard golf ball weighs 1.62 oz. A professional golfer with a smooth "grooved" swing drives off the tee with a follow-through that allows clubhead contact with the ball for 0.45 ms. The impact force averages 1800 lb during that interval. A heavy-swinging amateur hits the ball with an impact force of 2000 lb, but his "hacker" swing allows clubhead-ball contact for only 0.32 ms. Compare the speeds of the two balls as they leave the tee.

Solution Remember to use the form of the momentum equation for *weight units,* that is, Eq. (6.11), $Ft = w\,v/g$. Solving for v, $v = F\,t\,g/w$.

For the pro golfer:

$$v = \frac{1800\ \text{lb} \times 0.45 \times 10^{-3}\ \text{s} \times 32.2\ \text{ft/s}^2}{1.62\ \text{oz}/16\ \text{oz/lb}}$$

$$= 258\ \text{ft/s or about } 176\ \text{mi/h} \qquad \textit{answer}$$

For the amateur:

$$v = \frac{2000\ \text{lb} \times 0.32 \times 10^{-3}\ \text{s} \times 32.2\ \text{ft/s}^2}{1.62\ \text{oz}/16\ \text{oz/lb}}$$

$$= 203\ \text{ft/s or about } 138\ \text{mi/h} \qquad \textit{answer}$$

Illustrative Problem 6.8 A 6-lb sledge hammer is moving at 40 ft/s as it strikes a railroad spike. It is brought to rest in 0.025 s. Find the average driving force on the spike.

Solution Use the momentum equation, alternate form, Eq. (6.11):

$$Ft = \frac{w(v_2 - v_1)}{g}$$

Divide both sides by t and solve for F:

$$F = \frac{w(v_2 - v_1)}{gt}$$

Substitute:

$$F = \frac{6\ \text{lb} \times (0\ \text{ft/s} - 40\ \text{ft/s})}{32.2\ \text{ft/s}^2 \times 0.025\ \text{s}} = -298\ \text{lb} \qquad \textit{answer}$$

The minus sign merely means that the force reacting on the sledge hammer is in a direction opposite to its motion.

Since *direction* is a factor in impulse considerations, impulse (like momentum) is a *vector quantity*.

Impulse is also a vector quantity

Illustrative Problem 6.9 A 15,000-lb rocket is to be fired vertically from rest. It is desired to give a velocity of 21,000 mi/h during the burning time of the first-stage rocket motor, which is 70 s. Calculate the average total thrust which the rocket motor must develop, in pounds. Assume that g is constant at 32.2 ft/s^2 and ignore the loss of mass due to fuel consumed.

Solution Using Eq. (6.11),

$$Ft = \frac{w(v_2 - v_1)}{g}$$

solve for F:

$$F = \frac{w(v_2 - v_1)}{gt}$$

Now,

$$21{,}000 \text{ mi/h} = 30{,}800 \text{ ft/s} \left(\text{using the } \frac{44 \text{ ft/s}}{30 \text{ mi/h}} \text{ ratio}\right)$$

Substitute:

$$F = \frac{15{,}000 \text{ lb} \times 30{,}800 \text{ ft/s}}{32.2 \text{ ft/s}^2 \times 70 \text{ s}}$$

$$= 205{,}000 \text{ lb } \textit{thrust to accelerate}$$

$$= 205{,}000 \text{ lb} + 15{,}000 \text{ lb}$$

$$\text{Total thrust} = 220{,}000 \text{ lb} \qquad \textit{answer}$$

Illustrative Problem 6.10 The coach of a softball team is hitting fly balls to the outfielders for practice. The ball weighs 0.44 lb and it leaves the bat with a velocity of 65 ft/s (see Fig. 6.13). If the bat and ball are in contact for 2.7 milliseconds (ms), calculate the average force (lb) exerted on the ball by the bat (and on the bat by the ball—remember Newton's third law) during the contact time interval.

Solution: Since data are in gravitational units, the use of Eq. (6.11) is necessary.

$$Ft = \frac{w(v_2 - v_1)}{g}$$

Substituting,

$$F = \frac{0.44 \text{ lb} \times (65 \text{ ft/s} - 0 \text{ ft/s})}{32.2 \text{ ft/s}^2 \times 2.7 \times 10^{-3} \text{s}}$$

$$= 330 \text{ lb} \qquad \textit{answer}$$

Momentum has a direct relationship to Newton's second law, which we earlier described as the *law of acceleration* (Sec. 5.6).

From

$$F = ma \tag{5.1}$$

we can write

$$F = m\,\frac{v_2 - v_1}{t}$$

since acceleration is change in velocity per unit time. By simply transposing, and letting $\Delta v (= v_2 - v_1)$ equal the change in velocity,

$$Ft = \Delta mv$$

$$\text{Impulse} = \text{change in momentum}$$

This is the impulse-momentum equation, Eq. (6.10).

An alternate form of the second law of motion can now be stated:

Newton's preferred expression for the second law of motion

Newton's second law (Momentum form):

When a body is acted upon by an unbalanced force for a time t, the momentum change produced is equal to the product of the force and the time that it acts.

As a matter of fact, Newton himself favored this interpretation of the second law. His own statement was: "Change of motion is proportional to the force acting and takes place in the line of the force." Newton's "change of motion" is the product of mass and velocity, which we call *momentum*. The *magnitude* of the change in momentum is, of course, dependent on the time that the force acts.

6.10 ■ The Impact of Moving Fluids

The force of impact of a moving stream of water, or of a jet of steam, is of especial importance in the field of hydraulic- and steam-turbine design. Consider the following problem in this connection.

Moving fluids impact against objects in their path with great force

Illustrative Problem 6.11 If 250 ft^3/s of water is flowing through the penstock of a turbine at a velocity of 25 ft/s, calculate the average force (pounds) against the blades of the turbine rotor. Assume that the initial velocity of the water in the penstock (25 ft/s) is reduced to zero by the turbine.

Solution Use Eq. (6.11):

$$Ft = \frac{w(v_2 - v_1)}{g}$$

and rewrite in the form

$$F = \frac{w(0 - v_1)}{tg}$$

You will note that w/t = weight of water per second striking the rotor blades. Now 1 ft^3 water weighs 62.4 lb. Substituting given values, we obtain

$$F = \frac{(250\ \text{ft}^3/\text{s} \times 62.4\ \text{lb/ft}^3) \times (-25\ \text{ft/s})}{1\ \text{s} \times 32.2\ \text{ft/s}^2}$$

$$= -12{,}100\ \text{lb}$$

the force acting *back* on the water column. The equal and opposite force on the rotor blades is +12,100 lb. *answer*

A similar analysis can be applied to the problem of the force of impact of high-velocity streams of water from fire hoses as they strike the walls of burning buildings. Anyone who has observed wave action at the beach during a storm is aware of the massive impact force of moving fluids.

6.11 ■ Momentum and Newton's Third Law—Conservation of Momentum

Momentum, impulse, and the third law of motion

The third law of motion, the law of action and reaction, is directly involved with impulse and momentum. When a rifle is fired, the force of the rapidly expanding gases from the burning gunpowder accelerates the projectile and gives it *momentum* (see Fig. 6.14). During the same time that the bullet is acquiring momentum forward, the force of the expanding gases is *reacting* against the gun backward, and the momentum acquired by the bullet forward equals that acquired by the gun backward, since *equal and opposite forces act on each for the same length of time*. As an equation,

$$\underset{\text{(bullet)}}{Ft} = \underset{\text{(rifle)}}{-Ft}$$

therefore

$$m_B v_B = -m_G v_G \qquad (6.12)$$

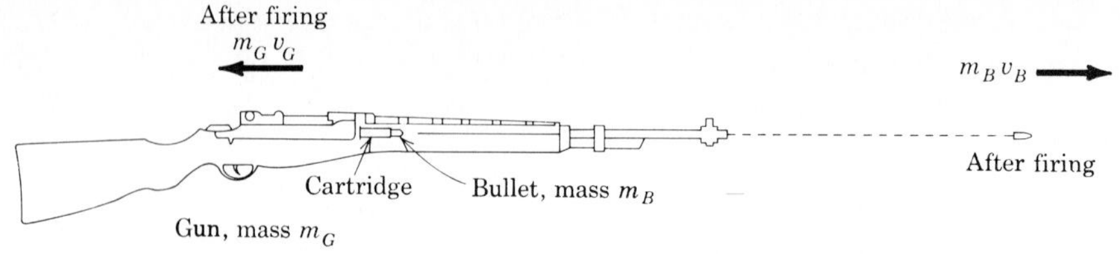

Fig. 6.14 Conservation of momentum. Consider the rifle and the bullet as a *system* whose total momentum with respect to the earth is zero before firing. Immediately after firing, the bullet has acquired momentum $m_B v_B$ to the right, and the gun has acquired momentum $m_G v_G$ to the left. Newton's laws require that $m_B v_B = -m_G v_G$ after firing, or $m_B v_B + m_G v_G = 0$. Both before and after firing, the momentum of the *gun-bullet system* is zero.

or

$$\text{Bullet momentum} = -\text{rifle momentum}$$

These momentums will be in opposite directions, of course, as the *minus signs* indicate. Consider now a problem related to gunnery.

Illustrative Problem 6.12 An antiaircraft gun fires a 54-lb projectile with a muzzle velocity of 2650 ft/s. The recoiling portion of the gun weighs 1450 lb. Find the theoretical velocity of recoil.

Solution *Velocity of recoil.* Making use of subscripts, let

$$\begin{aligned} w_p &= \text{weight of projectile} \\ v_p &= \text{velocity of projectile} \\ w_G &= \text{weight of recoiling portion of gun} \\ -v_G &= \text{velocity of recoil of gun} \end{aligned}$$

Using these subscripts and Eq. (6.12), and recalling that $m = w/g$

$$-\frac{w_G}{g} v_G = +\frac{w_p}{g} v_p$$

$$v_G = -\frac{w_p}{w_G} v_p \quad (g\text{'s cancel out})$$

Substituting,

$$v_G = -\frac{54 \text{ lb} \times 2650 \text{ ft/s}}{1450 \text{ lb}} = -98.7 \text{ ft/s} \qquad \textit{answer}$$

The negative sign indicates the direction of the recoil. Actually, the recoil would be controlled by strong springs or a hydraulic mechanism and its velocity would be much less than this theoretical value.

Internal components of a system can exchange momentums without any change in the total momentum of the system

The example above (without any recoil control features) is one case of the more general statement known as *the law of conservation of momentum*. Each body, that is, the projectile and the gun, undergoes exactly the same change in momentum, but in *opposite directions*. Consequently there is no net gain or loss of momentum involved in firing the gun-projectile system. Different internal components of the system exchange momentums, but the total momentum of the system is unchanged.

Momentum relations in the coupling of freight cars

In like manner, consider a moving freight car (Fig. 6.15) in a railway marshaling yard. Suppose it rolls slowly up to another car of *equal mass* which is at rest and collides with it. They lock together and both move forward, but with only half the speed of the original moving one. Since the same force magnitude acts on each car and they have the same mass, one will gain in velocity the amount which the other loses. Since their final common velocity is the same, it is therefore equal to one-half the velocity of the moving car before impact. Mathematically, we could state the momentum relations of the freight cars as follows, using symbols as indicated. Let

$$\begin{aligned} v_1 &= \text{velocity of first car before impact} \\ v_2 &= 0 = \text{velocity of second car before impact} \\ m &= \text{mass of each car} \\ v_1/2 &= \text{velocity of cars (locked together)} \\ &\quad\ \text{after impact} \end{aligned}$$

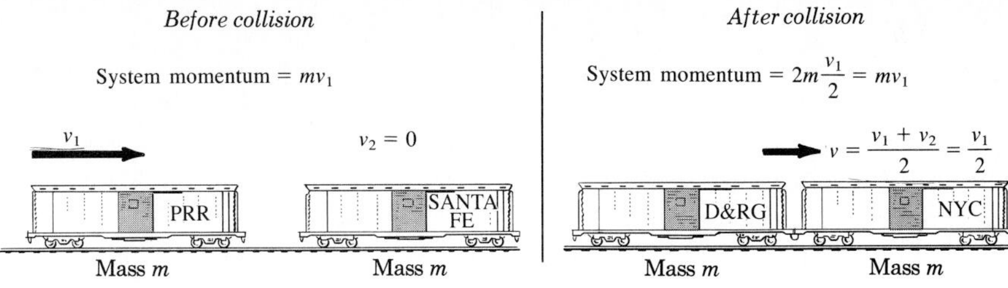

Fig. 6.15 Momentum is conserved at collision, whether the collision is elastic or inelastic. Railway marshaling yards freely make use of this principle as long trains are made up from individual cars. Kinetic energy is dissipated as one car collides with and "hooks on" to another, but momentum is conserved.

Then

$$\underset{\text{(first car)}}{mv_1} + \underset{\text{(second car)}}{mv_2} = \underset{\text{(total mass)}}{(m + m)} \times \underset{\left(\text{velocity after impact}\right)}{\frac{v_1}{2}}$$

(Before | After)

Simplifying yields

$$mv_1 + 0 = \frac{2mv_1}{2}$$

$$mv_1 = mv_1$$

or

Total momentum of the two-car system before impact
= total momentum of the system after impact

The foregoing examples illustrate the fact that when no *external forces* act on a system of particles or bodies, then no net impulse is imparted to the system, and no change in the momentum of the system will occur [Eq. (6.10)].

If the system consists of two or more particles or bodies, these bodies may exert forces on one another, but the force on either of two interacting bodies is equal in magnitude and opposite in direction to that on the other (Newton's third law), and the resulting *impulses* are also equal in magnitude and opposite in direction. Since it is impulse that causes change in momentum, the change in momentum of either interacting body is equal in magnitude and opposite in direction to that of the other body. For a system of bodies or particles,

$$\underset{\text{(system)}}{MV} = \underset{\text{(body 1)}}{m_1v_1} + \underset{\text{(body 2)}}{m_2v_2} + \underset{\text{(body 3)}}{m_3v_3} + \cdots \underset{\text{etc.}}{m_nv_n} \tag{6.13}$$

Total system linear momentum is the algebraic sum of the momentums of constituent bodies of the system

Equation (6.13) states that for a system of bodies, total system momentum is the vector sum of the momentums of the constituent bodies. If no external forces act on the system, its total momentum remains unchanged, regardless of exchanges of momentum among the constituent bodies of the system. A good example is that of gun recoil, as discussed in Illustrative Problem 6.12, and illustrated in Fig. 6.15.

The discussion of momentum thus far involves only those forces that cause *linear* motion, motion in a straight line in a specified direction. The momentum of rotating bodies (called *angular momentum*) will be dealt with in a later chapter.

The foregoing observations and examples lead to a statement of a general law of conservation of momentum.

The Law of Conservation of Momentum

The total linear momentum of a system is conserved (remains constant) if the vector sum of all external forces acting on the system is zero.

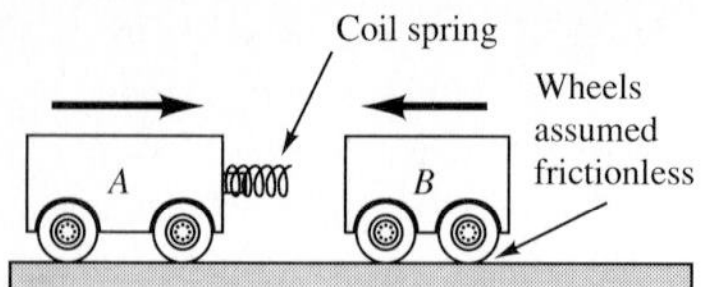

Fig. 6.16 Representation of a perfectly elastic collision. Bodies A and B experience a head-on collision, but a perfectly elastic (assumed) coil spring absorbs the kinetic energy of the rolling bodies and stores it momentarily as potential energy of compression. As the spring expands in recoil, the stored potential energy is reconverted to kinetic energy of the two-body system. There is no loss of kinetic energy in perfectly elastic collisions.

6.12 ■ Collisions

Elastic Collisions A perfectly elastic collision between two bodies is one in which there is no loss of kinetic energy of the system. Two billiard balls colliding approximate an elastic collision. During an elastic collision there will be energy transfer from one body to the other, but this energy is momentarily stored as potential energy of elasticity, and almost immediately given back as kinetic energy.

A simple example of a perfectly elastic collision is diagrammed in Fig. 6.16. A and B are two bodies moving toward a head-on collision. A coil spring on A absorbs the impulse of the collision and it is momentarily compressed, as some of the system's kinetic energy is converted to potential energy of elasticity in the spring. The coil spring immediately expands again and, as the bodies separate, the potential energy stored in the spring is reconverted to kinetic energy of the rebounding bodies.

In perfectly elastic collisions both momentum and kinetic energy are conserved. Both colliding bodies may be temporarily deformed, but their elasticity reconverts the deforming energy into kinetic energy as the bodies regain their original shape.

Inelastic Collisions A completely *inelastic collision* is one in which the colliding bodies move off as a single unit after the impact. There is no rebound—the bodies stick together. The freight-car problem is a good example. *Kinetic energy is not conserved in an inelastic collision*—it always decreases. Much of the kinetic energy in inelastic collisions is converted to energy forms such as heat, sound, light, internal molecular motion, wave energy, and in doing work that results in permanent deformation of the colliding bodies. Auto accidents are prime examples of inelastic collisions (see Fig. 6.17).

On the other hand, it must be remembered that *momentum is always conserved, in both elastic and inelastic collisions.*

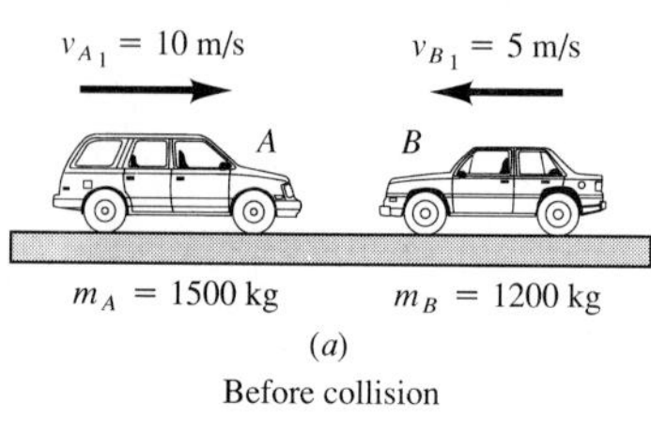

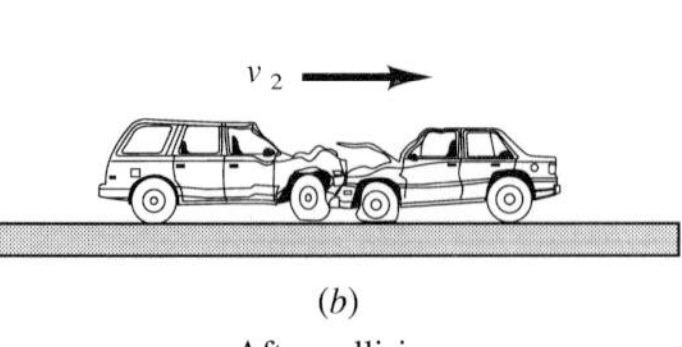

Fig. 6.17 Kinetic energy is dissipated in inelastic collisions. Autos A and B are shown in (*a*) just before a head-on collision. Each has its own momentum and its own kinetic energy. After collision (*b*), the two-car system moves to the right with momentum (instantaneous) undiminished. Kinetic energy of the system has markedly decreased however, being converted into work of deformation of the auto bodies, sound energy, and heat energy.

6.13 ■ Rockets and Jet Propulsion

The propulsion of jet aircraft, rockets, and missile-type weapons depends on the application of impulse and momentum principles and the law of action and reaction. True rockets and spacecraft carry both fuel and the oxidizing agent necessary to burn the fuel. They are therefore not limited in their operation to the earth's atmosphere. They can travel as far into space as their fuel and oxidizing agent supply permits. Jet aircraft, on the other hand, depend on the earth's atmosphere for their oxygen supply, either scooping in sufficient quantities due to their high speed or pumping it in with turbines driven by the jet engine (*turbojet* and *turbofan*). The forward thrust on the rocket or aircraft is obtained from the reaction force of the hot gases which are ejected backward at terrific speeds through the jet or *orifice* at the rear of the engine or rocket motor. A rocket or a space ship is accelerated *forward* by throwing some of its *mass* out *backward*. It is to be emphasized that the forward momentum gained by the aircraft or spacecraft can be acquired only at the expense of backward momentum gained (which is forward momentum *lost*) by gases escaping to the rear. A familiar example may serve to illustrate this theory of propulsion. If you stand in a small rowboat and heave an anchor astern, what happens? The boat moves forward, of course, and its *momentum* forward equals that of the anchor backward. If you attempt to jump from a small boat to a dock you may find that the boat's backward velocity (the v part of $-mv$) is such that you land in the water and not on the dock.

Rocket propulsion is based on momentum and impulse principles and on the third law of motion

Elementary Principles of Rocket Propulsion Rockets of the type currently in use as vehicles for space exploration are powered by chemical energy. The fuel is usually a liquid like kerosene or hydrogen, and liquid oxygen or other oxidizing agent is carried by the rocket to oxidize (burn) the fuel. In some rocket engines the fuel and oxidizer are combined in a solid form. A high-temperature, high-velocity jet of gases roars out backward from the rocket motor, (Figs. 6.18 and 6.19) and the momentum of these gases rearward exerts an impulse forward on the rocket. The force of this impulse is called *thrust*. These relationships can be analyzed as follows.

Fig. 6.18 Test firing of rocket-propelled missile from a launching pad at Cape Canaveral in Florida. The force which lifts and accelerates the missile, giving it upward momentum, is equal and opposite to the reaction force (thrust) of the hot gases thrown violently to the rear by the energy of combustion. The upward momentum of the missile is the result of the reaction force multiplied by the time that it acts ($mv = -Ft$). (Courtesy of NASA and U.S. Navy)

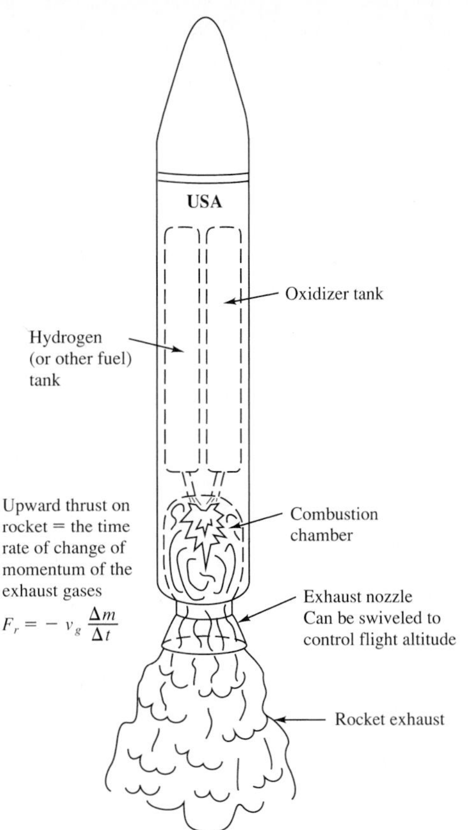

Fig. 6.19 Schematic representation of rocket propulsion. The sketch shows a liquid-fueled rocket engine. Solid propellants are also used to a great extent. The thrust on the rocket due to the rate of combustion gas ejection to the rear must be sufficient to lift the rocket and to accelerate it to a very high speed. (Details shown are schematic and do not depict any actual rocket.)

A rocket is a system in which momentum is conserved (air resistance assumed zero). The rocket is propelled forward by the forcible ejection of some of its mass (products of combustion) rearward.

The force propelling a rocket is called *thrust*

Let

v_g = rearward velocity of gases relative to rocket motor
Δm = very small increment of mass of escaping gas
Δt = very small increment of time required for mass Δm to escape

Then

$$\frac{\Delta m}{\Delta t} = \text{rate at which mass is discharged (assumed constant)}$$

and

$$v_g \frac{\Delta m}{\Delta t} = \text{time rate of change of the momentum of the escaping gases}$$

But (by Newton's second law) the time rate of change of momentum equals the force F given to the gases by the rocket motor, and (by Newton's third law) this force is equal and opposite to the force imparted to the rocket by the jet of escaping gas. Let F_R be the force (thrust) imparted to the rocket. Then

Calculating rocket thrust—a simplified analysis

$$F_R = -v_g \frac{\Delta m}{\Delta t} \tag{6.14}$$

The minus sign indicates that the velocity of the hot gas is opposite in direction to the force on the rocket body.

Other forces, such as the gravitational weight of the rocket ($w = mg$) and air resistance, will be acting on the rocket during launch. The vector sum of these would be opposite in direction to F_R, if the launch is vertically upward. Let F_G equal the aggregate of these forces. If M is the total mass of the rocket, Newton's second law gives

$$F_R - F_G = Ma \tag{6.15}$$

Calculating the vertical acceleration of a rocket

where a is the vertical acceleration of the rocket. M will vary during the flight as fuel is burned or as parts of the rocket are separated or jettisoned. If thrust remains constant, the rocket's acceleration will increase as mass decreases due to expended fuel and oxidizer.

Illustrative Problem 6.13 A Saturn rocket is being used to put a space lab in orbit. The total gross weight is 6,200,000 lb. Its five engines produce a total thrust of 7,500,000 lb. The velocity of the hot gases shooting out of the jets is 6800 ft/s. Find (*a*) the rate of ejection of hot gases in slugs per second, and (*b*) the acceleration of the rocket after 70 s of flight.

Solution

(***a***) From Eq. (6.14)

$$\frac{\Delta m}{\Delta t} = \frac{F_R}{-v_g} = \frac{7{,}500{,}000 \text{ lb}}{-(-6800 \text{ ft/s})}$$
$$= 1.1 \times 10^3 \text{ lb} \cdot \text{s/ft}$$
$$= 1.1 \times 10^3 \text{ sl/s} \qquad \textit{answer}$$

(***b***) The loss of mass after 70 s of firing is

$$70 \text{ s} \times 1.1 \times 10^3 \text{ sl/s} = 77{,}000 \text{ sl}$$

The remaining net mass of the vehicle at 70 s is

$$M_{70} = \frac{6.2 \times 10^6 \text{ lb}}{32.2 \text{ ft/s}^2} (\text{sl}) - 77{,}000 \text{ sl}$$
$$= 115{,}500 \text{ sl}$$

To calculate the acceleration 70 s after firing, neglect air resistance and assume the rocket is still close enough to the earth to equate $g = 32.2$ ft/s^2. The weight at 70 s is

$$w_{70} = mg = 115{,}500 \text{ sl} \times 32.2 \text{ ft/s}^2$$
$$= 3{,}720{,}000 \text{ lb}$$

The unbalanced force to cause acceleration is

$$F_{70} = 7{,}500{,}000 \text{ lb force} - 3{,}720{,}000 \text{ lb weight}$$
$$= 3{,}780{,}000 \text{ lb force}$$

From $F = ma$,

$$3{,}780{,}000 \text{ lb force} = 115{,}500 \text{ sl} \times a$$

from which

$$a = 32.7 \text{ ft/s}^2 \qquad \text{or a little more than 1 } g \qquad \textit{answer}$$

As a rocket rises far above the earth the value of g decreases, and in *outer space* it reduces to near zero. The rocket would then be weightless, but it must be remembered that its mass will remain unchanged, except for such losses as result from burning (ejected) fuel and jettisoning of rocket components.

Illustrative Problem 6.14 A submarine-launched missile has a thrust of 600,000 N under "full-burn" conditions, when combustion gases are being exhausted from its rocket engine at a velocity of 2200 m/s. What is the rate of ejection of fuel and oxidizer in kg/s?

Solution Newton's third law requires that the forward thrust on the missile be equal to the rearward force on the exhaust gases. Since m is the mass of exhaust gas *per second,* and v_1 is its initial velocity ($v_1 = 0$ in this case), we substitute in the impulse equation, $Ft = m\,\Delta v$, as follows:

$$600{,}000 \text{ N} \times 1 \text{ s} = \text{m } (2200 \text{ m/s} - 0 \text{ m/s})$$

Recalling that the units of N are kg · m/s^2, solve for mass/s, m/s.

$$m/s = \frac{600{,}000 \text{ kg} \cdot \text{m/s}^2}{2200 \text{ m/s}}$$

$$= 273 \text{ kg/s} \qquad \textit{answer}$$

POWER

Industry has to be concerned with work accomplished and energy expended. But fully as important are the *time* required to do a job and the *rate* at which energy is transformed. *Power* is the technical term which takes the time element into consideration.

Power is defined as the rate at which work is done or the rate at which energy is transformed or expended.

Power is the time rate of doing work

$$\text{Power} = \frac{\text{work}}{\text{time}} = \frac{\text{force} \times \text{distance}}{\text{time}}$$

$$P = \frac{W}{t} = \frac{Fs}{t} \tag{6.16}$$

If one machine does a certain amount of work in half the time required by a second machine, the first has twice the power of the second (see Fig. 6.20). Time is often all important in production, and consequently engineers and technicians are continually interested, not only in what a machine or device can do but also in the length of time it takes to do it.

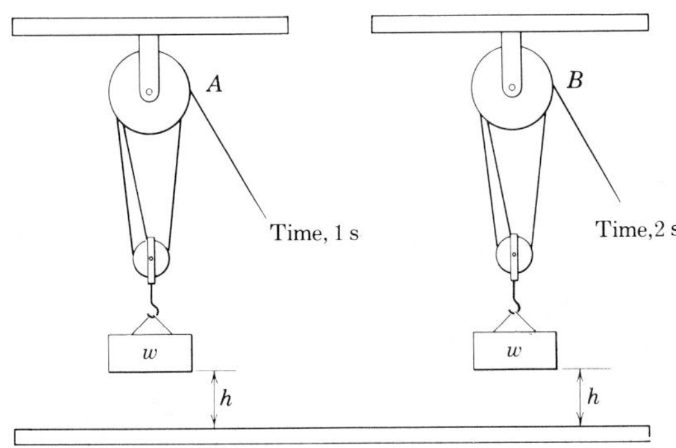

Fig. 6.20 Power equals work done per unit time. Hoist A lifts weight w a distance h in 1 s. It takes hoist B 2 s to accomplish the same lift. Hoist A has twice the power of hoist B.

6.14 ■ Units of Power

In the SI-metric system, forces are measured in *newtons,* distances in *meters,* and time in *seconds*. The basic power unit is then

$$P_{(\text{SI units})} = \frac{F\ (\text{N}) \times s\ (\text{m})}{t\ (\text{s})} = \frac{\text{newton} \cdot \text{meter}}{\text{second}} = \frac{\text{joule}}{\text{s}} \left(\frac{\text{J}}{\text{s}}\right)$$

Units of power in the SI and English (engineering) systems

The unit *joule per second* has been given the name *watt* (symbol W), in honor of James Watt (1763–1819), who built the first commercially successful steam engine. The watt is a rather small unit of power, and the *kilowatt* (kW, 1000 watts) is the practical power unit in the metric system. The *megawatt* (MW, 1 million watts) and the *gigawatt* (GW, 1 billion watts) are also in common use (see Table 6.2).

In the English (engineering) system, the basic power unit is defined as follows:

$$P_{(\text{engineering units})} = \frac{F\ (\text{lb}) \times s\ (\text{ft})}{t\ (\text{s})} = \frac{\text{ft} \cdot \text{lb}}{\text{s}}$$

The *kilowatt* and the *horsepower* defined

This unit is read as *foot-pounds per second*. It is too small for practical purposes, however, and the *horsepower* (hp) is the unit commonly used to express the power rating of most types of machinery in the United States. Although in its original determination the horsepower was intended to indicate the rate at which the average horse could work, it is today defined in more exact terms. One horsepower equals 550 ft · lb/s or 33,000 ft · lb/min by legal definition in the United States. In computing the horsepower ratings of machines either of these equivalents may be used.

$$1\ \text{hp} = 550\ \text{ft} \cdot \text{lb/s} = 33{,}000\ \text{ft} \cdot \text{lb/min}$$

Table 6.2 Units of Power

English System	*Metric System*
ft · lb/s	joule/s = 1 watt
Horsepower (hp) = 550 ft · lb/s	1 kilowatt (kW) = 1000 watts
= 33,000 ft · lb/min	1 megawatt (MW) = 1,000,000 watts
	1 gigawatt (GW) = 1 billion watts
Conversion	
1 hp = 746 W = 0.746 kW 1 kW = 1.34 hp	(memorize these)

Power outputs of typical engines and motors

Modern automobile and truck engines are in the 25- to 500-hp range. Large reciprocating-type aircraft engines develop as much as 3500 hp. Fan-jet engines (see Fig. 17.1) for modern aircraft develop up to 20,000 hp, each horsepower delivering about 1 lb of thrust force. Huge rocket engines like the Saturn can develop several million horsepower at peak thrust. Steam turbines for marine and stationary use develop 100,000 hp or more. Electric motors can be made to develop 20,000 hp and more, but the trend in industrial plants is to use unit installations of small electric motors rather than a few large, powerful ones. Fractional-horsepower motors and those up to 20 hp far outnumber the larger installations in modern industrial plants. Farm tractors range from 15 to 150 hp, rated at 2000 revolutions per minute (abbreviated rev/min or rpm), depending on function and auxiliary equipment to be powered by the tractor engine.

Illustrative Problem 6.15 A 50-hp motor is installed to operate a water pump which lifts water from a depth of 85 ft (see Fig. 6.21). Assuming no power losses, how many gallons per minute (gal/min) will be pumped at full power?

Fig. 6.21 Water pump driven by an electric motor for an orchard irrigation system. The pump itself is a turbine type (see Chap. 12) of several stages, with the first stage set at the level of the underground water (aquifer). The motor shown is 50 hp, but 800- to 1500-hp motors are often used when large flows from deep wells are required. (Pacific Gas and Electric Co.)

Solution First find how many pounds per minute will be pumped, and then obtain gallons per minute from the relation, 1 gal water weighs 8.34 lb. Assuming no losses, write the equation which sets

$$\text{Input power} = \text{output power}$$
$$\text{ft} \cdot \text{lb/min input} = \text{ft} \cdot \text{lb/min output}$$

Let Q = pounds of water per minute that must be lifted 85 ft:

$$50 \text{ hp} \times \frac{33{,}000 \text{ ft} \cdot \text{lb/min}}{1 \text{ hp}} = \frac{Q \text{ lb}}{\text{min}} \times 85 \text{ ft}$$

(Power input = power output)

Solve for Q:

$$Q = \frac{50 \text{ hp} \times \dfrac{33{,}000 \text{ ft} \cdot \text{lb/min}}{1 \text{ hp}}}{85 \text{ ft}} = 19{,}400 \text{ lb/min}$$

Therefore the flow (gal/min) is

$$\frac{19{,}400 \text{ lb/min}}{8.34 \text{ lb/gal}} = 2325 \text{ gal/min} \qquad \textit{answer}$$

if 100 percent efficiency is assumed.

Illustrative Problem 6.16 An electric motor creates a tension of 3000 N in a hoisting cable and reels it in at the rate of 5 m/s. What power is the motor supplying?

Solution From Eq. (6.16)

$$\text{P} = \frac{\text{Work}}{t} = \frac{Fs}{t}$$

$$= 3000 \text{ N} \times \frac{5 \text{ m}}{1 \text{ s}}$$

$$= 15{,}000 \text{ J/s} = 15{,}000 \text{ W} = 15 \text{ kW} \qquad \textit{answer}$$

6.15 ■ Power and Velocity

Force and velocity (speed) related to power

Power has been defined as the rate at which work is done, or the rate at which energy is expended. From Eq. (6.16),

$$P = \frac{W}{t} = \frac{Fs}{t}$$

Since from Eq.(4.1), $\dfrac{s}{t} = \bar{v}$, we can write

$$P = F\bar{v} \qquad (6.17)$$

Power = Force × velocity (speed)

Equation (6.17) is a convenient formula for the solution of many power problems. The term $\bar{v}$ is the average velocity (speed) of the moving force, F. Examples follow.

Illustrative Problem 6.17 A sled dog exerts a continuous force of 7 lb and pulls a sled across the snow at a speed of 6 ft/s. What horsepower is the dog generating?

Solution From Eq. (6.17),

$$P = F\overline{v} = 7 \text{ lb} \times 6 \text{ ft/s} = 42 \text{ ft} \cdot \text{lb/s}$$

$$P_{\text{hp}} = \frac{42 \text{ ft} \cdot \text{lb/s}}{\dfrac{550 \text{ ft} \cdot \text{lb/s}}{1 \text{ hp}}} = 0.076 \text{ hp} \qquad \textit{answer}$$

Illustrative Problem 6.18 The two jet engines of an aircraft produce a total thrust of 9000 lb, resulting in an air speed of 550 mi/h. What is the combined horsepower output of the two engines? The kilowatt output?

Solution From Eq. (6.17),

$$P = F\overline{v} = 9000 \text{ lb} \times 550 \text{ mi/h} \times \frac{44 \text{ ft/s}}{30 \text{ mi/h}}$$

$$= 7{,}260{,}000 \text{ ft} \cdot \text{lb/s}$$

$$P_{\text{hp}} = \frac{7{,}260{,}000 \text{ ft} \cdot \text{lb/s}}{550 \text{ ft} \cdot \text{lb/s} \cdot \text{hp}} = 13{,}200 \text{ hp} \qquad \textit{answer}$$

$$P_{\text{kW}} = \frac{13{,}200 \text{ hp}}{1.34 \text{ hp/kW}} = 9850 \text{ kW} \qquad \textit{answer}$$

6.16 ■ Efficiency

Machine output cannot be greater than the total energy input

It must not be assumed that the entire power output of an engine or motor is going into useful work accomplished. As a matter of fact, this is never the case, since in every actual machine, friction (see Chap. 7) causes some energy to be wasted in the form of heat. Any machine has certain bearings, gears, pulleys, and levers, whose frictional resistance causes a waste of energy. Consequently, *the useful work output of a machine is always less than the energy input to the machine*. Stated as an equation, the ***principle of work*** for actual machines is

Energy input − losses (due to friction and other causes) = work output

Efficiency of a machine, defined

> ***The ratio of the work output of a machine to the energy input is called its efficiency.***

Since no real machine has a work output as great as the energy input to it, efficiencies are always less than 1 and are commonly expressed as percentages.

$$\%\text{Eff} = \frac{\text{work output of machine}}{\text{energy input to machine}} \times 100$$

$$\%\text{Eff} = \frac{\text{power output}}{\text{power input}} \times 100 \tag{6.18}$$

or, stated another way,

$$\text{Output} = \frac{\%\text{eff} \times \text{input}}{100} \tag{6.19}$$

Illustrative Problem 6.19 A farm pump motor (see Fig. 6.21) uses electrical energy at the rate of 50 kW. The water level in the well is 200 ft below ground level. The actual water flow produced by the pump at ground level is 1000 gal/min. What is the efficiency of the motor-and-pump system?

Solution First, calculate the power output.

$$Q = \text{lb water/min} = 1000 \text{ gal/min} \times 8.34 \text{ lb/gal}$$
$$= 8340 \text{ lb/min}$$

$$\text{Power output} = \frac{8340 \text{ lb/min} \times 200 \text{ ft (lift)}}{\dfrac{33{,}000 \text{ ft} \cdot \text{lb/min}}{1 \text{ hp}}}$$

$$= 50.5 \text{ hp}$$

$$\text{Power input} = 50 \text{ kW} \times 1.34 \text{ hp/kW}$$
$$= 67 \text{ hp}$$

From Eq. (6.18),

$$\text{Eff} = \frac{50.5 \text{ hp}}{67 \text{ hp}} \times 100 = 75 \text{ percent} \qquad \textit{answer}$$

6.17 ■ Measuring Power

Manufacturers of all types of motors, engines, and turbines must subject their products to frequent tests to be certain that they will actually perform up to the rated power on the nameplate. Power actually delivered to a drive shaft, flywheel, gearhead, pulley, or propeller is called *brake power* (bp) because it can be measured by a braking device such as that suggested by Fig. 6.22. Such a test device is called a *Prony brake*. The engine or motor under test is coupled directly to the revolving drum (diameter d) of the Prony brake. The belt B is held on the drum D under tension by adjusting the spring balances at A and C. As the test engine turns at an observed speed in revolutions per second (rev/s) the frictional force of the belt on the drum is measured by the difference between the readings of the balances. This difference between the readings is the frictional force being applied by the test engine, and this force acts at the circumference of the drum during each revolution. The work done per second (power) is therefore

$$\text{Power} = \text{Force} \times \text{distance per second}$$
$$P = (F_1 - F_2) \times \pi d \times \text{rev/s} \tag{6.20}$$

If F_1 and F_2 are in newtons and d is in meters, power will be in joules/s = watts.

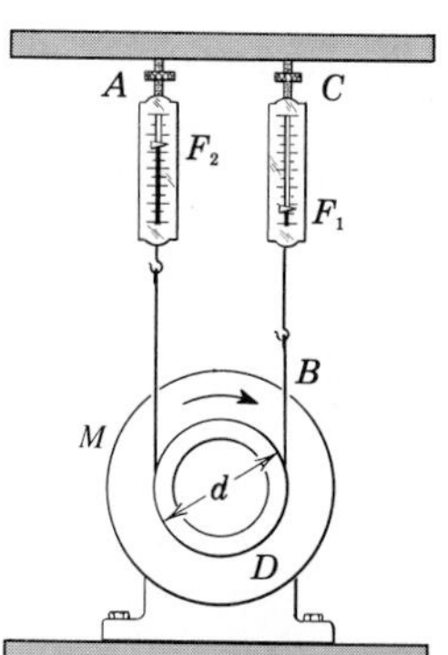

Fig. 6.22 Diagram of a simple Prony brake. An electric motor M is being tested for its horsepower rating. (For illustration of theory only. Modern industry makes use of far more sophisticated methods.)

Illustrative Problem 6.20 An electric motor is under test on a Prony brake. The spring balances read 840 N and 150 N, at 30 rev/s. The drum diameter is 20 cm. Find the brake power (bp) of the motor, in kW.

Solution From Eq. (6.20),

$$P = (840 \text{ N} - 150 \text{ N}) \times \pi$$
$$\times 0.2 \text{ m} \times 30 \text{ rev/s}$$
$$= 13{,}000 \text{ J/s (watts)}$$
$$= 13 \text{ kW} \qquad \textit{answer}$$

When F_1 and F_2 are measured in pounds and d is in feet, P will be in foot-pounds per second. Therefore brake horsepower (bhp) will be given by the formula

$$\text{bhp} = \frac{(F_1 - F_2) \times \pi d \times \text{rev/s}}{550} \tag{6.21}$$

Illustrative Problem 6.21 A newly developed high-compression truck engine is on the test block. At 3500 rev/min by the tachometer, the spring balances of a Prony brake read 525 lb and 75 lb. The diameter of the drum is 2.0 ft. Find the brake horsepower of the engine.

Solution Using Eq. (6.21)

$$\text{bhp} = \frac{(F_1 - F_2) \times \pi d \times \text{rev/s}}{550}$$

and substituting, changing rev/min to rev/s,

$$\text{bhp} = \frac{(525 \text{ lb} - 75 \text{ lb}) \times \pi \times 2 \text{ ft/rev} \times \dfrac{3500 \text{ rev/min}}{60 \text{ s/min}}}{\dfrac{550 \text{ ft} \cdot \text{lb/s}}{1 \text{ hp}}}$$

$$= 300 \text{ hp}$$ *answer*

Power output measurement with a dynamometer

The Dynamometer A much more common test device for measuring power output is the *dynamometer*. It is actually a heavy-duty electric generator whose magnetic field can be varied at the will of the operator, so that it can absorb all the power output of a test engine. The engine or turbine being tested is connected directly to the generator. The generator transforms mechanical energy into electric energy at a rate which can be measured by a suitable electric meter, in units of electric power (*watts*). Since 1 hp = 746 watts, and 1000 watts = 1 kW, it is easy to convert the measured output in kilowatts into horsepower. In actual practice, dynamometer dials are often calibrated to read horsepower directly. Figure 6.23 shows a dynamometer in use.

Fig. 6.23 An engine dynamometer in use in a test laboratory. The technician is running the engine at certain stipulated speeds (rpm) and determining its horsepower by reading the electric power output of the large generator (center of picture) that the engine is driving. The test equipment, including the computer and tape machine at left, can be programmed to simulate a wide variety of operating conditions. (Oldsmobile Division, General Motors Corp.)

QUESTIONS AND EXERCISES

1. Define the following in technical terms, and after each definition give two examples chosen from actual practice in industrial, engineering, agricultural or medical fields: (*a*) work, (*b*) kinetic energy, (*c*) potential energy other than gravitational, (*d*) gravitational potential energy, (*e*) momentum, (*f*) impulse, (*g*) power, (*h*) efficiency.

2. Based on some library reading, write a short paper on ''perpetual motion'' machines. Describe and sketch one model of such a machine different from the one shown in Fig. 6.12. What happens when such a machine is applied to a load to do ''useful'' work? Could such a machine work on Jupiter where the force of gravity is many times greater than it is on earth?

3. Suppose your home is heated by electric radiant panels. Trace the energy transformations involved if the electricity is generated (*a*) at a steam plant with coal-fired boilers; (*b*) at a hydroelectric plant in the mountains. Start with the sun in each case.

4. Using the formula for kinetic energy, show that the SI-metric unit of energy is the newton-meter, or joule.

5. A right fielder makes a long, arching throw to home plate. The ball arrives a bit high, at the same level above the ground as his throwing hand. From energy principles, prove that its speed at home plate is the same as when it left his hand. (Neglect air friction.)

6. Team up with a friend and, with a good stop watch, measure each other's horsepower output, by running up a flight of stairs whose vertical rise can be measured. Weigh yourselves just before the trials.

7. What are some other forms of potential energy besides GPE? How can the amount of PE be determined? Describe ways in which machines can use these forms of PE.

8. In Fig. 6.15 the colliding railway cars absorbed the impact of collision without appreciable deformation. Momentum of the system was conserved. What about the kinetic energy of the two-car system? Was KE conserved at collision? If the cars were damaged at collision, would momentum be conserved? Explain your answers fully.

9. Explain in detail, with sketches, why the concept of ''follow-through'' is important in sports like golf, baseball, and tennis. What specific factor in the impulse equation is being maximized by a full follow-through?

10. A tremendous force (thrust) is required to lift off and accelerate a space vehicle as it leaves the earth. When it is far out in space, with the earth's gravitational force at or near zero, is force (thrust) needed to accelerate it? Explain.

11. Billiard balls are nearly ''perfectly'' elastic. As two balls collide on the table what is the only evidence available to the human senses that there is indeed some small loss of kinetic energy in the two-ball system?

12. Explain why kinetic energy is not lost when work is done on an elastic body in a collision, but is lost if the body is inelastic.

PROBLEMS

NOTE: *Assume frictional, air resistance, and other losses are zero unless stipulated otherwise in the problem.*

Group One

1. A warehouse worker pushes a loaded dolly across the floor to a truck at the loading dock. The distance is 80 ft and the horizontal force exerted on the dolly is constant at 30 lb. How much work is done for each trip to the truck?

2. A switch engine in a railroad yard exerts a force of 4 tons (drawbar pull) to overcome the rolling friction of a string of freight cars and move them across the yard. How much work is done in moving them a distance of 400 yd along a level track?

3. A 60-lb casting is lifted to the bed of a milling machine through a vertical distance of 4 ft. Express the work done in foot-pounds.

4. A hoist slowly lifts an engine whose mass is 250 kg to a mezzanine level 7 m from the shop floor. Calculate the work done in joules.

5. A 1000-kg elevator rises from the first floor to the thirtieth floor of a hotel, a vertical distance of 130 m. How much has its GPE increased? Answer in joules.

6. Using the work-energy theorem, calculate the velocity of a wrecking ball that has fallen from rest through a vertical distance of 12 m.

7. Find the momentum of a 2000-kg auto when it is moving along at 80 km/h.

8. A kilowatt-hour (kW · h) is a unit of energy. Express this quantity of energy in joules, in foot-pounds.

9. If a horse actually worked at the rate of 1 hp, how many pounds of coal could be lifted out of a mine shaft 300 ft deep in 1 min with suitable hoisting equipment?

10. A 150-lb student runs up several flights of stairs through a vertical distance of 120 ft in 40 s. What average horsepower is being developed?

Group Two

11. Calculate the potential energy per cubic meter of water just before it goes over the edge of a fall whose vertical drop is 200 m. [One liter (1000 cm^3) of water has a mass of 1 kg.]

12. An industrial plant has a shop whose average power requirements are 500 kW. Assuming that it would be possible for human muscular energy to furnish this power, how many workers would have to be employed if each could work at a steady rate of 0.2 hp? (Contrast your answer with the 10 to 15 persons which such a shop would probably employ to operate its modern machines.)

13. A 2-kg projectile is fired from an antiaircraft gun at a muzzle velocity of 500 m/s. If the recoiling portion of the gun mechanism has a mass of 210 kg, what would be the velocity of uncontrolled recoil in meters per second?

14. If 1000 m^3 of water is pumped from a reservoir into a tank 100 m higher than the reservoir in 2 min, find (*a*) the increase in potential energy and (*b*) the power delivered by the pump. One liter (1000 cm^3) of water has a mass of 1 kg.

15. A construction elevator hauls a hopper of ready-mix concrete, total mass 1800 kg, to the pouring site 150 m above the ground in 30 s. What is the power required?

16. An escalator in a department store is designed to carry 300 passengers per minute from the ground floor to the mezzanine, 10 m vertically higher. The design assumes an average mass per person of 70 kg. Allowing 30 percent loss for friction, calculate the power output which the driving motor must have. (Answer in kW.)

17. An electric motor on a Prony brake-horsepower test turned at 1750 rpm. Spring balances attached to the belt read 54 and 16 lb. The diameter of the drum was 6 in. What horsepower did the motor develop?

18. A 65-ton freight car rolls along at 6 mi/h and collides with a stationary car weighing 50 tons, and the two lock and move off together. What is the velocity of the two-car train after the impact, in miles per hour?

19. A pendulum is 1.00 m long. If the bob is pulled aside until the cord makes an angle of 35° with the vertical and then released, with what speed does the bob pass through the rest position?

20. Assuming frictional forces are zero, determine the horsepower required to accelerate a 2500-lb auto from rest to 60 mi/h in 10 s, on a level road.

21. A force (thrust) of 4000 N is required to overcome road friction and air resistance in propelling an automobile at 80 km/h. What power (kilowatts) must the engine develop?

22. An electric motor is found to have an output of 25 hp. An electric wattmeter is placed in the circuit, and reads 19.5 kW. What is the motor's efficiency?

23. A disabled truck is being towed at 25 mi/h. The tension (force) in the towing bar is known to be 350 lb. What horsepower is being expended to tow the truck?

24. A rocket engine ejects hot gases at a speed of 2000 m/s. The rate of loss of mass, i.e., hot gases ejected, is 400 kg/s. Find the thrust exerted on the rocket, in newtons.

25. A ball, mass 150 g, moving at 30 m/s, is struck by the bat and reverses direction, leaving the bat at 50 m/s. What was the impulse applied to the ball?

26. The roller coaster of Fig. 6.24 is assumed to roll without friction. If it passes point T at 5 m/s, find its speed at point B and at point X.

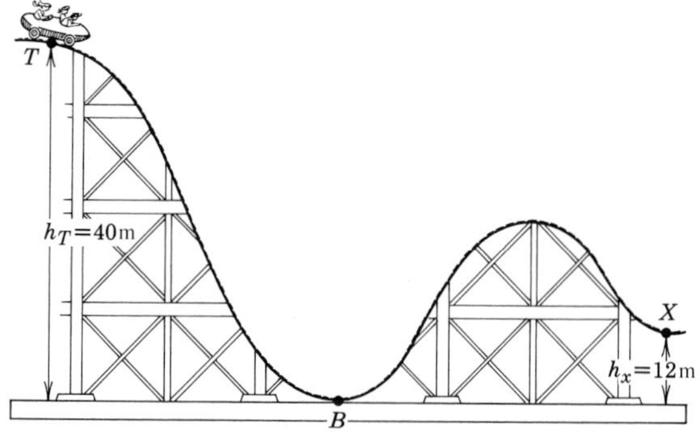

Fig. 6.24 Roller coaster sketch for Problem 26.

27. A farm tractor pulls a gang plow at a steady speed of 3 mi/h. If the engine is developing 80 hp, what is the "drawbar pull" on the plow, in pounds?

28. A space vehicle is on its way to Mars, at a point where it is still subject to some earth gravity; at a steady speed of 42,000 km/h. Instruments aboard indicate that the thrust being delivered by the rocket motor is 5000 N. What power is the rocket motor delivering, in megawatts?

29. A waterfall is 85 ft high, and 3000 ft^3/s flow over it driving a water wheel connected to an electric generator. If the overall efficiency is 22 percent, how many kilowatts does the generator develop? (Water weighs 62.4 lb/ft^3.)

30. An elevator is powered by a motor whose power input is 25 kW. The mass of the elevator and its load is 1000 kg. If the overall efficiency of the entire hoisting system is 75 percent, how long does it take the elevator to rise to the fourteenth floor, a vertical distance of 90 m?

Group Three

31. A major league pitcher can throw a 0.145-kg baseball with a speed of 100 mi/h (161 km/h). How fast would a 2.5 g table tennis ball have to move to have the same kinetic energy as the baseball?

32. By what factor must the speed of a moving object be increased to cause its kinetic energy to increase by (*a*) a factor of 2, (*b*) by a factor of 3?

33. A pocket computer uses energy at a rate of 0.03 J/s. The battery for the computer is capable of providing power at this rate for 300 h of operation. (*a*) How much energy must be stored in the battery when it is new? (*b*) If all the energy in a new battery could be converted to kinetic energy of a 0.145-kg baseball, initially at rest, how fast would the baseball move?

34. The hammer of a pile driver weighs 750 lb and falls a distance of 20 ft before it hits the top of the pile. The blow drives the pile 8 in. deeper into the ground. What was the average force driving the pile, in pounds? (Solve from energy-work considerations.)

35. A ballistic missile has a total weight of 65,000 lb. Its solid-propellant engine gives a full-power thrust of 165,000 lb. The velocity of hot gases ejecting from the rear of the rocket motor is 6700 ft/s. Find (*a*) the rate of gas ejection (in slugs per second) and (*b*) the acceleration 20 s after firing. Assume air resistance negligible and g still = 32.2 ft/s^2 at the 20-s position. Rocket fired vertically.

36. A stream of water 2 in. in diameter, squirting from a fire hose, hits the side of a building nearby at a velocity of 25 ft/s. What steady force does it apply to the wall?

37. A 15-g bullet fired from a rifle at 750 m/s embeds itself in a wooden block that is free to slide on the floor. The block's mass is 1 kg and the constant frictional force between block and floor is 5 N. How far will the block slide?

38. A rocket carrying a communications satellite is launched vertically from the earth. The lift-off weight, including fuel, is 60,000 lb. Just after lift-off the acceleration is measured as 0.25 g. If the speed of the hot exhaust gases is 8000 ft/s: (*a*) determine the rate of fuel usage in pounds per second; (*b*) at 3 min into the flight with flight path still vertical, what is the acceleration? (Neglect the slight decrease in earth's gravity at the 3-min altitude.)

39. The diagram of Fig. 6.25 represents an apparatus known as a ballistic pendulum. Prior to the advent of modern electronic and high-speed flash photographic methods, such an arrangement was often used to determine the muzzle velocity of bullets. The wooden block B acts as a pendulum bob. The projectile P strikes, penetrates, and lodges in the wood, causing B to swing to position B' with a net increase in GPE represented by the height h. From the data given on the diagram, calculate the speed v_b of the bullet just before impact in meters per second.

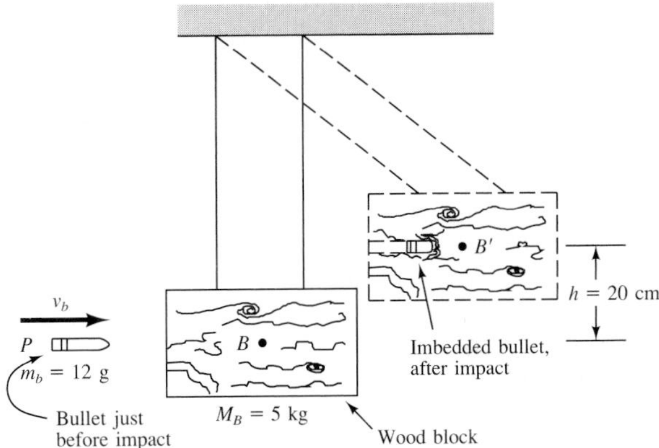

Fig. 6.25 Diagram of a ballistic pendulum for Problem 39. This is an inelastic collision. Kinetic energy is not conserved but momentum is conserved.

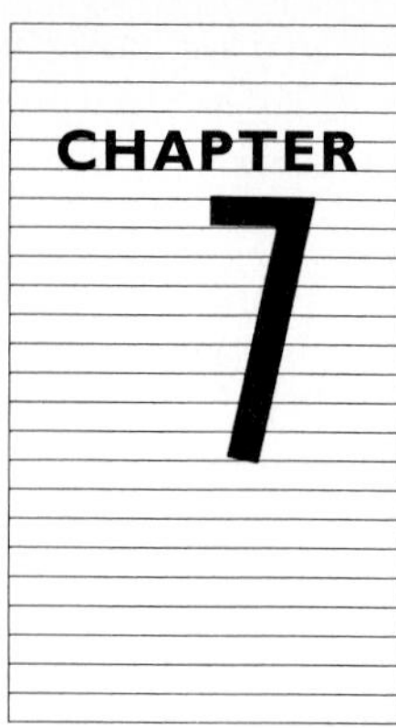

CHAPTER 7 BASIC MACHINES—FRICTION

Television, magazines, and newspapers continually portray the present as an age of microchips, computers, lasers, satellites, and industrial robots. To be sure, large segments of industry and commerce are involved with or dependent upon these somewhat glamorous developments. But just as surely, industrial production and the everyday lives of all of us are still heavily dependent on ordinary machines as well—machines like levers, gears, pulleys, pumps, cams, hoists, and screws.

Definition of a machine—a means to apply force to accomplish useful work.

The word *machine* means different things to different people. In physics we have a very specific definition:

A machine is a device for the advantageous application of a force.

A machine is a device for accomplishing useful work as an *output*, when there is an energy *input* to the machine. In doing useful work all machines also do some *useless* or *wasted* work. *Friction* is often the cause of this wasted work.

BASIC MACHINES

All complex machines are ingenious combinations of three *basic machines.*

Most business, industrial, and agricultural machines are merely combinations of what we call *basic machines*. In all the thousands of different kinds of complex machines in daily use, there are actually only three basic machines, the *lever*, the *inclined plane*, and the *hydraulic press*.

All the basic and simple machines have certain common properties. It takes a force input to operate them, and they yield an output force. There is an *energy input* to them, and they yield an *energy output*. Also, there are *energy losses* within the machine due to friction or other causes. In analyzing basic machines we will sometimes *assume* that these losses are zero, but actually they never are. In some machines they are so small, however, that for practical purposes they can be neglected without introducing serious error into the machine design.

All machines, including basic machines, obey the law of conservation of energy (see page 124). A restatement of the *law of conservation of energy for machines* is:

Energy input to a machine = energy output from the machine + energy losses in the machine (including "useless work")

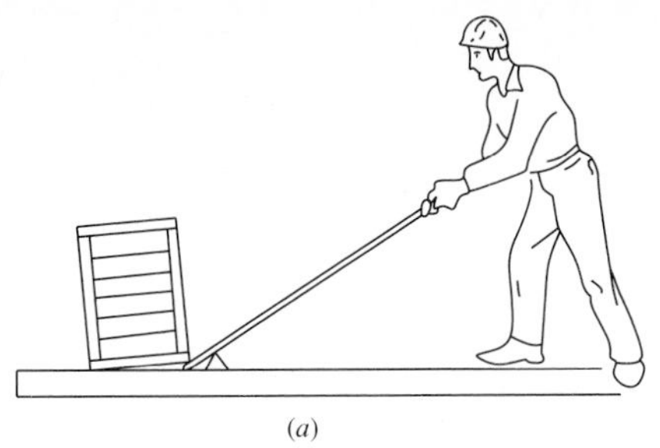

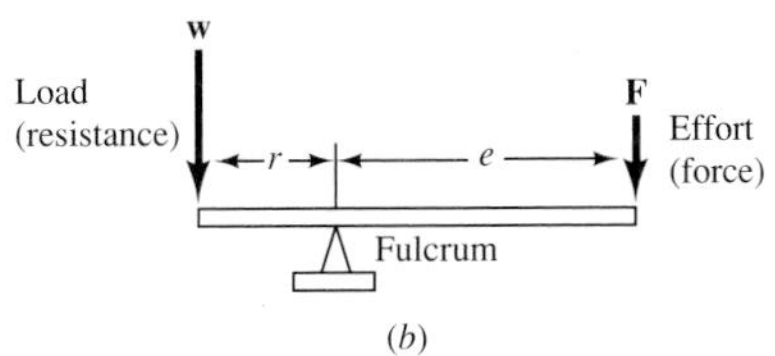

(a) (b)

Fig. 7.1 The principle of the lever. (*a*) Worker uses a lever and fulcrum to move a heavy crate. (*b*) The law of the lever is: $Fe = wr$.

7.1 ■ The Lever

A lever consists of any rigid bar or rod (it may be straight or bent) arranged in such a way that it can pivot about some definite point, so that a resistance w at one point along the bar can be overcome by a force F applied at some other point on the bar. Figure 7.1*a* illustrates one such arrangement in which a lever in the form of a *crowbar* is being used to tip a heavy packing crate. The diagram of (*b*) shows the relationships involved, along with the pivot point, which is called the *fulcrum*. If we designate the distance from the fulcrum to the resistance (load) as the *resistance arm* r and the distance from the fulcrum to the effort as the *effort arm* e, then the *law of the lever* can be stated thus:

The law of the lever.

Effort × effort arm = resistance × resistance arm

Assuming no internal losses, the law of the lever, stated as an equation, is

$$Fe = wr \tag{7.1}$$

Another way of stating the law of the lever is evident from analyzing Eq. (7.1). *When a lever is in balance, the forces applied are inversely proportional to their distances from the fulcrum,* or mathematically

$$\frac{F}{w} = \frac{r}{e} \tag{7.1$'$}$$

7.2 ■ Mechanical Advantage

Actual Mechanical Advantage (AMA) One of the important reasons for using a lever is to overcome a large resistance (load) by exerting a relatively small effort (force). When used for this purpose, a lever *multiplies force,* and the ratio

$$\frac{\text{force overcome (resistance)}}{\text{force applied (effort)}}$$

is defined as the *actual mechanical advantage* of the lever. In fact, for *any machine* the *actual mechanical advantage* is defined as *the number of times which the machine multiplies the force applied;* or, in other words,

Actual mechanical advantage (AMA) and *theoretical* mechanical advantage (TMA).

$$\text{AMA} = \frac{\text{force output of the machine}}{\text{force input to the machine}} \tag{7.2}$$

Theoretical Mechanical Advantage (Any machine). For the lever, Eq. (7.1) can be rewritten in the form

$$\frac{w}{F} = \frac{e}{r} \tag{7.1$''$}$$

The left side of this equation is seen to be the ratio of (force output)/(force input) and is therefore (from Eq. 7.2) the actual mechanical advantage. The right side is the ratio of the *effort arm* of the lever to the *resistance arm*.

$$\text{TMA (Lever)} = \frac{e}{r} = \frac{\text{effort}}{\text{resistance}} \tag{7.3}$$

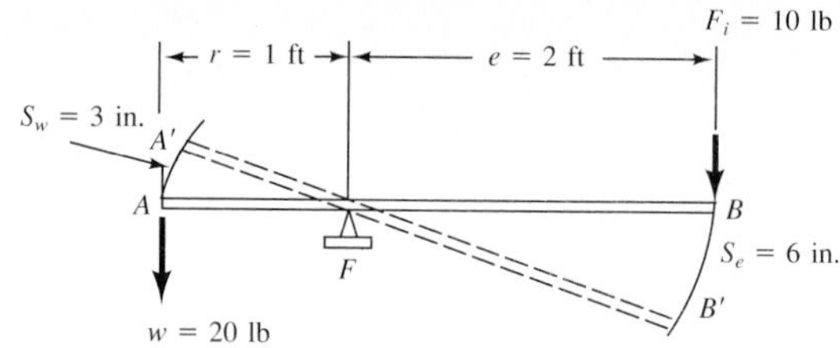

Fig. 7.2 The lever and the principle of work.

The ratio (e/r) is called the *theoretical mechanical advantage* (TMA). Note that Eq. (7.1″) states, in effect, that for levers, theoretical mechanical advantage equals actual mechanical advantage. This is true only in situations where frictional and other losses are so small that they can be considered negligible. If there is friction, the effort force F will have to be enough larger than its theoretical value to overcome the frictional resistance.

The idea of mechanical advantage has been introduced here in connection with the lever, but it is applicable to all simple machines. Remember that mechanical advantage is a *ratio of forces,* not a ratio of *energy output* to *energy input*. The latter ratio measures *efficiency,* and we will deal with that later in this chapter.

7.3 ■ The Lever and the Principle of Work

Although *force* can be multiplied by levers such as the crowbar, and *speed* or *distance* can be multiplied by such lever arrangements as valve rocker arms (Fig. 7.6), it is important to realize that in no case can *work* or *energy* be multiplied by a lever or any other known device! The principle of work was discussed in Chap. 6, and it holds true for levers and all other machines. Reference to Fig. 7.2 will show this to be so. The lever AB is 3 ft long, the effort arm e is 2 ft, and the resistance arm r is 1 ft. An input force F_i of 10 lb should therefore balance a resistance w of 20 lb since from Eq. (7.1)

$$10 \text{ lb} \times 2 \text{ ft} = 20 \text{ lb} \times 1 \text{ ft} = 20 \text{ lb} \cdot \text{ft}$$

The TMA of this lever (from Eq. 7.3) would be 2. Suppose we push down on the end B through a small arc ($S_e = 6$ in.) to the point B'. If the arc S_e is small compared to the length of the lever, BB' approximates a straight line. From simple geometry it is seen that the end A will have moved upward along its arc a distance $S_w = 3$ in. to the point A'. Since

$$\text{Work} = \text{force} \times \text{distance}$$

the work done by the input force F_i is

$$\text{Work input} = 10 \text{ lb} \times 6 \text{ in.} = 60 \text{ in.} \cdot \text{lb}$$

The work accomplished at the left (A) end is

$$\text{Work output} = 20 \text{ lb} \times 3 \text{ in.} = 60 \text{ in.} \cdot \text{lb}$$

The operation of a simple lever conforms to the *principle of work.*

Work output equals work input (assuming no losses), and we see that the principle of work (work-energy theorem) holds true for the lever. *Force* was multiplied, but *distance* was sacrificed, so there was no net gain or loss of energy.

In contrast to the lever of Fig. 7.2, some levers are designed to multiply distance and speed at the sacrifice of force.

7.4 ■ Examples and Types of Levers

The illustration of Fig. 7.1 shows the fulcrum located between the effort and the resistance. This arrangement is called a *Class I* lever. Other examples of Class I levers are oars, valve rocker arms, scissors, pliers, and "walking beams" of petroleum-pumping units (see Fig. 7.3). It should be evident that there are two other

Fig. 7.3 A Class I lever, the "walking beam" of a petroleum pumping unit. See also Fig. 7.6, valve rocker arm.

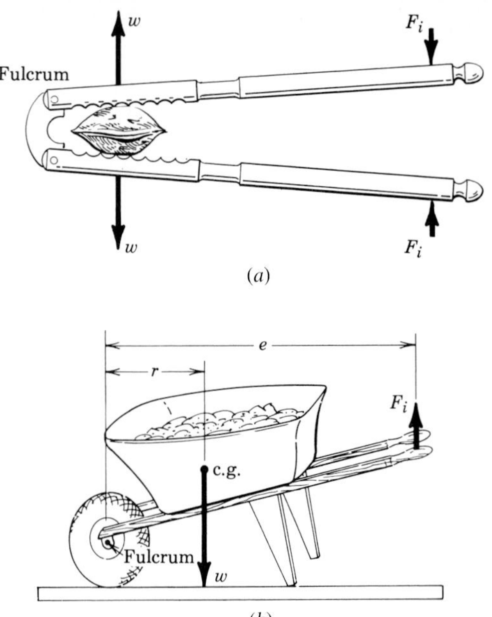

Fig. 7.4 Class II levers. (*a*) A common nut cracker. (*b*) a wheelbarrow.

possible combinations, i.e., resistance between effort and fulcrum (*Class II* levers), and effort between resistance and fulcrum (*Class III* levers). Figure 7.4 illustrates Class II levers, and Fig. 7.5 a Class III lever. See how many more of each type you can name, particularly those that are found in industrial machinery.

Illustrative Problem 7.1 The valve spring of a certain engine holds the valve closed with a force of 70 lb. The rocker-arm arrangement is shown in the diagram of Fig. 7.6. What input force must the push rod exert to crack open the valve?

Solution Using Eq. (7.1) and noting that it is to be solved for the effort F_i, we obtain

$$F_i = \frac{wr}{e}$$

Substituting,

$$F_i = \frac{70 \text{ lb} \times 4.5 \text{ in.}}{2.1 \text{ in.}}$$

$$= 150 \text{ lb} \qquad \textit{answer}$$

Note that this lever multiplies distance moved, and requires an input force *larger* than the output (resistance) force.

Other Levers The bars or rods of many levers are not perpendicular to the forces applied but, owing to a design requirement, they may be inclined or curved. They are, of course, still *rigid* members. Consider the wheelbarrow of Fig. 7.4*b*. The load w is assumed concentrated at the center of gravity, (c.g.). The effort arm is the

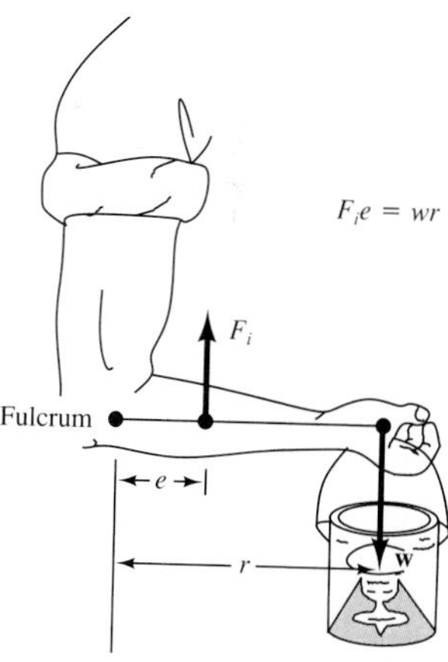

Fig. 7.5 A Class III lever, the human forearm.

Levers are found as component parts of many industrial machines. They are also present in the human arm and ankle.

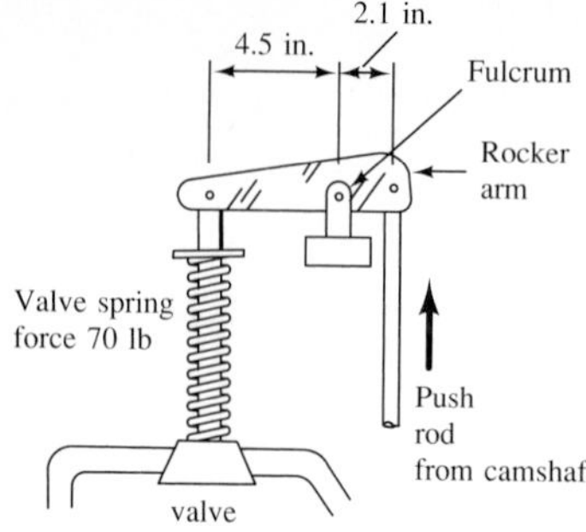

Fig. 7.6 Schematic diagram of a rocker-arm arrangement for an overhead-valve engine. A Class I lever (Illustrative Problem 7.1).

perpendicular distance from the fulcrum to the *line of action* of the input force, F_i. The shape of the wheelbarrow handles does not affect the "leverage" obtained, and $F_i e = wr$ as before.

Levers are in use in nearly all types of industrial machinery. Tractors, hoists, cranes, engines, pumps, presses, textile looms, bulldozers—the list of machines in which levers are involved is nearly endless. The human arm and ankle also contain levers.

Illustrative Problem 7.2 A wheelbarrow is loaded with 180 lb of wet concrete mix. The entire load is considered to act downward at the center of gravity, c.g. (see Fig. 7.4*b*). The resistance arm r is 15 in. and the effort arm e is 48 in. What must be the value of F_i, the input force (effort), to lift the load?

Solution From Eq. (7.1),

$$F_i = \frac{wr}{e}$$

Substituting,

$$F_i = \frac{180 \text{ lb} \times 15 \text{ in.}}{48 \text{ in.}}$$

$$= 56.3 \text{ lb} \qquad \textit{answer}$$

7.5 ■ The Inclined Plane

Analysis of the forces acting on an inclined plane.

If you had to move a heavy safe from the ground to a truck bed h ft high without the aid of hoisting equipment, you might obtain some long planks and roll the safe up the incline, as shown in Fig. 7.7. If the planks available were L ft long, the inclined plane would be at an angle $\theta = \sin^{-1}(h/L)$, as shown in the diagram. The weight of the safe is indicated by the vector w, a force acting vertically downward as shown. This force has two measurable effects: (1) It tends to cause the safe to roll back down the planks, and (2) it causes a force to be exerted perpendicularly against the planks. These two *component* forces are shown in the diagram by the vectors F_g and N, respectively. Note that, considered as a parallelogram of forces, w is the resultant of N and F_g, and N and F_g are mutually perpendicular components of w. From trigonometry, the magnitude of N, the force pressing normally (perpendicularly) against the planks, is

$$N = w \cos \theta \tag{7.4}$$

and the force with which the safe tends to roll back down the plane is

$$F_g = w \sin \theta \tag{7.5}$$

If friction is neglected (we shall consider its effects later), the force F required to pull the safe up the plane *at uniform speed* is equal and opposite to the force with which it tends to roll down the plane. That is,

$$F = F_g = w \sin \theta$$

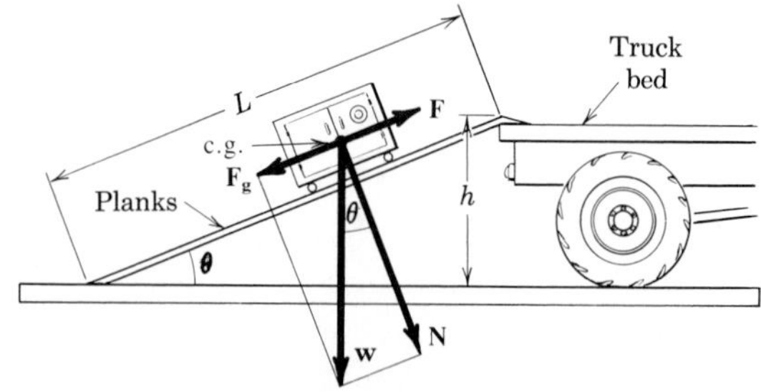

Fig. 7.7 The inclined plane is a basic machine. Here the use of an inclined plane to load a heavy safe on a truck is shown. The vector diagram is superimposed on the sketch.

But $\sin \theta = h/L$, and therefore

$$F = w \frac{h}{L} \tag{7.6}$$

The *theoretical* mechanical advantage (TMA) of the inclined plane is therefore

$$\underset{\text{(inclined plane)}}{\text{TMA}} = \frac{L}{h} \tag{7.7}$$

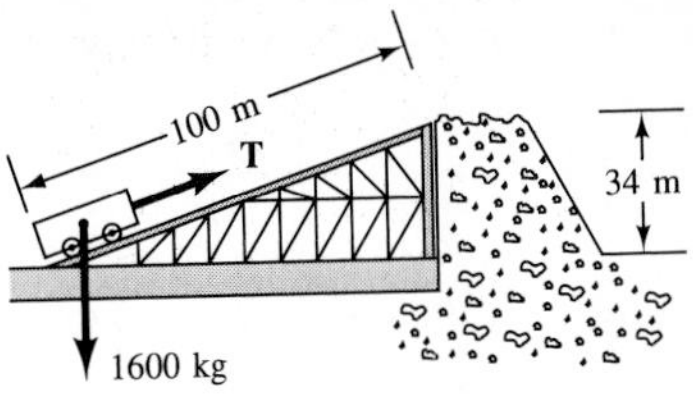

Fig. 7.8 An inclined-plane with tracks for hauling concrete up to the working level at a dam (Illustrative Problem 7.3).

The same result can also be obtained from the *principle of work*. Neglecting friction, the work done in pushing the safe up the plane is equal to the work which would be required if the safe were merely lifted through the vertical distance h. Stated mathematically,

$$FL = wh$$

The TMA of an inclined plane is the ratio of the length of the plane to its vertical rise.

Putting this equation in the form of a proportion, we obtain

$$\frac{w}{F} = \frac{L}{h}$$

which is the same result for theoretical mechanical advantage as obtained above from the vector analysis of the problem.

The inclined plane and the principle of work.

Illustrative Problem 7.3 In constructing a dam, heavy hoppers of concrete mix are slowly pulled up an inclined track (assumed frictionless), which rises 34 m for every 100 m along the incline. If the loaded hoppers have a mass of 1600 kg, what is the tension T in the hoisting cable (see Fig. 7.8).

Solution Note that $h/L = \frac{34}{100}$. Substitution in Eq. (7.6) gives

$$F = w\frac{h}{L} = \frac{mgh}{L} \qquad (\text{NOTE: } w = mg)$$

$$= \frac{1600 \text{ kg} \times 9.81 \text{ m/s}^2 \times 34 \text{ m}}{100 \text{ m}} = 5340 \text{ N} \qquad \textit{answer}$$

FRICTION

7.6 ■ Friction and Machines

Rarely, if ever, is friction absent when one object moves or slides over another. The surface of any material, no matter how smooth it may look or feel, is actually full of irregularities which oppose the sliding of some other surface. This force of opposition to the motion of bodies in contact is called *friction*.

Friction is a force which always acts to oppose the relative motion of two sliding surfaces in contact.

The definition of *friction*—a force that always acts to impede the relative motion of two bodies in contact.

In the light of this definition of friction, and reasoning from *Newton's third law—action and reaction*—it follows that if there is no force tending to cause relative motion, then there is no force of friction. Consider two bodies, A and B (Fig. 7.9a), pressed together by a force N. [The letter N is used to indicate a force *normal* (i.e., perpendicular) to the surfaces in contact.] As a force F is applied to body A, *tending* to make it slide to the right over body B, the force of *static friction* F_s, acting opposite to F and equal to F, increases as F increases, until motion actually impends (Fig. 7.9*b*). F_s (*static friction*) is zero when F is zero, and F_s has its maximum or limiting value ($F_{s_{max}}$), just before A begins to slide on B. Immediately after motion begins, frictional resistance drops somewhat below its maximum (*static*) value, as shown by the curve of Fig. 7.9*c*. This lower frictional resistance after sliding begins is called *sliding friction* or *kinetic friction* (*kinetic* means motion), and is designated F_k.

Static friction distinguished from *kinetic* (sliding) *friction*.

A possible explanation for the reduction in frictional resistance after motion begins is that the motion prevents the rough indentations from "settling in" or "locking," on the two surfaces. Also motion could create a "bounce" effect, allowing a thin cushion of air to infiltrate between the sliding surfaces to lower the friction. The fact that F_k is less than $F_{s,max}$ has important implications for the braking of automobiles, the "stopping power" of the brakes being much more effective just *before* skidding begins than it is if the wheels are locked, initiating skidding.

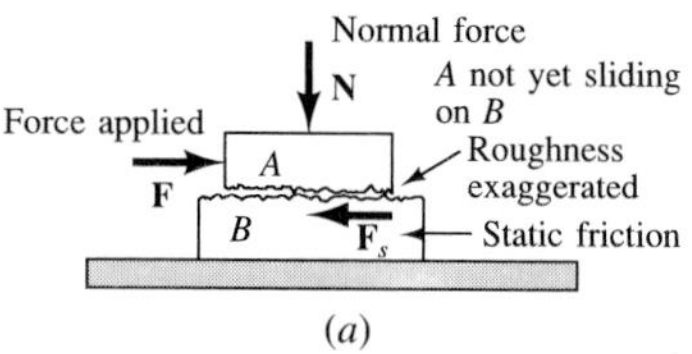

(*a*)

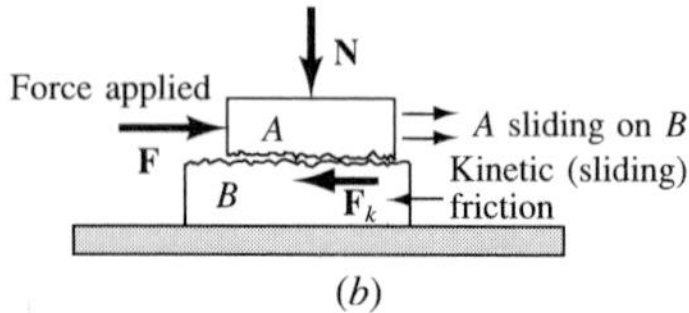

(*b*)

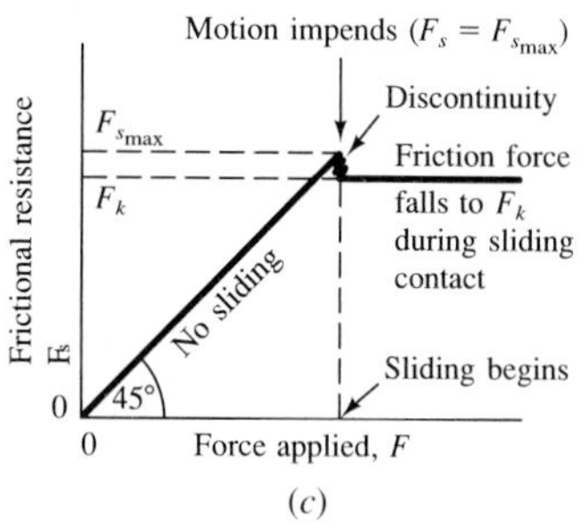

(*c*)

Fig. 7.9 Static versus kinetic, or sliding, friction. In (*a*) force F is insufficient to overcome the force of friction for the two bodies as shown. The force of static friction F_s increases as the force F applied increases. F_s reaches its limiting value $F_{s,max}$ just as sliding impends. (*b*) Once sliding has begun, the force of kinetic friction F_k has a somewhat lower value than $F_{s,max}$, and sliding continues. (*c*) Curve depicting the relationships among static friction F_s, kinetic friction F_k, and the force applied F, before and after sliding begins (not to any scale). Note that F_s increases as F increases (45° angle) until $F_{s,max}$ is reached. Then the force of retarding friction falls back to a value F_k during sliding contact.

7.7 ■ The Nature of Friction

Friction has been investigated for more than 200 years, but there still remain many uncertainties and problems for current research. For *dry surfaces* the following statements about the behavior of friction are applicable within wide limits:

1. The force of static friction (F_s) is not directly related to the area of contact but is directly related to the force pressing the two sliding surfaces together. Since this force acts perpendicularly on both surfaces, it is called the *normal force*. Within reasonable limits of sliding-contact area, the force required to start an object sliding on a floor is the same, regardless of the area in contact with the floor.

2. When motion begins, the force of kinetic friction is usually somewhat less than the maximum value of static friction just before motion begins. In other words, *sliding friction is less than limiting or static friction*. The relative velocity of the sliding surfaces does not seem to affect the value of sliding or kinetic friction for a considerable range of relatively low velocities. However, for high velocities, friction *decreases* as speed *increases*.

3. The force of friction is definitely related to the nature of the two surfaces sliding across each other. This constant of the particular surfaces is called the *coefficient of friction*. Values for almost any combination of surfaces can be found in engineering handbooks (see Table 7.1).

4. In addition to irregularities and roughness of the sliding surfaces, friction is due, in part, to molecular attraction (cohesion) between the surface molecules of the bodies in close contact (see Sec. 10.7).

7.8 ■ Analyzing Frictional Forces—The Coefficient of Friction

We have defined three quantities involved in friction problems, the force of static friction F_s, the force of sliding (kinetic) friction F_k, and the normal (perpendicular) force pressing the surfaces together, designated by the letter N. The constant of the two surfaces, referred to above as the *coefficient of friction*, is a fourth factor. The coefficient of friction is designated by the Greek letter μ (mu) and it requires two definitions—one for the static case and one for the sliding case.

$$\mu_s = \frac{\text{Force of static friction}}{\text{Normal force}} = \frac{F_{s,max}}{N}$$

or

$$F_{s,max} = \mu_s N \tag{7.8}$$

$$\mu_k = \frac{\text{Force of kinetic friction}}{\text{Normal force}} = \frac{F_k}{N}$$

or

$$F_k = \mu_k N \tag{7.9}$$

Since they are ratios of two forces, μ_s and μ_k have no units.

Table 7.1 Approximate coefficients of sliding friction μ_k and static friction, μ_s (Average values for dry surfaces unless otherwise noted). All values vary with surface conditions.

Materials	μ_k	μ_s	*Materials*	μ_k	μ_s
Hardwood on hardwood	0.25	0.5	Steel on concrete (smooth)	0.30	0.5
Rubber on dry concrete	0.75	1.0	Steel on babbit metal (dry)	0.14	0.20
Rubber on wet concrete	0.50	0.75	Steel on steel (oiled)	0.06	0.09
Metal on hardwood (varies)	0.40	0.6	Steel runners on ice	0.04	0.10 (depends on temperature)

Illustrative Problem 7.4 A heavy machine is being skidded across a factory floor. If the machine weighs 1800 lb and the coefficient of sliding friction (steel on greased concrete) is 0.120, find the horizontal force required to slide it.

Solution Use Eq. (7.9) and substitute given values.

$$F_k = \mu_k N = 0.120 \times 1800 \text{ lb}$$
$$F = 216 \text{ lb} \qquad \textit{answer}$$

Illustrative Problem 7.5 A generator weighing 2600 lb is mounted on skids and is being pulled at constant velocity up a ramp which rises 26 ft for every 100 ft along the incline. $\mu_k = 0.27$. What is the tension T in the hoisting cable if the cable is parallel to the incline?

Solution
First method (using vectors, trigonometry, and the principles of equilibrium). Sketch the relations of the problem as shown in Fig. 7.10, showing the forces as vectors. Applying the principles of equilibrium, it is evident that the algebraic sum of the forces on the skid in any direction must be equal to zero. Consequently, in the line of the ramp,

$$\Sigma F = T - F_g - F_k = 0$$

From the vector diagram,

$$N = w \cos \theta$$

And from Eq. 7.9,

$$F_k = \mu_k N = \mu_k w \cos \theta$$

From the diagram,

$$F_g = w \sin \theta$$

Therefore

$$T = w \sin \theta + \mu_k w \cos \theta$$

From trig tables or pocket computer

$$\theta = \sin^{-1} \tfrac{26}{100} = 15.1° \text{ and } \cos \theta = 0.966$$

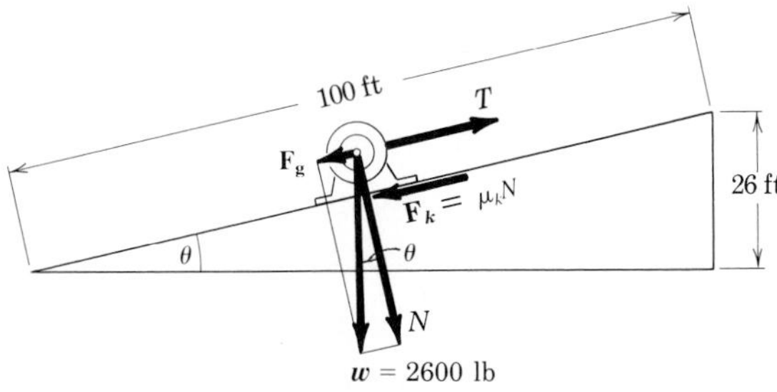

Fig. 7.10 Vector analysis of the forces acting while a heavy object is being skidded up an inclined plane (Illustrative Problem 7.5).

Transposing and substituting numerical values,

$$T = 2600 \text{ lb} \times \tfrac{26}{100} + 0.27 \times 2600 \text{ lb} \times 0.966$$
$$= 676 \text{ lb} + 678 \text{ lb} = 1354 \text{ lb} \qquad \textit{answer}$$

Second method (using the principle of work). The principle of work implies that the total work done by the hoisting cable tension equals the net work done against the force of gravity (increasing the GPE) plus the "wasted" work done against friction. Writing this relation in equation form, we have

$$100 \text{ ft} \times T = 2600 \text{ lb} \times 26 \text{ ft} + 100 \text{ ft} \times F_k$$

or
$$T = \frac{(2600 \text{ lb} \times 26 \text{ ft}) + (100 \text{ ft} \times F_k)}{100 \text{ ft}} = 2600 \text{ lb} \times 0.26 + 1 \times F_k$$

But
$$F_k = \mu_k w \cos \theta$$
$$= 0.27 \times 2600 \text{ lb} \times 0.966$$
$$= 678 \text{ lb}$$

And
$$T = 2600 \text{ lb} \times 0.26 + 678 \text{ lb}$$
$$= 676 \text{ lb} + 678 \text{ lb}$$
$$= 1354 \text{ lb, as before} \qquad \textit{answer}$$

It is again pointed out that the principle of work constitutes a powerful tool in the analysis and solution of problems in mechanics.

Table 7.1 gives some average values of μ_k, the coefficient of kinetic (sliding) friction, and of μ_s, the coefficient of static friction.

7.9 ■ Friction–Good or Bad

Friction is a problem that must be minimized in many mechanisms; but it is an absolutely necessary element in land transportation.

Friction is advantageous or disadvantageous depending entirely on the result desired. Certainly friction between rubber and concrete is desirable. And it would be very difficult even to walk if friction between the ground and shoe soles were absent. All automotive and rail transportation is based on friction, and tires and roads are designed with a view to assure its presence. On the other hand, and again using the automobile as an example, the friction between pistons and cylinder walls is undesirable, and every effort is made to minimize these effects.

For motor vehicles, on the average, more than one-third of the power output of the engine is used up in doing work against friction forces in the engine and the drive train. A good bit of the remaining engine power is used in overcoming air friction on the auto's body surface. The lubrication industry spends millions of dollars yearly in research to develop better lubricants to minimize friction in machines. The theory of lubrication is quite complex, but it is sufficient to say here that the presence of a film of oil or grease between two surfaces prevents them from sliding on themselves (if they are reasonably smooth to begin with) and substitutes fluid friction for the relatively greater friction which exists between solid surfaces. Lubricating oil is usually supplied to the bearings of engines, turbines, and other rotating machinery under pressure, and this fluid pressure minimizes metal-to-metal contact, thus reducing friction markedly.

Illustrative Problem 7.6 The total weight of a loaded snowmobile, including driver and passenger, is 840 lb. The coefficient of sliding friction between the runners and the snow crust is 0.045. If the vehicle is traveling at a steady speed of 30 mi/h, calculate the horsepower being used to overcome the friction between the runners and the snow.

Solution The force of friction against which work is being done is

$$F_k = \mu_k N = 0.045 \times 840 \text{ lb}$$
$$= 37.8 \text{ lb}$$

$$30 \text{ mi/h} = 44 \text{ ft/s}$$

Using Eq. (6.17), $P = Fv$ (Power = force × velocity)

Substituting values,

$$P = 37.8 \text{ lb} \times 44 \text{ ft/s}$$
$$= 1660 \text{ ft} \cdot \text{lb/s}$$

From which,

$$P_{\text{hp}} = \frac{1660 \text{ ft} \cdot \text{lb/s}}{550 \text{ ft} \cdot \text{lb/s-hp}}$$
$$= 3.02 \text{ hp} \qquad \textit{answer}$$

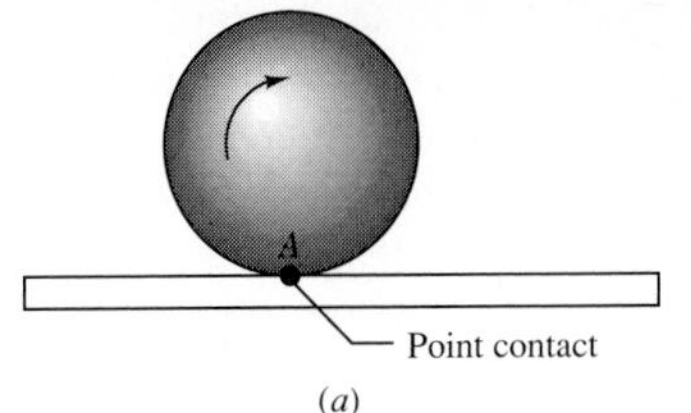

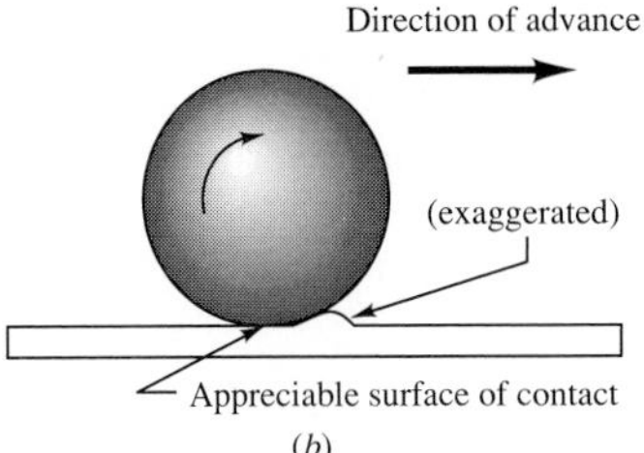

Fig. 7.11 Rolling friction. (*a*) Theoretical point contact; (*b*) actual situation (somewhat exaggerated) in rolling contact. There is some temporary deformation of both of the contact surfaces.

7.10 ■ Rolling Friction

The invention of the wheel was an important event in history. Humans have known for centuries that *rolling friction* is much less than *sliding friction*. In addition to providing wheels on which to roll vehicles, modern industry provides ball and roller bearings for machine components in place of the older type *sleeve* bearings, whenever the factors of long life, efficiency, and cost justify the practice.

Analysis of the rolling of a wheel or ball on a hard surface (Fig. 7.11*a*) reveals that theoretically at least, the contact is a single point. Further, if the wheel or ball rolls without slipping, there is no relative motion between the point *A* (on the wheel) and the point *A* (on the surface) at the instant of contact. Consequently there is, theoretically, *no sliding friction*. Practically, however, either the wheel or the surface (or both) flattens a bit under the force of contact, and a situation which is shown in exaggerated proportions in Fig. 7.11*b* occurs. The surface indentation makes it necessary for the wheel (or ball) to roll slightly uphill all the time, and the flattening of the wheel itself implies that there is an appreciable *surface* of contact rather than a *point* of contact. The deformation is, of course, extremely small and only temporary. Consequently, there is in reality some relative motion between wheel and surface, and therefore some sliding friction involved.

The Coefficient of Rolling Friction Coefficients of rolling friction (μ_r) are subject to many variable conditions of load, wheel (or ball) diameter, rotation, speed, etc., but the student can form some idea of their relative magnitude from the values in Table 7.2. As before, $F_f = \mu_r N$.

Anyone who roller skates or performs on a skateboard knows about rolling friction.

Table 7.2 Coefficient of rolling friction μ_r for dry surfaces

Situation	μ_r
Car wheels on rails	0.0045
Steel ball bearings on steel	0.0025
Steel roller bearings on steel	0.0003
Rubber tires on concrete	0.04

7.11 ■ Sleeve Bearings vs. Ball (or Roller) Bearings

Bearings may be classified according to their use as (1) *guide bearings*, such as crankshaft main bearings or line-shaft bearings; (2) *journal bearings*, such as the bearings on a crankshaft "throw" or on a locomotive driving wheel; and (3) *thrust bearings*. A thrust bearing is one that is mounted in such a way that the bearing has to support the weight of a pump or rotor or counteract the thrust of a ship's screw or an airplane propeller. Depending on the factors previously referred to (load, speed, cost, etc.), all three of these classes of bearings are made in both *sleeve* form and in *ball* (or *roller*) form. The basic difference in performance can be seen from an analysis of Fig. 7.12, which shows a guide bearing in sleeve form (*a*) and in ball (or roller) form (*b*).

Without ball bearings and roller bearings the engines and machines of modern industry would literally grind to a halt.

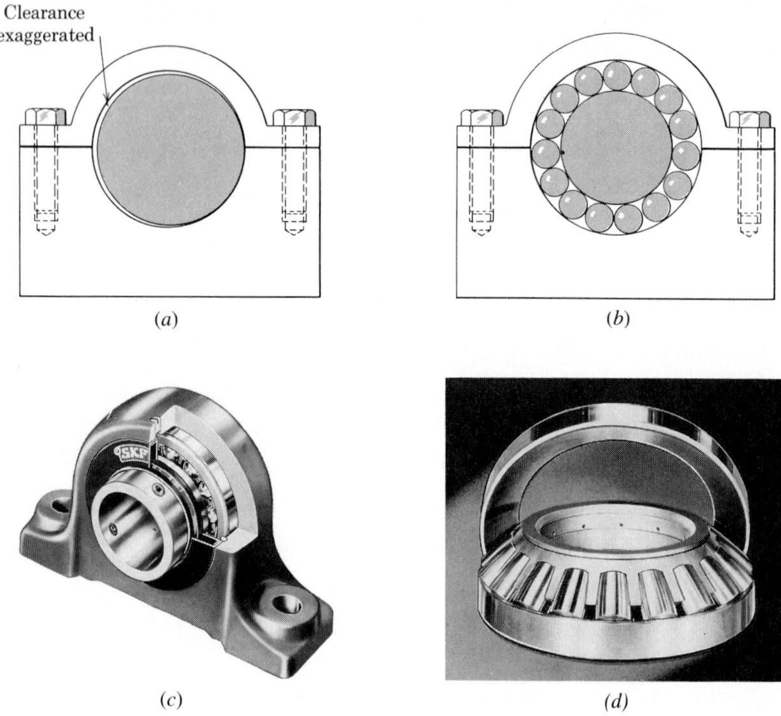

Fig. 7.12 Guide bearings. (*a*) Sleeve bearing, showing shaft touching bearing over a considerable area. Sliding friction occurs. (*b*) Ball (or roller) bearing, showing shaft rolling on balls and balls rolling in the outer "race." Only rolling friction occurs. (*c*) A commercial ball-type guide bearing, and (*d*) a roller thrust bearing (SKF Industries). All of these bearings are lubricated with oil or other special lubricant, often pumped under pressure.

The ball (or roller) form will have less friction, is easier to lubricate, and keeps the shaft more nearly centered with less wear and greater efficiency. The initial cost of ball or roller bearings is several times the cost of sleeve bearings. Parts (*c*) and (*d*) of Fig. 7.12 show two types of commercially manufactured bearings.

7.12 ■ Fluid Friction

Fluid friction, though of much smaller magnitude than friction between solids, is always present when fluids flow; and at high flow velocities, it is a very troublesome factor.

Liquids flowing through pipes are subject to frictional forces between themselves and the pipe wall and to frictional forces (called *viscosity*) between the molecules of the liquid itself. Air or gases flowing in pipes or ducts also encounter frictional forces. Ships moving through or over water and aircraft moving through air experience frictional forces which we call *drag*. Both air resistance and liquid resistance are examples of *fluid friction*. Fluid friction tends to be less pronounced than friction between solids. Pulling a raft on water requires less force than sliding the same raft over even a very smooth solid surface. And, frictional forces caused by air or gases are less than those caused by liquids. Objects moving through air encounter very little friction at low speeds, but as speed increases markedly, air friction becomes a powerful retarding force. Friction with atmospheric air, for example, generates such heat that meteorites burn up before they hit the earth, causing the shooting stars often seen in the August sky. Spacecraft are equipped with a special heat shield of ceramic tiles to prevent their being consumed by the heat of friction as they return to the earth's atmosphere at speeds in excess of 30,000 mi/h.

7.13 ■ The Effect of Friction on Machines

Friction is a major cause of decreased efficiency in machines. Research and development outlays to improve lubricants to decrease friction total millions of dollars annually.

Force is required to overcome friction. When a force acts through a distance, work is done and energy must be expended to do this work. Therefore friction always results in energy expended, which is "wasted" either in the form of "useless work" or in the form of heat within the machine.

For the purposes of the present discussion it will be assumed that the only losses within machines are frictional losses, since we are not concerned with heat engines at this point.

7.14 ■ The Inclined Plane and Efficiency

Friction on inclined planes makes the AMA appreciably less than the TMA.

Where frictional losses are small, as in most levers, the actual mechanical advantage is nearly equal to the theoretical mechanical advantage to be expected from the physical design of the machine. On the inclined plane, however, and in some other machines, friction is a factor of considerable importance and cannot be neglected. In such cases, actual mechanical advantage is considerably less than theoretical mechanical advantage. These relationships are clarified in Fig. 7.13. *Theoretically* (by Eq. 7.6), we should be able to pull block B up the plane at constant velocity with a force $F = wh/L$. However, a greater force than F is actually required because of the friction between the block and the plane. This greater force actually required to pull the block up the plane is shown as F' in the diagram. The *actual* mechanical advantage is then

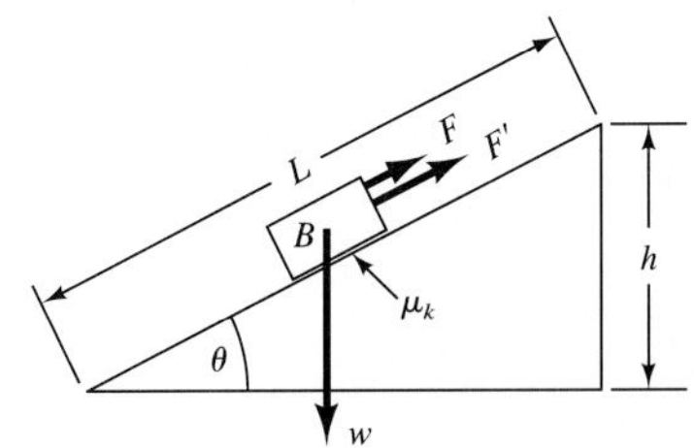

Fig. 7.13 Diagram for analyzing the efficiency of an inclined plane.

$$\underset{\text{(Inclined plane)}}{\text{AMA}} = \frac{w}{F'} \tag{7.10}$$

where F' is the actual force required to move an object up an inclined plane at constant velocity. The *net useful work output* in moving the block B through the vertical height h is, regardless of the path used, merely the product of the weight of the block w and the height h, or

$$\text{Net work output} = w \times h$$

The *work input* in moving the block up the plane is equal to the force applied F' times the distance up the plane L, or

$$\text{Work input} = F' \times L$$

Therefore the *efficiency of an inclined plane* is given by

The efficiency of an inclined plane (or any other machine) is the ratio of work output to work input.

$$\text{Eff} = \frac{\text{work output}}{\text{work input}} = \frac{wh}{F'L} \tag{7.11}$$

Now, noting that

$$\frac{\text{AMA}}{\text{TMA}} = \frac{w/F'}{L/h} = \frac{wh}{F'L} \tag{7.12}$$

we can relate efficiency and mechanical advantage, from Eqs. (7.10) and (7.11) to obtain the basic relation

$$\text{Eff} = \frac{\text{AMA}}{\text{TMA}} \tag{7.13}$$

(Efficiency of basic or simple machines)

Though derived from inclined-plane considerations, Eq. (7.13) gives the efficiency of any of the basic or simple machines.

The efficiency of any basic or simple machine is the ratio AMA/TMA.

Illustrative Problem 7.7 An inclined plane made of smooth planks is used to load heavy boxes onto a truck. The height of the truck bed is 4 ft, and the length of the planks is 10 ft. The coefficient of sliding friction μ_k between the boxes and plank surface is 0.30. Find (*a*) the theoretical mechanical advantage, (*b*) the actual mechanical advantage, and (*c*) the efficiency of this inclined plane.

Solution
(*a*) From Eq. (7.7),

$$\text{TMA} = \frac{L}{h} = \frac{10 \text{ ft}}{4 \text{ ft}} = 2.5 \qquad \textit{answer}$$

(*b*) From Eq. (7.10),

$$\text{AMA} = \frac{w}{F'}$$

But, (see Fig. 7.10),

$$F' = F_g + F_k = W \sin \theta + \mu_k w \cos \theta$$
$$= w(\sin \theta + 0.3 \cos \theta)$$

From the given dimensions, $\sin \theta = 0.40$, from which $\theta = 23.6°$ and $\cos \theta = 0.917$. Therefore,

$$F' = w(0.40 + 0.3 \times 0.917) = 0.642w$$

$$\text{AMA} = \frac{w}{F'} = \frac{w}{0.675w} = 1.48 \qquad \textit{answer}$$

(*c*) From Eq. (7.13),

$$\text{Eff} = \frac{\text{AMA}}{\text{TMA}} = \frac{1.48}{2.5} = 0.59 = 59\% \qquad \textit{answer}$$

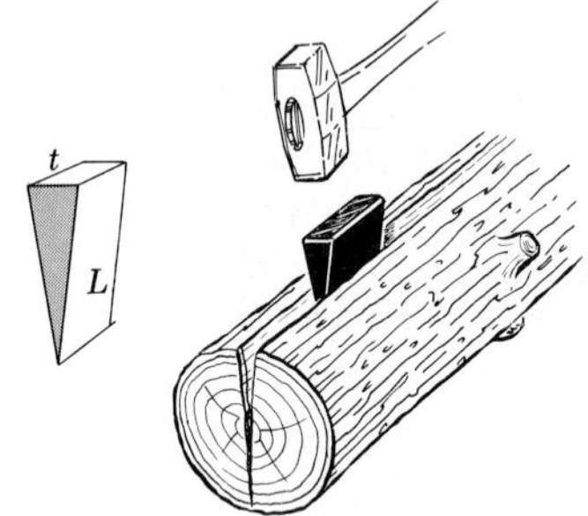

Fig. 7.14 The wedge is a double inclined plane.

MODIFICATIONS OF THE INCLINED PLANE

7.15 ■ The Wedge

The *wedge* (Fig. 7.14) is actually *a double inclined plane*. Its mechanical advantage is, in terms of the dimensions shown (and neglecting friction),

$$\underset{\text{(wedge)}}{\text{TMA}} = \frac{\text{L}}{\text{t}}$$

Friction is considerable and desirably so, else the wedge might be forced back out by the pressures acting on it. The actual mechanical advantage of a wedge is much less than the theoretical mechanical advantage and depends on the friction between the contacting surfaces. Wedges are used in splitting wood and in leveling heavy machinery before cementing the base (''grouting in'') to the foundation. The wedge principle is the basis for many cutting tools—knives, axes, chisels, lathe tools, and the cutting tools of planers and milling machines. The sharper the tool, the greater the ratio *L*/*t* (TMA). The rotating cams of internal combustion engines (see Fig. 7.15) are based on the wedge principle.

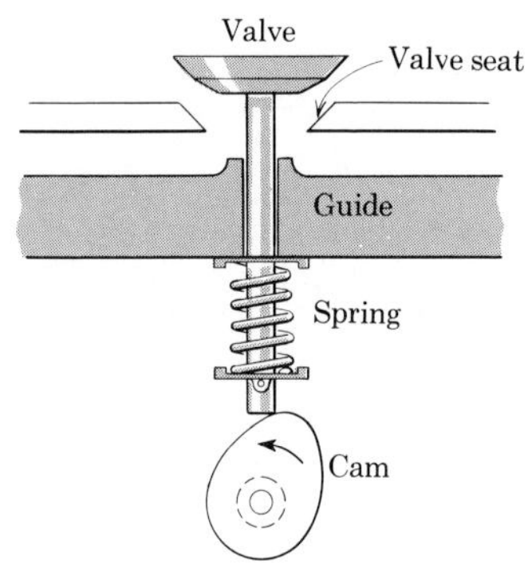

Fig. 7.15 The cam is basically a rotating wedge with rounded instead of sharp edges. The diagram shows one arrangement (schematic) for a cam-operated engine valve.

7.16 ■ Screws

Another application of the inclined plane is the *screw*. It is possible to obtain extremely high mechanical advantages with screws. Actually, a screw is merely an inclined plane wrapped around a cylinder so that a helical form results. Figure 7.16*a* shows the basic nomenclature applied to screws. The alternate crests and valleys are called *threads,* and the distance from one crest straight across to the next crest is called the *pitch* (*p* in the diagram) of the screw. This is the distance the screw advances as it is turned through one full revolution. Most screws are *right-hand threaded*; i.e., upon being turned to the right (clockwise) they advance away from the operator. Some bolts and screws are left-hand threaded. This arrangement is used if the normal rotation of a machine would tend to loosen a right-hand-threaded bolt.

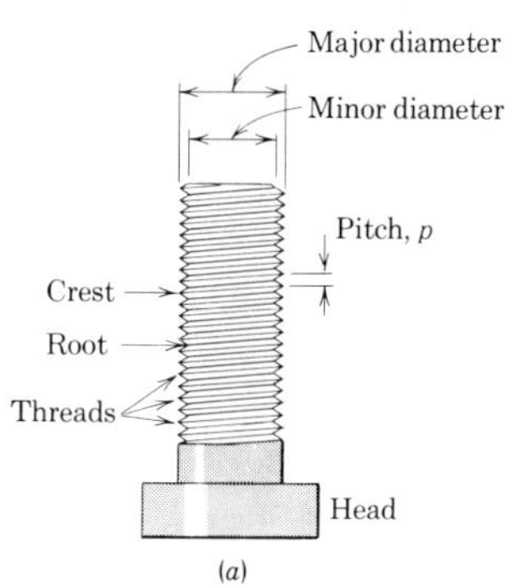

(a)

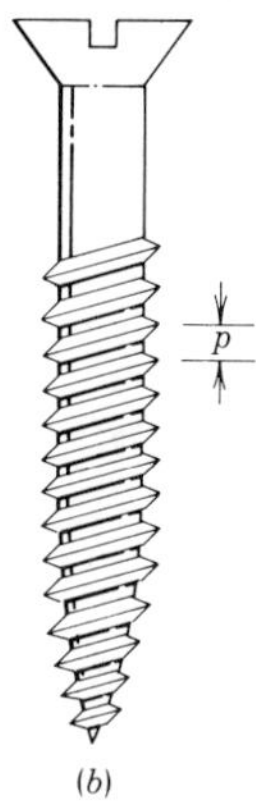

(b)

Fig. 7.16 Screws and their nomenclature. (*a*) A right-hand-threaded bolt, (*b*) a right-hand-threaded wood screw, (*c*) the common screw jack, for applying large forces, as in building construction, (*d*) one of the huge screws (propellers) on a large ship. (Official U.S. Navy photo)

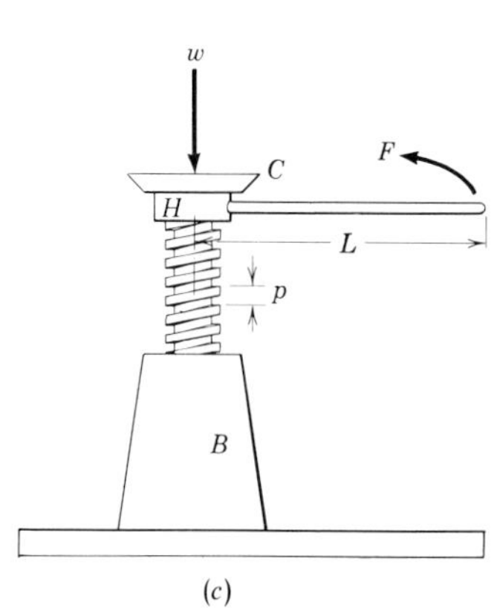

(c)

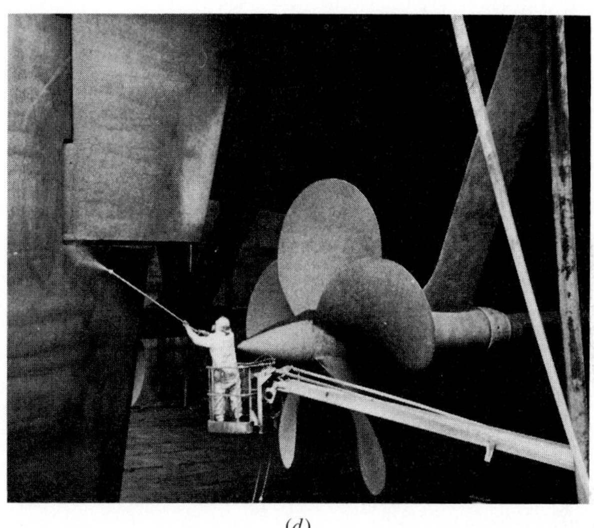
(d)

Bolts and screws have large TMA's, but are purposely designed to have a great deal of friction. Since they are fastening devices, the reaction forces of the materials they are compressing must not be sufficient to "back them off."

Ships' propellers are also screws (Fig. 7.16*d*). They "bite" or advance into water and throw the water to the rear.

Screws and screw jacks are purposely designed to have a great deal of friction. What is the reason for such a design?

7.17 ■ The Screw Jack

The screw principle is used in the design of the common screw jack used for jacking up houses, autos, trucks, heavy machinery, etc. The common form is illustrated in Fig. 7.16*c*. The weight w being lifted acts down on the cap C, which does not rotate, although the head H does. Lubrication between C and H reduces friction. The effort F is applied at the end of the handle, whose length, to the center of the screw, is L. The screw itself is usually of the square-thread type, lubricated to reduce friction somewhat, but designed so that friction will always be sufficient to keep the jack from "running down" under the load. In practice this means that screw jacks are less than 50 percent efficient.

Historians credit Archimedes (287–212 B.C.) with using the screw principle to raise water from a lower to a higher level. Archimedes was one of the foremost *natural philosophers* of ancient Greece.

7.18 ■ Mechanical Advantage and Efficiency of the Screw Jack

Just as for any machine, the *actual* mechanical advantage of a screw jack is the ratio resistance/effort, or, in terms of the diagram of Fig. 7.16*c*,

$$\underset{\text{(screw jack)}}{\text{AMA}} = \frac{w}{F} \tag{7.14}$$

The *theoretical* mechanical advantage is derived in terms of distances moved by the effort and the resistance. If the effort moves through one revolution, the resistance is raised a distance p, equal to the pitch of the screw. Defining the theoretical mechanical advantage as

$$\text{TMA} = \frac{\text{distance effort moves}}{\text{distance resistance moves}}$$

we have

$$\underset{\text{(screw jack)}}{\text{TMA}} = \frac{2\pi L}{p} \tag{7.15}$$

Owing to the necessarily large friction, screw jacks rarely have *efficiencies* higher than 30 to 40 percent. They might well have theoretical mechanical advantages of 500 or more, however. The efficiency of a screw jack can be calculated from the formula

$$\underset{\text{(screw jack)}}{\text{Eff}} = \frac{\text{AMA}}{\text{TMA}} = \frac{w/F}{2\pi L/p}$$

$$= \frac{wp}{2\pi LF} \tag{7.16}$$

Illustrative Problem 7.8 A screw jack has a screw whose pitch is 0.25 in. Its handle length is 18 in., and a load of 1.5 tons is being lifted. (*a*) If friction is neglected, what effort is required at the end of the handle? (*b*) What is the theoretical mechanical advantage of this jack? (*c*) If the force actually required at the end of the jack handle is 65 lb, what are the actual mechanical advantage and the efficiency of the jack?

Solution

(*a*) If friction is neglected,

$$\text{TMA} = \text{AMA}$$

or

$$\frac{2\pi L}{p} = \frac{w}{F}$$

from which

$$F = \frac{pw}{2\pi L}$$

$$= \frac{0.25 \text{ in.} \times 1.5 \text{ tons} \times 2000 \text{ lb/ton}}{2\pi \times 18 \text{ in.}}$$

$$= 6.63 \text{ lb} \qquad \textit{answer}$$

(*b*)

$$\text{TMA} = \frac{2\pi L}{p} = \frac{2\pi \times 18 \text{ in.}}{0.25 \text{ in.}} = 450 \qquad \textit{answer}$$

(*c*)

$$\text{AMA} = \frac{w}{F} = \frac{1.5 \text{ tons} \times 2000 \text{ lb/ton}}{65 \text{ lb}}$$

$$= 46 \qquad \textit{answer}$$

$$\text{Eff} = \frac{\text{AMA}}{\text{TMA}} = \frac{46}{450} = 0.102 = 10.2\% \qquad \textit{answer}$$

7.19 ■ Other Applications of the Screw Principle

The most familiar screws are the ordinary *carpenter's wood screw* and the *bolts* used in the assembly of machinery. *Vises*, screw-type *presses* and electromechanical garage door openers also utilize the large mechanical advantages possible with the screw principle. The cutting edges of drills and carpenters' bits are wedges, but the helical form which the metal chips or wood cuttings take is in the form of a screw. Turbine-type pumps also utilize the principle of the screw.

Ships' propellers are technically called *screws* (Fig. 7.16*d*). In their rotation they "bite" into and throw water to the rear in the same way that a screw or bit advances into wood as cuttings move up the spiral. In the case of the ship the force which accelerates the water to the rear has a reaction force which pushes the ship forward. The same principle applies to propeller-driven aircraft.

SIMPLE MACHINES BASED ON THE LEVER PRINCIPLE

7.20 ■ The Wheel and Axle

The simple lever, as discussed previously, has the limitation of a restricted arc of motion. This limitation can be removed by designing a lever system capable of continuous rotation. Such a device is called a *wheel and axle*. As shown in Fig. 7.17, a heavy load w is the resistance whose lever arm is the relatively small radius r of the *axle*. The force F is applied to the circumference of the attached *wheel,* whose radius is the larger distance R. The *actual* mechanical advantage of such a device is AMA = w/F. To obtain the *theoretical* mechanical advantage apply the principle of work. If the force F moves down through the distance s, the load w will be lifted through a smaller distance h. The principle of work requires that $Fs = wh$, or work input equals work output (losses assumed negligible). In one revolution of the wheel and axle the work input is $Fs = 2\pi RF$ and the work output is $wh = 2\pi rw$. Equating these,

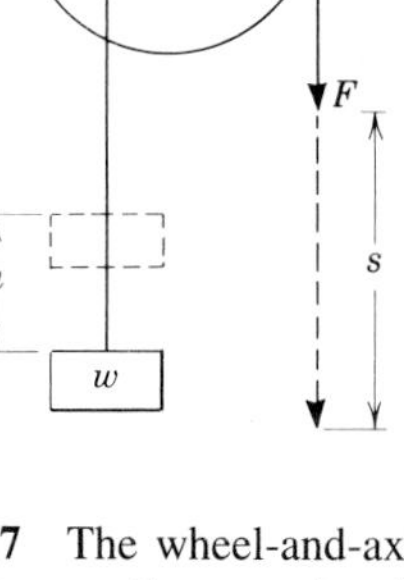

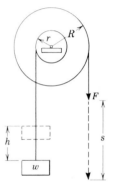

Fig. 7.17 The wheel-and-axle system is actually a rotating lever.

$$2\pi RF = 2\pi rw$$

and, rearranging,

$$\frac{w}{F} = \frac{R}{r} \qquad (7.17)$$

or

$$\underset{\text{(wheel and axle)}}{\text{TMA}} = \frac{R}{r} = \frac{D}{d} \qquad (7.18)$$

or, in words:

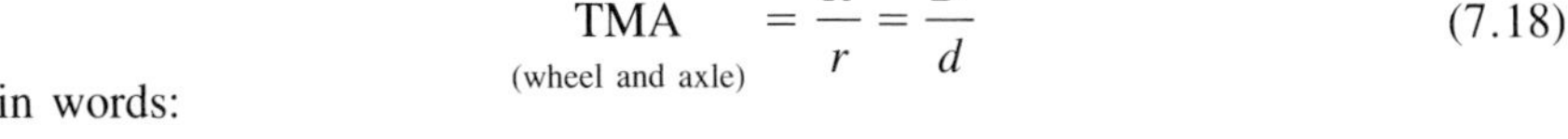
The theoretical mechanical advantage of a wheel and axle is the ratio of the radius (or diameter) of the wheel to the radius (or diameter) of the axle.

The larger the wheel is, compared to the axle, the greater the *TMA*.

Illustrative Problem 7.9 A wheel and axle system is used to lift a 2200-kg load. The wheel diameter is 150 cm and the axle diameter is 15 cm. The operator has to exert a force of 2250 N. Find the following: (*a*) TMA, (*b*) AMA, (*c*) efficiency.

The wheel-and-axle is the basis of many types of hoisting equipment.

Solution

(*a*)

$$\text{TMA} = \frac{R}{r} = \frac{D}{d} = \frac{150\text{ cm}}{15\text{ cm}} = 10 \qquad \textit{answer}$$

(*b*)
$$\text{AMA} = \frac{w}{F} = \frac{2200\ \text{kg} \times 9.81\ \text{m/s}^2}{2250\ (\text{kg}\cdot\text{m})/\text{s}^2} = 9.59 \qquad \textit{answer}$$

(*c*)
$$\text{Eff} = \frac{\text{AMA}}{\text{TMA}} = \frac{9.59}{10} = 0.959 = 95.9\% \qquad \textit{answer}$$

Examples of the wheel and axle include: the windlass or drum hoist, the pedal-and-sprocket system of a bicycle, automobile steering wheels, and hand drills.

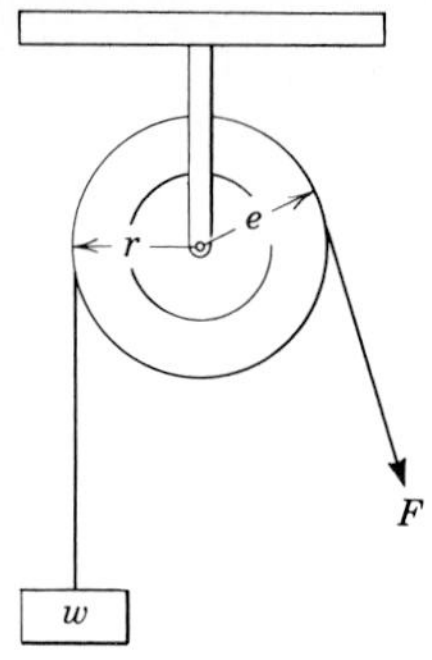

Fig. 7.18 The single fixed pulley affords a change in direction of the applied force, but does not change the mechanical advantage.

7.21 ■ Pulleys and Pulley Systems

A pulley is merely a continuous lever which has equal lever arms. As Fig. 7.18 shows, the only advantage of a *single fixed pulley* is whatever benefit might result from a change of direction. Since the effort arm *e* equals the resistance arm *r*, the theoretical mechanical advantage is

$$\underset{\text{(single fixed pulley)}}{\text{TMA}} = \frac{e}{r} = 1 \tag{7.19}$$

Even though they multiply neither force nor speed, single *fixed* pulleys are frequently used for the convenience of direction change as pointed out above.

The only advantage of a single fixed pulley is to change the direction of the effort force.

A *single movable pulley,* diagramed in Fig. 7.19 is frequently inconvenient as concerns the direction of application of the effort, but the following analysis shows it to have a TMA of 2. To move the load *w* up a distance *h*, both the supporting cords must be shortened by the amount *h*. This result is obtained by moving the applied force *F* a distance equal to 2*h*. Applying the principle of work, we write

$$Fs = wh$$
$$F \times 2h = wh$$

from which
$$\frac{w}{F} = \frac{2h}{h} = 2$$

Consequently
$$\underset{\text{(single movable pulley)}}{\text{TMA}} = 2 \tag{7.20}$$

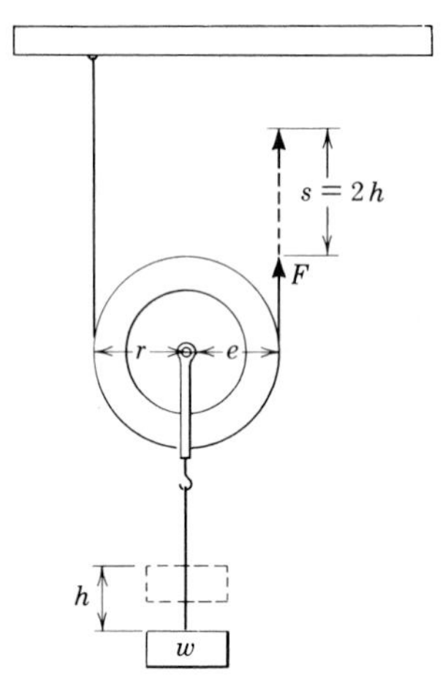

Fig. 7.19 The single movable pulley provides a TMA of 2. However, by itself it sometimes results in an inconvenient direction for the applied force.

7.22 ■ Compound Pulley Arrangements—Block and Tackle

In order to secure both the advantage of multiplying force or speed and the convenience of a selected direction of application of the force, pulleys are usually arranged in combinations called a *block and tackle*. The pulleys, which are commonly called *sheaves* (pronounced ''shivs''), are arranged two or more to a block. The blocks may be made of steel or wood and the sheaves are fitted to turn on a common lubricated shaft. Standard types are fitted with ordinary sleeve bearings, but those subjected to heavy-duty or high-speed operation have ball or roller bearings. Figure 7.20*a* is a diagram of a block and tackle with four supporting ropes to the movable block.

In order to illustrate the action of a block and tackle, Fig. 7.20*b* shows the blocks separated. Here there are three sheaves on each block and you can trace the path of the rope or cable over each one. Six strands support the movable block and the load. Assuming the pulleys to be frictionless, the tension throughout the rope or cable is constant, and each pulley permits a reversal of direction of the rope. With the arrangement shown (Fig. 7.20*b*) it can be seen that a force *F* on the ''hauling part'' contributes an equal force *F* in *each of the six strands which support the movable block*. Consequently the ratio $w/F = 6$, and in general,

The block-and-tackle has a centuries-old history, dating back to the earliest days of sailing ships.

$$\underset{\text{(block and tackle)}}{\text{TMA}} = \begin{array}{l}\text{No. of strands supporting or}\\ \text{pulling on the movable block}\\ \text{(and the load)}\end{array} \tag{7.21}$$

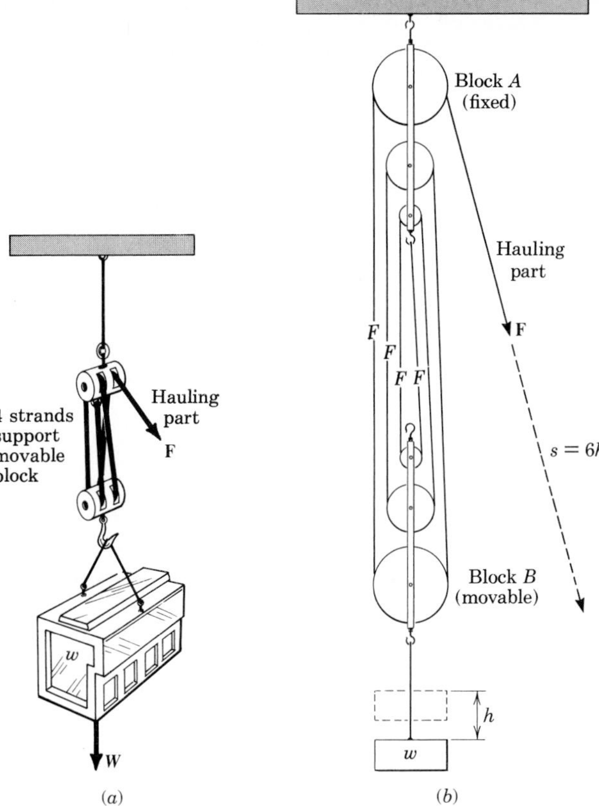

Fig. 7.20 Block and tackle. (*a*) Sketch showing a block and tackle in use. (*b*) Schematic, "exploded view" showing all the strands of rope and the forces acting in order to calculate the TMA. In this case, TMA = 6.

The same result can be obtained using the principle of work.

It should be noted that it would be possible to obtain a TMA of 7 from the block and tackle of Fig. 7.20*b* if the rope were attached initially to block *B* instead of to block *A*. What might be the disadvantage of this arrangement?

7.23 ■ The Chain Fall, or Differential Pulley

Wherever heavy machinery has to be handled in the machine shop, auto shop, or boat yard, there is likely to be found suspended from an overhead beam or monorail a device called a *chain fall* or *differential pulley*. Reference to the diagram of Fig. 7.21 reveals that a chain fall is actually a combination of a wheel and axle and a single movable pulley. An endless chain runs over the wheel and axle (R, r) and the movable pulley P, which is attached to the load at the bottom. The wheels of radii R and r turn together on the same axis and are slotted to fit the chain links so that no slippage will occur. Such an arrangement will cause friction, of course, so that the AMA of such a device is considerably less than its TMA. However, even with frictional losses, the AMA of chain falls can be made quite large, as will be shown.

Let the force F pull the chain down far enough to turn the fixed wheel and axle through one revolution. This will be a distance of $2\pi R$, and the work done in one revolution will be $2\pi RF$. The length of the chain unwound off the small wheel of radius r in one revolution will be $2\pi r$. The load w will therefore be raised a distance equal to $\frac{1}{2}(2\pi R - 2\pi r)$ or $\pi(R - r)$. The work done on the load will be $w\pi(R - r)$. Neglecting friction and applying the principle of work, we write

$$2\pi RF = w\pi(R - r)$$
$$\text{work input} = \text{work output}$$

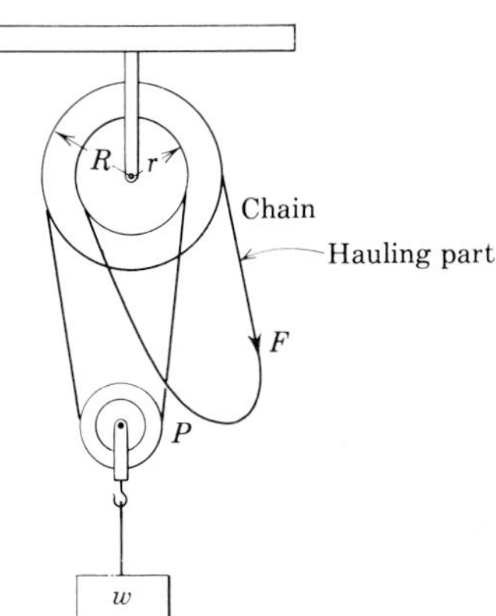

Fig. 7.21 The chain fall, or differential pulley.

The chain-fall, or differential pulley, is found in machine shops and auto shops. Today, however, most of them are energized by electric motors.

Rearranging gives

$$\frac{w}{F} = \frac{2\pi R}{\pi(R - r)} = \frac{2R}{R - r}$$

or

$$\underset{\text{(chain fall)}}{\text{TMA}} = \frac{2R}{R - r} = \frac{2D}{D - d} \tag{7.22}$$

The smaller the difference between the radii R and r, the greater the TMA becomes. In some models, friction is purposely planned to be great enough to prevent the load from running down when the force F is released. In other forms of the device a ratchet or catch holds the chain from running backward.

7.24 ■ Pulleys, Belts, and Chains Used to Transmit Torque and Power

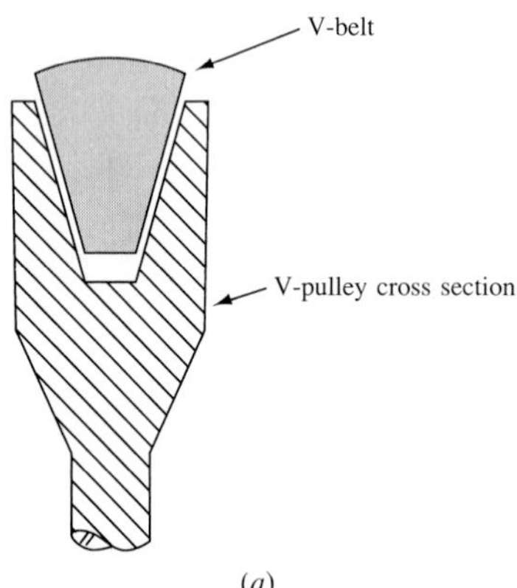

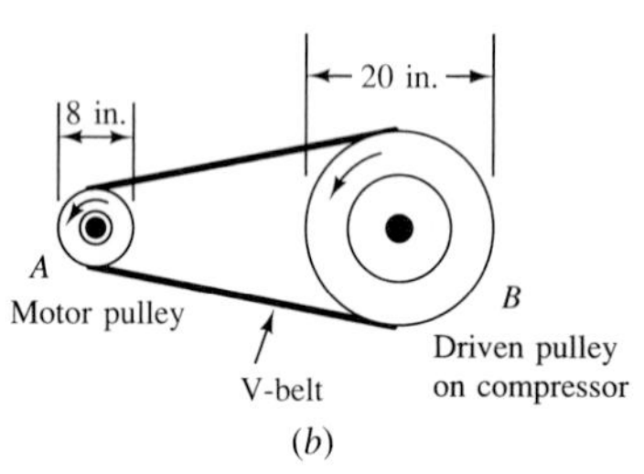

Fig. 7.22 V-belt drives. (*a*) Cross-sectional view of a V-belt in the groove of a V-pulley. Note that the belt contacts the pulley groove on the slanting sides, not on the bottom of the groove. (*b*) Diagram of a pulley-and-V-belt system suitable for driving an air compressor.

In many machines the driving motor or engine is connected to the machine to be driven by means of *pulleys* and *V-belts,* or by a chain-and-sprocket drive. The designation *V-belt* comes from the shape of the cross section of the belt (see Fig. 7.22*a*). The relative sizes of belt and pulley groove are so selected that the belt does not touch the bottom of the groove. The greater area of contact afforded by the sides of the V is utilized for the friction required to prevent slipping.

A V-belt drive arrangement such as might be used to drive an air compressor is diagramed in Fig. 7.22*b*. Note that both pulleys will turn *in the same direction*. The motor (driving) pulley A turns at motor speed, usually either 1150 or 1750 rev/min, depending on the type of motor and the nature of the electric power being used. Pulley B, on the air compressor, has a diameter (or radius) 2.5 times that of pulley A, and therefore the TMA of the system is 2.5. The *speed reduction* is therefore 2.5:1. This arrangement would be used when it is desired to multiply turning moment (technically called *torque* and discussed in detail in Chap. 9) at the sacrifice of speed. The smaller the driving pulley, relative to the driven pulley, the greater the torque multiplication. Reversing the procedure and making B the driving pulley would result in a *speed multiplication* of 2.5:1 at the sacrifice of torque. *Rotational speed* is expressed in revolutions per minute (rpm).

In general, for pulley and belt drives, in terms of pulley diameters,

$$\underset{\text{(belt-pulley drives)}}{\text{Speed reduction ratio}} = \frac{d}{D} \tag{7.23}$$

where D is the diameter of the driven pulley.

For heavy machinery which requires a great deal of torque for its operation, pulleys may have multiple grooves, each groove on the driving pulley being connected to a corresponding groove on the driven pulley by a V-belt (see Fig. 7.23). On lathes, drill presses, and similar machines where it is frequently desired to change speed ratios, a set of *step pulleys* is provided. These commonly have three or four matched pairs of pulleys, permitting changing the belt from one pair to another to get the desired speed ratio (see Fig. 7.24).

Illustrative Problem 7.10 A small air compressor is being driven by a belt-and-pulley arrangement like that shown in Fig. 7.22*b*. The motor pulley at A has a diameter of 6 in., and compressor pulley at B has a diameter of 16 in. If the motor speed is 1750 rev/min, find the rotational speed of the compressor.

Power transmission is often accomplished by means of pulleys, belts, and flexible chains.

Solution From Eq. (7.23), the speed reduction ratio in this case is

$$\frac{d}{D} = \frac{6 \text{ in.}}{16 \text{ in.}} = 0.375$$

The rotational speed of the compressor is therefore,

$$0.375 \times 1750 \text{ rev/min} = 656 \text{ rev/min} \qquad \textit{answer}$$

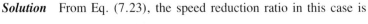

Fig. 7.23 V-belt drives for heavy machinery feature pulleys with multiple grooves to accommodate several V-belts in order to transmit the large torques required. (General Motors and Gardner-Denver Co.)

Chain drives If, in Fig. 7.22*b*, *sprockets* were used instead of pulleys, and a *link chain* instead of a V-belt, a chain drive would result. In terms of the number of teeth on the sprockets

$$\underset{\text{(chain drive)}}{\text{Speed ratio}} = \frac{N}{n} \tag{7.24}$$

where N and n refer to the large and the small sprockets, respectively. Chain drives have the advantage of being able to transmit greater torque without slippage. Their disadvantages include higher cost and somewhat noisier operation. Most bicycles and motorized cycles are equipped with sprocket-and-chain drives.

Link-Belt Drives These combine the flexibility and low noise level of V-belts with the nonslip characteristics of chain drives.

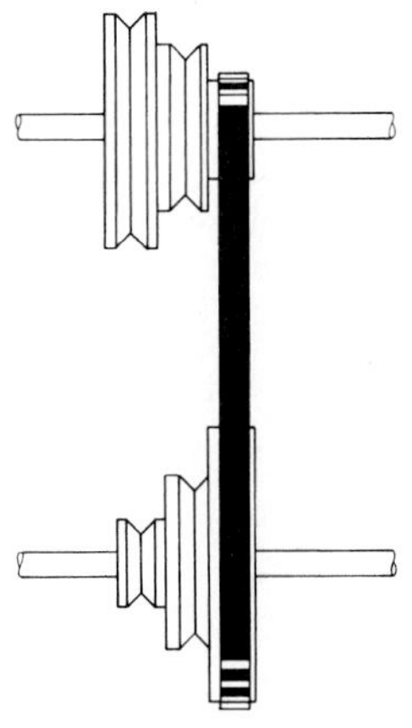

Fig. 7.24 A step-pulley and V-belt arrangement for operating a lathe or a drill press. This type of drive is being replaced by variable-speed motors in modern machine shops.

7.25 ■ Gears and Gear Combinations

In cases where absolutely no slippage can be tolerated and in situations where large torque must be transmitted at low speeds, pulleys and belts are replaced by gears and gear combinations. There is a tremendous variety of gears of all sizes and shapes, but they are ordinarily classified under four general headings: *spur gears, bevel gears, worm gears,* and *helical gears* (see Fig. 7.25). As a basic machine, a gear is just a wheel or pulley with teeth (that will mesh with teeth on a similar wheel) cut on the outer rim.

(a)

(b)

Fig. 7.25 Gears and gear boxes. (a) Huge spur-and-pinion gear combination for a very large press. (b) Bevel gear and helical gears in a speed reducer. (c) Worm gear combination in a speed reducer. Note the roller thrust bearings. (The Falk Corporation)

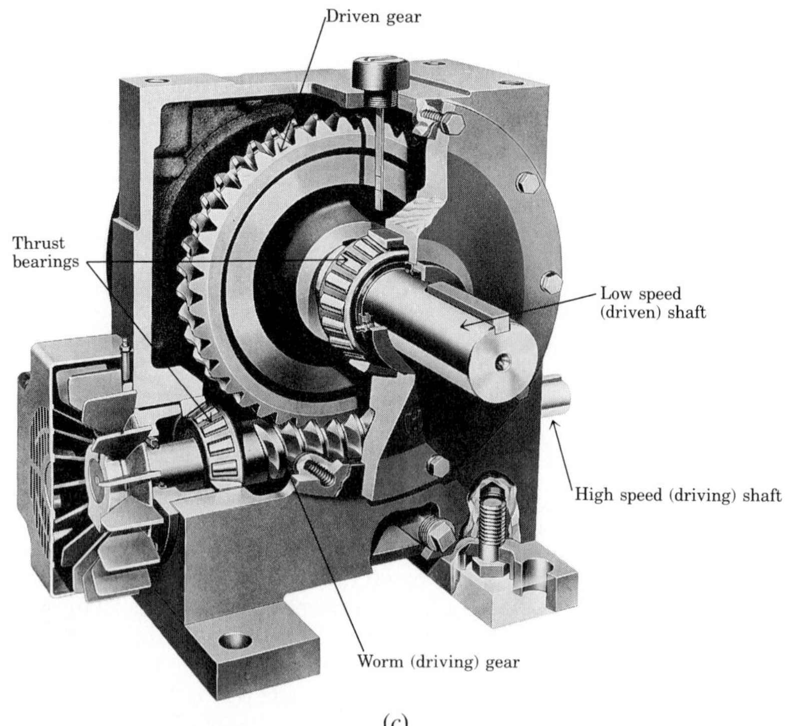

(c)

Spur Gears Spur gears are used to transmit rotary motion between parallel shafts. The teeth are cut parallel to the axis. Timing gears of engines and drive gears for lathes and presses are examples. When two spur gears are in mesh, a situation very similar to that of two pulleys with V-belt drive results. The significant difference is that the gears rotate in opposite directions, while belt-connected pulleys rotate in the same direction. The speed-reduction ratio (called *gear ratio* in this case) can be expressed either as the ratio of the diameters of the gear wheels or more commonly as the ratio of the number of gear teeth on one gear to the number on the other. Let N be the number of teeth on the large (driven) gear (see Fig. 7.26) and n the number on the small (driving) gear. The speed ratio is as follows:

Industries, and in fact, our daily lives, are so dependent on gears that the word itself is in everyone's vocabulary. We get "geared up" for a difficult task. We say, "I'm in high gear today"; or, in reference to an over-garrulous person, "His mouth and his brain are not in gear".

$$\underset{\text{(spur gears)}}{\text{Speed ratio}} = \frac{\text{rev/min driving gear}}{\text{rev/min driven gear}} = \frac{N}{n} \qquad (7.24')$$

With all types of gearing arrangements a rotational *speed increase* is accompanied by a *torque loss*; and a rotational *speed decrease* yields a *torque gain*.

Or, the rotational speeds are *inversely proportional* to the number of teeth on the gears. It is emphasized that, just as with pulleys, *if input power is constant,* a speed gain means a corresponding torque loss; and conversely, a speed reduction results in a torque increase. Both "speed gears" and "reduction gears" are common in industrial practice.

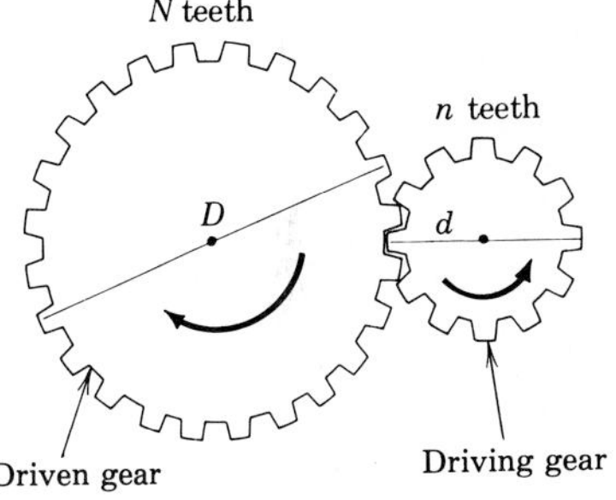

Fig. 7.26 Gear ratio of spur gears.

Gear Trains If it is desired to effect a speed reduction (or torque multiplication) greater than can be conveniently obtained by one pair of gears, a *gear train* is used. Figure 7.27 shows one simple arrangement. The overall gear ratio of such a train can be obtained by analysis as follows. Consider gears A and B as constituting one system. The gear ratio of this system is N_1/n_1. Gears C and D constitute a second system whose gear ratio is N_2/n_2. Since gear C turns at the same speed as gear B (both are keyed to the same shaft), the overall speed ratio formula is

$$\underset{\text{(gear train)}}{\text{Speed ratio}} = \frac{N_1}{n_1} \times \frac{N_2}{n_2} \times \cdots \times \frac{N_n}{n_n} \qquad (7.25)$$

As above, Eq. (7.25) yields the ratio

$$\underset{\text{(gear train)}}{\text{Speed ratio}} = \frac{\text{rev/min driving gear}}{\text{rev/min driven gear}}$$

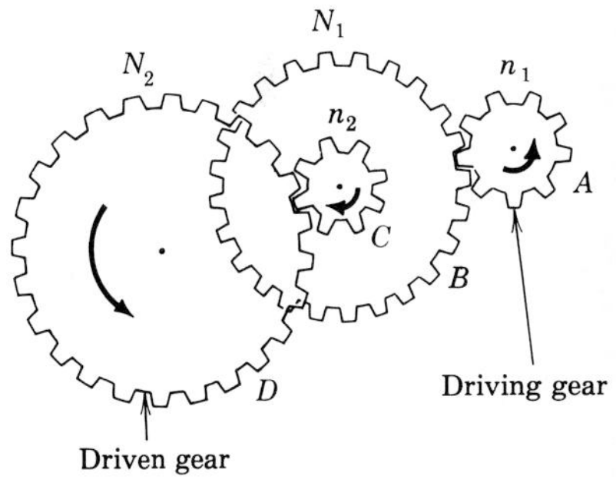

Fig. 7.27 Gear ratio of a gear train of spur gears for speed reduction (torque multiplication). If speed multiplication is desired, D could be made the driving gear and A the driven gear.

Worm Gears Many "prime movers" (engines, motors, turbines) are efficient only at high rotational speeds. For example, standard electric motor speeds range from 1750 to 3450 rev/min; internal combustion (gasoline and diesel) engines operate best in the range from 1500 to 6000 rev/min; and steam and hot-gas turbines turn at speeds of 3600 rev/min and higher. On the other hand, many *driven* machines must be operated at much slower speeds, from a few hundred revolutions per minute down to as little as 30 or 40 rev/min. *Speed-reducer* mechanisms, often featuring *worm gears*, are used for bringing high rotational speeds of "prime movers" down to the range required by the driven machines. Figure 7.25*c* is a photo of a speed reducer, with housing partially cut away, showing the worm gear, the driven gear, the thrust bearings, and other details.

From Fig. 7.25*c*, it can be seen that for each revolution of the driving (input) shaft, the "worm" gear will advance the driven gear by just one tooth. If the driven (output) gear has N teeth, N revolutions of the input shaft are required to produce one revolution of the output shaft. Therefore, in terms of rotational speed,

$$\underset{\text{(worm gear/speed reducer)}}{\text{rev/min}_{\text{output}}} = \frac{\text{rev/min}_{\text{input}}}{N} \qquad (7.26)$$

Why are thrust bearings required in worm-gear speed reducers?

Again, N is the number of teeth on the driven gear.

The number of helical grooves or threads on the input shaft (worm gear) does not affect the gear ratio. They (the input-shaft threads) must, of course, be spaced and machined for an exact fit with the helical cut of the teeth on the driven (output) gear.

7.26 ■ The Compound Machines of Industry

The principle of work (work-energy theorem) governs all machine operation. No machine can multiply *energy*. In fact, all machines exhibit energy losses, and the work output never equals the energy input.

The basic law that "you can't get something for nothing" certainly holds true for machines. If speed is to be multiplied, torque must be sacrificed, and conversely. The power output of a machine can never be greater than the power input; in fact, because of friction, the output is always less than the input. Machines may also have losses other than those due to friction, such as those from heat, magnetism, and similar effects.

The Principle of Work for machines can be stated as follows:

Energy (or power) output = energy (or power) input − energy (or power) losses

On actual industrial jobs the basic and simple machines analyzed in this chapter are rather infrequently used alone. Rather, they are combined in many ingenious and complex ways to produce the *compound machines* of industry. It is beyond the scope of a physics book to treat compound machines in detail, but in closing this discussion let us consider the power shovel of Fig. 7.28 as a typical example of a compound

Fig. 7.28 A complex industrial machine. The huge power shovel shown here is used in surface (strip) mining. Complex as it is, it is merely a combination of the basic and simple machines studied in this chapter. Note the relative size of the person in front of the truck cab.

machine. Notice the horns on the shovel itself. They are *wedges*—a form of inclined plane. The boom is a huge *lever*, operating with tremendous mechanical advantage. The forces which operate the boom are carried by the cables which, with the sheaves over which they pass, constitute a huge *block-and-tackle* system. Inside the cab the drum on which the cable is wound constitutes a *wheel and axle*. The wheel is a large *gear*, driven by a *gear train* from the source of power. The operator controls the machine by a set of *levers* at his seat. The cab and boom are rotated by a ring-and-pinion gear arrangement.

You can see from this example how the design, construction, and operation of complex machines depend on a thorough understanding of the principles of basic and simple machines.

QUESTIONS AND EXERCISES

1. Diagram the three classes of levers and list five practical applications (some from the human body) of each.

2. If a simple machine such as a lever is operated "backward," what change occurs in the AMA? in the efficiency? (Friction negligible.)

3. A screw jack is rusty and corroded. Before starting to use it you decide to lubricate it thoroughly. How does this affect its efficiency? its theoretical mechanical advantage (TMA)? its actual mechanical advantage (AMA)?

4. Based on some library research write a brief paper on the theory of friction in machine parts. Touch on all the following: (*a*) the effect of smoothness of surface, (*b*) the choice of alloys for sleeve bearings, (*c*) the effect of relative velocity of the rubbing parts, (*d*) the general theory of lubrication in reducing friction, (*e*) the "oil-wedge" theory of lubrication of sleeve-type bearings, and (*f*) the advantages of ball and roller bearings over sleeve-type bearings.

5. Some models of automatic transmissions for autos have small radiators (heat exchangers) through which the transmission oil is circulated to cool it. Why does the oil get so hot that cooling is necessary?

6. *A* and *B* are two bicycle riders. *A* has a 10-speed bike which he shifts into low gear as they approach a steep hill. *B* keeps hers in "regular" coaster-brake gear. They start even and race to the top of the hill, finishing in a dead heat. (*a*) Who pedaled faster? (*b*) Who applied more force on each down stroke of the pedal? (*c*) Who generated the greater power output? Who did the most work? (Assume same friction for both, and equal total weights.)

7. A block-and-tackle system is to be used to pull a stalled pickup truck out of the mud. There are two blocks of three sheaves each and plenty of strong wire rope. Diagram the arrangement which (*a*) would give the greatest TMA; (*b*) would probably be used in practice because of convenience. (There is a convenient tree to "anchor" one end of the equipment.)

8. Refer to the diagram of the differential pulley on p. 161. As force F pulls on the hauling part for one full revolution of the differential pulley, a length of chain $2\pi R$ passes over the large pulley. The small pulley is fixed to the large pulley so that the two turn as one, and in one full revolution a much shorter length of chain, $2\pi r$, goes up and over this pulley. Explain the "mystery" of the "ever-lengthening" chain.

9. The sketch of Fig. 7.29 shows a device called a turnbuckle. One bolt is right-hand threaded and the other is left-hand threaded. The yoke can be turned by hand or by any convenient tool, pulling the two bolts into the center at the same rate and thus putting tension on the cable, C. If the pitch of the screw(s) is p, and the diameter of the yoke (to which one's hand applies the forces $F - F$) is D, what is the TMA of the turnbuckle?

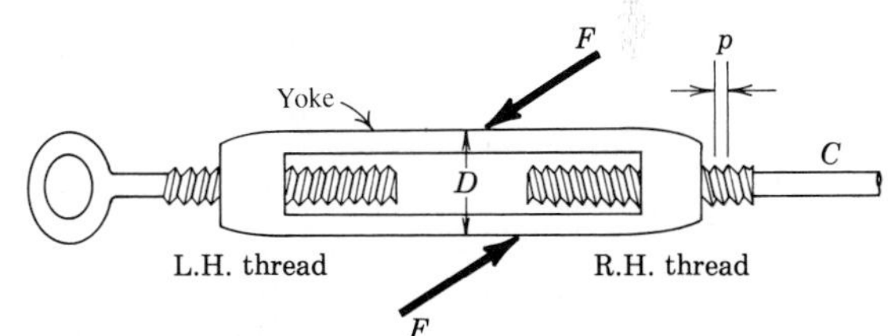

Fig. 7.29 Diagram of a turnbuckle (Exercise 9).

10. Much use is made in modern factories of industrial robots. Although they are electronically controlled, they are actually mechanical devices, incorporating many of the basic and simple machines discussed in this chapter. In an auto assembly-line robot, where and for what purpose might levers be used? Pulleys? Cams? Gears and gear trains?

11. Why are the rollers in thrust bearings in the form of truncated cones? In what kinds of machinery are thrust bearings used? Specifically, what is their purpose? Use a sketch in your explanation.

PROBLEMS

NOTE *Drawing a careful diagram for each problem and showing the various forces as vectors will assist you in understanding and solving the problem.*

Group One

1. A 185-lb man sits on a seesaw 4 ft from the fulcrum. Where must an 85-lb boy sit to balance the seesaw?

2. A resistance of 800 N must be overcome through a Class I lever whose resistance arm is 18 cm long and whose effort arm is 5 cm long. (*a*) What is the magnitude of the effort force? (*b*) For what purpose would such a lever be used?

3. A shovel is used to lift 20 kg of concrete mix. The worker's right hand is at the end of the shovel handle 1.3 m from the center of the load. The left hand acts as a fulcrum 45 cm from the center of the load. Neglect the weight of the shovel and find the force exerted by each hand.

4. The wheelbarrow of Fig. 7.4*b* contains 200 lb of gravel. If r is 18 in. and e is 4 ft, what total force F must be applied at the handles?

5. A pair of wire-cutting pliers is gripped in a person's hand so that the force applied is 20 cm from the pivot point of the jaws. A wire to be cut is 1.6 cm from the pivot axis. If a force of 300 N is exerted by the gripping hand as the wire is cut, what is the resistance offered by the wire?

6. If the human forearm acts as a Class III lever, as in Fig. 7.5, calculate the force F which must be applied by the muscles in order to lift a bucket of paint weighing 18 lb. Assume the arm muscles apply the force at a point 2.25 in. from the elbow, and that the distance from elbow to the bucket bail in the hand is 14 in.

7. What force is required to pull a loaded sled, mass 2500 kg, across hard-packed snow (level) if the coefficient of sliding friction between runners and snow is 0.035?

8. A windlass is one form of a wheel and axle. With the conditions diagramed in Fig. 7.30, find the force F that must be applied tangent to the outer rim of the wheel in order to hoist at constant speed a load whose mass is 250 kg, with a rope wound around the axle (drum). Wheel diameter is 0.95 m; drum diameter is 30 cm (answer in newtons).

9. A gear train like that shown in Fig. 7.27 is being actuated by gear A turning at 1750 rpm. Gear A has 10 teeth; gear B, 25 teeth; gear C, 12 teeth; and gear D, 30 teeth. What is the rotational speed of the shaft to which gear D is attached?

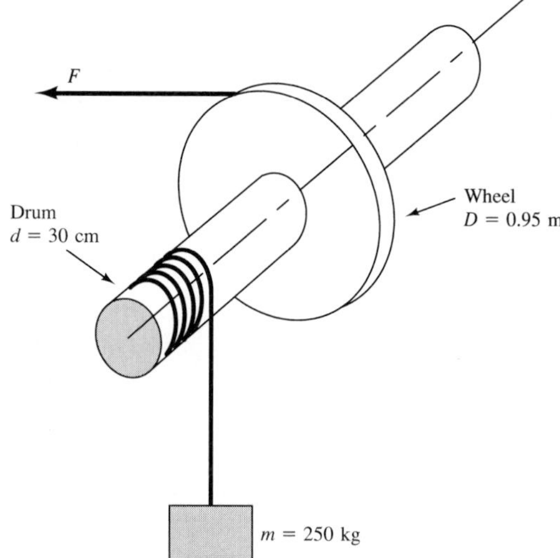

Fig. 7.30 Diagram of a windlass (wheel and axle) (Problem 8).

10. The coefficient of rolling friction, as well-inflated rubber tires roll on a macadam (blacktop) road, is known to be 0.052. On a level road at 60 mi/h, what horsepower is required from an auto's drivetrain just to cope with the tires' rolling friction? The auto weighs 3000 lb.

Group Two

11. Four sailors are 6 ft from the center of a capstan 18 in. in diameter, pushing with a force of 90 lb each. If the efficiency of the capstan is 60 percent and the sailors just weigh the anchor, how heavy is it? (Assume anchor is out of water.)

12. A windlass has a drum 15 cm in diameter and a crank 60 cm long. A man pulls up a bucket of concrete whose mass is 100 kg. What force must he exert on the crank if the device is 90 percent efficient?

13. A loaded dolly weighing 450 lb is rolled up a ramp 10 ft long, which rises 2.5 ft vertically. Neglecting friction, (*a*) what force parallel to the ramp is required? (*b*) Find the *normal* force pressing against the ramp.

14. The TMA of an inclined plane is 5.6. What percent grade is this incline?

15. The cable of a ski lift pulls up a 25° slope at uniform speed. If the total mass of chairs, skiers, and equipment is 4000 kg, what is the tension in the cable? (Neglect friction.)

16. A screw jack is being used on a house-moving job. The pitch of the screw is 0.25 in., and the handle is 3 ft long. (*a*) What is the theoretical mechanical advantage? (*b*) If the jack is only 46 percent efficient, what load could be lifted by a force of 125 lb applied to the end of the handle?

17. A small drive gear with 17 teeth, turning at 1700 rev/min, drives a larger gear with 96 teeth. On the same shaft with the large gear is a V-pulley 6 in. in diameter. Find the linear speed (in feet per second) of the V-belt which this pulley drives. Assume no belt slippage.

18. An 1800-lb sled is being pulled across level snow at uniform speed by Arctic explorers. The drag rope makes an angle (at the sled) of 28° above the horizontal. The tension in the rope is 105 lb. Find the coefficient of kinetic friction between sled runners and snow.

19. A force of 500 N is applied to the hauling part of a block-and-tackle system, and it moves through a distance of 5 m. The load, meanwhile, has moved 1 m. (*a*) At 100 percent efficiency, what load (kg) could be lifted by the 500-N force? (*b*) If the given force will lift only 200 kg, what is the efficiency of the system?

20. A block-and-tackle system is shown in schematic form in Fig. 7-31. (*a*) What is its TMA? (*b*) If it is desired to lift the load 6 ft, how far will the force F have to move? (*c*) If it is noted that a force F of 70 lb is required to lift a load w of 250 lb, what is the efficiency of the system?

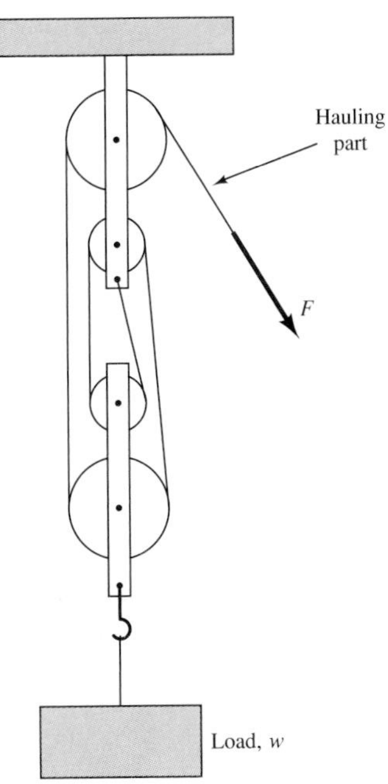

Fig. 7.31 Schematic block and tackle system (Problem 20).

21. An electric motor rated at 0.5 hp output, rotates at 3400 rev/min. It drives a V-belt from its motor pulley, the diameter of which is 4 in. Assuming no belt slippage, what is the linear speed of the belt in ft/s?

22. A mechanical automobile jack of the lever-and-ratchet type requires a force of 150 N pushing straight down for a distance of 25 cm, to cause the auto to be lifted a distance of 0.6 cm. If the jack has an efficiency of 75 percent (the ratchet device prevents run-down under load in this type of jack), how large a load (kg) can the 150 N effort lift?

23. A bicycle has a pedal sprocket with an effective diameter of 9 in., and the diameter of the driven sprocket on the rear wheel is 4 in. The bike wheels have diameters of 28 in. to the outer edge of the inflated tires. If a bike racer wants to make 30 mi/h, what must be the rotational speed (rpm) of the pedal sprocket?

24. The hoisting mechanism of an oil-well drilling rig (called the *draw works*) has an outer chain-driven sprocket of 3-ft diameter. The hoisting cable is wound around the drum, which is 10 in. in diameter. If the drive chain applies a force of 15,000 lb to the rim of the sprocket, what tension will be created in the hoisting cable?

25. The "crown block" (at the top of the derrick) and "traveling block" arrangement used in pulling oil-well casing has six sheaves per block. If it is desired to pull casing out of the hole at a linear speed of 10 ft/s, at what linear velocity must the cable be pulled in by the draw works? (NOTE: The "hauling part" comes down from the crown block to the draw works.)

26. A chain fall is being used to lift a 1750-lb aircraft engine from the overhaul stand to the nacelle for installation. The radii of the large and small wheels are, respectively, 18 and 17 in. If the efficiency is 40 percent, what force F must be applied to the hauling part of the chain?

27. A refrigeration compressor comes equipped with a 30-cm diameter pulley and is designed to operate at 500 rpm. What should be the diameter of the motor pulley if the motor speed is 1750 rpm?

28. An auto differential (at the rear axle) has a gear ratio of 3.6:1. The transmission is in low gear, where the ratio is 7.6:1. How many turns does the engine make for each revolution of the rear wheel?

29. A 1500-kg auto is moving at 80 km/h on a level concrete road when the driver slams on the brakes, seeing danger ahead. Assuming a straight skid on dry pavement, how far will the car skid before coming to a stop?

30. A 20-ton boat, mounted on its hull sled (weight 1.5 tons), is ready for launching at a boat yard. The coefficient of limiting (static) friction between the hull sled runners and the (greased) launching ways is 0.25. At what downward angle with the horizontal must the launching ways be slanted so that sliding will just begin?

Group Three

31. A 150-kilogram cake of ice is being pulled steadily up a steel ramp to the loading dock. The slope is inclined 25° to the horizontal. The force required (parallel to the ramp) is noted to be 820 N. Find μ between ice and steel.

32. A truck engine is delivering 195 hp. Assuming a 25 percent loss between engine crankshaft and rear wheels, find the coefficient of rolling friction for rubber tires on concrete if the loaded truck (total weight 15 tons) is making 16 mi/h up a 6 percent grade. (Neglect air resistance.)

33. A steel loading chute slants downward at 45° from a warehouse to a freight siding. It is 60 ft long. The coefficient of sliding friction between cardboard and steel is 0.45. What is the speed of sliding cardboard cartons as they arrive at the bottom of the chute? Work the problem two ways: (*a*) from acceleration considerations, and (*b*) using energy principles.

34. An auto engine is propelling a 2000 kg car up a 20 percent grade at 20 km/h. Assuming an overall efficiency of 65 percent from crankshaft to rear wheels and pavement, (*a*) what power (kW) is the engine developing? (*b*) What is the thrust at each rear wheel in newtons?

35. The system shown in Fig. 7.32 consists of a 10-kg mass acted on by the force of gravity, a frictionless pulley and cord, and a 4-kg hardwood block on a hardwood ramp inclined upward at an angle of 35°. What is the acceleration of the system? (See Table 7.1 for μ_k.)

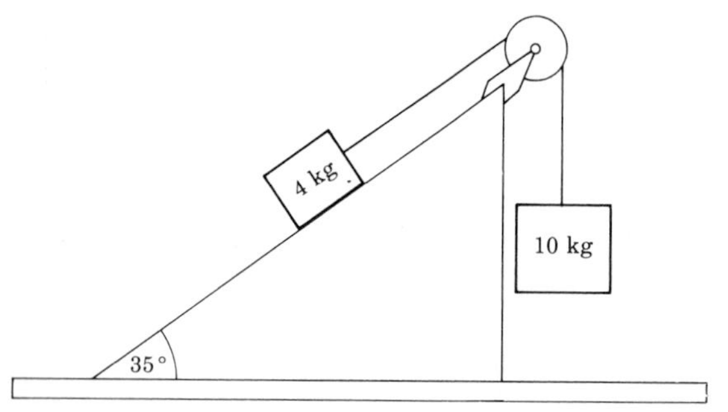

Fig. 7.32 Acceleration on an inclined plane, when friction is present (Problem 35).

36. The diagram of Fig. 7.33 shows a rather simple compound machine, in schematic form, consisting of an inclined plane and a block-and-tackle. (*a*) What is the TMA of the inclined plane, taken alone? (*b*) of the block-and-tackle, taken alone? (*c*) of the combination? (*d*) What force F, in newtons, is required to pull the load up the incline at a steady speed? (*e*) What is the AMA of the combination, given the data on the diagram?

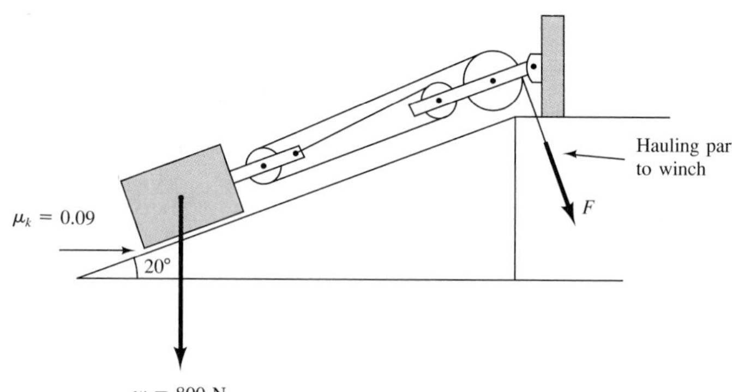

Fig. 7.33 A simple compound machine combining an inclined plane with a block-and-tackle system. (Problem 36)

37. The desired *linear speed* for a conveyor belt is 5 ft/s. It is driven by an 8-in.-diameter roller, over and around which the belt runs without slippage. The roller itself is turned by a V-belt drive from an electric motor rotating at 1750 rev/min. The motor pulley has an effective diameter of 4 in. What should be the diameter of the pulley on the belt roller?

38. A screw jack is designed for an efficiency of 40 percent. The pitch of the screw is 6 mm. The jack handle provides a radius of 70 cm out to the point of application of the effort force. What load (kg) can be lifted by applying a force of 300 N at the end of the jack handle?

39. A speed-reducer worm gear (see Fig. 7.25*c*) has an output shaft gear with 44 curved teeth. The input shaft gear is turned by a steam turbine rotating at 3600 rev/min. What is the rotational speed of the output shaft of the speed reducer?

40. A crate of weight w is to be dragged across a level floor at constant velocity by a force F, acting at an angle θ above the horizontal. If the coefficient of kinetic friction between crate and floor is μ_k, derive an expression for F in terms of the other variables.

CHAPTER

8 CIRCULAR MOTION AND SATELLITE MECHANICS

Nearly all complex machines have one or more parts or systems that move with *rotational motion*—motion in which all the particles of a rigid body move in circular paths around a fixed axis. Wheels, gears, pulleys, ships' screws and aircraft propellers, computer disks, engine crankshafts, centrifuges, clothes washers and dryers, cement mixers, and ferris wheels are just a few examples of machines that use rotational motion, with major or minor components spinning on a fixed axis.

Most machines have one or more components in rotational motion. Some machine components may experience both rotational and translational motion simultaneously.

This chapter and the one following will deal in considerable detail with particles, objects, and machine components moving in curved, mostly circular, paths.

At the outset we identify two kinds of circular motion: (1) Pure *rotation*, in which a body spins or rotates around an axis through the body itself, as does a phonograph record or a turbine rotor (Fig. 8.1); and (2) circular motion of a body *revolving*

Distinction between a *rotating* body and a *revolving* body.

Fig. 8.1 Rotor from a large steam turbine. When the turbine rotates, every stage, every vane, and every particle is in motion in a circular path (*rotation*) around the fixed axis. (Westinghouse Corporation)

Fig. 8.2 The earth *revolves* around the sun, describing a (nearly) circular path around a point or axis outside of itself. (The orbit of the earth is actually an ellipse of slight eccentricity.)

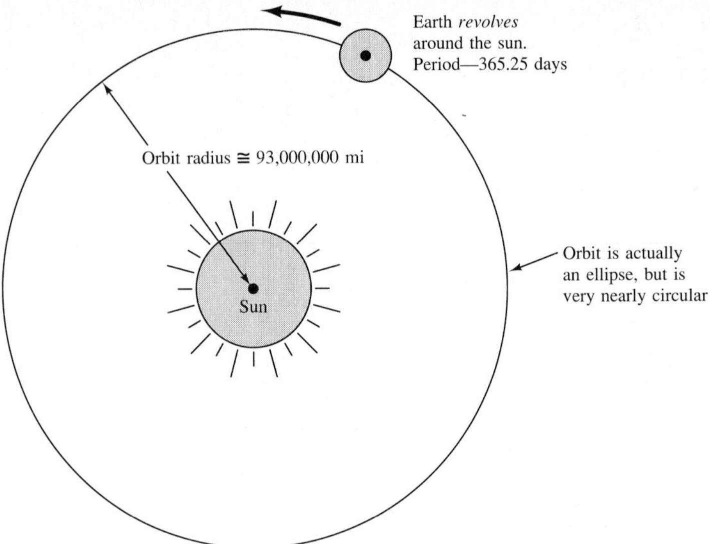

Earth *revolves* around the sun.
Period—365.25 days
Orbit radius ≅ 93,000,000 mi
Sun
Orbit is actually an ellipse, but is very nearly circular

around a point or an axis outside the body, as a ball being whirled at the end of a string or the earth revolving around the sun in its nearly circular orbit (Fig. 8.2).

CIRCULAR MOTION

8.1 ■ Rotation Contrasted With Translation

For a body to have rotary motion, two requirements must be met:

1. Every particle of the body must move in a circular path.
2. All these circles must have their centers on the same straight line. The straight line is said to be the *axis of rotation*.

In contrast, motion in which no imaginary straight line in a body changes direction as the body moves from one place to another is called *translation* (see Fig. 8.3 and Sec. 4.1).

Translation may be either *straight-line* or *curvilinear*, depending on the path of the body. The straight-line distance between the starting point and the end point of the translation is called the *displacement*.

Objects often have complex combinations of rotary motion and translation. This is true of the rotating parts of an auto engine when the car is in motion along the highway or a point on the rim of a Frisbee in flight.

8.2 ■ Angular Measurement

Since rotary motion is motion in a circular path, you should recall the geometry of the circle. The linear distance around a circle is the *circumference*, and the distance from the center to the circle is the *radius*. The angular displacement through which a radius turns if pivoted at the center is measured in *degrees* (°), and there are 360° in the total angular distance around a point (see Fig. 8.4). A portion of the circumference of a circle is called an *arc*.

Another measure of angular displacement is the *revolution*. One revolution is one complete circular trip around the center or axis of rotation. One revolution is thus equal to an angular distance of 360°. In science and technology there is still another unit of angular measure. It is larger than the degree but smaller than the revolution, and is called the *radian* (abbreviated rad).

Fig. 8.3 Translation and rotation compared. (*a*) Diagram of an engine cylinder, piston, and connecting rod showing how the connecting rod and crankshaft convert the back-and-forth (*reciprocating*) motion of the piston into the rotary motion of the crankshaft. (*b*) Cutaway view of a four-cylinder auto engine, showing various parts, some of which are in rotation when the engine is running, and some of which are in translational motion. Some, like the connecting rod, have complex motions. (Ford Motor Co.)

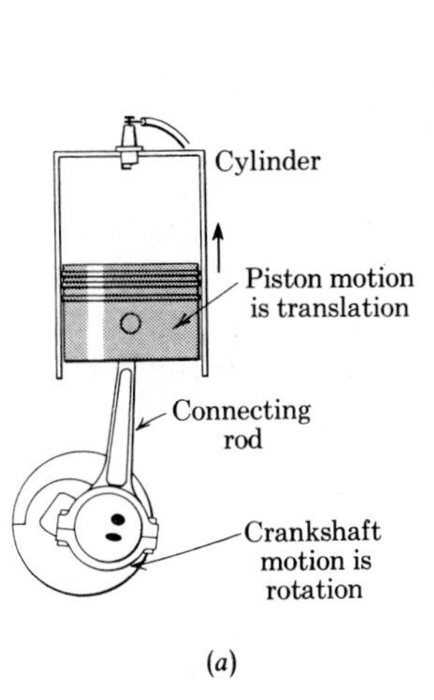

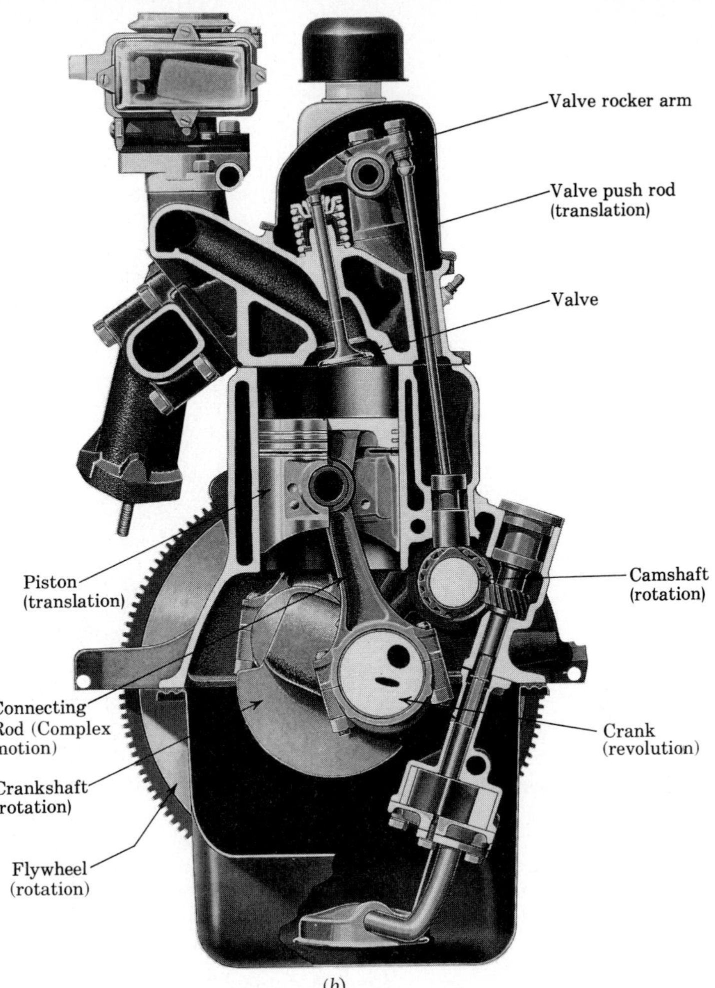

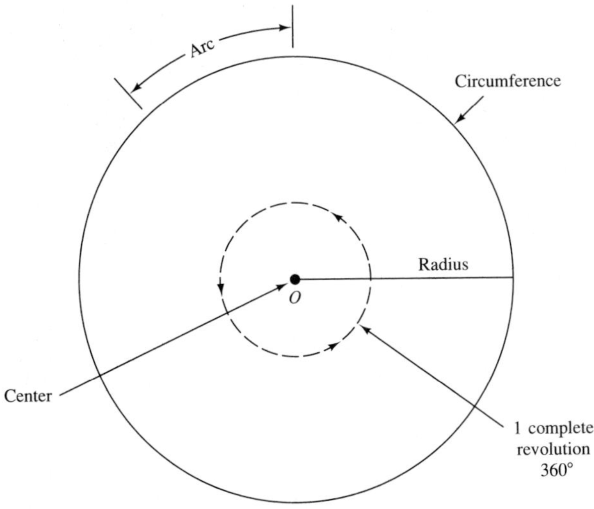

Fig. 8.4 Circle relationships. The circumference c = $2\pi \times$ radius. There are 360 degrees (°) in the angular distance around a point, or in one revolution.

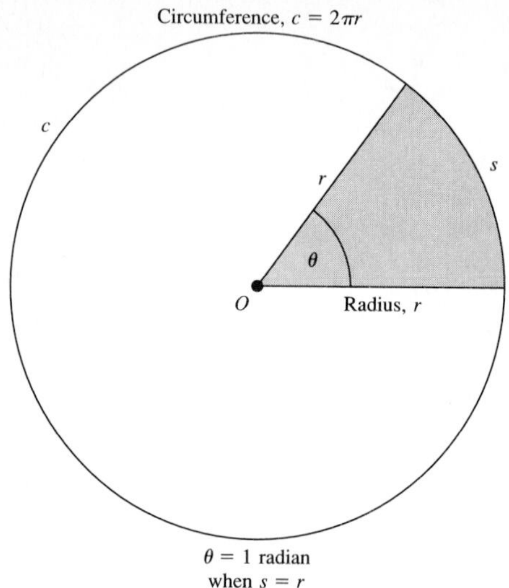

Fig. 8.5 Diagram to define the *radian*. $\theta = 1$ radian when s, the arc subtended by the angle θ, equals the radius r. 2π radians = 360°; π radians = 180°.

The radian is the measure of an angle whose arc along the circumference is just equal to the radius.

Definition of the *radian,* the scientific and engineering unit of angular measure.

The angle θ (Fig. 8.5) is defined as one radian when s, the arc length, is equal to r, the radius. Since $c = 2\pi r$, we have 2π rad = 360°, and 1 rad = 360°/2π = 57.3°

$$
\begin{aligned}
1 \text{ rev} &= 360 \text{ deg } (^\circ) \\
&= 2\pi \text{ rad} \\
1 \text{ rad} &= 57.3^\circ
\end{aligned}
$$

Since $\theta = 1$ rad when $s = r$ (by definition), $\theta = 2$ rad when $s = 2r$, and in general,

$$\theta = \frac{s}{r}$$

or,

$$s = r\theta \tag{8.1}$$

$$\text{Arc length} = \text{radius} \times \text{angular displacement in rad}$$

This is the basic equation for radian measure. It is one of the fundamental equations for motion in a circular path. The radian, as a unit of angular measurement, is a dimensionless quantity, since it is the ratio of two lengths. Of the three units of angular measure—the revolution (rev), the degree (°), and the radian (rad), the radian is the standard measure used by scientists. Mechanical engineers often use the revolution as a preferred unit.

8.3 ■ Angular Velocity

If rotary motion is *uniform*, i.e., at constant angular speed, it is often described in terms of revolutions per minute (rev/min or rpm) or revolutions per second (rev/s or rps). For many industrial purposes, these measures of angular velocity are perfectly satisfactory. However, we often need to know the *linear* (*tangential*) *speeds* of points that are in rotary motion, e.g., tip speeds of airplane propellers or rim speeds of high-speed turbine rotors. Angular velocities in revolutions per minute give no direct information on tip speed. Engineers and technicians therefore often express angular velocities in *radians per second*, denoted by the symbol ω (the Greek letter omega). Since $\theta = s/r$ from Eq. (8.1), if we divide both sides by t, we obtain

$$\frac{\theta}{t} = \frac{s}{rt}$$

Now θ/t is angular displacement per unit time, or the *angular velocity*, ω, measured in rad/s. And s/t is the linear speed, v, of a point on the circle. Consequently,

$$\text{Angular velocity} = \frac{\text{linear speed}}{\text{radius}}$$

As an equation,

Relationship between angular velocity in rad/s and linear (tangential) speed of a point or particle.

$$\omega = v/r$$

or

$$v = \omega r \tag{8.2}$$

Equation (8.2) is the basic relationship between the linear (tangential) speed of a point on a rotating object and the angular velocity of the object. Either angular velocity (ω) or linear speed (v) can be obtained if the other is known and if the radius can be measured.

In SI-metric units,

$$\omega\left(\frac{\text{rad}}{\text{s}}\right) = \frac{\text{m/s}}{\text{radius (m)}}$$

In English units,

$$\omega\left(\frac{\text{rad}}{\text{s}}\right) = \frac{\text{ft/s}}{\text{radius (ft)}}$$

You should note that angular velocity, ω, has the dimension of time^{-1} (that is, s^{-1}), since the length units on the right side of these equations cancel out.

The radian, being a ratio of two lengths, is a dimensionless quantity. Angular velocity is expressed in rad/s, and therefore it (ω) has the dimension of 1/s or s^{-1}.

Angular velocity in radians per second is related to angular velocity in revolutions per minute as follows:

$$\omega\left(\frac{\text{rad}}{\text{s}}\right) = \omega\left(\frac{\text{rev}}{\text{min}}\right) \times \frac{2\pi\ \text{rad}}{\text{rev}} \times \frac{1\ \text{min}}{60\ \text{s}}$$

or, as a formula

$$\omega\left(\frac{\text{rad}}{\text{s}}\right) = \frac{2\pi\ \text{rad/rev}}{60\ \text{s/min}} \times \frac{\text{rev}}{\text{min}} \tag{8.3}$$

Angular velocity in rev/min is often expressed as rpm, but this designation *should not be used* in problem solutions, since dimensional analysis would not be possible.

8.4 ■ Angular Displacement Related to Angular Velocity

If a rotating object turns through an angle of θ rad in a time t seconds, its *average* angular velocity is given by the relation $\omega = \theta/t$. This can be written in the form

$$\theta = \omega t \tag{8.4}$$

Equation (8.4) expresses *angular displacement* in rotary motion in terms of angular velocity and time. It is very much like Eq. (4.1″), which gives *linear displacement* in terms of average linear velocity and time ($s = \bar{v}t$), for translation.

Illustrative Problem 8.1 A bicycle wheel has a diameter of 26 in. The rider is pedaling with a translational speed of 30 ft/s. What is the wheel's angular velocity in (a) radians per second, and (b) revolutions per minute?

Solution (***a***) Assume no slippage of tire on pavement. Then the linear rim speed (tangential speed) of the wheel must be 30 ft/s. Select Eq. (8.2) which relates linear velocity, angular velocity, and radius, and write

$$v = \omega r$$

Substituting,

$$30 \text{ ft/s} = \omega \times \tfrac{13}{12} \text{ ft}$$

Solving,

$$\omega = \frac{30 \text{ ft/s}}{\tfrac{13}{12} \text{ ft}} = 27.7 \text{ rad/s} \qquad \textit{answer}$$

(***b***) To solve for the angular velocity in revolutions per minute, write Eq. (8.3) in this form:

$$\text{rpm} = \frac{\text{rev}}{\text{min}} = \frac{60}{2\pi}\left[\omega\left(\frac{\text{rad}}{\text{s}}\right)\right] \tag{8.3'}$$

Substitute the value of ω from (***a***) above, and solve:

$$\text{rpm} = \frac{\text{rev}}{\text{min}} = \frac{60 \text{ s/min} \times 27.7 \text{ rad/s}}{6.28 \text{ rad/rev}} = 265 \frac{\text{rev}}{\text{min}} \qquad \textit{answer}$$

Illustrative Problem 8.2 A cargo plane has propellers that describe arcs 4 m in diameter. What is the propeller-tip speed in meters per second when the shaft is turning 1200 rev/min?

Solution Use Eq. (8.3) and substitute known values.

$$\omega = \frac{2\pi \text{ rad/rev} \times 1200 \text{ rev/min}}{60 \text{ s/min}} = 126 \text{ rad/s}$$

Substitute this value of ω in Eq. (8.2), recalling that the radius $r = \dfrac{4 \text{ m}}{2} = 2$ m. This gives

$$v = \omega r = 126 \text{ rad/s} \times 2 \text{ m} = 252 \text{ m/s} \qquad \textit{answer}$$

Tip speeds of large-diameter rotors in fan-jet engines, steam turbines, and wind-energized generators are often a critical design problem.

NOTE: *Propeller and turbine tip speeds are often critically important. As the speed of sound (approximately 330 m/s or 1130 ft/s at standard sea-level conditions) is approached, shock waves are produced which materially decrease propeller efficiency.*

Illustrative Problem 8.3 An auto is being driven 50 mi/h. Its wheels, with tires fully inflated, have a diameter of 25 in. Find the angular velocity of the wheels in (*a*) rad/s and (*b*) rpm.

Solution

$$v = 50 \text{ mi/h} \times \left(\frac{44 \text{ ft/s}}{30 \text{ mi/h}}\right) = 73.3 \text{ ft/s}$$

(See *Note*, Sec. 4.6)

The linear rim speed (tangential speed) of the wheels (assuming no slippage) is then $v = 73.3$ ft/s.

(***a***) From Eq. (8.2), ($\omega = v/r$), substitute values and solve.

$$\omega = \frac{73.3 \text{ ft/s}}{25/(2 \times 12) \text{ ft}} = 70.4 \text{ rad/s} \qquad \textit{answer}$$

(*b*) From Eq. (8.3′), $\text{rpm} = \frac{\text{rev}}{\text{min}} = \frac{60 \text{ s/min} \times 70.4 \text{ rad/s}}{6.28 \text{ rad/rev}}$

$= 673 \text{ rev/min (rpm)}$ *answer*

ROTATIONAL KINEMATICS

The term *kinematics of rotation* is used to describe angular velocity and angular acceleration without reference to force, mass, and inertia considerations. When force, mass, and inertia *are* considered, the term *dynamics of rotation* is used. We have introduced the concepts of angular *displacement* and angular *velocity* in Secs. 8.2 to 8.4, and will now discuss some aspects of angular *acceleration*.

Kinematics, as applied to rotation, implies the study of circular motion without regard to any of the causes of the motion.

8.5 ■ Angular Acceleration

A rotating body may have a changing angular velocity ω, just as a body in linear motion may have a varying velocity v. If the angular velocity of a body changes from ω_1 to a new value ω_2 in a time interval t, the angular acceleration α is given by the expression

$$\alpha = \frac{\omega_2 - \omega_1}{t} \tag{8.5}$$

$$\text{Angular acceleration} = \frac{\text{change in angular velocity}}{\text{time}}$$

This equation is just as often seen in an alternative form,

$$\omega_2 = \omega_1 + \alpha t \tag{8.6}$$

Rotational motion is subject to angular acceleration, just as translational motion is subject to linear acceleration.

Illustrative Problem 8.4 A hydraulic turbine is turning at 310 rpm (rev/min) when the water supply to it is shut off. It comes to a complete stop in 90 s. Find the negative acceleration of the turbine rotor, assuming it to be uniform.

Solution From Eq. (8.3),

$$\omega_1 = 2\pi \frac{\text{rad}}{\text{rev}} \times \frac{310 \text{ rev/min}}{60 \text{ s/min}} = 32.45 \text{ rad/s}$$

$\omega_2 = 0$, the final angular velocity of the rotor.

Substituting in Eq. (8.5),

$$\alpha = \frac{0 \text{ rad/s} - 32.45 \text{ rad/s}}{90 \text{ s}}$$

$= -0.36 \text{ rad/s}^2$ *answer*

(The minus sign indicates negative angular acceleration)

Note the units of angular acceleration as: *angular distance* (radians) *divided by time squared*; or rad/s^2, or rad s^{-2}.

8.6 ■ Uniform Circular Motion and Centripetal Acceleration

When a body or particle moves in a circular path at constant speed, it is said to be in *uniform circular motion*. An example is a tooth on a gear which is revolving about a fixed axis at uniform angular velocity. In the diagram of Fig. 8.6*a*, consider a particle at position P on a rotating wheel, at which instant its linear velocity is represented by the vector $\mathbf{v}$, whose magnitude is ωr. A short time (t sec) later the particle is at P', and the *magnitude* of its velocity vector is still ωr. However, as the diagram shows, the *direction* of the velocity has changed. Therefore, since velocity is a quantity which has both magnitude and direction, the *linear velocity* of the particle is continually changing, even though its *linear speed* remains constant and equal to ωr. A changing velocity means *acceleration*. In this section we determine the magnitude and direction of the acceleration in uniform circular motion.

In Fig. 8.6*b* the two velocity vectors of Fig. 8.6*a* have been laid out from a common point A, parallel to their original directions. The vector $\mathbf{pp}'$, which closes the vector triangle, represents the small change in velocity in the short time t sec, and can be designated as $\Delta\mathbf{v}$.

As a vector equation,

$$\mathbf{v} + \Delta\mathbf{v} = \mathbf{v}'$$

NOTE: *Remember that the symbol Δ means "a very small amount of," and* $\Delta\mathbf{v}$ *is read "delta* v*." It does not mean delta multiplied by* v.

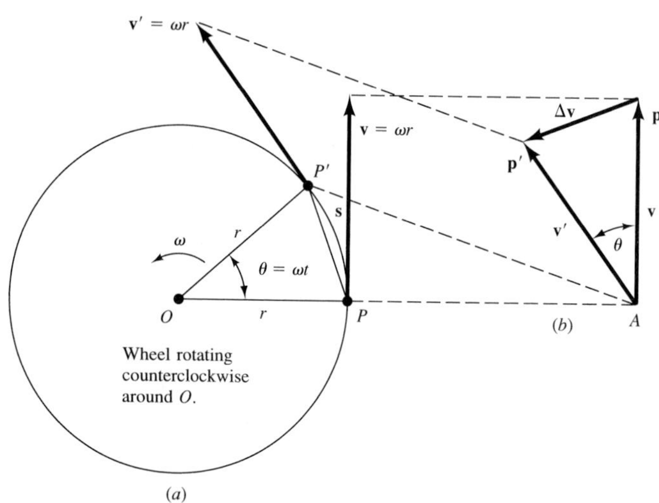

Fig. 8.6 Relationships involved in uniform circular motion. (*a*) Space diagram showing particle P on the rim of a wheel, rotating around the center O at a distance r, with constant angular velocity ω. P moves arc distance s and angular distance θ in time t sec. (*b*) Vector diagram, showing vector $\mathbf{pp}' = \Delta\mathbf{v}$ as the vector difference between $\mathbf{v}'$ and $\mathbf{v}$. As a vector equation, $\mathbf{v} + \Delta\mathbf{v} = \mathbf{v}'$, or $\Delta\mathbf{v} = \mathbf{v}' - \mathbf{v}$.

Vector diagrams like Fig. 8.6*b* can be drawn for smaller and smaller values of t. As t approaches zero as a limit, it is found that the direction of the vector $\Delta\mathbf{v}$ is always toward the center of the circle. And also, as t approaches zero as a limit, the arc distance s (Fig. 8.6*a*) becomes essentially equal to the chord PP', and triangle OPP' is similar to triangle App'. Consequently, since corresponding sides of similar triangles are proportional,

$$\frac{s}{r} = \frac{\Delta v}{v}$$

Acceleration in uniform circular motion is always directed toward the center of rotation. It is "center-seeking", or *centripetal* acceleration.

Now Δv represents a small change in velocity, and is therefore the result of an acceleration over a small time interval Δt. It can therefore be represented by the equation $\Delta v = a\,\Delta t$. Further, the arc distance moved by the body in time Δt is $s = v\,\Delta t$. Making these substitutions,

$$\frac{v\,\Delta t}{r} = \frac{a\,\Delta t}{v}$$

Solving for a gives

$$a = \frac{v^2}{r}$$

Since the acceleration of a point describing uniform circular motion *is always toward the center,* it is called *centripetal* ("center-seeking") *acceleration*. It is defined by the equation

$$a_c = \frac{v^2}{r} \tag{8.7}$$

Or, since $v = \omega r$,

$$a_c = r\omega^2 \tag{8.7$'$}$$

8.7 ■ Total Linear Acceleration in Circular Motion

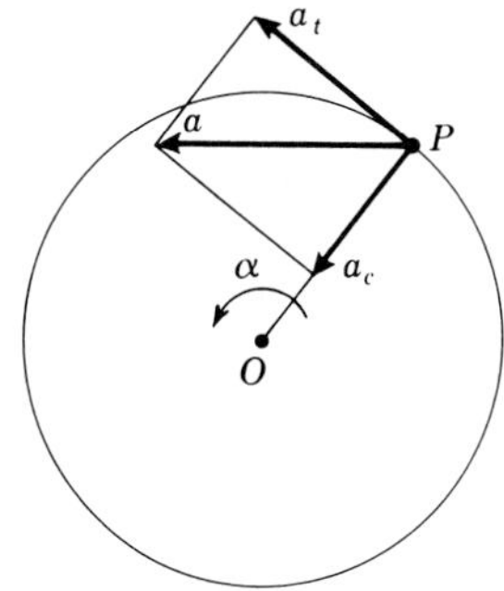

Fig. 8.7 Angular, tangential, and centripetal accelerations of a particle in circular motion and undergoing angular acceleration.

We have seen that a rotating body can have a changing angular velocity, which results in *angular acceleration* (Sec. 8.5). This angular acceleration is accompanied by linear or tangential acceleration of the particles of the body (see Fig. 8.7). Also, we have seen (Sec. 8.6) that a particle on the rim of a rotating body has an acceleration toward the center (*centripetal acceleration*) even when the circular motion is uniform, that is, when angular acceleration is zero. The combination of tangential acceleration and centripetal acceleration is called *total linear acceleration*.

In order to derive an expression for the *total acceleration* of a body in *accelerated angular motion*, we can begin by noting that for a particle rotating at distance r from the center of rotation, from Eq. (8.2),

$$\omega_1 = \frac{v_1}{r} \qquad \text{and} \qquad \omega_2 = \frac{v_2}{r}$$

Subtracting the first of these expressions from the second, gives

$$\omega_2 - \omega_1 = \frac{v_2 - v_1}{r}$$

Dividing both sides of this equation by t gives

$$\frac{\omega_2 - \omega_1}{t} = \frac{v_2 - v_1}{rt}$$

But $(\omega_2 - \omega_1)/t$ is *angular acceleration* α; and $(v_2 - v_1)/t$ is the *linear acceleration* a_t of the particle, *tangent to its momentary path* at P (see Fig. 8.7). This linear acceleration is the result of the angular acceleration of the rotating body. It is also called *tangential acceleration*. Combining these relationships results in the following expression:

$$\alpha = \frac{a_t}{r}$$

or, tangential acceleration,

$$a_t = \alpha r \tag{8.8}$$

Equation (8.8) shows that tangential acceleration is equal to the product of angular acceleration and the radius of the rotary motion.

Total linear acceleration of a particle in circular motion is the vector sum of tangential acceleration and centripetal acceleration.

The point whose tangential acceleration at a given instant is a_t also has an instantaneous *centripetal* acceleration $a_c = r\omega^2$ (Eq. 8.7$'$). From Fig. 8.7 it should be noted that the tangential acceleration a_t and the centripetal acceleration a_c of a particle P undergoing rotational acceleration are perpendicular to each other and that they have a resultant a, called *total linear acceleration*. From the geometry of Fig. 8.7 the *total instantaneous linear acceleration* is

$$a = \sqrt{a_t^2 + a_c^2} \tag{8.9}$$

Illustrative Problem 8.5 The angular velocity of a lathe chuck is 20 rad/s at time $t_1 = 0$. At time $t_2 = 50$ s the angular velocity is 100 rad/s. Find the angular acceleration, assuming it to be uniform.

Solution From Eq. (8.5), angular acceleration is

$$\alpha = \frac{\omega_2 - \omega_1}{t}$$

substituting values gives

$$\alpha = \frac{(100 - 20)\ \text{rad/s}}{50\ \text{s}}$$

$$= 1.6\ \text{rad/s}^2 \qquad \textit{answer}$$

Illustrative Problem 8.6 A particle on the rim of a high-speed grinding wheel (see Fig. 8.9) is 40 cm from the center of rotation. If the instantaneous angular velocity is 5π rad/s and the angular acceleration of the wheel is 50 rad/s^2, find the tangential and total linear accelerations of the particle.

Solution From Eq. (8.8) the tangential acceleration is

$$a_t = \alpha r$$
$$= 50\ \text{rad/s}^2 \times 0.4\ \text{m}$$
$$= 20\ \text{m/s}^2 \qquad \textit{answer}$$

Centripetal acceleration is

$$a_c = r\omega^2\ (\text{Eq. } 8.7')$$
$$= 0.4\ \text{m} \times (5\pi\ \text{rad/s})^2$$
$$= 98.7\ \text{m/s}^2$$

Total linear acceleration is, from Eq. (8.9),

$$a = \sqrt{a_t^2 + a_c^2}$$
$$= \sqrt{(20\ \text{m/s}^2)^2 + (98.7\ \text{m/s}^2)^2}$$
$$= 101\ \text{m/s}^2 \qquad \textit{answer}$$

8.8 ■ Generalized Equations of Angular Acceleration

The discussion of Sec. 8.4 related angular distance to angular velocity by the equation

$$\theta = \omega t \qquad (8.4)$$

The analogous equation relating linear distance to linear velocity is, from Sec. 4.2,

$$s = \bar{v}t \qquad (4.1'')$$

An entire set of equations describing angular acceleration have the same general form as those developed in Sec. 4.9 to describe linear acceleration. Compare the following:

Linear acceleration equations compared to angular acceleration equations.

Linear acceleration equations (From Sec. 4.9)		**Angular acceleration equations**	
$v_2 = v_1 + at$	(4.6)	$\omega_2 = \omega_1 + \alpha t$	(8.6)
$s = v_1 t + \frac{1}{2}at^2$	(4.10)	$\theta = \omega_1 t + \frac{1}{2}\alpha t^2$	(8.10)
$v_2^2 = v_1^2 + 2as$	(4.11)	$\omega_2^2 = \omega_1^2 + 2\alpha\theta$	(8.11)

Angular acceleration, like linear acceleration, may be either positive or negative—positive if ω is increasing, negative if ω is decreasing.

Illustrative Problem 8.7 A ship's propeller shaft is turning at 100 rev/min. The order is given to the engine room to increase the angular speed at a uniform rate to

400 rev/min and to reach that angular speed in exactly 1 min. Find the angular acceleration required.

Solution

$$\omega_1 = 100 \text{ rev/min} = 10.47 \text{ rad/s}$$

(Recall that 1 rev = 2π rad.)

$$\omega_2 = 400 \text{ rev/min} = 41.88 \text{ rad/s}$$

From Eq. (8.5),

$$\alpha = \frac{\omega_2 - \omega_1}{t} = \frac{41.88 \text{ rad/s} - 10.47 \text{ rad/s}}{60 \text{ s}}$$

$$= 0.52 \text{ rad/s}^2 = 0.083 \text{ rev/s}^2 \qquad \textit{answer}$$

Illustrative Problem 8.8 The rear wheels of an auto are turning at 60 rad/s. The brakes are applied for 5 s, giving a uniform (negative) angular acceleration of 8 rad/s^2. Find (*a*) the final angular velocity and (*b*) the number of revolutions turned by the rear wheels during the braking period.

Solution
(***a***) From Eq. (8.6),

$$\begin{aligned}\omega_2 &= \omega_1 + \alpha t \\ &= 60 \text{ rad/s} + (-8 \text{ rad/s}^2) \times 5 \text{ s} \\ &= 20 \text{ rad/s} \qquad \textit{answer}\end{aligned}$$

(***b***) Angular displacement,

$$\begin{aligned}\theta &= \omega_1 t + \tfrac{1}{2}\alpha t^2 \qquad (8.10) \\ &= 60 \text{ rad/s} \times 5 \text{ s} \\ &\quad + \tfrac{1}{2} \times (-8 \text{ rad/s}^2) \times (5 \text{ s})^2 \\ &= 200 \text{ rad}\end{aligned}$$

$$\text{from which, revs} = \frac{200 \text{ rad}}{6.28 \text{ rad/rev}} = 31.8 \text{ rev} \qquad \textit{answer}$$

FORCES INVOLVED IN CIRCULAR MOTION

8.9 ■ Centripetal Force

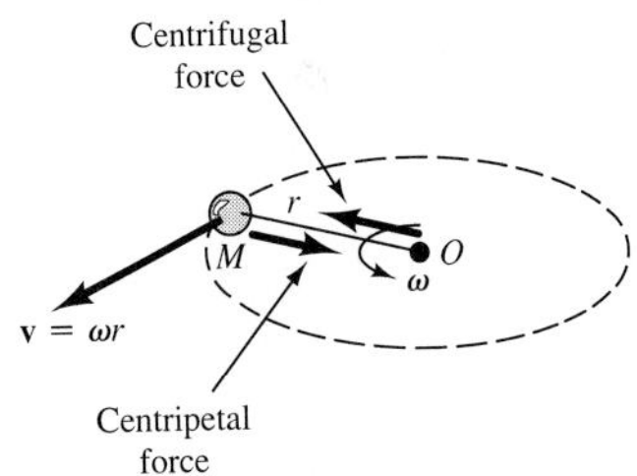

Fig. 8.8 Centripetal vs. centrifugal force. Centripetal force holds the ball in orbit. Centrifugal force exerts an outward "pull" on the hand of the person whirling the ball. The string is in tension, supplying the centripetal force.

Since objects or particles in circular motion undergo a continual change in direction of their motion, they are being constantly accelerated. This acceleration is always toward the center of rotation.

According to Newton's *first law of motion,* the particle on the rim of the wheel of Fig. 8.6*a* must be acted upon by an unbalanced force before it will change either its speed or its direction. The force causing centripetal acceleration in this case is supplied by the molecular forces of cohesion within the material of which the wheel is made. This force which causes centripetal acceleration is called *centripetal force*. If the molecular forces of cohesion were destroyed, the centripetal force would become zero. The particle would then immediately *move off at a tangent,* in the line of its motion at the instant the centripetal force was destroyed.

Newton's laws of motion apply to circular motion as well as to translational motion.

Using a simple illustration, let us look at the factors that determine centripetal force. Consider a ball of mass M being whirled in a circle about a center O, attached to a string whose length is r, as diagrammed in Fig. 8.8. Nearly everyone has, at one time or another, performed this simple trick. The tension in the string supplies the force necessary to hold the ball in its circular orbit, i.e., to keep it accelerating toward the center. Again, this is *centripetal force*.

Newton's second law for circular motion—centripetal force equals mass times centripetal acceleration. ($F_c = ma_c$).

It is Newton's *second law of motion* ($F = ma$) which relates force to mass and acceleration. Remember that a *force of* 1 N is necessary to give *a mass of* 1 *kg an acceleration of* 1 *m/s*^{2}. Applying the *second law* to uniform circular motion gives

$$F_c = ma_c \tag{8.12}$$

Centripetal force = mass × centripetal acceleration

But

$$a_c = \frac{v^2}{r} \tag{8.7}$$

and therefore

$$F_c = \frac{mv^2}{r} \tag{8.13}$$

Units for expressing centripetal force.

This equation expresses centripetal force in newtons when m is in kilograms, linear velocity v is in meters per second, and r is in meters. Or, in the English system, F_c is in pounds when m is in slugs, linear velocity v is in feet per second, and r is in feet.

If *weight* is used instead of *mass,* the equation becomes

$$F_c = \frac{wv^2}{gr} \tag{8.13$'$}$$

Equation (8.13) is often considered the basic formula for centripetal force, but a second form, equally useful, may be written when we recall that $v = \omega r$. Substituting and simplifying, we obtain

$$F_c = mr\omega^2 \tag{8.14}$$

or, with weight units,

$$F_c = \frac{w}{g}r\omega^2 \tag{8.14$'$}$$

Equations (8.14) and (8.14′) are the more convenient forms when the angular velocity in rad/s is known.

The force required to hold an object or a particle in a circular path increases as mass and linear speed increase. It should be noted, however, that centripetal force varies *directly* (i.e., with the first power) of mass, and is related to the *square* of the linear speed. Speed changes therefore have a much greater effect on centripetal force than do mass changes. Note also that an increase in the radius of the circular motion results in a lessened force requirement if mass and linear speed are held constant.

8.10 ■ Centrifugal Force

For every *action force* there is an equal and opposite *reaction force*. In the example of Fig. 8.8, the centripetal (action) force is exerted by the string on the ball, and it provides the "center-seeking" acceleration which holds the ball in a circular orbit. The reaction force is the outward force which the ball exerts on the string. The string transmits this tension force to point O, or to a person's hand. The inertia of the ball, its resistance to a change in direction, is the cause of this reaction force. Since this force is directed away from the center it is called *centrifugal* ("center-fleeing") *force*. It is important to note that centripetal force and centrifugal force, like any action-reaction pair of forces, act on *different* bodies, not on the same body—in this case centripetal force acts on the ball, and centrifugal force acts eventually on the hand.

Centripetal force and *centrifugal force* clarified and differentiated. Newton's third law applies.

What happens if the ball is whirled too fast? The string's tensile strength will eventually be exceeded and it will break. Since centripetal force can then no longer exist, acceleration toward the center ceases, and the ball flies off in straight-line motion tangent to the arc at the instant of the break, with the linear velocity v at that instant. You should be able to show, from Newton's laws, why the ball moves off tangentially.

Illustrative Problem 8.9 A 16-lb ball is being whirled in uniform circular motion at the end of a cable 3 ft long. If the angular velocity is 0.6 rev/s, find the tension (centripetal force) supplied by the cable.

Solution Since weight units are involved, use Eq. (8.14′),

$$F_c = \frac{w}{g} r\omega^2$$

From Eq. (8.3), solve for ω:

$$\begin{aligned} \omega &= 2\pi \text{ rad/rev} \times 0.6 \text{ rev/s} \\ &= 3.77 \text{ rad/s} \end{aligned}$$

Substituting into Eq. (8.14′),

$$\begin{aligned} F_c &= \frac{16 \text{ lb}}{32.2 \text{ ft/s}^2} \times 3 \text{ ft} \times (3.77 \text{ rad/s})^2 \\ &= 21.19 \text{ lb} \qquad \textit{answer} \end{aligned}$$

(Remember—the radian is a dimensionless quantity)

8.11 ■ Applications of Centripetal Force

Any rotating machine part, such as a propeller, turbine rotor, motor armature, engine flywheel, or grinding wheel (Fig. 8.9) is subject to internal stresses caused by the rotation. These stresses are, as shown by Eq. (8.14), directly proportional to the mass

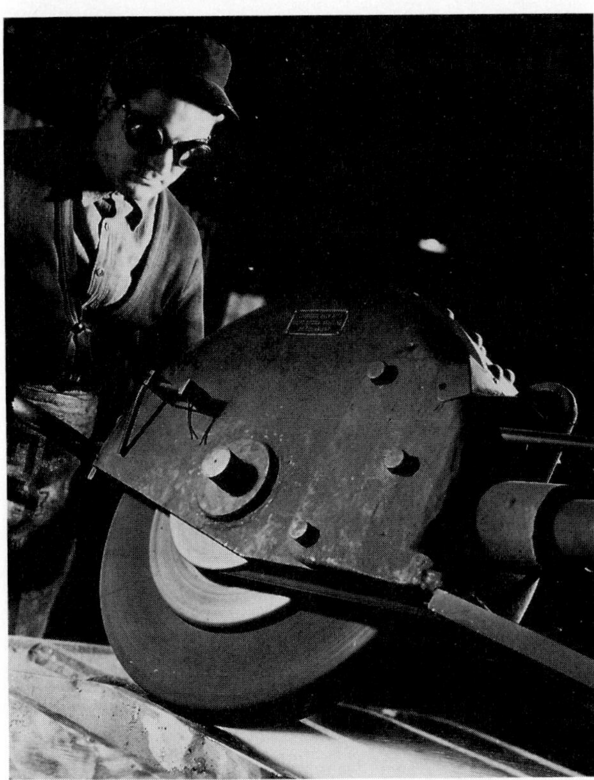

Fig. 8.9 High-speed grinding wheels must be able to supply large centripetal forces. They have been known to fly apart at high rotational speeds. Note the protective cover over the wheel. (*Steelways Magazine*)

Stresses in rapidly rotating machine parts can be tremendous, and machine design must predict them correctly or disaster will be the result.

of the rotating object and to the square of the angular velocity. All materials which go into the manufacture of rotating parts must be carefully tested to make sure that their molecular structure can cope with the stresses required for centripetal force. Rotational speeds of 10,000 to 100,000 rev/min are not uncommon for centrifuges and for spin-stabilized instruments such as gyrocompasses and inertial navigation systems. Slower rotational speeds, for example, 300 to 3600 rev/min, are required of ponderous equipment like steam-turbine rotors, wind-powered generators, and aircraft propellers, but their large diameters (8 to 16 ft) and huge masses set up tremendous internal stresses within the material.

8.12 ■ The Centrifuge

The centrifuge is indispensable to medical research and laboratory testing. It is also used in many commercial operations.

The centrifuge is a high-speed rotating device for the separation of particles of differing densities (Fig. 8.10). Liquids containing microscopic particles are placed in tubes in a small vessel and whirled at incredibly high speeds. The tremendous centrifugal forces involved have the result that the denser particles collect or ''sediment'' to the outside of the vessel, resulting in a separation of dense particles from lighter, or less dense, particles. In medical research, centrifuges are used in determining the sedimentation rates of blood samples. They are also used for concentrating solutions thought to contain viruses, hormones, etc., and for separating impurities from pharmaceuticals. The dairy industry uses the centrifuge principle both in making laboratory butterfat tests and in the separation of cream from milk. Spin-dry washers use the principle to separate water from wet clothes.

8.13 ■ Highway and Rail Travel Depend on Centripetal Force

Centripetal force is an essential factor to be considered in designing curves on highways and railroads.

When high-speed trains, autos, and trucks travel around curves, tremendous forces must be supplied by the roadbed to hold them in their curving paths. For example,

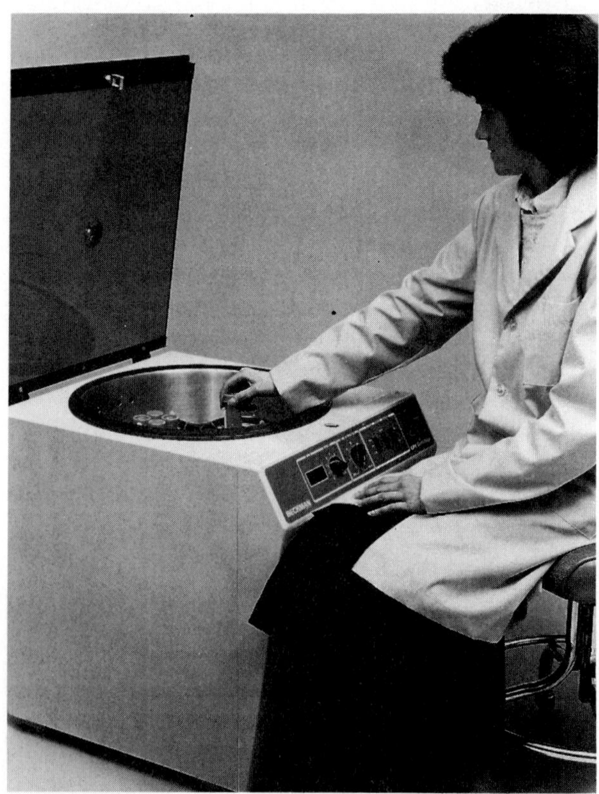

Fig. 8.10 A centrifuge used in medical research and testing. The rotor of this centrifuge can operate at speeds up to 3750 rev/min and set up centripetal accelerations equal to 3200 times the acceleration of gravity (3200 g's). (Beckman Instruments, Inc.)

Fig. 8.11 diagrams an automobile traveling on a curve of radius r at speed v. If the roadbed is level, the only force which can supply the required centripetal force is the force of friction between tires and pavement. Recall the analysis of frictional forces in Chapter 7. We write an equation for the instant when skidding is impending, i.e., at the instant when the force of static friction F_s has reached its maximum or limiting value, $F_{s,\max}$. Setting friction force equal to centripetal force, we have

$$F_{s,\max} = F_c$$

from which

$$\mu_s mg = mv^2/r$$

or

$$\mu_s = v^2/rg \qquad (8.15)$$

Note that the mass of the auto makes no difference, since m's cancel out in obtaining Eq. (8.15).

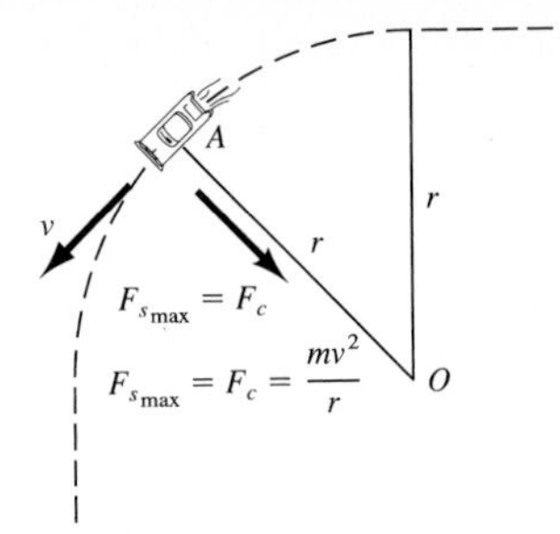

Fig. 8.11 Skidding on highway turns. Road friction can supply centripetal force to reduce the tendency to skid while going around corners. Skidding impends when the required centripetal force is equal to the maximum value of the force of static friction ($F_{s,\max}$) between tires and the road surface.

Illustrative Problem 8.10 What is the maximum speed permissible for an automobile rounding a level curve of 200-ft radius if μ_s between tires and roadbed is 1.0. Answer in miles per hour.

Solution Use Eq. (8.15) and solve it for v:

$v = \sqrt{\mu_s gr}$. Substitute known values, giving

$$v = \sqrt{1.0 \times 32.2 \text{ ft/s}^2 \times 200 \text{ ft}} = 80 \text{ ft/s}$$

$$= 80 \text{ ft/s} \times \left(\frac{30 \text{ mi/h}}{44 \text{ ft/s}}\right) = 54.5 \text{ mi/h} \qquad \textit{answer}$$

This seems a reasonable speed for a curve of 200-ft radius, but dry pavement and good tire treads are not always present. On a slick or icy pavement μ_s might well be less than 0.2 and a safe value of v would then be only 24 mi/h!

8.14 ■ Banking Roadbeds

The value of μ_s is a variable, being dependent on such factors as road surface, weather, and tire condition. Highway engineers therefore attempt to provide another and more constant factor to supply the required centripetal force on curves of short radius. This is done by inclining, or "banking," the roadbed. Suppose the auto in Fig. 8.12*a* is coming toward you at speed v around a curve of radius r. The center of the curve O is toward the left. Neglecting frictional forces entirely, we analyze the two forces, $w = mg$ (the weight of the car acting vertically downward) and N (the normal reaction force of the roadbed on the car). The value of the force N is not dependent on friction. Since the road is *banked* at an angle θ with the horizontal, N is inclined to the vertical by this same angle. Its horizontal and vertical components are, respectively, $N \sin \theta$ and $N \cos \theta$, as shown in Fig. 8.12*a*. The vertical component, $N \cos \theta$, is equal and opposite to the weight w. If it were not, the car would have acceleration in a vertical plane. The horizontal component, $N \sin \theta$, must supply the centripetal force F_c to hold the car in its circular path, if there is to be no reliance on friction to do so. Consequently (see vector diagram, Fig. 8.12*b*),

Banking a roadbed has the effect of allowing the roadbed itself to assist centripetal force in holding vehicles "on the curve," without having to rely only on friction between tires and pavement to do so.

$$w = mg = N \cos \theta \qquad (8.16)$$

and

$$F_c = mv^2/r = N \sin \theta \qquad (8.17)$$

Dividing Eq. (8.16) by Eq. (8.17) results in

$$\tan \theta = v^2/rg$$

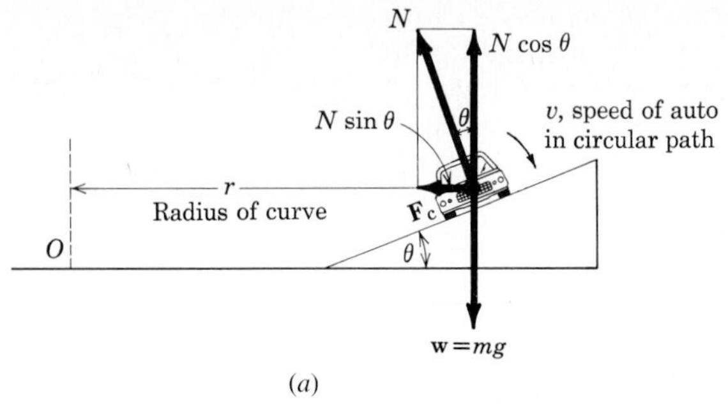

(a)

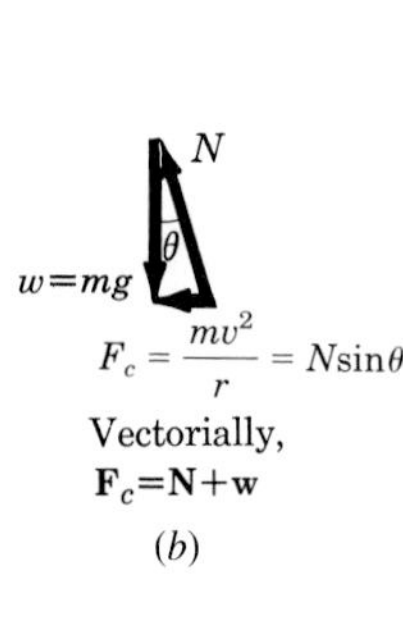

(b)

(c)

Fig. 8.12 Banking of highway curves. The angle θ in the figure is called the *banking angle*. (*a*) Geometric relationships of the banking angle. (*b*) Vector relationships, assuming that frictional forces are zero. (*c*) Banking of speedway on auto testing grounds. (Ford Motor Co.)

The *banking angle* is, then, for any speed v,

$$\theta = \tan^{-1} \frac{v^2}{rg} \tag{8.18}$$

where v = linear speed of vehicle, m/s or ft/s
r = radius of curve, m or ft
g = acceleration of gravity = 9.81 m/s^2 or 32.2 ft/s^2.

Note that w, the weight of the car, does not affect the value of the banking angle. However θ does depend on the value of the velocity v. Consequently, a turn can be banked correctly only for a speed chosen in advance. Engineers try to make the best estimate of the speed at which motorists may drive the curve, or they may bank it for the legal speed limit. Friction will also be present, even though it is assumed zero when determining the banking angle. This adds a factor of safety for the fast driver except when the road is icy. Figure 8.12*c* shows the banking of a speedway curve at an automotive proving ground.

Illustrative Problem 8.11 A roadway whose radius of curvature is 200 m is to be "banked." Traffic is expected to move around the curve at 90 km/h. What should be the value of the banking angle if no dependence is to be placed on friction?

Solution Using Eq. (8.18),

$$\theta = \tan^{-1}(v^2/rg)$$

Substitute the given data, changing km/h into m/s:

$$\theta = \tan^{-1} \frac{\left(\dfrac{90 \text{ km/h} \times 1000 \text{ m/km}}{3600 \text{ s/h}}\right)^2}{200 \text{ m} \times 9.81 \text{ m/s}^2}$$

$$= \tan^{-1} 0.319$$

By calculator, or from trig tables,

$$\theta = 17.7° \qquad \textit{answer}$$

GRAVITATION AND AEROSPACE PHYSICS

8.15 ■ Newton's Law of Universal Gravitation

In the century before Newton (sixteenth century) many scientists and philosophers contributed to knowledge of natural phenomena. Galileo studied freely falling bodies and formulated some of the general laws of accelerated motion. He also devised the first telescope for the magnification of celestial bodies. Tycho Brahe made determinations of the daily motions of the planets and some of the stars over a period of 30 years. Copernicus formulated the heliocentric theory of the solar system (the ancients had believed that the sun and the planets all revolved about the earth).

Using the solar system as a starting point, Newton generalized a theory of gravitational attraction which he felt would apply anywhere in the universe. It can be stated as follows:

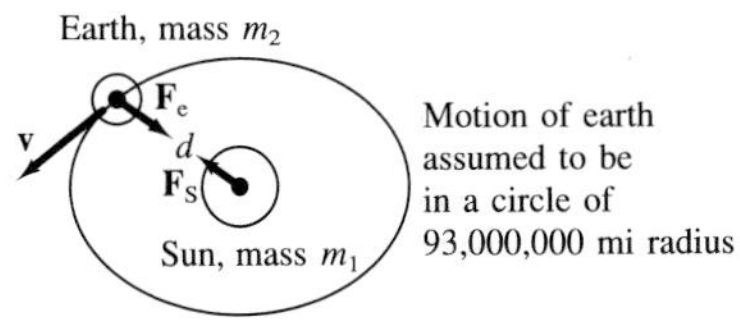

Fig. 8.13 The force of gravitational attraction between the sun and the earth. F_e is the force the sun exerts on the earth, and the equal and opposite force F_s is the force exerted on the sun by the earth. The vector **v** represents the earth's orbital speed, which is about 66,700 mi/h.

Newton's Law of Universal Gravitation

Any two bodies or particles in the universe attract each other with a force which is directly proportional to the product of their masses and inversely proportional to the square of the distance between them.

As a mathematical expression, this law can be written

$$F \propto \frac{m_1 m_2}{d^2}$$

For the case involving the earth's revolution around the sun, Fig. 8.13 diagrams the essential relationships.

The period of the earth's revolution around the sun is $365\frac{1}{4}$ earth days. The *average* value of d (distance from earth to sun, center to center) is about 93,000,000 mi. The vector **v** represents the earth's *linear* velocity in orbit, relative to the sun. (The magnitude of the earth's velocity in orbit is about 18.5 mi/s.) $\mathbf{F}_e$ is the centripetal force that keeps the earth in orbit. This force is supplied by the gravitational attraction between the earth and the sun. Newton's third law (action and reaction) requires that $\mathbf{F}_e = -\mathbf{F}_S$. Stated in words, the sun attracts the earth with the same force (in magnitude) that the earth attracts the sun.

The expression $F \propto m_1 m_2/d^2$ describes the nature of gravitational attraction, but it cannot evaluate the force F between two bodies. In order to do that the proportionality sign must be replaced with an equals sign and a constant which will validate the equals sign. Making use of such a constant, the force of gravitational attraction F between two bodies (anywhere in the universe) with masses m_1 and m_2 whose centers are at a distance d apart (see Fig. 8.14) is:

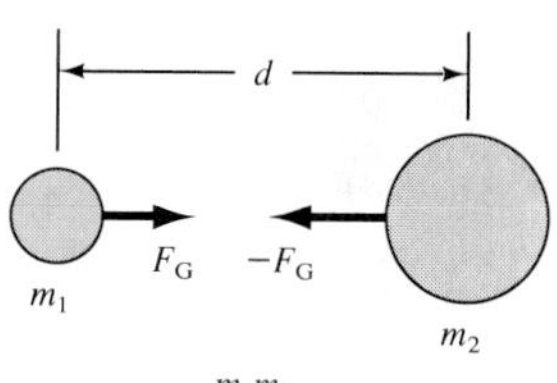

Fig. 8.14 The force of gravitation between two bodies is a universal property. The bodies attract each other with oppositely directed forces of equal magnitude. The force is proportional to the product of the masses of the bodies and inversely proportional to the square of the distance between their centers. **G**, the proportionality constant, is the *universal gravitation constant*.

$$F = \mathbf{G}\frac{m_1 m_2}{d^2} \tag{8.19}$$

The constant **G** is called the *universal gravitation constant*. It must not be confused with g, the acceleration of gravity. Scientists have been measuring the value of **G** for over 200 years, and the following values are now agreed upon.

G is the universal gravitational constant. It applies *anywhere in the universe.* It must not be confused with g, the acceleration of earth's gravity.

SI-Metric Units of G If F is in newtons, m_1 and m_2 in kilograms and d in meters, **G** has the value

$$\mathbf{G} = 6.67 \times 10^{-11}\ \mathbf{N} \cdot \text{m}^2/\text{kg}^2 \tag{8.20a}$$

or, since the dimensions of the newton are kg · m/s²,

$$\mathbf{G} = 6.67 \times 10^{-11}\ \text{m}^3/(\text{kg})(\text{s}^2) \tag{8.20b}$$

English-Engineering Units of G If F is in pounds, m_1 and m_2 in slugs, and d in feet, **G** has the value

$$\mathbf{G} = 3.44 \times 10^{-8}\ \text{lb} \cdot \text{ft}^2/\text{sl}^2 \tag{8.20c}$$

The distance d in Eq. (8.19) is the *straight-line distance between the centers of mass* of the two masses m_1 and m_2.

Between the planets and the sun gravitational forces are, of course, tremendous. But, for objects of our ordinary experience, the forces are small indeed, as the following problem illustrates.

Illustrative Problem 8.12 Two 16-lb shot spheres (as used in track meets) are held with their centers 2 ft apart. What is the force of attraction between them?

Solution In engineering units.

$$m_1 = m_2 = \frac{w}{g} = \frac{16\ \text{lb}}{32.2\ \text{ft/s}^2} = 0.497\ \text{sl}$$

Substituting in Eq. (8.19) using the value of **G** from Eq. (8.20*c*),

$$F = 3.44 \times 10^{-8}\ \frac{\text{lb} \cdot \text{ft}^2}{\text{sl}^2} \left[\frac{0.497\ \text{sl} \times 0.497\ \text{sl}}{(2\ \text{ft})^2} \right]$$

$$= 2.12 \times 10^{-9}\ \text{lb force} \qquad \textit{answer}$$

This force is perhaps about the weight of a mosquito's wing, and would be detectable only by precision scientific apparatus in a research laboratory.

G and g Related It is again emphasized that the *universal gravitation constant* **G** is a completely different concept and quantity from the *acceleration of earth's gravity* g. However, the two constants can be related mathematically as follows. Let

G and g are different concepts, but they are related by the simple mathematical expression of Eq. (8.22).

M = mass of earth
w = force of gravitation on a small mass m at or near the *surface* of earth (its weight)
r_e = the distance to the center of the earth from the small mass m (the earth's radius for an object on or near the earth's surface)

The substitution of these symbols in Eq. (8.19) gives

$$w = \mathbf{G}\frac{Mm}{r_e^2} \tag{8.21}$$

But from Newton's second law, $g = w/m$, and consequently

$$g = \frac{\mathbf{G}M}{r_e^2} \tag{8.22}$$

Equation (8.22) shows that g *decreases* as the square of the distance to the center of the earth *increases*. Since the earth's radius is not everywhere the same, g varies slightly over the earth, being less at the equator than it is at the poles.

Variation in Weight and in the Value of g with Distance from the Earth's Surface We have learned that the weight of an object is given by the relationship $w = mg$ [Eq. (5.2)]. At the earth's surface g has the value 32.2 ft/s² or 9.81 m/s². However, Eq. (8.22) requires that the value of g, and therefore the weight of any object, decrease if it is moved into near-earth space and on into outer space, making its distance from the center of the earth greater. The curve of Fig. 8.15 shows how the weight of a 100-lb sack of wheat (at the earth's surface) would vary with increasing distance from the center of the earth. For example, at 12,000 mi from the earth's surface (16,000 mi from the center of the earth), the 100-lb sack would weigh only about 6.25 lb. This is equivalent to saying that, at 16,000 mi from the center of the earth, g is only a little more than 6 percent of its value at the earth's surface, or about 2 ft/s² or 0.25 m/s².

The weight of an object, or of a human being, gets less and less as the distance from the earth's surface increases. Why? Because g decreases as distance from the earth increases; and $w = mg$.

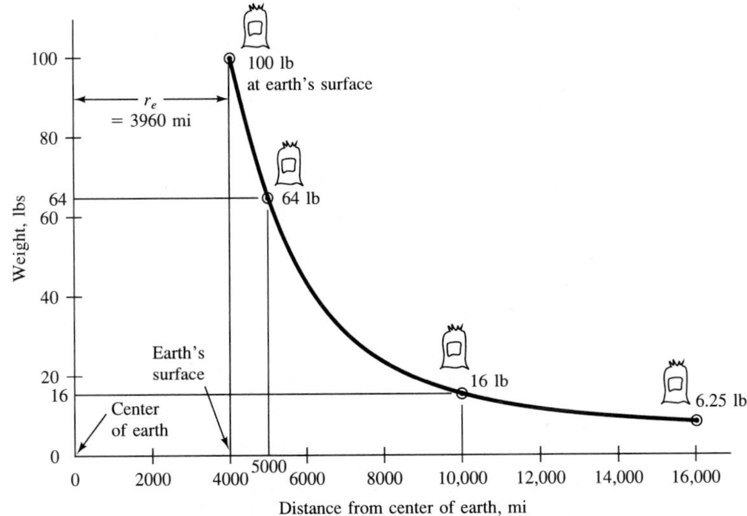

Fig. 8.15 Variation in weight with distance above the earth's surface.

8.16 ■ Mass of the Earth

The diagram of Fig. 8.16 shows the relationships involved in the mutual attraction between the earth and a 1-kg mass on its surface. We will use it to calculate the mass of the earth. We know that the earth imparts an acceleration $g = 9.81$ m/s² to any small mass at sea level. The radius of the earth is 6.37×10^6 m. **G** is given by Eq. (8.20*b*) as 6.67×10^{-11} m³/(kg)(s²). Solving Eq. (8.22) for M, the mass of the earth, gives

$$M = \frac{gr_e^2}{\mathbf{G}} \tag{8.23}$$

Substituting,

$$M = \frac{9.81 \text{ m/s}^2 \times (6.37 \times 10^6 \text{ m})^2}{6.67 \times 10^{-11} \text{ m}^3/(\text{kg})(\text{s}^2)}$$

$$\text{Mass of earth} = M = 5.96 \times 10^{24} \text{ kg}$$

which is equivalent to a weight of about 6.6×10^{21} tons (at 1 g).

Illustrative Problem 8.13 The moon's mass is 7.37×10^{22} kg (see Table 8.1) and its radius is 1740 km. Find (*a*) the acceleration of gravity on the surface of the moon and (*b*) the weight on the moon of a person whose mass is 80 kg.

Solution The equations above apply to *universal* gravitation, and they can therefore be used anywhere in the universe.

Fig. 8.16 Mutual attraction between the earth and a 1-kg mass on the earth's surface.

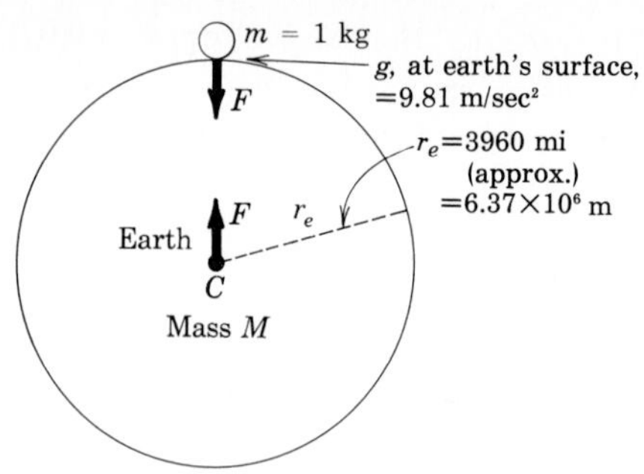

(*a*) Using Eq. (8.22),

$$g_{\text{moon}} = \frac{\mathbf{G}M_{\text{moon}}}{r_{\text{moon}}^2}$$

and substituting values for the moon,

$$g_{\text{moon}} = \frac{6.67 \times 10^{-11}\ \text{m}^3/(\text{kg})(\text{s}^2) \times 7.37 \times 10^{22}\ \text{kg}}{(1.74 \times 10^6\ \text{m})^2}$$

$$= 1.62\ \text{m/s}^2 \qquad \textit{answer}$$

or roughly one-sixth that at the earth's surface (g_{earth}).

(*b*) The weight of a person on the moon equals his or her mass times g_{moon}, or $w_{\text{moon}} = mg_{\text{moon}}$

Substituting,

$$w_{\text{moon}} = 80\ \text{kg} \times 1.62\ \text{m/s}^2 = 130\ \mathbf{N} \qquad \textit{answer}$$

Table 8.1 presents selected data for some of the bodies in the solar system.

Table 8.1 Selected data for the solar system

Celestial Body	*Mass, kg*	*Mean Radius, km*	*g at Surface, m/s²*
Earth	5.97×10^{24}	6,370	9.81
Moon	7.37×10^{22}	1,740	1.62
Sun	1.97×10^{30}	695,000	274.0
Mars	6.37×10^{23}	3,400	3.90
Jupiter	1.88×10^{27}	71,800	26.45

NOTES: $\mathbf{G} = 6.67 \times 10^{-11}\ \mathbf{N} \cdot \text{m}^2/\text{kg}^2$ or $6.67 \times 10^{-11}\ \text{m}^3/\text{kg} \cdot \text{s}^2$; Earth-moon distance = 385,000 km (239,000 mi) (approx.); Earth-sun distance = 150,000,000 km (93,000,000 mi) (approx.). (Distances are center-to-center.)

8.17 ■ Orbiting Satellites

The orbits of the planets around the sun are ellipses, but the elliptical eccentricity is small enough that we will treat them as circles. In the reality of space exploration, however, orbital eccentricity must be taken into account.

The orbits of the planets around the sun in our solar system are ellipses. Circular orbits are possible for man-made satellites, but they tend to be unstable, changing to elliptical form or ''decaying'' if gravitational forces vary or if orbital speed slows for any reason. In general, an elliptical orbit is stable—the satellite is *bound* to the central body. A hyperbolic orbit is an *escaping* orbit, in which the ''satellite'' will make one pass around the central body and then escape into space (see Fig. 8.17).

Launching and putting into orbit an earth satellite is a very complex operation. For successful orbiting, the final-stage rocket motor must impart exactly the right velocity

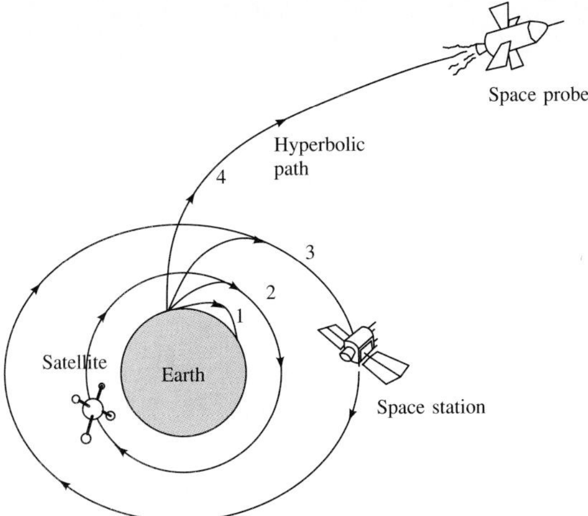

Legend: 1. A projectile—parabolic trajectory—falls to earth
2. Very low circular orbit—unstable, will decay
v_0 = 7900 m/s; 17,500 mi/h
3. Stable, elliptical orbit—400 mi + above
earth. v = 9500⁺ m/s; 21,000 mi/h
4. Hyperbolic path for outer-space probe
$v_{escape} = \sqrt{2}\, v_0$
= 11,200 m/s; 25,000 mi/h

Diagram not to scale. All data are approximate.

Fig. 8.17 The path described by objects launched from the earth's surface depends on the speed they attain. Illustrated are a projectile, which returns to earth; a small satellite in a low-earth orbit; a space station in a stable elliptical orbit; and an outer-space probe with a hyperbolic path.

for the intended radius of the orbit. Orbital paths of planets around the sun are actually ellipses. The orbits of man-made satellites can be adjusted by firing small propulsion rockets. These adjustments enable the satellite to be maneuvered into a nearly circular orbit at a selected distance from the surface of the earth, moon, or planet. The discussions to follow will assume circular orbits.

When orbits are assumed circular in satellite mechanics, the centripetal force needed for uniform circular motion is provided by gravitational attraction. We neglect, in the following treatment, the *drag* forces due to air resistance, which, although satellites operate hundreds of miles above the earth's surface, are sometimes of sufficient magnitude to cause the orbit to "decay" over a period of weeks or months, as happened with the U.S. *Skylab* in 1979, and with a U.S.S.R. satellite that disintegrated and crashed in northern Canada in 1978.

Man-Made Satellites In Fig. 8.18 a man-made satellite is shown circling a planet of mass M (it could be the earth) in an orbit of radius r_s. The distance r_s is measured to the center of the planet.

r_p = radius of planet
F_s = force of gravitational attraction between planet and satellite (the weight of the satellite)
g_s = value of the acceleration of gravity at distance of orbit from the center of the planet
v_s = linear speed of satellite in orbit

From Newton's second law, $F = ma$. But, in this case, $F = F_s$ and $a = g_s$, so $F_s = mg_s$. But from centripetal-force considerations (Eq. 8.13), the gravitational force of attraction must supply the centripetal force to hold the satellite in orbit, or

$$F_s = mg_s = \frac{mv_s^2}{r_s}$$

Gravitational force = centripetal force

Linear speed and orbital radius must be in accord with Eq. (8.24) for a satellite to remain in its orbit.

Therefore if the satellite is to remain in its orbit, its linear speed must be (solving for v_s)

$$v_s = \sqrt{g_s r_s} \tag{8.24}$$

Note that the *mass of the orbiting object does not affect the value of the required orbital speed.*

Any Satellite Orbiting Around Any Celestial Body The analysis above, leading to Eq. (8.24), was based on recognizing the *weight* of the satellite as $F_s = mg_s$. An alternate analysis, using the universal gravitational constant **G** follows.

Again, the starting point is Newton's second law, $F = ma$. The centripetal acceleration of the satellite is $a = v_s^2/r_s$. Now, from Eq. (8.19), the force of gravitational attraction between this satellite and its planet (Fig. 8.18) is $F_s = -F_p = \mathbf{G}mM/r_s^2$.

Required relationships for *any satellite* orbiting around *any celestial body.*

From which

$$\frac{\mathbf{G}mM}{r_s^2} = \frac{mv_s^2}{r_s}$$

and

$$v_s = \sqrt{\frac{\mathbf{G}M}{r_s}} \tag{8.25}$$

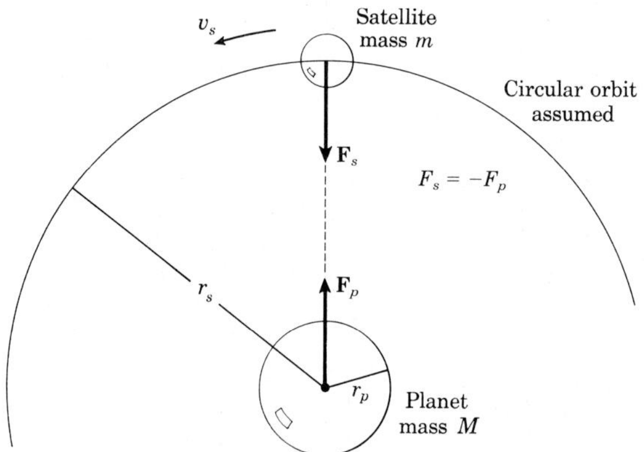

Fig. 8.18 Diagram of a man-made satellite of mass m revolving around a planet of mass M in a circular orbit of radius r.

Note that Eqs. (8.24) and (8.25) are equivalent—they both give the required orbital speed for a satellite to remain in a stable orbit around a planet. Equation (8.24) gives the required speed in terms of the value of g_s, the *acceleration of gravity at the satellite,* and the radius of the orbit; while Eq. (8.25) gives the required speed in terms of the *mass of the planet* and the radius of the orbit, making use of the universal gravitational constant. It is worth noting that Eq. (8.25) shows at a glance that the larger the orbital radius, the slower is the required orbital speed for a stable orbit. Note also that the *mass of the satellite* itself does not enter into the calculations at all.

For a satellite orbiting about any given planet, e.g., the earth, $g = \mathbf{G}M/r^2$ (Eq. 8.22). Note that **G** and M are both constants (**G** is a universal constant, and M is a constant for any one celestial body). Consequently, the relationship between the value of g at the satellite altitude and the value of g at the planet's surface is given by

$$\frac{g_s}{g_p} = \frac{r_p^2}{r_s^2} \tag{8.26}$$

where g_s = value of g at satellite orbit distance
g_p = value of g on surface of planet
r_s = radius of satellite orbit
r_p = radius of the planet

In other words, for a satellite orbiting any celestial body, g is inversely proportional to r^2, where r is the distance from the center of the celestial body to the satellite's orbital path.

Orbital Speed for Low Orbits For *very low orbits* around any planet, r_s is very nearly equal to r_p, and the required speed of an orbiting satellite becomes

$$v_s = \sqrt{g_p r_p} \tag{8.27}$$

Thus, if we substitute values for a satellite in a *very low orbit around the earth,* and solve Eq. (8.27) we obtain $v = 7900$ m/s or about 17,700 mi/h. This orbital speed gives a period of revolution around the earth, $p = 2\pi r_e/v_s = 1.4$ h or 1 h 24 min. Satellites in higher orbits will have *lower* orbital speeds and longer periods of revolution, as Eq. (8.25) indicates.

At a muzzle velocity of 7900 m/s (if that were possible) a rifle bullet fired horizontally from shoulder height would become an earth satellite briefly, until air resistance caused rapid orbital decay.

Illustrative Problem 8.14 A communications satellite is to be put into a circular orbit 400 mi above the earth's surface (see Fig. 8.19). If the mean radius of the earth is 3960 mi and g at the earth's surface is 32.2 ft/s², find (*a*) the required speed for the satellite to stay in orbit and (*b*) the time required for a complete "pass" around the earth.

Solution

(*a*) $r_s = 3960 + 400 = 4360$ mi. From Eq. (8.26),

$$g_s = \frac{g_p r_p^2}{r_s^2} = 32.2\ \text{ft/s}^2 \times \frac{(3960\ \text{mi})^2}{(4360\ \text{mi})^2}$$
$$= 26.6\ \text{ft/s}^2$$

Equation (8.24) then gives the necessary orbital velocity:

$$v_s = \sqrt{g_s r_s}$$
$$= \sqrt{\frac{(26.6\ \text{ft/s}^2) \times 4360\ \text{mi}}{5280\ \text{ft/mi}}}$$
$$= 4.69\ \text{mi/s} = 16{,}900\ \text{mi/h} \qquad \textit{answer}$$

Once in orbit, the various operational and communications systems of many satellites can be energized by photovoltaic (solar) cells mounted in large arrays or panels.

(*b*) From Eq. (8.2), we find the satellite's angular velocity:

$$\omega = v/r_s$$

Substituting from **(*a*)** gives

$$\omega = \frac{16{,}900\ \text{mi/h}}{4360\ \text{mi}} = 3.87\ \text{rad/h}$$

Since there are 2π rad/rev, the angular velocity in revolutions per hour is

$$\frac{3.87\ \text{rad/h}}{6.28\ \text{rad/rev}} = 0.616\ \text{rev/h}$$

The period, or time for one revolution, is

$$\frac{1}{0.616\ \text{rev/h}} = 1.623\ \text{h/rev} = 97.4\ \text{min} \qquad \textit{answer}$$

Figure 8.19 shows a satellite during final testing operations. Figure 8.20 illustrates some of the stages of putting a satellite into orbit around the earth. Final linear speed and orbital height adjustments are made by firing small rocket motors during the last stage before satellite separation.

Fig. 8.19 Intelsat VI communications satellite (artist's rendering). This satellite is nearing completion and undergoing tests. It is scheduled for deployment in a geosynchronous orbit, 22,300 mi above the earth, in 1989. It is 39 ft high in its operating mode, and weighs nearly 4000 lb. Six antennas will receive and transmit communications on an international network. It will be solar-powered, from panels of photovoltaic cells (PVCs) covering most of its outer surface, and producing 2600 watts of power. (Hughes Aircraft Company)

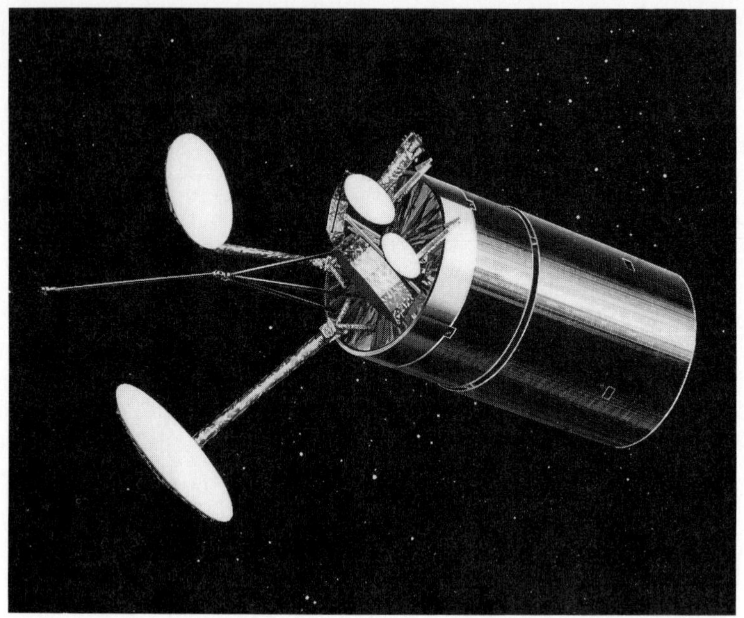

Illustrative Problem 8.15 Table 8.1 gives the mass of the earth M_e as 5.97×10^{24} kg. Calculate the value of g at the earth's surface, in SI-metric units.

Solution From Eq. (8.22),

$$g = \frac{\mathbf{G}M_e}{r_e^2}$$

Substituting,

$$g = \frac{6.67 \times 10^{-11}\ \text{m}^3/\text{kg}\cdot\text{s}^2 \times 5.97 \times 10^{24}\ \text{kg}}{(6.37 \times 10^6\ \text{m})^2}$$

$$= 9.81\ \text{m/s}^2$$

answer

Fig. 8.20 Artist's conception of the launch sequence of a three-stage rocket with satellite payload, showing the various stages of lift-off, flight to designated altitude, and satellite separation.

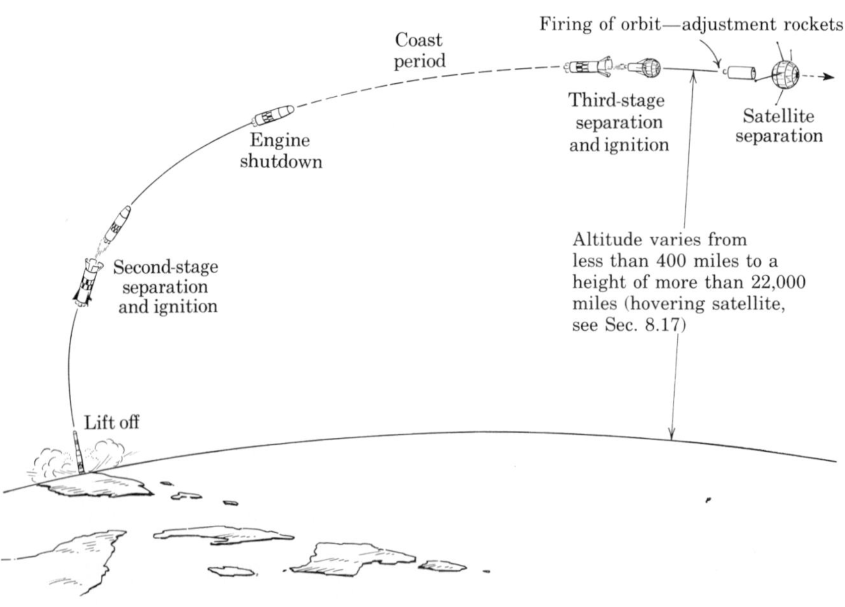

Illustrative Problem 8.16 Find the period of revolution of an earth satellite in a circular orbit 600 mi above the surface of the earth.

Solution First, determine the value of g_s, the acceleration of gravity at the satellite orbit distance from earth. From Eq. (8.26), substituting values,

$$g_s = 32.2\ \text{ft/s}^2 \times \frac{(3960\ \text{mi})^2}{(3960\ \text{mi} + 600\ \text{mi})^2}$$

$$= 24.3\ \text{ft/s}^2$$

Now use Eq. (8.24) to find the linear speed, v_s.

Substituting, $v_s = \sqrt{g_s r_s} = \sqrt{24.3\ \text{ft/s}^2 \times 4560\ \text{mi} \times 5280\ \text{ft/mi}}$
$= 24.19 \times 10^3$ ft/s
$= 16.5 \times 10^3$ mi/h

The period, $$T = \frac{\text{circumference of orbit}}{\text{linear speed in orbit}}$$

$$= \frac{2\pi \times 4560\ \text{mi}}{16.5 \times 10^3\ \text{mi/h}}$$

$= 1.74$ h or 1 h 44 min *answer*

The manufacture, orbiting, and operating of earth satellites is currently a mammoth scientific-engineering-industrial enterprise, involving government agencies and private corporations in America, Europe, and Asia.

8.18 ■ Practical Uses of Satellites

Space probes and earth satellites have many valuable uses. *Weather satellites* bring us worldwide meteorological information; *scientific satellites* collect information on magnetic fields, radiation from outer space, earth resources, volcanoes, earthquake faults, and the sphericity of the earth. *Surveillance satellites* have been used to gather information on military installations. A system of satellites at an altitude of about 750 mi services a worldwide *navigation network. Communications satellites* have made a significant contribution to worldwide communications. Such early satellites as Telstar and Early Bird revolutionized intercontinental television in 1962, when Telstar I was launched. Early Bird was placed in commercial service in 1965 and was reported to have a capability equivalent to that of 240 transatlantic telephone channels. At present, international communications systems (including television) are heavily dependent on satellites. ''Landsats'' gather information on the earth's crust and other physiographic features. ''Comsats'' make worldwide communications (including television) possible. ''Navsats'' guide ships and aircraft all over the globe.

Mammoth space laboratories are possibilities for the future. They would orbit the earth at altitudes of several hundred miles, and might house research laboratories or specialized manufacturing operations that must take place in the absence of an atmosphere or under ''zero-gravity'' conditions. Space shuttles (Fig. 8.21) can be used to commute between these satellites and the earth. There is even talk of ''space colonization,'' but the line between practical reality and science fiction is difficult to draw in these matters. Energy shortages may force cancellation of some of the space exploration and space utilization projects currently envisioned.

It is estimated that there are now (1988) some 4500 satellites in orbit around the earth. The orbits range from 100 to 22,500 mi above the earth's surface.

Fig. 8.21 The Space Shuttle Columbia clears the gantry and climbs into the early morning sky for a scientific mission into space. Note the powerful thrust of the rocket engines necessary to overcome inertia and the force of the earth's gravity. After a major disaster in 1986, followed by a two-year period of re-analysis and redesign of rocket components, the Shuttle program is currently (1989) in full operation again. (National Aeronautics and Space Adminsitration)

Hovering Satellites Most satellites are in orbits which require relative motion between them and the earth's surface. However, satellites can be placed in orbits where the required angular speed is exactly equal to the angular speed of the rotating earth, and thus the satellite ''hovers'' over one location *on the earth's equator* (see Fig. 8.22). Such satellites are said to be in a *geosynchronous* orbit, since their motion is synchronized with the earth's daily rotation. These satellites must be at altitudes of about 22,250 mi to satisfy Eq. (8.24), and they are used for international communi-

Fig. 8.22 Diagram of the relationship of a hovering satellite and its geosynchronous orbit to the earth. The view is from a point above the North Pole. The period of rotation of the earth and the period of revolution of the satellite are identical. The required altitude above the earth's surface is (about) 22,250 mi.

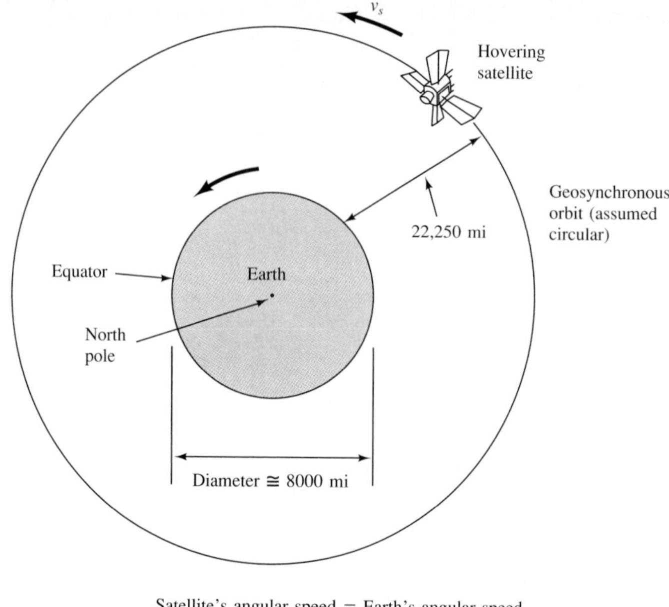

"Hovering" satellites are stationed directly over the earth's equator at an altitude of 22,250 mi. They are *geosynchronous* (*geo*—of or pertaining to the earth) since their period of revolution is synchronized with the earth's rotation. They revolve around the earth in exactly one day, remaining directly above the same point on the earth at all times.

cations, space program communications, and for television programs and telephone traffic between the nations of the world (Fig. 8.22).

Since they are in geostationary orbits, that is ''parked'' directly above the same spot on the earth's equator at all times, their period of revolution is exactly one (earth) day. The geometry of their location (22,250 mi distant from the earth's sphere, the diameter of which is about 8000 mi) allows any one satellite to be effective over nearly one-third of the earth's surface.

The use of hovering satellites as solar-power-generating stations, beaming fantastic quantities of electric power to earth, has been proposed, but a multitude of technical problems will have to be solved before this somewhat visionary possibility becomes a reality.

8.19 ■ Some Problems of Aeronautics and Space Flight

True weightlessness occurs only under conditions of zero gravity.

Weightlessness—Real and Apparent Astronauts during space flights may encounter conditions of *near-zero gravity* or weightlessness. They ''float around'' in the spacecraft cabin, and nothing that is not bolted down will stay put. To be *truly weightless* means that negligible gravity forces are acting, and that, in turn, implies a location far away from earth or any other celestial body. Even at 12,000 mi from the earth's surface the *g* force is still 6 percent of its value on earth. *Absolute zero gravity* would not be attainable anywhere in our solar system, but conditions of *near-zero gravity* can result when a space vehicle reaches distances of 50,000 mi or more from earth.

Apparent weightlessness occurs whenever a downward (i.e., directed toward the earth) acceleration is equal to *g*, the acceleration of earth's gravity. For example, a freely falling body is apparently weightless.

At zero gravity a spaceship or a human being, or any other mass would be truly weightless. *Apparent weightlessness* can, however, be readily experienced without zero-gravity conditions. One common example of apparent weightlessness is the situation in a rapidly downward-accelerating elevator, if the acceleration approaches or equals g—9.81 m/s^2 or 32.2 ft/s^2 (see Fig. 8.23). With a downward acceleration equal to g, the elevator and its occupants are actually freely falling bodies with respect to the earth and with respect to the building in which the elevator is installed. And, with respect to the elevator floor, the occupants are for the moment ''weightless,'' since they exert no downward force on the floor.

Fig. 8.23 Apparent weightlessness in an elevator. The elevator's downward acceleration is momentarily equal to or greater than g.

Orbital Weightlessness Astronauts in orbiting satellites experience apparent weightlessness also, since they and the orbiting vehicle are constantly in free fall with respect to the earth (or any other central body they are orbiting). In order to prepare space crews for the unusual problems of real or apparent weightlessness, extended periods of training are provided under simulated zero-gravity conditions. Trainees practice all the maneuvers and operations that are expected to be performed under the weightless conditions of the space flight.

Astronauts, and all other objects in orbiting satellites and space stations, experience *apparent* weightlessness, since they and the orbiting vehicle are freely falling bodies.

During astronaut training on earth, zero-gravity conditions must be created artificially. One convenient method is to use the principles of centripetal force. Astronaut trainees and their equipment are taken aloft in an airplane with a large cabin space which has been properly prepared in advance. At a suitable altitude the aircraft executes an outside loop (see Fig. 8.24) with a proper combination of curvature of path and speed which allows everybody and everything in the aircraft to be a freely falling body, that is, unsupported by the airframe.

Flying an "outside loop" with an aircraft can create conditions of *apparent weightlessness* in the plane's cabin. The same effect can be experienced when "going over the top" on a roller coaster in an amusement park.

Let A (Fig. 8.24) be an astronaut of mass m sitting on the floor of the aircraft. Two forces will act on him or her: (1) the force of gravity, $F_g = mg$ (the astronaut's

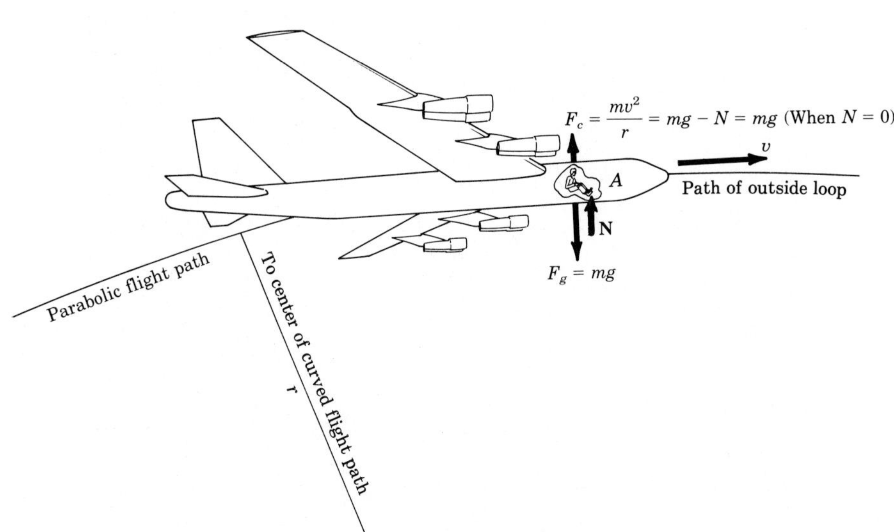

Fig. 8.24 Diagram of an aircraft flying an "outside loop" (near-parabolic path) to produce a situation of apparent zero gravity or "weightlessness." As weightlessness occurs, N = zero and $F_c = mg$.

weight), acting downward, and (2) the force N, with which the floor supports the astronaut. The condition for weightlessness is that $N =$ zero.

Now, the resultant of N and mg is the centripetal force F_c that ''holds'' the aircraft (and the astronaut) in the curved path. In equation form,

$$F_c = mg - N$$

With $N = 0$,

$$F_c = mg$$

But

$$F_c = mv^2/r$$

and we write

$$mv^2/r = mg$$

from which

$$v = \sqrt{gr} \tag{8.28}$$

which is the condition for apparent zero gravity. Note that the m's cancel out (all persons in the plane will be weightless at the same time regardless of their differences in size).

It is reemphasized that the crew and objects are not *really* weightless. They still have mass and the force of earth's gravity still acts on them ($w = mg$). They are actually in free fall and are unsupported, just like the person in the elevator of Fig. 8.23.

Illustrative Problem 8.17 Some astronauts are undergoing zero-gravity training in a jet transport plane, as illustrated in Fig. 8.25. The pilot is instructed to fly at 450 mi/h. What is the radius of curvature of the outside loop when ''weightlessness'' occurs?

Solution Square both sides of Eq. (8.28), obtaining

$$v^2 = gr \qquad \text{or} \qquad r = \frac{v^2}{g}$$

Since 450 mi/h = 660 ft/s,

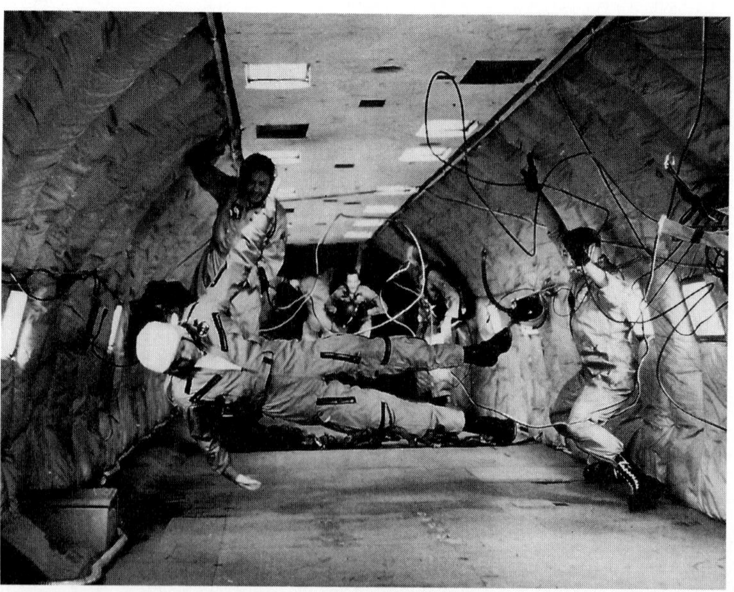

Fig. 8.25 Astronauts participating in zero-gravity training. The aircraft is flying an outside loop at just the right speed to make the plane and its occupants freely falling bodies. (National Aeronautics and Space Administration)

$$r = \frac{(660 \text{ ft/s})^2}{32.2 \text{ ft/s}^2}$$

$$= 1.35 \times 10^4 \text{ ft} \qquad \textit{answer}$$

If the curvature of the loop being flown were more gradual, the people would remain sitting on the floor; if a tighter loop were negotiated, they would appear to ''float upward'' and would bump against the ceiling of the aircraft.

8.20 ■ Landing on the Moon or on Another Planet

The problem of landing on the moon or on another planet is far more complex than that of orbiting the earth with a satellite. First, the vehicle must ''escape'' from the earth along a flight path which will result in its intercepting in space, perhaps many days or months later, the planet intended. This interception is in itself a complex problem of relative motion. If the ''target'' is one of the planets of our solar system, the sun may be used as the frame of reference for solving the relative-velocity problem of earth, rocket, and target planet.

Escape Velocity As a *projectile* climbs upward against the force of gravity, the work-energy principle requires that its kinetic energy diminish as the potential energy increases. In order to ''escape'' the earth's gravity, the initial velocity must provide an initial kinetic energy equal to or greater than the potential energy the projectile-earth system will have when it reaches a height where the earth's gravity is zero. This required initial velocity is called *escape velocity*. Since the derivation of a formula for escape velocity requires the use of the calculus, the relationship is given here without derivation. If r_e is the radius of the earth and g the acceleration of gravity *at the earth's surface,* the velocity which must be imparted to a *projectile* at the earth's surface in order for it to escape from the earth's gravity is given by

"Escaping" from the earth means reaching a point where the earth's gravity is essentially zero. The escape velocity from earth is of the order of 25,000 mi/h or about 7 mi/s.

$$v_{\text{escape}} = \sqrt{2gr_e} \qquad (8.29)$$

It is interesting to note that Eq. (8.29) gives the escape velocity for *any direction* of launching. (Air resistance is neglected throughout this discussion.)

It should be noted that we have been describing the escape velocity for a *projectile*. Much different principles apply to rocket-propelled vehicles. A rocket *starts* from zero velocity but it has a continued thrust, and it is common practice to accelerate a rocket-propelled vehicle to reach escape velocity soon after it leaves the earth's atmosphere, or, if it has orbited the earth preparatory to embarking on its space flight, soon after leaving the ''parking orbit.'' The continued thrust of rocket propulsion also prevents air resistance from slowing the vehicle.

Escape velocity from earth is of the order of 25,000 mi/h as compared to the velocity required to stay in a very low-altitude orbit around the earth, which is of the order of 17,500 mi/h. The required *orbital velocity* for a low-altitude earth satellite is, from (Eq. 8.27) $v_s = \sqrt{g_e r_e}$. The *escape velocity*, v_{escape}, from earth (Eq. 8.29), is

$$v_{\text{escape}} = \sqrt{2g_e r_e} \qquad \text{(from earth)}$$

Therefore escape velocity equals $\sqrt{2}$ times the orbital velocity for a very low altitude orbit.

As the probe or spaceship nears its intended target, one of three outcomes is possible:

1. Its speed and direction with respect to the target may be such that the centripetal force supplied by the target's gravitational attraction never becomes great enough to result in orbital ''capture.'' In this case the probe or spaceship will make one pass in a curved path near the planet, and then fly off into space with a hyperbolic path.

Some of the problems (greatly oversimplified) to be solved as a spacecraft nears an intended target planet or moon for a landing.

2. The speed and approach path may be such that the resulting centripetal acceleration (g_s) just satisfies Eq. (8.27), and the spaceship will be ''captured'' by the planet and remain in a circular or slightly elliptical orbit around the planet at a distance r_s from the center of the planet.

3. The speed and path may have been such as to bring the vehicle so close to the target planet (or moon) that the force of gravitational attraction is greater than that required as centripetal force for an orbit, and the spaceship will be "captured" in the very real sense that it crashes on the planet.

These three possibilities can be enhanced or prevented by firing additional guidance and control rockets on the spaceship, thus making course corrections in flight. Such mid-course corrections have been common practice on both manned and unmanned space missions.

Landing on another celestial body is usually preceded by going into orbit around it for several "passes" in order to accomplish such objectives as visual and photographic surveillance; measurement of magnetic fields, gravitational forces, radiation, and meteorite intensity; and making sure that all is in readiness for the descent to the surface. Landing on the planet (Fig. 8.26) requires the firing of retrorockets and course-correction rockets. Finally, the spaceship (or a part of it), if manned, must return to earth, and the return trip is no simpler than the outbound trip. It, too, requires escape, interception, orbit around the earth, and a safe landing.

Guidance rockets for mid-course corrections put the returning vehicle into the proper "window" for a return through the earth's atmosphere to the preselected landing area. The atmosphere supplies the braking force to slow the vehicle from its 30,000 mi/h (plus) speed to a manageable speed for a landing. The Apollo Program vehicles were landed by parachute, but the Space Shuttle (see Fig. 8.27) is an air-

Fig. 8.26 Artist's conception, based on preliminary design data, of a Mars landing vehicle. The next step in exploring Mars is to land an unmanned vehicle which can obtain soil and related samples, then return to its spacecraft for the trip back to earth. (National Aeronautics and Space Administration)

Fig. 8.27 The Space Shuttle is sometimes assigned duty as a space cargo carrier. Here a SATCOM K-1 communications satellite is being "spun out" of Columbia's cargo bay. It will then be maneuvered into its desired orbital position by signals sent to its own on-board rocket engines and control systems. (National Aeronautics and Space Administration)

craft, and it is piloted to its landing. The heat generated as the vehicle streaks through the atmosphere is tremendous, and a special heat shield is necessary to dissipate the heat and prevent a temperature buildup which might consume the vehicle like a fiery meteor.

QUESTIONS AND EXERCISES

1. Explain carefully the difference between *rotation* and *revolution*. Is uniform circular motion possible in both? Can centripetal force and centrifugal force be present in both of these kinds of circular motion?

2. State and define the three commonly used units of angular displacement. Use diagrams to define them.

3. The equation relating linear (tangential) speed of a point on a rotating object to the angular velocity of the object is $v = \omega r$. Prove that this equation is dimensionally correct.

4. A ship's propeller rotates on the propeller shaft as its axis. But the ship also moves forward. Describe what kind of motion the propeller has with respect to the earth. What technical term describes the path of a point on a propeller blade tip?

5. Analyze three common machines (for example, an auto, a lawn mower, a bulldozer—but you pick any three you like) and classify the movement of their main parts as to whether it is translation, rotation, or a combination of both. Simplify by neglecting any forward motion of the machine itself.

6. What is the essential difference between a projectile and a rocket-powered vehicle? Does the rocket ever become a projectile?

7. Explain carefully the distinction between **G** and g. Is either ever zero? Does **G** vary over the surface of the earth? Does g? Explain.

8. A manned landing on Mars is being planned. Explain each phase of the operation, step by step, including the process of bringing the space vehicle (or a part of it) with the astronauts back to earth safely.

9. Orbiting satellites revolve around the earth (like a stone being whirled around on a string) with the earth's gravitational pull supplying the centripetal force. If they "slow down," their orbits "decay" and they eventually crash. How, then, is it possible to have "hovering satellites"—satellites that stay "parked" over one spot on the earth's equator?

10. Everyone who drives an auto is aware of the danger of speeding around a banked curve on a highway. Under what conditions might it be dangerous to drive too slowly around such a curve?

11. Equation (8.22) relates g and **G** as follows: $g = \mathbf{G}M/r_e^2$. **G** and M are constants (**G** is a universal constant, and M is a constant for earth problems). Is it correct to deduce that g diminishes with the square of the altitude above the earth's surface?

12. Look at Eq. (8.13), $F_c = mv^2/r$, and at Eq. (8.14), $F_c = mr\omega^2$. They are equivalent expressions for centripetal force, one in terms of linear (orbital) speed, and the other in terms of angular velocity. How is it possible that in Eq. (8.13) centripetal force is *inversely* proportional to the radius of revolution, while in Eq. (8.14) it is *directly* proportional to the radius of revolution?

13. The term "weightlessness" is sometimes used to describe the apparent "zero gravity" caused by orbital motion around the earth. Explain why both of these terms are incorrect. Use the basic relation, $w = mg$ in your explanation. When an object is apparently "weightless" at a point above the earth where g is definitely not zero, what is really the state of motion of the body?

14. Centripetal force and centrifugal force were explained in terms of whirling a ball in a circular path at the end of a string (Fig. 8.8). Both forces, one acting on the ball and the other acting on the hand of the person doing the whirling, are supplied by tension in the string. Now take the case of a planet circling the sun or a satellite circling the earth. What supplies the *centripetal* force? Is there a *centrifugal* force in these cases? If so, what causes it?

15. Define very rigorously the meaning of *escape velocity*. Why is the situation different for a projectile than for a rocket? Why does Eq. (8.29) not have any term or factor in it to specify the direction or angle of elevation of the firing or launching?

PROBLEMS

Group One

1. The proper rotational speed for a certain lathe operation is 220 rev/min (rpm). Express this speed in rad/s.

2. The roulette wheel at a casino turned $7\frac{1}{4}$ revolutions before coming to a stop on the winning number. How many radians did it turn?

3. A bicyclist rides 3 mi on wheels that have a diameter of 26 in. to the outer edge of the tires. If no slippage occurs, how many radians do the wheels turn during the ride?

4. An engine flywheel is turning at 400 rev/min. What angular displacement does it undergo in 20 s?

5. Standard industrial electric motor speed is 1750 rev/min. How many radians per second is this?

6. A steam turbine rotor is 12 ft in diameter and turns at 1200 rev/min. How many miles does a point on its rim travel in a 24-h day?

7. To start an outboard motor, a rope which is wound on a 10-cm drum is given a strong pull. At what linear speed is the rope pulled if the engine is rotated at 600 rev/min? (Answer in meters per second.)

8. A shaft with a diameter of 6 in. is being turned down on a lathe at 200 rev/min. The cutting tool takes off a continuous ribbon of metal. How many linear feet of the metal ribbon will be cut off in 5 min?

9. A ball of mass 1 kg is being whirled on the end of a 75 cm cord at a rotational rate of 1.5 rev/s. Disregard gravitational effects and calculate the tension in the cord.

10. A friend demonstrates that he can whirl a pail half full of water in a vertical plane without any water falling out at the top of the arc. If the effective radius of the vertical circle his whirling describes is 3 ft, what is the *minimum* required tangential speed of the bucket at the top of the arc? (Draw a diagram and see Eq. (8.7)). (HINT To what constant must a_c be equal?).

Group Two

11. A ship's propeller shaft is turning at 100 rev/min when the order comes from the bridge to increase speed *uniformly* to a speed which requires 350 shaft rev/min, and accomplish the speed change in 2 min. Calculate the uniform angular acceleration required.

12. For the conditions of Problem 11, calculate the total angular displacement of the propeller shaft, in radians, during the 2-min period.

13. In a centrifuge, liquid suspected of containing a virus is being rotated at high speed in a circular path of 4-in. radius. If a centripetal acceleration of 100,000 g's is required for separation of the virus, at what speed (rev/min) must the centrifuge be driven?

14. An auto rounds an unbanked curve of 300-ft radius at 30 mi/h. What is the minimum value of μ_s required between tires and pavement to prevent skidding?

15. Standard-gauge rails are 4 ft $8\frac{1}{2}$ in. apart in the United States. On a curve of 1700-ft radius, how far (in inches) above the inner rail should the outer rail be elevated to allow trains to negotiate the curve at 50 mi/h without relying on friction or wheel flanges to hold the train on the tracks?

16. How many g's does a pilot experience in pulling out at the bottom of a vertical dive at 400 mi/h if the loop radius is 3000 ft?

17. What is the minimum speed (km/h) at which a pilot can execute an inside vertical loop of a 700-m radius and still be pressing against his seat at the top of the loop when his head is hanging down toward the earth?

18. An oval auto race track has semicircular ends designed on a 400-ft radius. If these turns are to be driven at 150 mi/h, what must be the banking angle if no reliance is to be placed on friction?

19. A small earth satellite is to orbit at a height of 300 km. If the orbit is assumed circular, how many minutes are required for a complete trip around the earth? Would a very large satellite require more or less time?

20. It is desired to place in circular orbit around the earth a 20,000-lb satellite whose path (assumed circular) will be 300 mi from the earth's surface. What must be the terminal velocity of the satellite vehicle in miles per hour as the last-stage rocket motor disengages itself?

21. If the command module of a lunar expedition is in a circular orbit around the moon, 20 km from the moon's surface, find (*a*) the module's linear speed in orbit and (*b*) the length of time required for one complete orbit.

22. It is desired to orbit a manned vehicle around Mars at a distance of 80 km from the surface. Find (*a*) the required orbital speed and (*b*) the period of the orbit, in hours.

23. Calculate the force of mutual attraction, in newtons, between the earth and the moon, given the average distance between their centers as 384,500 km (see Table 8.1).

24. Calculate the centripetal acceleration of the moon as it revolves around the earth, making a complete revolution in 27.3 days. Its mean distance from the earth is 384,500 km (answer in m/s^2).

25. For a "zero-gravity" exercise to be conducted in a military jet transport aircraft, it is decided to fly an outside loop (assumed circular) whose radius at the point of "zero gravity" is 5000 m. Find the required speed of the aircraft, in kilometers per hour.

26. If an astronaut weighs 120 lb on earth, what would be her apparent weight on Mars? (See Table 8.1.)

27. In order to remain in a circular orbit 500 km above the surface of the planet Jupiter, what orbital speed would be required? (See Table 8.1.)

28. Calculate the escape velocity in miles per hour from the earth's surface for a space vehicle whose total weight is 2.4 million lb; for one whose weight is 5 million lb.

29. Find the escape velocity in miles per hour from the surface of the moon.

30. A highway curve is banked in the expectation that drivers will not exceed 50 mi/h under conditions of dry pavement, which is concrete. The radius of the curve is 300 ft. (*a*) What should the banking angle be if no reliance is to be placed on friction? (*b*) At what maximum speed could this curve be driven if normal friction between rubber tires and concrete pavement is assumed?

Group Three

31. A fan-jet engine rotor 1.25 m in diameter is spinning at an *instantaneous* angular velocity of 20π rad/s, when there is a uniform angular acceleration of the rotor of 100 rad/s^2. Find the tangential and total linear accelerations of a point on the rim of the rotor at that instant.

32. Calculate the force of gravitational attraction that exists between the earth and the sun, given the data in Table 8.1 (see Fig. 8.15). Answer in newtons.

33. A surveillance satellite is orbiting the earth at an altitude of 800 mi. Below what linear speed in orbit would its orbit begin to "decay"?

34. It is desired to put a satellite into a stable orbit at an altitude of 400 mi above the surface of Mars. Calculate (*a*) its required orbital speed, and (*b*) the period of its revolution around Mars.

35. How much total force (newtons) must a set of bolts be able to withstand if they are to hold a 30-kg counterweight in place on the rim of a flywheel 3 m in diameter which is turning at 250 rev/min? (See Fig. 8.28.) (Neglect gravitational forces.)

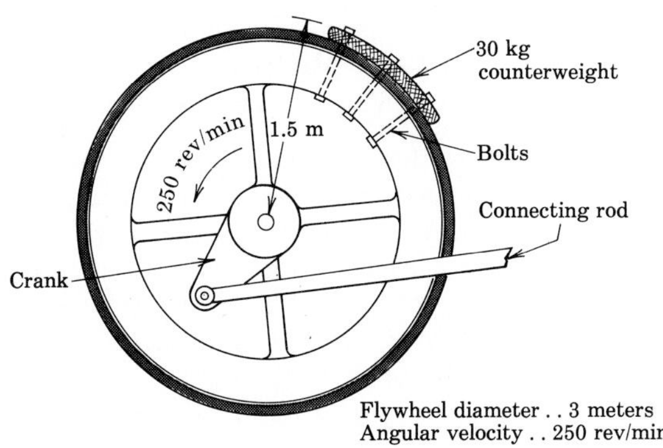

Fig. 8.28 Stresses in rotating machinery provide centripetal force (Problem 35).

36. A motorcycle and rider perform the stunt of circling in a vertically walled cylindrical track 30 ft in diameter. If the machine makes one revolution every 3 s and the combined weight of the machine and rider is 580 lb, with what force (pounds) does the wall push in on the cycle's tires? (Neglect the earth's gravitational force.)

37. At what altitude (km) above the earth's surface would the numerical value of g be half that at the earth's surface?

38. In an amusement park a roller coaster runs over the crest (top) of a curve whose radius of curvature is 10 m at a speed of 7.0 m/s. If a passenger weighs 55 kilos under normal conditions, what is her *apparent* weight at the top of the ride?

39. Prove mathematically that a satellite can hover over the same spot on the equator if it is in orbit at an altitude of 22,250 mi above the earth's surface.

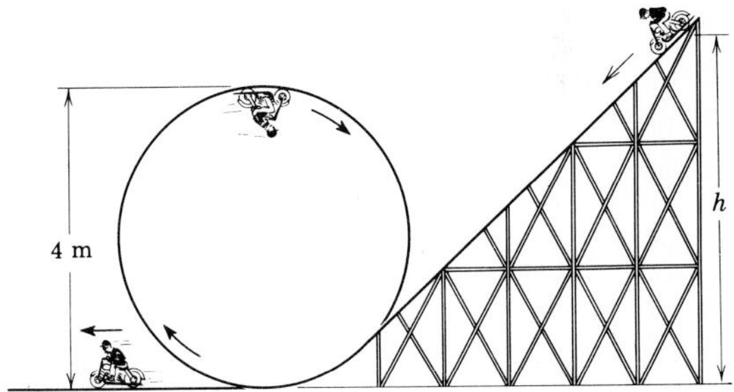

Fig. 8.29 "Looping the loop" on a two-wheeled cycle (Problem 40).

40. A stunt bicyclist "loops the loop" as shown in the diagram of Fig. 8.29. From what minimum height h must he start in order to roll down the incline and around the loop successfully? (Neglect friction. Weight of bike and rider not required. Why?)

41. A 20-ton space vehicle of the future is launched from the earth's surface, and by the time it reaches a distance of 50,000 mi (earth's gravity near zero at this distance), its speed has accelerated uniformly to 100,000 mi/h. Assume an *average value* of g of 1.8 ft/s^2 throughout this portion of the flight. Calculate the total energy that would have to be expended in bringing the vehicle to this speed at that distance from the earth's surface. Answer in kilowatt-hours. (Neglect the air resistance that would be encountered in the early part of the flight.)

CHAPTER 9

TORQUE, POWER TRANSMISSION, AND ROTATIONAL DYNAMICS

Having investigated the nature of circular motion in the last chapter, along with some of its applications to industry, medicine, transportation, and aerospace, we turn now to an inquiry into the *causes* and *effects* of rotational motion about a fixed axis. Since we will be dealing with rotating bodies (masses) and the forces that cause or retard their motion, the term *rotational dynamics* is used, and Newton's laws of motion are again involved.

Torque is the turning effect of an applied force.

In translational motion, *force* is what initiates or retards motion. The counterpart of force in rotational motion is *torque*. Torque is the *turning effect* in rotary motion; it is torque that causes a mass to start rotating about a fixed axis, and torque is required to slow down or stop rotation.

TORQUE

It was pointed out in previous discussions that force is the cause of a change in motion. This is true not only of translation, but of rotation as well. It is a matter of common experience that it takes force to start a body rotating and also to slow its rotation if it is already spinning. However, in rotational motion an additional factor besides force is involved. This additional factor is called the *moment arm*. The combined effect of force and moment arm is called *torque,* and the basic relationships are illustrated in Figs. 9.1 and 9.2.

9.1 ■ Torque Is the Cause of Rotary Motion

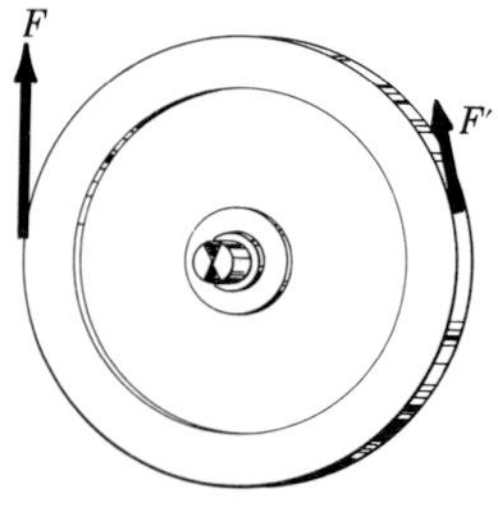

F

F'

Fig. 9.1 Torque causes a change in rotational motion. The result of an applied torque is rotational acceleration. The flywheel possesses rotational inertia, and torque is necessary to overcome that inertia.

If the flywheel of Fig. 9.1 is at rest, a force F applied at the rim will start it rotating clockwise around its axis. Or, if it is already rotating clockwise, a force F' applied in the opposite direction will gradually bring it to rest. However, force alone is not the only factor affecting the rotation of the wheel. The manner of applying the force is just as important as the magnitude of the force itself. A force is most effective in producing rotation of a body when (1) *its line of application is perpendicular to a radius at the point of application* and (2) *the force is applied as far as possible from the center of rotation*. Stated another way, the *turning effect of a given force is proportional to the perpendicular distance from the line of application of the force to the center of rotation*. Thus in Fig. 9.2*a*, AB is the line of application of force F, Q is its point of application, and OP is the perpendicular distance from the center of rotation to the line of application of the force. The distance OP is called the *moment arm* of the force. The *turning effect* produced by the force is defined as *the product of the force itself and its moment arm*. The turning effect is called *torque*. Torque is a vector quantity, considered positive if the turning effect is counterclockwise, negative if the turning effect is clockwise.

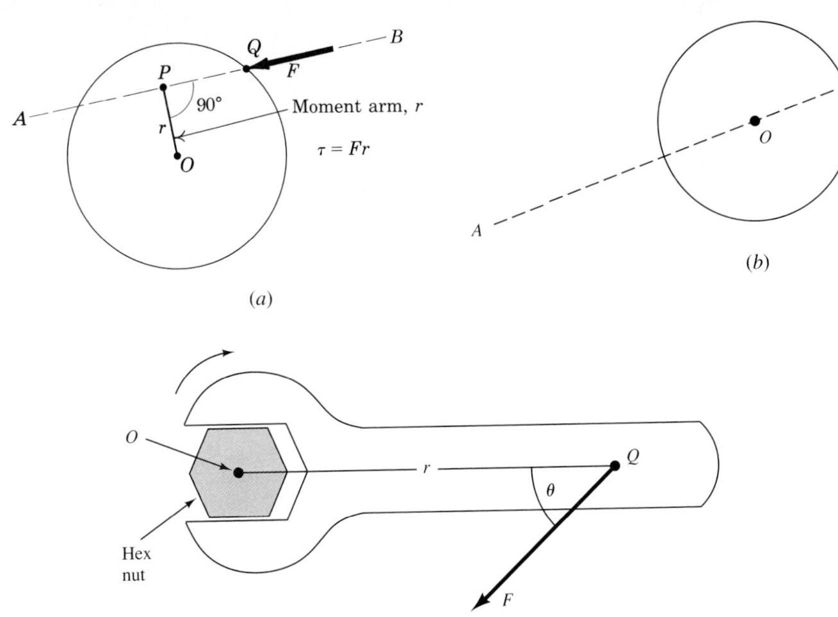

Fig. 9.2 *Moment arm* and torque. (*a*) *OP* is the moment arm of the force *F*. The product of force and its moment arm is torque: $\tau = Fr$. (*b*) If the line of action of an applied force passes through the axis of rotation of a rotatable body, the moment arm is zero, and no torque results. (*c*) When the applied force is not perpendicular to the moment arm *r*, the effective force is $F \cos \theta$, and the torque is $\tau = Fr \cos \theta$.

If we let the Greek letter τ (tau) stand for torque and designate the moment arm by r, we have

Torque defined as an equation.

$$\tau = Fr \tag{9.1}$$

Torque = force × moment arm

Note (Fig. 9.2*b*) that if the line of action of the force passes through the axis of rotation, r (the moment arm) = 0, and no torque results ($\tau = 0$).

Applied Force Not Perpendicular to Radius Figure 9.2*c* shows a situation in which the applied force acts at an angle other than 90° with the radius from the point of application. In this case the *effective* force perpendicular to the radius is $F \cos \theta$, and the torque is

$$\tau = Fr \cos \theta \tag{9.1'}$$

Any body, whether capable of rotating about a fixed axis or not, has inertia, and just as its inertia would resist a change in *translational* motion, so does its inertia (called *moment of inertia* for rotating bodies—see Sec. 9.4) resist a change in *rotational* motion. **Newton's first law** for rotating bodies can be stated as follows:

> ***A body at rest, but capable of rotating about a fixed axis, will remain at rest, and a body which is already rotating will remain in rotation at the same angular velocity, unless acted upon by an unbalanced torque.***

Newton's first law of motion (inertia) for rotating bodies.

Units of Torque F is most commonly measured in newtons (SI-metric), or in pounds (English-engineering). Moment arm r is expressed in meters or feet. The unit of torque in the metric system is the *newton-meter* (N · m) and the common engineering-industrial unit of torque in the U.S. is the *pound-foot* (lb · ft). The *pound-inch* (lb · in.) is also sometimes used. Note that although these English-system units involve the same quantities as the units of *work*, previously defined, the order of the quantities is reversed to avoid confusing one unit with the other. Torque and work are not the same concept by any means. Occasionally you may see the SI unit of torque

Metric and English system units of torque.

Fig. 9.3 Mechanic's torque wrench being used in the assembly of electrical apparatus. (P.A. Sturtevant Co.)

expressed as the meter-newton (m · N) to differentiate it from the N · m, a unit of work and energy (the joule).

The mechanic's torque wrench (Fig. 9.3) is a common industrial example of the use of torque in assembly operations. Nuts and bolts must be made up tight enough to hold the assembled parts against design stresses, but "strong-arm" methods often strip threads, shear bolts, or set up unequal stresses in the machine parts which cause trouble later. A torque wrench allows the mechanic to apply the specified amount of torque to every bolt or nut involved in an assembly operation. Some torque wrenches have dials and pointers which read directly in pound-inches, or in newton-meters. Others (especially motorized or compressed-air wrenches) have a ratchet which begins slipping when the torque reaches a value preset for the particular assembly operation.

Illustrative Problem 9.1 Specifications call for the head bolts on a refrigerating compressor to be set up to a torque of 510 lb · in. during final assembly. The torque wrench (see Fig. 9.3) measures 22 in. from the center of its bolt socket to the midpoint of the handle. What force (lb) at right angles to the handle must be applied?

Solution From Eq. (9.1),

$$\tau = Fr$$

Rearranging and substituting, $F = \dfrac{\tau}{r} = \dfrac{510 \text{ lb} \cdot \text{in.}}{22 \text{ in.}}$

$$= 23.2 \text{ lb} \qquad \textit{answer}$$

POWER TRANSMISSION

Power transmission in industry is accomplished largely through rotary motion.

In the early days of steam power, the first steam engines were capable only of back-and-forth or *reciprocating motion*. It soon became evident, however, that this new source of power from steam would be of greater value if the reciprocating motion of the piston could somehow be converted to *rotary motion*. The invention of the *connecting rod* (see Fig. 8.3) made this conversion to rotary motion possible. Today almost all mechanical power is converted into rotary motion, whether its source is steam, electricity, water, solar, or nuclear energy.

9.2 ■ Work and Power in Rotary Motion

The transmission of power is one of the most important applications of rotary motion. Consider the force F acting tangentially on the shaft whose cross section is shown in

Fig. 9.4. Let F turn the shaft counterclockwise through an arc distance s. The work done is

$$W = Fs$$

But (see Sec. 8.2)

$$s = r\theta \tag{8.1}$$

and therefore

$$W = Fr\theta$$

Also, by definition from Eq. (9.1), $Fr = \tau$

Substituting, we obtain

$$W = \tau\theta \tag{9.2}$$

or, in words,

Work = torque × angular displacement (in radians)

Now, dividing both sides of Eq. (9.2) by t, we obtain

$$\frac{W}{t} = \tau \frac{\theta}{t}$$

But W/t is the rate of doing work, or *power*, and θ/t is *angular velocity*. There results the very useful formula

$$P = \tau\omega \tag{9.3}$$

Power = torque × angular velocity

Power and torque, related by angular velocity.

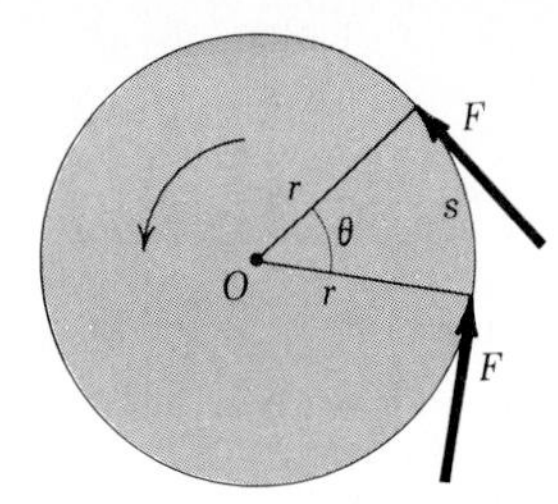

Fig. 9.4 Work done in rotary motion. The work done by torque, in turning a shaft through an angular distance θ, is $W = Fr\theta = \tau\theta$ (θ in radians).

This is the fundamental formula for power transmission by shafts, pulleys, and gears. An analysis of the formula reveals that for a given value of the power P, the torque τ is inversely proportional to the angular velocity. In other words, to transmit a given amount of power at a high rotational speed requires relatively little torque, while at a low speed high torque is necessary, if the same power is to be transmitted.

The units for Eq. (9.3) can be clarified by recalling that the unit of power is the *watt,* which is a joule/second, or a newton-meter/second. On the right-hand side of the equation, the unit of torque is the *newton-meter,* and the unit of angular velocity is s^{-1} since the radian is dimensionless. (See Sec. 8.3.) A similar analysis can readily be done for English units.

Illustrative Problem 9.2 An industrial engine is delivering 25 hp driving a small generator (see Fig. 9.5). The V-belt pulley on the generator has a mean diameter of 14 in., and it is turning at 1800 rev/min. Assuming no belt slippage, find the difference between the tension of the belt on the drive side of the pulley and that on the slack side (answer in pounds).

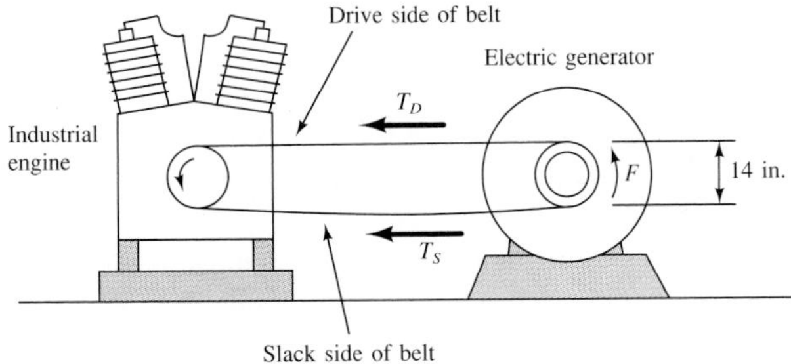

Fig. 9.5 Transmitting power by torque applied from a V-belt (Illustrative Problem 9.2).

Solution The difference in belt tension on the two sides of the driven pulley is actually the net force F applied to the rim of the pulley. It is applied with a moment arm of the given radius of 7 in. Using Eq. (9.3),

$$P = \tau\omega = Fr\omega$$

Force applied by a belt to a driven pulley is equal to the difference in belt tension between the drive side and the slack side.

Solve for F:

$$F = \frac{P}{r\omega}$$

Since the answer is to be in pounds, substitute as follows, and remember that the *radian is dimensionless.*

$$F = \frac{25 \text{ hp} \times \dfrac{550 \text{ ft} \cdot \text{lb/s}}{1 \text{ hp}}}{\dfrac{7}{12} \text{ ft} \times \dfrac{1800 \text{ rev/min} \times 2\pi \text{ rad/rev}}{60 \text{ s/min}}}$$

$$= 125 \text{ lb} \qquad \textit{answer}$$

Torque-Horsepower Formula Engineers use a handy approximate formula to relate torque, horsepower, and shaft angular velocity in revolutions per minute (rev/min). If you pay careful attention to units, you can derive this formula:

Shaft horsepower depends on the torque acting and on the shaft angular velocity.

$$\text{hp} = \frac{\text{torque (lb} \cdot \text{ft)} \times \text{rev/min}}{5250} \tag{9.4}$$

Such a formula enables design engineers to estimate the approximate horsepower requirements of rotating machinery if shaft revolutions per minute and the required torque are known.

Illustrative Problem 9.3 A hay mower is to operate from the power takeoff shaft of a farm tractor. If the required torque is 80 lb · ft when the drive shaft to the mower is turning at 500 rev/min, what horsepower must the tractor engine furnish to operate the mower?

Solution Solve Eq. (9.4) for horsepower:

$$\text{hp} = \frac{\text{torque} \times \text{(rev/min)}}{5250}$$

Substitute values:

$$\text{hp} = \frac{80 \times 500}{5250} = 7.62 \text{ hp} \qquad \textit{answer}$$

9.3 ■ Power Transmission by Drive Shafts

There is no such thing as an absolutely rigid body. Cables stretch under load; beams bend or are compressed, depending on the nature of the stresses they are subjected to; and shafts twist as torques are applied. In every case, the amount of stretch, bending, or twist is proportional to the stress applied. In other words, *stress is proportional to strain* within the limits of elasticity defined by Hooke's law (see Chap. 12).

An unbalanced torque causes angular acceleration.

ROTATIONAL DYNAMICS

Just as an *unbalanced force* is the cause of linear acceleration, so an *unbalanced torque* is the cause of angular acceleration. And just as the magnitude of the linear acceleration produced by a force depends on the mass of the object being accelerated, the angular acceleration due to a given torque depends on mass considerations. When angular acceleration, torque, and mass are all involved, the term *rotational dynamics* is used. Newton's second law relates angular acceleration and the *torque* which causes it, just as that law relates translational acceleration and the *force* causing it. With translational motion it is *mass* that relates force and acceleration, because mass is the inertial factor in that kind of motion. In rotation, however, the resistance to a change in angular velocity is not inherent in the *mass* of the rotating body alone, but depends on the *distribution of the mass*. The inertial factor in rotational motion is called *moment of inertia*.

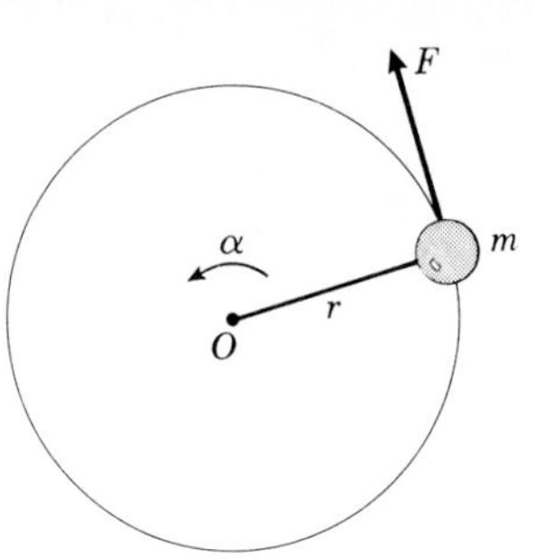

Fig. 9.6 Moment of inertia I of a mass m *revolving* around a point O is given by the relation $I = mr^2$, if the entire mass can be considered as effectively located at the distance r from the center of revolution O.

9.4 ■ Moment of Inertia

Figure 9.6 shows a small mass m at the end of a slim rod, revolving with angular acceleration α about point O at a distance r. It is being accelerated by a force F acting perpendicularly to r. Let the rod of length r be of such small mass that it is negligible compared to m. We can write Newton's second law for the linear (tangential) acceleration a_t, as

$$F_t = ma_t$$

Multiplying both sides by r gives

$$Fr = ma_t r$$

But $Fr = \tau$, the torque, from Eq. (9.1); and

from Eq. (8.8), $$a_t = \alpha r$$

the tangential acceleration. Therefore, torque

$$\tau = mr^2\alpha$$

But m and r^2 are constants for any given rotating body, and their product can be replaced by a single constant. The quantity mr^2 is called the *moment of inertia* (symbol I) of a rotating body or particle in which all the mass can be considered as being concentrated at a distance r from the center of rotation. It is also assumed here, and in the following discussions, that all parts of the rotating body have the *same* angular acceleration.

Moment of inertia is the inertial factor in rotational motion, just as *mass* is the inertial factor in translational motion.

Writing I for the quantity mr^2, we obtain:

$$\tau = I\alpha \qquad (9.5)$$

Torque = moment of inertia × angular acceleration

This is Newton's second law for rotational motion.

It is apparent from Eq. (9.5) that rotating bodies with large moments of inertia (I) require large torques to change the rate of angular motion.

Moment of inertia is a rather complex quantity in mechanics. The case referred to above, in which all the mass of a rotating body can be considered to be located at the same distance from the center of rotation, is rarely, if ever, encountered in practical industrial and engineering situations. Moment of inertia is actually a concept which involves not only the total mass rotating, but the manner in which that mass *is distributed* with respect to the center of rotation (see Fig. 9.7). In general, as mass is distributed farther from the axis of rotation, rotational inertia (i.e., moment of inertia) increases.

Moment of inertia of a rotating body depends not just on the mass of the body, but also on the *distribution of the mass* with respect to the axis or center of rotation.

In order to "visualize" the moment of inertia of a rigid body rotating about an axis, we can imagine the body being composed of a very large number of small mass

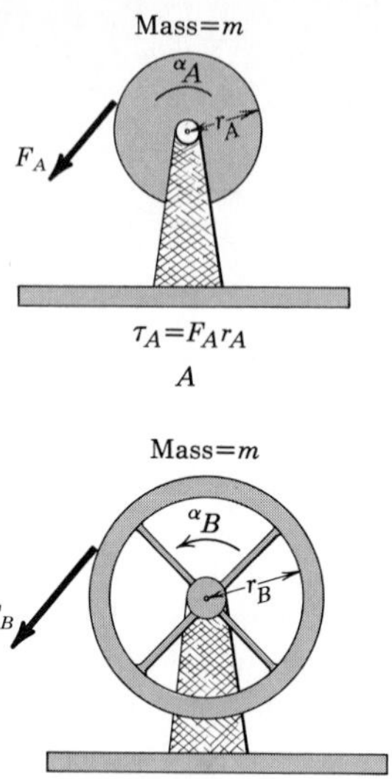

Fig. 9.7 Moment of inertia involves not just total mass, but the manner in which the mass is distributed with respect to the axis of rotation. Two flywheels of equal mass are shown. *A* has its mass concentrated near the axis of rotation. *B* has spokes so that most of its mass can be located in the rim. A much greater torque ($\tau = F_B r_B$) is required to set *B* in motion (or to stop it if it is already in motion) than is required for *A* ($\tau = F_A r_A$). Since $\tau = I\alpha$ in each case, I_B is greater than I_A, even though their masses are equal.

elements, each at its own distance from the center of rotation. Each, then, has its own small but finite moment of inertia mr^2. We can obtain the total moment of inertia of the entire body by summing up these small and discrete moments of inertia, as follows:

$$I = \Sigma\, mr^2 = m_1 r_1^2 + m_2 r_2^2 + m_3 r_3^2 + \cdots + m_n r_n^2 \tag{9.6}$$

where the Greek Σ (sigma) means "the sum of."

For rotors, shafts, flywheels, propellers, and other complex shapes, the use of the calculus is required to evaluate moments of inertia. However, for a simple thin ring whose total mass is *M* and whose entire mass can effectively be considered to be at a distance *r* from the center, we can derive an expression for the moment of inertia by the summing-up method of Eq. (9.6).

In Fig. 9.8, let *M* be made up of a very large number of small masses, m_1, m_2, $m_3, \cdots m_n$. Since each of the small masses is essentially at the same distance from *O* (that is, $r_1 = r_2 = r_3 = r$), we can sum up as follows:

$$\begin{aligned} I &= m_1 r_1^2 + m_2 r_2^2 + m_3 r_3^2 + \cdots \\ &= m_1 r^2 + m_2 r^2 + m_3 r^2 + \cdots \\ &= (m_1 + m_2 + m_3 + \cdots) r^2 \end{aligned}$$

But,

$$\Sigma[m_1 + m_2 + m_3 + \cdots] = M$$

the total mass of the ring. Therefore the moment of inertia of a thin ring is given by

$$I = Mr^2 \tag{9.7}$$
(Thin ring)

Units of Moment of Inertia The units of moment of inertia *I* are kg · m² or sl · ft². Occasionally, in industrial work where the pound (lb) is still used as a unit of mass, *I* may be expressed in lb · ft · s², which unit is the equivalent of sl · ft². The usual care must be taken when using the equation $\tau = I\alpha$, since Newton's second law requires *absolute units* of force. The *newton* of force (SI-metric) and the *slug* of mass (English-engineering system) are the units to use when computing $\tau = Fr = I\alpha$.

Moments of inertia of some common shapes are given in Fig. 9.9 without derivation, since calculus methods are necessary for developing these formulas.

Illustrative Problem 9.4 A flywheel approximates the shape of a solid disk. Its mass is 6 sl and its diameter is 4 ft. Find its moment of inertia.

Solution From Fig. 9.9, for a solid disk,

$$\begin{aligned} I &= \tfrac{1}{2} Mr^2 \\ &= \tfrac{6}{2}\ \text{sl} \times (2\ \text{ft})^2 \\ &= 12\ \text{sl} \cdot \text{ft}^2 \end{aligned}$$
answer

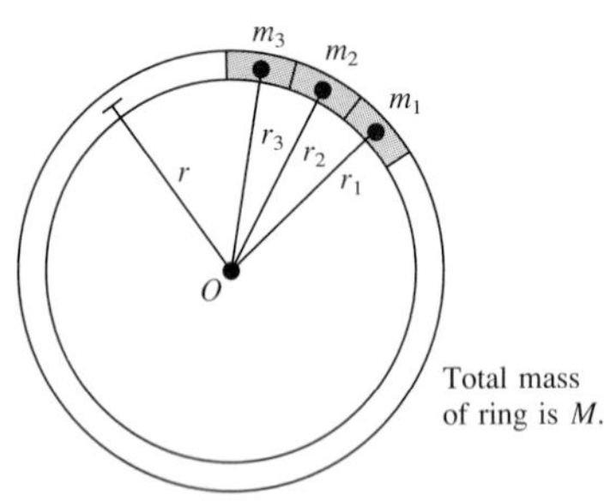

Fig. 9.8 Division of a thin-walled ring into small masses for the determination of moment of inertia. $I = Mr^2$ for a thin-walled ring.

Illustrative Problem 9.5 Find the moment of inertia of a 450-lb flywheel if all of its weight can be considered as being in an annular cylinder at the periphery of a set of spokes (assumed weightless). The distance from the axis of rotation to the outer edge of the annular cylinder is 3 ft, and the distance to the inner edge is 2.6 ft.

Solution From Fig. 9.9, the moment of inertia of an annular cylinder about its axis is given by

$$I = \tfrac{1}{2} M\,(r_1^2 + r_2^2)$$

Converting weight units to mass units,

$$m = \frac{w}{g} = \frac{450\ \text{lb}}{32.2\ \text{ft/s}^2} = 13.98\ \text{sl}$$

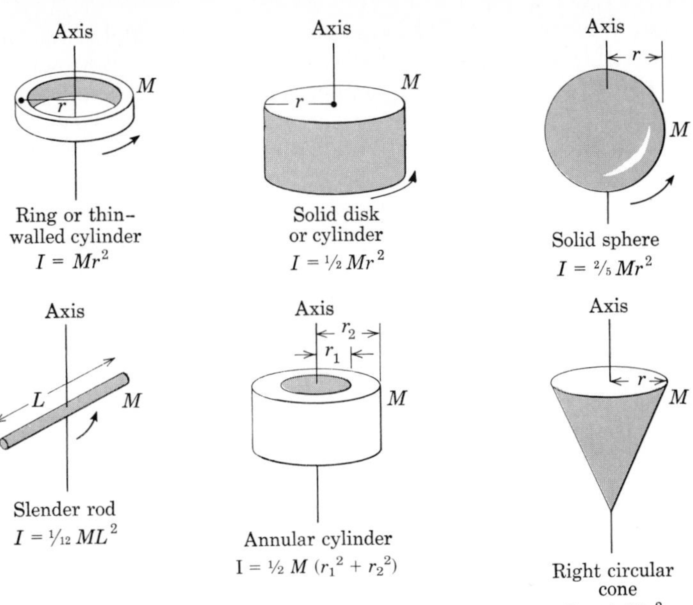

Fig. 9.9 Moments of inertia of some common bodies (all of mass M) around the axes indicated.

Substituting
$$I = \frac{13.98 \text{ sl} \times (\overline{2.6\text{ft}}^2 + \overline{3.0 \text{ ft}}^2)}{2}$$
$$= 110 \text{ sl} \cdot \text{ft}^2 \qquad \textit{answer}$$

Moment of inertia is a critical factor in flywheel design.

Illustrative Problem 9.6 If the flywheel of Illustrative Problem 9.5 is rotating at 400 rev/min (rpm), what torque (lb · ft) is required to bring it to rest in 2.0 min?

Solution From Eq. (8.6),
$$\omega_2 = \omega_1 + \alpha t$$
$$\omega_2 - \omega_1 = \alpha\ t \qquad (1)$$

But $\omega_2 = 0$
$$\omega_1 = 400 \text{ rev/min} = 6.67 \text{ rev/s}$$
$$= 13.34\ \pi \text{ rad/s (since there are } 2\ \pi \text{ rad per revolution)}$$
$$t = 2 \text{ min} = 120 \text{ s}$$

Substituting in (1) above,
$$0 - 13.34\ \pi \text{ rad/s} = \alpha \times 120 \text{ s}$$

from which,
$$\alpha = -\frac{13.34\pi}{\ } \text{rad/s}^2$$

From Eq. (9.5), torque $\tau = I\alpha$. From Illustration Problem 9.5, $I = 110 \text{ sl} \cdot \text{ft}^2$

Substituting,
$$\tau = (110 \text{ sl} \cdot \text{ft}^2)(-\frac{13.34\pi}{120} \text{ rad/s}^2)$$
$$= -38.4 \text{ lb} \cdot \text{ft} \qquad \textit{answer}$$

The negative answer indicates a retarding torque.

NOTE *To check the dimensional analysis, recall that the dimensions of the slug are lb · s^2/ft (Table 5.1).*

Torques inevitably produce "reverse torques," in accord with Newton's third law—action and reaction. Motor and engine mounts must be designed to absorb reverse torque.

Torque and Newton's Third Law For any torque applied to one rotating body, a torque equal in magnitude and opposite in direction will be applied to some other body. For example, as an aircraft engine exerts torque to turn the propeller, the aircraft is subjected to an equal but opposite torque, which would actually rotate the airframe if it were not compensated for by the force from air pressure on the airfoil surfaces. As a motor or engine applies torque to a shaft, the shaft exerts an opposite torque on the motor or engine. The motor or engine mounts must be designed to absorb this "reverse" torque. The "reverse" torque is of such magnitude on helicopters that an extra propeller, usually mounted toward the rear of the aircraft, is required to counteract the tendency for the aircraft to rotate in the opposite direction from that of the rotor blades.

9.5 ■ Angular Momentum

The rotational counterpart of linear momentum is *angular momentum*. It will be recalled that linear momentum was defined by Newton as "the quantity of motion" possessed by a body. It is given by the expression

$$\boldsymbol{p} = mv \qquad (6.8)$$

Linear momentum is the product of mass (the inertia factor) times linear velocity. It therefore follows logically that *angular momentum* is the product of *moment of inertia* times *angular velocity*. Denoting angular momentum by $\boldsymbol{L}$, we have

$$\boldsymbol{L} = I\omega \qquad (9.8)$$

$$\text{Angular momentum} = \text{moment of inertia} \times \text{angular velocity}$$

Linear momentum of a body in translation is a measure of the impact or impulse which it will impart if it acts on some other object. And angular momentum is a measure of the angular impulse which a rotating body can deliver to a shaft or a brake. Practical problems of angular momentum and angular impulse are encountered in automotive transmissions, spin-dry laundry equipment, impulse wrenches, and torque converters for heavy industrial and marine machinery, to list only a few examples.

Units of angular momentum.

The units of angular momentum are:

In the SI-metric system: $\dfrac{\text{kg} \cdot \text{m}^2}{\text{s}}$

In the English (engineering) system: $\dfrac{\text{sl} \cdot \text{ft}^2}{\text{s}}$

Illustrative Problem 9.7 The flywheel of a stationary diesel engine is in the form of a flat circular disk whose radius is 2 ft and whose weight is 200 lb. It is rotating at 450 rev/min. Find its angular momentum.

Solution Angular velocity is

$$\omega = \frac{2\pi \text{ rad}}{\text{rev}} \times 450 \frac{\text{rev}}{\text{min}} \times \frac{1}{60 \text{ s}} = 47.1 \frac{\text{rad}}{\text{s}}$$

$$I_{\text{disk}} = \frac{1}{2} \times \frac{200 \text{ lb}}{32.2 \text{ ft/s}^2} \times (2 \text{ ft})^2$$

$$= 12.42 \text{ sl} \cdot \text{ft}^2$$

$$\boldsymbol{L} = I\omega = 12.42 \text{ sl} \cdot \text{ft}^2 \times 47.1 \text{ rad/s}$$

$$= 585 \frac{\text{sl} \cdot \text{ft}^2}{\text{s}} \qquad \textit{answer}$$

Conservation of Angular Momentum It will be recalled that in interactions between bodies which make up a system in translatory motion, the total momentum of the system of interacting bodies is conserved (see Sec. 6.11). It is only in the case where an *external force* acts on the system that momentum is changed. When Newton's second law, $Ft = mv_2 - mv_1$, is expressed in the form

$$F = \frac{mv_2 - mv_1}{t}$$

it states, in effect, that a force acting on a body is measured by the time rate of change of momentum of the body. In like manner, *torque* is measured by the *time rate of change of angular momentum* of a *rotating* body. As an equation

$$\tau = \frac{I(\omega_2 - \omega_1)}{t}$$

or

$$\tau t = I(\omega_2 - \omega_1) \tag{9.9}$$

Angular impulse = Change in angular momentum

Equation (9.9) is the rotational counterpart of Eq. (6.10).

> **The law of conservation of angular momentum is: The total angular momentum of a body or system of bodies is a constant unless the body or system of bodies is acted upon by an external torque.**

Angular momentum is conserved in the absence of external torques.

There are a number of interesting applications of conservation of angular momentum in the fields of sports and entertainment. Figure skaters, ballet dancers, gymnasts, and high-divers often start into a spin at slow speed with arms and legs outstretched (I at a maximum); then, pulling in arms and legs (making I smaller), they spin at a much greater angular velocity, with *angular momentum unchanged* (Fig. 9.10*a*). The phenomenon can be illustrated in the laboratory or lecture room with the arrangement shown in Fig. 9.10*b*.

Discus throwers wind up in a tight, high-speed spin (small I, large ω), then allow the discus and their arms to spiral out, resulting in a larger value of I (and increased *linear* speed of the discus) at the sacrifice of *angular velocity* ω (Fig. 9.10*c*). In the final stages of their throws the inertial reaction force of the accelerating discus, acting backward through hand and arm, exerts a torque which stops the throwers' angular motion entirely.

The quantity τt in Eq. (9.9) is called *angular impulse*; it equals the *change in angular momentum* of the body about the same axis.

Angular impulse τt is equal to change in angular momentum.

Angular Momentum Is a Vector Quantity Earlier, we found that linear velocity, acceleration, and momentum are vector quantities. So are angular velocity, angular acceleration, and angular momentum. *Angular velocity* can be represented by a vector whose length to scale is equal to the rotational speed in rad/s, and whose direction is along the axis of rotation in the direction of advance of a right-hand-threaded screw (see Fig. 9.11*a*).

Angular momentum is represented by a vector whose magnitude is equal to $I\omega$, and whose direction is also given by the advance of a right-hand-threaded screw. The fingers and thumb of the right hand can also be used, as in Fig. 9.11*b*, for a "right-hand-thumb rule" to determine the direction of the vector.

Since angular momentum is a *vector quantity,* an externally applied torque is required to change the *direction* of the axis of rotation of a spinning object, whether or not the speed of rotation is changed. The greater the magnitude of **L** (angular momentum), the greater the torque required to change the direction of the axis of

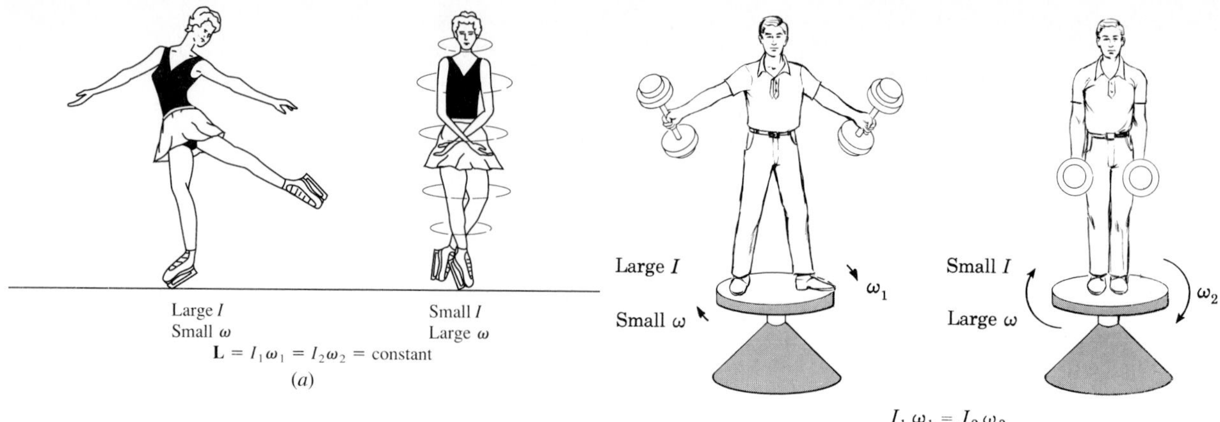

Fig. 9.10 Examples of conservation of angular momentum. (*a*) Figure skaters make frequent use of conservation of angular momentum as they go from a ''wide arc'' position into a tight spin. (*b*) A laboratory demonstration of conservation of angular momentum. Changing the distribution of mass of a rotating body results in a change of angular *velocity,* but the angular *momentum* is constant. (*c*) The discus thrower first builds angular momentum in a tight spin, then allows the discus to spiral outward to a much larger arc in which its *linear* velocity increases greatly. The inertia of the discus itself exerts a backward torque on the spinning athlete, and his spin stops as the athlete applies force to accelerate the discus.

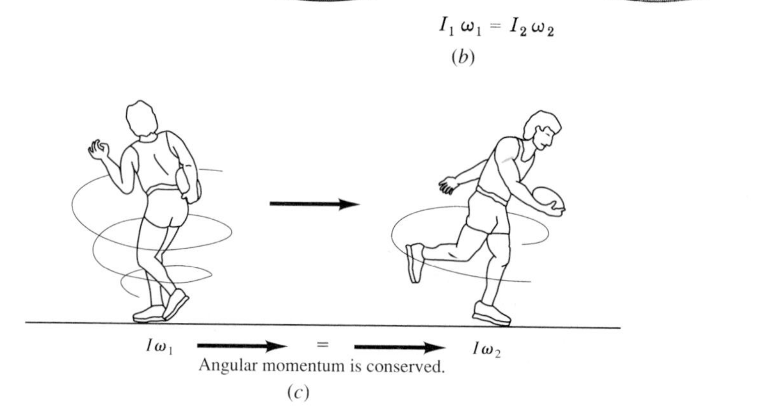

rotation. Another way of stating this gives a rotational counterpart of **Newton's first law** of motion:

> ***A rigid body rotating about a fixed axis will continue rotating about that axis with constant angular speed, unless acted upon by an unbalanced torque.***

Note that this is not true for nonrigid bodies, i.e. bodies where mass distribution can be changed while the body is rotating.

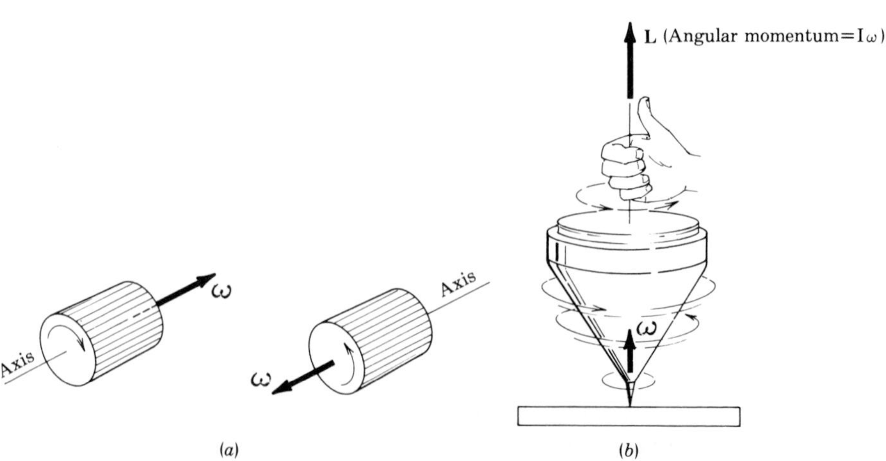

Fig. 9.11 Vector considerations in rotational motion. (*a*) Vector representation of angular velocity, ω. The vector direction is that in which a right-hand-threaded screw would advance. (*b*) The angular momentum vector of a spinning top. If one imagines the fingers of the right hand wrapped around the axis of rotation in the direction of the rotation, the extended thumb shows the direction of the angular momentum vector **L.**

Spin Stabilization and the Gyroscope The spinning top is a simple example of angular momentum. Its "spin stabilization" keeps its axis from wobbling until, due to friction, the angular velocity and the angular momentum fall to near zero. Spin is purposely imparted to gun projectiles and to footballs to keep them from wobbling in flight. Two-wheel cycles are partially spin-stabilized by the angular momentum of their rotating wheels (Fig. 9.12).

Objects that spin or rotate while in translational motion tend to retain their original direction of motion because of their angular momentum. This phenomenon is called *spin stabilization.*

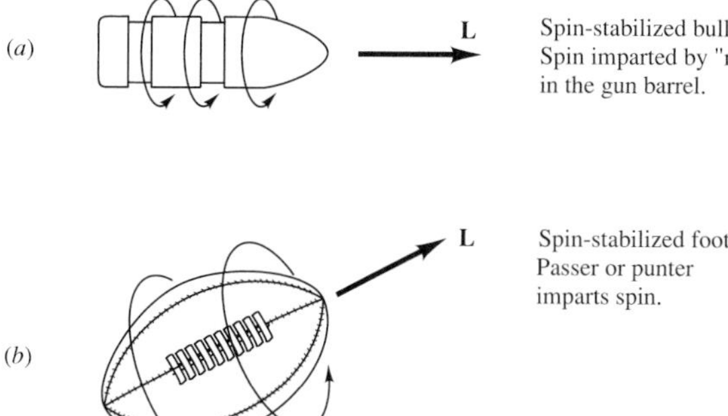

(a)
L
Spin-stabilized bullet.
Spin imparted by "rifling" in the gun barrel.
(b)
L
Spin-stabilized football.
Passer or punter imparts spin.

ACK

(c)

Fig. 9.12 Spin stabilization. (*a*) Gun projectiles would have very little accuracy if they were not spin-stabilized. (*b*) Both accuracy and distance are increased by spin-stabilizing a football. (*c*) The angular momentum of bicycle wheels helps to stabilize the cycle in an upright position. (Alan Degasis photo)

The gyroscope (Fig. 9.13) is a device especially designed to utilize the vector properties of angular momentum. Gyroscopes are used in spin-stabilized navigation equipment for aircraft, submarines, and aerospace vehicles, and also in some ballistic missile guidance systems.

A gyroscope is basically a rotating disk or wheel with most of its mass concentrated in the rim so that it will have a large moment of inertia. The rotor axis is mounted in an assembly of rings and bearings called gimbals (see Fig. 9.13). The mounting arrangement is such that the gyro rotor is free to turn about any one of three mutually perpendicular axes. With its large moment of inertia, and when it is spinning with high angular velocity, the angular momentum **L** of the gyro rotor is very great, and it possesses a high degree of space-directional stability, in accord with the law of conservation of angular momentum. (Remember that **L** is a vector quantity.) In other words, if the rotor is set spinning with the axis pointed toward the North Star (for example), the axis will continue to point in that direction. The three-dimensional gimbal mounting allows the base of the gyro (or the aircraft, ship, torpedo, rocket, or space vehicle in which the gyro is installed) to turn or change direction at will without transmitting any torques* to the rotor axis. The gyro rotor, then, once spinning at a high angular velocity, will remain in a fixed plane in space regardless of the maneu-

* Except for the extremely small torques resulting from friction in the gimbal bearings. These bearings are finely crafted to be as nearly frictionless as possible.

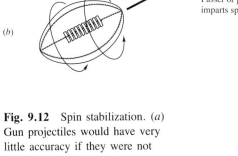

Fig. 9.13 A gimbal-mounted gyroscope. Note that most of the rotor's mass is concentrated in the outer rim, to maximize moment of intertia. The arrangement of gimbal rings and bearings allows three degrees of freedom for the gyro rotor axis. The gyro rotor axis z can maintain any initially chosen orientation in three-dimensional space, while the support S is moved at will. The center of mass of the gyro wheel is located at the intersection of the space axes, x, y, and z, and therefore no torque is exerted on it by forces resulting from the motion of its support system. (Minor torques exist at the gyro rotor bearings, resulting from friction, but these are kept as near zero as possible by precision bearing design and by special lubricants.) (Central Scientific Company)

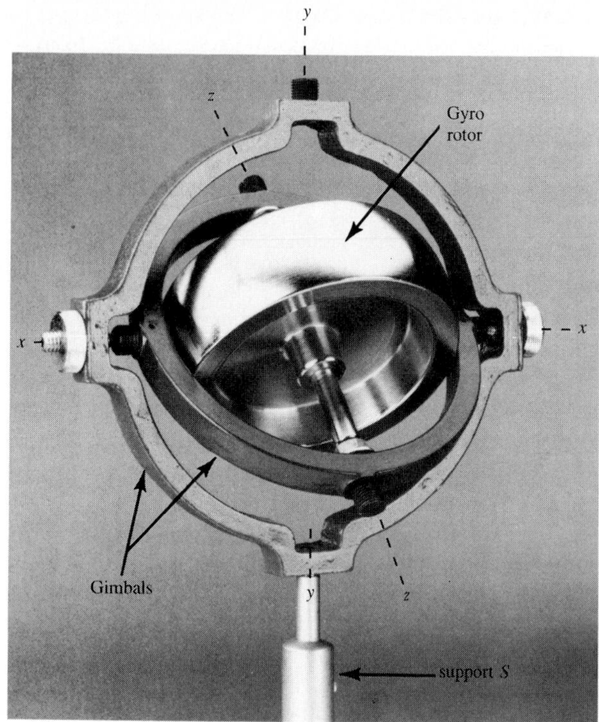

A gyro rotor is a spin-stabilized mass. Its gimbal-and-ring mounting allows it to turn freely about any one of three mutually perpendicular axes, and three mutually perpendicular axes describe space. Consequently, a gyro rotor possesses space-directional stability.

vers of the craft or vehicle in which it is installed. Instruments based on gyro-rotor space-directional stability include the *gyrohorizon indicator* and the *gyropilot*. Instruments utilizing the gyroscope, but incorporating additional features for the *intentional* application of external torques, include the *gyrocompass*, the *gyrostabilizer* for passenger ships, and *inertial guidance systems* in missiles and aerospace vehicles.

A spinning top turns about the sharpened point at the end of its axis, this point touching the table, floor, or ground (Fig. 9.11*b*). Consequently, if and when the top leans the slightest bit from the vertical, its center of gravity is no longer directly above the point of support and the force of gravity produces a torque acting on the axis of rotation. A gyro, on the other hand, is mounted in the gimbals in such a manner that it turns freely in any plane, *around its own center of gravity* (Fig. 9.13). There is, therefore, no moment of force (torque) exerted on it by gravity, and it is this absence of gravity-induced torques that allows the rotor axis to continue to point toward a fixed direction in space.

As mentioned above, in some gyroscopic applications, torques are intentionally applied. In such cases the gyro axis is said to *precess*. Predicting the exact nature (rate and direction) of the precession that will result from an intentionally applied torque involves a level of vector analysis and mathematical treatment well beyond the scope of an introductory text.

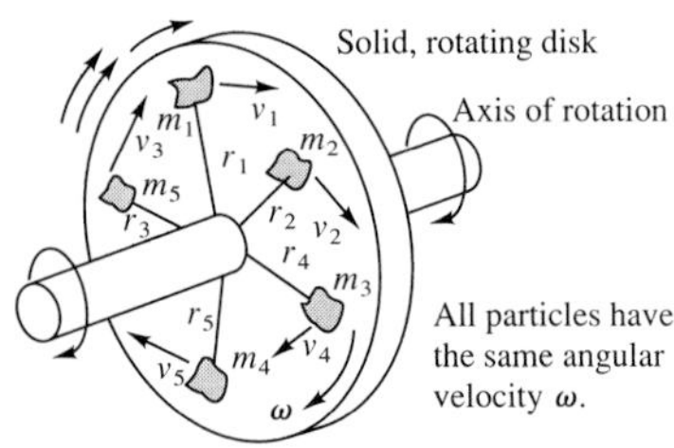

Fig. 9.14 Particles making up a solid rotating disk. Let m_1, m_2, m_3, m_4, and m_5 represent an infinite number of particles making up the total disk. All of the masses have the same *angular* velocity ω, but they may have different distances from the axis of rotation and different instantaneous linear velocities.

9.6 ■ Rotational Kinetic Energy

In order to express rotational kinetic energy (KE_{rot}) in mathematical terms, it is helpful to visualize the rigid rotating body as being composed of a very large number of small particles at various distances from the axis of rotation. These particles may have different masses, and they have different linear (tangential) speeds at any instant; but they all move with *the same angular velocity* ω. The diagram of Fig. 9.14 depicts a rotating disk and shows a few of these hypothetical particles. It serves as a basis for the following analysis.

The kinetic energy of the particle whose mass is m_1 is given by KE $= \frac{1}{2}m_1 v_1^2$, where v_1 is its instantaneous tangential velocity. But m_1 is at a distance r_1 from the axis of rotation, and we know from Sec. 8.3 and Eq. (8.2) that the relation between linear speed and angular velocity is $v = \omega r$. Therefore, the kinetic energy of mass m_1 can be expressed as

$$\begin{aligned} KE_{m_1} &= \tfrac{1}{2}m_1 v_1^2 \\ &= \tfrac{1}{2}m_1 r_1^2 \omega^2 \end{aligned}$$

The summation of the kinetic energies of all such particles will give the rotational kinetic energy of the entire wheel, as follows:

$$\begin{aligned} KE_{rot} &= \tfrac{1}{2}m_1 r_1^2\omega^2 + \tfrac{1}{2}m_2 r_2^2 \omega^2 + \tfrac{1}{2}m_3 r_3^2 \omega^2 + \cdots + \tfrac{1}{2}m_n r_n^2 \omega^2 \\ &= \Sigma\tfrac{1}{2}mr^2 \omega^2 = \tfrac{1}{2}(\Sigma mr^2)\omega^2 \end{aligned}$$

where the symbol Σ stands for the summation.

But (see Sec. 9.4), Σmr^2 is the moment of inertia I of a rotating body, and consequently,

Rotational kinetic energy (KE_{rot}) is directly proportional to the moment of inertia I of the rotating body and to *the square* of its angular velocity ω.

$$KE_{rot} = \tfrac{1}{2}I\omega^2 \tag{9.10}$$

The same result can also be obtained from an application of the *work-energy theorem* (principle of work) to rotational motion. The work-energy theorem requires that the kinetic energy of a rotating body be equal to the work done on it to accelerate it from rest (zero angular velocity) to its present angular velocity. It will be recalled that to cause angular acceleration, torque must be applied. From Sec. 9.4, an applied torque produces an angular acceleration given by

$$\tau = I\alpha \tag{9.5}$$

Multiplying both sides of Eq. (9.5) by θ, the angular distance turned during the acceleration period, gives

$$\tau\theta = I\alpha\theta \tag{9.11}$$

But, from Eq. (9.2), $\tau\theta$ is the *work done to bring the spinning body from rest up to its present angular velocity*. Now, since the initial angular velocity is zero, we can rewrite the acceleration equation, Eq. (8.11), in the form

$$2\alpha\theta = \omega^2$$

from which $$\alpha\theta = \frac{\omega^2}{2}$$

Therefore, the work done (and the rotational kinetic energy acquired) is:

$$\text{Work done} = \tau\theta = I\alpha\theta = I\frac{\omega^2}{2} = \tfrac{1}{2}I\omega^2 = KE_{rot}$$

the same expression as Eq. (9.10) above.

Illustrative Problem 9.8 A solid cylinder with a mass of 200 kg and radius 0.5 m is rotating about its axis at an angular velocity of 9 rad/s. Find (*a*) its rotational kinetic energy; and (*b*) the uniform torque that must be applied to stop it from rotating in one minute.

Solution The moment of inertia of a solid cylinder is (see Fig. 9.9) $I = \frac{1}{2}Mr^2$.

Substituting given values, $$\begin{aligned} I &= \frac{200 \text{ kg} \times (0.5 \text{ m})^2}{2} \\ &= 25 \text{ kg} \cdot \text{m}^2 \end{aligned}$$

(*a*) From Eq. (9.10),

$$KE_{rot} = \tfrac{1}{2}I\omega^2 = \frac{25\ \text{kg}\cdot\text{m}^2 \times (9\ \text{rad/s})^2}{2}$$

$$= 1012\ \frac{\text{kg}\cdot\text{m}^2}{\text{s}^2} \quad \text{(since the radian has no dimensions)}$$

$$= 1012\left(\frac{\text{kg}\cdot\text{m}}{\text{s}^2}\right)\text{m} = 1012\ \text{N}\cdot\text{m} = 1012\ \text{J} \qquad \textit{answer}$$

(*b*) The uniform negative acceleration required to stop the cylinder from rotating in exactly one minute is obtained from Eq. (8.5) as follows:

$$\alpha = \frac{\omega_2 - \omega_1}{t} = \frac{0\ \text{rad/s} - 9\ \text{rad/s}}{60\ \text{s}} = 0.15\ \text{rad/s}^2$$

The required torque is obtained from Eq. (9.5), $\tau = I\alpha$. Substituting values,

$$\tau = 25\ \text{kg}\cdot\text{m}^2\left(-0.15\ \frac{\text{rad}}{\text{s}^2}\right) = -3.75\ \frac{\text{kg}\cdot\text{m}^2}{\text{s}^2}$$

$$= -3.75\left(\frac{\text{kg}\cdot\text{m}}{\text{s}^2}\right)\text{m} = -3.75\ \text{N}\cdot\text{m} \qquad \textit{answer}$$

(The minus sign merely indicates that the required torque is oppositely directed to the existing rotation.)

Flywheels for Energy Storage The use of heavy flywheels to ''smooth out'' the flow of mechanical energy from reciprocating engines has been common for more than a century. The flywheel gains kinetic energy as the engine supplies torque, ''stores'' the energy temporarily as rotational KE, and then provides it to the load during the interim when the engine is getting ready for its next power stroke.

There is currently some interest in the energy storage capabilities of really mammoth flywheels. As solar power, wind power, and tidal power—all of them intermittent in nature—are being developed, the problem of energy storage will become very important. Huge flywheels could possibly provide one means of ''stockpiling'' energy as it is produced, and storing it in the form of rotational KE for delivery at another time of day or even later in the week.

The following problem is illustrative of this energy-storage concept.

Illustrative Problem 9.9 A mammoth flywheel, weight 20 tons, is used to ''store'' mechanical energy. It is in the form of an annular cylinder with $r_1 = 6$ ft and $r_2 = 7.3$ ft. The effect of spokes is to be neglected. The flywheel is brought from rest up to an angular velocity of 300 rev/min (rpm) and left spinning until its energy is needed. (*a*) Find its rotational kinetic energy in foot-pounds. (*b*) For what period of time could energy be drawn from it at the rate of 5 hp?

Solution

(*a*) From Fig. 9.9, the moment of inertia of an annular cylinder is

$$I = \tfrac{1}{2}M\,(r_1^2 + r_2^2)$$

Substituting, remembering to express the mass M in slugs (20 tons = 1242 sl)

$$I = \frac{1242\ \text{sl}\ (\overline{6\ \text{ft}}^2 + \overline{7.3\ \text{ft}}^2)}{2} = 621\ \text{sl}\ (36 + 53.3)\ \text{ft}^2$$

$$= 55{,}450\ \text{sl}\cdot\text{ft}^2 \quad \text{(the moment of inertia)}$$

$$\text{KE}_{\text{rot}} = \tfrac{1}{2}I\omega^2 = \frac{55{,}450\ \text{sl}\cdot\text{ft}^2}{2} \times \left(\frac{300 \times 2\pi}{60}\ \frac{\text{rad}}{\text{s}}\right)^2$$

$$= 27.3 \times 10^6\ \text{sl}\cdot\text{ft}^2/\text{s}^2 \quad \text{(the radian has no dimensions)}$$

Recalling that a slug is 1 lb · s²/ft,

$$\text{KE}_{\text{rot}} = 27.3 \times 10^6 \frac{\left(\dfrac{\text{lb} \cdot \text{s}^2}{\text{ft}}\right)\text{ft}^2}{\text{s}^2} = 27.3 \times 10^6 \text{ ft} \cdot \text{lb} \qquad \textit{answer}$$

(*b*) Since 1 hp = 33,000 ft · lb/min, we write, for the time period the flywheel will provide 5 hp,

$$T = \frac{27.3 \times 10^6 \text{ ft} \cdot \text{lb}}{5 \text{ hp} \times 3.3 \times 10^4 \text{ ft} \cdot \text{lb/min} \cdot \text{hp}} = 165 \text{ min or } 2.75 \text{ h} \qquad \textit{answer}$$

9.7 ■ Rotational and Translational Kinetic Energies Combined

As a ball, cylinder, or wheel rolls along a surface, it has both kinetic energy of translation and kinetic energy of rotation. The rolling cylinder of Fig. 9.15, for example, rotates around C and has a rotational kinetic energy given by

$$\begin{aligned}\text{KE}_{\text{rot}} &= \tfrac{1}{2} I\omega^2 \\ &= \tfrac{1}{2}(\tfrac{1}{2}mr^2)\omega^2 \\ &= \frac{mr^2\omega^2}{4} \qquad \text{(for a solid cylinder)}\end{aligned}$$

Rotating bodies can also be moving with translational motion, in which case they possess both KE_{rot} and KE_{trans}.

In addition, its center of mass C moves to the right with *linear velocity* v, and the cylinder has $\text{KE}_{\text{trans}} = \frac{1}{2}mv^2$. The *total kinetic energy* of a body which has both angular velocity and linear velocity (for example, any rolling object) is then

$$\text{KE}_{\text{tot}} = \tfrac{1}{2} I\omega^2 + \tfrac{1}{2}mv^2 \qquad (9.12)$$

Illustrative Problem 9.10 A light cord is wrapped around a solid cylinder of mass 10 kg and radius 10 cm. The cylinder rotates without friction about a horizontal axis, as illustrated in Fig. 9.16. A mass of 1 kg is attached to the free end of the cord and the system is released to the force of gravity with the small mass at a height of 2 m above the floor. Find (*a*) the linear speed of the 1-kg mass and (*b*) the angular velocity of the cylinder, at the instant that the 1-kg mass hits the floor.

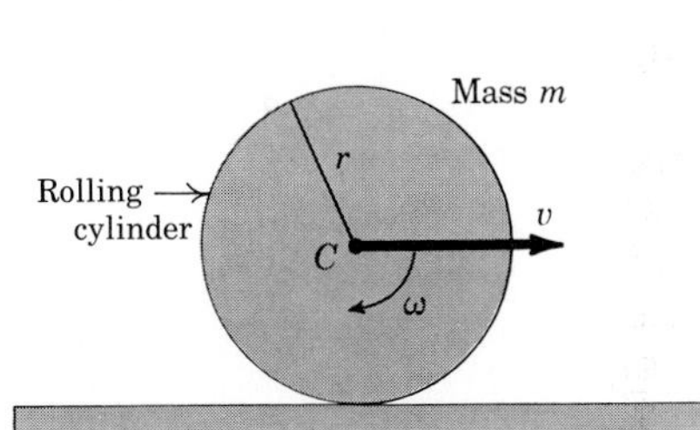

Fig. 9.15 A rolling cylinder of mass m and radius r is rolling on a level surface with linear velocity v and angular velocity ω. It has both kinetic energy of rotation and kinetic energy of translation. $\text{KE}_{\text{tot}} = \text{KE}_{\text{rot}} + \text{KE}_{\text{trans}} = \frac{1}{2}I\omega^2 + \frac{1}{2}mv^2$.

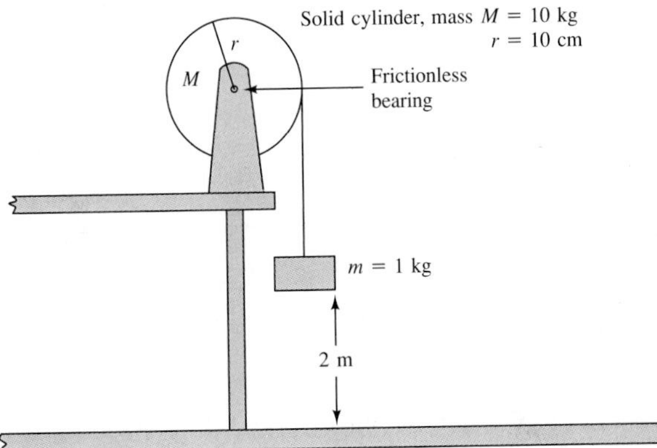

Fig. 9.16 The rotational dynamics of a solid cylinder set in rotation by a torque provided by the force of gravity (Illustrative Problem 9.10).

Solution The law of conservation of energy requires that the initial potential energy of the 1-kg mass (with respect to the floor) be equal to the total kinetic energy of the system as the mass m hits the floor. The initial potential energy is PE $= mgh$; and the final kinetic energy is $KE_{tot} = \frac{1}{2}mv^2 + \frac{1}{2}I\omega^2$, where m refers to the descending mass, and I and ω refer to the rotating cylinder. For a solid cylinder, $I = \frac{1}{2}Mr^2$. As usual, $\omega = v/r$.

(*a*) Setting up the equation for conservation of energy,

$$mgh = \tfrac{1}{2}mv^2 + \tfrac{1}{2}(\tfrac{1}{2}Mr^2)\left(\frac{v}{r}\right)^2$$

Factoring and simplifying, (Note that r's cancel, indicating that the translational velocity is independent of the radius of the cylinder.)

$$mgh = \tfrac{1}{2}(m + \tfrac{1}{2}M)\ v^2$$

$$v^2 = \frac{mgh}{\frac{1}{2}m + \frac{1}{4}M}$$

Substituting and evaluating,

$$v^2 = \frac{19.6\ \text{kg} \cdot \text{m}^2/\text{s}^2}{3\ \text{kg}}$$

$$v = \sqrt{\frac{19.6}{3}\ \frac{\text{m}^2}{\text{s}^2}} = 2.56\ \text{m/s} \qquad \textit{answer}$$

(*b*)

$$\omega = \frac{v}{r} = \frac{2.56\ \text{m/s}}{0.1\ \text{m}} = 25.6\ \text{rad/s} \qquad \textit{answer}$$

Energy of a Body Rolling Down an Inclined Plane In Chapter 6 (Secs. 6.4 to 6.8), energy transformations were discussed in detail. The relationships between gravitational potential energy (GPE) and kinetic energy (KE) were given special emphasis. The *work-energy theorem* (principle of work) was introduced and applied to a variety of situations involving energy of position and energy of motion. Among the important principles derived from those discussions was the interesting finding that work done against gravity as a mass is lifted from a lower elevation to a higher one is independent of the path used (Sec. 6.7). A somewhat related and equally interesting phenomenon is the fact that the translational speeds of circular disks or cylindrical bodies *rolling* (without friction) down smooth surfaces are *independent of their masses and their radii*, as the following problem shows.

Illustrative Problem 9.11 A solid circular cylinder (or disk) of radius r and mass m is released at point P on an inclined plane (Fig. 9.17), and it rolls without slipping

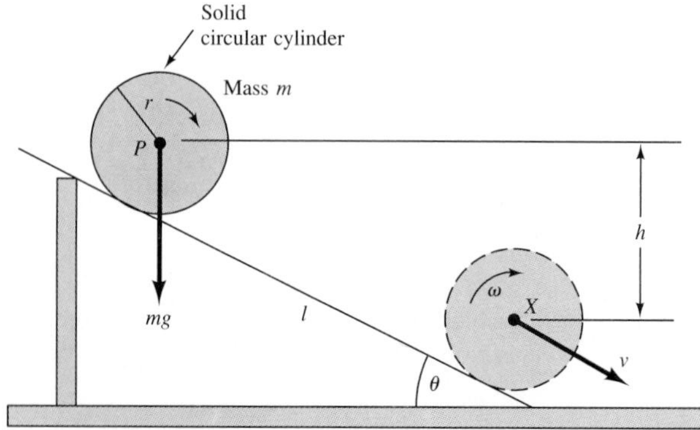

Fig. 9.17 Cylinder rolling without friction or slippage down an inclined plane (Illustrative Problem 9.11).

and without friction to the bottom of the plane at X. Find its linear (translational) speed at X. (Points P and X refer to the center of mass of the cylinder, on its axis of rotation.)

Solution Since no work is done against friction, the work-energy theorem requires that the sum of the kinetic energy change and the gravitational potential energy change be equal to zero. We therefore write

$$(\mathrm{KE}_{\mathrm{trans}} + \mathrm{KE}_{\mathrm{rot}})_X - (\mathrm{KE}_{\mathrm{trans}} + \mathrm{KE}_{\mathrm{rot}})_P + (mgh_X - mgh_P) = 0$$

But $(\mathrm{KE}_{\mathrm{trans}} + \mathrm{KE}_{\mathrm{rot}})_P = 0$ and $mgh_X = 0$. Recalling that $\mathrm{KE}_{\mathrm{trans}} = \frac{1}{2}mv^2$ and $\mathrm{KE}_{\mathrm{rot}} = \frac{1}{2}I\omega^2$, we write

$$[(\tfrac{1}{2}mv_X^2 + \tfrac{1}{2}I\omega_X^2) - 0] - mgh_P = 0$$

or,

$$mgh = \tfrac{1}{2}I\omega_X^2 + \tfrac{1}{2}mv_X^2 \tag{1}$$

A large-diameter heavy wheel or hoop and a small-diameter lightweight wheel (hoop) will roll down a hill (friction neglected) with the same translational velocity. The translational velocity attained depends only on the vertical height (difference in elevation).

The moment of inertia for a solid cylinder (see Fig. 9.9) is $I = \frac{1}{2}mr^2$, and the angular velocity at X is $\omega_X = v_X/r$. Substituting these relationships in (1) above gives

$$\left[\tfrac{1}{2}mv_X^2 + \tfrac{1}{2}(\tfrac{1}{2}mr^2)\left(\frac{v_X}{r}\right)^2\right] - mgh_P = 0 \tag{2}$$

Simplifying (2) by cancelling m's and r's yields

$$\tfrac{1}{2}v_X^2 + \tfrac{1}{4}v_X^2 = gh$$

or

$$\frac{3}{4}\,v_X^2 = gh$$

from which,

$$v_X = \sqrt{\frac{4\,gh}{3}} \qquad \textit{answer}$$

Note the following conclusions from the above Illustrative Problem:

1. Neither the radius of the cylinder nor its mass has any effect on the translational speed attained. All solid cylinders will attain the same speed on a given inclined plane.
2. The *slope* of the plane does not affect the speed attained either, as long as the cylinder *rolls without slipping* and *without friction*.
3. See Sec. 6.6 to recall that the speed of an object *falling* from a height h (without friction), or *sliding* down a frictionless incline through a difference in elevation h, is $v = \sqrt{2gh}$.
4. For a *rolling* object, v_{trans} is less than it is for a falling or a sliding object, if friction is absent in all three cases. Why?

From Eq. (1) in the above problem, one can calculate the linear velocity that *any round body* will attain as it rolls without friction down an inclined plane. The results of such calculations give: For a hoop, $v = \sqrt{gh}$; for a sphere, $v = \sqrt{10/7\ gh}$. All bodies of the same *kind* will attain the same velocity regardless of their masses or radii; but bodies of *different* geometry roll at different speeds. If a solid cylinder (or disk), a thin-walled ring (hoop), and a solid sphere are all released simultaneously at the top of an inclined plane, the sphere will arrive at the bottom first, the cylinder next, and the hoop last. A block *sliding without friction* would beat them all to the bottom!

9.8 ■ Linear and Angular Motion Compared

In concluding the chapter, Table 9.1 is provided, to bring together, for ready reference, the principal equations and symbols of linear and angular motion.

Table 9.1 Comparison of symbols and equations for translational and rotational motion

Concept	*Linear (Translation)*	*Angular (Rotation)*
Distance (displacement)	s	θ
Velocity	$\bar{v} = \frac{s}{t}$	$\omega = \frac{\theta}{t}$
Acceleration	$a = \frac{v_2 - v_1}{t}$	$\alpha = \frac{\omega_2 - \omega_1}{t}$
Uniformly accelerated motion	$v_2 = v_1 + at$ $s = v_1 t + \frac{1}{2}at^2$ $v_2^2 = v_1^2 + 2as$	$\omega_2 = \omega_1 + \alpha t$ $\theta = \omega_1 t + \frac{1}{2}\alpha t^2$ $\omega_2^2 = \omega_1^2 + 2\alpha\theta$
Mass (inertia)	m	I (Moment of inertia)
Force (torque)	$F = ma$	Torque, $\tau = I\alpha$
Work	$W = Fs$	$W = \tau\theta$
Power	$P = Fv$	$P = \tau\omega$
Momentum	$p = mv$	$L = I\omega$
Impulse	$Ft = m(v_2 - v_1)$	$\tau t = I(\omega_2 - \omega_1)$
Newton's second law	$F = ma$	$\tau = I\alpha$
Kinetic energy	$KE_{trans} = \frac{1}{2}mv^2$	$KE_{rot} = \frac{1}{2}I\omega^2$

QUESTIONS AND EXERCISES

1. $F = ma$ is the mathematical statement of Newton's second law for translatory motion. What is the equivalent equation for rotation? Define each term in the equation.

2. Work and torque can both be measured in units of newton-meters (N·m). Explain the difference between the work unit (N·m) and the torque unit (N·m).

3. If the power being transmitted by a shaft has to be calculated, what two factors must be known or measured?

4. How is moment of inertia different from mass? In what way(s) is it similar to mass?

5. If you were building a coaster car for the "roller derby" (the cars start from rest and coast down a hill to the finish line) would you use relatively heavy, large-diameter wheels and tires or lighter, smaller ones? (Total mass of car same in either case. Assume friction equal also.)

6. In designing flywheels, as much as possible of the total mass is placed in the outer rim. Why?

7. Explain why the rifling of a gun barrel tends to increase the accuracy of the bullet's flight. Use the vector property of angular momentum in your explanation.

8. A bowling ball has several finger holes drilled in it. As a result of these, would you expect the moment of inertia of a bowling ball to be greater than or less than $\frac{2}{5}Mr^2$?

9. If all the ice near the north and south poles of the earth were to melt, a very large volume of water would be slowly distributed over the world's oceans. If such an event occurred, would you predict that the length of the day would increase, decrease, or stay the same? Explain your answer in detail.

10. A hoop (a thin ring with all the mass distributed around the rim) of mass m rolls down a smooth inclined plane. Prove that when it arrives at the bottom, half its KE is rotational and half translational.

11. A solid sphere, a solid cylinder, and a hollow cylinder all start simultaneously from rest and roll down a smooth incline. Assuming no friction and no slippage, in what order will they reach the bottom? Explain.

12. If a mass m is hung from a string secured to the ceiling, the tension in the string is mg. If the string is wound around the rim of a wheel also of mass m which is free to turn and which is then released, will the tension still be mg? Why? When the mass hits the floor, what has happened to the potential energy it had just before it was released to unwind?

13. A solid cylinder and a solid block, both made of steel and both of the same mass, are released simultaneously at the top of a smooth incline. Assume no friction for the sliding block and no slippage of the rolling cylinder. Which arrives at the bottom first? What if the cylinder had only half the mass of the block? What if a bicycle wheel of the same mass were substituted for the cylinder?

14. A thin-walled cylinder and a rough-surface rectangular block are positioned at the top of a smooth incline. Both are released at the same instant, and the cylinder rolls down without friction. The block slides down, and half of its original potential energy is used in doing work against friction. Which arrives at the bottom first? Prove your answer.

15. From an examination of the cases of Fig. 9.9 explain why flywheels are designed with as much mass as possible in the outer rim.

16. If you built a helicopter with a single blade rotating horizontally around a vertical axis, and then took it up for a flight, what unsatisfactory flight characteristic would it have? Assuming a safe landing, how would you alter the design before your next flight? Explain in detail how your modification would correct the problem encountered on the initial flight.

17. Explain in detail how Galileo could predict the behavior of *falling bodies* from observations made on bodies sliding or rolling down inclined planes.

18. An experienced bike rider, when starting or negotiating a turn, will shift some body weight to "lean into" the turn, instead of or in addition to merely turning the handlebars. Explain this technique in terms of the gyroscopic action of the cycle wheels.

PROBLEMS

Group One

Note: For English-system problems treat pounds as masses.

1. A floppy disk of the type used with personal computers accelerates from rest to 300 rpm in $\frac{1}{8}$ s. Assume the angular acceleration is constant during the $\frac{1}{8}$-s time interval. What is the magnitude of the angular acceleration of the disk in rad/s^2?

2. A wrench handle is 30 cm long from the center of the bolt being turned to the point of application of the force on the handle. If a force of 90 N is applied at an angle of 20° with a perpendicular to the wrench handle, what torque is being exerted? (See Fig. 9.2*c*.)

3. A force of 50 N is applied at the rim of a wheel 1 m in diameter. The line of action of the force is perpendicular to the radius of the wheel at that point. What torque is exerted on the wheel?

4. Wrenches used by final-assembly personnel in industry have built-in torquemeters. If the wrench handle is 18 in. long and the torquemeter reads 500 lb·in., what force is the mechanic applying (*a*) when he pulls perpendicular to the wrench handle? (*b*) when the position is such that he has to pull at an angle of 50° with the wrench handle?

5. Find the horsepower being delivered by an engine which exerts a torque of 4000 lb·ft on a rotating kiln in a cement factory and turns it at 30 rev/min.

6. What is the power transmitted by a shaft turning at 3000 rev/min if the torque is found to be 1600 N · m?

7. A steam turbine in a power-generating plant starts from rest and accelerates at a uniform rate of 3.4 rad/s^2. Find its angular velocity in rev/min after 70 s.

8. If the allowed angular acceleration of a turbojet engine is 60 rad/s^2, and if idling speed is 600 rev/min, how long does it take to attain an angular speed of 5000 rev/min? (Assume uniform acceleration.)

9. The torque being delivered to a shaft turning at 2500 rpm is measured as 250 lb · ft. What is the horsepower output of the driving engine?

10. Find the moment of inertia of a hoop (thin ring) whose mass is 40 kg and whose radius is 75 cm.

Group Two

11. A long-playing phonograph record has the form of an annular cylinder with $r_1 = 3.75$ mm, $r_2 = 15$ cm, and mass = 130 g. (*a*) What is its moment of inertia? (*b*) What is its kinetic energy (KE_{rot}) when it is rotating at a constant angular speed of $33\frac{1}{3}$ rpm?

12. An engine governor is rotating at 2500 rpm and is slowed to 500 rpm in 40 s. Assuming uniform deceleration, calculate (*a*) the deceleration in rad/s^2, and (*b*) the number of revolutions turned by the governor during the 40-s period.

13. An elevator is designed to lift a total unbalanced load of 2500 kg at a *steady rate* of 300 m/min. The hoisting drum around which the elevator cable is wound is 1 m in diameter. (*a*) What is the angular velocity of the drum during lift? (*b*) What power (kW) is required?

14. Find the moment of inertia of a large cylindrical roller in a steel-rolling mill from the following data: diameter, 4 ft; length, 6 ft; weight, 2800 lb.

15. What is the moment of inertia of a squirrel-cage rotor of a large electric motor if it is in the form of an annular cylinder whose outer radius is 80 cm and whose inner radius is 50 cm. Its mass is 50 kg. Neglect weight of spokes.

16. Consult Table 8.1, pg. 190 for data and calculate (*a*) the moment of inertia of the earth and (*b*) the force in newtons required to stop the earth from rotating in a 24-h

period if it were applied along the equator and effectively used as a brake. Consider the earth as a sphere of uniform density.

17. A flywheel in the form of a solid disk of diameter 0.8 m has a mass of 100 kg. In order to accelerate it from an initial speed of 50 to 2000 rev/min in 10 s, what torque must be applied?

18. An engine has a flywheel whose moment of inertia is 50 sl · ft^2. What unbalanced torque will accelerate the flywheel from rest to 700 rev/min in 20 s?

19. A heavy wheel, weight 400 lb, is in the form of an annular cylinder, $r_2 = 1.8$ ft, $r_1 = 1.2$ ft. (Neglect spokes.) If it is spinning at 500 rev/min and a load is applied which brings it uniformly to a stop in 30 s, what torque is exerted?

20. A small airplane propeller approximates a slender rod and rotates around an axis as shown in Fig. 9.9. It is 7 ft long and has a mass of 2.5 sl. What uniform torque must be exerted to bring its angular velocity from zero to 1800 rev/min in an elapsed time of 10 s?

21. A bicycle wheel is assumed to be in the form of a thin ring (see Fig. 9.9) with the mass of spokes and hub neglected. It weighs 6 lb and has an effective radius of 13 in. What torque must be applied to give it a uniform acceleration of 25 rad/s^2?

22. An electric motor turns 1750 rev/min as it delivers 2 hp to drive an air compressor by means of pulleys and a V-belt. The diameter of the motor pulley is 6 in. and the diameter of the air compressor pulley is 14 in. Assume no belt slippage and find the difference between the tension on the drive side of the compressor pulley and that on the slack side. (Answer in lb.)

23. A heavy flywheel (mass 1000 kg) is in the form of an annular cylinder, with $r_1 = 1.2$ m and $r_2 = 1.8$ m. (Mass of spokes and hub to be neglected.) What torque applied to this flywheel would give it an acceleration of 9 rad/s^2?

24. Find the angular momentum of a spinning shaft (solid cylinder) 3 m long and 15 cm in diameter, mass 750 kg, if it is rotating at 400 rpm.

25. How rapidly would a 7.4-cm-diameter baseball have to spin to have a rotational kinetic energy equal to the translational kinetic energy of an identical nonspinning ball thrown at a speed of 80 mi/h? (Ans. in rpm.)

Group Three

26. A carousel (merry-go-round) with a full load of riders has a moment of inertia of 30×10^4 sl·ft^2. An electric motor must be selected that can cause the merry-go-round to accelerate from rest to a rotational speed of 3.8 rev/min in one minute. A technician hired as a consultant for the project performs some calculations and recommends the installation of a 10-hp motor on the assumption that the associated chain-belt and gear drive system is 20 percent efficient. Do the necessary calculations yourself and evaluate the consultant's recommendation. Was it correct? If not, what size motor should be specified?

27. A floppy disk for a personal computer has the form of an annular cylinder with $r_1 = 1.42$ cm, $r_2 = 6.5$ cm, and $M = 1.2$ g. In order to have a suitably short access time for retrieving information, the disk must be accelerated from rest to a rotational speed of 300 rev/min in $\frac{1}{8}$ s. (*a*) What is the kinetic energy of the disk when its rotational speed is 300 rev/min? (*b*) What torque is required to accelerate the disk from rest to 300 rev/min in $\frac{1}{8}$ s? (*c*) What magnitude force applied at the radius r_1 and directed tangent to the inner circular opening, will produce the required torque?

28. The braking process in a typical automobile converts the kinetic energy of the moving auto into heat energy as a result of work done against friction at the brake drums. As a research experiment, a 650-kg prototype electric auto is stopped by converting its kinetic energy to rotational motion of a 1-m-diameter cylindrical flywheel whose moment of inertia is 40 kg · m^2. Assume the angular speed of the flywheel is zero at an instant when the translational speed of the auto is 55 mi/h (24.6 m/s) and the "brakes" are applied (that is, the flywheel is activated). (*a*) What will be the angular speed of the flywheel just as the auto comes to a stop—that is, when the flywheel has absorbed all the auto's translational kinetic energy? (*b*) What will be the linear (tangential) speed of a point on the circumference of the flywheel just as the automobile stops?

29. A figure skater has just gone into a pirouette on one skate with both arms and the other leg and foot extended. Her angular velocity in this position is 1 rev/s, and her moment of inertia is 4.5 kg·m^2. Neglecting friction, what will her angular velocity be when she pulls in to the fast-spin position, where her moment of inertia is only 1.2 kg · m^2? (See Fig. 9.10.)

30. The earth's linear speed in its orbit around the sun is 9.85×10^4 ft/s. In addition to this translational motion, the earth spins on its polar axis at a rate of one revolution every 24 h. Find (*a*) the earth's rotational kinetic energy and (*b*) its kinetic energy of translation. Compare the two. (See Table 8.1 for data.)

31. A bowling ball has a mass of 7.5 kg and a diameter of 20 cm. It is rolling without slipping along the floor with a translational velocity of 6.5 m/s. Find its translational, rotational, and total KE.

32. An "energy-storage" flywheel is being designed. It will be in the form of a huge ring, with all the mass assumed to be located in the ring at a mean distance of 5 m from the axis of rotation. The design maximum angular velocity is 380 rev/min. At 20 rev/min, drawing additional power from the device is impractical. If the mass of the wheel is to be 50,000 kg, (*a*) how much energy (joules) will be available between the above angular velocity limits? (*b*) If this energy were drawn from the wheel at a steady rate over a 4-h period, what power (kW) would be available? Neglect all friction and other losses.

33. A flywheel in the form of a solid disk has a radius of 30 cm and a mass of 10 kg. It is rotating on a shaft that is connected through a clutch to an internal-combustion engine, which is to be started by the impulse furnished by the flywheel as it slows from 500 to 100 rev/min in 2.5 s. Calculate the magnitude of the torque delivered to the engine crankshaft, assuming no losses in the clutch system.

34. Find the magnitude and direction of the angular momentum of a bicycle wheel whose weight is 7 lb and radius is 13 in., when the bicycle is headed due north at 20 mi/h. Assume the bike is in a vertical plane, and treat the wheel as a thin ring, neglecting spokes and hub.

35. A punch press in a machine shop has a heavy flywheel that supplies all the energy for each punching or metal-forming operation. A small motor brings the flywheel back up to a speed of 400 rev/min in the interval between "punches." The moment of inertia of the flywheel is 30 kg · m^2. (*a*) What is the angular speed of the flywheel at the instant of completing a metal-forming operation requiring 5500 J of work? (*b*) What steady power must the motor supply to the flywheel in order to bring it back up to 400 rev/min in a time interval of 6 s?

36. A hoop (thin ring) rolls down the inclined plane shown in Fig. 9.18 with no slippage and no friction. What is the instantaneous translational speed (m/s) of the hoop at the bottom of the incline? (Note that mass and diameter are not involved. Why? Review Illustrative Problem 9.10.)

37. In Fig. 9.19 a 5 kg mass hangs by a cord which is wrapped around a solid disk as shown. The disk can rotate around its supporting axis *C* without friction. The mass of the disk is 40 kg and its radius is 15 cm. When the hanging mass is released, find (a) the acceleration with which it descends; and (b) the distance it moves downward in the first 2.0 seconds.

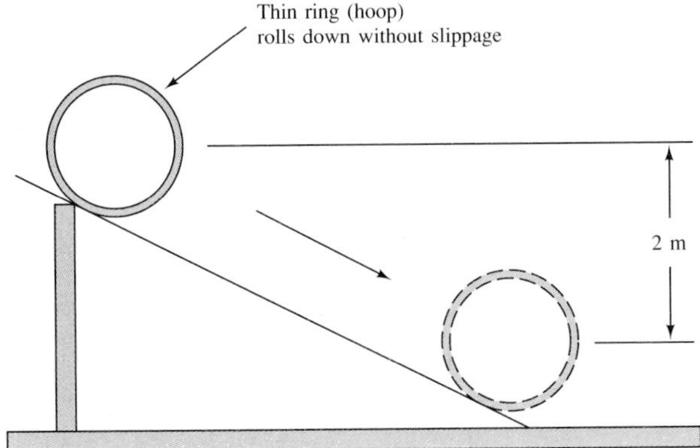

Fig. 9.18 Thin ring (hoop) rolling without friction or slippage down an inclined plane (Problem 36).

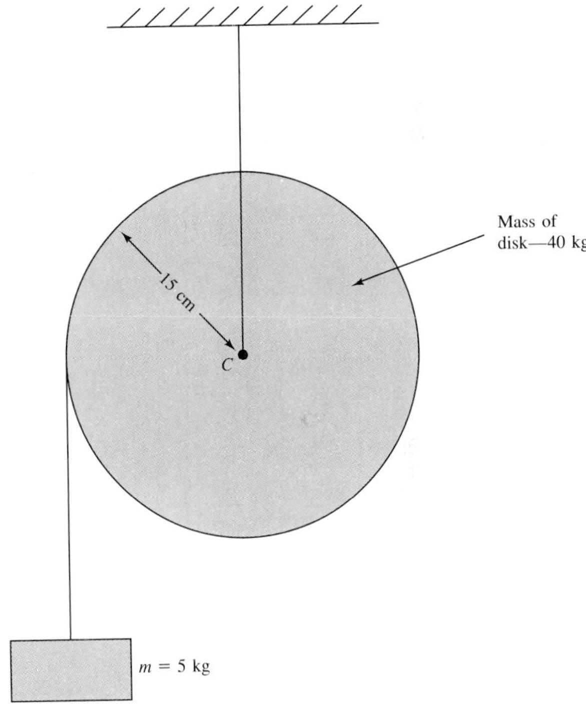

Fig. 9.19 A solid disk is rotated by the torque resulting from a hanging mass (Problem 37).

PART 3
MECHANICAL PROPERTIES OF MATTER

Steel is formed into sheets in an automated rolling mill. (Bethlehem Steel Corp.)

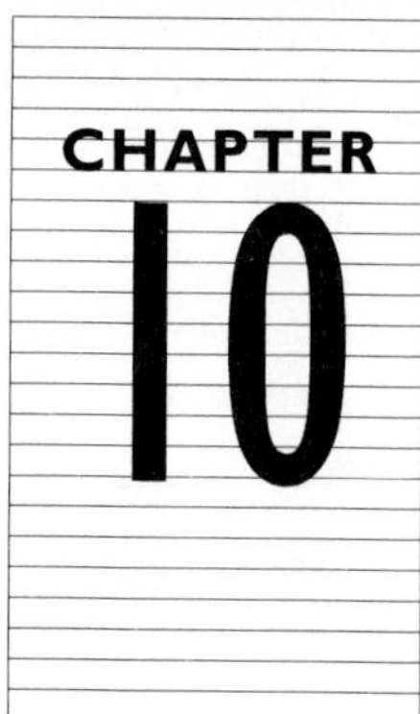

THE STRUCTURE OF MATTER—ATOMS, MOLECULES, ELEMENTS

All physical objects are composed of a substance called *matter*. We have studied various properties of matter that can be observed or measured by the naked eye. Matter has *mass*. Matter in motion possesses *kinetic energy* and *momentum*. Its *potential energy* is determined by its position or internal configuration. We know that material objects attract each other with forces called *gravitational* forces.

Matter is composed of tiny particles called *atoms* and *molecules*.

We will now deal with particles so small that they are invisible even when viewed through an optical microscope. Their masses are too minute to be detected by the most sensitive analytical balance. Most of the conclusions that have been made regarding the atomic and molecular nature of matter are the result of indirect physical and chemical observations.

10.1 ■ Elements and Their Symbols

Elements are the building blocks of all matter. They cannot be decomposed by ordinary chemical change.

Chemists generally subdivide matter into three categories: *elements*, *compounds*, and *mixtures*. An *element* is a substance that cannot be decomposed by ordinary types of chemical change; nor can it be made by chemical union. At present there are only 92 different kinds of elements found in nature. However, atomic physicists have artificially produced a number of additional elements in the laboratory by nuclear bombardment. Some common *natural* elements are carbon, oxygen, nitrogen, mercury, iron, sulfur, and calcium. An alphabetical listing of the known elements is given in Appendix VIII.

Symbols identify elements.

Each element is given a symbol which identifies it. Usually the symbol used for the element is taken from the first one or two letters of the name for the element. For example, consider the following elements and their symbols; carbon, C; hydrogen, H; nitrogen, N; oxygen, O; aluminum, Al; and bismuth, Bi. Some of the symbols may seem unrelated to the names of the elements. Usually those elements have been known for many centuries and have retained the symbols of their Latin names. For example, the symbol for iron is Fe, taken from the Latin word *ferrum;* copper is identified by the symbol Cu, taken from *cuprum;* gold is recognized as Au, from the Latin word *aurum*.

At ordinary room temperatures and atmospheric pressures only ten of the elements are gases and two are liquids. The rest are found in the solid state. Most of these are metals.

Elements combine to form *compounds*. The compound can be separated into two or more simpler substances.

Two or more elements may combine chemically to form a new substance called a *compound*. The compound will have properties that differ from the elements that compose it. For example, water is a compound of the elements hydrogen and oxygen. Alcohol is a compound of carbon, hydrogen, and oxygen—the same elements found in the compound sugar, but in different proportions and combined in a different way. More than two million compounds are known today, each possessing definite chemical and physical properties that enable the chemist to identify it.

Not all materials, even though they may seem to be homogeneous, are compounds. Most of them are mixtures of compounds. A *mixture* consists of two or more substances (elements or compounds) with no constant percentage composition and with each component retaining its original properties. In mixtures, the amounts of the constituent parts can be varied in any proportion. Examples of mixtures are soil, air, milk, and such combinations as iron and sulfur and sand and water.

The ratio of elements or compounds in a *mixture* may vary. The elements and compounds can be separated by ordinary physical means.

10.2 ■ Atoms Are the Smallest Particles of Elements

Consider a hypothetical experiment of dividing a piece of pure iron into two pieces. You would then have two pieces with identical properties. If it were possible to keep dividing each part into smaller particles, eventually there would result a particle which could not be divided again and still have the properties of the original piece of iron. The name *atom* has been given to this smallest ''indivisible'' portion of the element.

All elements are composed of small, indestructible (by ordinary means) particles called *atoms*. The atoms of a given element are almost identical in mass,* size, and shape. Atoms are extremely stable particles in chemical reactions. They cannot be broken up even by the hottest furnace or by any ordinary means used by chemists. The atom is the smallest part of an element that enters into a chemical change to form a compound. Thus, the atoms of the elements can be considered as building materials from which all compounds are made; but the atoms lose their original identities when they unite to form compounds. *Molecules* of a compound are produced by the union of two or more elements. In general, molecules of the most stable compounds are formed when one atom of each of two elements combine. Ordinary table salt, for example, is formed when one atom of sodium and one atom of chlorine are chemically combined. Water is formed by the chemical union of two gases, hydrogen and oxygen.

Atoms of an element are identical in mass, structure, and size.

Atoms combine chemically to form *molecules.*

10.3 ■ Molecule and Formula

A *molecule* is the smallest particle of a substance that has all the chemical and physical properties of the substance. All molecules of one substance are alike chemically, while those of different substances are never alike. Molecules are made up of atoms.

A molecule is the smallest unit quantity of matter which can exist by itself and retain all the properties of the substance.

Atoms of elements always combine in the same ratio to form a molecule of a given compound. The ratio of the number of atoms in a molecule of a compound is represented by a *formula* for the compound. The formula for ordinary table salt, for example, is NaCl, indicating that each molecule of the salt has one atom of sodium chemically united with one atom of chlorine. Two atoms of hydrogen and one atom of oxygen combine to form a molecule of water. The formula for water is H_2O. The formula for cane sugar is $C_{12}H_{22}O_{11}$, which indicates that each molecule of the sugar is made up of 12 atoms of carbon, 22 atoms of hydrogen, and 11 atoms of oxygen (a total of 45 atoms). The small numbers following each symbol in a formula are called *subscripts* and represent the ratio in which atoms combine to form a particular molecule of a compound.

Formulas indicate the kinds and numbers of atoms in a molecule.

Most molecules have two or more atoms. However, in some instances, molecules consist of single atoms. For example, molecules of metallic elements have only one atom. Also, there are a few gases, called *noble* (or inert) *gases,* that have one atom in each molecule. Examples of inert gases are helium, neon, argon, krypton, xenon, and radon. When the molecule contains only one atom, it is called *monatomic*. The molecules of the gaseous elements hydrogen, oxygen, nitrogen, chlorine, and fluorine are *diatomic,* consisting of two atoms.

Monatomic and diatomic molecules.

* Actually, there is a slight variation in the masses of the atoms of some elements. These atoms of an element that differ in mass are called *isotopes*.

The masses of atoms and molecules are extremely small. For example, the mass of an atom of oxygen has been calculated to be 26.57×10^{-27} kg. Scientists avoid using such small numbers by expressing the masses of atoms and molecules in *atomic mass units* (u), where

$$\text{One atomic mass unit} = 1\text{u} = 1.6605 \times 10^{-27} \text{ kg}$$

Atomic mass.

One atomic mass unit represents one-twelfth the mass of the most abundant form of the carbon atom. Thus, the mass of an atom of ordinary carbon is 12.0000 u. A list of the atomic masses of the elements is found in Appendix VIII. The table lists the relative masses of the 92 elements found in nature. For example, it gives the average atomic mass of iron as 55.847 u. Thus, each average atom of iron has a mass of $55.847 \times 1.6605 \times 10^{-27}$ kg.* Chemists use atomic masses in solving quantitative relations between the reactants in a chemical reaction.

Illustrative Problem 10.1 A molecule of ordinary table salt (NaCl) is made up of one atom of sodium and one atom of chlorine. How many molecules are there in 1 kg of table salt?

Solution From Appendix VIII

$$\begin{aligned}\text{Mass of one atom of sodium (Na)} &= 22.990 \text{ u}\\ \text{Mass of one average atom of chlorine (Cl)}^{\dagger} &= 35.453 \text{ u}\\ \text{Mass of one average molecule of salt (NaCl)} &= 58.443 \text{ u}\end{aligned}$$

$$\begin{aligned}\text{Then the number of molecules of NaCl} &= \frac{\text{mass of NaCl}}{\text{mass of 1 molecule of NaCl}}\\ &= \frac{1 \text{ kg}}{58.443 \text{ u} \times 1.6606 \times 10^{-27} \text{ kg/u}}\\ &= 1.030 \times 10^{25} \text{ molecules}\end{aligned}$$

10.4 ■ The Kinetic-Molecular Theory

The *kinetic-molecular theory* explains many properties of matter.

The kinetic-molecular theory helps to explain the properties of matter. In the kinetic-molecular theory, it is assumed that molecules are constantly in motion with varying speeds. One of the most direct evidences of this motion can be observed by the student in the laboratory. A device illustrated in Fig. 10.1 draws smoke from a

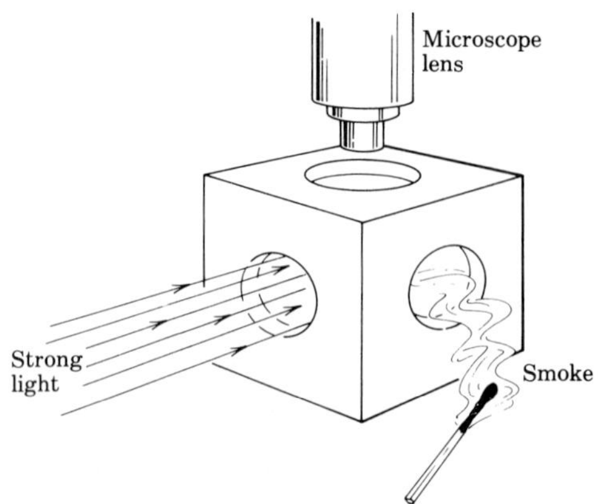

Fig. 10.1 Brownian movement apparatus. The ceaseless, zigzag motions of tiny smoke particles are observed under the microscope.

* Iron has four stable isotopes.

† Chlorine has two stable isotopes.

burning match into a small box by the compression and release of a rubber bulb. A strong ray of light entering through a hole from one side of the smoke box will illuminate the smoke particles in the box. The motion of the particles can be observed by a high-powered microscope. The tiny particles will appear to dart first one way and then another. A typical (random) movement pattern of one of the particles is illustrated in Fig. 10.2. Such movements are called *Brownian movements,* after an English botanist, Robert Brown, the man who first observed them in 1827.

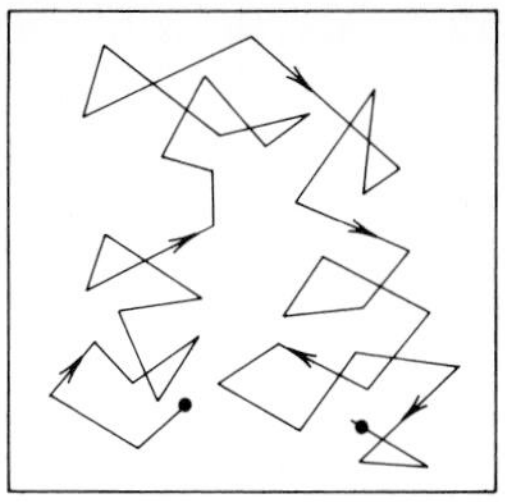

The particles of smoke are very much larger than the molecules of air in which they are suspended. The motion of the particles is caused by millions of collisions per second with the air molecules. When more of these collisions occur on one side of the particle than on the other sides, the smoke particle will dart away from the excess collisions.

Fig. 10.2 Typical random movement of a smoke particle under bombardment from air molecules.

Molecules are assumed to be perfectly elastic. When two molecules interact, they rebound without any net loss of energy or momentum for the two-molecule system. This means that, for a given equilibrium state, the total amount of kinetic energy and the total momentum of the molecules in a mass of matter does not change.

Molecules attract each other very little when they are far apart. When they are close together they attract each other very strongly. But when the molecules are too close together, they repel each other, often with tremendous forces. These forces of attraction and repulsion determine many properties of matter. One of these is the *state* of matter.

10.5 ■ The Three States of Matter

Matter is classified into *solids, liquids,* and *gases.*

In dealing with the physical properties of matter it is convenient to divide matter into three general classes, or states: *solids, liquids,* and *gases.* (A "fourth state"—*plasmas*—is suggested by some modern physicists.) The same kind of matter may exist in each of the three states. As an example, liquid water may be frozen to a solid (ice) or vaporized to a gaseous state. In each of these forms it is composed of identical molecules. The different states of matter are explained by the relative positions of these molecules and the kinds of freedom of their motions.

Matter is said to be in a solid state if it retains a definite shape and a definite volume. This volume and shape, however, may be subjected to change under varying external conditions such as temperature or the application of tensional or compressional forces.

In the solid state it is probable that the molecules oscillate or vibrate about certain fixed points. The molecules are held together by relatively strong molecular forces, called *cohesive* forces, in such a way that they cannot move far from their positions with reference to other molecules in the body. It is the strong molecular forces of cohesion that cause solids to retain a definite shape and volume.

Liquids have a definite volume but an indefinite shape. For example, a gallon of water maintains a constant volume under constant temperature conditions but varies its shape to conform to that of the containing vessel. Liquids are nearly incompressible.

Evidence of molecular motion in liquids.

In liquids the molecules are not held together in rigid patterns. Liquid molecules are, in general, almost as close together as the molecules of solids, but they slip over each other with ease and hence have no fixed positions. Molecular motion in liquids can be evidenced by the fact that a drop of ink released in a glass of water can be observed to diffuse quite rapidly until a uniform mixture results. The existence of empty spaces between molecules of a liquid can be easily demonstrated by mixing 0.5 liters (L) of alcohol with 0.5 L of water (see Fig. 10.3). The mixture occupies a volume considerably less than 1 L. Apparently the molecules of one liquid fill in some of the spaces between the molecules of the other.

Gases differ from solids in that they have neither a definite shape nor a definite volume. Regardless of the size or shape of the containing vessel, a gas will completely fill the container. The term *fluid* is used to include both liquids and gases.

Molecular model of a gas.

When matter exists in the gaseous state, its molecules possess relatively high velocities and great freedom of motion. The molecules are far apart compared to

Fig. 10.3 Adding 0.5 L of alcohol to 0.5 L of water produces less than 1 L of mixture because the molecules of one liquid fill in some of the spaces between the molecules of the other.

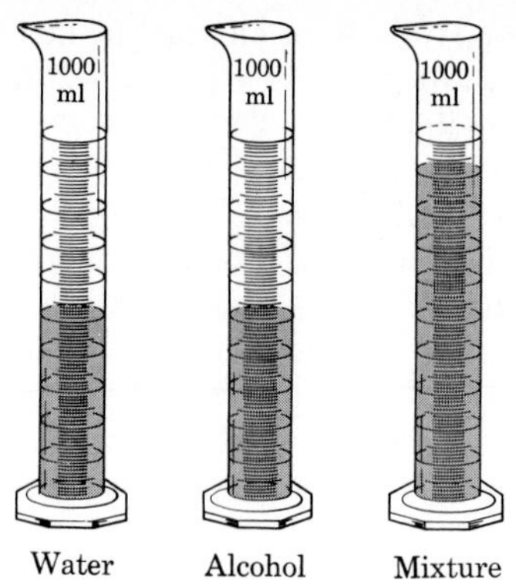

those in solids and liquids. This is evidenced by the fact that the molecules of 1 L of water separate to occupy about 1600 L when the liquid is changed to saturated steam at 100°C and atmospheric pressure.

Illustrative Problem 10.2 Assume that a gas is made up of spherical atoms with a diameter of 10^{-10} m. At room temperature and atmospheric pressure there will be 2.45×10^{19} atoms/cm^3. What percent of a 1 cm^3 volume will be occupied by the atoms?

Solution The volume of 1 (spherical) atom of the gas is, from $V_{\text{sphere}} = \frac{4}{3}\pi r^3$,

$$\frac{4}{3}\pi\left(\frac{10^{-10}\text{ m}}{2} \times \frac{100\text{ cm}}{1\text{ m}}\right)^3 = \frac{\pi}{6} \times 10^{-24}\text{ cm}^3$$

Therefore the volume occupied by 2.45×10^{19} atoms equals

$$\frac{\pi}{6} \times 10^{-24}\text{ cm}^3 \times 2.45 \times 10^{19} = 1.28 \times 10^{-5}\text{ cm}^3$$

The percentage of a 1 cm^3 volume actually occupied by the atoms is

$$\frac{1.28 \times 10^{-5}\text{ cm}^3}{1\text{ cm}^3} \times 100 = 1.28 \times 10^{-3}\% \qquad \textit{answer}$$

Stated another way, only 13/10,000 of the volume of the gas is actually made up of the atoms. The rest is space between the atoms.

In a *change of state,* the size of the molecules remains the same, but the space between the molecules changes. While the molecules of a solid or liquid are closely spaced, the molecules of a gas are relatively far apart, separated by a void. Molecules of a gas exert almost no cohesive force upon one another. This lack of mutual attraction of gaseous molecules, coupled with the fact that they may have a velocity exceeding in many cases that of the fastest rifle bullet, helps explain the relatively rapid *diffusion* of gases.

Diffusion in gases.

One of the most striking properties of a gas is that of unlimited expansibility. No matter how small an amount of gas is placed in a vessel, the gas will expand until it completely fills the vessel. If only half as much gas had been placed in the vessel, the vessel would still be *full,* but at a lower pressure. For this reason it is impossible to get a complete vacuum in a container. No matter how much air is "pumped" from a container, the remaining air will almost instantly redistribute itself throughout the entire vessel. The molecules tend to expand indefinitely if unconfined.

Gases are assumed to be perfectly elastic and can readily be compressed. This combination of perfect *elasticity* and *compressibility* gives gases the property called *resiliency,* i.e., the ability to yield to a force and to return promptly to original condition when that force is removed. It is this characteristic property of air that is utilized in the pneumatic tire, tennis balls, basketballs, and the like.

10.6 ■ Molecular Motion and Gas Pressure

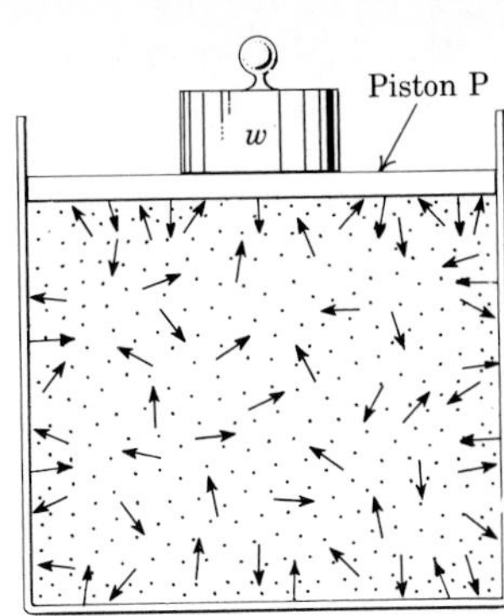

Fig. 10.4 Molecular bombardment creates the force which holds up the weight w. As a molecule collides elastically with the piston (or the walls of the cylinder) it bounces back with its speed unchanged. The changes in momentum of the molecules as they collide and bounce back provide the force on the piston which sustains the weight w.

Daniel Bernoulli (1700–1782), a Swiss scientist, first proposed the hypothesis that has now been generally accepted as the *kinetic theory of gases*. In his attempt to account for the compressibility peculiar to gases, he conceived of gas pressure as being caused by countless molecular impacts against the walls of the container.

In order to discuss this theory more fully, let us assume an imaginary cylindrical vessel filled with a gas (Fig. 10.4), which we shall show by a great number of small dots, each of which represents a molecule moving with extremely high velocity. Let us imagine further that the vessel is fitted with a movable frictionless piston P of negligible weight, upon which is placed a weight w. It is the continual bombardment against the piston by the moving molecules that sustains the weight w. In order to compress the gas, a greater weight would have to be added to compensate for the resultant greater number of impulses imparted to the piston by more frequent impacts of the compressed gas. In like manner, by diminishing the weight w, the bombardment on the piston would force it to rise, letting the gas expand until a state of equilibrium is reached in which the decreased pressure due to fewer impacts of the less dense gas on the piston is balanced by the lighter weight.

According to the kinetic theory, pressures exerted by gases on the walls of the vessels which contain them are due to the continuous bombardment of the walls by the gas molecules. The greater the number of molecules, the greater the pressure. Thus, by pumping more (molecules of) air into a container, we increase the pressure in the container.

MOLECULAR FORCES IN SOLIDS

10.7 ■ Molecular Forces of Attraction

Forces of attraction between *like* molecules are called *cohesive forces*. It is this molecular attraction which gives solids rigidity and strength. For example, it requires approximately 50 tons of force to pull apart a 1-in-diameter round bar of hard steel. Since various substances differ in the magnitude of the force required to pull them apart, it is evident that the forces of mutual attraction between some kinds of molecules are greater than those between other kinds of molecules.

Cohesion and adhesion.

Cohesion is greatest in solid substances. These forces not only hold the object together, but they maintain the definite shapes of solid objects. The cohesive forces in liquids are weaker than those in solids, and while appreciable, they are not sufficient to give the liquid a shape of its own. The cohesive forces among the molecules of a gas are negligible.

Relative cohesive and adhesive strengths determine the effectiveness of glues.

Forces of attraction between *unlike* molecules are called *adhesive forces*. Such forces are utilized in binding surfaces together with glue. In many cases, adhesive forces are greater than cohesive forces. These relative strengths are important in industry. In gluing together pieces of furniture, the adhesive forces between the glue and the wood and the cohesive attraction of glue for glue is a vital problem. Since, in this case, the adhesive forces exceed the cohesive forces, the layer of glue should be made as thin as possible. Adhesive forces come into play in the use of such common objects as pencils, ink, chalk, cement, paint, shellac, and tar.

10.8 ■ Molecular Forces of Repulsion

To compress a solid such as a steel beam or a liquid such as oil or water is not easy. It can be done, but even large forces applied to a solid or a liquid produce relatively

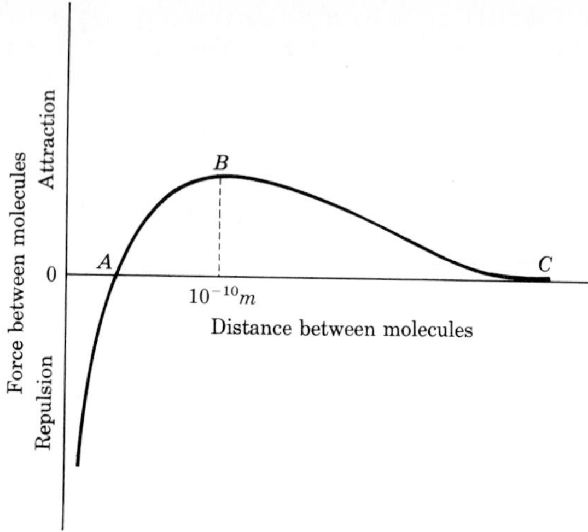

Fig. 10.5 Variation of forces with distance between the molecules. When the molecules are far apart for gases (as at C), there are almost no net forces between them.

Molecules exert forces of attraction and repulsion.

small compressions. The molecules of solids and liquids apparently resist being pressed closer together by repelling each other.

Molecules have their greatest cohesive attraction when they are about 10^{-10} m apart.

The force exerted by one molecule upon another depends upon the distance between them. Experiments have shown that molecules have maximum attraction for each other when they are about 10^{-10} m apart. They repel each other when they are closer than that, and their mutual attraction decreases rapidly when they are more than 10^{-10} m apart. A graph illustrating how the force between molecules varies with the distance between them is shown in Fig. 10.5. At C molecules are so far apart that forces between them are essentially zero. As the molecules come closer together, the attractive forces increase until they reach a maximum, shown at B. As the molecules move still closer, repulsive forces appear which oppose the forces of attraction. At point A no net force acts between the molecules. As molecules move still closer, the net repulsion forces between the molecules rise sharply.

The forces between molecules are primarily electrical.

At first one might be led to believe that forces of attraction between molecules are gravitational, but physicists have concluded that attractive and repulsive molecular forces are primarily a result of electric fields associated with the atoms which make up the molecules.

QUESTIONS AND EXERCISES

Indicate which of the following fifteen statements are true and which are false. Rewrite the false statements so they will be true.

1. Neon is an element.

2. Table salt is an element.

3. When two or more elements combine chemically a compound is formed.

4. The smallest particle of a compound that has all the chemical properties of the compound is called a molecule.

5. A chemical change takes place when water boils.

6. Only through the most powerful optical microscope can an atom be seen.

7. The composition of a pure compound always remains the same except in chemical changes.

8. Pure water is a mixture.

9. Each molecule of the compound NH_4NO_3 (ammonium nitrate) contains one atom of nitrogen.

10. Nitrogen is the lightest of all known elements.

11. All molecules of the same substance have the same weight.

12. The molecular weight of a compound is equal to the sum of the weights of all the atoms of which it is composed.

13. A mixture is composed of two or more elements united chemically.

14. Fluid and liquid are two words having the same meaning.

15. Two or more elements can not be chemically combined to make a compound.

16. Distinguish between a compound and an element.

17. What is the difference between an atom of oxygen and a molecule of oxygen?

18. What weight of an element is represented by its symbol? What weight of a compound is represented by its formula?

19. List three evidences that support the theory that molecules are in continual motion.

20. If gases possess the property of unlimited expansibility, why does the atmosphere not leave the earth?

21. Left uncovered, fish stored in the refrigerator will soon impart its flavor to other food in the refrigerator. Explain this in the light of molecular motion theory.

22. Why does a damp piece of cloth absorb water more quickly than a dry one?

23. A silk umbrella is more effective in shedding rain than one made of cotton. Give a possible explanation in terms of forces of adhesion and cohesion.

24. Explain why oil poured on water spreads into a very thin film on the water surface.

25. When a patch is removed from a bicycle inner tube the rubber is often torn before the patch comes off. Explain.

PROBLEMS

Group One

1. How many atoms of hydrogen are there in one molecule of ammonium hydroxide? The formula for ammonium hydroxide is NH_4OH.

2. How many hydrogen atoms are there in one molecule of ammonium sulfate whose formula is $(NH_4)_2SO_4$?

3. The formula for aluminum sulfate is $Al_2(SO_4)_3$. In one molecule of $Al_2(SO_4)_3$ there are how many (*a*) aluminum atoms, (*b*) sulfur atoms, and (*c*) oxygen atoms?

4. What is the mass of 6.02×10^{26} atoms of copper (Cu)?

5. What is (roughly) the ratio of the mass of a copper atom and the mass of a sulfur atom?

6. What is the molecular weight of acetic acid whose formula is $HC_2H_3O_2$?

7. If aluminum sulfate has the formula $Al_2(SO_4)_3$, what is its molecular weight?

Group Two

8. What is the percentage (by mass) of nitrogen (*N*) in the compound NH_4NO_3?

9. How many ounces of carbon are present in a 2-lb bag of sugar whose formula is $C_{12}H_{22}O_{11}$?

10. How many grams of sulfur are in 2 kg of ammonium sulfate? The formula for ammonium sulfate is $(NH_4)_2SO_4$.

11. Copper atoms are spherical with a diameter in the order of 10^{-10} m. In solid copper the distance between adjacent atoms has been determined to be 3.6×10^{-10} m and there are 2.14 atoms/cm^3. Find the volume percentage of copper atoms in solid copper.

CHAPTER 11 PROPERTIES OF SOLIDS

In the chapters on mechanics we have studied the action of forces on objects in equilibrium. We have dealt with the transmission of forces from one body to another. Throughout these discussions we have treated bodies as being rigid and have assumed that under the action of applied forces a body moves as a whole with translation, rotation, or combinations of both.

Density equals mass per unit volume.

In reality, there is no such thing as a rigid body. All so-called rigid bodies deform or yield to some extent under the influence of applied forces, although the deformations are usually quite small and do not affect the conditions of equilibrium. However, the study of such deformations is an important consideration in the understanding of mechanical materials and structural design. Some of the properties of solids which we shall study in this chapter are *density, elasticity, tensile strength, hardness,* and the ability to withstand extreme temperatures. Engineering handbooks tabulate the results of many studies that have been made of these properties.

DENSITY AND SPECIFIC GRAVITY

11.1 ■ Two Kinds of Density

Gold is often described as a *heavy* metal, while aluminum is referred to as one of the *lighter* metals. What is really meant is that the gold has a greater density than aluminum. The *density* of matter is defined in metric units as the *mass per unit volume*. Algebraically, this can be expressed as

$$\text{Mass density} = \frac{\text{mass}}{\text{volume}}$$

$$\rho = \frac{m}{V} \tag{11.1}$$

Engineers and technicians often employ units based on *weight*.

Mass densities, which will be represented in this text by the Greek letter ρ (rho), are usually expressed in kilograms per cubic meter or in grams per cubic centimeter.

Mass density in the English system is expressed in slugs per cubic foot. Engineers and technicians also often employ units of density based upon weight units. *Weight density* is defined as the *weight per unit volume*. We shall use the Roman letter D to represent weight densities. The unit of weight density in the English system is the *pound per cubic foot*. In the metric system it is the *newton per cubic meter*. Thus,

$$\text{Weight density} = \frac{\text{weight}}{\text{volume}}$$

$$D = \frac{w}{V} \tag{11.2}$$

Since weight is equal to the product of mass and the acceleration of gravity ($w = mg$), weight density is related to mass density by the formula

$$D = \frac{w}{V} = \frac{mg}{V} = \frac{m}{V}g$$

or

$$D = \rho g \tag{11.3}$$

or

weight density = mass density × acceleration of gravity.

Weight densities can be determined by multiplying mass units by units for the acceleration of gravity.

It should be noted that since weight may vary from place to place, weight density also varies, but mass density does not. The weight density of iron on the surface of the earth is about 77,000 N/m^3 (490 lb/ft^3), while it would be about 13,000 N/m^3 (82 lb/ft^3) on the surface of the moon.

Weight units vary with changes in altitude, but mass densities do not.

Table 11.1 Densities of common substances

Material	*Mass Density* ρ		*Weight Density D, lb/ft^3 at sea level*
	kg/m^3	*g/cm^3 (and specific gravity)*	
Solids:			
Aluminum	2,700	2.7	168.5
Bone	1700–2000	1.7–2.0	106–125
Brass	8,600	8.6	540
Copper	8,890	8.89	555
Cork	540	0.54	34
Glass	2400–2800	2.4–2.8	150–175
Gold	19,300	19.3	1,205
Ice	910	0.91	57.2
Iron	7,900	7.9	493
Lead	11,400	11.4	712
Paper	700–1150	0.7–1.15	44–72
Platinum	21,400	21.4	1,330
Silver	10,500	10.5	655
Steel	7,830	7.83	489
Liquids:			
Alcohol, ethyl (20°C)	790	0.79	49.4
Benzene	900	0.90	56.1
Gasoline	690	0.69	42
Kerosene	820	0.82	51
Linseed oil	940	0.94	59
Mercury	13,600	13.6	849
Turpentine	870	0.87	54
Water at 0°C	1,000	1.00	62.4
Ocean	1,025	1.02	64
Gases (32°F, 1 atm):			
Air	1.29	0.00129	0.0807
Ammonia	0.76	0.00076	0.0481
Carbon dioxide	1.96	0.00196	0.1234
Carbon monoxide	1.25	0.00125	0.0781
Helium	0.18	0.00018	0.0111
Hydrogen	0.090	0.00009	0.0056
Nitrogen	1.25	0.00125	0.0781
Oxygen	1.43	0.00143	0.0892
Propane	2.02	0.00202	0.1254

Weight density is used in computing the effects of forces on materials. Mass density is used when mass is considered.

The definition for density of matter is the same whether its state is solid, liquid, or gas. Weight density is commonly used when we are interested in the effects of forces on material, while mass density is used when mass is to be considered.

Table 11.1 gives the densities of some common substances.

Illustrative Problem 11.1 What is the mass density of a block of granite whose dimensions are 2 m by 75 cm by 25 cm if the block has a mass of 1020 kg?

Solution Converting the dimensions of the block to meters and using the formula V = length × width × height, we have

$$V = 2\text{ m} \times 75\text{ cm} \times \frac{1\text{ m}}{100\text{ cm}} \times 25\text{ cm} \times \frac{1\text{ m}}{100\text{ cm}}$$

$$= 0.375\text{ m}^3$$

Substituting values in Eq. (11.1) yields

$$\rho = \frac{m}{V} = \frac{1020\text{ kg}}{0.375\text{ m}^3} = 2720\text{ kg/m}^3 \qquad \textit{answer}$$

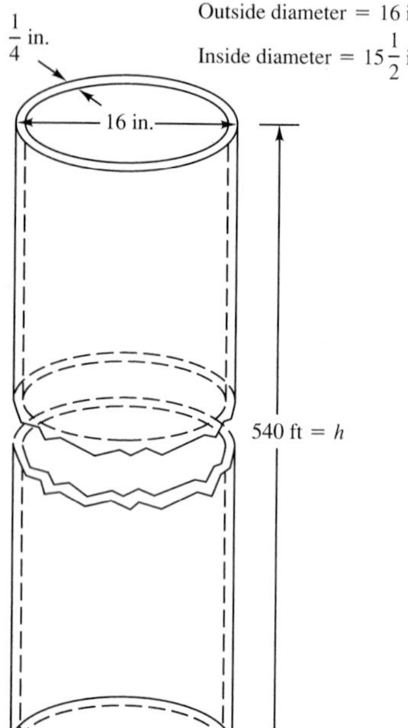

Fig. 11.1 Dimensions of a typical well casing used for agricultural water pumps.

Illustrative Problem 11.2 An agricultural well casing (see Fig. 11.1) is formed by welding together 30-ft sections of steel pipe. The pipe has an outside diameter of 16 in. and is $\frac{1}{4}$ in. thick. What would be the weight of such a casing if it is 540 ft long? Neglect weight loss due to the perforations at the bottom of the well.

Solution Solve Eq. (11.2) for w, to get $w = DV$. The formula for the volume of the steel in the hollow cylinder is $V = \pi(r_1^2 - r_2^2)h$. In this case $r_1 = 8$ in. $= \frac{2}{3}$ ft and $r_2 = 7\frac{3}{4}$ in. $= \frac{31}{48}$ ft. Table 11.1 gives $D_{\text{steel}} = 489\text{ lb/ft}^3$.

Thus,

$$\begin{aligned} w &= DV \\ &= 489\text{ lb/ft}^3 \times \pi[(\tfrac{2}{3}\text{ ft})^2 - (\tfrac{31}{48}\text{ ft})^2] \times 540\text{ ft} \\ &= 22{,}680\text{ lb} = 11.34\text{ tons} \end{aligned} \qquad \textit{answer}$$

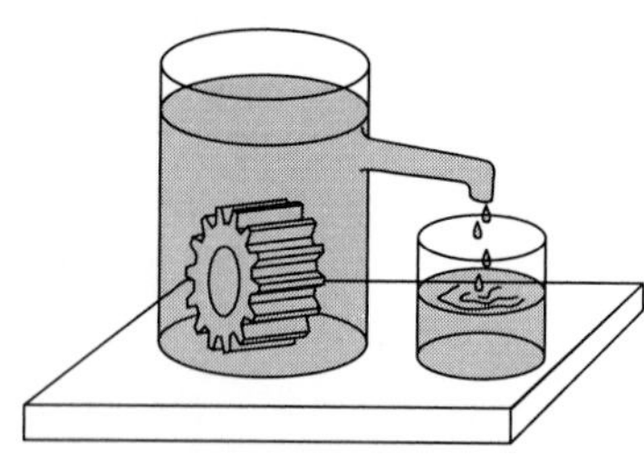

Fig. 11.2 Determining the density of a casting. The casting will displace a volume of water equal to that of the solid. Then by weighing the casting and the displaced water, its density can be determined.

Illustrative Problem 11.3 A helical metal gear casting weighing 11.7 lb is lowered into a vessel full of water (Fig. 11.2). The displaced (overflow) water is carefully collected and weighed. If the overflow water weighs 1.59 lb, what is the weight density of the casting?

Solution It is apparent that the volume of the casting will equal the volume of the water it displaces. The volume of the overflow water must first be determined. From Table 11.1 we observe that the weight density of water is 62.4 lb/ft³. Solving Eq. (11.1) for V gives

$$V = \frac{w}{D}$$

Now, for the overflow water (and for the casting),

Substituting values, $$V = \frac{1.59\text{ lb}}{62.4\text{ lb/ft}^3} = 0.0255\text{ ft}^3$$

Substitute in Eq. (11.2), and solve for the density of the casting:

$$D = \frac{w}{V} = \frac{11.7\text{ lb}}{0.255\text{ ft}^3} = 459\text{ lb/ft}^3 \qquad \textit{answer}$$

11.2 ■ Density and Specific Gravity

Specific gravity is density relative to that of water.

The statement "ice is lighter than water" is meaningless unless it implies that weights of equal volumes of ice and water are compared. A cubic centimeter is a convenient volume for comparison. Thus 1 cm^3 of water has a mass of 1 g, and 1 cm^3 of ice has a mass of 0.91 g. Since the weights are proportional to the masses, we can say that for equal volumes ice is 0.91 times as heavy as water.

It is customary to compare weights or masses of substances with the weight or mass of an equal volume of water, since water is the most abundant common liquid. The concept of *specific gravity* comes from such comparisons.

The *specific gravity* (sp gr) of a substance is defined as the ratio of its density to that of water at 4°C (39.2°F). Water has its maximum density at this temperature (see Sec. 14.19).

If ρ_s (or D_s) is the density of a substance and ρ_w (or D_w) is the density of water, the specific gravity of the substance is found by the formula

$$\text{sp gr} = \frac{\rho_s}{\rho_w} = \frac{D_s}{D_w} \tag{11.4}$$

A sometimes more convenient, but equivalent, way of defining specific gravity is

$$\text{sp gr} = \frac{\text{mass (or weight) of substance}}{\text{mass (or weight) of an equal volume of water}} \tag{11.5}$$

The distinction between density and specific gravity must be clearly understood. *Density* refers to *mass or weight per unit volume* of a given substance. The numerical value of the density of a substance will vary with the units of weight and volume used. Aluminum, for example, has densities of 2700 kg/m^3, 2.7 g/cm^3, 168.5 lb/ft^3, and 0.097 $lb/in.^3$.

Specific gravity is numerically equal to the mass in grams per cubic centimeter (g/cm^3).

The *specific gravity* of a substance is a pure number that has no units. Since, in the metric system, the mass density of water is 1 g/cm^3, *the specific gravity of any substance is equal numerically to the mass density of that substance expressed in grams per cubic centimeter*. Hence in the previous example we would know that aluminum has a specific gravity of 2.7, or that it is 2.7 times as heavy as an equal volume of water. Other specific gravities can be found by referring to Table 11.1. The column labeled "g/cm^3" can also be read "specific gravity."

Solving Eq. (11.4) for densities, we get

$$\rho_s = (\text{sp gr}) \times \rho_w \tag{11.4'}$$

and

$$D_s = (\text{sp gr}) \times D_w \tag{11.4''}$$

For mass density, therefore,

$$\text{Density (kg/m}^3\text{)} = \text{sp gr} \times 1000\ \text{kg/m}^3$$

or

$$\text{Density (g/cm}^3\text{)} = \text{sp gr} \times 1\ \text{g/cm}^3 \tag{11.4'''}$$

For weight density, since the density of water is 62.4 lb/ft^3,

$$\text{Density (lb/ft}^3\text{)} = \text{sp gr} \times 62.4\ \text{lb/ft}^3 \tag{11.4^{IV}}$$

Illustrative Problem 11.4 An irregularly shaped metal casting whose mass is 35 kg displaces 13 L of water. What is (*a*) the density, and (*b*) the specific gravity of the casting?

Solution The volume V of the casting is 13 L since it displaces 13 L of water. Expressing this value in cubic meters,

$$V = 13\ \text{L}\left(\frac{1000\ \text{cm}^3}{1\ \text{L}}\right)\left(\frac{1\ \text{m}}{100\ \text{cm}}\right)^3 = 0.013\ \text{m}^3$$

(***a***) Using Eq. (11.1),

$$\rho = \frac{m}{V} = \frac{35 \text{ kg}}{0.013 \text{ m}^3} = 2700 \text{ kg/m}^3$$

(*b*) Then

$$\text{sp gr} = \frac{\rho_s}{\rho_w} = \frac{2700 \text{ kg/m}^3}{1000 \text{ kg/m}^3} = 2.7 \qquad \textit{answer}$$

11.3 ■ Elasticity

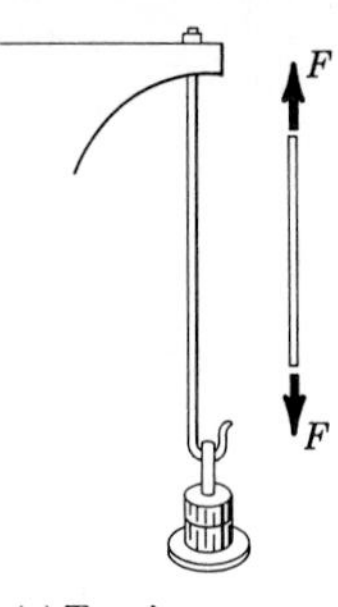

(*a*) Tension

Solid bodies tend to maintain their shape and volume. However, whenever a body is acted upon by an external force, a change in shape, or *distortion,* of the body results. That property of matter which, upon the removal of these distorting forces, will cause the body to resume its original shape or volume is called *elasticity*. Every substance has some degree of elasticity, but the term *elastic* is usually applied to those bodies that return *readily* to their original shape or volume when the distorting forces are removed.

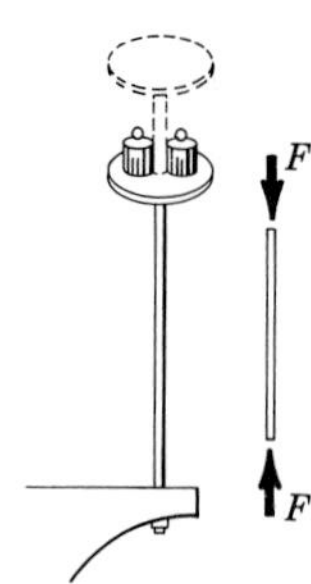

(*b*) Compression

Whenever the shape of a solid is distorted, the molecules are rearranged: some are separated, and others are pushed closer together. When the forces of distortion are removed, the forces of attraction and repulsion between the molecules tend to cause the original shape to be restored. It can be shown that the forces between the molecules are primarily electric. Materials such as putty, sealing wax, and lead are called *inelastic*.

Industry is continually attempting to increase personal comfort and safety by utilizing the property of elasticity. The innerspring mattress, the various springs in autos and trains, the balls used in many sports, and the tires on automobiles are just a few examples where this property of elasticity is of major importance.

STRESS AND STRAIN

11.4 ■ Stresses in Solids

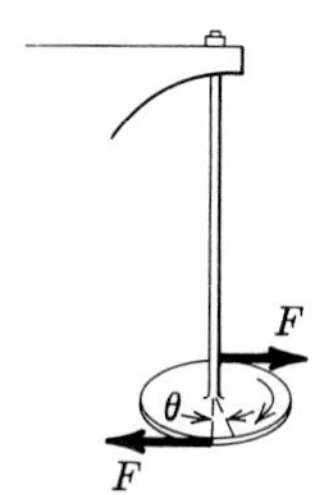

(*c*) Twisting (Shear)

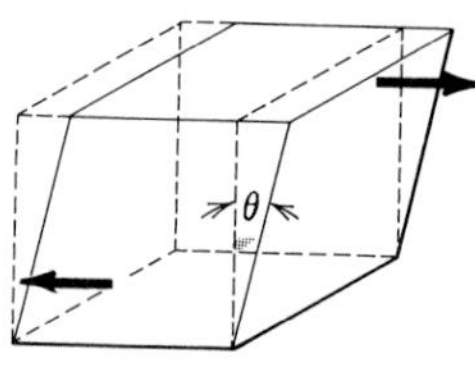

(*d*) Shear

Fig. 11.3 Different kinds of stress. (*a*) *Tension*: Equal and opposite forces act away from each other along the same line of action. (*b*) *Compression*: Equal and opposite forces act toward each other along the same line of action. (*c*) *Shear*: Equal and opposite forces are exerted along different lines of action. Volume stresses will be discussed later in connection with liquids and gases in this and subsequent chapters.

Whenever a solid body is deformed by the action of external forces, there are set up within that body internal molecular forces which tend to resist changes in shape and/or volume. These internal forces are called forces of *stress*. In specifying a girder, a column, or a machine part, engineers must know the forces it is expected to resist. There are four kinds of stresses: *tension, compression, volume* and *shear* (see Fig. 11.3). The stress can be expressed as the ratio of the applied external force creating a distortion to the area over which the force acts.

$$\text{Stress} = \frac{\text{external force acting}}{\text{area over which force acts}}$$

$$S = \frac{F}{A} \qquad (11.6)$$

The unit for stress in the English system is pounds per square inch (lb/in^2 or psi). The SI unit for stress is newton per square meter (N/m^2) and this unit is called the *pascal* (Pa).

$$1 \text{ Pa} = 1 \text{ N/m}^2$$
$$1 \text{ kPa (kilopascal)} = 10^3 \text{ Pa}$$
$$1 \text{ GPa (gigapascal)} = 10^9 \text{ Pa}$$

When the actions of an external force tend to stretch or pull the parts of a body apart, the body is said to be under *tension*. Tension, for example, occurs when a rope or cable is tied to a rigid support and sustains a weight. The belt that is transmitting power from one pulley to another is in a state of tension. The column of a building, and the foundation of a house serve a different function. They resist external forces which tend to push their parts together, thereby shortening them. These members are said to be under *compression* (see Fig. 11.3*b*).

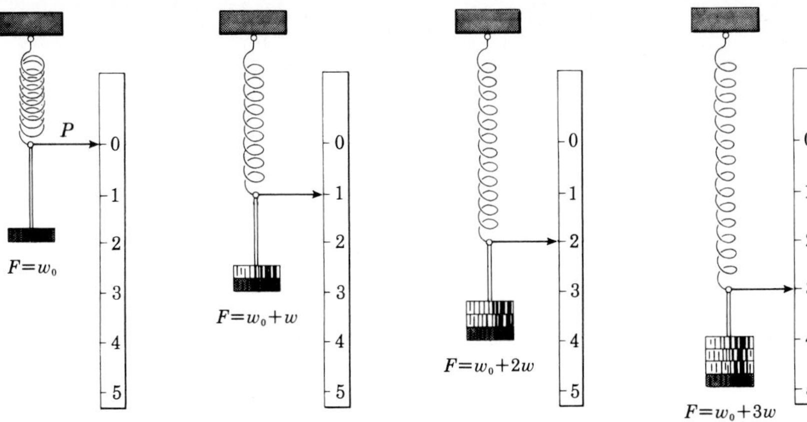

Fig. 11.4 An experiment which illustrates that the deformation is proportional to the applied force (Hooke's law).

11.5 ■ Stress Causes Strain

No matter how small a stress is applied to a body, the body will yield a little. The yielding of bodies under stress may be in the form of elongation, shortening, twisting (torsion), or change in volume and is called *strain*. Strain is a necessary consequence of stress.

The mathematical relationship between stress and strain can be discovered by performing a simple experiment. A helical spring is suspended from a rigid support (Fig. 11.4). A weight holder is attached to the bottom end of the vertical spring, and a weight is placed on the holder so that its weight combined with that of the holder is w_0. A vertical scale is placed next to a pointer P on the holder. Next, weights are added one by one to the holder and corresponding elongations are recorded. Finally, the weights are removed in reverse order and the elongations are recorded.

Elastic deformation is proportional to applied stress.

Results of this experiment show that upon removal of the weights the pointer P will return to its original position. If we then compute the corresponding elongation e for each weight w applied, we shall find that all the values of w/e are approximately equal. In other words w/e is a constant, or $w = ke$. That is, the elongation is proportional to the force (weight) producing it, and k is the constant of proportionality.

A similar experiment can be set up in which the shortening or compression of a beam or spring can be measured for various loads or forces. It can be shown that *within certain limits* the compression (strain) is proportional to the force (stress) applied. Likewise, the amount of twisting of a rod or bar is proportional to the stress causing it. These facts were first expressed as a scientific truth in 1660 by Robert Hooke (1635–1703), an English physicist, and the law is now named after him.

Hooke's Law

If the applied forces on a body are not too large, the deformations resulting are directly proportional to the forces producing them.

Hooke's law states that the strain is proportional to the stress.

Or, more simply, *strain is proportional to stress*. Hooke's law applies to all kinds of strains if the stresses are not too great. (What happens when the forces *are* too great is discussed in Sec. 11.8.)

11.6 ■ Potential Energy of Spring under Stress

Work is done on a spring whenever it is stretched or compressed. The energy expended in doing this work is stored in the spring as potential energy. If k is the spring constant and s is the displacement, it can be shown (see Sec. 19.2 for derivation) that the work W is equal to $\frac{1}{2}ks^2$

$$PE = W = \tfrac{1}{2}ks^2 \quad (19.2)$$

Work done in elongation or compression is stored as potential energy.

(Potential energy of spring under tension or compression)

Illustrative Problem 11.5 If a spring has a spring constant k of 4,260 N/m, how much must it be stretched to store 37.4 J of energy in the spring?

Solution

$$W = PE = \tfrac{1}{2}ks^2 \quad (19.2)$$

Solve Eq. (19.2) for s.

$$s = \sqrt{\frac{2W}{k}}$$

$$= \sqrt{\frac{2 \times 37.4 \text{ J} \times 1 \text{ N} \cdot \text{m}/1 \text{ J}}{4260 \text{ N/m}}}$$

$$= 0.1325 \text{ m} = 13.25 \text{ cm} \qquad \textit{answer}$$

11.7 ■ Modulus of Elasticity

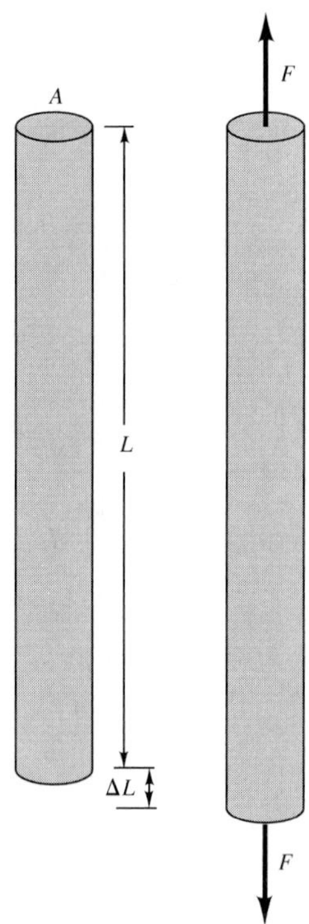

Fig. 11.5 For an object under tension, the strain ($\Delta L/L$) is proportional to the stress (F/A). The proportionality constant Y is called the modulus of elasticity.

Hooke's law states that, within certain limits, the strain produced in a body is proportional to the stress, or stress/strain = a constant k. This constant for a particular substance is called its *modulus of elasticity*. In order to arrive at a value of k, we must first express stress and strain in terms of some common measurable units. The *stress,* which is the internal resistance of a body to a change of shape, is expressed by Eq. (11.6):

$$\text{Stress} = \frac{\text{force applied}}{\text{area of cross section}}$$

$$\text{Stress} = \frac{F}{A} \quad (11.6)$$

The *strain* is the amount of change in size of the material per unit original size. The size may be expressed in terms of length, area, or volume units.

Tension and Compression Tensile stresses produce tensile strains, expressed as

$$\text{Tensile strain} = \frac{\text{elongation}}{\text{original length}}$$

$$\text{Tensile strain} = \frac{e}{L} = \frac{\Delta L}{L} \quad (11.7)$$

where ΔL is called the *increment of* L and is the change in L as it passes from one value to a second. ΔL is found by subtracting the first value from the second (see Fig. 11.5). Remember that the increment of L is read ''delta L.'' Thus, ΔL signifies a small change in L. Note also that ΔL may be either positive or negative.

The *ratio of stress to strain* for cases of tension or compression in a given material is called *Young's modulus* of elasticity for that material. It is usually denoted by the letter Y.

$$Y = \frac{\text{stress}}{\text{strain}} = \frac{\Delta F/A}{\Delta L/L}$$

from which Young's modulus becomes

$$Y = \frac{\Delta F\ L}{A\ \Delta L} \quad (11.8)$$

In Eq. (11.8) ΔL is the increment of L which corresponds to an increment ΔF of F. Young's modulus is expressed in newtons per square meter (Pa, or pascal) or in pounds per square inch (lb/in.^{2}). Table 11.2 gives values of Y for some common materials.

The ratio Y of the tensile (or compressional) stress F/A to the tensile (or compressional) strain $\Delta L/L$ is called the *Young's modulus.*

As can be seen from Table 11.2, the numerical value for Young's modulus varies for each material and may vary for the same kind of substance under differing conditions of physical structure. The heat treatment given a substance affects this modulus. This constant is also affected by the temperature of the material. For example, the modulus for a rubber band will increase with a rise in temperature, while that for steel will decrease as the temperature rises.

Table 11.2 Approximate Values of Young's Modulus

	Y	
Material	*N/m² (Pa)*	*lb/in.²*
Aluminum	7.0×10^{10}	10×10^6
Brass, cast	9×10^{10}	13.1×10^6
drawn	13×10^{10}	18.9×10^6
Copper	12.5×10^{10}	18×10^6
Iron, cast	9.1×10^{10}	13×10^6
wrought	18.3×10^{10}	26×10^6
Lead, rolled	1.6×10^{10}	2.3×10^6
Steel, drawn	20.0×10^{10}	29×10^6
mild	17.2×10^{10}	25×10^6
Rubber, vulcanized	14.0×10^4	20
Tungsten, drawn	35.0×10^{10}	51×10^6

Values of Young's modulus.

Illustrative Problem 11.6 A surveyor uses a (drawn) steel tape that is 0.300 in. wide and 0.015 in. thick. The tape is 100.000 ft long when supported throughout its length and is pulled with a force of 15 lb. What will be the length of the tape if it is pulled with a force of 50 lb?

Solution

$$Y_{\text{drawn steel}} = 29{,}000{,}000 \text{ lb/in}^2 \qquad \text{from Table 11.2}$$
$$\Delta F = 50 \text{ lb} - 15 \text{ lb} = 35 \text{ lb}$$
$$L = 100.000 \text{ ft}$$

Use Eq. (11.8) to find ΔL

$$\Delta L = \frac{\Delta FL}{AY}$$
$$= \frac{35 \text{ lb} \times 100.000 \text{ ft}}{(0.300 \text{ in.} \times 0.015 \text{ in.}) \times (29{,}000{,}000 \text{ lb/in.}^2)}$$
$$= 0.0268 \text{ ft}$$

Hence $L + \Delta L = 100.000 \text{ ft} + 0.0268 \text{ ft} = 100.027 \text{ ft}$ *answer*

The modulus of elasticity in cases of compression can be computed in a similar way as long as no bending is involved. In this case, ΔL is the amount the member is shortened by the increment ΔF of the compressing force.

Illustrative Problem 11.7 How much will a wrought-iron bar 5 by 12 cm in cross section and 2 m long shorten under a compression load of 2500 N? (Assume $Y = 18.3 \times 10^{10}\ \text{N/m}^2$.)

Solution From Eq. (11.8),

$$\Delta L = \frac{\Delta FL}{AY}$$

Substitution yields

$$\Delta L = \frac{2500\ \text{N} \times 2\ \text{m}}{0.05\ \text{m} \times 0.12\ \text{m} \times 18.3 \times 10^{10}\ \text{N/m}^2}$$
$$= 4.6 \times 10^{-6}\ \text{m} = 4.6 \times 10^{-4}\ \text{cm}$$
$$= 4.6\ \mu\text{m}$$ *answer*

11.8 ■ Elastic Limit, Ultimate Strength

Hooke's law holds only if the stresses are not too large.

In the discussion to this point we have assumed that upon removal of the forces which produced the various strains the bodies would resume their original shapes or volumes. The expression "within certain limits" was used in connection with Hooke's law. However, if in the experiment on tensile stress (Fig. 11.4) a sufficiently heavy load is hung from the spring, the spring *will not return* to the original length when the load is removed. The same situation can be made to develop in each of the other three types of elasticity mentioned if the stresses are made large enough. *Hooke's law,* then, *holds only for values of stresses less than the elastic limit.*

The relation between stress and strain for a body under tension can be represented by a *stress-strain diagram.* Figure 11.6 represents the stress-strain diagram for mild steel. A body that has been deformed under stress will regain its original size and shape on removal of the stress unless the stress has exceeded a certain limit. Thus, starting at O, the strain increases in direct proportion to the stress. That is, the graph is a straight line. The maximum stress from which the material will completely recover is called the *elastic limit* (point A on the graph). If the stress is not increased beyond the elastic limit, the curve will be retraced as the strain becomes zero when the stress is removed.

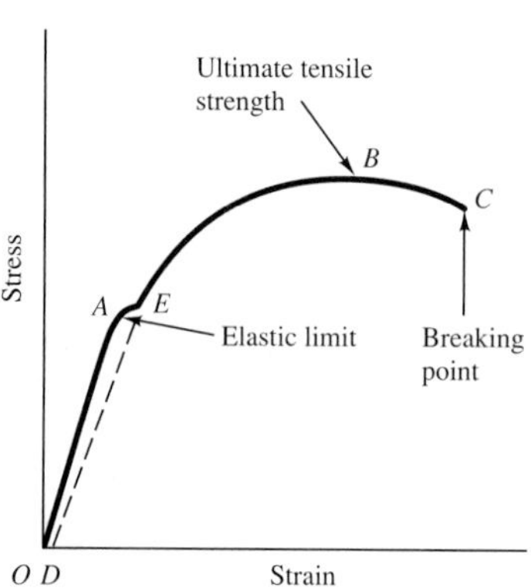

Fig. 11.6 Stress-strain diagram for mild steel. Segment OA represents the *elastic region*. Up to a point A the metal will return to its original length if the applied stress is removed.

If the stress exceeds the elastic limit, the strain will increase more rapidly (the curve flattens out). The region represented by AE is called the region of *plastic flow*. The elastic properties of the metal will have been changed. If the stress is then decreased, the strain will decrease as shown by the dashed line ED. The strain will not return to zero when the stress is reduced to zero. The material will have acquired a *permanent set,* represented by OD in the graph.

Table 11.3 Elastic Limit and Ultimate Tensile Strength

	Elastic Limit		Ultimate Tensile Strength	
Material	$\text{N/m}^2 \times 10^8$	$\text{lb/in.}^2 \times 10^3$	$\text{N/m}^2 \times 10^8$	$\text{lb/in.}^2 \times 10^3$
Aluminum, cold rolled	1.3	19	1.4	21
Brass, cast	3.8	55	4.7	67
Copper-aluminum-nickel, rolled	7.9	115	8.2	120
Iron, annealed	1.4	20.3	2.9	42
drawn	7.7	111	7.8	114
wrought	1.7	25	2.4	50
Steel, hard	2.7	40	5.5	80
mild	2.1	30	4.1	60
spring-tempered	11.7	170	13.8	200
Tungsten, drawn			41.2	600

The elastic limit refers to the maximum stress that can be applied before a material ruptures. For brittle materials the ultimate strength is the breaking stress.

Fig. 11.7 A compression molding press rated at 4000 tons is used to produce truck hood units. (Erie Press Systems.)

If the stress had been continually increased beyond the elastic limit, eventually a point B would have been reached where the strain would increase with no increase in stress. Indeed, beyond this point loads may even be lightened without halting the continuing deformation. The greatest stress that can be applied is called the *ultimate strength* of the material. Ultimately a point will be reached where rupture occurs. This point is called the *breaking strength*.

Each substance will have a different stress-strain diagram. But, in general, each graph will have the same characteristics. A substance for which EC is relatively long is called *ductile* (see Sec. 11.16). Such a substance can be subjected to a large increase in length before breaking. Substances for which EC is quite short are called *brittle* (see Sec. 11.15).

Ductile metals can be deformed far beyond their elastic limits.

Table 11.3 lists values of elastic limit and ultimate tensile strength for a number of common metals.

Illustrative Problem 11.8 A hard steel rod 12 ft long has a diameter of $\frac{3}{4}$ in. and is carrying a load of 3 tons. (*a*) What is the total elongation in the rod caused by this load? (*b*) What is the load required to cause the rod to just reach its elastic limit?

Solution

(*a*) $F = 3$ tons $= 6000$ lb; $L = 12$ ft; diameter $= \frac{3}{4}$ in.; $Y = 29{,}000{,}000$ lb/in^2. Use Eq. (11.8) to solve for ΔL

$$\Delta L = \frac{\Delta F \times L}{AY}$$

$$= \frac{6000 \text{ lb} \times 12 \text{ ft} \times 12 \text{ in./1 ft}}{\pi/4(\frac{3}{4} \text{ in.})^2 \times 29{,}000{,}000 \text{ lb/in.}^2}$$

$$= 0.067 \text{ in.} \qquad \textit{answer}$$

(*b*) The load to cause a stress of 40×10^3 lb/in^2 (see Table 11.3) is

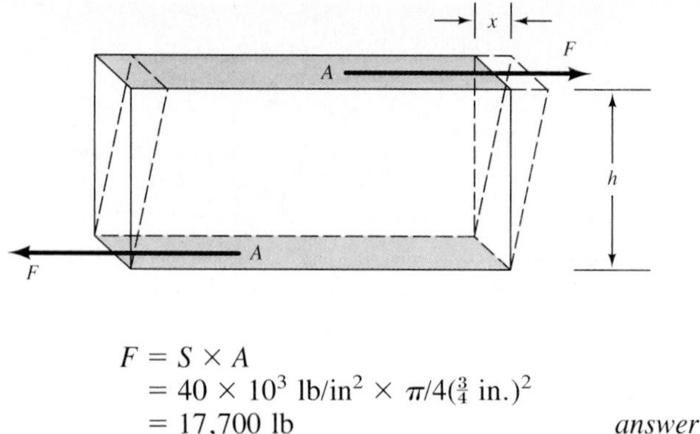

Fig. 11.8 The shearing stress on the block is the ratio of force F to A. Note that A is the area of the surface parallel to the applied force, not perpendicular, as it is for stresses of tension or compression. Note also that x is perpendicular to h.

$$
\begin{aligned}
F &= S \times A \\
&= 40 \times 10^3 \text{ lb/in}^2 \times \pi/4(\tfrac{3}{4} \text{ in.})^2 \\
&= 17{,}700 \text{ lb} \qquad \textit{answer}
\end{aligned}
$$

11.9 ■ Shearing Stress

A shearing stress occurs when equal and opposite forces are applied to an object along different lines of force.

A third form of stress is that of *shear* in which certain particles of the body tend to slide over the particles in that body adjacent to them. A *shearing stress* is applied to an object when equal and opposite forces F, F are exerted on an object along different lines of action. This type of stress is illustrated in Fig. 11.8. The tangential forces F, F acting on a rectangular block will change the shape of the block without changing the volume. The shearing stress is defined as the force per unit surface area across which the shearing action occurs. Common units of shearing stress are newtons per square meter (N/m^2) and pounds per square inch (lb/in.^{2}). Shear elasticity of shape is characteristic of solids and not of liquids and gases.

An example of shearing stress is that set up in rivets (see Fig. 11.9) of a steel structure. A rivet has to keep one plate from sliding over another, as well as hold the plates together (see Fig. 11.9). This same type of stress is set up when a piece of sheet metal is cut with tin shears, or when steel plates are cut with heavy power shears in steel-fabricating plants.

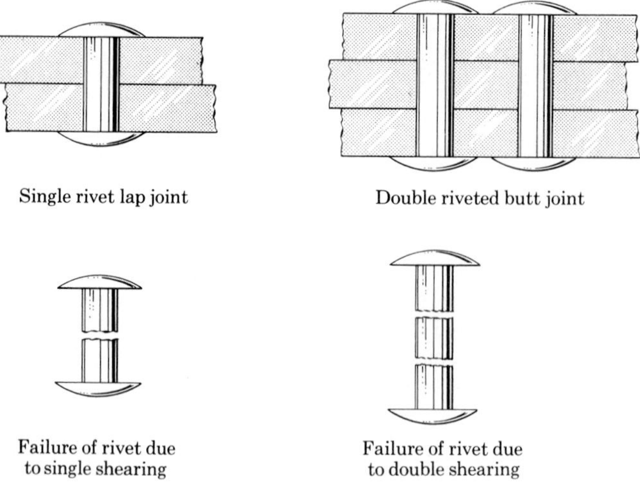

Fig. 11.9 Single and double shear of a rivet.

Illustrative Problem 11.9 What force must be applied in punching out $\frac{1}{2}$-in.-diameter holes in a steel plate $\frac{3}{8}$ in. thick if the shearing strength (stress) of the material is 38,000 lb/in.^{2}?

Solution The force must act over an area equal to the lateral surface area of a cylindrical hole punched out. Use the formula for the lateral surface area of a cylinder:

$$A = 2\pi rh$$
$$= 2(3.14) \times \tfrac{1}{4}\text{ in.} \times \tfrac{3}{8}\text{ in.} = 0.589\text{ in.}^2$$

the lateral area over which the shearing force acts. Now, selecting Eq. (11.6) which relates force, stress, and area,

$$\text{Stress} = \frac{F}{A}$$

Solving for F,

$$F = \text{stress} \times A$$
$$= 38{,}000\text{ lb/in.}^2 \times 0.589\text{ in.}^2 = 22{,}400\text{ lb} \qquad \textit{answer}$$

Illustrative Problem 11.10 Suppose two of the plates of Prob. 11.10 are riveted together; how much force per rivet, applied at right angles, would be necessary to shear off the rivets? Assume the shearing strength of the rivet material to be 30,000 lb/in.^{2}. Neglect the friction between the plates.

Solution Here the area to be sheared is the cross-sectional area of the rivet, which is assumed to have the same diameter as the holes ($\tfrac{1}{2}$ in.). Selecting Eq. (11.6) again,

$$F = \text{stress} \times A$$
$$= S \times \pi r^2$$
$$= 30{,}000\text{ lb/in.}^2 \times 3.14 \times (\tfrac{1}{4}\text{ in.})^2 = 5900\text{ lb} \qquad \textit{answer}$$

11.10 ■ Shear Modulus or Torsion Modulus

When an object is subjected to a pair of equal forces which act in opposite directions but *not along the same line,* the shape of the body will be changed. Consider a pair of forces F, F acting on the upper and lower faces of a rectangular body, as in Fig. 11.10. Assume the pair of forces shears the block so that face $BCDE$ becomes the parallelogram $BCD'E'$, with edges BE and CD turning through the small angle ϕ. ϕ is called the *angle of shear*.

Upon removal of the shearing forces, the body will return to its original shape if the elastic limit of the material has not been exceeded. This type of elasticity is called *elasticity of shear*. It is of great significance in structural design and in power transmission.

The block may be considered as made up of many layers of horizontal sheets of area A, each of which under shearing forces slides slightly with respect to the adjacent layers. The *shearing stress* is defined as the tangential force per unit area producing the sidewise motion. Hence, the shearing stress is $\Delta F/A$, where A is the area of the upper or lower face of the block. The *strain* is defined as the amount of the lateral displacement x per unit height. Thus, in Fig. 11.10, if h is the height of the rectangular block and x is the lateral displacement of the top with respect to the lower face, the strain is x/h. The shear modulus n is the ratio of shearing stress to shearing strain.

Note that A is the area of the surface *parallel* to the applied force F (and not perpendicular as for tension and compression) and x is perpendicular to h.

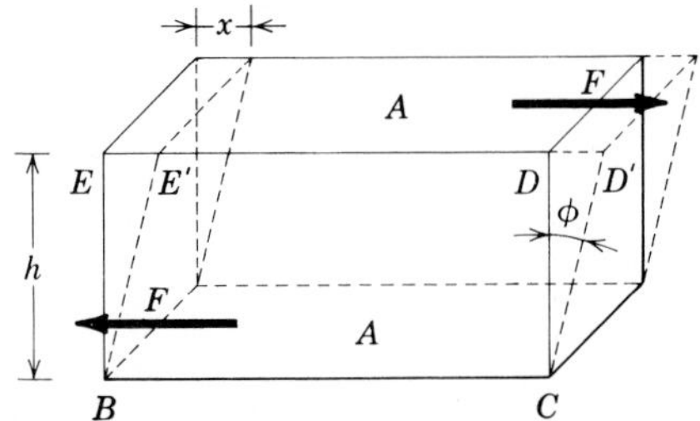

Fig. 11.10 Effect of shearing forces on a rectangular body. Shearing forces tend to change the shape of the body, but not its volume.

$$n = \frac{\text{stress}}{\text{strain}} = \frac{\Delta F/A}{x/h} = \frac{h\,\Delta F}{xA} \tag{11.9}$$

Shear modulus.

For very small angles $\tan \phi = x/h \approx \phi$, where ϕ is in radians. Hence,

$$n = \frac{\Delta F/A}{\phi} = \frac{\Delta F}{A\phi} \tag{11.9'}$$

Illustrative Problem 11.11 A 30-cm cubical solid of rubber has two parallel and opposite forces of 110 N each applied to opposite faces (see Fig. 11.10). If the angle of shear is 1.23×10^{-3} radians, calculate (*a*) the lateral displacement x between the two faces, and (*b*) the shear modulus of the rubber, in gigapascals, GPa.

Solution

(***a***)

$$\phi = \frac{x}{30\text{ cm}} = 0.00123 \text{ (approx.)}$$

$$x = 0.0369\text{ cm} \qquad \textit{answer}$$

(***b***)

$$n = \frac{\Delta F}{A\phi} \tag{11.9'}$$

$$= \frac{110\text{ N}}{(30\text{ cm})^2 \times 0.00123\text{ rad}}$$

$$= 99.4\ \mathbf{N}/\text{cm}^2 = 99.4\ \frac{\mathbf{N}}{\text{cm}^2} \times \left(100\ \frac{\text{cm}}{1\text{ m}}\right)^2 \times \frac{1\text{ GPa}}{10^9 \frac{\mathbf{N}}{\text{m}^2}}$$

$$\approx 0.001\text{ GPa} \qquad \textit{answer}$$

Shear is also involved in the twisting of a straight rod with circular cross section. The drive shaft of a car or any shaft which transmits rotary motion throughout its length is subjected to twisting forces when under load (see Fig. 11.11). Within the shaft, *twisting stresses* are set up to counteract the external forces applied.

If one end of a cylindrical solid rod or shaft is clamped in a fixed position and the other is turned by an applied torque τ (see Fig. 11.12), successive sections of the rod move over each other. The resultant twist θ produced by the applied torque τ depends on the length L and the radius r of the rod. It can be shown, by using calculus, that for a solid rod,

The angle of shear θ is a measure of shear strain.

$$\theta = \frac{2\tau L}{\pi n r^4} \tag{11.10}$$

where θ is in radians, τ is in newton-meters, L in meters, r in meters, and n in newtons per square meter. In English units, τ is in pound-feet, L in feet, r in feet, and n in pounds per square foot.

The angle of twist θ for a hollow cylinder of radius r and wall thickness t is given by

$$\theta = \frac{\tau L}{2\pi n r^3 t} \tag{11.11}$$

Values of the shear modulus n for some common materials are given in Table 11.4.

Fig. 11.11 Oil-field rotary drilling table in operation. Strings of drill pipe, often more than a mile long, are subjected to tremendous twisting stresses throughout their entire length. The heavy hose at the left carries a special drilling "mud" into the well. (Chevron Corporation)

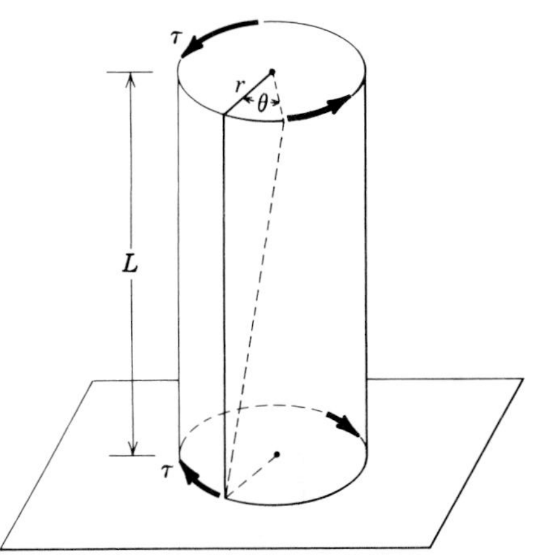

Fig. 11.12 Shearing stress effect on a cylindrical body. The stress increases from zero at the center to a maximum at the outer surface. The angle of twist θ depends upon the length L of the cylinder, the applied torque τ, the fourth power of the radius r, and the shear modulus n of the material.

Illustrative Problem 11.12 Through what angle will the free end of a copper rod be twisted if it is 110 cm long and has a diameter of 0.68 cm, if a torque of 320 N · cm is applied to it? (Use Table 11.4.)

Solution Convert the given data to units consistent with Eq. (11.10) and substitute:

Table 11.4 Shear modulus for selected solids

Material	*Shear Modulus n*	
	$N/m^2 \times 10^{10}$	$lb/in.^2 \times 10^6$
Aluminum, rolled	2.37	3.44
Brass, cold rolled	3.53	5.12
Copper	4.24	6.14
Lead, rolled	0.54	0.78
Nickel	7.30	10.6
Platinum, pure drawn	6.42	9.32
Steel	8.04	11.7
Tungsten, drawn	14.8	21.5

$$\theta = \frac{2(320 \text{ N}\cdot\text{cm}) \times (110 \text{ cm})}{3.14(4.24 \times 10^{10} \text{ N/m}^2)(1 \text{ m}/100 \text{ cm})^2(0.34 \text{ cm})^4}$$

$$= 0.40 \text{ rad} \approx 23°$$ *answer*

A typical problem faced by the technician.

Illustrative Problem 11.13 A technician is given the task of designing a drive-shaft coupling and is given the following parameters. He must use 1-in.-diameter bolts that are to be placed at a distance of 5 in. from the center of the shaft (see Fig. 11.13). The shaft is to transmit 4750 hp at a speed of 1200 rev/min. What is the least number of 1-in. bolts he should use for the coupling? The assembly is to be built for the bolts to withstand a shearing stress of 13,000 lb/in.^{2}

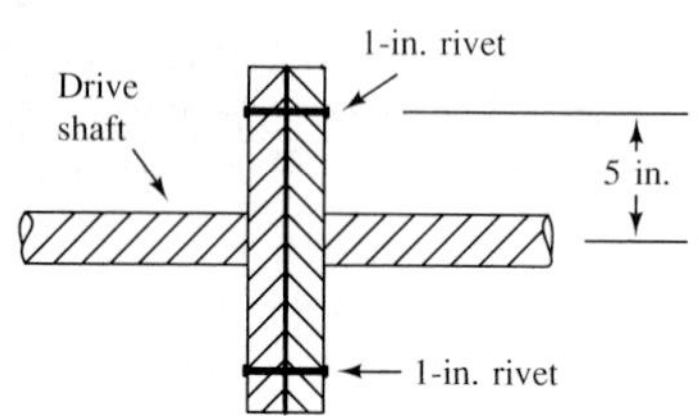

Fig. 11.13 Illustrative Problem 11.13.

Solution The power P transmitted is 4750 hp which is equivalent to 2,612,500 ft · lb/s. The angular speed ω is 1200 rev/min which is equivalent to 40π rad/s. The cross-sectional area A of each bolt is $(\pi/4)(1 \text{ in.})^2 = \pi/4 \text{ in.}^2$ The radius r of the bolt circle (moment arm) $= \frac{5}{12}$ ft. The allowable shearing stress S per bolt = 13,000 lb/in.^{2}

Let n be the number of bolts required. Then $\tau = Fr$ Eq. (9.1)

$$\tau = (nAS)r \tag{1}$$

$$= n(\pi/4 \text{ in.}^2)(13{,}000 \text{ lb/in.}^2)(\tfrac{5}{12} \text{ ft})$$

$$= \frac{65{,}000\pi}{48} n \text{ lb}\cdot\text{ft}$$

Using Eq. (9.3), $\quad P = \tau\omega$, we get $\tau = P/\omega \qquad (2)$

or

$$\tau = \frac{2{,}612{,}500 \text{ ft}\cdot\text{lb/s}}{40\pi \text{ rad/s}}$$

$$= \frac{65{,}300}{\pi} \text{ lb}\cdot\text{ft}$$

From Eq. (1) and Eq. (2), we can solve for n.

$$\frac{65{,}000\pi}{48} n \text{ lb}\cdot\text{ft} = \frac{65{,}300}{\pi} \text{ lb}\cdot\text{ft}$$

From which, $\qquad n = 4.89 \qquad$ *answer*

Thus, five bolts should be used.

11.11 ■ Bulk Modulus and Compressibility

All substances can be diminished in size under sufficient pressure. The decrease in volume of a solid or liquid is proportional to the pressure exerted on its outer surface. Uniform pressure can be placed on a solid if it is submerged in a liquid, for the liquid pressure will be transmitted undiminished and will act perpendicularly to each surface. (Such hydrostatic pressures will be discussed in Chap. 12.) The pressure applied will be the force acting on a unit area.

Consider a cube and a sphere of some material on which forces are applied normal to each surface and uniformly distributed over the surface (Fig. 11.14). The solids will be in equilibrium under the action of these external forces, but the volumes will be reduced. If the original volume of a solid V decreases by an increment ΔV, the *volume strain* will be $\Delta V/V$. The *volume stress* is the normal force per unit area; i.e., volume stress is $\Delta F/A = \Delta P$, where ΔP is the increase in pressure on the surface.

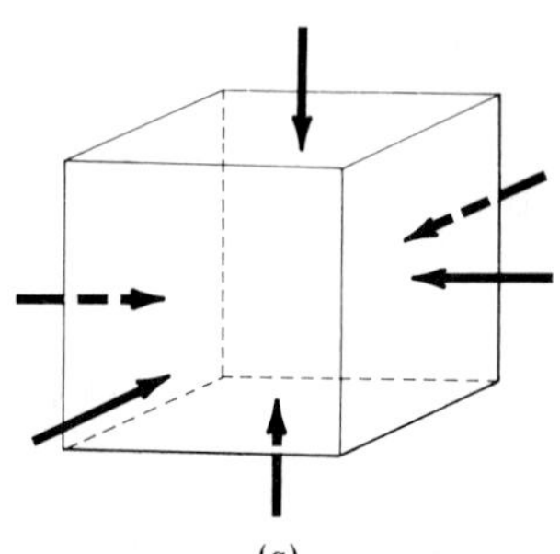

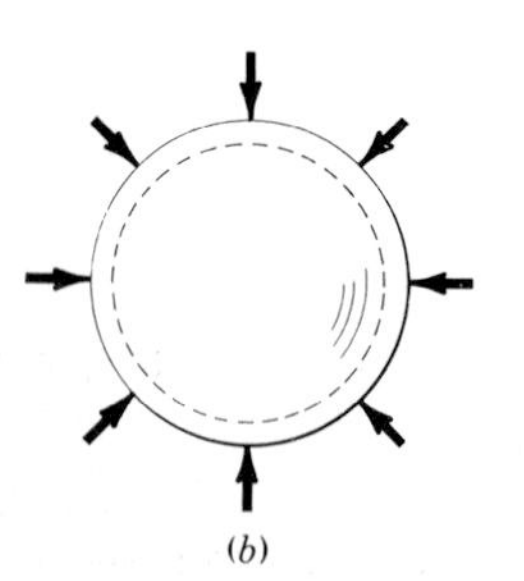

Fig. 11.14 Volume stresses. Forces are normal to the surfaces of (*a*) a cube and (*b*) a sphere.

The ratio of stress to strain in these cases is called the *bulk modulus* (B) or the *coefficient of volume elasticity*. Thus

$$B = \frac{\text{stress}}{\text{strain}} = -\frac{\Delta F/A}{\Delta V/V} - \frac{\Delta P}{\Delta V/V} = -\frac{\Delta P\ V}{\Delta V} \qquad (11.12)$$

Bulk modulus is a measure of the force required to compress a material.

The negative sign is used here because an increase in pressure causes a corresponding decrease in volume.

The type of deformation which involves only volume changes applies particularly to fluids, since they offer resistance only to change of volume. They can have only bulk moduli of elasticity. The bulk moduli of solids and liquids (see Table 11.5) are relatively large numbers, indicating that large forces are needed to produce even minute changes in volume. Gases are more easily compressed and have correspondingly smaller bulk moduli.

Table 11.5 Bulk Modulus for Solids and Liquids

	Bulk Modulus B	
Material	*N/m² × 10¹⁰*	*lb/in.² × 10⁶*
Solids:		
Aluminum	7.0	10
Brass	6.1	8.5
Copper	14	20
Glass	3.7	5.3
Iron, cast	9.6	14
Lead	0.77	1.1
Steel	16	23
Liquids:		
Ethyl alcohol	0.110	0.16
Kerosene	0.13	0.19
Lubricating oil	0.17	0.25
Mercury	2.8	4.0
Water	0.21	0.31

The *compressibility* c of a material is the reciprocal of its bulk modulus.

The reciprocal of the bulk modulus B is called the compressibility c.

$$\text{Compressibility } c = \frac{1}{B} = \frac{-\Delta V}{\Delta P V} \qquad (11.13)$$

Illustrative Problem 11.14 Find the decrease in volume of 7.5 liters of water under a pressure of 2.3×10^4 N/m² (use Table 11.5). Answer in cubic centimeters.

Solution Solve Eq. (11.12) for ΔV:

$$\Delta V = -\frac{\Delta P V}{B}$$

From Table 11.5, $B = 0.21 \times 10^{10}$ N/m² for water. Substituting proper converted units in the above equation gives

$$\Delta V = -\frac{2.3 \times 10^4 \text{ N/m}^2 \times 7.5\text{L} \times (10^3 \text{ cm}^3/1 \text{ L})}{0.2\text{L} \times 10^{10} \text{ N/m}^2}$$

$$= -\frac{2.3 \times 7.5 \times 10^7}{0.2\text{L} \times 10^{10}} \text{ cm}^3$$

$$= -0.082 \text{ cm}^3 \qquad \textit{answer}$$

The negative sign merely means that volume has decreased.

Illustrative Problem 11.15 Determine the compressibility of steel (use Table 11.5): What does the answer actually mean?

Solution The bulk modulus of steel is 23×10^6 lb/in.^{2}. Substituting this value for B in Eq. (11.13) gives for compressibility

$$c = \frac{1}{B} = \frac{1}{23 \times 10^6 \text{ lb/in.}^2} = 4.35 \times 10^{-8} \text{ in.}^2\text{/lb} \qquad \textit{answer}$$

This means that a hydrostatic pressure of 1 lb/in.^{2} will cause an original volume V_0 to decrease by 0.0000000435 V_0; that is ΔV_0 (the decrease in volume) = 0.0000000435 V_0 for each lb/in.^{2} increase in the pressure.

11.12 ■ Factor of Safety

The factor of safety inherent in an engineering design will depend upon the dangers involved in case of structural failure.

Engineers in designing a bridge or machine always plan to make each member strong enough to withstand several times as much load as will ever be imposed upon it. This is to allow for any unexpected overloading of the structure or for possible overlooked or imperceptible flaws in the material. The *ratio of the ultimate strength of any material to its maximum expected stress* is called the *factor of safety* of that material. Stated in another way, the factor of safety of a structure is the ratio of the maximum load the structure will maintain to the maximum load that it is built to maintain. For example, if an elevator is built to withstand a load of 5 tons but at no time is expected to carry more than a load of 1 ton, its factor of safety is 5.

The factor of safety varies with the material. For example, a factor of safety of 10 is common practice in brick structures, while steel structures use a factor of about 4. The nature of the load also affects the factor of safety to be recommended, as do wind and flood conditions and the possibility of earthquakes.

Illustrative Problem 11.16 Determine the factor of safety involved in using a cable 0.64 cm in diameter to support a load of 540 kg. The ultimate tensile strength of the cable is 5.5×10^8 N/m^2.

Solution Use Eq. (11.6),

$$F = 540 \text{ kg} = 540 \text{ kg}\left(\frac{9.81 \text{ N}}{1 \text{ kg}}\right) = 5300 \text{ N}$$

and

$$A = \frac{\pi d^2}{4} = 3.14\frac{[0.64 \text{ cm} \times (1 \text{ m}/100 \text{ cm})]^2}{4}$$
$$= 3.2 \times 10^{-5} \text{ m}^2$$

Applied stress,

$$S = F/A = \frac{5300 \text{ N}}{3.2 \times 10^{-5} \text{ m}^2}$$
$$= 1.65 \times 10^8 \text{ N/m}^2$$

The ultimate (breaking) stress is 5.5×10^8 N/m^2. Hence the factor of safety is

$$\frac{5.5 \times 10^8 \text{ N/m}^2}{1.65 \times 10^8 \text{ N/m}^2} = 3.3 \qquad \textit{answer}$$

11.13 ■ Metal Fatigue

When metallic material is subjected to repeated stresses over long periods of time, the internal structure of the material will change. As a result, certain regions are weakened and the metal may rupture under the repeated applications of stresses. Failure usually occurs around areas where there is a minute defect (a flaw) in the internal structure of the material. These flaws grow a little each time a stress acts, and eventually may lead to failure. This loss of strength due to repeated applications of stress is known as *fatigue*.

Metal fatigue is used to describe the failure or rupture of a metal part under repeated application of a stress which is well below the elastic limit. It is particularly important in the design of parts which are subjected to cyclic stresses.

Fatigue is a common cause of failure in machinery that is subjected to cyclic stresses. Thus, it is important to discover flaws in a machine part before it is installed. In many manufacturing plants x-rays are used to detect hidden flaws in parts of complicated machinery.

For most steels the endurance limit, or fatigue stress, is that stress which will withstand 10 million cycles without breaking. However, the number of cycles which will produce failure decreases rapidly with an increase in the stress applied. For example, doubling the stress on a metal bar may reduce by half the number of cycles needed for failure.

Any discontinuity in a section of material, e.g., a bend, groove, or hole, will increase the chance of failure. Stresses at these sections of discontinuity will reach values three or more times the value of the average stress in the member. Such weakening is used sometimes by the person who wants to cut a piece of wire but has no wire cutters. Whereas the person may be unable to rupture the wire by pulling on it, he or she can produce a break in the wire by repeated bendings at the desired point of rupture.

11.14 ■ Hardness, Malleability, and Ductility

Additional properties of matter.

Engineers, in selecting suitable materials, must consider many other properties in addition to those discussed up to this point. Among them are *hardness, malleability,* and *ductility*. One of the methods employed in determining the relative hardness of various substances is the *scratch test*. At slow speeds, the harder of two substances can always scratch the less hard material. An arbitrary scale showing the positions of 10 substances in the scratch test (see Table 11.6) has been made, ranging from soft talc to the hardest of all known substances, diamond.

Other materials are classified for hardness by comparing, by scratching, with these known substances. Mohs' scale of hardness of some common structural materials is as follows: aluminum, 2 to 2.9; brass, 3 to 4; emery, 7 to 9; iron, 4 to 5; lead, 1.5; marble, 3 to 4; steel, 5 to 8.5; tin, 1.5 to 8; wax (0°C), 0.2.

Table 11.6 Mohs' Scale of Hardness

1. Talc
2. Rock salt or gypsum
3. Calcite
4. Fluorite
5. Apatite
6. Feldspar
7. Quartz
8. Topaz
9. Corundum
10. Diamond

11.15 ■ Hardness by the Brinell Scale

The property of hardness is commonly measured by either the Brinell number or the Rockwell number.

One industrial method of measuring the degree of hardness of metals involves pressing a 10-mm hardened chrome-steel ball into the metal to be tested, with a force exerted by a mass of 3000 kg. The diameter of the resulting indentation is used as the measure of hardness. This is called the *Brinell method* (Fig. 11.15). Each solid is given a *Brinell number* (see Table 11.7 for examples). It is found by dividing the load (in kilograms-force) by the surface area of the impression (in square millimeters). The larger the Brinell number, the harder the material.

Fig. 11.15 Metal hardness tester with a system of digital readout of Brinell values. (Tinius Olsen Testing Machine Co.)

Fig. 11.16 Coils of aluminum foil feedstock are shown being checked before entry into an annealing oven. The annealing process eliminates work hardness and softens the metal to allow further thickness reduction to foil gauges. These large 30,000-lb coils, when rolled on a mill to foil gauge, would stretch about 265 mi. (Reynold's Metals Co.)

The Rockwell hardness machine is designed to test materials of widely varying hardness. This is done by changing indenters on the machine. For very hard materials a diamond cone is used. Hard steel balls are used for the softer materials.

Table 11.7 Brinell Hardness of Materials

Material	*Brinell Number*
Aluminum, annealed	16
Chromium	91
Copper-aluminum (11.73%), hard-rolled	269
Iron, annealed	77
Lead, cast	4.2
Platinum, drawn	64
Carbon steel	460
Alloy steel-nickel-vanadium-carbon-manganese-silicon	627

Hardness is dependent on the elastic moduli, the elastic limit, and the hardening produced by "working" (as annealing) the metal.

Steel can be hardened by heating it to a high temperature and then cooling suddenly by plunging it into water or oil. However, such steel becomes too brittle for most uses. It can then be tempered (toughened) by reheating and cooling slowly. As the metal loses hardness, it gains toughness. Thus, if the metal were allowed to cool slowly and completely, it would be left soft and tough but not brittle. This process is called *annealing* (Fig. 11.16).

Hardness must not be confused with brittleness. For example, steel is hard and tough, while glass is hard but brittle.

11.16 ■ Ductility

The *ductility* of a material is that property which permits it to be drawn into a wire. The smaller the diameter of the wire to which it can be drawn, in general, the greater

Fig. 11.17 Glowing ingot takes on new shape as it is worked by rolls of a blooming mill. The ingot eventually will be shaped into a bloom or billet. From this semifinished piece of steel, finished structural shapes will be rolled. Rolling not only shapes the steel but also improves its mechanical properties. (Bethlehem Steel Corporation)

the ductility. Many common metals are tenacious enough to allow rods or large wires to be drawn through a hole in a hard steel plate (*die*) that has a smaller diameter than that of the original rod or wire. Gold, silver, and copper are common metals with a high degree of ductility. Platinum can be drawn into wires 0.00003 in. in diameter. Glass, when heated to softness, can be drawn into such fine threads that garments and curtains can be woven from them. Iron and steel are also ductile, but not to the extent that characterizes the above materials. In *wiredrawing* steel, it is common practice to resort to heat treatment (annealing) between successive passes through the die.

Ordinary wire for fencing and utility use is *hot-drawn*. Piano wire and other special wires are *cold-drawn* so that their hardness will be retained.

11.17 ■ Malleability

Malleability is the property of a material which enables it to be hammered or rolled into a desired shape. The process usually requires the operation to be carried on at high temperatures.

The property of a material that makes it capable of being rolled or hammered into thin sheets of various shapes is called *malleability*. Gold is extremely malleable and has been rolled into sheets that are 1/300,000 in. thick—so thin that they will actually transmit diffused light. Copper, aluminum, and tin are quite malleable—one reason why we have so many pots, pans, and appliances made from them. Lead is extremely malleable and was the first metal used for pipes. In fact, our word "plumbing" comes from the Latin *plumbum,* meaning lead. Aluminum foil is sheet aluminum that has been rolled very thin. It is an excellent wrap for food, candy, gum, cigarettes, and other packaged products.

The capability of industry to roll steel (Fig. 11.17) into girders of different shapes or cross sections is important to the engineer. The malleability of various steels is a matter of extreme importance in the automobile industry because of the *streamlined* shapes of auto tops, bodies, and fenders demanded by modern design.

QUESTIONS AND EXERCISES

1. What is the name given to the force of attraction between (*a*) like molecules, (*b*) unlike molecules?

2. Discuss the difference in behavior of the molecules of a solid when in tension and compression.

3. Why are molecular collisions assumed perfectly elastic?

4. Molecular forces of tension and compression are due primarily to what phenomena associated with the atoms which make up the molecule?

5. Which of the following is the most elastic: rubber, steel, tungsten? (HINT: See Table 11.2.)

6. Describe the motion of the particles of a solid.

7. List the various types of distortion. What does Hooke's law imply in each?

8. In what type problem situations are weight densities important? What equation relates mass density and weight density?

9. How are mass density and specific gravity related?

10. Explain in molecular terms why when you stretch a spring it will return to its original shape when you "let go."

11. What are the relative elongations of two steel rods under the same tension if the first is twice as long, but with a diameter twice that of the second rod?

12. Why do you think scissors are often referred to as shears?

13. What property of a substance will permit it to be drawn into fine wires?

14. Give several properties of aluminum which make the metal effective in the production of pots, pans, and cans.

15. Which stress is the larger, the pascal or the kilogram/meter^2? How many times as large?

16. What determines the factor of safety of a structure? Why should a brick structure require a considerably higher factor of safety than a steel structure?

17. Two identical springs are under tension. What are the relative potential energies stored in the springs if the first is stretched twice as much as the second? Assume the stresses are within the elastic limit.

18. Do some research reading on prestressed concrete. Why are steel rods often embedded in concrete used for buildings and bridges? What does "prestressed" mean in this application?

19. Select the correct answer: The Brinnell number of a given material is a measure of (*a*) specific gravity, (*b*) specific heat, (*c*) weight, (*d*) hardness, (*e*) density.

20. Which answer is correct? The ultimate strength of a beam divided by the allowable stress is the (*a*) yield point, (*b*) percentage of elongation, (*c*) percentage of reduction in area, (*d*) factor of safety, (*e*) working stress.

PROBLEMS

Group One

1. A 100-g weight will stretch a spring whose original length is 15 cm, to 18 cm. What will be the length of the spring if a total weight of 400 g is applied and the spring remains within its elastic limit?

2. What is the approximate mass of the air in a room with dimensions 8.6 m × 4.3 m × 2.7 m? Assume 0°C, 1 atm conditions.

3. What is the mass of a solid brass ball if its diameter is 5.0 cm?

4. A spring has a spring constant of 3750 N/m. What energy will be stored in the spring if it is stretched 11.25 cm?

5. If a rope with a diameter of 0.50 in. will break under a tension of 6700 lb, what will be the breaking strength of a rope of the same material with a diameter of (*a*) 0.25 in., (*b*) 0.75 in.?

6. How many grams of gasoline could you put in a paper cup that has an interior volume of 170 cm^3?

7. The volume of a 1-gal container is 231 in.^{3}. What would be the weight of 1 gal of (*a*) water? (*b*) gasoline?

8. A metal garbage can has a volume of 30 gal. There are 231 in.^{3} in a gallon. (*a*) What is the volume in cubic feet? (*b*) What weight of water can the garbage can hold?

9. A 1-ft-diameter cylindrical copper disk weighs 100 lb. What is the thickness of the disk in inches?

10. A 1-ft-diameter Styrofoam cylinder weighs 100 lb. If the weight density of Styrofoam is 1.25 lb/ft^3, how long is the cylinder?

Group Two

11. What is the density of a proton if it has a mass of 1.67×10^{-27} kg and a radius of about 10^{-15} m?

12. What is the weight density of a material whose mass density is 8900 kg/m^3?

13. Find the stress developed in a steel rod which has a diameter of 0.75 in. and which is hung vertically and supports a load of 8000 lb.

14. A steel plate $\frac{1}{2}$ in. thick has an ultimate shearing strength of 40,000 lb/in.^{2}. What force must be applied to punch a $\frac{3}{8}$-in. hole through the plate?

15. Suppose four $\frac{1}{4}$-in. wrought-iron rivets were used to rivet together two of the plates of Problem 14. What force would have to be applied before the four rivets would shear off at right angles? Let the shearing strength of the rivets be 35,000 lb/in.^{2}. (Neglect friction between the plates.)

16. A drawn steel wire is 175 cm long and has a diameter of 0.214 cm. What is the tension in the wire if it is tightened and stretched 0.307 cm?

17. A lead ball with a radius of 1.2 cm is placed in oil under a compressional stress of 3.0×10^9 Pa. What is the fractional change in its volume due to this pressure?

18. If atmospheric pressure is 1.013×10^5 Pa, by what percentage will a volume of a brass ball increase when placed in a vacuum?

19. What load must be applied to a 2.5 cm cylindrical aluminum bar 2.0 m long to stretch the bar 0.13 cm?

20. How much will the volume of 250 cm^3 of mercury diminish under a pressure of 1.38×10^6 N/m^2?

21. How much will a spring with a spring constant of 25.0 lb/in. need to be stretched to store 5.16 ft · lb of energy?

22. Aluminum ruptures when a sheer stress of 1.4×10^6 N/m^2 is applied. Find the force needed to punch a 12.5 cm diameter hole in an aluminum plate 0.80 cm thick.

23. A wrought-iron member of a small highway-bridge truss is built to support expected tensile loads of 250,000 *N*. The factor of safety used in the truss is 5. What should be the cross-sectional area of the member? The ultimate tensile strength of wrought iron is 3.2×10^8 N/m^2.

24. What is the factor of safety in effect if the maximum tensile load to be supported by a steel rod $\frac{3}{4}$ in. in diameter is 8000 lb? Ultimate tensile strength of the steel $= 160 \times 10^3$ lb/in.^{2}.

25. Rubber used in a certain shock absorber has a cross-sectional area of 13 cm^2 and is 6.4 cm long. The rubber has a Young modulus of 2.4×10^8 N/m^2. How much will it support without being compressed more than 0.48 cm?

26. A wire that is 4 m long and has a diameter of 1.0 mm is elongated 1.2 mm when a 5-kg mass is hung from its end. What is the value of Young's modulus for the material of the wire?

27. Four 2-in. diameter cast-iron rods support under compression a load of 3 tons. Each rod has a pre-loaded length of 6 ft. Find the decrease in length of each rod produced by this load.

28. What decrease in length of a 6.2-m steel girder whose cross-sectional area is 62 cm^2 will result when it is subjected to a compressional load of 48,000 kg? (Assume $Y = 21 \times 10^{10}$ N/m^2.)

29. Two parallel and opposite forces, each of 4450 N, are applied tangentially to opposite faces of a 25.4 cm cubical block of steel. Find the angle of shear and the relative displacement of the faces.

30. Calculate the bulk modulus in gigapascals of glycerine if a liter of the liquid contracts 0.63 cm^3 when subjected to a pressure of 30 kg/cm^2.

31. A drawn-brass wire with diameter of 0.162 in. is 6 ft long when supporting a load of 40 lb. How much longer will the wire be when the load is increased to 120 lb?

32. A hydraulic press contains 0.30 m^3 of oil when under no load. What is the decrease in volume of the oil when it is subjected to a pressure of 3.5×10^7 N/m^2 if the compressibility of the oil is 2.0×10^{-10} m^2/N?

Group Three

33. A drawn-steel wire 300 m long whose cross-sectional area is 0.65 cm^2 hangs vertically in a deep well. How much does it stretch under its own weight? (Take the density of steel as 7800 kg/m^3. Find the average force acting throughout the length of the wire.)

34. How many steel cables of 0.75-in.^{2} cross section should there be in the hoisting mechanism of a 6-ton elevator (no counterweights) which will have a maximum upward acceleration of 12 ft/s^2? Assume that the factor of safety is 5 and that the tensile strength of the cable steel is 40,000 lb/in.^{2}

35. A marine engine delivers 250 hp at 1000 rev/min to a solid steel drive shaft 3 m long and 3 cm in diameter. Through what angle is the shaft twisted?

36. A drawn steel wire 1.00 mm in diameter and 2.50 m long supports a mass of 88 kg at its midpoint as shown in Fig. 11.18. How much is the wire elongated?

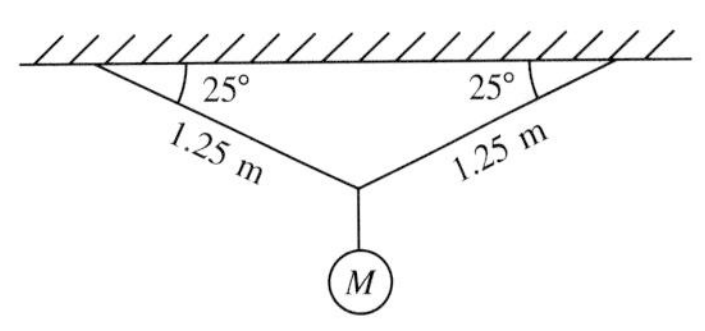

Fig. 11.18 Problem 36.

37. Two equal-length parallel wires 20 cm apart, one copper and one drawn steel, are used to support a load W as shown in Fig. 11.19. The cross-sectional area of the copper wire is twice that of the steel wire. Determine the distance x from the steel wire for applying the load so the wires will remain with equal lengths.

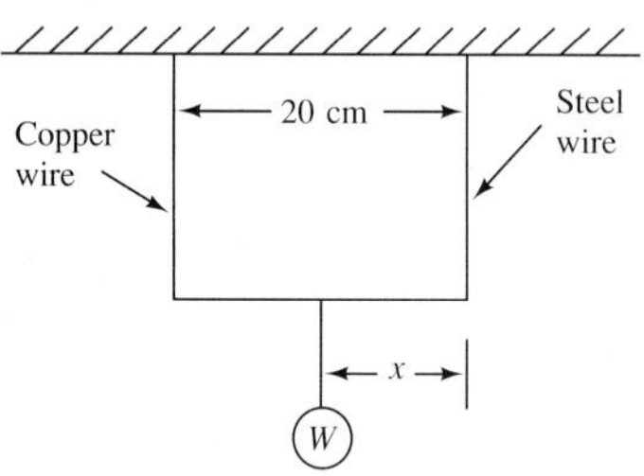

Fig. 11.19 Problem 37.

38. A steel ball 15.2 cm in diameter weighing 14.2 kg is suspended from a point 3.07 m above the floor by a mild steel wire of unstretched length of 2.90 m. The diameter of the wire is 0.914 mm. If the ball is set swinging so that its center passes through the lowest point at 4.88 m/s, by how much does it clear the floor?

39. Each of the two alloy-steel drive shafts of a Navy ship is 60 ft long and 8 in. in diameter. If the modulus of rigidity $n = 11.7 \times 10^6$ lb/in.^{2}, and the shaft is delivering 25,000 hp at 400 rev/min, what is the total twist produced in the shaft, in radians?

40. A coupling connecting a motor to a pump transmits 5.4 hp at 1750 rev/min. Two $\frac{1}{4}$-in.-diameter pins are used to hold the couplings together (see Fig. 11.20). When the coupling is transmitting 5.4 hp at 1750 rev/min what is the shearing stress, in lb/in.^{2}, on each pin?

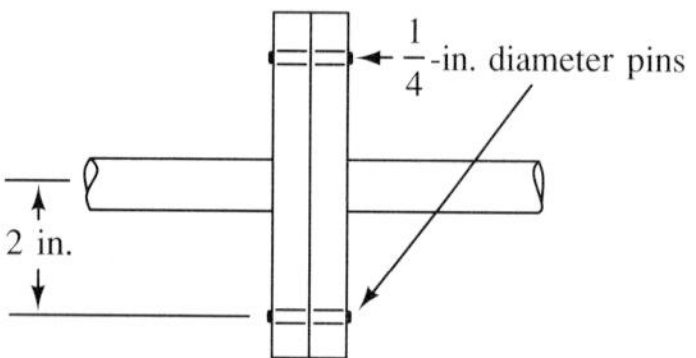

Fig. 11.20 Problem 40.

41. A $\frac{1}{4}$-in.-diameter shear pin, as shown in Fig. 11.21, is used to protect a mechanism at the end of a drive shaft. The shaft has a diameter of 2 in. and turns 4.0 rev/min. It is driven by a $\frac{1}{8}$-hp motor through a reduction gear having an efficiency of 82 percent. What will be the shearing stress, in pounds per square inch, on the shear pin when the motor is delivering its rated horsepower?

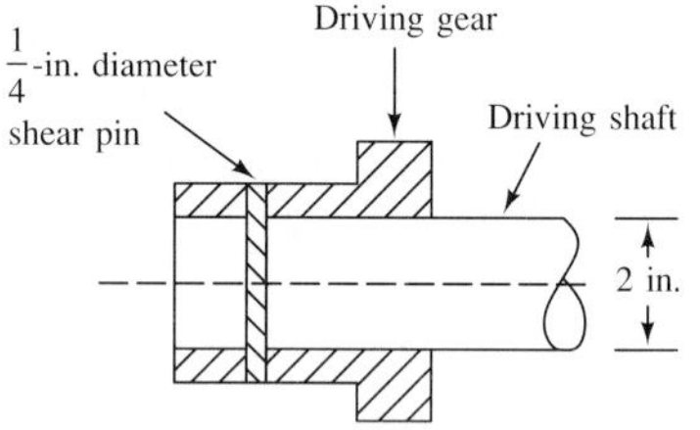

Fig. 11.21 Problem 41.

42. A beam 20.0 m long and weighing 1500 kg is supported by two 6.0-m-long circular pillars at the ends of the beam (see Fig. 11.22). A 20,000 kg load is added to the beam as shown. (*a*) What are the forces F_1 and F_2 at the supports? (*b*) What minimum diameter should the cylindrical supports have to support the beam and load if the supports are made of concrete with an ultimate compressive strength of 20×10^6 N/m^2. Use a factor of safety of 6. (*c*) How much will the support carrying the heavier load compress under the load if Young's modulus for concrete under compression is 2.0×10^{10} N/m^2?

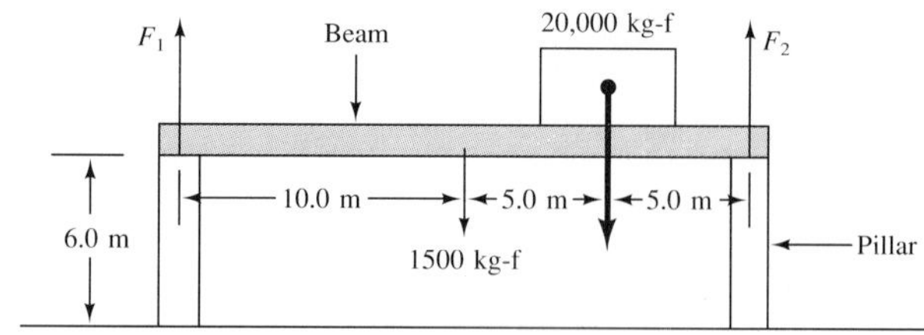

Fig. 11.22 Problem 42.

CHAPTER 12 PROPERTIES OF LIQUIDS

Molecules of liquids are, on the average, farther apart than the molecules of solids. The average distance a molecule moves before colliding with another molecule is generally greater in a liquid than in a solid. This permits the molecules of the liquid to move faster, resulting in an increase in momentum. The net result of the greater momentum of liquid molecules is that the motion of liquid molecules is translational rather than vibrational as in solids.

A liquid, like a gas, cannot sustain a shear stress and hence does not maintain a fixed shape. Liquid molecules have the ability to flow. Thus, while maintaining a fixed volume, a liquid will assume the shape of the container.

12.1 ■ Liquids

Hydrostatics describes the characteristics of liquids at rest.

The study of liquids is divided, for convenience, into two main parts: *hydrostatics*, or liquids at rest, and *hydraulics*, or liquids in motion. The properties of liquids at rest can very often be expressed by means of simple relationships. However, liquids in motion present problems which are usually more difficult to solve than those experienced in hydrostatics. This is due to the presence of frictional and other disturbances whose actions often cannot be expressed in simple mathematical terms.

Liquids are almost incompressible. For example, a pressure of 16,400 lb/in.^{2} (1.13×10^8 N/m^2) (Fig. 12.1) will cause a given volume of water to decrease by only 5 percent from its volume at atmospheric pressure. Practically the only type of elasticity possessed by a liquid is that of volume elasticity. The volume elasticity of

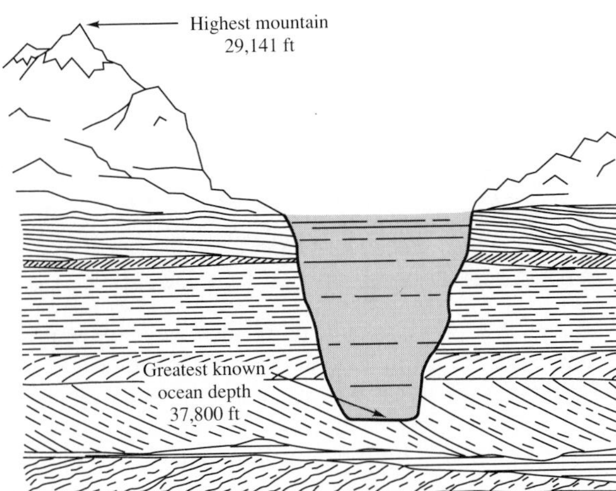

Fig. 12.1 At the greatest known ocean depths, pressure is equal to 16,400 lb/in.^{2} (1120 atm), but the volume of a given amount of water decreases by only 5 percent.

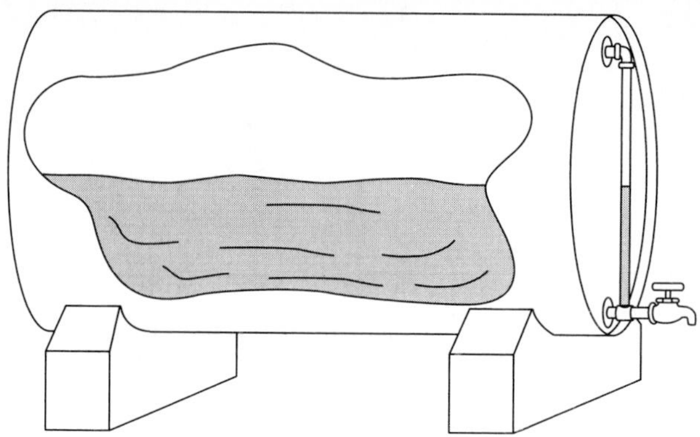

Fig. 12.2 The water in the gauge stands at the same level as that in the tank.

water is almost perfect. If a tremendous pressure is applied to a volume of water, the water will be compressed slightly, but on removal of the pressure the volume will almost instantly return to its original value.

HYDROSTATICS

12.2 ■ A Liquid Seeks Its Own Level

When a liquid is poured into a vessel, it flows until it takes the shape of the vessel. The height of the liquid in the vessel will depend on the volume of the liquid and on the shape of the vessel.

A liquid will tend to keep its surface level. When the surface is not level, the liquid will flow in whatever direction will make the surface level. "Water runs downhill" is a common expression for this property. "Water seeks it own level" is another (see Fig. 12.2).

12.3 ■ Pressure Is Force per Unit Area

A fluid at rest will exert forces on the walls of the container which are perpendicular to the containing surface. The normal force per unit area is called *pressure:*

Pressure is the normal force per unit area.

$$\text{Pressure} = \frac{\text{total force}}{\text{area}}$$

$$P = \frac{F}{A} \tag{12.1}$$

Units of pressure are obtained as ratios of force units and area units. Common units for pressure are the newton per square meter (N/m^2), called a *pascal* (Pa), in the SI-metric system; and the pound per square inch in the English-engineering system.

Units of pressure.

Units of pressure

SI-metric unit	pascal (Pa) = 1 N/m^2
English-engineering unit	lb/in.^{2} (psi)
	1 psi = 6895 Pa

The *atmosphere*, another pressure unit, represents the average pressure exerted by the earth's atmosphere at sea level. One atmosphere is equal to 1.013×10^5 Pa or 14.7 lb/in.^{2} The *millibar* (mb) is used in meteorology and is equal to 100 N/m^2 or 100 Pa.

Pressures are also measured in millimeters of mercury (mm Hg) and in in. Hg. The unit mm Hg is called the *torr* (see Sec. 13.14).

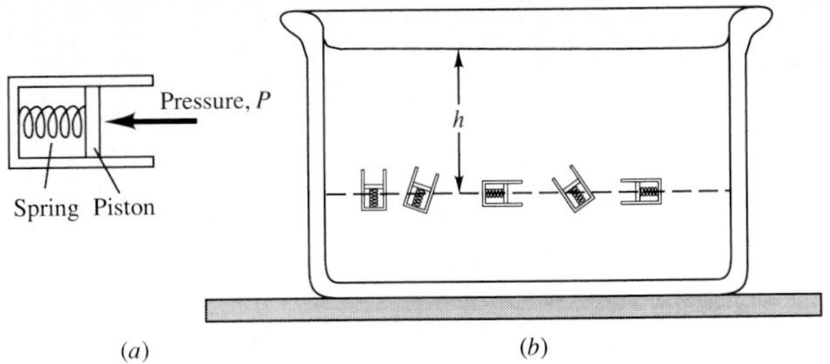

Fig. 12.3 The fluid pressure is measured by the degree of compression on a spring. The compression (fluid pressure) will be the same in all directions at any given depth h.

Equation (12.1) can be solved for F to give

$$F = P \times A \tag{12.1'}$$

Total force equals the product of pressure and area.

12.4 ■ Hydrostatic Pressure Increases with Depth of Liquid

Hydrostatic pressure is directly proportional to the depth of the liquid.

Anyone who dives under the surface of water notices that the pressure on the eardrums at a depth of even a few feet is quite noticeable. Deep-sea divers wear special gear to protect them against extreme pressures.

In January 1960 Jacques Piccard and Lt. Donald Walsh, USN, descended in a specially constructed underwater vessel to a depth of 37,800 ft off the coast of Guam. Pressures on the vessel reached 16,400 lb/in.^{2}, requiring steel walls 5 in. thick for the underwater craft.

Careful measurements show that the *pressure of a liquid* is directly proportional to the depth and for a given depth the liquid exerts the same pressure in all directions (see Fig. 12.3).

Thus, the force on a dam is not determined by how much water there is impounded in the lake, but by how deep the water is directly behind the dam.

12.5 ■ The Pressure-Force Formulas for Liquids

Consider a cylindrical container with vertical sides filled with a liquid. Let A be any horizontal cross-sectional area of a column of the liquid and h the height (depth) of the column. A can be the cross-sectional area of the bottom of the container (Fig. 12.4*a*) or the cross-sectional area of any (imaginary) column within the liquid (Fig. 12.4*b*). The entire weight of the column of liquid above A will be pressing down on the area A. The volume V of the column is Ah. The weight of the column of liquid can be found if we know the density and volume. The density is found from Eq. (11.2), which states that

$$D = w/V$$

$$\text{Weight density} = \text{weight per unit volume}$$

From Eq. (11.2) and the formula $V = Ah$, we get

$$w = DV = DAh$$

Equation (11.3) relates weight density and mass density: $D = \rho g$. Substituting ρg for D in the equation for w, gives

$$w = DAh = \rho gAh$$

Since the force F acting on area A is equal to the weight w of the column of liquid,

$$F = DAh = \rho gAh \tag{12.2}$$

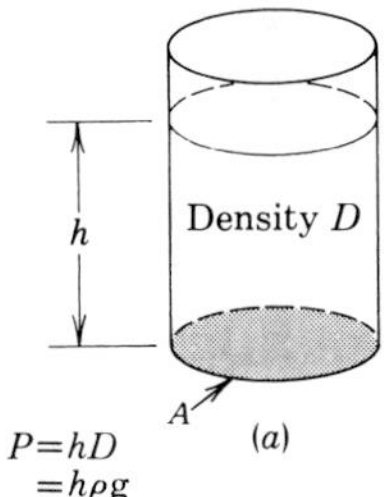

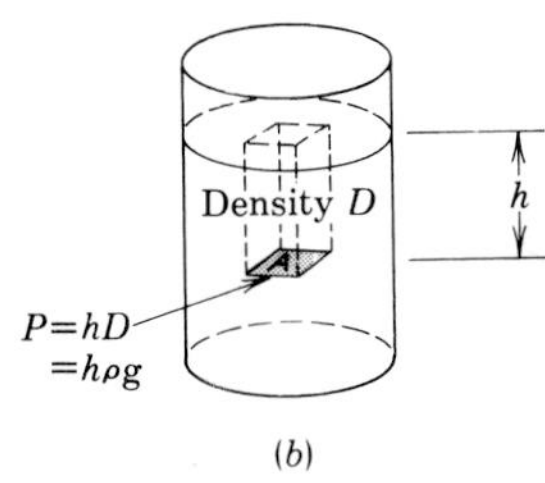

Fig. 12.4 The formula $P = hD = h\rho g$ applies at (*a*) the bottom of the tank as well as at (*b*) any depth h below the liquid surface.

which gives the force of a liquid on a horizontal area below a free surface.

If we divide each member of Eq. (12.2) by A, we get

$$\frac{F}{A} = \frac{DAh}{A} = \frac{\rho gAh}{A}$$

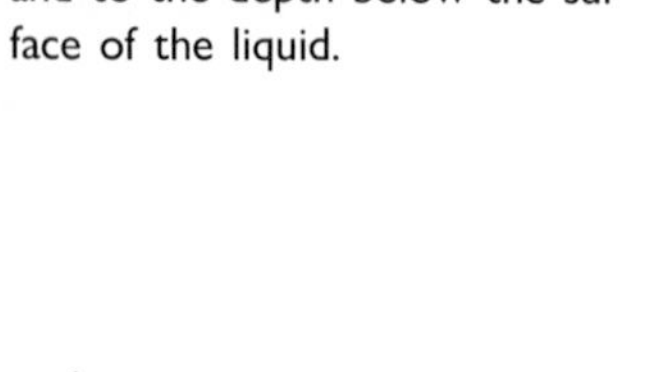

Liquid pressure is directly proportional to the density of the liquid and to the depth below the surface of the liquid.

We can now substitute pressure P for F/A to get the formula for *liquid pressure* at a given depth below the surface of the liquid.

$$P = hD = h\rho g$$
$$\text{Pressure} = \text{depth} \times \text{weight density} \tag{12.3}$$

It is essential that depths and densities be expressed in consistent units when using Eq. (12.3). When D is in pounds per cubic foot and h is in feet, P will be in pounds per square foot. When ρ is in kilograms per cubic meter, h in meters, and g is 9.81 m/s^2, P will be in newtons per square meter or pascals.

Since $D = 62.4$ lb/ft^3 (from Table 11.1) for water, it follows that if height h is 1 ft,

$$\begin{aligned} P &= hD \\ &= 1 \text{ ft } (62.4 \text{ lb/ft}^3) = 62.4 \text{ lb/ft}^2 \\ &= 62.4 \text{ lb/ft}^2 \, (1 \text{ ft/12 in.})^2 = 0.433 \text{ lb/in.}^2 \end{aligned}$$

This is equivalent to saying that the weight of a column of water 1 ft high and 1 in.^{2} in cross-sectional area is 0.433 lb.

Using the same approach for metric units we find that, since ρ for water is 1000 kg/m^3, if height h is one meter, then

$$\begin{aligned} P &= h\rho g \\ &= 1 \text{ m } (1000 \text{ kg/m}^3)\,(9.81 \text{ m/s}^2) \\ &= 9.81 \times 10^3 \frac{\text{kg}}{\text{m} \cdot \text{s}^2}\left(\frac{1 \text{ N}}{1 \text{ kg} \cdot \text{m/s}^2}\right) \\ &= 9.81 \times 10^3 \text{ N/m}^2 = 9.81 \times 10^3 \text{ Pa} \end{aligned}$$

This, in turn, is equivalent to saying that the weight of a column of water 1 m high and 1 m^2 in cross-sectional area is 9.81×10^3 N.

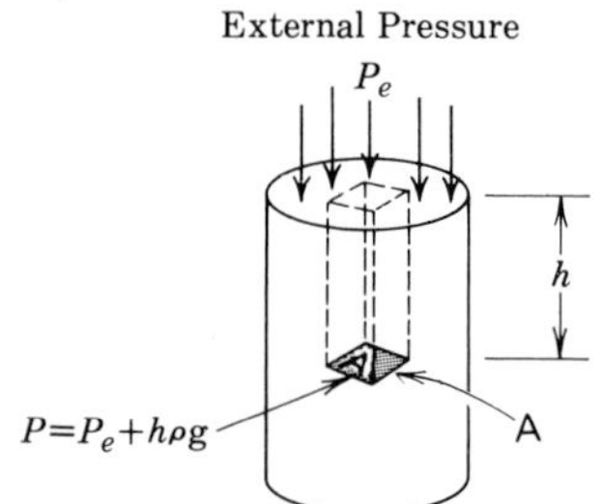

Fig. 12.5 The total pressure at A is the sum of the liquid pressure due to the liquid depth h and the external pressure P_e.

Total Pressure It should be noted that Eq. (12.3) is used to determine the pressure due to the liquid alone. The *total pressure* within a fluid also depends upon the external pressure P_e at the surface (see Fig. 12.5). To obtain the total pressure P, the external pressure P_e must be added to the fluid pressure. The total pressure P is often called the *absolute pressure*.

$$P = P_e + hD = P_e + h\rho g \tag{12.4}$$

The external pressure on the surface of a liquid may be the pressure of the atmosphere or pressure caused by a piston. In general, at sea level, the atmospheric pressure is equal to 14.7 psi or 1.013×10^5 N/m^2 (pascal).

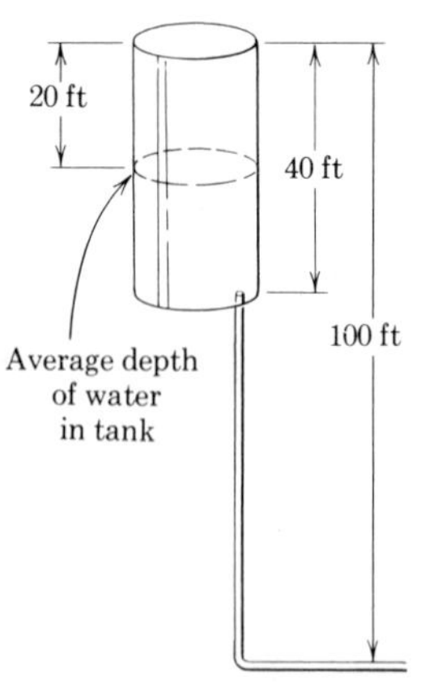

Fig. 12.6 Illustrative Problem 12.1.

Illustrative Problem 12.1 A cylindrical water tank 40 ft high and 20 ft in diameter is filled with water (Fig. 12.6). (*a*) What is the water pressure on the bottom of the tank? (*b*) What is the total force on the bottom? (*c*) On the vertical wall? (*d*) What is the pressure (pounds per square inch) in a water pipe at street level which is 100 ft below the water surface?

Solution

(*a*)

$$\begin{aligned} h &= 40 \text{ ft} \\ D &= 62.4 \text{ lb/ft}^3 \\ P &= h \times D \end{aligned}$$

$$= 40 \times 62.4$$
$$= 2500 \text{ lb/ft}^2 \qquad \textit{answer}$$

(b)

$$F = P \times A$$
$$= 2500 \text{ lb/ft}^2 \times 3.14 \times (10 \text{ ft})^2$$
$$= 785{,}000 \text{ lb (on bottom)} \qquad \textit{answer}$$

(Note that this is equal to the weight of the water in the cylindrical tank.)

(c) In calculating the total outward force against the tank wall, it must be realized that the pressure along the sides varies with the depth. Since the pressure varies directly with the depth, we can use the pressure at the midpoint of the tank as an average value with which to compute force against the vertical wall. The lateral surface area of a cylinder is given by substituting

$$S = 2\pi rh$$
$$= 2 \times 3.14 \times 10 \text{ ft} \times 40 \text{ ft}$$
$$= 2510 \text{ ft}^2$$

Then, substituting in Eq. (12.2), and taking h as *average depth*, or 20 ft:

$$F = 2510 \text{ ft}^2 \times 20 \text{ ft} \times 62.4 \text{ lb/ft}^3$$
$$= 3{,}130{,}000 \text{ lb} \qquad \textit{answer}$$

(d)

$$h = 100 \text{ ft}$$

$$P = \frac{100 \text{ ft} \times 62.4 \text{ lb/ft}^3}{144 \text{ in.}^2/\text{ft}^2}$$

$$= 43.3 \text{ lb/in.}^2 \qquad \textit{answer}$$

Illustrative Problem 12.2 Water is poured into one side of a U-tube, while oil is poured into the other until they both balance in the exact middle of the bottom section of the tube (see Fig. 12.7). How much oil with a density of 820 kg/m^3 must be poured in the one side to exactly balance 24.0 cm of water in the other side?

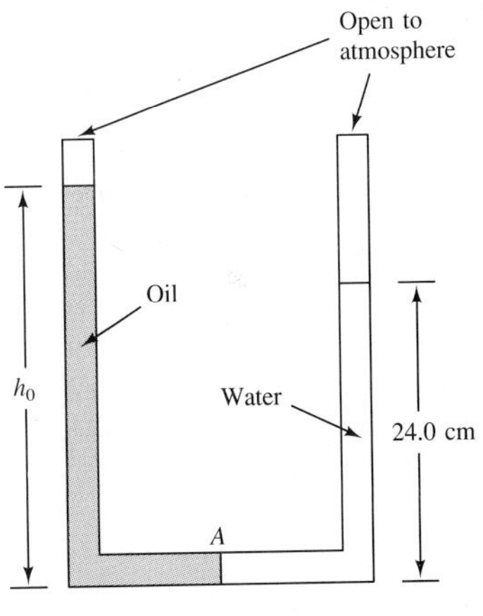

Fig. 12.7 Illustrative Problem 12.2.

Solution The oil pressure should exactly equal (but be in the opposite direction to) the water pressure at A. (We use the subscripts o for oil and w for water.)

$$P_o = P_w$$
$$h_o \rho_o g = h_w \rho_w g$$

Solve for h_o:

$$h_o = \frac{h_w \rho_w g}{\rho_o g}$$

$$= \frac{h_w \rho_w}{\rho_o}$$

$$= 24.0 \text{ cm} \left(\frac{1000 \text{ kg/m}^3}{820 \text{ kg/m}^3}\right)$$

$$= 29.3 \text{ cm} \qquad \textit{answer}$$

Illustrative Problem 12.3 What horsepower motor will be needed to lift water 100 ft and deliver it at a rate of 120 gal/min at a discharge pressure of 60 psi (see Fig. 12.8)? (A review of Secs. 5.16 and 5.17 may be helpful in solving this problem.)

Solution Let P_1 be the pressure equal to a 100-ft column of water and P_2 equal the discharge pressure. Then

$$P_1 = hD = 100 \text{ ft}(62.4 \text{ lb/ft}^3) = 6240 \text{ lb/ft}^2$$

$$P_2 = 60 \frac{\text{lb}}{\text{in.}^2}\left(\frac{12 \text{ in.}}{1 \text{ ft}}\right)^2 + 8640 \text{ lb/ft}^2$$

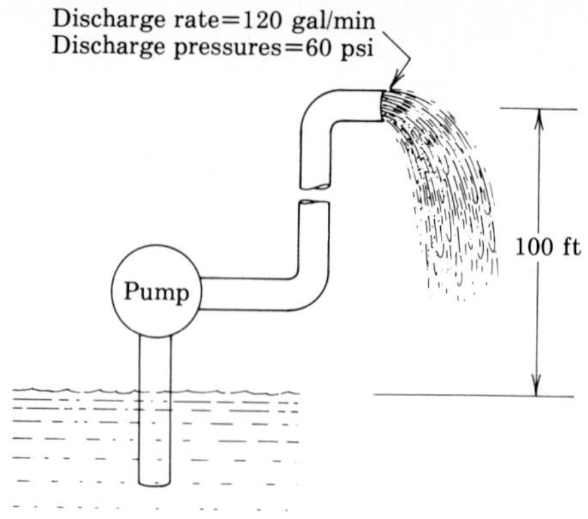

Fig. 12.8 Illustrative Problem 12.3.

The total water pressure P at the pump will be

$$P = P_1 + P_2 = 6240 \text{ lb/ft}^2 + 8640 \text{ lb/ft}^2$$
$$= 14{,}880 \text{ lb/ft}^2$$

Water is lifted at the rate of

$$120 \frac{\text{gal}}{\text{min}} = 120 \frac{\text{gal}}{\text{min}} \left(\frac{1 \text{ ft}^3}{7.48 \text{ gal}} \right)$$
$$= 16.0 \text{ ft}^3/\text{min}$$

Eq. (6.16) states that

$$\text{Power} = \frac{\text{force} \times \text{distance}}{\text{time}} = \frac{Fs}{t}$$

Eq. (12.1′) states that $F = P A$. By proper substitution, we can get

$$\text{Power} = \frac{P As}{t}$$

But As = volume V. Hence

$$\text{Power} = \frac{PV}{t}$$

and the horsepower needed is

$$\text{HP} = 14{,}880 \frac{\text{lb}}{\text{ft}^2} \left(16.0 \frac{\text{ft}^3}{\text{min}} \right) \left(\frac{1 \text{ hp}}{33{,}000 \text{ ft} \cdot \text{lb/min}} \right)$$
$$= 7.21 \text{ hp}$$

answer

12.6 ■ Shape and Size of Container Related to Liquid Pressure

Blaise Pascal was a French mathematician, scientist, and philosopher. He is credited with being the first person to invent a mechanical calculating machine, as well as for propounding a number of principles on the mechanics of fluids.

Equation (12.3) indicates that the pressure due to the weight of a liquid is independent of the size and shape of the reservoir or container; the pressure depends only upon the depth below the free surface and on the density of the liquid.

Blaise Pascal was the first to prove experimentally that shape and volume of a container do not affect pressure. Figure 12.9 shows the essential parts of an apparatus to illustrate this principle. Three vases with different shapes but with equal cross-sectional openings at the bottom have a flexible diaphragm stretched across the bottom openings. These containers are then placed on a solid plate which by its vertical motion imparts rotary motion to a pointer pivoted about a fixed point. When each of

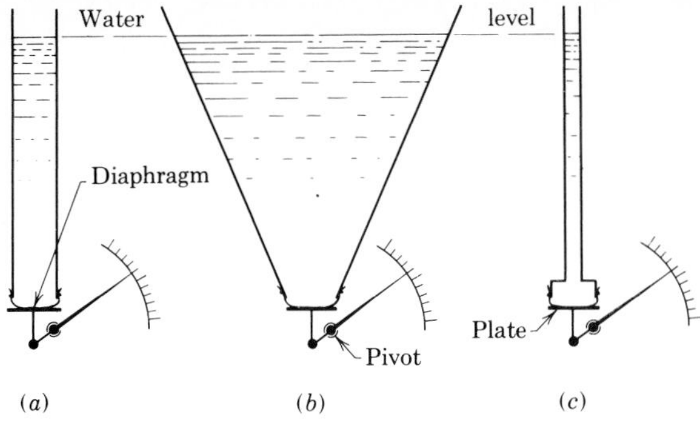

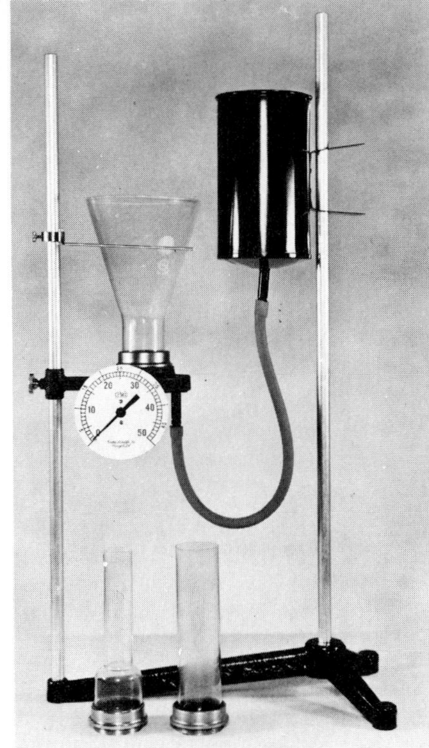

Fig. 12.9 Pascal's vases. (*a*) to (*c*) Despite varying quantities of water in the vessels, the pressure at the bottom is the same. (*d*) A typical arrangement of Pascal's vases used in the physics laboratory.

(*d*)

the vases is filled with water to the same height, the water forces acting on the diaphragm are found to be equal by identical readings on the scale—this despite the fact that the weight of water in vase *b* may exceed by many times that in either *a* or *c*. Since the forces at the bottom in each of the vases were shown to be the same and since the cross-sectional areas are the same, the quotient F/A (the pressure) must in consequence be the same. Figure 12.9*d* shows a typical arrangement of Pascal's vases often used in the physics laboratory.

12.7 ■ Transmission of Liquid Pressure—Pascal's Principle

Since liquids are nearly incompressible, an external force produces a pressure in a confined liquid that is transmitted throughout the liquid. This property is a special case of the general law known as *Pascal's principle*.

> ***Pascal's Principle***
>
> ***Pressure applied to a confined fluid at rest is transmitted undiminished throughout the confining vessel or system.***

Illustrative Problem 12.4 A cubical container (see Fig. 12.10) is completely filled with 8 L of water. (*a*) What is the upward force on the top lid *BC*? (*b*) On one of the sides? Next, a vertical pipe with internal diameter of 2.5 cm is connected to the lid. One liter of water is then poured into the pipe. (*c*) What will be the resulting forces on sides *BC* and *AF*?

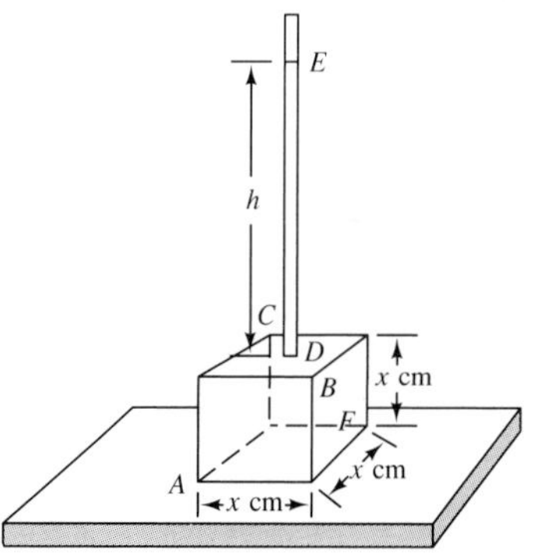

Fig. 12.10 Illustrative Problem 12.4.

Solution

(*a*) $F_{BC} = 0$

(*b*) Since the water pressure varies directly as the depth below the free surface, we know the average pressure will occur halfway down the side *AB*. Since the containing vessel has a volume of 8 L (full), we can find the length of each side from the formula for the volume of a cube.

$$V = (\text{length } x \text{ of side})^3 = 8 \text{ L}$$

$$x^3 = 8 \text{ L} \times \frac{1000 \text{ cm}^3}{1 \text{ L}}$$

$$x = 20 \text{ cm} = 0.20 \text{ m}$$

$$F_{AB} = \rho g A h \qquad (12.2)$$

$$= 1.000 \text{ kg/m}^3 \times 9.81 \text{ m/s}^2 \times (0.20 \text{ m})^2 \times \frac{0.20 \text{ m}}{2}$$

$$= 39 \text{ N}$$ *answer*

(*c*) The diameter of the pipe = 2.5 cm
The volume of water in the pipe = 1 L. Solve for *h*.

$$\frac{\pi}{4}(2.5 \text{ cm})^2 h = 1000 \text{ cm}^3$$

$$h = 203.7 \text{ cm} = 2.037 \text{ m}$$

Now,

$$F_{BC} = \rho g A h$$
$$= 1{,}000 \text{ kg/m}^3 \times 9.81 \text{ m/s}^2 \times 0.04 \text{ m}^2 \times 2.037 \text{ m}$$
$$= 800 \text{ N}$$ *answer*

$$F_{AB} = 1{,}000 \text{ kg/m}^3 \times 9.81 \text{ m/s}^2 \times 0.04 \text{ m}^2 \times (2.037 \text{ m} + 0.10 \text{ m})$$
$$= 839 \text{ N}$$ *answer*

Of interest is the fact that by adding only 1 L to 8 L of water, the resultant force on the sides has increased more than twentyfold (Also note: Application of Pascal's principle: 839 N = 800 N + 39 N).

12.8 ■ The Hydraulic Press

An important industrial application of Pascal's principle is found in the *hydraulic press*. This machine makes it possible to obtain an enormous force by exerting a relatively small effort. In Fig. 12.11*a* if a small force F_1 is applied to the smaller piston of area A_1, the pressure F_1/A_1 is transmitted undiminished throughout the confined liquid. This pressure acting on the larger-area piston A_2 will exert a total force on it equal to the product of the area A_2 times the pressure. Hence,

$$F_2 = P\,A_2 = \frac{F_1}{A_1}A_2$$

or, rearranging,

$$F_2 = \frac{A_2}{A_1}F_1 \qquad (12.5)$$

The force that can be exerted by the large piston equals the force exerted on the small piston times the ratio of the areas of the pistons.

Thus by simply changing the ratio of A_2 to A_1, the force exerted on the larger piston for a given effort on the smaller piston can be made as large as desired. It should be emphasized that where force is gained, distance is lost. For example, if the force obtained at the larger piston is 1000 times the effort applied on the smaller one, the distance through which the larger piston will move is only 1/1000 the distance the smaller one travels. That is, the product of the force and the distance is the same for both pistons (neglecting friction):

Fig. 12.11 (*a*) Principle of the hydraulic press. (*b*) The crosshead of a hydraulic press shapes the workpiece at the forge shop of a steel plant. The press is guided by press columns, each weighing 92,000 pounds. (Bethlehem Steel Photo)

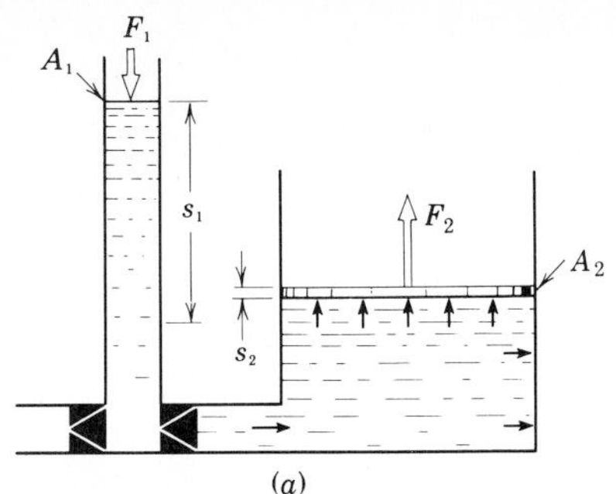

(*a*)

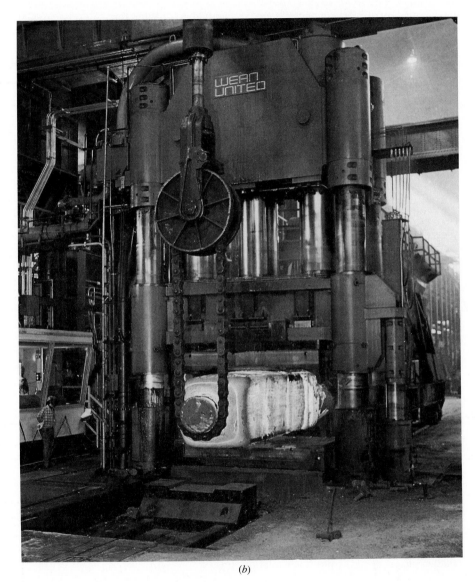

(*b*)

$$F_1 s_1 = F_2 s_2 \tag{12.6}$$

Equation (12.6) illustrates the principle of work which was discussed at length in Chap. 5.

Representing the forces by F_1 for the small piston and F_2 for the large piston, the areas by A_1 and A_2, respectively, and the diameters by d_1 and d_2, we may write the following relation for the hydraulic press.

$$\frac{F_1}{F_2} = \frac{A_1}{A_2} = \frac{d_1^2}{d_2^2} \tag{12.7}$$

Hydraulic presses have many industrial uses. They are used for baling paper and cotton; stamping out crankcases, tops, and other sheet metal parts for automobiles; for forcing lead through dies; and punching holes through iron plates. Other common hydraulic applications occur in barbers' or dentists' chairs, in the brake systems of autos and airplanes, and in presses for extracting oil from seeds.

Illustrative Problem 12.5 The smaller and larger pistons of a hydraulic press have diameters of 7.50 cm and 100 cm, respectively. (*a*) What force must be applied by the smaller piston in order to develop a compressive force of 1.11×10^6 N at the larger piston? (*b*) How far must the small piston travel to move the large piston 2.54 cm? Ignore friction.

Solution
(*a*) $d_1 = 7.50$ cm, $d_2 = 100$ cm, F_1 = force on smaller piston, $F_2 = 1.11 \times 10^6$ N. Using Eq. (12.7), we have

$$\frac{F_1}{F_2} = \frac{d_1^2}{d_2^2}$$

$$\frac{F}{1.11 \times 10^6 \text{ N}} = \left[\frac{7.50 \text{ cm}}{100 \text{ cm}}\right]^2$$

Then

$$F_1 = \left[\frac{7.50}{100}\right]^2 (1.11 \times 10^6 \text{ N})$$

$$= 6.24 \times 10^3 \text{ N} \qquad \textit{answer}$$

(*b*) Letting s_1 = distance small piston travels and $s_2 = 2.54$ cm, we can use Eq. (12.6):

$$F_1 s_1 = F_2 s_2$$

$$6.24 \times 10^3 \text{ N} \times s_1 \text{ cm} = 1.11 \times 10^6 \text{ N} \times 2.54 \text{ cm}$$

$$s_1 = \frac{1.11 \times 10^6 \text{ N} \times 2.54 \text{ cm}}{6.24 \times 10^3 \text{ N}}$$

$$= 452 \text{ cm} = 4.52 \text{ m} \qquad \textit{answer}$$

The smaller piston moves a total distance of 4.52 m in a series of short strokes.

The hydraulic jack is an important application of Pascal's principle.

Illustrative Problem 12.6 The hydraulic jack illustrated in Fig. 12.12 has dimensions as shown. What force F on the large piston will balance the 25-lb force applied at the end of the handle? Assume 100 percent efficiency.

Solution The 25-lb force has a lever arm of 10.5 in. Let F_s indicate the force on the smaller piston. Then, from lever relationships,

$$F_s \times 1.5 \text{ in.} = 25 \text{ lb} \times 10.5 \text{ in.} \tag{7.1}$$

$$F_s = 175 \text{ lb}$$

The pressure P_l on the large piston = the pressure P_s on the smaller piston (Pascal's principle). The forces are therefore proportional to the areas.

$$\frac{F_{\text{large}}}{F_{\text{small}}} = \frac{A_{\text{large}}}{A_{\text{small}}}$$

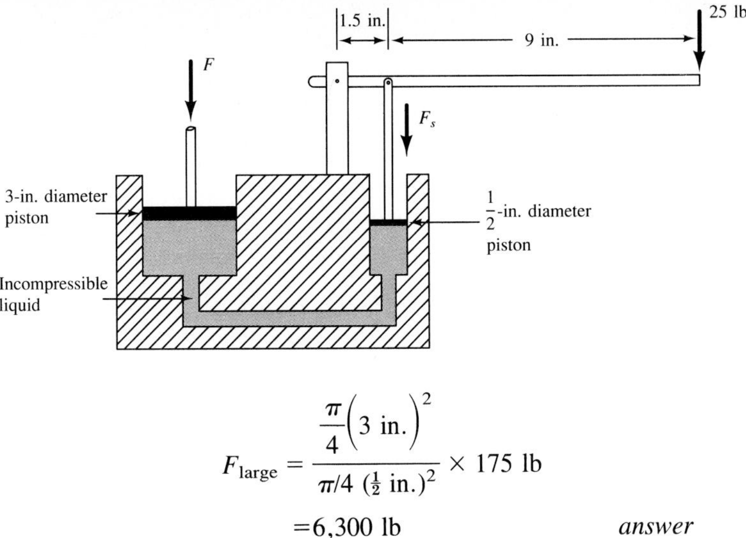

Fig. 12.12 Schematic of a hydraulic jack.

$$F_{\text{large}} = \frac{\frac{\pi}{4}(3\text{ in.})^2}{\pi/4\ (\frac{1}{2}\text{ in.})^2} \times 175\text{ lb}$$
$$=6{,}300\text{ lb} \qquad \textit{answer}$$

12.9 ■ Archimedes' Principle

Buoyancy We all have had numerous opportunities for observing the buoyant effect of liquids. When we go swimming, our bodies are held up almost entirely by the water. Wood, ice, and cork float on water. When we lift a rock from a stream bed, it suddenly seems heavier on emerging from the water. Boats rely on this buoyant force to stay afloat.

The amount of this buoyant effect was first computed and stated by the Greek philosopher Archimedes (287–212 B.C.).

Archimedes was perhaps the greatest physicist, engineer, inventor, and mathematician of antiquity. He invented the compound pulley, the spiral screw, the auger, and burning glasses to set fire to besieging enemy Roman ships. However, he prided himself most on his work as a mathematician. He developed ideas which many years later have been incorporated into modern calculus.

Archimedes' Principle

When a body is placed in a fluid, it is buoyed up by a force equal to the weight of the displaced fluid.

If the body weighs more than the liquid it displaces, it will obviously sink but will *appear* to lose an amount of weight equal to that of the displaced liquid. If the body weighs less than that of the displaced liquid, the body will rise to the surface, eventually floating at such a depth that it will displace a volume of liquid whose weight will just equal its own weight. *A floating body displaces its own weight of the fluid in which it floats*. This is an alternative statement of Archimedes' principle, sometimes called the *law of flotation*.

Archimedes' principle can easily be verified experimentally in the laboratory (see Fig. 12.13). (1) Weigh a solid object in air. (2) Fill an overflow vessel to the brim with water and weigh a small can to be used to catch overflow water. (3) Slowly submerge the solid object in the water and catch the overflow water in the small can. Read from the scale the apparent weight of the solid when it is submerged in the water. A definite decrease in weight of the solid will be observed. Now, weigh the small can with the overflow water in it. It will be noted that the increase in weight of the small can (the weight of the displaced water) will be equal to the apparent weight "loss" of the solid.

12.10 ■ Theoretical Proof of Archimedes' Principle

Archimedes' principle can be derived from the laws of liquid pressure as follows. Consider a rectangular block of height h and cross-sectional area A completely im-

Fig. 12.13 Experimental verification of Archimedes' principle.

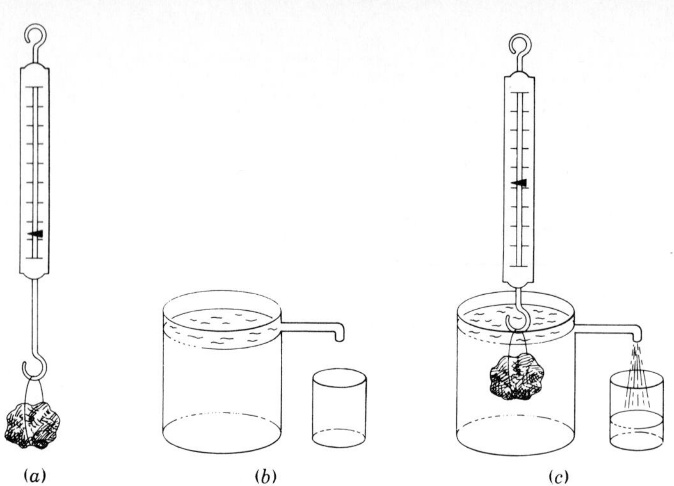

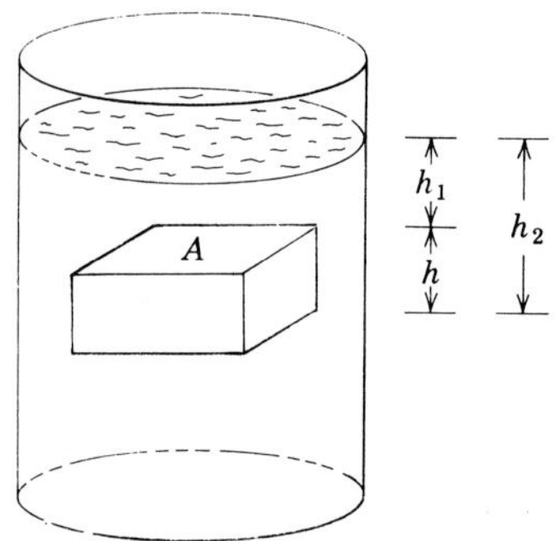

Fig. 12.14 Diagram for theoretical proof of Archimedes' principle.

Buoyant force equals weight of displaced liquid.

mersed in a liquid of density D (or ρg) so that its top is h_1 units below the surface (see Fig. 12.14).

The block has six rectangular surfaces. On the vertical faces the liquid exerts horizontal forces which are in equilibrium (why?). On the top face the liquid exerts a downward force Ah_1D and on the bottom face an upward force Ah_2D. Since h_2 is greater than h_1, the liquid will exert a net force upward equal to

$$\begin{aligned} F &= Ah_2D - Ah_1D \\ &= AD(h_2 - h_1) \\ &= AhD \text{ (or } Ah\rho g) \end{aligned}$$

But Ah is the volume of the block and hence also equal to the volume of the displaced liquid. Substituting V for Ah, we get, for the net upward force,

$$F = VD = V\rho g = \text{weight of the liquid displaced}$$

Although we have proved Archimedes' principle only for a regularly shaped block, the proof is equally valid for an irregular one, since we can think of an irregular solid as being made up of a large number of very small rectangular blocks.

Archimedes' principle has many applications. Although made of steel, which is much heavier than water, ocean liners float because the steel ship encloses a large volume and the *average density* of the entire volume of the ship fully loaded is less than that of water.

The size of a ship is generally described in terms of the weight of the seawater it will displace when afloat. One of the world's largest merchant ships is an oil tanker constructed in France in 1977 with a seawater displacement of 555,031 tons. It is 1359 ft long and 206′ 9″ wide.

Illustrative Problem 12.7 A rectangular barge 25 ft wide, 48 ft long, and 8 ft deep floats in a canal lock which is 32 ft wide, 60 ft long, and 12 ft deep. When there is no load on the barge other than its own weight, the bottom of the barge is 3 ft beneath the water surface. The depth of the water in the lock is 7.5 ft. If a load of steel which weighs 75 tons is added to the barge, what will be the new water depth h in the lock?

Solution First make a sketch of the barge and lock relationships. The load of steel will cause the barge to lower, relative to the water surface, a distance Δh. Archimedes' principle states, in effect, that the weight of steel added (since it is held up by water), equals the weight of water displaced. The weight of the displaced liquid is

$$\begin{aligned} DV &= 62.4 \text{ lb/ft}^3 \times (25 \text{ ft} \times 48 \text{ ft} \times \Delta h \text{ ft}) \\ &= 75 \text{ tons} = 150{,}000 \text{ lb} \end{aligned}$$

Solving for Δh, we get

$$\Delta h = 2.0 \text{ ft}$$

the distance the barge sank as the steel was loaded.

The bottom of the barge will now be 3 ft + 2.0 ft = 5.0 ft beneath the water surface.

The volume of water in the lock before and after the steel load is added is constant. Thus,

$$\underset{\text{(before load)}}{32 \text{ ft} \times 60 \text{ ft} \times 7.5 \text{ ft} - 25 \text{ ft} \times 48 \text{ ft} \times 3 \text{ ft}} = \underset{\text{(after load)}}{32 \text{ ft} \times 60 \text{ ft} \times h \text{ ft} - 25 \text{ ft} \times 48 \text{ ft} \times 5.0 \text{ ft}}$$

Solve for h. $$h = 8.8 \text{ ft} \qquad \textit{answer}$$

Submarines and deep-sea vessels apply Archimedes' principle in their design. The submarine submerges by admitting seawater into tanks, thus counteracting the net upward force of buoyancy on the ship. To raise the submarine, compressed air or pumps are used to force the water out of ("blow") the tanks.

Archimedes' principle holds for all fluids. A gas balloon will ascend if the combined weight of bag, gas contained in the bag, and the balloon's load is less than the weight of the air it displaces. As the balloon rises into the atmosphere, the external air pressure decreases, causing the balloon to expand. However, at the same time the density of the air displaced diminishes, so that the buoyancy per unit volume is decreased. The balloon will continue to rise until the weight of the air it displaces is no longer any greater than the total weight of the balloon and its contents.

Archimedes' principle holds for both liquids and gases.

12.11 ■ Specific Gravity of Solids

There are several methods used in finding the specific gravity of solids. Each method employs the same general rule: (1) Weigh the object. (2) Find the weight of an equal volume of water. (3) Divide the weight of the object by the weight of an equal volume of water. Archimedes' principle must often be utilized in obtaining the weight of an equal volume of water.

Illustrative Problem 12.8 A piece of cast iron weighs 23.7 lb in air and 20.6 lb when submerged in water. (*a*) What is the specific gravity of the iron? (*b*) What is the weight density of the iron in pounds per cubic foot?

Specific gravity of solids can be found by applying Archimedes' principle.

Solution

(***a***) By Archimedes' principle, we know that the weight of water equal in volume to that of the cast iron is 23.7 − 20.6 or 3.1 lb, since this is the apparent loss in weight.

Hence from Eq. (11.4), the specific gravity of the iron is,

$$\text{sp gr} = \frac{\text{wt of iron}}{\text{wt of equal volume of water}} = \frac{23.7 \text{ lb}}{3.1 \text{ lb}} = 7.65$$

(***b***) Using Eq. (11.4), the density of the iron is

$$\begin{aligned} D &= 62.4 \text{ lb/ft}^3 \times (\text{sp gr of iron}) \\ &= 62.4 \text{ lb/ft}^3 \times 7.65 \\ &= 480 \text{ lb/ft}^3 \qquad \textit{answer} \end{aligned}$$

12.12 ■ Specific Gravity of Liquids

There are several methods of determining the specific gravity of liquids. Three of these will be discussed here.

The comparison of the weight of a liquid with that of an equal volume of water is made with moderate accuracy and facility by the use of special specific-gravity bottles, called *pycnometers*. The bottle is carefully weighed, first empty, then full of the liquid, and then full of water. By subtracting the weight of the empty bottle in each case, the weights of equal volumes of the liquid and water are obtained. Specific gravity is then obtained from Eq. (11.5):

$$\text{sp gr} = \frac{\text{weight of liquid}}{\text{weight of equal volume of water}}$$

The second method can be explained by describing an imaginary laboratory problem (see Fig. 12.15). Weigh a solid (1) in air, (2) in water, and (3) in the solution whose specific gravity is desired. Assume the solid weighs 0.85 **N** in air, 0.49 **N** in water and 0.61 **N** in the solution of unknown specific gravity.

The difference, 0.85 **N** − 0.49 **N** = 0.36 **N**, is the buoyant force exerted by the water on the submerged cylinder and therefore (by Archimedes' principle) the weight of the displaced water. Thus, the specific gravity of the solid is

$$\text{sp gr of solid} = \frac{0.85\ \mathbf{N}}{0.36\ \mathbf{N}} = 2.4$$

In like manner, the difference 0.85 **N** − 0.61 **N** = 0.24 **N**, is the weight of the unknown solution which is equal in volume to that of the cylinder and hence also to that of the displaced water.

The specific gravity of the unknown liquid is equal to the weight of the unknown liquid displaced by the solid, divided by the weight of the water displaced by the solid. For this example,

$$\text{sp gr of unknown liquid} = \frac{0.24\ \mathbf{N}}{0.36\ \mathbf{N}} = 0.67$$

The quickest and most common method employed in determining specific gravities of liquids is by using a *hydrometer*. The hydrometer is usually a sealed glass tube with an enclosed scale in a cylindrical stem and a bulb weighted with mercury or shot at the bottom (Fig. 12.16). The depth to which a given hydrometer will float in a

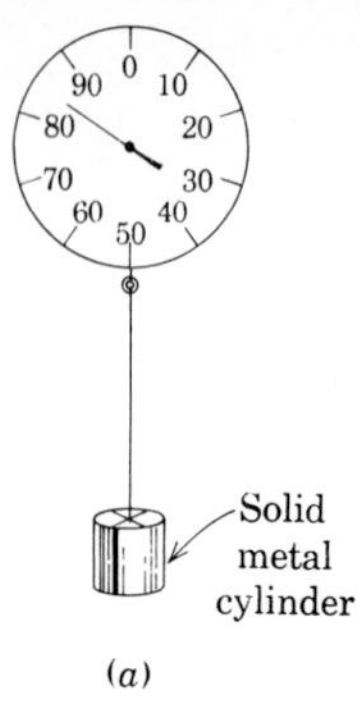

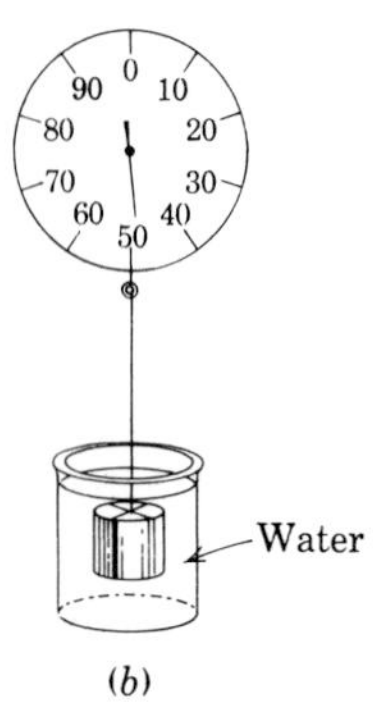

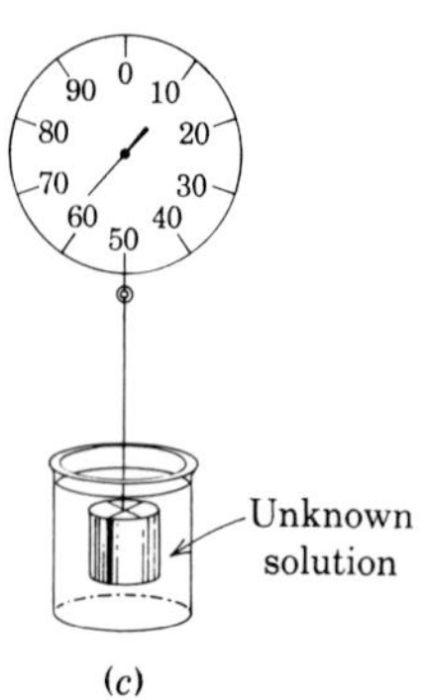

Fig. 12.15 Determining specific gravity of solids and liquids using Archimedes' principle.

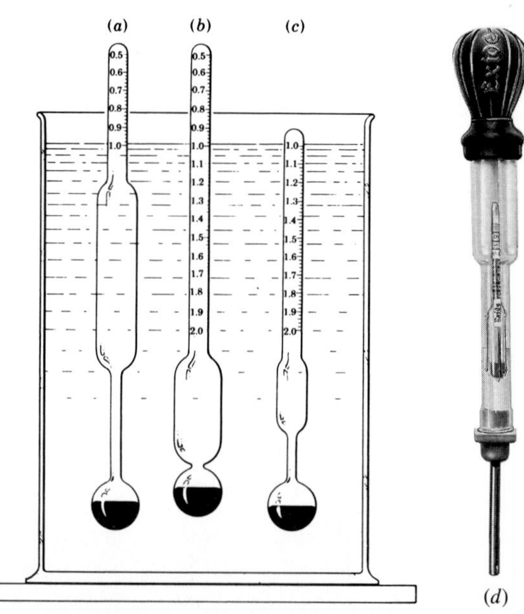

Fig. 12.16 Hydrometers for (*a*) light liquids, (*b*) light or dense liquids, (*c*) dense liquids; all three are shown in water. (*d*) Battery tester hydrometer. The rubber bulb and outer glass are for drawing a sample of the electrolyte out of the battery being tested. (Exide Sales, Electric Storage Battery Co., Automotive Division)

liquid is determined by the specific gravity of the liquid. In every case the hydrometer will sink to that level which will cause it to displace an amount of the liquid whose weight equals that of the instrument. The greater the specific gravity of the liquid, the higher out of the liquid the hydrometer will float. Because the depth at which the instrument floats is inversely related to the specific gravity of the liquid, a scale which reads specific gravity directly can be mounted vertically along the instrument.

The higher the hydrometer floats, the more dense the liquid.

Hydrometers have numerous commercial uses. They are used to measure the specific gravities of alcohols, gasoline, other petroleum products, acids, antifreeze mixtures, sugar solutions, and the like. A common form of hydrometer is the battery-acid tester (Fig. 12.16*d*).

HYDRAULICS

12.13 ■ Fluids in Motion

Fluids that are in motion exhibit several characteristic differences from fluids at rest. Frictional resistances within the fluid itself (called *viscosity*) and inertia contribute to these differences. Inertia, it will be recalled, means the resistance which a mass offers to a change in its motion. When a fluid is at rest, the pressure is the same at all points of the same elevation. However, this is not true when the liquid is in motion. Figure 12.17*a* represents the condition of a liquid at rest. When valve *E* is closed, the levels of liquid in *A*, *B*, *C*, and *D* are the same. When the valve is opened and liquid flows (Fig. 12.17*b*), the liquid level in each tube will fall in such a way as to be progressively lower than that in the preceding one. Should the valve be opened wider to permit more liquid to flow, the difference in levels will be further increased. This drop in pressure is therefore seen to be caused in some manner by the *motion* of the liquid.

Hydraulics includes the study of flow of liquids, Bernoulli's equation, and pump analyses.

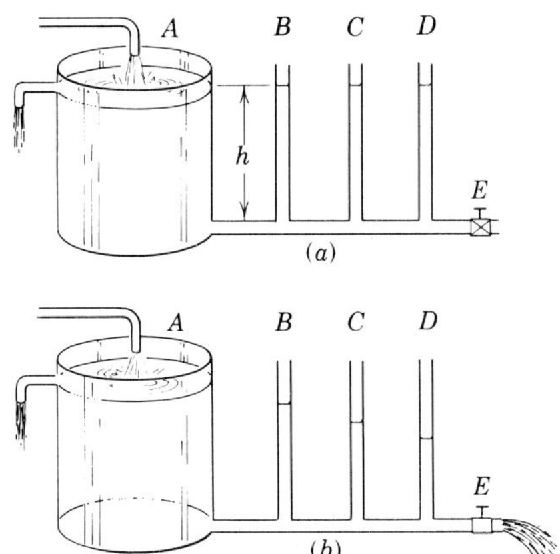

Fig. 12.17 Fluid friction reduces pressure head. (*a*) Valve *E* is closed, liquid is not flowing, and gauges *B*, *C*, and *D* all indicate the same pressure. (*b*) Steady flow is assumed and friction causes a drop in pressure along the pipe.

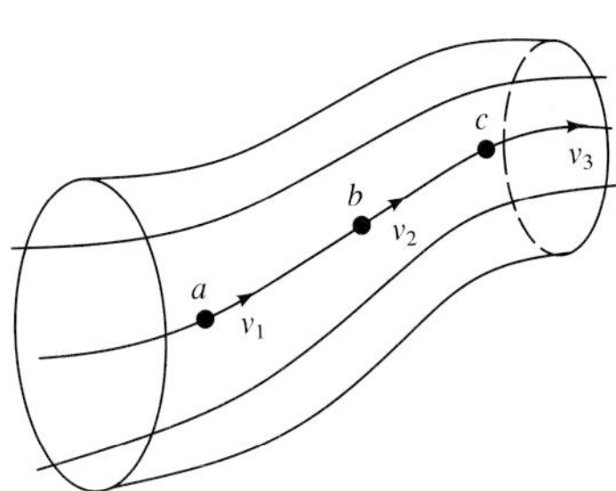

Fig. 12.18 In laminar flow the path *abc* is fixed and streamline. The current paths do not cross one another and are essentially parallel.

Characteristics of Flow In general there are two types of fluid flow. A flow is said to be *streamline* or *laminar* if each particle follows a smooth path. Consider a fluid flowing in a flexible tube (Fig. 12.18). Let the velocities of a given particle as it passes *a*, *b*, and *c* be v_1, v_2, and v_3, respectively. In laminar flow subsequent particles passing *a* will also have a velocity v_1 and will pass *b* with a velocity v_2 and *c* with velocity v_3. All particles passing through *a* will pass through *b* and *c*. A line, such as *abc*, which represents the path followed by a particle of the fluid is called a *streamline*. Streamlines in a pipe or tube will not cross.

Laminar flow and turbulent flow.

Turbulent flow, on the other hand, is unsteady and characterized by the presence of eddies and whirlpools in motion. The paths of particles in turbulent flow are continually changing and crossing each other.

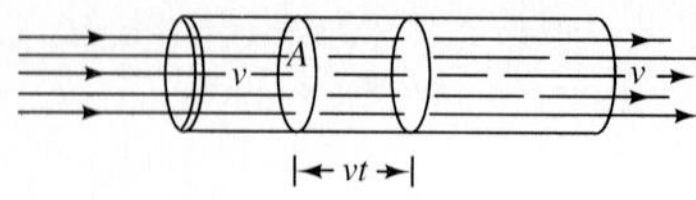

Fig. 12.19 The rate of flow Q of liquid through a pipe is Av. $Q = Av$ is the flow equation.

The Flow Equation When an incompressible liquid flows in a pipe or channel under steady-state flow conditions, the quantity of flow past all points in the pipe or channel is a constant. If v is the average speed of the liquid in a section of the pipe in Fig. 12.19, the distance through which the liquid will move in time t is vt. If A is the cross-sectional area of the pipe, then the volume of flow V in time t is equal to the volume of the cylinder with base area A and height equal to vt. The volume equals $A(vt)$. The rate of flow Q of the liquid is given by

$$Q = \frac{V}{t} = \frac{Avt}{t} = Av \qquad (12.8)$$

$$\text{Flow rate} = \text{area} \times \text{velocity}$$

This is the basic *flow equation* for fluids. Q is commonly expressed in m^3/min, ft^3/min or ft^3/s units.

Illustrative Problem 12.9 A jet of water 1 in. in diameter, flowing at the rate of 0.325 ft^3/s, impinges on a surface as shown in Fig. 12.20. What is the magnitude of the steady force F provided by the wall to stop the jet of water?

Solution Newton's second law of motion can be written

$$Ft = \frac{w(v_2 - v_1)}{g} \qquad \text{(Impulse = change in momentum)}$$

The rate of flow Q is given by

$$Q = Av \qquad (12.8)$$

Solving for v,

$$v = \frac{Q}{A}$$

The *initial* velocity

$$v_1 = \frac{0.325 \text{ ft}^3/\text{s}}{\pi/4\ (\frac{1}{12}\text{ ft})^2} \qquad v_2 = 0 \quad \text{(assumed)}$$

Hence,

$$F = \frac{w(v_2 - v_1)}{g\,t} = D\left(\frac{V}{t}\right)\left(\frac{v_2 - v_1}{g}\right)$$

$$= 62.4 \text{ lb/ft}^3 \times 0.325 \text{ ft}^3/\text{s} \left(\frac{0 - \dfrac{0.325 \text{ ft}^3/\text{s}}{\pi/4(\frac{1}{12}\text{ ft})^2}}{32.2 \text{ ft/s}^2}\right)$$

$$= -37.5 \text{ lb}$$ *answer*

The minus sign indicates that the force is acting opposite to the direction of flow.

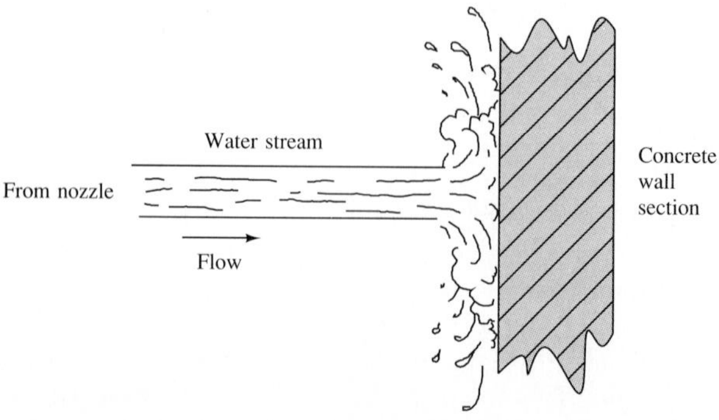

Fig. 12.20 Illustrative Problem 12.9.

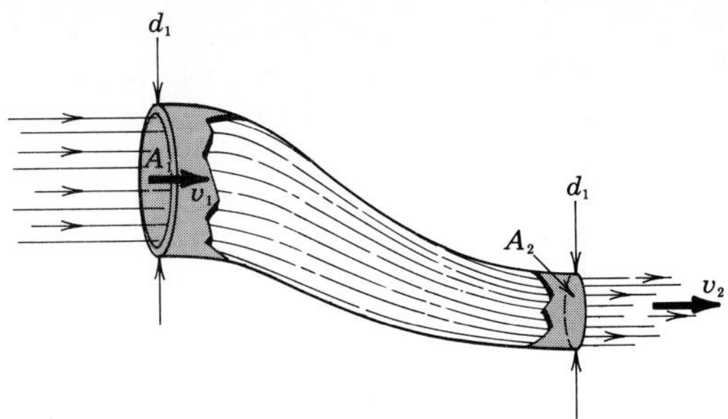

Fig. 12.21 Fluid flowing in a tube of variable diameter. $A_1v_1 = A_2v_2$.

Under steady-state flow conditions the product of the cross-sectional area and the fluid velocity at all points along the pipe is a constant.

The steady-flow equation for an incompressible liquid flowing in a pipe with varying cross-sectional areas (see Fig. 12.21) is

$$Q = A_1v_1 = A_2v_2 \tag{12.9}$$

or since $A = \pi d^2/4$,

$$\frac{v_1}{v_2} = \frac{d_2^2}{d_1^2} \tag{12.10}$$

It will be noted that the velocity varies *inversely* as the cross-sectional area of the pipe or inversely as the square of the diameter if it is a round pipe.

Illustrative Problem 12.10 Water in a pipe 12 cm in diameter flows with a velocity of 8 m/s. This pipe is joined to a second pipe that is constricted down to 3 cm in diameter. What is the flow velocity in the 3-cm pipe?

Solution Substitute in Eq. (12.10) to get

$$\begin{aligned} v_1 &= \left(\frac{d_2^2}{d_1^2}\right)v_2 \\ &= (12\text{ cm})^2/(3\text{ cm})^2 \times 8\text{ m/s} \\ &= 130\text{ m/s} \qquad \textit{answer} \end{aligned}$$

12.14 ■ Liquid Heads in Fluid Flow

In the study of hydraulics the vertical depth h_p (in feet) of any point below the free surface of the liquid is called the *pressure head* or *hydrostatic head*. In engineering practice, the total head p is commonly expressed in *feet of water* or in pounds per square inch. *If the liquid is water*, using Eq. (12.3), we find that if h_p is measured in feet, the pressure

$$\begin{aligned} p &= h_pD \\ &= \left(h_p\text{ ft} \times \frac{12\text{ in.}}{1\text{ ft}}\right) \times \left(62.4\text{ lb/ft}^3 \times \frac{1\text{ ft}^3}{1728\text{ in.}^3}\right) \\ &= 0.433\ h_p\text{ lb/in.}^2 \end{aligned}$$

$$P\ (\text{lb/in.}^2) = 0.433(h_p\text{ ft}) \tag{12.11}$$

Drop in pressure due to friction is called the *friction head* (see Fig. 12.17). This resistance to flow due to friction is (approximately) directly proportional to the total area of the rubbing surface, i.e., the product of the length of the channel and its wetted circumference, for round pipes. The resistance also increases with increased roughness of the internal channel surface and with increased velocities of the liquid. Abrupt changes in the cross-sectional area of the channel and changes of direction of flow increase the resistance to flow. With extreme velocities, the frictional resistance

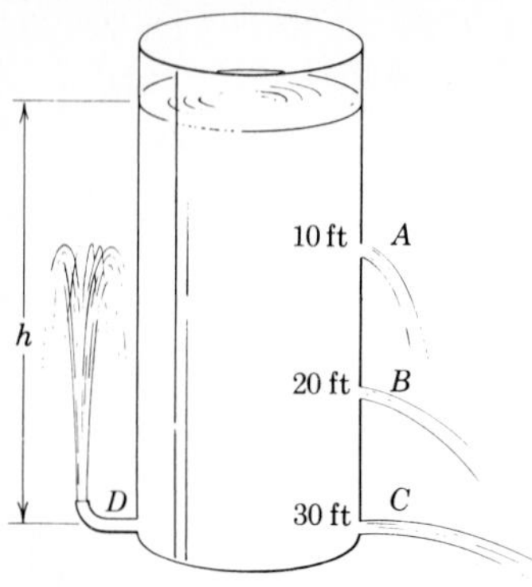

Fig. 12.22 Velocity of water flow from an orifice varies with pressure head.

The liquid leaves the pipe with the same velocity v that a freely falling object would attain falling from height h.

set up is increased many times because of the setting up of surging and eddying currents of the fluid, called *turbulent flow*. Friction head results in a drop in pressure head.

The pressure in a pipe or channel due to the velocity of a liquid is called the *velocity head*. It can be shown that if v is the flow velocity and g the acceleration of gravity, the velocity head is determined by the formula

$$h_v = \frac{v^2}{2g} \tag{12.12}$$

The value of h_v in Eq. (12.12) indicates a pressure due to the velocity of the liquid particles which is equivalent to a hydrostatic head of h_v ft of the liquid. To convert this to pounds per square inch use Eq. (12.11).

Let us consider a tank (Fig. 12.22) filled with water. Openings are placed at depths below the water surface of 10, 20, and 30 ft as shown. The *pressure head* at A is 10 ft or, using Eq. (12.11),

$$P_{10} = 0.433 \times 10 \text{ ft} = 4.33 \text{ lb/in.}^2$$

In like manner, the pressure at B is 8.66 lb/in.^{2} and at C, 12.99 lb/in.^{2}

If Eq. (12.12) is solved for v, the formula for the velocity of flow at the various openings is obtained:

$$\text{Velocity of flow} = \sqrt{2g \times \text{pressure head}}$$
$$v = \sqrt{2gh} \tag{12.13}$$

The relationship expressed by Eq. (12.13) is known as *Torricelli's theorem*.

Substituting values of 10, 20, and 30 ft for h in Eq. (12.13), we get values for velocities of flow at A, B, and C of 25.3, 35.9, and 44.0 ft/s, respectively. The direction of flow does not affect the velocity. Hence if C and D are at the same depth, the velocity of flow will be the same. It must be made clear, however, that Eqs. (12.12) and (12.13) are theoretical and would be strictly true only if frictional forces within the liquid and at the walls and openings could be neglected. If there were no friction, the velocity head at D would be just the exact amount required to force a jet of water back up to the level of the water surface in the tank. The diagram shows it falling short, since friction is always present.

12.16 ■ Bernoulli's Principle

The foregoing paragraphs have shown that several factors affect the flow of liquids in pipes. The diagram of Fig. 12.23 will serve to relate these factors and to introduce a basic law of liquid flow.

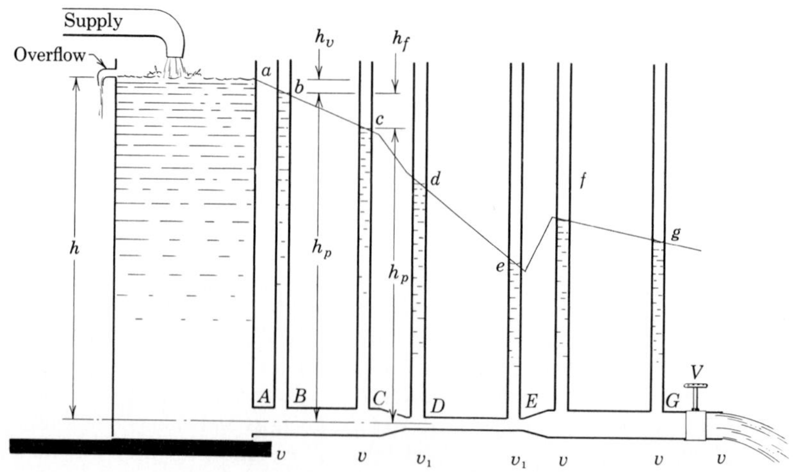

Fig. 12.23 The total head related to velocity head, friction head, and pressure head. Note that where velocity increases, the pressure head decreases, and where velocity decreases, the pressure head increases.

Let water be maintained in a reservoir at a constant level a when valve V is open with steady-state flow conditions in effect. The *total head* is indicated by h. Water leaves the reservoir at A and acquires velocity v as it reaches B. The drop in level from a to b is due to this acquisition of velocity. The drop in head h_v is equal to $v^2/2g$ in accordance with Eq. (12.12). Note that *where friction is neglected,* total head h is equal to *pressure head* h_p plus *velocity head* h_v.

There is no change in velocity from B to C, but a further drop in pressure occurs from level b to level c. This drop is caused by friction in the pipe length BC and is indicated by h_f in the diagram. It should be noted that *where friction in pipes is taken into account,* total head equals pressure head plus velocity head plus friction head.

$$h = h_p + h_v + h_f \tag{12.14}$$

If a smaller pipe or constriction (section DE) is placed in the pipe, flow velocity must of necessity increase, since the flow *rate* is a constant. This increased velocity v_1 results in a further pressure drop, down to level d, and friction in the narrow pipe results in another drop to e. As the water enters the section FG, whose diameter is the same as that of the original pipe, the flow velocity reduces (since flow *rate* is constant) and pressure rises to level f. The drop from f to g is caused by friction in the section FG.

A study of these flow considerations by Daniel Bernoulli resulted in a fundamental theorem of fluid flow known as *Bernoulli's principle*. One way to state Bernoulli's principle is:

Daniel Bernoulli (1700–1782) was a Swiss mathematician and physicist. In his *Hydrodynamica* he developed fluid flow principles and his version of the theory of gases and liquids. He also worked on problems in acoustics and differential equations which he used in developing equations of hydrodynamic flow.

> **When fluids flow in pipes or channels under steady-state conditions, where the velocity is high the pressure is low, and where the velocity is low the pressure is high.**

Actually, Eq. (12.14) is also a statement of Bernoulli's principle. (Another statement will be given shortly.)

The path traced by a particular molecule of a fluid flowing with steady motion (without turbulence) is called the *streamline* of flow. Consider a volume V of fluid with density p flowing between planes aa and bb (Fig. 12.24). Assume that (1) the liquid is incompressible, (2) the flow is streamlined, and (3) there is no fluid friction (zero viscosity). As the liquid passes plane a it has potential energy $h_a\rho Vg$ and kinetic energy $\frac{1}{2}mv_a^2 = \frac{1}{2}\rho Vv_a^2$. The relationship between V, A_a, and s_a is $V = A_a s_a$, or $s_a = V/A_a$. In like manner, $s_b = V/A_b$. The force pushing the volume V of the fluid past plane a is P_aA_a, and that pushing the volume V past plane b is P_bA_b.

Thus the work done to push volume V a distance s_a past a is $P_aA_a(V/A_a) = P_aV$. Similarly, at plane b the volume V has potential energy $h_b\rho Vg$ and kinetic energy

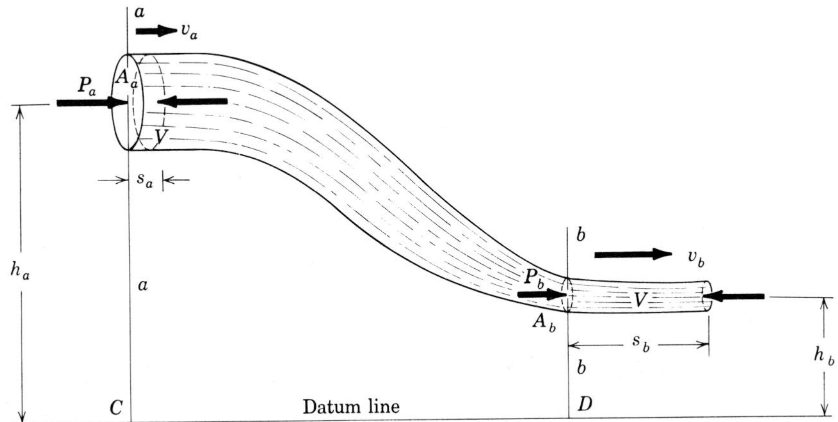

Fig. 12.24 Liquid flow through a tube of variable cross section at different heights.

$\frac{1}{2}\rho V v_b^2$, and $P_b V$ units of work are required to push it a distance s_b units past plane b.

If we can neglect viscous losses of energy (we assume streamline flow), the sum of the *three energies* for a unit volume V is the same at a as at b:

Bernoulli's equation is based on the principle of conservation of energy in fluid flow.

$$P_a V + \tfrac{1}{2}\rho V v_a^2 + h_a \rho V g = P_b V + \tfrac{1}{2}\rho V v_b^2 + h_b \rho V g \tag{12.15}$$

Dividing each term of Eq. (12.5) by V, we get

$$P_a + \tfrac{1}{2}\rho v_a^2 + h_a \rho g = P_b + \tfrac{1}{2}\rho v_b^2 + h_b \rho g \tag{12.16}$$

You will note that the potential energies were determined by measuring h_a *and* h_b from an arbitrarily chosen datum line CD. Another form of Bernoulli's equation can be developed by dividing each term of Eq. (12.16) by ρg, yielding an equation with dimensions of a length:

$$\frac{P_a}{\rho g} + \frac{v_a^2}{2g} + h_a = \frac{P_b}{\rho g} + \frac{v_b^2}{2g} + h_b \tag{12.17}$$

Each term of Eq. (12.17) is called a *head:* $P/\rho g$ is the *pressure head,* $v^2/2g$ the *velocity head*, and h the *elevation head.*

In terms of the three heads discussed above (neglecting friction head), Bernoulli's principle may then be stated another way:

An alternative statement of Bernoulli's principle.

At any section of a pipe in which a fluid is flowing under steady-state conditions without friction, the total head is constant; whatever pressure head is lost appears as a gain in velocity head.

Illustrative Problem 12.11 Water flows in the pipe illustrated in Fig. 12.24 at the rate of 8.50 m^3/min. The diameter at a is 30.4 cm. The diameter at b is 15.2 cm. The pressure at a is 1.03×10^5 **N/m**^{2}. What is the pressure at b if the center at b is 60.4 cm lower than the center at a?

Solution

$$\begin{aligned} A_a v_a &= A_b v_b \\ &= 8.50\ \text{m}^3/\text{min}\ (1\ \text{min}/60\ \text{s}) \\ &= 0.142\ \text{m}^3/\text{s} \end{aligned}$$

$$A_a = \pi(0.304\ \text{m}/2)^2 = 7.26 \times 10^{-2}\ \text{m}^2$$
$$A_b = \pi(0.152\ \text{m}/2)^2 = 1.81 \times 10^{-2}\ \text{m}^2$$

Thus

$$v_a = \frac{0.142\ \text{m}^3/\text{s}}{A_a} = \frac{0.142\ \text{m}^3/\text{s}}{7.26 \times 10^{-2}\ \text{m}^2} = 1.96\ \text{m/s}$$

and

$$v_b = \frac{0.142\ \text{m}^3/\text{s}}{A_b} = \frac{0.142\ \text{m}^3/\text{s}}{1.81 \times 10^{-2}\ \text{m}^2} = 7.85\ \text{m/s}$$

For water, $\rho = 1000$ kg/m^3. From Eq. (12.16),

$$P_b = P_a + \tfrac{1}{2}\rho(v_a^2 - v_b^2) + \rho g(h_a - h_b)$$

Then

$$\begin{aligned} P_b = {} & 1.03 \times 10^5\ \text{N/m}^2 + \tfrac{1}{2}(1000\ \text{kg/m}^3) \\ & \times [(1.96\ \text{m/s})^2 - (7.85\ \text{m/s})^2] \\ & + 1000\ \text{kg/m}^3(9.81\ \text{m/s}^2)(0.604\ \text{m}) \end{aligned}$$

$$= 1.03 \times 10^5 \text{ N/m}^2 - 0.289 \times 10^5 \frac{\text{kg} \cdot \text{m/s}^2}{\text{m}^2}$$

$$+ 0.0593 \times 10^5 \frac{\text{kg} \cdot \text{m/s}^2}{\text{m}^2}$$

$$= 8.00 \times 10^4 \text{ N/m}^2 \text{ (Pa)}$$ *answer*

Illustrative Problem 12.12 Water discharges from a tank through a 6-in. (internal diameter) pipe under the conditions shown in Fig. 12.25. The water travels a horizontal distance of 3.45 ft from *A* before falling into an open channel 3.68 ft below.
(*a*) What is the flow rate, in ft³/s through the pipe?
(*b*) What is the magnitude of the velocity head?
(*c*) What is the head loss due to friction in the 6-in. pipe, including entrance and exit losses?

Solution
(*a*) After leaving the pipe the water will travel as a projectile 3.42 ft horizontally, while dropping 3.68 ft.

We can find the time of flight by using Eq. 4.12, $s = \frac{1}{2}gt^2$. Thus, the time to drop 3.68 ft (and travel 3.45 ft horizontally) is

$$t = \sqrt{\frac{2\ s}{g}} = \sqrt{\frac{2 \times 3.68 \text{ ft}}{32.2 \text{ ft/s}^2}} = 0.478 \text{ s}$$

The velocity of efflux $$v = \frac{3.45 \text{ ft}}{0.478 \text{ s}} = 7.22 \text{ ft/s}$$

The volumetric flow rate,

$$Q = Av \qquad (12.8)$$

$$= \frac{\pi}{4}(6 \text{ in.})^2 \times \frac{1 \text{ ft}}{144 \text{ in.}^2} \times 7.22 \text{ ft/s}$$

$$= 1.42 \text{ ft}^3\text{/s}$$ *answer*

(*b*) The velocity head,

$$h_v = \frac{v^2}{2g} \qquad (12.12)$$

$$= \frac{(7.22 \text{ ft/s})^2}{2(32.2 \text{ ft/s}^2)}$$

$$= 0.81 \text{ ft}$$ *answer*

(*c*) The friction-head loss is calculated as follows:

$$h_1 + \frac{P_1}{\rho g} + \frac{v_1^2}{2g} = h_2 + \frac{P_2}{\rho g} + \frac{v_2^2}{2g} + h_{\text{loss}} \qquad (12.17)$$

$$h_{\text{loss}} = h_1 - h_2 + \frac{P_1 - P_2}{\rho g} + \frac{v_1^2 - v_2^2}{2g}$$

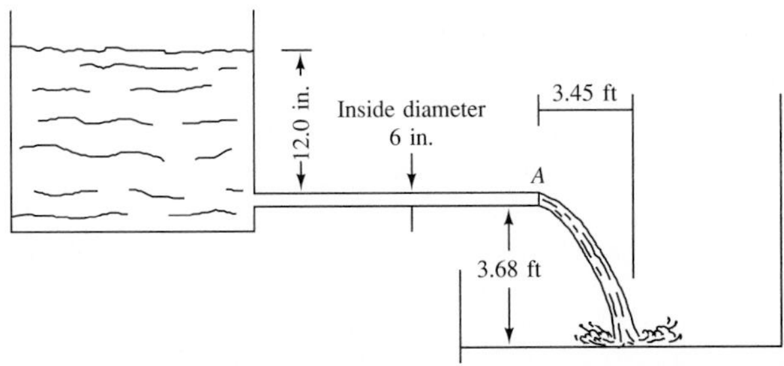

Fig. 12.25 Illustrative Problem 12.12.

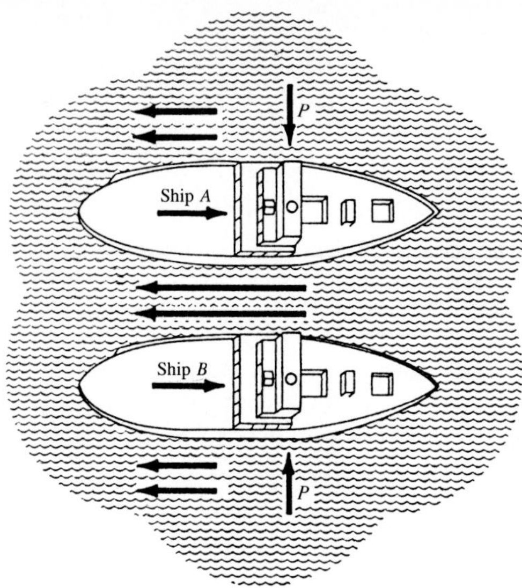

Fig. 12.26 Bernoulli's principle in a practical situation. Pressure is greater on the outboard side of the two ships.

By taking the center of the pipe as the datum line:

$$P_1 = P_2 = 1 \text{ atm}$$
$$v_1 = 0 \text{ and } h_2 = 0$$

Then

$$h_{\text{loss}} = 12 \text{ ft} - 0 + 0 - \frac{(7.22 \text{ ft/s})^2}{2 \times 32.2 \text{ ft/s}^2}$$
$$= 12 \text{ ft} - 0.81 \text{ ft}$$
$$= 11.19 \text{ ft} \qquad \textit{answer}$$

Bernoulli's theorem explains many phenomena which at first may seem strange to the layman. Suppose two ships steaming on a parallel course in still water decide to travel in close formation (see Fig. 12.26). The relative motion of the ships with respect to the water would be such as to cause the water between the ships to "speed up" relative to the ships because of the narrowed space between them. The pressure in the water between the ships will therefore be diminished and will become less than the water pressure on the far sides of the ships. This difference will cause the ships to come closer and closer together, with danger of collision if care in steering is not taken.

The Bernoulli effect is even more pronounced with gases than it is with liquids (see Sec. 13.25).

12.16 ■ Measurement of Liquid Flow—The Venturi Meter

Fig. 12.27 The venturi meter can be used to determine the velocity of fluid flow within a pipe. The liquid height in the constriction is lower because the pressure there is lower. The height difference h is proportional to the speed difference, which in turn depends on the flow speed difference. Once the velocity of the fluid is known, the flow rate can be calculated.

Bernoulli's theorem provides a ready means for measuring the flow of a liquid through a pipe. The *venturi meter,* illustrated in Fig. 12.27, consists of a horizontal section of pipe containing a constriction, or throat, having properly designed tapers to avoid turbulence and assure streamline flow. Bernoulli's equation applied to the wide (a) and constricted (b) portions of the pipe becomes

$$P_a + \tfrac{1}{2}\rho v_a^2 = P_b + \tfrac{1}{2}\rho v_b^2$$

The h_a and h_b terms of Eq. (12.16) drop out since the meter is level.

Since velocity v_b is greater than v_a, the pressure P_b is less than pressure P_a. Consequently the liquid in the throat manometer will not rise as high as in the pipe manometer. The difference in manometer heights is a measure of the difference in pressures. Thus, if h is the difference in manometer heights,

$$P_a - P_b = \rho g h$$

Assuming the liquid to be incompressible, we know

$$A_a v_a = A_b v_b$$

or

$$v_b = \frac{A_a}{A_b} v_a$$

Bernoulli's equation then becomes

$$P_a - P_b = \rho g h = \tfrac{1}{2}\rho(v_b^2 - v_a^2)$$
$$= \tfrac{1}{2}\rho\left(\frac{A_a^2}{A_b^2} v_a^2 - v_a^2\right)$$

Solving for v_a^2,

$$v_a^2 = \frac{\rho g h}{\tfrac{1}{2}\rho(A_a^2/A_b^2 - 1)}$$

and then for v_a,

$$v_a = \sqrt{\frac{2gh}{A_a^2/A_b^2 - 1}}$$

Since the discharge rate $Q = A_a v_a$, we get

$$Q = A_a \sqrt{\frac{2gh}{A_a^2/A_b^2 - 1}} \tag{12.18}$$

Hence, if the diameters of the pipe and the throat are known, one need only read the difference in the heights of the liquid in the two tubes to have sufficient data to determine the volume flow rate.

The difference h between the manometer levels of a venturi meter is the only measurement needed to find the flow rate if pipe and throat areas are known.

The reduction of fluid pressure at a constriction finds application in such devices as the carburetor of an auto, the aspirator or spray nozzle (see Sec. 13.24), and the filter pump.

Illustrative Problem 12.13 Compute the discharge rate of water through a venturi meter having a pipe diameter of 10 cm and a throat diameter of 5.0 cm if the difference of the liquid (water) heights in the manometer tubes is 6 cm.

Solution

$$A_a = 25\pi \text{ cm}^2 \qquad A_b = 6.25\pi \text{ cm}^2 \qquad h = 6 \text{ cm}$$

$$\frac{A_a^2}{A_b^2} = \left(\frac{25\pi \text{ cm}^2}{6.25\pi \text{ cm}^2}\right)^2 = 16$$

Substitute known values in Eq. (12.18):

$$Q = 25\pi \text{ cm}^2 \sqrt{\frac{2 \times 981 \text{ cm/s}^2 \times 6 \text{ cm}}{16 - 1}}$$

$$= 2200 \text{ cm}^3/\text{s} \qquad \textit{answer}$$

SURFACE PROPERTIES OF LIQUIDS

12.17 ■ Surface Tension

The kinetic-molecular theory (Sec. 10.7) states that although molecules of a substance are in continual motion, they possess mutual attractions for other neighboring molecules. Forces between like molecules are termed *cohesive* forces. Forces between unlike molecules are called *adhesive* forces.

Fig. 12.28 The cohesive forces are greater than the adhesive forces at the surface.

Cohesive and adhesive forces at the interface between two different materials give rise to interesting phenomena. One of these is that of surface tension for liquids. The forces acting on the molecules constituting the surface layer of a liquid are different from those acting on the remaining molecules. No net average cohesive force can act on a molecule in the interior of a body of liquid (Fig. 12.28) because other molecules are symmetrically arranged around it. Consequently, the cohesive forces have the same magnitude in all directions.

However, at the boundary with air, the surface molecules experience a net cohesive force from the other liquid molecules below them which is greater than the net adhesive forces between water molecules and air molecules. The net effect is a greater compacting of molecules at the surface, which causes the surface to act like a stretched membrane. This effect is called *surface tension*. The inward pull on the surface layer tends to make the surface as small as possible. It is this tension that causes dewdrops, raindrops, soap bubbles, etc., to assume the shape of spheres, since for a given volume the shape having minimum surface area is the sphere. Lead shot is manufactured by passing molten lead through a sieve from a high tower and allowing it to fall into water. The descending molten-lead particles assume a spherical form and solidify in this shape before hitting the water.

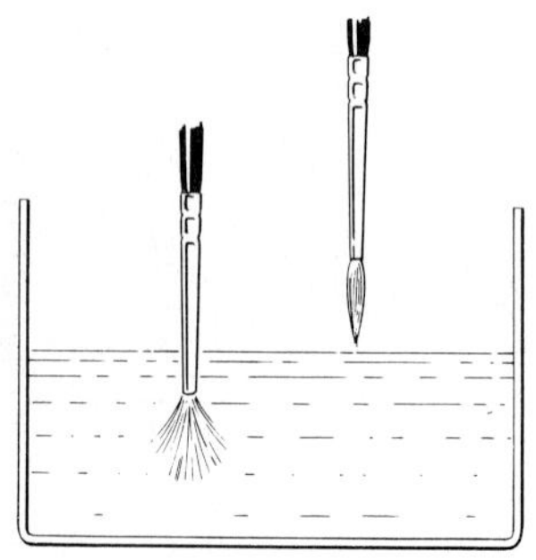

Fig. 12.29 Surface tension acting on a paintbrush.

When a paintbrush is placed in paint, the bristles spread out, but when it is lifted out, the film of liquid on the bristles contracts because of surface tension, and that pulls the bristles together, making a fine point (Fig. 12.29).

Industry has manufactured many products designed to overcome surface tension. The effectiveness of soap and *detergents* depends upon the ability of these compounds to break down the cohesion and surface tension of the liquid and thus permit the liquid to penetrate under dirt and grease to dissolve it. These substances are commercially used in washing powders, shampoos, dentifrices, in many germicides, and in agricultural sprays.

12.18 ■ Measurement of Surface Tension

A common sewing needle or razor blade lowered gently to the surface of water in an open vessel can be made to float, as shown in Fig. 12.30. The metal will float on the surface even though it may be 7 times as dense as water.

The presence of the needle (Fig. 12.30*b*) causes the water surface to be depressed. According to Newton's third law, if the needle causes the water to experience a downward force, the water must cause the needle to experience an upward force.

If next the needle is dipped into the water, cohesive and adhesive forces will cause a thin film of water to adhere to the needle. To pull the needle upward through the water surface layer requires a force in excess of the weight of the needle. This phenomenon is due to surface tension. All liquids have the property of exerting a tension upon adjacent portions of the liquid.

Surface tension is defined as the force per unit length which must be applied to overcome the molecular forces of attraction of the liquid. It can be measured by

(*a*)

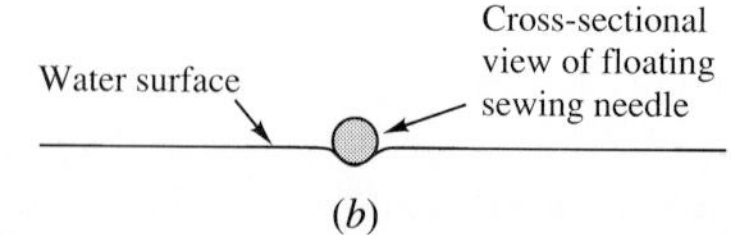

(*b*)

Fig. 12.30 (*a*) A razor blade and a needle can be made to float on the water surface. (*b*) Cross-sectional view of the floating needle.

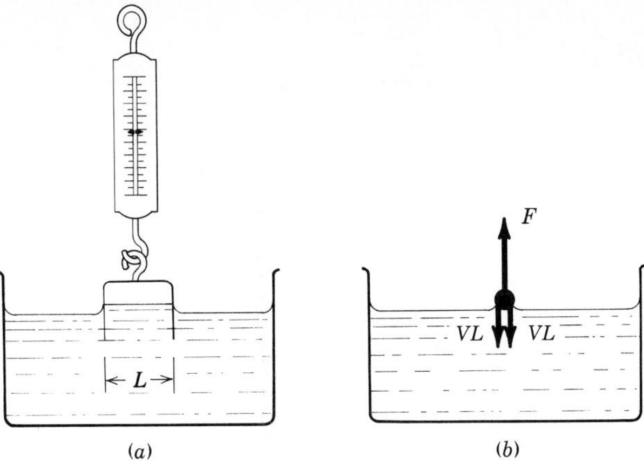

Fig. 12.31 Measurement of surface tension: (*a*) plan view; (*b*) force vectors involved.

observing the force needed to pull an inverted U-shaped wire (Fig. 12.31) upward through the surface of the liquid.

Suppose a force F (not including weight of the wire) is required to pull a wire of length L through the surface layer. As the wire leaves the surface, two surfaces of one film of liquid are attached to it, making the effective length $2L$. The two surfaces exert a downward force on the wire equal to $2\gamma L$, where γ(*gamma*) represents the surface tension of the liquid in newtons/meter of contact between the liquid and the wire. Thus

$$F = \gamma L + \gamma L = 2\gamma L \tag{12.19}$$

or

$$\gamma = \frac{F}{2L} \tag{12.20}$$

In general, the surface tension of liquids decreases with rise in temperature. Values of the surface tension of a few common liquids measured in air are given in Table 12.1.

Table 12.1 Surface tension of some common liquids in contact with air

Liquid	*γ, newtons/m*
Alcohol at 0°	0.024
at 20°	0.023
Benzene at 20°	0.029
Glycerin	0.063
Mercury at 20°	0.480
Olive oil	0.035
Soap solution	0.026
Water at 20°	0.073
at 100°	0.059

F
Olive oil
D = 3.0 cm (diameter of ring)

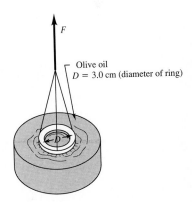

Fig. 12.32 The force F that is required to lift the ring from the liquid is equal to the sum of the force of surface tension and the weight of the ring.

Illustrative Problem 12.14 A fine wire in the shape of a ring of 3.0 cm in diameter is placed horizontally in a vessel containing olive oil (Fig. 12.32). What force in addition to the weight of the wire will be required to pull the wire from the oil surface?

Solution Use Eq. (12.19) for F:

$$F = 2\gamma L$$

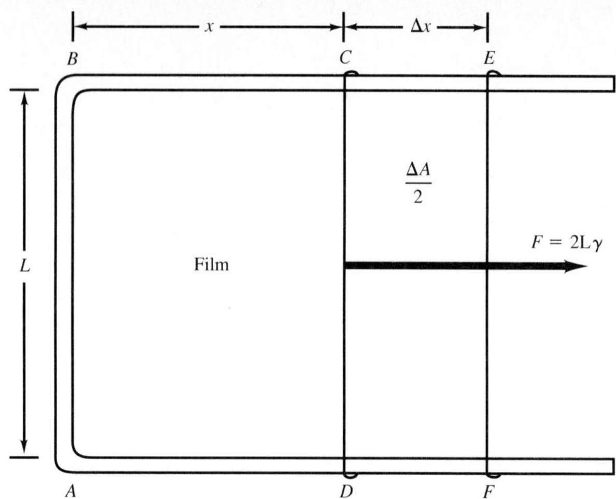

Fig. 12.33 Apparatus for measuring surface tension. The force $F = 2L\gamma$ is required to hold the movable wire CD in equilibrium against the surface tension.

where L is the circumference of the ring.

$$L = \pi d = 3.14 \times 3.0 \text{ cm} = 9.42 \text{ cm}$$

Then

$$F = 2 \times 0.035 \text{ N/m} \times 9.42 \text{ cm} \times \frac{1 \text{ m}}{100 \text{ cm}}$$

$$= 6.6 \times 10^{-3} \text{ N} \qquad \textit{answer}$$

Surface tension can also be defined as the work done per unit increase of surface area as film surface area is increased against molecular forces. Consider the U-shaped apparatus shown in Fig. 12.33 which is a rectangular wire whose segment CD is free to move without friction. If the apparatus is dipped in a soap solution there will be formed a film represented by $ABCD$. In order to prevent the film from contracting, pulling the movable side CD toward AB, a force F must be applied. Equation (12.19) tells us the force F is equal to $2L\gamma$. If the wire CD is slowly pulled a distance Δx to a new position EF, work must be done. This work $W = F\Delta x$. Thus,

$$W = F\ \Delta x$$
$$= \gamma(2\text{L}\Delta\text{x})$$

But

$$2L\ \Delta x = \Delta A$$

Therefore

$$W = \gamma \Delta A \tag{12.21}$$

where Δx is the change in x and ΔA is the *total increase in area* of the soap film. (Remember the film has two surfaces.) And so we can solve for

$$\gamma = \frac{W}{\Delta A} \tag{12.22}$$

Thus, the surface tension γ is not only the force per unit length, but also the work done per unit increase in surface area. Hence, γ can be expressed in N/m or J/m^2 units. It will be left as an exercise for the student to prove the two units are dimensionally equivalent.

Illustrative Problem 12.15 A soap film is formed in a U-shaped wire similar to that in Fig. 12.33. The length of the movable slide is 8.0 cm. How much work will be required to move the slide 4 cm against the surface tension? Assume the surface tension of the soap solution is 0.026 N/m.

Solution $L = 8.0$ cm; $x = 4.0$ cm; $\gamma = 0.026$ N/m.

The work done,

$$
\begin{aligned}
W &= F\ \Delta x \\
&= 2\gamma L \Delta x \\
&= 2(0.026\ \text{N/m})\left(8.0\ \text{cm} \times \frac{1\ \text{m}}{100\ \text{cm}}\right)\left(4.0\ \text{cm} \times \frac{1\ \text{m}}{100\ \text{cm}}\right) \\
&= 1.7 \times 10^{-4}\ \text{N} \cdot \text{m} \\
&= 1.7 \times 10^{-4}\ \text{J}
\end{aligned}
$$

answer

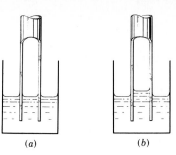

12.19 ■ Capillary Action

Adhesive, cohesive, and surface-tension forces combine to give a liquid the characteristics which, depending upon the liquid used, will cause it to be raised or depressed in a tube of small bore. These tubes are called *capillary tubes*.

Consider a clean glass tube placed in a vessel of water, as shown in Fig. 12.34. Since the water wets the walls of the tube, the forces of adhesion between the molecules of water and glass must be greater than the forces of cohesion between the water molecules. Consequently a thin film of water will quickly be drawn up the side of the tube (Fig. 12.34*a*). Surface tension causes this film to contract (Fig. 12.34*b*), with the net result that the water column rises in the tube. As the water rises in the tube, adhesive forces continue to wet more surface inside the tube. The water will continue to rise in the tube until the upward pull of the adhesive and surface-tension forces just balances the downward pull of gravity on the column of water lifted (Fig. 12.34*c*).

When a liquid does not wet the surface of the tube (cohesion greater than adhesion), such as mercury in glass, the cohesion of the particles of mercury for one another makes it appear as if there were repulsion between the glass and mercury. The surface tension pulls downward on the surface, and the liquid inside the tube is depressed until it stands at a lower level (Fig. 12.35). When the liquid does wet the tube surface, the upward force is equal to the product of the length of the film (the circumference $2\pi r$ of the tube) and the upward component of the tension γ of the surface. This upward component is $\gamma \cos \theta$ (see Fig. 12.36).

$$F = 2\pi r\ \gamma \cos \theta$$

The weight W of the liquid is the product of weight density ρg and the volume.

$$W = \rho g\ (\pi r^2 h)$$

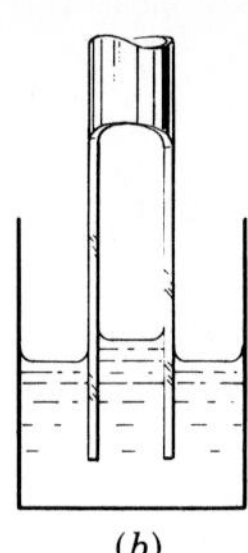

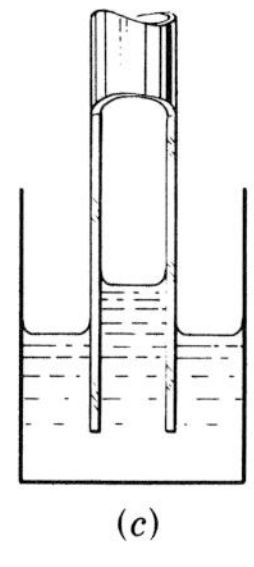

Fig. 12.34 Capillary action with glass tube in water.

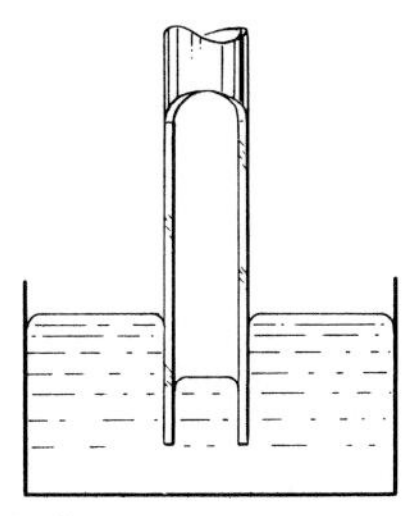

Fig. 12.35 Capillary tube with a glass tube in a container of mercury.

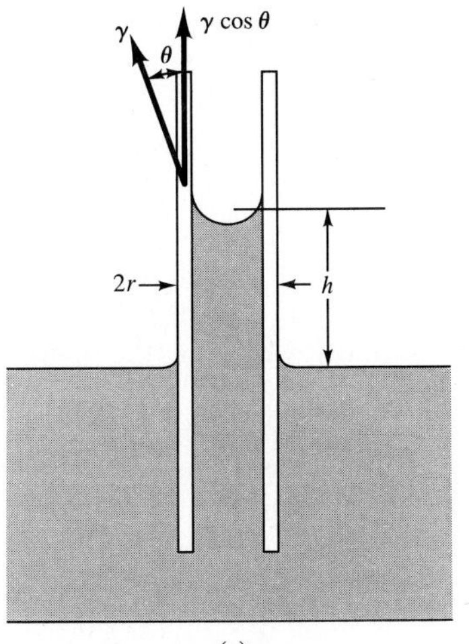

(*a*)
Glass tube in water.
Water wets glass. $\theta < 90°$

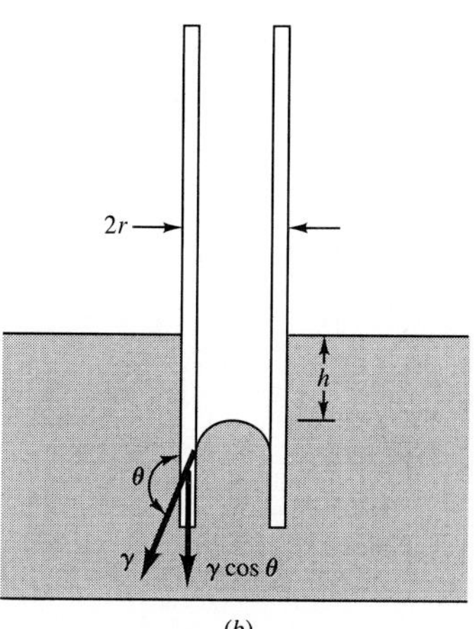

(*b*)
Glass tube in mercury.
Mercury does not wet glass. $90° < \theta < 180°$

Fig. 12.36 Liquid in a capillary tube. (*a*) Glass tube in water. The water wets the glass. $\theta < 90°$. (*b*) Glass tube in mercury. The mercury does not wet the glass. $90° < \theta < 180°$.

Thus,

$$2\pi r\gamma \cos \theta = \rho g(\pi r^2 h)$$

$$h = \frac{2\gamma \cos \theta}{r\rho g} \qquad (12.23)$$

For many liquids, such as water in glass, the value of the angle θ is nearly zero. Since cosine $0° = 1$, Eq. (12.23) would then reduce to $h = 2\gamma/r\rho g$.

Illustrative Problem 12.16 A clean capillary tube is placed in a vessel of mercury. Find how much the mercury will rise in the tube if the diameter of the tube is 2 mm and $\theta = 130°$.

Solution From Table 12.1, we find that the surface tension γ for mercury is 0.480 N/m. From Table 11.1 ρ for mercury is 13.6 g/cm^3. Substituting in Eq. (12.23) we get

$$h = \frac{2\gamma \cos \theta}{r\rho g}$$

$$= \frac{2 \times 0.48 \text{ N/m} \times 1 \text{ m/100 cm} \times \dfrac{1 \text{ kg} \cdot \text{m/s}^2}{1 \text{ N}} \times 0.643}{1 \text{ mm} \times \dfrac{1 \text{ cm}}{10 \text{ mm}} \times \dfrac{13.6 \text{ g}}{\text{cm}^3} \times \dfrac{1 \text{ kg}}{1000 \text{ g}} \times 9.81\dfrac{\text{m}}{\text{s}^2}}$$

$$= -0.463 \text{ cm}$$ *answer*

The negative sign for the result indicates that the mercury column does not rise but is depressed below the surface of the mercury in the vessel.

Many common phenomena are due in part to capillary action. The rise of oil in lampwicks, the effectiveness of a blotter or towel, the rise of water in the soil, the flow of sap in a plant are examples of this action.

12.20 ■ Volatility

Another important property of liquids is that of *volatility*. A liquid is said to have a high degree of volatility if it can be readily changed to a vapor state. Perfumes and many food flavors depend upon this property. The effectiveness of gasoline as a fuel in an internal-combustion engine depends to a great extent upon its volatility, since the liquid gasoline must be quickly converted to a vapor state before proper combustion can occur. On the other hand, too high a volatility can be a disadvantage in an internal-combustion engine. If the volatility is too high, the combustion is not smooth throughout the stroke of the piston and the engine *pings* or *knocks*. Refinery chemists are constantly striving to produce a properly balanced fuel, one that has enough volatile *lighter fractions* to assure proper and quick starting but enough *heavier components* to result in slow burning and smooth engine performance.

12.21 ■ Viscosity

The internal friction, or resistance to flow, set up within a liquid is called *viscosity*. The water on the surface of a river moves more rapidly than the water near the bottom or sides, and the water in the center of a pipe also moves with greater velocity than the water next to the pipe surface. This difference in velocity is due in part to the friction between the water and the river bed or the pipe, which causes the adjacent layers of water to move more slowly. These more slowly moving layers of water in turn, because of internal friction (viscosity), tend to retard the motion of adjacent layers. Hence viscosity, or internal friction, is a very important factor in fluid flow.

Viscosity is associated with the resistance of two adjacent layers of fluid to any motion relative to each other.

This property of viscosity is especially significant in the study of oils. In industry, oils with the higher viscosity are the ''heavier'' oils. It should be pointed out that the term heavier as here used does not refer to density. An increase in temperature of the oil is accompanied by a decrease in its viscosity, and vice versa. Viscosity is responsible for most of the dissipation of energy, and hence for most of the expense, in transporting liquids through pipelines. Thus it is found that a great deal of energy is expended in transporting crude oil in pipelines—the colder the weather, the greater the energy expended. Sometimes the oils must be heated before they can be pumped.

The effectiveness of lubricating oils depends, among other things, upon their viscosity. Lubricating oils should be viscous enough so that they will not be squeezed out of the bearing, yet not so viscous as to offer needless resistance to the motion of the parts being lubricated.

The viscosity of a fluid can be determined by measuring the time required for a given quantity of the fluid to escape through a long tube of small diameter. For determining the viscosity of paints, varnishes, lacquers, and similar liquids, the test often consists in timing the flow of the liquid through a standard orifice. The orifice is screwed into the conical bottom of a heavy polished-metal cup. The instruments used in these tests are called *viscosimeters*.

QUESTIONS AND EXERCISES

1. If a rock loses 1 lb weight when submerged 1 ft in water, how much will it lose when submerged 5 ft deep in water?

2. A brass ball is weighed, first in kerosene (sp gr 0.82) and then in benzene (sp. gr. 0.90). In which will its apparent weight be the greatest? Explain.

3. Will increasing the diameter of a capillary tube cause water to rise or fall in the tube?

4. An ice cube floats in a glass brimful of water. When the ice melts will the water overflow? Give reasons for your answer.

5. When a ship sails from the ocean into a freshwater river, does it sink deeper or ride higher out of the water? Give reasons for your answer.

6. Upon what factors does rate of liquid flow through a hole in the bottom of a tank depend?

7. How is it possible in a hydraulic press to eventually obtain a greater linear displacement at the output piston than the linear displacement of one stroke at the input piston?

8. Explain how Bernoulli's equation is an example of the law of conservation of energy.

9. An overflow can full to the brim with water is placed on a weighing scale. After a small block of wood is placed in the water, what will happen to the balance reading?

10. A glass is brimful of water. What will happen when a cube of ice is gently placed on the water surface. Give reasons for your answer.

11. A steel ball is in an empty cup that is floating in a pail partly full of water. If the ball is removed from the cup and dropped in the water, will the water level rise, fall, or remain the same? Explain your reasoning.

12. Why are drops of mercury on glass more nearly spherical than drops of water on glass?

13. Explain why alcohol will cling to a glass rod while mercury does not.

14. Is it possible to "sip" a liquid through a straw on the moon? Give reason for your answer.

15. Explain why, despite varying quantities of water in Pascal's vases (Fig. 12.10), the force is the same at the bottom of each vase. Then, why does one vase full of water weigh more than another?

16. Give two reasons why it is impractical to use water as the working fluid of a barometer.

17. Each of the pipes from *A, B, C* in Fig. 12.37 are identical in size. If we neglect internal friction, discuss the relative rates of flow in each pipe.

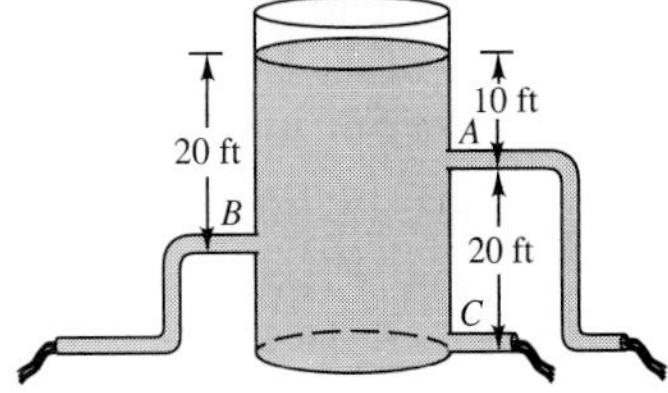

Fig. 12.37 Question 17.

18. An empty sealed aluminum can is floating in a vessel of water. What happens to the water level in the vessel if after holes are punched in the can it is returned to the vessel, and after filling with water, sinks to the bottom?

19. A 3.75-cm-diameter table tennis ball and a 3.75-cm-diameter steel ball are held underwater. According to Archimedes' principle, how do the magnitudes of the buoyant forces experienced by the two balls compare?

20. Legend has it that Archimedes, by applying his principles of buoyancy, was able to prove that a goldsmith gave the king of Syracuse a "gold" crown adulterated with silver. Do you suppose Archimedes was able to prove the fraud?

PROBLEMS

Group One

1. The height of water impounded behind two identical dams is the same. Dam *A* holds back a lake which stretches 1320 m from the dam. Dam *B* holds back a lake which stretches 5280 m from the dam. What is the ratio of the forces acting on the dams?

2. What is the force exerted on an area of 4.0 m^2 at a level in a liquid where the average pressure is 3 Pa?

3. If the specific gravity of brick is 1.8, what is its density in kilograms per cubic meter?

4. Convert a pressure of 1050 mb to lb/in.^{2}.

5. What is the pressure (psi) in a city water main that is 120 ft below the surface of the water in the reservoir? (Assume water not flowing.)

6. What is the difference in water pressure in two offices on different floors of a building if one office is 20 m higher than the other? Answer in pascals.

7. Cotton from a gin is passed to a baler, where it is compressed by hydraulic presses into 500-lb bales. If the diameters of the pistons of the cotton press are 2 and 30 in., respectively, and the force required to bale the cotton is 12,000 lb, how much force must be applied on the small piston?

8. What air pressure will be necessary to lift a hydraulic auto hoist in a service station if the air acts against a cylinder 8 in. in diameter and is required to lift a 4000-lb car? (Assume the weight of the hoist to be 500 lb.)

9. A block of wood with dimensions 20 by 10 by 5 cm floats with its largest surface horizontal. If its specific gravity is 0.6, how deep will it sink in water?

10. A rectangular block of wood floats in water with four-tenths its volume out of water. What is the specific gravity of the wood?

11. An aluminum block with dimensions of 15 in. × 12 in. × 24 in. is securely bolted to the bottom of a tank of water in such a way that no water is between the block and the bottom of the tank. What is the buoyant force of the water on the block?

12. What is the theoretical pressure head required to push water through a round orifice one inch in diameter with an exit velocity of 45 ft/s?

13. A U-shaped wire (see Fig. 12.31) having a horizontal length of 8.4 cm is lowered into benzine at 20°. What force, in addition to the weight of the wire, is needed to break the film that forms as the wire is slowly raised through the surface?

14. A waterfall has a drop of 43.7 ft and a flow rate of 52.5 ft^3/s. What horsepower is available if the power is harnessed at 100 percent?

Group Two

15. The water company of a residential district has drilled a new well and installed pumps to force the water into storage tanks. How high above ground must the storage tanks be placed to obtain a hydrostatic pressure of 2.62×10^5 Pa on the water main at ground level?

16. A cylindrical tank 50 cm high is filled two-thirds with water and one-third with linseed oil. What is the pressure at the bottom of the tank? (Use Table 11.1.)

17. A piece of metal weighs 41.0 lb in air. Completely submerged in water it weighs 25.4 lb, and submerged in an unknown liquid it weighs 28.2 lb. What is the density of the unknown liquid?

18. A solid was weighed in water, kerosene and alcohol. Its loss of weight in water was 5.06 N, in kerosene 4.15 N, and in alcohol 4.00 N. What is the specific gravity of (*a*) kerosene and (*b*) alcohol?

19. An iron casting weighs 8.35 lb in air and 7.16 lb when submerged in water. (*a*) What is the volume of the casting? (*b*) What is the weight density of the casting?

20. With what speed will water escape from a hole 2.54 cm in diameter at the bottom of a water storage tank 11.0 m high? How many cubic meters will be discharged each second?

21. An auto ferry is 30 ft long and 15 ft wide and has vertical sides. When six cars are driven on board, the barge sinks 8 in. deeper into water. How much do the six autos weigh?

22. The tube of a mercury barometer has an inside diameter of 3.0 mm. What error is introduced in the reading because of capillary action? Assume the angle of contact between the mercury and glass is 128°.

23. An empty bottle weighs 46.35 g. When it is filled with water the full bottle weighs 103.07 g and when filled with an unknown liquid the full bottle weighs 92.64 g. What is the specific gravity of the unknown liquid?

24. Gasoline flows from a one-inch diameter pipe at the bottom of a vertical cylindrical tank at the rate of 50 gal/min. How high is the gasoline in the tank?

25. A capillary tube whose inside diameter is 0.75 mm is dipped in a solution whose density is 1.25 g/cm^3. The solution rises 2.75 cm in the tube. What is the surface tension of the solution? Assume the angle of contact is 0°.

26. A liter flask "weighs" 225.0 g when empty and 1.057 kg when full of turpentine. Find the specific gravity of turpentine.

27. A hydraulic jack has a ram diameter of 3.8 cm and a pump piston diameter of 1.6 cm. The jack handle has a mechanical advantage of 15 to 1. (*a*) Neglecting friction, what weight will the jack lift when 13.5 kg is applied to the handle? (*b*) What is the pressure of the hydraulic fluid under these conditions?

28. A cylindrical water tank has a hemispherical dome. The dimensions are shown in Fig. 12.38. Calculate the total force exerted by the water on the bottom of the tank when the tank is full.

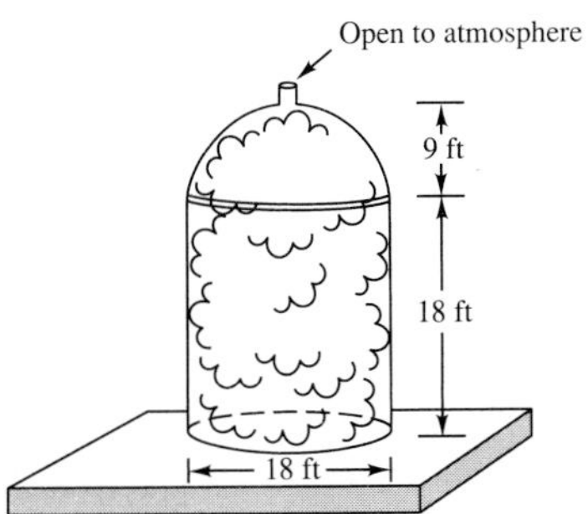

Fig. 12.38 Problem 28.

29. A pump requires 100 hp to deliver water (sp gr = 1) at a certain capacity and to a given elevation. What horsepower would be required to pump liquid which has a sp gr of 0.87 if the capacity and elevation are not changed?

30. An object weighs 2.49 N in air and has a volume of 49.2 cm^3. How much will it weigh when totally submerged in water?

31. One kg of water is mixed with 2 kg of a liquid whose sp gr is 0.80. What is the specific gravity of the mixture?

32. What is the minimum water pressure that is needed at the main (street level) if it is to deliver water to the twentieth story 70 m above?

33. A rock weighs 15.4 lb in air and has an apparent weight of 12.7 lb when submerged in water. What is the density of the rock?

34. How high will water at 20°C rise in a glass capillary tube 0.30 mm in diameter. Assume the angle of contact is zero.

35. A circular tank with height h ft and radius r ft is filled with liquid having a density D lb/ft^3. What is the formula for the total force on its vertical wall?

36. A cylindrical tank 7 m high and full of water develops a hole 3 m below the top. How far from the base of the tank will the escaping water strike the ground? (Neglect friction.)

37. A hydraulic jack is shown schematically in Fig. 12.39. What force F should be applied at A to lift a 907-kg weight at B? (Neglect the friction forces of the piston and the linkage.)

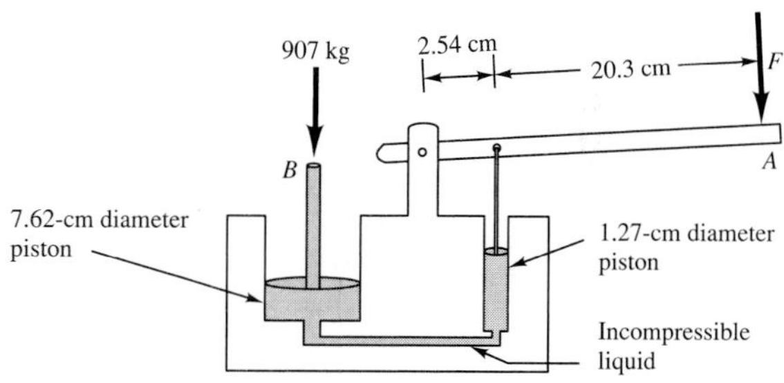

Fig. 12.39 Problem 37.

38. The water level of a dam is 420 ft above the valve of a 6-ft-diameter outlet pipe (called the *penstock*). (*a*) What is the hydrostatic pressure at the center of the valve? (*b*) When the valve is opened, water flows into the penstock at the rate of 1600 ft^3/s. Find the velocity of flow through the penstock. (*c*) Find the velocity head and the pressure head (in feet of water) in the penstock when water is flowing.

Group Three

39. Water flows at a velocity of 1.52 m/s in a pipe with a 7.62-cm diameter. It then passes through a 5.08-cm pipe and finally through a 10.16-cm pipe. Find (*a*) the velocity in the last two pipes and (*b*) the velocity head in each of the three pipes.

40. Salt water (density 64 lb/ft^3) flows through a tube of variable cross section and height as shown in Fig. 12.24. Find the velocity and pressure at b if P_a = 24 lb/in.^{2}, v_a = 25 ft/s, h_a = 32 ft, h_b = 22 ft, and the cross-sectional area at b is one-half that at a. (Neglect frictional losses.)

41. A 2-in.-diameter nozzle used in hydraulic mining, projects a stream of water to a height of 225 ft when the inclination of the nozzle is 60° from the horizontal. (*a*)

What is the velocity of the water as it leaves the nozzle? (*b*) At what rate, in ft^3/s, is the water delivered from the nozzle? (*c*) How much horsepower is available at the nozzle?

42. A bowl is made from an alloy of gold and silver. The bowl weighs 1000 g in air and 940 g in pure water. How much gold and how much silver did the bowl contain?

43. A 75-g bracelet is made by goldplating a copper ring. When it is dropped in a full vessel of water, 6.5 cm^3 of water overflows. What is the mass of the gold in the bracelet?

44. A tank, open at the top (Fig. 12.40) is filled to a depth h with water. Water spurts from a hole in the side a distance y below the free water surface. Find the distance x, in terms of h and y, from the tank to the spot where the water strikes the ground.

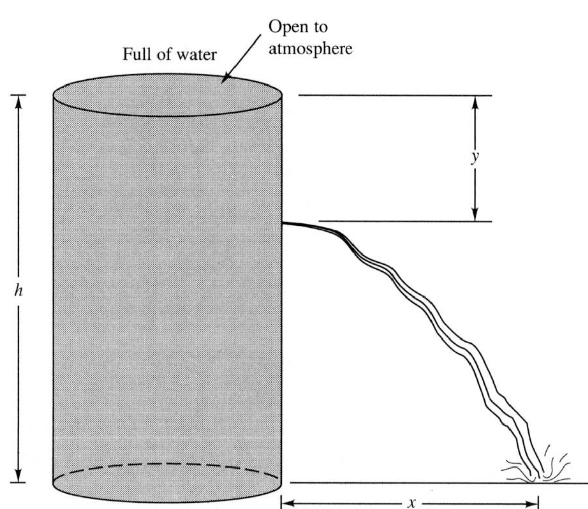

Fig. 12.40 Problem 44.

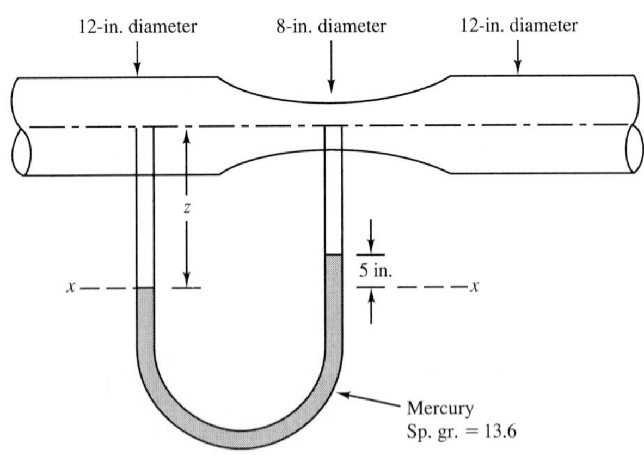

Fig. 12.41 Problem 45.

45. A venturi meter (manometer type) has an 8-in.-diameter throat and is installed in a 12-in.-diameter water pipeline (Fig. 12.41). What is the water flow in ft^3/s in the pipeline if a mercury manometer registers a 5-in. differential in the mercury columns? (Neglect fluid friction effects.)

46. Water flows in a pipe of variable cross section at different heights at a uniform rate as shown in Fig. 12.42. At A the diameter is 10.0 cm and the pressure is 2.4×10^4 N/m^2. At B the diameter is 5.0 cm and the pressure is 1.4×10^4 N/m^2. The center of the pipe at B is 0.6 m higher than the center of the pipe at A. Determine (*a*) the velocity of flow at A and B, and (*b*) the rate of flow through the pipe.

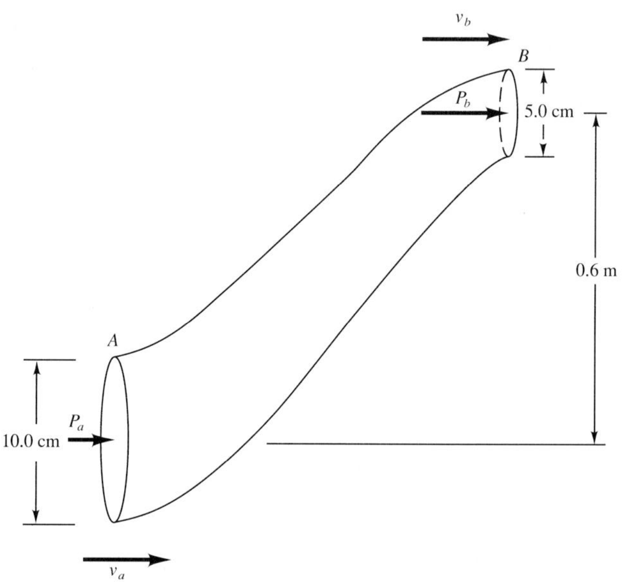

Fig. 12.42 Problem 46.

47. A cube with sides equal to 15 cm and a density ρ_0 of 0.70 g/cm^3 floats in two liquids of different densities as shown in Fig. 12.43. The heavier liquid has a density ρ_1 of 1.00 g/cm^3 and a depth of 12 cm. The lighter liquid has a density ρ_2 of 0.80 g/cm^3 and forms a layer of thickness 5 cm. What is the distance x that the cube projects above the lighter liquid surface?

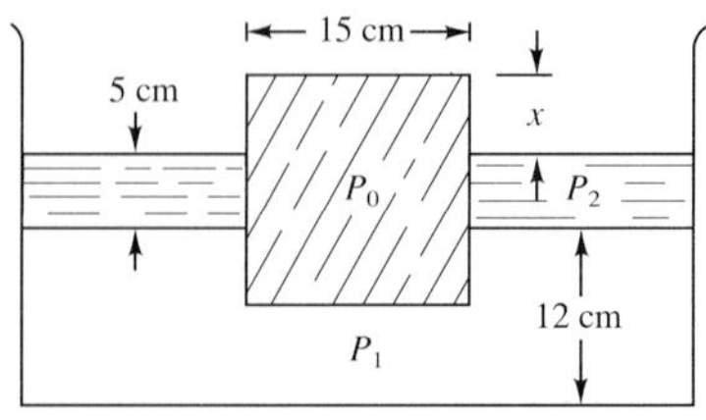

Fig. 12.43 Problem 47.

PROPERTIES OF GASES

In many respects, gases resemble liquids: both consist of many tiny molecules that move about in constant random motion. Both are capable of flowing and are therefore designated by the common term *fluid*. Each applies pressure on the surfaces within which it is confined, and each exerts upward buoyant forces in accordance with Archimedes' principle. The discharge velocities through an orifice can be determined for gases as they are determined for liquids. Gases, like liquids, conform to the shape of the containing vessel. Neither is able to exert shearing stresses, except those due to viscosity.

Gases, however, differ from liquids in two respects: gases are very compressible, whereas liquids are nearly incompressible. A liquid has a fixed volume, while a gas has an unlimited volume, determined only by the volume of the containing vessel.

13.1 ■ Characteristics of Gases

We have seen in Chap. 10 that when matter exists in the gaseous state its molecules possess high velocities and such a great freedom of motion that any given mass of gas will completely fill any container, regardless of its shape. Gases obey the laws of Pascal and Archimedes. They are perfectly elastic, they can readily be compressed, and they provide buoyancy.

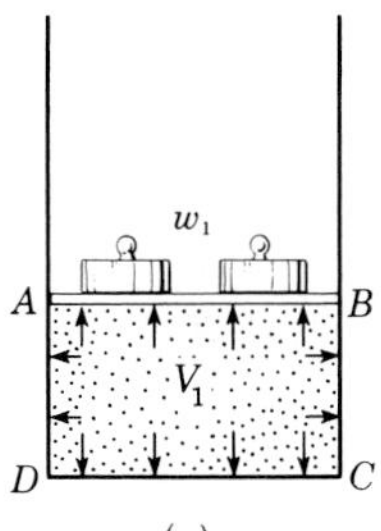

(*a*)

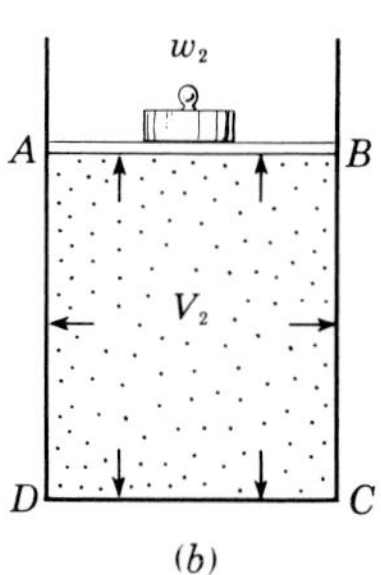

(*b*)

Fig. 13.1 Bombardment of molecules on piston *AB* supports weight *w*.

13.2 ■ Variation of Volume of a Gas with Pressure—Boyle's Law

It can be proved experimentally that whenever the pressure on a gas is increased, the volume of the gas is decreased. On the other hand, a gas which is allowed to expand to a greater volume will exert a lesser pressure on the container. The kinetic theory of matter offers a simple and adequate explanation of this pressure-volume relationship.

Let us assume that a certain mass of air is confined in a cylindrical container and that the pressure P_1 of the air is adequate to sustain a weight w_1 on the piston *AB* (Fig. 13.1*a*). The molecules of the air are continually bombarding each other, the sides of the vessel, and the bottom of the piston *AB* (of negligible weight) supporting the weight w_1. The impact of these molecules on *AB* develops a total force equal to the area of the piston times the pressure. Suppose the volume of the air under this condition to be V_1. When the weight w_1 is replaced by a weight w_2 equal to one-half of w_1, the pressure P_2 necessary to support w_2 is only half the original pressure. The piston *AB* is forced up until the volume V_2 of the confined gas is doubled (Fig. 13.1*b*) and the consequent total impact of the molecular bombardment on *AB* is halved.

This relationship between pressure and volume was first stated as a law in 1662 by Robert Boyle, after whom it is named.

Robert Boyle (1627–1691) was born in Dublin, Ireland. Most of his scientific work was done in England, where he became one of the most brilliant scientists of the seventeenth century.

Boyle's Law

When the temperature of a gas is kept constant, the volume of an enclosed mass of gas varies inversely as the absolute pressure upon it.

Boyle's law holds quite accurately for a wide range of pressures, but deviations from the law must be considered when dealing with extreme pressures.

This relationship can be conveniently expressed as an equation:

$$\frac{V_1 \text{ (original volume)}}{V_2 \text{ (new volume)}} = \frac{P_2 \text{ (new absolute pressure)}}{P_1 \text{ (original absolute pressure)}}$$

or

$$P_1V_1 = P_2V_2 \tag{13.1}$$

Absolute pressure will be explained in Sec. 13.20.

Illustrative Problem 13.1 The oxygen to be used in an oxyacetylene welding outfit is stored in a cylinder 1.40 m long, and of 17.8-cm internal diameter. The gas is at an absolute pressure of 1.55×10^6 Pa. To what volume, in cubic meters, would the gas expand if released to atmospheric pressure? (Assume normal atmospheric pressure = 1.013×10^5 Pa.)

Solution Recall the formula for the volume of a right circular cylinder and substitute for known values.

$$\begin{aligned} V_1 &= \pi r^2 h \\ &= 3.14\left(\frac{0.178\ m}{2}\right)^2 (1.40 \text{ m}) \\ &= 3.48 \times 10^{-2} \text{ m}^3 \end{aligned}$$

Now, it is given that

$$P_1 = 1.55 \times 10^6 \text{ Pa}$$
$$P_2 = 1.013 \times 10^5 \text{ Pa}$$

Next solve Eq. (13.1) for V_2 and substitute for known values:

$$\begin{aligned} V_2 &= \frac{1.55 \times 10^6 \text{ Pa}(3.48 \times 10^{-2} \text{ m}^3)}{1.013 \times 10^5 \text{ Pa}} \\ &= 0.532 \text{ m}^3 \qquad \textit{answer} \end{aligned}$$

13.3 ■ Variation of Volume of a Gas with Temperature

Jacques Alexander Cesar Charles (1746–1823) was a French scientist. He researched the work done by other scientists relating to the volume of a given mass of gas with temperature. He is given credit for first putting the relation into an equation, since known as **Charles' Law**. He was the first to experiment with gas balloons using hydrogen.

Charles' Law Boyle's law assumes conditions of constant temperature, and in actual industrial situations this is rarely the case. Temperature changes continually occur, and they affect the volume of a given mass of gas. Until we study the subject of heat and temperature in Chap. 14, the effect of adding heat to a gas will have to be treated superficially. It can be shown that if pressure is unchanged, an increase in temperature of a gas is accompanied by an increase in its volume.

Charles' Law

If constant pressure is maintained, the volume of a gas is proportional to its absolute temperature.

("Absolute temperature" is more rigorously explained in Chap. 14.)* *Charles' law* is given in algebraic form by

$$\frac{V_1 \text{ (original volume)}}{V_2 \text{ (new volume)}} = \frac{T_1 \text{ (original abs temp)}}{T_2 \text{ (new abs temp)}}$$

$$\frac{V_1}{V_2} = \frac{T_1}{T_2} \tag{13.2}$$

Illustrative Problem 13.2 A thin balloon is filled with 220 in.^{3} of air at a temperature of 280° abs. To what volume will the air expand if the temperature rises to 320° abs? (Neglect changing forces of constriction of balloon.)

Solution Original volume = 200 in.^{3}; new volume = V_2 in.^{3}; original abs temp = 280°; new abs temp = 320°. Use Eq. (13.2), and write

$$\frac{220 \text{ in.}^3}{V_2 \text{ in.}^3} = \frac{280°}{320°}$$

$$280V_2 = 220 \times 320$$

$$V_2 = 250 \text{ in.}^3 \qquad \textit{answer}$$

13.4 ■ Variation of Pressure of a Gas with Temperature—Gay-Lussac's Law

Since an increase in the temperature of a gas will cause it to expand if the pressure is kept constant, it is reasonable to expect that if a certain mass of gas were heated in a closed container so that its volume had to stay constant, there would be a consequent increase in the gas pressure. Experiments have found this to be true.

The exact relationship between pressure, temperature, and volume of gases is developed at greater length in Chap. 14. The formula to be developed in Chap. 14 takes the form

$$\frac{P_1 \text{ (original absolute pressure)}}{T_1 \text{ (original absolute temperature)}} = \frac{P_2 \text{ (new absolute pressure)}}{T_2 \text{ (new absolute temperature)}}$$

$$\frac{P_1}{T_1} = \frac{P_2}{T_2} \tag{13.3}$$

Equation (13.3) is known as *Gay-Lussac's law*. In the next chapter it will be noted that for every increase of 1 C° in the temperature of a given mass of gas at constant pressure there is an accompanying volume increase equal to approximately $\frac{1}{273}$ of the gas volume at 0°C. For each decrease of 1 C° the volume decreases by approximately $\frac{1}{273}$ of the volume at 0°C. Although this generalization breaks down at conditions of extremely low and extremely high temperatures, it is a satisfactory approximation for conditions normally encountered.

Joseph Louis Gay-Lussac (1778–1850) was a French chemist and physicist. His research on combining volumes of gases resulted in the law that bears his name. His work on the relation between the volumes of gases and their temperatures won him enduring fame. He was a pioneer in scientific balloon observations. He and Louis Thenard were the first to isolate the element boron.

13.5 ■ Law of Combining Gas Volumes

In 1808 the French scientist Joseph Gay-Lussac pointed out that when combining with one another gases under the same conditions of temperature and pressure will always combine in simple volume relations (if the product is a gas). This relationship is known as the *law of combining volumes*. The volumes of gaseous substances that are involved in a chemical change are in the ratio of small integer numbers. For

* Other names for absolute temperature are *Kelvin temperature* (SI-system); and *Rankine temperature* (English system).

A generalization such as this about the volumes of reactants and products when they are solids or liquids cannot be made.

example, when hydrogen and chlorine gases combine to form hydrogen chloride gas (at constant conditions of temperature and pressure):

$$1 \text{ L hydrogen} + 1 \text{ L chlorine} \xrightarrow[\text{(combine chemically)}]{} 2 \text{ L hydrogen chloride}$$

When hydrogen and oxygen gases combine chemically at a temperature above 100° C to form steam (water vapor), the volume ratios are:

$$2 \text{ L hydrogen} + 1 \text{ L oxygen} \longrightarrow 2 \text{ L steam}$$

Amadeo Avogadro (1776–1856) was an Italian physicist. He was a professor at the University of Turin.

The Italian scientist Amadeo Avogadro reasoned that in order for gases to combine in simple ratios, it must be true that under the same conditions of temperature and pressure equal volumes of *all gases* contain the same number of molecules. That is to say, a liter of hydrogen contains just as many molecules as a liter of chlorine, or oxygen, or any other gas if they are measured under the same conditions of temperature and pressure. This fact has been thoroughly substantiated by scientific evidence.

13.6 ■ Diatomic Nature of Some Gas Molecules

The two examples just cited illustrate why some molecules of gases must contain 2 atoms. Such molecules are called *diatomic*. One liter of hydrogen and one liter of chlorine produce two liters of hydrogen chloride.

This fact can readily be explained by assuming that each molecule of hydrogen and of chlorine has two atoms and the reaction occurs between the atoms. Thus,

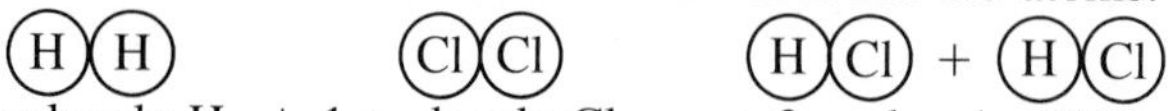

(H)(H) (Cl)(Cl) (H)(Cl) + (H)(Cl)

$$1 \text{ molecule } H_2 + 1 \text{ molecule } Cl_2 \longrightarrow 2 \text{ molecules HCl}$$

The symbols for each of the molecules of the gases in the reaction are H_2, Cl_2, and HCl. The chemical reaction is written

$$H_2 + Cl_2 \rightarrow 2\ HCl$$

When hydrogen and oxygen combine to form water vapor the reaction can be expressed by the chemical equation

$$2\ H_2 + O_2 \rightarrow 2\ H_2O$$

Symbolically,

(H)(H) (H)(H) (O)(O) (H)(O)(H) + (H)(O)(H)

$$2 \text{ molecules hydrogen} + 1 \text{ molecule oxygen} \longrightarrow 2 \text{ molecules water}$$

13.7 ■ Gram-Molecular Weight of Gases

Avogadro's theory enabled scientists to calculate the relative weights of single molecules (and atoms) of gases. By weighing equal volumes of gases under similar conditions, scientists have been able to compare relative weights (and thus the masses) of individual molecules. Conversely, when the ratio of the weights of two gases is the same as the ratio of the individual weights (or masses) of the individual molecules, then there must be the same number of molecules of the two gases present. The symbol for an element not only identifies the element but also represents one atom of it. Thus, H represents one atom of hydrogen. Ne represents one atom of neon. H_2 represents a *molecule* of hydrogen which is made up of two atoms. In like manner, NH_3 indicates a molecule of ammonia gas, which is formed by the union of one atom of nitrogen and three atoms of hydrogen.

In order to solve quantitative relations involving atoms and molecules, some convenient weight (mass) unit needs to be selected for the various atoms. The atomic mass unit (μ) commonly used is the gram. One atomic mass unit is defined as one-twelfth the mass of a carbon 12 atom. It is equal to 1.660×10^{-24} g. Atomic weights of most of the elements have now been determined with great accuracy. They can be found in Appendix VIII.

The atomic weights listed in Appendix VIII represent *relative weights* of the same number of atoms of the elements. Thus, we find that since the atomic weight of hydrogen is 1.00797 and the atomic weight of oxygen is 15.9994 then an atom of oxygen is 15.9994/1.00797 (approximately 16) times as heavy as an atom of hydrogen. The *gram-atomic weight* (abbreviated *gram-atom*) represents the relative weight of an element expressed in gram units. The gram-atomic weight of hydrogen is 1.00797 g.

Once the atomic weights of the various elements were determined, it became a simple matter to calculate the relative weights (masses) of molecules of the elements or compounds. These relative weights (masses) are called *molecular weights* or *gram-molecular weights* (abbreviated GMW). All that is needed to calculate gram-molecular weights is to know the number and kind of atoms in the molecule and the atomic weights of those atoms.

Illustrative Problem 13.3 A molecule of ammonia gas has the formula NH_3. (*a*) How much (mass) hydrogen gas is needed to unite with 7.38 g of nitrogen to form ammonia gas? (*b*) What is the mass of the ammonia gas?

Solution (***a***) The atomic weight of nitrogen is 14.0067 and hydrogen has an atomic weight of 1.00797 (from Appendix VIII). Thus, each hydrogen atom has a mass 1.00797/14.0067 that of an atom of nitrogen. But each molecule of ammonia gas has three atoms of hydrogen and one atom of nitrogen. Therefore the mass of hydrogen required to combine with 7.38 g of nitrogen to form the ammonia gas is

A problem on combining masses of gases.

$$\text{Mass of hydrogen} = 3\left(\frac{1.00797}{14.0067}\right) \times 7.38 \text{ g}$$

$$= 1.59 \text{ g} \qquad \textit{answer}$$

(***b***)

$$\text{Mass of ammonia gas} = 7.38 \text{ g} + 1.59 \text{ g}$$

$$= 8.97 \text{ g} \qquad \textit{answer}$$

Since the gram-molecular weights of pure substances are in the same ratio as the weights (masses) of their molecules we find, for example, that

$$\frac{2(1.00797) \text{ g hydrogen}}{\text{weight of 1 molecule hydrogen}} = \frac{2(15.9994) \text{ g oxygen}}{\text{weight of 1 molecule oxygen}} = N(\text{a constant})$$

The constant N is called *Avogadro's number*.

Avogadro's number = 6.023×10^{23}

Thus, Avogadro's number represents the number of molecules in one gram-molecular weight of a given substance. In the SI system Avogadro's number is 6.023×10^{26}, the number of molecules in one kilogram-mole (kilomole) of the substance.

13.8 ■ The Mole

One mole represents 1 g formula weight of a pure substance, whether element or compound.

A more convenient and more widely used term than gram-molecular weight of a substance is the mole. *A mole of any substance contains Avogadro's number of molecules of the substance*. Or, stated another way, the number of molecules in 1 mole of any substance *is Avogadro's number*. A mole of hydrogen gas (H_2) means 2(1.00797) g = 2.01594 g of hydrogen. A mole of oxygen gas (O_2) means 31.9988 g of it; and a mole of ammonia gas (NH_3) means (14.0067 + 3 × 1.00797) g = 17.03061 g of the gas. Under the same conditions of temperature and pressure equal mole quantities of gases occupy the same volume. One mole of any gas under standard conditions of temperature and pressure contains Avogadro's number of molecules and occupies a volume of approximately 22.4 L. This volume is called the *gram-molecular volume* (abbreviated GMV) of gases. Thus 2(1.00797) g of hydrogen at standard temperature and pressure conditions (STP) occupies a volume of 22.4 L.

The term "mole" has assumed a much wider meaning to chemists than merely its applications to gases. In the broad sense one mole is equal to the gram-molecular weight of any pure substance. The physical state of the substance, whether solid, liquid, or gas, does not affect the statement. One gram-molecular weight of any pure substance contains 6.023×10^{23} molecules and one kilogram-molecular weight of the substance contains 6.023×10^{26} molecules.

Illustrative Problem 13.4 The formula for ordinary table salt is NaCl. How many grams in a mole of NaCl?

Solution Each molecule of NaCl contains one atom of sodium (Na) and one atom of chlorine (Cl). From Appendix VIII, the atomic weights of the elements are:

$$\begin{aligned} \text{Na} &= 22.9898 \\ \text{Cl} &= \underline{35.453} \\ \text{NaCl} &= 58.4428 \end{aligned}$$

A mole equals the gram-molecular weight of NaCl = 58.443 g *answer*

Illustrative Problem 13.5 How many molecules are there in 74.8 L of carbon dioxide (CO_2) gas? Assume STP conditions.

Solution One mole of CO_2 gas contains $N = 6.023 \times 10^{23}$ molecules. Then, 22.4 L of CO_2 contains 6.023×10^{23} molecules, and 74.8 L CO_2 contains

$$\frac{74.8\ \text{L}}{22.4\ \text{L}} \times 6.023 \times 10^{23}\ \text{molecules} = 2.01 \times 10^{22}\ \text{molecules} \qquad \textit{answer}$$

Illustrative Problem 13.6 Use Avogadro's number to determine the mass of one molecule of ammonia (NH_3) gas.

Solution One molecule of ammonia gas is made up of one atom of nitrogen and three atoms of hydrogen. The gram-molecular weight of one molecule of NH_3 can be determined from the atomic weights given in Appendix VIII.

Finding the mass of one molecule of a gas.

$$\begin{aligned} \text{1 atom nitrogen} &= 14.0067\ \text{g} &&= 14.0067\ \text{g} \\ \text{3 atoms of hydrogen} &= 3(1.00797\ \text{g}) &&= \underline{3.0239\ \text{g}} \\ \text{Gram-molecular weight of NH}_3 &= && \ 17.0306\ \text{g} \end{aligned}$$

One mole of NH_3 gas has a mass of 17.0306 g and contains $N = 6.023 \times 10^{23}$ molecules. Then one molecule of NH_3 has a mass of

$$\begin{aligned} m &= \frac{17.0306\ \text{g}}{6.023 \times 10^{23}} = 2.83 \times 10^{-23}\ \text{g} \\ &= 2.83 \times 10^{-26}\ \text{kg} \qquad \textit{answer} \end{aligned}$$

13.9 ■ Vapors and Gases

A vapor is the gaseous form of any substance. The term *vapor*, however, is commonly used to describe only those gases which exist as liquids or solids at ordinary temperatures or pressures. Examples of vapors are water vapor, gasoline vapor, alcohol vapor, or naphthalene vapor. The ability of solids and liquids to change to a vapor

Table 13.1 Density and specific gravity of gases and vapors

Substance	Density: kg/m³ or g/L at 0°C and 760 mm	Density: lb/ft³ at 32°F and 1 atm	Specific Gravity: Air = 1.000	Specific Gravity: Oxygen = 1.000
Acetylene	1.173	0.0732	0.9073	0.8208
Air	1.293	0.0807	1.0000	0.9047
Ammonia	0.7710	0.0481	0.5963	0.5395
Carbon dioxide	1.977	0.1234	1.529	1.383
Carbon monoxide	1.250	0.0781	0.9671	0.8750
Freon-12	5.39	0.337	4.21	3.77
Helium	0.1785	0.0111	1.1380	0.1249
Hydrogen	0.0899	0.0056	0.0695	0.0629
Isobutane	2.673	0.1669	2.067	1.870
Methyl chloride	2.3043	0.1434	1.7824	1.6123
Neon	0.9003	0.0562	0.6964	0.6300
Oxygen	1.429	0.0892	1.105	1.000
Sulfur dioxide	2.020	0.1261	1.562	1.414

state is essential to our everyday comfort and convenience—indeed to our very existence. The evaporation of water to the vapor state makes possible the formation of clouds, rain, and snow. The operation of the steam engine depends upon this same property. Before gasoline can effectively be used in an engine, it must first be converted to the vapor state. Perfumes, liquefied petroleum gas (LPG), refrigerants, and mothballs are only a few examples of substances whose effectiveness depends upon their being changed to the vapor state.

Vapors play an important role in our daily comfort and in commerce and industry.

13.10 ■ Density of Gases and Vapors

Just as different solids and liquids vary in density, so do gases and vapors. In determining the specific gravities of gases and vapors, the weights of equal volumes of the gases or vapors are compared with that of either oxygen or air. Table 13.1 lists the densities and specific gravities of a few of the more common gases and vapors.

Specific gravity of a gas is its density relative to air (or oxygen).

$$\text{sp gr}_{\text{gas}} = \frac{\text{weight of a volume of gas}}{\text{weight of an equal volume of oxygen (or air)}} \tag{13.4}$$

13.11 ■ Industrial Gases

There are hundreds of gases used commercially throughout the world today. Each gas has its own peculiar chemical and physical properties. These properties make many of them suitable for specific industrial purposes. Some of the more common gases and a few of the uses to which they have been put by industry are listed in Table 13.2.

13.12 ■ The Atmosphere

All creatures on the surface of the earth live at the bottom of a deep ocean of air. The total force (equal to the weight of the air) that this atmosphere exerts on the earth is almost incomprehensible. We gain a notion of this tremendous force when we realize that the total force exerted by the atmosphere on an acre of the earth's surface at sea level is 46,100 tons. Each square meter of surface is acted upon by a force equal to the weight of a column of air 1 m² in cross section, with an altitude equal to the height of the atmosphere. This is roughly 1.013×10^5 N. The force on each square

The weight of air causes air pressure.

Air pressure exerts a force that is the same in all directions (Pascal's law).

Industrial gases, constitute an entire industry.

Table 13.2 Some industrial gases and their uses

Gas	*Use*
Acetylene	High-temperature welding and cutting steel
Ammonia	Refrigerant, preparation of fertilizers, soap, and glass
Argon	Filler for electric-light bulbs and signs (blue color)
Carbon dioxide	Carbonated beverages, fire extinguishers, baking powders
Chlorine	Bleaching agents, germicides, anesthetics, solvents, chemical-warfare agents
Coal gas	Fuels
Coke-oven gas	Fuels
Natural gas	Fuels
Water gas	Fuels
Producer gas	Fuels
Freon (12, 21, 22, etc.)	Refrigerants, aerosol propellant for spray cans
Helium	Lighter-than-air craft, low-temperature physics research
Hydrogen	Balloons, conversion of cottonseed oil and peanut oil into solid edible fats, welding, fuel for rockets (in liquid form)
Methyl chloride	Refrigerant
Nitrogen	Filler for incandescent lamps, component of rocket fuel (liquid)
Oxygen	Welding and cutting, component of rocket fuel (in liquid form)
Sulfur dioxide	Bleaching agent, preservative, preparation of sulfuric acid (which enters into the manufacture of a multitude of products), refrigerant
Water vapor	Steam, for engines and turbines, heating plants, etc.

inch of surface at sea level is roughly 14.7 lb. You may wonder why we are not conscious of these terrific forces or even crushed by them. Pascal's law offers a simple, adequate explanation. Air pressure is transmitted equally in all directions; consequently, though the external force on the human body due to air pressure is several tons, the body does not suffer, for an equal pressure on the inside of the body balances it.

The effect of atmospheric pressure can be effectively demonstrated by a simple experiment, diagramed in Fig. 13.2. Tie a thin sheet of rubber over the top end of a jar (Fig. 13.2*a*). Then seal (with grease) the mouth of the jar over a flat disk connected to a vacuum pump. As air is removed from the jar, the pressure inside the jar will be reduced. As a consequence, the outside (atmospheric) pressure will push the rubber down inside the jar as shown in Fig. 13.2*b*.

Houses with windows and doors closed in the path of a tornado frequently burst their walls outward. The pressure in the eye of a tornado is very low, and when the tornado passes by, the tremendous forces of atmospheric pressure within the house may easily push the walls out.

The breathing of humans and animals is an application of atmospheric pressure. A person inhales by reducing the pressure in the lung cavity, expanding it with the diaphragm muscles. The greater pressure of the outside air forces a fresh supply to flow into the lungs. Air is forced out of the lungs when the internal pressure is made greater than the pressure of the atmosphere.

Many animals use atmospheric pressure to aid them in drinking (''sucking'') liquids. By reducing the pressure in the mouth, water is forced into the mouth by the outside atmospheric pressure.

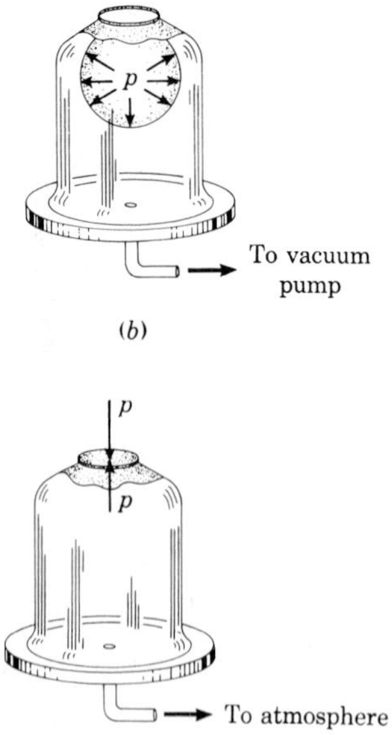

Fig. 13.2 Effect of atmospheric pressure on rubber diaphragm.

13.13 ■ Variations in Pressure of the Atmosphere

Because air has weight, the force of gravity of the earth on the atmosphere causes about half the air in the atmosphere to lie within 3.5 mi (5.6 km) of the surface of the earth. Since the atmospheric pressure at any point is a measure of the weight of the column of air of unit cross-sectional area above that point, it is evident that the pressure will vary with elevation. The greatest atmospheric pressure is found at the lowest levels. Atmospheric pressure on the top of Mt. Everest (8882 m or 29,141 ft) is only about 2.8×10^4 Pa or 4 lb/in.^{2}. At 30 mi, the pressure has been calculated to be only about $\frac{1}{6000}$ that at sea level. (See Fig. 13.3 for variation of atmospheric pressure with height.)

Atmospheric pressure varies with altitude.

13.14 ■ The Barometer

Instruments which are employed to measure atmospheric pressure are called *barometers*. The earliest and still the simplest barometer is the *mercury barometer*. Evangelista Torricelli (1608–1647) filled a long glass tube, closed at one end, with mercury so as to permit no air in the tube. Then, keeping the open end sealed with his finger, he inverted the tube and placed it with the open end immersed in a vessel containing mercury (Fig. 13.4*a*). When he removed his finger, the mercury fell in the column until a definite height of mercury remained. He found that no matter what the length of the tube, as long as it had a length greater than 30 in. (76 cm), the mercury always leveled off at the same height. Torricelli reasoned that the pressure of the atmosphere on the free surface of the mercury in the vessel was transmitted (*Pascal's law*) to the mercury in the tube, thereby counterbalancing the downward pressure due to the weight of the column of mercury. At sea level the height of the mercury sustained was about 76 cm (30 in.).

Measurement of atmospheric pressure.

If a similar experiment were conducted using water, one might expect (since mercury is about 13.6 times as heavy as water) the atmosphere at sea level to support a column of water 13.6 × 760 mm (30 in.), which is about 10.4 m (34 ft). Experiment reveals that this is the case. Pascal, when he read of Torricelli's experiment, reasoned that if the air were less dense at higher altitudes, the height of the mercury

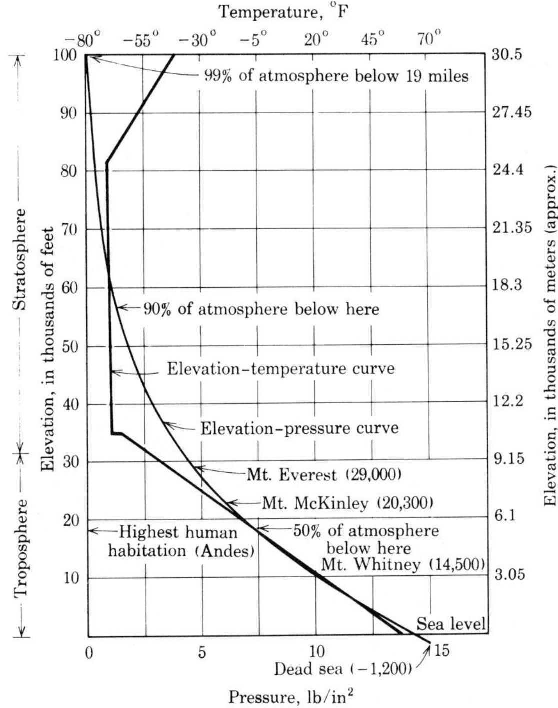

Fig. 13.3 Atmospheric temperatures, pressures, and densities diminish with increasing heights.

column supported on the top of a mountain should be appreciably less than it is at sea level. This theory was tested and proved by him.

The simple torricellian tube has been highly refined and perfected into the mercurial barometer (Fig. 13.4*b*). With it even the slightest changes in atmospheric pressure can be detected.

13.15 ■ Atmospheric Pressure

Atmospheric pressure is usually expressed in terms of millimeters of mercury (mm Hg) or inches of mercury (in. Hg), not in pascals or pounds per square inch. Pressure in millimeters or inches of mercury is read directly from the mercurial barometer, which usually has a vernier attachment for precision reading. The barometer represented in Fig. 13.4*b* is designed to measure pressures accurate to 0.01 in. Hg. Since a column of mercury 29.92 in. high with cross-sectional area of 1 in.^{2} weighs 14.7 lb, it follows that a pressure of 29.92 in. Hg is equivalent to 14.7 lb/in.^{2}.

Often a unit called the *bar* is employed for pressure. One *bar* is equivalent to 10^5 Pa. One *millibar* (mb) is equal to 10^2 Pa. The millibar is routinely used in meteorology. It can be shown that it is also equivalent to 100 N/m^2 (100 Pa).

Quite frequently—especially for large pressures—the unit used for pressure is the atmosphere (atm). The *atmosphere* represents the average atmospheric pressure at sea level on the earth.

$$\text{One atmosphere (atm)} = 29.92 \text{ in. Hg} = 760 \text{ mm Hg}$$

The unit mm Hg is so frequently used that it has been given the name *torr*, in honor of Torricelli: 1 atm = 760 torrs.

Here are several equivalents of the *atmosphere* (pressure):

$$\begin{aligned} 1 \text{ atm} &= 14.7 \text{ lb/in.}^2 \\ &= 1.013 \times 10^5 \text{ N/m}^2 \\ &= 1.013 \times 10^5 \text{ Pa} \\ &= 1013 \text{ millibars (mb)} \\ &= 760 \text{ mm Hg} \\ &= 760 \text{ torr} \\ &= 29.92 \text{ in. Hg} \\ &= 10.36 \text{ m of water} \\ &= 34.0 \text{ ft of water} \end{aligned}$$

Illustrative Problem 13.7 The barometer at the foot of a mountain reads 1006 mb. At the summit it reads 837 mb. What is the height of the mountain if the average density of the air between the two places is 0.081 lb/ft^3?

Solution

$$1013 \text{ mb} = 14.7 \text{ lb/in.}^2$$

$$P = hD \qquad \text{or} \qquad \Delta P = \Delta h \times D$$

A change Δh will produce a change ΔP. Solving for Δh,

$$\Delta h = \frac{\Delta P}{D} = \frac{(1006 - 837) \text{ mb}/1013 \text{ mb} \times 14.7 \text{ lb/in.}^2 \times 144 \text{ in.}^2/1 \text{ ft}^2}{0.081 \text{ lb/ft}^3}$$

$$= 4360 \text{ ft}$$ *answer*

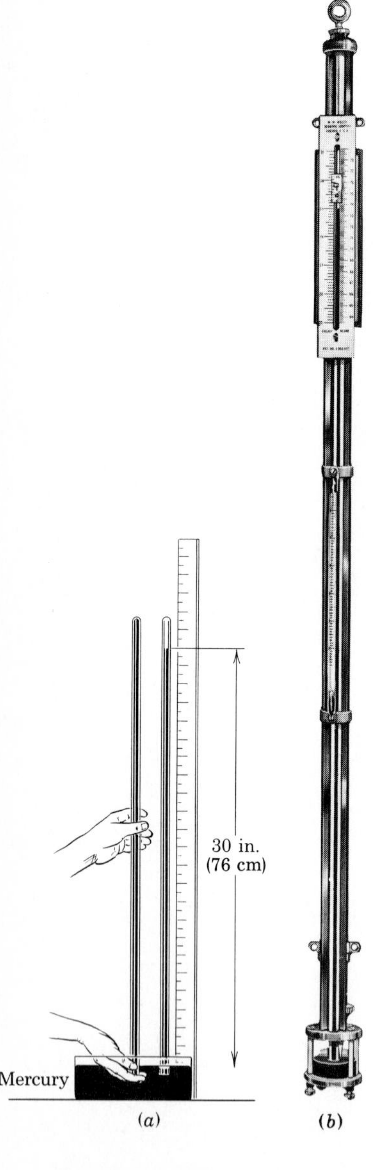

Fig. 13.4 (*a*) Torricelli's experiment for measuring atmospheric pressure. (*b*) Mercury barometer. (Central Scientific Co.)

13.16 ■ The Aneroid Barometer

The mercury barometer is inconvenient in many cases because of its length and the use of a free liquid. For purposes where a more rugged instrument is desired, the mercurial barometer is replaced by the *aneroid barometer* (Fig. 13.5). It consists principally of one or more small disk-shaped, accordion-pleated metal boxes (called *sylphons*) which have been partially evacuated and sealed. As the pressure of the air

Fig. 13.5 The barograph is an aneroid barometer linked to a pen which traces changes in atmospheric pressure on a rotating record sheet. (Courtesy Taylor Instrument Co.)

changes, the sylphon moves slightly up or down or sideways. This small motion is multiplied manifold by a delicate system of levers and is communicated to a pointer which moves over a scale to correspond to the readings of a mercury barometer or to be read directly in pounds per square inch. Large aneroid barometers have been made so sensitive that they will show the difference in atmospheric pressure between the floor level and the top of a table. Mercury barometers have been made which are even more sensitive than the most sensitive aneroid.

Household barometers are usually of the aneroid type.

13.17 ■ The Altimeter

Aviators use a form of aneroid barometer called the *altimeter,* in which the pointer moves along a scale which is calibrated to read directly in feet above sea level. Table 13.3 shows how the pressure of the atmosphere will change as one ascends above sea level.

The barometric altimeter has limited use, partly because atmospheric pressure is quite variable, and because the instrument does not take into account the terrain below.

Table 13.3 Standard atmospheric pressure at various heights above sea level

Altitude		*Pressure*			
Kilometers	*Miles*	*Millibars*	*mm Hg (torr)*	*in. Hg*	*lb/in.*^{2}
0	0	1013	760	29.9	14.7
2	1.2	795	596	23.5	11.5
4	2.5	616	462	18.2	8.9
6	3.7	472	354	13.9	6.8
8	5.0	356	267	10.5	5.2
10	6.2	264	198	7.8	3.8
12	7.5	193	145	5.7	2.8
14	8.7	141	106	4.2	2.0
16	9.9	103	77	3.0	1.5
18	11.2	75	56	2.2	1.1
20	12.4	55	41	1.6	0.8

The barometer is not entirely satisfactory as an altimeter, because even at a single elevation the pressure of the atmosphere changes continually because of changing weather conditions. It is imperative that the altimeter be constantly adjusted to the atmospheric pressure prevailing at the time of use. Also, the aviator is interested

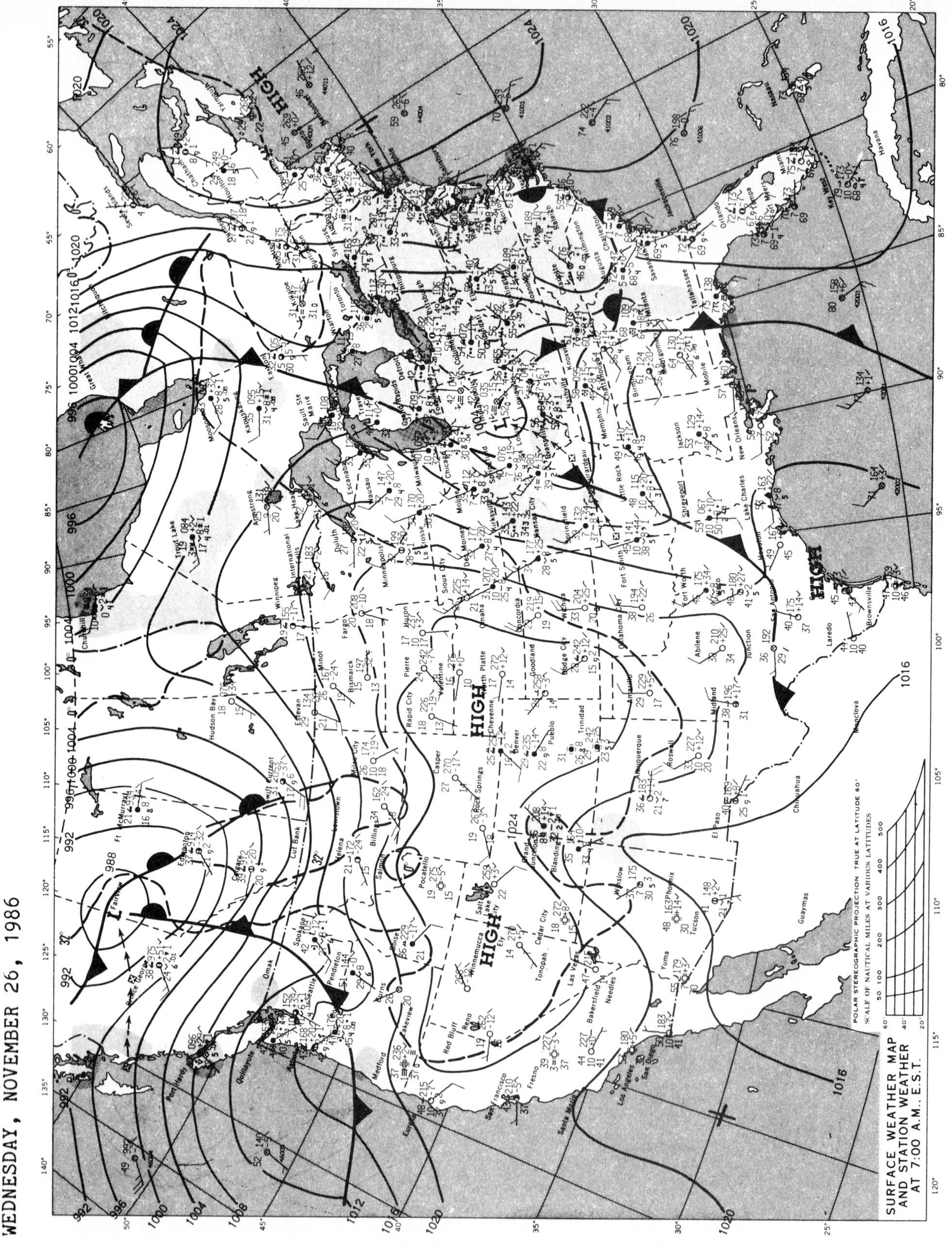

Fig. 13.6 A U.S. daily surface weather map. The map shows station data and the analysis for 7:00 a.m. EST. Tracks of well-defined low-pressure areas are indicated by a chain of arrows; locations of these at 6, 12, and 18 hours preceding maptime are indicated by small white crosses in black squares. (Courtesy NOAA/National Weather Service)

primarily in his height above the terrain below him, and not in his height above sea level. Modern altimeters, using radar or other electronic devices, measure height above terrain rather than height above sea level.

13.18 ■ Weather Forecasting

Although the fluctuating barometer pressure due to changing weather conditions makes a barometer unreliable as an altimeter, it nevertheless renders it a valuable instrument in predicting weather conditions. Experience shows that a falling barometer often indicates an approaching storm, while a rising barometer usually indicates fair weather. These changes are a fair guide to the amateur forecaster in predicting weather conditions for short periods of time. However, the science of weather forecasting is a very complex one, and barometer readings alone cannot be relied on for accurate weather forecasting.

The United States government has established the *National Weather Service*, whose function it is to give weather reports and forecasts to all those concerned. Weather stations have been set up at strategic locations throughout the nation and at many places throughout the world. Weather satellites keep track of major storm movements. Simultaneous instrument readings at the many different places are radioed in to central stations, or obtained from orbiting satellites. Weather maps are prepared and distributed by video recorders and television. Wind velocities, temperatures, and relative humidities all play an important part in these forecasts. Figure 13.6 shows a typical weather map prepared by the National Weather Service. On the map can be found the low-pressure and high-pressure areas. The direction and velocities of winds are usually included. Generally speaking, the areas of low barometric pressure are the storm centers. These usually move in an easterly direction across the United States. Lines joining places of equal barometric pressure (expressed in millibars) on a weather map are called *isobars*.

The science of weather forecasting is complex. Weather satellites and weather maps aid in forecasts.

Illustrative Problem 13.8 The atmospheric pressure suddenly dropped 135 mb as a tornado passed by a house. If the house was airtight (all windows, doors, etc., tightly shut), what additional (outward) force would this cause on a 30 in. × 30 in. window?

Solution

$$\Delta F = \Delta PA \qquad (12.1')$$

$$= 135 \text{ mb} \times \frac{14.7 \text{ lb/in.}^2}{1013 \text{ mb}} \times (30 \text{ in.})^2$$

$$= 1760 \text{ lb} \qquad \textit{answer}$$

In all likelihood the window would be shattered. It is clearly not wise to seal a building when a large sudden drop in air pressure is imminent.

13.19 ■ The Bourdon Gauge

Various types of gauges are used in industry for measuring gas and steam pressure. One of the most common type is called the *Bourdon gauge*. This device (Fig. 13.7) consists essentially of a hollow bronze tube G bent into a nearly complete ring and closed at one end B. When fluids enter the tube under pressure, the ring tends to straighten out. This motion at B is transmitted by a linkage D to a rack R (which is pivoted at C). The rack actuates a pinion attached to shaft E, to which a pointer is attached. By proper calibration, Bourdon gauges can be made to read pressures in pounds per square inch (psi), newtons per square meter (Pa), in. Hg, or mm Hg (torrs).

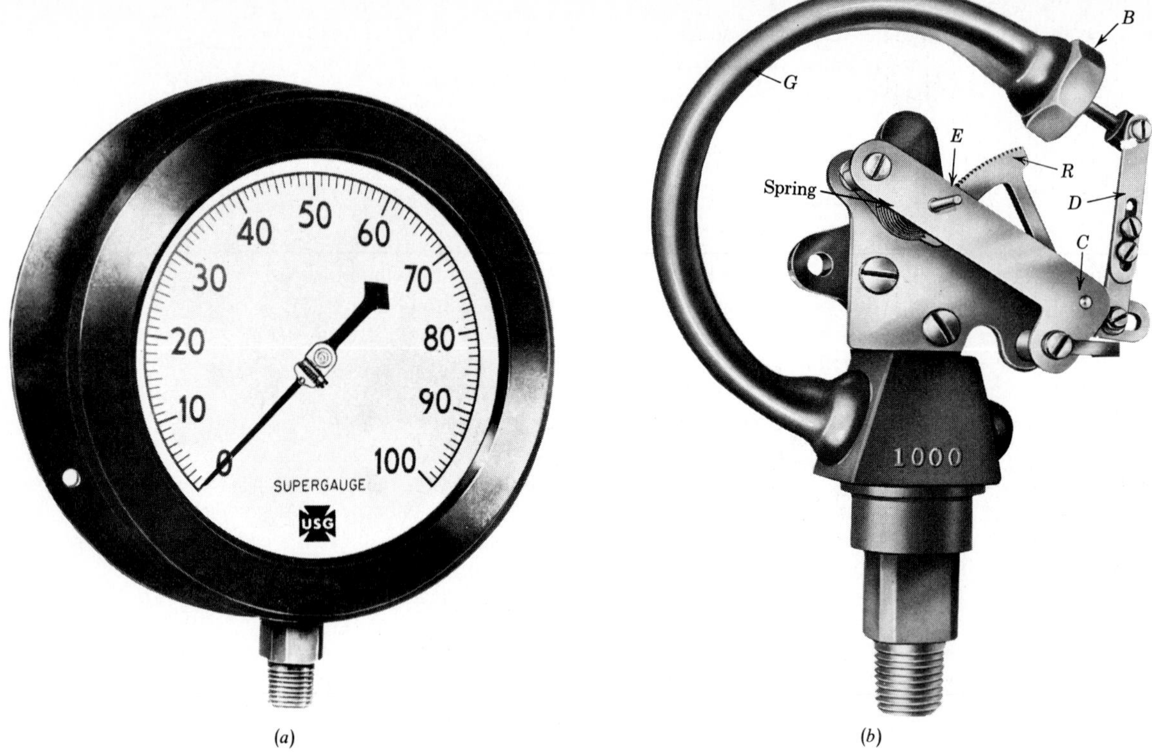

Fig. 13.7 (*a*) Bourdon gauge. (*b*) Pressure straightens out the curved tube, actuating a linkage which moves the pointer along a calibrated scale. (United Gauge Division, American Machine and Metals, Inc.)

13.20 ■ Gauge Pressure and Absolute Pressure

Gauge pressures cannot be used with Boyle's law. Atmospheric pressure must be added to the gauge pressure to get the absolute pressure.

In practice we may be interested in two kinds of pressure readings. We may want to know the *difference* between the actual pressure being measured and the pressure of the atmosphere. This pressure is called *gauge pressure*. Or we may wish to know the actual *total pressure* on the confined gas, including that of the atmosphere on the outside. This pressure is called the *absolute pressure*.

In predicting the behavior of gases under different conditions of temperature and pressure, it is essential to use *absolute pressures*. However, in most practical applications the pressure desired is the gauge pressure. The following relationship is used to determine either pressure if the other is known.

$$\text{Absolute pressure} = \text{gauge pressure} + \text{atmospheric pressure}$$
$$P_{abs} = P_g + P_{atm} \tag{13.5}$$

13.21 ■ Vacuum Gauge

When pressures less than 1 atm are to be measured, the instrument is calibrated to read from 0 to 14.7 lb/in.^{2} (1.013×10^5 Pa) or from 0 to 30 in. (76 cm) Hg, and the scale reading is interpreted as the degree of vacuum. A reading of 14.7 lb/in.^{2} or 30 in. (76 cm) Hg on such a gauge would indicate a perfect vacuum, whereas a reading of zero on either scale would indicate (normal) atmospheric pressure. Vacuum gauges find frequent application in the testing of reciprocating engines, vacuum pumps and vacuum systems, and refrigeration compressors and connecting lines.

13.22 ■ The Manometer Gauge

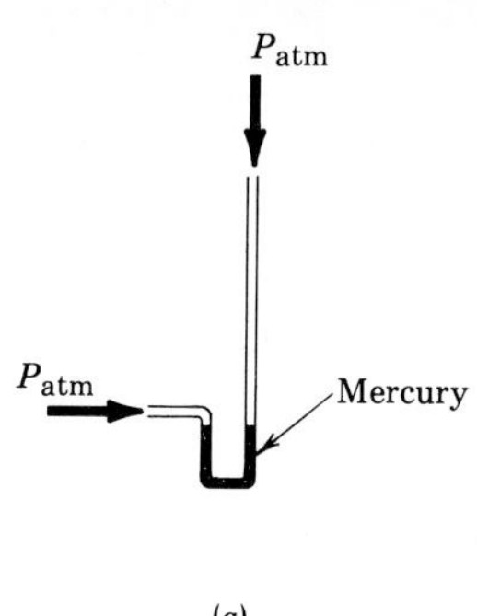

(a)

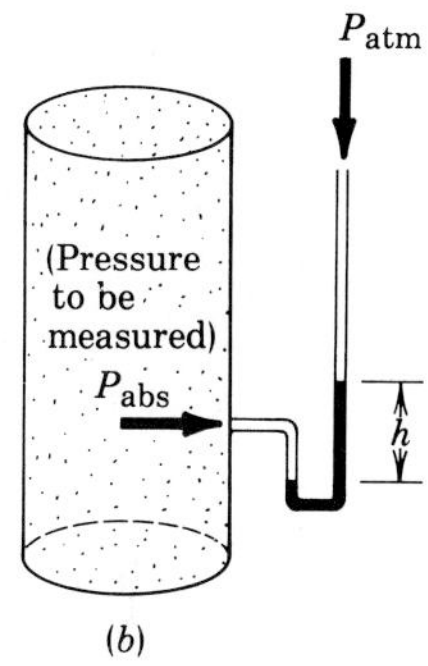

(b)

Fig. 13.8 The open-tube manometer (not drawn to scale).

Another common device for measuring the gauge pressure of a confined body of gas is the *open-tube manometer* (Fig. 13.8). The manometer is a U-shaped tube and contains a liquid of known density, usually mercury, but some contain water or oil. When both ends of the tube are open to the atmosphere (Fig. 13.8*a*), the level of the mercury will be the same in both arms of the manometer. (Why?)

In practice, one end of the manometer is connected to the container in which the gas pressure is to be measured; the other is open to the atmosphere (Fig. 13.8*b*). The mercury will rise in the open tube end until the pressures are equalized. The difference h between the two levels of mercury (or whatever fluid is in the tube) is a measure of the difference between the pressure in the chamber and the atmospheric pressure at the open end. Thus, the absolute pressure in the container,

$$P_{abs} = P_{atm} + hD_{mercury} \quad (13.6)$$

$$= P_{atm} + h\rho_{mercury}g \quad (13.6')$$

and

$$\text{Gauge pressure} = hD = h\rho g \quad (13.7)$$

Illustrative Problem 13.9 An open-tube mercury manometer is used to measure the pressure inside a pressurized tank of gas. The difference in the mercury levels is 32 cm. What is the absolute pressure (in pascals) of the confined gas in the tank?

Solution The gauge pressure is 32 cm of mercury. The atmospheric pressure is 76 cm of mercury. Hence, the absolute pressure (Eq. 13.5) = 32 cm + 76 cm = 108 cm Hg. The pressure exerted by a column of mercury 108 cm high is equal to the pressure of the confined gas.

$$\begin{aligned} P_{abs} &= \rho g h \\ &= 13600 \text{ kg/m}^3 \, (9.81 \text{ m/s}^2) \\ &\quad \times (108 \text{ cm})(1 \text{ m}/100 \text{ cm}) \\ &= 1.44 \times 10^5 \frac{\text{kg}\cdot\text{m}}{\text{m}^2\cdot\text{s}^2}\left(\frac{1 \text{ N}}{1 \text{ kg}\cdot\text{m/s}^2}\right) \\ &= 1.44 \times 10^5 \text{ N/m}^2 \text{ (Pa)} \qquad \textit{answer} \end{aligned}$$

The open-tube manometer is inconvenient for accurate measurements because barometric pressures are not constant and must therefore be measured frequently. The closed-tube manometer (Fig. 13.9) eliminates this problem by sealing one arm of the tube and highly evacuating the space above the mercury in that arm. With this device, the difference in the mercury columns is supported entirely by the pressure of the gas confined in the tank. Thus, the closed-tube manometer is used to measure directly the total (*absolute*) pressure of the gas. The pressure in pascals would be calculated from $P = h\rho g$.

The closed-tube manometer measures directly the total pressure of the gas.

Illustrative Problem 13.10 A closed-tube manometer is used to measure the pressure inside the tank of gas in Fig. 13.9. The difference in the mercury levels is 32 cm. What is the gas pressure?

Solution The difference in the mercury levels is a direct measure of the pressure inside the tank, since the pressurized gas from the tank is the only factor contributing to the difference. Thus,

$$\begin{aligned} P &= \rho g h \\ &= 13600 \text{ kg/m}^3 \, (9.81 \text{ m/s}^2) \\ &\quad \times (32 \text{ cm})(1 \text{ m}/100 \text{ cm}) \\ &= 4.27 \times 10^4 \text{ N/m}^2 \\ &= 4.27 \times 10^4 \text{ Pa} \qquad \textit{answer} \end{aligned}$$

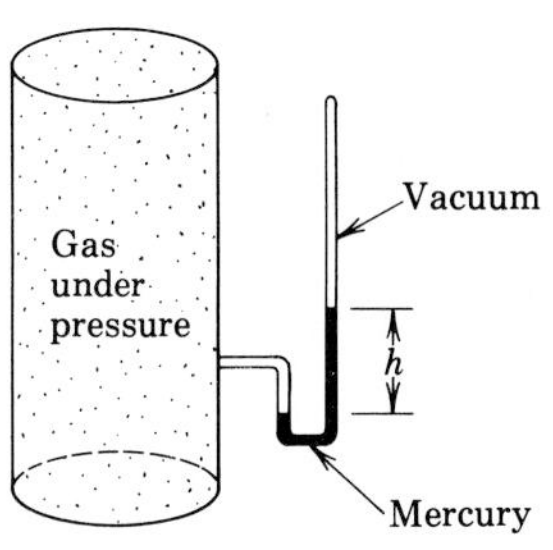

Fig. 13.9 The closed-tube manometer (not drawn to scale).

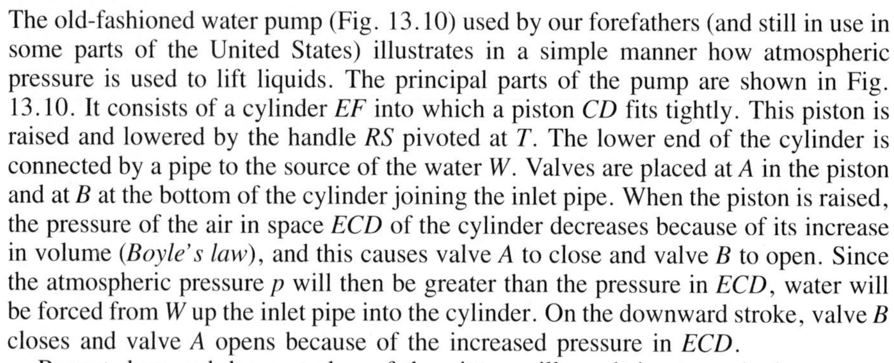

DEVICES OPERATED BY ATMOSPHERIC PRESSURE

13.23 ■ The Lift Pump

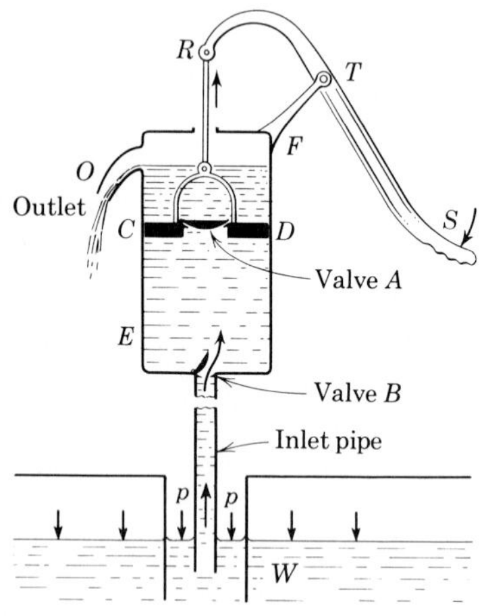

Fig. 13.10 The lift pump.

The old-fashioned water pump (Fig. 13.10) used by our forefathers (and still in use in some parts of the United States) illustrates in a simple manner how atmospheric pressure is used to lift liquids. The principal parts of the pump are shown in Fig. 13.10. It consists of a cylinder *EF* into which a piston *CD* fits tightly. This piston is raised and lowered by the handle *RS* pivoted at *T*. The lower end of the cylinder is connected by a pipe to the source of the water *W*. Valves are placed at *A* in the piston and at *B* at the bottom of the cylinder joining the inlet pipe. When the piston is raised, the pressure of the air in space *ECD* of the cylinder decreases because of its increase in volume (*Boyle's law*), and this causes valve *A* to close and valve *B* to open. Since the atmospheric pressure *p* will then be greater than the pressure in *ECD*, water will be forced from *W* up the inlet pipe into the cylinder. On the downward stroke, valve *B* closes and valve *A* opens because of the increased pressure in *ECD*.

Repeated up-and-down strokes of the piston will result in atmospheric pressure forcing the water up until it passes through valve *A*. The water above *CD* is then lifted and flows out the spout *O* of the pump. But as the water above *CD* is lifted, more water is forced by atmospheric pressure from *W* into *ECD*. And so the process continues. It should be apparent that since the atmospheric pressure at sea level is approximately 34 ft of water, the most perfect pump of this type could not lift water more than 34 ft from the water level in the well. Usually the maximum height lifted is about 28 ft, since most such pumps do not attain a high degree of vacuum in *ECD*. However, liquids can be lifted many hundreds of feet *above the piston*, as long as the piston is not more than about 28 ft above the water level in the well or cistern.

13.24 ■ The Siphon

At first glance the siphon seems to be a device which defies gravity by causing water to "run uphill."

The siphon is commonly used in irrigating row crops.

The siphon is an interesting application of the laws of physics, in which water can be made to "run uphill," utilizing the force of atmospheric pressure. It is a bent tube of rubber, glass, metal, or plastic which is used to carry water from one level over a small elevation to a lower level. To operate the siphon, one has only to fill the tube full of water, or other liquid, and then, holding closed both ends of the tube, lower it until one end is immersed in the vessel from which the liquid is to be drained and the other end is at a lower elevation. On opening the ends, liquid will flow continuously from the higher to the lower elevation. The explanation of the siphon is quite simple. The atmospheric pressures at levels *A* and *D* (Fig. 13.11*a*) are essentially the same. Thus the atmosphere attempts to push water up the two arms with equal force. However, the downward force of the column of water in the shorter column at *A*, because of its weight, is less than that of the longer column of water at *D*; thus a greater resultant upward force acts on the shorter side *A*. Hence the water will move from the point of the greater force to that of the lesser—from *A* across *C* and down to *D* in the diagram.

The siphon will cease to function (1) when the elevation of *D* comes up to the elevation of *A* or (2) when the water level at *A* falls below the tube opening on the shorter side. It should be pointed out that the siphon will not work at all if the difference in elevation between *A* and *C* exceeds 34 ft. For practical purposes, siphons are quite ineffective if the difference in elevation is more than about 27 ft.

Illustrative Problem 13.11 The 2-in. diameter siphons shown in Fig. 13.11*b* siphon water from a ditch and deliver it to furrows. The free end of the siphon hangs in air at a level 18 in. below the level of water in the ditch. What is the water flow, in ft^3/min, from each siphon, delivered to the ditch?

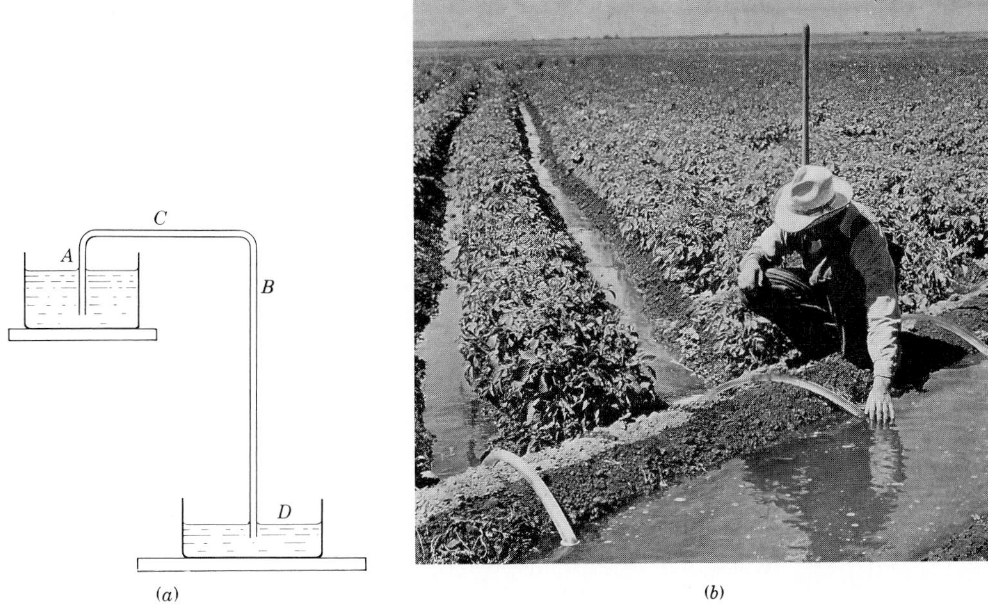

Fig. 13.11 (*a*) The siphon. (*b*) Irrigating by siphoning water from an open ditch. This method is preferred by many farmers over the more permanent "pipeline-and-stand" method. It gives a more uniform rate of flow into every crop row. It is cheaper to install, and the control of weeds near the ditch is no problem. (U.S. Department of Agriculture)

Solution Use Bernoulli's theorem. Let subscript d represent conditions at the water surface in the ditch and f represent the conditions at the free end of the siphon.

$$P_d + \tfrac{1}{2}\rho v_d^2 + h_d\rho g = P_f + \tfrac{1}{2}\rho v_f^2 + h_f\rho g \qquad (12.16)$$

$$P_d = P_f = \text{the atmospheric pressure}$$

$$v_d = 0$$

Then,

$$\tfrac{1}{2}\rho v_f^2 = h_f\rho g - h_d\rho g$$

$$v_f^2 = 2g(h_f - h_d)$$

$$= 2(32.2\ \text{ft/s}^2)(18/12\ \text{ft})$$

$$= 9.8\ \text{ft/s}$$

$$Q = v_f A_f$$

$$= 9.8\ \text{ft/s} \times \pi(1/12\ \text{ft})^2 \times 60\ \text{s/min}$$

$$= 13\ \text{ft}^3/\text{min} \qquad \textit{answer}$$

13.25 ■ Applications of Bernoulli's Principle to Gases

Bernoulli's principle (see Sec. 12.15) applies to all kinds of fluid motion. There are many interesting examples of its application associated with gases of which we shall cite a few.

Sports enthusiasts have for years argued pro and con on the question of whether a baseball pitcher can really throw a curve ball. Tests with research laboratory equipment and high-speed photography show that the ball really does curve a great deal. Anyone who is a beginner at golf knows for sure that his "slice" or "hook" is not imaginary. Bernoulli's principle contributes to our understanding of the forces which cause a ball to curve.

Figure 13.12 shows a spinning ball thrown from left to right. This situation is equivalent to the ball standing still, but spinning, while the air moves from *right to left*. The spinning ball drags some air along with it, causing the air on one side of the ball to rush past the ball faster than the air on the other side. As a consequence, the

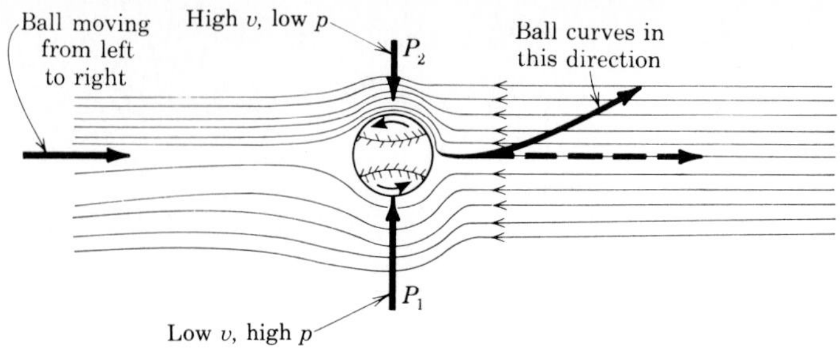

Fig. 13.12 A pitched spinning baseball travels in a curved path.

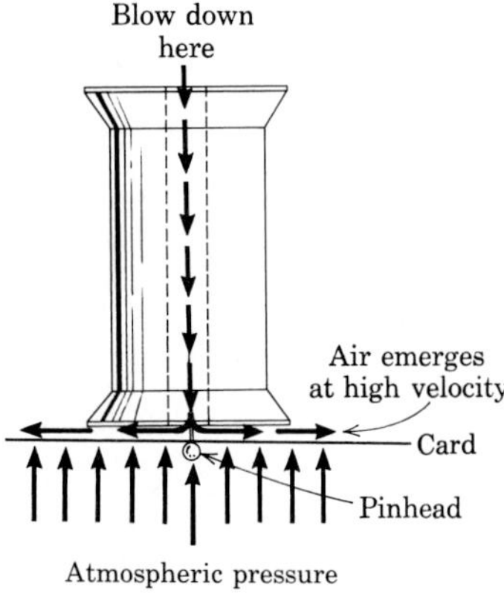

Fig. 13.13 Living room demonstration of Bernoulli's principle. Increasing the velocity of the air blown in the spool will increase the pressure holding the card to the spool.

pressure P_1 is greater than the pressure P_2. To produce a downward curve (a ''drop ball'' or ''sinker''), the ball is given ''overspin'' around a horizontal axis. The ''in'' or ''out'' curve ball is achieved by spinning the ball on a vertical axis. Roughing the surface of the ball increases the effectiveness of the ''curve.'' (Why?)

Put a pin through a light card and then place a sewing-thread spool over the pin as shown in Fig. 13.13. Next blow through the spool and try to blow the card off. The harder you blow, the tighter the card will be pressed to the spool by atmospheric pressure. Can you explain why the card will not fall from the spool?

When a moving airfoil is inclined at a slight angle with the horizontal, air will be deflected from the lower surface to create a reacting force which will tend to lift the foil. At the same time air rushes over the longer upper wing surface at a higher velocity than over the shorter (lower) wing surface. Thus (Bernoulli's principle) the pressure below the foil is greater than that above the foil (Fig. 13.14*a*). This difference in pressure provides about 85 percent of the lifting force which causes an airplane to rise and maintain altitude in flight. When the plane flies with constant speed, the lift L, the weight $\mathbf{w}$ of the plane, the drag D due to air friction, and the thrust T due to the engine are in equilibrium. Vectors representing the forces form a closed polygon, as shown in Fig. 13.14*b*. Thus, $\mathbf{L} + \mathbf{w} + \mathbf{D} + \mathbf{T} = 0$, as a vector equation.

The action of a hand held garden sprayer (Fig. 13.15) or a paint gun depends upon Bernoulli's principle. In the paint gun a stream of high-velocity air passes over the top of a vertical tube dipped in the paint in a jar. The high-velocity air creates a lower pressure at the top of the tube than the pressure at the surface of the paint in the jar. The unbalanced pressure causes a force which pushes the paint up the tube. As the liquid spills over the top of the tube, the airstream blows it out of the gun in a fine misty (''atomized'') spray.

Gasoline undergoes atomizer action in the carburetor of an automobile and is quickly changed to a vapor state on mixing with the air before reaching the cylinders of the engine.

Many modern auto engines use *fuel injectors,* however, rather than carburetors.

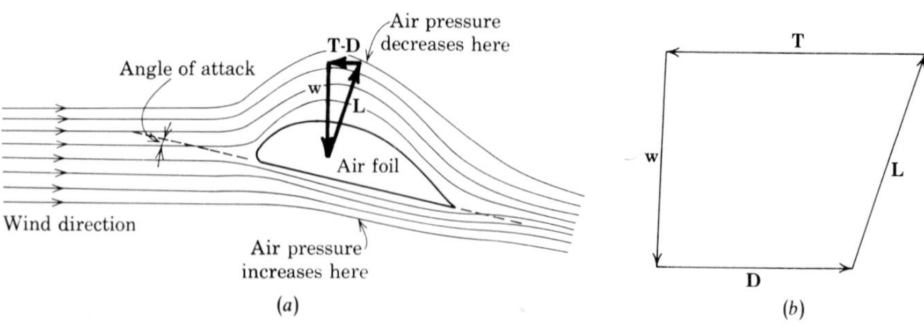

Fig. 13.14 (*a*) Lift produced on an airfoil is an example of Bernoulli's principle. (*b*) When the airfoil moves with constant velocity, $\mathbf{L} + \mathbf{w} + \mathbf{D} + \mathbf{T} = 0$ as a vector equation.

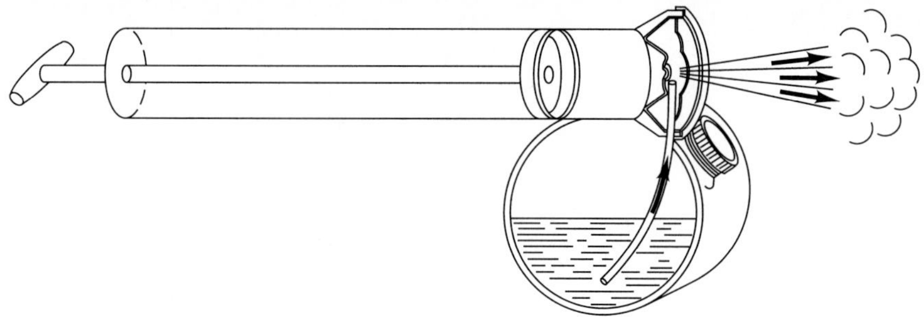

Fig. 13.15 Cross-sectional view of a hand garden sprayer.

QUESTIONS AND EXERCISES

1. What is the formula for the steam molecule?

2. What is meant by the atomic weight of an element?

3. What is meant by *molecular weight*? How are molecular weights obtained from atomic weights?

4. Why are gases more easily compressed than liquids?

5. On the basis of the kinetic theory, describe what happens when one gas mixes with another.

6. Since the molecules of different gases vary in size, how is it possible for equal volumes of different gases, under similar conditions, to contain the same number of molecules?

7. How is it possible to measure gas pressures in inches and millimeters, which are units of length, not pressure?

8. Explain why a dental plate of false teeth clings to the roof of the mouth.

9. Is it possible to change both the temperature and the pressure of a given mass of gas without changing its volume? Explain.

10. Why would a siphon fail to operate over a hill more than 34 ft high?

11. Two objects of different volumes have the same apparent weight when submerged in water. Compare their weights when weighed in air. Give reasons for your answer.

12. Discuss the advantages and disadvantages of the closed-type and the open-type manometers.

13. Will smoke from a fireplace rise faster or slower in the flue when a breeze is blowing across the top of the chimney? Give reason for your answer.

14. Why is it wise to keep the windows of a building open during a tornado?

15. A balloon with helium will rise. Why? Will the balloon continually rise or is there a limit to its height? If there is a limit what determines how high the balloon will rise?

16. Explain how in Fig. 13.11*b* water can be made to flow ''uphill'' during part of its flow. Why doesn't the liquid in each side flow back down the tube?

17. How does the density of a gas vary with pressure exerted on it?

18. A lead ball and an iron ball with the same diameter are dropped from the top of a building. Which will strike the ground first?

19. Why is a rock easier to lift under water than in air?

20. How is the ascent and descent of a submarine controlled?

PROBLEMS

Group One

1. In 2500 molecules of sulfuric acid (formula, H_2SO_4) how many atoms of (*a*) hydrogen, (*b*) sulfur, and (*c*) oxygen are there?

2. Convert a barometric pressure of 720 mm Hg to an equivalent pressure in Psi units.

3. Find the greatest theoretical height that water can be raised over a siphon in the mountains where the barometer reads 22 in. Hg.

4. If a cylindrical can 8 cm in diameter and 10 cm tall were completely evacuated, what force would the atmosphere exert on each end of the can?

5. What is the atmospheric pressure in lb/in.^{2} if the mercury barometer stands at 74.5 cm?

6. Express normal atmospheric pressure in kg/cm^2 units.

7. What is the mass of one molecule of sugar if its formula is $C_{12}H_{22}O_{11}$?

8. The system shown in Fig. 13.16 is filled with oil, closed by a piston P, and open to the atmosphere in pipes A and B. The diameter of pipe A = one-half the diameter of pipe B. The piston P is lowered 2 in. Compare how much the oil will rise in the two pipes?

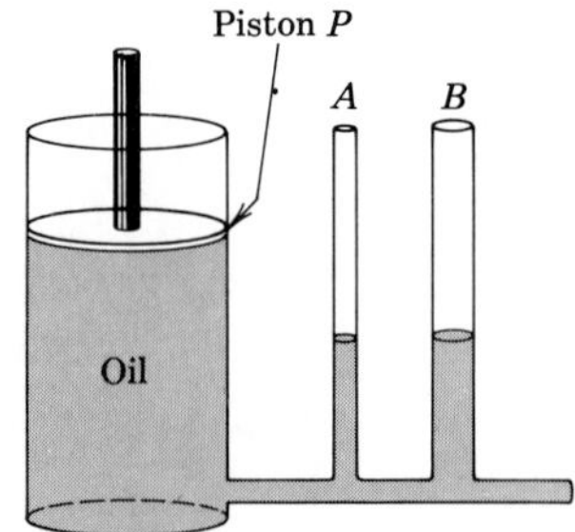

Fig. 13.16 Problem 8.

9. Two manometers are used to measure gas pressures P_A and P_B (Fig. 13.17). h_A is greater than h_B. (*a*) Compare P_A and P_B. (*b*) What is the gauge pressure at A? (*c*) What is the absolute pressure at B?

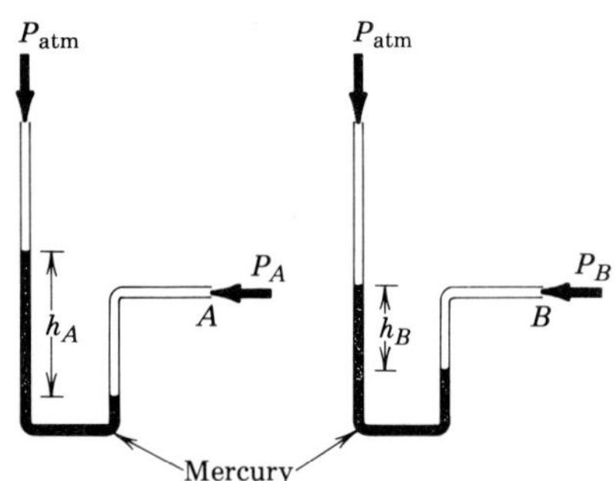

Fig. 13.17 Problem 9.

10. Describe the situation which would cause a closed-tube manometer to appear as shown in the drawing of Fig. 13.18.

11. If the gauge pressure acting on 2 ft^3 of a gas increases from 15 lb/in.^{2} to 30 lb/in.^{2} what happens to the volume? (Give approximate answer in ft^3.)

12. Five liters of a gas at an absolute pressure of 760 torrs is compressed to 1.0 liter. If the temperature remains the same, what is the new pressure?

13. When hydrogen (H) burns in the presence of oxygen (O) water (HOH) is formed. If 4 g of hydrogen is burned

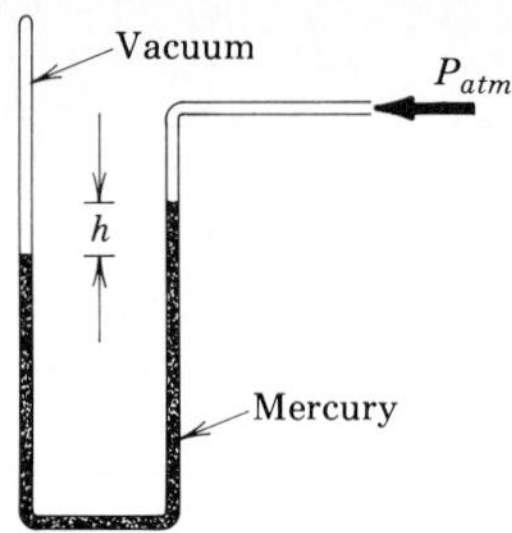

Fig. 13.18 Problem 10.

in a vessel containing 34 g of oxygen, how much oxygen is left over (does not combine chemically with the hydrogen)?

14. An open-tube, a closed-tube, and a mercury barometer are placed side by side in a room (see Fig. 13.19). (*a*) What is the absolute pressure in flask A? (*b*) in flask B?

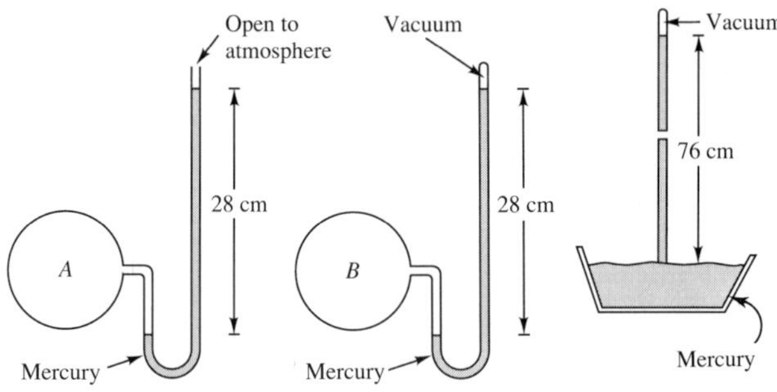

Fig. 13.19 Problem 14.

Group Two

15. What is the mass of an atom of oxygen?

16. Seven hundred eighty-five ml of a gas at STP conditions weigh 1.308 g. What is the molecular weight of the gas?

17. A woman can exert a force of 200 N on the handle of an air pump. To what absolute pressure can she pump an automobile tire if she uses a single barrel pump (without levers), and the diameter of the pump piston is 6.0 cm?

18. How many atoms are there in 60 g of copper?

19. How many molecules are there in 5 L of ammonia gas (formula, NH_3)? Use STP conditions.

20. How many molecules are there in 5 lb of lead sulfate? The formula for lead sulfate is $PbSO_4$.

21. The formula for ammonium sulfate is $(NH_4)_2SO_4$. How many molecules are there in 2.50 lb of ammonium sulfate?

22. In the closed-tube manometer illustrated in Fig. 13.20, $h = 4.5$ in. Find P_{abs} in lb/in.^{2}

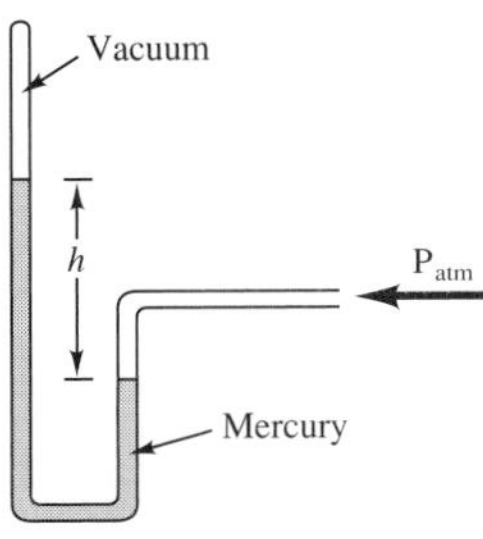

Fig. 13.20 Problem 22.

23. In the open-tube manometer of Fig. 13.21, $h = 25$ mm, $P_{atm} = 760$ mm. Find P_{abs} in pascals.

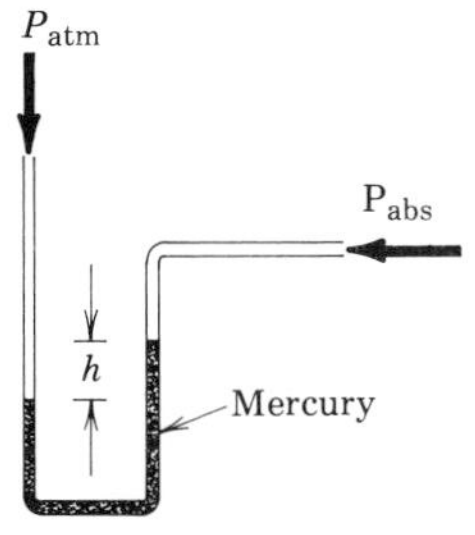

Fig. 13.21 Problem 23.

24. In Fig. 13.11*a*, the difference in the elevations of *A* and *D* is 12 cm and the tube has a cross-sectional area of 1.4 cm^2. Find (*a*) the velocity of efflux and (*b*) the quantity of water (in cubic meters per minute) transferred at *D*.

25. An open-tube manometer is used to measure the pressure of a tank of gas. The meter reads 37 cm Hg. Atmospheric pressure is 74 cm Hg. What is the absolute gas pressure (*a*) in centimeters of mercury; (*b*) in pounds per square inch; (*c*) in pascals?

26. A closed-tube manometer is used to measure the pressure of a gas system. Find the gas pressure (*a*) in pounds per square inch and (*b*) in newtons per square meter given that the difference of the mercury levels in the two arms is 12.5 cm.

27. Generally, when making a rapid descent in an airplane whose cabin is not pressurized, the human ear will "pop" as the air pressure behind the ear drum is equalized to the outside pressure. What would be the approximate force on the ear drum (area 7.8×10^{-2}in.^{2}) if this did not happen and the plane suddenly dropped from a 10,000 ft elevation to sea level? (Use Fig. 13.3.)

28. A blower-type fan delivers air at the rate of 2500 ft^3/min against a static pressure of 2.2 in. of water. How much air horsepower is being delivered by the fan if air has a density of 0.0807 lb/ft^3?

29. A compressed air tank has a volume of 3.75 ft^3 and holds air at a gauge pressure of 60.0 lb/in.^{2}. How many cubic feet of air can be drawn from the tank at normal atmospheric pressure? Assume no temperature change.

30. A helium balloon, its trappings, and payload weigh 1540 lb. How large in volume must the balloon be if it is to lift the total load of 1540 lb from a sea-level location? The weight density of helium is 0.0111 lb/ft^3. Note: The 1540 lb does not include the weight of the helium. Assume STP conditions in the balloon.

31. The flat roof of a house is 30 ft long and 24 ft wide. It weighs 8.25 tons. On hearing of the approach of a tornado, the owner of the house closed all the windows and doors of the house. The air pressure in the house remained at 1013 mb. When the tornado hit the house the outside pressure dropped to 975 mb. What was the net upward force (in tons) on the roof caused by this drop in pressure?

32. The density of propane gas is 0.1254 lb/ft^3 at 32°F and 1 atm pressure. A 2.500 ft^3 tank contains 5 lb of propane gas at 32°F. Calculate the gauge pressure of the gas within the tank.

33. A block of cork has a specific gravity of 0.24. By what percentage will the apparent weight of the block increase when placed in a vacuum?

34. When sugar is sufficiently heated to char it, water (H_2O) is given off and only carbon is left. The formula for sugar is $C_{12}H_{22}O_{11}$. If 4.00 kg of sugar is charred: (*a*) What is the mass of carbon left? (*b*) What mass of water is given off?

35. The standard tank of oxygen used in industry has compressed into it gas which would occupy 244 ft^3 at atmospheric pressure. The compressed gas has a gauge pressure (70°F) of 2200 lb/in.^{2}. Determine the inside volume of the tank.

36. The 150 gallon pressure tank of a home watersupply system is full of air at atmospheric pressure. 100 gal of water is then forced in, trapping and compressing all the air. What is the resultant water pressure delivered from the tank?

37. An air bubble released at the bottom of a lake will expand to three times its original volume by the time it reaches the surface of the lake. If atmospheric pressure is 76.0 cm Hg, how deep is the lake? (Assume no temperature change.)

38. A compressed air tank has a capacity of 4.1 ft^3 and is filled with air at normal atmospheric pressure. How many additional cubic feet of atmospheric air must be pumped into the tank in order to raise the gauge pressure to 90 lb/in.^{2}?

39. A caisson used in constructing a tunnel under a river is lowered to the bed of the river 50 ft under water. Atmospheric pressure is 14.7 lb/in.^{2}, and the density of air is 0.0805 lb/ft^3. (*a*) What is the pressure (due to water pressure head plus atmospheric pressure) at the bottom of the river? (*b*) When air is pumped to workers in the caisson at the bottom of the river, to what fraction of the original volume has the air been reduced? (*c*) What is the density of the air within the caisson? (Assume no temperature change.)

40. The volume of a certain automobile tire is 1600 in.^{3}. How much air in cubic inches at standard pressure must be forced in to raise the gauge pressure from 24 to 32 lb/in.^{2}? (Assume no temperature change.)

Group Three

41. An open U-tube (Fig. 13.22) is partially filled with mercury. Then another liquid of unknown density is poured in the left arm until the levels of the free liquid surfaces are stabilized as shown. Find the density of the unknown liquid.

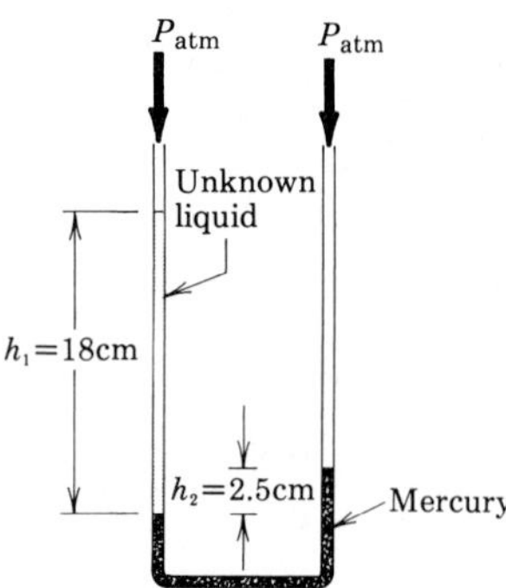

Fig. 13.22 Problem 41.

42. An auto has a mass of 1930 kg and is supported by four identical tires. Each tire is filled with air at a gauge pressure of 1.655×10^5 Pa. Find the theoretical area of contact with the ground (in square centimeters) for each tire. Neglect the tensile stresses within the fabric of the tires.

43. A gas balloon, its trappings, and gondola weigh 3000 N. It is filled with 1500 m^3 of helium at atmospheric pressure. What will be its maximum possible payload near the earth's surface?

44. "Bottled gas" (butane) in a steel tank is under a gauge pressure of 125 lb/in.^{2}. After some of the gas has been used, the gauge pressure drops to 50 lb/in.^{2}. What fractional part of the original amount of gas still remains in the tank?

45. A cylindrical diving bell, open at the bottom and filled with air at 1 atm pressure, has a diameter of 1 m and a height of 3 m. How high will the water rise in the bell when the bottom of the bell is immersed to a depth of 12 m? Assume atmospheric pressure of 1.013×10^5 N/m^2.

46. A 6-in.-diameter piston weighing 200 lb is placed in the top of a 14-in.-high cylinder containing air. How far down will the piston move in compressing the air before coming to rest? Assume no leakage takes place, and no temperature change.

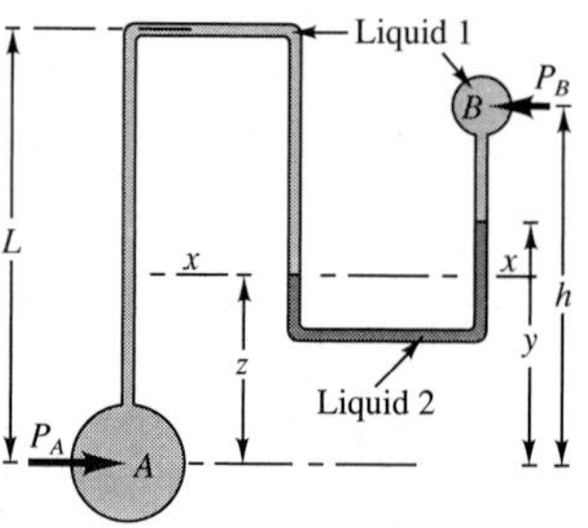

Fig. 13.23 Problem 47.

47. In Fig. 13.23, liquid No. 1 has a sp gr of 1.0 and liquid no. 2 has a sp gr of 0.800. $L = 32$ cm, $z = 9$ cm, $h = 28$ cm, and $y = 18$ cm. What is the difference in pressure between points A and B?

48. Two cubical tanks, one measuring 3 ft on each edge and the other measuring 2 ft on each edge, are joined together by a pipe of negligible length in which there is a closed valve. The smaller tank contains oxygen under gauge pressure of 325 lb/in.^{2} The large tank contains air at atmospheric pressure. What will be the resulting gauge pressure in both tanks if the valve is opened and gas is permitted to flow from the smaller tank to the larger tank?

49. A compound was analyzed and found to contain 52.14% carbon, 13.13% hydrogen, and 34.73% oxygen (by weight). What would be its simplest formula?

PART 4

HEAT AND THERMODYNAMICS

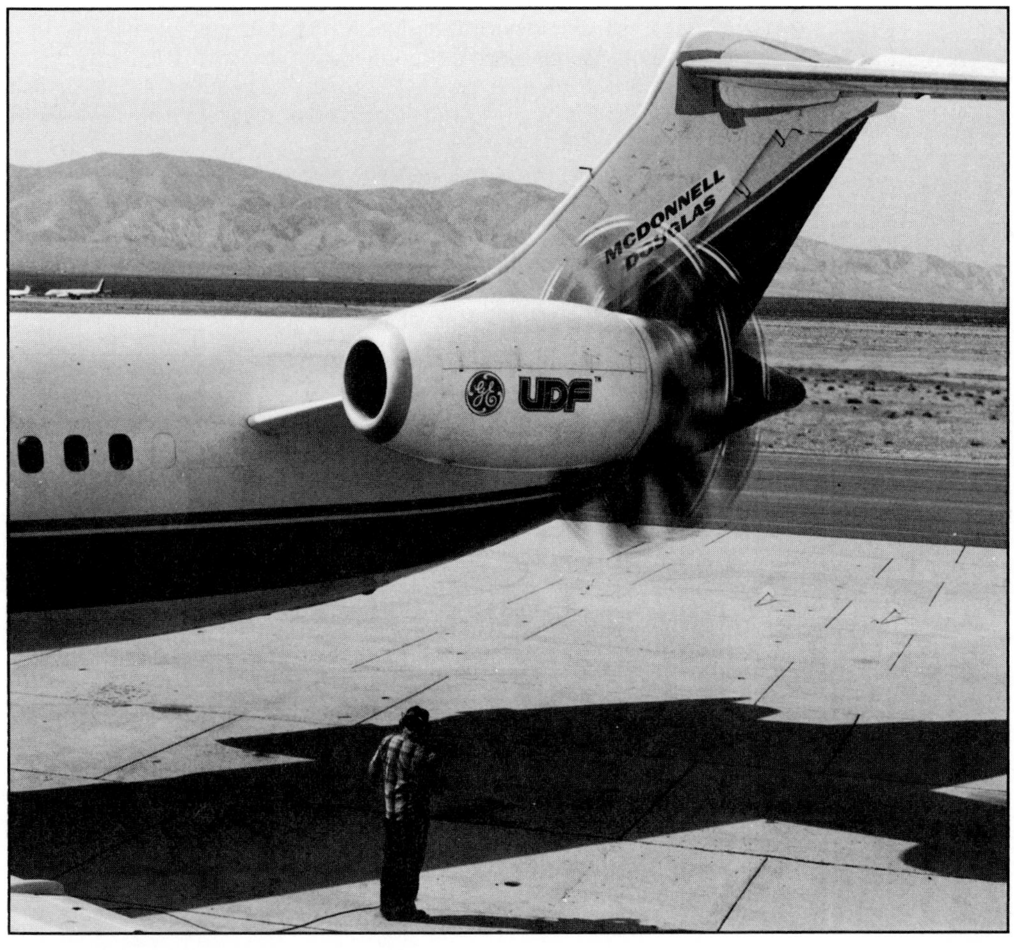

New heat engine in test trials. General Electric unducted fan (UDF) engine for commercial aircraft. (McDonnell Douglas Aircraft Company)

TEMPERATURE AND HEAT

With this chapter *heat energy,* or *thermal energy,* becomes the major subject of study. Heat energy can be converted into mechanical energy quite readily by the use of suitable devices; and the reverse process—conversion of mechanical energy into heat energy—occurs continuously whenever friction is present. Energy transformations involving heat and electricity are commonplace, as are those involving heat and light. The energy of the atom and of the atomic nucleus can also be converted to heat energy.

Temperature is a measure of the "hotness" or "coldness" of a body or a substance. It can be considered as an indication of the *intensity* of the heat energy of a substance. It is expressed in *degrees,* and is measured by instruments called *thermometers*. In this chapter we will explain temperature and its measurement, discuss the nature of heat, and examine some of the effects of heat on matter. In subsequent chapters heat transfer, change of state, heat engines, and refrigeration and air conditioning will be considered in detail.

TEMPERATURE AND ITS MEASUREMENT

14.1 ■ Temperature

Our ideas of temperature are related to the sense of feel of the human body. If a substance feels *hot,* we say that its temperature is high. If it feels *cold,* it is said to have a low temperature. The terms *warm* and *cool* describe mild degrees of hotness and coldness, respectively. Bodily sensations however, are not very accurate indicators of temperature. You can do a simple experiment to demonstrate this fact. After a few minutes with one hand in hot water and the other in ice-cold water, place both hands in lukewarm water. The "cold" hand tells you that the lukewarm water is hot, and the "hot" hand tells you it is cold.

Temperature difference *depends on the rate at which heat is transferred to or from the body*. Lukewarm water transfers heat *into* the "cold" hand, and the "hot" hand transfers heat into the lukewarm water. Sensations of "hot" and "cold" are really not very accurate indicators of temperature. Later, a definition of temperature will be proposed which relates temperature to the velocity (and kinetic energy) of moving molecules.

14.2 ■ The Liquid-in-Glass Thermometer

In 1714, Gabriel Daniel Fahrenheit (1686–1736) designed the first successful liquid-in-glass thermometer and calibrated it in terms of a temperature scale (the *Fahrenheit scale*) which is still used throughout the English-speaking world. The typical liquid-

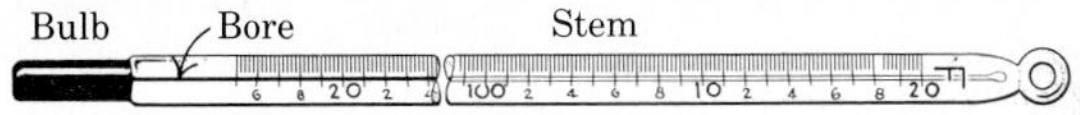

Fig. 14.1 A standard mercury-in-glass laboratory thermometer for use with the Fahrenheit temperature scale.

in-glass thermometer (see Fig. 14.1) consists of a glass bulb and a stem in which a slender bore or tube is filled with either mercury or alcohol. The space above the liquid in the tube is evacuated. Changes in temperature cause the liquid to rise or fall in the narrow stem because of expansion or contraction of the liquid in the bulb. The expansion of the glass bulb is almost negligible when compared with that of either mercury or alcohol, and since the stem is sealed off, variations in atmospheric pressure have no effect on the readings. The stem bore must be uniform throughout if accuracy is expected at all readings. The sensitivity of thermometers can be increased by making the bore smaller with respect to the volume of the bulb.

Common thermometers sense temperature changes by means of the expansion and contraction of a liquid (mercury or alcohol) in a small-bore tube or stem.

Calibrating a Thermometer Common thermometers, whether intended for use with the SI-metric or English system of measurement, are calibrated using the freezing point and the boiling point of water (at standard atmospheric pressure, 760 mm Hg) as *fixed points*. The process is shown schematically in Fig. 14.2. The thermometer is first placed in a mixture of ice and pure water which has been sitting (with occasional stirring) for some time. This mixture is in the *equilibrium state* between liquid water and ice. A mark is placed on the thermometer tube where the mercury level stabilizes after a few minutes in the water-ice mixture. The process is then repeated with the thermometer in a boiling water and steam mixture, and a mark is again made on the thermometer tube at the point where the mercury comes to rest. The space between these two fixed points is then divided into a number of equal parts, each part representing *one degree* (°) of temperature. The method of division depends on the thermometer *scale* being used.

The freezing point and the boiling point of pure water at atmospheric pressure are the *fixed points* used in calibrating common thermometers.

14.3 ■ The Fahrenheit Temperature Scale

The Fahrenheit temperature scale is now defined as having its fixed points at the freezing point of water (32°F) and at the boiling point of water (212°F) at standard atmospheric pressure. The range between these fixed points is divided into 180 equal

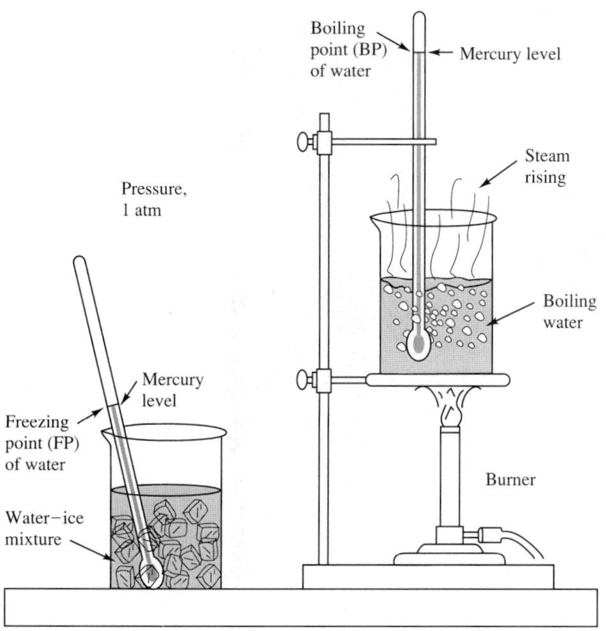

Fig. 14.2 Schematic representation of the process of calibrating mercury-in-glass thermometers. The thermometer is first placed in an ice-and-water mixture that has been standing for some time to attain temperature equilibrium. The resultant mercury level is marked as the *ice-point*, or freezing point of water (0°C or 32°F). The thermometer is then placed in vigorously boiling water and its final mercury level is marked as the *steam point*, or boiling point of water (100°C or 212°F). The calibration process must be carried out at a pressure of 1 atmosphere, 760 mm Hg, or 29.92 in. of mercury.

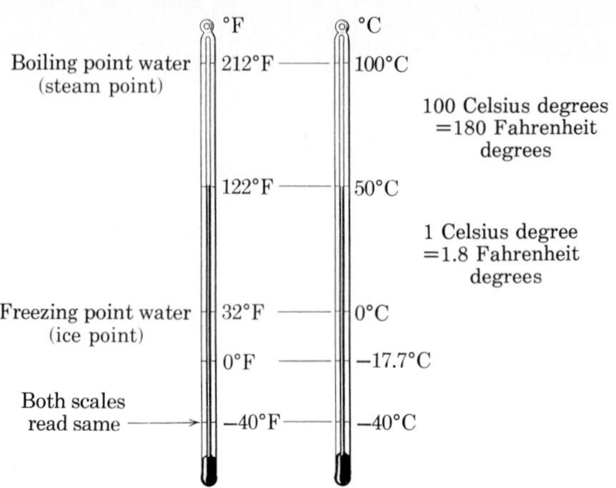

Fig. 14.3 Fahrenheit and Celsius temperature scales compared. The reference points for both scales are the freezing point of water (ice-point), and the boiling point of water (steam-point), both referred to a pressure of 1 atm.

divisions, each corresponding to 1° of temperature (see Fig. 14.3). The scale is then extrapolated above 212° and below 32° as far as desired, subject to the temperature limitations of the liquid with which the particular thermometer is filled. The Fahrenheit scale is still used in U.S. industry and commerce and in meteorology, engineering, agriculture and medicine.

14.4 ■ The Celsius Temperature Scale

About 1742 Anders Celsius (1701–1744) proposed the "centigrade" scale, with the freezing point of water to be pegged at 0°C and the boiling point of water at 100°C (Fig. 14.3). Such a scale had the advantage of simplicity and ease of decimal calculation, and it soon became the standard throughout the world for scientific work. Although called the *centigrade scale* for over two centuries, it has been officially renamed the *Celsius scale*. The Celsius scale is used in all countries where the SI-metric system has been adopted. It is also used in the U.S., for scientific work.

14.5 ■ Comparison of Fahrenheit and Celsius Scales

The two scales are compared in Fig. 14.3. Analysis reveals that 100 Celsius degrees are equivalent to 180 Fahrenheit degrees ($212 - 32 = 180$). Therefore, 1 Celsius degree represents a temperature change equivalent to 1.8 Fahrenheit degrees. Also note that the freezing point of water is at 0°C and at 32°F. In comparing a temperature on one scale with that on another, the freezing point (fp) of water is always a reference point to be used.

To convert from one scale to the other, suppose it is desired to obtain the Fahrenheit reading which corresponds to a 50°C reading. Fifty Celsius degrees above the freezing point of water equal 50×1.8, or 90 Fahrenheit degrees *above the freezing point*. But the freezing point on the Fahrenheit scale is 32°F, and we are 90°F above that. Adding 90 and 32, we obtain 122°F as the Fahrenheit equivalent of 50°C. As another example, what is the Celsius reading that corresponds to a Fahrenheit reading of 20° below zero (−20°F)? Note that −20°F is 52 Fahrenheit degrees below the freezing point of water; and 52 Fahrenheit degrees = 52/1.8, or 28.8 Celsius degrees below the freezing point of water. The Celsius reading is therefore −28.8°C.

The following rules for converting temperatures from one scale to another may be used, but they need not be memorized if the student visualizes the scales as shown in Fig. 14.3.

1. To convert Celsius to Fahrenheit, multiply the Celsius reading by 1.8 and add 32.

2. To convert Fahrenheit to Celsius, subtract 32 from the Fahrenheit reading and divide by 1.8.

In applying these rules the correct *algebraic sign* of the readings must be used.

Stated as mathematical formulas the rules are:

$$F = 1.8C + 32 \tag{14.1}$$

$$C = \frac{F - 32}{1.8} \tag{14.2}$$

where F and C represent Fahrenheit and Celsius readings of the same actual temperature. Note from Fig. 14.3 that at −40° both scales have the same reading.

Attention is here called to the difference between a temperature *reading* and a temperature *range*. A temperature reading is recorded as 62°F, for example; or as 28°C. For temperature *readings,* the ° (degree) symbol comes immediately after the number, and *before* the *F* or the *C* as 26°F or 40°C. On the other hand, if a temperature *range* is to be indicated, we say, "40 Fahrenheit degrees," or "66 Celsius degrees" and write these as 40 F° and 66 C°, respectively, the degree symbol *following* the F or the C.

Temperature *readings* and temperature *ranges* are written differently.

14.6 ■ Construction of Common Thermometers

Ordinary liquid-in-glass thermometers are made in many grades of quality for a wide range of purposes. The glass bulb is made as thin as possible consistent with sturdiness, because glass is a relatively poor conductor of heat and thickness of the bulb would cause poor sensitivity. The stem bore above the liquid is evacuated and sealed off, so there is no air pressure to act down on the liquid surface as it rises and falls because of changing temperature.

Mercury freezes at about −38°F, and consequently, for temperatures below this level, alcohol-filled thermometers are used. (Alcohol freezes at about −200°F.) Mercury is best for higher temperatures, since the boiling point of alcohol is about 171°F. For extremely high and extremely low temperatures, special types of thermometers, to be described later, must be used. There are many industrial, clinical, and home uses for liquid-in-glass thermometers. Figure 14.4 shows a type used by weather stations, and Fig. 14.5 illustrates two types of clinical thermometers to measure the human body temperature (average 98.6°F).

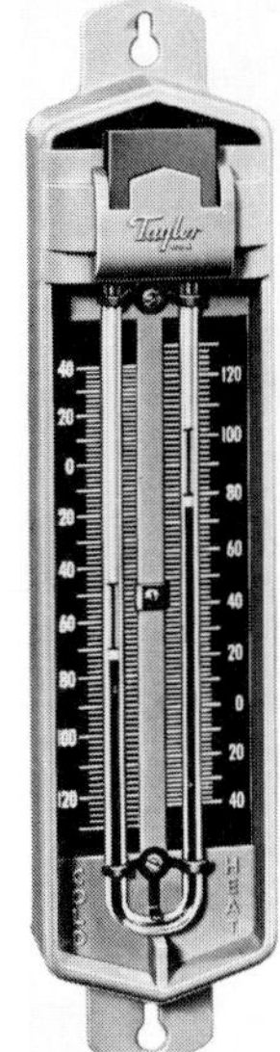

Fig. 14.4 Maximum and minimum thermometer for use at weather stations or by orchardists to determine suitable locations for planting trees and vines. The instrument is alcohol-filled, with a range from −40 to 130°F. (Taylor Instrument Co.)

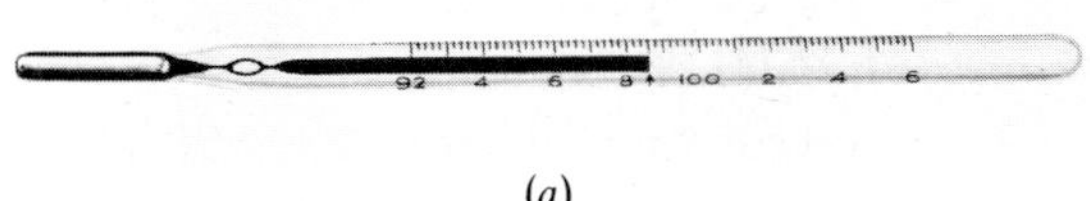

(*a*)

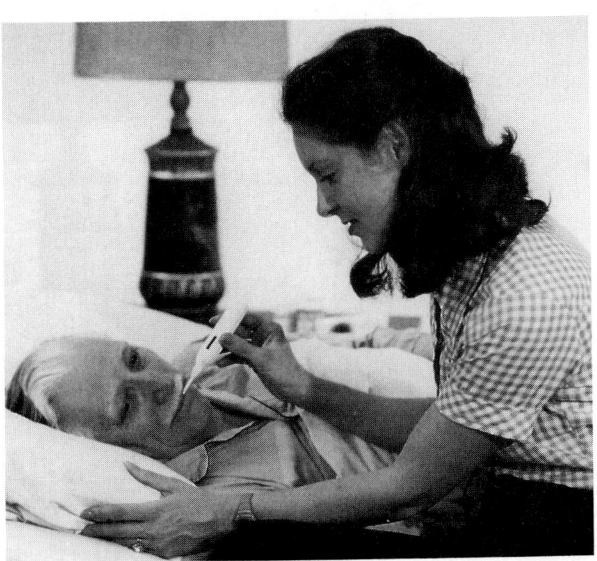

(*b*)

Fig. 14.5 Clinical thermometers are used to determine human body temperature. The average "normal" temperature, taken orally, is 98.6°F. (*a*) Standard mercury-in-glass clinical thermometer. (*b*) Nurse takes patient's temperature with a battery-powered digital-display, solid-state electronic thermometer. (Electromedics, Inc.)

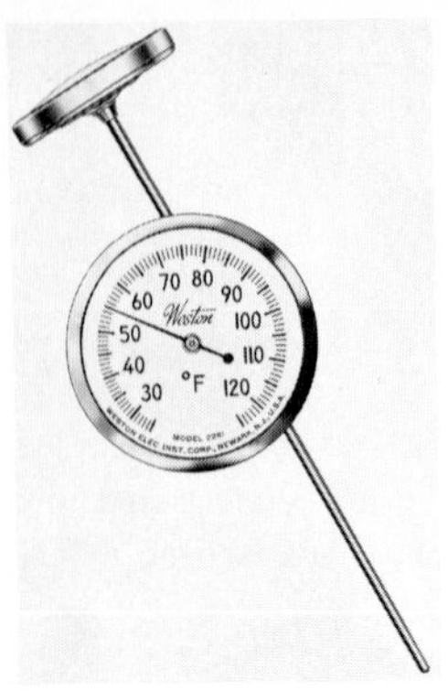

Fig. 14.6 Dial thermometer for the Fahrenheit scale. The pointer rotates across the dial as a result of the differential expansion of a bimetallic element. (Central Scientific Co.)

Dial thermometers and other remote-reading thermometers may be actuated by a *bimetallic element* undergoing *differential expansion*; or by the variation in *electrical resistance* with temperature of selected metals; or by the electric current generated by the *thermoelectric* (Seebeck) *effect*.

14.7 ■ Dial Thermometers

Many industrial thermometers register the temperature reading by means of a pointer on a dial. These thermometers do not contain any liquid. They may operate on the principle of thermal expansion, or as a result of the thermoelectric effect (see Sec. 14.9). Different metals expand at different rates (see Sec. 14.14), and some dial thermometers make use of this principle of unequal expansion. A *bimetallic element* (brass bonded to iron) bends one way on being heated, the opposite way on being cooled (see Sec. 14.16). This motion is transmitted by a suitable linkage to a pointer which moves across a calibrated scale (see Fig. 14.6). Although dial thermometers are not as accurate as liquid-in-glass thermometers, their more rugged construction makes them ideal for many industrial purposes.

14.8 ■ Resistance Thermometers

Metals, which are usually good conductors of electricity, vary in the amount of *resistance* they offer to electric current. Also, any one metal shows a remarkable variation in *electric resistivity* with variations in temperature. The electric resistivity of pure metals decreases with falling temperature and increases with rising temperature. For some metals, notably platinum, this rise and fall in resistivity is such a direct and constant variation that the temperature can be very accurately determined by measuring the resistance, in *ohms*, of a given sample of the metal. One form of such a resistance thermometer is shown schematically in the sketch of Fig. 14.7. The electrical leads from the thermometer are connected to a source of electric current and to a very sensitive *ohmmeter*, which measures the resistance of the platinum resistance coil at the temperature being measured. The resistance may be read and converted into a temperature reading from a prepared table or graph, or the ohmmeter may be calibrated to read temperature directly. The quartz tube of the instrument and the platinum coil both have melting points of over 1650°C, and so the instrument is well suited for measuring reasonably high temperatures (up to 1350°C in actual practice) for heat-treating and annealing processes in the metallurgical industry.

Another type of resistance thermometer is a two-terminal semiconductor device called a *thermistor*. (See Chap. 33.)

14.9 ■ Thermocouples

A very important discovery was made by T. J. Seebeck (1770-1831) about 1820. He found that if two different metals, such as copper and iron, are joined together to form a closed loop and one junction is kept at a different temperature from the other, an electric current can be observed in the closed loop. This discovery is known as the *thermoelectric effect*, or sometimes as the *Seebeck effect*. Figure 14.8 shows, in diagram form, the essentials of a temperature-measuring device based on the thermoelectric effect. It is called a *thermocouple*. In the diagram, T_h is the temperature of the hot junction and T_c the temperature of the cold junction (the *reference temperature*).

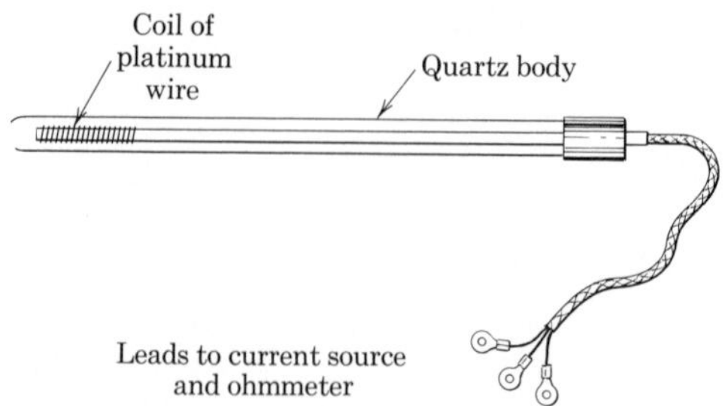

Fig. 14.7 Schematic diagram of one form of platinum resistance thermometer.

Careful experiments by Seebeck and others have shown that the electron flow is related in a predictable manner to the difference in temperature ($T_h - T_c$) between the junctions. If the temperature of one junction is kept at a constant known value—say 0°C—the temperature of the other junction (which may be many feet away—inside a furnace or refinery tank, for example) can be readily and accurately determined by noting the electric voltage produced, and then consulting a previously prepared temperature-voltage graph for the particular pair of metals being used. Figure 14.9 shows such a graph for a copper-iron thermocouple. A chromel-alumel combination is often used in thermocouples. Its temperature-voltage curve approximates a straight line. A multirange microthermometer with remote-sensor thermocouple is illustrated in Fig. 14.10. Dials of such instruments are calibrated so that voltages produced are read as degrees F or degrees C.

The thermocouple is convenient because of the small size of the sensing junction. It can be sealed in pipes, ducts, stacks, furnaces, cylinder heads, etc., wherever a temperature measurement is desired. Even a single junction is quite sensitive, and if additional sensitivity is desired, multiple junctions may be connected in series, resulting in an instrument of great sensitivity called a *thermopile*.

The copper-iron thermocouple is not much used in actual practice because of corrosion and the relatively low sensitivity and low melting point of these metals. Combinations of metals most often used for thermoelectric thermometers (called *thermels*) are copper and constantan, chromel and constantan, chromel and alumel, and platinum and platinum-rhodium alloy.

NOTE: *Constantan is an alloy composed of 55 percent copper, 45 percent nickel; chromel is 90 percent nickel, 10 percent chromium. In addition to extreme sensitivity, the platinum-rhodium combination has a high melting point and can be used for temperatures up to 1480°C.*

The extreme sensitivity of thermopiles is difficult to appreciate without observing them in operation. As an example, however, a thermopile composed of just eight copper-iron thermocouples, series-connected, is capable of detecting temperature differences of 0.005 C°. The use of the more sensitive metal combinations together with an increase in the number of junctions will increase the sensitivity even further.

14.10 ■ Optical Pyrometers

To measure temperatures beyond the range of resistance thermometers and thermocouples, *optical pyrometers* are often used. Both the brightness and the color of an incandescent surface change in a predictable way as the temperature increases. An

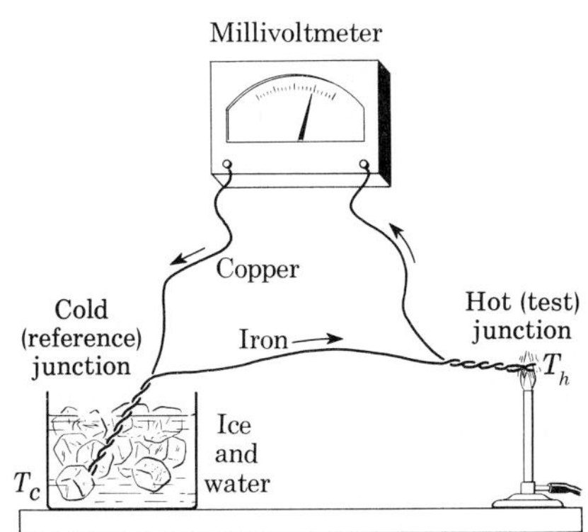

Fig. 14.8 Schematic diagram of apparatus arrangement to demonstrate the *Seebeck effect* or *thermoelectric effect*—the operating principle of the thermocouple.

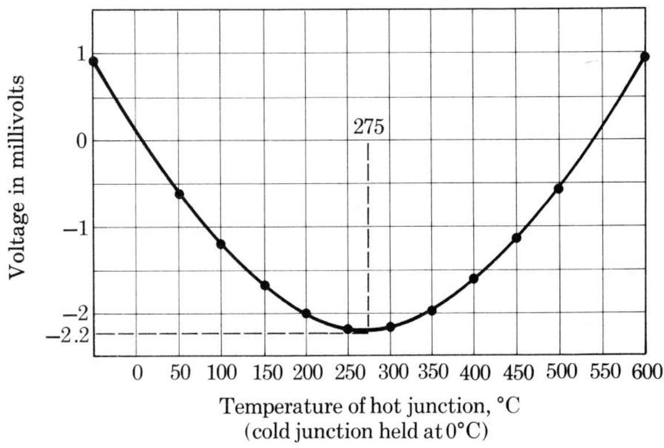

Fig. 14.9 Temperature–voltage graph for a copper-iron thermocouple. The curve shows the voltage produced for a range of hot-junction temperatures, if the cold junction is maintained at 0°C.

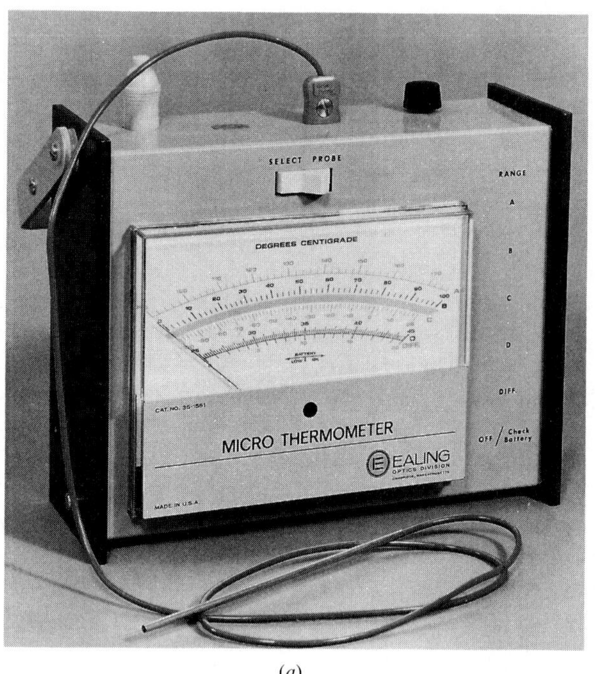

(*a*)

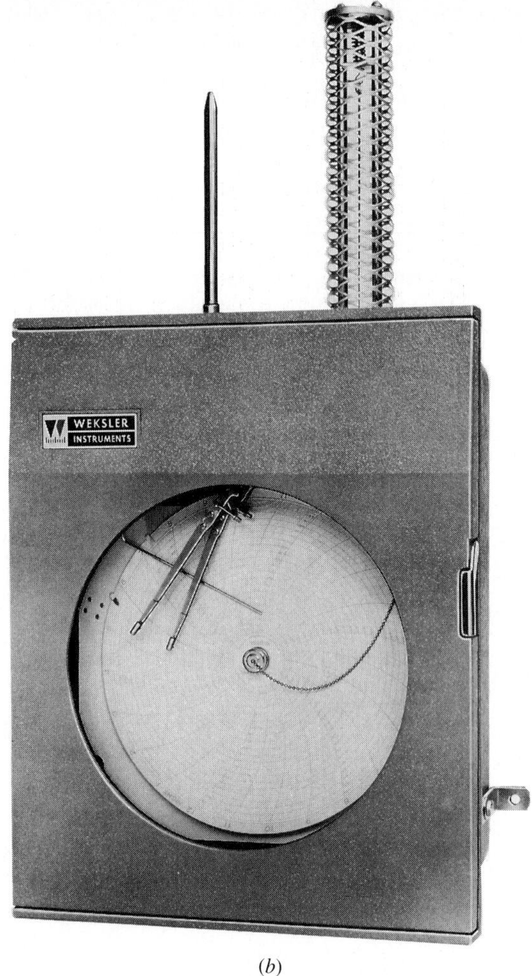

(*b*)

Fig. 14.10 Thermometers for reading and recording temperatures in remote locations. (*a*) Multirange microthermometer for Celsius temperature readings in remote or difficult-to-reach locations. (The Ealing Corp.) (*b*) Remote-sensor recording thermograph. (Weksler Instruments, Inc.)

The temperature of hot metals and molten material is related in a predictable manner to the color of the "glow" they radiate. Optical *pyrometers* (from the Greek, literally "fire meter") use this property to determine approximate temperatures in the metals industry, and the temperature of lava in volcano eruptions.

optical pyrometer enables the operator to estimate rather accurately the temperatures of glowing surfaces by their brightness and color. The temperature is judged visually (see Fig. 14.11*a*) by comparing the brightness and color of the molten or glowing mass with the brightness and color of a heated filament (whose temperature is known) within the instrument. The filament is heated by a controlled electric current until it is judged by the operator to be of the same brightness as that of the incandescent body whose temperature is desired. The optical arrangement of the instrument is shown diagrammatically in Fig. 14.11*b*. The operator views the glowing filament against the background of the incandescent material and then adjusts the current which heats the filament until color and brightness of the one are judged to be the same as those of the other. The meter which measures the current supplied to the filament is calibrated to read in degrees of filament temperature rather than in units of electric current. Tungsten filaments are ordinarily used since the melting point of this metal is over 3300°C. Temperatures up to about 3000°C are measured directly. Above this range, the instrument can be modified by the introduction of wedge-type filters to bring the intensity and color of the source down within the limits of the tungsten filament. A dial on the instrument reads corrected temperatures for the introduction of varying amounts of

(a)

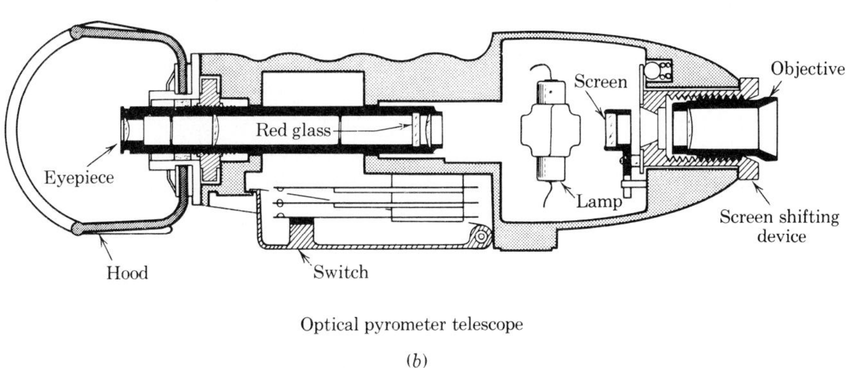

Optical pyrometer telescope

(b)

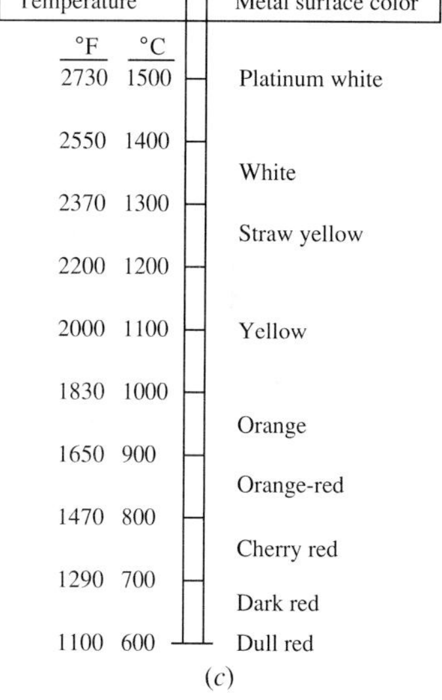

Temperature		Metal surface color
°F	°C	
2730	1500	Platinum white
2550	1400	
		White
2370	1300	
		Straw yellow
2200	1200	
2000	1100	Yellow
1830	1000	
		Orange
1650	900	
		Orange-red
1470	800	
		Cherry red
1290	700	
		Dark red
1100	600	Dull red

(c)

Fig. 14.11 The optical pyrometer is used to measure the temperature of glowing-hot surfaces. (*a*) The instrument in use, as a technician determines the temperature of molten steel. (*b*) Diagram of the construction details of an optical pyrometer (Leeds and Northrup Co.). (*c*) Approximate relationships between the temperature of glowing materials and the colors they emit.

wedge thickness. Fig 14.11*c* provides some approximate color-temperature equivalencies for objects hot enough to glow.

Optical pyrometers have their greatest use in estimating the temperature of molten metals in the iron and steel industry. They are also used by volcanologists in estimating the temperature within an active volcano.

Work with very low (cryogenic) temperatures (below −250°C) requires specialized thermometers, one form of which is shown in Fig. 14.12. Further discussion of very low temperatures will be found in Chapter 18.

Fig. 14.12 The ultrasonic thermometer. Measurement of low temperatures in cryogenic research can be readily accomplished by the use of high-frequency sound waves (see Chap. 21). The tall Dewar (insulated) flask (center) contains the cryogenic substance (liquid hydrogen, helium, or nitrogen, for example) and also a quartz crystal for producing high-frequency sound of known frequency. The exact temperature in the cryogenic range (below about −200°C) is a function of the speed of sound in the cold substance. With such a thermometer, temperature determinations accurate to a few hundredths of a degree Celsius are possible in the range −250 to −270°C. (National Bureau of Standards)

THE NATURE OF HEAT

Several hundred years ago scientists thought of heat as being a material substance which could be made to flow in and out of a body, with properties similar to those of a fluid. We still speak of heat "flowing" from one object to another. The French chemist Lavoisier (1743-1794) gave the name *caloric* to this hypothetical fluid, taking it from the Latin word for heat. From this same root comes our word *calorie* for a unit of heat.

14.11 ■ Heat as a Form of Energy

The "caloric theory" of heat was effectively challenged by the experiments of Count Rumford, as he observed the relationship of work and heat in boring mill operations.

Benjamin Thompson (1753-1814), born in colonial America but destined to spend most of his life in Europe, was the first to demonstrate that heat was not a fluid substance but a *form of energy*. While superintendent of a Bavarian ordnance works (about 1790), Thompson (known in Europe as Count Rumford) was in charge of boring brass cannon. Then, even as now, one of the basic problems of machine-shop work was to provide the proper coolant liquid to carry away the heat generated by machining operations. Thompson was struck by the fact that whenever work was done on the boring mill, heat was generated. He kept records on the change in temperature of known amounts of coolant water and also on the amount of mechanical work being done by the horses which were the factory "engines" of that day. He saw that as long as work was done by the drilling bit on the brass, heat was produced, and when work stopped, heat flow stopped. His conclusion was that heat was not a fluid, not *caloric*, not a substance at all, but a *form of energy* which appeared to flow into the metal and the coolant water, as a result of mechanical work.

14.12 ■ The Concept of Internal Energy

Any given quantity of matter possesses a certain amount of *internal energy*. This energy is in addition to whatever potential energy the body may have with respect to the earth or any other body. Internal energy is also in addition to any kinetic energy the body as a whole may have due to its motion.

The *internal energy* of a body or system is that energy which is the sum total of: (1) the translational kinetic energy of the body's or system's molecules, electrons, and nucleons in random motion; (2) the rotational or vibrational energy of its atoms; (3) the potential energy of action-at-a-distance forces between molecules, and of the binding forces between and among atoms and subatomic particles. The total amount of internal energy possessed by a body or system depends on its temperature, its mass, its physical state (i.e. solid, liquid, or gas), and on such other variables as pressure, volume, magnetic and electric and binding-force properties, and chemical characteristics. Of all these variables, it is temperature that concerns us here, since it is *temperature* that is directly related to the *translational kinetic energy of molecules*. Temperature and heat are related in that they both are thought to be manifestations of molecular kinetic energy.

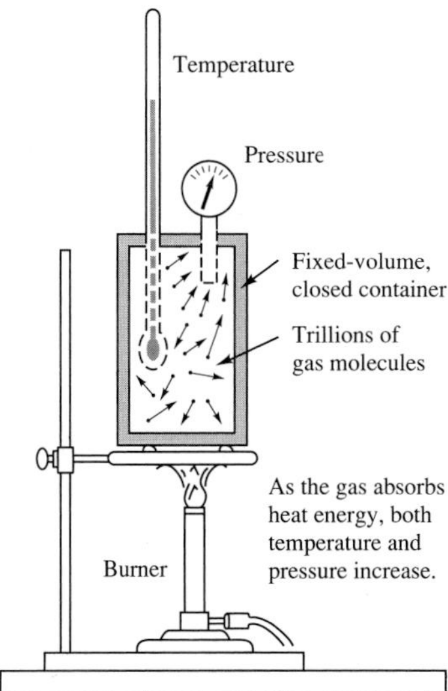

Fig. 14.13 A closed container (constant volume) containing a given quantity (mass) of gas. At room temperature, according to the kinetic theory, the molecules of the gas are in constant motion at varying speeds, colliding with each other and with the walls of the container. Their average speed depends on the temperature. As the temperature rises, average molecular speed rises and, since the volume is constant, the pressure increases.

14.13 ■ Temperature and Heat

The *kinetic theory of gases* was introduced and briefly discussed in Chap. 10. A further extension of that discussion is given here. Suppose a quantity of gas is enclosed (trapped) in a container equipped with a pressure gauge and a thermometer (Fig. 14.13). The kinetic theory assumes that the gas molecules are in constant random motion in different directions at different speeds. For any given condition of pressure, volume, and temperature, we can conceive of some *average molecular speed* which causes the molecules to have a certain number of impacts per second against the walls of the container. The frequency and kinetic energy of these impacts are what cause the pressure, which is read on the gauge. If the container and enclosed gas are heated, what happens? The thermometer registers an increased temperature, and the gauge registers an increased pressure. Adding heat is therefore associated with a temperature increase. Since the number and total mass of the molecules is unchanged (closed container) the increase in pressure can come only from more frequent impacts of the same number of molecules on the walls of the container. But more frequent impacts can result only from an increase in the *average molecular speed*. And molecular speed (velocity) is directly related to molecular kinetic energy ($\text{KE} = \frac{1}{2}mv^2$).

The **kinetic theory** then, leads to a new concept of temperature:

The (absolute) temperature T of a gas is related to the average translational kinetic energy of the molecules of the gas.

A "hot" gas has molecules with high average kinetic energy; a "cold" gas has molecules of relatively low kinetic energy. These same generalizations hold also for liquids and solids. (The meaning of *absolute* temperature in terms of two new temperature scales will be explained in a later section.)

The kinetic theory of gases relates the absolute (Kelvin or Rankine) temperature of a gas to the translational kinetic energy of its molecules.

Consider two bodies or material systems in the same physical state (i.e. gas, liquid, or solid) at different temperatures T_1 and T_2 (T_1 hotter than T_2) (Fig. 14.14*a*). If the two systems are brought together (gases or liquids will mix, solids will attain good surface contact), it is observed that T_1 decreases and T_2 increases until some final temperature T_3 (intermediate between T_1 and T_2) is reached (Fig. 14.14*b*). Kinetic theory explains this phenomenon by supposing that the high-energy molecules of the T_1 (hot) system transfer energy by collision to the low-energy molecules of the T_2 (cold) system, until finally the average translational KE of the molecules of

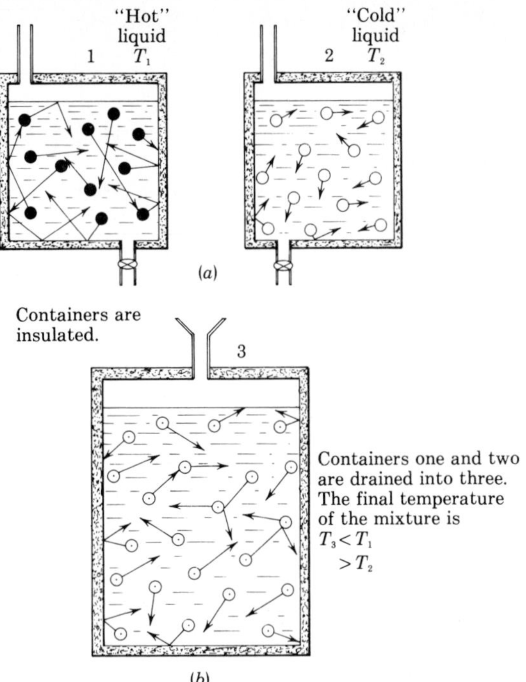
"Hot" liquid
1 T_1
"Cold" liquid
2 T_2
(a)
Containers are insulated.
3
Containers one and two are drained into three. The final temperature of the mixture is
$T_3 < T_1$
$> T_2$
(b)

Fig. 14.14 Kinetic theory and thermal equilibrium. High-energy molecules transfer energy to low-energy molecules by collision. Heat flows from a high-temperature substance into a low-temperature substance until thermal equilibrium is reached.

both systems is the same. At this point the two systems are said to be in *thermal equilibrium*—their temperatures are equal.

For substances that do not undergo a change of state in the process (i.e. they do not boil or freeze or melt or condense), the kinetic theory leads to the conclusion that energy always flows from a hotter body to a colder body. This energy which flows, or is transferred, from a hot body to a cold body is *molecular kinetic energy*. We call it ***heat energy***.

Heat energy, defined.

Heat energy is molecular kinetic energy which is transferred from one body or system to another as a result of temperature difference.

Going back to the idea of *internal energy*, we now see that heat energy is that part of internal energy that is contributed by the translational kinetic energy of the moving molecules.

Heat energy may be added to a body or it may flow out of a body. For heat to flow from one body or substance to another, *there must be a temperature difference*.

Heat will not "flow" unless there is a temperature difference; but heat exchanges do not always result in a temperature change. For example, temperature remains constant during a change of state.

However, heat exchange does not always result in a temperature change. Instead, heat gain or loss may result in a change of state (melting, freezing, evaporating, condensing) with no change in temperature, or heat flow may cause work to be done (as in an engine cylinder). These relationships among internal energy, change of state (phase), and work constitute the science of *thermodynamics*, which we will study in later chapters.

SOME EFFECTS OF HEAT

14.14 ■ Expansion Due to Heat—Thermal Expansion

Steel rails are laid with gaps between the ends to allow for expansion due to temperature changes. Piping systems must be provided with flexible (expansion) joints to allow for expansion and contraction caused by temperature changes, either from the surroundings or from the fluids inside the pipes. Pistons are fitted in engine cylinders, and valve-rod clearances are adjusted to allow for expansion under operating conditions. Tanks full of a cool liquid will overflow as the liquid temperature rises. A toy balloon partially inflated in a cool room may expand to full size and burst if placed out in the hot sun. All these examples illustrate the general law that almost all substances (solids, liquids, and gases) expand on being heated and contract on being cooled.

Thermal expansion of a substance occurs in all directions simultaneously, but for convenience the discussion of expansion is treated under three separate headings:

1. *Linear expansion*, applicable to such objects as wires, cables, and rods, where change in length is the factor of primary importance (solids only)
2. *Area expansion*, applicable to flat sheets, where variation in thickness is immaterial (solids only)
3. *Volume expansion*, applicable to solids, liquids, and gases

14.15 ■ Linear Expansion

Consider a metal rail as it warms up from an early-morning temperature of 3°C to an afternoon temperature of 18°C. Each unit length of the rail becomes a bit longer as the temperature rises degree by degree. The amount of increase of each unit length for each degree rise in temperature is a property of the particular material, in this case, steel, and is called the *coefficient of linear expansion*.

The **coefficient of linear expansion** is defined as *the change in length per unit of original length per degree change in temperature*. As a mathematical formula (see Fig. 14.15),

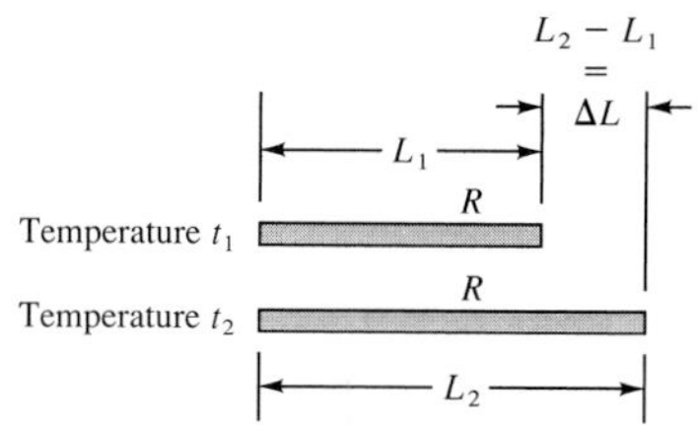

Fig. 14.15 Linear expansion caused by temperature change. A metal rod R has a measured length L_1 at temperature t_1. On being heated to temperature t_2, it expands (exaggerated in the diagram) to a length L_2. The net elongation due to expansion is $L_2 - L_1 = \Delta L$.

$$\alpha = \frac{L_2 - L_1}{L_1(t_2 - t_1)} \qquad (14.3)$$

where α = coefficient of linear expansion
L_1 = length at some lower temperature t_1
L_2 = length at some higher temperature t_2

Or, if ΔL represents the change in length caused by a temperature change Δt,

$$\alpha = \frac{\Delta L}{L_1\,\Delta t} \qquad (14.3')$$

Letting Δt stand for the change in temperature, and solving Eq. (14.3) for L_2, we obtain

$$L_2 = L_1(1 + \alpha\,\Delta t) \qquad (14.4)$$

a convenient formula for determining the new length if the original length and change in temperature are known.

Temperatures on the Celsius and Fahrenheit scales are denoted by t (lowercase). *Absolute temperatures* (to be introduced later in Sec. 14.20) are denoted by T (capital).

The coefficient α has different values for different substances and varies somewhat over different temperature ranges even for the same substance. As might be expected, the values are very small, but nevertheless measurable and significant in almost every manufacturing and assembling operation. For steel, as an example, the coefficient is 0.0000061 ft/(ft) (F°). Note that this would also be 0.0000061 in./(in.) (F°), but that it would be 0.000011 cm/cm, or m/m per Celsius degree, since the Celsius degree represents a change in temperature 1.8 times that of a Fahrenheit degree.

Linear expansion and contraction caused by temperature variations must be taken into account in all construction work and in manufacturing and assembly operations.

Values of α for some common substances, reasonably accurate for ordinary temperature ranges, are given in Table 14.1

Table 14.1 Coefficients of Linear Expansion for Some Common Materials

Material	Per C° cm/cm or m/m	Per F° in./in. or ft/ft
Aluminum	24×10^{-6}	13×10^{-6}
Brass	18×10^{-6}	10×10^{-6}
Brick	10×10^{-6}	5.6×10^{-6}
Concrete	10×10^{-6}	5.6×10^{-6}
Copper	17×10^{-6}	9.5×10^{-6}
Glass, ordinary	$7\text{-}9.5 \times 10^{-6}$	$4\text{-}5 \times 10^{-6}$
borosilicate (Pyrex)	3×10^{-6}	1.7×10^{-6}
Lead	30×10^{-6}	17×10^{-6}
Invar-steel alloy (average)	0.8×10^{-6}	0.44×10^{-6}
Iron (wrought)	12×10^{-6}	6.7×10^{-6}
Silver	20×10^{-6}	11×10^{-6}
Steel (mild)	12×10^{-6}	6.7×10^{-6}

Illustrative Problem 14.1 A valve pushrod of mild steel measures 28.00 cm as the engine is assembled at a temperature of 20°C. What clearance should be allowed if the operating temperature is expected to be 100°C?

Solution We want to solve for the maximum *elongation* $\Delta L = L_2 - L_1$. Write Eq. (14.3) in the form

$$\Delta L = L_2 - L_1 = \alpha L_1(t_2 - t_1)$$

Substituting numerical values (see Table 14.1),

$$\begin{aligned}\Delta L &= [(12 \times 10^{-6})/°\text{C}] \times 28.00 \text{ cm} \times 80°\text{C} \\ &= 0.0269 \text{ cm} \qquad \textit{answer}\end{aligned}$$

Thermal Stresses Experience shows that tremendous forces are exerted by expansion. Buckling of pipes and girders, "freezing" of pistons in cylinders, and the force of explosions are all examples of expansion due to heat.

In designing any structure or machine part that is subject to appreciable temperature variations, adequate allowance must be made for thermal expansion and contraction, else deformation or even complete failure of the structure or part may result from thermal stresses. (See Fig. 14.17.)

14.16 ■ Differential Expansion—Bimetallic Elements

If two strips of different metals, such as iron and brass, are riveted or bonded together at a certain initial temperature, as long as the temperature remains at that level the *compound bar* will remain flat and unbent. However, when the temperature changes, since the metals will expand at different rates with great force, the bar will have to bend. For a given increase in temperature, brass expands more than iron and the compound bar, or *bimetallic element*, as it is called, will bend as shown in Fig. 14.16*a*.

Bimetallic elements are the sensing and actuating devices in most types of common thermostats.

Such bimetallic elements are the basis for dial thermometers and also form the working element in thermostats for heating and cooling controls. The essentials of a typical thermostat are diagramed in Fig. 14.16*b*. In actual practice the contact point *P* is usually in the form of a small magnet so that a "snap action" results. Thus the gap opens and closes with a positiveness which prevents the contact points from burning and pitting as a result of repeated arcing in the electric circuit. Heating thermostats close the electric circuit as temperature falls, while cooling thermostats are set to close the circuit as temperature rises. Temperature control of rooms to a tolerance of plus or minus 2°F can easily be achieved. Research laboratories, final assembly rooms for sensitive instruments, and electronic computer laboratories are just a few of

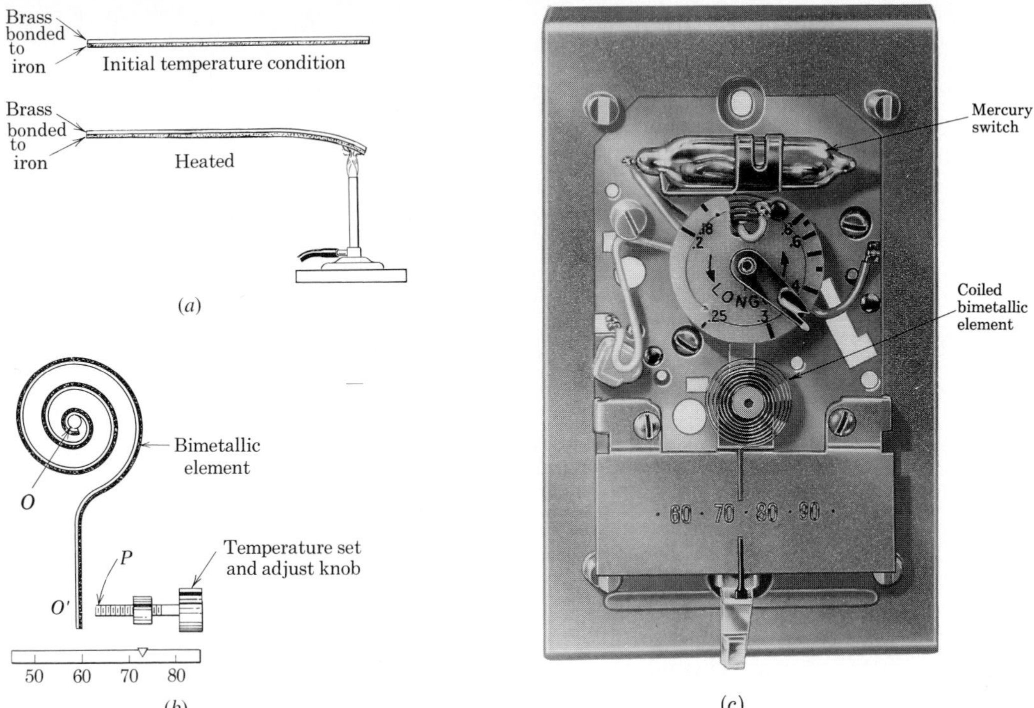

Fig. 14.16 Differential expansion. A bimetallic strip will bend when it undergoes a temperature change, because the two metals of which it is made have different rates of expansion. (*a*) Laboratory demonstration. (*b*) Schematic diagram of the essential components of a simple thermostat. *O* and *P* are connected to a low-voltage electric circuit. Closing the circuit at $O'P$ can actuate the burner of a furnace or boiler or the motor starter for refrigeration and air conditioning equipment. (*c*) An air-conditioning cooling thermostat with cover removed, showing the bimetallic element (coil) and mercury switch. The action of the bimetallic coil tips the glass tube containing the mercury and the mercury flows to a sealed contact in the glass and closes the circuit.

the industrial facilities which need thermostatically controlled temperatures. (See Fig. 14.16*c*.)

The principle of differential expansion is used also in the construction of clocks and watches. Pendulums of clocks and balance wheels of watches expand and contract with temperature changes, and time-measurement errors result. Suitable compensation is effected by using steel and brass so combined and distributed in the pendulum or balance wheel that the expansion of the one works against, or *compensates* for, the expansion of the other.

Industrial Applications of Linear Expansion The effects of heat expansion must be reckoned with constantly throughout industry. In bridge and building construction, expansion joints must be provided for each major section. In oil-field piping and in the plumbing, refrigeration, and steam-fitting trades careful attention must be given to providing expansion joints at the right places (see Fig. 14.17). The importance of expansion to the fitting of engine parts has already been referred to. Surveyors take careful account of the effects of temperature variations on the steel tapes with which they measure the earth's surface.

14.17 ■ Area Expansion

When heated, a surface expands in both length and width, and consequently in area. The **area coefficient of expansion** of a material is defined as *the change in area per*

Fig. 14.17 Expansion joints at a compressor station on a transcontinental pipeline carrying natural gas. Boosting the pressure on the gas raises its temperature and the hot gas causes significant expansion of the steel piping. The large U-bends provide flexibility to prevent pipe breakage.

unit of original area per degree change in temperature. Mathematically,

$$\alpha_A = \frac{A_2 - A_1}{A_1(t_2 - t_1)} \tag{14.5}$$

where

α_A = area coefficient of expansion
$A_2 - A_1$ = change in area = ΔA
$t_2 - t_1$ = change in temperature = Δt

An alternative form is

$$\alpha_A = \frac{\Delta A}{A_1\,\Delta t} \tag{14.5$'$}$$

The predicted new area A_2 is

$$A_2 = A_1(1 + \alpha_A\,\Delta t) \tag{14.6}$$

The *area* coefficient of expansion α_A is equal to 2α, where α is the *linear* coefficient of expansion.

Values of α_A are not usually listed in tables, since it turns out that α_A is in every case very approximately equal to 2α, where α is the *linear coefficient* of expansion. For purposes of ready calculation, Eq. (14.6) therefore becomes

$$A_2 = A_1(1 + 2\alpha\,\Delta t) \tag{14.7}$$

The area of a hole cut in a sheet of metal expands just as if it were a sheet of the same material. In calculating the increase of area of the hole, the value of α to be used is that of the material in which the hole was cut.

Illustrative Problem 14.2 A round flat plate or disk of copper has a diameter of exactly 30 cm at a temperature of 15°C. What is its area when heated to 300°C?

Solution From Eq. (14.7),

$$A_2 = A_1(1 + 2\alpha\,\Delta t)$$

From Table 14.1,

$$\alpha_{(\text{cooper})} = (17 \times 10^{-6})/\text{C}^\circ$$

$$\text{Substituting,} \quad A_2 = \pi \times (15\ \text{cm})^2 \times \frac{1 + 2\,(17 \times 10^{-6})}{\text{C}^\circ} \times 285\ \text{C}^\circ$$

$$= 707\ \text{cm}^2 \times 1.00969 = 714\ \text{cm}^2 \qquad \textit{answer}$$

an increase of about 7 cm^2.

A 30-cm *hole* in a copper sheet subjected to the same temperature increase would have the same area increase.

14.18 ■ Volume, or Cubical, Expansion

Volume Expansion of Solids The **volume coefficient of expansion** β is defined as *the change in volume per unit of original volume per degree change in temperature*. As a formula,

$$\beta = \frac{V_2 - V_1}{V_1(t_2 - t_1)} \tag{14.8}$$

Or, alternatively,

$$\beta = \frac{\Delta V}{V_1\ \Delta t} \tag{14.8'}$$

Therefore

$$V_2 = V_1(1 + \beta\ \Delta t) \tag{14.9}$$

Experimental results and mathematical analysis agree that the value of β very closely approximates 3α, *for solids*. We can therefore write

$$V_2 = V_1(1 + 3\alpha\ \Delta t) \tag{14.10}$$

The *volume* coefficient of expansion of solids β is equal to 3 α, where α is the *linear* coefficient of expansion.

A cavity in a hollow object expands just as if it were a solid block of the same material. Thus a steel gasoline tank which holds 22 gal at 60°F will hold 22(1 + 3 × 0.0000067 × 50) = 22.02 gal at 110°F. Note that β for the hollow space inside is the value of β for the steel of which the container is made ($\beta = 3\alpha$).

The inner volume of a metal tank expands and contracts with temperature changes just as if it were a solid block of the tank material.

Volume Expansion of Liquids Liquids follow the same law for volume expansion as solids, except that in general the coefficients are larger for liquids. β for liquids is however, determined by actual experiment. A list of values of β for some common liquids is given in Table 14.2.

In the experimental determination of values of β for liquids, it is of course necessary to have the liquid in some sort of container. The container will also change in volume during the experiment. An *apparent* value of β for the liquid will therefore be obtained. To find the true value of the expansion coefficient β_T, the coefficient for the container, $\beta_C = 3\alpha_C$, must be added to the apparent value β_A. As an equation,

$$\beta_T = \beta_A + \beta_C \tag{14.11}$$

The values in Table 14.2 are *true* values.

Table 14.2 Volume coefficients of expansion for common liquids

Liquid	*β Per C°*	*β Per F°*
Alcohol, methyl	122×10^{-5}	68×10^{-5}
Gasoline	96×10^{-5}	53×10^{-5}
Glycerin	53×10^{-5}	29×10^{-5}
Mercury	18.2×10^{-5}	10.1×10^{-5}
Petroleum	89.9×10^{-5}	50×10^{-5}
Sulfuric acid	58×10^{-5}	32×10^{-5}
Turpentine	94×10^{-5}	52×10^{-5}
Water (at 20°C)	20.7×10^{-5}	11.5×10^{-5}

Table 14.3 Density of water at various temperatures

Temperature, °C	*Density g/cm³*
0	0.99987
2	0.99997
4	1.0000
6	0.99997
10	0.99988
15	0.99912
20	0.99823
25	0.99705

Illustrative Problem 14.3 A gasoline retailer accepts delivery of 10,000 gal of gasoline, as measured by the flow meter on the delivery truck. In the heat of a summer day the gasoline temperature was 92°F. If the average temperature of the gasoline when pumped into customers' cars from the dealer's underground tank is 64°F, what volume loss will the dealer sustain?

Solution The temperature difference Δt is 92° − 64° = 28 F°. Table 14.2 gives $\beta_{\text{gasoline}} = 53 \times 10^{-5}$ per F°. From Eq. (14.8),

$$V_2 - V_1 = \beta V_1(t_2 - t_1)$$

Substituting

$$V_2 - V_1 = \frac{53 \times 10^{-5}}{\text{F}^\circ} \times 10{,}000 \text{ gal} \times 28 \text{ F}^\circ$$

$$= 148.4 \text{ gal} \qquad \textit{answer}$$

14.19 ■ The Unusual Expansion of Water

When water is cooled, its volume contracts steadily, as expected, until its temperature reaches 4°C. At this temperature, water has its greatest density. If cooled further, water *expands* slightly until the freezing point (0°C) is reached. As it freezes, it expands considerably. The bursting of water pipes in cold winter climates is a common result of this expansion due to freezing.

Table 14.3 gives the density of water at selected temperatures, and the curve of Fig. 14.18 shows graphically the peculiar expansion of water.

It is this peculiar behavior of water which causes lakes and oceans to freeze on top first, rather than at lower levels. Consider a body of water as winter sets in. The

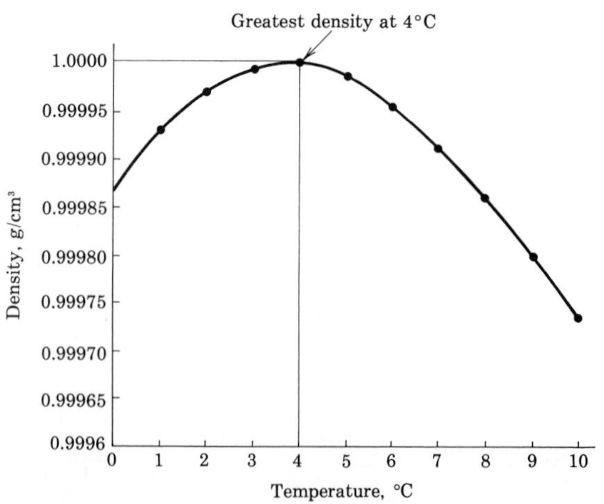

Fig. 14.18 The unique density-temperature curve for water. Water has its greatest density at 4°C. From there down to 0°C (the freezing point) it expands. And, as it changes state to ice, it expands again. Exposed water pipes may burst as water freezes in them, and foundation footings of buildings can be "heaved" upward as soil water freezes in cold climates.

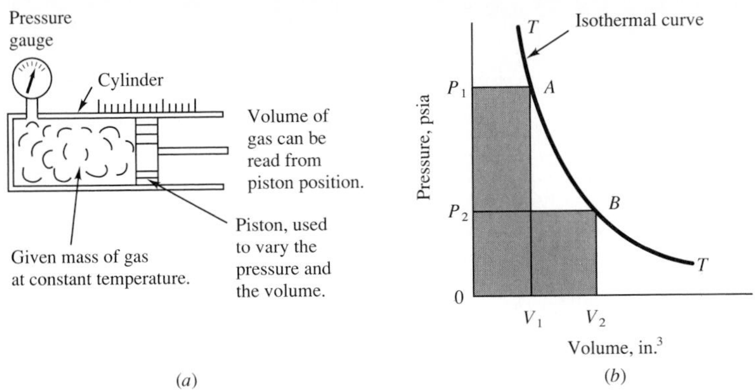

Fig. 14.19 Pressure-volume (*P-V*) relations of an ideal gas at constant temperature, a graphic presentation of Boyle's law: (*a*) A cylinder with a freely moving (frictionless) piston contains a quantity of gas. The piston can be moved to increase or decrease the volume occupied by the gas. (*b*) Plot showing the variation of absolute pressure with changes in volume. (Remember that temperature is assumed constant.)

warmer water, being less dense, continually rises to the top, and a steady circulation is maintained until the entire body of water reaches a temperature of 4°C. Below this temperature the *colder water is less dense and stays on top*, where it quickly freezes. Extremely cold weather is required to freeze a layer as much as a foot thick, since ice is a poor conductor of heat. What might the consequences be if water did not have this reversal in its expansion curve? What would happen to lakes and ponds in very cold climates?

If it were not for the unusual expansion characteristics of water, lakes, streams, and bays might freeze all the way to the bottom in extremely cold winters. Given that both ice and water are very poor *conductors* of heat, what could be predicted for the following summer?

THE LAWS OF GASES

14.20 ■ The Thermal Behavior of Gases—Pressure-Volume-Temperature Relationships—Boyle's, Charles', and Gay-Lussac's Laws

In Chap. 13 some properties of gases were discussed briefly, among them *Boyle's law* and *Charles' law* which, taken together, describe the pressure-volume-temperature behavior of ideal gases. The present discussion will explore these *P-V-T* relations of gases in greater detail.

The *condition* of a gas is described by three factors—its *pressure*, its *temperature*, and its *volume*. Within reasonable limits of temperature and pressure, i.e., not close to the liquefaction points, *all gases* have the same *P-V-T* behavior.

All of the so-called *gas laws* to be discussed below presume an "ideal gas." In practice, such a gas does not exist, but the behavior of real gases is close enough to that of the hypothesized ideal gas that the gas laws are extremely useful.

Gases at Constant Temperature—Boyle's Law A given mass of (an ideal) gas is enclosed in a cylinder, trapped behind a piston which is free to move without friction (see Fig. 14.19*a*). It will occupy a volume which is inversely proportional to the *absolute pressure* applied on the gas by the piston. If a series of different pressures are applied (slowly, so the temperature does not rise), the corresponding volumes can be noted from a scale on the side of the cylinder. If *absolute pressure* readings are then plotted against volume readings, a curve like that of Fig. 14.19*b* results. (It should be recalled that *absolute pressure* equals *gauge pressure* plus *atmospheric pressure*.)

Since this pressure-volume relationship occurs without any temperature change, it is called a *constant-temperature process*. Curves obtained by plotting constant-temperature processes are called *isothermals* (from the Greek *iso*—"equal", and *therm*—"warm"). Careful analysis of such curves reveals that for all combinations of pressure and volume, the areas of the rectangles of which *P* and *V* are the sides, *are equal*. In Fig. 14.19*b*, for example, the area P_1AV_10 equals the area P_2BV_20. Since the area of a rectangle equals the product of length and width, for the present case, with *temperature constant*,

$$P_1V_1 = P_2V_2 = k \quad \text{(a constant)} \tag{14.12}$$

or

$$\frac{V_1}{V_2} = \frac{P_2}{P_1} \quad \text{Boyle's law (temperature constant)}$$

Boyle's law relates the volume of a given mass of gas to the absolute pressure, when the temperature is held constant.

Boyle's law is restated here for convenience:

The volume of a given mass of any gas is inversely proportional to the absolute pressure, if temperature is held constant.

Illustrative Problem 14.4 An air compressor takes atmospheric air ($P_1 =$ 14.7 lb/in.^{2} abs) and compresses it into a pressure tank whose volume is 20 ft^3. The initial pressure in the tank before the compressor starts is 1 atm. After the temperature of the tank cools to room temperature, the gauge reads 600 lb/in.^{2}. What volume of atmospheric air was forced into the tank?

Solution Temperature is constant, and the problem is a pressure-volume, or Boyle's law, problem.

$$\begin{aligned} P_1 &= 14.7 \text{ lb/in.}^2 \text{ abs} \\ &\quad \text{(in tank and atmospheric air)} \\ P_2 &= \text{gauge pressure} + \text{atmospheric pressure} \\ &= 600 \text{ lb/in.}^2 + 14.7 \text{ lb/in.}^2 = 614.7 \text{ lb/in.}^2 \end{aligned}$$

From Boyle's law, the total volume of air at pressure P_1 is given by

$$\frac{V_1}{V_2} = \frac{P_2}{P_1} \quad \text{or} \quad V_1 = \frac{V_2P_2}{P_1}$$

Substituting,

$$V_1 = \frac{20 \text{ ft}^3 \times 614.7 \text{ lb/in.}^2}{14.7 \text{ lb/in.}^2} = 836 \text{ ft}^3$$

Volume of atmospheric air added is

$$836 \text{ ft}^3 - 20 \text{ ft}^3 = 816 \text{ ft}^3 \qquad \textit{answer}$$

Fig. 14.20 Sketch of the essential details of an apparatus for verifying Boyle's law in the laboratory—simple U-tube form.

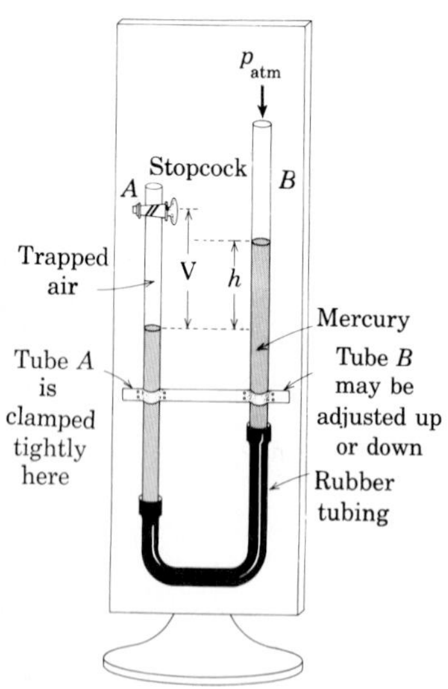

Verifying Boyle's Law in the Laboratory Boyle's law can be verified in the laboratory with the simple apparatus shown in diagram form in Fig. 14.20. Two glass tubes, one open and the other capable of being tightly closed, are connected at the bottom with a U-tube loop of rubber hose. Tube A is rigidly mounted on a laboratory stand, and tube B can be adjusted vertically at the will of the experimenter. Mercury is poured into the open tube B, with the stopcock at A open, until the mercury rises about halfway up tube A. The stopcock is then tightly closed, trapping air in tube A at atmospheric pressure, and the apparatus is ready for use. Tube B is moved vertically with respect to tube A, and a series of readings of V (volume of trapped air) and h (difference in levels of mercury in the two tubes) is taken. Since the glass tubing is of uniform bore, the distance from the mercury level to the stopcock in tube A can be taken as the measure of volume of the trapped air, and the difference h in levels of mercury in the two tubes is the *gauge pressure* applied *on the trapped air*.

At each trial, the *absolute pressure* on the trapped air is equal to the pressure contributed by the mercury column of height h plus the atmospheric pressure:

$$P_{abs}(\text{cm Hg}) = h \text{ (cm Hg)} + 76 \text{ cm Hg (atm press.)}$$

Proof of Boyle's law can be obtained from experimental data, by multiplying P for each trial by V for that trial. If all the PV products are essentially the same (that is, if

PV = a constant), Boyle's law has been verified (see Eq. 14.12). Results, even from such a crude apparatus, are usually in error by no more than 2 to 3 percent.

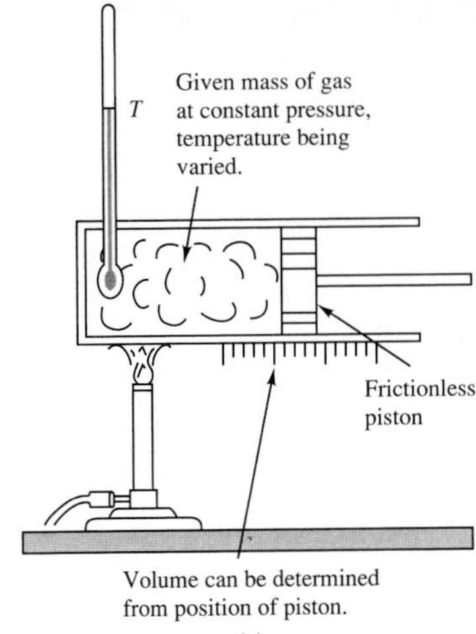

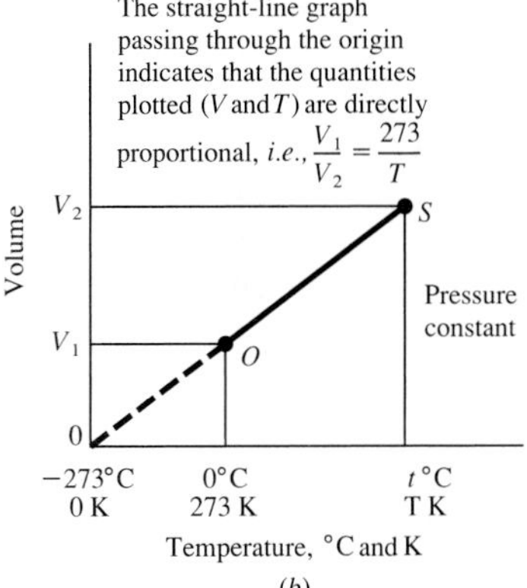

Fig. 14.21 Volume-temperature relations of a gas at constant pressure, a graphic portrayal of Charles' law. (*a*) Diagram of process to study the variation of volume with changing temperature. The freely sliding (frictionless) piston allows the pressure to remain constant at 1 atm. (*b*) Plot of data from trials taken in (*a*). The volume of a gas at constant pressure is directly proportional to its absolute temperature.

Gases at Constant Pressure—Charles' Law If a given mass of any gas is enclosed in a chamber or cylinder fitted with a freely sliding piston, the pressure *will remain constant*, since any slight increase or decrease of gas pressure will cause the (frictionless) piston to move outward or inward until the pressure equalizes again. If, now, such a container of gas is heated (see Fig. 14.21*a*), the volume is found to increase with rising temperature in a manner depicted by the graph of Fig. 14.21*b*.

In a series of careful experiments the noted French physicist Jacques Charles (1746–1823) discovered that the volume of gas varies with temperature (if the pressure is constant) in accordance with the relation

$$V_t = V_0(1 + \beta\,\Delta t) \tag{14.13}$$

where β is the coefficient of volume expansion, and V_0 is the volume at 0°C. Charles and subsequent workers found β *for all gases* to be about $\frac{1}{273}$ per Celsius degree throughout normal temperature ranges (well above liquefaction). This discovery led to the following hypothesis: *If any mass of gas at 0°C is cooled at constant pressure, its volume decreases by* $\frac{1}{273}$ *for each Celsius degree, and therefore at −273°C it should have zero volume*! (Actually, all gases liquefy before reaching −273°C, and so the theory is not subject to experimental verification.)

The coefficient of volume expansion β of *any gas* at constant pressure is 1/273 per C°, with V_0 at 0°C.

Absolute Temperature—The Kelvin Scale This line of reasoning led to the proposal by Lord Kelvin (William Thomson, 1824–1907) of a new temperature scale, whose zero would be 273° below zero Celsius.* This scale, of tremendous importance in all heat-engineering work, is referred to as the *Kelvin* or *absolute temperature scale*. A similar absolute scale developed for use with the Fahrenheit degree is called the *Rankine scale* for William J. M. Rankine (1820–1872), a Scottish engineer who proposed it. Figure 14.22 shows these absolute scales lined up with the Fahrenheit and Celsius scales for comparison. Any Kelvin temperature is obtained by adding 273° to the Celsius temperature, and any Rankine temperature by adding 460°

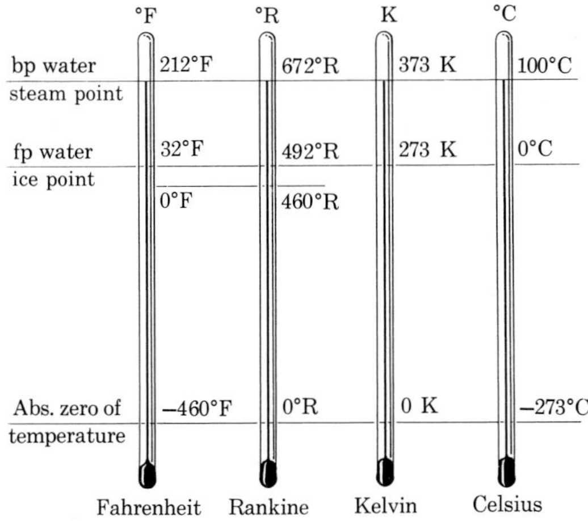

Fig. 14.22 The four basic temperature scales compared. The sketch is schematic only, since no liquid-in-glass thermometer could actually record such a range of temperatures. Temperatures are rounded to the nearest whole degree.

* The currently accepted value of "absolute zero" is −273.16°C.

The Kelvin (or absolute) temperature scale is based on the hypothesis of an *absolute zero* at −273°C. Degrees on the Kelvin scale are called *kelvins* (K), and 1 kelvin = 1 C°. Kelvin temperatures are written without a degree sign (°), as, for example, 345 K.

to the Fahrenheit temperature. *Kelvin temperatures use no degree symbol.* A temperature of 20°C is 293 kelvins, or 293 K.

$$T_K = t_C + 273$$
$$T_R = t_F + 460$$

The kelvin (K) is a fourth *fundamental unit*, a unit of temperature. Length (*L*), mass (*M*) and time (*T*) have been previously defined.

Referring again to Fig. 14.21*b*, suppose that the *V* and *T* values obtained in a range of temperatures from 0°C to t°C plot to a straight line, as shown. Using the same scale and extending the "curve" downward and to the left (shown dashed), we see that it will hit the zero volume line at −273°C, or 0 K. This point is called the *absolute zero* of temperature. The straight-line graph tells us that *with pressure constant, the volume of a gas is directly proportional to the absolute temperature*. This relationship may be stated mathematically either by Eq. (14.13) or, more commonly, by the following:

Charles' law relates the volume of a given mass of gas to the absolute (Kelvin) temperature, with pressure held constant.

$$\frac{V_1}{V_2} = \frac{T_1}{T_2} \qquad \text{Charles' law (pressure constant)} \qquad (14.14)$$

where the capital *T*'s indicate absolute temperatures, either Kelvin (K) or Rankine (R).

Charles' Law is restated here:

For a given mass of gas, if pressure is held constant, the volume is directly proportional to the absolute temperature.

Gases at Constant Volume—Gay-Lussac's Law If a given mass of gas is held in a container of fixed volume while the temperature is varied, the *pressure-temperature relations at constant volume* can be obtained. (Actually, the volume of the container will vary slightly with temperature changes, but a correction factor can be applied.) With an apparatus like that shown schematically in Fig. 14.23*a*, a series of pressure readings at different temperatures can be obtained. Data from such experiments, when plotted, give results as shown in Fig. 14.23*b*. Note that the locus of *P-T* points is a straight line, indicating that *at constant volume,*

$$\frac{P_1}{P_2} = \frac{T_1}{T_2} \qquad \text{Gay-Lussac's law (volume constant)} \qquad (14.15)$$

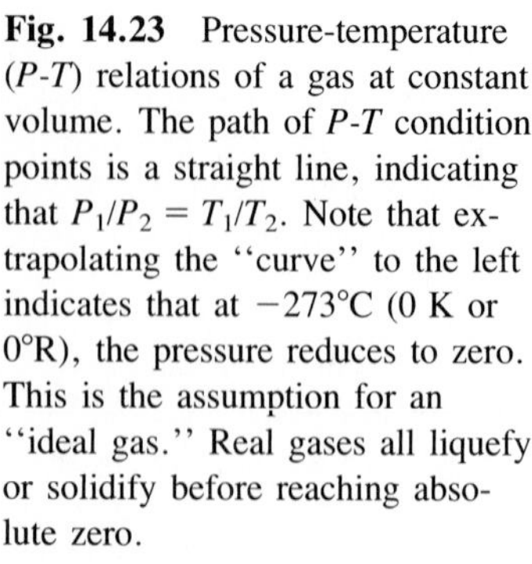

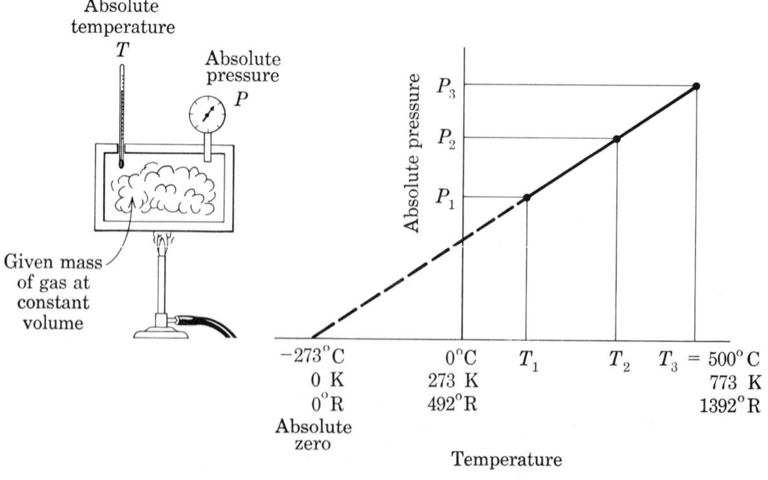

Fig. 14.23 Pressure-temperature (*P-T*) relations of a gas at constant volume. The path of *P-T* condition points is a straight line, indicating that $P_1/P_2 = T_1/T_2$. Note that extrapolating the "curve" to the left indicates that at −273°C (0 K or 0°R), the pressure reduces to zero. This is the assumption for an "ideal gas." Real gases all liquefy or solidify before reaching absolute zero.

The behavior of gases indicated by Eq. (14.15) was initially discovered by Joseph Gay-Lussac (1778–1850), a French experimental scientist.

Gay-Lussac's law relates the absolute pressure of a given mass of gas to the absolute temperature, with volume held constant.

Gay-Lussac's Law

For a given mass of any gas held at constant volume, the absolute pressure is directly proportional to the absolute temperature.

Illustrative Problem 14.5 A tank whose volume is assumed to remain constant is filled with carbon dioxide at a *gauge pressure* of 375 N/cm² when the temperature is 18°C. While being transported under a hot sun its temperature rises to 60°C. What is the new gauge pressure?

Solution *Absolute pressures and temperatures* must be used.

$$P_1 = 375 \text{ N/cm}^2 \text{ gauge} + 10.1 \text{ N/cm}^2 \text{ (1 atm)}$$
$$= 385.1 \text{ N/cm}^2 \text{ abs.}$$
$$T_1 = 18°\text{C} = 291 \text{ K} \qquad T_2 = 60°\text{C} = 333 \text{ K}$$

From Eq. (14.15),

$$P_2 = \frac{P_1 T_2}{T_1}$$
$$= \frac{385.1 \text{ N/cm}^2 \times 333 \text{ K}}{291 \text{ K}}$$
$$= 440.7 \text{ N/cm}^2 \text{ abs} = 430.6 \text{ N/cm}^2 \text{ gauge} \qquad \textit{answer}$$

14.21 ■ The General Gas Law (Ideal-gas Law)

We have found that *Boyle's law* describes the pressure-volume conditions of a gas held at constant temperature; that *Charles' law* governs the volume-temperature conditions of a gas maintained at constant pressure; and that, when volume is held constant, *Gay-Lussac's law* describes the pressure-temperature relationships. However, these limitations, i.e., constant pressure, or constant temperature, rarely occur in actual practice, since it is usually not possible to control the pressure or the temperature of a gas very closely under actual conditions of commercial and industrial use (see Fig. 14.24).

The laws of Boyle and Charles and Gay-Lussac are combined to form the *general gas law* (also called the *ideal-gas law*):

$$\frac{P_1 V_1}{T_1} = \frac{P_2 V_2}{T_2} = k \qquad \text{(a constant for a given quantity of gas)} \qquad (14.16)$$

This equation relates *absolute* pressures, *absolute* temperatures, and volumes of a given mass of gas. The term *ideal* is used since real gases liquefy at low temperatures and at high pressures, and the "law" is therefore not subject to verification at these extremes.

The *General Gas Law* combines Boyle's, Charles', and Gay-Lussac's laws into a single equation. Like the three laws which it combines, it is based on the concept of an ideal gas. Note the assumptions that are inherent in the ideal gas theory.

An *ideal gas* is a theoretical concept. No such "perfect" gas actually exists, although helium meets many of the criteria. The concept involves a number of assumptions, of which the following are typical:

1. A large number (trillions) of molecules are in completely random motion within a finite (sizable) volume.

2. No gravitational, magnetic, or other "action-at-a-distance" forces exist be-

Fig. 14.24 Butane tanks at a petroleum refinery. Industrial gases are usually stored under high pressure so the volume can be minimized. Since the volume of these tanks is a constant, their daily *P-T* (pressure-temperature) behavior could be predicted by Gay-Lussac's law. (Standard Oil Co. of New Jersey)

tween molecules—only the forces of impact and reaction during collisions. This assumption is reasonably true only when the molecules are not too close together.

3. Collisions between molecules occur very frequently, and they are assumed to be perfectly elastic. Both KE and momentum are conserved at collision.

4. The molecules are extremely small compared to the spacing between them.

It will be noted from Eq. (14.16) above, that when $T_1 = T_2$ (temperature constant), *Boyle's law* results:

$$P_1V_1 = P_2V_2 \qquad \text{or} \qquad PV = k$$

When $P_1 = P_2$ (pressure constant), *Charles' law* is obtained:

$$V_1/T_1 = V_2/T_2$$

And, when $V_1 = V_2$, Gay-Lussac's law results:

$$\frac{P_1}{P_2} = \frac{T_1}{T_2}$$

A general statement of the *ideal gas law* is

$$PV/T = k \tag{14.17}$$

Eq. (14.17) for an ideal gas shows the ratio PV/T equal to a constant k. The nature of this constant is perhaps of more interest to the chemist and theoretical physicist than it is to the technician or engineer, but an explanation of it and of its relationship to the gas laws is essential, in any event.

Decades of carefully controlled experiments, coupled with detailed theoretical analyses, show that the constant k is actually the product of two factors, n and R, where

n = the number of moles (review Sec. 13.8) of gas in the volume V of Eq. (14.17)

R = the *universal gas constant*. The value of R has been determined experimentally and is *the same for all gases*. The actual numerical value depends on the units being used, as follows:

The *universal gas constant R*. Its numerical value depends on the units being used. It has the *same value for all gases*.

Units of P (Absolute pressures)	*Units of V*	*Value of R*
N/m^2 or Pa (pascals)	m^3	$8.314 \dfrac{J}{g \cdot mol \cdot K}$
Atm	liters (L)	$0.0822 \dfrac{L \cdot atm}{g \cdot mol \cdot K}$
N/m^2 or Pa	m^3	$8314 \dfrac{J}{kg \cdot mol \cdot K}$
$lb/in.^2$ (psi)	ft^3	$0.0236 \dfrac{psi \cdot ft^3}{g \cdot mol \cdot °R}$

In addition, then, to Eqs. (14.16) and (14.17) there is a third formulation of the ideal gas law that involves the actual quantity (moles) of gas being subjected to the *P-V-T* conditions specified. This formulation, called the *equation of state* for an ideal gas, is

$$PV = nRT \tag{14.18}$$

The equation of state for an ideal gas.

where
P = the absolute pressure on the gas
V = the volume of the gas
n = the number of moles of the gas in volume V
R = the universal gas constant, whose value depends on the units used (see above)

It should be noted that n (the number of moles of the gas) is in fact the actual mass of the gas divided by the *molecular weight* of the gas, or

$$n = \frac{m}{M} \tag{14.19}$$

Interpreting the meaning of Eqs. (14.18) and (14.19) necessitates recalling from Sec. 13.8, that

1. A *mole* is the gram-molecular weight (or, if SI units are being used, the kilogram-molecular weight) of a substance; that is, a mass in grams (or kilograms) equal numerically to its molecular weight. (Abbreviations are g · mol and kg · mol.)
2. A mole of any substance contains the same number of molecules as a mole of any other substance. The number of molecules in a mole is *Avogadro's number*—$N_A = 6.025 \times 10^{23}$ molecules per gram-mole, or 6.025×10^{26} molecules per kilogram-mole.
3. At standard conditions of pressure and temperature (STP) (1.013×10^5 Pa, and 273 K), a mole of *any gas* occupies 22.4 L; a kilogram-mole, 22,400 L.

One mole of any substance (element or compound) has the same number of molecules as a mole of any other substance. That number is Avogadro's number, 6.025×10^{23} molecules per gram-mole.

Some typical problems involving the ideal gas law follow.

Illustrative Problem 14.6 A cylinder contains a mass of 500 g of carbon dioxide (CO_2). How many carbon dioxide molecules are there in the cylinder?

Solution The gram-molecular weight of CO_2 is 12 + 32 = 44
The number of g · mols of CO_2 is [from Eq. (14.19)],

$$n = \frac{m}{M} = \frac{500 \text{ g}}{\dfrac{44 \text{ g}}{\text{g} \cdot \text{mol}}} = 11.36 \text{ g} \cdot \text{mol}$$

Multiplying by Avogadro's number (the number of molecules in a gram-mole),

$$N = n \times N_A = 11.36 \text{ g} \cdot \text{mol} \times 6.025 \times 10^{23} \text{ molecules/g} \cdot \text{mol}$$
$$= 68.44 \times 10^{23} \text{ molecules} \qquad \textit{answer}$$

Illustrative Problem 14.7 A small spherical tank whose volume is 12 ft^3 is to be evacuated and then filled with nitrogen at a gauge pressure of 3000 psig and a temperature of 95°F. What weight of nitrogen is forced into the tank?

Solution Convert 95°F to 555°R, and 3000 psig to 3014.7 psia.

Solve Eq. (14.18) for n,

$$n = \frac{PV}{RT}$$

Substituting

$$n = \frac{3014.7 \text{ lb/in.}^2 \times 12 \text{ ft}^3}{\dfrac{0.0236 \text{ lb/in.}^2 \cdot \text{ft}^3}{\text{g} \cdot \text{mol} \cdot {}^\circ\text{R}} \times 555^\circ\text{R}}$$

$$= 2760 \text{ g} \cdot \text{mol of } N_2$$

$$\text{The mass of } N_2 = \text{g} \cdot \text{mol of } N_2 \times \text{molec. wt. of } N_2$$
$$= 2760 \times 28 = 77{,}280 \text{ g}$$

Converting to pounds weight,

$$\frac{77{,}280 \text{ g}}{454 \text{ g/lb-wt}} = 170 \text{ lb} \qquad \textit{answer}$$

Illustrative Problem 14.8 A helium cylinder has a volume of exactly 3.0 ft^3. It is filled to an *absolute* pressure (gauge plus atmospheric) of 2400 lb/in.^{2} during the night at a temperature of 60°F. What will be the pressure in the tank at the fairgrounds in direct sun, when the temperature of the gas in the tank is 90°F?

Solution First convert temperatures to the Rankine (abs.) scale.

$$60^\circ\text{F} = 520^\circ\text{R}, \qquad \text{and } 90^\circ\text{F} = 550^\circ\text{R}$$

Making use of Gay-Lussac's law (constant volume), solving for P_2,

$$P_2 = P_1\left(\frac{T_2}{T_1}\right)$$

Substituting,

$$P_2 = 2400 \text{ lb/in.}^2 \text{ abs.} \times \frac{550^\circ\text{R}}{520^\circ\text{R}}$$

$$= 2540 \text{ lb/in.}^2 \text{ abs.} \qquad \textit{answer}$$

14.22 ■ International Standards for Temperature Measurement

The four temperature scales discussed above are very satisfactory for scientific and engineering work in normal temperature ranges. Based as they are on the boiling and freezing points of water and on the behavior of gases, they are not always useful for industrial purposes. Industry has a need for a more comprehensive scale for higher temperatures, utilizing certain of the practical industrial thermometers discussed in previous pages.

Such a practical temperature scale has been worked out and adopted by most nations of the world. It is called the International Practical Celsius Temperature Scale (IPCTS). A series of fixed points based on certain standard metals is the central idea of the scheme. Some of these fixed points are listed in Table 14.4.

Internationally agreed-upon standards for temperature measurement.

It was agreed that certain thermometric methods would be used for appropriate temperature ranges. A few of these are given in Table 14.5.

Down to −260°C (13 K) a platinum resistance thermometer can be used with fair accuracy. Between this point and absolute zero, temperature-measuring methods have not been standardized, but low-temperature physicists are using methods based on the variations in magnetic susceptibility, with temperature, of certain crystalline salts (see Sec. 18.10 on *cryogenics*). Ultrasonic thermometers, which send very short, high-frequency vibrations through the substance in question, are also used. The smaller the wave velocity, the lower the temperature of the substance.

Table 14.4 Some fixed points on the International Practical Celsius Temperature Scale (IPCTS)

Standard	*Temperature, °C*
Normal boiling point of liquid hydrogen	−252.9
Normal boiling point of liquid oxygen	−182.96
Normal melting point of ice	0.00
Normal boiling point of water	100.00
Boiling point of liquid sulfur	444.60
Melting point of silver	961.9
Melting point of gold	1064

Table 14.5 Thermometric methods for stated temperature ranges

Range	*Standard Method of Measurement*
From −190 to 630°C	Platinum resistance thermometer
From 630°C to the gold melting point	Thermocouple of platinum and platinum-rhodium alloy
Above the gold point	Monochromatic optical pyrometer

QUESTIONS AND EXERCISES

1. Diagram and explain the operation of (*a*) a dial thermometer, (*b*) a simple thermocouple.

2. If a service station owner takes delivery of gasoline from a tank truck on a hot summer day, and stores it in his cool undergound tanks, does he lose or gain gallonage when he pumps it into his customers' cars?

3. On a graph paper, plot a neat graph which will allow you to convert Fahrenheit temperatures to Celsius temperatures all the way from −50 to 250°F. Use the vertical axis for Fahrenheit readings and the horizontal axis for Celsius readings. What happens at −40°?

4. Temperature readings from the freezing point of water down to zero degrees Fahrenheit are positive readings. When the same temperature range is observed with a Celsius thermometer, the readings are all negative. Explain.

5. Make use of a diagram or sketch, and explain why the top surface of a lake or pond freezes first as winter sets in. What might be the result if freezing occurred from the bottom up?

6. Explain the relationship between *heat* and *temperature*. Let your explanation show that you are familiar with the kinetic-molecular theory of heat energy.

7. Use a sketch to show how a bimetallic element works. If possible, examine a heating or cooling thermostat and diagram the basic elements which make it function.

8. Do some library reading about Benjamin Thompson (Count Rumford). Describe his experiments that led to the abandonment of the old *caloric theory* of heat.

9. The temperature of molten steel is to be measured inside an electric furnace. What methods and instruments would you recommend?

10. Show mathematically that the general gas law (Eq. 14.16) represents a combination of Boyle's, Charles', and Gay-Lussac's laws.

PROBLEMS

Group One

1. The freezing point of mercury is −38.9°C. Express this as (*a*) a Fahrenheit reading; (*b*) a Kelvin reading.

2. A nurse announces that your temperature is 38.6° Celsius. If you ask her to express this in Fahrenheit, what reading should she give?

3. A Fahrenheit thermometer indicates a room temperature of 68°F. What Celsius temperature is this?

4. An influenza patient ran a temperature of 104.2°F. In three days of treatment the temperature was down to normal, 98.6°F. Express this range for a Celsius thermometer.

5. The temperature gauge on the instrument panel of a sports car registers 85°C. What Fahrenheit temperature is this?

6. The directions on an aerosol spray can say it is not to be stored at temperatures higher than 95°F. Express this in degrees Celsius.

7. On the radio you hear the temperature announced as −12°C. What Fahrenheit temperature is this?

8. A metal rod begins to glow red when heated to about 700°C. Express this temperature in degrees Rankine.

9. The boiling point of alcohol is 171°F. How many kelvins is this?

10. Superheated steam enters the first stage of a turbine at 440°C and emerges from the final stage at 110°C. Express these as Rankine temperatures.

Group Two

11. A 110-ft invar-steel tape was calibrated as correct at a temperature of 72°F. On a winter day when the temperature was 48°F, the tape was used to mark off a 1-mi race course. The value obtained and recorded was precisely 5,280.00 ft. What should be the correct value?

12. Steel rails are 40 ft long and are laid when the temperature is 40°F. What minimum gap would be allowed between rails (inches) if the maximum temperature expected is 130°F?

13. Steel cables on the suspension section of the San Francisco Bay Bridge are about 9000 ft long. If the temperature range is from 25°F (winter) to 90°F (summer), what is the variation in cable length?

14. A metal rod 1.0 m long expands exactly 1.20 mm when heated from 10°C to 100°C. Calculate the coefficient of linear expansion of the material.

15. A valve pushrod in an internal combustion engine is made of mild steel and is exactly 12.000 cm long at 20°C. If the clearance between the rod and the rocker arm is to be 0.038 mm at the engine operating temperature of 95°C, what must be the clearance at 20°C? (Answer to the nearest 0.001 mm.)

16. A brass hot-water pipe is 40 ft long when installed at a temperature of 50°F. By what length (in inches) will it expand when filled with hot water at 200°F?

17. A survey line is run on the desert (120°F) and measures exactly 18 mi. The 100-ft invar-steel alloy tape used was standardized at 60°F. By how many feet (too long or too short?) was the 18-mi measurement in error?

18. A Pyrex glass flask holds exactly 1000 cm^3 of methyl alcohol at 20°C. How much will overflow when the alcohol is heated to 70°C?

19. A 100-gal steel tank is completely filled with gasoline at 30°F. If the temperature rises to 100°F, how many gallons of gasoline will overflow?

20. The density of aluminum at 20°C is 2.70 g/cm^3. Calculate its density at 400°C, accurate to three significant digits.

21. The density of mercury at 0°C is 13.596 g/cm^3. Find its density at 100°C, accurate to 4 significant digits.

22. A certain mass of gas occupies 3.5 m^3 at 150°C. If pressure remains constant, calculate the volume at 0°C.

23. A copper pipe is 85 ft long when installed outdoors at a temperature of 40°C. How many centimeters will it contract with a nighttime temperature of 6°C?

24. A stainless steel gauge block (see Chap. 2) is calibrated at exactly 1.0000 in. at 60°F. What would be the percent error of using it in a warm machine room at 95°F?

25. Find the approximate weight of all the air in a room measuring 12 ft × 15 ft × 9 ft. Assume standard conditions of temperature and pressure (STP). (See Table 13.1.)

26. A round hole 10.00 cm in diameter is cut in a copper sheet at a temperature of 22°C. Find the diameter of the hole when the copper sheet is heated to 150°C.

27. A section of a concrete bridge is 250 ft long. What is the change in length of this section from a cold morning at 20°F to a sunny afternoon temperature of 70°F?

28. The tires on an auto are inflated to a gauge pressure of 28 psig at a temperature of 45°F. After an hour of fast

driving the tire temperature is 110°F. Assume no change in tire volume and calculate the new gauge pressure.

29. A combustible mixture of gases is confined in an engine cylinder under the following conditions: $P_1 = 1$ atm; $V_1 = 350$ in.^{3}, $t = 80$°F. The piston moves up on the compression stroke, at the end of which the conditions are $P_2 = 8$ atm; $V_2 = 45$ in.^{3} (Pressures given are absolute.) Find the final temperature of the gas mixture just before the cylinder fires.

30. A gas is confined in a cylinder with a *freely sliding* piston. The gas temperature is initially 10°C. To what temperature (°C) must the gas be heated to have it expand and move the piston out enough to triple the gas volume? Assume the pressure remains constant at 1 atm.

Group Three

31. Physics students for half a century have been assigned this puzzler: A copper band is placed completely around the earth at the equator, with a snug fit at 80°F. (Assume earth is a smooth sphere.) Then if the temperature of the copper band rises by 5°F, and it is readjusted so that there will be equal clearance (space) between it and the earth's surface all the way around, what is the magnitude of the space? (Assume the earth does not expand.)

32. It was pointed out (Sec. 14.5) that at −40°, a Fahrenheit thermometer and a Celsius thermometer have the same reading. There is another point where they have the same *numerical* reading, but one is below zero (minus) and the other is above zero (plus). Determine these two readings by a graphical plot.

33. A piston compresses gas in a cylinder. Before the compression, the volume of gas was 0.55 L, pressure was 1 atm, and temperature was 80°C. At maximum compression, the volume is reduced to 0.08 L, and the pressure is read from a gauge as 11×10^5 Pa. What is the temperature of the compressed gas in degrees Celsius?

34. A truck tire has a volume of 2.5 ft^3 when filled with air at a gauge pressure of 85 lb/in.^{2}. How many cubic feet of atmospheric air is required to inflate this tire? (Assume no temperature change.)

35. An auto tire is inflated to a pressure of 26 lb/in.^{2} gauge on a cool morning when the temperature is 40°F. After a high-speed run the temperature of the tire is 150°F. Assuming no volume change, what is the new gauge pressure in lb/in.^{2}? In pascals?

36. Cold outside air at −20°C is drawn into a furnace heat exchanger and heated to a discharge temperature of 60°C. If pressure is assumed constant, by what factor is the volume multiplied?

37. An air pump with a piston whose diameter is 4 in. and whose stroke is 12 in. takes atmospheric air and compresses it into a tank whose volume is 5 ft^3. Assuming 100 percent volumetric efficiency of the pump, and no temperature change (neither assumption is true in actual practice), find the gauge pressure in the tank after 100 strokes of the pump. The tank has air at atmospheric pressure to start.

38. A refrigeration compressor takes in 20 in.^{3} of cold refrigerant "gas" at each stroke of the piston and compresses it to a volume of 9.0 in.^{3} (mass constant) before discharging it to the condenser. The "cold gas" enters the compressor cylinder at 50°F and 37 lb/in.^{2} gauge pressure. The "hot gas" leaves the compressor at 120 lb/in.^{2} gauge pressure. Find the Fahrenheit temperature of the hot gas, assuming no heat losses to surroundings.

39. A high-compression auto engine takes in 8.5 volumes of fuel-air mixture at 40°C and atmospheric pressure and compresses it to 1 volume at 245°C. Find the absolute pressure in the cylinder at the end of the compression stroke, in newtons per square meter (Pa).

40. A steel I-beam has a cross-section area of 12 in.^{2} It is riveted and welded in place on a "skyscraper" construction project when the temperature is 50°F. What is the least force that will prevent it from expanding in length as its temperature rises to 84°F? Why does its original length not enter into the problem? (HINT: Review Young's modulus.)

41. Show mathematically that the volume of 1 g · mol of an ideal gas at STP is 22.4 L.

42. A quantity of 50 mol of CO_2 is forced into an initially evacuated cylinder whose volume is 6 L. The temperature at the conclusion of the process is 30°C. Calculate the final pressure in the cylinder, in Pa.

CHAPTER 15 HEAT AND CHANGE OF STATE

According to the kinetic theory, heat is molecular kinetic energy transferred from one body or substance to another as a result of temperature difference. In general, when heat flows into a body the temperature of the body rises; and when heat flows out of a body, its temperature drops. However, under certain prescribed conditions, heat can be exchanged between bodies or substances *without any temperature change*. Such heat exchanges occur during a reorganization of the structure of a body or substance called a *change of state*.

Three states of matter—solids, liquids, and gases—were defined in Chap. 10. The "state" (or "phase") of a substance depends on its molecular arrangement and also on its temperature and its internal energy at the time. Pressure is also a factor in determining the state of a substance.

Many change-of-state processes occur naturally, but we are more concerned here with the changes of state that are made to take place for scientific, engineering, and industrial purposes. Common examples of forced changes of state are found in steam-generating plants, in chemical plants and petroleum refineries, in heating and refrigerating systems, in foundries, in steel making, and in the smelting and refining of ores.

HEAT QUANTITY AND HEAT MEASUREMENT

Heat quantities cannot be measured by thermometers, because heat is energy in transition. Thermometers measure *heat intensity*, not *heat quantity*.

Thermometers are useful instruments for measuring *temperature*, but a thermometer alone cannot measure the *amount* of heat *energy* that flows in or out of a substance. The total quantity of heat required to warm up, or melt, or vaporize a substance, or the quantity of heat a substance will give off as it cools (or as it condenses or solidifies) cannot be directly measured by a thermometer. What is needed is a unit of *heat measurement*, and a method of determining heat flow that involves temperature change as one of the factors in the calculation.

If two bodies or substances at different temperatures are placed in contact, heat energy will flow from the hotter one to the cooler one. The temperature of the hotter one falls, and the temperature of the cooler one rises. The heat exchange will continue until *thermal equilibrium* (both bodies reach the same temperature) is attained. Heat energy always flows from the hotter object to the colder object. And, as we have emphasized before, heat will *not flow* unless there is a temperature difference.

15.1 ■ Heat Quantity

In a strictly rigorous definition, the term *heat* has the sense of *energy transfer due to a temperature difference*. In the light of the present kinetic-molecular explanation of heat energy, to say that a given body or substance *has*, or *contains* a stipulated quantity of heat is incorrect, since heat is energy in transit.

Fig. 15.1 Molten (liquid) steel at about 2900° Fahrenheit gushes from an electric furnace into a huge 150-ton ladle. Heat energy and change of state are basic to the entire metals industry. (Bethlehem Steel Corporation)

15.2 ■ Units of Heat

Since heat is a form of energy it can only be measured in terms of the effect it has on some material substance. The substance chosen to standardize the heat unit is water. As usual, we define two basic units, one for the metric system, and one for the English (engineering) system.

The old "caloric theory" of the nature of heat has been replaced by the kinetic-molecular hypothesis, but the "old" units of heat-quantity measurement are still very much in use.

Metric Unit of Heat

The basic unit of heat in the metric system is that amount which will raise the temperature of one gram of water one Celsius degree. This quantity is called a calorie (cal).

1 cal of heat raises the temperature of 1 g of water 1 C°. The actual dimensions of the calorie are g · C°.

In order to be compatible with SI-metric units, the kilogram-calorie, or *kilocalorie* (kcal), is the favored unit for scientific and engineering work.

1 kcal (1000 cal) raises the temperature of 1 kg of water 1 C°

The basic heat unit in the English (engineering) system is the *British thermal unit* (Btu). Its definition follows.

English System Unit of Heat:

1 Btu is the amount of heat necessary to raise the temperature of 1 lb of water 1 F°

One Btu of heat raises the temperature of 1 lb of water 1 F°. The dimensions of the Btu are lb · F°.

Since there are approximately 454 g in the mass of a standard English pound, and since 1 Celsius degree is equivalent to 1.8 Fahrenheit degrees, relationships between the metric and English units of heat are derived as follows:

$$1 \text{ Btu raises } 454 \text{ g water } 1/1.8 \text{ C}^\circ$$

or,

$$1 \text{ Btu} = 1 \text{ lb} \cdot \text{F}^\circ \times \frac{1 \text{ C}^\circ}{1.8 \text{ F}^\circ} \times \frac{454 \text{ g}}{1 \text{ lb}} \times \frac{1 \text{ cal}}{1 \text{ g} \cdot \text{C}^\circ} = 252 \text{ cal}$$

$$1 \text{ Btu} = 0.252 \text{ kcal}$$
$$1 \text{ kcal} = 3.97 \text{ Btu}$$

The phrase ''mass of a standard English pound'' is necessary for a rigorous definition, since the pound is a unit of weight or force. However, in industrial and commercial *practice* in the U.S. the pound is often used without the above distinction. We shall not repeat the ''mass of a pound'' phrase in these chapters on heat and heat engines, since it is in the study of heat and its engineering applications that the pound is most often treated as a mass unit.

Strictly speaking, the heat-absorbing capacity of water is not a constant over a broad temperature range. As a result, the calorie is rigorously defined as the amount of heat required to raise the temperature of 1 g of water from 14.5°C to 15.5°C. Similarly, a rigorous definition of the British thermal unit stipulates that it is the amount of heat required to raise the temperature of 1 lb of water from 63°F to 64°F. The variation with temperature is so small, however, that in engineering, industry, agriculture, and medicine, the value of the calorie and the British thermal unit are considered constant throughout the range of temperatures in ordinary daily use.

A Calorie (with the capital C) is actually a kilocalorie.

It should also be noted that the *kilocalorie* is called a *Calorie* (capitalized) by physicians and nutritionists. This measure of the energy content of foods is actually 1000 times as large as the metric *calorie*.

$$1 \text{ Calorie} = 1000 \text{ calories} = 1 \text{ kcal}$$

Heat quantity units can also be expressed in terms of energy units—joules and foot-pounds. The use of joules to express heat energy is becoming quite common in science and engineering research, but the foot-pound is not being used as an English-system heat quantity.

SI or MKS Heat Units Since heat is a form of energy, it really could be measured in *energy units*, and there is currently a definite trend in both science and engineering to express heat quantities in standard energy units, namely joules (J).

The relationships between the ''older'' heat units and standard energy units are given here:

1 kilocalorie (kcal) = 4186 joules (J)
1 calorie (cal) = 4.186 J
1 J = 2.39×10^{-4} kcal

Also:
1 Btu = 778 ft · lb = 1055 J
1 ft · lb = 0.00129 Btu = 1.356 J

15.3 ■ Measuring the Heating Values of Fuels—Heat of Combustion

On the basis of the definition of the unit heat quantity, the amount of heat supplied to a *given mass of water* can be determined as follows:

Let
H = heat quantity supplied
m = mass of water being heated
t_1 = initial temperature
t_2 = final temperature

The heat absorbed by the water is then

$$H = m(t_2 - t_1) \qquad (15.1)$$

If m is in grams and t_2 and t_1 in degrees Celsius, H will be in calories; or, if m is in kilograms and t_2 and t_1 in degrees Celsius, H will be in kilocalories.

For English-system heat measurements, Eq. (15.1) becomes

$$H = w(t_2 - t_1) \tag{15.1'}$$

where w is the *weight* in pounds on earth. The *slug*, as a unit of mass, is not used in heat calculations, because the Btu (based on the pound) has been and still is the basic unit of heat quantity in the English (engineering) system.

Units for extremely large quantities of heat—the *therm* and the *Quad*.

The *therm* (1 therm = 10^5 Btu) is often used as a unit to express large quantities of heat. An even larger unit is the *Quad* (1 Quad = 1 quadrillion, or 1000 trillion, or 10^{15} Btu). In a recent year the total energy demand of the United States was about 120 Quads.

In this age of petroleum, another unit of heat energy is sometimes encountered, expressed as a daily rate. *One million barrels per day oil equivalent* (MB/DOE) is often used in industrial reports and economic surveys. Depending on the fuel value of the oil being used, it takes about 170 million barrels of oil to produce one Quad of heat energy. The current usage of petroleum in the U.S. (1988) is in excess of 27 MB/day with the balance of total energy demand coming from such sources as coal, natural gas, nuclear energy, hydroelectric plants, solar energy, wind energy, and geothermal energy.

In order to measure the heating value (or *heat of combustion*) of fuels, all the heat given off by the fuel is absorbed by a given quantity of water. The container in which the water for heat-measurement experiments is heated is called a *calorimeter*. Figure 15.2 shows the elements of a laboratory calorimeter suitable for student use. In actual fuel value tests a modification called a *continuous-flow calorimeter* is used, one form of which (for gaseous fuels) is diagramed in Fig. 15.3.

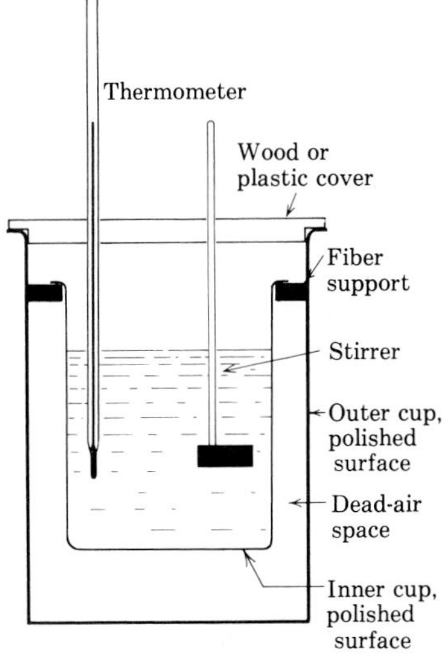

Fig. 15.2 Cross-section diagram of a simple laboratory calorimeter.

The hot gases of combustion go up through the central opening C (Fig. 15.3), and then down through a number of flues D. Water flows in at A and out at B. The rate of water flow is carefully controlled in order to obtain a steady but rather small difference (10 to 20 F°) between water-in and water-out temperatures. The test is run for a definite time, during which the fuel consumed and the water flow through the calorimeter are carefully measured.

From the amount of fuel used and from the number of kilocalories or Btu required to heat the measured amount of water to a fixed change in temperature (computed from Eq. 15.1 or Eq. 15.1′), the fuel value can be determined. Before the actual test trial is made, the apparatus must operate long enough for a steady-state condition of water flow and temperature difference to be obtained.

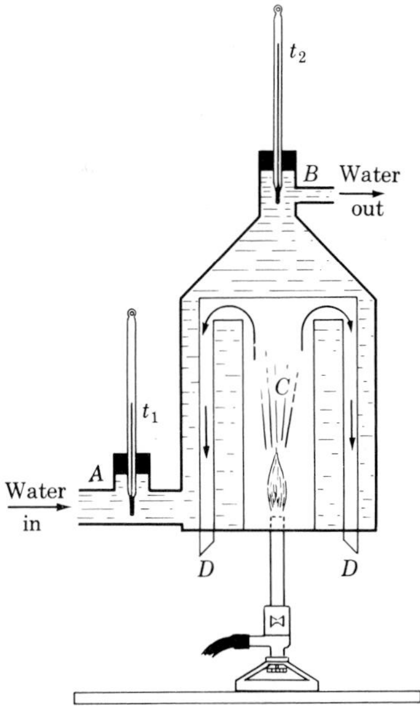

Fig. 15.3 One form of a continuous-flow calorimeter for laboratory use in determining the heating value (heat of combustion) of a gaseous fuel. Solid or liquid fuels would require a different kind of burner.

Illustrative Problem 15.1 During a fuel-value test on natural gas, 4.62 ft³ of gas were consumed while 195 lb of water passed through the continuous-flow calorimeter. Throughout the test run the water-in temperature was 62°F, and the water-out temperature was 85.5°F. Assuming all the heat was absorbed by the water and that the burner used was 100 percent efficient, what is the fuel value of the gas in Btu per cubic foot?

Solution Substituting in Eq. (15.1′),

$$\begin{aligned} H &= w(t_2 - t_1) \\ &= 195 \text{ lb} \times 23.5 \text{ F}^\circ = 4580 \text{ lb} \cdot \text{F}^\circ \\ &= 4580 \text{ Btu} \quad \text{(heat absorbed by the water)} \end{aligned}$$

Dividing,

$$\text{Fuel value} = \frac{H(\text{Btu})}{V(\text{ft}^3)} = \frac{4580 \text{ Btu}}{4.62 \text{ ft}^3}$$

$$= 990 \text{ Btu/ft}^3 \qquad \textit{answer}$$

The heating values (heats of combustion) of all common fuels have been determined by this and similar methods. Table 15.1 gives some average values for typical fuels.

Table 15.1 **Heating value, or heat of combustion, of some common fuels**

Fuel	Heating Value Btu/lb	Btu/ft³ at STP	kcal/kg or cal/g
Coal (best grade)	14,000		7,800
Coke	13,000		7,200
Diesel oil	19,500		10,700
Fuel oil (varies)	19,000		10,800
Gasoline	20,000		11,300
Kerosene	19,800		11,000
TNT (for comparison)	6,520		3,600
Gases:			
Natural	24,000	1,100	13,300
Manufactured (varies)	10,000–15,000	450–700	5,500–8,300
Hydrogen	61,000	2,800	30,800
Wood	5,000–8,000		2,800–4,450

Heats of combustion of fuels are measured in a continuous-flow calorimeter, using water as the heat absorber.

It is emphasized that the fuel values listed in Table 15.1 represent the *maximum possible* heat output of the fuels. In actual practice, every furnace, boiler, engine, smelter, or stove has efficiency losses ranging from 10 percent up to 65 or 70 percent.

Illustrative Problem 15.2 An industrial furnace burns natural gas at the rate of 500 ft^3/h. The furnace has an overall efficiency of 78 percent. Calculate the net heat *output* of the furnace in (*a*) Btu/h, (*b*) kcal/h, and (*c*) kJ/h.

Solution From Table 15.1, the heat of combustion of natural gas is 1100 Btu/ft^3.

(***a***) $$\text{Heat output} = 500\frac{\text{ft}^3}{\text{h}} \times 1100\frac{\text{Btu}}{\text{ft}^3} \times 0.78$$

$$= 429{,}000 \text{ Btu/h} \qquad \textit{answer}$$

(***b***) 1 Btu = 0.252 kcal (Sec. 15.2)

$$429{,}000\frac{\text{Btu}}{\text{h}} \times 0.252\frac{\text{kcal}}{\text{Btu}} = 108{,}100 \text{ kcal/h} \qquad \textit{answer}$$

(***c***) 1 Btu = 1055 J (Sec. 15.2)

$$429{,}000\frac{\text{Btu}}{\text{h}} \times 1055\frac{\text{J}}{\text{Btu}} = 4.53 \times 10^8 \text{ J}$$

$$= 4.53 \times 10^5 \text{ kJ} \qquad \textit{answer}$$

15.4 ■ Different Substances Absorb Heat at Different Rates—Specific Heat Capacity

A slab of iron in the direct rays of the sun becomes unbearably hot to the touch while a pan of water alongside it merely becomes lukewarm. Many everyday experiences tell us that some substances heat up more rapidly than others. If equal masses of copper and water are heated over the same flame for equal periods of time, the temperature of the copper rises about 10 times as fast as the temperature of the water. To cause the same temperature change, the water would have to be heated 10 times as long. It is thus seen that water has far greater ability to absorb heat than copper does. We say it has greater *heat capacity*. The *heat capacity* of a substance is *the quantity of heat required to produce unit temperature change*. Water has the highest heat capacity of any common substance. By definition, the *heat capacity of water* is 1 kcal/kg · C°, or 1 cal/g · C°, or 1 Btu/lb · F°.

Specific Heat Capacity

That quantity of heat whose addition or removal produces unit temperature change in unit mass (or weight in the English system) of any substance is termed specific heat capacity.

Specific heat capacity, defined.

Specific heat capacity is designated by the letter c. It is sometimes called, simply *specific heat*. It is the heat capacity per unit mass (or weight).

Since water has the highest heat capacity of any common substance, the specific heat capacity of water has been given the value of 1.00 kcal/kg · C° or 1.00 Btu/lb · F° ($c = 1.00$ for water). The specific heat capacities of all other substances are less than 1.00. It is important to note that, when the specific heat capacity of water is said to be 1.00, this means 1 kcal/kg · C°, or 1 cal/g · C°, or 1 Btu/lb · F°.

Water has the highest heat capacity of any substance. The specific heat capacity of water is 1.00 kcal/kg · C°, and all other common substances have specific heats less than 1.00.

Specific heat may be looked upon as a measure of the "thermal inertia" of a substance.

In industrial heating processes it is desirable to know in advance how much heat a material will absorb before reaching a certain temperature. Knowing the specific heat of a substance, we can predict in advance how much heat will be required to raise its temperature by any given amount. If H is the heat required, m the mass of the substance, c the specific heat of the substance, and $t_2 - t_1$ the temperature change, we can write *the fundamental heat equation*

$$H = mc(t_2 - t_1) = mc\,\Delta t \text{ (SI units)} \tag{15.2}$$

or

$$H = wc(t_2 - t_1) = wc\,\Delta t \text{ (English units)} \tag{15.2$'$}$$

The fundamental heat equation:

$H = mc\Delta t$ (C°) (SI)
$H = wc\Delta t$ (F°) (English)

If m is in grams and temperatures are in Celsius degrees, H will be in calories; and if m is in kilograms and temperatures are in Celsius degrees, H will be in kilocalories. For the English system, remember to use weight in pounds, and Fahrenheit degrees. H will then be in Btu.

15.5 ■ Measuring the Specific Heat Capacity of Liquids

The measurement of specific heats of liquid substances is carried out in a calorimeter similar to that pictured in Fig. 15.2. Specific heats of liquids are determined by using the *method of mixtures*, in which a known mass of the test liquid at a known temperature is poured into a known mass of hot water whose exact temperature is also known. Any two liquids at different temperatures will, when mixed, soon come to some common intermediate temperature (thermal equilibrium). The heat given up by the hotter liquid will warm up the cooler liquid until the final temperature is the same throughout the mixture. *If there is no heat lost or gained in the process* (and the calorimeter is designed to keep losses near zero), the heat units given up by the hot liquid as it cools will be exactly equal to the heat units absorbed by the cold liquid as it warms up. This *law of mixtures*, or *law of heat exchange* results in the general equation

The specific heat capacity of substances is determined from measurements with calorimeters, making use of the *law of heat exchange*.

$$m_h c_h (t_h - t_f) = m_c c_c (t_f - t_c) \tag{15.3}$$

$$\text{Heat lost by hot substance} = \text{heat gained by cold substance}$$

where the subscript h refers to the hot substance, the subscript c refers to the cold substance, and t_f stands for the final temperature of the mixture.

The c which is *not* a subscript is the *specific heat capacity*.

As stressed above, Eq. (15.3) assumes no heat lost or gained in the mixing process. Of necessity, however, the mixing occurs in a container (the inner cup of the calorimeter), and it too will absorb heat from the hot substance and warm up to the final temperature of the mixture. (The dead-air space, fiber support ring, and polished surfaces tend to prevent heat transfer and radiation losses outside this inner cup.) We must therefore modify Eq. (15.3) to include the amount of heat absorbed by the inner calorimeter cup and its associated equipment. Writing m_{cal} and c_{cal} for the mass and specific heat, respectively, of the calorimeter's inner cup, we find that Eq. (15.3) becomes

$$m_h c_h(t_h - t_f) = m_c c_c(t_f - t_c) + m_{cal} c_{cal}(t_f - t_c)$$

$$\begin{matrix}\text{Heat lost by} \\ \text{hot substance}\end{matrix} = \begin{matrix}\text{heat gained by} \\ \text{cold substance}\end{matrix} + \begin{matrix}\text{heat gained by} \\ \text{calorimeter}\end{matrix}$$

which is usually rearranged and written in this form:

$$m_h c_h(t_h - t_f) = (m_c c_c + m_{cal} c_{cal})(t_f - t_c) \tag{15.4}$$

The quantity $m_{cal}c_{cal}$ represents the heat absorbed by the calorimeter cup per degree temperature change. The term $m_{cal}c_{cal}$ is sometimes referred to as the *water equivalent* of the calorimeter cup.

$$\text{WE}_{cal} = m_{cal} c_{cal} \tag{15.5}$$

Again, remember to use weight (w) for English-system heat-exchange calculations.

Illustrative Problem 15.3 One hundred fifty grams of ethyl alcohol ($c = 0.60\ \text{cal/g} \cdot \text{C}°$) at a temperature of 20°C is poured into a calorimeter that already contains 100 g of water at 85°C. The liquid mixture is stirred until thermal equilibrium is reached. The mass of the aluminum calorimeter cup is 60 g, and c for aluminum is 0.22 cal/g · C°. Find the final temperature of the mixture.

Solution Use Eq. (15.4), noting that the calorimeter cup is, in this case, one of two "hot" substances.

Substituting values,

$$\left(100\ \text{g} \times 1.00\frac{\text{cal}}{\text{g} \cdot \text{C}°} + 60\ \text{g} \times 0.22\frac{\text{cal}}{\text{g} \cdot \text{C}°}\right)(85 - t_f)\text{C}°$$

$$= \left(150\ \text{g} \times 0.60\frac{\text{cal}}{\text{g} \cdot \text{C}°}\right)(t_f - 20)\text{C}°$$

$$\begin{aligned} 113.2\ (85 - t_f) &= 90(t_f - 20) \\ 9620 - 113.2\ t_f &= 90\ t_f - 1800 \\ 203.2\ t_f &= 11{,}420 \\ t_f &= 56.2°\text{C} \qquad \textit{answer} \end{aligned}$$

15.6 ■ Measuring the Specific Heat Capacity of Solids

Suppose it is desired to measure the specific heat of a certain metal sample whose mass is 125 g. Arrange a calorimeter as shown in Fig. 15.4. Place a known amount (say 300 g) of water in the calorimeter. Let its temperature (and that of the calorimeter cup) be 15.5°C. Heat the metal sample in an open steam bath for several minutes until it can be assumed to be at the steam temperature throughout. Record the temperature of the steam and also that of the water and calorimeter. Now lower the hot sample into the calorimeter water, cover with the wooden lid and stir or shake gently until the temperature as indicated by the thermometer reaches a steady final value. If the steam temperature (and the hot-metal temperature) is 100°C, the final temperature is 19.5°C, and the mass of the aluminum ($c = 0.22\ \text{cal/g} \cdot \text{C}°$) calorimeter cup is 110 g, we can substitute in Eq. (15.4) and obtain the specific heat of the metal sample as follows:

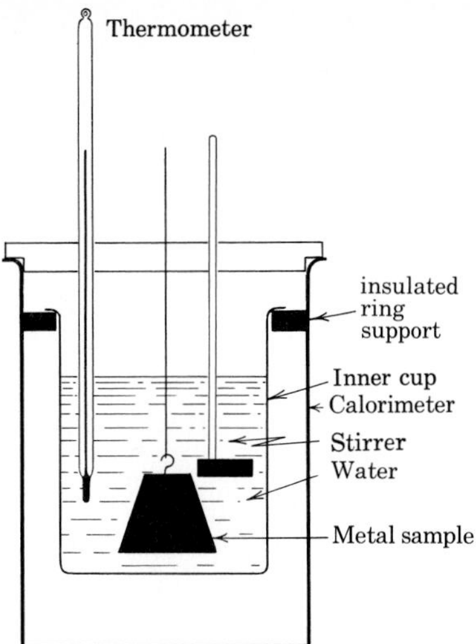

Fig. 15.4 A laboratory calorimeter fitted with items of equipment needed to measure the specific heat of a metallic solid.

$$125 \text{ g} \times c_h(100°\text{C} - 19.5°\text{C}) = (300 \text{ g} \times 1 \text{ cal/g} \cdot \text{C}° + 110 \text{ g} \times 0.22 \text{ cal/g} \cdot \text{C}°)(19.5°\text{C} - 15.5°\text{C})$$

from which

$$c_h = 0.129 \text{ cal/g} \cdot \text{C}°$$

Through carefully conducted experiments of this sort, specific heats for many common substances have been determined. Values for some common substances are given in Table 15.2.

Table 15.2 Specific heat capacities of common substances at 20°C

Substance	*Specific Heat Capacity*	*Substance*	*Specific Heat Capacity*
Air	0.24	Lead	0.03
Alcohol, ethyl	0.60	Masonry	0.20
Aluminum	0.22	Mercury	0.033
Brass	0.091	Silver	0.056
Copper	0.093	Steam (100°C)	0.48
Earth (dry soil)	0.20	Stone (avg)	0.192
Glass	0.21	Tin	0.055
Gold	0.032	Water	1.00
Ice	0.50	Wood (avg)	0.42
Iron (steel)	0.115	Zinc	0.093

NOTE: Specific heats have the same magnitudes in either metric or engineering (English) units. The values in the table can be read as cal/g · C°, kcal/kg · C°, or Btu/lb · F°.

Illustrative Problem 15.4 If 15 kg of steel ball bearings at 100°C are immersed in a bath of 25 kg of water at 20°C, assuming no losses of heat to or outside the container, what is the final temperature?

Solution Letting t_f represent the final temperature, the law of heat exchange, Eq. (15.3), gives

$$m_h c_h(t_h - t_f) = m_c c_c(t_f - t_c)$$

Multiplying, we have

$$m_h c_h t_h - m_h c_h t_f = m_c c_c t_f - m_c c_c t_c$$

Collecting terms and factoring to facilitate solution gives

$$m_h c_h t_h + m_c c_c t_c = t_f(m_c c_c + m_h c_h)$$

and

$$t_f = \frac{m_h c_h t_h + m_c c_c t_c}{m_c c_c + m_h c_h}$$

Substituting,

$$t_f = \frac{15 \text{ kg} \times 0.115 \text{ kcal/kg} \cdot \text{C}^\circ \times 100 \text{ C}^\circ + 25 \text{ kg} \times 1.0 \text{ kcal/kg} \cdot \text{C}^\circ \times 20 \text{ C}^\circ}{25 \text{ kg} \times 1.0 \text{ kcal/kg} \cdot \text{C}^\circ + 15 \text{ kg} \times 0.115 \text{ kcal/kg} \cdot \text{C}^\circ}$$

$$= \frac{672 \text{ kg} \cdot \text{C}^\circ}{26.7 \text{ kg}} = 25.2^\circ\text{C}$$ *answer*

Illustrative Problem 15.5 How much heat in Btu's is required to raise the temperature of a one-ton block of steel from 90°F to 800°F?

Solution Use the fundamental heat equation (15.2′), with the weight in pounds. The specific heat of steel (Table 15.2) is 0.115 Btu/lb · F°. Substituting values gives

$$H = 2000 \text{ lb} \times 0.115 \text{ Btu/lb} \cdot \text{F}^\circ \times (800^\circ\text{F} - 90^\circ\text{F}) = 163{,}000 \text{ Btu}$$ *answer*

15.7 ■ Water and Heat

Inspection of Table 15.2 shows how remarkably high the heat capacity of water is compared to other common substances. The amount of heat absorbed by a substance when it is heated is exactly equal to the amount it will give off as it cools through the same temperature range. Therefore large quantities of water can act as excellent storehouses for heat energy. In lakes and oceans, water even acts as a controlling factor, or brake, on seasonal temperature changes. In summer, large bodies of water absorb vast quantities of the sun's heat which would otherwise cause unbearably hot temperatures on the adjacent land; and in winter this heat in the water is slowly given up to the surrounding air, thus tending to moderate the winter weather. This accounts for the "even" climate of islands and narrow strips of land surrounded by the sea.

Water is a frequently used heat-transfer medium for industrial and engineering projects.

Water as a Heat-Transfer Medium For the same reason (high thermal capacity), water is a universally used substance for transferring heat in industry. Water is heated in boilers and pumped through pipes to radiators perhaps hundreds of feet away, there to give up heat as it warms homes, apartments, or shops. Cooled, it is pumped back to be heated again and the cycle continues, water serving as the carrier of heat, or the *heat-transfer medium*.

Illustrative Problem 15.6 A hot-water boiler and associated pump circulates 20 gal/h of hot water to the baseboard finned-tube radiators in a small apartment. The water-in temperature at the radiators is 180°F and the water-out temperature is 90°F. How much heat (Btu/h) is given off to the rooms?

Solution Use Eq. (15.2′), the fundamental heat equation. Let t = the time period = 1 h.

$$\frac{H}{t} = \frac{w}{t} c\,(t_h - t_c)$$

Substituting

$$H_{\text{Btu/h}} = 20 \text{ gal/h} \times 8.3 \text{ lb/gal} \times 1.00 \text{ Btu/1b} \cdot \text{F}^\circ \times (180 - 90)\text{F}^\circ = 14{,}900 \text{ Btu/h}$$ *answer*

Besides being used to carry heat to spaces where heat is desired, water is used to carry heat away from spaces where it is not desired. Water circulates in the *jackets* around cylinders and combustion chambers of engines and picks up engine heat, preventing temperatures which would damage the engine parts. The hot water is pumped to a radiator and the heat is dissipated there to the surrounding air. Chilled water is circulated through air-conditioning cooling coils, and as warm air is forced over and through the chilled-water coil, the air gives up its heat to the water. As the water is warmed, the air is cooled, a heat-transfer process. The cooled air is then forced by a blower into the spaces to be air-conditioned (see Chap. 18).

CHANGE OF STATE (PHASE)

The fact that all matter exists in one of three states—*solid, liquid,* or *gas*—has been discussed at some length. Whether a substance exists in the solid state, the liquid state, or the gaseous state depends on its temperature, its heat content, and on the pressure to which it is subjected. The temperature of a given mass of a substance depends on the heat content and on its thermal capacity. As an example let us investigate the effects of changing heat content on the temperature of water at standard atmospheric pressure (76 cm Hg or 14.7 lb/in.^{2}).

The state (or phase) of a substance at any given time depends on its pressure and temperature conditions and on its total heat content.

As mentioned earlier, the addition or removal of heat from a body or substance does not always result in a change of temperature. Changes in internal energy can also result in changes of state or phase, under conditions which vary for every different substance. These conditions for *water* are discussed in detail in the following section.

15.8 ■ The Temperature-Heat (*T-H*) Diagram for Water—Pressure Constant

As an example, take 1 lb of ice at -20°F and add heat to it slowly. Its temperature is found to rise 1 F° for each 0.5 Btu of heat added. (This means that the specific heat c of ice is 0.5 Btu/lb · F°.) The line *AB* of the diagram of Fig. 15.5 shows how the temperature of ice rises as heat is added through a temperature rise of 52 F°.

We see from the figure that 26 Btu of heat will raise the temperature of the pound of ice to 32°F, the melting point of ice. This is also the *freezing point* of water—a temperature at which liquid and solid exist together in thermal equilibrium. Here it will be observed that the addition of more heat does not result in any further tempera-

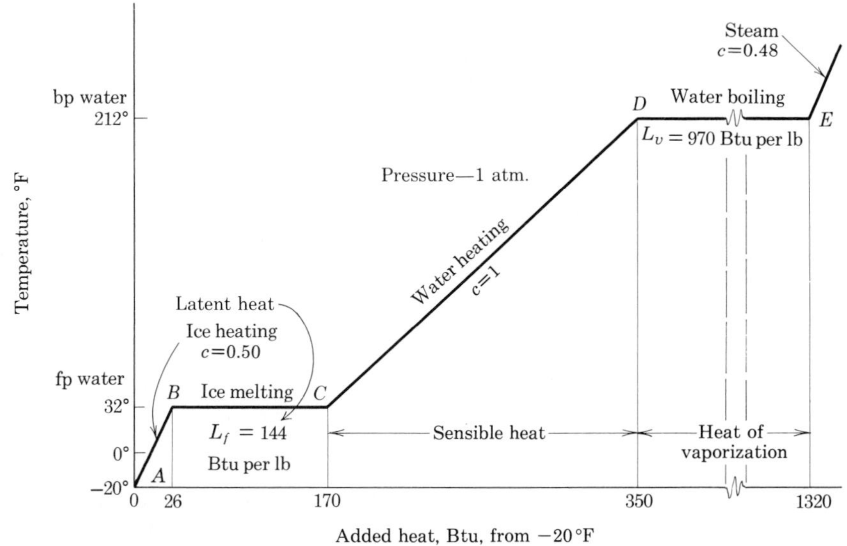

Fig. 15.5 Approximate temperature-heat (*T-H*) diagram for 1 lb of pure water (initially ice at -20°F) at atmospheric pressure. As heat is added to the ice, the temperature and the phase (or state) of the water are described by the plot shown. Note the discontinuity or break in the horizontal (heat-content) scale. (English-units).

The behavior of a substance as it changes phase from solid to liquid to vapor is well illustrated by its temperature—heat diagram (pressure constant).

ture increase but that the water begins to melt from solid to liquid form—a *change of state*. Careful measurements show that it takes 144 Btu to change (melt) 1 lb of ice at 32°F into liquid water at 32°F. Since this heat does not show up in increased temperature, we say it is hidden or *latent heat*. The process of melting is technically called *fusion*.

Latent heat of fusion, defined.

The latent heat of fusion of any substance is the amount of heat required to melt unit mass of the substance without any change of temperature or pressure.

Latent heat of fusion of water

$$L_f = 144 \text{ Btu/lb} \quad \text{(English system)}$$

The melting (fusion) process is shown by line *BC* of the diagram. By the time all the ice has melted (point *C* of the diagram) a total of 170 Btu (26 + 144) of heat has been added since the process was started at −20°F.

The further addition of heat results in increased temperature at the rate of 1 F° rise for each British thermal unit added, since the specific heat of liquid water is 1.00 Btu/lb · F°. (See line *CD* of the diagram.) After having absorbed another 180 Btu the water will be at *state point D*, at a temperature of 212°F, ready to boil. Total heat added (above −20°F) to this point is 350 Btu. The heat added between point *C* and point *D* is called *sensible heat*, since a change in heat added can be "sensed" as a temperature change. At the boiling point, water will change to vapor, or conversely, vapor can condense to liquid water, depending on the *latent heat* situation. To boil away the entire pound of water requires the addition of another 970 Btu of heat, and again this process occurs (line *DE*) *with no change in temperature*. Note that *DE* is a broken line, since the original scale cannot be maintained on the textbook page size. The *DE* process is technically called *vaporization*.

Latent heat of vaporization, defined.

The latent heat of vaporization of any substance is the amount of heat required to vaporize unit mass of the substance without any change of temperature or pressure.

Latent heat of vaporization of water (at 212°F)

$$L_v = 970 \text{ Btu/lb} \quad \text{(English system)}$$

This brings us to *state point E* of the diagram with 1 lb of steam at 212°F. This steam is said to be *saturated* or *wet steam*. At point *E* 1320 Btu of heat has been added to the pound of water, measured from the starting point at −20°F.

The same process for one kilogram of water is diagramed in Fig. 15.6. Using it and the explanation above for the English system, you should be able to follow each step of the metric *T-H* diagram for water and understand what is happening in every stage of the process.

The *metric* values for *latent heat of fusion* and *latent heat of vaporization* of water are

$$L_f = 80 \text{ kcal/kg} = 80 \text{ cal/g} \quad \text{(metric)}$$
$$L_v = 540 \text{ kcal/kg} = 540 \text{ cal/g} \quad \text{(metric) (at 100°C)}$$

In a manner similar to that just described for water, but with markedly different ranges of temperature, pressure, and heat content most common substances will

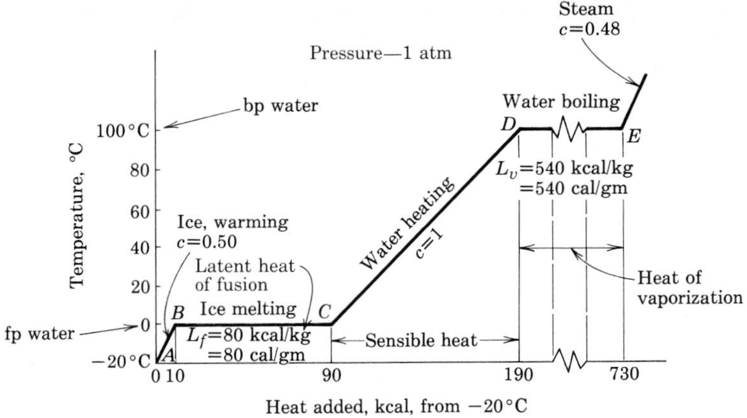

Fig. 15.6 Temperature-heat (*T-H*) diagram for 1 kg of pure water (initially ice at −20°C) at atmospheric pressure. (SI-metric units).

undergo changes of state as heat is supplied or taken away. Substances which are normally solids (metals, glass, etc.) can be melted and even vaporized if sufficient heat is added. Substances like air, oxygen, and nitrogen—normally gases—can be liquefied and finally frozen solid if enough heat is taken away from them. Table 15.3 gives melting points and boiling points for some common substances, and Table 15.4 the latent heats of fusion and vaporization of various substances.

Illustrative Problem 15.7 How much heat must be removed by a refrigerator to produce ice cubes whose temperature is 15°F, from 20 lb of water whose initial temperature is 70°F?

Solution Heat removal occurs in three steps: (1) cooling the water to the freezing point, (2) removing the latent heat of fusion L_f, and (3) further cooling the ice from 32 to 15°F.

Step 1: $H_1 = 20 \text{ lb} \times 1 \text{ Btu/lb} \cdot \text{F}^\circ \times (70 - 32)\text{F}^\circ =$ 760 Btu
Step 2: $H_2 = 20 \text{ lb} \times 144 \text{ Btu/lb} =$ 2880 Btu
Step 3: $H_3 = 20 \text{ lb} \times 0.50 \text{ Btu/lb} \cdot \text{F}^\circ \times (32 - 15)\text{F}^\circ =$ 170 Btu
Total heat removed 3810 Btu *answer*

Table 15.3 Approximate melting points and boiling points of some common substances at atmospheric pressure

Substance	*Melting Point, °C*	*°F*	*Boiling Point, °C*	*°F*
Air	(about)−215	−353	−195 to −185	−321 to −297
Ammonia	−75	−102	−33	−28
Carbon dioxide	−58	−70	−60	−76
Copper	1080	1980	2300	4170
Helium	−273	−458	−269	−452
Hydrogen	−259	−434	−253	−423
Iron (steel)	1530	2790	2980	5400
Lead	325	620	1620	2950
Mercury	−39	−38	356	675
Nitrogen	−210	−346	−196	−321
Oxygen	−218	−362	−182	−297
Platinum	1770	3220	4300	7770
Tin	232	450	2260	4100
Tungsten	3365	6090	5900	10,650

Table 15.4 Latent heats of fusion and vaporization of some common substances at atmospheric pressure

	Latent Heat of Fusion L_f			*Latent Heat of Vaporization L_v*		
Substance	*kcal/kg*	*kJ/kg*	*Btu/lb*	*kcal/kg*	*kJ/kg*	*Btu/lb*
Alcohol, ethyl	24.9	105	45	204	853	367
Ammonia	108	453	195	327	1370	465
Copper	32	134	58	1211	5070	2170
Lead	5.9	25	10.6	210	880	315
Mercury	2.8	12	5.0	71	297	128
Nitrogen	6.2	26	11	47.8	200	85
Oxygen	3.3	14	5.9	51	212	92
Water	80	335	144	540	2260	970

15.9 ■ Melting and Freezing

Crystalline substances like ice, and most metals, melt at fairly definite temperatures. The melting point of these substances is generally at the same temperature as the freezing point. Many substances, on the other hand, are noncrystalline, or *amorphous*, like glass, waxes, and fats, and their softening or hardening is a gradual process which takes place not at a fixed temperature but over an appreciable temperature range.

The exact melting points and heats of fusion of crystalline substances are of great importance in the metallurgical industry. In metallurgy, almost every process depends on the ability to predict in advance what will happen when heat is added or taken away. The properties of toughness, hardness, malleability, tensile strength, and elasticity are determined by the molecular and crystalline structure of the finished metal. This molecular arrangement is, in turn, tailored to specifications by the metallurgist using accumulated knowledge of the effects of heat on the molecular and crystalline structure of the particular metal.

Melting Is Actually a Cooling Process When a substance such as ice melts without heat being intentionally supplied, the heat necessary to melt it is absorbed from its surroundings, with a consequent drop in temperature. Thus, *melting* can be thought of as a *cooling process*. For example, each pound of ice that melts in an icebox will absorb 144 Btu of heat from the food products in the space. If ice did not melt in an "icebox," it would do very little cooling. Conversely, *freezing* can be looked upon as a *heating process*, since as a liquid freezes, it gives up to its surroundings an amount of heat equal to its latent heat of fusion. This is *heat of solidification*, L_s, equal to L_f at the same temperature and pressure. When water freezes to ice, 144 Btu (latent heat of fusion) is given to the surroundings for each pound of ice frozen; or 80 kcal for each kilogram of ice frozen. Ranchers take advantage of this property of water by irrigating or spraying citrus trees and vineyards during freezing and subfreezing weather. The latent heat of fusion given off as the water freezes, can "warm" the trees, vines, fruit, and the immediately surrounding air, by several degrees.

At its melting point a substance can be either melting or freezing, depending on the direction of heat flow. If heat is being supplied to the substance, *melting* (a change of phase) begins and continues; whereas if heat is being removed, *freezing* (also a change of phase) results. As pointed out in Sec. 15.8, the melting (freezing) point is that temperature at which the liquid and the solid phases of the substance can exist together.

Change of Volume on Freezing Most substances contract on freezing, resulting in a greater density of the solid. Water and some alloys are notable exceptions to this rule. Water expands markedly as its freezes, and ice has a specific gravity of only about 0.9. Ice therefore floats in water—a fact that is indeed fortunate, else the polar ice caps would spread well into what we now call the temperate zones.

Type metal (an alloy of lead, antimony, and copper) expands considerably on solidifying, a property which is essential for clear type with sharp, well-defined edges. So-called *gray iron* also expands slightly on solidifying, making it an ideal metal for making castings, since it expands against the mold and reproduces every detail that the design calls for.

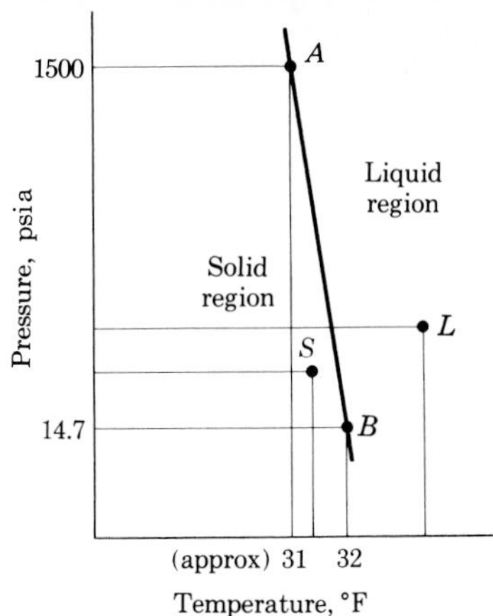

Fig. 15.7 Diagram showing the effect of pressure on the freezing point of water. A *P-T* curve (at constant heat) for the solid-liquid boundary. Pressure causes ice to melt at a lower temperature than it normally would. (Not to scale.)

Effect of Pressure on Freezing Everything said about melting and freezing thus far has been based on the assumption that normal atmospheric pressure prevails. But every ice skater has observed that the pressure of the skate blades actually melts the ice under the blade. This melting forms a slippery film of water between the ice and the blade which facilitates skating. Part of the explanation for this melting is that *pressure causes it by a sudden lowering of the freezing point*. Since the actual contact area between skate blades and ice is very small, the *pressure*, even for a small child, is very great. However (see Fig. 15.7), only a small reduction in freezing point is possible, and on *very cold* days, no melting occurs under the skate blades and skating becomes difficult if not impossible. Carefully conducted experiments have shown that for liquids which expand on freezing, such as water, increasing the pressure lowers the freezing point; and for liquids which contract on freezing a pressure increase raises the freezing point. The effect is very slight, however, because it takes a pressure of about 1500 lb/in.^{2} to lower the freezing point of water 1°F. An increase in pressure of 1 atm lowers the freezing point by only about 0.007°C.

The graph of Fig. 15.7 shows a pressure-temperature diagram for water in the vicinity of 32°F. The line *AB* represents a series of *P-T* points for which the solid and liquid phases are in equilibrium. *As long as no heat is added or taken away*, any point to the right of *AB*, for example, *L*, will represent a *P-T* situation for a *liquid*, and any point to the left of *AB*, for example, *S*, will represent a *P-T* situation for the *solid* phase.

An important recreational and professional sport (ice skating) would be impossible were it not for the effect of pressure on the freezing point of water.

We have been investigating some of the factors which cause a substance to undergo a change of state at the solid-liquid boundary. Let us now discuss change of state at the liquid-vapor boundary. We shall begin with an elementary analysis of the boiling of water.

15.10 ■ What Makes Water Boil?

Fill a flask half full of water (Fig. 15.8), support it over a burner, and place a thermometer and a steam outlet with a pinch clamp in the stopper. As the water heats, the first effect noticed is that air which was dissolved in the water is driven out as small bubbles. Later, bubbles of steam will form at the bottom but as they rise to the cooler water near the top they condense and disappear. Finally, as the temperature of the entire mass of water reaches 100°C or 212°F (at standard sea-level pressure), bubbles of steam reach the surface and *boiling* is said to occur. The steam in the flask is invisible, but as it shoots out into the room, it condenses into tiny fog droplets, which can be seen and which are ordinarily called "steam" in everyday usage. Liquid and vapor exist together.

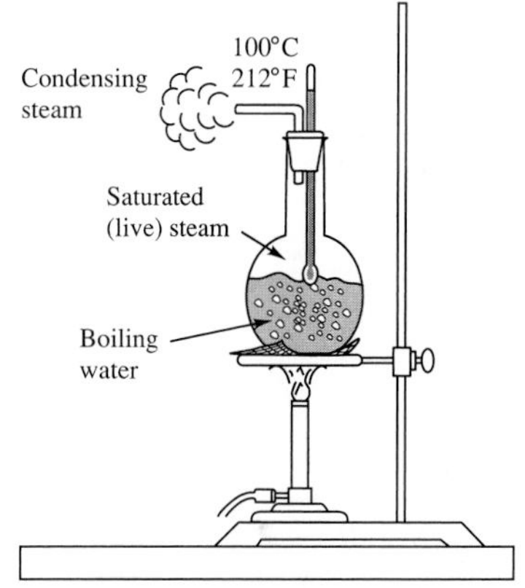

Fig. 15.8 The standard way to boil water is to add heat.

If the outlet valve is now *partly* closed by the pinch clamp (*careful!*), to create some back pressure in the flask, it will be noted that boiling stops until more heat is absorbed, the temperature rising perhaps 2 or 3°F (depending on the amount the pressure is increased) before full boiling begins again.

As a final experiment, remove the burner, allow the water to cool down well below the boiling point (say to 150°F), and connect a vacuum pump to the flask. As the pressure is reduced on the water surface, the water will burst into a vigorous "boil" even at this reduced temperature. (Pump very briefly, as moisture contaminates the vacuum pump oil rapidly.)

Boiling under a vacuum to lower the boiling-point temperature is common in many industrial processes—the refining of sugar and the processing of milk being two examples. The opposite process—boiling under increased pressure to raise the boiling point—is exemplified by the common kitchen pressure cooker. Anyone who has tried to cook beans, peas, or other foods at a high-altitude mountain camp, knows how the boiling point of water is lowered by reduced (atmospheric) pressure.

15.11 ■ The True Nature of Boiling

Boiling point—the temperature at which the vapor pressure in a liquid becomes equal to the pressure acting down on the surface of the liquid. "Normal boiling point" means that the pressure acting on the liquid surface is 1 atm.

We have seen that boiling depends not only on the temperature but on the pressure above the liquid. Heat energy added increases the temperature (see "sensible heat" portion of Figs. 15.5 and 15.6), which means an increased molecular speed of the liquid molecules. This causes some of the molecules of liquid to possess energies appropriate to the vapor state; i.e., bubbles of vapor are formed within the volume of the liquid. The formation of these bubbles is possible only when the pressure created by the vapor is equal to, or greater than, the pressure acting on the surface of the liquid. In accordance with these principles, *boiling point* is defined as *that temperature at which the vapor pressure within the liquid becomes equal to the pressure on the liquid surface* (see Fig. 15.9).

Vaporization of a liquid, on the other hand, occurs at all temperatures at which the substance exists as a liquid. [Actually, there is direct vaporization of some solids: ice, naphthalene (mothballs), solid carbon dioxide (dry ice) are examples. This vaporization of solids is called *sublimation*. Ice kept for months at a temperature well below its melting point will slowly lose weight due to sublimation.] Even at extremely cold temperatures, some molecules possess energies sufficient to "fly off" from a liquid surface. The escape of these high-energy molecules into the atmosphere above the liquid creates a pressure called *vapor pressure*. If a liquid is placed in a container and the space above it is evacuated of air, molecules of the liquid will leave the surface (evaporate) and collect in the space above the liquid (see Fig. 15.10). Because of

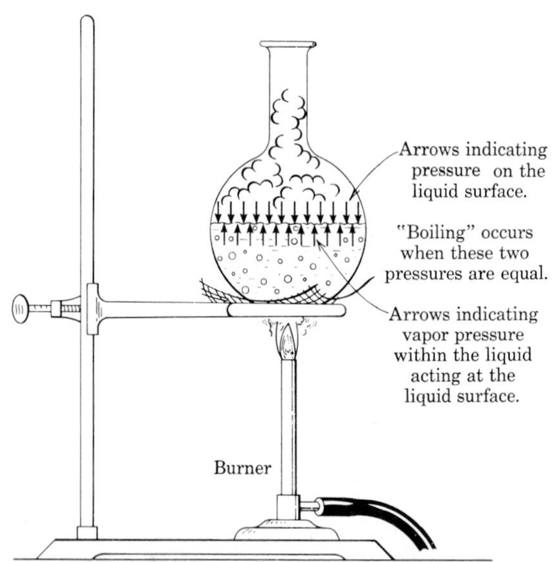

Fig. 15.9 Boiling point, defined. The boiling point of a liquid is reached when the vapor pressure within the liquid becomes equal to the pressure directly above the surface of the liquid. Pressure equilibrium can result either from adding heat to the liquid and thus increasing its internal vapor pressure or by reducing the pressure acting on the surface until it becomes equal to the *existing* vapor pressure in the liquid.

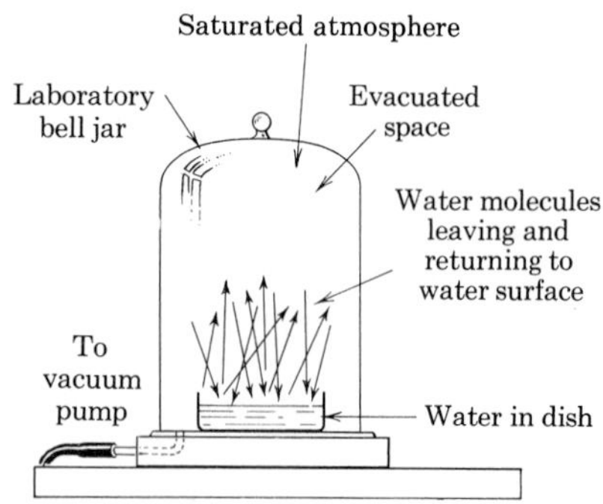

Fig. 15.10 Saturated vapor pressure. When the space above a liquid surface contains all the molecules of the liquid that it can hold, it is said to be *saturated*. Under the bell jar the space above a dish of water is evacuated so that it will contain only evaporated water molecules and no air. Water (or any other liquid) will evaporate into this space until the pressure above the liquid surface is equal to the internal vapor pressure of the liquid. This pressure is called the *saturated vapor pressure* for that temperature.

Table 15.5 Saturated vapor pressure of some common liquids

Liquid	Temperature °C	Temperature °F	Pressure in. Hg	Pressure mm Hg (torrs)
Alcohol, ethyl	20	68	1.73	43.9
	60	122	8.74	223
	78.3	173	29.92 (1 atm)	760
Mercury	20	68	0.000047	0.0019
	100	212	0.0107	0.272
	357	675	29.92 (1 atm)	760
Water	0	32	0.18	4.57
	20	68	0.69	17.5
	60	122	3.64	149
	100	212	29.92 (1 atm)	760

collisions with each other, some molecules fall back in. When the number leaving the surface is just equal to the number returning to the surface, an equilibrium condition is reached and the vapor above the liquid is said to be at its *saturated* vapor pressure for that temperature.

The higher the temperature of the liquid, the more high-energy molecules there are and the more will escape from the surface, creating a higher vapor pressure. The value of the vapor pressure depends on the temperature, and *only on the temperature,* for any given liquid. Table 15.5 gives values of the saturated vapor pressure of some common liquids for a normal range of temperatures. Table 15.6 gives some properties of saturated steam for selected Fahrenheit temperatures.

Illustrative Problem 15.8 The heating requirements of a residence have been calculated at 180,000 Btu/h for the winter conditions at the site. The house has a radiator system operating on saturated steam (1 atm pressure, 212°F) from a low-pressure steam boiler. How many pounds of steam per hour must be circulated through the baseboard radiators of the house to balance the "heating load" if the return to the boiler is hot water (condensed steam) at 140°F?

Solution See Table 15.6. Note that it sums up heat content of saturated steam from a base at 32°F. Each pound of saturated steam will give off 970 Btu as it condenses to hot water at 212°F. Further, each pound of the condensate (hot water) will give off 1 Btu/F° as it cools from 212°F to 140°F, providing 72 Btu/lb of additional heating for the rooms.

Let $\frac{w}{t}$ = the weight of steam required, lb per hour

Table 15.6 Properties of saturated steam at selected Fahrenheit temperatures

Absolute Pressure (Gauge + Atmospheric), lb/in.^{2} abs (psia)	Temperature, °F	Heat Btu/lb above 32°F: Sensible Heat (in Liquid)	Heat of Vaporization	Total Heat Content
1.0	102.2	70	1034	1104
14.7 (1 atm)	212 (bp)	180	970	1150
25	240	209	950	1159
50	281	250	923	1173
100	328	298	888	1186
250	401	376	824	1200
500	467	450	754	1204

Then $$\frac{w}{t}(970 + 72) \text{ Btu/lb} = 180{,}000 \text{ Btu/h}$$

$$\frac{w}{t} = \frac{180{,}000 \text{ Btu/h}}{1042 \text{ Btu/lb}} = 173 \text{ lb/h} \qquad \textit{answer}$$

15.12 ■ The Kinetic-Molecular Hypothesis

Many phenomena of physics cannot be satisfactorily explained in the sense that *physical proof* is readily available. Sometimes *theories* or *hypotheses* are formulated which satisfactorily explain observed behavior. Change-of-state phenomena, for example, can best be explained by a pattern of reasoning known as the *kinetic-molecular hypothesis*. This hypothesis assumes that in any quantity of a substance there is a somewhat random distribution of molecular velocities (energies). For example, in a quantity of gas (whose temperature as read from a thermometer is 65°F), it might be assumed that the 65°F temperature is only an *average* energy indication. The actual energies (speeds) of the trillions of gas molecules are assumed to vary all the way from those with near zero energy to those with energies associated with a temperature of hundreds of degrees. It would be logical to assume, however, that most of the molecules would have speeds (energies) at any given time which would cluster around the *average* value and that speeds near either extreme would be much less common.

James Clerk Maxwell (1831–1879) was able to show (in 1859) that statistically, molecules do have a definite (though not entirely random) probability distribution of speeds. He assumed that in a random assembly of (ideal) gas molecules in motion, all energy values are theoretically possible. Under his assumptions, the distribution of molecular speeds can be represented by a curve resulting from a plot of the relative numbers of molecules at various speeds, against the speeds themselves. Figure 15.11 gives the general form of the curve of a *maxwellian distribution*. Indicated on the curve are: (1) The predicted *average* speed; (2) the most probable speed; and (3) the root-mean-square (rms) speed.

If such a probability distribution of molecular speeds actually occurs, it is interesting to think about what would happen if the higher-energy molecules of a substance could be drawn off in some way. Would the temperature of the remaining mass change? Could a change of state result? The paragraphs that follow attempt to shed some light on these questions and to lend support to the kinetic-molecular hypothesis.

The kinetic-molecular hypothesis includes the assumption that in any quantity of a substance a full range of molecular speeds is probable, and that there will be a nearly random distribution of speeds. The Maxwell-Boltzmann distribution curve portrays the relationships for an ideal gas.

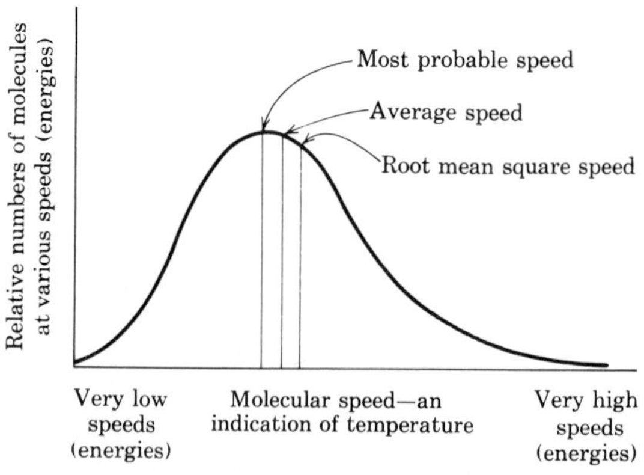

Fig. 15.11 The Maxwell-Boltzmann distribution curve of molecular velocities for a gas at a specified temperature, T. The curve, which approximates that for a normal distribution, but departs from it somewhat, illustrates one possible distribution of molecular speeds with respect to average speed. The average speed, in turn, is a measure of molecular kinetic energy and therefore of the absolute temperature of the gas. (Not to any scale.)

15.13 ■ Evaporation and Condensation

Evaporation is a phenomenon that occurs at all temperatures, as pointed out above. Since in general, it is the higher-energy molecules that leave the liquid, the average molecular energy (and speed) of the remaining molecules is lowered. This results in a drop in temperature of the remaining liquid. *Evaporation is therefore a cooling process.* There is direct, sensible evidence of this, since perspiration cools the body as moisture evaporates from the skin. Some factors which affect the rate of evaporation of liquids into the atmosphere are:

Evaporation is a *cooling process*, and condensation is a *warming process.*

1. The *volatility* of the liquid, which is measured by the vapor pressure it will create at a given temperature. Note from Table 15.5 that alcohol has a vapor pressure about 2.5 times that of water at room temperature. Alcohol is said to be more *volatile*. Thus, volatile liquids like alcohol, ether, and gasoline have a marked cooling effect on the skin, since they evaporate quickly at normal temperatures.

2. The *temperature*. At higher temperatures liquids evaporate faster, since more molecules have energies sufficient to escape from the liquid surface. Processes requiring evaporation, e.g., crystallizing salts and sugars from solutions and making pure water from seawater, are carried on at high temperatures.

3. The *pressure above the liquid*. The lower the pressure above the liquid, the higher the evaporation rate. The crystallization processes just referred to are usually carried out under a partial vacuum.

4. The *surface area* of liquid exposed. Greater surface area gives more molecules a chance to escape without falling back into the liquid.

5. *Air currents* blowing over the surface. These aid evaporation by removing those molecules already evaporated, thus preventing an equilibrium state from being reached above the liquid surface. The cooling effect of a breeze is thus explained, since the air movement allows faster evaporation.

6. The amount of *liquid vapor already present in the surrounding atmosphere*. Damp (high humidity) air will allow little evaporation of water, for example. Evaporation occurs readily into an atmosphere with low humidity. On the other hand, if the surrounding atmosphere is already saturated, for every molecule that leaves the liquid surface, one reenters from the saturated atmosphere above.

Condensation Saturated vapors condense to liquids at any given pressure if they lose an amount of heat equal to the latent heat of vaporization, L_v, at that pressure. For example (Table 15.4), at atmospheric pressure, one pound of saturated ammonia vapor will condense to saturated liquid ammonia if 465 Btu of heat is removed. Heat removal to cause a change of state is done commercially by a piece of equipment called a *condenser*, as we shall see in a later chapter. When vapors condense, the *latent heat* they possessed as vapors is transferred to and absorbed by the surroundings, where it shows up as *sensible heat*. This explains why, as clouds form and it begins to rain, the temperature of the atmosphere usually rises. *Condensation,* in this sense, *is a heating process*. *After* the rain, as water evaporates again, the temperature falls, confirming the observation made previously that *evaporation is a cooling process*.

The amount of heat given off as unit mass of a vapor condenses to a liquid is its *latent heat of condensation* L_c. $L_c = L_v$ at the same pressure and temperature.

15.14 ■ Some Interesting Demonstrations

The definition of boiling as the condition which exists when the vapor pressure of the liquid equals the pressure above the liquid will explain the following interesting demonstrations, which at first may seem contrary to everyday experience.

These demonstrations all offer support to the kinetic-molecular hypothesis.

1. *Boiling by cooling*. First bring some water to a boil in an open flask. Take away the burner and *working rapidly* and *carefully*(!) insert a tight stopper, fitted with a thermometer, in the flask. (This arrangement must be "airtight.") Invert the flask

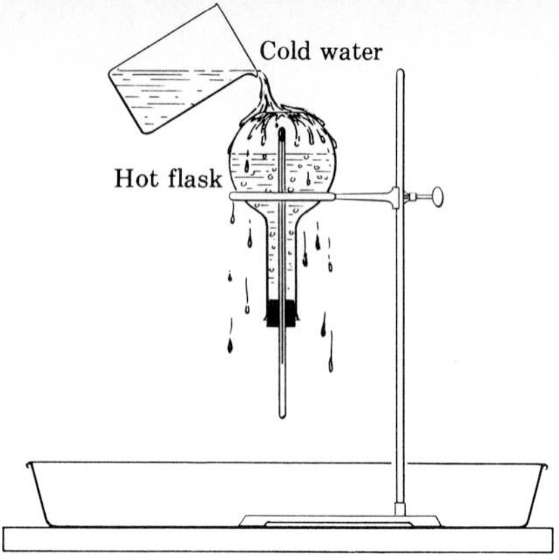

Fig. 15.12 Boiling water by cooling hot water—a seeming paradox.

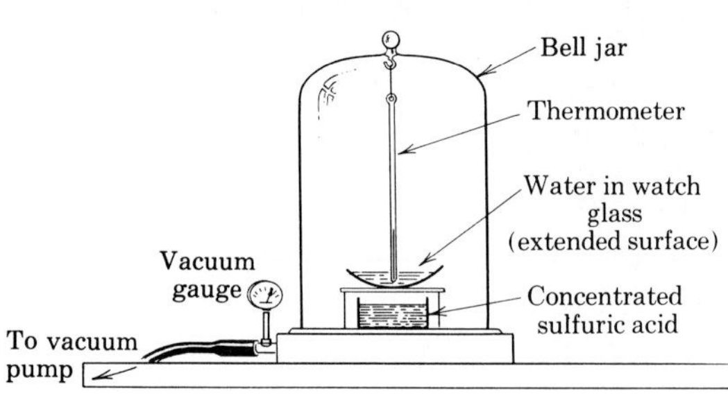

Fig. 15.13 Freezing water by boiling it without adding any heat. You have to see it to believe it. After you believe it, do you understand it?

The two experiments suggested here are best done as instructor demonstrations, since care and skill are required, both for the sake of safety and to prevent leaks, which will spoil the results.

and *quickly* pour cold water over it to prevent steam pressure from forcing the stopper out (see Fig. 15.12). As cold water is poured over the hot flask, the temperature drops but vigorous boiling continues. Cool the contents of the flask to about 150°F, and let stand. The boiling stops. Pour more cold water on; as the temperature drops further, boiling begins again!

Apparently we have boiled water by cooling it. The real explanation is left for the student to think through, keeping in mind the vapor-pressure explanation of boiling given above.

2. *Freezing by boiling under reduced pressure*. Place a small quantity of water in a watch glass under a bell jar which can be evacuated by a high-capacity vacuum pump. Suspend a thermometer in the water in such a way that its scale can be read through the bell jar. Under the water dish place a crystallizing dish of *concentrated* sulfuric acid, which helps the pump by absorbing water vapor rapidly. Seal all joints with vacuum wax, for even the slightest leak will prevent the demonstration from being successful. The arrangement is diagramed in Fig. 15.13 (NOTE: Large amounts of water vapor can damage a vacuum pump, since the vapor will condense and contaminate the oil. Change the pump oil after these demonstrations are completed.)

Turn on the pump and watch the vacuum gauge. When the pressure has fallen to about 5 cm Hg pressure, bubbles will begin to form in the water. These are dissolved air coming out of solution. At about 2 cm pressure, *real* boiling begins. As the pressure drops further, boiling continues, and the falling thermometer indicates that the water is being rapidly cooled. At about 7 mm pressure, boiling becomes intermittent, the bubbles appearing rather infrequently but bursting violently when they do appear. At about 5 mm pressure boiling almost ceases, and as the gauge drops toward 4 mm pressure, it will be observed that the thermometer registers 0°C. A few minutes' more pumping results in the sudden freezing of the water. Ice crystals form first around the edges; then needlelike crystals suddenly form across the entire dish. Watch closely or you may miss the sudden ice-crystal formation.

We have boiled water, not by heating it but by reducing the pressure on it. We have cooled water by boiling it, and continued boiling has frozen it into ice! If you can explain these phenomena in terms of the kinetic-molecular theory you are beginning to understand change of state.

15.15 ■ The Liquid-Vapor Boundary

Change of state at the liquid-vapor boundary can be portrayed by a *P-T* curve in much the way that Fig. 15.7 illustrates conditions at the solid-liquid boundary. The curve of Fig. 15.14 shows the *P-T* diagram (vapor-pressure curve) for water from about 230 to 32°F. The curve itself represents a series of *P-T* points for which the liquid and vapor phases are in equilibrium. Any point to the right of and below the curve, such as *V*,

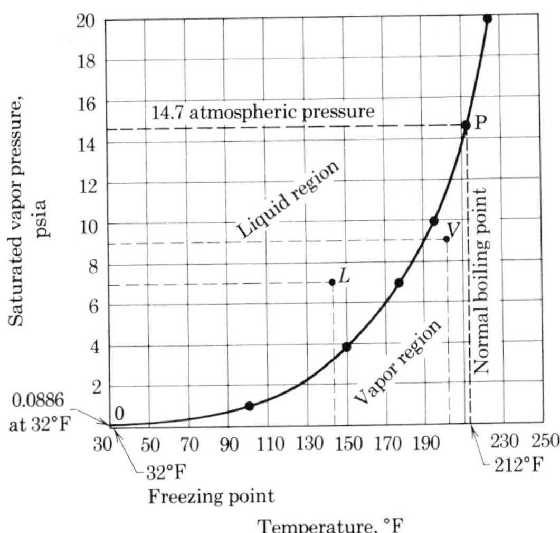

Fig. 15.14 The effect of pressure on the boiling point of water. This is a pressure-temperature (*P-T*) curve for water at the liquid-vapor boundary.

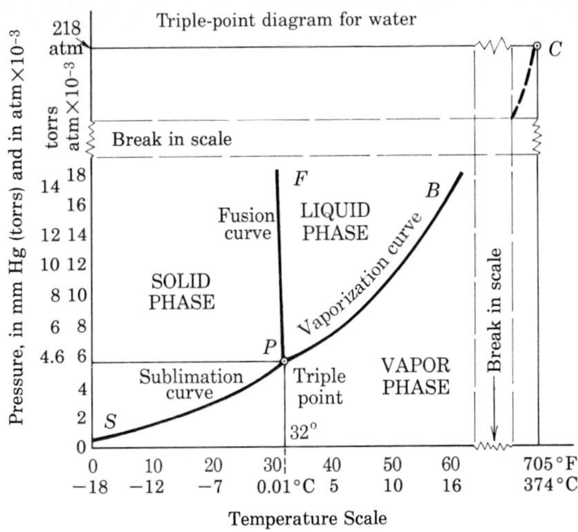

Fig. 15.15 The triple point diagram for water. The triple point is a *P-T* condition where all three phases (states) of a substance can exist simultaneously. Also shown (with a scale break in the curve) is the *critical point* for water. The critical point *C* marks the upper limit of the vaporization curve.

represents a *P-T* combination which will result in vapor; and any point to the left of and above the curve, such as *L*, is a *P-T* situation for liquid. Only the curve for water is shown here. Different *P-T* conditions would apply to other substances, but the general nature of the curve is similar.

15.16 ■ The Triple Point

In the experiment illustrated by Fig. 15.13 (freezing by boiling under reduced pressure), water existed in a very unusual situation as the process reached its climax. Ice was forming, liquid water was still present, and vapor was being given off (boiling was occurring) all at the same time. The *P-T* condition for water at this point, if accurate pressure and temperature readings are taken, is found to be $P = 4.6$ mm Hg (torrs) = 0.18 in. Hg; $t = 0.01°C = 32.002°F = 492°R = 273.17$ K. This *condition point,* at which all three states of matter (solid, liquid, gas) can momentarily exist simultaneously, is called the *triple point*. Many common substances exhibit triple-point behavior, but we shall consider only water in this discussion.

The *triple point* is a condition point on a pressure-temperature diagram of a substance at which all three states (phases) of the substance can exist simultaneously.

By choosing a somewhat different scale than that used in Figs. 15.7 and 15.14, it is possible to portray on a single diagram the relationships involved in the triple-point condition for water. In Fig. 15.15, *BP* is a section of the *vapor-pressure curve,* or boiling-point curve, for water. The *fusion curve FP,* the locus of state points in phase equilibrium between solid and liquid, is also drawn in, as is the *sublimation curve SP,* the locus of state points in phase equilibrium between solid and vapor. Pressures and temperatures shown on the diagram are only approximate. Note that at any temperature higher than about 0.01°C, no matter what the pressure, water will not exist as a solid, and that at any pressure less than about 4.6 torrs (0.006 atm), water will not exist as a liquid, no matter what the temperature.

Freeze Drying If the temperature of a substance is below that at its triple point, and if the then solid substance is subjected to a very low pressure, it will change phase *directly* from a solid to a vapor, without passing through the liquid phase. This property is often used to dehydrate substances which would be damaged if dehydration were accomplished by heating or other means. "Freeze drying" is accomplished in a vacuum chamber by first lowering the temperature of the substance below the freezing point of water, whereupon its water content freezes. A vacuum is then produced, which (see Fig. 15.15) moves the state point of the (now frozen) water content to the right of and below the sublimation curve. The resultant water vapor is pulled out of the chamber by the vacuum pump. Some examples of products produced by freeze drying are instant coffee, certain pharmaceuticals, and blood plasma.

The *critical point* is the upper end point of the vapor pressure curve of any substance. For water the critical point occurs at 218 atm and a temperature of 374°C. No substance can exist as a liquid at a temperature above its critical point, no matter what pressure is applied.

The Critical Point There is an upper end point to the vapor-pressure curve of any substance. Called the *critical point,* it occurs for water at a pressure of 218 atm and a temperature of about 374°C. No substance can exist as a liquid at a temperature above its critical point. No matter how great a pressure may be exerted upon it, a substance will still be in vapor form if its temperature exceeds the critical temperature for that substance. As the pressure gets greater its density increases, but it will not liquefy.

15.17 ■ The Importance of Change of State to Industry

Scores of industrial processes depend on change of state and on heat transfer.

Although much of the foregoing discussion of change of state has concerned itself with water, the student should not gather the impression that water is the only important substance in which industry uses change-of-state processes. Here are some other applications.

Refining Metals Most metals are mined in the form of ores, which contain many impurities, and frequently the metal is present in the form of an oxide. The basic procedure in obtaining the metal itself may be to heat the ore to the melting point, where impurities can be removed by chemical reduction, physical separation, or electrolytic means. Melting (a change of state) is the first step in the refining of many metals.

Fig. 15.16 A fractionating tower in a petroleum refinery. Pressure-temperature (*P-T*) conditions are carefully controlled over the entire height of the tower. Vaporized crude petroleum is separated into many components or *fractions* by such towers. Asphalt and heavy waxes and tars remain at the bottom. Heavy oils come off next, then fuel oil, distillate, kerosene, and gasoline in that order. Nearer the top the most volatile fractions, such as butane, propane, and methane are collected. (Chevron Corporation)

Refining Petroleum Petroleum is an exceedingly complex mixture of chemical compounds called hydrocarbons. The problem of the refiner is to separate out one useful compound from another, i.e., to separate butane, propane, gasoline, naphtha, kerosene, distillate, diesel fuel, fuel oil, etc., from the crude oil. This separation is effected by heating under carefully controlled temperature and pressure conditions in petroleum stills called *fractionating towers* (see Fig. 15.16). The more volatile hydrocarbons undergo a change of state first and are drawn off. Then come the next most volatile, and so on. These vapors are condensed by cooling water and are drawn off at the proper point on the fractionating tower. Pressures are controlled as needed, by pressure pumps or vacuum pumps.

Liquefaction of Gases Many gases are of great commercial importance today. Oxygen, hydrogen, acetylene, nitrogen, propane, carbon dioxide, and ammonia are just a few examples. Liquefied petroleum gas (LPG) is a major product on world markets. Oxygen and nitrogen are used by hospitals in the treatment of postoperative and other very ill patients. Oxygen and hydrogen serve as a team to provide rocket fuel for the space age. Propane and LPG serve as fuels on thousands of farms and in rural communities. Ammonia is the refrigerant at ice plants, cold-storage warehouses, and skating rinks; and huge quantities of it are used in the manufacture of fertilizers. Carbon dioxide is the ''fizz'' agent in a billion bottles of soft drinks consumed every year.

We think of most of these substances as *gases*. But for transport and use, nearly all of them are first liquefied. Each has its own change-of-state conditions, and these must be carefully observed in transit and storage. At the point of use, tanks of the liquid material, usually under high pressure, are fitted with precision valves and pressure regulators to assure safe and economical use of the ''bottled gas.''

Heat Engines and Change of State Internal-combustion engines, though the fuel may be a liquid, receive their energy from the sudden expansion of hot gas in the cylinders. The fuel-air mixture must be supplied to the cylinder in the form of a vapor. *Carburetion* is one process of vaporizing the liquid fuel and mixing it in proper proportions with air. Many gasoline and diesel engines are equipped with *fuel injectors* (one for each cylinder) to inject the proper amount of fuel as an "atomized" spray at high pressure. Steam engines, both reciprocating and turbine types, use water as the heat-transfer medium and cycle it back and forth from the liquid phase to the vapor phase (steam), using it over and over again. Heat engines are discussed in detail in Chap. 17.

Change of State and Refrigeration The entire refrigeration and air-conditioning industry is based on the principles of change of state. The process of refrigeration consists in removing heat from some space or substance where its presence is not wanted and rejecting it to some space or substance where its presence is immaterial. The medium for this heat removal is an easily liquefied vapor called a *refrigerant*. Refrigeration and air conditioning are the subject of Chap. 18.

These few examples drawn at random from the metallurgical, mechanical-power, and refrigeration industries are illustrative of the extremely wide applications of change-of-state phenomena to industry.

QUESTIONS AND EXERCISES

1. Explain what is meant by *heat capacity* and *specific heat capacity*. How do they differ in meaning?

2. Why does a large body of water tend to modify the climate in its immediate vicinity?

3. Explain why *melting* is a cooling process and *freezing* is a heating process. Also why *evaporation* is a cooling process and *condensation* is a heating process.

4. Give a technical definition of *boiling*. Use it to explain why it is possible to "bring water to a boil" without adding any heat to it.

5. Define L_f and L_v. What conditions must be stated in each case? How are these two related to L_s and L_c?

6. Define *fractional distillation*. Do some library research and explain the approximate *P-T* conditions required in a fractionating tower to obtain distillate, kerosene, gasoline, and butane.

7. Under what conditions can heat be added to, or taken away from, a substance without any temperature change occurring?

8. In cold climates, in addition to heating the air in their houses, many people also humidify the air. What are some reasons for doing this?

9. By what approximate multiple is saturated steam (212°F, 1 atm pressure) a more effective heat-transfer medium than 212-degree water, if the base temperature for both is 100°F?

10. Explain why, as fog rolls in or as rain commences, the air temperature usually rises. And why is it usually cooler soon after a rain?

11. Explain the mechanism by which perspiration cools the human body. Why does it not "work" in humid or "sticky" weather?

12. Does water boil at the same temperature all over the earth's surface? What factors affect the boiling point of water? By what techniques could you raise the boiling point? Lower it?

13. Citrus growers and growers of winter vegetables often face severe frosts that would damage or completely destroy their crops. One frost protection technique used is to run the sprinkler irrigation system and allow the water to freeze to ice all over the trees and plants. Explain the rationale for this operation.

14. Describe exactly what the "triple point" means for water. What significance does it have for science and industry?

15. Explain why water condenses on the outer surface of a glass containing an iced drink.

16. A manufacturer of iceboxes for camp and trailer use claims that the *ice compartment* is so well insulated that the ice will not melt for 10 days. Would this box be a good buy for protecting perishable foods?

17. It is a common experience to boil (i.e. evaporate) water and other liquids by adding heat to them. Many liquids with low boiling points (e.g., ammonia, carbon dioxide) evaporate or "boil" if the pressure on them is lowered suddenly. Where does the heat to boil them come from?

PROBLEMS

Group One

1. A swimming pool measuring 30 × 40 ft and averaging 5 ft deep is filled with water at a temperature of 66°F. How many Btus will the pool heater have to provide to warm the entire pool to a temperature of 82°F? Assume no losses during the heating operation.

2. A slab of hot pig iron at 800°F weighs 1.5 tons. How much heat does it lose as it cools to a temperature of 80°F?

3. Five thousand pounds of warm (100°F) air passes through a finned-tube cooling coil per hour. The air has negligible moisture content. How much cooling (Btu/h) is done by the coil if the air emerges from the coil at 57°F? (See Table 15.2.)

4. A 150-kg sheet of aluminum is heated from 21 to 450°C. How much heat was required? (*a*) In kcal, (*b*) in joules.

5. How many Btu are required to raise the temperature of 1 gal of water from 32 to 212°F?

6. A slice of whole wheat bread is rated by dietitians at 100 Calories. If the bread were used as a fuel and converted 100 percent into heat, how many grams of water would it heat from 15 to 75°C?

7. How many kilocalories of heat are required to raise the temperature of 700 kg of copper from 18°C to its melting point at 1082°C? How many joules?

8. It is desired to heat a steel ingot weighing 450 lb from 60 to 850°F. What is the minimum possible amount of heat (Btu) required?

9. Ten pounds of molten lead at 630°F are dropped through a sieve to form shot for shotgun shells. The shot cool to 600°F and solidify while falling. They land in 10 lb of water whose temperature is 50°F. What is the final temperature of the water?

10. An 80-kg person consumes 3000 kcal of food energy per day. If the same thermal energy were used to heat an 80-kg block of iron originally at 20°C, what would be the final temperature of the iron?

Group Two

11. A BLT (bacon-lettuce-tomato) sandwich is rated at 450 Calories, when provided with the mayonnaise and condiments. If the entire heat energy content of the sandwich could be converted to human work by climbing up a mountain, how high could a 150-lb person climb on the energy from the sandwich? (The energy equivalent of 1 Btu is 778 ft · lb.)

12. An auto whose mass is 1000 kg is travelling at 24 m/s. The brakes are slammed on in a panic stop and the car skids to a halt. (*a*) Find the work done against friction, in joules. (*b*) Assuming that all of this kinetic energy was dissipated in heat, how many kcal does this amount to?

13. A 60-gal water heater has a rating of 75,000 Btu/h. If it is 85 percent efficient, how long will it take to heat the 60 gal of water from 60°F to 135°F?

14. How many gallons of fuel oil (sp gr 0.76) are required to raise the temperature of 100,000 gal of water from 40°F to the boiling point? (Assume 100% efficiency of water heater.)

15. During a test run with a continuous-flow calorimeter, 5 kg of high-grade coal were used. The water-in temperature was 12°C, the water-out temperature 60°C. A volume of 690 L of water ran through during the test. The burner is known to be only 85 percent efficient. Calculate the fuel value of the coal in kilocalories per kilogram.

16. An oil burner for a home furnace must supply 40,000 kcal/h to the furnace heat exchanger. The burner is 60 percent efficient, and the fuel oil density is 0.72 kg/L. Find the consumption of fuel in liters per day (see Table 15.1).

17. Air at 32°F has a density of 0.081 lb/ft^3. If a furnace must heat 5000 ft^3/min of this air to a temperature of 140°F, using natural gas as the fuel, what will be the consumption of gas per day if the burner and heat-exchanger system is 65 percent efficient?

18. How much heat (Btu) is required to convert 1 gal of water at 50°F into saturated steam at 212°F?

19. How much heat is required to convert 500 kg of water at 10°C into saturated steam at 120°C?

20. Mercury in the laboratory is initially at 20°C. How much heat (kcal) must be removed from 10 kg of it to just freeze the entire quantity?

21. A 650-g iron stew pot, with 1 L of water in it is placed directly on the ''Cal-Rod'' heating element of an electric range. The heating element is rated at 1.8 kW. The initial temperature of the pot and water is 22°C. How long (min) will it take for the water to come to the boiling point? Neglect all losses to the surroundings.

22. If a natural gas–fired furnace has an overall efficiency of 80 percent, how many cubic feet of gas would be required to heat all the air in a house whose dimensions are 35 × 60 × 9 ft ceiling height, from 40° to 72°F? Take density of air as 0.08 lb/ft^3 and c for air as 0.24 Btu/lb · F°. Neglect other losses.

23. An electric hot-water heater has a heating element rated at 5 kW. Assuming no losses, how long will it take to heat 50 gal of water from 55 to 140°F?

24. How much heat is required to raise the temperature of 500 kg of lead, initially at 20°C, to its melting point and then melt it? (See Tables 15.2, 15.3, and 15.4.) Answer in joules.

25. Ships at sea make fresh water using "evaporators" to boil (i.e., evaporate) sea water, and then condense the steam. The condensed steam is salt-free "fresh" water. Assume c, L_v, b.p., and density are the same for sea water as for fresh water (not actually true). How much heat energy must be supplied (Btu) to produce 40,000 gal of fresh water from sea water at 60°F. Neglect all equipment losses.

26. An electric heating coil is installed in an air duct. Air at 40°F is blown through the coil at a rate of 500 ft^3/min (cfm). The density of air at this temperature is 0.08 lb/ft^3. If the air emerges from the coil at 120°F, at what rate (kW) is electric energy being used?

27. If the engine and drive train of an automobile have an overall efficiency of 14 percent, how far (miles) will 1 gal of gasoline propel a 2000-lb car up a 6 percent grade? (Be careful with units. Neglect losses such as road friction, air resistance, etc. Given that 1 Btu is equivalent to 778 ft · lb.)

28. What is the end result of removing 1250 Btu from 1 lb of saturated steam at 212°F?

29. After 10 gal of water at 70°F is poured slowly over a 50-lb cake of ice at 32°F, the ice is dried and weighed. What is its weight? (Assume all the water is cooled to ice temperature.)

30. If 50 kg of ice at −10°C is placed in an icebox, how much heat (in kilocalories) will be removed from the contents of the box by the time all the ice has melted and drained out as a liquid at 0°C? (Assume no heat gains or losses from or to the outside of the box.)

Group Three

31. Solar collectors commonly use circulating water as a heat-transfer medium. If the water-in temperature to a bank of roof-mounted solar collectors is 75°F and the water-out temperature is 140°F, what would be the necessary water flow rate (gal/h) through the collectors to remove 150,000 Btu/h from them? (Assume water density is 8.34 lb/gal.)

32. A low-pressure steam boiler is designed to produce 30,000 lb of steam (212°F, 1 atm pressure) per hour. The feed water enters the boiler at 60°F. The heat of combustion of the fuel oil is 19,000 Btu/lb, and its specific gravity is 0.70. If the burner-boiler system has an efficiency of 58 percent, how many gal/h of fuel oil are used?

33. A copper calorimeter cup (m = 100 g) contains 100 g of ice, floating in 200 g of water. The entire system is initially at 0°C. A steam line slowly bubbles 50 g of steam (100°C, 1 atm pressure) into the water, which is being gently stirred. What is the final "condition" of the calorimeter and its contents?

34. A 1-kg sample of an unknown metal is being tested for its specific heat. It is placed in an aluminum calorimeter whose mass is 227 g. Room temperature is 22°C. A mass of 0.45 kg of water at 95°C is poured into the calorimeter, covering the test sample. The final (equilibrium) temperature is 81°C. What is the specific heat of the sample, and (probably) what metal is it?

35. A 100-g aluminum calorimeter cup contains 200 g of water at 20°C. Fifty grams of ice at 0°C and 300 g of iron at 100°C are placed in the cup, and the contents are stirred until a steady state is reached. Compute the mass of ice and water and the final (equilibrium) temperature.

36. A 200-kg steel forging is taken from a soaking pit at 900°C and immersed in a cooling tank containing 100 kg of water at 50°C. How much water is left when temperature equilibrium is reached?

37. Saturated steam is supplied to a radiator at 25 lb/in.^{2} abs, and the condensed steam (water) leaves at 160°F and 1 atm pressure. How much heat is furnished to the room by each pound of steam? (Use Table 15.6.)

38. What capacity, in Btu per hour, must a refrigerating plant have to freeze twenty 300-lb cakes of ice in 4 h and take their temperature down to 0°F? The water is at 60°F to start, and the cans in which the water is contained are of steel, each weighing 65 lb.

39. Calculate the cost of heating 50,000 gal of water from 60°F and converting it into saturated steam under 500 lb/in.^{2} abs pressure if the fuel is natural gas costing $2.00 per thousand cubic feet. Assume the burner-boiler combination to be 65 percent efficient. (Use Table 15.6).

40. A low-pressure steam engine is operating on 10,000 lb/h of saturated steam, supplied from a boiler at 250 psia. (See Table 15.6.) The boiler "feed water" is supplied at 62°F. The boiler burners use commercial fuel oil, whose weight density is 6 lb/gal. The overall boiler/burner efficiency is 55 percent. Find the fuel oil demand in gallons per hour.

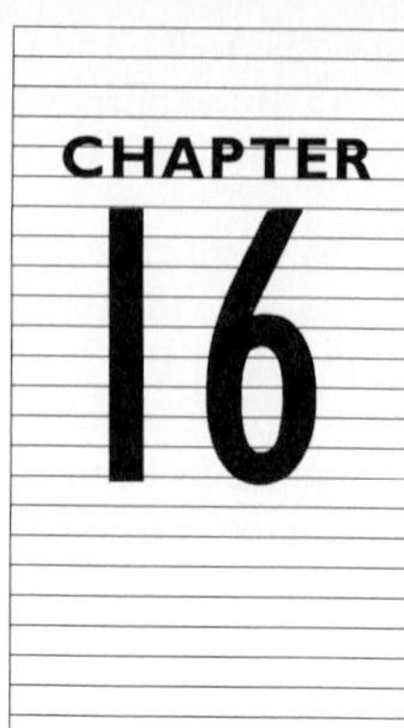

HEAT TRANSFER

We have made the point that heat is energy in transition, but not much has yet been said about the ways in which heat flows, or about the media through which it is transferred.

Heat flows "down a temperature hill."

No matter what the heat *quantity*, heat will always flow from a substance or source at a higher temperature to one at a lower temperature. In other words, heat flows from a substance at higher *temperature* not necessarily from a substance possessing a larger *quantity* of heat energy. It is said that "heat flows down a temperature hill," just as water flows downhill to lower elevations (see Fig. 16.1).

HEAT-TRANSFER PROCESSES

The three heat-transfer processes—conduction, convection, and radiation.

Heat can be transferred, or moved, by three different processes, all of which are common to everyday experience. These heat-transfer methods are known as *conduction, convection,* and *radiation*. We shall be able to give only the barest of introductions to each of these branches of the study of heat transfer. Each is a science in itself. Engineers and researchers spend entire lifetimes making new discoveries and developing new applications within each of these fields of heat transfer. We will present the important fundamentals in each field and indicate some of the more common practical applications.

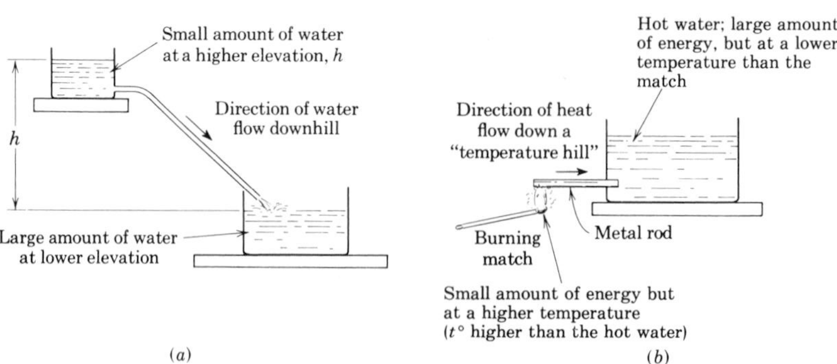

Fig. 16.1 Heat flow is analogous to water flow. (*a*) Regardless of the relative amounts of water at two locations, water will always flow from the location at a higher elevation to the location at a lower elevation. "Water flows downhill." (*b*) Heat also flows "down a temperature hill," to a substance at a lower temperature, regardless of the relative amount or quantity of heat available at the two different temperature levels.

16.1 ■ Conduction

Conduction is the primary method of heat transfer in solids. When a metal rod is held with one end in a flame, the other end soon gets hot. A pan set on a stove conducts the heat through to the substance contained in the pan. Heat from engine cylinders passes through the cast-iron cylinder walls to the cooling liquid in the "jackets" surrounding them. No motion of the heated metal can be observed by the eye, but the kinetic-molecular hypothesis would predict that the hot molecules are in violent motion. These higher-energy molecules and atoms in contact with the flame collide with their neighbors, striking them hard blows and thus passing on energy to them. These in turn strike others, and heat energy is thus passed on through the metal (see Fig. 16.2). The molecules themselves do not actually move along the rod; they vibrate more or less in place and pass the energy along, much as early fire-fighting companies passed water buckets along the line of volunteers from the town pump to the blaze. *Heat energy is transferred* by conduction, but the molecules of the medium through which it moves remain in place.

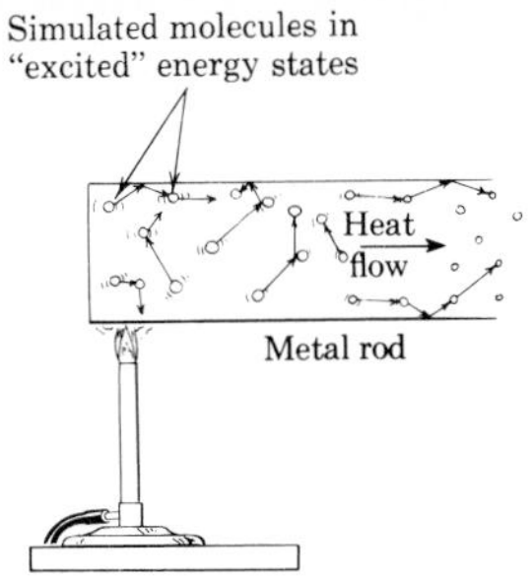

Simulated molecules in "excited" energy states

Heat flow

Metal rod

Fig. 16.2 A schematic explanation of heat conduction. Heat energy adds to the internal energy of the molecules and atoms of a substance. Molecules in "excited" energy states go into violent motion, colliding with adjacent molecules and thus passing the energy along throughout the substance.

> ***In conduction, heat energy is passed along or transferred by molecular collisions, without measurable movement of particles of the conducting medium.***

16.2 ■ Thermal Conductivity

All solid materials conduct heat to some extent, some being excellent conductors, others very poor conductors. (Liquids and gases also conduct heat but very poorly compared to most solids. Recalling the molecular theory, can you suggest a possible reason for this?) Metals, in general, are very good conductors; wood, glass, and brick are fair conductors; and porous substances with many tiny air pockets are poor conductors. It is an interesting fact that the heat conductivity of substances seems to parallel quite closely their electric conductivity. Good conductors of heat are generally good conductors of electricity, and conversely. Atoms of metals generally have "free electrons," and these play a part in both electrical conductivity and heat conductivity. The ability of a given substance to conduct heat is called its *thermal conductivity,* denoted by the letter k. The rate at which heat will flow through a slab or rod or wall depends on a number of factors, as follows:

Metals are generally good conductors of heat.

Thermal conductivity is the ability of a substance to conduct heat.

Conduction is the result of molecular, atomic, and electron vibratory motions of very small magnitude.

1. The *thermal conductivity* k of the material
2. The *cross-sectional area* A of the material through which the heat is flowing
3. The *temperature difference* ($t_2 - t_1 = \Delta t$) between the hot side and the cold side of the material
4. The *thickness* L of the material

Figure 16.3 shows the factors involved in heat conduction through solids.

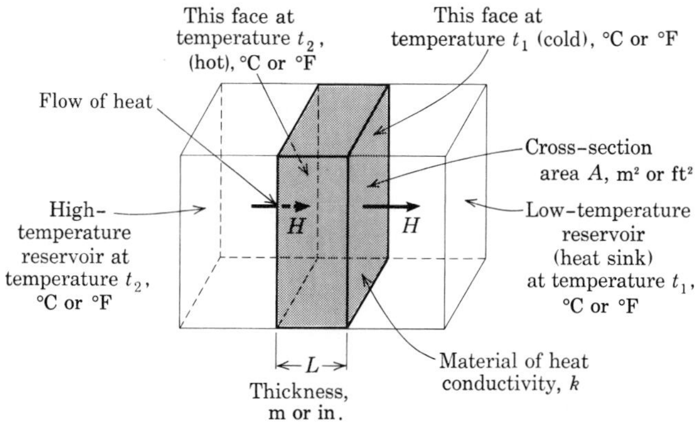

Fig. 16.3 Heat conduction through solids. The diagram shows the factors involved in the basic equation for heat conduction, Eq. (16.1).

The total amount of heat conducted through a material also depends on the length of time T the heat flows. Careful experiments have verified the following equation for heat transfer by conduction.

$$H = \frac{kAT(t_2 - t_1)}{L} = \frac{kAT\ \Delta t}{L} \tag{16.1}$$

In SI-units, H is the heat in kilocalories (or joules) conducted through a material whose cross-sectional area is A (in square meters) in time T (seconds), when the temperature difference Δt is $(t_2 - t_1)$C°. The thickness of the material L is measured in meters.

In English-engineering units, H is the heat (in Btu) conducted through a substance of cross-sectional area A (in square feet) in time T (hours) when the difference in temperature Δt of the two sides is $(t_2 - t_1)$F°. L is the thickness of the material in inches.

The units of k depend, of course, on what system of measurement is being used with regard to the other factors in Eq. (16.1).

Units for expressing heat conductivity.

Units for Heat (Thermal) Conductivity In the English system, k has the units Btu/(h)(ft^2)(F°/in.).

In the SI-metric system, two sets of units are in common use—the "older" form, in terms of kcal/s; and the now often-encountered form in terms of joules/s or *watts*. The SI-metric units of k are

1. kcal/(s)(m^2)(C°/m)
2. J/(s)(m^2)(C°/m) or, W/(m^2)(C°/m)

When k-values in watts are used, the time factor T in Eq. (16.1) is already included, since 1 J/s = 1 watt.

You may have noted that these two expressions for units of k (SI-metric) can be simplified by cancelling units of length, resulting in the following:

1′. kcal/(s)(m)(C°)
2′. W/(m)(C°)

Although these do represent *mathematical* simplifications, they result in units that are not easily related to the cross-section area and the thickness of the material for which k is the thermal conductivity. In this book, therefore, we shall use the somewhat more cumbersome but more understandable forms (1) and (2) above.

As the metric system comes into more common use in the U.S., heat flow rates expressed in watts will be encountered more frequently.

It should also be noted here that, although metrication is progressing very slowly in the United States, there is nevertheless some movement toward use of the SI units, expressing heat flow rates in watts. How rapidly this change will affect industry and commerce in the United States, only time will tell. Both the kcal/s and the watts units should be familiar to students.

Heat conductivities for all common substances have been determined by careful experiment. Table 16.1 lists values of k for a number of common materials and for some insulating materials.

Illustrative Problem 16.1 A plate-glass window in an office building measures 4 by 8 m and is 1.2 cm thick. When its outer *surface* is at a temperature of -4C° and the inner *surface* at 0°C, how much heat is transferred through the glass in 1 h?

Solution Use Eq. (16.1):

$$H = \frac{kAT(t_2 - t_1)}{L}$$

From Table 16.1, for glass

$$k = 1.9 \times 10^{-4}\ \text{kcal/(s)(m}^2\text{)(C°/m)}$$

Table 16.1 Thermal conductivities of various materials (ordinary temperature ranges); and conversion factors for heat units.

Material	k		
	kcal/(s)(m²)(C°/m)	W/(m²)(C°/m)	Btu/(h)(ft²)(F°/in.)
Air, at rest	0.055×10^{-4}	0.024	0.168
Aluminum	5.0×10^{-2}	212	1480
Brick	1.7×10^{-4}	0.7	5.0
Concrete	2×10^{-4}	0.8	6.0
Copper	9.2×10^{-2}	378	2640
Cork board	0.1×10^{-4}	0.05	0.25
Window glass, plate	1.9×10^{-4}	0.79	5.5
Insulating board, fiber	0.14×10^{-4}	0.04–0.05	0.35
Iron, cast	1.1×10^{-2}	50.5	350
Kapok	0.08×10^{-4}	0.034	0.24
Fiberglass (loose batt)	0.09×10^{-4}	0.037	0.27
Sawdust	0.14×10^{-4}	0.057	0.41
Silver	10×10^{-2}	412	2880
Steel	1.1×10^{-2}	44.6	312
Water, liquid	1.4×10^{-4}	0.61	4.28
Water, ice	5.2×10^{-4}	2.23	15.6
Wood, oak	0.4×10^{-4}	0.16	1.10
pine	0.29×10^{-4}	0.12	0.84

Conversion Factors: 1 kcal = 3.97 Btu = 4186 J

$1 \text{ Btu/h} = 6.998 \times 10^{-5} \text{ kcal/s} = 2.93 \times 10^{-4} \text{ kW}$

$1 \text{ kW} = 0.2389 \text{ kcal/s} = 3.413 \times 10^{3} \text{ Btu/h}$

$1 \text{ kcal/s} = 4.186 \text{ kW} = 1.429 \times 10^{4} \text{ Btu/h}$

Substituting gives

$$H = \frac{1.9 \times 10^{-4} \text{ kcal/(s)(m}^2\text{)(C°/m)} \times 32 \text{ m}^2 \times 3600 \text{ s} \times 4\text{C°}}{1.2 \times 10^{-2} \text{ m} \times 1 \text{ h}}$$

$$= 7300 \text{ kcal/h} = 8.49 \text{ kW}$$
$$= 28{,}980 \text{ Btu/h} \qquad \textit{answer}$$

Illustrative Problem 16.2 The outside surface temperature of a brick wall in direct sunshine is 110°F. The wall is 40 × 9 ft × 12 in. thick. Find the steady-state flow of heat (Btu/h) through the wall into the interior space when the inner surface of the wall is 80°F.

Solution From Table 16.1, note that k for brick is 5 Btu/(h)(ft²)(F°/in.). Substituting values in Eq. (16.1),

$$H = \frac{5 \text{ Btu/(h)(ft}^2\text{)(F°/in.)} \times 360 \text{ ft}^2 \times (110 - 80)\text{F°}}{12 \text{ in.}}$$

$$= 4500 \text{ Btu/h} \qquad \textit{answer}$$

16.3 ■ Thermal Resistance

Materials with high thermal resistance are good *insulators.*

Thermal resistance (R) is a measure of a material's resistance to heat flow. Materials that have high thermal resistance are classed as *insulating* materials. The greater the value of R, the greater is the resistance to heat flow, and the better insulator the material will be. Thermal resistances are called "R-values" by engineers. R-values can be expressed either *per inch of thickness* or for the *thickness as stated* of the material. Defined,

Thermal resistance is the reciprocal of *thermal conductivity.* Thermal resistances are called *R-values* by engineers. R-values are a measure of resistance to heat flow.

$$R_{\text{(per inch)}} = \frac{1}{k} \tag{16.2}$$

$$R_{\text{(thickness stated)}} = \frac{L}{k} \tag{16.2'}$$

$$\text{Thermal resistance} = \frac{\text{thickness}}{\text{thermal conductivity}}$$

Labeling of insulating materials with R-values. The greater the R-value, the better the insulating value.

R-values are now a (U.S.) federally mandated designation that must be stated by the manufacturer on insulating materials. For example, a 1-in.-thick pressed-fiber insulating board ($k = 0.35$, Table 16.1) has an R-value of about 3, stamped "R-3" on each board. Or, 8 in. of blown fiberglass insulation in a ceiling ($k = 0.27$) has a total R-value of about 29.6, or R-30 (Fig. 16.4). Since R (in general) $= L/k$, the English system units of R are (h)(ft^2)(F°/Btu). This is a cumbersome unit to handle, and for that reason R-values are usually stated without units, in industrial and engineering practice. We will use these units only in calculations where keeping track of them is important in solving the problem.

In terms of thermal resistance R, heat transfer through a homogeneous material is directly proportional to the temperature difference between the two surfaces of the material and inversely proportional to the thermal resistance of the material.

The basic heat transfer equation.

$$H = \frac{A(t_2 - t_1)}{R} = \frac{A\,\Delta t}{R} \tag{16.3}$$

Fig. 16.4 Applying thermal insulating blankets in stud-frame walls of a residence under construction. A $3\frac{1}{2}$-in.-thick blanket of fiberglass ($k = 0.27$) gives an insulating value (thermal resistance) of R-11. (Owens-Corning Fiberglas)

16.4 ■ Conduction through Composite Materials

So far we have dealt only with homogeneous materials, where both k and R retain their same values all the way through the given substance. In actual practice, however, heat transmission more often than not occurs through structural components made of different materials. Walls, floors, ceilings, doors, and even windows are usually built up of several different layers. Also, trapped air is a poor conductor (i.e., it has high thermal resistance) and walls often contain "dead air" spaces sandwiched in between the wall surface materials.

For a structural member such as a wall that is made up of a number of different materials, the total R-value is equal to the sum of the R-values of the components (Fig. 16.5). In other words, the *thermal resistance* of the wall is just the sum total of the thermal resistances of its component parts.

$$R_T = R_1 + R_2 + R_3 + \cdots + R_n \tag{16.4}$$

Illustrative Problem 16.3 Compare the heat lost from an indoor space by conduction through a wall of dimensions 30 × 12 ft, whose cross section is indicated in Fig. 16.5, with the heat lost through the same wall except that it has no insulating board under the lath and plaster. The indoor wall surface temperature is 70°F and the outdoor wall surface temperature is 15°F, steady state. (NOTE: The R-values shown in Fig. 16.5 are for the thickness stated.)

Solution

1. For the wall as shown in Fig. 16.5,

$$R_T = R_1 + R_2 + R_3 = 0.66 + 6.0 + 2.0 = 8.66$$

$$H = \frac{360 \text{ ft}^2 \times 55 \text{ F}^\circ}{8.66 \text{ (h)(ft}^2\text{)(F}^\circ\text{/Btu)}}$$

$$= 2290 \text{ Btu/h} \qquad \textit{answer}$$

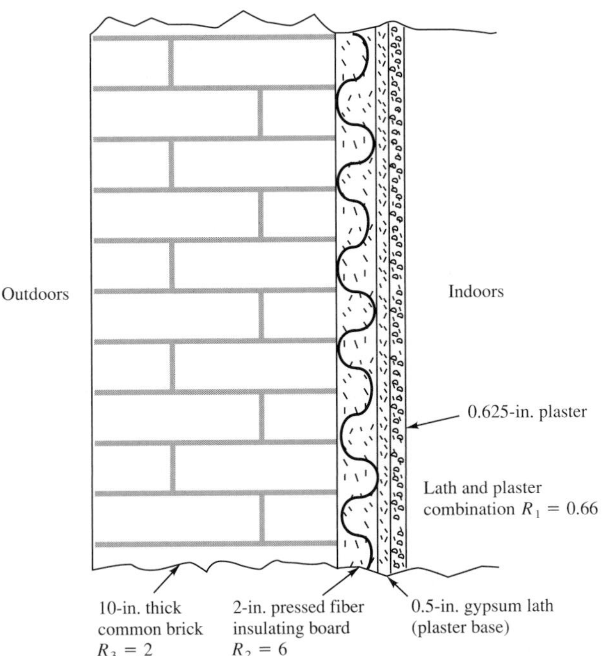

Fig. 16.5 Cross section of a portion of an exterior wall of a building, showing composite construction from materials with different R-values (Illustrative Problem 16.3).

2. For the same wall but without the insulation,

$$R_T = 0.66 + 2.0 = 2.66$$

(The 2 in. of insulating board more than triples the thermal resistance of the wall)

$$H = \frac{360 \text{ ft}^2 \times 55 \text{ F}^\circ}{2.66 \text{ (h)(ft}^2\text{)(F}^\circ\text{/Btu)}}$$

$$= 7445 \text{ Btu/h} \qquad \textit{answer}$$

16.5 ■ Overall Heat Transmission Coefficients—U-values

A composite wall can be looked upon as a *series resistance*, for heat flow calculations.

For structures made up of different materials, *overall heat transmission coefficients*, called *U-values*, have been adopted. A composite wall or structure, as indicated above, can be looked upon as a *series resistance* in terms of heat flowing through it. The resistances of the different materials are merely added in series, as indicated in Eq. (16.4). The overall heat transmission of such a wall (*thermal transmittance*) is just the reciprocal of the thermal resistance, or

$$U = \frac{1}{R_T} = \frac{1}{R_1 + R_2 + \cdots + R_n} \tag{16.5}$$

U-values are a measure of the overall heat transmission of a composite wall or structural component.

For a composite wall, since U is the reciprocal of the total thermal resistance, Eq. (16.3) can be rewritten as follows:

$$\underset{\text{(composite wall)}}{H} = UA\,\Delta t \tag{16.6}$$

U-values have the units Btu/(h)(ft^2)(F°). They are calculated from computer simulations, or the composite wall or structure being studied can be set up in a laboratory and its U-value measured under controlled conditions. U-values and R-values are compiled in engineering handbooks for use in air-conditioning, heating, and construction work. Space does not allow their inclusion in a basic physics text. However, a few examples will be given.

1. The insulated wall of Fig. 16.5, which has an R-value of 8.66, would have an overall transmission coefficient

$$U = \frac{1}{8.66} = 0.115 \text{ Btu/(h)(ft}^2\text{)(F}^\circ\text{)}$$

The uninsulated version of that wall would have

$$U = \frac{1}{2.66} = 0.376 \text{ Btu/(h)(ft}^2\text{)(F}^\circ\text{)}$$

2. The ''standard'' stud-frame exterior wall for residence construction, with 2 × 4 studs on 16-in. centers, with exterior wood siding, and interior lath and plaster (or finished gypsum ''drywall''), has transmission coefficients as follows:

With 3.5-in. blanket insulation, U ≅ 0.07 Btu/(h)(ft^2)(F°).
Without insulation, U ≅ 0.26 Btu/(h)(ft^2)(F°).
In other words, the heat loss (or gain, in summer) will only be one-third as much with 3.5 in. of insulation between the studs as it will be without the insulation.

Many industrial and everyday processes require *good* conduction.

From the above discussions, one might conclude that all conduction is undesirable, that we are always seeking ways to reduce and minimize it. Such is definitely not the case. Many industrial, research, commercial, and home practices require *good* conduction. Furnaces, boilers, kitchen utensils, sterilizers, heat exchangers of all types, including auto radiators, are just a few of the applications for which materials, usually metals, are chosen specifically for their high rates of thermal conductivity.

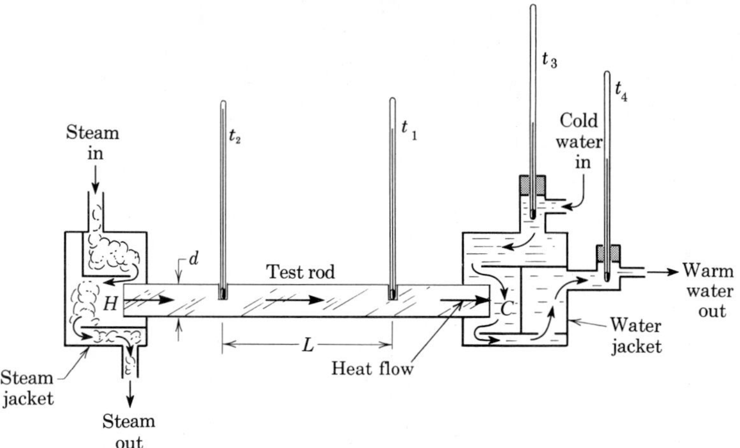

Fig. 16.6 Diagram of Searle's apparatus for measuring the heat conductivity of metals and other solid materials. Heat from live steam enters the test rod at *H* and flows along the rod to *C* where it flows into and heats the circulating water in the water jacket.

16.6 ■ Measuring Thermal Conductivity

In every phase of industrial and engineering work in which heat transfer is involved, it is of extreme importance to use the material which possesses the desired conductivity. Millions of dollars are saved annually by controlling heat losses (or gains) in industrial processes. Most test laboratories have equipment for determining the heat conductivity of materials. Figure 16.6 illustrates, in diagram form, a laboratory apparatus for determining k for metals.

A rod of the test material (*HC* in the diagram) has a steam jacket fitted at one end and a water jacket at the other. The entire assembly is enclosed in a heavily insulated case (not diagramed) to prevent heat losses to the atmosphere. Live steam is supplied to the *H* end of the test rod, and the amount of heat transmitted to the *C* end is measured by the continuous-flow calorimeter arrangement provided. Cold water enters the jacket at the *C* end at a temperature t_3, is warmed by the heat transmitted by the test rod, and leaves the apparatus at temperature t_4. If a known mass of water m passes through the *C* jacket, the total heat *H* which the water picks up from the test rod can be calculated from the basic heat equation $H = mc(t_4 - t_3)$, or $H = wc(t_4 - t_3)$ where $c = 1$ for water.

In a regular test run, the material to be tested is placed in the apparatus, steam is turned in to the *H* end, and a small stream of water is circulated through the cold end. After 15 or 20 min, the water flow can be adjusted until a steady-state condition is obtained, i.e., thermometer readings t_2, t_1, t_3, and t_4 remain at constant levels. When this condition is obtained, the *water out* is caught in a catch bucket for a measured time T and weighed. The four thermometer readings are recorded, and the length L and diameter d of the test rod are measured. From these data, k can be calculated, as in the following sample problem.

Illustrative Problem 16.4 A newly developed alloy is being tested for its heat conductivity by the apparatus and method described above. The following data are taken (English system):

$$\begin{array}{ll} L = 6 \text{ in.} & t_3 = 60°\text{F} \\ d = 1 \text{ in.} & t_4 = 68°\text{F} \\ t_2 = 185°\text{F} & T = 25 \text{ min} \\ t_1 = 102°\text{F} & w = 3.5 \text{ lb water} \end{array}$$

Determine the heat conductivity of the alloy from these data.

Solution First calculate the heat supplied to the water at the *C* end. Substituting $H = wc(t_4 - t_3)$, (English system) gives

$$H = 3.5(68 - 60) = 28 \text{ Btu in 25 min}$$

Now solve Eq. (16.1) for k:

$$k = \frac{HL}{AT(t_2 - t_1)}$$

Substituting the known values gives

$$k = \frac{28 \text{ Btu} \times 6 \text{ in.}}{\left(\dfrac{\pi \times 1^2}{144 \times 4}\right) \text{ft}^2 \times \frac{25}{60} \text{ h} \times (185 - 102)\text{F}°}$$

$$= 890 \text{ Btu/(h)(ft}^2\text{)(F°/in.)} \qquad \textit{answer}$$

Where strength is a factor, steel or iron is used in heat transfer equipment, even though its conductivity is not as high as that of some other metals. Boilers and furnaces are examples of equipment where steel is necessary. Copper is an excellent metal for heat transfer. It combines high heat conductivity [k = 2640 Btu/(h)(ft^2)(F°/in.)], with reasonably good strength, resistance to corrosion, and relatively low price. As new developments on the energy front—solar energy, geothermal energy, and nuclear energy—come into full use, heat transfer will be an even more important industry, and new metals and alloys with high values of k and great strength will be in demand.

16.7 ■ Convection

Convection heat transfer takes place by means of a moving fluid.

Heat is not *conducted* very well by liquids and gases, but heat transfer is easily accomplished in both liquids and gases by the process known as *convection*.

Warm winds carry heat from one place to another on the earth. Wind is a convection current which results from unequal heating of the earth's surface. Warm ocean currents carry heat from the tropics to warm what would otherwise be uninhabitable northern shores. Warm air and cold air (depending on the season) are blown (forced convection) through ducts and registers to heat and cool homes, offices, and factories. Hot water rises through pipes from boilers to remote radiators in other parts of the building or plant, carrying heat along with it. These are all examples of *heat transfer by convection*.

Convection Explained Note that in all the examples cited above the heat-transfer medium (water or air) moved from one place to another and *carried the heat with it*. Here we have the distinguishing characteristic of *convection—heat transfer by motion of the medium*. Conduction takes place by motion and collision of the atoms and molecules of the medium, whereas convection necessitates the *actual movement of the medium as a whole*. Recall the analogy of the volunteer fire brigade: If instead of passing the buckets along the line, each person carried a bucket from pump to blaze, this would be comparable to the manner in which heat is carried by convection.

In convection, heat energy is transferred by "hot" or high-energy molecules moving as a fluid and carrying heat energy with them.

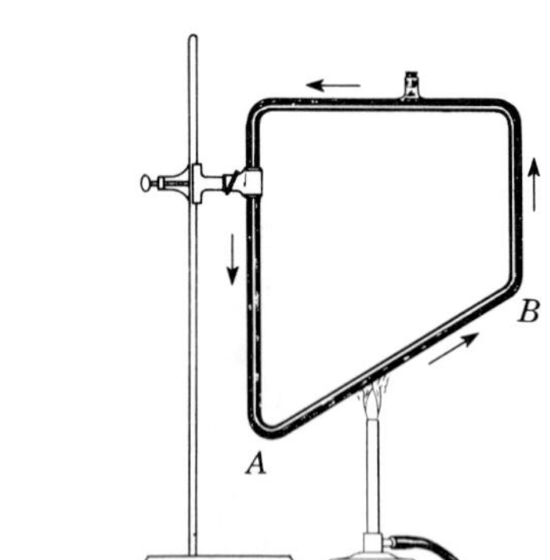

Fig. 16.7 Simple laboratory setup to demonstrate convection currents in water. Heat transfer by convection occurs in all fluids.

16.8 ■ Forced and Natural Convection

Some of the examples of convection cited above occur naturally while others depend upon mechanically operated blowers or pumps. Two simple experiments, one for liquids, one for gases, will serve to illustrate natural convection, frequently called a *convection current*.

Arrange a bent glass tube in the form shown in Fig. 16.7. Fill it with water in which is placed some sawdust or dye crystals. If the section AB is gently heated, a circulation begins almost immediately in the direction of the arrows.

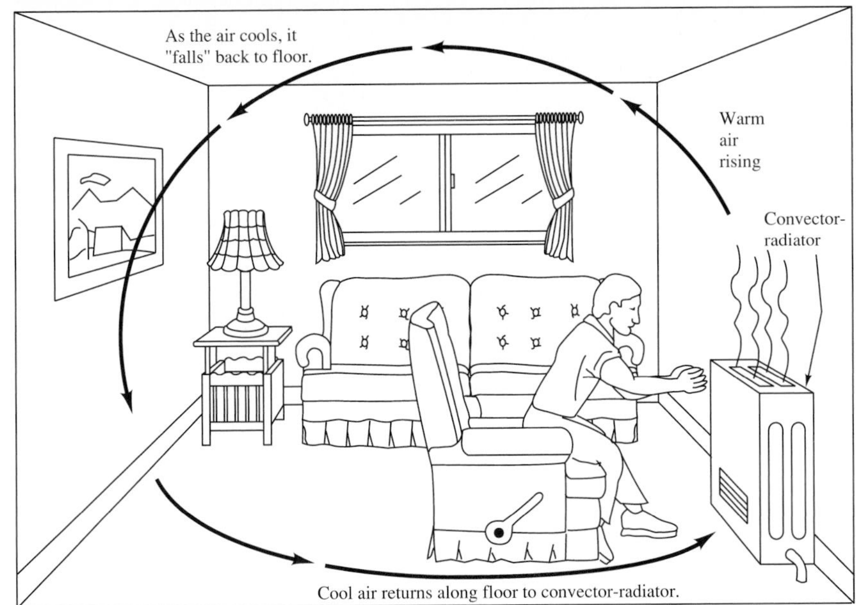

Fig. 16.8 Near any source of heat a convection current of air will be established. In the illustration, heated air rises from the convector because, having expanded, its density is less than that of the surrounding (ambient) air. As it slowly cools again it ''falls'' to the floor and moves across to the source of heat again. The action of a convector-radiator is akin to that of a pump or blower. It ''pumps'' out warm air and receives cold air, setting up a flow called a convection current.

Anyone sitting near a fireplace or a hot radiator (convector) is aware almost equally of the warmth from the fire or radiator in front, and the gentle but cold air movement coming from the rear. This air movement can be easily observed if someone near a fireplace is smoking. The tobacco smoke drifts to the hearth, into the fireplace and up the chimney.

These examples of natural convection (see Fig. 16.8) show that convection currents are established in liquids and gases. These convection currents are in turn caused by the expansion of the liquid or gas due to heat. The rapid expansion creates a quantity of liquid or gas of decreased density, and this lighter fluid rises, being pushed (buoyed) up by the surrounding heavier fluid. As hot gases rise in a chimney or over a bonfire, cold surrounding air flows in to take their place. This current continues as long as the fire burns, and in stoves or fireplaces it is called the *draft*. Draft is of considerable importance in the design of a fireplace, a furnace firebox, or an industrial chimney, since an adequate supply of oxygen to support combustion depends upon how good a draft is obtained. The design and location of the chimney are the most important factors in obtaining a good draft.

Convection can be either *natural* (winds, ocean currents, fluid motion because of density differences) or *forced* (by pumps, fans, or blowers).

Many home heating systems are based on natural convection, while others utilize forced convection (see Chap. 18).

Forced convection uses a moving medium (liquid or gas or air) to carry heat, but the motion of the medium is caused (or aided) by mechanical means—pumps or fans. The cooling system of an automobile is a good example of a *convector* (it is mistakenly called a *radiator*), since the liquid (water or a coolant specified by the manufacturer) is force-circulated from the engine block to the heat exchanger (''radiator'') and back again by means of a V-belt-driven water pump (see Fig. 16.9).

16.9 ■ Surface Films—the Convection Coefficient

Convection occurs only in fluids, and fluids (liquids or gases) are often heated or cooled as a result of being in contact with a hot or cold solid. Two cases of such contact are very common: (1) a liquid or a gas (often air) moving along a flat surface like a wall or a metallic plate, and (2) fluids moving in pipes or ducts where the contact surfaces are curved. In either case there is a temperature difference Δt between the surface of the wall or pipe and the body of the fluid. Furthermore, although the fluid moves along the wall or pipe surface, there is a thin *boundary film* of fluid in direct contact with the solid surface that either moves very slowly or not at all, because of friction and/or adhesive forces. This region of the fluid is called a *stagnant layer*. (See Fig. 16.10.)

Heat transfer at the boundary between a solid surface and a fluid is affected by a *boundary film* or *stagnant layer* of the fluid. The effect of the boundary film is expressed as a *convection coefficient*.

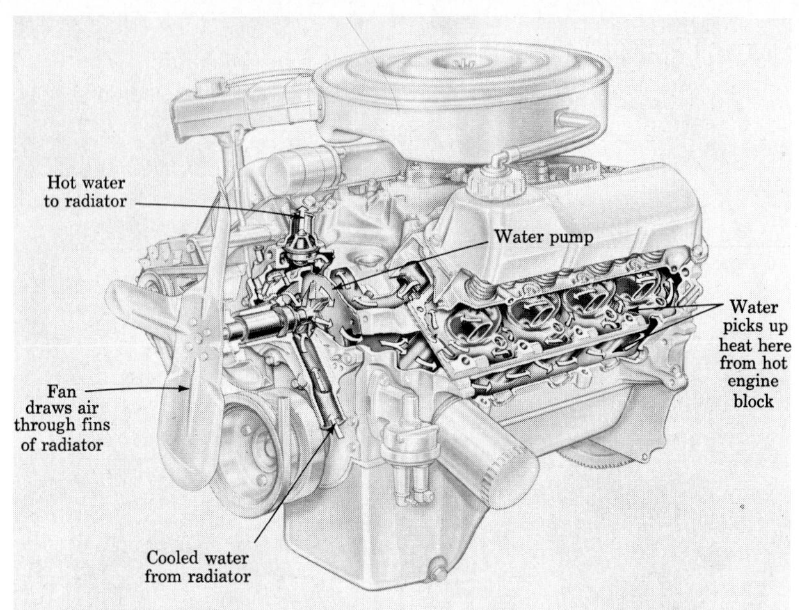

Fig. 16.9 An auto engine cooling system. The "radiator" is actually a combination radiator, conductor, and convector. It does radiate some heat, but much of the heat it must dissipate is transferred to the stream of air that passes through and around the fins and tubes of the radiator core. A fan (cycled by a thermostat on many modern autos) maintains a high volume of air flow through the radiator core. (Ford Motor Company)

Figure 16.10 shows a cross section of an office-building exterior wall, not receiving any direct solar radiation but absorbing heat from hot summer air. Its outer surface temperature is designated by t_h (hot). Heat is transmitted by *conduction* to the inner surface of the wall, where the temperature is t_w (warm). There is a rising current of air along the warm interior wall surface, and in that rising current, next to the wall, is a *stagnant layer* or *boundary film* of thickness L (exaggerated in Fig. 16.10). There will be some conduction to the room air through the boundary film, and there will

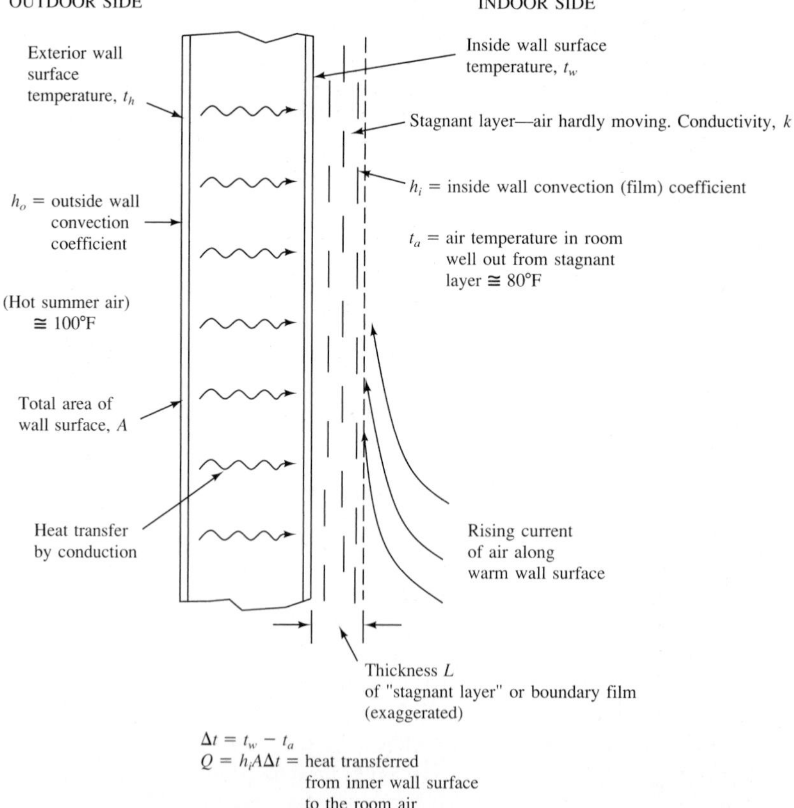

Fig. 16.10 Cross-section of an exterior wall showing heat transmission through the wall by conduction, and heat transfer from the interior surface of the wall to the indoor air by convection. The *film coefficient* (or *convection coefficient*) is illustrated.

also be heat transfer by *convection* as the air currents move along the wall. Heat therefore flows into the room from the warm inner surface of the wall by a combination of conduction and convection. (There will also be some heat transfer by *radiation,* but we neglect that for the present. See Sec. 16.10 below.) Let t_a be the temperature of the indoor air well away from the boundary film.

Conductive-convective heat transfer of this kind takes place in accordance with the following equation:

$$Q = hA\ \Delta t \tag{16.7}$$

The basic equation for the rate of convective heat transfer from a wall surface to surrounding air.

where Q = the rate of heat transfer, Btu/h
A = the area of the wall surface, ft^2
$\Delta t = t_w - t_a$, the difference in temperature between the (inner) wall surface and the ambient air in the room
h = a factor called the *film coefficient* or *convection coefficient*. The term *surface conductance* is sometimes used also. h has the dimensions k/L and it is therefore the *conductivity* of the fluid divided by the *thickness* of the boundary film.

Convection (film) coefficients are not susceptible to ready calculation, but are determined experimentally. Tables and charts are available in engineering handbooks giving values of film coefficients for a wide variety of conditions and fluids encountered in actual practice.

For smooth interior surfaces (plaster, dry-wall, wood paneling), and for air velocities of 0 to 3 mi/h, $h_i \approx 1.7$ Btu/(h)(ft^2)(F°), or 9.64 W/(m^2)(C°). For rougher (outdoor) surfaces such as stucco, brick, concrete, or rough-sawn wood siding, and an air velocity of 15 mi/h, $h_o \approx 6.5$ Btu/(h)(ft^2)(F°), or 31.5 W/(m^2)(C°).

In a manner similar to that just described for the interior surface of a wall, the heat exchange between the exterior surface of a wall and the outdoor air is also the result of conductive-convective heat transfer.

Convection effects are quite important in calculating the rate of heat transfer through walls and windows. Neglecting convection can cause large errors in heating and cooling load estimates if the U-value used for the wall or window does not incorporate the thermal resistance of boundary films. U-values, as used for heating and air-conditioning load calculations, must express the heat transfer rate from *outdoor air* to *indoor air,* not just from *outer wall surface* to *inner wall surface*. Consequently, in order to obtain a U-value that will yield correct heating and cooling load estimates, the thermal resistances of the exterior and interior boundary films must be introduced into the computation. Recall Eq. (16.5), which gives

Gross errors in determining heat losses or gains through building structure components can result if convection effects are not taken into account.

$$U = \frac{1}{R_T} = \frac{1}{R_1 + R_2 + R_3 + \cdots R_n}$$

with convection coefficients *not* involved. The R's, it will be recalled, are the thermal resistances (R-values) of the several wall components.

The thermal resistances of the outdoor and interior boundary films are respectively: $R_o = 1/h_o$ and $R_i = 1/h_i$. These are reciprocals because the dimensions of h are k/L and the dimensions of R are L/k.

For example, Fig. 16.11 portrays a cross section through a typical concrete wall of an industrial office building, with convection data given.

Now, adding convection (boundary film) effects to Eq. (16.5), and using the R-values given in Fig. 16.11 for 10-in. concrete and plaster, we can write, for the wall of Fig. 16.11,

$$U = \frac{1}{\dfrac{1}{6.5} + 0.833 + 0.47 + \dfrac{1}{1.7}}$$

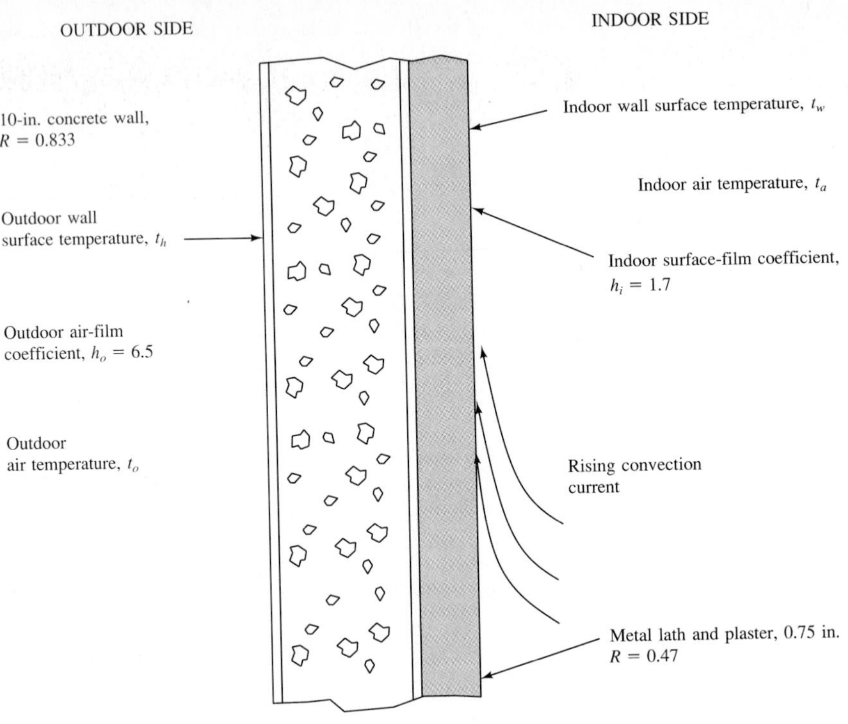

Fig. 16.11 Cross-section of a poured-concrete exterior wall of an industrial building, illustrating the calculation of *overall* U-values that take into account convection heat transfer at the outer and inner wall surfaces.

$$= \frac{1}{0.154 + 0.833 + 0.47 + 0.588}$$

$= 0.49$ Btu/(h)(ft^2)(F°) when convection effects are included.

If convection effects are *not* incorporated, the calculation gives

$$U = \frac{1}{0.833 + 0.47} = 0.77 \text{ Btu/(h)(ft}^2\text{)(F°)}$$

The consequences of such an error in the U-value are made evident in the following problem.

Illustrative Problem 16.5 Calculate the heat transmission in Btu/h (the *cooling load*) through the wall of Fig. 16.11 to *indoor air* under the following summer conditions:

Outdoor air temperature	100°F
Indoor air temperature	80°F
Wall area	1200 ft^2

(*a*) Use the U-value corrected for convection effects, $U = 0.49$
(*b*) Use the U-value uncorrected for convection effects, $U = 0.77$

Solution (***a***) $H = UA\,\Delta t$
$= 0.49$ Btu/(h)(ft^2)(F°) $\times$ 1200 ft^2 $\times$ 20 F°
$= 11{,}760$ Btu/h *answer*

(***b***) $H = UA\,\Delta t$
$= 0.77$ Btu/(h)(ft^2)(F°) $\times$ 1200 ft^2 $\times$ 20 F°
$= 18{,}480$ Btu/h *answer*

Overestimating the load calculation by an error of this magnitude would be an inexcusable engineering blunder. The result, for the contractor involved, would be the loss of the job to some other firm whose correct load estimate led to a bid price for the job markedly lower than that of the "mistaken" firm.

Conduction and convection, as we have seen, often operate in tandem. Their effects are present in many different industries, and the ability to determine the magnitude of these effects is often crucial to the process or the design being undertaken. In this basic text we have been able to give only the barest introduction to these heat-transfer methods. We turn now to the third method of heat transfer—radiation.

16.10 ■ Radiation

The third method of heat transfer (*radiation*) cannot be explained either by the molecular theory or by familiar analogies such as that of the fire brigade. We can easily observe the process of radiation and measure its effects; we can create and use sources of radiant heat for many industrial processes; but the cause of radiant energy and the method by which it travels through space cannot be explained by common, everyday events.

We saw in the previous discussion that air currents flow from a room toward a burning fireplace. Yet, seated halfway across the room, we can "feel" the heat of the glowing embers on our faces, shins, and outstretched hands. Quite obviously, this heat is not carried to us by the air, for the air is moving *away from us toward the fireplace*. Neither is it *conducted* to us by the air, for air is a very poor *conductor* of heat.

As another example of radiation, consider the following: A steel bar, cherry-red from the forge, is held a few inches from and somewhat above the face. The heat is intense, but it cannot be due either to convection (hot air will rise from the bar) or to conduction. Now place a pane of transparent glass between the glowing bar and the face, and note the results: (1) You still see the cherry-red bar, and (2) despite the fact that glass is a very poor *conductor* of heat, you feel the burning almost as intensely as before. Now replace the pane of glass by a thin board, a sheet of paper, or anything else opaque to light. The immediate results noted are (1) you no longer see the bar, since light cannot travel from it to your eyes, and (2) you no longer feel its heat.

This simple experiment would seem to justify the conclusion that radiant heat is like light. The process of heat transfer by radiation seems to be similar to the passage of light through space. This conclusion is further borne out by the manner in which heat reaches us from the sun. Certainly convection and conduction are not involved, for some 93 million miles of nearly empty space separate us from the sun. When the sun sets or is temporarily obscured by a dense cloud, both its light and its heat diminish simultaneously.

Scientists believe that heat energy as well as light energy is transmitted by electromagnetic waves. The process is known as *radiation*. Radiant heat, like light, seems to travel in straight lines. It is reflected and absorbed by the same surfaces which reflect and absorb light. It can be brought to a focus by lenses and mirrors in the same manner as light (see Fig. 16.12). Radiant heat travels with the same speed as light, 3×10^8 m/s, or 186,000 mi/s in space or in a vacuum. Radiant heat and light both seem to be the *energy of wave motion,* and the difference between them is in the frequency and length of the waves. Heat energy for the most part is a longer-wavelength radiation than light energy. Heat waves include, but are not limited to, the region of the electromagnetic spectrum known as the "infrared."

Both light and heat are transmitted through space by electromagnetic waves—a process known as *radiation.*

Low-temperature heat waves are of a wavelength that places them in the *infrared* region of the electromagnetic spectrum.

Radiation is heat transfer by means of electromagnetic waves, which travel with the speed of light. "Heat waves" occur in the visible and infrared regions of the electromagnetic spectrum.

Fig. 16.12 Radiant heat, like light, can be brought to a focus by (*a*) lenses, and (*b*) mirrors.

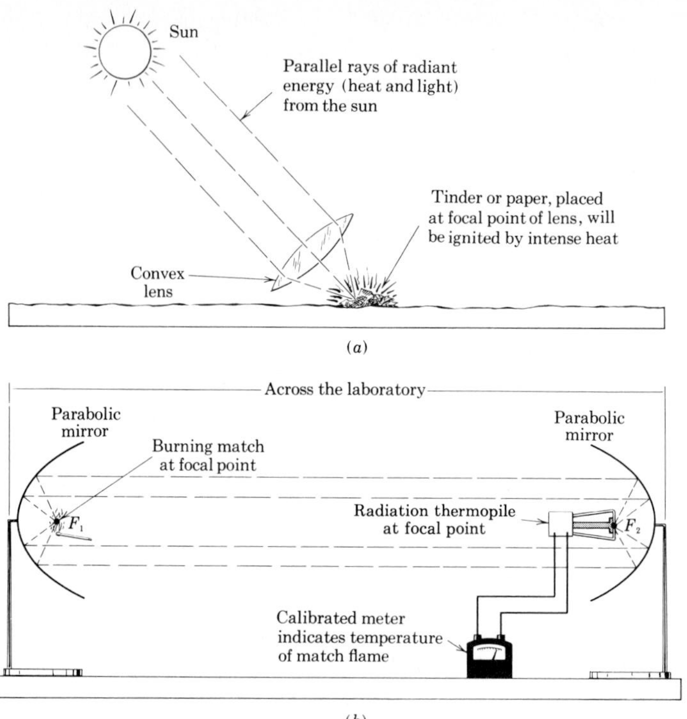

The Solar Constant Though least understood, radiation is perhaps the most important of the three methods of heat transfer, for it is the only method by which heat energy reaches the earth from the sun. In full sunlight and perpendicular to the sun's rays, the earth (at the top of its atmosphere) receives about 19.4 kcal/(min)(m^2) or 7.16 Btu/(min)(ft^2) in the form of radiant energy. This value is often referred to as the *solar constant,* I_0.

Table 16.2 Values of the Solar Constant, Various Units

I_o = 429 Btu/(h)(ft^2)
1165 kcal/(h)(m^2)
1.94 cal/(min)(cm^2)
4870 kJ/(h)(m^2)
1.353 kW/m^2

Radiation and Absorption Substances which absorb radiation are heated by it. This suggests that the absorbed radiation in some manner causes an increase in internal energy (molecular, atomic, and particle motion) within the body.

Another theory of radiant energy is that it consists of high-energy "bundles" or particles called *photons.*

Radiation energy must strike *and be absorbed by* a material substance before any heat is produced.

A completely satisfying explanation of electromagnetic phenomena like light and radiant heat is as yet unavailable; consequently scientists resort to theories and hypotheses. Another theory (in addition to the electromagnetic wave theory) is that light and radiant heat are composed of bundles of energy called *quanta* (singular *quantum*). The term *photons* is also used (see Chap. 33). According to the *quantum theory,* proposed by Max Planck (1858–1947) (see Chap. 33), the energy of each quantum is proportional to the frequency of the electromagnetic radiation which is the source of the energy. When light and/or radiant heat impinges on a surface such as the earth, the quanta impart their energies to the substance they strike. These "bundles" of energy increase the excitation of molecules and atoms on the absorbing surface, with consequent increases of both temperature and thermal energy. Transparent substances, through which radiation travels with little loss, are not appreciably heated, and neither are substances whose surfaces reflect most of the radiant energy falling upon them. In brief, then, before radiant energy in the form of waves or quanta can be converted into sensible heat, it must strike *and be absorbed* by some material substance.

All bodies radiate energy; the hotter the body the greater the amount of radiation. As a body radiates, its temperature falls, unless heat is supplied to keep the temperature up. All bodies also absorb radiant energy in varying amounts, and a continual exchange of energy is thus going on among bodies at different temperatures. Just as with other methods of heat transfer, net radiation heat transfer occurs from a body at a higher temperature to a body at a lower temperature.

Gustav Kirchhoff (1824–1887) found that the emitting or radiating ability of a surface is directly proportional to its absorbing ability. In other words, good radiators are also good absorbers, and *vice versa*.

Good radiators are good absorbers, and vice versa.

A surface that could absorb all radiant energy that falls on it (a perfect absorber) would appear to be dead black. Such a surface would radiate, at any given temperature, better than any other surface. This "perfect" radiator is called an *ideal blackbody,* and we use the term "blackbody radiation" to describe this theoretical condition. Actually, there are no perfect absorbers. Black velvet and carbon black (soot) come closest, with absorption of about 97–98 percent of incident radiation. Polished, light-colored surfaces are poor radiators, and poor absorbers as well, since they reflect most of the radiant energy that falls on them.

The "perfect" radiator (or absorber) is called an *ideal blackbody.*

The *ideal blackbody* cannot be attained in practice, and the best approach to it is a small hole leading into a cavity with rough, dark interior walls (Fig. 16.13). Such a cavity then radiates to the outside, through the hole, at a rate which depends solely on the absolute (Kelvin) temperature of the interior.

The Stefan-Boltzmann Law of Radiation In 1879, Josef Stefan (1835–1893), an Austrian physicist, was able to show experimentally that the *rate of emission* of radiant energy from an ideal blackbody (cavity radiation) is proportional to the fourth power of the Kelvin temperature. Ludwig Boltzmann (1844–1906) subsequently confirmed Stefan's findings from theoretical calculations. Their work resulted in one of the basic laws of radiation.

Total radiation rate per unit area of radiating surface is proportional to the *fourth power* of the Kelvin temperature.

Stefan-Boltzmann Radiation Law

$$R = \epsilon\sigma T^4 \tag{16.8}$$

where R = the rate of total radiation emittance per unit surface area
T = the absolute temperature of the radiating surface, in kelvins
σ = the radiation constant
and ϵ = the *emissivity* of the radiating surface (see below)

If R is in kcal/(s)(m^2),

$$\sigma = 1.36 \times 10^{-11} \text{ kcal/(s)(m}^2\text{)(K}^4\text{)}$$

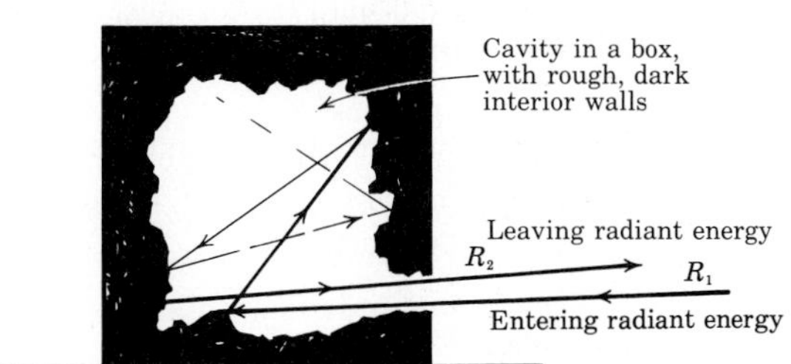

Fig. 16.13 Theoretical *blackbody* (or *cavity*) radiation. Such a cavity is a near-perfect absorber. As any ray of radiant energy, R_1 enters the cavity it undergoes some absorption and some reflection at each incidence with the cavity walls. Near-total absorption is the eventual result, and virtually none of the *entering* radiant energy finds its way back out through the small entry hole. The only radiation to emerge from the cavity (R_2) is that which results in photons (or quanta) whose source is heat energy *inside* the cavity. The combustion chamber of a steam boiler or the interior of a blast furnace are approximations to blackbody radiation.

If R is in W/m^2,

$$\sigma = 5.67 \times 10^{-8}\ \text{W/(m}^2\text{)(K}^4\text{)}$$

The symbol R, as used here for *radiation emittance,* must not be confused with R for thermal resistance (Sec. 16.3).

The critical effect of temperature on radiation is readily evident from Eq. (16.8) since doubling the Kelvin temperature increases the rate of radiation sixteenfold. As the temperature of a radiating body increases, the wavelength of the radiation becomes shorter. White-hot or blue-hot surfaces are much hotter than red-hot surfaces.

Emissivity of a surface is the ratio of its radiant emittance per unit area to that of an ideal blackbody at the same temperature.

Emissivity of a Radiating Surface The *emissivity* ϵ of any surface is the ratio of the radiant emittance R of that surface, to the radiant emittance of an *ideal blackbody* of the same surface area and surface temperature. By definition, $\epsilon_{\text{blackbody}} = 1.00$. For comparison, the emissivity of common brick is about 0.93; of flatblack paint about 0.97; of the human body surface about 0.96; for glossy white-painted surfaces, about 0.91; and for polished metals about 0.10.

Illustrative Problem 16.6 The temperature of a cherry-red ingot from the soaking pit of a steel mill is 600°C. Assuming it to radiate like a blackbody, how much heat energy is radiated from all its surfaces per second if its dimensions are 0.5 by 1 by 3 m?

Solution Use Eq. (16.8) with $\epsilon = 1.00$

$$R = \epsilon\sigma T^4$$
$$\sigma = 1.36 \times 10^{-11}\ \text{kcal/(s)(m}^2\text{)(K}^4\text{)}$$

Substituting, and changing 600°C to 873 K,

$$R = 1.00 \times 1.36 \times 10^{-11}\ \frac{\text{kcal/s}}{(\text{m}^2)(\text{K}^4)} \times (873\ \text{K})^4$$
$$= 7.9\ \text{kcal/(s)(m}^2\text{)}$$

$$\text{Total surface area} = 10\ \text{m}^2$$
$$\text{Total heat radiated} = 79\ \text{kcal/s}$$
$$= 331\ \text{kW}$$
$$= 1.13 \times 10^6\ \text{Btu/h} \qquad \textit{answer}$$

At very high temperatures quite a few actual surfaces (the sun, for example, or the surfaces of molten metals) approach the characteristics of ideal blackbodies. If the total radiant emittance of such bodies per unit surface area can be determined by measurement, then the surface temperature can be calculated from Eq. (16.7). This is one of the methods by which the sun's surface temperature (about 6200 K) has been estimated.

16.11 ■ Solar Energy

The possibilities for solar energy development can be appreciated by performing a simple calculation with the solar constant. In full sunlight, with the sun directly overhead, and a clear atmosphere, the amount of radiant heat striking the flat roof of a 1500-ft^2 house would be

The solar power (radiation intensity) available at a given location and time on the earth's surface is called *insolation.* This term must not be confused with *insulation.*

$$H = 1500\ \text{ft}^2 \times 7.16\ \text{Btu/(min)(ft}^2\text{)}$$
$$\times\ 60\ \text{min/h}$$
$$= 640{,}000\ \text{Btu/h}$$

This is roughly four times the heat output rating of the furnace that would be needed to heat such a home on cold winter days. However, the practical everyday reality is far less than the above theoretical value would indicate. The earth's atmosphere, with its clouds, haze, and dust, blocks a significant amount of solar radiation. Solar power *available* at a given location and time is called *insolation*. (Note the spelling and do not confuse this word with *insulation*, a material that minimizes heat transfer by conduction). Solar radiation intensity (i.e., solar power per unit area) varies with the latitude, the season of the year, the time of day (falling to zero at night), and the amount of cloud cover and other atmospheric contaminants. Solar power is often expressed as *average daily insolation*, symbol I, and values given in tables are ordinarily those for "clear-day" conditions. Extensive tabular data are available in solar engineering handbooks for all locations on earth, for every day of the year, and for any angle of incidence of the sun's rays. We provide here just a few average values for clear-day conditions, to illustrate the potential for intensive development of solar energy.

Average January *daily* (clear-day) *insolation* across much of central United States is about 650 Btu/day·ft^2 of surface area. For locations in the southwest United States, the insolation increases to a range of 1200 to 1500 Btu/day·ft^2. At solar noon on a clear day in July at 35° to 40°N latitude, the insolation on a horizontal surface is of the order of 320 Btu/h·ft^2. The thermal energy available from solar power is illustrated by the following problem.

Illustrative Problem 16.7 A commercial laundry has a flat roof that accommodates 5000 ft^2 of solar collectors. The average daily insolation is known to be 1600 Btu/day·ft^2 in July. Feed water is supplied to the collectors at 60°F and the water flow is controlled so that the water-off temperature from the collectors is 205°F. Based on a nine-hour operating day, how many gallons per hour of hot (205°F) water can be supplied? (Collector efficiency is 48 percent.)

Solution The total heat supplied to the water is equal to the total insolation on the collectors multiplied by the collector efficiency.

$$\begin{aligned} H &= 1600 \text{ Btu/day·ft}^2 \times 5000 \text{ ft}^2 \times 0.48 \\ &= 3{,}840{,}000 \text{ Btu/day or } 426{,}700 \text{ Btu/h for 9 h} \end{aligned}$$

Let V = the water volume in gallons (1 gal = 8.34 lb). Specific heat c of water is 1.00 Btu/lb·F°. Δt = (205 − 60)°F.

Recalling that $$H = wc\,\Delta t, \qquad w = \frac{H}{c\,\Delta t}$$

Therefore, $$V_{\text{gal}} = \frac{H}{8.34c\,\Delta t}$$

Substituting, $$V = \frac{426{,}700 \text{ Btu/h}}{8.34 \text{ lb/gal} \times 1.00 \text{ Btu/lb·F°} \times 145 \text{ F°}}$$

$$= 353 \text{ gal/h} \qquad \textit{answer}$$

It could also be noted that if natural gas were the fuel used in a hot water boiler to produce the same amount of heat per day (3,840,000 Btu) with a boiler efficiency of 80 percent, the gas consumption would be about 4400 ft^3/day.

Vast amounts of energy are available from solar radiation, but large collection areas are required.

Much of the incident solar radiation on buildings is reflected. There is also the problem of storing the heat absorbed during periods of sunlight for use at night or during prolonged cloudy periods. To collect significant amounts of solar energy, large collection areas are required. Also, factors of geography and climate are such that in the regions needing the most heat during the winter months, the sun is not "directly overhead," and the radiant energy received is far less than the given value of the solar constant. With slanting rays the solar constant is decreased by a factor equal to the cosine of the angle made with the perpendicular. Solar collector panels can be mechanized to "track" the sun, but this is an expensive installation. Solar heating designs are sometimes architecturally undesirable, and furthermore, present costs are quite high. However, dozens of different designs are under development,

As the supplies of fossil fuels dwindle and their cost rises dramatically, solar energy will become an important component of the world's energy supply.

and in a few more years the state of the art will have advanced to the point where solar heating of homes, commercial buildings, and factories may be commonplace.

Energy from Photovoltaic Cells When solar radiation strikes certain metallic elements and compounds, the radiant energy of the sun is converted directly into electric current. The theory of the *photoelectric effect* is dealt with in Chap. 33, and here we are concerned only with some practical applications for the present and future. (See Sec. 33.5 for an explanation of how a solar cell produces electric current.)

Direct conversion of solar radiation into electric power is accomplished by photovoltaic cells (PVCs). Each 4-in.-diameter silicon cell in full sun (perpendicular incidence) will produce about 1 W of electric power.

Silicon has proved to be a good material for photovoltaic (solar) cells (PVCs) (Fig. 16.14). Cadmium/sulfur and gallium/arsenic compounds have also been used. The efficiencies of solar cells are quite low, with silicon cells converting only about 10 to 12 percent of the sun's energy that falls on them into electric energy. Some newly developed gallium-arsenide cells are reported to be over 20 percent efficient, and 30-percent efficiency is hoped for in the future. The cost of solar-cell electric power is quite high at present. Compared to the approximate 7 to 10 cents per kilowatt-hour charged by many electric utility companies for power generated by steam, hydro, and nuclear plants, home-generated photovoltaic power has a current (1988) cost of from 20 to 30 cents per kilowatt-hour when all costs are properly charged against the system. However the cost of PVCs and PV systems decreases each year.

Each solar cell is quite small (Fig. 16.14), on the order of four inches in diameter (they are not necessarily round), and nearly 1000 silicon cells are required to produce 1 kW of electrical output (about 1 W per 4-in. cell in full sun at perpendicular incidence). Installations of massive size are required to produce significant amounts of electric power. Using silicon solar cells, a 6.5-megawatt plant, now in operation (small by steam plant standards) requires in excess of 150 acres for the solar arrays and associated equipment (Fig. 16.15).

Solar-cell arrays are used to power satellites of all kinds (Skylab, for example, had an installation of nearly 150,000 cells, producing 22 kW), and they are frequently used to generate small amounts of power at remote locations on earth. Preliminary conceptual schemes for huge solar-powered, hovering-satellite generating stations have been proposed by the National Aeronautics and Space Administration (NASA), but it will probably be many years before such a station is in operation. The sheer size and mass of such a plant would involve hundreds of space missions in order to get it assembled and operating in a suitable orbit (Fig. 16.16). Furthermore, the problems of transmitting that amount of power through space and of receiving it at one or more ground stations are by no means simple of solution.

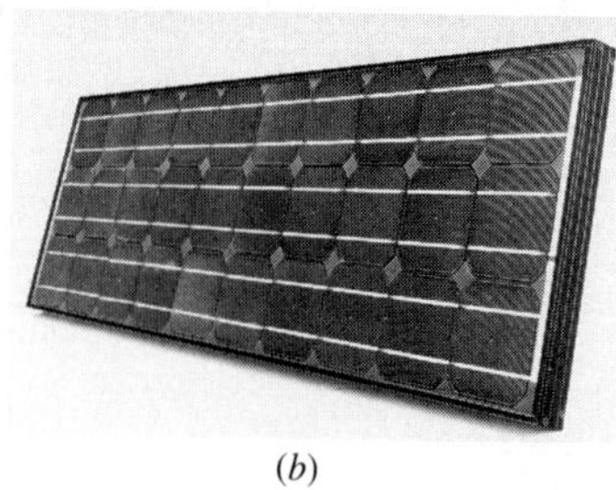

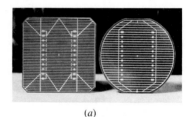

(*a*)

(*b*)

Fig. 16.14 Typical silicon solar cells. (*a*) Four-inch round and 4-in. square photovoltaic cells, each with an electric power output of about 1-W in full sun at perpendicular incidence. (*b*) A solar module of larger square cells, with a total output of 42 W in full sun. (ARCO Solar, Inc.)

16.12 ■ Applications of Heat-Transfer Processes

We take advantage of the known facts about radiation in many ways. For example, white clothing will keep us cooler in summer; in tropical climates houses are commonly painted white to reflect much of the sun's radiant energy. Boilers and steam and hot-water pipes are painted either white or aluminum color to reduce heat losses from radiation. Heating-system radiators should be (but for esthetic reasons seldom are) painted a dark color so they will readily radiate heat to the space surrounding them.

Thermos bottles utilize the principles of all three methods of heat transfer in their construction (see Fig. 16.17). The inner container is usually of glass, which is a poor conductor. Its outer surface is silvered to reduce radiation losses. Between this surface and the outer container, the space is evacuated to prevent convection (and conduction) losses. The outer container is usually lined with cork or another heat insulator and the outer surface is frequently of shiny metal to reduce the radiation losses further. If the bottle contains a cold liquid, the same principles apply, but heat gains are minimized instead of heat losses.

Fig. 16.15 Solar cells for the commercial production of electric power. The huge array shown here in close-up is one of 756 identical arrays installed on 160 acres on the Carrisa Plains in Central California. Each array is mounted on a dual-axis, computer-controlled "tracker," that tracks the sun all day for every day in the year. The reflectors shown increase efficiency by directing additional sunlight onto the solar cells. The yearly energy output of the present 6.5-MW installation is about 14 million kilowatt hours, enough for 2300 average homes. A total power output of 16 MW is the planned capacity for this plant, as further expansion takes place in the future. (ARCO Solar, Inc., and Pacific Gas and Electric Co.)

Fig. 16.16 Artist's rendering of one concept for a Satellite Solar power station (SSPS). The mammoth satellite with arrays of solar cells would be placed in a geosynchronous orbit, 22,500 mi above the earth's equator. Such concepts as this are only "visionary" at the present time. (National Aeronautics and Space Administration)

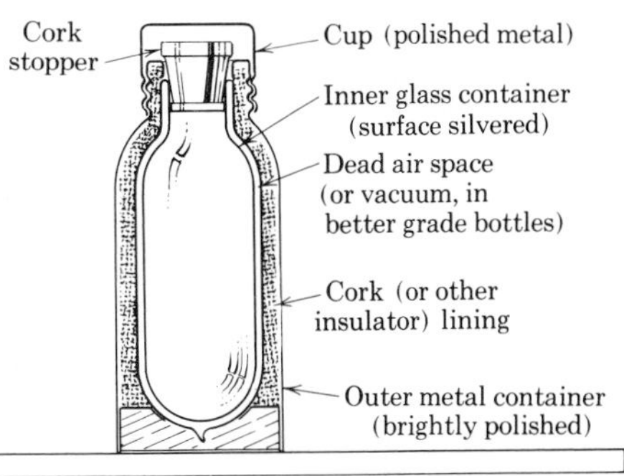

Fig. 16.17 Cross-sectional view through a typical Thermos bottle. It is so constructed that heat losses and gains of all three types (conduction, convection, and radiation) are minimized.

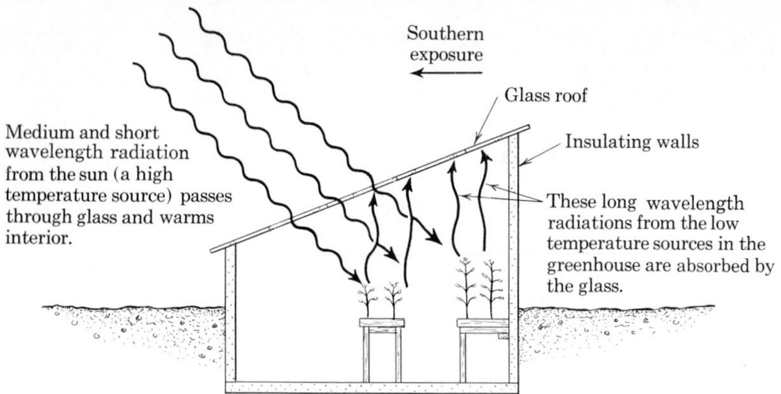

Fig. 16.18 A greenhouse is constructed in such a manner as to receive as much solar radiation as possible and lose as little heat as possible. Glass transmits most of the medium- and short-wavelength radiation (light waves) from the sun, absorbing very little of it. Heat (infrared) waves from the high-temperature sun, although the glass absorbs them to some extent, are mostly transmitted also. As the greenhouse interior warms up from absorbing the sun's rays, objects and surfaces begin reradiating, but this is low-temperature, long-wavelength radiation and most of it is absorbed by the glass and is not lost to the cold outdoors; consequently, the term ''heat trap'' for a greenhouse.

Greenhouses and cold frames used by gardeners for growing vegetables and flowers in cold climates take advantage of the fact that radiation from hot bodies, e.g., the sun, is in general composed of short-wavelength high-frequency energy, while low-temperature radiation is of longer wavelength and lower frequency. Glass, although transparent to medium- and medium-short-wavelength radiation, is relatively opaque to extremely short (ultraviolet) radiation. Also, glass does not transmit the longer wavelengths (from low-temperature sources in the greenhouse) of infrared waves very well, and reradiation is minimized. The sun's rays pass through the glass roof and warm the interior. The plants and objects within radiate in turn, but their radiation is of longer wavelengths (low frequency), which is characteristic of low-temperature sources, and much of it is absorbed by the glass. The greenhouse thus acts as a one-way valve for heat energy, and the structure itself becomes a ''heat trap'' (see Fig. 16.18). Crop production in commercial greenhouses is an important industry in north temperate-zone climates. Part of the necessary heat is solar heat ''trapped'' in this manner. Solar-heated homes and office buildings are also designed to maximize the ''trapping'' of solar radiation.

Heat transfer and heat exchange are such important processes that, in economic terms, they represent a multibillion industry worldwide.

Many of the practical applications of heat transfer occur in, or are representative of, the heating, refrigeration, and air-conditioning industries. We shall examine some of these applications later in Chap. 18.

Heat exchange plays an essential role in many chemical-plant and refinery processes. Superheated steam is usually the heat-transfer medium, and the process of transferring the heat from the steam to the product being circulated is carried on in units of equipment called heat exchangers (see Fig. 16.19). The tubes of heat exchangers are ordinarily made of copper because of its excellent conductivity.

Fig. 16.19 A heat-exchanger bundle from a petroleum refinery. Heat exchangers in one form or another are essential to nearly all heavy industry and to heating and cooling equipment in all applications. Heat transfer in a heat exchanger is mainly by conduction and convection. Radiation losses are kept to a minimum by giving the exterior shell (not shown) a coating of aluminum or white paint. (Chevron Corporation)

QUESTIONS AND EXERCISES

1. What is the determining factor in the direction of net heat flow by conduction? convection? radiation?

2. Outdoors on a day when the temperature is well below freezing, one's bare hand will "freeze" to a metal tool handle. This does not occur with a wooden-handled tool. Explain.

3. Discuss in detail how the kinetic-molecular hypothesis explains the process of conduction in solids.

4. Trace every heat-transfer process involved in dissipating the heat from an operating automobile engine. Specify each material that takes part in the process and state whether conduction, convection, or radiation is involved at each stage of the process.

5. Pour a small quantity of water in a paper cup and then heat the bottom of the cup *gently* over an open flame until the water boils. Why does the paper not catch fire from the flame?

6. Why is it necessary for a fireplace or a furnace combustion chamber to have a proper chimney or flue for efficient operation? What is the purpose of the draft?

7. A pond will very quickly acquire a thin layer of ice when the air temperature falls below 32°F. Why is a much lower temperature required for several days before the ice layer thickens appreciably and is strong enough for skating?

8. On a winter day the outside air temperature is −15°C and the air inside a house is +20°C. Would the outside and inside surface temperatures of a plate glass window be −15°C and +20°C respectively? Which surface (outside or inside) is likely to be closest to the temperature of the air immediately adjacent to it? Explain.

9. A house has a solar collector array on the roof made up of many coils of copper pipe, encased in glass-covered housings, the insides of which are painted a dead black. Water is circulated in the pipes in a closed system which includes a pump and the "radiators" in the house. Trace every heat transfer process from the sun itself to a person sitting by one of the radiators who feels warmed by it.

10. At 40°N latitude on December 21st at solar noon, what would be the approximate angle that the sun's rays make with the perpendicular? Using a diagram see if you can calculate the approximate value of the solar constant at this latitude at midday in bright sunlight. (Neglect atmospheric effects.)

11. During the redecorating of her house a woman decided to paint all the room radiators a high-gloss white. They had been a dark gray color. What effect, if any, is this change likely to have on the heating system?

12. The total thermal resistance R of an exterior wall has been determined as R-30. The design outdoor winter air temperature is 5°F, and the desired indoor air temperature is 72°F. Can the heat loss through the wall in Btu/h be calculated from these data? If not, what other factors must be known and applied?

13. Why is more heat lost from a warm room through a wall to the cold outdoors on a windy day than on a day with the same value of Δt, but with no wind?

14. The spaces between the studs in typical frame walls are filled with blanket insulation (fiberglass) about 3.5 in. thick, to reduce conduction and convection heat transfer. However, air itself has a conductivity k only about half that of fiberglass (see Table 16.1). Why not leave the spaces uninsulated, just as air spaces?

15. Suppose the daily routine in your home is that everybody leaves for work at 7:30 a.m. and nobody returns before 5:30 p.m. The daytime outdoor temperature gets as low as 35°F but never goes below 30°F. In order to economize on heating bills in the winter, would you turn the furnace off entirely when you leave in the morning? Turn the thermostat down to, say, about 55°F? Or, set the thermostat at or very near the desired indoor temperature, keeping the house warm all day? Discuss the alternatives, with respect to economy of system operation.

16. In hot summer climates people wear white clothing, since it reflects, rather than absorbs the sun's rays. If you live in a cold winter climate and are concerned only with conserving body heat, what color clothing should you wear? Explain your answer.

17. In order to obtain good air distribution in a living space heated and cooled with a forced-air system, where should the registers supplying warm air in winter be placed? Those supplying chilled air in summer?

PROBLEMS

NOTE: *Except where metric units are specifically stated, all problems are to be solved with English-system units.*

Group One

1. Radiation in the infrared region of the electromagnetic spectrum is found to have a frequency of 6×10^{12} hertz (Hz). (See Sec. 22.7). Find the wavelength of these rays in meters.

2. Heat waves 1 μ in length are in the *near infrared* region of the electromagnetic spectrum. What is the frequency of these waves?

3. A glass view window is 7 mm thick and measures

3 × 4 m. When its outside surface temperature is 42°C and the inside surface temperature is 36°C, what is the rate of conduction heat gain into the living space, in kilowatts?

4. The thermal conductivity k of mineral wool insulation is 0.25 Btu/(h)(ft^2)(F°/in.). Find the R-value of a 10-in.-thick blanket of this material.

5. "Four inches of fiberglass," usually in the form of batts or "blankets," is a common specification for insulating the walls of homes. How thick would the walls have to be to provide the same insulation effect if they were constructed of solid (*a*) pine, (*b*) brick, (*c*) concrete?

6. If one side of a 5 cm-thick corkboard wall is at a temperature of 30°C and the other at −10°C, how many kilocalories per hour will be conducted through 1 m^2 of the wall?

7. A glass window measures 8 by 5 ft and is $\frac{1}{4}$ in. thick. The outside glass surface temperature is +5°F, and the inside temperature of the glass is 28°F. How many Btu/h are lost through this window?

8. A large household refrigerator has a cabinet which is equivalent to a box made of corkboard, 4 in. thick and 65 ft^2 in area. If the interior is to be maintained at 38°F while the kitchen temperature is 76°F, what is the conduction heat gain in Btu/h into the cabinet? (Assume door is not opened during the hour of the test.)

9. A shoe store has a total wall surface 100 ft long by 14 ft high. The wall is of common brick, 8 in. thick. If the outdoor wall temperature is 10°F on a winter day and the inside wall temperature is being maintained at 54°F, find the heat loss in Btu/h through the wall.

10. The ceiling of a building is insulated with 6 in. of fiberglass and measures 30 by 50 ft. If the inside ceiling temperature is 80°F and the attic temperature is 30°F, how many Btu/h are lost through the ceiling?

Group Two

11. In a test run with a continuous-flow calorimeter, cold water enters at 55°F and leaves after it has been warmed to 120°F. The flow rate is measured as 20 gal/h. Calculate the heat transferred to and carried away by the water (*a*) in Btu/h, (*b*) in watts.

12. How thick would a brick wall have to be to have the same R-value as 4 in. of rock wool or fiberglass insulation?

13. A composite wall is known to have an R-value of 11. Calculate its overall coefficient of heat transmission U, taking into account the inside and outside film coefficients, as given in Sec. 16.9.

14. An iron steam pipe has an outside diameter of 20 cm and is 0.8 cm thick. The steam temperature inside the pipe is 130°C, and the outside of the pipe is at 70°C. How many kcal/h are lost by conduction from a 5-m section of this pipe? (HINT: Assume effective diameter of 19.2 cm.)

15. The inside dimensions of an icebox are 3 × 2 × 2 ft. The walls are of corkboard 4 in. thick. The inside temperature is 38°F, and the outside temperature is 95°F. How many pounds of ice are melted per 24-h day by heat leakage through the walls?

16. A sample of metal is being tested to determine the value of k. With an apparatus like that of Fig. 16.6, the following data are recorded:

$L = 5$ in.	$t_1 = 130°F$
$d = 2$ in.	$t_2 = 205°F$
$w = 15.3$ lb water	$t_3 = 50°F$
$T = 18$ min	$t_4 = 60°F$

What is the value of k for this sample?

17. A furnace door has a round hole in it 6 cm in diameter. The interior of the furnace is at a temperature of 750°C. How many kilocalories will be radiated from this hole in 1 min? (Assume blackbody radiation.)

18. Find the rate of radiation of a blackbody whose temperature is 50°C and compare it with the rate from the same body when its temperature is raised to 450°C.

19. A spherical satellite is in geocentric orbit around the earth. It is 5 m in diameter and its surface emissivity $\epsilon = 0.55$. If its average surface temperature is 40°C, find the rate of energy radiation (radiated power) from its surface in watts. (Assume that it gains no radiant energy from the earth or any other source.)

20. What is the heat loss in joules/h through a plate glass window 0.65 cm thick by 3 m long by 1 m high. The inner surface temperature is 6°C and the outer surface temperature is 0°C.

21. An exterior wall of a building has the following structure from the outside in: 8-in. brick; a 1-in. air space; a 1-in.-thick fiber insulating board; and finally, oak panelling 0.75 in. thick. Take data from Table 16.1 and calculate the R-value of the wall.

22. One type of double-glazed insulating glass has two $\frac{1}{8}$ in.-thick glass sheets separated by a $\frac{1}{16}$ in. air space. (*a*) Calculate its U-value in metric units, surface to surface (convection effects neglected). (*b*) Determine the corrected U-value, taking into account the coefficients for film boundaries as given in Sec. 16.9.

23. A round copper rod is in an arrangement like that of Fig. 16.6. It is 14 in. long and has a diameter of 2 in. Its "hot end" is maintained at 212°F by saturated steam. The "cold end" of the rod is surrounded by water in a continuous-flow calorimeter and is maintained at 150°F by a flow of water. The water-on temperature is 55°F and the water-off temperature is 105°F. Calculate the water flow rate through the calorimeter in gal/h.

24. The value of k, the thermal conductivity for a certain concrete, is 2.1×10^{-4} kcal/(s)(m^2)(C°/m). What thickness of this concrete is required in order to have the same R-value as 5 cm of corkboard?

25. Calculate the rate at which the sun loses energy by radiation. Take its surface temperature as 6000 K, its emissivity as 0.98, and its diameter as 14×10^5 km. Assume that it receives no radiant energy from any other source. Answer in kilowatts.

26. A walk-in commercial freezer (cold storage room) has 8-in.-thick concrete walls lined on the inside with 2-in.-thick corkboard. The air temperature inside the freezer is kept at 15°F. On a hot summer day the outside air temperature is 95°F. Take convection effects into account and calculate the heat loss rate per square foot of wall surface, in Btu/h.

27. The sun's surface temperature is about 6200 K. At solar noon on the earth's equator (sun directly overhead) the radiation intensity (solar power per unit area) striking the earth is about 1.03 kW/m^2. If a major sun-spot disturbance reduced the sun's surface temperature to 5800 K, what would be the radiation intensity at the same location on the earth?

Group Three

28. A picnic cooler has inside dimensions of $1 \times 1 \times 2$ ft. The walls and top and bottom all have R-values equivalent to 1 in. of corkboard. A large chunk of ice has been in the cooler long enough so that the entire inside of the cooler, including the ice itself, is at a temperature of 32°F. The temperature of the outer surface of the cooler is 80°F. What is the steady-state rate of melting of the ice, in lb/h?

29. A 6×10 ft view window is double-glazed, the glazing being two panes of glass, each 0.25 in. thick, separated by an air space 0.30 in. thick. Outdoor air temperature is 12°F, and indoor air temperature is 72°F. Taking convection effects into account, calculate the heat loss rate through the window, in Btu/h.

30. The tungsten filament in an electric lamp operates at 2600 K, with a power input of 30 W. Assume tungsten emissivity is 0.35 and calculate the surface area of the filament.

31. A room has a total heat transmission loss of 4.2 kW. If a glowing cherry-red iron sphere (surface temperature 800°C) were placed on a pedestal in the middle of the room, what would have to be its diameter in order that it radiate enough heat to just balance the room heat loss and maintain the temperature of the room surfaces at 22°C? (Assume the iron ball is an ideal blackbody.)

32. The operating temperature of the filament in a tungsten lamp is 2900 K and the emissivity $\epsilon = 0.38$. Calculate the required surface area for a 40-W lamp.

33. A room for a computer laboratory is 30 by 40 by 9 ft high. The walls and ceiling are equivalent to 4 in. of fiberglass insulation. Assume no heat gains or losses through the floor. The computers dissipate heat energy into the space from their use of electric power at a rate of 8 kW. If a temperature of 70°F is to be maintained within the space against an outside wall temperature of 90°F and an attic temperature of 110°F, what is the required cooling capacity of an air conditioner in Btu/h to balance the combined equipment and heat-transmission loads? (NOTE: 1 kW of electric energy dissipates 3400 Btu/h of heat.)

34. A steam radiator (painted black) is operating with a surface temperature of 210°F. Its effective surface area is 20 ft^2. If saturated steam enters at 240°F and leaves as hot water at 212°F, how many pounds of steam per hour must be supplied? [HINT: Convert problem conditions to SI-metric units and use Eq. (16.8). See Table 15.6 for steam properties.]

35. A copper kettle whose bottom surface is 0.5 cm thick and 50 cm in diameter rests on a burner which maintains the bottom surface of the kettle at 110°C. A steady-state heat flow occurs through the copper bottom into the vessel, where water is boiling at atmospheric pressure. The actual temperature of the inside surface of the kettle bottom is 105°C. How much water (in kilograms) boils away in 1 h?

36. A layer of ice on a lake is 6 cm thick. If the top surface of the ice is at −10°C and the temperature of the water just below the ice layer is 0°C, at what rate does the ice become thicker? (Answer in centimeters per hour.)

37. A radiation pyrometer is used to measure the rate of energy emission from a hole of area 1 cm^2 in the wall of a furnace for making alloy steel. If the pyrometer reading is 140 W, what is the temperature (degrees Celsius) of the furnace interior? [HINT: Recall that 1 W = 1 J/s. (Given: 1 kcal = 4186 J.) Assume blackbody radiation.]

CHAPTER 17 THERMODYNAMICS—HEAT ENGINES

Heat warms our homes, offices, and factories; it smelts ores and refines metals; it processes foods, textiles, paper, and medicines; it provides hot water for a thousand uses in manufacturing and in the home; and, of special concern for the present chapter, it powers autos, trucks, trains, farm equipment, stationary power plants of all kinds, ships, aircraft (Fig. 17.1), and spacecraft.

Machines that use heat to produce energy of motion are called *heat engines*. The science of converting heat energy into mechanical energy is called *thermodynamics*. This science is based on two fundamental laws called the *laws of thermodynamics* and much of this chapter will be concerned with these laws.

Fig. 17.1 A new "turbofan" jet engine ready to begin service with commercial air lines. This PW4000 engine will develop from 50,000 to 60,000 lb of thrust, and is 7 percent more fuel efficient than previous models. It features digital electronic engine control. (United Technologies—Pratt and Whitney)

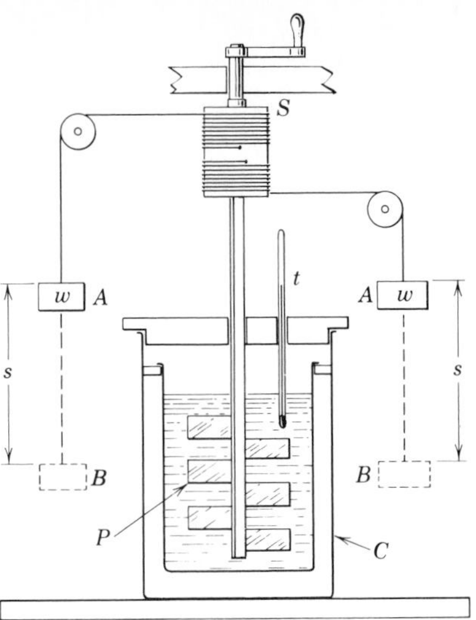

Fig. 17.2 Joule's apparatus (schematic representation) for determining the mechanical equivalent of heat.

BASIC PRINCIPLES

17.1 ■ The Mechanical Equivalent of Heat

In an earlier chapter, the studies of Count Rumford on the nature of heat were mentioned. He found that heat could be produced indefinitely from mechanical work and suggested that heat was a form of energy. His studies, though carefully performed and well interpreted, did not go far enough to determine the quantitative relationship which exists between heat and mechanical work. It remained for James Prescott Joule (1818–1889), an English scientist, to determine the heat equivalent of mechanical work. Joule performed a famous experiment in 1843 with an apparatus (similar to that shown schematically in Fig. 17.2) which determined the equivalence between heat and work with considerable precision.

Joule was primarily a man of industry, the nineteenth-century equivalent of today's engineer or technologist. His avocation or hobby was scientific investigation.

He fitted a calorimeter C with a paddle arrangement P which could be turned by falling weights w, unwinding cords from spindle S. As the weights move from A to B through a distance s, an amount of work $W = 2ws$ is done. This energy is expended by the paddle in churning the water in the calorimeter cup. Joule was successful in showing that as mechanical work was done on the water, its temperature increased. Knowing the mass of water in the calorimeter and its temperature rise, he calculated the amount of heat which resulted from a known amount of mechanical work. His findings, which have been verified by many subsequent workers, resulted in a determination of the *heat equivalent of mechanical work.*

The Mechanical Equivalent of Heat

The mechanical equivalent of heat is the ratio of mechanical work done to the heat produced. This ratio is known as Joule's constant, symbol J.

Joule's constant—The heat equivalent of mechanical work.

Joule's constant
$$J = \frac{W}{H} \tag{17.1}$$

The presently accepted values for the constant J are

$$J = 4.186 \text{ J/cal}$$

or $$4186 \text{ J/kcal} \quad \text{(metric units)}$$

or $$J = 778 \text{ ft·lb/Btu} \quad \text{(English-engineering units)}$$

When friction is present, mechanical energy is continually converted into heat.

A *heat engine* is required to convert heat energy into mechanical energy.

This is equivalent to saying that if an energy transformation could be effected without loss, 1 kcal would accomplish 4186 J of work, and 1 Btu of heat would accomplish 778 ft·lb of mechanical work. The conversion of work into heat is continually going on, since friction is never absent when work is being done. Also, heat is continually being transformed into mechanical energy by natural processes and, in modern times, by man-made machines.

NOTE: *Joule's constant* is denoted by J, while the symbol for the energy unit *joule* is J.

Illustrative Problem 17.1 A sledgehammer, mass 2 kg, is moving at 25 m/s when it strikes a 150-g steel spike, driving it into a railroad tie. If 40 percent of the impact energy is converted into heat in the spike, calculate its rise in temperature.

Solution

$$\begin{aligned}\text{KE of hammer} &= \tfrac{1}{2}mv^2 = 1\text{ kg} \times (25\text{ m/s})^2 \\ &= 625\text{ kg}\cdot\text{m}^2/\text{s}^2 \\ &= 625\text{ J}\end{aligned}$$

(Review units of the joule, Sec. 6.2.) The energy which goes into heat is

$$W = 0.40 \times 625\text{ J} = 250\text{ J}$$

From Eq. (17.1), the heat produced is

$$H = \frac{W}{J} = \frac{250\text{ J}}{4.186\text{ J/cal}}$$

from which

$$H = 59.7\text{ cal}$$

From Eq. (15.2)

$$H = mc(t_2 - t_1)$$

$$\text{Change in temperature} = t_2 - t_1 = \frac{H}{mc}$$

From Table 15.2, the specific heat c for steel is 0.115 cal/(g)(C°). Substituting gives

$$\begin{aligned}t_2 - t_1 &= \frac{59.7\text{ cal}}{150\text{ g} \times 0.115\text{ cal/(g)(C°)}} \\ &= 3.46\text{C°}\end{aligned}$$

answer

17.2 ■ Conservation in Energy Transformations

The conversion of mechanical energy into heat is easily accomplished—in fact, it is inescapable. This conversion is usually an undesirable process, and the heat produced is generally regarded as "wasted" for the purpose in mind. Friction is often the cause of "wasted" energy.

Converting heat to mechanical energy, on the other hand, is not so easily accomplished. It will be recalled that heat energy is the internal energy or molecular kinetic energy of a body or substance which is being transferred from one body or system to another as a result of temperature difference. Since internal or molecular kinetic energy is the result of an almost infinite number of collisions, vibrations, and oscillations, a device, which we call a *heat engine,* is required to transform this kind of energy into mechanical energy. And, as we have noted and emphasized repeatedly, a *temperature difference* is an absolute requisite. Heat will do work only as it flows from high temperature to a lower temperature.

There is no heat engine that can carry out the transformation from heat energy to mechanical energy at anywhere near 100 percent efficiency. Some heat engines are more efficient than others. It is emphasized, however, that actually no energy is *lost* in either of these processes. The *law of conservation of energy* holds here, as in all other energy transformations. When producing heat from mechanical work, we can write

$$W = JH + W_w$$

Mechanical-energy input = energy converted to heat + "wasted" work (17.2)

The law of conservation of energy for heat-to-work transformations.

and when work is obtained from heat, energy conservation is expressed by

$$H = \frac{W}{J} + H_w$$

Heat-energy input = heat converted to work + "wasted" heat (17.3)

There is a high degree of certainty that it is impossible to create or destroy energy: Any increase in one form of energy will be accompanied by an equal decrease in some other form of energy. If electric energy is dissipated, for example, heat and/or mechanical energy will result. As nuclear energy is released by fission, thermal energy is the result; and when heat energy is applied to junctions of certain dissimilar metals, electric energy is produced. Countless experiments meticulously performed by the best scientists of the past 300 years lead to the following general principle:

Law of Conservation of Energy

Energy can neither be created nor destroyed, but it can be converted or transformed from one form to another.

The law of conservation of energy was perhaps the most comprehensive generalization in all of science prior to Einstein's special theory of relativity and his principle of mass-energy equivalence, $E = mc^2$ (see Chap. 34). Even so, the principle of conservation of energy still applies, by broadening it to express the *principle of conservation of mass-energy*.

The following problem illustrates the conversion of heat to mechanical work.

Illustrative Problem 17.2 A steam turbine generates 50,000 hp and has an overall efficiency of 22 percent. If the boiler fuel oil used has a fuel value of 19,000 Btu/lb, calculate the fuel-oil consumption per hour.

Solution Determine the amount of mechanical work done in 1 h, in foot-pounds, as follows:

$$W = 50{,}000 \text{ hp} \times \frac{33{,}000 \text{ ft}\cdot\text{lb/min}}{1 \text{ hp}} \times \frac{60 \text{ min}}{1 \text{ h}}$$

$$= 9.9 \times 10^{10} \text{ ft}\cdot\text{lb/h}$$

The mechanical equivalent of heat is 778 ft·lb = 1 Btu. Thus,

$$W = 9.9 \times 10^{10}\ \frac{\text{ft·lb}}{\text{h}} \times \frac{1\ \text{Btu}}{778\ \text{ft·lb}}$$

$$= 12.7 \times 10^7\ \text{Btu/h}$$

which would be the heat input quantity required if the boiler-turbine system were 100 percent efficient. The oil *actually* consumed, however, amounts to

$$\frac{12.7 \times 10^7\ \text{Btu/h}}{19{,}000\ \text{Btu/lb} \times 0.22} = 30{,}400\ \text{lb/h} \qquad \textit{answer}$$

17.3 ■ The Laws of Thermodynamics

Such transformations of energy involve producing heat from motion or the reverse—motion from heat. Consequently, the name *thermodynamics* (literally, "motion from heat") is the technical term applied to this branch of physics. The science of thermodynamics is based on two fundamental laws, both of which have tremendous significance for industry and for everyday living.

The First Law of Thermodynamics This law is actually a restatement of the law of conservation of energy for the special case involving heat energy.

The First Law of thermodynamics.

In any thermodynamic process, total energy remains constant—none is created or destroyed.

The meaning of the first law of thermodynamics for *a cyclical process in a heat engine* can be illustrated by a schematic diagram. In Fig. 17.3 a heat engine containing a working substance (usually a gas, such as steam, or the products of internal combustion) is shown operating between a high-temperature reservoir and a low-temperature reservoir. It is assumed that a supply Q_1 of heat energy is available at temperature T_1 from the high-temperature reservoir for each cycle. After an amount of work ΔW is done by the engine in that cycle, an amount of (unused) heat Q_2 is rejected to the low-temperature reservoir at temperature T_2. The law of conservation of energy, in this case the *first law of thermodynamics for a heat engine* states that

First Law of Thermodynamics—Heat Engine

$$\Delta W = Q_1 - Q_2 = \Delta Q \qquad \textbf{(17.4)}$$

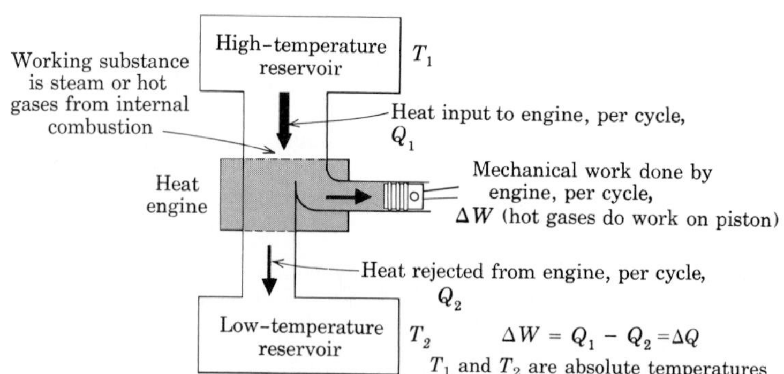

Fig. 17.3 Block diagram illustrating the heat and energy relationships of the first law of thermodynamics.

Or, stated another way, the mechanical work done by a heat engine in one complete cycle is equal to the difference between the heat energy taken in at the higher temperature and the heat energy rejected at the lower temperature. T_1 and T_2 are *absolute temperatures*—Kelvin or Rankine.

Note that Eq. (17.4) applies only to a *complete* cycle of a heat engine, not to single processes (i.e., expansion or compression) within a cycle. A more general treatment follows.

Internal Energy and the First Law It should be noted that, in the above formulation of the first law (Eq. 17.4), the assumption of a cyclic process involving a heat engine was made. That is, work is done as heat energy in transit flows through the system (engine). The source of the heat is the high-temperature reservoir. Heat that is not converted to work is rejected to the low-temperature reservoir. But what about the *internal energy* of the molecules of the working substance? Equation (17.4) contains no information about this repository of energy. Suppose that the process were not cyclical and that heat energy were not being continually supplied from the high-temperature source—could any work be done by the system? The answer is yes, on a one-time basis the working substance could expand and do work *at the expense of its store of internal energy*.

Internal energy and the First Law.

For general applicability to all kinds of thermodynamic processes, the first law must take into account the internal energy of the system. Let U be the energy of the molecules of a thermodynamic system, that is the *internal energy*. Q_1, as before, is the heat input, Q_2 is the heat rejected, and ΔQ is the *net heat* added. ΔQ can result in either a change in internal energy ΔU, or in the accomplishment of external work, ΔW.

A generalized statement of the first law of thermodynamics takes into account the internal energy of the system:

Heat energy added to and in transit through a system is equal to the algebraic sum of the change in internal energy of the system and the external work done.

As an equation,

Generalized Statement—First Law of Thermodynamics

$$Q_1 - Q_2 = \Delta Q = \Delta U + \Delta W \qquad (17.5)$$

Heat added to a system = change in internal energy + work done by the system

It should be noted that in all of the above discussions, it is assumed that the "system" is perfectly insulated (except for the entry of heat from the high-temperature source), preventing any undesired heat losses or gains from conduction, convection, or radiation.

In interpreting Eq. (17.5) it will be helpful to keep in mind that it involves these four important concepts:

1. The law of conservation of energy
2. The idea of internal energy
3. The realization that heat is not a substance, but is energy in transit
4. The equivalence, on a two-way basis, between heat and work

Adiabatic, or constant-heat, processes.

Constant Heat Processes In many thermodynamic processes there is no gain or loss of heat (ΔQ is zero). These are called *adiabatic* processes. In such processes the internal energy U of the system may undergo rapid and significant changes, as work is being done either *by* the system or *on* the system. Adiabatic processes will be described later in this chapter and in Chap. 18.

The Second Law of thermodynamics—when and how heat energy can do useful work.

The Second Law of Thermodynamics This law has to do with the *availability* of heat energy for the purpose desired.

Heat will not flow "up a temperature hill" unless mechanical work is expended to force it to do so.

An alternate statement of the second law is:

Mechanical energy cannot be obtained from any source of heat unless that source undergoes a drop in temperature.

Several other interpretations of the second law are given below. Thus, in an engine, a hot substance flows from a hot cylinder to a cooler exhaust and does mechanical work as the temperature drops. (A refrigerator, which is a heat engine in reverse, takes heat from a low-temperature source and transfers it to a place of higher temperature, but *mechanical work must be done on the working substance by the compressor in the process*.) (See Chap. 18.)

The *first law* of thermodynamics tells us that energy is conserved in all processes *that actually take place*, but it does not tell us *whether or not a specific process will take place*. When a baseball's kinetic energy (at about 90 mi/h velocity) is absorbed by a catcher's mitt, much of the energy of motion shows up as heat, which raises the temperature of the mitt. The first law is obeyed. But no one has ever observed the reverse process, which, *if it did happen*, would also be consistent with the first law. Sudden cooling of a catcher's mitt does not expel a baseball at 90 mi/h. By the same token, a jar of hot water does not rise into the air gaining potential energy as it cools and freezes, even though there is sufficient energy in the water to supply the mechanical work needed for such a feat (Fig. 17.4). An automobile is brought to a stop from high speed by the absorption of its kinetic energy by the braking system, with a consequent sudden heating of the brake drums and brake pads or shoes. Luckily for us the reverse does not happen: we do not sharply accelerate in the auto as a result of the braking system's sudden surrender of heat energy with consequent frost formation on the brake drums.

There is an incalculable amount of heat energy in warm summer air. Why not run an engine on this plentiful and cheap "hot gas"? The second law tells us why not—there is no "cold reservoir" to which the exhaust from the engine can be rejected. Look at Fig. 17.3 and note that the work done per cycle would be zero, since the engine would exhaust into a "reservoir" whose temperature is the same as that of the heat energy source.

These examples of what *does not happen* seem to say that there is some kind of natural law which stipulates the *direction* in which thermodynamic processes can take place. Analysis of the many processes theoretically possible for isolated systems under the first law reveals that only those which move in a direction *from an orderly state toward a more disorderly state* will *actually* occur. Steam or gas molecules in a confined chamber at high pressure represent an orderly arrangement of high-speed molecules, under control, and available to do work. The exhaust steam or the spent gases from an internal-combustion engine and the "wasted heat" which heats up the engine parts—these are disorderly, more random manifestations of thermal energy at lower temperatures, and we cannot put these lower levels of energy back through the engine and get useful work.

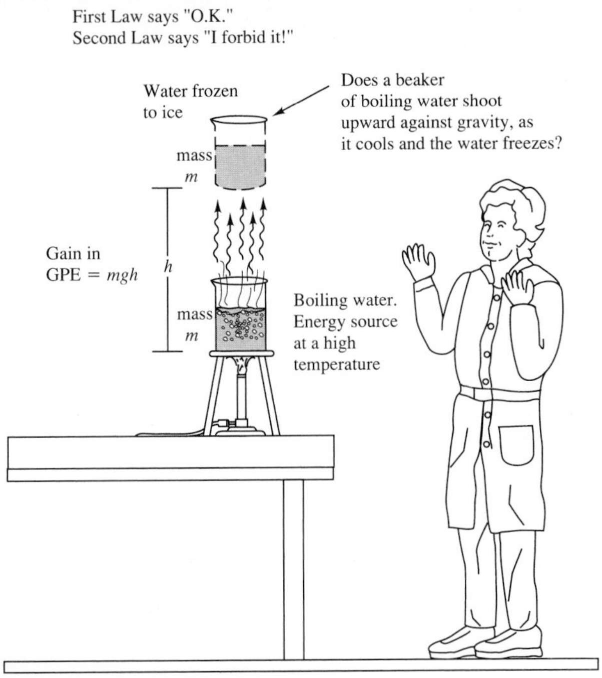

First Law says "O.K."
Second Law says "I forbid it!"

Water frozen to ice

Does a beaker of boiling water shoot upward against gravity, as it cools and the water freezes?

mass m

Gain in GPE = mgh

h

mass m

Boiling water. Energy source at a high temperature

Fig. 17.4 A hot, high-energy mass *does not* use its energy to rise against gravity, losing heat and gaining gravitational potential energy as its mass cools and perhaps freezes! The First Law does not prohibit such an event, but the Second Law does. If it *did happen,* all the First Law would require is that the gain in GPE equal the loss in heat energy.

The second law of thermodynamics is independent of the first law, and it says that energy processes always move from order to disorder, and that heat flows of its own accord from high temperature to low temperature. Three men of science who devoted much of their lives to the study of heat and thermodynamics have made separate but equivalent statements of the second law, which are summarized here:

Three different but equivalent interpretations of the Second Law.

The Second Law of Thermodynamics—Three Formulations

As formulated by Sadi Carnot (1796–1832):

It is impossible to construct an engine which will convert a given quantity of heat energy into an equivalent amount of work.

As formulated by Rudolph Clausius (1822–1888):

It is impossible for any self-operating device to take heat continuously from a reservoir at one temperature and deliver it to a reservoir at a higher temperature.

As formulated by Lord Kelvin (1824–1907):

As a result of natural processes, the world's supply of energy available to do work is continually decreasing.

17.4 ■ Entropy

Entropy—a measure of disorder; or of the degree of energy unavailability.

The thermodynamic property of a system which measures its degree of disorder is called *entropy* (from a Greek word meaning "transformation"). Entropy may be thought of as a measure of the degree of energy degradation in any process. The second law of thermodynamics is sometimes called the *law of entropy*. Clausius defined entropy mathematically as follows: If ΔQ is a small reversible change in the heat content of a body, and if the change takes place at temperature T, then the *change in entropy* ΔS is:

$$\Delta S = \frac{\Delta Q}{T} \tag{17.6}$$

Change in entropy = heat gain (or loss) divided by absolute temperature

Natural phenomena always occur (in isolated systems) in such a direction that entropy (the state of disorder or *unavailability* of energy) *increases*. Only in an idealized reversible process can ΔS be zero. Referring again to Fig. 17.3, we see that heat from a reservoir of hot gas (temperature T_1) is allowed to flow into a reservoir of "cold" gas (temperature T_2), with an engine in the path of the heat flow converting some of the heat to mechanical energy. Suppose no engine had been there. The high-temperature gas would then merely get cooler, and the low-temperature gas would get warmer until the mixture reached some equilibrium temperature. There is no decrease in *total energy* in this second process (first law), but the opportunity to convert some of the heat in the "hot" reservoir to mechanical work has been lost, not temporarily, but forever (second law). Energy itself has not been lost, but the opportunity to get some work done has been irretrievably lost. The mixture will not, by itself, separate back into hot gas and cold gas. In this process, entropy will have increased in accordance with Eq. (17.6), and *availability* of energy for conversion to mechanical work will have decreased.

The natural condition throughout the universe seems to be one which tends toward "running down," toward some final state at which the entire universe will finally come to the same temperature. Entropy will have increased to a maximum, and theoretically all thermodynamic processes (including life processes, of course) would end. This so-called heat death of the universe is assumed to be many billions of years in the future, however.

17.5 ■ Basic Principles of Heat Engines

Operating principles of heat engines.

A material substance must be employed in the transformation of heat to work accomplished by a heat engine. Ordinarily, a gas is the material substance. It is heated and allowed to do work as it expands against a piston or in a turbine casing. In reciprocating steam engines, superheated steam expands and drives the piston down the cylinder. As the volume increases, the steam pressure decreases and its temperature drops. In steam turbines, high-pressure steam shoots through nozzles against blades on the turbine rotor, and the impulse and reaction forces whirl the rotor at high speed. The steam expands with great force and moves on to the next stage of rotor blades. (See Fig. 17.14.) In internal-combustion engines, hot gases are produced by high-temperature burning within the cylinder, and again the low-temperature exhaust indicates that mechanical work has been done. In rocket engines the hot gases from combustion exit rearward at high speed causing the reactive force (momentum principles) that propels the vehicle. All heat engines which do work absorb heat at a high temperature, turn some of it into mechanical work, and exhaust or reject the rest at a lower temperature. In any case, $\Delta W = Q_1 - Q_2$ (Eq. 17.4).

Reciprocating engines must return the piston to the starting point. As the piston returns, the gas in the cylinder must be compressed. Work has to be done *by the piston* on the gas to accomplish this compression. The sequence of expansion and compression is called a *cycle*. During a cycle, work is done *by the hot gas* during expansion, and a smaller amount of work has to be done *on the gas* during compression. For an ideal engine, the difference between the amount of work done on the piston by the hot expanding gas and the amount of work done by the piston on the gas to compress it is the net work accomplished.

17.6 ■ Work of Expansion and Compression

As a hot gas expands in a cylinder it will do work as its expansion force moves a piston through the length of the engine stroke. And, in order to complete a cycle, the piston must do work *on the gas* as it reverses direction and moves back to the starting position in the cylinder, compressing the gas. Figure 17.5*a* shows a cylinder fitted with a movable piston and containing a hot-gas working substance. The cross-sectional area of the piston is A, and the expanding gas moves the piston from position 1 to position 2. Let us first assume that the pressure of the gas P remains constant throughout the expansion stroke. We call this an *isobaric* (constant-pressure) expansion. Isobaric expansions are not characteristic of real engines but are most nearly approximated in old-style single-acting steam engines.

As a gas expands it can do work.

An *isobaric* expansion is one that occurs at constant pressure.

Now force equals pressure times area, so the force on the piston throughout an isobaric expansion is $F = PA$. If the piston is pushed a distance s, the work done is $W = Fs$. But the area of the piston times the distance moved is $V_2 - V_1 = V$, the volume swept out in one stroke. Consequently,

$$W = P(V_2 - V_1) = PV \tag{17.7}$$

Verification of this equation is readily obtained from an analysis of Fig. 17.5*b*. The *P-V* curve for a constant-pressure expansion (*E*) is seen to be a straight horizontal line. From the geometry of the figure, the work done during expansion is verified as $W = P(V_2 - V_1) = PV$. Note that the product PV is the area (shaded) under the *P-V* curve.

The isobaric case so far discussed is not typical of actual practice. In most engines and turbines the pressure drops off markedly during expansion in the cylinder or in the turbine stages, and the *P-V* curve has a configuration more like that shown in Fig. 17.5*c*, moving down and to the right, toward lower pressure as the volume increases. Mathematical analysis of such curves by integral calculus shows that the work done by the gas in expanding is again *equal to the area under the P-V curve*, that is, equal to the shaded area of Fig. 17.5*c*.

Work and the *P-V* diagram.

The work done on a piston by an expanding gas is equal to the area under the P-V curve for the expansion.

For the compression (return) stroke, when work must be done *on the gas* by the piston, Fig. 17.5*d* illustrates the piston arrangement. Figure 17.5*e* shows the compression curve C moving upward and to the left, toward higher pressure as the volume is reduced. The (input) work of compression is also indicated by the area under the *P-V* curve, as shown by the shaded area.

Work must be done *on a gas* to compress it.

Finally, the net work accomplished in this two-stroke cycle is equal to the difference between the expansion work (output) and the compression work (input). This difference is shown graphically in Fig. 17.5*f* as the shaded area *EYCX* enclosed by the expansion curve and the compression curve.

P-V curves for heat engine analysis are often called *indicator diagrams*. Areas under the curves can either be calculated by computers or measured by instruments called *planimeters*.

Treating the cylinder and piston arrangement as a closed system, as we have done in the discussions of this section, is a simplification that enables us to look at expansion, and later, compression as discrete events, uncomplicated by other factors and conditions that exist in real engines. We now turn to an explanation of the cyclic processes more or less typical of real heat engines, preparatory to an analysis of some of the engines themselves.

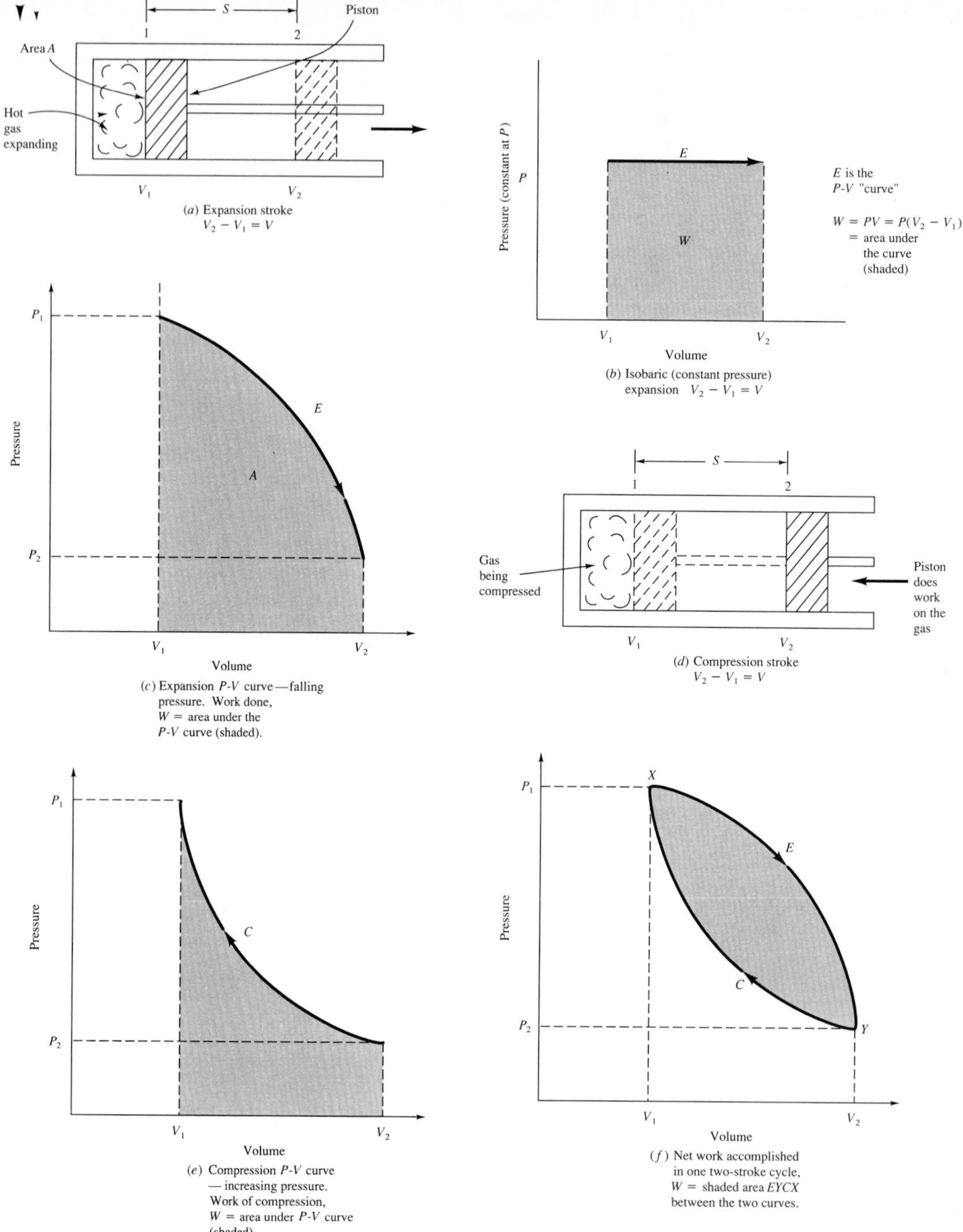
Area A
Hot gas expanding
S
1
2
Piston
V_1
V_2
(a) Expansion stroke $V_2 - V_1 = V$
Pressure (constant at P)
P
E
W
V_1
V_2
Volume
(b) Isobaric (constant pressure) expansion $V_2 - V_1 = V$
E is the P-V "curve"
$W = PV = P(V_2 - V_1)$ = area under the curve (shaded)
Pressure
P_1
P_2
E
A
V_1
V_2
Volume
(c) Expansion P-V curve—falling pressure. Work done, W = area under the P-V curve (shaded).
S
1
2
Gas being compressed
Piston does work on the gas
V_1
V_2
(d) Compression stroke $V_2 - V_1 = V$
Pressure
P_1
P_2
C
V_1
V_2
Volume
(e) Compression P-V curve — increasing pressure. Work of compression, W = area under P-V curve (shaded).
Pressure
P_1
P_2
X
E
C
Y
V_1
V_2
Volume
(f) Net work accomplished in one two-stroke cycle, W = shaded area EYCX between the two curves.

Fig. 17.5 The work of expansion and compression. (*a*) Cylinder and piston arrangement as a hot gas begins to expand, pushing the freely moving piston to the right. (*b*) The pressure-volume (*P-V*) "curve" for a constant-pressure (*isobaric*) expansion. The *P-V* "curve" is a straight line, and the work done on the piston is equal to the area of the shaded rectangle under the "curve." (*c*) The *P-V* curve for an expansion as pressure falls. Again, the work done on the piston is equal to the shaded area under the *P-V* curve. (*d*) The cylinder and piston situation for a compression stroke. The gas is compressed as the piston moves to the left. (*e*) The *P-V* curve for a compression, as pressure increases. The work done on the gas by the piston is equal to the area (shaded) under the *P-V* curve. (*f*) Net work accomplished by successive expansion and compression in a two-stroke cycle. The net work accomplished is equal to the shaded area between the two *P-V* curves.

17.7 ■ Cyclic Processes Involved in Heat Engines—Isothermals and Adiabatics

***An isothermal process** is one in which the pressure and volume of a gas vary without change in the temperature.*

Isothermal processes occur at constant temperature. Boyle's law is obeyed. Heat must be allowed to flow in and out during isothermals.

In order to keep temperature constant, heat must be allowed to flow in and out of the gas as its pressure and volume vary. Under these conditions, Boyle's law of gases is obeyed. (See Fig. 14.19 for a graphical interpretation of an isothermal process.)

Boyle's law is expressed mathematically as

$$PV = k \text{ (a constant)} \tag{14.12}$$

for an ideal gas. In an isothermal process, therefore, P varies inversely with V, and the temperature remains constant, as heat energy flows in and out of the cylinder.

If a gas expands or is compressed (P and V changing) *without permitting heat to flow* from or to the gas, the temperature *will not stay constant,* and Boyle's law does not hold true. This kind of process, where there is no flow of heat into or out of the system, is known as a constant-heat, or *adiabatic,* process.

***An adiabatic process** is one in which pressure and volume vary in such a way that there is no heat flow into or out of the system. Temperature is not constant in an adiabatic process.*

Adiabatic processes occur at constant heat. The general gas law is obeyed. Temperature is not constant.

Adiabatic expansion and compression are illustrated in Fig. 17.6.

In actual practice, a perfectly adiabatic process is impossible, since a perfectly insulated system is impossible. By definition, in an idealized adiabatic process, $\Delta Q = 0$.

An insulated cylinder (theoretically insulated so well that no heat is lost or gained during a complete cycle) is fitted with an insulated piston and a sensitive pressure gauge and thermometer (Fig. 17.6*a*). A scale along the cylinder enables the volume to be determined for any piston position. Let the cylinder contain a given mass of gas at temperature T_1 (degrees Rankine), pressure P_1 (pounds per square inch absolute, psia), and volume V_1 (cubic inches) when the piston is in position 1. This situation corresponds to point 1 on the graph of Fig. 17.6*b*. (NOTE: Heat-power engineering in the United States still makes much use of the English-engineering system of units, and Rankine temperatures.)

As the hot gas pushes the piston to position 2, an adiabatic (constant-heat) expansion takes place, the pressure drops to P_2, and the temperature drops to a lower value T_2. Moving the piston to position 3 lowers the pressure and temperature still further and results in the plotted point marked 3 on the graph. A series of such points would plot to the smooth curve shown. This curve is known as an *adiabatic*.

An *adiabatic process* for an ideal gas obeys the *general gas law* (Sec. 14.21), which says that the pressure of a given quantity of a gas is proportional to the absolute temperature, and inversely proportional to the volume. As an equation, this relation may be expressed as

$$\frac{PV}{T} = k \tag{14.17}$$

Note the contrast with the equation for an *isothermal* in which the product of pressure and volume is a constant ($PV = k$).

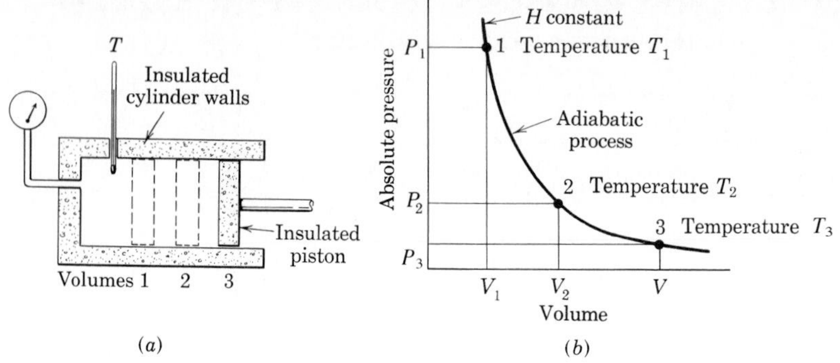

Fig. 17.6 Diagrams illustrating an adiabatic (constant-heat) process. (*a*) The piston-and-cylinder arrangement. Positions 1, 2, and 3 represent gas volumes for which corresponding absolute pressure values are obtained. (*b*) The *P-V* diagram for the process, showing temperature changes, but indicating no heat flow into or out of the cylinder.

Heat engines make use of processes which roughly approximate isothermals and adiabatics in the compression and expansion of gases. The two processes will now be compared and plotted with respect to the same *P-V* coordinates.

In Fig. 17.7 let the *ic* arrow indicate the direction of an isothermal compression (*V* decreases as *P* increases, temperature being held constant). The work of compression will produce heat, and the only way a constant temperature can be held is for the heat to flow out as process *ic* occurs.

Isothermal compression is a process in which mechanical work is done on the gas and heat is given off, while temperature remains constant.

Isothermal *expansion* is shown by the curve in the *ie* direction, and heat must be supplied.

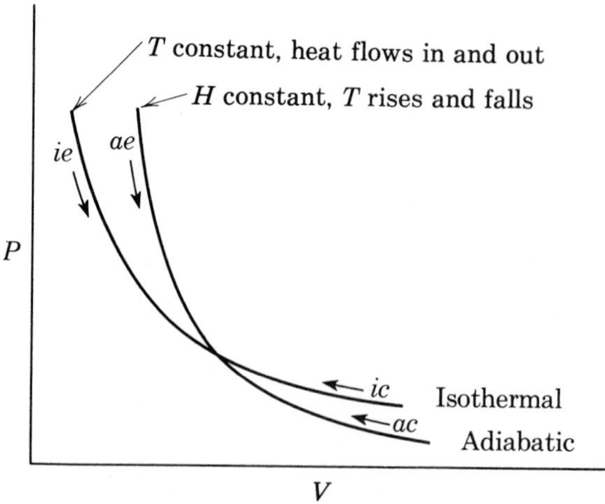

Fig. 17.7 Isothermal and adiabatic curves plotted with respect to the same set of *P-V* coordinate axes.

In isothermal expansion, mechanical work is done by the gas on the piston, and heat must be supplied if temperature is to remain constant.

Referring again to the adiabatic curve of Fig. 17.7 we note that *ac* indicates the direction of adiabatic compression. As work is done *by the piston* on the gas to compress it, the heat equivalent of mechanical work evidences itself as an increase in temperature.

In adiabatic compression the work done on the gas increases the temperature of the gas. In adiabatic expansion work is done by the gas and the temperature of the gas decreases. Any adiabatic process occurs at constant heat.

17.8 ■ The Ideal Heat Engine—Carnot's Cycle

Carnot's ideal (four-stroke) cycle.

The French physicist Sadi Carnot, in 1824, proposed an ideal engine operating on a cycle (since called the *Carnot cycle*) which has the highest possible efficiency for the temperature limits within which it operates. His ideal engine was to operate on a cycle of four heat processes, in the following manner.

A theoretically frictionless one-cylinder engine (see Fig. 17.8) is assumed to have perfectly *nonconducting* walls, a *nonconducting* piston, and a *perfectly conducting* cylinder head. The cylinder is assumed to contain an ideal gas as the *working substance*. The box S_1 represents a source of heat from which heat at a high temperature T_1 (°R or K) can be obtained, and the box S_2 is a receiver of heat, or *sink*, to which heat at a low temperature T_2 can be rejected. The box I is an insulating stand. The four diagrams of Fig. 17.8 show the four stages in Carnot's ideal cycle, and the graph of Fig. 17.9 shows the resulting P-V-T conditions for a complete cycle.

An isothermal expansion is illustrated in Fig. 17.8*a*. At the start the cylinder is assumed to be standing on the heat source S_1, with the piston at A near the bottom of the cylinder. The gas in the cylinder is at temperature T_1 and it occupies a small volume at high pressure, as indicated by point A on the P-V diagram of Fig. 17.9. A quantity of heat Q_1 flows into the cylinder and the gas expands while the piston moves to B. The temperature stays constant at T_1 during this stage, since heat is flowing in from source S_1 to maintain the temperature at the same level. The *isothermal* expansion curve AB of Fig. 17.9 represents this stage of the cycle.

Next (Fig. 17.8*b*) the cylinder is placed on top of the insulator box I, and the gas is allowed to expand further, moving the piston to "top dead center" at C. During this second stage of the expansion no heat leaves or enters the cylinder and work is done

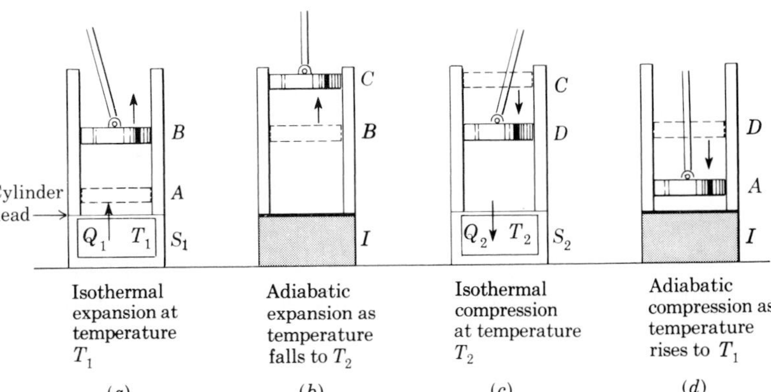

Fig. 17.8 Schematic diagram of the four processes of Carnot's ideal heat-engine cycle.

Fig. 17.9 The Carnot cycle represented by two isothermals and two adiabatics. The area enclosed by the four curves is the net work accomplished per cycle, ΔW. This concept is for an "ideal gas."

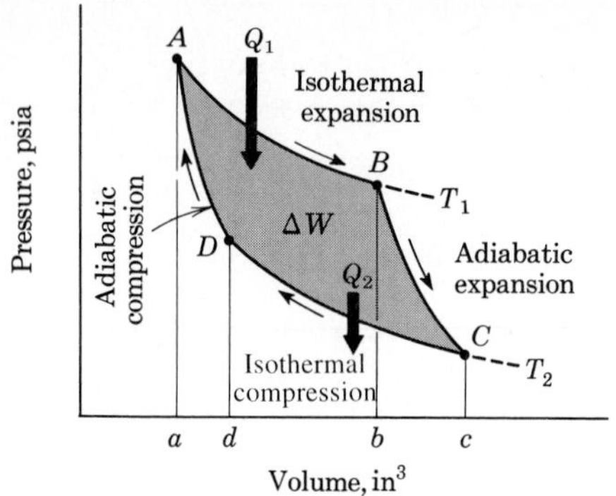

on the piston as the temperature of the gas in the cylinder falls to a value T_2. The process is illustrated on the *P-V* diagram by the *adiabatic* expansion curve *BC* of Fig. 17.9.

The cylinder is now placed on the heat sink S_2 (Fig. 17.8*c*), whose temperature is cold enough so that heat will flow into it from the cylinder. The piston is now thrust downward to *D*, and the heat equivalent of this mechanical work Q_2 flows out into the heat sink. The temperature of the gas in the cylinder remains constant at T_2. This first phase of the compression is illustrated by the *isothermal* compression curve *CD* on the *P-V* diagram.

Finally (Fig. 17.8*d*), the cylinder is placed on the insulating stand once more, and the piston is thrust downward to *A*. More work is thus done on the gas, and since no heat can escape, a rise in temperature is the result. The final *P-V-T* conditions are the same as those at the start, and the second stage of the compression is illustrated by the *adiabatic* curve *DA* on the *P-V* diagram of Fig. 17.9.

Work done in a complete cycle of a heat engine is equal to the area bounded by the *P-V* curves of the cycle.

During the complete cycle just described, an amount of heat Q_1 flows into the cylinder at the high temperature T_1, and an amount Q_2 is exhausted at the lower temperature T_2. External work is done on the piston by the expanding gas during the expansion phase *ABC*, and some work is done on the gas by the piston during the compression *CDA*. The external work accomplished during expansion is represented by the area *ABCca*, and the energy given back during compression by the area *ADCca*. The net work ΔW accomplished in one complete cycle is given by *ABCca* − *ADCca*, which is the shaded area *ABCD* within the closed curves. This is equivalent to writing $\Delta W = Q_1 - Q_2$.

The foregoing discussion leads to the following summary, with respect to heat engines:

1. A source of heat at high temperature (a steam boiler or burning fuels) called a *heat source* or *reservoir*, supplies an amount of heat per cycle, Q_1 at temperature T_1.
2. The engine uses a part of this input heat to do mechanical work on the basis of $W = JH$. (J is the mechanical equivalent of heat, 4.186×10^3 J/kcal, or 778 ft · lb/Btu).
3. The remaining heat Q_2 leaves the engine in the "exhaust" at temperature T_2. The exhaust heat is rejected or discarded to a low-temperature reservoir or *sink*, (often the atmosphere).
4. The theoretical mechanical work accomplished by the engine, per cycle, is $\Delta W = Q_1 - Q_2$.

Illustrative Problem 17.3 The *P-V* diagram of an isothermal expansion of a hot gas in an engine cylinder has the form shown in Fig. 17.10. (*a*) How much work (ft · lb) is done on the piston per stroke? (*b*) How much heat (Btu) has to be supplied to keep the gas temperature constant during the stroke?

Solution (***a***) The net work done per stroke is expressed by the area under the *P-V*

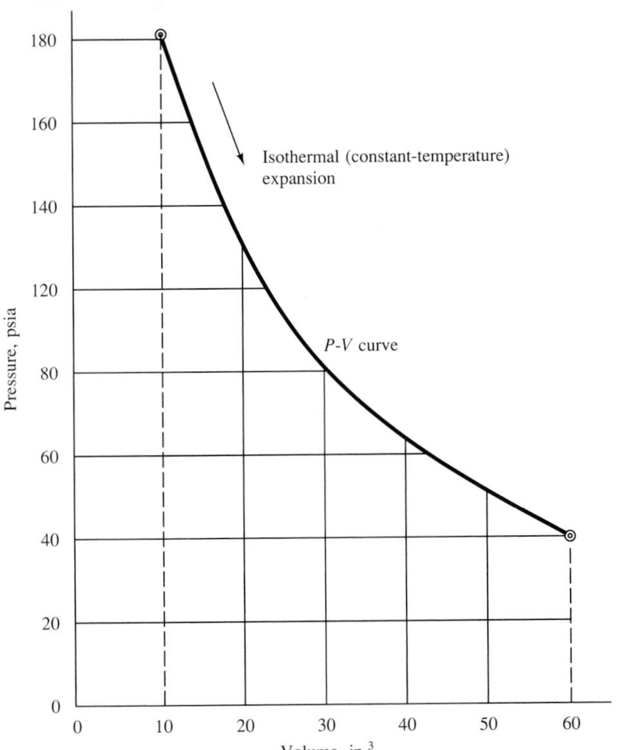

Fig. 17.10 *P-V* diagram for an isothermal expansion of a hot gas in an engine cylinder (Illustrative Problem 17.3).

curve. Counting squares under the curve yields approximately 20.3 (large) squares. Each large square represents 20 lb/in.^{2} × 10 in.^{3} = 200 lb · in.

$$\Delta W = \frac{200 \text{ lb} \cdot \text{in.}}{12 \text{ in./ft}} = 16.67 \text{ ft} \cdot \text{lb per stroke} \qquad \textit{answer}$$

(***b***) Since temperature remains constant, the internal energy of the working substance is constant and $\Delta U = 0$. Therefore, from Eq. (17.5), $\Delta Q = \Delta W$. Then $Q =$ 16.67 ft · lb/stroke. Since 778 ft · lb = 1 Btu (see Sec. 17.1) we know that

$$Q = \frac{16.67 \text{ ft} \cdot \text{lb/stroke}}{778 \text{ ft} \cdot \text{lb/Btu}} = 0.0214 \text{ Btu/stroke} \qquad \textit{answer}$$

17.9 ■ Efficiency of the Ideal Engine

On the basis of these considerations, and using the concept of absolute temperature, Carnot was able to derive an expression for the ideal efficiency of any heat engine operating between the temperature limits T_1 and T_2.

The efficiency of *any machine* was given in Chap. 6, in percent, as

$$\% \text{ Eff} = \frac{\text{work output}}{\text{energy input}} \times 100 \tag{6.18}$$

For a heat engine

$$\% \text{ Eff} = \frac{\text{work output}}{\text{mech. equiv. of heat input}} = \frac{\Delta W}{Q_1} \times 100$$

From the first law of thermodynamics

$$\Delta W = Q_1 - Q_2$$

Consequently

$$\text{Heat engine eff} = \frac{Q_1 - Q_2}{Q_1} = 1 - \frac{Q_2}{Q_1} \tag{17.8}$$

In the ideal (Carnot) engine, the heat transferred to or from the working substance is directly proportional to the absolute temperature of the reservoir which supplies or receives the heat, or

$$\frac{Q_2}{Q_1} = \frac{T_2}{T_1}$$

Efficiency of an ideal (Carnot) engine.

Therefore Carnot efficiency (ideal engine) is

$$\text{Carnot eff} = 1 - \frac{T_2}{T_1} = \frac{T_1 - T_2}{T_1} \tag{17.9}$$

where T_1 = absolute temperature of hot gas supplied to engine
T_2 = absolute temperature of (cooler) gas exhausted from engine

Analysis of Eqs. (17.8) and (17.9) shows that: (1) For an engine to be 100 percent efficient, no heat could be exhausted—Q_2 must be zero, an impossible event. (2) The less heat exhausted, the greater the efficiency. (3) For a given temperature of the hot gas supplied to the engine, the greatest efficiency results when the exhaust gas is at the lowest possible temperature.

A heat engine is most efficient when the difference between the operating temperature and the exhaust temperature is a maximum.

A real engine cannot possibly approach Carnot's theoretical cycle of operation, and actual operating engines are always much less efficient than Eq. (17.9) would indicate. Friction is always present, for one thing, and actual cycles only approximate the theoretical isothermal and adiabatic processes described above.

Equation (17.9) shows that the efficiency of a heat engine increases as the difference between T_1 and T_2 increases; or, an engine is the more efficient, the greater the difference between the temperature of the hot gas supplied (steam or gases from internal combustion) and the temperature of the exhausted gas. Exhaust temperature cannot be lower than that of the medium to which the working substance is exhausted (the atmosphere, or perhaps water). Since T_2 cannot be lowered at will by the engine designer or operator, the most feasible way to get higher efficiency is to make T_1 as high as possible. (T_2 *can be lowered* within limits by the use of condensers with steam engines—see below.)

Illustrative Problem 17.4 A steam engine is supplied saturated steam at a pressure of 250 lb/in.^{2} abs and exhausts it into the atmosphere. What is the "ideal" Carnot efficiency of this engine?

Solution From Table 15.6 note that *saturated* steam at 250 lb/in.^{2} abs is at a temperature of 401°F, and at 14.7 lb/in.^{2} abs its temperature is 212°F. Changing these to Rankine temperatures and substituting in Eq. (17.9), we have

$$\text{Carnot eff} = 1 - \frac{672}{861} = 1 - 0.78 = 0.22 \text{ or } 22\% \qquad \textit{answer}$$

Operating between these temperature limits, even the ideal engine could convert into mechanical energy no more than 22 percent of the heat energy supplied. Real engines, as pointed out above, have efficiencies well below the theoretical Carnot efficiency. However, *real* steam engines operate on *superheated* steam rather than saturated steam, so that T_1 can be very high, and efficiency can therefore be increased somewhat.

STEAM ENGINES

Heat engines designed to use steam as the working substance are of two types, *reciprocating engines* and *turbines*. In both cases the complete steam plant involves a

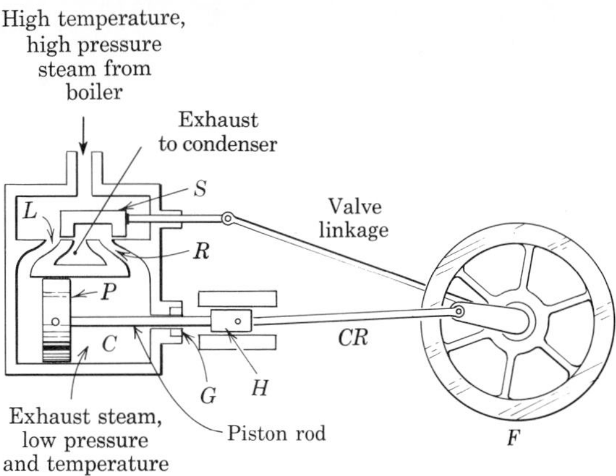

Fig. 17.11 Diagram (schematic) of a slide-valve, double-acting, reciprocating steam engine.

boiler in which to produce steam, the engine itself, and a *condenser* into which the exhaust steam is discharged to save both the water* and the heat which still remain in the spent steam. Condensers can be operated in such a manner that they create a partial vacuum at the exhaust exit and they lower the temperature T_2 appreciably. This increases efficiency.

17.10 ■ Reciprocating Steam Engines

The *basic principles* of operation of the single-cylinder slide-valve reciprocating steam engine have not changed a great deal since the time of James Watt, although many design improvements have been made. The engine can be represented diagrammatically, as in Fig. 17.11. Hot high-pressure steam enters the cylinder through port L, which is opened at the proper instant by the slide valve S. The hot steam expands, pushing the piston P toward the right end of the cylinder C. As the piston moves to the right, the port R is also open, and the spent steam from the previous stroke exhausts to the condenser. The piston is joined to the crosshead H by the piston rod and thence to the crank of the flywheel by the connecting rod CR. The flywheel F is mounted on the main shaft of the engine, and so also is an eccentric, or cam, which operates the slide-valve linkage. The slide valve closes the intake port L about one-third of the way through the stroke, and the stroke is completed by the expansion of the steam which is already in the cylinder. As the expanding steam does work on the piston, the steam temperature falls. At the end of the stroke, the slide valve interchanges the connections so that port R is now open to incoming live steam and port L is connected to the exhaust. Thus, a continual reciprocating motion of the piston is maintained, which is converted into rotary motion of the flywheel and shaft by the crankshaft and crank.

Such engines are frequently called *double-acting* engines, since work is done on the piston by the expanding steam from both ends of the cylinder. The piston rod enters the cylinder through a *packing gland G,* which must be carefully made and frequently inspected to ensure that steam does not leak out around the piston rod and that the packing is not tight enough to cause undue friction on the rod.

Superheated steam is supplied to such an engine from a *boiler* or *steam generator*. The spent steam exhausts to a *condenser* (usually water-cooled) which is operated at a sufficiently low temperature that steam condenses rapidly enough to create a partial vacuum at the exhaust valve ports. Both the partial vacuum and the low exhaust temperature tend to increase engine efficiency.

Modern steam engines run on superheated steam.

* Boiler feedwater is expensive since it must be quite pure. Impurities precipitate out and clog up boiler tubes, steam lines, and engine and turbine parts. Expensive chemical treatment is required to maintain purity, so as much water as possible is condensed and reused.

Fig. 17.12 The idealized steam cycle, or Rankine cycle, for a condenser-equipped reciprocating steam engine. Such representations are sometimes called *indicator diagrams*.

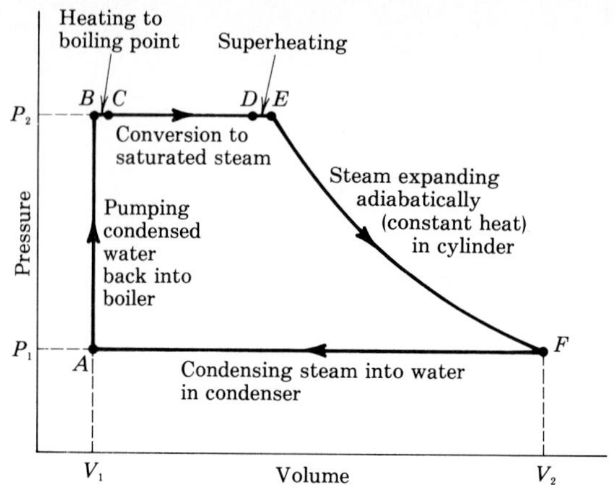

These processes are illustrated in simplified form by the *P-V* diagram of Fig. 17.12. This idealized cycle (the Rankine cycle) begins with liquid (boiler-feed) water coming from the condenser at point *A* on the diagram—low pressure and temperature. Line *AB* represents the water being pumped, under pressure, into the boiler. Line *BC* represents heating the water at constant pressure to the boiling point; line *CD* represents further heating and evaporation to saturated steam; the line *DE*, further heating and expansion to superheat the steam. The superheated steam then enters the cylinder and expands adiabatically, doing work on the piston (*EF*). The line *FA* represents cooling and condensation at reduced pressure in the condenser, to complete the cycle. Water is the working substance.

The advantage of a condenser can be shown by recalculating Illustrative Problem 17.4, this time assuming that the steam exhausts from the cylinder at point *F* into a condenser at 110°F, instead of blowing to the atmosphere as saturated steam at 212°F. For the condenser-equipped engine (all other conditions as in Illustrative Problem 17.4),

$$\text{Carnot eff} = 1 - \frac{570}{861}$$

$$= 1 - 0.662$$

$$= 0.338, \quad \text{or 34 percent} \qquad \textit{answer}$$

The condenser thus increases the efficiency by over 10 percent.

It is again pointed out that real engines do not attain the efficiencies predicted from the theoretical Carnot cycle.

17.11 ■ Horsepower Developed by Reciprocating Steam Engines

We have seen from the analysis of the Rankine cycle that pressure is not constant during the stroke. However, an average or *mean effective pressure* (MEP) for the stroke can be determined. If MEP is multiplied by piston area, the average force on the piston is obtained. This force times the length of stroke gives the work done per stroke. Recalling that *power is work per unit time* and that 1 hp equals 33,000 ft·lb/min, we write

Calculating the horsepower output of a reciprocating steam engine.

$$\underset{\text{(reciprocating engine)}}{\text{hp}} = \frac{PLAN}{33{,}000} \tag{17.10}$$

where

P = MEP, lb/in.^{2}
L = length of stroke, ft
A = piston area, in.^{2}
N = number of power strokes per minute

Illustrative Problem 17.5 Calculate the horsepower being developed by a double-acting reciprocating steam engine making 250 rev/min if piston diameter is 10 in., stroke is 22 in., and MEP is 240 lb/in.^{2}.

Solution Substitute in Eq. (17.10) being careful to use the units indicated and remembering that there are two power strokes per revolution in a double-acting engine.

$$\text{Power} = \frac{240 \text{ lb/in.}^2 \times \frac{22}{12} \text{ ft} \times \pi \times 25 \text{ in.}^2 \times 2 \times 250 \times 1/\text{min}}{33{,}000 \text{ ft} \cdot \text{lb/min/hp}}$$

$$= 523 \text{ hp} \qquad \textit{answer}$$

The reciprocating steam engine is inefficient, heavy, and generally unsuited to most modern uses for "prime movers." It has been largely replaced by the steam turbine.

17.12 ■ The Steam Turbine

Since ancient times it has been known that continuous rotational motion from steam was possible. At some time between 150 B.C. and A.D. 250 (scholars disagree on the dates of his life) a Greek scientist of Alexandria, Hero (or Heron), made and demonstrated a toy steam engine operating as a *reaction turbine* (see Fig. 17.13). A simple laboratory demonstration will serve to show the basic principles of another type of steam turbine—the *impulse turbine*. Boil water in a flask and allow the steam to issue at high velocity from a nozzle made from a bent glass tube. Let the steam jet strike a bladed wheel made of cardboard. The heat contained by the steam makes for extremely high-speed, high-energy molecules, which hit the blades of the rotor. The loss in momentum of the steam molecules on impact with the vanes is imparted to the vanes of the rotor as *impulse,* in accordance with the impulse equation $Ft = mv_2 - mv_1$.

An English engineer, Sir Charles Parsons (1854–1931), was the first to construct a commercially successful engine based on reaction principles (about 1884). Gustaf de Laval and Charles Curtis, beginning about 1896, made a series of improvements on Parsons' original design, the most important of which was to combine the impulse principle with the reaction principle in the same turbine. As a result of this and other refinements, the steam turbine is today among the most efficient of heat engines, as Table 17.1 shows.

Figure 17.14 shows some of the construction features of high-speed, multi-stage steam turbines.

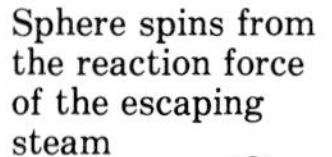

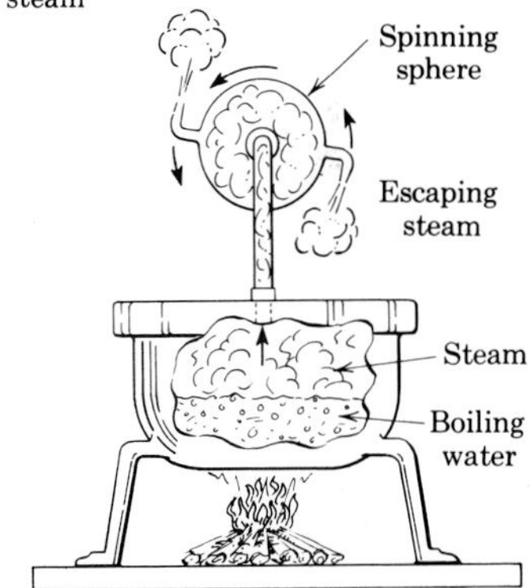

Fig. 17.13 A schematic representation of Hero's engine. This first steam turbine was merely a toy. As far as is known, it was not developed into a practical engine. It operated on reaction principles. (General Electric Co.)

Table 17.1 Efficiencies of typical heat engines

Engine	*Percent Efficiency*
Reciprocating steam engine, noncondensing	6–8
Condenser-equipped	12–16
Steam turbine	Up to 35
Gasoline engine	Up to 28
Industrial gasoline engine (heavy)	30–35
Industrial engine (natural gas)	35
Hot-gas turbine	20
Diesel engine	30–38

INTERNAL–COMBUSTION ENGINES

For steam engines of all types the *working substance* (steam) is heated in a boiler and carried to the cylinder by pipes called *steam lines*. The products of combustion are vented up the flue, or "stack," as it is called. In the *internal-combustion engine,*

Fig. 17.14 A high-speed, multistage steam turbine for use as a prime mover in an electric power plant. Such turbines operate with superheated steam at 1000°F or more. Sixteen stages allow for full expansion of the steam. Rotors and stators are designed for both impulse and reaction effects. (General Electric Co.)

however, the burning takes place within the engine itself, and the products of combustion, i.e., the hot gases, are the actual working substance. Atmospheric air usually provides the oxygen for combustion, and it becomes a major constituent of the working substance. The chemical energy of the fuel is thus converted into kinetic energy of a moving piston or turbine rotor. Internal-combustion engines may be classified as (1) *reciprocating,* (2) *turbine,* and (3) *jet,* or *rocket.* The reciprocating engine will be discussed first.

17.13 ■ Reciprocating Engines

Such engines may have one or many cylinders, and there are at least four common arrangements of the cylinders. They are designed to burn many different fuels, but gasoline, diesel oil, butane, and natural gas are the most common. Two different cycles of operation are possible, and many specialized types of engines are manufactured for widely differing demands. We shall begin by analyzing the operation of a simple one-cylinder, four-stroke-cycle gasoline engine.

The Four-Stroke Cycle The four-stroke cycle is called the *Otto cycle,* for the German engineer who designed the first successful engine of this type. Figure 17.15 shows in diagrammatic form the basic components and cycle of operation of an

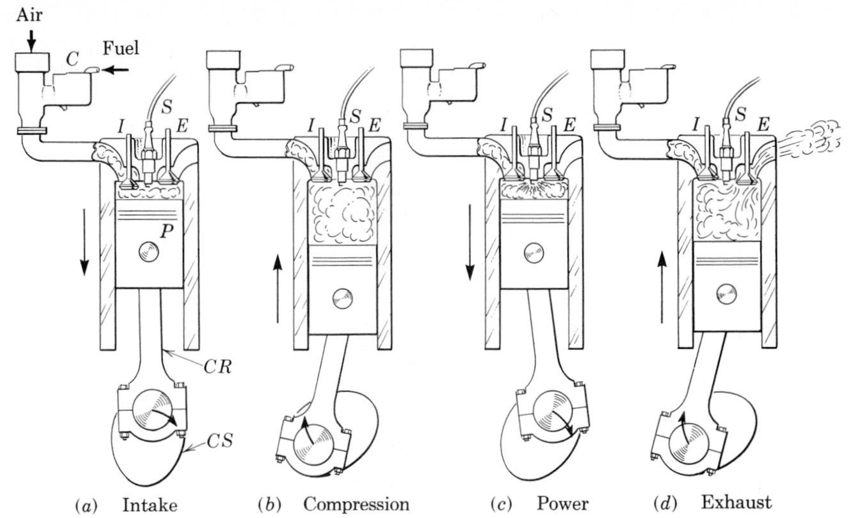

Fig. 17.15 Diagrams of the four-stroke cycle of an internal-combustion engine with fuel mixture supplied by a carburetor, *C*. *I*, intake valve; *E*, exhaust valve; *S*, spark plug; *P*, piston; *CR*, connecting rod; *CS*, crankshaft. (General Motors Corp.)

engine operating on a four-stroke cycle, with a carburetor for mixing air and gasoline vapor. The fuel and air mixture enters the cylinder through the intake valve *I*, shown open, in diagram (*a*). The exhaust valve *E* is closed. The piston *P* is starting down in the cylinder and is connected to the crankshaft *CS* by the connecting rod *CR*. Note that there is no crosshead or piston rod and that the cylinder is open to the crankcase at the bottom. This downward stroke with the intake valve open is called the *intake stroke*.

The most common type of reciprocating internal-combustion engine operates on the four-stroke *Otto cycle*.

Diagram (*b*) shows both valves closed and the piston starting upward. On this stroke the fuel-air mixture in the cylinder is compressed to about one-eighth its full-cylinder volume, depending on the design of the particular engine, and the sudden pressure increase causes an appreciable temperature rise. This stroke is the *compression stroke*. In diagram (*c*) a spark from the ignition system has jumped the gap of the spark plug *S*, and combustion has begun. The fuel does not *explode* but burns progressively through most of the stroke. The rapidly expanding gases force the piston down on the *power stroke*. Both valves are closed. Diagram (*d*) shows the piston starting back up again after passing bottom dead center. The exhaust valve has opened, and the rising piston will force out most of the spent gases. This is the *exhaust stroke,* and, upon its completion, the cycle will begin again. The cycle is, then, *intake, compression, power, exhaust*. Only a part of the heat of combustion of the fuel goes into mechanical work. Much of it is "wasted" in heat conducted through the cylinder walls, and much of it is "rejected" in the exhaust gases. The ideal Carnot efficiency is not closely approached by such engines.

A schematic indicator (*P-V*) diagram for a *gasoline-fueled* Otto-cycle engine is shown in Fig. 17.16. Note that the intake stroke (denoted by *ia*) takes place at

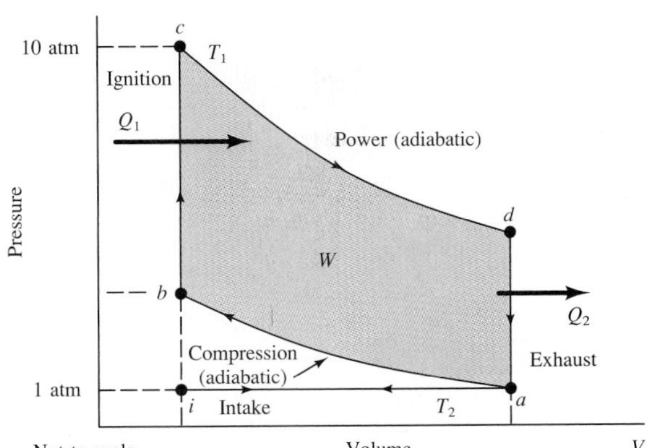

Fig. 17.16 Indicator diagram for a gasoline-fueled, four-stroke, Otto-cycle engine. Starting at *i* with the intake stroke:
ia—intake stroke provides fuel-air mixture at 1 atm pressure (assumed isobaric).
ab—adiabatic compression of fuel-air mixture—work done on gas by the piston.
bc—constant-volume (isochoric) process—ignition occurs and heat Q_1 flows into system. Pressure and temperature both increase.
cd—power stroke—adiabatic expansion—work done on piston by the expanding gas.
da—exhaust valve opens, pressure and temperature drop as heat Q_2 is rejected in the exhaust stroke.
ai—further exhaust at atmospheric pressure—temperature constant.
W—net work per cycle equals area bounded by *abcd*.

atmospheric pressure, as the air-fuel mixture is drawn into the cylinder. The compression stroke *ab* occurs adiabatically as both pressure and temperature increase and volume decreases. During ignition (process *bc*) a quantity of heat Q_1 flows into the system as pressure increases from 1 atm to about 10 atm. The temperature during ignition increases to T_1 at about 800 K. The power stroke *cd* is an adiabatic expansion, with both pressure and temperature dropping sharply. The exhaust valve opens at *d* and, as pressure and temperature drop still further from *d* to *a*, a quantity of heat Q_2 is exhausted to the atmosphere. Temperature in the cylinder falls to T_2 at about 400 K. The exhaust stroke of the piston begins at *a* and is portrayed by *ai*, with volume being reduced at constant temperature T_2 and at 1 atm pressure. As the exhaust stroke concludes, the intake stroke begins again, producing process *ia*, and the cycle repeats itself. The net work per cycle *W* is indicated by the area bounded by *abcd*.

Figure 17.17 shows a one-cylinder, gasoline-fueled, air-cooled engine for four-cycle operation.

For decades the carburetor has been the standard device for vaporizing fuel and mixing it with air. The current trend for automobile engines is toward the use of fuel injectors, which spray the atomized fuel into the cylinder (or manifold) at the proper instant. Such systems are controlled by on-board solid-state electronic devices (''black boxes'').

The two-stroke cycle can improve the horsepower-per-pound ratio of internal combustion engines.

The Two-Stroke Cycle The four-stroke cycle has the disadvantage of producing only one power stroke per cylinder for every two revolutions of the engine shaft. This deficiency is serious where a lightweight engine is desired. To increase the horsepower-per-pound ratio, the two-stroke-cycle engine has been developed. This cycle gives a power stroke every time the piston goes down, or one power stroke per revolution. The two-stroke cycle has only a *power stroke* and a *compression stroke*. In between these, the functions of *exhaust* and *intake* must be performed. There are several different methods in common use for achieving the two-stroke cycle. Two-stroke engines are used for outboard motors, motorcycles, small electric generators, chainsaws, pumps, truck refrigerators, and the like. Many diesel engines are two-cycle engines (see Fig. 17.18).

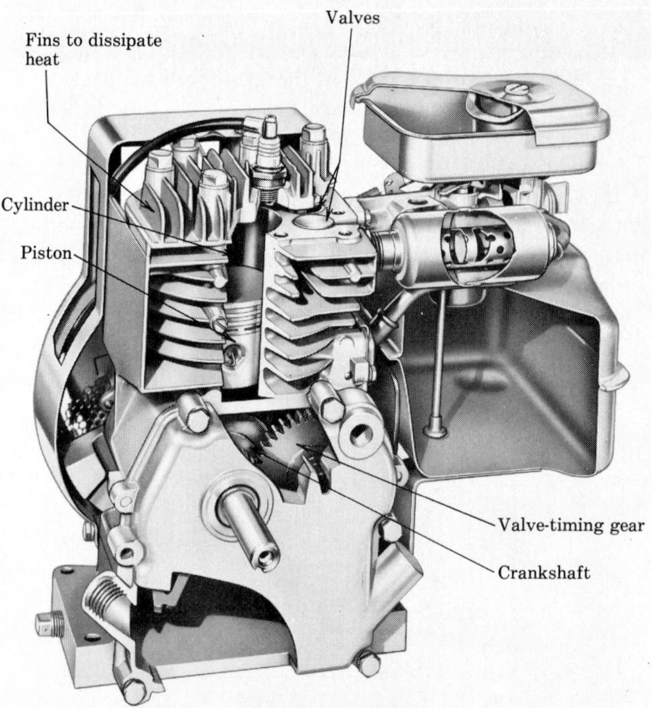

Fig. 17.17 Cutaway view of a one-cylinder, air-cooled, gasoline-fueled engine for four-cycle operation. Such small engines are used by the millions to power small tractors, electric generating plants, garden tillers, etc. (Briggs and Stratton Corp.)

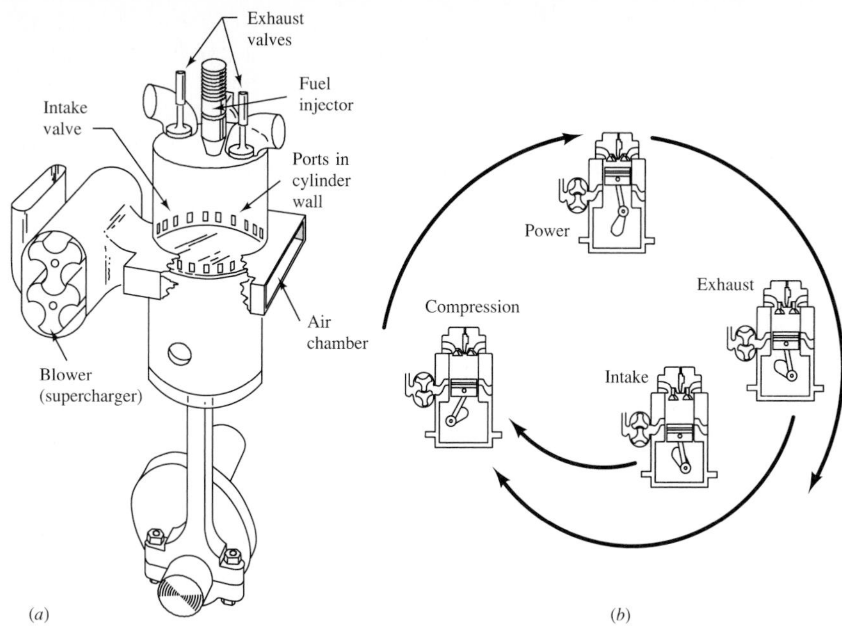

Fig. 17.18 Schematic representation of the two-stroke cycle of a diesel engine.

17.14 ■ The Diesel Engine

In 1892, Rudolf Diesel (1858–1913), a young German engineer, built a new type of internal-combustion engine based on the principle that the heat of compression in a cylinder could flash the fuel and thus dispense with the spark plug and the associated components of the ignition system.

Early diesel engines were massive machines, sometimes weighing as much as 250 lb for each horsepower developed, and were thus suitable only for stationary installations. Improvements over the years have brought this ratio down to as low as 5 lb/hp in smaller, higher-speed engines.

Principles of Operation In any internal-combustion engine, the higher the compression ratio, the greater the efficiency, provided the fuel used will burn slowly enough to prevent "knocking." The diesel engine uses compression ratios up to 20:1, instead of the 6:1 to 10:1 common for gasoline engines. The fuel is classified as *diesel fuel,* a rather heavy type of distillate, which, though it has a high Btu content, burns slowly in the cylinder without "pinging."

Diesel engines are generally more efficient than gasoline-fueled engines, because diesels operate at a higher temperature.

The diesel engine "fires" as the heat of compression in the cylinder raises the temperature of the air/oil vapor mixture to its flash point.

For a four-stroke cycle engine, air is forced into the cylinder by atmospheric pressure or by a supercharger as the piston moves down on the intake stroke. On the upstroke, the air is compressed to less than one-sixteenth its full-cylinder volume, and this sudden compression raises the temperature in the cylinder to about 1000°F, which is well above the flash point of the fuel oil. At this instant, fuel is sprayed into the cylinder in a fine mist by a fuel injector, whose pump builds up pressure of over 2100 lb/in.^{2}. The oil begins to burn immediately and burns evenly throughout the stroke. This power stroke is followed by an exhaust stroke, and the four-stroke cycle is complete. Diesel engines are often built to operate on the two-stroke cycle (see Fig. 17.18). Two-cycle diesels *must* be supplied with a supercharger in order to get enough air in the cylinder for proper combustion.

17.15 ■ Hot Gas Engines

Another type of heat engine, now used in many industrial applications, is the hot gas turbine. Its design and principles of operation are similar to those of the steam turbine. There are two major differences: (1) the working substance is the mixture of hot gases from the combustion, rather than steam; and (2) the combustion takes place within the engine itself. The fuel may be kerosene or petroleum distillate. Operating temperatures are very high, 2000°F and more in some engines. These extreme temperatures require the development of special heat-resisting alloys. Rotational speeds are also high, and gear reduction is necessary. A few locomotives are powered by gas turbines. Some electric generating plants use gas turbines as the "prime mover." "Turboprop" aircraft also use these engines.

17.16 ■ Jet Engines

Jet aircraft engines provide forward thrust by expelling the products of combustion rearward at extremely high velocity. These engines rely on the atmosphere for combustion oxygen.

The workhorse of modern commercial and military aviation is the *turbojet engine*. The term *turbofan* is also used for some models of this engine. Its combustion chambers and general design are somewhat similar to the gas turbine. There is a turbine rotor or "fan" which is driven by the hot gases, and its main function is to drive a rotary air compressor which is required to force in enough air for combustion (see Figs. 17.1 and 17.19). Forward thrust is obtained not from a propeller but from the reaction force of the hot gases escaping through the "jet" to the rear at fantastic speeds. The discussions of Chaps. 5 and 6, which explained the basic principle of jet propulsion should be recalled here. The diameter and shape of the orifice through which the hot gases escape are matters of tremendous importance in jet-engine design. It is only with these engines that airplane speeds in excess of 600 mi/h have been attained. Speeds in the 2000-mi/h range are common for some military aircraft, and supersonic commercial transport aircraft (SSTs) capable of 1900 mi/h are operating on scheduled transoceanic flights.

17.17 ■ Rocket Engines

All the internal-combustion engines so far discussed have had one feature in common. In order to sustain combustion, oxygen must be obtained from the atmosphere. An engine that does *not* depend on atmospheric air, but has self-sustaining combustion with built-in oxidizers, is called a *rocket engine*. Though successful rocket motors are of relatively recent development (about 50 years), humans have experimented with rockets for several centuries. The principle of flight used by a rocket is jet propulsion, as with the jet engines discussed above. There are no moving parts to the rocket motor proper, and it is the fuel and the oxidizer which make the difference between rockets and jet engines. Rocket fuel may be in either liquid or solid form, but in either case the rocket *carries its own oxygen* or other oxidizing agent. Solid propellant fuels (with oxidizer) are shaped to burn evenly and progressively, and their chemical composition is such that just the right amount of oxygen is present to assure good combustion. Liquid-fueled rockets may use liquid hydrogen or alcohol as the

Rocket engines also get their thrust from the expulsion of combustion products to the rear, but they carry their own oxidizing agent with them.

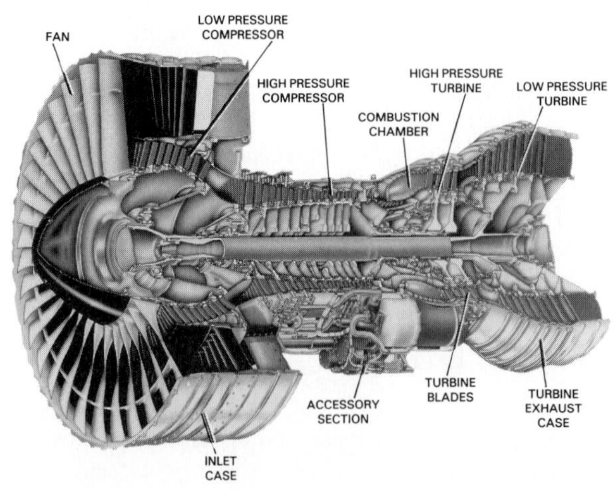

Fig. 17.19 Modern "Turbofan" jet aircraft engine in cutaway view. Turbines drive the "fans" that pack in huge volumes of air to support combustion. (United Technologies—Pratt and Whitney)

fuel, and the oxygen supply may be liquid oxygen (LOX), or some other oxidizer. The proper mixture of fuel and oxidizer is regulated by a complex valve system. Computer-control of the valves, pumps, and sensors is often essential to a satisfactory "burn."

Most of the large rocket engines for space exploration operate on liquid fuels. To date, greater thrusts and longer "burn" times have been obtained with liquid than with solid propellants. However, the difficulties and dangers involved in handling liquid fuels and oxidizers have spurred the development of solid-fuel rockets. Antiaircraft rockets and submarine-fired ballistic missiles are examples of currently produced solid-propellant rocket engines designed for military use.

Rocket Engines for Spacecraft The huge rocket engines that were used for the Apollo space program (moon exploration) were liquid-fueled and developed thrusts of over 7 million pounds.

The Saturn 5 vehicle (Fig. 17.20) was the workhorse of the Apollo program of the late 1960s. Its first-stage booster is 138 ft long and 33 ft in diameter. Its motor consists of five liquid-fueled engines each developing 1.5 million pounds of thrust. It weighs nearly 4.5 million pounds when fully loaded. The second stage is 82 ft long and also has five engines, each developing 200,000 pounds of thrust. On top of the

Fig. 17.20 The rocket engine for the Saturn-5 booster, which was the first-stage, lift-off engine. Each of the five jets visible in the photograph delivered 1.5 million pounds of thrust. This was the first-stage vehicle for the U.S. Apollo series of flights to the moon in the late 1960s and early 1970s.

Fig. 17.21 A predawn launch of Atlas/Centaur 65 carrying the *IntellSat VA-712* satellite into orbit around the earth. The Atlas/Centaur vehicle can carry a "payload" of nearly three tons. (National Aeronautics and Space Administration)

S-2 stage is the S-4B stage, which is 58 ft long and has a 200,000-lb-thrust rocket motor. It was able to place 240,000 lb in an earth orbit and 90,000 lb in the vicinity of the moon.

The failure of the rocket engine for the Space Shuttle Challenger in early 1986 posed a serious challenge to thermodynamics scientists and engineers to place more emphasis on reliability and safety in future designs. New designs for the Shuttle engines have resulted in resumption of Space Shuttle operations early in 1989. In addition some of the dependable rocket vehicles of the recent past, such as the Atlas Centaur, shown in Fig. 17.21, are being used to put satellites in orbit.

QUESTIONS AND EXERCISES

1. Why is a relatively small quantity of heat energy at a high temperature more useful for thermodynamic purposes than a vast quantity of heat energy at a low temperature?

2. Which of the gas laws is obeyed by an adiabatic process? By an isothermal process? Prove your statements by the equations for these gas laws.

3. The first law of thermodynamics tells us that energy is conserved in all heat-to-work and work-to-heat processes *that actually take place*. It does not tell us whether or not a specific process *will take place*. In general, what kinds of processes will occur and what kinds will not? Suggest several unique and interesting consequences for everyday life if the second law could be "repealed." (See Sec. 17.3.)

4. A refrigerator seems to violate the second law of thermodynamics, in that it takes heat away from bodies already quite cold and rejects this heat to bodies (or water or air) which may be very warm. How can this process be consistent with the second law?

5. Starting with the definition of work, $W = Fs$, show that the amount of work done by a gas expanding at constant pressure P pushing a piston through a swept volume V is $W = PV$.

6. An inventor claims to have built an engine which will operate between 800 and 400 K and which will give a work output of 2500 J for every kilocalorie of heat input. He wants financing from you to build and market his engine. Would you invest? Explain your answer with a mathematical analysis.

7. Make a list of relative advantages and disadvantages for steam turbines vs. reciprocating steam engines.

8. Diagram and explain in detail the four-stroke operating cycle of an automotive-type gasoline engine.

9. Compare the advantages and disadvantages of diesel engines and gasoline engines for automotive and truck use, and do the same for diesel engines and steam turbines for marine use.

10. Use diagrams and explain briefly the principle of operation of (*a*) the gas turbine, (*b*) the turbojet engine, (*c*) the rocket motor.

11. What do you think were some of the most difficult problems encountered by Joule in his attempts to measure the mechanical equivalent of heat? Recall the approximate date of his work.

12. Exactly what does a heat engine accomplish? What is meant by the term *working substance?* Why is it necessary? How do heat engines illustrate the law of conservation of energy (the first law of thermodynamics)?

13. As work is done in an adiabatic expansion where does the energy come from? Interpret Eq. (17.5) in this connection.

14. Explain the concept of *entropy* in the light of the second law of thermodynamics.

15. Explain how it is possible to express the work done by an expanding gas in terms of the *area* under a curve on an indicator diagram.

16. As a gas expands or is compressed isothermally, what happens to the *internal energy* in the gas?

17. A two-stroke cycle engine has a power stroke every revolution of the engine, while a four-stroke engine only has a power stroke every two revolutions. Would Carnot predict that the two-stroke is twice as efficient? More efficient? Rationalize your answer.

18. A gasoline engine is operating efficiently while exhausting its ''waste''gases to the outdoor atmosphere. What would Clausius and Carnot predict if the exhaust pipe were fed into a blast furnace where the temperature is 2000°F?

19. What are the operational advantages to be expected from adding a water-cooled condenser to a steam turbine or reciprocating steam engine? (Use Fig. 17.12 in your discussion.)

20. Why is an impulse-reaction steam turbine more efficient than one based on impulse principles alone?

21. From Carnot's efficiency equation, prove that the efficiency of a heat engine is greatest when the difference between the temperature of the working substance and the temperature of the exhaust ''sink'' is a maximum.

22. Recalling Joule's constant and Carnot's ideal cycle and its maximum efficiency, show that a high-compression engine will be more efficient than a low-compression engine, assuming that all other conditions are equal.

23. The same sample of hot gas expands in a cylinder from a volume V_1 to a volume V_2, in two separate expansions. First it is expanded *isothermally,* then *adiabatically*. In which expansion does the gas perform more work? Which process is more efficient? Explain your answers.

PROBLEMS

Group One

1. A steam turbine operates between 950 and 180°F. Its actual efficiency is known to be 32 percent. What is its Carnot efficiency?

2. An old single-acting steam engine has a cylinder bore of 18 in. The piston stroke length is 36 in. and the mean effective pressure (MEP) is 20 lb/in.^{2}. Find the work done on each power stroke (ft·lb).

3. What is the Carnot efficiency of a hypothetical engine that operates between 100°C (boiling point of water) and 0°C (freezing point of water)?

4. An experimental ocean-thermal power plant uses surface water at 82°F as its source of heat. Cold water (40°F), pumped from the ocean depths, is the cold reservoir. Calculate the maximum possible (Carnot) efficiency of such an engine.

5. How much work (ft·lb) will a four-cycle gasoline engine accomplish from each gallon of gasoline, if an overall efficiency of 25 percent is assumed? (See Table 15.1, and use 5.75 lb as the weight of 1 gal of gasoline.)

6. A single-cylinder, double-acting steam engine has a cylinder diameter of 14 in. and a stroke of 28 in. The mean effective pressure (MEP) throughout each stroke is 315 lb/in.^{2}, and the engine speed is 260 rev/min. Find its horsepower output.

7. The temperature of steam entering a steam turbine is 1350°R. It exits into the condenser at 665°R. Find the theoretical maximum efficiency of the turbine.

8. While resting, the average adult produces heat at a rate of about 65 kcal/h (basal metabolism rate). How many watts of power is this?

9. A coal-fired steam plant burns 0.5 kg/h of coal per kW of electric power produced. Calculate the overall efficiency of the burner-boiler steam-turbine generator system.

10. A steam-turbine generator plant burns 8300 gal/h of fuel oil and delivers 50,000 kW of electric power. What is the overall efficiency of this plant? Take specific gravity of fuel oil as 0.76.

11. A 1-hp motor drives a paddle in 10 gal of water for 1 h. If the device is 100 percent efficient (no heat lost), how much of a temperature rise in Fahrenheit degrees will occur?

12. What is the maximum possible (Carnot) efficiency of an engine which is supplied heat at 350°C and rejects heat at 110°C?

Group Two

13. How many Btu are required to melt a 1-lb ice cube (32°F) and then heat the resulting water to the boiling point? If an equivalent amount of work were done on the ice cube against the force of gravity, how high (feet) would it be raised?

14. An engine is assumed to be operating at Carnot efficiency between the temperatures of 800 and 450 K, and it rejects 1 kcal of thermal energy during each cycle. How much useful mechanical work (joules) does it accomplish during each cycle?

15. A Carnot (ideal) engine has an efficiency of 45 percent. If hot gas enters the cylinder at 650°C, what is the exhaust temperature?

16. A hot-gas turbine uses kerosene as fuel. It is rated at 15,000 hp and operates a 10-MW electric generator at full load. The engine burns kerosene at 4000 lb/h. Find the overall efficiency of the engine-generator system.

17. While an engine is idling, it expends energy at the rate of 1.5 hp. If 25 percent of this energy goes into heat development by friction of the moving parts, how many Btu/min are dissipated due to friction?

18. A jet aircraft uses 75,000 lb of kerosene for a five-hour flight. How many foot-pounds of energy are expended? If the engines provide an average of 40,000 hp during the flight, what is their efficiency?

19. Forty kilocalories of heat flows into a quantity of gas trapped in a cylinder. If volume remains constant, how much does the internal energy of the gas increase? (Answer in kilojoules.)

20. A cylinder 12 cm in diameter, fitted with a freely moving piston, contains a gas at a pressure of 8 atm. (*a*) What force (newtons) is exerted on the piston? (*b*) If the piston moves a distance of 1 cm, how much work is done? (Assume constant pressure.)

21. A piston does 1 kJ of work in the adiabatic compression of an ideal gas in a cylinder. (*a*) How much heat flowed out of the gas? (*b*) Did the internal energy of the gas change?

22. The latent heat of vaporization of water is 540 kcal/kg. How much does the internal energy of 1 kg of water at 100°C increase as it is vaporized to saturated steam at 100°C? Assume no external work is done. (Answer in kJ.)

23. An ideal (Carnot) engine absorbs 500 kJ of heat per cycle from a hot reservoir at 400°C, and exhausts to a reservoir at 110°C. What is the work done per cycle, in joules?

24. A Carnot engine receives 2000 Btu of heat at 1200°R and exhausts 1400 Btu to a low-temperature reservoir, in each cycle. Find the exhaust temperature.

25. A gas expands in a cylinder under a constant pressure of 1000 kPa. The volume increases from 1 to 2.5 L. How much work is done on the piston, in kilojoules?

26. The cold reservoir of a Carnot engine is at 24°C. If its efficiency is to be 35 percent, what must be the temperature of the hot reservoir?

27. An engine operating on the Carnot cycle is observed to exhaust, or "waste," 65 percent of its input heat. Find its efficiency.

28. A Carnot engine whose hot reservoir is at 380 K uses 320 cal in accomplishing useful work per cycle, and rejects 250 cal to the cold reservoir. (*a*) What is the efficiency of this cycle? (*b*) Find the temperature of the cold reservoir.

29. An "ideal" engine operates from a hot reservoir at 1340°R. It uses 80 Btu per cycle in accomplishing useful work and rejects 180 Btu to the cold reservoir. (*a*) Find its efficiency. (*b*) What is the temperature of the cold reservoir?

30. Calculate the ideal, or Carnot, efficiency of a steam turbine if saturated steam is supplied to stage 1 at 2000 lb/in.^{2} abs (900°F) and exhausts to the condenser at 1 lb/in.^{2} abs (see Table 15.6).

31. An "ideal" Carnot engine operates between reservoirs at 400°C and 20°C. (*a*) Find its theoretical efficiency. (*b*) If the work output per cycle is 9750 J, how many kilocalories of heat must the hot reservoir supply for each cycle? (*c*) How much heat does the low-temperature reservoir absorb per cycle?

Group Three

32. An ideal gas maintains a constant gauge pressure of 1 atm on a piston in a cylinder, while the piston sweeps out a volume of 1 L. (*a*) How much work is done (joules)? (*b*) Explain how the unit L·atm is equivalent to a unit of energy or work. (*c*) What is the energy equivalent of 1 L·atm in kJ? In kcal?

33. The *P-V* diagram of Fig. 17.22 illustrates schematically a hypothetical cycle for a certain piston–cylinder–ideal gas system. (*a*) How much work is done on the gas in the compression process *A* to *B*? (*b*) In the further compression *BC*? (*c*) How much work is done *on the piston* in the expansion *CD*? (*d*) What happened during the process *DA*? Was any work done? (*e*) Estimate the total output work done in one entire cycle. (All answers in joules.)

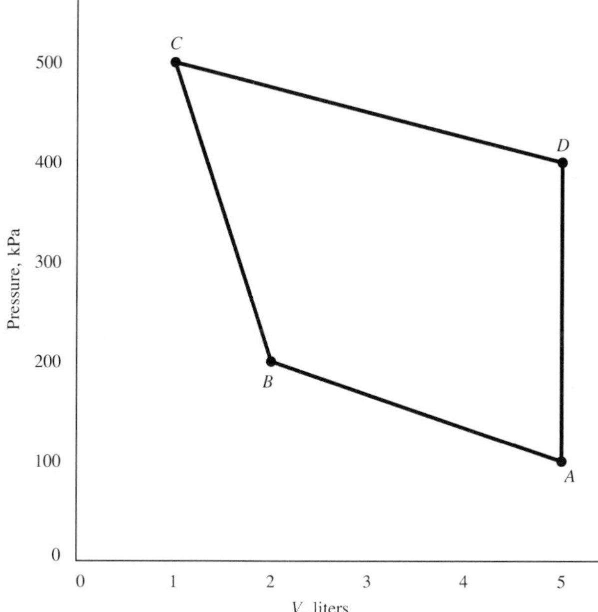

Fig. 17.22 Pressure-volume (*P-V*) diagram for a hypothetical heat engine cycle (Problem 17.32).

34. A coal-fired, steam turbine–driven electric power generating plant produces 1 MW · h of electric energy for every 600 kg of coal burned to fire the boilers. What is the overall efficiency of this plant?

35. A new V-6 auto engine operates on a four-stroke cycle. The cylinders have a diameter of 3 in. and the length of the stroke is 3.75 in. If the mean effective pressure (MEP) on the pistons during each power stroke is 200 lb/in.^{2}, find the horsepower output of the engine at 4500 rpm.

36. A certain gasoline engine has an efficiency of 25 percent. For each Btu of heat energy input, find (*a*) the work output of the engine in ft·lb; and (*b*) the amount of heat exhausted.

37. Compute the velocity with which an ice cube (at 0°C) would have to strike against a concrete wall in order to melt it to water at 0°C. Assume all the kinetic energy is converted to heat and that all the heat generated goes into melting the ice cube.

38. A truck engine burns 7.5 gal/h of diesel fuel.On a dynamometer test it shows 110 hp output at 2400 rev/min. What is its efficiency at this speed? Specific gravity of diesel fuel is 0.72.

39. A jet aircraft flew 1500 statute miles at an average speed of 600 mi/h. Fuel consumption was 4500 gal of kerosene, density 6.8 lb/gal. The jet's engines produced an average power of 20,000 hp during the flight. Find the efficiency of the engines.

40. What is the energy equivalent of 1 atm·in.^{3} in ft·lb?

41. A spacecraft rocket engine produces 400 lb of thrust while the vehicle is traveling at 16,000 mi/h. Calculate the horsepower output of the engine.

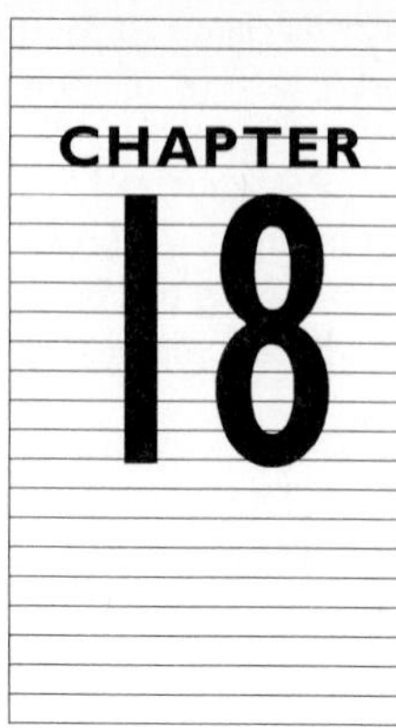

REFRIGERATION AND AIR CONDITIONING

Refrigeration is based on thermodynamic principles dating back to the work of Lord Kelvin (1824–1907). Air conditioning too dates back several centuries in terms of simple heating and cooling of living spaces. Modern air conditioning, however, implies far more than just heating and cooling. Year-round systems heat and humidify indoor air in winter; cool and dehumidify air in summer, and filter out impurities and odors all year long, under automatic control.

The refrigeration and air-conditioning industries are directly related to human health and comfort, including food preservation.

Both refrigeration and air conditioning are major industries in all industrialized nations of the world. Without them, the food supply and the health and well being of a major share of the world's population would be in jeopardy much of the time.

The refrigeration and air-conditioning industries use the SI-metric system of measurement in most countries of the world, and there is currently a slow trend toward metrication in the United States. However, the major emphasis in the United States is still on the English (engineering) system of units in these industries, and that system will be used exclusively in this chapter.

REFRIGERATION

Refrigeration is a process of heat removal.

Refrigeration is defined as the process of removing heat from a space or a product in order to lower its temperature to some specified value, or to freeze it to a solid. People sometimes speak loosely of refrigeration as the "production of cold," as if *heat* and *cold* were two separate conditions capable of independent existence. But cold is merely the *absence* of heat. Some heat is present in all substances whose temperature is above absolute zero, and refrigeration is the process of removing some of this heat and dissipating it to some other substance or space where its presence will be unobjectionable. *Heat removal* and *heat dissipation* are the two processes which are always involved in refrigeration. The idea of "cold production" is completely inaccurate.

18.1 ■ The Unit of Refrigeration Capacity

Units of refrigeration capacity are The Btu/h and the ton of refrigeration (12,000 Btu/h).

Since refrigeration is a process of heat removal, refrigeration machines are rated as to capacity in terms of the quantity of heat which they are capable of removing from a given space or product in unit time. The standard unit of refrigeration capacity is the *Btu per hour*. A much larger unit is the *ton of refrigeration*. This is the quantity of heat removal required to freeze 1 ton of ice from water which is already at 32°F, in a 24-h day. The latent heat of fusion of ice is L_f = 144 Btu/lb, and therefore the ton of refrigeration is the equivalent of 144 Btu/lb × 2000 lb/ton · day or 288,000 Btu/

ton of heat removal per day. It is more common however to rate the ton of refrigeration on an hourly basis, and the ton of refrigeration capacity is practically defined as

$$\begin{aligned} 1 \textit{ ton of refrigeration} &= 288{,}000 \text{ Btu/24 hour day} \\ &= 12{,}000 \text{ Btu/h} \\ &= 200 \text{ Btu/min} \qquad (15.2') \end{aligned}$$

This amount of heat removal will freeze about 10 gal/h of water (at 32°F) into ice at 32°F.

Illustrative Problem 18.1 Ten thousand pounds of freshly picked grapes are brought in from the field for precooling. Their initial temperature is 90°F. They are placed in the precooler to be cooled to a temperature of 40°F in exactly 3 hours. The specific heat c of grapes is 0.86 Btu/lb · F°. Assume a steady-state cooling rate and calculate the refrigeration capacity required in Btu/h and in tons of refrigeration.

Solution The heat that must be removed is calculated from the basic heat equation

$$H = wc\,\Delta t \qquad (15.2')$$

Substituting,

$$\begin{aligned} H &= 10{,}000 \text{ lb} \times 0.86 \text{ Btu/lb} \cdot \text{F}^\circ \times (90 - 40)^\circ\text{F} \\ &= 430{,}000 \text{ Btu} \end{aligned}$$

Since this amount of heat is to be removed as a steady heat flow in 3 h, the *rate* of heat removal and the refrigeration capacity required is:

$$\frac{430{,}000 \text{ Btu}}{3 \text{ h}} = 143{,}300 \text{ Btu/h} \qquad \textit{answer}$$

or,

$$\frac{143{,}300 \text{ Btu/h}}{12{,}000 \text{ Btu/h/ton}} = 11.94 \text{ tons of refrigeration} \qquad \textit{answer}$$

18.2 ■ Basic Thermodynamics of Refrigeration

It was emphasized in the previous chapter that in any thermodynamic process total energy is conserved (First Law). Also, heat energy can accomplish mechanical work only when the circumstances are such that the heat can "flow down a temperature hill" (Second Law). As examples of the *second law of thermodynamics,* we studied in detail several heat engines which perform mechanical work as hot, high-pressure gas *does work on a piston,* and low-pressure gas at reduced temperature is exhausted.

The refrigeration process at first glance seems to violate the second law of thermodynamics, since a refrigerating machine takes heat from a low-temperature source and rejects it to a space or substance at a higher temperature. A refrigerator makes heat *run up a "temperature hill"!* The explanation is, of course, that work must be done *on the refrigerant gas* by the piston in a refrigerating machine. A heat engine *accomplishes* work as heat flows "downhill." A refrigerating machine *must have work done on it* to make heat flow "uphill." In this sense, then, a refrigerating machine is a *reversed heat engine*. Ordinarily the work is done on the refrigerant by a compressor which is driven by an electric motor. The compressor may be either a piston-in-cylinder, reciprocating type; or a turbo compressor with rotating blades or vanes.

A refrigerating machine is a reversed heat engine. Heat is made to flow "up a temperature hill" by doing work on the refrigerant vapor.

18.3 ■ Mechanical Refrigeration

Almost all refrigeration (including the manufacture of ice) is today based upon the use of certain low-boiling-point working substances called *refrigerants* as heat-transfer media. *Ammonia* (bp −28°F at 1 atm pressure), *methyl chloride* (−10.6°F), *sulfur dioxide* (13.8°F), *Freon-12* (−21.6°F), *Freon-22* (−41.4°F), and *Carrene-1* (105.2°F), are examples of commonly used refrigerants.

Refrigerants are low-boiling-point substances that transfer heat by means of changes of state.

Fig. 18.1 The pressure-temperature (P-T) curve for the refrigerant Freon-12. Also called the saturated vapor-pressure curve.

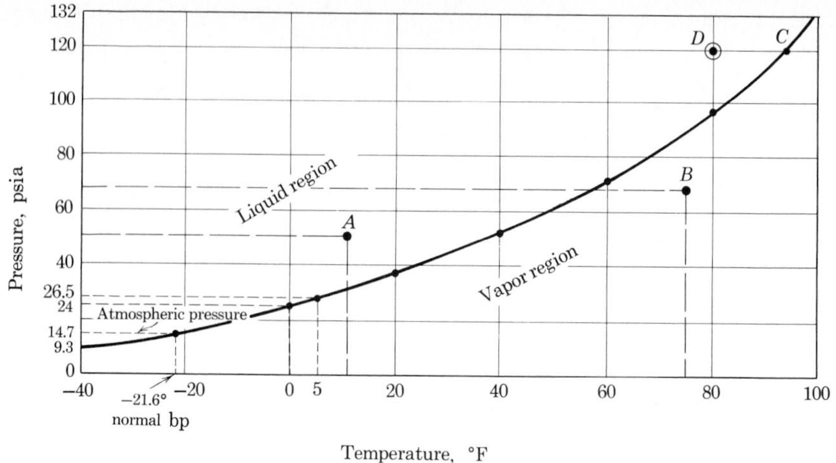

The saturated vapor pressure curve of a substance is plotted on a pressure-temperature grid.

These substances are all vapors (gases) at normal temperatures and atmospheric pressure but are easily liquefied by an increase in pressure and/or a drop in temperature. The curve of Fig. 18.1 illustrates the pressure-temperature behavior of a typical refrigerant. The diagram shows the pressure-temperature (P-T) curve (*vaporization curve*) for saturated Freon-12* vapor for a range of temperatures typical of a refrigerating machine using this gas as a refrigerant. The curve can be considered as the path of all pressure-temperature condition points which represent equilibrium between the liquid and the vapor state. Any P-T combination which would result in a point plotted to the left of and above the curve (such as A) would result in the liquid phase. A P-T combination resulting in any point B, to the right and below the curve, indicates a vapor condition.

Each refrigerant has its own pressure-temperature (vaporization) curve, and the choice of a refrigerant depends to a great extent on how it reacts to changes in temperature and pressure, and on the desired operating temperature of the refrigerator's coils. The boiling point of the refrigerant must be low enough to attain the desired temperature without having to pull a high vacuum on it. On the other hand, it should not require excessively high pressures to liquefy the refrigerant at the temperature of the surrounding air or that of the cooling water available. Note from the curve of Fig. 18.1 that Freon-12 satisfies both these conditions for most cooling applications. The liquid phase boils at −21.6°F without any vacuum. Pulling a slight vacuum will lower the boiling point still further and make colder temperatures possible. At the other end of the curve note that gaseous Freon-12 will begin condensing to a liquid at 132 lb/in.^{2} abs pressure at the temperature of the air on a hot summer day, 100°F. If cool water were used to condense it, the condenser pressure could be correspondingly less.

NOTE: *In mechanical engineering the notation lb/in.^{2} abs is commonly written psia, and we shall use this notation in this chapter. The student will recall that absolute pressure equals gauge pressure plus atmospheric (14.7 lb/in.^{2}) pressure.*

18.4 ■ Refrigerating Systems

Two basic, and fundamentally different, systems account for almost all mechanical refrigeration. They are (1) *compression refrigeration systems* and (2) *absorption refrigeration systems*. Compression systems account for a major share of installed

* Freon-12 is the patented trade name for the refrigerant *dichlorodifluoromethane*. It is now more commonly referred to as, simply, R-12. Fluorocarbon refrigerants are currently being phased out of production, since there is some evidence that they create ''holes'' in the earth's ''ozone layer.''

tonnage in commercial refrigeration, but absorption systems are increasingly popular for large-capacity air-conditioning systems. We will describe the operation of both types. *Thermoelectric refrigeration,* a nonmechanical system, should be mentioned but space limitations do not permit a treatment of it in this text.

COMPRESSION REFRIGERATION SYSTEMS

Refrigerating machines take many forms, depending on the uses for which they are designed. The relative location of the component parts of a commercial refrigeration system will differ, for example, from the arrangement of components in a household refrigerator. The following discussion considers some basic principles common to all compression refrigerating systems.

18.5 ■ Compression-System Components

Figure 18.2 is a schematic diagram showing the basic components of a compression refrigerating system using Freon-12 as the refrigerant.

A compression refrigeration cycle involves sequential *evaporation*, *compression*, *condensation* (liquefaction), and adiabatic *expansion* through a throttling valve.

In such a system, a cycle consists of alternate *compression, liquefaction, expansion,* and *evaporation* of the refrigerant. The first heat transfer occurs as the space and products being cooled give up heat to the refrigerant through the walls of a copper coil called the *evaporator*. This heat boils the liquid refrigerant contained therein and is "carried away" by the gaseous refrigerant as latent heat of evaporation. Evaporators for refrigeration machines are of many different sizes, shapes, and designs,

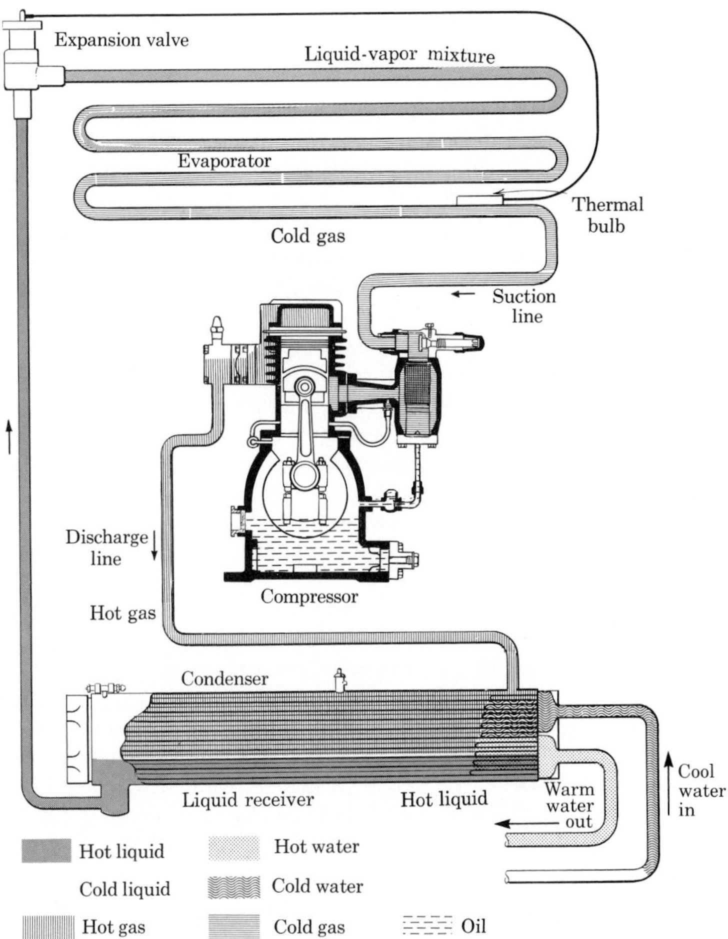

Fig. 18.2 Schematic diagram of the basic components of a compression refrigerating system. The flow of the various fluids is indicated. The compressor maintains a low pressure on its "suction" side, and a high pressure on the discharge side. The hot refrigerant gas, under high pressure, is cooled in the condenser by (in this version) a flow of cooling water. (Carrier Corp.)

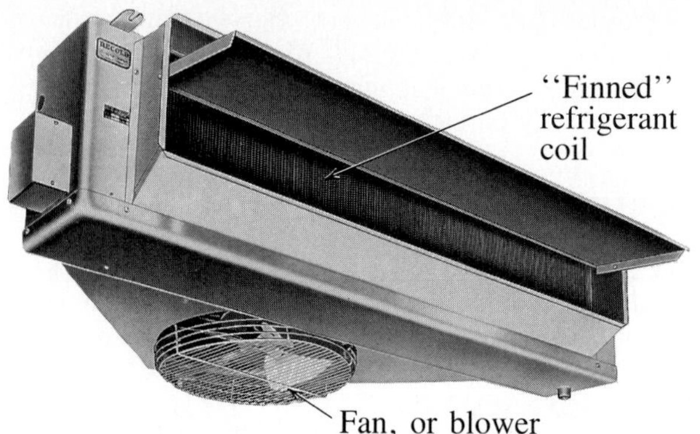

Fig. 18.3 A common type of blower-equipped evaporator. This model is ordinarily suspended from the ceiling of walk-in freezer boxes or cold-storage rooms. (Recold Corp.)

depending on the cooling job to be done. The evaporator is located within the space to be cooled and may be at a considerable distance from the other components of the system. It may be a simple coil of pipe for a "walk-in" freezer room, as in the diagram of Fig. 18.2, or it may take the form of a blower-equipped and "finned" evaporator as shown in Fig. 18.3.

The *compressor* does work on the refrigerant vapor, raising its temperature and pressure so that it can transfer heat to the coolant in the condenser.

The *compressor* is a pump for removing vapor from the evaporator on the *suction stroke* and compressing it on the *compression stroke*.

The compressor shown in the diagram of Fig. 18.2 is a reciprocating type, with cylinders and pistons. However, rotary and turbine-type compressors are also used, as later illustrations will show.

Vapor "suction" from the evaporator is essential to keep the evaporator pressure (and therefore the temperature) low (see the curve of Fig. 18.1). Vapor *compression* is necessary to boost the gas temperature high enough (superheated) to permit heat rejection from the hot vapor along a descending temperature gradient into the cooling water or air at the condenser.

The final heat rejection to cooling water or air is effected by the *condenser*. The typical water-cooled condenser is a steel shell containing a coil of copper pipe in which the refrigerant vapor circulates while being cooled by the water flowing through the shell. Air-cooled condensers are often used, and they look somewhat like the radiator of an automobile. In air-cooled condensers the refrigerant gas circulates in coiled copper tubing while air is blown across the "fins" and around the tubes by a fan. "Fin-and-tube" construction gives a large surface area for heat transfer (Fig. 18.4).

Cooling the superheated refrigerant gas in the condenser while it is at the high pressure built up by the compressor brings the gas to a *P-T* condition representing saturation, i.e., to some point such as *C* (120 psia, 95°F) on the curve of Fig. 18.1. Further cooling results in condensation, and the warm liquid refrigerant, whose condition (pressure-temperature) is now at some such point as *D* on Fig. 18.1, collects in the *liquid receiver,* still at high pressure, ready for another cycle.

Liquid refrigerant is supplied to the evaporator by the *expansion valve,* or "throttling valve," a device which automatically regulates the quantity of refrigerant supplied. The expansion valve (see Fig. 18.5) is usually operated by a bellows arrangement containing a gas, the pressure of which is controlled by the evaporator temperature. When evaporator temperature is low, the gas pressure in the bellows is low, allowing the valve to close. As evaporator temperature rises, gas pressure in the bellows increases, opening the expansion valve. Liquid refrigerant then flows through the valve from the high-pressure side of the system to the low-pressure side, that is, to the evaporator. This sudden drop in pressure results in an adiabatic expansion in which some of the liquid refrigerant "flashes" to vapor. Each pound of liquid refrigerant that vaporizes results in significant (adiabatic) cooling of the remaining liquid. The latent heat of vaporization of R-12 at 5°F is about 68 Btu/lb, and for every pound that becomes vapor, 68 Btu leaves the liquid to become associated with the vapor. The temperature of the liquid drops sharply as it continues to boil in the low-pressure evaporator.

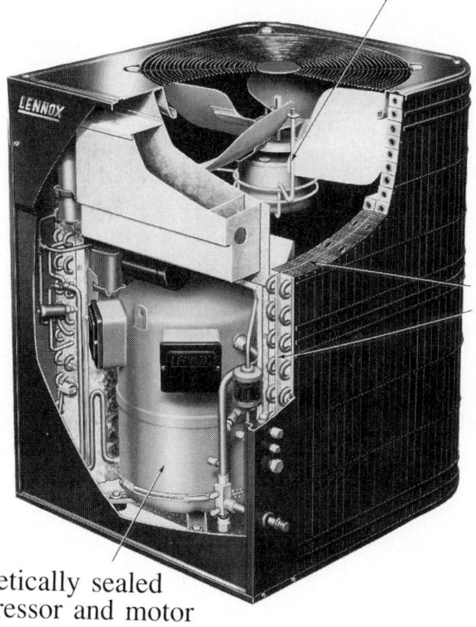

Fig 18.4 A finned-tube condenser forms the outer casing of this small-capacity refrigerating unit for home air-conditioning service. The fan pulls atmospheric air through the fins and tubes of the condenser and discharges the now-heated air vertically upward. (Lennox Industries, Inc.)

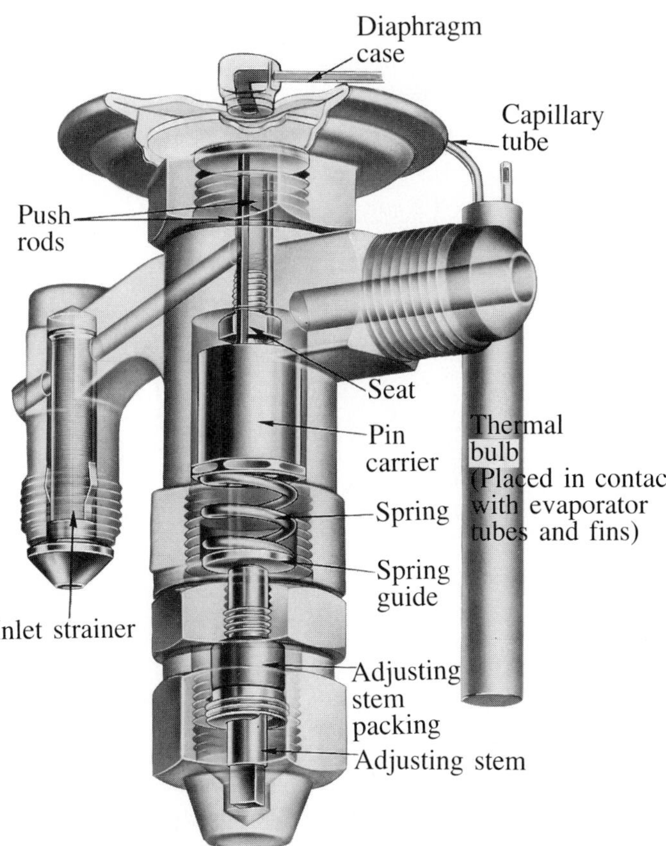

Fig 18.5 Cutaway rendering of a typical refrigeration expansion valve. The thermal bulb (right) is strapped to the suction line near the exit end of the evaporator as shown in Fig. 18.2. Gas pressure in the thermal bulb actuates the bellows arrangement of the valve and controls the flow of the refrigerant to the evaporator. (Sporlan Valve Co.)

Fig 18.6 Refrigeration condensing units. (*a*) Typical air-cooled commercial refrigeration condensing unit, with components labeled. (Tecumseh Products Company) (*b*) Outdoor air-cooled condensing unit for home air-conditioning system. (Carrier Corporation)

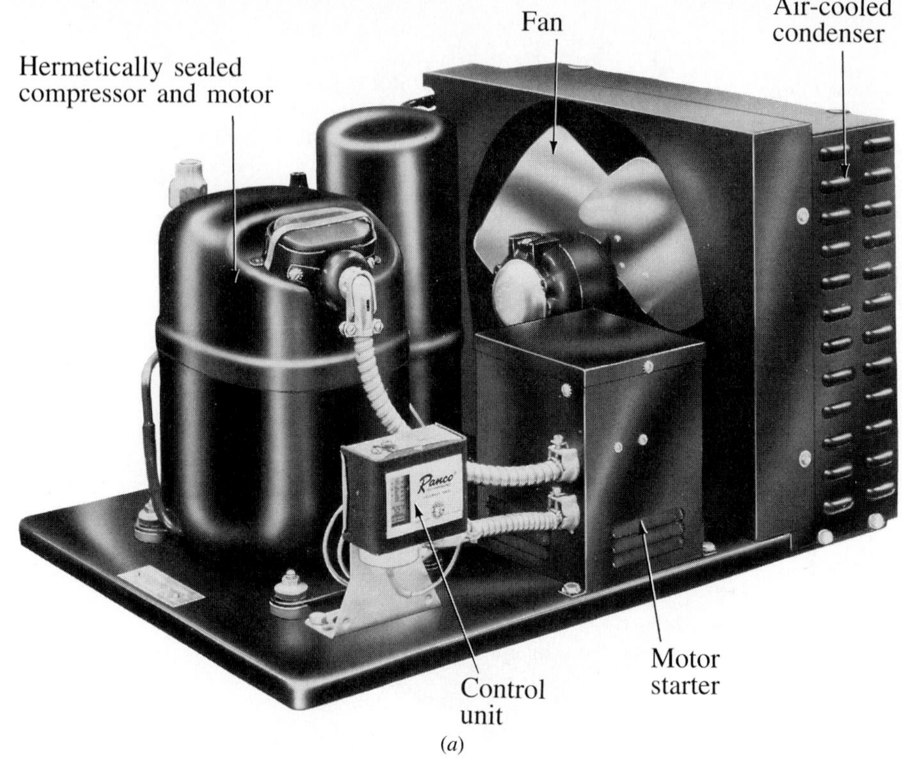

(*a*)

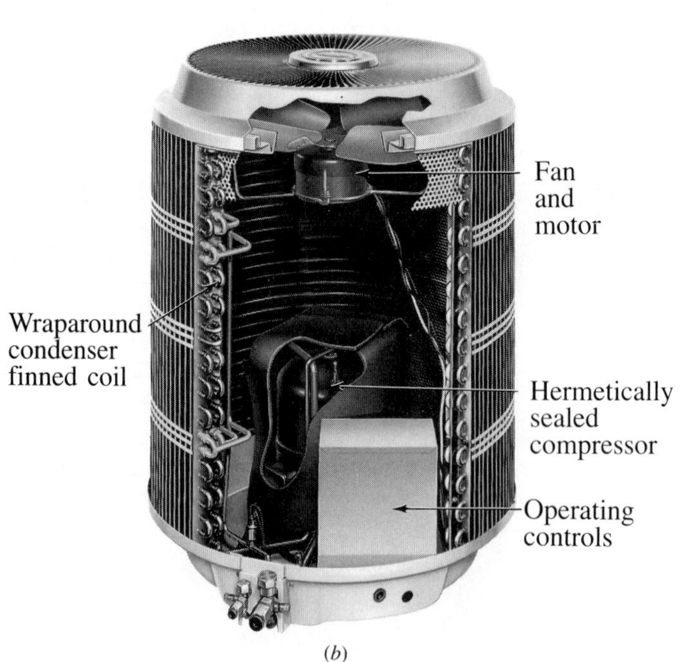

(*b*)

Refrigerating machines are rated on a standard cycle—5°F evaporating, 86°F condensing.

A ''standard'' cycle has been agreed upon for rating refrigerating machines. This cycle presumes a 5°F evaporator temperature and an 86°F condenser temperature. For R-12, a pressure of 26.5 psia is required in the evaporator for a boiling point of 5°F. This pressure must be maintained by the ''suction line'' from the compressor, in order to stabilize the 5°F evaporator temperature (for R-12). Any other refrigerant would have a different set of pressure-temperature requirements.

There are other methods than expansion valves for controlling refrigerant flow.

Small-capacity systems like those in household refrigerators, window air conditioners, dehumidifiers, etc., ordinarily use capillary-tube refrigerant-flow control.

This elementary analysis of refrigerating system components serves to illustrate several basic concepts of a compression refrigeration system:

Basic concepts of compression refrigeration—summarized.

1. The net effect is *heat transfer*—from the space or product being cooled, along a descending temperature gradient to the refrigerant in the evaporator; and after compression, from the hot gas along a second descending temperature gradient to the condenser cooling medium.

2. The purpose of the compressor is to work as a *reversed heat engine*. Work is done *by the piston on a low-pressure, low-temperature gas* to compress it and raise its temperature to a point high enough *to make heat rejection possible*. This is just the reverse of a conventional heat engine in which a *high-pressure, high-temperature gas does work on a piston* and lower-temperature gas at reduced pressure is exhausted.

3. The expansion valve or capillary tube may be regarded as the balance system between the high side and the low side, furnishing refrigerant at an automatically controlled rate to meet the demands of the product- and space-cooling loads.

The combination of compressor, condenser, and associated pumps, fans, and controls is called a *condensing unit*. (See Fig. 18.6.)

18.6 ■ Thermodynamics of Compression Refrigeration Systems—Carnot's Idealized Refrigerator

A refrigerator is a heat engine which absorbs heat at a low temperature and rejects it at a higher temperature. This process at first thought seems at variance with the *second law of thermodynamics*. More careful consideration reveals, however, that the *availability-of-energy* (entropy) criterion demanded by the second law is met by the fact that work is done *on the refrigerant gas* (working substance) by an external source of energy. Heat can be made to "flow up a temperature hill" if it is "pushed up" by energy from an external source.

Carnot's *ideal* refrigerating machine—analysis of a cycle.

A compression refrigerating machine, then, acts as a reversed heat engine. We can use the analysis applied to the *ideal* heat engine (see Chap. 17) and apply the same analysis to an *idealized* refrigerator. Carnot's theoretical treatment, applied to a compression refrigeration system (see Fig. 18.7), follows. Let

Q_1 = heat rejected in one cycle to a high-temperature reservoir (the condenser) at temperature T_1

Q_2 = heat absorbed in one cycle from a low-temperature reservoir (the evaporator) at temperature T_2 (Q_2 is the *refrigerating effect*)

ΔW = work done in one cycle *on the refrigerant gas* by an *external* source of energy

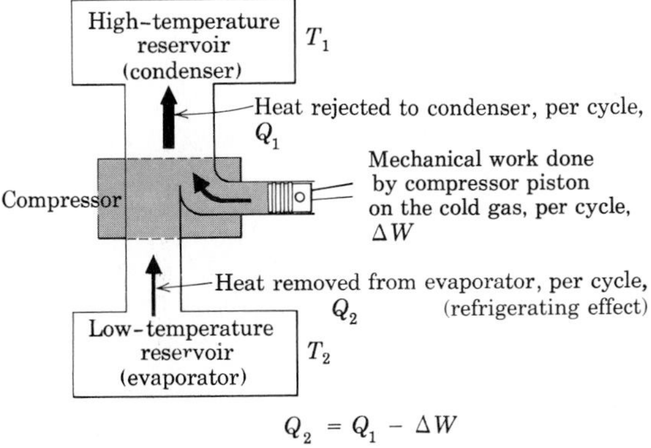

Fig. 18.7 Block diagram illustrating the first law of thermodynamics for a refrigerating machine. Note that, for the *ideal* cycle, the heat rejected to the condenser on each cycle is the sum of the heat removed from the product by the evaporator (that is, the net refrigerating effect) plus the heat equivalent of compressor work for the cycle.

Recalling Eq. (17.4) and keeping in mind that the refrigerator is acting as a reversed heat engine, the work ΔW done in one cycle on the refrigerant gas is

$$\Delta W = Q_1 - Q_2 = \Delta Q \tag{18.1}$$

(*First law of thermodynamics for a refrigerating machine*)

Stated another way,

$$Q_2 = Q_1 - \Delta W \tag{18.1'}$$

which says that the refrigerating effect per cycle, Q_2, is equal to the heat rejected to the condenser minus the heat equivalent of the mechanical work done by the compressor.

Stated still another way,

$$Q_1 = Q_2 + \Delta W \tag{18.1''}$$

which says that the heat rejected to the condenser in one cycle is equal to the heat picked up in the evaporator (the refrigerating effect) plus the heat equivalent of compressor work.

Figure 18.7 illustrates the energy balance for a refrigerating machine. Compare it with Fig. 17.4, which shows the energy balance for a heat engine.

It appears that Eqs. (18.1) and (17.4) are identical in form, but it must be remembered that the term ΔW in Eq. (17.4) represents the *work output of the engine* in one cycle, while in Eq. (18.1), ΔW is the *work which must be done on the refrigerator* in one cycle by some external source of power.

The efficiency of a refrigerating machine is expressed as its *coefficient of performance*, c.o.p.

Efficiency of a Refrigerator—Coefficient of Performance In evaluating the efficiency of a refrigerating machine we compare the amount of *heat removed* from the evaporator per cycle with the amount of external work which must be done *on the refrigerator* in order to obtain that amount of heat removal. In this case, the *output* is the heat extracted from the evaporator Q_2, and the *input* is the mechanical work ΔW supplied from an external source. The ratio of output to input for a refrigerator is not called efficiency; the more descriptive term *coefficient of performance* (c.o.p.) is used.

$$\underset{\text{(refrigerator)}}{\text{c.o.p.}} = \frac{Q_2}{\Delta W} = \frac{Q_2}{Q_1 - Q_2} \tag{18.2}$$

$$= \frac{\text{the refrigerating effect per cycle}}{\text{the heat equivalent of compressor work per cycle}}$$

In a refrigerator, as in a heat engine (see Sec. 17.9), heat transfer to or from the working substance in an *ideal* machine is directly proportional to the absolute temperature T at which the heat flow occurs. In mathematical terms, for a refrigerator,

$$\frac{Q_2}{Q_1} = \frac{T_2}{T_1} \quad \text{or} \quad Q_2 = \frac{Q_1 T_2}{T_1} \tag{18.3}$$

Substituting the value of Q_2 from Eq. (18.3) in Eq. (18.2) and simplifying gives

$$\underset{\text{(ideal refrigerator)}}{\text{Carnot c.o.p.}} = \frac{T_2}{T_1 - T_2} \tag{18.4}$$

where T_2 = absolute temperature of the evaporator (cold reservoir), °R or K

T_1 = absolute temperature of the hot gas as it enters the condenser (hot reservoir)

Analysis of Eq. (18.4) indicates that the colder the required evaporator temperature T_2 (°R or K), the lower the coefficient of performance will be. The more nearly equal the temperatures of the evaporator and the condenser are, the higher the c.o.p. will be. But the *essential purpose of a refrigerating machine* is to produce low evaporator temperatures, and consequently the value of ΔW must be high and c.o.p.'s must be relatively low in order to obtain low evaporator temperatures.

Furthermore, heat transfer in the condenser from the hot gas to the cooling water (or air) is greater if T_1 is high compared to the temperature of the condenser cooling medium. Consequently, when engineers design for *overall performance* of a refrigerating system, it is not sufficient merely to attain a high value of the c.o.p., without giving attention also to other operating requirements and conditions. All factors (necessary evaporator temperature, cost of electric power, availability of suitable condenser-cooling media, space and noise considerations, etc.) must be taken into account by the designers of the components of a refrigerating system.

For *heat engines,* efficiency is greater when T_1 and T_2 are far apart; but for *refrigerating machines,* c.o.p. is greater when T_1 and T_2 are close together.

Unlike the *efficiencies* of heat engines, the *coefficients of performance* of refrigerating machines are ordinarily *greater than unity*. Consider the following illustrative problem.

Illustrative Problem 18.2 A compression refrigeration system is operating between 10°F evaporating and 130°F condensing. Find its Carnot coefficient of performance.

Solution

$$T_2 = 460 + 10 = 470°\text{R}$$
$$T_1 = 460 + 130 = 590°\text{R}$$

From Eq. (18.4)

$$\text{Carnot c.o.p.} = \frac{470}{590 - 470} = 3.92 \qquad \textit{answer}$$

For an *ideal* refrigerator operating between the temperatures given, the above result indicates that the *net refrigerating effect* is 3.92 times as great as the heat equivalent of the mechanical work done on the refrigerator by the external source of energy. In simpler terms, for the system described, 3.92 Btu of heat will be removed from the evaporator for every 778 ft · lb of work (the mechanical equivalent of 1.0 Btu) done *on the compressor* by the driving motor. From the *ideal* (Carnot) *cycle,* c.o.p.'s of from 2.0 to 7.0 are possible. Actual refrigerators do not give coefficients as high as those computed from Eq. (18.4), however, because there are many losses which the ideal cycle does not take into account. When the *actual net refrigerating effect* is measured by suitable instruments and compared to the *actual energy input* to the compressor from the driving motor (converted to heat units, using Joule's constant), the coefficient of performance is calculated from

$$\underset{\text{(actual)}}{\text{Refrig. c.o.p.}} = \frac{\text{net refrigerating effect (actual)}}{\text{heat equivalent of mechanical work}} \tag{18.5}$$

Coefficients of performance of refrigerating machines under actual operating conditions range from 1.5 to about 4.0, depending on the application and on the state of maintenance and repair of the equipment.

Figure 18.8 illustrates the arrangement of refrigerating machine components for a typical cold-storage plant.

The fact that a refrigerating machine in good operating condition can remove three times as much heat from products or air as the heat equivalent of the mechanical work done on the machine, might seem at first to be a violation of the first law of thermodynamics. The key word is *remove*. The machine does not *produce* heat—it takes heat already there in the products or air being cooled, and *removes* that heat by means of two changes of state, and finally rejects it to cooling water or to outdoor air. This is *heat transfer*, not *energy production*, and the first law is not transgressed in any way.

A refrigerating machine in good working order can accomplish about three times as much heat removal as the heat equivalent of compressor work. Are you sure you can explain this seeming paradox?

Fig 18.8 A typical refrigerating machine for a commercial installation. The components shown constitute the "high side", or condensing unit. The evaporator(s) will be located in the rooms or walk-in freezers to be cooled. This is a water-cooled machine (note shell-and-tube condenser). (Carrier Corporation)

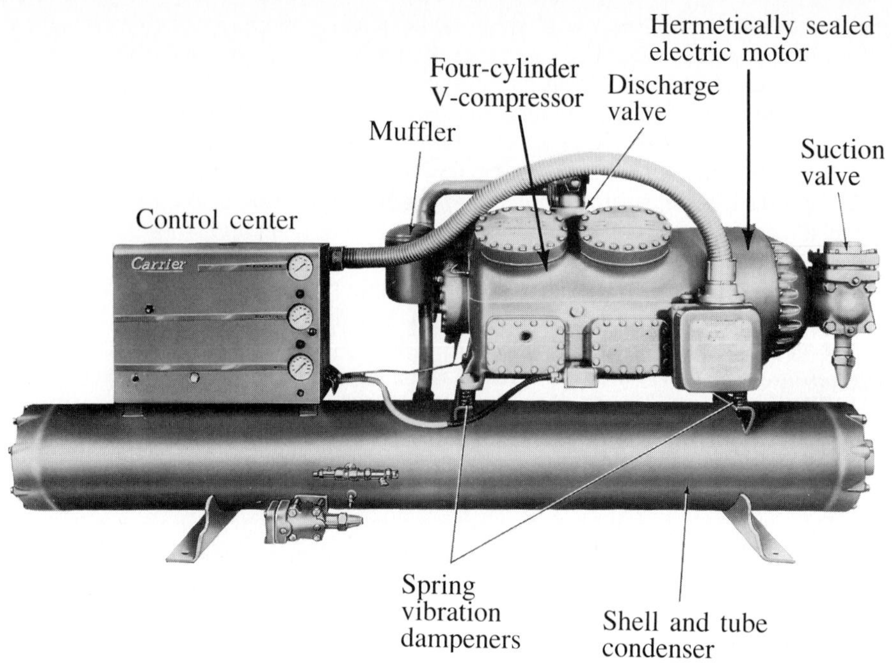

Another efficiency measure for refrigeration, of special interest to energy conservation, is the *energy efficiency ratio*, EER.

Energy Efficiency Ratio (EER) Coefficient of performance (c.o.p.) is the efficiency measure that is of primary concern to engineers and technicians that *design and test* refrigerating machines. At the application and commercial level, however, a different efficiency measure is commonly used. It is called the *energy efficiency ratio* (EER). The EER of a refrigerating unit or air-conditioning unit is the ratio of the Btu/h of heat removed to the total power in watts required to operate the entire machine, including controls, fans, pumps, and any other auxiliary equipment.

$$\text{EER} = \frac{\text{Heat removed per hour, or net refrigeration effect (Btu/h)}}{\text{Electrical power input, entire unit (watts)}}$$

$$= \frac{\text{Btu/h}}{\text{W}} \tag{18.6}$$

Illustrative Problem 18.3 A commercial refrigerating unit is actually producing 5 tons of refrigeration. The electric power input during a test run averages 6.1 kW. Find the EER of the unit.

Solution The net refrigerating effect is

$$5 \text{ tons} \times 12{,}000 \text{ Btu/h/ton} = 60{,}000 \text{ Btu/h}$$

$$\text{EER} = \frac{60{,}000 \text{ Btu/h}}{6100 \text{ W}} = 9.84 \qquad \textit{answer}$$

Illustrative Problem 18.4 A 7.5-ton air-conditioning unit is operating with an air-cooled condenser. On a very hot summer day, the *actual* heat removal (net refrigerating effect) is measured accurately as 81,000 Btu/h. The total electric power input to the machine (compressor, pumps, fans, etc.) is measured as 8.85 kW. Find the EER of the unit.

Solution

$$\text{EER} = \frac{81{,}000 \text{ Btu/h}}{8850 \text{ W}} = 9.15 \qquad \textit{answer}$$

Under average operating conditions, equipment of current manufacture that is operating with air-cooled condensers can often attain EERs of 11.0 or more. However, if the ambient air temperature goes much above 90°F, electric power input rises significantly and the EER decreases.

ABSORPTION REFRIGERATION SYSTEMS

We have seen that a refrigerant can be made to cycle from the low-pressure side to the high-pressure side of a refrigerator as a result of raising its temperature and pressure, using mechanical work as the energizer. By a proper selection of the refrigerant and associated system components, *heat itself can be the energizer*. The *absorption refrigerator* removes heat (refrigerates), and in the process the refrigerant, mixed with an absorbent, is heated by a flame or by hot steam. An elementary discussion of absorption refrigeration follows.

In absorption refrigeration, a *heat input* is used to refrigerate. Can you explain the thermodynamics of this process?

18.7 ■ Description of an Absorption System

The principles of absorption refrigeration were first discovered by Michael Faraday more than a hundred years ago. A detailed technical discussion of absorption refrigeration will not be given here, but the basic principle will be explained with the aid of the block diagram of Fig. 18.9. Both the explanation and the diagram are much simplified. There are no moving (rotating) parts to such systems, but the valve-and-control arrangement is highly complex.

All of the components indicated schematically in Fig. 18.9 are hermetically sealed inside a steel shell. A complex system of sensors, valves, pumps, and controls keeps the system in continuous operation, responding to the cooling demand. The input heat can be supplied by a gas or oil burner, by steam, by electric-resistance heaters, or by solar-heated hot water.

Two absorption systems are in common use today—the *ammonia-water system*, with ammonia the refrigerant and water the absorbent; and the *water–lithium bromide*

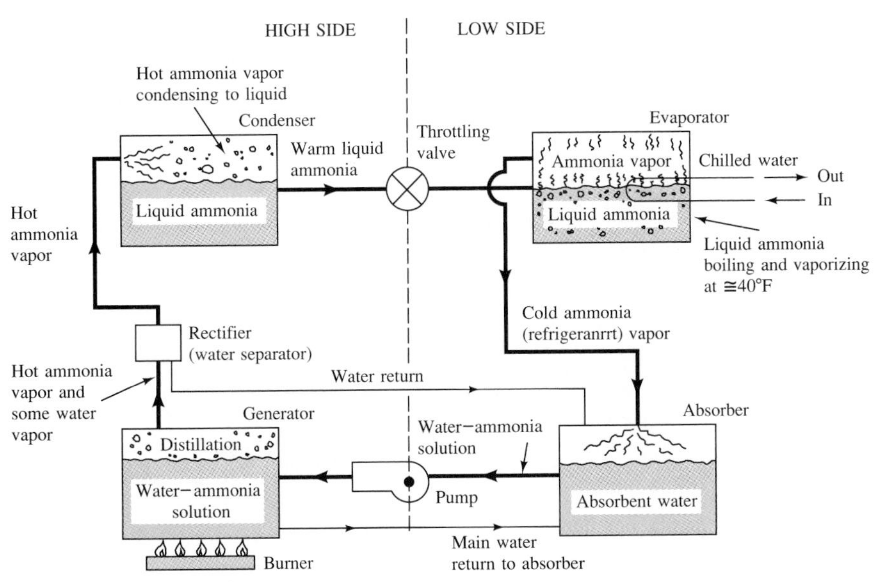

Fig 18.9 Block diagram of an ammonia-water absorption refrigerating system, operating as a water chiller for an air-conditioning system. Note carefully the components on the "low side" and those on the "high side."

system, with *water as the refrigerant* (!) and lithium bromide the absorbent. The *evaporator* and the *absorber* constitute the ''low side'' of the system, and the *generator* and the *condenser* the ''high side.''

The *evaporator* (Fig. 18.9) serves the same purpose that it does in a mechanical (compression) refrigerator. Here the liquid refrigerant flashes to a cold liquid-vapor mixture as it passes (adiabatically) through the throttling valve. The cold liquid continues to boil and removes heat (refrigerates) as it does so. Each pound of refrigerant that vaporizes removes an amount of heat from the product being cooled equal to its latent heat of vaporization (of ammonia in this case) at that temperature.

The refrigerant vapor is absorbed so rapidly in the *absorbent* that a low pressure is maintained in the evaporator. The *absorber* thus has the same effect as the "suction" stroke of a compressor.

The *generator* acts like the compression stroke of a compressor.

The cold refrigerant vapor now passes to the *absorber,* where it goes into solution or is mixed with the *absorbent*. Ammonia, for example, is extremely soluble in water. The refrigerant-absorbent mixture is then pumped to the *generator,* where the input heat energy drives the refrigerant vapor out of solution and heats it to a hot high-pressure vapor. The ''hot gas,'' now under the high-side pressure, goes to the condenser (water-cooled or air-cooled), where it condenses to a warm liquid still under high pressure, ready for another pass through the throttling valve to begin another cycle. The condenser-cooling medium removes and carries away the generator heat and also the heat removed in the evaporator from the product being cooled (the *refrigerating effect*).

A necessarily complex system of controls, flow valves, sensors, and pumps is not shown. All of these are installed, tested, and preset at the factory.

Absorption systems are nearly noiseless and vibrationless in operation. They operate over long periods without repairs, since there are almost no rotating parts to wear out. Their major disadvantage, and it is a serious one, is that they have much lower coefficients of performance than compression systems. In other words, per ton of refrigeration accomplished, absorption systems require a greater energy input than do compression systems.

Ammonia-water systems are commonly used with small tonnage applications such as home refrigerators and small air-conditioning systems, up to about 15 tons. They may be direct-fired by gas or oil flame, or energized by electricity or solar-heated hot water. Water-lithium bromide systems are favored for large-capacity installations, with single-unit machines capable of producing up to 1500 tons of refrigeration. These large machines are most often energized by steam from an associated steam boiler or by hot water from solar collectors. They are especially well-suited to air conditioning duty where hundreds of tons of cooling are required and evaporator temperatures are in the 38 to 42°F range. Figure 18.10 shows a large-capacity absorption machine. For a more detailed discussion of absorption refrigeration, the student may consult any standard textbook on refrigeration or air conditioning.

Fig 18.10 A large absorption refrigerating machine designed for water chilling service. All components diagrammed in Fig. 18.9 (evaporator and generator) are contained within the ''package'' illustrated, preassembled at the factory. This unit employs steam or hot water at 190 to 212°F as the energizer. It is manufactured in capacities up to 800 tons of refrigeration. (Carrier Corporation)

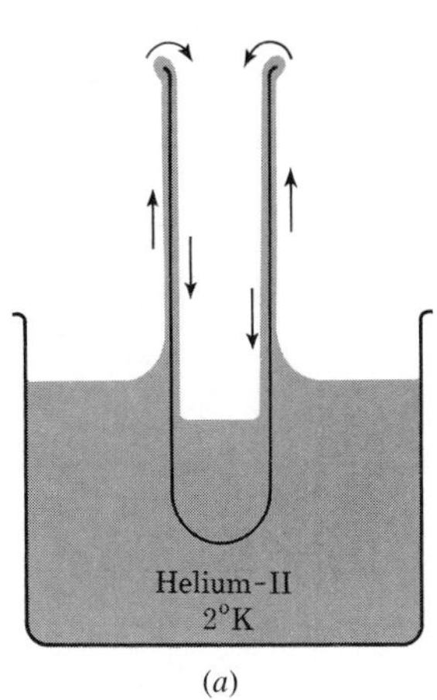

(*a*)

(*b*)

Fig 18.11 (*a*) An illustration of ''creep'' at cryogenic temperatures. ''Superfluid'' helium II climbing up and over the side of a test tube at a temperature of 2 K. (*b*) Transporting cryogenic liquids to hospitals and laboratories. The term ''cryogens'' is now used for such liquids. Liquid helium has a boiling point of −452.1°F (4.21 K or −269°C), at 1 atm. pressure. In other words it boils at just 4 degrees Celsius above absolute zero. Liquid nitrogen, at 1 atm, boils at 77 K or −196°C. Both of these cryogens are used in Magnetic Resonance Imaging systems in hospitals and clinics. The very low temperature of boiling helium creates a condition of ''superconductivity'' in the coils of the large electromagnets used in the MRI system. Liquid nitrogen, with its quite low boiling point temperature, serves as an ideal ''insulator'' for the process, thus reducing the consumption of the far more costly liquid helium. (Linde/Union Carbide Corp.)

18.8 ■ Cryogenics

There has been a great deal of interest in the past half-century in the physics of extremely low temperatures. The term *cryogenics* (from the Greek, meaning ''icy cold'') has been applied to the study of physical phenomena at temperatures below that of liquid air (86 K, or 130°R). Liquid air is considered to be at the ''warm end'' of the cryogenic region, and *low-temperature physics* usually implies temperatures below 10 K.

The science of low-temperature physics is called *cryogenics.* Many unusual things happen to substances as their temperatures drop below 20 K. Electrical superconductivity is one attribute of very low temperatures.

Many unusual things happen at very low temperatures, a few of which will be cited here as examples. Below about 2.2 K liquid helium behaves in a most peculiar manner. Its viscosity reduces to zero (*superfluidity*), and it climbs right up the side of a glass tube or container. If an empty test tube, previously cooled to about 2 K, is pushed partly below the surface of superfluid helium II, the helium will climb the outside of the test tube, go right over the top, and down the inside wall until the test tube is filled to the same level inside as the helium level is on the outside (see Fig. 18.11*a*). This phenomenon is aptly called *creep.*

Living tissue can be preserved and metabolic processes held in abeyance for long periods of time at cryogenic temperatures. Cryogenic surgery (destruction or excision of selected tissues or cells by freezing with liquid nitrogen) is now successful in some types of cancer therapy (especially in treating precancerous skin conditions) and in brain surgery.

Cryogenic Liquids A number of substances which are gases at normal temperatures exist as liquids in the cryogenic region, e.g., air, carbon dioxide, methane, oxygen, nitrogen, hydrogen, and helium. The boiling temperatures of some cryogenic liquids are listed in Table 18.1, with a few more common substances also listed for comparison. (See Fig. 18.11*b*.)

Table 18.1 Boiling points of some cryogenic and common substances

	Boiling Point			
Substance	°F	°R	°C	K
Water	212	672	100	373
Ether	95	555	35	308
Ammonia	−28	432	−33	240
Methane	−259	201	−161	112
Oxygen	−297	163	−183	90
Nitrogen	−320	140	−196	77
Air	−330	130	−187	86
Hydrogen	−423	37	−253	20.3
Helium IV	−452	8	−269	4.2
Helium III	−454	6	−270	3.2

Liquid oxygen (LOX) is used in large quantities as an oxidizing agent for liquid-fuel rocket motors. Liquid hydrogen serves as the fuel in some rockets and it is also manufactured in large quantities. Liquid air is the intermediate stage in obtaining pure oxygen and pure nitrogen. Liquid helium is of extreme value in low-temperature laboratory research, since the lowest temperatures obtainable are reached by using liquid helium as the cooling medium. A temperature of 0.71 K has been reached by pulling a vacuum on liquid helium and thus cooling it adiabatically through its own evaporation.

AIR CONDITIONING

Complete air conditioning implies all of the following: Temperature control, humidity control, air-purity control, noise control, and fresh air circulation.

Air conditioning can be classified under two basic headings: (1) *winter air conditioning* and (2) *summer air conditioning*. As might be expected, *heating* is the most important aspect of the winter phase, and *cooling* that of the summer phase. Besides *temperature control* however, both winter and summer air-conditioning systems should provide other essentials to human comfort, namely, *humidity control; control of dust, smoke, odors, and noise;* and *proper air circulation,* including a supply of fresh outside air for ventilation.

Year-round air conditioning maintains conditions in the human "comfort zone" during the entire year, under automatic control.

Industrial air conditioning maintains an indoor climate that is conducive to optimum industrial production.

Water vapor is the most variable constituent of air. The amount of water vapor present in air ranges all the way from a mere trace to as much as 0.03 lb of water in a pound of air. Air is said to be *saturated* with water vapor when it contains as much water as it possibly can hold at that temperature—when any drop in temperature causes condensation or *dew*. The higher the temperature, the greater the amount of water vapor that can be mixed with air before saturation is reached. *Humidity* is the term used to describe the amount of water vapor in the air.

Atmospheric factors which affect human comfort are the air temperature, the amount of water vapor in the air, and the relative cleanliness of the air, that is, its freedom from odors, dust, and smoke. Air which is too dry irritates the nasal passages and chaps lips and skin, while air containing too much water vapor feels dank and "heavy," slows the normal perspiration rate, and aggravates certain respiratory difficulties.

18.9 ■ Humidity

The amount of water vapor contained in the air is measured in two ways. *Specific humidity* (W) gives *the actual weight of moisture contained in any given amount of air*. *Relative humidity* (RH) is the *ratio of the amount of water vapor actually present*

Table 18.2 Water vapor contained in saturated air at various temperatures

Air Temperature, °F	*Weight of Water Vapor to Saturate gr/lb of Air (1 lb = 7000 gr)*
50	53.5
60	77.3
70	110.5
80	155.8
90	217.6
100	301.3
110	415.0
120	569.0
130	780.0

in the atmosphere *to that amount which would be present if the air were saturated* at that temperature. Specific humidity is often expressed in *grains* of water per pound of dry air, the *grain* (gr) being a unit of weight equal to 1/7000 lb. Some tables and charts give specific humidity in *pounds of water vapor per pound of dry air*. Relative humidity is expressed as a percentage of saturation. At saturation the relative humidity (RH) is 100 percent, while absolutely dry air has a relative humidity of 0 percent. Table 18.2 shows the maximum amount of water vapor that 1 lb of air can contain at various Fahrenheit temperatures.

Air contains a great deal of water vapor. On a humid summer day a home living room 24 ft x 20 ft x 9 ft could have nearly 5 lb of water vapor mixed with the air in the room.

Human comfort is directly related to the *relative humidity* of the surrounding air.

Relative humidity is a ratio or a percent. It expresses the percentage of saturation (with moisture) of the air at a specified location.

It has been found by experiment that human comfort is much more closely related to *relative humidity* than to *specific humidity*. This is reasonable, for it is the nearness to saturation which interferes with normal bodily cooling processes by evaporation of moisture from the skin. Most people feel comfortable when the indoor condition is in the range 70 to 72°F with RH at 40 to 50 percent (winter); and at 76 to 80°F, 50 percent RH (summer). The actual amount of water vapor which air can hold is determined solely by the temperature, and the temperature is an important factor in determining *relative humidity,* also. Note from Table 18.2 that at 80°F, 156 gr of water vapor per pound of air results in saturation, or 100 percent relative humidity. At 110°F, however, this same amount (156 gr) of water vapor would give a RH of only $\frac{156}{415} = 0.376$ or 37.6 percent.

18.10 ■ Measuring Relative Humidity

The most convenient means for measuring relative humidity is the *sling psychrometer*. Figure 18.12 shows a common type. The instrument consists of two identical thermometers mounted on a light frame which can be whirled in the air. One thermometer, the *wet bulb* (WB), is covered with a wick which is saturated with water before taking a reading. The other thermometer, the *dry bulb* (DB), has no wick. As the instrument is whirled or ''slung'' through the air, evaporation from the wet wick occurs, cooling the bulb of the wet-bulb thermometer. Its temperature reading falls below that of the dry-bulb thermometer, and the difference between the two readings is called the *wet-bulb depression*. Since the cooling effect on the ''wet bulb'' depends on the evaporation rate from the wick, and the evaporation, in turn, depends on the degree of saturation of the surrounding air, it follows that the wet-bulb depression is a measure of relative humidity.

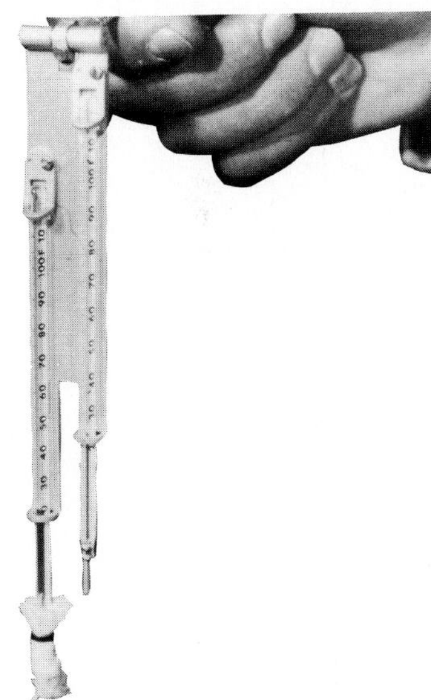

Fig 18.12 A sling psychrometer for measuring the relative humidity of air. The wet-bulb reading and the dry-bulb reading are used with a psychrometric chart to obtain the relative humidity (RH).

The properties of air are conveniently presented in graphical form by means of a *psychrometric chart*. Many air-conditioning problems can be successfully solved by reference to such a chart. A portion of a psychrometric chart for normal temperatures is shown in Fig. 18.13.

Illustrative Problem 18.5 The RH of the air in an auditorium is being determined with a sling psychrometer. The instrument gives these readings: dry bulb, 85°F; wet bulb, 77°F. What is the RH in percent?

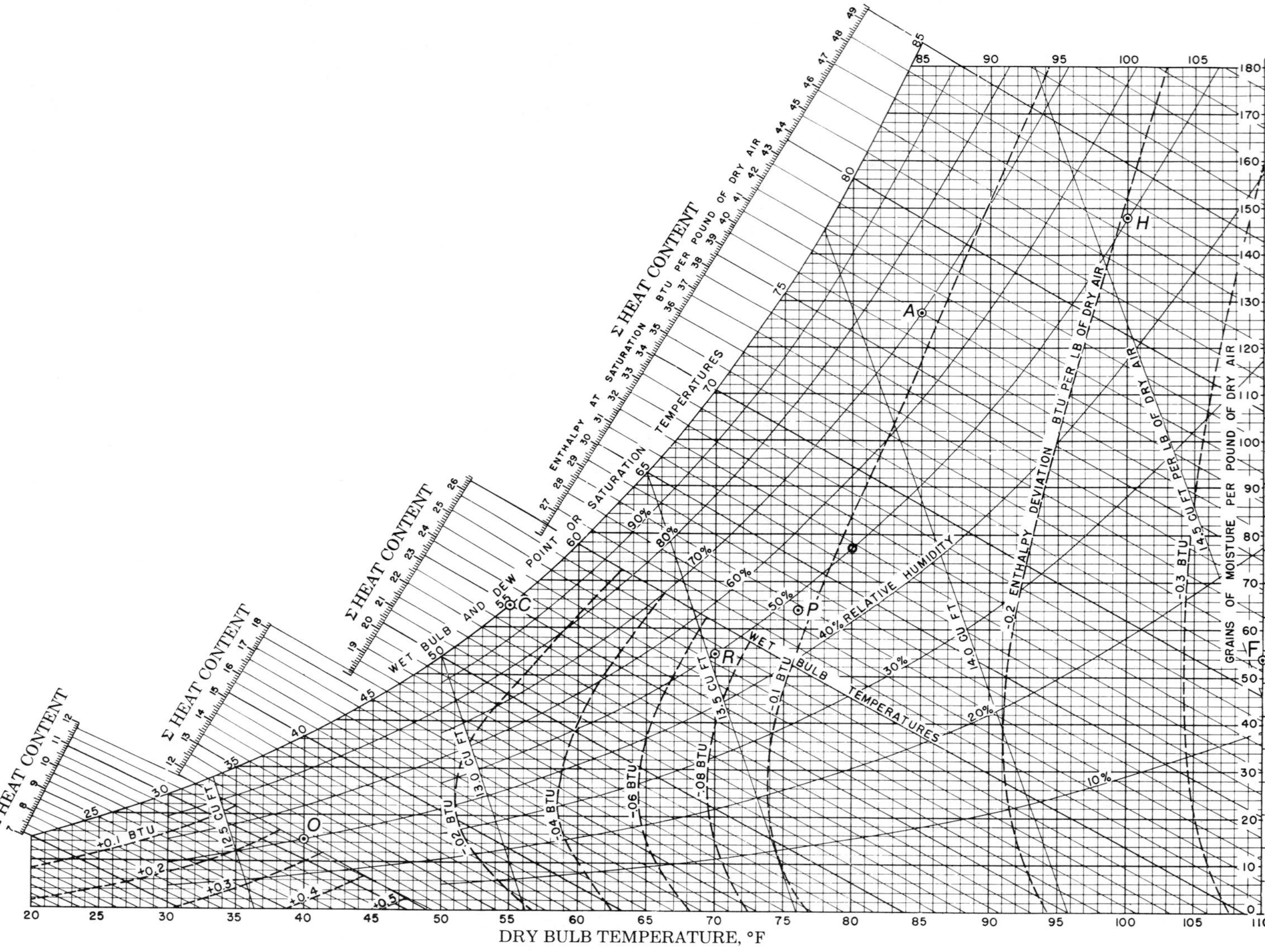

Fig 18.13 Psychrometric chart (normal temperatures) for comfort air-conditioning use. In addition to values of relative humidity, specific humidity, and specific volume, the chart provides values for total heat content in air (sensible and latent). The Σ Heat Content scale (left of the saturation line) provides the total heat (enthalpy) of air measured from an arbitrary base of 0°F, dry air. (Carrier Corporation.)

Solution Enter the psychrometric chart (Fig. 18.13) along the DB temperature scale at 85° (dry bulb). Follow the vertical line upward until it intersects the 77° (wet bulb) line slanting downward and to the right. Note the intersection (point *A* on the chart) is on the 70 percent RH line.

70% RH *answer*

WINTER AIR-CONDITIONING SYSTEMS

There are three basic types of systems for winter air conditioning: (1) *steam or hot-water systems,* (2) *warm-air systems,* and (3) *electric heating systems*. Each has several variations, and we shall treat only the more typical arrangements in a very elementary way. Some are true air-conditioning systems, and others do only part of the job.

18.11 ■ Steam and Hot-Water Systems

The typical steam (or hot-water) heating system for homes and factories comprises a furnace and boiler usually located in a basement or separate boiler room, with radiators located in the several spaces to be heated. Figure 18.14 shows a typical hot-water system in diagram form. Coal, oil, or gas may be used as the fuel. Solar energy is readily applicable to hot-water systems.

Steam and hot-water heating systems are dependable and usually quiet in operation, and temperature can be effectively controlled. They usually have no provision for humidifying the air in the heated spaces, and for two reasons (no ventilation and no humidity control) steam and hot-water radiator systems are not true air-conditioning systems but merely *heating systems*. The tendency of these systems is to produce an inadequately ventilated and low-humidity condition not conducive to optimum human comfort. Console-type humidifiers can, of course, be provided in each space to be heated.

Steam and hot-water heating systems provide a steady, even temperature, but humidity control is often not provided.

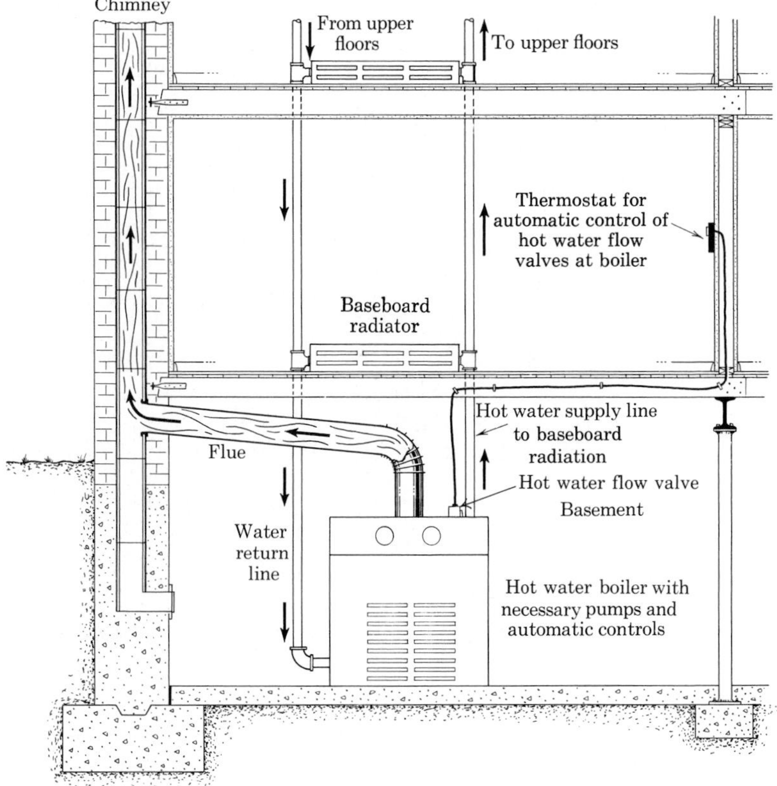

Fig 18.14 Simplified diagram of a hot-water heating system for a home. Only basic components are indicated. Zone control (not shown) can be provided by multiple thermostats and modulating water-flow valves.

18.12 ■ Warm-Air Systems

Forced-air heating systems provide quick response to thermostat signals. Ventilation and humidity control are usually provided also. Care must be taken with the installation of these systems or noise, drafts, and uneven temperature control may result.

There are several different types of warm-air systems, but the *forced-air system* is the most common. Figure 18.15 shows a typical, small forced-air furnace. The furnace may be in a basement, but can be located at any level since the air circulation is forced by the blower and does not depend on convection currents. It may be gas-, oil-, or coal-fired. It includes a large air-heating chamber, called the *heat exchanger*, through which air is circulated in close contact with the hot iron *firebox*. Fresh outside air is pulled into the system at all times, the ratio of fresh to recirculated air being controlled by damper settings. In the furnace bonnet, there is often a spray-type humidifier, which may be automatically controlled from *humidistats* in the conditioned spaces. Relative humidity can be controlled within limits of ± 10 percent. *Filters* for dust control are installed in the path of the return air and fresh air.

Electrically Heated Systems In areas where electric energy is competitive in cost, electric heat is very common. One typical electric-heat installation is similar in most respects to the forced-air system described above. The furnace unit is fitted with electric strip (resistance) heaters instead of facilities for the burning of a fuel. A blower forces the air through the strip heaters and circulates it to the rooms through ducts and registers in the same manner as described above. Thermostats and electric controls cycle the operation of the strip heaters and the blowers to regulate room temperatures. Another system features electric-resistance baseboard heaters in each room. Electrically energized radiant panels are also quite common.

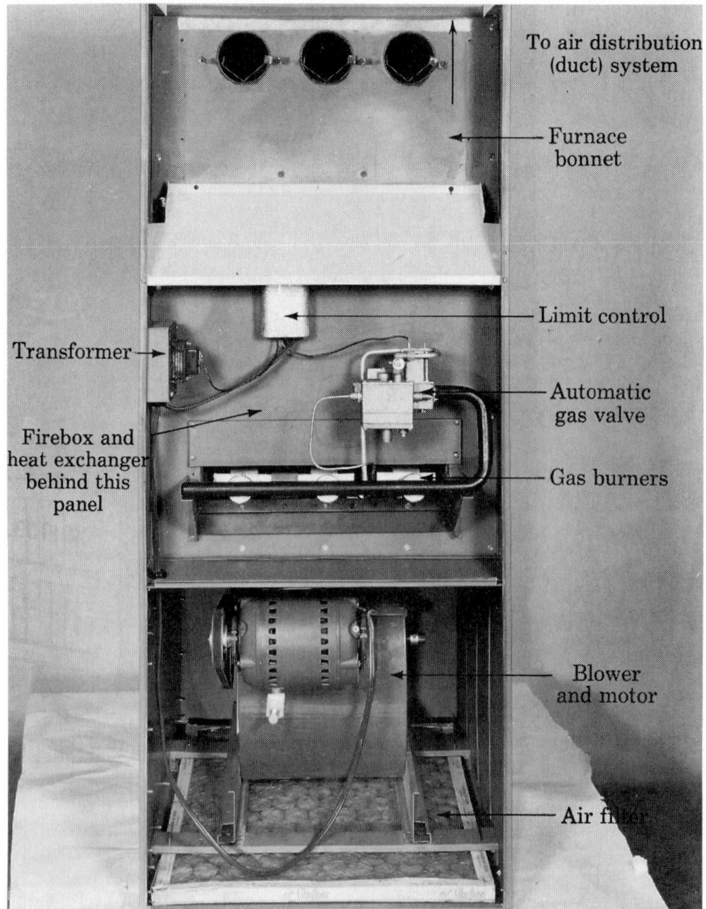

Fig. 18.15 Gas-fired forced-air furnace with front panel removed, showing essential operating parts and controls. In many modern furnaces, the "pilot light" has been replaced by an electronic ignition system. Furnaces of this type can be installed in a basement, a mechanical equipment room, or even in a specially prepared closet. Adequate ventilation for combustion must be provided. The air distribution system (ducts) is not shown. (York Division, Borg-Warner Corporation)

As an approximation, the heat output from electric strip (resistance) heaters can be calculated from the following:

$$\underset{\text{(electric power)}}{1\text{ kW}} = \underset{\text{(heating rate)}}{3410\text{ Btu/h}} \tag{18.7}$$

Humidifiers and filters are installed in electric heat systems in the same manner as in other forced-air systems.

SUMMER AIR-CONDITIONING SYSTEMS

It will be recalled that the basic problem of a summer air-conditioning system is to *cool and dehumidify* the air. Both cooling the air and dehumidifying it are problems of *heat removal*. To cool hot, dry air, we remove its *sensible* heat, which is that heat whose effect is noted as an increase in temperature, so called because the body ''senses'' the fact that the air is hotter. The sensible heat is removed in accordance with the basic heat equation $H = wc(\Delta t)$, where c is the specific heat of dry air (0.24 Btu/lb·F°), w is the weight of air being cooled, and Δt the temperature drop desired. (*Latent heat,* on the other hand, causes no dry-bulb temperature increase but is the result of water vapor in the air.)

A thermometer reading gives an indication of the *sensible heat* in air, but no indication whatever of the *latent heat* (moisture content) of the air.

In dehumidifying air, the problem is to remove the latent heat of vaporization L_v of the water vapor in the air. The vapor then condenses and drains off, carrying latent heat away with it.

A summer air-conditioning system must accomplish the following: (1) Drop the dry-bulb temperature by removing sensible heat, and (2) remove the moisture in the air by condensing the vapor on a cold coil surface, so that it will drain away.

18.13 ■ Basic Components of Summer Air-Conditioning Systems

Many different arrangements of equipment for summer air conditioning are possible. Any system, however, must contain the following basic components:

1. The *air-conditioning unit* itself, consisting of the cooling and dehumidifying coil, the blower, and filters. This unit is ordinarily housed in a sheet-metal casing and is installed in the duct system in such a way that it can handle both the recirculated air and the ventilation air from the outside. Figure 18.16 shows a common type of air-conditioning unit.

Basic components of a summer air-conditioning system.

Fig 18.16 Fan-coil unit, or air-conditioning unit, for installation in a mechanical room or directly in a main duct serving one ''zone'' of an air-conditioning system. (Carrier Corporation)

Fig 18.17 The refrigerating equipment for a large air-conditioning system. At center right is a large centrifugal compressor and water chiller. Hot-water boilers for winter heating are in the background. Chilled water and hot water are piped from this central mechanical room to "zone" air-conditioning units throughout the building. (The Trane Company)

2. The *refrigerating equipment,* which may be at a remote location from the air-conditioning unit. Several air-conditioning units may operate from a single refrigerating plant. (See Fig. 18.17.)

3. The *duct-and-register system,* properly sized and insulated to deliver conditioned air to the spaces without objectionable noise or drafts. The flow of air in ducts is governed by the equation

Calculating air flow in ducts.

$$Q = Av \tag{18.8}$$

where Q = quantity of air to be supplied, ft^3/min
A = cross-sectional area of duct, ft^2
v = air velocity in duct, ft/min

4. The *control system,* which cycles the refrigerating equipment, controls the air-conditioning unit, and operates mixing dampers to control airflow.

18.14 ■ The Heat Pump

The heat pump is an unusual device. It removes heat from a cold source (air or water), "pumps it up" to a high temperature by the mechanical work of compression, and delivers it to a warmer place.

Thermodynamics is a fascinating science, full of seeming paradoxes. One of these is the refrigerator itself, which at first thought seems to defy the second law of thermodynamics. Another is the absorption refrigerating machine, which operates by *adding heat* from a flame to the refrigerant, and in which water is often used as the refrigerant. A third is the *heat pump,* a compression refrigeration air conditioner that both heats and cools air, cycling automatically from cooling to heating on demand from a room thermostat.

Heat pumps operate equally well for summer cooling. In the cooling mode they function just like any other summer air-conditioning unit.

A commercial heat pump for air conditioning operates like any air-cooled compression refrigeration air conditioner when summer cooling is desired (Fig. 18.18*a*). Heat is absorbed from the room air by the *cooling coil* (evaporator) and is rejected (along with the heat of compressor work) to outdoor air by the condenser. The heat pump has a complex system of valves and controls which allows the evaporator and the condenser to exchange functions. When the machine is switched to the *heating cycle* (Fig. 18.18*b*), the outdoor coil becomes the evaporator. It must operate at a temperature well below that of the cold winter air, so that heat flow *from the cold winter air into the very cold refrigerant* will occur. This heat, together with *the heat equivalent of compressor work,* is rejected to room air by the indoor coil, now operating as the condenser at a high temperature. The usual blower-and-duct system forces

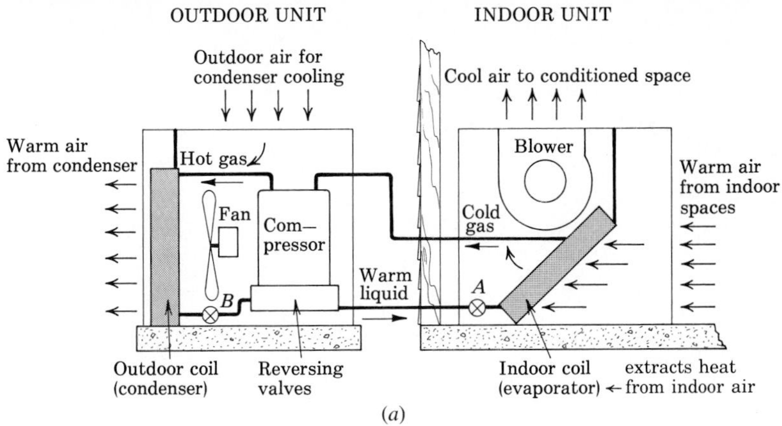

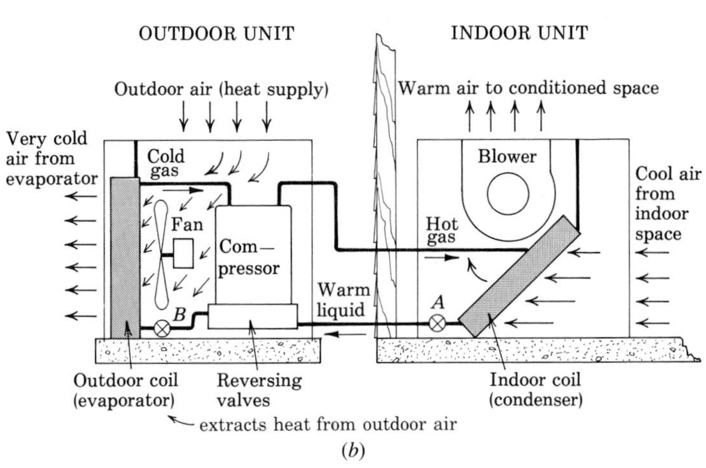

Fig 18.18 An air-to-air heat pump for year-round air conditioning. Diagrams and flowcharts illustrate the principles of operation of (*a*) the cooling cycle, and (*b*) the heating cycle. The reversing valves and their controls (location only indicated in the diagrams) are key elements in cycling the operation from heating to cooling and back again.

air over the indoor coil to be heated, and circulates the warm air to the conditioned spaces. Figure 18.19 shows a single-package heat pump for homes and small commercial applications.

Heat pumps have quite interesting thermodynamic properties. The coefficient of performance for a heat pump on the winter cycle is greater than that for a refrigerator operating between the same temperature limits, because in the heat pump, the heat equivalent of compressor work is *added* to the useful heat output of the machine. In contrast, for summer cooling, the heat pump is on a regular refrigeration cycle, and the heat of compressor work is a loss factor which must be removed from the refrigerant by the condenser and rejected to the outside air.

Coefficient of performance of a heat pump

The theoretical (i.e., Carnot) coefficient of performance of a heat pump on *winter cycle* is given by

$$\underset{\text{(ideal–Carnot)}}{\text{Heat pump c.o.p.}} = \frac{T_1}{T_1 - T_2} \tag{18.9}$$

where T_1 = absolute temperature of the heating (indoor) coil (the condenser on winter cycle)

T_2 = absolute temperature of the evaporator, which is the outdoor coil on winter cycle

Heat pumps are quite effective heating devices in climates where mild winters are the rule. In very cold climates however, their efficiencies drop off rapidly, as the following calculations show.

Fig 18.19 A "packaged" heat pump for year-round residential and light commercial air-conditioning use. Outer panels have been removed to show the essential components. (Lennox Industries, Inc.)

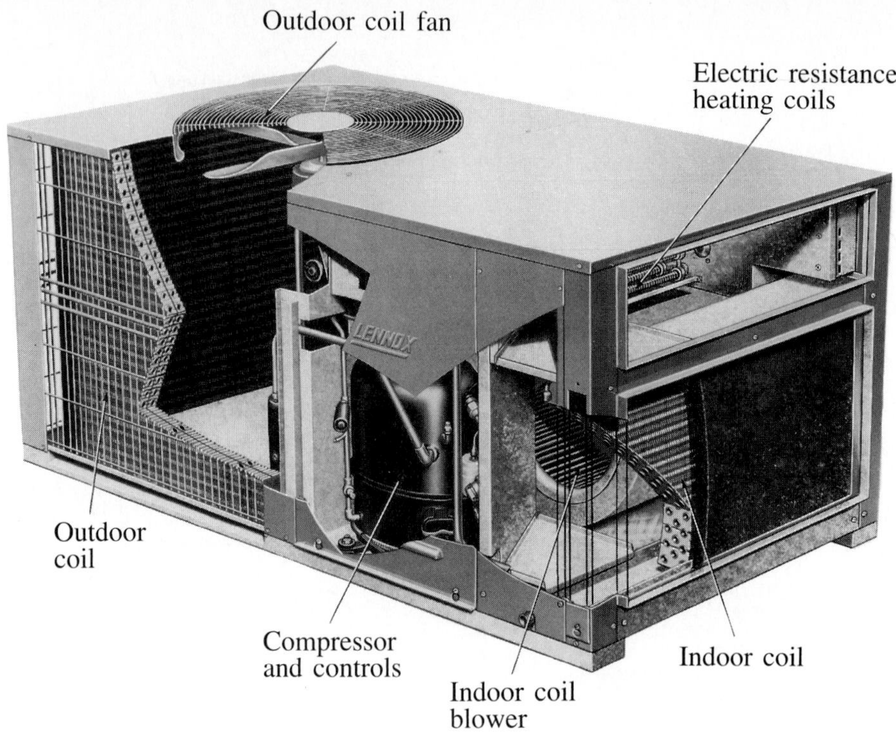

Illustrative Problem 18.6 A heat pump is set to produce an indoor coil temperature of 120°F. What is its *theoretical* (Carnot) *coefficient of performance* when the outdoor coil operates at (*a*) 30°F? (*b*) −10°F?

Solution

(***a***)

$$\underset{\text{(Carnot)}}{\text{c.o.p.}} = \frac{580°\text{R}}{(580 - 490)°\text{R}}$$

$$= 6.44 \quad \text{at} + 30°\text{F} \qquad \textit{answer}$$

(***b***)

$$\underset{\text{(Carnot)}}{\text{c.o.p.}} = \frac{580°\text{R}}{(580 - 450)°\text{R}}$$

$$= 4.46 \quad \text{at} - 10°\text{F} \qquad \textit{answer}$$

From these idealized-cycle Carnot coefficients, the decrease in heating effectiveness in cold climates is readily apparent. The *actual coefficient of performance* of a heat pump on winter cycle is given by

$$\underset{\text{(actual)}}{\text{Heat pump c.o.p.}} = \frac{\text{heat obtained from condenser (indoor coil)}}{\text{heat equivalent of electric energy input to compressor motor}} \qquad (18.10)$$

Using this indicator of *actual* performance, and substituting data obtained under operating conditions, the coefficients of performance for air-to-air heat pumps on winter cycle range from a high of around 4.0 in very mild climates down to about 1.5 in cold winter conditions. Figure 18.20 gives an indication of the typical performance of "packaged" heat pumps operating on an air-to-air cycle. It is common practice to supplement the heat output of heat pumps with electric strip (resistance) heaters for use whenever the outdoor temperature drops much below 20°F. A 5-kW electric strip heater would provide supplementary heat in the amount of 17,050 Btu/h (5kW × 3410 Btu/h · kW).

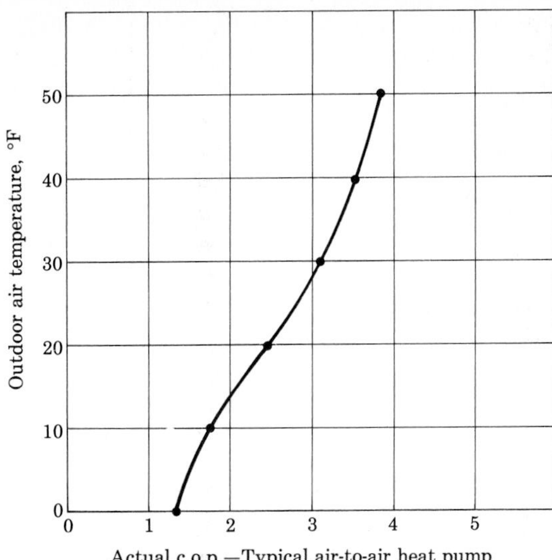

Fig 18.20 Performance curve for air-to-air heat pumps on heating cycle showing the approximate *actual* c.o.p. attained at various outdoor temperatures. The curve indicates average performance of many models from several manufacturers.

Water-to-Air Heat Pumps The discussion of heat pumps thus far has been limited to air-to-air cycles, in which heat is extracted from already cold winter air and "boosted" to a high enough temperature to heat interior spaces. Well water is a much better source of heat than cold outdoor air. The temperature of well water in winter is usually above 55°F, compared to winter air temperatures in some regions of 20°F and on down to below zero. Furthermore, each pound of water will provide 1 Btu/lb · F° of heat, while air provides only 0.24 Btu/lb · F°.

The well water is pumped into and through the evaporator (outdoor coil) and the water, now chilled, is returned to the earth via a second well. The heat picked up from the well water in the evaporator, along with the heat equivalent of mechanical work done by the compressor, is extracted from the hot condenser (the indoor coil) by a blower-energized airstream that is then distributed by a duct-and-register system to the rooms to be heated.

In this chapter only the most basic and elementary principles of refrigeration and air conditioning have been presented. Such a brief treatment can provide only a limited understanding of the complexities of the subject and of the scope of the industry. Students interested in further study are encouraged to consult any of the several standard textbooks on refrigeration and/or air conditioning.

QUESTIONS AND EXERCISES

1. Explain how it is possible for a "warm" liquid refrigerant in the condenser to enter the evaporator and immediately begin absorbing heat from products that are much colder than the condenser temperature. [HINT: What happens to the "warm" liquid as it goes through the expansion valve from high pressure (condenser) to low pressure (evaporator)?]

2. Compressing a refrigerant gas adds the heat equivalent of mechanical work to the gas and raises its temperature. This would seem to be the wrong thing to do since later in the cycle the refrigerant in the evaporator must be very cold to absorb heat from the product being cooled. Explain this seeming contradiction.

3. Starting with the basic definition of a ton of refrigeration, show that 1 ton of refrigerating capacity equals 12,000 Btu/h.

4. Explain in detail how both the first and second laws of thermodynamics apply to the compression refrigerating cycle.

5. Diagram a compression refrigerating system and discuss in detail the function of each of the basic components? What is included in the *high side?* The *low side?*

6. Discuss all the essential purposes of the compressor in a compression refrigerating system? Explain in terms of the second law of thermodynamics.

7. Does the fact that refrigerating machines have coefficients of performance greater than unity mean that such machines do not obey the first law of thermodynamics? (The net refrigerating effect per cycle may be 3 or 4 times the thermal equivalent of compressor work.) Explain in detail how it is possible to get more cooling from a machine than the heat equivalent of mechanical work (input) to the machine.

8. Which essential process on the "low side" in the refrigeration cycle is an adiabatic process? What does it accomplish?

9. Hold a small can of (liquid) Freon in your hands. It is at room temperature. Open the valve and let some liquid escape into a beaker. (What escaped into the air at the same time?) The liquid in the beaker will start to boil and soon reach a temperature of −21.6°F. Explain the sudden temperature drop. Why does a mere release of pressure produce this effect? What name is given to this cooling process?

10. Why is it possible for a refrigerating unit equipped with a water-cooled condenser to accomplish more refrigeration on a given summer day than one of the same design and horsepower equipped with an air-cooled condenser?

11. In an absorption refrigerating system, what element of the system is the counterpart of the high-pressure side of the compressor in a mechanical system? Of the low-pressure ("suction") side of the compressor?

12. In a lithium bromide absorption refrigeration system, *water is the refrigerant*. Explain how it is possible to obtain low temperatures (that is, refrigerate) with ordinary water as a refrigerant.

13. Why is humidity an important factor in human comfort? Why is relative humidity of more importance than specific humidity?

14. List the relative advantages and disadvantages of steam or hot-water vs. forced-air heating systems.

15. What are the four basic components which any air-conditioning system must have?

16. Explain how it is possible to extract heat energy from cold outdoor air at 10°F and use it to heat indoor spaces with warm air at 90°F. For a given amount of electric energy used, why will a heat pump provide more heat than an electric resistance heater?

PROBLEMS

Group One

1. A refrigerated water cooler is designed to deliver 40 gal/h of chilled water at 50°F. The water-in temperature from the water main is 75°F. What minimum capacity of refrigeration (Btu/h) is required?

2. A 10-ton refrigerating machine has an *actual* c.o.p. of 3.4. What is the minimum required horsepower of the compressor motor?

3. An ice plant freezes 500 tons of ice per day from water at 72°F and subcools the ice to 18°F. Neglecting all losses, what is the necessary refrigerating capacity of the plant in Btu/h?

4. An ideal (Carnot) refrigerator operates between −10°C and +22°C. How much work must be done by the compressor per joule of heat removed in the evaporator?

5. An actual refrigerator is operating between 5°F (evaporator) and 95°F (condenser). It is found to have a coefficient of performance of 3.2. If a Carnot refrigerator operated under these conditions, what would be its c.o.p.?

6. From the psychrometric chart determine the relative humidity RH and specific humidity W for each of the following combinations of dry-bulb and wet-bulb temperatures: (*a*) 105° DB, 69° WB, (*b*) 95° DB, 72° WB, (*c*) 92° DB, 86° WB, (*d*) 40° DB, 35° WB.

7. A 100-lb cake of ice at 32°F is placed in a refrigerator so well insulated that gains and losses to the exterior are negligible. In exactly 8 hours, heat from the products being cooled has melted all the ice and the 32°F water has drained off. What was the cooling capacity of the ice in tons of refrigeration?

8. An "ideal" refrigerator is operating at a rate of exactly one ton of refrigeration (net refrigerating effect). Tests show that in one hour, 15,000 Btu are rejected to the condenser. How much mechanical work (foot-pounds) does the compressor do on the refrigerant gas in one hour?

9. An ice plant manufactures 100 tons of ice daily (24 h) from supply water whose temperature is 65°F. The ice cakes are at a temperature of 10°F. What is the capacity of the refrigeration plant in tons of refrigeration? (Assume no losses.)

10. A round duct has a diameter of 18 in. What flow of air (cubic feet per minute) can this duct deliver at a velocity of 600 ft/min?

11. A rectangular duct measures 34 by 20 in. If it must deliver 4000 ft^3/min, what velocity will be required?

12. If a rectangular duct is to carry 10,000 ft^3/min at a velocity of 800 ft/min, what vertical dimension must it have if its horizontal dimension is limited to 38 in.?

13. The condition of the air in an office is known to be 70°F DB and 40 percent RH. From the data given in Table 18.2, calculate the specific humidity W of the room air.

Group Two

14. An ideal (Carnot) refrigerator has a coefficient of performance of 3.6. Its low-temperature reservoir (evaporator) is operating at −15°C. (a) What is the condenser temperature? (b) How much work is required per kilojoule of heat removed in the evaporator?

15. A soda pop cooler is filled with 800 aluminum cans of carbonated beverages. Each can contains 12 oz of beverage, and the cans themselves weigh 1 oz each. Take the specific heat of the liquid beverages to be 0.97 Btu/lb · F° and the specific heat of aluminum as 0.22 Btu/lb · F°. Find the refrigeration capacity required to cool the beverages and cans from 85°F to 38°F in an elapsed time of 2 h.

16. An ice plant freezes 100 rectangular cakes per 24-h day, each weighing 300 lb. The starting point is water at 65°F, and the ending point is ice at 10°F. Each cake is frozen in a stainless steel can whose weight is 75 lb. Assuming no other losses, what must be the refrigerating capacity (tons of refrigeration) of the plant?

17. A 500-lb batch of ice-cream mix whose initial temperature is 70°F is to be frozen (freezing point is 28°F) and then subcooled to 10°F in 20 min. The specific heat of ice-cream mix is 0.78 before freezing and 0.45 after freezing. Its latent heat of fusion L_f is 126 Btu/lb. What refrigeration capacity (tons) must the freezer have?

18. An inventor has submitted plans and specifications for a new compressor-operated, air-cooled refrigerator. It is to operate between 10°F and 120°F. He claims his design will produce one ton of refrigeration with exactly 1 hp of mechanical input. Evaluate this design thoroughly. Show mathematically whether it is feasible or not.

19. Calculate (from Table 18.2) the approximate relative humidity if the specific humidity is found to be 128 gr/lb of air when the temperature is 105°F.

20. A commercial air-to-air heat pump is operating on an outdoor air temperature of 30°F. (*a*) From the curve of Fig. 18.20, what is its approximate *actual* c.o.p.? (*b*) If the total heating requirement of your house at that outdoor temperature is 140,000 Btu/h, what must be the kilowatt input to the compressor motor to balance that heating load on a steady-state basis?

21. A blower is handling 7500 ft^3/min (cfm) of outside air at 35°F dry-bulb, 60 percent RH. How many pounds of air per hour are being circulated? (See the specific volume lines on the psychrometric chart.)

22. A Carnot-cycle machine operates between the temperature limits of 35 and 140°F. Find (*a*) its theoretical efficiency when operated as a heat engine, (*b*) its c.o.p. as a refrigerator, and (*c*) as a heat pump.

23. A home air conditioner on summer cycle removes heat from a house at a rate of 76,000 Btu/h. The overall c.o.p. under the existing operating conditions is known to be 2.65. Find: (*a*) The electric power input requirement in kW; and (*b*) the EER.

24. Find the relative humidity of air whose DB temperature is 85°F, if it is known to contain 172 gr of moisture per pound of air (use Table 18.2 and interpolate between listed values).

25. An air-to-air heat pump is operating at 25°F evaporating temperature, 130°F condensing temperature. The electric power input is measured as 14 kW and the actual heat output from the indoor coil (the condenser) is 135,000 Btu/h. Find (*a*) the Carnot-cycle c.o.p., (*b*) the actual c.o.p., (*c*) the EER.

26. An air-to-air heat pump is powered by a 7.5-hp electric motor and is extracting heat from outside air whose temperature is 20°F. (*a*) What is the approximate *actual* c.o.p. at this operating temperature? (See Fig. 18.20.) (*b*) Calculate the heat output of the machine in Btu/h.

27. A water-to-air heat pump extracts heat from the cold reservoir (well water) at 58°F and maintains a 115°F indoor coil (condenser) temperature. (*a*) Calculate the ideal (Carnot) c.o.p. of the machine. (*b*) What would be the Carnot c.o.p. if the unit were operating off outdoor air at 20°F for the cold reservoir?

28. A refrigeration compressor motor uses power at a rate of 4.2 kW. Calculate the heat equivalent of compressor work in Btu/h. If the cooling capacity of this machine is measured at 4.0 tons, what is its EER? Its *actual* c.o.p.?

Group Three

29. An air conditioner is operating with an EER of 10.5. It is maintaining a uniform temperature of 78°F in an office building whose cooling load at that temperature is known to be 120,000 Btu/h. What is the total input power demand of the unit in kW? What is the actual c.o.p. of the machine?

30. An actual performance test on a refrigeration machine was carried out with a continuous-flow calorimeter. The following data were obtained:

Water-in temperature to the evaporator	68°F
Water-out temperature from the evaporator	40°F
Water flow quantity	15 gal/min
Power consumption of the driving motor	17 kW

Find (*a*) the refrigeration capacity, tons, (*b*) the horsepower per ton of refrigeration, (*c*) the actual c.o.p., (*d*) the EER.

31. An air-to-air heat pump has an overall c.o.p. of 2.35. Electric energy costs 8.2 cents per kilowatthour. Fuel oil for a home furnace costs $1.04 per gallon, and the overall efficiency of the furnace is 65 percent. Calculate the comparative operating costs of these two installations in Btu delivered per dollar of energy cost.

PART 5

WAVE MOTION AND SOUND

Air waves, pipes, sound, and music. (Harris Organs)

VIBRATORY MOTION AND WAVES

One of the most important phenomena in nature is the transmission of energy from one point to another. There are various ways in which this is done. Energy can be transferred by the movement of materials or objects from one place to another. The kinetic energy of a speeding bullet, for example, is transferred to the target.

The transfer of heat energy through metal from a region of high temperature to one of lower temperature was considered in Chap. 16. In later chapters the transfer of electric energy in a metal conductor will be considered. The transfer of heat energy and electric energy through metals depends upon the motion of molecular and atomic particles composing the metal.

ENERGY TRANSFER BY WAVES

In the next six chapters we will consider yet another method of energy transfer. In this method energy is transferred without the movement of matter from a source of energy to a receiver, but by means of waves. Examples of energy transfer by waves are everywhere around us: water waves, sound waves, earthquake waves, waves along a stretched wire, radio waves, microwaves, and light waves, to cite a few.

Waves such as water waves, sound waves, and waves in a spring require a material medium (solid, liquid, or gas) to transfer energy. Such waves are called *mechanical waves*. The material substance is called the *medium* for the wave. In later chapters we will study waves that do not require a material medium, such as light waves, and infrared, ultraviolet, x-ray, gamma ray and television waves. These are called *electromagnetic* waves. Electromagnetic waves are propagated through a vacuum (space), or through air or other gases.

Even though many kinds of waves exist, they all have some common properties. By observing and studying visible waves in ropes, springs, and water, we can discover some characteristics that are common to all waves, including even the invisible ones.

19.1 ■ Vibrations and Periodic Motion

Energy can be transmitted by the movement of material particles from one place to another or by wave motion in a medium.

Mechanical waves originate in the vibration of a finite mass.

Wave motion and vibrations are intimately related. Water waves, sound waves, earthquake waves, and waves in a wire or rope all have as their source of energy a vibration. A vibrating tuning fork, a piano wire, and an air column in an organ pipe produce sound waves. A rock thrown in a quiet pool of water supplies the energy that results in water waves.

Most mechanical waves originate from objects vibrating so rapidly that their motion is difficult to analyze with the naked eye. We will study the properties of these waves by observing slowly vibrating objects, such as a mass oscillating on the end of a spring, and waves that form along a rope or cord.

When an object vibrates or oscillates, the motion of the object repeats itself, back and forth, over the same path. Any motion which repeats itself in equal intervals of time is called *periodic motion*. Many phenomena in everyday experiences are examples of periodic motion—a swinging pendulum, the turning balance wheel of a watch, the reciprocating pistons in an automobile engine, the vibrating string of a violin, the vibrating air columns in musical instruments, and a vibrating mass on the end of a coiled spring. Satellites (including the moon) in orbit about the earth, and the planets in orbit around the sun also exhibit periodic motion.

19.2 ■ Elastic Potential Energy

Suppose we attach a mass m at the end of a spring supported on a frictionless surface (Fig. 19.1). When we stretch the spring, it resists being lengthened. If we release the stretching force, the spring returns to its original length (zero gravitational force is also assumed).

When the spring is stretched by a force F, the magnitude of the force required is proportional to the elongation s (see Sec. 11.5). Thus,

$$F = ks \qquad \text{(Hooke's law)} \tag{19.1}$$

where k is a proportionality constant that depends upon the units used for F and s.

In like manner, if we compress the spring, it resists being shortened and will return to its original length upon removal of the compressional force. In each case, the restoring force is always equal in magnitude, but opposite in direction, to the deforming force. Hence the restoring force F is $-ks$; that is it acts in a direction opposite to the deforming force and toward the neutral position 0.

The work done in stretching (or compressing) the spring is the product of the force F and the distance through which it acts. In these instances, we will need to use the average force $\bar{F}$ to find the work:

Work done in stretching or compressing a spring is stored in the spring as potential energy.

$$\bar{F} = \frac{F_{\text{initial}} + F_{\text{final}}}{2}$$

$$= \frac{0 + ks}{2}$$

$$= \tfrac{1}{2}ks$$

Thus the work done on the spring is

$$W = \tfrac{1}{2}ks(s) = \tfrac{1}{2}ks^2 \tag{19.2}$$

This work is stored in the spring as *elastic potential* energy.

When a stretched or compressed spring is released, the potential energy stored in the spring is converted to kinetic energy of the object attached to the spring.

When the restraining forces are released, the potential energy $\tfrac{1}{2}ks^2$ is converted into kinetic energy of the mass (or into work done on some other object). If we can neglect the mass of the spring (assumed small compared to m) and friction losses, we can readily determine the velocity of m as it passes the equilibrium point O. At the equilibrium position, $s = 0$, and all the potential energy of the spring will have been

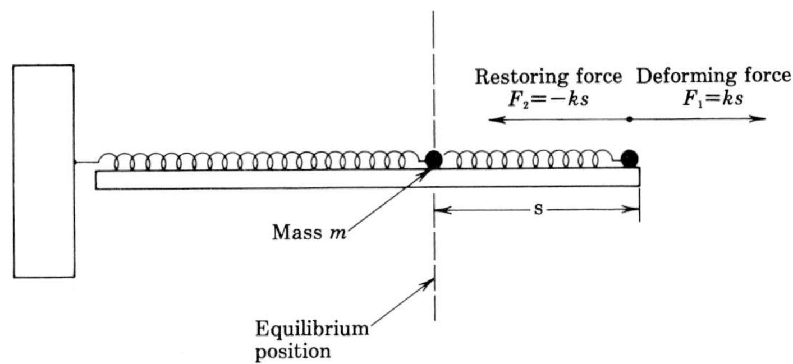

Fig. 19.1 Within limits of perfect elasticity, the stretching force F is proportional to the elongation (Hooke's law).

converted to the kinetic energy of the mass m:

$$\tfrac{1}{2} m v_{max}^2 = \tfrac{1}{2} k s^2$$

or

$$v_{max} = \sqrt{(k/m)} s \tag{19.3}$$

Illustrative Problem 19.1 The spring shown in Fig. 19.1 has a spring constant k for elongation of 15 N/m, and $m = 0.75$ kg. The spring is pulled by a force which will produce an elongation of 50 cm and is then released. What will be the speed of the mass when it passes through its equilibrium position?

Solution At the equilibrium position the velocity of m will be at a maximum.

$$\begin{aligned} v_{max} &= \sqrt{\frac{k}{m}}\, s \\ &= \sqrt{\frac{15 \text{ N/m}}{0.75 \text{ kg}}}\,(50 \text{ cm}) \\ &= \sqrt{20 \frac{\text{N/m}}{\text{kg}} \frac{(1 \text{ kg} \cdot \text{m/s}^2)}{1 \text{ N}}}\,(0.50 \text{ m}) \\ &= 4.47 \text{ s}^{-1}(0.50 \text{ m}) \\ &= 2.2 \text{ m/s} \end{aligned}$$

answer

19.3 ■ Simple Harmonic Motion

One form of periodic motion can be illustrated by a mass m oscillating at the end of a spring. Consider a spring with one end fixed to a rigid support and a mass m attached at the other end (Fig. 19.2). Idealize that the mass slides without friction on

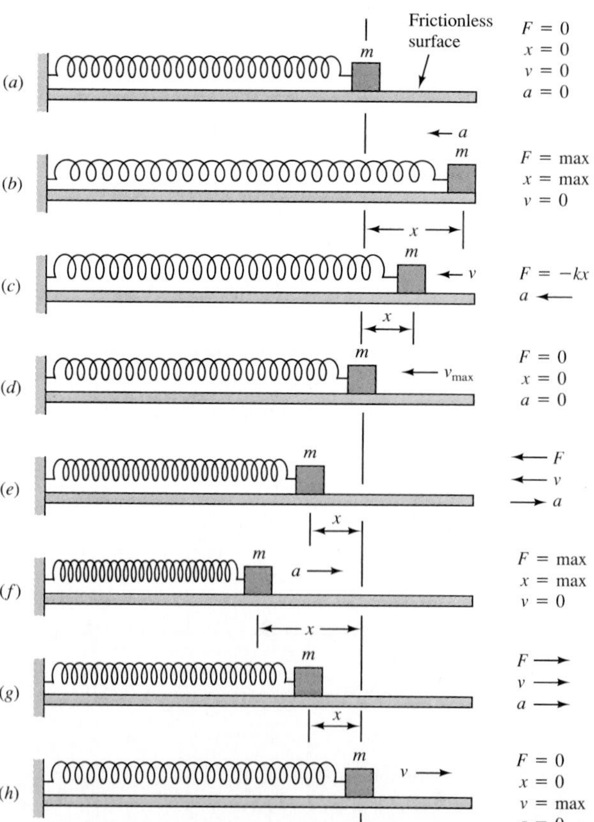

Fig. 19.2 Characteristics of SHM of a mass oscillating at the end of a spring.

the horizontal surface. Assume further that the mass of the spring is not a factor in the motion.

The spring has a natural length or equilibrium position at which it exerts no force on m. Let this *equilibrium position* be as shown in Fig. 19.2*a*. If the mass is moved either to the right, which stretches the spring, or to the left, which compresses it, the spring exerts a force on the mass which acts in the direction of returning it to the equilibrium position. This force is called the *restoring force*. The magnitude of the restoring force, by Hooke's law (Chap. 11), is directly proportional to the distance x the spring has been stretched or compressed.

The energy in a harmonic motion shifts back and forth between potential and kinetic energy.

If the spring is initially elongated a distance $x = A$ (Fig. 19.2*b*) and then released, the spring will exert a force F that pulls the mass m toward the equilibrium position. When $x = A$, and m is held in place, $v = 0$ (Fig. 19.2*b*). Upon release, the restoring force F will accelerate the mass toward the equilibrium position. In Fig. 19.2*c* the force is decreasing, the acceleration is decreasing, and the velocity is increasing to the left. When the mass reaches the equilibrium position (Fig. 19.2*d*), the force on it is zero, its speed is maximum, and the acceleration is zero. (Remember that we assumed that m slides without friction.)

As the mass continues moving to the left, the force on it changes direction, the acceleration will be directed to the right (negative) and the speed will diminish (Fig. 19.2*e*). The mass will stop when $x = -A$ (Fig. 19.2*f*), $v = 0$ and F is a maximum again. It then begins moving back toward the equilibrium position, with acceleration a directed to the right. In the interval between maximum compression and the equilibrium positions, F and a will decrease and v will increase (toward the right).

When x again equals zero (Fig. 19.2*h*) the force F and the acceleration a will be zero and v will be maximum. The spring will begin to be stretched again, F and a will change direction and increase until $x = A$ and $v = 0$. The motion then will be repeated (if no friction is present).

All simple harmonic motions are periodic, but not all periodic motions are harmonic.

The motion just described is termed *simple harmonic motion* (SHM). When the mass m is displaced a small distance x from the equilibrium position, the spring exerts a restoring force $F = -kx$, where k is the force constant of the spring. Applying Newton's second law to the motion of m in the x-direction, we get

$$F = -kx = ma$$

and

$$a = -\frac{k}{m}x \tag{19.4}$$

that is, the acceleration is inversely proportional to the mass of the moving body, and is directly proportional to its displacement from the equilibrium position.

Simple harmonic motion (SHM) is motion described by a particle (or mass) about an equilibrium position such that the acceleration is always directed toward the equilibrium position and is proportional to the displacement.

19.4 ■ Amplitude, Period, Frequency

Some terms that must be defined to fully describe SHM.

A few terms must be precisely defined in order to deal with periodic motion and wave motion. The distance x from the equilibrium position of a particle undergoing periodic motion is called the *displacement*. The maximum displacement of the particle is called the *amplitude* (A). One *cycle* refers to the complete to-and-fro motion from some initial point back to the same point *going in the same direction*. The *period* (T) is the time required for one complete cycle. The *frequency* (f) of the motion is the number of complete cycles per second. Frequency is the reciprocal of period.

Simple harmonic motions have a constant frequency.

$$f\left(\frac{\text{cycles}}{\text{second}}\right) = \frac{1}{T\left(\dfrac{\text{second}}{\text{cycle}}\right)} \tag{19.5}$$

The frequency of wave motion is the number of vibrations that pass a point every second.

The unit of frequency in vibratory motion is the hertz (Hz).

The standard unit of frequency is the *hertz* (Hz), named after the German physicist Heinrich Hertz (1857–1894).

$$\text{One hertz} = 1 \text{ cycle/s}$$

Multiples of cycles/second (Hz) are used for high frequencies. Thus,

$$1 \text{ kilocycle/s (kc/s)} = 10^3 \text{ cycles/s} = 1 \text{ kilohertz (kHz)}$$

$$1 \text{ megacycle/s (Mc/s)} = 10^6 \text{ cycles/s} = 1 \text{ megahertz (MHz)}$$

Obviously, the system shown in Fig. 19.2 has energy when it is oscillating. When the spring is stretched or compressed to its maximum (Fig. 19.2*b* and *f*) the system possesses maximum potential energy. As the spring begins to return to its equilibrium position, it will gain kinetic energy and lose potential energy. The spring, in its back-and-forth motion, possesses only kinetic energy when it passes the equilibrium position. It has zero potential (stored) energy at that point in the cycle.

If there is no appreciable dissipation of energy for the system of Fig. 19.1 or Fig. 19.2, the sum of the potential energy and the kinetic energy is a constant at every instant during the oscillation. The total energy is equal to the initial energy given the system. Thus

The total mechanical energy of a vibrating system is proportional to the square of the amplitude.

$$E_{PE} + E_{KE} = \text{constant} = \tfrac{1}{2}kA^2$$

or

$$\tfrac{1}{2}ks^2 + \tfrac{1}{2}mv^2 = \tfrac{1}{2}kA^2 \qquad (19.6)$$

Illustrative Problem 19.2 A 4.0-kg body vibrates in SHM with an amplitude of 5.0 cm. Find the speed of the body (*a*) at the midpoint of the vibration and (*b*) at a point 2.0 cm from the midpoint. Assume $k = 0.75$ N/m, and frictionless horizontal motion.

Solution

(***a***) At the midpoint of the vibration, $s = 0$. Then

$$\tfrac{1}{2}mv^2 = \tfrac{1}{2}kA^2 \qquad \text{or} \qquad v^2 = \frac{kA^2}{m}$$

$$v^2 = \frac{0.75 \text{ N/m } (0.05 \text{ m})^2}{4 \text{ kg}}$$

$$= 4.69 \times 10^{-4} \frac{\text{N}\cdot\text{m}}{\text{kg}}\left(\frac{1 \text{ kg}\cdot\text{m/s}^2}{1 \text{ N}}\right)$$

$$= 4.69 \times 10^{-4} \text{ m}^2/\text{s}^2$$

$$v_{max} = 0.022 \text{ m/s} \qquad \text{or} \qquad 2.2 \text{ cm/s} \qquad \textit{answer}$$

(***b***) At $s = 2.0$ cm,

$$\tfrac{1}{2}ks^2 + \tfrac{1}{2}mv^2 = \tfrac{1}{2}kA^2$$

$$\tfrac{1}{2}(0.75 \text{ N/m})(0.02 \text{ m})^2 + \tfrac{1}{2}(4 \text{ kg})v^2 = \tfrac{1}{2}(0.75 \text{ N/m})(0.05 \text{ m})^2$$

$$(2 \text{ kg})v^2 = 7.875 \times 10^{-4} \text{ N}\cdot\text{m}$$

$$v^2 = \frac{7.875 \times 10^{-4}\text{N}\cdot\text{m}\left(\dfrac{1 \text{ kg}\cdot\text{m/s}^2}{1 \text{ N}}\right)}{2 \text{ kg}}$$

$$= 3.9375 \times 10^{-4} \text{ m}^2/\text{s}^2$$

$$v = 1.98 \times 10^{-2} \text{ m/s}^2 \approx 2.0 \text{ cm/s} \qquad \textit{answer}$$

19.5 ■ Simple Harmonic Motion and the Sine Curve

Often it is desirable to represent periodic oscillations by a graph. One way of doing this is shown in Fig. 19.3. A pen or stylus attached to a vertical oscillating mass *m*

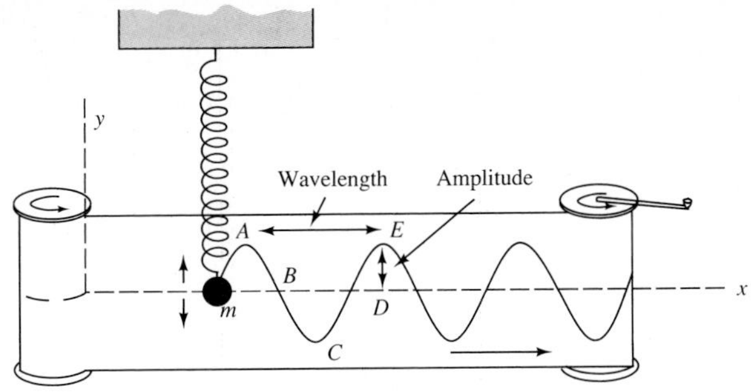

Fig. 19.3 The oscillating spring illustrates the sinusoidal nature of SHM as a function of time. The roll of paper is pulled from left to right at constant speed while mass m oscillates up and down in SHM.

traces out a wave on a roll of paper moving horizontally at constant speed. The coordinates of the graph are y, the vertical displacement from the equilibrium position, and x in units of time. The curve traced on the roll of paper is the familiar sine curve.

To have a complete description of a particle undergoing simple harmonic motion, the position, velocity, and acceleration for any given time must be known.

19.6 ■ Simple Harmonic Motion and Uniform Circular Motion

Simple harmonic motion has a simple relationship to a particle rotating in a circle with uniform speed. Consider a particle P moving with uniform linear speed v_r and uniform angular speed ω around a circle (called the *reference circle*) of radius r with center at 0 (Fig. 19.4). The particle P is in uniform circular motion (UCM). Imagine rays of light parallel to the X axis producing a moving shadow of P on a line MN lying in the XY plane and perpendicular to the X axis at E. Let Q represent the shadow of P on line MN. As P moves with uniform circular motion, point Q will oscillate back and forth along MN with SHM in the same manner as the oscillating mass of Fig. 19.3.

The projection of uniform circular motion on a straight line describes SHM.

The displacement EQ of point Q at any instant is

$$y = r \sin \theta \tag{19.7}$$

The velocity of Q is the component of the *instantaneous* velocity v_r of point P along a line parallel to YY', or

$$v = v_r \cos \theta \tag{19.8}$$

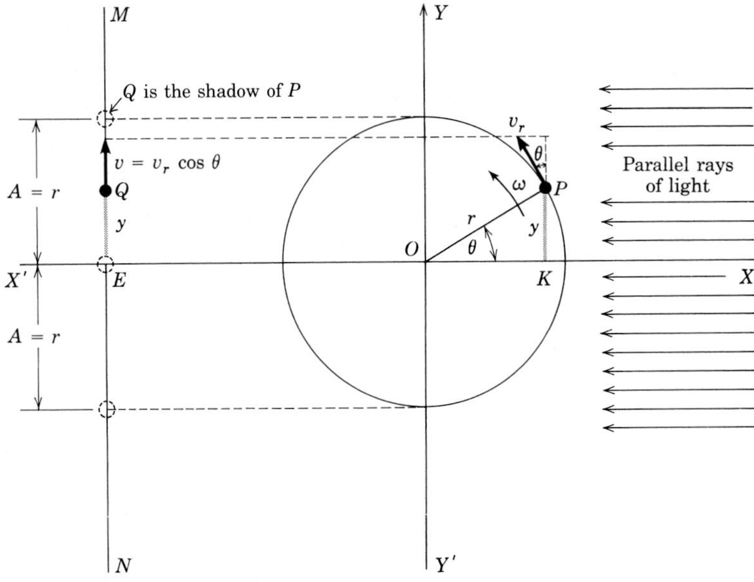

Fig. 19.4 Circle of reference for SHM.

Fig. 19.5 Velocity and acceleration for UCM and SHM.

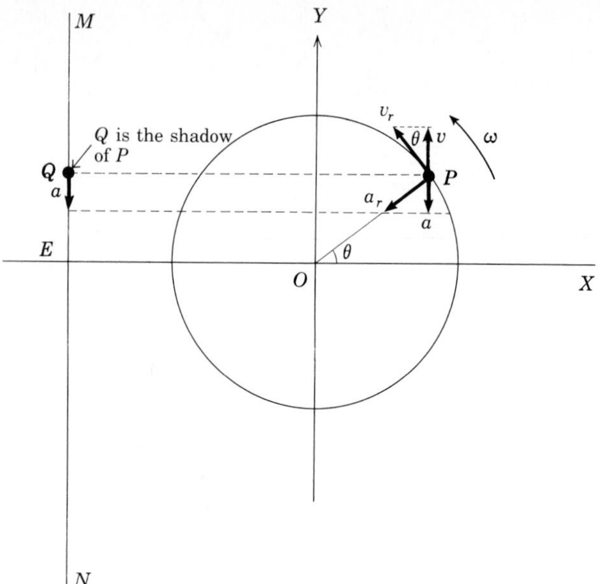

The acceleration at each instant of a particle in SHM is proportional to the negative value of the displacement at that instant.

The centripetal acceleration a_r (see Fig. 19.5 and Sec. 8-6) of particle P toward the center of the circle is v^2/r or $\omega^2 r$. The acceleration of Q is the component of the centripetal acceleration of P along a line parallel to the Y axis and may be written

$$a = -a_r \sin\theta = -\omega^2 r \sin\theta = -\omega^2 y \tag{19.9}$$

The negative sign is introduced since ω is constant and positive and a is directed downward when y is positive, and vice versa. Thus we see that the acceleration of Q is always directed toward the neutral position E and is proportional to the displacement y, as required in SHM.

19.7 ■ Period of Simple Harmonic Motion

It is usual to express the period T for a particle with SHM in terms of the acceleration and displacement of the particle. Consider a mass m suspended from a spring undergoing SHM (see Fig. 19.6). The period T of mass m will be the same as that of a similar mass P in UCM of the same amplitude (shown dotted). When f is measured in hertz, and T in seconds, the value of the angular velocity in radians per second is

Fig. 19.6 The mass at the end of a spring executes SHM. The dotted circle shows that Q moves as the shadow of P would move if P were in UCM.

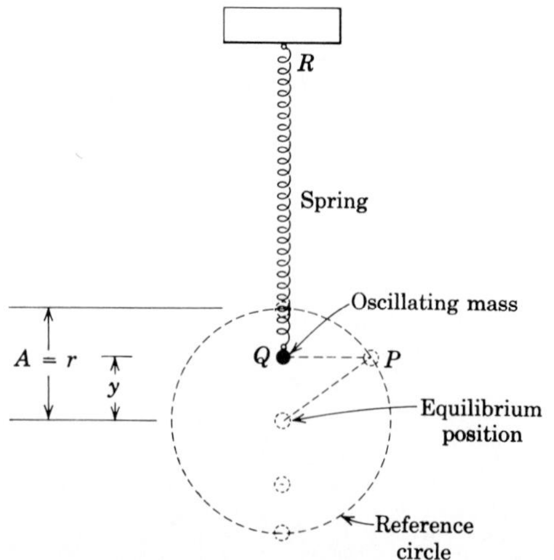

found by the equation

$$\omega = 2\pi f = \frac{2\pi}{T} \tag{19.10}$$

From Eq. (19.9)

$$a = -\omega^2 y = -\frac{4\pi^2}{T^2}\,y \tag{19.11}$$

Solving first for T^2 and then for T, we get

$$T^2 = -4\pi^2\frac{y}{a}$$

$$T = 2\pi\sqrt{-y/a} \tag{19.12}$$

where the acceleration is that of the body when it is displaced y units from the equilibrium position. It is evident that the term under the radical sign will always be positive since the displacement of a body with SHM will always be of opposite sign to the acceleration.

The period can also be expressed in terms of the spring constant (or other measure of elasticity) and the mass of the body.

From Hooke's law we have the relation $F = -ky$, where F is the restoring force, y is the displacement, and k is the (spring constant) measure of elasticity. From Newton's second law we know also that the force F to accelerate the mass m is equal to ma. Equating the two expressions for F,

$$ma = -ky \qquad \text{or} \qquad -\frac{y}{a} = \frac{m}{k}$$

and substituting m/k for $-y/a$ in Eq. (19.12), we get

$$T = 2\pi\sqrt{m/k} \tag{19.13}$$

Consistent units to use for T, m, and k in Eq. (19.13) are seconds, kilograms, and newtons per meter; or seconds, slugs, and pounds per foot.

Note that increasing the mass increases the period, while increasing the stiffness (elasticity) of a spring decreases the period. It should also be evident, since amplitude A does not appear in Eq. (19.13), that *the period does not depend upon the amplitude of motion*.

Changing the amplitude of an oscillating spring does not affect the frequency of the motion.

Illustrative Problem 19.3 A 0.100-kg mass is hung from a spring. It is noted that when an additional 0.400-kg mass is added to the spring, the spring elongates 80.0 cm. If the 0.500-kg mass is then set vibrating, what should the period of oscillation be? What is the frequency?

Solution

$$m = 0.500\text{ kg} \qquad F = -ky \qquad \text{or} \qquad k = -F/y$$

$$k = -\frac{-0.400\text{ kg}}{80.0\text{ cm}} \times \frac{9.81\text{ N}}{1\text{ kg}} \times \frac{100\text{ cm}}{1\text{ m}}$$

$$= 4.90\text{ N/m}$$

By Eq. (19.13),

$$\begin{aligned} T &= 2\pi\sqrt{m/k} \\ &= 2\pi\sqrt{0.500\text{ kg}/(4.90\text{ N/m})} \\ &= 2.01\text{ s} \qquad \textit{answer} \\ f = 1/T &= 1/2.01\text{ s} \\ &= 0.50\text{ cycles/s} \\ &= 0.50\text{ Hz} \qquad \textit{answer} \end{aligned}$$

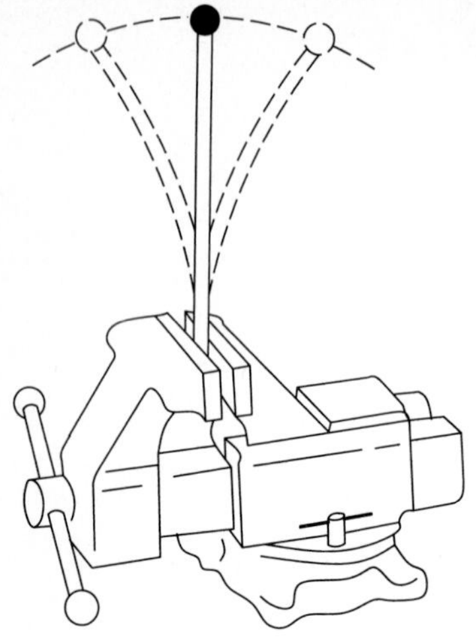

Fig. 19.7 Illustrative Problem 19.4.

Illustrative Problem 19.4 A 2.8-lb ball is fastened to a metal strip clamped tightly in a vise at one end (Fig. 19.7). A force of 3.4 lb will pull the ball 6 in. to one side. Find the period of vibration of the weighted strip. Assume SHM.

Solution

$$k = \frac{F}{y} = \frac{3.4 \text{ lb}}{6 \text{ in.}} \times \frac{12 \text{ in.}}{1 \text{ ft}}$$
$$= 6.8 \text{ lb/ft}$$
$$m = \frac{w}{g} = \frac{2.8 \text{ lb}}{32 \text{ ft/s}^2}$$
$$= 0.0875 \text{ lb·s}^2\text{/ft}$$
$$T = 2\pi\sqrt{m/k}$$
$$= 2\pi\sqrt{\frac{0.0875 \text{ lb·s}^2\text{/ft}}{6.8 \text{ lb/ft}}}$$
$$= 0.71 \text{ s} \qquad \textit{answer}$$

Illustrative Problem 19.5 A spring stretches 0.250 m when a 0.450 kg mass is hung from it. The spring is stretched an additional 0.150 m from its equilibrium position and then released. Determine (*a*) the frequency of oscillation, and (*b*) an equation for the displacement as a function of time. Assume SHM.

Solution
(***a***) The spring constant is equal to

$$k = \frac{F}{y} = \frac{0.450 \text{ kg} \times 9.81 \text{ N /1 kg}}{0.250 \text{ m}} = 17.66 \text{ N/m}$$

Use Eq. (19.13) to find T.

$$\mathrm{T} = 2\pi\sqrt{\frac{\mathrm{m}}{\mathrm{k}}}$$
$$= 2\pi\sqrt{\frac{0.450 \text{ kg} \times \dfrac{1 \text{ N}}{1 \text{ kg·m/s}^2}}{17.66 \text{ N/m}}}$$
$$= 2\pi\sqrt{0.02458 \text{ s}^2}$$
$$= 1.003 \text{ s}$$

Then $$f = \frac{1}{T} = \frac{1}{1.003 \text{ s}} = 1.00 \text{ s}^{-1} \qquad \textit{answer}$$

(***b***) $y = -0.150\ m$ when $t = 0$. Use Eq. (19.8) with a 90° phase difference.

$$= \mathrm{r} \cos \theta = \mathrm{r} \cos 2\pi \mathrm{ft}$$
$$= -0.150 \cos 2\pi\ (1.00)\mathrm{t}$$
$$= -0.150 \cos 6.28\ \mathrm{t} \qquad \textit{answer}$$

19.8 ■ Speed of a Particle in Simple Harmonic Motion

The principle of conservation of energy enables us to determine the speed of a particle in SHM in terms of its frequency f, amplitude A, and displacement s. Solving Eq. (19.6) for v,

$$\tfrac{1}{2}ks^2 + \tfrac{1}{2}mv^2 = \tfrac{1}{2}kA^2$$

Then $$mv^2 = k(A^2 - s^2)$$

and $$v = \sqrt{\frac{k}{m}}\sqrt{A^2 - s^2}$$

Eq. (19.13), ($T = 2\pi\sqrt{m/k}$) can be rearranged as

$$\sqrt{\frac{m}{k}} = \frac{T}{2\pi}$$

or $$\sqrt{\frac{k}{m}} = \frac{2\pi}{T} = 2\pi f$$

The speed of the particle, thus, is equal to

$$v = \pm 2\pi f\sqrt{A^2 - s^2} \tag{19.14}$$

The sign of v will depend upon whether s is positive or negative and upon whether it is moving toward or away from the equilibrium position.

The speed of a particle in SHM can be determined if its frequency, amplitude, and displacement are known.

Illustrative Problem 19.6 The crankshaft of an auto engine turns 4800 rev/min. The stroke of the piston is 10 cm. How fast is the piston travelling (*a*) at the middle of its stroke, (*b*) when it is 2 cm from the bottom of its stroke? Assume that the piston has simple harmonic motion.

Solution The amplitude A of the motion of the piston is $\frac{1}{2}$(10 cm) = 0.05 m. The frequency of oscillation f is 4800 rev/min = 4800/60 rev/s = 80 Hz.
(***a***) Use Eq. (19.14). s = 0 at the middle of its stroke. Hence,

$$\begin{aligned} v &= 2\pi f\sqrt{A^2 - s^2} \\ &= 2\pi(80\ \text{s}^{-1})\sqrt{(0.05\ \text{m})^2} \\ &= 25\ \text{m/s} \qquad \textit{answer} \end{aligned}$$

(***b***) $s = 0.03$ m when the piston is 2 cm from the bottom of a stroke. Then

$$\begin{aligned} v &= 2\pi(80\ \text{s}^{-1})\sqrt{(0.05\ \text{m})^2 - (0.03\ \text{m})^2} \\ &= 20\ \text{m/s} \qquad \textit{answer} \end{aligned}$$

19.9 ■ The Simple Pendulum

The *simple pendulum* consists of a concentrated mass at the end of a cord or wire of negligible weight. The motion of the simple pendulum closely approximates SHM if θ in Fig. 19.8 is small. Consider a small, relatively heavy bob at the end of a very light rod or string. If such a bob of mass m is displaced from its equilibrium position as shown in Fig. 19.8, two forces act upon the bob: the attraction of gravity mg vertically downward and the pull P of the string acting toward O.

The restoring force is the component of the weight perpendicular to the string. This component is

$$-mg \sin\theta = -mg \sin \angle COA$$

The negative sign is used to indicate that the force F is directed toward the neutral position C. (Angles measured in a counterclockwise direction are given positive signs, and those measured in a clockwise direction are given negative signs.) The displacement along the arc is $s = l\theta$ (if θ is in radians). We see that the force is proportional to $\sin\theta$ while the displacement is proportional to θ. However, if θ is small, $\sin\theta$ may be replaced by θ (in radians) without serious error. Hence, to the degree of this approximation, the restoring force F is directly proportional to the displacement, and we can apply the equations of SHM to the motion of the pendulum:

$$\begin{aligned} F &= ma \\ &= -mg\sin\theta = -mg\frac{x}{l} = -mg\frac{s}{l} \quad \text{(approx)} \end{aligned}$$

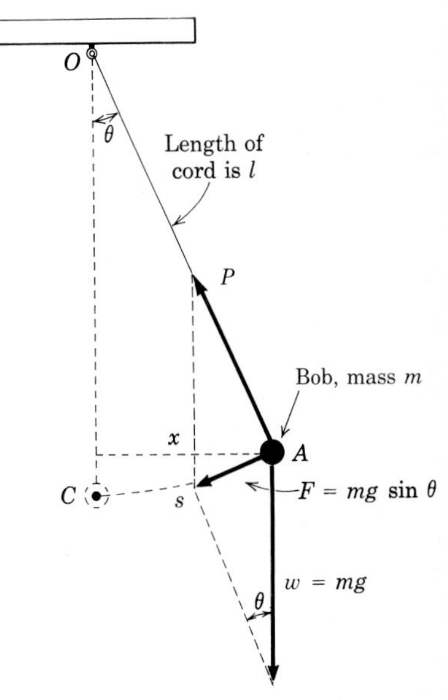

Fig. 19.8 Forces acting on a simple pendulum.

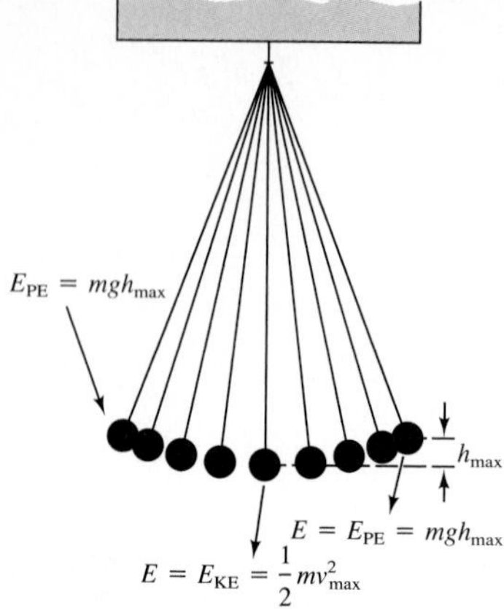

Fig. 19.9 As the pendulum swings, there is a continuous conversion between potential and kinetic energy. In the ideal pendulum, the sum of the energies is constant. (Small amplitude is assumed.)

The period of a simple pendulum is independent of the mass of the bob.

Equating the two forces, we get

$$-mg\frac{s}{l} = ma$$

and

$$-s/a = l/g$$

Substitute in Eq. (19.12) where s is equivalent to y:

$$T = 2\pi\sqrt{l/g} \quad \text{period of a pendulum} \tag{19.15}$$

The swinging pendulum is another example of the continuous conversion of potential and kinetic energy of a system. At the pendulum's maximum arc swing (Fig. 19.9), or maximum height h_{max}, the mass stops momentarily ($v = 0$) and all the mechanical energy of the system is now potential energy. At the bottom of the swing ($h = 0$, and $s = 0$) the velocity of the mass is maximum and all the mechanical energy is kinetic energy. In between these positions the total mechanical energy is made up of both kinetic and potential energies. At any instant, the total energy equals

$$E = mgh + \tfrac{1}{2}mv^2 = mgh_{max} = \tfrac{1}{2}mv^2_{max} \tag{19.16}$$

The period of a pendulum is seen (Eq. 19.15) to depend only upon the length of the pendulum and the acceleration due to gravity at the location of the pendulum. The period does not depend on the mass of the bob and is independent of the amplitude *so long as the amplitudes are small* (remember that for small θ's, $\theta \approx \sin\theta$). It can be shown that an angle of 10° will produce an error of about 0.2 percent if Eq. (19.15) is used.

Equation (19.15) can be applied in determining the value of g. By using a pendulum of fixed length and measuring its period of vibration, the value of g at a particular location can readily be determined, as the following problem demonstrates.

Illustrative Problem 19.7 A simple pendulum has a length of 4.05 m and executes 20 complete vibrations of small amplitude in 80.8 s. What is the value of the acceleration of gravity g where the pendulum is located?

Solution Solve Eq. (19.15) for g:

$$\begin{aligned} T &= 2\pi\sqrt{l/g} \\ T^2 &= 4\pi^2(l/g) \\ g &= 4\pi^2 l/T^2 \end{aligned}$$

Substituting $l = 4.05$ m and $T = 80.8/20 = 4.04$ s in the equation for g, we get

$$g = \frac{4\pi^2 \times 4.05 \text{ m}}{(4.04 \text{ s})^2} = 9.79 \text{ m/s}^2 \qquad \textit{answer}$$

Illustrative Problem 19.8 The simple pendulum shown in Fig. 19.9 has a mass of 75.0 g and is released from a height h of 15.0 cm above its lowest position (where $h = 0$). Determine (a) the velocity of the mass at the bottom of its swing, and (b) the tangential velocity when the pendulum is at a height of 6.0 cm above its lowest point.

Solution (***a***) At the top of the swing all the energy is potential.

$$\begin{aligned} E_{PE} &= mgh_{max} \\ &= 0.075 \text{ kg } (9.81 \text{ m/s}^2)(0.150 \text{ m}) \\ &= 0.1104 \text{ kg·m}^2/\text{s}^2 \text{ (joules)} \end{aligned}$$

At the bottom of the swing all the energy is kinetic

$$\begin{aligned} E_{KE} &= \tfrac{1}{2}mv^2_{max} = E_{PE} \\ \tfrac{1}{2}(0.075 \text{ kg})(v^2) &= 0.1104 \text{ kg·m/s}^2 \\ v^2 &= 2.944 \text{ m}^2/\text{s}^2 \\ v &= 1.72 \text{ m/s} \qquad \textit{answer} \end{aligned}$$

(*b*) At height $h = 6.0$ cm

$$\tfrac{1}{2}mv^2 + mgh = mgh_{max}$$

Then

$$\tfrac{1}{2}v^2 = gh_{max} - gh$$

$$\begin{aligned} v^2 &= 2g(h_{max} - h) \\ &= 2 \times 9.81 \text{ m/s}^2 \times (0.15 \text{ m} - 0.061 \text{ m}) \\ &= 1.74618 \text{ m}^2/\text{s}^2 \\ v &= 1.32 \text{ m/s} \qquad \textit{answer} \end{aligned}$$

19.10 ■ The Torsion Pendulum

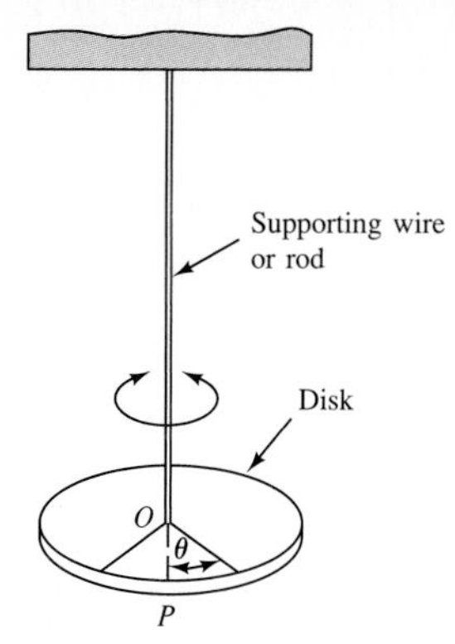

Fig. 19.10 The torsion pendulum oscillates with simple harmonic motion.

The *torsion pendulum* consists of a solid object (usually a circular disk or cylinder) suspended by a wire, ribbon, or thin rod (Fig. 19.10). When the object is turned through a small angle θ, the twisted wire, ribbon, or rod will exert a restoring torque on the body proportional to the angular displacement θ. From Hooke's law, the restoring torque is

$$\tau = \kappa\theta \tag{19.17}$$

where κ (the Greek letter kappa) is called the *torsion constant* of the supporting member. The value of κ depends on the material and dimensions of the supporting wire, rod, or ribbon. The torsion constant is measured in newton-meters per radian (N · m/rad) (see Sec. 8.1).

The torque in a torsional pendulum is proportional to the angular displacement.

The period of a torsion pendulum is found by the equation (given here without derivation)

$$T = 2\pi\sqrt{\frac{I}{\kappa}} \tag{19.18}$$

where κ is the torsion constant and I is the moment of inertia of the object about the axis of rotation (see Sec. 9.4). I is measured in units of kg · m^2.

The similarity between the equation for the period of vibratory motions and that for rotational oscillations should be readily apparent. I and κ in Eq. (19.18) are analogous to m and κ, respectively in Eq. (19.13). Equation (19.18) makes no restriction on the size of angle θ. The period of oscillation is independent of the size of θ, as long as the elastic limit of the supporting member is not exceeded.

Illustrative Problem 19.9 A circular cylinder has a mass of 5.0 kg and a diameter of 56 cm. It is suspended by a wire that has a torsion constant of 0.36 N · m/rad. Determine its frequency of oscillation.

Solution In Sec. 9.4 the moment of inertia I of a circular cylinder is given by the formula $I = \tfrac{1}{2}mr.^2$ Hence, substituting values,

$$\begin{aligned} I &= \tfrac{1}{2}(5.0 \text{ kg})\left(\frac{0.56 \text{ m}}{2}\right)^2 \\ &= 0.196 \text{ kg} \cdot \text{m}^2 \end{aligned}$$

From Eqs. (19.5) and (19.19),

$$\begin{aligned} f &= \frac{1}{T} \\ &= \frac{1}{2\pi}\sqrt{\frac{\kappa}{I}} \\ &= \frac{1}{2\pi}\sqrt{\frac{0.36 \text{ N} \cdot \text{m/rad}}{0.196 \text{ kg} \cdot \text{m}^2}} \\ &= \frac{1}{2\pi}\sqrt{1.837 \frac{\text{N}}{\text{kg} \cdot \text{m}} \times \frac{\text{kg} \cdot \text{m/s}^2}{\text{N}}} \\ &= 0.216 \text{ s}^{-1} \text{ or } 0.216 \text{ Hz} \qquad \textit{answer} \end{aligned}$$

19.11 ■ Equations for Simple Harmonic Motion

In Sec. 19.6 it was determined that the displacement of a particle undergoing SHM is a function of the sine of θ, where θ is the angular displacement of the related particle undergoing UCM; that is, $y = r \sin \theta$. But since the maximum amplitude A of the SHM of Q in Fig. 19.4 is equal to the radius r of the related reference circle, we have the equation for the displacement, $y = A \sin \theta$.

The frequency f of SHM is the same as the frequency, or number of revolutions per unit time, in the reference circle. Since there are 2π radians per complete revolution, the angular velocity ω of P in Fig. 19.4, in radians per second, is related to f by the equation

$$\omega = 2\pi f \qquad (19.19)$$

Furthermore, the point Q makes one complete vibration for each revolution of P. Therefore, f is also the number of vibrations per second (or *frequency of vibration*) of point Q.

Since $f = 1/T$ (Eq. 19.5), we can derive the equation

$$\omega = \frac{2\pi}{T} \qquad (19.20)$$

We know from Eq. (8.4) that $\theta = \omega t$. Then

$$\theta = 2\pi f t \qquad (19.21)$$

and

$$\theta = \frac{2\pi}{T}\,t \qquad (19.22)$$

Thus, if particle Q passes through its equilibrium position ($\theta = 0$) when $t = 0$ (that is, $y = 0$ when $t = 0$) the particle has a displacement equal to

Sinusoidal equations for displacement, velocity, and acceleration for SHM.

$$y = A \sin(2\pi f t) \qquad (19.23)$$

or

$$y = A \sin\left(\frac{2\pi}{T}\,t\right) \qquad (19.24)$$

where f is the frequency (and T the period) of the motion and t is the time elapsed after passing the equilibrium position.

From Eq. (8.2) we know that v, the tangential velocity of the particle P, is equal to ωr (or ωA). Hence, Eq. (19.8) becomes

$$v = A \cos \theta$$

or

$$v = 2\pi f A \cos(2\pi f t) \qquad (19.25)$$

and

$$v = \frac{2\pi}{T} A \cos\left(\frac{2\pi}{T}\,\kappa\right) \qquad (19.26)$$

Equation (8.7′) states that $a_r = r\omega^2$. But in the present case, $r = A$ and $r\omega^2 = A\omega^2$. Hence Eq. (19.9) becomes

Maximum speed of a particle in SHM occurs when the particle is in its equilibrium position.

$$a = -\omega^2 A \sin \theta$$

or

$$a = -4\pi^2 f^2 A \sin(2\pi \mathrm{ft}) \qquad (19.27)$$

and

$$a = -\frac{4\pi^2}{T^2} A \sin\left(\frac{2\pi}{T}\,t\right) \qquad (19.28)$$

Maximum acceleration of a particle in SHM occurs when the particle is at its maximum displacement.

Note that y and a are both zero and v has a maximum magnitude when $t = 0, \pi, 2\pi, 3\pi$, etc., and the acceleration is maximum when $t = \pi/2, 3\pi/2, 5\pi/2$, etc. Summarizing,

$$s_{max} = \pm A \tag{19.29}$$

$$v_{max} = \pm 2\pi f A = \pm \frac{2\pi}{T} A \tag{19.30}$$

$$a_{max} = \mp 4\pi^2 f^2 A = \mp \frac{4\pi^2}{T^2} A \tag{19.31}$$

Illustrative Problem 19.10 The displacement of a particle which oscillates with simple harmonic motion is given by the equation $y = 8.0 \sin 0.50\pi t$, where y is measured in centimeters and t in seconds. Determine (*a*) the amplitude, (*b*) the frequency, and (*c*) the period of the motion. What are (*d*) the displacement, (*e*) the velocity, and (*f*) the acceleration of the particle when $t = 0.75$ s?

Solution Given $y = 8.0 \sin 0.50\pi t$

(***a***) Since $y = A \sin 2\pi ft$, A must $= 8.0$ cm *answer*

(***b***) $2\pi f = 0.50\pi$. Then $f = 0.25\ \text{s}^{-1}$ *answer*

(***c***) $T = \dfrac{1}{f} = \dfrac{1}{0.25\ \text{s}^{-1}} = 4$ s *answer*

(***d***)
$$\begin{aligned} y &= 8.0 \sin 0.50\pi(0.75) \\ &= 8.0 \sin \left[0.375\pi(\text{rad}) \times \frac{360°}{2\pi\ \text{rad}}\right] \\ &= 8 \sin 67.5° \\ &= 7.4\ \text{cm} \end{aligned}$$
answer

(***e***)
$$\begin{aligned} y &= 2\pi f A \cos(2\pi ft) \\ &= 0.50\pi \times 8.0 \cos 67.5° \\ &= 4.0\pi(0.383) \\ &= 4.8\ \text{cm/s} \end{aligned}$$
answer

(***f***)
$$\begin{aligned} a &= -4\pi^2 f^2 A \sin(2\pi ft) \\ &= -4\pi^2(0.25)^2(8)(0.924) \\ &= -18.2\ \text{cm/s}^2 \end{aligned}$$
answer

Illustrative Problem 19.11 The displacement of a particle undergoing SHM is given by the equation $s = 10.0 \sin 1.57t$, where s is measured in centimeters and t in seconds. Determine the magnitudes of (*a*) the amplitude, (*b*) the maximum velocity, and (*c*) the maximum acceleration of the particle.

Solution Since $s = 10.0 \sin 1.57t$ in this instance and since $s = A \sin 2\pi ft$ for all SHM, we know that $A = 10.0$ cm and $2\pi f = 1.57\ \text{s}^{-1}$. Hence $f = (1.57/2\pi)\ \text{s}^{-1} = 0.250\ \text{s}^{-1}$. Then

(***a***) $s_{max} = 10.0$ cm *answer*

(***b***)
$$\begin{aligned} v_{max} &= 2\pi f A \\ &= 2\pi(0.250\ \text{s}^{-1})(10.0\ \text{cm}) \\ &= 15.7\ \text{cm/s} \end{aligned}$$
answer

(***c***)
$$\begin{aligned} a_{max} &= 4\pi^2 f^2 A \\ &= 4\pi^2(0.250\ \text{s}^{-1})^2(10.0) \\ &= 24.7\ \text{cm/s}^2 \end{aligned}$$
answer

The graphs of Eqs. (19.23), (19.24), and (19.27) are shown in Fig. 19.11. Thus, we see that the velocity v reaches its maximum and minimum 90° ($\pi/2$ rad) ahead of the displacement y. We say that the velocity *leads* the displacement by 90°. Similarly,

Fig. 19.11 Graphical representation of displacement, velocity, and acceleration in SHM, as a function of time. (*a*) Plot of $A \sin(2\pi ft)$, (*b*) plot of $v = 2\pi fA \cos(2\pi ft)$, (*c*) plot of $a = -4\pi^2 f^2 A \sin(2\pi ft)$.

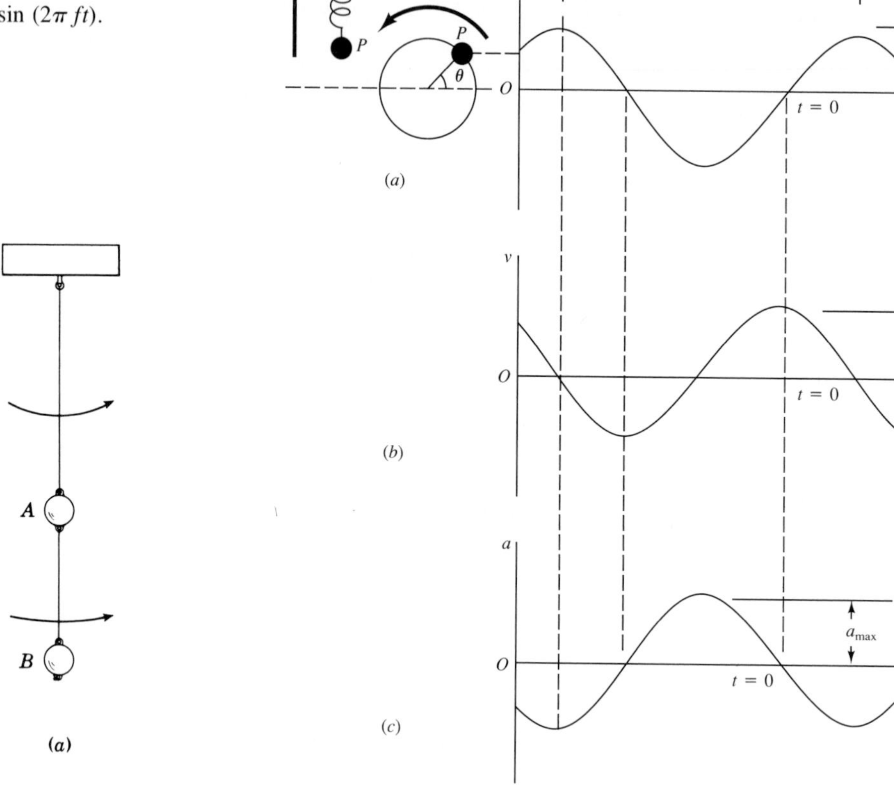

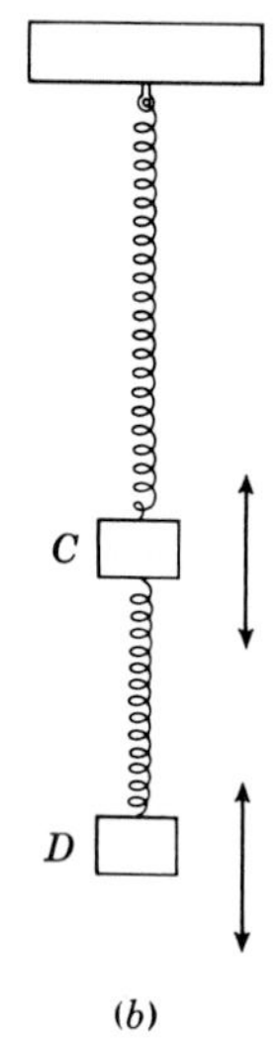

Fig. 19.12 Sources of complex vibrations.

note that the acceleration leads the velocity by 90° and the displacement by 180°. The displacement, velocity, and acceleration can each be represented by sine (or cosine) curves of the same frequency but differing in *phase*. Two variables that are represented by sine waves are said to be *in phase* if they reach maxima, zeros, and minima at the same instant.

19.12 ■ Complex Periodic Vibrations Can Be Resolved into Simple Harmonic Motions

If we hang a pendulum *B* from the bob *A* of a first swinging pendulum (Fig. 19.12*a*), the bob *B* of the second pendulum will undergo a complex vibration which will be the resultant of the simpler motions of the individual pendulums. In like manner, we might support a mass *C* from a spring and then from mass *C* hang a second weighted spring *D* (Fig. 19.12*b*). When both masses are set in motion, the motion of the lower mass will be complex. Suppose the period of the spring with mass *C* is 2 s and the period of the spring with mass *D* is 1 s. The dashed black line of Fig. 19.13 is the graph of the motion of *C* with only the top spring-mass combination. The solid black line represents the graph of the motion of *D* with only the bottom spring-mass combination. The dashed-dotted line is the graph of the complicated motion of *D* when the two are set into motion.

If can be shown that any periodic complex motion is the vector sum of a number of simple harmonic motions. We shall make use of this fact in our study of sound waves and electric currents in later chapters.

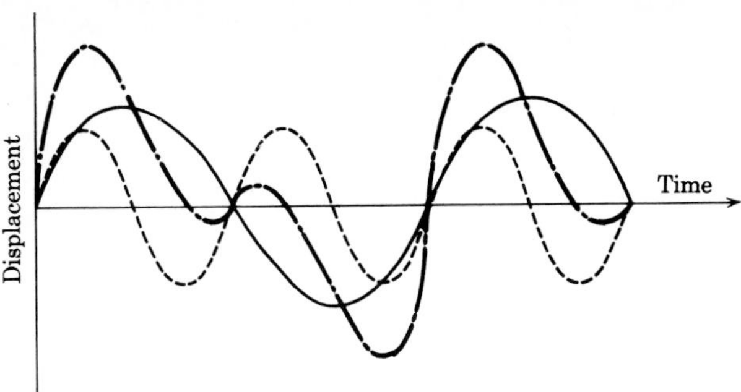

Fig. 19.13 Complex periodic motions are combinations of several simple harmonic motions.

19.13 ■ Damped Harmonic Oscillations

The systems we have dealt with so far in this chapter were assumed to operate under ideal conditions in which no friction was present. In such systems a particle oscillates indefinitely with no change in amplitude and frequency. However, all real vibrating systems (oscillators) are subject to dissipative forces, such as friction, which retard the motion of the system. Consequently, the mechanical energy of the system will decrease in time.

The potential energy of a stretched spring is not completely converted to kinetic energy when $s = 0$, because some of it goes into work done against the frictional forces that are acting. Thus, at the end of each vibration the loss in energy leads to a shorter elongation of the spring, until finally the amplitude of vibration drops to zero. This effect is called *damping*. Figure 19.14 shows a typical graph of displacement as a function of time in a *damped harmonic motion*. The dotted line in the figure is an exponential curve which shows that the amplitude decays exponentially with time.

Increasing the rate of decay can be accomplished by increasing the dissipative force (friction). One example of a damping oscillator is shown in Fig. 19.15. The damping of the system is accomplished by a *dashpot* made by a perforated piston that moves inside a cylinder filled with a viscous liquid. When the dashpot is disconnected, the spring and mass m execute simple harmonic motion (with only very slight damping) at its *natural frequency*. But when the dashpot is connected to the system, the resistive dissipative force of the liquid causes the SHM to be damped more rapidly. The degree of damping can be controlled by varying the viscosity of the liquid in the dashpot.

If the liquid is so viscous that the resistive force is greater than the restoring force of the spring, the system will be *overdamped*. The displaced system will not oscillate but will *slowly* return to its equilibrium position. The loss in the mechanical energy of the system dissipates into heat energy in the resistive liquid.

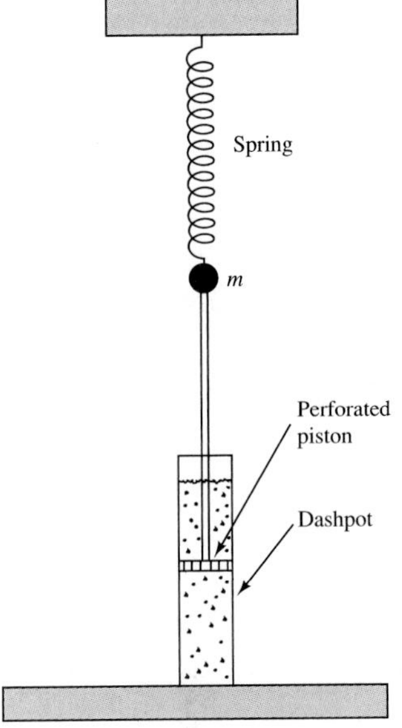

Fig. 19.15 A damped oscillator. The degree of damping is controlled by alternating the viscosity of the liquid in the dashpot.

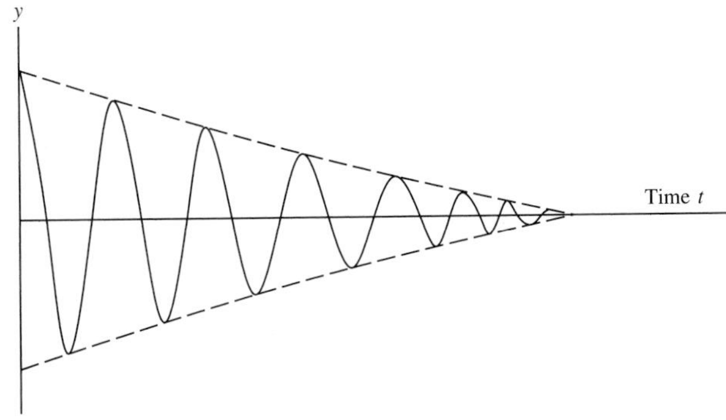

Fig. 19.14 A damped oscillation.

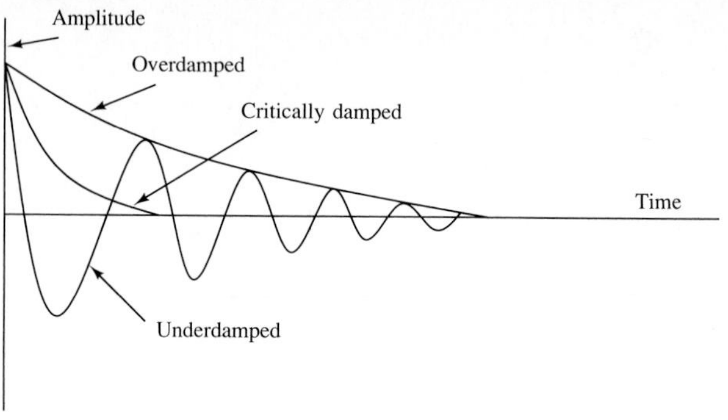

Fig. 19.16 Types of motion that occur when an oscillating system is underdamped, critically damped, and overdamped.

There is a degree of damping which will cause an oscillating mass to reach its equilibrium position in a mimimum time and come to rest in that position. Such a system is *critically damped*. Figure 19.16 shows graphs of the displacement versus time curves which represent the three types of damping just defined.

The use of damping devices is common in industry. Autos are fitted with shock absorbers to decrease the number of "bounce" cycles when the car hits a rut or obstacle on the road. Electric instruments, such as voltmeters and ammeters, are equipped with devices which cause critical damping in the moving vanes. If electric meters are underdamped, the measuring needles on the instruments will oscillate back and forth excessively before arriving at a stable reading. On the other hand, if the meters are overdamped they will take too long to reach equilibrium, negating the ability to record rapid changes in the quantity to be measured. Weighing scales are slightly underdamped, but are damped enough to give relatively quick readings.

19.14 ■ Forced Vibrations. Resonance.

The energy of a damped oscillator decreases with time because of dissipative frictional forces. This energy loss can not only be restored, but the energy of the system can be increased by applying an external force that does work on the system. For example, suppose a heavy ball is mounted on a flat steel spring and clamped in a vise, as shown in Fig. 19.17. Assume a *natural frequency* of vibration of the system to be 2 vib/s (Hz). Next imagine some means of applying a to-and-fro force with a frequency of 5 vib/s to the bottom of the spring. As a consequence, the system will be forced to vibrate at 5 vib/s. During part of each cycle the system will oppose the

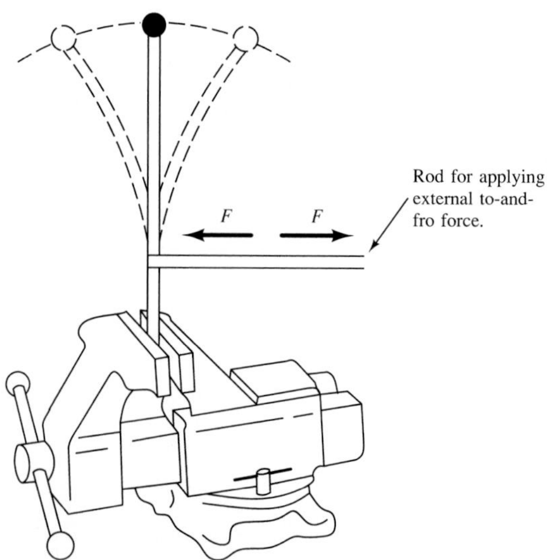

Fig. 19.17 An example of forced vibrations.

driving force. Hence the amplitude of vibration will be quite small. The system will be undergoing a *forced vibration*.

Now suppose the frequency of the to-and-fro *forced motion* is gradually slowed until it is equal to the natural frequency of the system, 2 vib/s. When this happens, the amplitude of vibration becomes relatively large and the system is said to be in *resonance* with the driving force.

If the frequency of the driving force is reduced still more, the amplitude of vibration of the spring-ball system will diminish until it is again quite small.

The principle of resonance can be demonstrated by a simple experiment illustrated in Fig. 19.18. Hang three pendulums on the same light rod. Let the lengths of pendulums B and C be fixed and that of A be variable. Next set pendulum A to swinging. If no two pendulum lengths are equal, pendulums B and C will jiggle a little, but they will not swing appreciably.

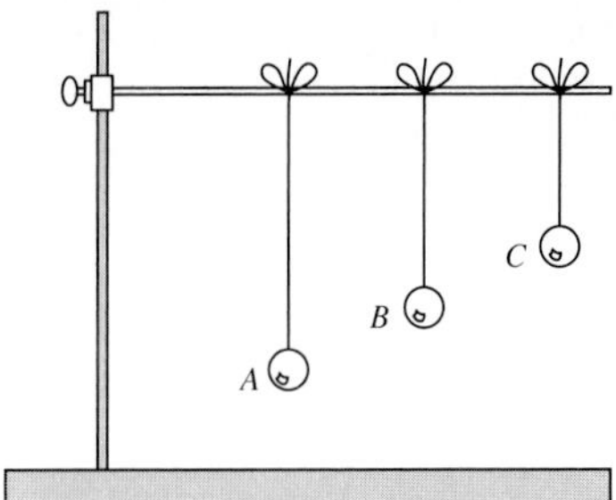

Fig. 19.18 Illustration of resonance.

Now if the length of the swinging pendulum A is shortened to equal that of pendulum B, the motion of pendulum A will be transferred to B in greater measure and B will start to swing with an appreciable amplitude while C continues to jiggle slightly. As the length of pendulum A is shortened more until it equals that of pendulum C, C will start to swing and the amplitude of B will diminish until it is negligible.

One vibrating object with the same natural frequency as another may cause the second object to vibrate with increased amplitude.

The ability of a small periodic driving external force of proper frequency to build up a very large amplitude of motion in another system of the same natural frequency is evident when the child in the swing "pumps" at the right time to increase the amplitude of motion. The rattle of a jalopy, the vibrations set up in dishes, a swinging chandelier, a tuned radio circuit, and the vibrations of the sounding box of a violin are all examples of a resonant response to external vibrations.

Resonance plays a very important role in mechanical and electrical systems as well as in sound. Soldiers marching across a bridge will break step to avoid a regular rhythm of marching feet that might set the bridge structure into violent and dangerous vibrations if the rhythm corresponded to the natural frequency of the structure. Vibrations can be set up mechanically with a small force if the frequency of the applied force is set to equal the natural frequency of the body. We are all familiar with the appearance of annoying rattles or sounds in an automobile at certain running speeds and their disappearance with a slight speed change.

In heavy machinery and structures it is important that the distribution of mass be such that the natural frequencies of the various parts will not be the same as the possible frequencies of its moving parts. Otherwise the vibrations of the smaller moving parts create a condition of resonance with heavier components and may build up a very large amplitude of motion, with serious consequences.

WAVE MOTION

19.15 ■ Types of Waves

Although mechanical waves may travel considerable distances, the particles of the medium themselves oscillate only over a limited region of space. There are many kinds of waves. They generally can be classified as one of two types, transverse and longitudinal.

Transverse Waves If an ordinary rope, fastened at one end, is stretched (under tension) across a room and given a sudden up-and-down motion at the free end, a portion of a wave, called a *pulse*, will travel along the rope (Fig. 19.19) with a fixed velocity v. The wave velocity depends upon the tension T and the mass m per unit length L. As an equation

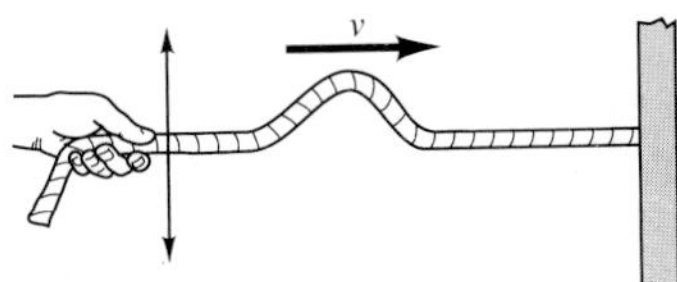

Fig. 19.19 A single pulse is traveling along a taut rope.

$$v = \sqrt{\frac{T}{m/L}} \qquad (19.32)$$

(speed of a transverse wave in a rope or string)

where T is in newtons, m in kilograms, and L in meters.

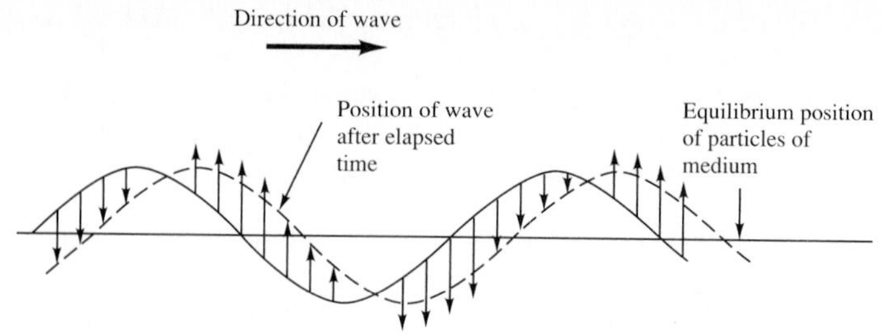

Fig. 19.20 A transverse wave. The arrow indicates the direction in which the particles move in the medium as the wave passes.

If the up-and-down motion of the hand is continuous, a series of waves travel along the rope, in which each portion of the rope will move up and down perpendicular (*transverse*) to the direction of the wave motion (see Fig. 19.20). Such waves are called *transverse waves*

> **In a transverse wave the direction of oscillation of the particles of the medium is perpendicular to the direction of motion of the wave.**

Illustrative Problem 19.12 One end of a rope is fastened to a fixed support. The other end passes over a pulley at a distance of 12 m from the fixed end. A load of 3.0 kg is attached to the free end of the rope. The mass of the rope between the fixed end and the pulley is 800 g. What would be the speed of a transverse wave set up in the rope?

Solution The tension T in the rope is caused by the 3.0-kg mass.

$$T = 3 \text{ kg} \times \frac{9.81 \text{ N}}{1 \text{ kg-force}} = 29.43 \text{ N}$$

$L = 12$ m; $m = 0.800$ kg
Hence,

$$v = \sqrt{\frac{T}{m/L}}$$

$$= \sqrt{\frac{29.43 \text{ N}}{0.800 \text{ kg/12 m}}} \qquad (1 \text{ N} = 1 \text{ kg} \cdot \text{m/s}^2)$$

$$= 21 \text{ m/s}$$ *answer*

Light waves, radio and television waves, heat waves, and X-rays are examples of transverse waves.

Definition of *compression* and *rarefaction* in wave motion.

Longitudinal Waves Not all waves are transverse. Consider a spiral spring under tension with its ends firmly attached to supports (Fig. 19.21). When a few coils at the left end are pinched closer together, the distortion is called a *compression*. When the compressed coils are suddenly released, they will attempt to spread out to their equilibrium positions. In so doing, they will compress the coils immediately adjacent to them. In this way the compression will move (to the right, in Fig. 19.21) along the spring.

If the coils at the left end of the spring were initially stretched apart (Fig. 19.22), the distortion is called a *rarefaction*. When released, the rarefaction will travel to the right along the spring, just as the compression did. In each case kinetic energy is transferred from particle to particle along the medium without motion of the medium as a whole. Such a wave is called a *longitudinal wave*.

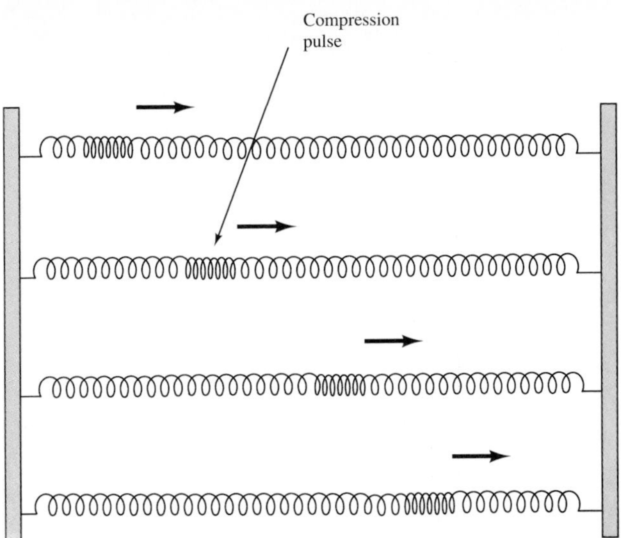

Fig. 19.21 A compression pulse traveling along an elastic medium.

In a longitudinal wave the direction of oscillation of particles of the medium is parallel to the direction of wave propagation.

If the spring of Fig. 19.21 were alternately compressed and expanded with periodic regularity, a continuous longitudinal wave would develop in the spring.

An important example of a longitudinal wave is a *sound wave*. A sound wave in air is a longitudinal wave that consists of alternating regions of high and low pressures (compressions and rarefactions). The loudspeaker of a radio, for example, alternately compresses and rarefies the air for each note it produces. Longitudinal waves in air are depicted schematically in Fig. 19.23.

Sound waves in gases and liquids are longitudinal waves.

It is common practice, however, to represent a longitudinal wave *graphically* by a curve similar to that for a transverse wave (Fig. 19.24). Wavelength, frequency, and wave velocity each have meaning for a longitudinal wave as well as for transverse waves. The wavelength is the distance between successive compressions (or rarefac-

The characteristics of both transverse and longitudinal waves can be shown by graphs.

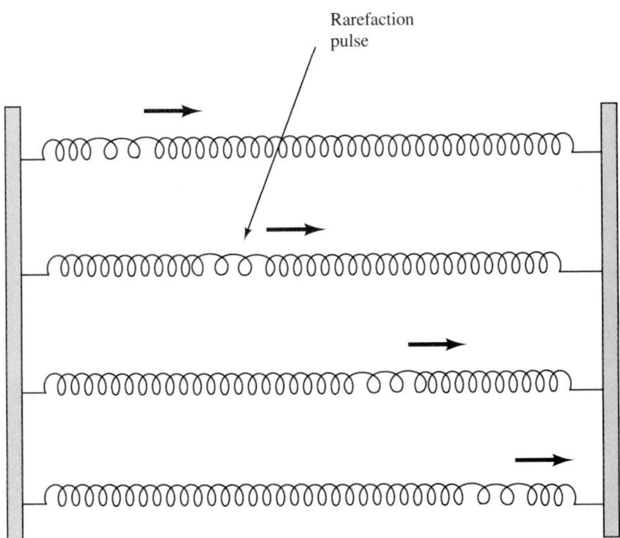

Fig. 19.22 A rarefaction pulse traveling along an elastic medium.

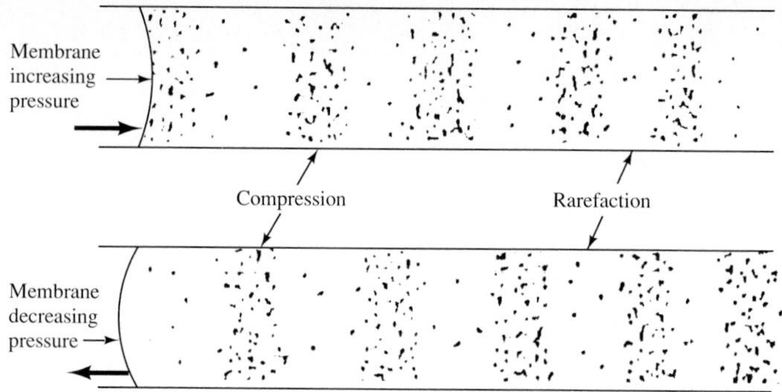

Fig. 19.23 The diaphragm of a loudspeaker produces longitudinal sound waves that are made up of compressions and rarefactions in air.

tions) and frequency is the number of compressions (or rarefactions) that pass a given point per second. The wave velocity v is the velocity with which each compression (or rarefaction) appears to move.

The equation for the velocity of a longitudinal wave has a form similar to that for a transverse wave.

Longitudinal Waves in Fluids Fluids transmit longitudinal waves. Their velocity in a fluid is given by

Speed of longitudinal waves in a fluid.

$$v = \sqrt{\frac{B}{\rho}} \tag{19.33}$$

(speed of longitudinal wave in a fluid)

where B is the bulk modulus in pascals of the fluid and ρ is the mass density in kg/m^3.

Longitudinal Waves in Solids When a longitudinal wave is propagated in a solid rod, it can be shown that the velocity of the longitudinal wave in the rod is given by

Speed of a longitudinal wave in a solid.

$$v = \sqrt{\frac{Y}{\rho}} \tag{19.34}$$

(speed of a longitudinal wave in a solid)

where Y is Young's modulus of the solid material in N/m^2 and ρ is its mass density in kg/m^3.

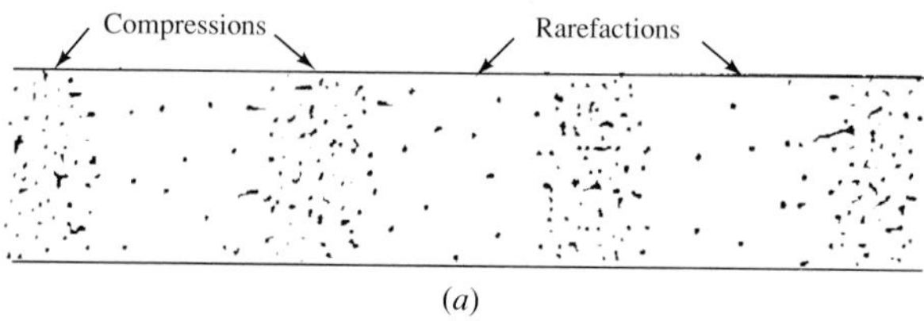

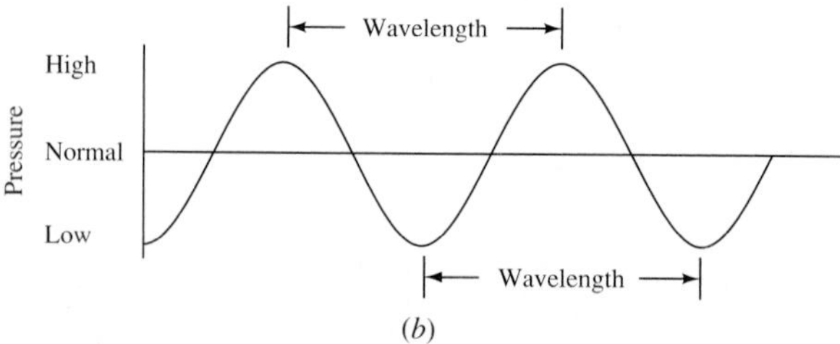

Fig. 19.24 (*a*) A longitudinal wave. (*b*) A conventional graphical representation of a longitudinal wave, as though it were a transverse wave.

Illustrative Problem 19.13 Determine the speed of sound waves in water.

Solution The bulk modulus B for water is 2.1×10^9 Pa (see Table 11.5). The mass density ρ of water is 1000 kg/m^3 (see Table 11.1). Using Eq. (19.33), we find that

$$v = \sqrt{\frac{B}{\rho}}$$

$$= \sqrt{\frac{2.1 \times 10^9 \text{ N/m}^2 \times \dfrac{1 \text{ kg} \cdot \text{m/s}^2}{1 \text{ N}}}{1000 \text{ kg/m}^3}}$$

$$= 1450 \text{ m/s} \qquad \textit{answer}$$

Illustrative Problem 19.14 How long does it take a sound wave to travel the length of a steel pipe that is 2.45 km long?

Solution Referring to Table 11.1 and Table 11.2, we find that $\rho = 7{,}830 \times 10^3$ kg/m^3 and $Y = 20.0 \times 10^{10}$ N/m^2 for steel. Then, from Eq. (19.34),

$$v = \sqrt{\frac{Y}{\rho}}$$

$$= \sqrt{\frac{20.0 \times 10^{10} \text{ N/m}^2}{7{,}830 \text{ kg/m}^3}}$$

$$= 5050 \text{ m/s}$$

and

$$t = \frac{s}{v}$$

$$= \frac{2450 \text{ m}}{5050 \text{ m/s}}$$

$$= 0.485 \text{ s} \qquad \textit{answer}$$

Water Waves When one watches the waves in a body of water from a distance, it appears that the water itself is traveling in the direction of the wave; but on observing an object floating on the water surface it becomes apparent that the water does not actually travel with the wave. The high points of waves are called *crests;* the low points are called *troughs*. The floating object can be seen to follow an elliptical or circular path, returning to its original position with each complete vibration. In like manner, the water molecules move in orbit about their original positions when a wave passes by (see Fig. 19.25).

A water wave moves horizontally across the body of water, but the particles of the medium themselves oscillate around more-or-less fixed positions.

It is apparent that what moves forward as successive waves pass over the water is not the water itself, but the energy which makes the masses of molecules of water rise

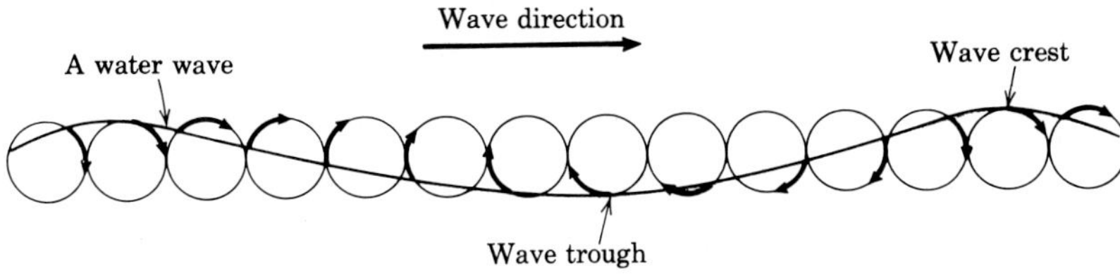

Fig. 19.25 Water molecules move in circular or elliptical orbits about their original positions as the wave passes by.

and fall. This energy is supplied by the agent that creates the disturbance in the water which produces the waves.

19.16 ■ Wavelength, Frequency and Speed of Waves

The distance (see Fig. 19.26) between two successive crests (or troughs) is called the *wavelength* λ (the Greek letter lambda). The distance which a crest (or trough) of a wave travels in 1 s is called its *speed*. The number of crests passing a given point in 1 s is called the *frequency*. The frequency of a wave is equal to the number of vibrations per second of the wave. Hence the total distance the wave will advance in 1 s (*speed*) *is equal to the product of the frequency times the wavelength*. As an equation

Wavelength and frequency for every type wave are related by the equation $v = f\lambda$.

$$v = f\lambda$$

***Speed* = frequency × wavelength** **(19.32)**

This is the basic equation of wave motion. It holds for any kind of wave motion, whether sound, light, heat, radio, water waves, or waves in solid materials.

Illustrative Problem 19.15 Waves roll up on a beach at the rate of five waves every 10 s. If the distance between the crests of the waves is 7.50 m, with what speed are the waves moving?

Solution A rate of five waves every 10 s is equivalent to a frequency of $\frac{1}{2}\ \mathrm{s}^{-1}$. Substituting the known values for f and λ in Eq. (19.35), we obtain

$$\begin{aligned} v &= f\lambda \\ &= \tfrac{1}{2}\ \mathrm{s}^{-1} \times 7.5\ \mathrm{m} \\ &= 3.75\ \mathrm{m/s} \qquad \textit{answer} \end{aligned}$$

Illustrative Problem 19.16 A sound wave which travels 1100 ft/s has a frequency of 256 Hz. What is the length of the wave?

Solution Solving Eq. (19.35) for λ and substituting known values, we get

$$\text{Wavelength} = \frac{\text{speed}}{\text{frequency}}$$

$$\begin{aligned} \lambda &= \frac{v}{f} \\ &= \frac{1100\ \mathrm{ft/s}}{256\ \mathrm{vib/s}} \\ &= 4.30\ \mathrm{ft} \qquad \textit{answer} \end{aligned}$$

This is the wavelength of middle C of the musical scale when $v = 1100$ ft/s.

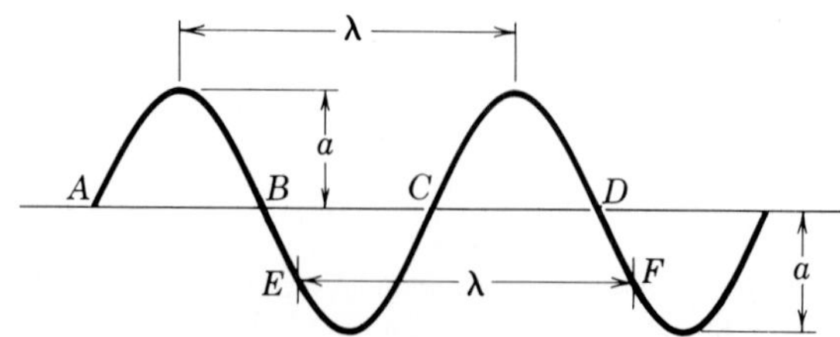

Fig. 19.26 Amplitude and length of wave.

19.17 ■ Reflection of Waves

When a traveling pulse reaches an obstacle or comes to the end of its transmitting medium, it may be completely reflected, or partially reflected and partially transmitted, at the boundary between the two media. The nature of the reflected pulse is determined by the type of boundary that exists. A pulse traveling along a rope is reflected as illustrated in Figs. 19.27 and 19.28.

When a pulse reaches the fixed end of a rope as shown in Fig. 19.27*b* it will exert an upward force on the support. Since the support is rigid, no part of the pulse energy is actually transferred to the support. The support will exert an equal but opposite force down on the rope (Newton's third law). This downward force on the rope will produce a reflected inverted pulse (Fig. 19.27*c*) that will be 180° out of phase (see Sec. 19.11) with the oncoming pulse.

In Fig. 19.28 the tension in the rope is maintained by connecting the end to a ring of negligible mass that is free to move vertically (assume no friction). As the pulse reaches the end of the rope (Fig. 19.28*b*), it exerts a force on the free end, causing the ring to accelerate upward. The ring will move upward to a height greater than that of the incoming pulse. The downward component of the increased tension on the rope will bring the ring back to its original position. The reflected pulse will not be inverted (no change in phase).

When the boundary of a traveling pulse is neither completely rigid nor completely free, part of the incident pulse is transmitted and part is reflected. Suppose, for example, that a light cord is attached to a heavier cord (Fig. 19.29). When the pulse reaches the boundary between the two sections, part of the pulse is reflected and part is transmitted. If the second section has a greater linear density than the first (*a*), the reflected pulse will undergo a phase change of 180° (the pulse will be inverted). If the second section has a lower linear density than the first (*b*), the reflected pulse will not undergo a phase change. In each case the reflected and transmitted pulses will have smaller amplitudes than that of the incident pulse.

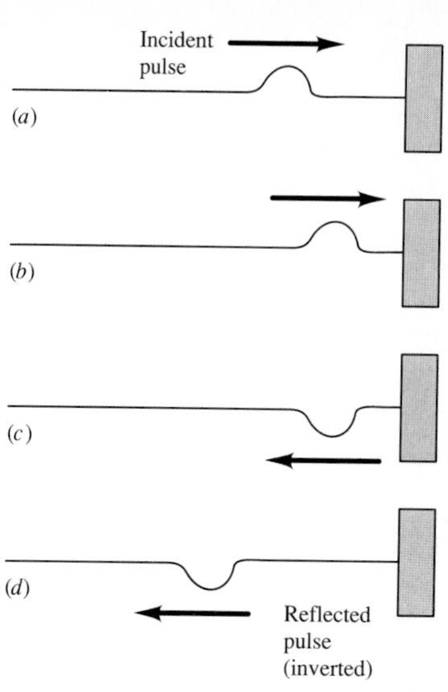

Fig. 19.27 Reflection of a wave pulse on a rope when the end is fixed. The reflected pulse is inverted, but the shape remains the same.

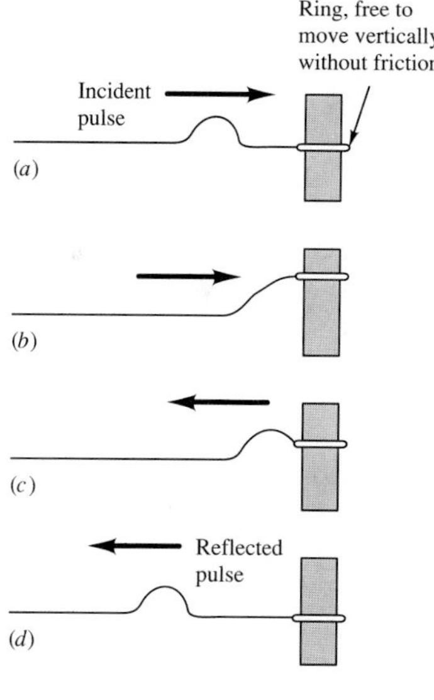

Fig. 19.28 Reflection of a wave pulse on a rope when the end of the rope is free to move in the plane of the wave pulse. The reflected pulse is not inverted, and the shape remains the same.

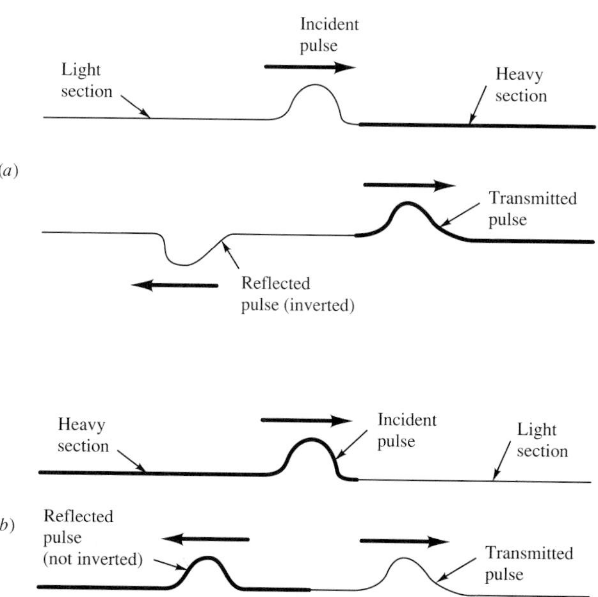

Fig. 19.29 (*a*) A pulse traveling from a lightweight cord to a heavier cord. Part of the incident pulse is reflected and inverted (in the light cord) and part is transmitted to the heavier cord. (*b*) A pulse traveling from a heavy cord to a lighter one. Part of the incident pulse is reflected and part is transmitted. The reflected pulse is not inverted.

19.18 ■ The Superposition Principle

Two or more waves passing through the same point at the same time add vectorially.

Suppose two pulses approach each other on a rope, one traveling to the right with an amplitude of 5 cm and the other traveling to the left with an amplitude of 3 cm. The series of sketches shown in Fig. 19.30 shows what would happen when the two pulses meet and cross each other. The resultant displacement of the medium at each instant as they cross is the *sum* of the displacements of the independent pulses at each instant. Thus, the maximum displacement of the rope (where the pulses pass each other) is 8 cm.

Each pulse proceeds along the rope making its own contribution to the rope's displacement no matter what any other pulse is doing. This property of pulses (and waves) is called the *superposition principle*. The superposition principle applies no matter how many separate pulses (or waves) are present in the medium. It is valid for all kinds of waves, provided the amplitudes are not too great.

Superposition Principle

When two or more pulses (or waves) travel past a point in a medium at the same time, the medium will have a displacement that is the algebraic sum of the displacements of the individual pulses (or waves).

Figure 19.31 shows what happens when two pulses of the same amplitude and 180° out of phase cross each other going in opposite directions. If each pulse has the same shape and size, there will clearly be a moment (when they pass each other) when the medium has zero displacement (see Fig. 19.31*c*). At the instant of complete cancellation, the total energy of both pulses is manifested in the kinetic energy of the medium segment where the cancellation occurs.

Figure 19.32 shows what happens when two periodic transverse waves of different amplitude and frequency travel in the same direction along a stretched cord. The

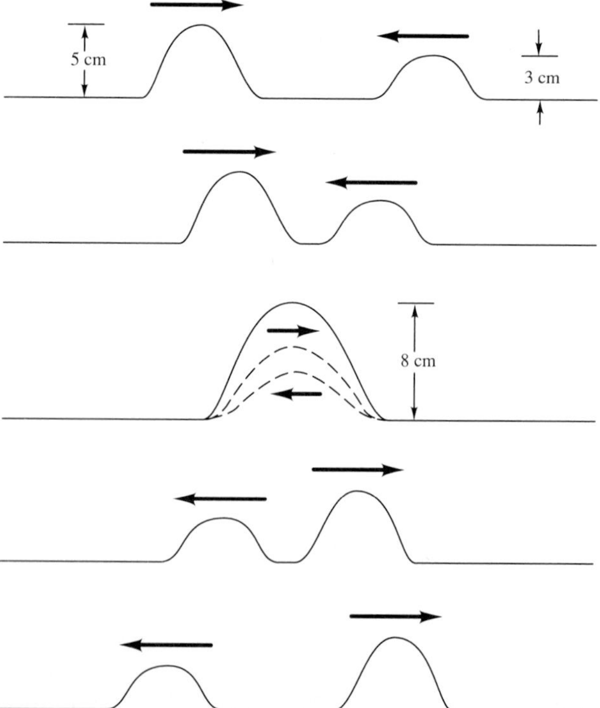

Fig. 19.30 The superposition of two pulses on a rope. The resultant amplitudes are additive. The movement of each pulse is independent of the other pulses present.

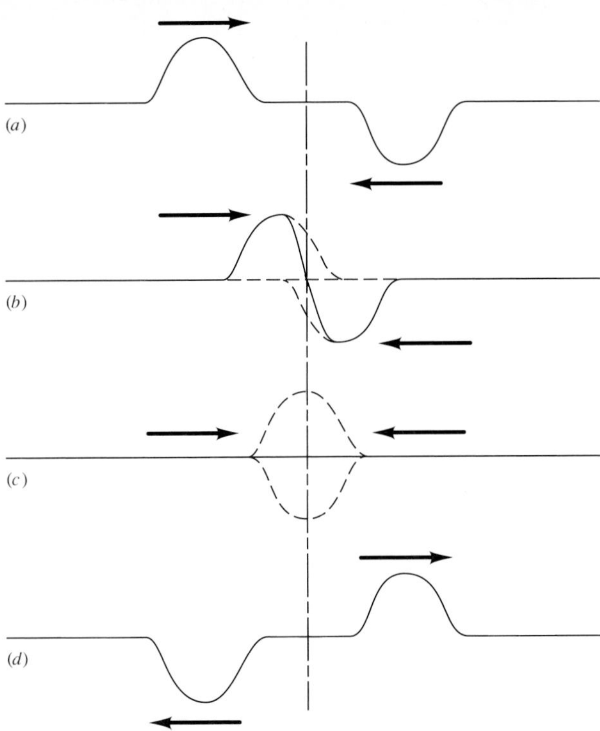

Fig. 19.31 The superposition of two equal and opposite pulses. (*a*) Before they meet; (*b*) before complete cancellation; (*c*) at complete cancellation; (*d*) after they meet and pass.

speed of the two wave trains will be the same since they are traveling in the same medium, the cord. The component waves are shown with dashed lines; the resultant wave is the full line. It can be seen that at each point along the cord the displacement of the resultant wave is equal to the algebraic sum of the displacements of the two waves at those points.

19.19 ■ Superposition and Interference

When two sets of waves travel in the same direction with the same frequency in such a way that crest meets crest and trough meets trough, the waves are said to be *in phase*. The displacement of the medium at all points, due to the two waves, is equal to the sum of the displacements of the individual waves (see Fig. 19.33*a*). These waves are said to undergo *constructive interference*. The waves interfere with each other constructively, that is, they build on each other. On the other hand, if the two waves travel in such a way that at the same points crest meets trough and trough

The effect of superposing two or more waves is called *interference*. Interference can be either *constructive* or *destructive*.

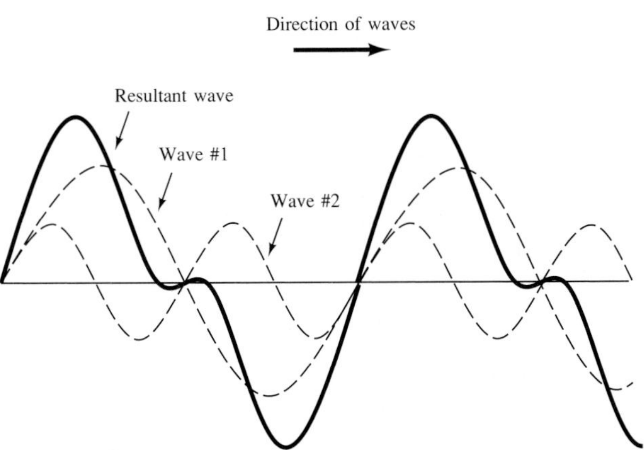

Fig. 19.32 The superposition of two periodic waves of different amplitudes and frequency traveling in the same direction.

Fig. 19.33 Interference of two periodic waves of the same frequency traveling in the same direction. The component waves are dashed lines and the resultant wave is drawn as a full line. (*a*) Constructive interference; (*b*) destructive interference; (*c*) complete destructive interference when the two waves have the same amplitude.

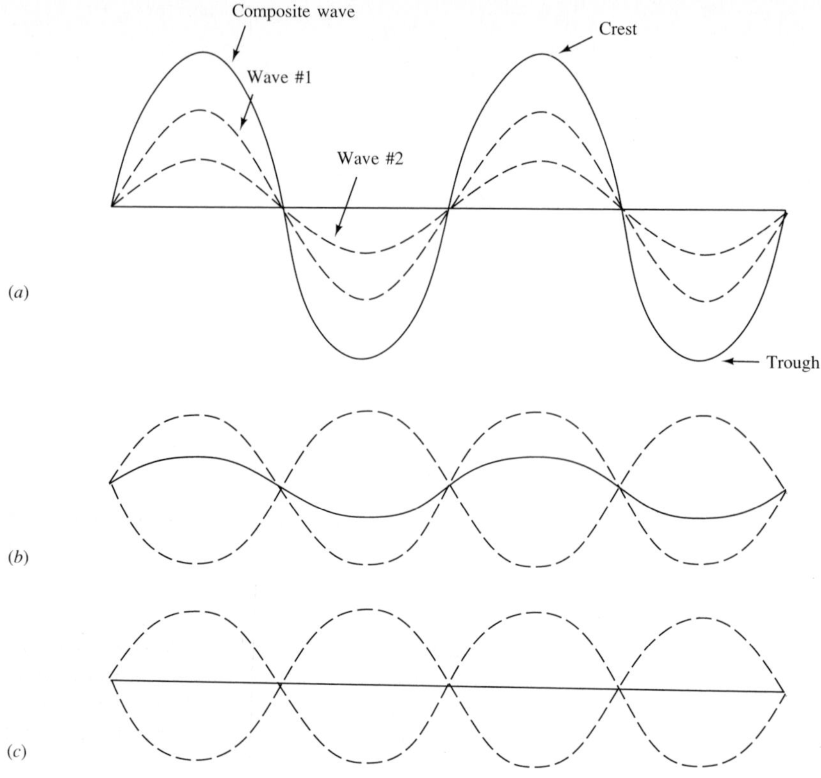

meets crest, the waves are 180° out of phase. In such a case the displacements of the two waves have opposite signs at each point and they interfere destructively (see Fig. 19.33*b*). If two such waves have the *same amplitude* and the same frequency, and are 180° out of phase, *complete destructive interference* will result (see Fig. 19.33*c*).

19.20 ■ Standing Waves

Standing waves are produced through interference between incident and reflected waves.

The interference between two sinusoidal waves of the same amplitude and frequency traveling in opposite directions through a medium creates patterns known as *standing waves* (also called *stationary waves*). In practice two such waves are produced by a wave and its reflection at a suitable barrier. Consider the two waves (dashed and dotted lines) shown in Fig. 19.34*a* and *b*. The incident wave (1) and its reflected wave (2) are in phase after reflection and travel with the same speed v. Figure 19.34*c* shows the two waves (1 and 2) plotted together as the same sine curve, since they are in phase and have identical amplitudes. (Note that both dashes and dots are used to plot this curve.) The displacement of the resultant wave (the solid black curve in *c*) at this instant is found by adding the displacements of the incident and the reflected waves. The two waves undergo constructive interference.

Figure 19.34*d* shows the waves when wave 1 has moved one-eighth wavelength ($\lambda/8$) or one-eighth period ($T/8$) to the right, and wave 2 has moved an eighth wavelength (or one-eighth period) to the left. The resultant wave has not moved at all. Note that the maximum, minimum, and zero points are exactly in the same locations as in Fig. 19.34*c*, but the resultant displacement at each point is reduced.

When the waves have moved another one-eighth wavelength ($t = T/4$), there will be complete destructive interference. At this instant the resultant is zero everywhere (see Fig. 19.34*e*).

Figures 19.34*f* and *g* shows the configurations as the two waves continue their motions to the right and left. Again, when $t = 3T/8$ and $T/2$ the resultant wave forms

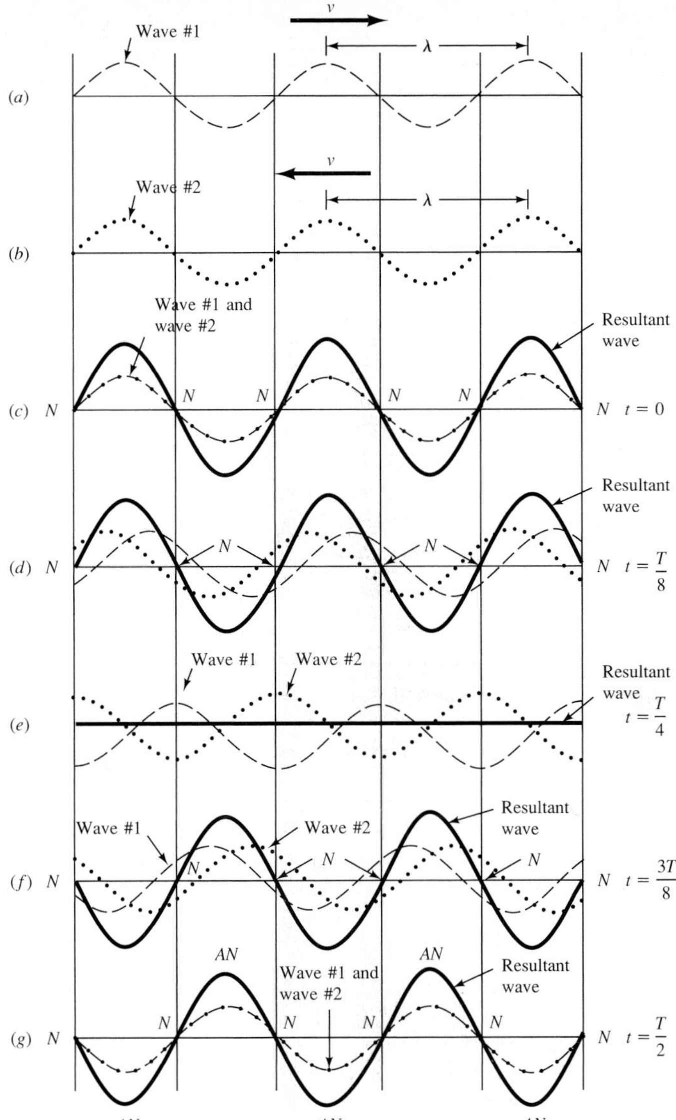

Fig. 19.34 Standing waves are produced by the superposition of two waves of the same amplitude and frequency. The stationary wave is shown as the solid line. The points marked N are *nodes*, and those marked AN are *antinodes*.

are stationary. In Fig. 19.34*g* each wave has moved a half wavelength ($\lambda/2$), and since $t = 0$, the waves are again coincident, but 180° out of phase from the waves of Fig. 19.34*c*.

Points of complete destructive interference are called *nodes*.

Points of constructive interference where the displacement is maximum are called *antinodes*.

The resultant waves in Fig. 19.34*c* to *f* have several interesting and important characteristics. The waves appear to stand still, with certain points (denoted by N) always remaining stationary and the other points (where destructive interference is complete) between them oscillating up and down. The stationary points are called *nodes*, and the points midway between nodes, oscillating with maximum amplitudes, are called *antinodes*. The distance between successive nodal points (and between antinodal points) is always one-half wavelength. No standing wave can be produced if the two waves have different wavelengths.

Standing waves are also set up in media that sustain compressional waves. Many objects vibrate in a manner that establishes standing waves. The strings of a piano, a guitar, or a violin and the air columns in other musical instruments establish various modes of standing waves.

QUESTIONS AND EXERCISES

1. In SHM, when the acceleration is greatest what is true about the (*a*) displacement and (*b*) the velocity?

2. In SHM, when the velocity is greatest, what is true about (*a*) displacement and (*b*) acceleration?

3. What is the numerical value for the product of the frequency and the period of periodic motion?

4. Which, if any, of the following will be doubled by doubling the amplitude of a swinging pendulum: (*a*) the period, (*b*) the frequency, (*c*) the maximum velocity?

5. A pendulum clock is found to run too fast so that it gains time. In trying to adjust the clock, should you make the length of the pendulum (*a*) longer or (*b*) shorter? Give the reasoning behind your answer.

6. How would the period of a pendulum be affected if the pendulum were moved from sea level to the top of a mountain?

7. When a particle undergoes SHM, its displacement at any time is always directly proportional to which of the following: (*a*) mass, (*b*) frequency, (*c*) acceleration, (*d*) velocity?

8. Which of the following properties of mechanical waves is independent of the others: (*a*) frequency, (*b*) wavelength, (*c*) amplitude, (*d*) velocity?

9. Which of the following will always change when two SHM waves are added: (*a*) phase, (*b*) wavelength, (*c*) amplitude, (*d*) frequency?

10. What is it that is transferred along the direction of wave motion?

11. Distinguish between transverse and longitudinal waves. Give examples of each.

12. Does the speed of propagation of a wave in a given medium depend on its amplitude? Give reasons for your answer.

13. What determines whether there will be a change of phase when a wave is reflected? Explain your answer.

14. Must a spring obey Hooke's law in order to oscillate with simple harmonic motion?

15. At what point in the motion of a simple pendulum is the tension in the string (or rod) (*a*) greatest, (*b*) the least? Give reasons for your answers.

16. What will happen to the frequency of a spring-mass system that oscillates with simple harmonic motion if the mass is doubled?

17. How will the period of a simple pendulum change if (*a*) the length is halved, (*b*) the length is tripled, (*c*) the amplitude is halved, and (*d*) the mass is doubled?

18. Give three common examples of damped oscillations.

19. The speed of sound waves in air depends on the air temperature, but the speed of light waves in a vacuum does not. Why is this true?

20. If a spring is cut in half, how is its spring constant affected? How would the frequency of SHM of the shorter spring differ from that of the entire spring, using the same tension in both cases?

PROBLEMS

Group One

1. Doubling both the length and the mass of the bob of a pendulum will multiply the period of the pendulum by what numerical factor?

2. Doubling both the mass and the spring constant of a vibrating system will multiply the frequency of vibration by what factor?

3. What is the period of vibration of a wave if its frequency is 400 Hz?

4. A train of waves moves along a cable with a speed of 12 m/s. The wavelength is equal to 3 cm. What are (*a*) the frequency and (*b*) the period of the source of the wave vibration?

5. Two pulses have maximum displacements of 3 cm and 4 cm in the same direction. What will be the maximum displacement as they pass each other?

6. The speed of radio waves in air is 3.00×10^8 m/s. What is the length of radio waves having a frequency of 1230 kHz?

7. What is the speed of a periodic wave disturbance if the wave has frequency of 7.2 Hz and a wavelength of 0.25 m?

8. Calculate the wavelength of water waves that have a frequency of 0.60 Hz and a speed of 3.0 m/s.

9. A waterwave machine in an amusement park produces waves that have a wavelength of 2 m and a frequency of 1.5 Hz. What is the speed of the waves?

10. The distance between two successive nodes in a vibrating cord is 16 cm. The frequency of the wave source is 4 Hz. Determine (*a*) the wavelength and (*b*) the velocity of the wave.

11. The speed of radio waves in a vacuum is 3.00×10^8m/s (the speed of light). Determine the wavelength for (*a*) an AM radio station that broadcasts on a frequency of 1440 kHz, (*b*) an FM radio station broadcasting at a frequency of 104 MHz.

12. A 2-m length of wire has a mass of 36.3 g. It is stretched with a tension of 44.5 N. What would be the speed of a transverse wave set up in the wire?

13. What is the period of a pendulum 1.00 m long on the moon, where the acceleration of gravity is 1.62 m/s^2?

14. What is the period of a pendulum that is 0.600 m long, located at sea level on earth?

15. Standing waves are set up in a wire by a source vibrating at 220 Hz. Six nodes are counted along a length of 45 cm (including one node at each end). Determine (*a*) the wavelength of the waves traveling in the wire, and (*b*) the speed of the waves.

16. What is the potential energy, in joules, of a spring if 0.3 N is being applied to compress the spring 3 cm?

Group Two

17. A particle undergoes SHM with a period of 0.75 s and an amplitude of 3.0 cm. What is its maximum speed?

18. A simple pendulum is 2.00 m long and oscillates in a place where g = 9.81 m/s^2. How many complete oscillations will the pendulum make in 10 min?

19. A 1-lb mass on the end of a spring describes SHM with a period of 2 s and amplitude of 3 in. Find (*a*) the maximum velocity of the mass and (*b*) the maximum acceleration of the mass. (HINT: Relate the SHM to a corresponding UCM.)

20. An object executes SHM with a period of 0.55 s and an amplitude of 2.5 cm. Calculate (*a*) its maximum velocity and (*b*) its maximum acceleration.

21. The maximum velocity of an object moving with SHM is 0.45 m/s. Its period of vibration is 0.8 s. Calculate (*a*) its amplitude and (*b*) is maximum acceleration.

22. An object moving with SHM has an amplitude of 5 cm and a period of 0.04 s. Find (*a*) its maximum velocity and (*b*) its maximum acceleration.

23. An object, at the end of a spring, vibrates with SHM. The shadow of the object is projected on a screen by a beam of parallel light. The length of the shadow's path from one end to the other is 6 cm. The time required for the shadow to travel this distance is 2 s. Find (*a*) the frequency of the SHM and (*b*) the velocity of the shadow at the center of its paths.

24. A bob weighing 2.5 lb is hung from the end of a spring whose Hooke's law constant is 0.65 lb/in. The bob is set to oscillating with an amplitude of 5 in. What is the speed in feet per second of the bob when it passes through its equilibrium position?

25. A body describing SHM has a maximum velocity of 4 m/s and a maximum acceleration of 16π m/s^2. Find (*a*) the period and (*b*) the amplitude of motion.

26. One end of a horizontal cord passes over a (frictionless) pulley and supports a mass of 2.5 kg. The other end is set oscillating at a constant frequency of 120 Hz. The linear mass density of the cord is 0.12 kg/m. Determine (*a*) the speed of the transverse wave in the cord, and (*b*) the length of the wave.

27. A spring is mounted as shown in Fig. 19.1. Using a spring balance and pulling horizontally it was determined that a force of 6 N caused a displacement of 3 cm. A 3-kg mass was then attached to the end, pulled aside a distance of 7.0 cm and released. Assume Hooke's law applies. Find (*a*) the period and frequency of vibration, (*b*) the maximum velocity of the mass, (*c*) the maximum acceleration of the mass, (*d*) the velocity of the mass when it is 3.5 cm from its equilibrium position and moving toward it, and (*c*) the time required for the mass to move from its initial position to a point 3.5 cm from its neutral position.

28. A body is moving horizontally without friction with SHM of amplitude 8.4 cm and frequency 5.0 Hz. Determine the acceleration and velocity when the body is 3.0 cm from its equilibrium position.

29. A force of 210 N is required to elongate a spring 12 cm long. When a mass is suspended from the spring and set into vertical motion, it will oscillate with a frequency of 2.5 Hz. What is the mass of the oscillating body? (Neglect the mass of the spring.)

30. A cord has a mass of 0.25 kg and a length of 6.0 m. Tension is maintained in the cord by suspending a 2.0-kg mass from on end as shown in Fig. 19.35. Determine

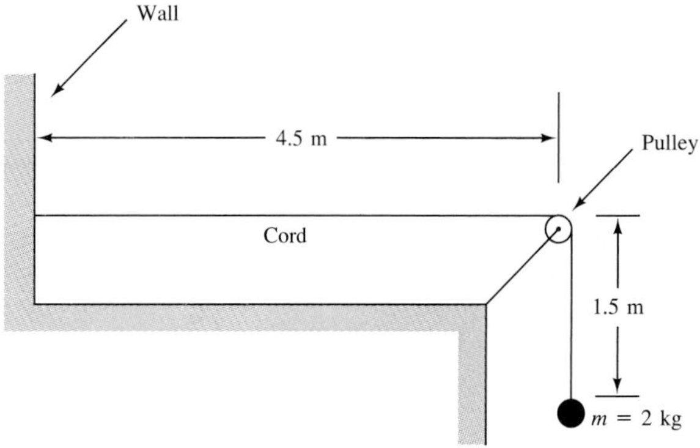

Fig. 19.35 Problem 30.

(*a*) the speed of a pulse in the horizontal portion of the cord, and (*b*) the time required for the pulse to travel the 4.5 m from the wall to the pulley.

31. A torsion pendulum (see Fig. 19.10) consists of a uniform disc that has a mass of 5.0 kg, an outer radius of 25 cm, and is supported by a wire with a torsion constant of 0.35 N · m/rad. Determine its period of oscillation.

32. A 60-g mass when suspended from a spring will stretch it a distance of 12 cm. If the 60-g mass is now replaced by a 100-g mass and then set to vibrating up and down, what will be the resultant frequency of vibration of the system?

33. A body whose mass is 0.30 kg is suspended from a spring whose elasticity constant is 0.15 N/m. It is set to vibrating with an amplitude of 2.0 cm. Find the speed of the mass when it passes through the midpoint of its vibration.

34. Determine the speed of the mass of Prob. 33 when it is 1.2 cm from the midpoint of its vibration.

35. A mass attached to a spring undergoes SHM with an amplitude of 6 cm. When the mass is 3 cm from the equilibrium position, what is the ratio of the kinetic energy to the potential energy of the system?

36. When a mass of 40 g is suspended from a spring, the spring stretches 1.0 cm. If an additional 160-g mass is suspended from the spring, what will be the natural frequency of the up-and-down motion of the system? (Assume Hooke's law holds.)

37. A 100-g mass is suspended from a spring. The mass is pulled down 2 cm by a force of 0.3 N and then released. Find (*a*) the frequency of the motion and (*b*) the maximum velocity.

38. A simple pendulum is to have a period of 1 s. How long, in inches, should it be if $g = 31.5$ ft/s^2 at that location?

39. A pendulum has a length of 4 m and executes 50 complete vibrations in 210 s. Find the acceleration of gravity at the location of the pendulum.

40. A pendulum has a length of 29.50 in. Find, accurate to four significant figures, its period at a location where $g = 32.16$ ft/s^2.

Group Three

41. Young's modulus for a stretched wire is 7.0×10^{10} N/m^2. What must the stress (F/A) be in the wire in order that the speed of longitudinal waves shall equal five times the speed of transverse waves along the wire?

42. A mass of 7.5 kg hangs from a spring and oscillates with a period of 1.5 s. How much will the spring shorten when the mass is removed from the spring?

43. A pure brass sphere is suspended from the ceiling by a wire whose torsion constant is 0.30 N · m/rad. The diameter of the sphere is 8.0 cm. Determine the period of rotational oscillation of the sphere.

44. A 12-lb weight suspended from a spring will temporarily stretch the spring a distance of 4 in. (the elastic limit is not exceeded). The 12-lb weight was replaced by a new weight and the system was set to vibrating up and down. The period of vibration was 1.5 s. How heavy is the new weight?

45. A clock pendulum has a period of 1.00 s where the acceleration of gravity is 980 cm/s^2. How many seconds will it lose in 24 h if it is taken to an altitude where $g =$ 976 cm/s^2?

46. A metal wire has the following properties: Coefficient of linear expansion, $\alpha = 15 \times 10^{-6}$ per C°; Young's modulus of elasticity, $Y = 20 \times 10^{10}$ N/m^2; mass density, $\rho = 7.83 \times 10^3$ kg/m^3. Each end of the (horizontal) wire is attached to rigid supports. If the tension is essentially zero at 20°C, what will be the speed of a transverse wave along the wire when the temperature is 5°C?

47. One end of a stretched cord is given a transverse motion with a frequency of 15 Hz. The cord is 75 m long, has a mass of 0.75 kg, and is stretched with a tension of 600 N. Determine (*a*) the wave speed and (*b*) the wave length. (*c*) What must be the frequency of motion to maintain the same wavelength when the tension is doubled?

48. A pendulum clock keeps perfect time at a place where $g = 9.807$ m/s^2. The pendulum is 1.000 m long. If the length of the pendulum were shortened by 1.93 cm, (*a*) how much time would it lose in a 24-h day? (*b*) If the clock were set for the correct time at noon of a given day, how long would it be before it would next indicate the correct time?

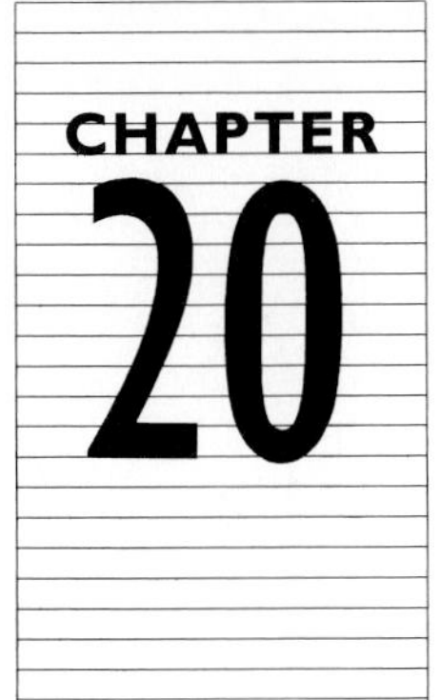

SOUND WAVES

Sound waves are probably the most important example of longitudinal waves. They can travel through any material medium (solid, liquid, or gas). The speed of sound will depend upon the properties of the medium. Sources of sound waves are present almost everywhere in our environment. Some sounds are quite pleasing and others may be extremely irritating. The production and control of sound has become a major problem in technology. It has been proven that excessive or prolonged sounds with very large intensities can produce drastic changes in a person's personality. Sounds that are painfully loud to the ear can rupture ear drums and, under continued exposure, can cause permanent hearing loss.

The science of *acoustics*, which is a term used to denote the production, transmission, reception, and control of sound, has developed into a major industry. In this chapter we will consider some of the characteristics and effects of sound.

The word "sound" usually means different things to the physiologist than it does to the physicist. For physiologists sound is an auditory sensation transmitted by the ear and interpreted by the brain. They use such terms as pitch, loudness, and tone quality. Physicists are more interested in the production and transmission of the disturbance which gives rise to the auditory sensation. They describe sound in terms of frequency, intensity (energy), and overtones. Each of these aspects of sound will be discussed in this chapter.

Sound is the name of the physiological sensation produced when a longitudinal wave train reaches the ear. It also denotes the physical cause of the wave train.

20.1 ■ Sources of Sound

Sound waves in air can be produced in many ways. Three sources of audible sound waves are (1) strings set into vibration by striking, plucking, or bowing—examples, respectively, are the piano, the guitar, and the violin; (2) vibrating membranes—such as those of the loudspeaker, the drum, the harmonica, and human vocal cords; and (3) vibrating air columns—such as those in the pipe organ, clarinet, trombone, and flute. Familiar examples of vibrating bodies emitting nonmusical sounds include such things as the clatter of dishes dropped on the floor, the warning whirr of a rattlesnake, the roar of busy traffic, and the clap of thunder.

The source of sound is a vibrating object.

Consider how a loudspeaker works. An electric voltage generator produces an oscillating current which operates a *transducer*, a device for changing electric signals to mechanical motion. The transducer causes the diaphragm of the speaker to oscillate with varying frequencies (see Fig. 20.1). The movement of the diaphragm causes a sound wave to move out from the loudspeaker. The alternate compressions and rarefactions, upon reaching the ear, will set up vibrations in the eardrum. The eardrum is sensitive to minute pressure changes produced by the mechanical vibrations of air in the sound wave. These pressure changes produce a chain of physiological events (in the normal ear) that carry information that is interpreted by the brain as sound.

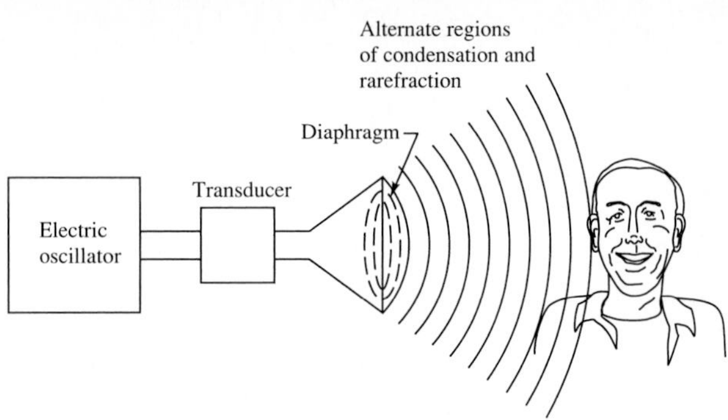

Fig. 20.1 The transducer (enclosed in the loudspeaker housing) converts electric oscillations of the generator to mechanical oscillations in the diaphragm of the speaker. The speaker sends out successive regions of condensation and rarefaction. The listener's ear and brain convert these to the sensation of sound.

20.2 ■ Transmission of Sound

Sound energy is mechanical energy transferred as a longitudinal wave by vibrating molecules.

In order for a sound to be heard, three conditions must be met: (1) there must be a source of wave energy consisting of a vibrating body; (2) there must be a material medium for transmitting the sound wave; and (3) there must be a device for receiving the sound, such as the ear. If any of these three conditions is not met, no sound will be heard. For example, when astronauts Neil Armstrong and Edward E. Aldrin journeyed to the moon, they were faced with the situation of needing to communicate with each other through the vacuum surrounding their space suits. The solution to the problem was accomplished by relying on radio waves for communication purposes. We will discuss in later chapters how radio waves—which are electromagnetic waves, needing no material medium—propagate in a vacuum.

Experimental proof that sound will not travel in a vacuum can be obtained by placing an alarm clock under a bell jar to which is attached a vacuum pump (Fig. 20.2). The alarm bell may be clearly heard when air is present in the jar. As the air is gradually evacuated from the jar, the sounds become fainter and fainter. If enough air is removed, the sound can hardly be heard. As air is once more permitted to enter the jar, the sound regains its original intensity.

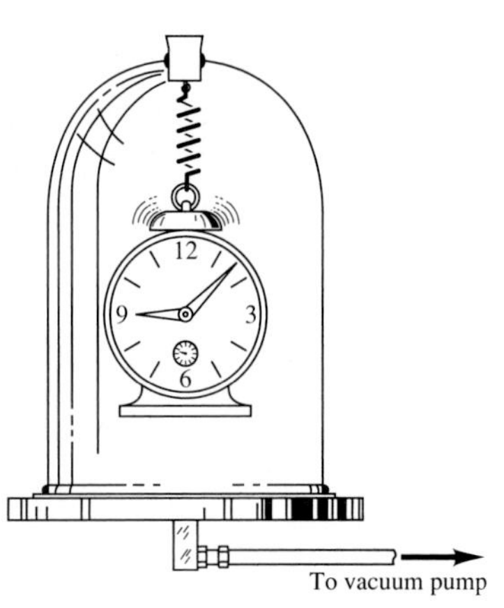

Fig. 20.2 A bell ringing in a vacuum can be seen but not heard.

20.3 ■ Speed of Sound

When lightning is seen from a distance, observers wait a considerable time before they hear the clap of thunder. When a band is followed by a long line of men marching in step to its music, the marching men are seen to be progressively more out of step as their distance from the band increases.

These and many more examples can be cited to illustrate the relatively slow speed of sound in air. Light on the other hand, travels at such a prodigious speed that over short distances its passage can be considered as instantaneous. Consequently, the speed of sound can be determined by dividing the distance between a gun and an observer by the time that elapses between the *observation* of the flash of the gun and *hearing* the report of the gun. By measuring the velocity, first in one direction and then in an opposite direction, errors due to the velocity and direction of the wind can be minimized by averaging.

The accepted values for the speed of sound in air are 331.4 m/s at 0°C or 1087 ft/s at 32°F. For small differences in temperature, it can be shown that the speed of sound in air increases about 0.61 m/s per degree Celsius increase in temperature. Written as an equation,

$$v_s = 331.4 + 0.61t \tag{20.1}$$

where v_s = speed, m/s at t°C, and where t may have either + or − values.

For each degree Fahrenheit increase in temperature, the speed of sound in air increases about 1.1 ft/s. As an equation,

$$v_s = 1087 + 1.1(t - 32) \tag{20.2}$$

where v_s = speed, ft/s at t°F (t can be either greater or smaller than 32).

This change in speed due to change in temperature may cause pipe organs and other wind instruments to be appreciably out of tune if the temperature change of the surrounding air is extreme.

It is important to realize that in speaking of the speed of sound we are referring to the speed with which energy of vibration is passed from molecule to molecule and not to the velocity of the individual molecules.

The speed of sound is the speed with which the vibration is passed from molecule to molecule, not the speed of the individual molecules.

The velocity of sound in a given medium is the same for all frequencies. The sounds from all the instruments of a band reach a listener's ear at the same time. Were this not true, and if the velocity depended on the frequency, the music would sound very different for listeners at varying distances from the band.

Illustrative Problem 20.1 The flash of lightning from an electrical storm is seen in the sky 7 s before the sound of thunder is heard. How far from the observer is the lightning if the temperature is 15°C?

Solution The speed of sound in air at 15°C is 331.4 + 0.61(15) or 341 m/s. Distance = speed × time, or

$$s = vt$$

Substituting, we have

$$\begin{aligned} s &= 341 \text{ m/s} \times 7 \text{ s} \\ &= 2390 \text{ m} \qquad \textit{answer} \end{aligned}$$

Sound waves are transmitted by gases, liquids, and solids. In general, the speed of sound is greater in liquids than in gases, and greater still in solids than in liquids. We discussed in Chap. 19 (see Sec. 19.15) that the speed of propagation of longitudinal waves (as well as of transverse waves) depends upon the density of the medium and upon its elastic properties; the greater the elasticity and the less the density, the greater the speed.

Sound waves can be transmitted by any medium having volume elasticity (i.e., gases, liquids, solids).

Sound travels through water more rapidly and with less energy loss than through air.

Table 20.1 gives the speed of sound in some common media. It is emphasized that neither the frequency nor the intensity of vibration of the sound source determines the sound speed in a given medium. The medium itself and the temperature determine the speed of sound.

Table 20.1 Speed of Sound in Various Media

	Velocity	
Medium	*m/s*	*ft/s*
Aluminum	5,104	16,740
Brass	3,500	11,480
Glass	5,010–6,010	16,410–19,690
Iron	4,720–5,210	15,480–17,390
Steel	5,103	16,730
Brick	3,660	11,980
Granite	6,010	19,685
Oak	3,850	12,620
Pine	3,320	10,900
Alcohol	1,240	4,070
Water (4°C)	1,420	4,700
Water (25°C)	1,500	4,920
Sea Water (25°C)	1,530	5,020
Air, dry (0°C)	331.4	1,087

The speed of sound is different in different materials. It depends upon the bulk modulus and the density of the material.

20.4 ■ Intensity and Loudness of Sound

Loudness is related to the intensity of the sound wave. The greater the intensity, the louder the sound.

The intensity of a sound wave is the rate at which energy flows through a unit area perpendicular to the direction of travel of the wave.

The human ear judges the magnitude of the auditory sensation produced by a sound as a measure of its *loudness*. Just as it is difficult to describe anything as being "twice as red," "twice as hot," or "twice as sweet," so it is that loudness is a characteristic of sound that is difficult to describe in exact (or simple) terms.

For a given frequency, the loudness of a sound is closely related to its intensity. The associated physical quality, *intensity*, is the rate at which sound energy flows through a unit area perpendicular to the direction of wave propagation. As an equation,

$$I = \frac{P}{A} \tag{20.3}$$

$$\text{Sound intensity} = \frac{\text{Sound power (watts)}}{\text{Unit area}}$$

where P is the sound power in watts (W), A is the area in square meters (m^2), and I is the sound intensity in watts per square meter (W/m^2). Thus, the intensity I does not depend upon the hearing acuity of a listener. The least audible sound, called the *threshold of hearing*, of the human ear has an intensity I_0 of about 10^{-12} W/m^2. The loudest sound that can be heard safely without damage to the ears has an intensity of about 1 W/m^2. It is interesting to note the great range of intensities over which the ear is sensitive. For example, a barely audible sound (the lower threshold of hearing) has an intensity which is a million-millionth that of the upper level (called the *threshold of pain* or the *upper limit of hearing*).

The S.I. unit for intensity of sound is the watt/meter^2.

Intensity of sound is measured with acoustical apparatus by comparing it with the intensity of the threshold of hearing. Because of the great range of intensities over which the ear is sensitive, a *logarithmic* intensity scale is used.

Carl Marie von Weber (1786–1826) was a German composer who created the vogue of romantic opera in Germany. He produced more than 250 compositions, such as Invitation to the Dance, and several operas, such as Oberon.

Gustav Theodor Fechner (1801–87) was a German philosopher and physicist.

The Weber-Fechner law states that the magnitude of the sensation of loudness is proportional to the logarithm of the sound intensity, or

$$\beta = 10 \log \frac{I}{I_0} \tag{20.4}$$

where β = magnitude of loudness in decibels (dB)
log = logarithm to base 10
I = intensity of sound in W/m^2
I_0 = intensity at the threshold of hearing (10^{-12} W/m^2)

The physical unit of loudness is the decibel.

The Weber-Fechner law is a fair approximation for more or less pure tones throughout a normal range of frequencies.

The unit of loudness is called the *decibel** in honor of Alexander Graham Bell. The definition of the decibel is quite technical, but for our purposes we can say that each time the power delivered to the ear is *multiplied* by 10, the loudness sensation is increased by 10 decibels (dB). Thus, when the loudness of a sound increases by 10 dB, the sound intensity is increased by a factor of 10. An increase of 20 dB increases the sound intensity by a factor of 100. An increase of 30 dB increases the sound intensity by a factor of 1000, and so on.

Alexander Graham Bell (1847–1922) was a Scottish-American scientist and inventor. He invented the telephone and introduced the deaf-mute system of "visible speech."

Tests have shown that the average ear can distinguish sound-intensity differences equal to about 1 dB. Intensity levels, in decibels, for some common sounds are indicated in Fig. 20.3.

An increase in intensity by a factor of 10 corresponds to a loudness level increase of 10 dB.

* The decibel is equivalent to 10^{-1} *bel*. The bel unit of loudness is generally too large for ordinary situations of loudness measurement. Ten decibels = 1 bel.

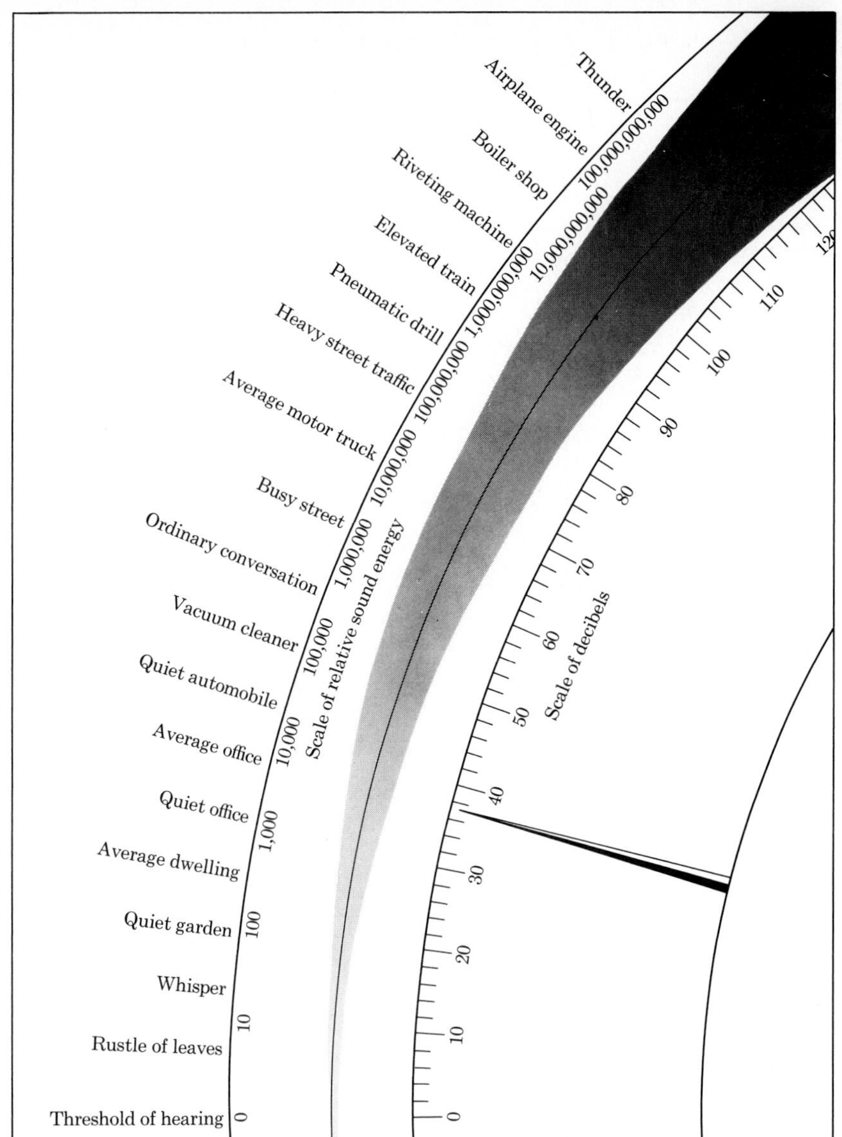

Fig. 20.3 Chart of sound level loudness and intensity, indicated in decibels and in units of relative sound energy. (*Electronics*)

Illustrative Problem 20.2 How do the intensity levels of two sounds compare if their loudness levels are 20 dB and 60 dB?

Solution For each 10-dB increase in loudness the intensity is increased tenfold. The increase in loudness is 60 dB − 20 dB = 40 dB. Hence the intensity will be increased by 10 × 10 × 10 × 10 = 10,000-fold. The second sound intensity will be 10,000 times as great as that of the first sound.

Illustrative Problem 20.3 Find the magnitude of loudness in decibels at the threshold of hearing.

Solution In this case $I_1 = I_0$ in Eq. (20.4). Thus

$$\beta(\text{dB}) = 10 \log\frac{I_1}{I_0} = 10 \log\frac{I_0}{I_0}$$

$$= 10 \log 1 \quad \text{(See log table in Appendix V)}$$
$$= 0 \text{ dB} \qquad \textit{answer}$$

Illustrative Problem 20.4 Street noises entering through an open window register a reading of 68 dB on a sound meter on the window sill. When the double window is closed, the meter registers 42 dB. What percentage of the sound is prevented from entering the room?

Solution Use Eq. (20.4):

$$68 \text{ (dB)} = 10 \log\frac{I_1(\text{W/m}^2)}{I_0(\text{W/m}^2)}$$

$$\log\frac{I_1}{I_0} = 6.8$$

Use Table V of the Appendix or a calculator to get

$$I_1/I_0 = 6.31 \times 10^6$$

Then

$$I_1 = 6.31 \times 10^6 \, I_0 \text{ W/m}^2$$

In like manner,

$$42 \text{ (dB)} = 10 \log\frac{I_2(\text{W/m}^2)}{I_0(\text{W/m}^2)}$$

and

$$I_2/I_0 = \text{antilog } 4.2$$
$$= 1.58 \times 10^4$$
$$I_2 = 1.58 \times 10^4 \, I_0$$

Then

$$\frac{I_1 - I_2}{I_1} \times 100 = \frac{(631 - 1.58)(10^4) \text{ W/m}^2}{6.31 \times 10^6 \text{ W/m}^2} \times 100 = 99.7\% \qquad \textit{answer}$$

It should be emphasized that as the power received by the ear is *multiplied* by 10, the loudness scale is increased (*added*) by 10 dB. Hence it should be clear that the wave-power scale increases very much faster than the corresponding decibel scale.

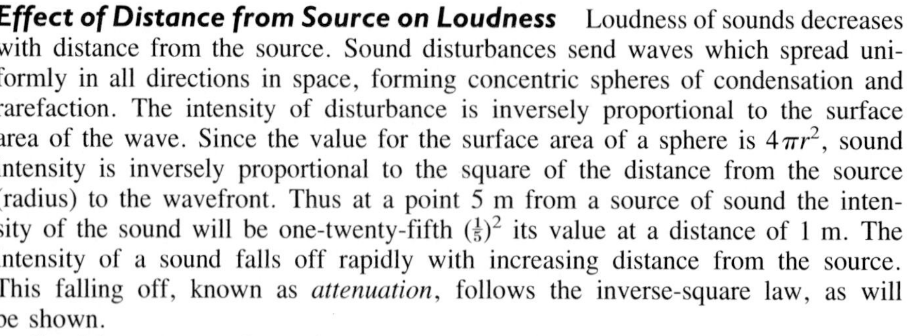

Effect of Distance from Source on Loudness Loudness of sounds decreases with distance from the source. Sound disturbances send waves which spread uniformly in all directions in space, forming concentric spheres of condensation and rarefaction. The intensity of disturbance is inversely proportional to the surface area of the wave. Since the value for the surface area of a sphere is $4\pi r^2$, sound intensity is inversely proportional to the square of the distance from the source (radius) to the wavefront. Thus at a point 5 m from a source of sound the intensity of the sound will be one-twenty-fifth $(\frac{1}{5})^2$ its value at a distance of 1 m. The intensity of a sound falls off rapidly with increasing distance from the source. This falling off, known as *attenuation*, follows the inverse-square law, as will be shown.

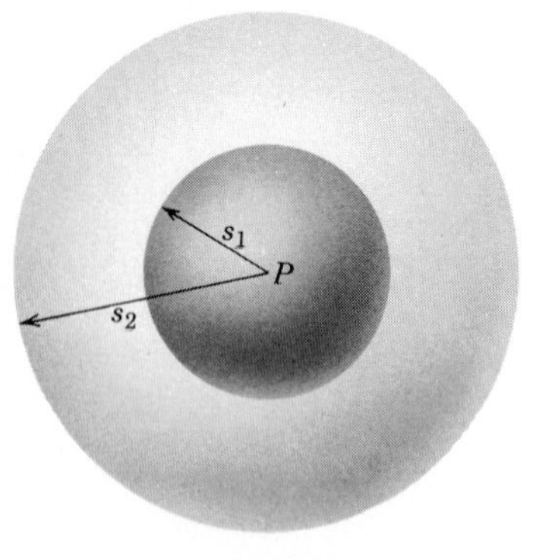

Fig. 20.4 Sound waves travel out from a source in alternate compression-rarefaction waves, which are concentric spheres.

Consider a source of sound at O (Fig. 20.4). At a distance s_1 the original sound energy is dissipated over a total area of $4\pi s_1^2$, and the sound intensity (energy per unit of area) could be expressed as

$$I_1 = E/4\pi s_1^2$$

When the sound wave arrives at the distance s_2, the original energy is dissipated over an area $4\pi s_2^2$, and

$$I_2 = E/4\pi s_2^2$$

Dividing, we obtain the formula which relates the intensity of sound and distance from the source,

$$I_1/I_2 = s_2^2/s_1^2 \tag{20.5}$$

where I_1 and I_2 are intensities of the sound at respective distances s_1 and s_2. This formula is strictly true, however, only if the distances involved are large compared to the size of the source of vibration, if no sound is *reflected* to the listener, and if no absorbing material is between the source and listener.

Illustrative Problem 20.5 The loudspeaker of a public address system produces sound waves with a power output of 120 W. (*a*) Determine the intensity at a distance 3 m from the source. (Assume the sound is a point source.) (*b*) How far can the sound travel and still give a loudness level of 40 dB (equivalent in intensity to that of ordinary conversation at 0.91 m distance)? (*c*) What will be the intensity level and the loudness level 1200 m distant from the speaker?

Solution

(*a*)

$$I = \frac{P}{4\pi r^2} = \frac{120\text{ W}}{4\pi(3\text{m})^2} = 1.06\text{ W/m}^2 \qquad \textit{answer}$$

In the next section it will be pointed out that this intensity borders on the threshold of pain.

(*b*)

$$\beta = 10 \log\frac{I}{I_0}$$

$$40\text{ dB} = 10 \log\frac{I}{I_0}$$

or

$$\log\frac{I}{I_0} = 4$$

Then

$$I = 10^4 I_0$$
$$= 10^4 \times 10^{-12}\text{W/m}^2$$

and

$$I = \frac{P}{4\pi r^2}$$

Substituting, we get

$$10^{-8}\text{W/m}^2 = \frac{120\text{ W}}{4\pi r^2}$$

From which $r = 3.1 \times 10^4$m (about 19 mi!) *answer*

(*c*)

$$\frac{I_1}{I_2} = \frac{s_2^2}{s_1^2}$$

Substituting known values gives

$$\frac{I_1}{1.06\text{ W/m}^2} = \frac{(3\text{m})^2}{(1200\text{m})^2}$$

and

$$I_1 = 6.6 \times 10^{-6}\text{ W/m}^2 \qquad \textit{answer}$$

$$\beta = 10 \log\frac{6.6 \times 10^{-6}}{10^{-12}}$$

$$= 10(6.8) = 68\text{ dB} \qquad \textit{answer}$$

Fig. 20.5 Limits of audible sound intensities. The audible range for the normal ear lies in the region enclosed by the two curves.

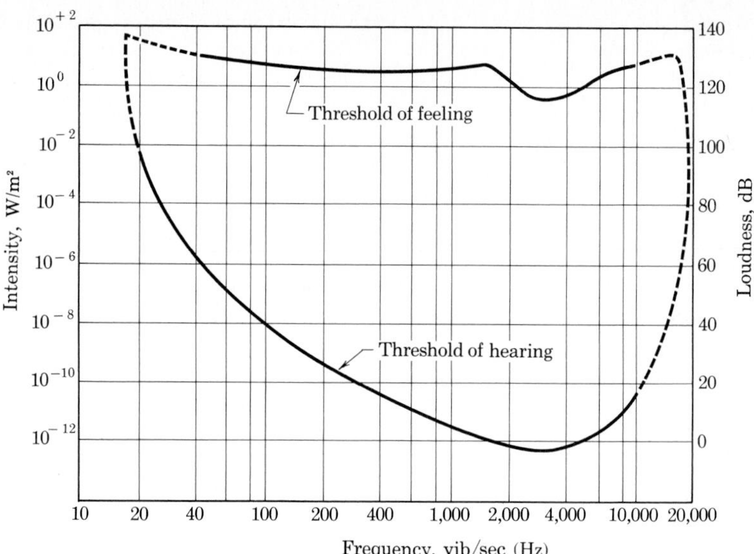

20.5 ■ Audible Frequencies

Two characteristics of sound common to the listener are *loudness* and *pitch*. Each produces a physiological sensation to the listener and to each of these sensations there corresponds a physically measureable quantity. We have seen how loudness is related to the intensity of the sound wave. Pitch refers to how "high" or "low" the sound seems. The lower the frequency, the lower the pitch; the higher the frequency, the higher the pitch. Pitch, however, also depends to a slight degree on loudness.

Not every human ear responds the same to a given sound. There is a considerable variation in the ability of individuals to hear sounds of high or low frequency and of high or low intensity (loudness).

Figure 20.5 shows graphically how the upper and lower limits of audible sound intensity (plotted vertically) vary as a function of frequency (plotted horizontally). The graph is a composite of results obtained by testing many persons. Therefore, it refers to the "average" ear. Intensities below the lower curve, the *threshold of hearing*, are generally insufficient to produce any sensation of hearing. Plots like that of Fig. 20.5 are called *audiograms*. It will be noted that the lowest intensity which will produce audible sound is for frequencies between 2000 and 4000 Hz. That is, in this range the ear requires the least energy to cause a sensation of sound. Between 2000 and 4000 Hz the loudness of a sound in order to be heard can be less than 0 dB. Intensities below the lower line are generally insufficient to produce a sensation of hearing. Sound waves whose frequencies are below the audible range (i.e., 20 Hz) are called *infrasonic*. Some sources of infrasonic waves are those produced by vibrating heavy machinery, and earthquakes.

Loudness intensity affects the frequency limits of audibility.

To illustrate how loudness (intensity) affects the frequency limits of audibility, consider where the 20-dB horizontal line in Fig. 20.5 cuts the threshold-of-hearing curve. Note, also, that at a frequency of 1600 Hz the threshold of hearing is 0 dB. Yet, when the loudness level is 20 dB (100 times the power intensity) frequencies below 300 and above 12,000 Hz cannot be heard. At a loudness of 40 dB, with a power intensity 10,000 times the threshold of hearing for 1600 Hz, the lowest frequency that can be heard is about 120 Hz, whereas the upper limit is near 16,000 Hz.

The normal human ear can detect sounds with an intensity as low as 10^{-12} W/m² and can tolerate sounds with intensities as high as 1 W/m².

The upper curve of Fig. 20.5 represents the intensity level of the loudest pure tone that can be tolerated without pain. Frequencies in excess of about 20,000 Hz are not heard by the normal ear but may produce a sensation of discomfort or pain if they are of high intensity. The upper curve is called the *threshold of feeling or pain*. If a sound is much louder than the upper level of hearing, the ear may be damaged by it. In fact, hearing may be impaired in a person subjected to long periods of sounds above 100 dB. Federal standards (OSHA) now set permissible noise exposure limits. In many occupations ear (hearing) protectors must be worn to protect workers from hearing loss.

Ultrasonic waves can be heard by dogs, birds, and many other species of animals.

Frequencies above 20,000 Hz are called *ultrasonic* (be careful not to confuse it with supersonic, which has to do with speeds greater than the speed of sound in air, about 344 m/s or 1130 ft/s). Experiments conducted with animals reveal that some animals can hear sounds with frequencies as high as 100,000 Hz. Dogs, for example, respond to whistles that emit frequencies as high as 50,000 Hz. Bats flying in total darkness emit a series of sharp squeaks with frequencies as high as 100,000 Hz. By listening to the echos of the squeaks from nearby objects, the bats can determine where the objects are and direct their flight accordingly. In the next chapter we shall study how ultrasonic waves have a number of applications in medicine and other fields.

20.6 ■ Vibrating Strings

In Chap. 19 we saw how standing waves are established on a string (see Fig. 19.34). Standing waves are involved in stringed musical instruments.

Consider a string of length L stretched between two fixed ends (see Fig. 20.6). By means of a tightening device at one end, the string can be drawn tight enough to vibrate when plucked. Standing waves will be set up in the string by a continuous superposition of waves incident on and reflected from both ends. For a stationary wave to exist in the string, each end (at $x = 0$ and $x = \text{L}$) must be a node, because the ends cannot move. When the string is plucked, energy is transferred to the string and causes it to vibrate in a transverse motion. The vibrating string transfers energy to the air molecules and causes them to vibrate with longitudinal motion. The frequency of the longitudinal (sound) wave is the same as the frequency of the transverse vibration of the string. The vibrating string consumes energy which must be replenished by continuous plucking or stroking in order to maintain a continuous sound.

When the string vibrates as a whole (Fig. 20.6*a*), it produces its lowest frequency f_1, called the *fundamental* or *first harmonic*. For the fundamental frequency, the length L of the string equals one-half the wavelength ($\lambda_1/2$) of the sound wave produced. Or, the wavelength λ_1 equals twice the length L.

$$\lambda_1 = 2L \tag{20.6}$$

When a finger is placed on the string at its midpoint ($x = L/2$), that point becomes a node. The string will vibrate in two segments (Fig. 20.6*b*). Because the effective length is halved, the wavelength will be halved and the frequency is doubled. In like manner, when the string vibrates in three segments (Fig. 20.6*c*), its frequency will be tripled. A string vibrating in four equal segments will have a frequency four times that of the fundamental. In general, nodes occur at points where $x = 0$, $x = \lambda/2$, $x = 2\lambda/2$, $x = 3\lambda/2$, etc. In other words, nodes occur at the end of the string and every half-wavelength thereafter to the end of the string. *Nodes* occur at

$$x = \frac{n\lambda}{2} \tag{20.7}$$

where $n = 0, 1, 2, 3, \ldots$ (any integer)

Antinodes occur at points where $x = \lambda/4, 3\lambda/4, 5\lambda/4, \ldots$, etc. Thus, *antinodes* occur at

$$x = \frac{(2n - 1)\lambda}{2} \tag{20.8}$$

where $n = 0, 1, 2, 3, \ldots$ (any integer)

Figure 20.6 shows four possible modes of vibration. In (*a*) $\lambda_1 = 2L$. Then, $f_1 = v\lambda_1 = v/2L$. In (*b*) $f_2 = 2v/2L$, in (*c*) $f_3 = 3v/2L$, and (*d*) $f_4 = 4v/2L$. Thus,

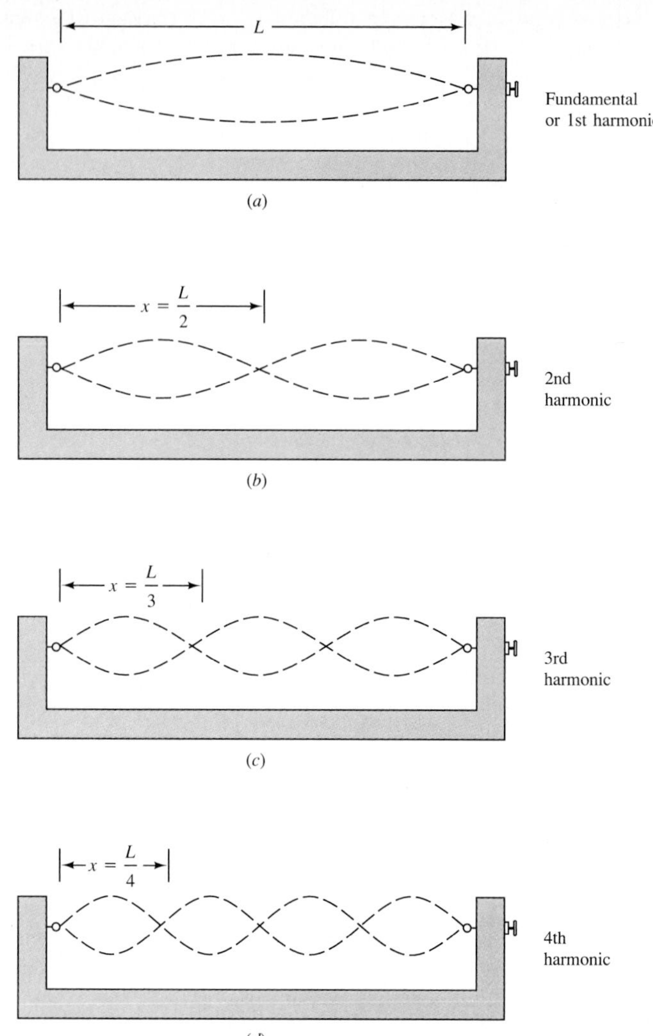

Fig. 20.6 Modes of standing waves in a string of length L, fixed at both ends.

$$f_n = \frac{nv}{2L} \qquad (20.9)$$

where f_n = the frequency of the vibrating string,
n = the number of segments in which the string is vibrating,
v = the speed of the sound wave in the surrounding air,
L = the length of the string

The lowest-frequency standing wave in a musical instrument is called the fundamental.

Modes of vibration whose frequencies are whole multiples of the lowest frequency are called *harmonics*. The *fundamental* is the first harmonic. The frequency that is double the frequency of the harmonic is the frequency of the second harmonic. The frequency that is triple that of the fundamental is the third harmonic, and so on. Harmonics whose frequencies are greater than the fundamental frequency are also called *overtones*. The second harmonic is the first overtone, the third harmonic is the second overtone, and so on.

A harmonic is a vibration whose frequency is an integral multiple of the frequency of the fundamental.

Harmonics with integral frequencies higher than that of the fundamental are called overtones.

Illustrative Problem 20.6 What will be the frequencies of the first three overtones of a wire in tension that has a fundamental frequency of 396 Hz?

Solution The first overtone will be the second harmonic. The first harmonic has a frequency of 396 Hz. Hence,

$$\text{Frequency of 1st overtone} = 2(396\ \text{Hz}) = 792\ \text{Hz} \qquad \textit{answer}$$

$$\text{Frequency of 2nd overtone} = 3(396\ \text{Hz}) = 1188\ \text{Hz} \qquad \textit{answer}$$

$$\text{Frequency of 3rd overtone} = 4(396\ \text{Hz}) = 1584\ \text{Hz} \qquad \textit{answer}$$

The fundamental frequency of the vibrating string is $f_1 = \dfrac{v}{2L}$. Therefore, it follows from Eq. (19.30) that

$$f_1 = \frac{1}{2L}\sqrt{\frac{T}{m/L}} \qquad (20.10)$$

where L = the length in meters (m)
T = the tension in newtons (N)
and m = the mass in kilograms (kg)

Stringed instruments can be analyzed by applying this equation. Thus, we see that the frequency of the fundamental depends on the length, the tension, and the mass per unit length. Increasing the tension on a string increases the frequency (raises the pitch), and vice versa. Adjusting the tension is the means for "tuning" the instrument. The longer the string, the lower the frequency; the shorter the string, the higher the frequency. The bass section of the piano has the longer wires. Pressing the finger against the fingerboard of a violin or guitar changes the length of the vibrating portion of the string, thus changing the pitch. Also, the lower notes from a piano or harp are produced by wires that, not only are longer, but are heavier—that is, their linear densities are greater.

The fundamental frequency of a vibrating string depends upon its length, linear density, and the tension.

Illustrative Problem 20.7 A guitar string is 87 cm long and has a mass of 3.5 g. The length of the string between the bridge and the support post is 60 cm. The string is under a tension of 510 **N**. What are the frequencies of (*a*) the fundamental and (*b*) the first two overtones?

Solution Using Eq. (20.10),

(***a***)

$$f_1 = \frac{1}{2L}\sqrt{\frac{T}{m/L}} = \frac{1}{2 \times 0.60\ \text{m}}\sqrt{\frac{510\ \text{N} \times \dfrac{1\ \text{kg}\cdot\text{m/s}^2}{1\ \text{N}}}{0.0035\ \text{kg}/0.87\ \text{m}}}$$

$$= 297\ \text{Hz} \qquad \textit{answer}$$

NOTE: The two L's are different in the equation.

(***b***)

$$\text{First overtone frequency} = 2(297\ \text{Hz}) = 584\ \text{Hz}$$
$$\text{Second overtone frequency} = 3(296\ \text{Hz}) = 891\ \text{Hz} \qquad \textit{answer}$$

Stringed instruments would not produce very loud tones if they relied solely on their vibrating strings to produce the sound waves. The strings are too thin to compress and expand much air. Consequently, they use a mechanical amplifier such as a sounding board (for the piano) and a sounding box (for the violin or viola or guitar). When the strings are set into vibration, the sounding board or box will resonate with the frequencies emitted by the instrument. Since the sounding board or box has a much greater area in contact with the air, it can produce a much stronger (amplified) sound wave. An effective sounding box is not as important in the electric guitar because the vibrations are amplified electrically.

Sounds in stringed instruments are amplified by sounding boards and sounding boxes. For resonance there is a transfer of energy and a resultant louder sound.

20.7 ■ Vibrating Air Columns

"Wind instruments" produce sound from vibrations of standing waves in a column of air within a tube or pipe. These instruments are generally divided into two subgroups, "woodwinds" and "brasses." In the woodwind group we find such instruments as the pipe organ, flute, piccolo, clarinet, oboe, saxophone, and basoon. In the brass group are the cornet, horn, trumpet, trombone, and tuba.

Vibrating columns of air are produced in the above instruments in various ways. For some a vibrating reed or the lip of the player helps set up vibrations of the air column. In the flute, piccolo, and the pipe organ a stream of air is directed against one edge of an opening in the instrument. This produces a turbulence which sets up the vibrations. Only vibrations of certain frequencies persist. The frequency will depend on the dimensions of the air column in the instrument. The length of the vibrating air column is controlled in instruments such as the flute, the clarinet, and the saxophone by covering and uncovering various holes in the instrument. The air columns in the trumpet, the French horn, and the tuba are controlled by pushing down various combinations of valves. The trombone uses a sliding sleeve which varies the length of the vibrating column.

The frequencies of musical instruments are difficult to calculate for most musical instruments, except for the long, narrow tube or pipe musical instrument. Organs make use of pipes of different lengths, some open at both ends and some closed at one end. The various ways in which air columns may vibrate are shown in Figs. 20.7 and 20.8. Standing waves in air are longitudinal in nature and as such are difficult to represent in any drawing. In the drawings of Fig. 20.7 and 20.8 the positions of the loops and nodes are shown as if the standing waves were transverse (dotted lines in the diagrams). However, the directions of the air particles in the longitudinal waves are indicated by full arrows inside the pipes.

The phase relationship between an incident wave and the wave reflected from one end of the tube will depend on whether or not the end is open or closed. The situation is analogous to the phase relationship between an incident and a reflected transverse wave at the end of a string. When the string is free to move at its end (see Sec. 19.28), the reflected and incident waves are in phase and the free end becomes an antinode. In like manner, there will be an antinode in the air column at the ends of a pipe open at both ends, since the air can move freely at the ends.

There is always an antinode at the end of an open pipe.

In an open pipe the fundamental frequency is $v/2L$ and all harmonics are present.

Modes of operation for a pipe open at both ends are shown in Fig. 20.7. The dotted lines represent the displacement amplitude of the vibrating air within the tube. The heavy arrows indicate the direction of motion of the air particles. In each of the three modes shown there must be at least one node. The antinodes are at the open ends.* Where there is only one node the wavelength of the longest standing wave (called the *fundamental*) is twice the length of the pipe (Fig. 20.7*a*). The next mode in which the pipe will vibrate (resonate) is shown in Fig. 20.7*b*. When there are two nodes and three antinodes, the wavelength equals the length of the pipe and the pipe will resonate with a frequency twice that of the fundamental. This frequency is of the second harmonic (first overtone). In Fig. 20.7*c* the mode has a *natural frequency* of $3f_1$ (three times the frequency of the fundamental). Its frequency is that of the third harmonic (second overtone). In a pipe open at both ends, all harmonics are present and the natural frequencies can be expressed by the equation

$$f_n = n\frac{v}{2L} \qquad n = 1, 2, 3, \ldots \tag{20.11}$$

where n = the speed of sound in air
L = the length of the pipe

Thus, the possible frequencies to which an open pipe may resonate are definite and fixed in value. The frequencies depend only upon the length of the pipe and the velocity of the sound in air at that time and place.

* Figure 20.7 shows the antinodes exactly at the open ends of the pipe. Actually, the antinodes lie slightly outside the open ends.

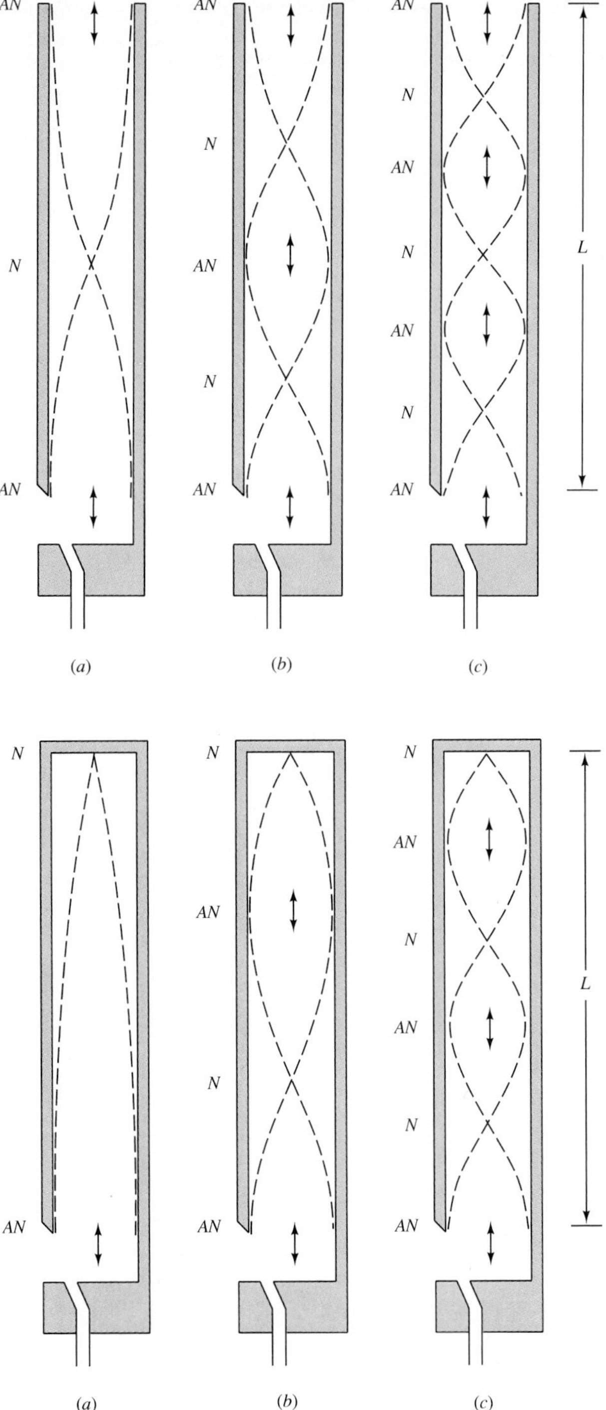

Fig. 20.7 Modes of vibration (standing longitudinal waves) in an organ pipe open at both ends. (*a*) The fundamental (first harmonic), where $\lambda_1 = 2L$ and $f_1 = v/\lambda_1 = v/2L$. (*b*) The second harmonic (first overtone), where $\lambda_2 = L$ and $f_2 = v/\lambda_2 = 2f_1$. (*c*) The third harmonic (second overtone), where $\lambda_3 = \frac{2}{3} L$ and $f_3 = \frac{3}{2} v/L = 3f_1$. Note that the natural frequencies include a series of frequencies f_1, $2f_1$, $3f_1$, ...

Fig. 20.8 Modes of vibration (standing longitudinal waves) in an organ pipe closed at one end. (*a*) The fundamental (first harmonic), where $\lambda_1 = 4L$ and $f_1 = v/\lambda_1 = v/4L$. (*b*) The third harmonic (first overtone), where $\lambda_3 = \frac{4}{3}L$ and $f_3 = \frac{3}{4}v/L = 3f_1$. (*c*) The fifth harmonic (second overtone), where $\lambda_s = \frac{4}{5} L$ and $f_5 = \frac{5}{4}v/L \leq 5f_1$. Note that only the odd harmonics are present in the closed pipe.

There usually are several modes of vibration in an organ pipe present at the same time. The number of harmonics will depend upon the dimension of the pipe, the ratio of the length to width, and whether the pipe is straight or crooked.

There is always a node at the closed end of a tube or pipe.

The first three modes of vibration for a pipe *closed at one end* are shown in Fig. 20.8. Note that there is always a displacement node at the closed end (just as a node occurs at the fixed end of a vibrating string) and an antinode at the open end. For the pipe closed at one end the wavelength of the fundamental is four times the length of

In a pipe closed at one end the fundamental frequency is $v/4L$ and only odd harmonics are present.

the pipe. However, there is another difference between the resonating frequencies of open and closed pipes. From the figures, it is evident that only odd harmonics can be produced in a pipe closed on one end. Thus, if the frequency of the fundamental is f_1, the overtones have frequencies of $3f_1$, $5f_1$, $7f_1$, ... Since it is impossible for waves having frequencies 2, 4, 6, . . . times the frequency of the fundamental to have a node at one end and an antinode at the other, even harmonics are not produced by a pipe closed at one end.

Therefore, in a pipe closed at one end, only odd harmonics are present, and they are given by the equation

$$f_n = n\frac{v}{4L} \quad (n = 1, 3, 5, \ldots) \tag{20.12}$$

where v = the speed of sound in air
L = the length of the pipe

Illustrative Problem 20.8 What will be the fundamental frequency and the frequencies of the first three overtones of a 1.26 m long organ pipe at 22°C if it is (*a*) open, (*b*) closed?

Solution The speed of sound in air can be found by using Eq. (20.1).

$$\begin{aligned} v &= 331.4 + 0.61t \\ &= 331.4 + 0.61\ (22) \\ &= 345 \text{ m/s} \end{aligned}$$

(***a***) Open pipe

$$f = n\frac{v}{2L}$$

$$f_1 = 1\ \frac{345 \text{ m/s}}{2(1.26 \text{ m})} = 137 \text{ Hz} \qquad \textit{answer}$$

$$f_2(\text{1st overtone}) = 2f_1 = 274 \text{ Hz} \qquad \textit{answer}$$
$$f_3(\text{2nd overtone}) = 3f_1 = 411 \text{ Hz} \qquad \textit{answer}$$
$$f_4(\text{3rd overtone}) = 4f_1 = 548 \text{ Hz} \qquad \textit{answer}$$

(***b***) Closed pipe

$$f_1 = n\frac{v}{4L} \quad (n = 1, 3, 4, 7, \ldots)$$

$$f_1(\text{fundamental}) = 1f_1\frac{345 \text{ m/s}}{4(1.26 \text{ m})} = 68 \text{ Hz} \qquad \textit{answer}$$

$$f_3(\text{1st overtone}) = 3f_1 = 205 \text{ Hz} \qquad \textit{answer}$$
$$f_5(\text{2nd overtone}) = 5f_1 = 342 \text{ Hz} \qquad \textit{answer}$$
$$f_7(\text{3rd overtone}) = 7f_1 = 479 \text{ Hz} \qquad \textit{answer}$$

CHARACTERISTICS OF MUSICAL SOUNDS

20.8 ■ Pitch of a Musical Sound

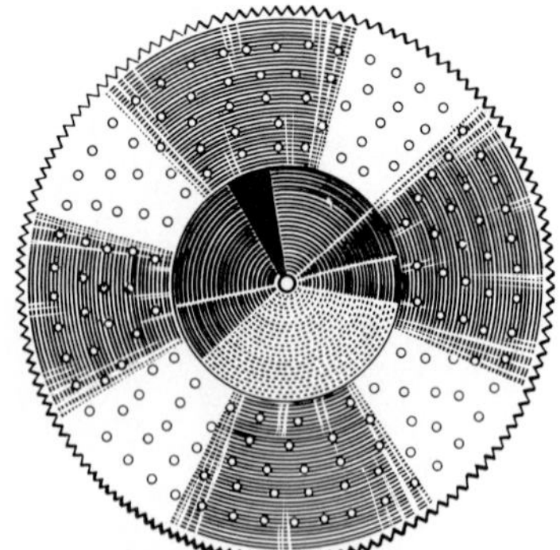

Fig. 20.9 Savart's wheel. (Central Scientific Co.)

The *pitch* of a musical sound depends upon the frequency of vibration of the source. We think of pitch in terms of the "highness" or "lowness" of the tone as interpreted by the ears and brain. The relation between the pitch of a tone and its frequency of vibration can be experimentally demonstrated by the use of Savart's wheel (Fig. 20.9). This is a wheel with pointed teeth equally distributed on its circumference and mounted on a shaft so that it can be rotated at varying speeds. When the wheel is rotating, a card is pressed against the teeth. One transverse vibration will be imparted to the card every time a tooth strikes it. The card in turn sends out a longitudinal wave throughout the surrounding air medium. When the frequency is about 20 or 30 vib/s (Hz), a recognizable musical tone of low pitch can be heard. Increasing the speed of

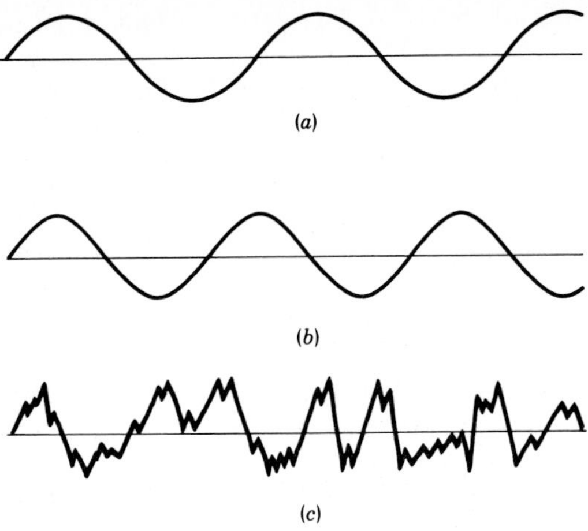

Fig. 20.10 Curves representing (*a*) a pure musical tone, (*b*) a pure musical tone of higher pitch, and (*c*) noise.

the wheel will increase the vibrations per second of the card and consequently the frequency of the forced vibrations of the air about it. This will be recognized as a higher pitch.

Thus it can be demonstrated that a sound has a definite pitch when the source is vibrating at a fixed frequency. Increasing the frequency of vibration of the source causes a rise in pitch, while decreasing the frequency will cause the pitch to fall.

If the teeth of the wheel were irregularly spaced, the sound would have no definite pitch; only a *noise* would be heard. We can represent the difference between a pure musical tone and a noise graphically (Fig. 20.10). A pure musical tone is distinguished by a regularity of the condensation-rarefaction pattern, while a noise is represented by a very irregular and haphazard curve.

A musical sound is always the result of regular vibrations, while noise is usually the result of irregular vibrations.

20.9 ■ Quality of Sound

Musical sounds differ in quality.

Sounds can be distinguished by three characteristics: *loudness, pitch,* and *quality*. We have seen how the physiological sense of the loudness of a sound is the ear's response to the intensity of the sound wave. It is quite subjective in the sense that two listeners may not perceive that two sounds are equally loud. In like manner, the frequency of vibration is the sound sensation which the ear judges as high or low. It, too, is a subjective quality and cannot be measured with instruments. Pitch is related to the objective quantity of frequency. For a pure tone (a single frequency), the pitch becomes higher as the frequency is increased and lower as the frequency is decreased. But, there is not an exact one-to-one correspondence between pitch and frequency. For example, the pitch of a pure tone of constant frequency seems to become lower as the intensity level is raised.

We all know that two people singing the same note do not sound exactly the same. The difference in the sound is due to the *quality* of the note. It is the difference in the quality of tone of a violin that distinguishes a good instrument from a poor one.

Most vibrating bodies vibrate with many modes of vibration. Thus, sounds have complex waveforms.

In order to understand the reasons for differences in quality, it must be realized that musical sounds do not usually result from single vibrations of a given frequency. When a tone is the result of a single frequency of vibration it is called a *pure* tone. However, most musical sounds are a blending of vibrations of different frequencies in a particular way. The different frequencies occur when the sounding body vibrates in parts at the same time that it is vibrating as a whole. For example, the string of a musical instrument may vibrate as a whole, but usually the whole is also vibrating in two or more equal parts, giving off frequencies which are exact multiples of the frequency of vibration of the whole. The result is a complex vibration which gives it a characteristic quality.

The normal ear is capable of supplying the fundamental tone from its higher harmonics.

The quality of a musical sound depends upon the number, the frequencies, and the relative intensity of the various overtones that blend with the fundamental. The rich tone of a violin may contain 10 or more overtones. In some instruments, the overtones are relatively strong, while in others they may be weak or missing completely. Figure 20.11 illustrates the blending of the fundamental (*a*) and the first two overtones (*b* and *c*) into the complex musical sound (*d*) which results. The pitch that is most easily recognized as a musical sound is that produced by the fundamental while the quality of the sound depends on the number and relative amplitude of its harmonics.

Many of the notes produced by musical instruments are rich in harmonics. In some instruments some of the harmonics are more prominent (greater intensity) than the fundamental. The human ear is capable of assigning a characteristic pitch (that of the fundamental) to an array of frequencies in a harmonic series, even though the overtones may be louder than the first harmonic. Thus, the ear supplies the fundamental, provided the correct harmonics are present. It is possible to eliminate the fundamental frequency entirely, by means of filters, without changing the pitch as registered by the ear. This property makes it possible to produce radios or audio amplifying systems with small speakers (which do not radiate low frequencies, but generate fairly good higher harmonics) that give the listener a sense of hearing the low frequencies, when actually he or she is hearing only multiples of those frequencies.

The physiological and physical properties of audible sound can be summarized as follows:

Sensory effects	*Physical property*
Loudness	Intensity
Pitch	Frequency
Quality	Waveform complexity

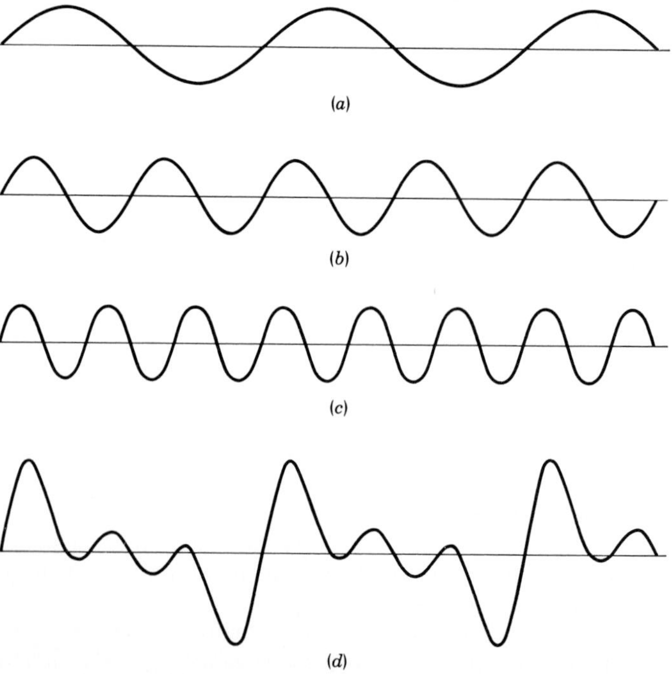

Fig. 20.11 Resolving the fundamental tone and overtones into one blended musical sound.

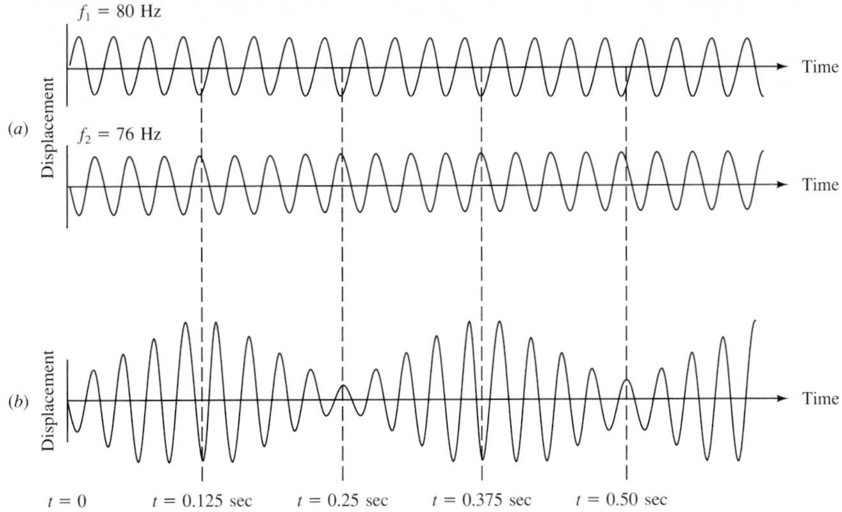

Fig. 20.12 Beats occur when two sound waves of slightly different frequency travel in the same direction. (*a*) Individual waves, (*b*) Resultant periodic variation in intensity as a result of superposition. The number of beats per second is equal to $f_1 - f_2$.

20.10 ■ Interference of Sound Waves—Beats

Interference of standing waves was mentioned in Sec. 19.19. One kind of interference results when two waves of the same amplitude and frequency travel through the same region in opposite directions. We now will consider another type of interference that results when two waves of equal magnitude but *slightly different frequencies* travel through the same medium.

Beats are variations in loudness of a sound due to the interference of two sound waves of nearly the same frequency.

Consider two tuning forks of slightly different frequencies sounding simultaneously. Sound waves from the two sources are shown in Fig. 20.12*a*. Figure 20.12*b* shows how the two waves interfere with each other and how the sound intensity level rises and falls. These alternating periods of increasing and decreasing volume of sound are called *beats*, and the number of times per second the sound builds to a maximum and dies away again is known as the *beat frequency*. An analysis of the waveforms of Fig. 20.12 shows that the beat frequency is equal to the difference in frequency between the two component sounds causing the beats.

The beat frequency is the difference between the frequencies of the two sources of sound.

The phenomenon of beats thus offers a ready method of tuning a string, horn, or other instrument against a standard. First the two tones are matched as closely as possible by ear. Then while the test string or instrument and the standard are sounded together, the musician listens for beats. The faster the beat frequency, the farther out of tune his instrument is. He makes adjustments until the beat frequency is reduced to one or fewer beats per second, the accuracy of his final tuning depending on the demands of the particular situation. When no beats are observed, the two tones have the same frequency.

20.11 ■ The Doppler Effect

When a source of sound is moving toward an observer, the pitch is higher than when the source is at rest; when the source is receding from the observer, the pitch is lower.

The frequency of a sound wave, usually equal to that of the source, can be changed by the motion of the sound source or the observer or both. It is a common experience to hear an approaching airplane appear to be speeding up as it approaches the listener and then suddenly appear to slow down as it passes over the listener. The pitch of the siren of a speeding police car (Fig. 20.13) is higher as it approaches the listener than when the car recedes from the listener.

This change of pitch is due to an apparent change in frequency, or *frequency shift*. The shift in frequency due to relative motion between the source and the receiver is

Christian Johann Doppler (1803–1853) was an Austrian physicist who discovered the shift in frequencies of light waves (electromagnetic waves) produced by the relative motion of source and observer.

called the *Doppler effect*. For sound, the change in pitch (due to the frequency shift) is quite marked, even when the relative speeds of source and listener are fairly low. Figure 20.13 will serve to clarify the change in frequency due to the Doppler effect. (In this discussion it will be assumed that the air or other medium is at rest in our reference frame.) Let v_s be the speed of the source of the sound waves (the police car) and v be the speed of sound in the medium. Let A represent an observer being approached by the sound source and B an observer from whom the sound source is receding.

The siren is moving away from the waves traveling to B and toward the waves traveling to A. The waves behind the siren are therefore stretched out, and those in front are compressed or crowded together. The actual speed of the sound waves is the same in all directions, but the net effect of the motion of the source is that observer A receives more waves per second and observer B receives fewer per second. The pitch at A is higher than the actual pitch of the siren and the pitch at B is lower than the actual pitch. Two cases of relative motion must be dealt with.

Source Moving If a source of sound of frequency f_s moves with speed v_s, the frequency f_O of the tone heard by a *stationary observer* may be obtained from

$$f_O = f_s \frac{v}{v \pm v_s} \tag{20.13}$$

where v is the speed of sound in air at that place. If the source is approaching the observer, the algebraic sign in the denominator is minus; if receding, the sign is plus.

Observer Moving If an observer moves with a speed v_o with respect to a *stationary source* of sound of frequency f_s, the frequency of the tone heard by the moving observer is

$$f_O = f_s \frac{v \pm v_O}{v} \tag{20.14}$$

where the plus sign is applied when the observer is approaching the source and the minus sign when the observer recedes from the source.

Equations (20.13) and (20.14) can be combined into one that includes all four cases. It is

$$f_O = f_s \left(\frac{v \pm v_O}{v \mp v_s} \right) \tag{20.15}$$

Fig. 20.13 A fast-moving police car sounding its siren illustrates the Doppler effect. Observer A will hear a higher-frequency sound, and observer B will hear a lower-frequency than that actually being emitted by the siren.

The upper sign is to be used if the velocity is one of approach. The lower sign is used if the velocity is one of receding motion.

Doppler's principle, though an interesting phenomenon in sound, has more practical applications in the fields of light and radar. Frequency shifts in light waves received from distant stars tell us whether the star is approaching our solar system or receding from it. The same principle is used by highway patrol officers to detect "speeders."

The Doppler effect has applications in fields other than sound.

An observed frequency shift in a returning radar wave reflected from a target airplane tells whether the target is approaching or receding, and the application of timed pulsing to this "doppler" gives the *range rate*, the actual rate at which the range to the target is closing or opening.

Applied to ultrasonic sounds, Doppler's principle is used in determining the range rate of enemy submarines and in the operation of sonar.

Illustrative Problem 20.9 A factory whistle is known to have a frequency of 480 Hz. An auto is approaching the source at 100 km/h when the air temperature is 32°C. What is the pitch heard by the auto's occupants?

Solution Since this is a case of observer moving toward a stationary sound, Eq. (20.14) is used, with the plus sign.

$$v = 331.4 + 0.61(32) = 350.9 \text{ m/s}$$
$$v_0 = 100 \text{ km/h} = 27.77 \text{ m/s}$$

Substituting known values in Eq. 20.14 gives

$$f_0 = 480 \text{ Hz}\frac{(350.9 + 27.8) \text{ m/s}}{350.9 \text{ m/s}} = 518 \text{ Hz} \qquad \textit{answer}$$

Illustrative Problem 20.10 An ambulance, traveling 88 km/h, is moving toward an approaching car which is traveling at a speed of 50 km/h. The air temperature is 32°C. If the siren is emitting a frequency of 1020 Hz, what is the frequency heard by the driver of the car (*a*) as the vehicles approach each other, and (*b*) after they pass each other? (Assume no wind.)

Solution Use Eq. (20.15) with $f_s = 1020$ Hz, $v_s = 88$ km/h, and $v_0 = 50$ km/h. The velocity of the sound in air is

$$v = 331.4 + 0.61(32) = 350.9 \text{ m/s}$$

$$v_s = 88\frac{\text{km}}{\text{h}}\left(\frac{1 \text{ h}}{3600 \text{ s}}\right)\left(\frac{1000 \text{ m}}{1 \text{ km}}\right) = 24.4 \text{ m/s}$$

$$v_0 = 50 \text{ km/h} = 13.9 \text{ m/s}$$

(***a***) Substituting known converted values and using proper signs, we get

$$f_0 = f_s\left(\frac{v + v_0}{v - v_s}\right) = 1020\left(\frac{350.9 + 13.9}{350.9 - 24.4}\right) = 1140 \text{ Hz} \qquad \textit{answer}$$

(***b***)

$$f_0 = f_s\left(\frac{v - v_o}{v + v_s}\right)$$

$$= 1020\left(\frac{350.9 - 13.9}{350.9 + 24.4}\right) = 916 \text{ Hz} \qquad \textit{answer}$$

QUESTIONS AND EXERCISES

1. Distinguish between the frequency and the pitch of a musical sound.

2. Give a fundamental difference between a sound wave and a light wave.

3. What effect does the air temperature have on the speed of sound in air?

4. Why is it impossible for a person to hear a gunshot 25 ft away on the surface of the moon?

5. Give an illustration which would show that the speed of sound is independent of frequency.

6. Distinguish between intensity and loudness of a sound.

7. What name is applied to sound vibrations (*a*) below the audio range, (*b*) above the audio range?

8. What is a fundamental tone?

9. How is it possible for a single vibrating musical string to produce several pitches at the same time?

10. What happens to the resonant frequencies of an organ pipe when the temperature of the air increases?

11. What will be the beat frequency produced by two tuning forks of 352 and 356 Hz vibrating simultaneously?

12. A 384-Hz sound wave travels down a steel pipe at a speed of 5100 m/s. What will be the speed in the pipe of a 768 Hz sound wave?

13. The wavelength of a 384-Hz sound wave traveling in a steel pipe is 13.2 m. What will be the wave length of a 768 Hz sound wave in the pipe?

14. Explain how a pure note of high frequency and high intensity is capable of shattering a fine glass tumbler.

15. Explain how the sounding box of a violin amplifies the sound produced by the vibrating string.

16. A pipe organ is ''tuned'' when the temperature is 20°C. The temperature in the room drops to 10°C, with a consequent decrease in the speed of sound in air. Equations (20.1) and (20.2) indicate that the frequencies of the pipes vary as the temperature varies. Yet the organ will sound in tune. Why?

17. Name the three characteristics of a musical sound. Tell what physical property of the wave determines the characteristics as heard by a person.

18. The speed of sound in a given medium depends upon which of the following? (*a*) the frequency of the sound wave, (*b*) the loudness of the sound, (*c*) the temperature of the medium, (*d*) the wavelength of the sound.

19. Which of the following are not correlated: (*a*) pitch and frequency, (*b*) quality and waveform, (*c*) loudness and intensity, (*d*) loudness and resonance?

20. Which of the following remains unchanged when a sound wave goes from air into water: (*a*) speed, (*b*) frequency, (*c*) wavelength, (*d*) amplitude, (*e*) none of these.

PROBLEMS

Group One

1. What must be the length of a pipe closed at one end to produce a fundamental tone with a frequency of 264 Hz (C note on the major scale)? Use 343 m/s for the velocity of sound in air.

2. If the fundamental frequency produced by an instrument is 400 Hz, what is the frequency of its third harmonic?

3. If the power of the sound in average conversation is about 10^{-5} W, how many people would expend, in average conversation, as much energy as would keep a 60-W light bulb glowing?

4. If middle C (264 Hz) has a wavelength of 5.65 m in a body of water, what is the speed of sound in the water?

5. What are the frequencies of the first four harmonics of a tone whose fundamental frequency is 330 Hz?

6. If the rumble of thunder is heard 7.52 s after the lightning flash is seen, how far from the observer did the lightning occur? Assume the temperature of the air is 22°C.

7. What is the level of sound in decibels of a sound whose intensity is 7.5×10^{-6} W/m^2?

8. What is the speed of sound in air at (*a*) -24°C, and (*b*) 40°C?

9. What is the speed of sound in air at (*a*) -75°F, and (*b*) 100°F?

10. A string is 120 cm long. When it is plucked at the middle it produces a fundamental note of A above middle C (440 Hz). If the string is held 30 cm from one end, what is the frequency of (*a*) the shorter part of the string, and (*b*) the longer part of the string?

11. If the intensity of a sound is doubled, by how many decibels will it be increased?

12. At what temperature is the speed of sound in air 1137 ft/s?

13. The equations for the speed of waves in a fluid, solid, and a vibrating string are respectively, given by: (*a*) $v = \sqrt{B/d}$, (*b*) $v = \sqrt{Y/d}$, and (*c*) $v = v\sqrt{T/m/L}$. Show that in each case the dimensions of v are length/time.

Group Two

14. A steel wire 120 cm long, whose mass is 12 g, is under a tension of 480 N. Determine (*a*) the wavelength of its fundamental mode of vibration, (*b*) the frequency of the fundamental tone, and (*c*) the wavelength of the fundamental sound wave.

15. What is the frequency heard by a listener who is leaving, at a speed of 25 m/s, a stationary sound source emitting a frequency of 980 Hz if the temperature is 24°C.

16. At 0°C and 760 mm Hg, the density of oxygen is 1.43 kg/m^3 and the bulk modulus is 1.418×10^5 N/m^2. Find the speed of sound in oxygen under those conditions.

17. A 3520-Hz sound source and an observer are approaching each other. The sound source is traveling 60 mi/h and the observer is traveling 45 mi/h. The air temperature is 82°F. What is the frequency of the sound heard by the observer?

18. One of the strings in a piano is only 7.5 cm long and produces a fundamental tone whose frequency is 4224 Hz. How long must a similar string be (same mass per unit length and under the same tension) to produce a fundamental whose frequency is 33 Hz? (Your answer should convince you of the necessity of using less tension and wires of greater linear density for the low notes.)

19. What will be the fundamental frequencies and the frequencies of the first two overtones for a 43.4-cm-long organ pipe at 20°C if it is (*a*) open, and (*b*) closed?

20. A piano wire is 55 cm long, has a mass of 5.2 g, and is under a tension of 450 N. (*a*) What is the frequency of its fundamental? (*b*) What is the number of the highest overtone that can be heard by a person who is capable of hearing only frequencies up to 12,000 Hz?

21. What must be the length of an organ pipe open at both ends to produce a fundamental note with a frequency of 264 Hz (middle C) if the temperature is 20°C? What will be the frequency of the fundamental if the temperature is 30°C?

22. A worker strikes the steel rail of a railroad track with a sledgehammer. The sound of the blow reaches an observer through the rail and through the air. The difference in time for the sound to reach the observer is 3 s. How far, in meters, is the observer from the worker? (Use Table 20.1.)

23. A blow struck on a steel rail was heard through the rail in 0.2 s and through the air in 2.9 s. If the temperature was 54°F, (*a*) how far from the observer was the blow struck and (*b*) what was the speed of sound in the rail?

24. A target shooter fires at a target with a rifle. The bullet travels at the average rate of 615 m/s, and 4 s after firing the gun, she hears the bullet strike the target. How far away is the target if the temperature is 25°C?

25. What is the intensity in watts per square centimeter of a sound which has an intensity of 50 dB?

26. What is the intensity level in decibels of a sound which has an intensity of 10^{-8} W/m^2?

27. Two sounds of the same frequency have intensities of 10^{-7} and 10^{-10} W/cm^2. How many decibels louder is the first sound than the second?

Group Three

28. How long will it take for sound waves to travel 450 yd in a copper tube?

29. Sound energy is radiated uniformly in all directions from a source at the rate of 40 W. (*a*) What is the sound intensity in watts per square meter at a point 35 m from the source? (*b*) What is the intensity level, in decibels, at that point?

30. What is the intensity level, in decibels, for Prob. 29 if there is a 15 percent absorption of acoustic energy in the 35-m path?

31. A 32-voice choir is singing at an intensity level of 78 dB. Each voice has the same intensity level. What is the level of each voice?

32. A steel piano wire is 42 in. long and has a diameter of 0.0254 in. What tension should be applied to the wire to make it produce a fundamental frequency of 264 Hz?

33. The frequency of the second overtone of an organ pipe open at both ends is equal to the frequency of the second overtone of an organ pipe closed at one end. Determine the ratio of the length of the pipe with open ends to that of the pipe closed at one end.

34. Find (*a*) the acoustic energy level and (*b*) the sound intensity level at a point 30 m from a source of sound if the sound radiates acoustic energy uniformly in all directions at the rate of 2.0 W.

TECHNICAL APPLICATIONS OF SOUND—ACOUSTICS

The science of controlling sound in buildings is called *acoustics.*

The science of the production, transmission, reception, and control of sound is called *acoustics*. The beginning study of acoustics can be traced in large measure to the work of W. C. Sabine (1868–1919) of Harvard University. It was he who first began to reduce the complexity of sound and hearing in buildings to relatively simple mathematical formulas with which the design of auditoriums and rooms could be greatly improved.

Sound production and reproduction constitute the basis of an entire industry today; and, by the opposite token, the control and suppression of unwanted sounds (noise) is equally important, especially in large urban centers. The term *sound pollution* has properly been applied to the impact of unwanted sound on the sensibilities of unwilling hearers.

It is now routinely expected that architects will specify satisfactory acoustical standards, as well as providing for building safety, convenience, air conditioning, and proper lighting.

21.1 ■ Reflection of Sound

We are all familiar with the rebounding of a rubber ball when it is thrown against a wall. In much the same manner sound waves are reflected from various surfaces. Thunder, for example, is due to successive reflections of sound from cloud to cloud or from cloud to land surface. Certain buildings, for example, Statuary Hall at the United States Capitol in Washington, D.C., and the Mormon Tabernacle in Salt Lake City, are so constructed that sounds will be reflected from their walls and ceilings to give a "whispering gallery" effect. Because of reflected sounds, a pin dropped in a certain place in the Tabernacle can be clearly heard at the other end of the room. In Statuary Hall it is possible to stand at a certain spot in the building and distinctly hear a whisper uttered by someone at certain spots on the other side of the room, although the whisper may not be heard at points in between.

Sound can be reinforced or focused by reflection from smooth surfaces that are properly shaped and placed.

A simple experiment can be performed to illustrate the phenomenon of sound reflection. Place a low-volume source of sound near a hard wall surface (Fig. 21.1*a*) and separate the hearer and the source by a panel *PR*. As the hearer *E* moves about the room, he will discover that it is difficult to hear the sound on the other side of the panel for all positions except those that lie in a given direction. It can be shown that the direction of best audibility is the one which makes angle *SOR* equal to angle *EOR*, which illustrates the following law:

The Law of Reflected Sound Waves

The angle of reflection r is equal to the angle of incidence i:

$$\textbf{Angle } r = \textbf{angle } i$$

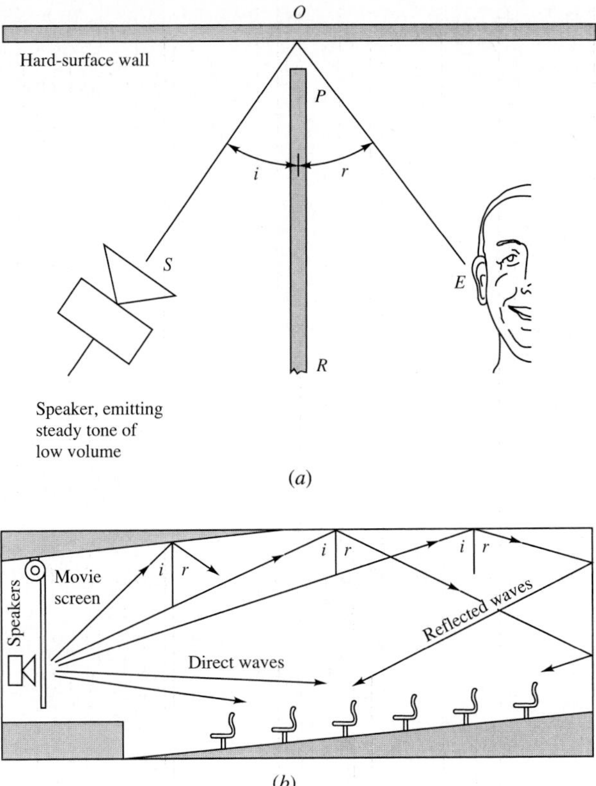

Fig. 21.1 Reflection of sound. (*a*) When a sound wave strikes a reflecting surface and is reflected, the angle of reflection *r* is equal to the angle of incidence *i*. An audio-oscillator with speaker, turned to low volume, is a good source for this simple experiment. (*b*) Auditoriums and lecture halls should be acoustically designed to make use of the law of sound reflection. The audience in a movie theater hears both direct and reflected sound.

Figure 21.1*b* shows how reflected sound can improve hearing in a movie theater.

An *echo* is a reflected sound wave returning to be heard at the place where it originated.

A sound that is reflected from a surface to the listener's ear is called an *echo*. The human ear is able to distinguish two sounds as being distinctly separate only if they reach the ear at least $\frac{1}{10}$ s apart. If the two sounds reach the ear with a shorter time interval between them, they will blend together and give the impression of a single sound. The distance sound must travel to be audible as an echo can readily be determined. Since sound travels at the rate of about 330 m/s (1100 ft/s), or 33 m (110 ft) each 0.1 s, the distance the reflected sound must travel to be distinguished from the original sound must be at least 33 m (110 ft) more than the distance the original sound travels directly to the ear. Consequently, echoes do not cause an interference hearing problem in small to medium-sized auditoriums. Echoes are commonly heard in long corridors and in deep canyons. The depth of a well can be determined by making a sound at the top of the well and noting the time required for the sound to reach the bottom and be reflected back to the ear.

Echoes within large enclosed spaces can cause sound interference and result in muffled and unintelligible speech and discordant music.

Megaphones and loudspeakers are shaped to concentrate sound waves in a desired direction. In various musical instruments, such as the trumpet, the shape of the horn is such as to direct the sound waves in a desired fashion. Proper directing or focusing of sound waves is important in large auditoriums and amphitheaters. Many of the older auditoriums and churches were provided with curved sounding boards placed behind the speaker or stage. The advent of electronically amplified public-address systems has made this practice less common. Many large amphitheaters, however, are still provided with curved bowls behind the stage and slanting baffles above to direct the sound, by reflection, to the audience (see Fig. 21.2).

21.2 ■ Refraction of Sound Waves

At the boundary between two media, the speed of sound changes, but the frequency does not. Sound travels generally faster in denser media than in less dense media.

Whenever a sound wave passes from one medium into another of different density or elasticity, its speed is changed. However, frequency remains unchanged as a new medium is encountered. Since $v = f\lambda$, with frequency constant, λ must increase or decrease in direct proportion to v. A speed increase means a longer wavelength, and

Fig. 21.2 The Sydney Myer Music Bowl in Melbourne, Australia. There is a partially excavated seating area of 2000 seats, above which is a large suspended canopy which stretches upward to a sloping lawn that will seat some 20,000 more people. The canopy and a reflecting shell at the back of the stage project sound effectively to very large audiences. (Bolt, Beranek, and Newman, Inc.)

The change in the speed of sound at a boundary between two media causes a bending of the sound wavefronts, called *refraction.*

vice versa. Speed changes at media boundaries may alter the direction of propagation of the wave. The bending of the direction of a wave due to a change of speed is called *refraction*. All sound waves traveling in the open air are refracted to some extent. This is true because the air is seldom at rest and varies throughout in temperature and density. These variations in temperature and density cause changes of sound speed, with consequent refraction of the sound wave. This phenomenon explains why sound usually does not appear to carry so far in the daytime as at night. The sun in the daytime warms the surface of the earth and the air immediately adjacent to the surface. Hence the air next to the surface is warmer than at higher levels. Since the speed of sound increases with an increase in temperature, the part of the wavefront next to the earth's surface travels faster than that at higher levels. The effect is a bending upward of the wavefront away from the observer on the ground (Fig. 21.3*a*). At night the reverse is true. The ground surface cools faster than the air above. Hence the layers of air next to the ground are the colder ones, and the sound waves are refracted downward (Fig. 21.3*b*). Often in the evening one can hear the sound of voices or music distinctly, even though they may be at a considerable distance from the listener.

Wind may also be a factor causing refraction of sound. The speed of wind is usually slower next to the surface of the earth than at higher levels, because of greater air friction at the surface. Hence if the sound travels in a direction against that of the wind, the top part of the wave is slowed more than the bottom; the result is a bending upward of the wavefront (Fig. 21.4*a*). Conversely, if the wave is traveling with the wind, the faster upper layers of wind bend the wavefront downward (Fig. 21.4*b*). Sound is thus seen to stay longer on the surface if it travels with the wind than if it travels against it.

Refraction of sound waves in water creates problems for underwater sound ranging (*sonar*). Currents and layers, or strata, in ocean water often exhibit significant temperature and density differences, with rather definite boundaries between one water mass and another where the density is appreciably greater or less. Sound signals and their echoes are refracted or bent at the boundaries between two such regions. The refraction makes a returning sonar echo appear to be coming from some other direc-

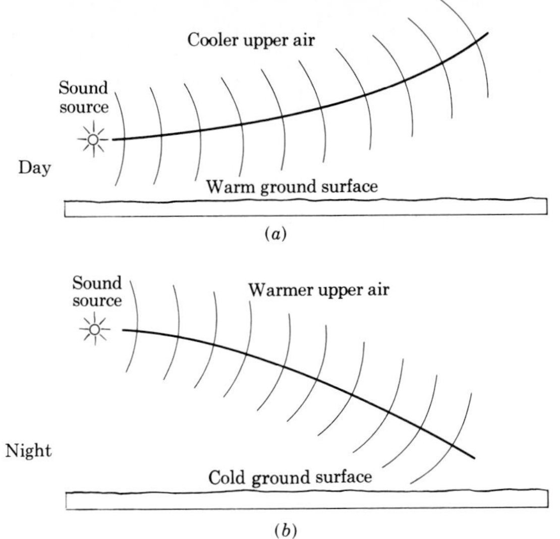

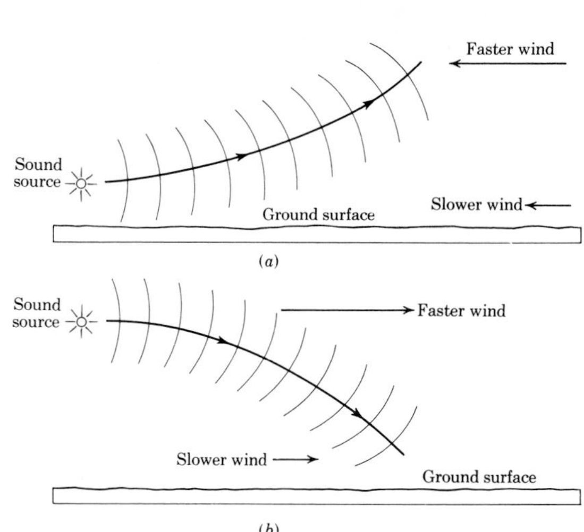

Fig. 21.3 Sound waves are refracted upward or downward according to whether the layers next to the ground are the warmer or the colder air.

Fig. 21.4 Sound waves are refracted upward traveling against the wind and downward traveling with the wind.

tion than the actual bearing to the source of the echo. (See Sec. 21.7 for a brief description of sonar.)

21.3 ■ Absorption of Sound

A sound wave will continue to recede from its source until it is converted into some other form of energy. When a sound wave passes into or through a given material, some of the sound energy is absorbed and converted into heat energy. That is, as the sound-wave energy strikes the absorbing material, it increases the motion of the molecules of the absorbent. This increase in molecular motion appears as added heat energy (see Secs. 14.11 and 14.12). Porous materials are effective sound absorbers because they contain many pockets of air whose molecules can readily be set into increased motion.

Sound is a form of mechanical energy, and like other forms of energy, it can be absorbed.

The greater the conversion to heat, the greater the *absorption coefficient*. The absorption coefficient of a given material is *the fraction of the incident sound energy that it will absorb at each reflection or transmission*. Some materials have low absorption coefficients, and sound waves pass through them or are reflected from them with little loss of energy. Other materials, such as sponge rubber, rugs, draperies, pressed plant fibers, and porous felt, are good absorbing materials and are used commercially for such purposes.

Sound energy is absorbed at a reflecting surface and also within a medium through which its waves are being propagated.

A perfect absorber of sound is a complete nonreflector of sound energy. For indoor conditions an *open window* is a perfect absorber since all the sound energy falling on it passes through it—all absorption and no reflection. The unit of absorbing power in the English system is that of one square foot of open window. This unit is called the *sabin*. The elementary principles of absorption just described are basic to the discussion of the acoustics of rooms and auditoriums, which follows.

The *sabin* is the unit of sound absorption. It is the absorbing power of one square foot of open window. The *metric sabin* is the absorbing power of one square meter of open window.

21.4 ■ Acoustics of Buildings

Materials with a high coefficient of absorption are of importance for the acoustical treatment of rooms and auditoriums. An auditorium is said to have good acoustics when speech or music can be heard almost equally well throughout the space, without troublesome echoes and reverberations. The podium and stage should be so designed

Multiple echoes in halls and auditoriums (and out of doors as well) are called *reverberations*.

Fig. 21.5 Diagram illustrating the units of absorbing power—the sabin and the metric sabin. Sound intensity levels are measured in watts per square meter by sound-level meters, often called *decibel meters*. Such a meter would indicate that there is no reflected sound energy from an open window. All the incident sound energy passes on through—the open window is a perfect sound absorber. One *sabin* is the absorbing power of 1 ft^2 of open window, and one *metric sabin* is the absorbing power of 1 m^2 of open window.

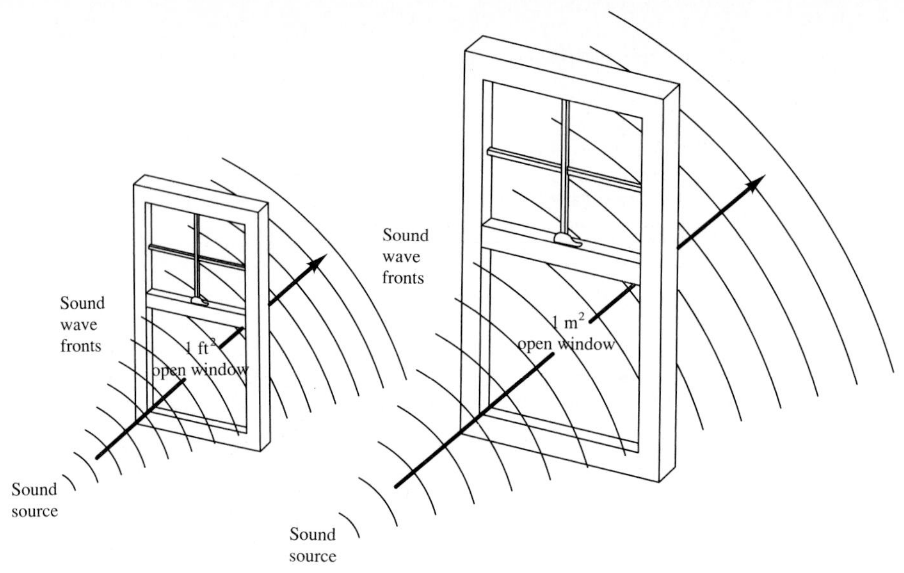

that speech sounds are projected out into the audience and not ''lost'' backstage. Multiple echoes from the ceiling and walls of the room should not be entirely absent or the room will be acoustically ''dead,'' as if the speaker were addressing a crowd in the open air. On the other hand, if multiple echoes (called *reverberations*) persist too long, the echoes from previous syllables uttered by the speaker will arrive at the listener's ear just in time to interfere with the hearing of the next syllable. Music, too, can be adversely affected by excessive reverberation.

Interference is another factor that must be considered in designing auditoriums. Interference will cause variations in intensity—loud spots in some places, and dead spots in others. Interference can be minimized by a proper choice of the dimensions and shape of the auditorium and by having ''clean'' lines, free from pillars, overhangs, and unnecessary architectural embellishments.

The time it takes for an average sound intensity in a room to diminish to one-millionth (10^{-6}) of its original value (a reduction of 60 decibels, db) at the instant the sound source is turned off is called the reverberation time.

Reverberation time is a critical factor in the design of auditoriums for music and speech.

"All-purpose" auditoriums are often ill-suited to either speech or music. Optimum reverberation time for speech (little or no reverberation desired) is about 1.2 s. Concert halls, however, in order to have a "live sound," should be designed for a reverberation time of about 1.6 s.

The reverberation time depends directly on the volume of the room and inversely on the total absorption of the sound by the various surfaces in the room.

The best reverberation time for sound in a room depends on its intended use. For satisfactory reception of speech, a relatively short reverberation time is essential. This is necessary because successive utterances will overlap if there is a long reverberation time. For moderately sized auditoriums (with volumes of between 100,000 ft^3 and 300,000 ft^3) the best reverberation time for speech should be about 1.2 s. In larger auditoriums, no speaker's voice can be heard without the use of a pubic address system, regardless of how short the reverberation time. Most rooms with volumes greater than 200,000 ft^3 require electronic amplification for the satisfactory hearing of speech. Music halls should have somewhat longer reverberation times, since multiple echoes, within limits, will reinforce orchestral sounds and result in a ''live'' auditorium with a certain feel of ''resonance.'' For moderately sized concert halls the optimum reverberation time is about 1.6 s.

As has been noted, Sabine made some of the earliest studies of acoustics for auditoriums, and although many researchers have since refined his techniques and supplemented his findings, his basic formula for reverberation time is still used. Sabine's formula is

$$t = 0.05 \frac{V}{a} \qquad \text{(English system)} \tag{21.1}$$

where

t = reverberation time, s, as defined above
V = volume of room, ft^3
a = absorbing power of the room, including all objects and materials in room, sabins. (See Table 21.1)

The use of the sabin in calculations will be illustrated below.

The experimentally determined factor 0.05 in Sabine's formula assumes an average indoor velocity of sound in air of 1130 ft/s when the temperature is 72°F.

For *metric* calculations, the reverberation time formula is

Sabine's formula is used in both the English and metric systems of measurement, but care must be taken with units.

$$t = 0.164 \frac{V}{a} \qquad \text{(metric system)} \tag{21.1'}$$

where

t = time, s
V = volume of room, m^3
a = total absorbing power of room, *metric sabins*

With the absorbing power of one square foot of open window being assigned a value of 1.00, other materials and surfaces can be rated accordingly, based on accurate measurements by sound-intensity meters of the residual sound energy reflected from known areas of these materials. For example, if 5 ft^2 of a certain carpet material results in the same decrease in residual sound energy (attenuation) as 1 ft^2 of open window, that carpet is absorbing one-fifth as much sound energy as an open window. The absorbing power of unit area of a surface or material is termed the *coefficient of absorption*. The coefficient of absorption of the carpet cited is then 0.2. In the English system coefficients of absorption have the units sabins/ft^2; and in the metric system, metric sabins/m^2. The symbol β (Greek *beta*) is used to denote the absorption coefficient.

The gradual diminution of audible sound, as sound energy is absorbed or dissipated, is known as *attenuation.*

Table 21.1 lists absorption coefficients (β) for several common materials and surfaces, plus some values of *total absorption* for objects, such as chairs and people, commonly found in auditoriums. Accurate determinations with modern sound-level meters show that the coefficients vary significantly with the frequency of the sound. The values given in the table are for a frequency of 512 Hz.

The total *absorbing power* of a surface or material is equal to its area multiplied by its absorption coefficient, that is, $a = \beta A$. The total absorbing power of all the surfaces, materials, and objects in a room is equal to the sum of all the separate absorbing powers of the room's surfaces, materials, and objects.

$$a_T = a_{\text{objects}} + \beta_1 A_1 + \beta_2 A_2 + \beta_3 A_3 + \cdots \tag{21.2}$$

where the A's are areas of the several room surfaces.

Illustrative Problem 21.1 Compute the reverberation time of a small auditorium 70 by 110 by 25 ft high. The following are the areas, in square feet, of the building surfaces: smooth plaster, 10,500; wood panel, 6000; linoleum, 6700; heavy carpet, 1000; velour draperies, 500; Acousti-Celotex, 1000. There are 500 people (seated) and 300 empty upholstered chairs in the auditorium.

Solution The calculation of absorbing power a is carried out in tabular form, as follows:

Table 21.1 Sound Absorption Coefficients—512 Hz

	Absorption coefficient β (sabins/ft^2 or metric sabins/m^2) β
Open window	1.00
Acousti-Celotex	0.82
Acoustic plaster	0.52
Brick, common	0.025
Carpet, heavy, with pad	0.40
Chalkboard	0.05
Draperies, cotton	0.40
Velour	0.55
Felt, heavy	0.70
Glass	0.025
Linoleum, asphalt tile	0.03
Marble	0.01
Plaster wall, smooth	0.03
Wood wall or floor, painted or varnished	0.04
Wood wall, floor, or paneling (unpainted)	0.08

	Absorbing power, sabins	
	English	Metric
Audience, each person, seated	4.4	0.44
Auditorium chairs, wood, each	0.3	0.03
Upholstered, each, empty	2.0	0.20
Occupied	1.5	0.15

$$
\begin{aligned}
&\text{Plaster, } 10{,}500 \times 0.03 &&= 315 \text{ sabins}\\
&\text{Wood panel, } 6000 \times 0.08 &&= 480\\
&\text{Linoleum, } 6700 \times 0.03 &&= 200\\
&\text{Heavy carpet, } 1000 \times 0.40 &&= 400\\
&\text{Velour draperies, } 500 \times 0.55 &&= 275\\
&\text{Acousti-Celotex, } 1000 \times 0.82 &&= 820\\
&\text{500 people, } 500 \times 4.4 &&= 2200\\
&\text{500 chairs, occupied, } 500 \times 1.5 &&= 750\\
&\text{300 chairs, empty, } 300 \times 2.0 &&= \underline{600}\\
& &a &= 6040 \text{ sabins}
\end{aligned}
$$

From Eq. (21.1), (English units),

$$t = 0.05 \times \frac{70 \times 110 \times 25}{6040} = 1.59 \text{ s} \qquad \textit{answer}$$

Illustrative Problem 21.2 An auditorium has basic dimensions of 40 m × 20 m × 12 m ceiling height. It is to be used primarily for musical performances and a "live" sound is desired. A reverberation time of 1.6 s is chosen as the acoustical specification. The following areas and surfaces are a part of the basic construction: hardwood floor, 700 m^2; floor covered with heavy carpet, 100 m^2; walls and ceiling, smooth plaster, 1800 m^2. At performances, the design calls for 700 people seated and 50 unoccupied upholstered chairs. Calculate the number of square meters of Acousti-Celotex that must be applied to the walls and ceiling to attain the design reverberation time of 1.6 s.

Solution From Eq. (21.1′) (metric units),

$$a = \frac{0.164\ V}{t}$$

Substituting $= \dfrac{0.164 \times 40 \text{ m} \times 20 \text{ m} \times 12 \text{ m}}{1.6 \text{ s}}$

$= 984$ metric sabins

Now calculate the absorbing power of the design elements that are fixed—the floor, the audience, and the seats. Call this a_1. (See Table 21.1 for values.)

Absorbing power of the fixed elements

Floor, wood	$700 \text{ m}^2 \times 0.08 =$	56 metric sabins
Floor, carpet	$100 \text{ m}^2 \times 0.40 =$	40
700 people	$\times 0.44 =$	308
700 chairs, occupied	$\times 0.15 =$	105
50 chairs, empty	$\times 0.20 =$	100
	$a_1 =$	609 metric sabins

The remaining wall and ceiling surfaces, then, must contribute a total absorbing power of $984 - 609 = 375$ metric sabins.

Two (simultaneous) equations, as follows, will be used:

Let x = remaining plaster area (m^2), and y = the required Acousti-Celotex area (m^2).

(1) $x + y = 1800$ (summing up the areas)
(2) $0.03x + 0.82y = 375$ metric sabins

From Eq. (1), $x = 1800 - y$

Substituting this value of x into Eq. (2),

$$0.03\,(1800 - y) + 0.82y = 375$$
$$54 - 0.03y + 0.82y = 375$$
$$0.79y = 321$$
$$y = 406 \text{ m}^2 \qquad \textit{answer}$$

It is best not to apply too much absorbing material to ceilings, since in long auditoriums the ceiling is needed as a reflector to carry sound into the back rows of seats (see Fig. 21.1*b*).

A modern tendency is to put too much reliance on electronic amplifier systems to "improve the acoustics" of halls and auditoriums. Merely amplifying sound energy, however, is often no aid to better hearing. If undesirable echoes result in interference, no amount of loudness will compensate for unintelligible speech or muffled music. If the reverberation time is too short, the spoken voice sounds too "thin" and music is nonresonant and without "fullness," even though the volume is satisfactory. Furthermore, amplifying systems, unless they are of exceptionally high quality (and high cost) and have been installed by experienced electronic technicians and acousticians, may exhibit poor fidelity; and the higher the gain is turned up, the greater the distortion. If a hall is large enough to need electronic amplification, attention should also be given to improving its basic (structural) acoustics as well.

If the acoustics of a hall or auditorium are unsatisfactory, at least as much effort and expense should go into improving the hall's structural elements as goes into designing and installing an electronic amplifying system.

Reflecting surfaces are often used to improve the acoustical properties of rooms and auditoriums. Parabolic shells are quite effective, with the sound source located as near as possible to the focal point of the parabolic shell. Large auditoriums, not originally designed for concerts or opera performances, can often be made acoustically acceptable by the proper use of baffles or shells projecting outward and upward from the stage. Outdoor "bowls" are nearly always fitted with such devices. (See Fig. 21.6.)

The science of acoustics is not limited to auditoriums. Other applications include the design of movie sound studios, recording studios, anechoic chambers for testing laboratories, sound barriers along expressways, and other devices for suppressing the "noise pollution" of everyday life.

(*a*)

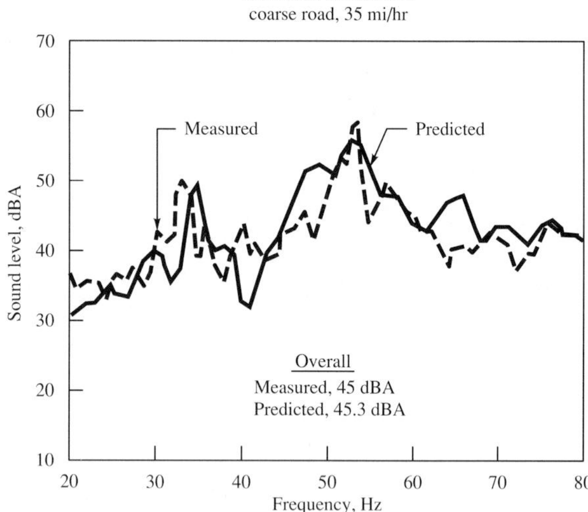

(*b*)

Fig. 21.6 (*a*) An example of good acoustical design. The Foellinger Great Hall in the Krannert Center for Performing Arts at the University of Illinois. The auditorium seats 2100 persons and is used primarily for music. The audience and seats provide most of the absorbing power. A reverberation time of 2.25 s results in a very ''live sound''—excellent for music. (University of Illinois at Urbana-Champagne) (*b*) Auto manufacturers test their new models for road noise with precision sound-level meters. Computer simulations of expected noise levels are checked against actual road tests. The actual road test, which is here compared to predicted results, was conducted with a new type of van body on a coarse-surface road at 35 mi/h. (General Motors Corp.)

21.5 ■ Supersonics—Traveling Faster Than the Speed of Sound

Any object travelling faster than the speed of sound in air is said to be *supersonic*.

Until fairly recently the only things that could travel through air faster than sound were light and certain gun-fired projectiles. In this era of jet aircraft and high-altitude rockets, however, objects moving through air at speeds in excess of the speed of sound are commonplace. Because the flow of air around such high-speed objects is a complex phenomenon, the study of *supersonic* speeds has a major place in research laboratory programs throughout the nation.

Sonic booms are caused by shock waves trailing behind aircraft or spacecraft flying at supersonic speeds.

When an object moves through air with a speed greater than that of sound at the same altitude, the particles of air are unable to get out of the way fast enough, causing a tremendous compression of air in front of the object. This highly compressed air, followed by a considerable rarefaction, forms a cone-shaped envelope of shock waves similar to the sketch of Fig. 21.7. If you are in the line of travel of such a shock wave, the great pressure change from condensation to rarefaction as the wave passes you causes a *sonic boom*. One reason the sonic boom startles a person is that the approach of a supersonic aircraft is absolutely silent, since it travels ahead of the noise it makes. Some booms that reach the earth can be quite destructive in terms of window breakage, as well as being annoying for their loud sound.

The shock wave that produces a sonic boom is a strong pulse whose formation can be explained as follows. First, note from Fig. 21.8*a,* the wavefront pattern that is established as an object moves through air with *less than the speed of sound.* As Doppler's principle (Sec. 20-11) predicts, the wave fronts ahead of the moving object still *move out ahead* of the object, even though they are compressed closer together.

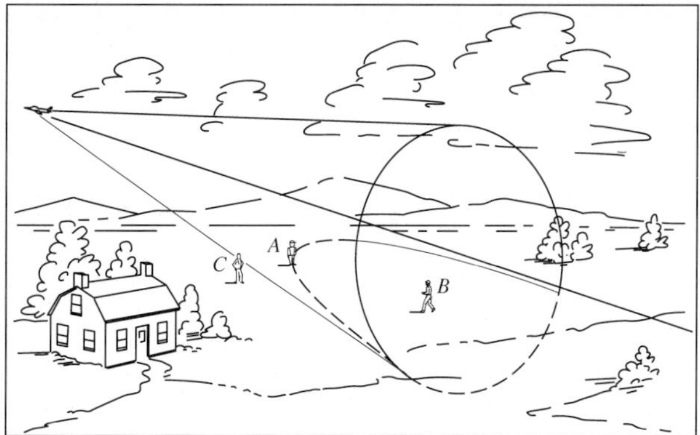

Fig. 21.7 Shock-wave cone produced when an aircraft travels at supersonic speeds. B has already heard the sonic boom, A is just sensing its arrival, and C will hear it very soon.

The wave fronts trailing the moving object are spaced farther apart behind the object. In other words, the wavelength is shorter in front of the moving object, and longer behind it. In contrast, consider Fig. 21.8*b*, depicting an object whose speed is greater than the speed of sound. The source is going faster than the sound it produces. All the spherical sound waves now trail the speeding object, and their ''envelope'' is conical in shape. The surface of this moving cone is the *shock wave front*. It is the result of constructive interference among the myriads of sound waves emanating from the leading edges of the moving source. The shock wave moves at the same speed as the aircraft, rocket, or bullet, but because of its conical shape, the resultant sonic boom is not apparent to an observer on the ground until several seconds after the supersonic object has passed overhead. The exact time lapse depends on the speed and altitude of the moving object.

A supersonic object is a source of sound, but the source is going faster than the sound waves it produces.

In reality, two shock waves are produced by a large aircraft or spacecraft such as the Space Shuttle—one at the leading edge of the craft and one at the rear. Consequently, a ''double boom'' is typical of the sound that is heard. Supersonic speeds are typical of many military aircraft today, and also of transport aircraft of the Concorde (SST) type.

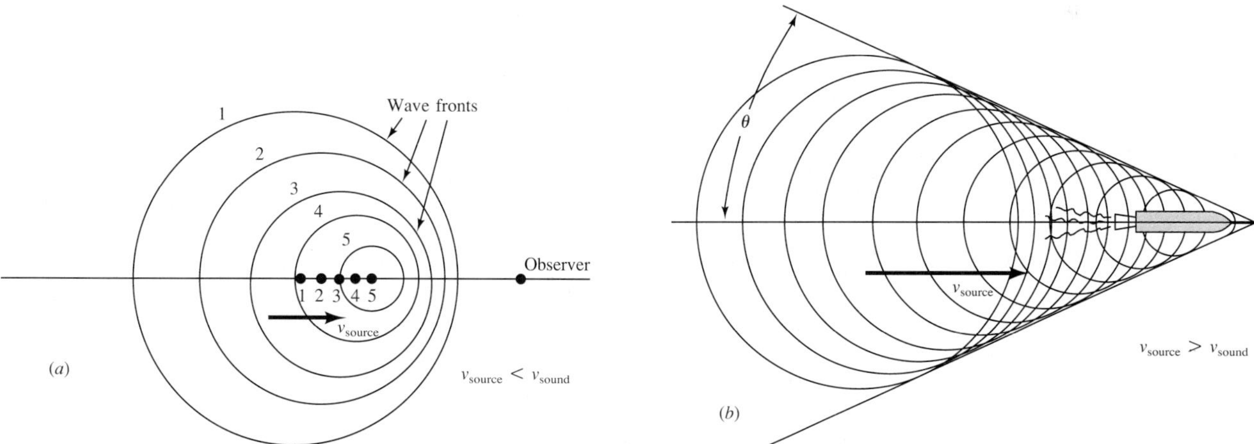

Fig. 21.8 Wave fronts and the Doppler effect. (*a*) Here the source of sound is flying overhead at less than the speed of sound. The wave fronts in front of the moving source are compressed—those trailing it are stretched out. Sound waves still move out ahead of the source. An observer on the ground will hear the sound *before* the source is directly overhead. The pitch of the sound will be higher than the source frequency justifies, however. (*b*) Sound waves from a flying source moving faster than the speed of sound. The spherical wavefronts emanating from the needle nose of the rocket never get ahead of the rocket. They all trail behind and their ''envelope'' is a conical surface or *shock wave* that moves with the source. The angle ϕ is a function of v_{source} and v_{sound} at that time and place (see Eq. 21.4).

Mach number is the ratio of the speed of a moving body to the speed of sound in air at that place and time. Mach 1.00 is about 770 mi/h, or 1130 ft/s, or 1240 km/h.

"Breaking the sound barrier" occurs as an aircraft reaches and exceeds Mach 1.00.

Mach Number Speeds which approach or exceed the speed of sound are described by a designation called *Mach number* (for Ernst Mach, 1838–1916, an Austrian scientist). Mach number is the ratio of the speed of the moving body to the speed of sound in air at that place and time:

$$\text{Mach number} = \frac{\text{speed of moving body}}{\text{speed of sound}} \tag{21.3}$$

Since the speed of sound varies appreciably with the temperature, the Mach number speed depends on where the object is flying—at a low or high altitude, or on a cold or warm day. For sea-level flight under normal daytime conditions, Mach 1.00 is about 344 m/s (1130 ft/s), or 1240 km/h (770 mi/h). The phrase ''breaking the sound barrier'' refers to an aircraft's accelerating from less than the speed of sound on through Mach 1.00.

As an aircraft or spacecraft approaches the speed of sound in air (Mach 1.00), the sound waves that have been moving out ahead of the vehicle become ''piled up'' or compressed. There is a tightly packed region of alternate compressions and rarefactions that constitutes what we call the *sound barrier*. As Mach 1.00 is reached and exceeded, there is a momentary period of great turbulence (breaking the sound barrier) and then a smooth ride at supersonic speeds. After Mach 1.00 is exceeded the barrier disappears, since the sound waves from the vehicle are now ''piling up'' and interfering with one another *behind* the vehicle.

High-speed military aircraft and SSTs like the Concorde fly at speeds well in excess of Mach 2.00.

Projectile Speeds Most high-speed projectiles exceed the speed of sound, and as they pass through air they form a *bow wave* that ''vees'' back from the tip, as shown in the shadowgram photograph of Fig. 21.9. The same effect occurs when a fast boat moves across the surface of water at a speed greater than the speed of the water waves themselves.

Let us represent the conditions shown in such a photograph by the diagram of Fig. 21.10*a*. Let the angle ϕ be the angle between the bow-wave front and the axis of the projectile. The wavefront represents a condensation traveling out from the projectile, and the wavefront is perpendicular to the direction of the propagation of the sound itself. The vector diagram of Fig. 21.10*b* shows v_s representing the velocity of the sound in a direction perpendicular to the bow-wave front. The value of v_s is known if the air temperature is known. The length of vector v_B, representing the velocity of the projectile, is obtained by erecting a perpendicular to v_s at C, which will cut v_B at B. From the figure,

$$v_s = v_B \sin \phi$$

from which

$$v_B = v_s/\sin \phi \tag{21.4}$$

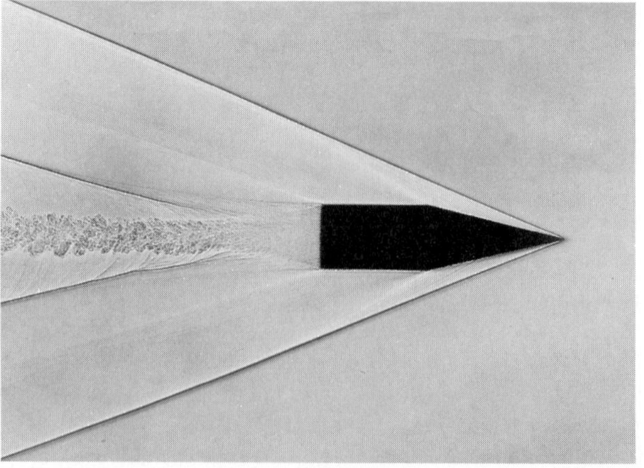

Fig. 21.9 A supersonic projectile produces a cone-shaped shock wave. By special photographic techniques, a *shadowgram* can be obtained. This is a test projectile fired from a 20-mm antiaircraft gun. The speed of the projectile can be determined from such a photograph (see Eq. 21.4).

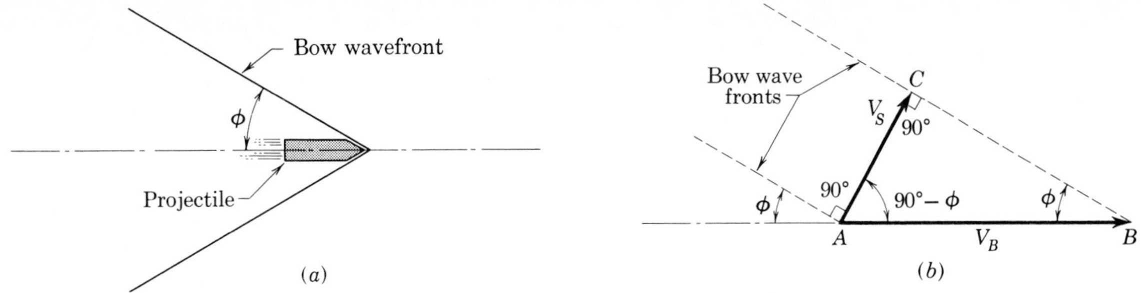

Fig. 21.10. Measuring the speed of a supersonic object. (*a*) A projectile with a bow wavefront that makes an angle ϕ with the flight axis. (*b*) Analysis of the angles and vectors to derive an expression for the velocity of the projectile (see Eq. 21.4).

Illustrative Problem 21.3 Ballistic test photographs to determine the speed of a projectile showed a bow-wave angle $\phi = 24°$. The air temperature was 23°C. What was the speed of the projectile.

Solution The speed of sound at the time was

$$v_s = 331.4 + 0.61(23) = 345.4 \text{ m/s}$$

From Eq. (21.4),

$$v_B = v_s/\sin \phi$$

we obtain

$$v_B = \frac{345.4}{\sin 24°} = \frac{345.4}{0.41} = 840 \text{ m/s} \qquad \textit{answer}$$

21.6 ■ Sound Waves and Seismography

Sound waves and other vibrations travel through the earth's crust, their velocity and waveform being affected by the nature of the formations through which they pass. *Seismology* (*seismo* is a Greek root meaning "earthquake" or "earth vibration") is the science that deals with earthquakes and artificially produced vibrations in the earth's crust. The technique of recording and measuring both natural and induced earth vibrations is called *seismography*.

The study of waves and vibrations in the earth is known as *seismology*—a science applicable to both earthquake analysis and explorations for oil.

Locating Earthquakes The approximate locations of earthquakes can be determined by the use of an instrument called a *seismograph*. When an earthquake occurs, the vibration source produces two distinct types of waves: longitudinal and transverse. These two types of waves travel with different velocities through the earth's crust. Hence it is possible to determine the distance to the source by recording with the seismograph the time interval between the arrival of the two types of waves. By recording the character and time pattern of the vibrations with seismographs at several points on the earth's surface, it is possible to locate fairly accurately the center (called *epicenter*) of the disturbance.

A study of earthquakes has given us our most reliable information about the makeup of the earth's interior. Engineers, by studying the character of the shocks, are now able to construct dams, canals, pipelines, and buildings which will resist or withstand all but the most severe quakes.

Seismographs are installed at known critical points along major earthquake "faults" for continual monitoring of earth movements in those regions. Predicting earthquakes is still not possible, but knowledge gained from seismology can minimize destruction of life and property.

Exploration for Oil Seismographs are also used in geophysical exploration for petroleum by determining underground structures. Waves set off by a buried explosive charge travel through the earth and are partially refracted and reflected to the surface through adjacent layers of formations of different densities (Fig. 21.11). The returning seismic waves are picked up by sets of sound detectors, called *geophones* or *transducers*, arrayed on the ground at equal intervals. The transducers convert the mechanical earth vibrations into weak electrical impulses. The impulses are amplified and recorded on tape or a film strip. The record is superimposed on grid lines, usually spaced at 0.01-s intervals. The instant of the firing of the explosive charge is also recorded on the tape.

The first impulses recorded by the geophones are results of the refracted waves. Impulses from the reflected waves will follow. The time for the reflected waves to appear will depend on their velocity within each of the layers penetrated and on the depths of the reflecting surfaces. The time intervals which elapse between the firing of the charge and the arrival of the reflected waves can be used to determine the depths of subterranean structures that may trap the oil and/or gas. Impulses are timed with an accuracy of 0.001 s. This produces an accuracy in depth determination of 1.0 to 1.5 m (3 to 5 ft) for every 300 m (1000 ft) of depth.

Seismographs are also used extensively by volcanologists to study the action of volcanoes. By keeping records over time of the earth movements around active volcanoes (or those suspected of becoming active) thermal and pressure conditions deep under the mountain can be estimated. As seismic activity builds to a danger point as shown on the monitoring seismographs, warnings are issued that may save lives and property.

21.7 ■ Ultrasonics—High-Frequency Sounds

Sounds with frequencies above about 20,000 Hz are not sensed by the average normal human ear. These high-frequency sounds are classified as *ultrasonic*.

The science and technology of high frequency sound vibrations is called *ultrasonics*. The normal human ear can hear vibrations up to approximately 20,000 Hz. Sound waves with vibrations greater than 20,000 Hz (beyond the limit of audibility) are called ultrasonic. Ultrasonic waves have been developed that vibrate at frequencies near 20 billion hertz. Ultrasonic waves can be generated from mechanical (such as whistles and sirens), electromagnetic, or thermal sources. Devices which convert

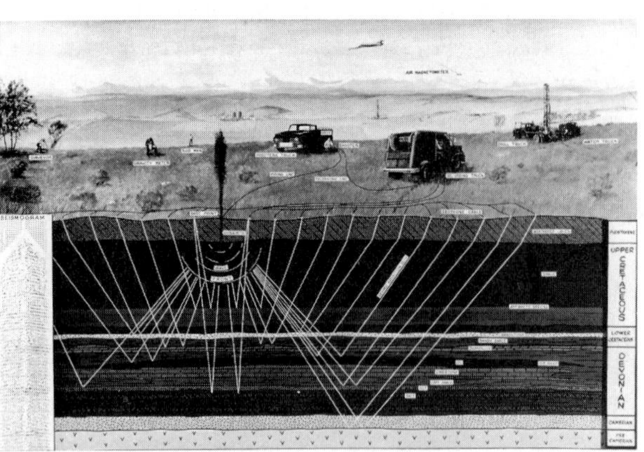

Fig. 21.11 A geophysical crew in the field using three primary geophysical methods—seismic, gravity, and magnetic. In seismic exploration, explosive charges are set off in holes drilled in the earth's surface. These create sound waves that, as they travel in the earth's crust, are refracted and reflected from formations of different composition and structure. By timing the return of the various portions of the wave, the geologist is able to draw accurate profiles of the subterranean strata using knowledge previously acquired from actual core-drilling operations. (Courtesy of the Society of Exploration Geophysicists)

electric, magnetic, mechanical, or heat energy into acoustic or ultrasonic energy (or vice versa) are called *transducers*.

The applications of ultrasonic vibrations are many, and the potential is great for more practical applications to be developed in the near future. We cite a few current uses of ultrasonics.

Sonar Although *audible* sound waves travel considerably farther in water than in air, their range in water is still quite limited. Ultrasonic waves however, can pass through many miles of water before the intensity drops to half its original value. The device for using ultrasonic waves for underwater observation, communication, and navigation is called *sonar*. The word is an abbreviation for Sound Navigation and Ranging. The frequencies most often used in submarine sonar systems range from 20,000 to 50,000 Hz (Fig. 21.12).

Ultrasonic sound is used extensively in underwater ranging and communications—an application called *sonar.*

Ultrasonic beams can be made that are narrow and highly directional. Thus, when an ultrasonic signal is sent through the water as a narrow beam, any obstacle in its path will reflect the beam back to a transducer-receiver, which converts the reflected signal into an audible sound and to a pattern on a cathode-ray tube (CRT) viewing screen.

Submarines are not alone in their dependence on ultrasound for navigation and/or communication. Dolphins are known to "navigate" and locate underwater objects, including food fish and other dolphins, by emitting short pulses of ultrasound at frequencies around 100 kHz. Bats also use ultrasound by sending out short pulses of up to 150 kHz and then using the returning echoes to avoid objects in the dark, or to locate small objects (insects, for example) for their evening meal. Dogs can hear sounds well above 20,000 Hz, and dog "whistles" operating in the ultrasonic region are often used to call dogs or signal sheepdogs to attend to their flocks.

Oceanography Ultrasonic waves are used in determining the depth of ship channels, the location of shoals, the exploration of the ocean and lake floors, underwater distance measurements, and in fishing. The operation is called *depth sounding* and is similar to that used in sonar ranging. The instrument used to detect the reflected waves is known as an *echo sounder*. The sounder (transducer) on the oceanographic

Fig. 21.12 Ship control personnel and sonar technicians on watch in a U.S. nuclear submarine. Sonar (underwater ultrasound) is essential to the mission of such vessels. (Official U.S. Navy photograph)

Sonar (often called *echo-sounding* by oceanographers) is used extensively to measure ocean depths, plot ship channels, chart shoals, and determine the underwater conformation of icebergs. Echo sounders are also used as fish finders by commercial fishermen.

vessel functions automatically. A signal is periodically directed vertically downward into the water, and the time interval between the transmission of the signal and the reception of its echo is a measure of the water depth. The method of depth sounding is also of value in locating schools of fish, submerged icebergs, derelicts, or other obstructions in the channels of ship travel. Similar equipment allows nuclear powered submarines to cruise safely just below the ceiling of the ice cap of the polar regions.

Illustrative Problem 21.4 A crew on a ship in the U.S. Coast and Geodetic Survey, doing depth-sounding work off the Alaskan coast, found that an interval of 1.90 s elapsed between the sending of a wave pulse to the ocean bottom and the return of the reflected wave. If the speed of sound in the water was 1460 m/s, what was the ocean depth?

Solution Using the formula $v = s/t$, solve for s and substitute the known values for v and t:

$$\begin{aligned} s &= vt \\ &= 1460 \text{ m/s} \times 1.90 \text{ s} \\ &= 2770 \text{ m} = \text{distance traveled by sound} \end{aligned}$$

Since the distance to the ocean floor is one-half the distance the sound travels, we get the ocean depth

$$\begin{aligned} D_{\text{ocean}} &= \tfrac{1}{2}(2770 \text{ m}) \\ &= 1385 \text{ m} \qquad \textit{answer} \end{aligned}$$

Ultrasound in Medicine The penetrating power of ultrasonic waves makes them valuable in certain types of diagnostic work in medicine. Transmission, reflection, and refraction of ultrasonic waves help the doctor locate solid particles, organs, and growths (e.g., gallstones, kidney stones, cysts, and tumors) in the human body.

One of the most useful new tools in diagnostic medicine is *sonography, or ultrasound imaging.* Body organs, tissues, tumors, arteries, gallstones, kidney stones, etc., can, with computer enhancement, be shown in color on a cathode-ray tube screen.

The rapidly developing use of ultrasound as an aid in medicine is known as *sonography* or *ultrasound imaging*. High-frequency sound waves are ''pulsed'' into the patient's body from a small transducer which is moved back and forth on the skin directly over the organ or other object being studied. The echoes that return to the transducer from the tissue, tumor, or fetus are ''digitized'' or reconstructed by a computer, and translated into a line-by-line graphical presentation, often in color, on a cathode-ray tube (CRT) viewing screen. Combined with a technique called *digital color Doppler*, sonography can utilize Doppler's principle to determine whether or not blood is flowing normally in an artery under study. Current versions of cardiac ultrasound equipment allow full-color representations on the screen of a beating heart, with the actions of the valves and the blood flow in the aorta readily visible. Ultrasound is at present the first choice among body-scanning techniques for the prenatal examinations of pregnant women (Fig. 21.13).

In order for sound waves to provide useful information about objects, tissues, or organs of the body, the waves themselves must be extremely short compared to the size of the ''target'' being examined. Short waves mean high frequencies, and sonography equipment (ultrasound scanners) often use frequencies above one megahertz (MHz).

In addition to diagnostic application, as above, ultrasound at high-intensity levels has been used in therapy. Heat is generated in bodily tissue, with reported beneficial effects in the treatment of arthritis and similar skeletal and muscular problems. Research is currently being conducted on the use of extremely high-energy ultrasound shockwaves for ''smashing'' gallstones. If the process is perfected, many major surgical operations can be avoided.

Nondestructive Testing The sonar echo reflection method is widely employed by industry in the inspection and detection of internal flaws or holes in many materials. By the proper adjustment of the transmission frequency of the waves, solid

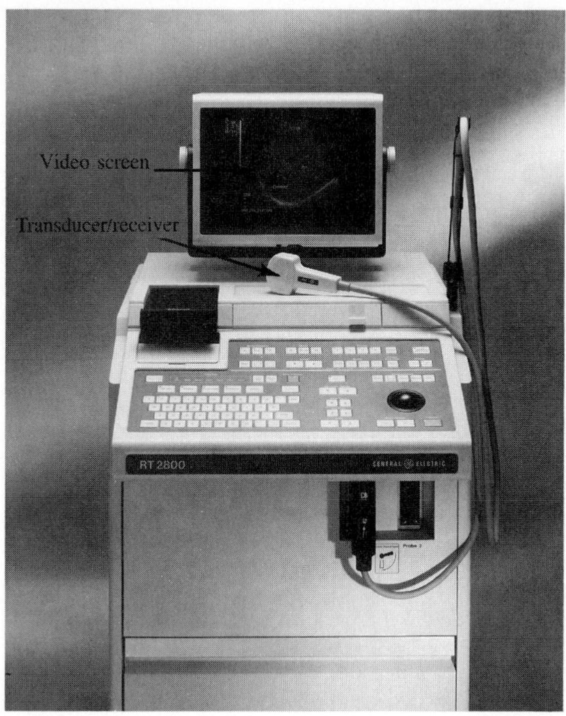

Fig. 21.13 Ultrasound scanner for use in medical diagnosis. As the transducer/receiver is rubbed across the body area immediately above the organ or suspected abnormality to be scanned, ultrasonic waves generated by the machine are reflected back from the organ or tissue under study. These "echoes" are computerized to provide a visual image on the face of the cathode ray tube (CRT). (General Electric Co.)

materials up to 20 ft thick can be penetrated by ultrasonic waves. Because of differences in densities and elastic properties, the holes, cracks, and impurities found in solid materials will reflect the ultrasonic waves. Ultrasonic waves can thus detect flaws in objects such as welds, castings, forgings, and machine parts without any damage to the object being tested.

Ultrasonic Cleaning Another commercial application of ultrasonic energy is that of ultrasonic cleaning. The process is used in the cleaning and degreasing of parts and assemblies where extreme cleanliness is required or where laborious hand-cleaning is to be avoided. It is used especially in cleaning small complicated parts used in the automotive, aircraft, and electronics industries. It is also widely used to clean optical, surgical, and other precision instruments. The object to be cleaned is subjected to high-intensity ultrasonic waves in a cleaning fluid. The waves tear the adhering foreign particles from the surface (Fig. 21.14).

Other Uses Ultrasonic drills are used in machining very hard materials. Since the drill does not rotate, it can produce holes of any shape.

Ultrasonic welding is used in bonding such metals as aluminum and tin. The intense vibratory action of the ultrasonic welding head removes the oxide scale from the aluminum, thus eliminating the need for fluxes.

By utilizing the properties of sound, engineers have made it possible to record and preserve famous voices, great music, and the sounds of important events. All sound-recording devices obey the same general principles. All sounds are vibrations. To reproduce the sounds, these vibrations must be recorded with as little distortion as possible on some sort of disk or tape or film. It then remains for the impressions on the record to be reconverted to the identical sound-wave patterns produced by the original source.

The more common storage systems for sound at present are: disk records, magnetic tapes, sound-on-film, and computer disks (floppy and hard-disk types).

By modulating both edges of the master disk groove, stereophonic sound signals can be etched in a single groove and picked up separately by a special stereophonic cartridge. Two electrical signals are picked up by a single transducer. These signals are amplified separately and are reproduced by two or more loudspeakers. When recording for stereophonic sound of a large sound source, e.g., a large orchestra, microphones are placed near each end of the ensemble. When these recordings are played back through a two-speaker system, the listener experiences an angular separation of sounds which approximates that which he would experience at the live concert.

Tape Recorders The familiar tape recorder impresses the sound waves on a thin plastic tape in the form of patterns of magnetized ferric oxide or chromium dioxide particles. The magnetic patterns on the tape correspond to the wave motion of the original sound. The tape cassette can be played back immediately, or the sound can be wiped out by demagnetization and the tape used over and over again. Both records and tapes can be used as components of high-fidelity and stereophonic sound systems.

Sound on Film The sound which accompanies a motion picture is sometimes recorded, by electro-optical devices, as a small variable-density (or variable-area) sound track on the edge of the film. The sound waves are converted into a small strip on the side of the film with variations in film density or width corresponding to the variations in the voice currents. Light that is allowed to shine through the sound track as the film runs through the projector will then have variations corresponding to the voice or music recorded on the film. These variations in light energy are converted into corresponding changes in electric energy by a photoelectric cell. These weak electrical variations are amplified and converted into sound by the loudspeaker. Dialogue for the film may be recorded as the pictures are being taken. However, other sounds are usually not recorded on the film at the same time the pictures are taken. They are recorded in the studio later and then synchronized to assure exact coordination with the action on the film.

With the introduction of stereophonic sound reproduction, magnetic sound tracks are fast replacing optical tracks. It is now possible to place five or six magnetic sound tracks in the space formerly occupied by one optical track.

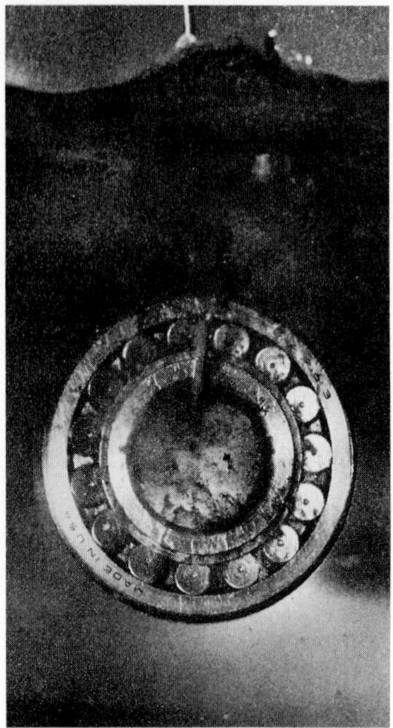

(*a*)

(*b*)

Fig. 21.14 Ultrasonic degreasing of machine parts. (*a*) Immersion in solvent through which high-frequency sound waves are transmitted. The high-intensity ultrasonic waves blast foreign particles from the surface of the metal. (*b*) Final vapor rinsing and drying. (Branson Cleaning Equipment Co.)

Microphones and Loudspeakers—Transducers The term *transducer* is used to indicate any device that converts sound waves into electrical variations (a *microphone* or *transmitter*); or, conversely, that converts electrical variations into sound waves (a *loudspeaker* or *receiver*). In some transducers both transmitting and receiving are accomplished by the same device.

A typical microphone of the *moving-coil type* is schematically portrayed in Fig. 21.15. The diaphragm *D* is small and cone-shaped, made of fluted or corrugated paper or plastic. It vibrates in exact accord with the sound waves (compressions and rarefactions) it intercepts. Attached to the cone's neck or ''funnel spout'' is a coil *C* of many turns of fine wire, wound on a cardboard spool that vibrates back and forth in a magnetic field supplied by a circular permanent magnet *M*. A weak alternating electric voltage, with properties which are the electrical counterpart of the original sound waves, is induced in the vibrating coil and is fed to an electronic amplifier, the output of which can energize loudspeakers or recording devices.

One form of loudspeaker has a similar construction and similar principles of operation, but the procedures are reversed. As depicted in Fig. 21.16, the diaphragm is quite large, since large volumes of air must be set in motion. Again, the diaphragm is fluted or corrugated to allow flexibility of movement. A *voice coil C*, of many turns, wound on a hollow cylindrical core as shown, is attached at the apex of the cone. It is free to vibrate back and forth in the field of the permanent magnet *M*. Fluctuating voltages from an amplifier are fed to the voice coil and these cut the lines of the magnetic field, resulting in forces that push the voice coil (and the diaphragm) back and forth in a faithful reproduction of the wave form from the amplifier. The diaphragm then produces alternate compressions and rarefactions in the surrounding air that constitute sound to the listening ear. The faithfulness with which a microphone-amplifier-loudspeaker system reproduces all the nuances of tone and quality of the original sound is called *fidelity*.

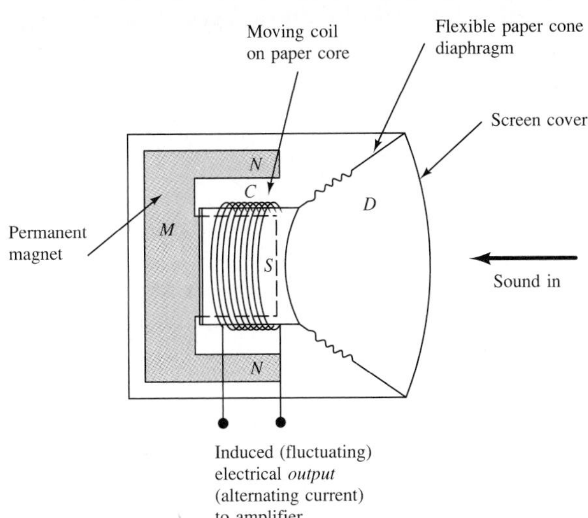

Fig. 21.15 Schematic cross-section diagram of the basic parts of a permanent-magnet moving-coil microphone. A microphone is a *transducer* (converts one form of energy into some other form). Microphones convert input sound energy into an output of alternating current electricity with the same waveform as the input sound.

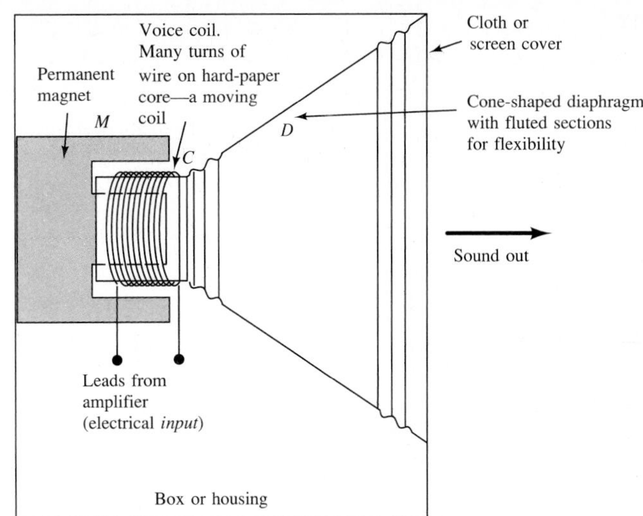

Fig. 21.16 Diagram of a typical permanent-magnet moving-coil loudspeaker (also a transducer). It converts an alternating current electrical energy input from an amplifier into sound energy with the same wave characteristics.

QUESTIONS AND EXERCISES

1. An old "rule of thumb" says that when you see a bolt of lightning, start counting seconds until the sound of thunder arrives. Then divide by 5, and the quotient is the approximate distance to the lightning, in miles. Prove that this rule is either (*a*) approximately correct, or (*b*) in error.

2. One of the age-old arguments of philosophers and cracker-barrel arguers is over whether or not "sound" exists if there is no creature on hand with ears to hear it. Present your own arguments on this question.

3. A sound of frequency 1000 Hz is traveling in air at 20°C. It encounters a layer of air whose temperature is 15°C. By means of a diagram, show how the sound is refracted at the boundary between the two air masses.

4. Suggest several ways in which sounds in an auditorium could be directed to or focused on distant parts of the room—without using amplification.

5. What is the general relationship of the velocity of sound to the density of the medium in which it travels?

6. Explain carefully the difference between *supersonics* and *ultrasonics*.

7. How can the *sabin,* a unit of sound absorption in the English system of measurement, be used with metric measurements?

8. Explain how *loudness* of a sound is different from its *intensity*.

9. A concert hall is described by an orchestra conductor as being "too dead." What is meant by this description? Does the hall need more or less sound-absorbing material? Explain what you would do to "liven up" the hall.

10. A large, flat, sound baffle is to be suspended from an auditorium ceiling out in front of the stage. Its purpose is to reflect sound to the rear seats in the auditorium. By a sketch, show the proper angle at which the baffle should be mounted, with respect to the speaker's podium on the stage.

11. Based on some library study, write a short paper on the current uses of ultrasound in medical diagnosis and treatment.

12. When sound energy is absorbed by a carpet, or drapes, or special acoustical materials, what happens to the energy? Exactly how is it dissipated?

13. In the same amplifying system both the microphone and the loudspeaker may feature a coil that is free to move in a magnetic field. What then is the essential difference in their operation? Explain why both are classified as *transducers*.

14. Interpret the physical significance of the term "sound barrier." Specifically, what happens as a high-speed object in air approaches, equals, and then exceeds Mach 1.00?

PROBLEMS

Group One

1. A gun is fired near a cliff, and 3 s later the person who fired the gun hears the echo. If the air temperature is 25°C, how far away is the cliff (meters)?

2. With a depth-sounding apparatus, an oceanographic survey vessel finds that 3.4 s elapse between the transmitting of a pulse and the reception of its echo. If the velocity of sound in sea water at that location is 5020 ft/s, what is the ocean depth?

3. If the air temperature is 23°C, how far away (minimum distance) from a wall must you be in order for you to hear a distinct echo of the sound as you shoot a pistol?

4. An oil exploration seismic crew detonates a test charge and receives an echo from an underground structure exactly 4.475 s later. Take the average speed of sound in the earth's crust to be 4100 m/s and calculate the depth of the reflecting stratum.

5. Find the total absorbing power (sabins) of 3000 ft^2 of Acousti-Celotex tiles installed on the walls and ceiling of an auditorium.

6. A cruise ship steaming in Glacier Bay, Alaska, sounds one short blast on the whistle. The echo returns from the vertical face of a nearby glacier in 1.85 s. The air temperature is 54°F. How far away is the glacier (yards)?

7. The musical note A has a frequency of 440 Hz. (*a*) What is its wavelength in air at 0°C? (*b*) in water at 4°C?

8. A violin string sounds a fundamental tone of 880 Hz. If the air temperature in the concert hall is 23°C, what is the wavelength of the musical tone (meters)?

9. A typing classroom has a hardwood floor, varnished, with dimensions 40 ft × 50 ft. (*a*) What is the absorbing power of the floor, in sabins? (*b*) What would be the new absorbing power of the floor if it is entirely covered with heavy carpet?

10. In some "Ultrasound" machines for medical diagnosis the frequency of the sound generated is as high as 10^6 Hz (1 MHz). What is the wavelength of these waves in human tissue if the speed of sound (average) is 1450 m/s in the human body?

Group Two

11. At night and in a dense fog a Coast Guard lighthouse blasts its foghorn and simultaneously sends an underwater sound signal. If the two signals arrive at a ship exactly 6.83 s apart, how far is the ship from the lighthouse (meters)? Take the speed of sound in air as 341 m/s; in ocean water as 1520 m/s.

12. A submarine has been "tracking" another submarine with sonar. An exact range is now desired, and a ping ($f = 45{,}000$ Hz) is sent out in a pulse with a duration of 30 μsec. (*a*) How many complete waves are there in the pulse? (*b*) If the echo returns in 1.955 s, what is the range to the target submarine, in yards? (Take 5000 ft/s as speed of sound in ocean water.)

13. How many persons must be present in an auditorium to contribute the same acoustic absorbing power as 5000 ft^2 of heavy carpet, with pad (see Table 21.1)?

14. The four walls of a gymnasium are wood paneling, painted, with a total area of 6000 ft^2. What area of Acousti-Celotex tiles should be installed on the walls to increase the absorbing power of the total wall area by a factor of 10?

15. An aircraft test pilot reports to his ground controller that he has just reached Mach 3.00. If the temperature of the air where the plane is flying is −10°F, what is the plane's speed in mi/h?

16. A window with an open area of 1 ft^2 faces a busy expressway. The noise level at the plane of the window is measured as 75 dB. Compute the value of the acoustic power entering the window, in watts.

17. A phonograph is momentarily producing a tone with a pitch of middle C (264 Hz). The speed of the record groove relative to the pick-up needle is 0.32 m/s. Calculate the distance in millimeters from one crest to the next crest in the wavy record groove.

18. A classroom has the dimensions 36 by 45 by 12 ft. There are 220 ft^2 of glass, 250 ft^2 of wood paneling, 300 ft^2 of chalkboard, 1000 ft^2 of smooth plaster, and 175 ft^2 of painted wood wainscot. The ceiling is covered with acoustic plaster, and the floor with linoleum tile.

There are 70 tablet armchairs (wood), 50 of which are occupied by students. Find the reverberation time in seconds.

19. It is desired to increase the reverberation time of the classroom of Problem 18 to obtain a more ''live'' sound. How many square feet of the ceiling's Acoustic-Celotex tile will have to be removed (exposing more smooth plaster) to result in a reverberation time of 1.3 s?

20. An echo fathometer is being used to measure ocean depths. If an ultrasonic signal of 40,000 Hz is emitted from a transducer on the hull of the ship and returns as an echo from the ocean bottom exactly 0.885 s later, how deep is the ocean at that point? Take the speed of sound in ocean water at that location as 4940 ft/s.

Group Three

21. A college's all-purpose room, as originally constructed, has 16,000 ft^2 of unpainted wood-panelled walls; 22,000 ft^2 of varnished hardwood floor; and 22,000 ft^2 of acoustic-plaster-covered ceiling. The ceiling height is 20 ft. (*a*) Calculate its reverberation time when empty and (*b*) when an audience of 800 persons is present, seated in wood chairs.

22. A submarine is stationary at a depth of 500 ft. Its sonar system is pinging with extremely short pulses whose frequency is 50,000 Hz. The velocity of sound in water at that location is known to be 5020 ft/s. (*a*) What is the wavelength in the ocean water of the ultrasound signal constituting the pings? (*b*) A target submarine is picked up and the elapsed time between an emitted pulse and its returned echo is 2.945 s. What is the distance (yards) to the target?

23. A small concert hall has the following conformation: 60 ft by 100 ft by 30 ft ceiling height. The aisles have 1600 ft^2 of heavy carpet, the balance of the floor being varnished wood. The ceiling is entirely coated with acoustic plaster ($\beta = 0.52$). The side and rear walls have 5200 ft^2 of acoustic tiles ($\beta = 0.82$), and 2600 ft^2 of wood panel wainscoting, unpainted. The front ''wall'' is mostly stage, with a reflecting parabolic shell. Assume it to be a flat surface of 1800 ft^2 area, with an average $\beta =$ 0.05. There are 500 upholstered seats. (*a*) Compute the reverberation time of the hall when all the seats are filled. (*b*) Musicians who perform in the hall, and patrons as well, complain vehemently about the acoustics. ''A dead hall'' is the consensus. Without any changes to the floor or ceiling, would removing the acoustic tiles from both side walls and rear wall, making them all painted wood surfaces, give the hall a ''live sound'' ($t = 1.5$ s or longer)? If not, how many square feet of the ceiling would have to be re-covered with smooth plaster to attain the 1.5-s reverberation time?

24. A particular groove in a disk phonograph record is moving past the needle at 0.24 m/s. At that instant the distance, crest to crest, on the wavy indentations in the groove is 0.545 mm. What is the frequency of the tone produced?

25. A shadowgraph of a test projectile (see Fig. 21.9) shows an angle $\phi = 28°$ between the bow wavefront and the flight axis of the projectile. If the air temperature at the time is 90°F, find the speed of the projectile.

26. A porpoise swimming in ocean water whose temperature is 25°C (see Table 20.1) at a speed of 5 m/s emits an ultrasonic pulse or ping of very short duration whose frequency is 90 kHz. A second porpoise happens to be swimming directly toward the first at a speed of 5.5 m/s. At the instant the ping is emitted by porpoise no. 1, porpoise no. 2 is 6000 m distant. (*a*) What is the elapsed time until porpoise no. 2 hears the ping? (*b*) What is the frequency it hears? (Hint: Review Doppler's principle, Sec. 20.11.)

27. A supersonic aircraft is in level flight at an altitude of 10,000 ft over level terrain at a speed of Mach 1.5. The air temperature (assumed constant all the way to the ground) is 30°F. When a person on the ground below directly under the flight path hears the sonic boom, how far past a vertical line from the observer's position has the aircraft traveled (miles)?

PART 6

LIGHT AND OPTICS

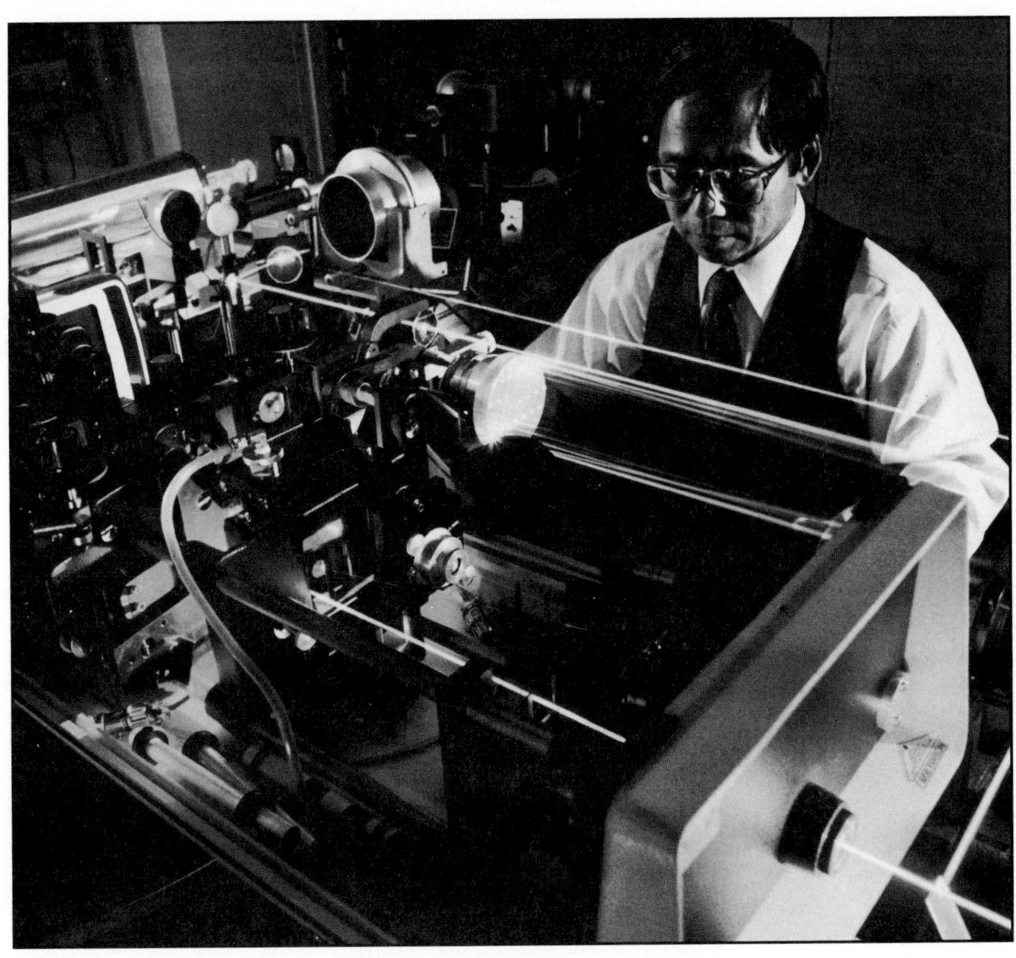

Laser light is used in research on atmospheric pollution. (Ford Motor Company)

THE NATURE OF LIGHT AND ILLUMINATION

During the day, we can see the beauty and richness of the world around us because the sun shines. At night we get some clues about the universe we live in because we can see light from the moon, the planets, and the stars. In past centuries, artificial light was provided at night by candles, lanterns, and torches that used some sort of flame to produce illumination, until Thomas Edison developed the electric incandescent bulb in 1879. Since then, the incandescent bulb has been improved, and engineers and inventors have worked to develop other types of convenient, effective electric lights. In addition to the technological questions that had to be answered to develop electric lights, there are other, more fundamental questions about light.

What is light? How is it transmitted? How can light travel through empty space? What causes the sensation of color?

From earliest recorded times people have sought answers to these questions. They have pried into the mysteries of light and speculated about its exact nature and its relationship to human vision. Today much is known about light and how it behaves. We know that light is a form of energy, and we know the laws that predict its behavior.

In this and the next two chapters we will study some of the properties of and laws developed for this form of wave energy called light.

22.1 ■ Early Theories of Light

Energy can be transmitted from one place to another by two means: (1) by the actual motion of matter from one point to the other or (2) by a wave disturbance traveling through the intervening medium. Both these methods of transmitting light energy have been advanced by scientists in the past. Sir Isaac Newton assumed light to be a stream of small particles or *corpuscles* originating from the light source and moving in straight lines through space. Christian Huygens (1629–1695), on the other hand, conceived of light as a form of wave motion transmitted through a vague, intangible medium which he called the *ether*. Each of these two seemingly incompatible theories has had its staunch supporters, and the relative merits of the two theories have been debated for three centuries.

The Dutch physicist and astronomer Christian Huygens was the first person to clearly see the rings of Saturn.

22.2 ■ Two Newer Theories for Light

In the latter part of the nineteenth century most scientists supported the *wave theory*, as a result of the *electromagnetic wave* studies of James Clerk Maxwell. He pictured light as transverse waves of progressively changing electric and magnetic field intensities. These fields were presumed to act at right angles to each other and to the

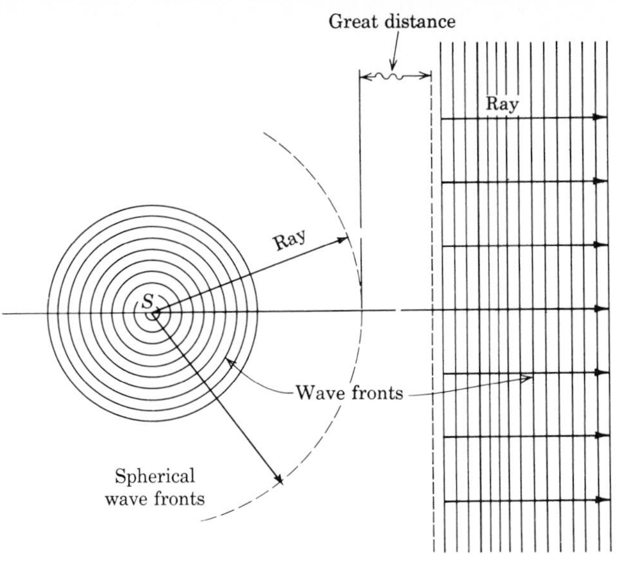

Fig. 22.1 Spherical wave fronts from a point source appear to be plane wave fronts at a great distance from the source, just as the earth's curved surface appears to be flat because of the earth's large diameter.

direction of propagation. The electromagnetic wave theory satisfactorily explains such phenomena as color, interference, polarization, and varying wavelengths in different media but fails to explain the easily observed *photoelectric effect* (explained later, in Sec. 34.3) and certain other phenomena.

The Scottish mathematician and physicist, James Clerk Maxwell (1831–1879), who is most famous for his contributions to the theory of electricity and magnetism, made a major contribution to astronomy by mathematically analyzing the rings of Saturn.

Evidence began to accumulate early in the twentieth century that when light interacts with matter, it behaves as if its energy were in the form of small particles. This theory, known as the *quantum theory*, was first advanced by Max Planck (1858–1947) and occupies an important place in the study of light. The small particles of light energy are called *photons* (see Sec. 34.3).

And so it is that we do not have a single, consistent theory of the nature of light. Our present view must be a combination of the two theories; we think of light as being dual in character.

Two theories for the nature of light—the wave theory and the photon (quantum) theory.

The lack of a single theory need not handicap us in the study of applied physics, for we are primarily interested in studying what light *does* and how it *behaves* under varying circumstances. For *the purposes of this chapter* we shall consider the properties of light waves to be similar to those of other transverse waves, as discussed in Chap. 19.

22.3 ■ Wavefronts and Rays

Light from a point source S (see Fig. 22.1) spreads out in a succession of spherical wavefronts. As they get farther and farther from the point source, the successive spherical wavefronts become more and more like planes. Imaginary lines perpendicular to the wavefronts, called *rays*, are used to show the direction in which the waves are moving. We shall find it convenient to use rays to describe the transmission, reflection, and refraction of light and the formation of images.

Light rays. A light ray is an imaginary line drawn perpendicular to the wave fronts in the direction of the wave propagation.

The rectilinear propagation of light rays can be quite clearly illustrated by placing an incandescent lamp on one side of a cardboard which has a small hole punched in it and placing a white screen on the other side (Fig. 22.2). Light will come from the lamp, pass through the hole, and form an inverted image of the bulb on the screen. Points on the image will lie on the straight lines joining the hole and the corresponding points on the object. Quite clear photographs can be taken with a *pinhole camera*, which employs this principle. A hole is punched in a closed box, and light traveling from the object is allowed to pass through the hole onto the photographic film at the back of the box. The hole must be made small in order to ensure a sharp image.

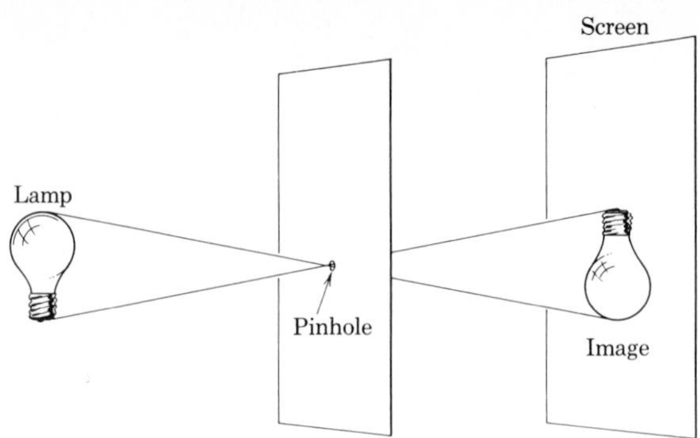

Fig. 22.2 Rectilinear propagation of light. An inverted image of a bright object is formed on a screen by rays of light from the object that pass through a pinhole in an opaque screen.

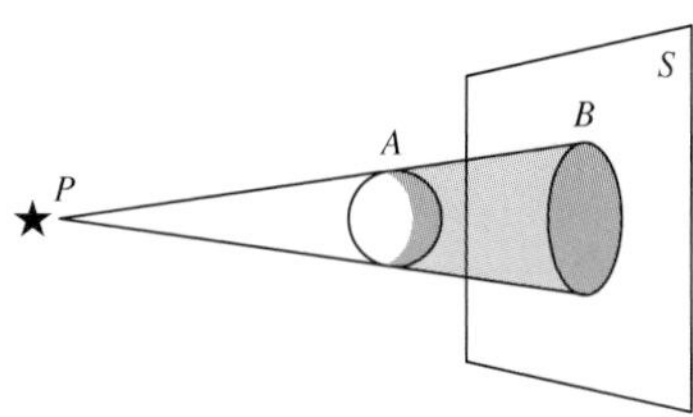

Fig 22.3 A sharp shadow is cast by an opaque object if the light source is small.

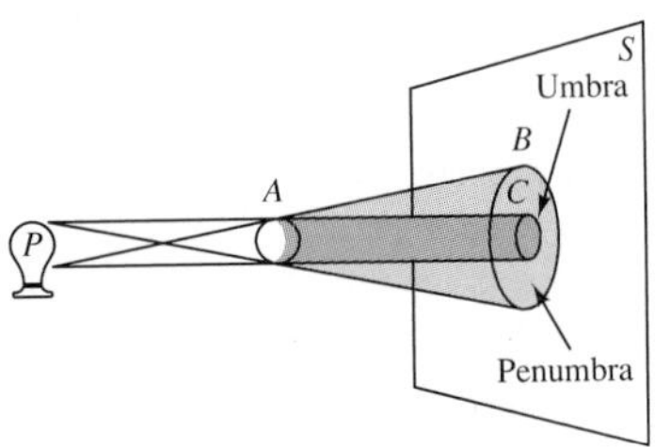

Fig. 22.4 Shadows cast by a light source larger than a point. The totally shadowed region is called the *umbra* and the partially shadowed region is called the *penumbra*.

22.4 ■ Shadows and Eclipses

Straight-line propagation of light from a point source P can be illustrated by showing that the light will cast a sharp shadow B of an opaque object A on a screen S (see Fig. 22.3). The fact that the shadow is sharp shows that the light from P does not bend appreciably in traveling from A to B.

If the light source is not small, the outer boundary of the shadow will not be sharp (see Fig. 22.4). Points on the screen outside circle B are fully illuminated by P. The points of the screen inside circle C receive almost no light. The part of the screen between C and the outer boundary of B will be partially illuminated by P. This outer shadow will gradually shade off from complete shadow at C to complete illumination outside of B.

The part from which all the rays of light are excluded (total shadow) is called the *umbra*; the partially shadowed region is called the *penumbra*.

Eclipses are caused by light from the sun producing shadows of the earth and the moon. An eclipse of the sun occurs when the moon appears directly between the sun and the earth. When this happens, if the moon is close enough to the earth, the moon's shadow falls upon the earth, forming both an umbra and a penumbra (Fig. 22.5*a*). Observers along the path of the umbra see a *total* eclipse, and observers along the path of the penumbra see a *partial* eclipse. For some possible relative positions of the sun, moon, and earth, the end of the umbra is located above the earth's surface as shown in Fig. 22.5*b*, and there is no total eclipse. In that case, observers along the center of the path of the penumbra can see a thin bright ring of the edge of the sun around the moon's disk. This type of partial eclipse is called an *annular eclipse*.

An eclipse of the moon occurs when the moon passes through the shadow of the earth. Since the earth is larger than the moon, it is possible for the entire moon to be

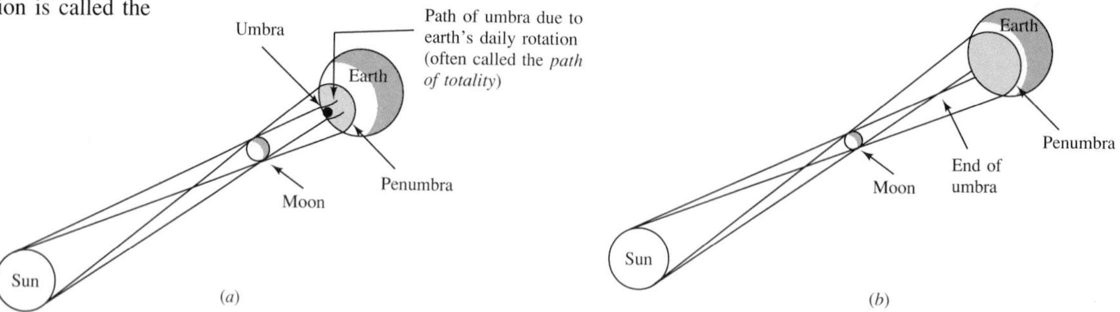

Fig. 22.5 (*a*) An eclipse of the sun occurs when the moon is located between the sun and the earth, and the moon casts its shadow on the earth. (*b*) It is possible for an eclipse to occur in which the full lunar shadow (the umbra) does not even touch the earth. In that case, there is no region where the eclipse is total.

in the earth's umbra, and for the lunar eclipse to be visible to all observers on the "night side" of the earth.

22.5 ■ Speed of Light

In 1926 to 1929, A. A. Michelson (1852–1931) measured the speed of light over a distance of about 44 mi between southern California's Mt. San Antonio and Mt. Wilson and return. The essential features of Michelson's method are illustrated in Fig. 22.6. Light from an arc lamp S was focused by a lens L onto face 1 of an octagonal mirror M at an angle of 45° and reflected to a distant concave mirror D and thence to a plane mirror m. From m the beam was reflected back to D, which returned it to face 3 of M. From there it was reflected into a telescope T. The mirror was then set in rotation and its speed gradually increased until face 2 moved to the point where it would just catch the beam of reflected light from D. It would then reflect the light to the telescope T. The speed of rotation of M was very accurately determined. Hence, the time for the light to travel the distance of $2MD$ was equal to the time for the mirror to make $\frac{1}{8}$ rev. The average of nearly 3000 such measurements gave the speed of light as 186,284 mi/s or 2.99796×10^8 m/s, only 0.0012 percent different from the best values obtained recently in the 1970s.

The American physicist Albert A. Michelson (1852–1931) performed a number of experiments over a period of approximately 40 years designed to measure accurately the speed of light.

Michelson later conducted similar experiments using an evacuated tube 1 mi long. He found that the speed of light is 0.03 percent greater in a vacuum than in air.

Other experiments using microwaves, with wavelengths much longer than light waves, have helped to confirm that in free space the speeds of waves *in all parts of the electromagnetic spectrum* have the same value.

In the early 1970s, four U.S. Bureau of Standards experiments that used helium-neon lasers to measure the speed of light gave results so precise that the accuracies of the values obtained for the speed of light were limited only by uncertainties involving the definition of the meter. Because of this in 1983 the meter was *redefined* to be the distance that light travels in a vacuum in 1/299,792,458 s. This definition of the meter also *defines* the speed of light to be 299,792,458 m/s.

The speed of light is now a *defined* constant.

For most practical purposes, we can take the speed of light in air or in a vacuum to be 186,000 mi/s or 3.00×10^8 m/s. The symbol c is used to represent this value for the speed of light. The relationship between the speed of light c, the distance d traveled by the light, and the time t required for the light to travel that distance in air or in a vacuum, is given by the equation

The symbol c is the scientific designation for the speed of light.

$$d = ct \tag{22.1}$$

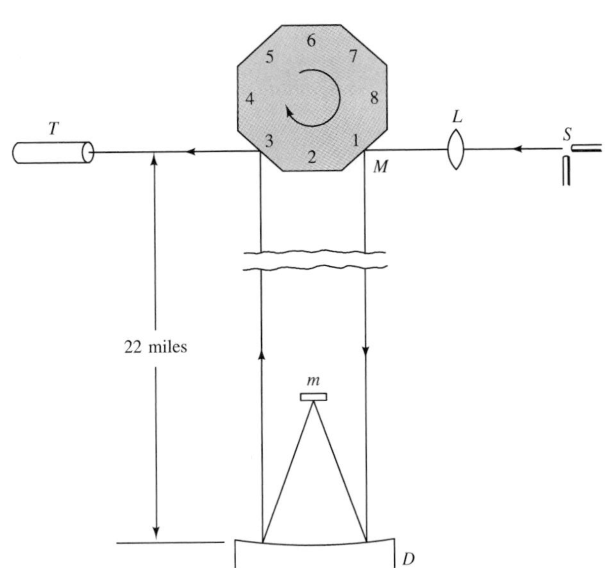

Fig. 22.6 Diagram of apparatus used by Michelson to determine the speed of light.

22.6 ■ Speed of Light in Other Media

Numerous experiments have shown that the speed of light in any medium is less than its speed in a vacuum. For example, the speed of light in water is about three-fourths its speed in air. Equation (22.1) can be used for calculations involving the speed of light in transparent liquids and solids, but the value for c must be replaced by the (smaller) value for the speed of light in the particular substance (see Illustrative Problem 22.2).

Illustrative Problem 22.1 How far can light travel in air in (*a*) one millisecond (ms), (*b*) one microsecond (μs)? [State your answer for part (*a*) in miles and for part (*b*) in feet.]

Solution Substitute the numerical values for the speed of light and for the times into Eq. (22.1). For both parts (*a*) and (*b*), $c = 186{,}000$ mi/s or 1.86×10^5 mi/s.
(***a***) Since 1 ms = 10^{-3} s,

$$\begin{aligned} d &= ct \\ &= (1.86 \times 10^5 \text{ mi/s})(10^{-3} \text{ s}) \\ &= 186 \text{ mi} \qquad \textit{answer} \end{aligned}$$

(***b***) Since 1 μs = 10^{-6} s,

$$\begin{aligned} d &= (1.86 \times 10^5 \text{ mi/s})(10^{-6} \text{ s}) \\ &= 0.186 \text{ mi} \left(\frac{5280 \text{ ft}}{1 \text{ mi}}\right) \\ &= 982 \text{ ft} \qquad \textit{answer} \end{aligned}$$

Illustrative Problem 22.2 How much time is required for a beam of light to travel one-half mile in water?

Solution Solve Eq. (22.1) for t and substitute the numerical values for d and for the speed of light in water, which is $0.75c$.

$$\begin{aligned} t &= \frac{d}{0.75c} \\ &= \frac{0.5 \text{ mi}}{(0.75)(1.86 \times 10^5 \text{ mi/s})} \\ &= 3.58 \times 10^{-6} \text{ s} \\ &= 3.58\ \mu\text{s} \qquad \textit{answer} \end{aligned}$$

RADIOMETRY AND PHOTOMETRY

The science of light measurement, called *photometry*, is a subdivision of a broader area called *radiometry*, which is concerned with the measurement of radiated electromagnetic waves of all wavelengths. Photometry is concerned with the measurement of *visible* electromagnetic waves associated with three quantities: the *luminous intensity I* of the source, the *luminous flux F* or flow of light from a source, and the *illumination E* on a surface. Photometric measurements are an important part of the process of designing effective, efficient, and economical indoor and outdoor lighting systems. All photometric measurements take into account the fact that the human eye is not equally sensitive to all wavelengths of light.

22.7 ■ The Electromagnetic Spectrum

Visible light is only a very small part of the *electromagnetic spectrum.*

The *visible spectrum* constitutes only a very small portion of the whole range of electromagnetic waves which are similar in nature to light, but which are *invisible*. In addition to visible light the electromagnetic spectrum (Fig. 22.7) includes waves such as radio, microwaves, infrared, ultraviolet, x-rays, and gamma rays that are too long or too short to affect the sensory nerves of the eye. Even though the chart in Fig. 22.7 has limits, there is no limit to how long electromagnetic waves can be, and it is probably safe to assume that still shorter waves exist and will someday be discovered.

The speed c, the frequency f, and the wavelength λ of all electromagnetic waves are related by the equation

$$c = f\lambda \qquad (22.2)$$

Speed of light = frequency × wavelength

Illustrative Problem 22.3 What is the frequency of blue light with a wavelength 475 nm? (1 nm = 10^{-9} m.)

Solution Solve Eq. (22.2) for the frequency f and substitute numerical values for c and λ.

$$f = \frac{c}{\lambda}$$

$c = 3.00 \times 10^8$ m/s, and 475 nm = 4.75×10^{-7} m.

$$f = \frac{3.00 \times 10^8 \text{ m/s}}{4.75 \times 10^{-7} \text{ m}}$$

$$= 6.32 \times 10^{14} \text{ s}^{-1}$$

$$= 6.32 \times 10^{14} \text{ Hz} \qquad \textit{answer}$$

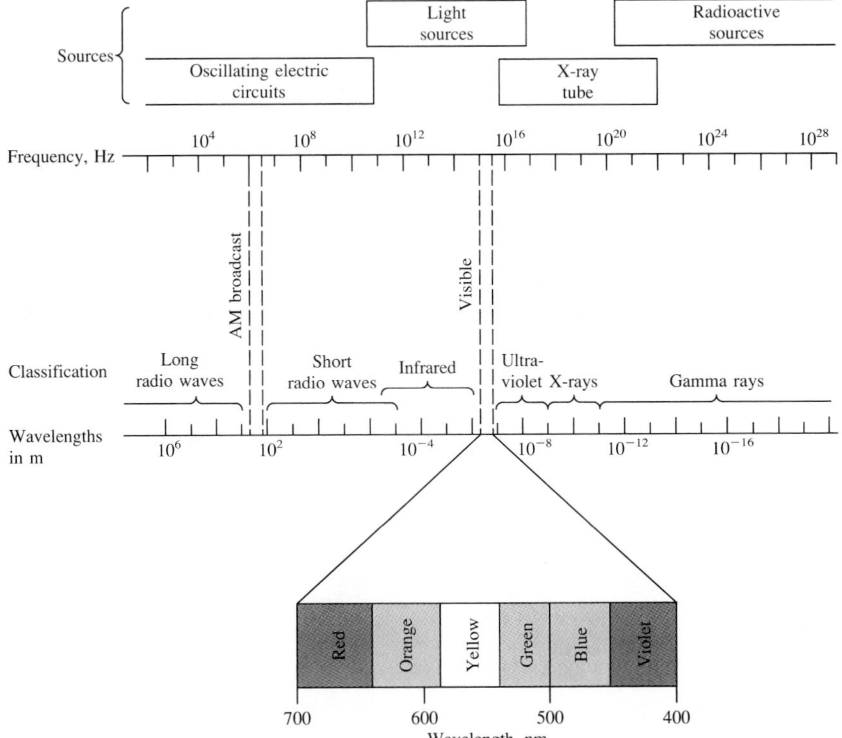

Fig. 22.7 Chart of the electromagnetic spectrum with an expanded view of the visible spectrum.

22.8 ■ Luminous Intensity of a Light Source

Although the ultimate concern in light-intensity measurements is usually the amount of light actually falling on a given surface, we will first consider the intensity of a light *source*. The intensity of illumination on a surface varies directly with the luminous intensity I of the light source.

The term *luminous intensity* has replaced the older term *candlepower.*

For years the unit of *luminous intensity* was called the *candle* (or candlepower, cp), and it was based on the early use of candles as standard sources of light. One *candle* was originally defined as the quantity of light given out by a certain candle that burned whale oil at the rate of 120 gr/h (1 grain (gr) = 1/7000 lb).

The international standard of luminous intensity is the candela.

In 1948, a new more highly reproducible *international standard light source* and a new unit of luminous intensity of a source called the *candela* were adopted by action of the International Committee on Weights and Measures. The international standard of luminous intensity for photometric measurements is a glowing blackbody radiator (see Sec. 16.9) maintained at the temperature of freezing platinum, 2046 K. The arrangement for such a primary standard is shown schematically in Fig. 22.8. The light from the glowing surface of the fused thorium (the radiating "blackbody") is visible through a 1.452-mm-diameter (area = 1/60 cm^2) hole at the top of the crucible. The light emitted through this hole has a luminous intensity of one candela (cd). A luminous intensity of one candela is essentially the same as that of the old standard candle. The *candela* (along with the *meter*, *kilogram*, *second*, *ampere*, *kelvin*, and *mole*) is one of the seven *fundamental units* in the International System of Units (SI).

The candela is one of the *fundamental units* of the International System of Units (SI).

One definition of the candela is that it is the luminous intensity of $\frac{1}{60}$ cm^2 of white-hot platinum at its melting (or freezing) point temperature of 2046 K.

Remember that the candela is a unit of *luminous intensity* of a *light source.*

22.9 ■ The Photometer—Measurement of Luminous Intensity

Robert Wilhelm Eberhard Bunsen was known as one of the founders of organic chemistry. He was one of the first scientists to organize the science of spectroscopy and to study the chemical effects of light.

The Bunsen grease-spot photometer can be used to compare the relative luminous intensities of two light sources.

A photometer is an instrument designed to measure the relative luminous intensities of light sources. A photometer capable of fair precision was devised by Robert Bunsen (1811–1899), a German chemist and inventor. It consists essentially of a flat screen of white paper with a central, translucent grease spot which transmits some light (Fig. 22.9). The grease-spot screen is mounted in a box with openings on three sides so that light entering from X and S will strike the grease spot. The intensity of illumination on the two sides of the grease spot can be viewed from the third open side of the box by means of tilted mirrors. If the illumination on one side of the screen is more intense than that on the other, the grease spot when viewed in the mirror on that side will appear darker (in contrast) than when viewed in the mirror of the other side. Conversely, it will appear lighter if viewed from the side of weaker illumina-

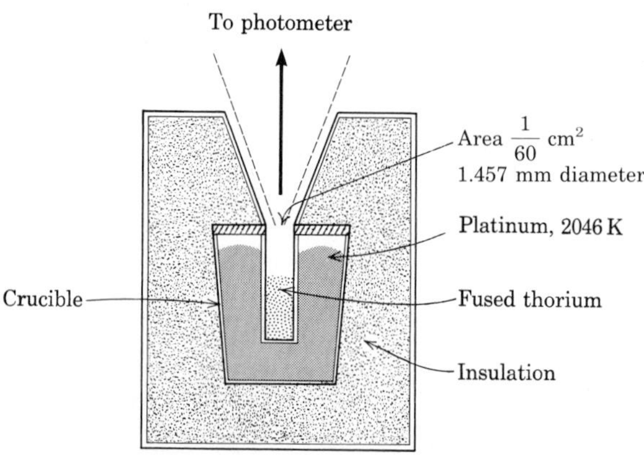

Fig. 22.8 Diagram of a primary standard of luminous intensity for photometric measurements.

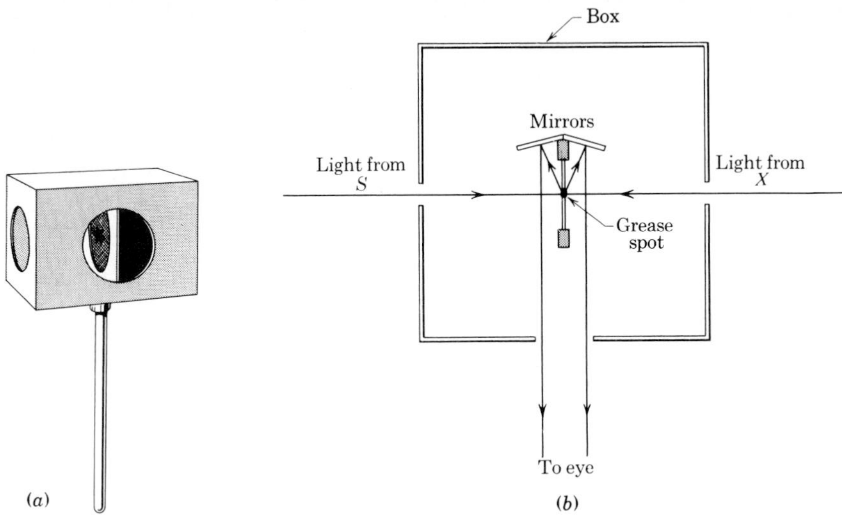

Fig. 22.9 Bunsen grease-spot photometer. (*a*) Pictorial view of box with grease-spot screen inside. (*b*) Ray diagram showing paths of light rays from standard light source S and from unknown source X.

tion. If the two sides are illuminated equally, the grease spot will appear to be the same neutral shade of gray when viewed in either mirror. If the two light sources and the photometer box are mounted on an optical bench as shown in Fig. 22.10 the distances d_x and d_s can be varied until the two sides of the screen appear equally bright. When they are equally bright, the luminous intensity of the unknown source can be found from the equation

$$\frac{\text{Intensity of } X}{\text{Intensity of } S} = \frac{(\text{distance of } X \text{ from screen})^2}{(\text{distance of } S \text{ from screen})^2}$$

$$I_x/I_s = d_x^2/d_s^2 \qquad (22.3)$$

This is known as the *photometric* equation.

Illustrative Problem 22.4 Equal illumination appeared on a photometer screen which was placed 80 cm from a 40-cd standard lamp and 120 cm from a second lamp. What was the luminous intensity of the second lamp?

Solution Substituting in Eq. (22.3) gives

$$\frac{I_x}{40 \text{ cd}} = \left(\frac{120 \text{ cm}}{80 \text{ cm}}\right)^2$$

$$I_x = 90 \text{ cd} \qquad \textit{answer}$$

Many other forms of photometers have been devised, but they all depend upon an adjustment which will make the illumination on two surfaces appear the same. In all the discussion on photometric measurement thus far we have assumed that both the

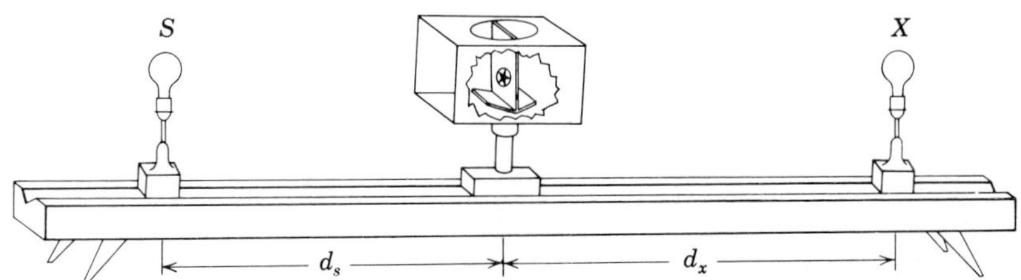

Fig. 22.10 Photometer used to compare lamp X of unknown luminous intensity with standard lamp S.

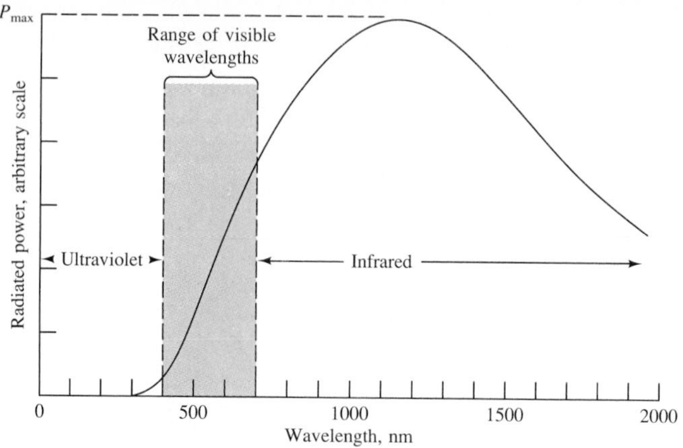

Fig. 22.11 Variation of radiated power with wavelength for 100-W incandescent bulb. Most of the radiation is emitted at invisible wavelengths in the infrared portion of the electromagnetic spectrum.

unknown and the known light sources have the same color. If they do not have the same color, the methods described cannot be used in determining equal illumination on a screen.

22.10 ■ Luminous Flux—Light Flowing Out from a Source

Only a small part of the radiation emitted by an incandescent bulb is within the range of wavelengths visible to the human eye.

Figure 22.11 shows how the power radiated by a common 100-W incandescent bulb depends on wavelength. The bulb exhibits maximum radiation at a wavelength of approximately 1200 nm, and most of the radiation is in the infrared range, at wavelengths longer than those to which the eye is sensitive. Only 2.5 percent of the total radiated power lies in the visible spectrum (see Illustrative Problem 22.5).

The *total* amount of electromagnetic *energy* radiated per unit time by a light source is called the *radiant flux P* of the source. *Radiant* flux is measured in watts. The *visible* light output of a light source, called *luminous flux* (*F*), is measured, not in watts, but in *lumens*. The lumen is basically defined in the following way:

The *lumen* is defined in terms of the brightness of yellow-green light having a radiant flux of $\frac{1}{683}$ W.

One lumen (lm) is the luminous flux emitted by a light source radiating 1/683 W at a wavelength of 555 nm.

The *relative visibility* of light depends on the wavelength (or color) of the light.

Stated differently, one watt of radiation at a wavelength of 555 nm is equal to a luminous flux of 683 lm. At any wavelength other than 555 nm, the conversion from watts to lumens must take into account the *relative visibility* of the light at that wavelength (see Fig. 22.12). For example, at 520 nm (green light) the relative visibility is 0.71 and one watt of light at 520 nm provides a luminous flux of (683 lm) (0.71) = 485 lm.

The reason that 555 nm is chosen as the defining wavelength is that this is the spectral region of yellow-green light, to which the eye is most sensitive.

At wavelengths shorter than about 400 nm, or longer than about 700 nm, the *luminous* flux flowing out of a light source is zero, regardless of how great the *radiant* flux is at those wavelengths. Most light sources, including common incandescent bulbs and fluorescent lamps, radiate light over a wide range of wavelengths, and the conversion from radiant flux in watts to luminous flux in lumens must take into account the power radiated at all the wavelengths in the visible spectrum. When this is done for typical 60-, 75-, and 100-W soft-white incandescent bulbs, luminous fluxes of 855, 1170, and 1710 lumens, respectively, are obtained.

Luminous efficiency expressed in lumens per watt.

The *luminous efficiency* of a light bulb is the luminous flux emitted per watt of power input to the bulb. The luminous efficiency in lumens per watt can be multiplied by the factor 1 W/683 lm to obtain a dimensionless ratio that is equal to the fraction

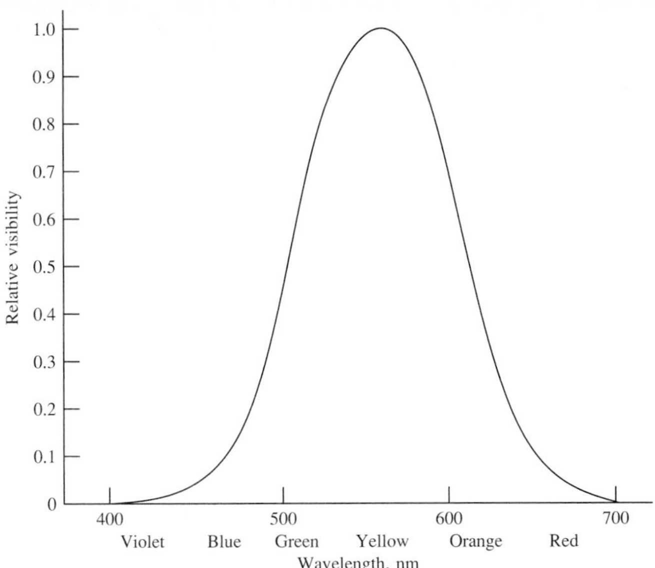

Fig. 22.12 Relative visibility of light of different wavelengths in the visible spectrum.

of the power of the light bulb that appears as visible radiation. Multiplying that fraction by 100 gives the percent of the power that appears as visible radiation.

Illustrative Problem 22.5 A 100-W incandescent bulb has a luminous output of 1710 lm, and a 40-W fluorescent tube has luminous output of 3000 lm. (*a*) What is the luminous efficiency for each of the two light sources? (*b*) What percent of the power of each light source appears as visible radiation?

Solution

(***a***) The luminous efficiency of a light source is equal to the luminous flux of the source divided by the power rating of the source.

$$\text{Luminous efficiency} = \frac{\text{luminous flux (lumens)}}{\text{power (watts)}}$$

For the 100-W bulb:

$$\text{Luminous efficiency} = \frac{1710 \text{ (lumens)}}{100 \text{ (watts)}}$$

$$= 17.1 \text{ lm/W} \qquad \textit{answer}$$

For the 40-W fluorescent tube:

$$\text{Luminous efficiency} = \frac{3000 \text{ (lumens)}}{40 \text{ (watts)}}$$

$$= 75 \text{ lm/W} \qquad \textit{answer}$$

The luminous efficiency of the fluorescent tube is more than four times as great as that of the incandescent bulb because most of the light emitted by the incandescent bulb lies in the invisible infrared portion of the electromagnetic spectrum, whereas much more of the light emitted by the fluorescent tube lies in the visible portion of the spectrum. The heat associated with a typical incandescent bulb is due to the infrared radiation emitted by the bulb. The operating temperature of a typical fluorescent tube is relatively low because it emits very little infrared radiation.

Incandescent bulbs radiate more infrared energy ("heat waves") than fluorescent lamps of the same power rating.

(***b***) For each light source the fraction of the power that appears as visible radiation can be determined by multiplying the luminous efficiency from part (*a*) above by the factor 1 W/683 lm.

For the 100-W incandescent bulb,

$$(17.1 \text{ lm/W})(1 \text{ W/683 lm}) = 0.0250.$$

Multiplying the above number by 100 shows that 2.5 percent of the radiated power appears as visible radiation. *answer*

For the 40-W fluorescent tube,

$$(75 \text{ lm/W})(1 \text{ W/683 lm}) = 0.110.$$

For the fluorescent tube, 11.0 percent of the radiated power appears as visible radiation. *answer*

22.11 ■ Luminous Flux Related to Luminous Intensity

Definition of a *solid angle* and of the *steradian.*

The luminous intensity of a *source* does not depend on distance from the source.

We have emphasized that luminous *intensity* and its unit, the *candela*, are visual attributes of light *sources*. Luminous *flux*, on the other hand, is a *flow of visible light* emanating *from* a source, and its unit is the *lumen*. If one visualizes light rays streaming out from a source in all directions (three-dimensional space), then luminous flux is a measure of the concentration of these rays in space. In order to quantify the concept of luminous flux and its unit, the lumen, it is necessary to develop some relationships of the *solid angle* that surrounds a point in space.

In Fig. 22.13*a*, C is the center of a sphere of radius r whose spherical surface is denoted by S. A point source of light is to be visualized at C. *Unit solid angle* is defined as that solid angle subtended at the center C by a portion of the spherical surface whose area A is equal to the square of the radius r of the sphere (see the right-hand sketch in the sphere of Fig. 22.13*a*). This unit solid angle is called one *steradian* (sr). (*Stere* is a Greek word meaning "solid," or "having three dimensions," and *radian* means an angle whose arc is equal to the radius). In general, a solid angle Ω is defined by the equation

$$\Omega = \frac{A}{r^2} \tag{22.4}$$

where Ω is the solid angle subtended at the center of a sphere of radius r by a portion or section of the sphere's surface whose area is A (see the left-hand sketch in the sphere of Fig. 22.13*a*).

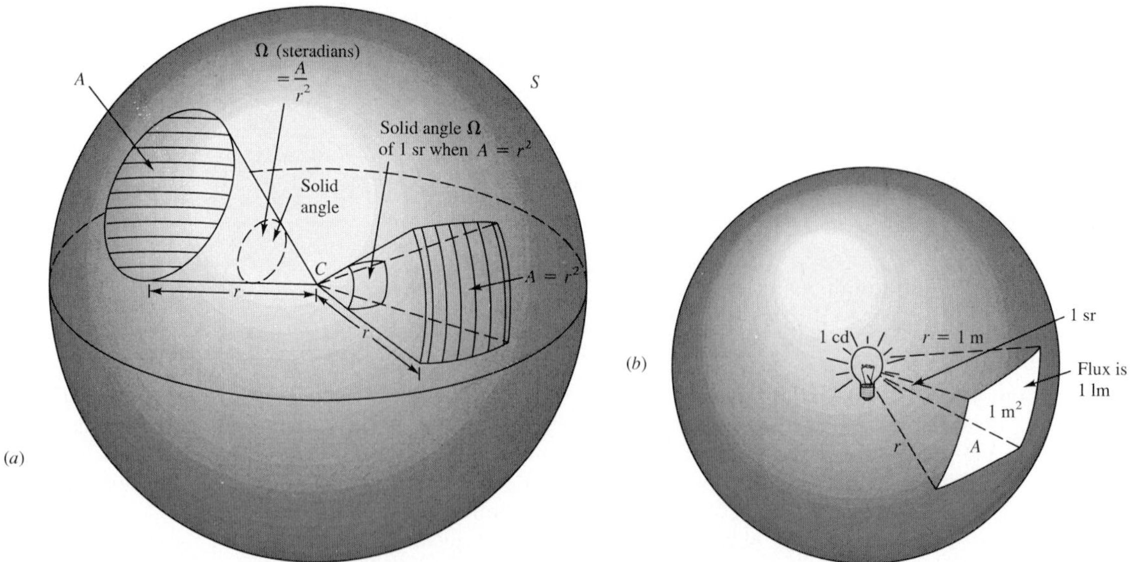

Fig. 22.13 (*a*) A solid angle Ω can be defined in terms of the area A of a segment of the surface of a sphere and the radius r of the sphere. In the diagram above Ω (in steradians) $= A/r^2$. If $A = r^2$, then Ω is equal to one steradian (sr). (*b*) One lumen of light flux through 1 m^2 of spherical area at a 1 m distance from a 1-cd source.

Now, think of an idealized 1 cd source sending out luminous flux in all directions, and located at C, the center of the sphere of Fig. 22. 13b. Let $r = 1$ m (or 1 ft), and imagine a hole A whose area is 1 m^2 (or 1 ft^2). This geometry leads to a second definition of the lumen (the basic definition in terms of a standard that can be tested in the laboratory was given on page 530):

One lumen is the quantity of light flux radiated through unit spherical area at unit distance from a source whose luminous intensity is l candela (cd).

This definition is equivalent to stating that a light source of intensity one candela, radiating uniformly in all directions, will emit one lumen of flux over a solid angle of one steradian. As an equation,

$$\begin{array}{ccc} F & = & I\Omega \\ \text{Luminous flux} & = & \text{Luminous intensity} \times \text{solid angle} \\ \text{(lumens)} & & \text{(candelas)} \quad\quad \text{(steradians)} \end{array} \tag{22.5}$$

In Eq. (22.5) the unit for luminous intensity I is the candela, and it is readily seen that the candela is equivalent to a lumen/steradian (lm/sr), or 1 cd = 1 lm/sr.

Since the surface area of a sphere of radius r is $4\pi r^2$, when the area A in Eq. (22.4) is equal to the entire surface area of the sphere, the maximum solid angle Ω is equal to 4π steradians. If 4π steradians is substituted for Ω in Eq. (22.5), an equation can be written that relates the intensity of the source (in cd or lm/sr) to the total luminous flux (in lm) emitted by the source.

The relationship between the luminous intensity of a source that radiates uniformly in all directions ("point source") and the total luminous flux emitted by the source.

$$F = 4\pi I \tag{22.6}$$

$$(\text{Total flux} = 4\pi \times \text{source intensity in cd})$$

This equation is valid only for sources that radiate uniformly in all directions.

The common incandescent bulbs in use today give a little more than 1 cd/W of electric power used and fluorescent lamps provide about 4 cd/W. For example, a 100-W bulb has a luminous intensity of about 135 cd. Thus the 100-W bulb will emit a *total* flux of about $4\pi \times 135 = 1700$ lm. Since luminous flux represents a *flow* of light energy, the same flow will occur through a 1-m^2 area at a distance of 1 m from a 1-cd source as will flow through a 1-ft^2 area at a distance of 1 ft from the same source (see Fig. 22.13b).

Illustrative Problem 22.6 The total luminous flux F is 490 lm for a light bulb that emits light uniformly in all directions. What is the luminous intensity of that bulb?

Solution Since the bulb emits light uniformly in all directions, the luminous intensity is the same in all directions. The value for the luminous intensity I can be calculated using Eq. (22.6):

$$F = 4\pi I$$

$$I = \frac{F}{4\pi}$$

Since a total luminous flux F of 490 lm is emitted uniformly in all directions, F = 490 lm and

$$I = \frac{490 \text{ lm}}{4\pi \text{ sr}}$$

$$= 39.0 \text{ lm/sr}$$

$$= 39.0 \text{ cd} \qquad \textit{answer}$$

Illustrative Problem 22.7 A ceiling light in a room uniformly emits light throughout the region below with a luminous intensity of 600-cd. What is the total luminous flux F emitted by the light? (Neglect ceiling reflection.)

Solution Since the light is uniformly emitted only into the region below the ceiling, the total luminous flux passes through a solid angle of only ½(4π) or 2π steradians. Since a candela is equal to a lumen/steradian, the use of Eq. (22.5) gives

$$\begin{aligned} F &= I\Omega \\ &= (600 \text{ lm/sr})(2\pi \text{ sr}) \\ &= 3770 \text{ lm} \qquad \textit{answer} \end{aligned}$$

22.12 ■ Illumination on a Surface

Illumination (illuminance) is equal to the luminous flux incident on a unit area of the surface.

When light strikes a surface, we say the surface is illuminated. *Illumination* (or *illuminance*) on a surface (see Fig. 22.14) is defined as the luminous flux per unit area of that surface. If the surface is uniformly illuminated,

$$E = F/A \tag{22.7}$$

where

F = luminous flux, lumens
A = area, square meters or square feet
E = illumination, luxes or footcandles

The two common units of illumination are the lumen per square meter (called *lux*) and the lumen per square foot (called *footcandle*, ft-c). One ft-c is equivalent (approx.) to 10.76 lux. Approximate values of the illumination from various sources are listed in Table 22.1.

Let us imagine a point source of light S emitting light energy uniformly in all directions, and consider two spheres of radii d_1 and d_2 having S as a common center (Fig. 22.15). Unless some of the light energy is absorbed, the amount of luminous

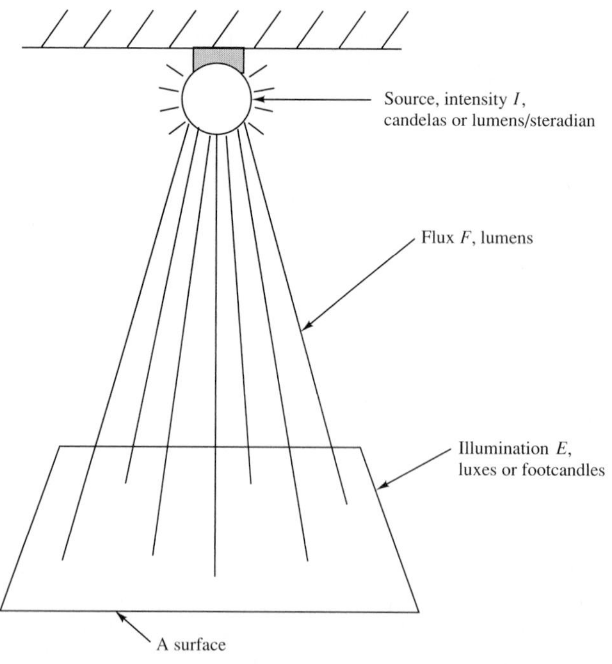

Fig. 22.14 Diagram showing flux F from a source of luminous intensity I providing illumination E on the surface below the light source.

Table 22.1 Approximate illumination from some light sources

Source	Illumination	
	Luxes	*Footcandles*
Starlight	0.0003	0.00003
Full moonlight	0.25	0.023
Clear sky	11,000	1000
Snow in sunlight	32,000	3000
Sun overhead	100,000	10,000

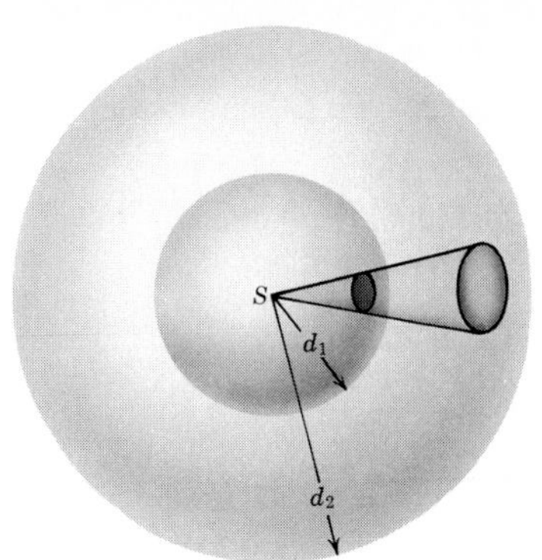

Fig. 22.15 Light travels from its source in concentric spheres.

flux F_1 passing through the spherical surface whose area is $4\pi d_1^2$ must equal the luminous flux F_2 through the surface whose area is $4\pi d_2^2$. Hence if we let E_1 and E_2 equal the intensity of illumination at d_1 and d_2, respectively, it must be true that

$$F_1 = F_2$$

and, from Eq. (22.6),

$$E_1(4\pi d_1^2) = E_2(4\pi d_2^2)$$

Dividing both sides of the equation by $4\pi E_2 d_1^2$, we get

$$E_1/E_2 = d_2^2/d_1^2 \qquad (22.8)$$

This is an example of the *inverse-square law*:

The illumination on a surface from a point source of light varies inversely as the square of the distance from the surface to the source.

The inverse square law for illumination.

(This law, it will be recalled, is also true for the intensity of sound.) It should be emphasized that Eq. (22.8) is true only if the illuminated surface is at right angles to the rays of light and if the source of light is a point source.

The illumination E on a surface can be found if the distance d between the surface and the light source and the intensity I of the light source are known. Thus

$$E = F/A = 4\pi I/4\pi d^2 = I/d^2 \qquad (22.9)$$

If d is measured in meters and I in candelas, E will be expressed in luxes; if d is measured in feet and I in candelas, E will be expressed in footcandles.

Illustrative Problem 22.8 What is the illumination, in footcandles, on a surface placed 5 ft directly below a 160-cd lamp if all light from reflecting surfaces is blanked out?

Solution Substituting known values in Eq. (22.9), we get

$$\begin{aligned}\text{Illumination } E &= I/d^2 \\ &= 160 \text{ cd}/(5 \text{ ft})^2 \\ &= 6.4 \text{ ft-c} \qquad \textit{answer}\end{aligned}$$

If the rays of light from the source strike a surface with an angle of incidence θ (measured from the normal to the surface) the intensity of illumination on the surface (see Fig. 22.16) is

$$E = \frac{I \cos \theta}{d^2} \qquad (22.10)$$

Illustrative Problem 22.9 A sheet of paper on a table is located 4.00 m directly below a 1650-lm lamp (Fig. 22.16). (*a*) What illumination is provided on the paper? (*b*) What is the illumination on the paper if it is moved 3 m horizontally from the original position?

Solution

(***a***)

$$E_1 = I/d^2 \qquad \text{and} \qquad F = 4\pi I$$

Solve for I:

$$I = F/4\pi$$

Then

$$E_1 = F/4\pi d^2 = 1650\ \text{lm}/4\pi(4.00\ \text{m})^2$$
$$= 8.21\ \text{lm/m}^2 = 8.21\ \text{luxes} \qquad \textit{answer}$$

(***b***)

$$d^2 = (4\ \text{m})^2 + (3\ \text{m})^2 = 25\ \text{m}^2$$
$$d = 5\ \text{m}$$
$$\cos\theta = 4\ \text{m}/5\text{m} = 0.800$$
$$E_2 = I\cos\theta/d^2$$
$$= \frac{(F/4\pi)\cos\theta}{d^2}$$
$$= \frac{(1650\ \text{lm})(0.800)}{4\pi(5\ \text{m})^2}$$
$$= 4.20\ \text{lm/m}^2 = 4.20\ \text{luxes} \qquad \textit{answer}$$

Table 22.2 below gives a summary of the radiometric and photometric quantities discussed in the sections above, together with the units in which they are measured.

Table 22.2 Summary of radiometric and photometric quantities

Name	*Symbol*	*Meaning*	*Unit (and abbreviation)*	*Definitions of units*
Radiant flux	P	Total quantity of electromagnetic radiation	watt (W)	W = joule/s
Luminous flux	F	Total quantity of visible light	lumen (lm)	1 lm = the light flux from a 555-nm source (yellow-green) radiating at the rate of 1/683 W.
Luminous intensity	I	Brightness of *source*	candela (cd) or lumen/steradian (lm/sr)	1 cd = 1 lm/sr
Luminous efficiency		Luminous flux emitted/input power	lumen/watt (lm/W)	
Illumination (illuminance)	E	Luminous flux per unit area	lux, footcandle (ft-c)	lux = lm/m^2 footcandle = lm/ft^2

22.13 ■ Proper Illumination

The amount and kind of illumination needed to produce optimum light conditions in a room vary greatly with the function or use of the room. Because the problem of *computing* the actual intensity of illumination on any given surface in a room is quite difficult, the intensity is usually obtained by direct measurement using instruments calibrated to read light intensities directly in luxes or footcandles (Fig. 22.17).

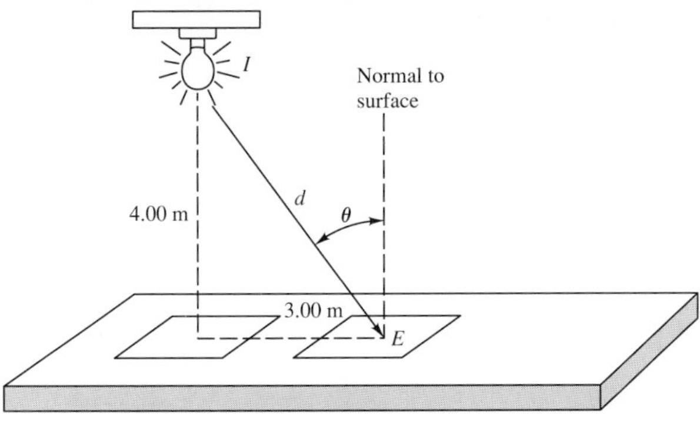

Fig. 22.16 Intensity of illumination on a surface at an angle θ to the incident rays of light.

Today we are growing more conscious of the need for proper illumination. Industrial plants find a marked improvement in the efficiency and output of employees working under proper illumination. A definite decrease in eyestrain, accidents, and poor workmanship follows the installation of proper lighting. Good lighting consists of not only enough light but light of the proper kind. Proper illumination should provide uniform, diffused, and shadowless light.

Table 22.3 suggests optimum values of intensity of illumination for various situations.

(*a*)

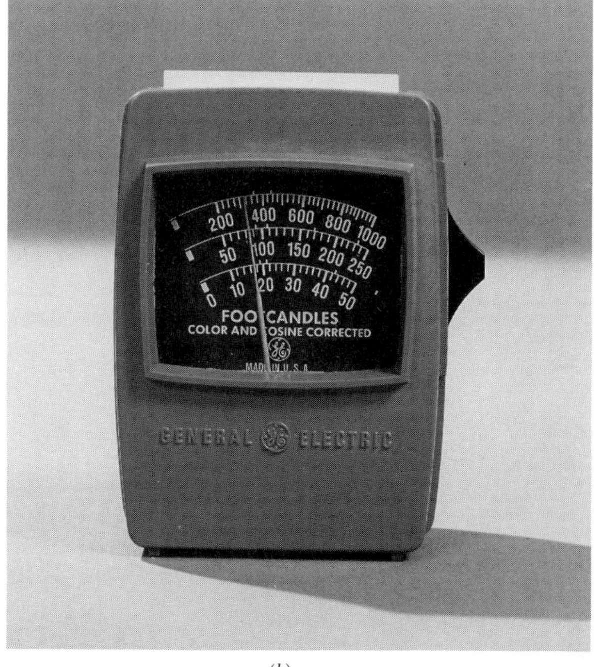

(*b*)

Fig. 22.17 (*a*) A *lux meter* for measuring illumination in-lumens per square meter (lm/m^2). Note the fiber-optic cable for use in monitoring illumination intensity at the precise point of interest. (Courtesy PASCO Scientific) (*b*) A pocket-sized *footcandle meter,* which is a convenient device for people concerned with proper industrial lighting. (Courtesy General Electric Co.)

Table 22.3 Optimum illumination requirements

Situation	*Luxes*	*Footcandles*
Hallways	54–110	5–10
Library, reading rooms	320–540	30–50
Kitchen	160–270	15–25
Office, classroom, laboratory	650–1100	60–100
Industrial assembly plant	430–1100	40–100
Stores (except jewelry stores)	220–540	20–50
Hospital operating rooms	540–11,000	500–1000
Night baseball	320–430	30–40
Continued close work (drafting, etc.)	540–1100	50–100
Jewelry manufacturing	1100–3200	100–300

22.14 ■ Color and the Length of Light Waves

Most of us have been impressed at times with beautiful displays of color produced by nature. Analysis of light has shown that the *different color sensations are mainly due to the varying wavelengths of light to which the eye is sensitive*. Of all the colors of the rainbow, it has been found that red light has the longest wavelength, orange the next longest, and so on down to violet with the shortest wavelength. Light from the sun is a combination of many wavelengths which produces the effect on the eye usually described as white light. Scientists have found it possible by the use of prisms (discussed in Chap. 24) and gratings to analyze sunlight or light from any other source into its constituent colors and to measure the wavelengths of the different colors of light. The wavelengths and frequencies of the central portion of each of the colored regions in the *visible spectrum* (Fig. 22.7) are given approximately in Table 22.4. See also the spectrum on the front endpapers.

Table 22.4 Approximate wavelengths and frequencies of certain colors of light—the visible spectrum

	Wavelengths			
Color	*m*	*in.*	*nm*	*Frequency, Hz*
Red	67×10^{-8}	0.000028	670	4.48×10^{14}
Orange	62×10^{-8}	0.000024	620	4.83×10^{14}
Yellow	57×10^{-8}	0.000022	570	5.25×10^{14}
Green	52×10^{-8}	0.000020	520	5.76×10^{14}
Blue	47×10^{-8}	0.000018	470	6.39×10^{14}
Violet	41×10^{-8}	0.000016	410	7.32×10^{14}

22.15 ■ Colors of Objects

Color transmission by *selective absorption*.

When white light strikes a colored *transparent* object, e.g., a piece of blue glass, only wavelengths in the blue region of the spectrum are transmitted. All or almost all the other colors are absorbed. Light passing through a transparent color filter owes its transmitted color to a process of *selective absorption*. For example, a blue filter placed in the path of a beam of white light will absorb the red, orange, and some of the yellow colors (Fig. 22.18*a*). If a yellow filter is placed in the path of white light, it will absorb the blue, violet, and some of the red and green colors (Fig. 22.18*b*). When both are inserted in the path of the beam, only green and a little yellow or blue are transmitted. To the eye the transmitted beam will appear bright green.

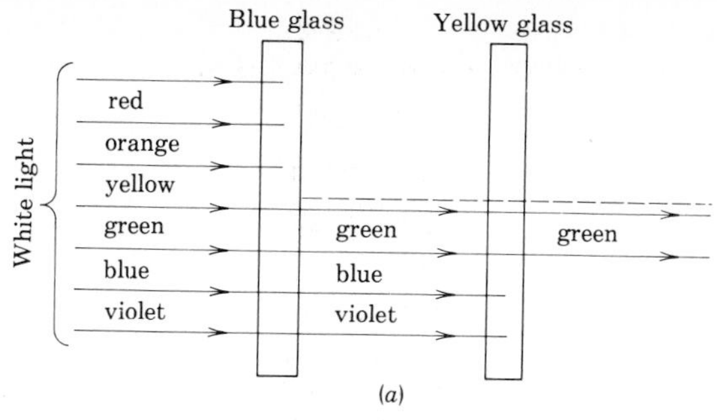

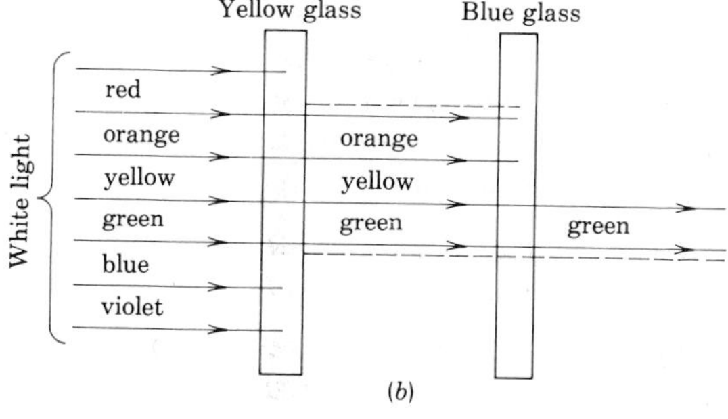

Fig. 22.18 Absorption and transmission of white light by blue and yellow filters.

Opaque objects, such as a piece of cloth, have characteristic colors because we see only the light that is *reflected* from them. For example, if white light shines on a piece of red cloth, it appears red because the dye in the cloth absorbs most of the wavelengths of the incident light except those near the red region of the visible spectrum. The red wavelengths are then *reflected,* giving a red appearance to the cloth. If an object reflects all the light it receives, we say it is *white*. If an object absorbs all the colors it receives, we say it is *black*. Since light is a form of energy, when some of it is absorbed, it must be converted to another form. It is converted largely to heat. This is one reason why dark clothing is worn in cold climates and light-colored or white suits in the tropics.

Many interesting effects can be obtained by using artificial light of one predominant color on fabrics or pigments of other colors. For example, a pure blue necktie will appear black when viewed under a red or yellow light source. This is because blue pigments reflect little or no yellow or red light. A car which appears blue in the daytime will appear black under yellow sodium street lamps.

22.16 ■ Complementary Colors

We have seen how colors may be subtracted from a beam of light when the light is reflected or passes through a filter. The reflecting surface and the filter remove certain frequencies from the light.

It would seem reasonable to suppose that simple colors can be combined to form new colors. Experiments with beams of different colored lights have shown that most colors and hues can be produced by adding varying amounts of three different colors. The three colors which can most effectively be used to produce other colors and hues are red, green, and blue. Consequently, these have been called the *primary colors*.

When these primary colors are projected on a white screen so they overlap as shown in Fig. 22.19 additive mixtures of the primary colors will be produced. You will note that the addition of the red and green colors will produce yellow; that red and blue colors produce magenta; and green and blue colors produce cyan. By varying the intensities of the red, green, and blue lights, combinations can be produced which to the human eye will match every color and hue of the spectrum.

Observe that proper combinations of the primary colors will produce what to the eye appears white. Any two colors which combine to form white light are said to be

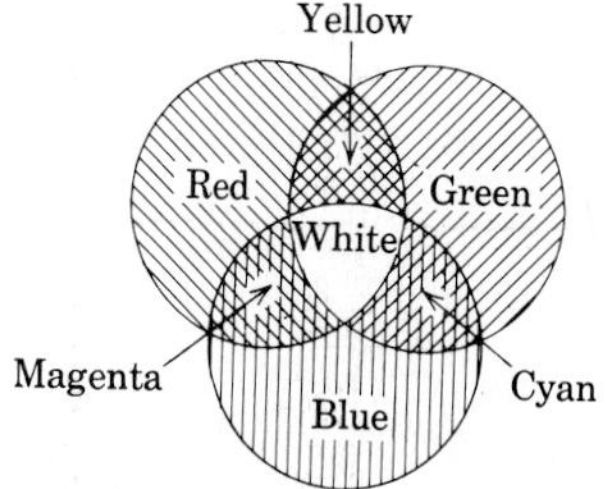

Fig. 22.19 Overlapping of the primary colors red, green, and blue produces white light. Any two primary colors combine to give the complement of the third. The complementary colors are cyan, magenta, and yellow.

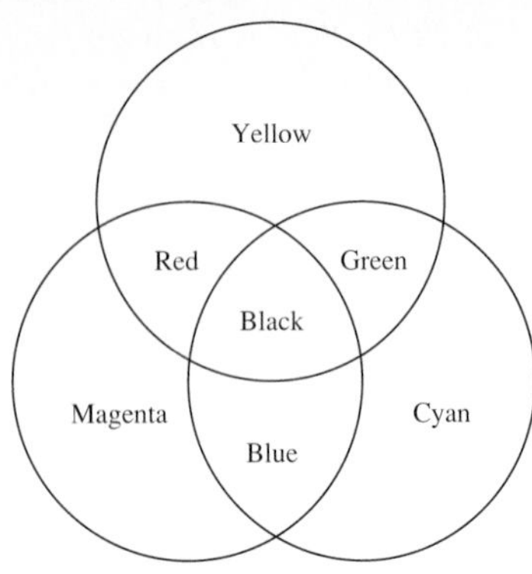

Fig. 22.20 Primary pigments cyan, magenta, and yellow, when mixed in proper proportions, appear as black. Combined in pairs, two primary pigments give the complement of the third pigment, forming red, green, and blue colors.

complementary. The complement of red is cyan; of green is magenta; of blue is yellow. However, many other pairs of colors can be complementary.

22.17 ■ Subtraction of Colors When Pigments Are Mixed

Most of us are familiar with the fact that if we mix yellow and blue *pigments*, we get a green mixture. This is because the yellow pigment absorbs most of the wavelengths of the visible spectrum except those which produce a sensation of yellow and some of the waves on either side of yellow, i.e., orange and green (see Table 22.4). In like manner, blue pigments absorb most wavelengths except blue, some green, and some indigo. The only color that is not completely absorbed by one or the other of the pigments is green. Hence green is reflected from the mixture. Similar explanations can be given for the resultant color of mixtures of all kinds of dyes and pigments.

When pigments are mixed, each one subtracts certain colors from the incident light and the resulting color depends upon the light waves that are not absorbed. The *primary pigments* are the complements of the primary colors. They are cyan, magenta, and yellow. When the three primary pigments are mixed in proper proportions, all wavelengths of the spectrum will be subtracted and the mixture will appear black (see Fig. 22.20).

Color mixing and analyzing has become an important science. Many industries depend on a thorough knowledge and skillful application of this science. Consider the variety of colors that are now available and used for clothing, automobile finishes, wallpaper, floor coverings, and paints for interior and exterior decorating.

22.18 ■ Penetration of Light Waves

Scattering of light by small particles in the atmosphere.

The beautiful blue of a midday sky and the gorgeous shades of red and yellow that frequently accompany a sunrise or sunset can be explained in terms of the *scattering* of the sun's rays. When white light passes through the atmosphere, the short-wavelength blue rays are scattered in all directions by the particles of the atmosphere, while the longer-wavelength red rays tend to continue straight through. The blue rays which are scattered by particles of air, water, and dust in the atmosphere are what we see when we look up in the sky. At sunrise and sunset we see only the longer red and yellow rays which have passed a long distance through the atmosphere. The shorter blue and green rays have been scattered and dissipated.

Red aircraft-beacon lights are used because of their greater penetrability in the fog and smoke of "heavy" weather. In like manner, yellow sodium lights are used on many bridges where foggy weather persists. However, merely covering a source of white light by a red or amber filter does not necessarily increase the penetrability of its rays, since by eliminating the shorter rays the intensity of the beam will be decreased.

22.19 ■ Uses of Infrared Rays

Infrared rays have an even greater penetrating power than red and yellow rays. Using camera film which is sensitive to infrared rays, one can take pictures of objects which are completely invisible to the eye. Infrared photography is employed in mapping from aircraft and satellites, since these rays can penetrate fog and haze and give a much sharper picture. Infrared photographic analysis of the earth's surface from Land Sat satellites is used for a multitude of purposes—for agriculture, forestry, geology, volcanology, oceanography, and many more. Infrared operations of this kind are one phase of a burgeoning new science called *remote sensing*.

Considerable use is made of infrared rays in gymnasiums and in sports medicine. These rays tend to penetrate the skin and warm the tissue of patients suffering from muscular aches and pains. In industry, freshly painted autos and refrigerators are dried under large batteries of infrared lamps. The penetrating rays afford a more

Fig. 22.21 An example of transmission of information over a short distance using infrared radiation. The results of calculations performed by the pocket computer are sent to the printer using infrared rays. (Courtesy, Hewlett-Packard Company)

uniform drying of many layers of paint at the same time. Many foods are dehydrated by using these rays—with a resultant saving of time as well as a saving of many of the vitamins in the food. Food warmers and ovens make use of infrared rays.

Although infrared rays are sometimes termed *heat rays*, the rays themselves are not warm. They *produce* heat by stimulating an increased motion of molecules in the objects they strike. It is the infrared radiation from the sun that is mainly responsible for the heating effect of the sun (see Sec. 16.11).

Infrared rays heat objects they strike.

Infrared radiation is sometimes used to transmit information over short distances. For example, coded pulses of infrared radiation are emitted by some hand-held remote-control units to control the operation of television sets and video cassette recorders (VCRs). In another application, infrared pulses are used to transmit information from a pocket computer to a printer for the computer, replacing the wire cable often used to connect such devices (see Fig. 22.21).

Other applications of infrared radiation include motion detectors for security systems for homes and businesses, heat detectors to indicate the presence of fire, analysis of infrared emissions from buildings to detect heat leaks (poor insulation), and photographic autofocus systems for cameras.

22.20 ■ Properties of Ultraviolet Radiation

Ultraviolet radiation is scattered by the atmosphere even more readily than the visible blue rays. Hence very few ultraviolet rays reach the earth from the sun. It is fortunate that most of these rays are scattered and absorbed by ozone in the stratosphere before reaching the earth, for if ultraviolet radiation from the sun reached the earth undiminished, it would prove disastrous to most life on the globe. These rays can produce serious burns of which "sunburn" is a painful example. Many mountain climbers have belatedly come to realize this fact. Since there is less atmosphere to penetrate at high altitudes, more ultraviolet rays reach the earth at high altitudes than at sea level. Hence sunburning takes place more rapidly on mountain peaks than at lower elevations.

The sun's ultraviolet radiation, acting on the human body, aids in the formation of vitamin D in the body, which helps to make healthy teeth and bone. Some food products are purposely irradiated with ultraviolet rays until the desired amount of vitamin D is developed in the food.

Carbon arcs and mercury arcs, which emit strong ultraviolet radiations, are used in sunlamps. Window glass is almost perfectly transparent to visible light, but it does

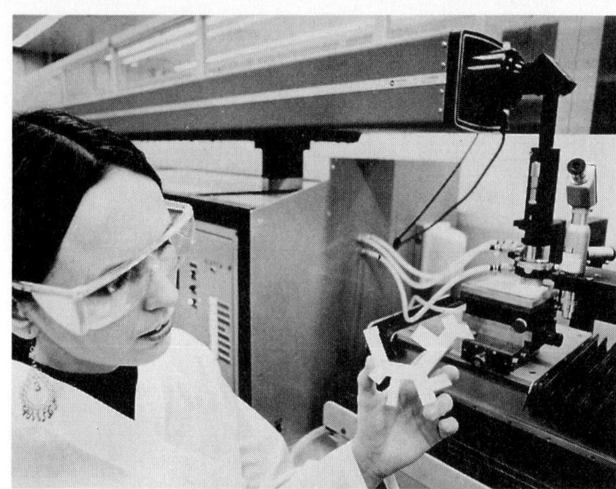

Fig. 22.22 A carbon dioxide laser (top of photo) is being used here to cut a very hard ceramic base into the irregular shape shown in the technician's hand. The ceramic base will be used for a thin-film electronic circuit in telecommunications equipment. (Western Electric Co.)

Although ultraviolet radiation is invisible, it is harmful to the eyes, and with overexposure, to the skin.

not transmit the ultraviolet light. Therefore it is almost impossible to get a sunburn from sunlight that has come through a closed window. Pure quartz glass, which will transmit ultraviolet rays, is used in sunlamps. It is important that the eyes be shielded whenever sunlamps are used. Likewise, since the radiation from a welding arc contains a dangerous amount of ultraviolet, the welder's eyes must be protected by a mask with a suitable filter.

22.21 ■ Lasers

The laser produces a very narrow beam of intense radiation.

Ordinary light sources are incoherent; i.e., they produce waves of light energy of many different wavelengths and with random phase relationships. As a result, there is little reinforcement among them, and their energy is soon diffused and lost. The *laser* is a device that produces an intense, highly concentrated, and almost parallel beam of light. The word laser is an acronym made up of the first letters of the words: Light Amplification by Stimulated Emission of Radiation. A discussion of the theoretical principles involved in laser production requires familiarity with the energy levels of the electrons in the structure of various atoms. Such an account is given in Chap. 34 but we shall discuss here some properties and uses of lasers.

Lasers produce an intense, narrow light beam of a very pure single color (monochromatic light). Laser beams can be intense enough to vaporize the hardest and most heat-resistant materials (Fig. 22.22). This property is useful in drilling holes in diamonds for wire-drawing dies and to destroy tumors or to weld a detached retina in modern eye surgery. It is used to perform microsurgery on cells of the body.

The helium-neon laser provides an ideal narrow straight line for all kinds of alignment purposes. Such a beam diverges by less than one part in a thousand for distances of millions of miles. An even smaller divergence of a laser beam directed to the surface of the moon, 240,000 mi distant, has produced a spot of light only a few feet in diameter. Laser beams are widely used to guide machines for drilling tunnels and to align jigs employed in producing large jet aircraft.

Lasers are used in medicine, engineering, construction, research, industry, and education.

Laser beams directed vertically downward from an airplane in flight can be used to determine the plane's height above ground or to map fine details of the land below.

Laser beams can be used for communications. Today one laser beam could carry the signals from all existing radio stations, and one beam can carry as many as seven television programs at one time. However, the beams can be blocked by fog or rain. So, to be effective under all weather conditions, the beams would need to be enclosed in protecting pipes or conducted along optic fibers. Laser beams, because of their directionality and use of small amounts of power, do find uses in long-distance communication through outer space.

Most lasers in use today are of low power and do not cause injury if carefully used. However, one should *never look directly* down a laser beam or along a reflected beam. For the 1-mW helium-neon lasers now often used in the physics laboratories of many colleges and technical institutes, the maximum safe exposure time for the eye is 0.1 s. Since this is less than the human reaction time, extreme caution must be used when working with a laser to prevent possible permanent eye damage. (See Sec. 34.14 for an additional discussion of lasers and applications of lasers.)

Exposure of the eye to a 1-mW laser beam for a fraction of a second can cause permanent damage to the eye.

QUESTIONS AND EXERCISES

1. List four ways in which light waves differ from sound waves. Then list four ways in which light waves are similar to sound waves.

2. Cut the bottom end out of an empty food can. Use rubber bands to cover one end of the can with a piece of translucent wax paper and the other end with aluminum foil. After making both end coverings taut, use a pin to punch a hole in the center of the aluminum foil. Use this simple apparatus to observe image formation by rectilinear propagation of light through a pinhole as described in Fig. 22.2. Note that an inverted image of an object, such as a bright light bulb, can be seen on the wax paper. Experiment to see how the image size and brightness depend on the distance between the object and the pinhole. Investigate also how the image brightness and sharpness depend on the diameter of the pinhole.

3. What objects cast the shadows that cause solar and lunar eclipses?

4. A student claims that light can travel farther in a hundredth of a second than sound (speed = 344 m/s) can travel in an hour. Do you believe the claim?

5. In what range of wavelengths is the eye most sensitive to light?

6. Does a 100-W incandescent bulb radiate mainly ultraviolet, visible, or infrared electromagnetic waves?

7. Discuss the difference between luminous flux and luminous intensity. Explain the units in which each is measured.

8. Which of the following does *not* properly pair with the unit of measurement following it: (*a*) luminous intensity—candela; (*b*) intensity of illumination—lux; (*c*) luminous flux—footcandle; (*d*) luminous flux—lumen?

9. Which has a greater luminous efficiency, a fluorescent lamp or an incandescent bulb? Discuss the reason for the difference.

10. Does the luminous intensity of a source decrease with distance from the source? How about luminous flux?

11. Distinguish between the meaning of the terms *light* and *illumination*.

12. Which is greater, an illumination of 5 lm/m^2 or 5 ft-c?

13. The following pairs of colors (light) are mixed additively. Name the resultant colors: (*a*) blue and green, (*b*) red or green, (*c*) blue and yellow, (*d*) red and blue, (*e*) magenta and green.

14. The following pairs of colors (pigments) are mixed subtractively. Name the resultant colors: (*a*) yellow and cyan, (*b*) yellow and blue, (*c*) cyan and magenta, (*d*) red and cyan, (*e*) magenta and yellow.

15. (*a*) What color *added* to red will give white? (*b*) What color *subtracted* from yellow will give black?

16. What color is the complement of (*a*) red, (*b*) yellow, (*c*) green, (*d*) cyan, (*e*) blue, (*f*) magenta?

17. Six lamps, each emitting one of the following colors—violet, blue, green, yellow, orange, and red—are viewed through a red glass filter. No other light is present. Describe the ''color'' you would see as you view each lamp through the filter.

18. Does the atmosphere scatter more red rays or more blue rays from the sun? Give reasons for your answer.

19. Why is the sky blue during full daylight and why is the sun red at sunset?

20. How effective will placing a yellow filter over the headlights of a car be in producing increased visibility in foggy weather? Why?

21. A hot iron can be photographed in a dark room only if which one of the following is true: (*a*) the lens aperture is opened wide; (*b*) a long exposure time is used; (*c*) infrared sensitive film is used; (*d*) the iron is painted with a luminous substance?

22. Discuss several ways in which laser light differs from the light produced by a common incandescent bulb.

23. What is the maximum length of time that light from a 1-mW laser can enter the eye without causing permanent damage?

PROBLEMS

Group One

1. How far (miles) can a light beam in air travel in 0.005 s?

2. What length of time would be required for a laser beam to travel from the surface of the earth to an orbiting satellite 22,250 mi above the earth's surface?

3. A camera is set with its shutter speed at $\frac{1}{500}$ s. How far (meters) will light travel during the time of exposure?

4. The North Star (Polaris) is 1.6×10^{15} mi from the earth. How many years elapse before light emitted from the star reaches the earth?

5. A helium-neon gas laser emits red light of wavelength 632.8 nm. What is the frequency of that light?

6. Equal illumination appeared on a photometer screen which was placed 75 cm from a 50-cd standard lamp and 45 cm from a second lamp. What was the luminous intensity of the second lamp?

7. A laboratory photometer (see Fig. 22.10) has a standard 32-cd lamp at one end and a lamp of unknown intensity at the other. The two sides of the screen appear equally illuminated when the screen is 3.5 ft from the standard lamp and 5.2 ft from the unknown. What is the luminous intensity of the unknown lamp?

8. The eye is not equally sensitive to all wavelengths of light. Is the relative visibility of light greater at a wavelength of 490 or 590 nm?

9. A certain tunable laser, which is adjusted so that it emits light only at 555 nm, emits a total luminous flux of 3.4 lm. What is the radiant power of the laser in watts?

10. The luminous flux of a 75-W soft-white incandescent bulb is 1170 lm. What would be the luminous flux of a 75-W source of light that emitted all its light at a wavelength of (*a*) 555 nm, (*b*) 610 nm?

11. The luminous flux for a 60-W soft-white incandescent bulb is 855 lm. What is the luminous efficiency for the bulb in lumens/watt?

12. The luminous efficiency of a 150-W incandescent bulb is 18.6 lm/W. What is the luminous flux for the bulb?

13. The stated luminous flux is 1170 lm for a 75-W incandescent bulb and 2790 lm for a 150-W incandescent bulb. For which bulb is the luminous efficiency (in lumens/watt) greater?

14. The total luminous flux F is 445 lm for a 40-W bulb that emits light uniformly in all directions. What is the luminous intensity of that bulb in candelas?

15. A light bulb that emits light uniformly in all directions has a luminous intensity of 15.1 lm/sr. What is the total luminous flux emitted by the bulb?

16. What is the illumination, in footcandles, on the surface of a desk located 4 ft directly below a 135-cd lamp? (Assume there are no nearby surfaces to send reflected light to the desk.)

Group Two

17. The width of a 100-W incandescent bulb at its widest point is 6 cm. An image of the bulb is formed on the film in a pinhole camera in which the distance from the pinhole to the film is 14 cm. What is the width of the image of the bulb on the film when the distance from the bulb to the pinhole is (*a*) 30 cm, (*b*) 100 cm?

18. A light beam carrying telephone signals in an optical fiber travels the distance from one repeater (amplifier) to another in 10^{-4} s (100 μs). The speed of light in the optical fiber is $0.67c$. What is the length of the optical fiber between repeaters?

19. The distance from the earth to the moon has been determined by measuring the time for a beam of laser light to travel from the earth to a mirror on the moon, and back to the earth again. If the time for the pulse of laser light to make the round trip is 2.510 s, what is the distance (in mi) from the earth to the moon?

20. The human eye is sensitive to electromagnetic waves with wavelengths between 400 and 700 nm (the visible spectrum). (*a*) What is the lowest *frequency* to which the eye is sensitive? (*b*) Is the eye sensitive to light of twice that frequency?

21. The frequency of radiation emitted by a certain laser is 3.22×10^{14} Hz. (*a*) What is the wavelength of the radiation emitted by the laser? (*b*) Does the laser radiation lie in the ultraviolet, visible, or infrared portion of the electromagnetic spectrum? (*c*) What is the relative visibility of radiation of the wavelength emitted by the laser? (*d*) Would the high intensity beam of laser radiation be visible? Why, or why not?

22. A photometer has a 32-cd standard lamp at one end 85 cm from the photometer screen. At what distance from the screen must a 40-cd standard lamp be placed at the other end of the photometer to cause equal illumination on the photometer screen from the two sources?

23. If 3 s is the proper time of exposure in printing a photograph using an incandescent lamp 70 cm from the printing frame, what length of exposure would be required in printing a picture from the same negative at a distance of 2 m from the same light source?

24. The relative visibility of two different wavelengths in the visible spectrum is 0.5. One of the wavelengths is 510 nm. What is the approximate value of the other wavelength?

25. A 40-W fluorescent lamp has a total luminous output of 3000 lm. The luminous efficiency of a 60-W incandescent bulb is 14.25 lm/W. What is the minimum number of 60-W incandescent bulbs that would provide a greater luminous flux than a single 40-W fluorescent lamp?

26. The luminous flux for a 200-W incandescent bulb is 3910 lumens. What percentage of the 200-W appears as visible radiation?

27. If 20 lux is correct for reading, how far from your reading material should a 64-cd light be placed in order to give proper illumination? (Assume no other light reaches the page.)

28. A screen is 5 m from an 80-cd source of light, and the screen surface makes an angle of 60° with the line drawn from the source to the center of the screen. What is the intensity of illumination on the screen?

29. What would be the intensity of illumination on a magazine held 5 ft from a 40-cd lamp? What would be the intensity of illumination 10 ft from the lamp? (Assume no reflection from the walls.)

30. The efficiency of a 300-W tungsten lamp is rated at 12 lm/W. What illumination in footcandles will be given by such a lamp on a surface 5 m from the lamp?

31. A standard 24-cd lamp at a distance of 40 cm from a screen gives the same illumination as a lamp of unknown intensity placed 120 cm from the screen. What is the intensity of the unknown lamp?

Group Three

32. What are the approximate values of the relative visibilities of light of the following frequencies: (*a*) 4.78×10^{14} Hz, (*b*) 7.69×10^{14} Hz, (*c*) 5.40×10^{14} Hz?

33. A 150-W incandescent bulb, assumed to radiate light uniformly in all directions, has a luminous intensity of 222 lm/sr. What is the illumination E in lm/m^2 (luxes) on a flat surface located 2 m directly below the bulb?

34. A 24-cd lamp is placed 0.36 m from a screen. How far from the screen must a 64-cd lamp be placed for the illumination on the screen to be three times as great as it was with the first lamp?

35. A football stadium is illuminated at night from eight towers, each with a bank of twenty-four 1000-W lamps with luminous efficiencies of 32 lm/W. One-third of the luminous flux reaches the playing field, which has dimensions of 120 by 75 m. Find the average intensity of illumination on the field.

36. If four lighted candles are placed at one end of a meter stick and one identical candle at the other end, where must a screen be placed on the meter stick in order to get equal illumination from both sources? (HINT: Let x = the distance in centimeters of the one candle to the screen and $100 - x$ the distance in centimeters of the four candles to the screen.)

37. The full moon provides an illumination of 0.023 lm/ft^2 at the surface of the earth. Consider a 40-W incandescent bulb with a total luminous flux of 490 lm. At what distance above the earth's surface would such a bulb provide the same illumination as the moon? (Assume the bulb emits light uniformly in all directions.)

38. At what angle must a surface be inclined so that light falling on it from an 80-cd luminous intensity source at a distance of 4 ft will give an illumination of 1.21 ft-c?

39. The surface of a long rectangular workbench is illuminated from above by three 300-cd incandescent bulbs. The bulbs, which are separated from one another by 6 ft, are located 4 ft above the surface of the workbench. (*a*) What is the illumination on the workbench surface directly below the center bulb? (*b*) What is the illumination on the workbench surface directly below the midpoint between the center bulb and either of the other two bulbs?

40. In measuring the speed of light by Michelson's method, a rotating mirror having 32 sides was used. The rotating mirror was 5.000 mi from the stationary mirror. What minimum rotational speed, in revolutions per second, was required of the rotating mirror to catch the reflected beam on the very next mirror surface?

CHAPTER 23 REFLECTION AND REFRACTION OF LIGHT

Reflection, refraction, and image formation by mirrors and lenses can be explained using geometric, ray optics.

The last chapter dealt with the measurement of the brightness of luminous and illuminated objects. The next chapter will include discussions of optical instruments, such as telescopes, microscopes, and cameras, that can be used to observe or photograph luminous or illuminated objects. In this chapter we will discuss, among other topics, the mirrors and lenses that are essential components of optical instruments. Since image formation is accomplished with mirrors by reflecting light, and with lenses by refracting light, the laws of reflection and refraction will be discussed in some detail.

Geometric, or ray, optics (Sec. 22.3) can be used to discuss the laws of reflection and refraction, to explain the process of image formation by mirrors and lenses, and to explain the operation of many optical instruments.

REFLECTION OF LIGHT

23.1 ■ Regular and Irregular Reflection

Diffuse scattering of light from the surface of this page allows you to read these words.

Common experience shows that when light strikes most objects, some of the light is reflected from the object. Indeed, it is by this reflected light that we are able to see the object. If the surface is rough, light striking it is reflected in many directions. This kind of *irregular* reflection (Fig. 23.1*a*) is called *diffuse reflection*. The distance between the "hills and valleys" and the depth of the "valleys" of the rough surface illustrated in Fig. 23.1*a* need only be large compared to the wavelength of light to cause diffuse reflection. The paper these words are printed on is an example of a surface sufficiently rough to reflect diffusely the light that strikes it.

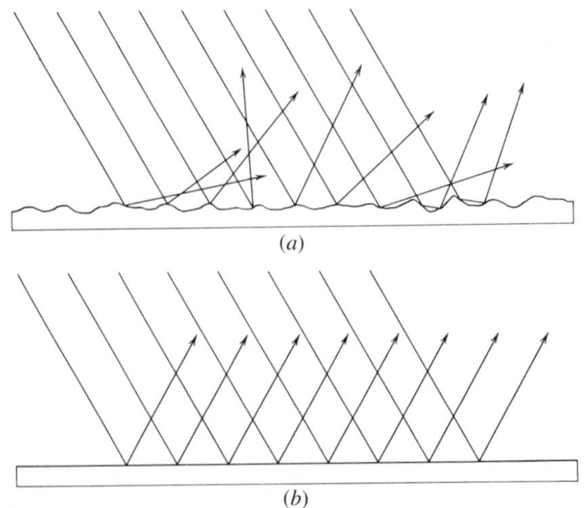

Fig. 23.1 (*a*) Irregular or diffuse reflection of light from a rough surface, and (*b*) regular or specular reflection of light from a smooth surface.

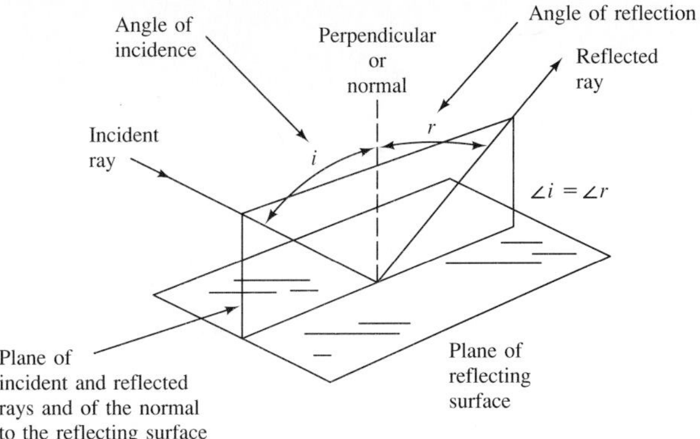

Fig. 23.2 The angle of incidence i is equal to the angle of reflection r, and the incident ray, the reflected ray, and the normal (perpendicular) to the reflecting surface all lie in the same plane.

If a surface is polished so that it is very smooth, parallel rays in a beam of light that strike the surface will, after reflection, still be parallel, as illustrated in Fig. 23.1*b*. This type of *regular* reflection, which is called *specular reflection*, occurs at smooth glass surfaces and smooth shiny metal surfaces. The result of specular reflection of light from an object is the appearance of an *image* of the object in the smooth reflecting surface. The best images are produced by smooth, shiny, metallic surfaces called *mirrors*.

23.2 ■ Laws of Reflection

The laws governing the direction of beams of light which are regularly reflected are the same for any smooth surface. When a beam of light enters a small opening in a dark room, the path of the beam can be clearly seen by the illuminated dust particles in the room. If this beam is allowed to strike a plane mirror (Fig. 23.2), the directions of the *incident* and *reflected* rays can easily be seen.

If a perpendicular (normal) is drawn to the reflecting surface at the point where the incident ray strikes the surface, the angle between the normal and the incident ray is called the angle of incidence, while the angle between the normal and the reflected ray is called the angle of reflection.

Careful experimental studies of specular (i.e., regular) reflection show that light rays obey the following laws:

The Laws of Reflection

1. ***The angle of incidence i is equal to the angle of reflection r.***
2. ***The incident ray, the reflected ray, and the normal to the reflecting surface all lie in the same plane.***

The equality of the angles of incidence and reflection can be stated mathematically.

$$\text{Angle } i = \text{angle } r \tag{23.1}$$

Solar Cooking Oven A solar cooking oven is an example of using the laws of reflection. Such an oven consists of an insulated chamber for the food with a glass cover plate to allow sunlight to enter the oven. The energy in an entering beam of bright sunlight is sufficient to heat the air in the oven to temperatures high enough to cook a variety of foods. Designers of solar cooking ovens usually include an arrangement of metallic mirrors to gather energy from a beam of sunlight with a cross-sectional area much greater than the area of the glass cover plate (see Fig. 23.3).

Fig. 23.3 Solar cooking oven. (*a*) Diagram showing how appropriately positioned metallic reflectors can act as a "funnel" to channel rays of sunlight (solar energy) into the cooking chamber. (*b*) A solar collector showing the four metallic reflectors. Note that the interior of the cooking chamber is painted black to allow maximum absorption of the incoming sunlight. (Photo courtesy Burns-Milwaukee, Inc.)

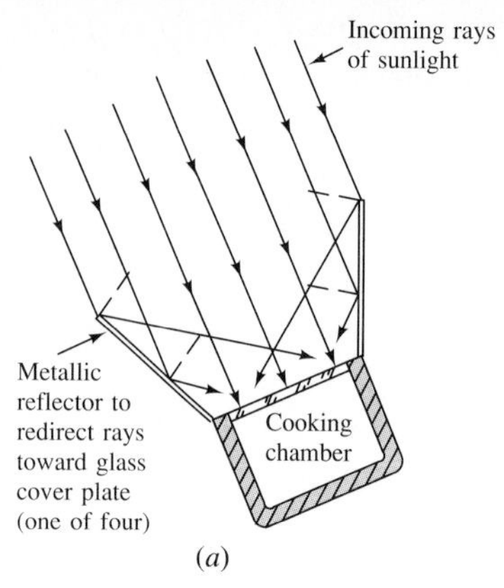

(*a*)

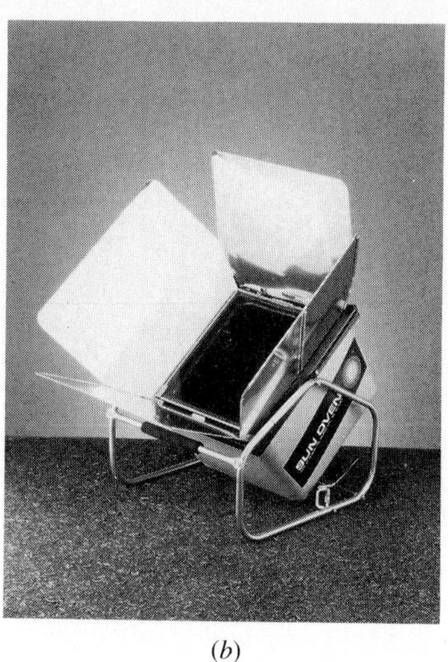

(*b*)

23.3 ■ The Optical Lever

A ray of light reflected from a rotating mirror rotates twice as rapidly as the mirror.

In some scientific instruments small rotational displacements are magnified by attaching a very small plane mirror to the rotating part. Let us consider an incident ray of light from a laser source striking the mirror MM' at O, and reflected as OR (Fig. 23.4). If we rotate the mirror through an angle ϕ, the normal ON is rotated through an equal angle to ON'. The new angle of incidence will be $i' = i + \phi$. By the law of reflection, the new angle of reflection r' also equals $i + \phi$. We see then that the increased deviation of the reflected ray

$$\text{Angle } ROR' = \text{angle } SOR' - \text{angle } SOR$$

or

$$\angle ROR' = 2(i + \phi) - 2i = 2\phi$$

Hence such an arrangement, called an *optical lever*, will turn a reflected ray of light through *twice* the angle through which the mirror is turned. By shining the reflected ray on a scale several feet from the mirror, extremely small angular displacements can be measured. Optical levers of this type are used in magnifying the motions of indicators in several types of measuring instruments, such as the sextant.

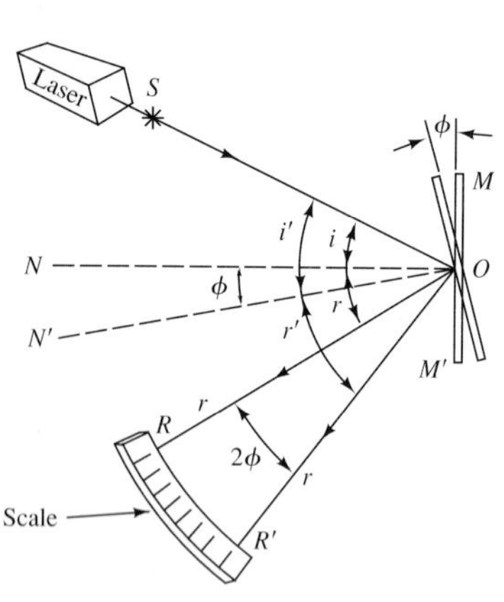

Fig. 23.4 In the optical lever arrangement illustrated, when mirror MM' rotates through an angle ø, a narrow beam reflected by the mirror rotates through an angle 2ø.

Illustrative Problem 23.1 An optical lever arrangement like that illustrated in Fig. 23.4 is set up with a mirror-to-scale distance OR of 2 m. If the mirror rotates through an angle of 0.5°, what will be the distance RR' between the initial spot R and the final spot R' where the laser beam strikes the scale?

Solution The distance RR' along the curved scale, the distance $OR = r$ from the mirror to the scale, and the angle 2ϕ between the initial and final directions of the reflected laser beam are related by an equation of the form $s = r\theta$. In the equation $s = r\theta$ (see Sec. 8.2), s represents the length of the arc of a circle of radius r that is subtended by an angle θ, where θ is in radians. For this problem, $s = RR'$, $r = OR$, and $\theta = 2\phi$, and the following equation can be used to calculate RR' if the 0.5° rotation angle of the mirror is converted to radians (πradians = 180°).

$$\begin{aligned} RR' &= (OR)(2\phi) \\ &= (2\text{ m})[(2)(0.5)(\tfrac{\pi}{180})] \\ &= 0.035\text{ m} \\ &= 3.5\text{ cm} \end{aligned}$$ *answer*

Illustrative Problem 23.2 An optical lever arrangement, like that illustrated in Fig. 23.4, is to be set up to meet the following condition: When the mirror is rotated through an angle ϕ of 0.2°, the laser beam spot should move a distance RR' of 1 cm on the scale. What should be the distance OR from the mirror to the scale?

Solution The same equation relating RR', OR, and 2ϕ used in Illustrative Problem 23.1 can be used for this problem.

$$RR' = (OR)(2\phi)$$

For this problem, the unknown is the mirror-to-scale distance OR. Solving the above equation for OR gives

$$OR = \frac{RR'}{2\phi}$$

Substituting the value 1 cm for RR', the angle 0.2° for ϕ, and converting the angle ϕ from degrees to radians, gives

$$OR = \frac{1 \text{ cm}}{(2)(0.2°)\left(\dfrac{\pi}{180°}\right)}$$

$$= 143 \text{ cm} \qquad \textit{answer}$$

23.4 ■ Images Formed by Plane Mirrors

In order to form the image of a self-luminous or illuminated object in a plane mirror, each point on the object must send out rays in all directions. One of these rays coming from each point of the object (represented by the pencil with ends marked AB in Fig. 23.5) will strike the surface of the mirror MM′ at just the right spot to be reflected to the eye E. For example, a light ray coming from point A strikes the mirror at P and is reflected, according to the law *angle i = angle r*, and comes to the eye E. The light *appears* to be coming from a point A' back of the mirror. Light from B is reflected by the mirror to the eye as if coming from B', lying on a line perpendicular to the mirror and the same distance behind it as B is in front of the mirror. By applying rules of geometry, it can be proved that *an image in a plane mirror is the same size as its object and that the image appears as far behind the mirror as the object is in front of it.*

The image formed by a plane mirror is called a *virtual image*.

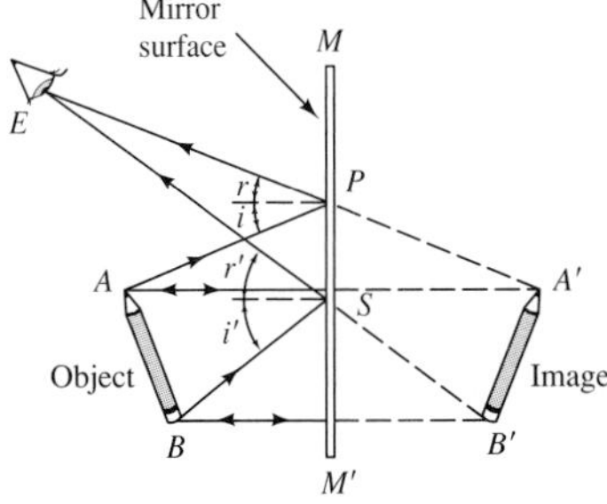

Fig. 23.5 Image of a pencil formed by a plane mirror. The image appears to be as far behind the mirror as the object actually is in front of it.

A virtual image is formed by rays which seem to radiate from the image, but which actually come from another source. A virtual image cannot be projected onto a screen.

The image of an object in a plane mirror appears laterally reversed with respect to the reflecting plane. For example, the reflection of the printed page of a book is reversed from right to left, and as you look at yourself in a mirror, what appears to be the right hand of your image is actually the image of your left hand.

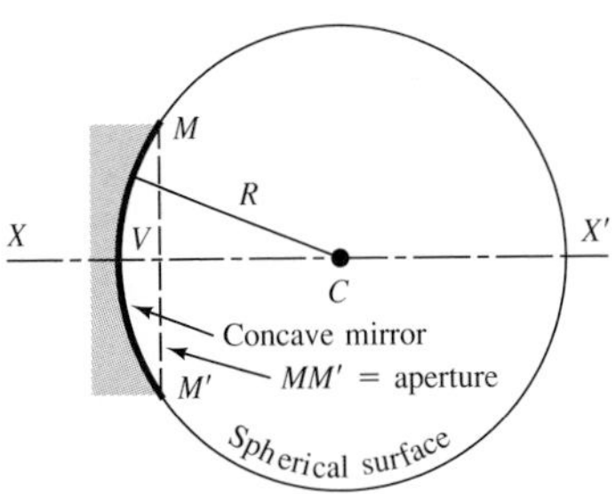

Fig. 23.6 A spherical mirror is made from a segment of a spherical surface.

23.5 ■ Spherical Mirrors

Spherical mirrors are classified as *concave* or *convex*, depending on whether the reflecting mirror is an inner or an outer segment of a sphere (Fig. 23.6 for example, is a diagram of a concave spherical mirror). The *center of curvature* C of the mirror is the center of the sphere. The *radius of curvature* R of the mirror is the radius of the

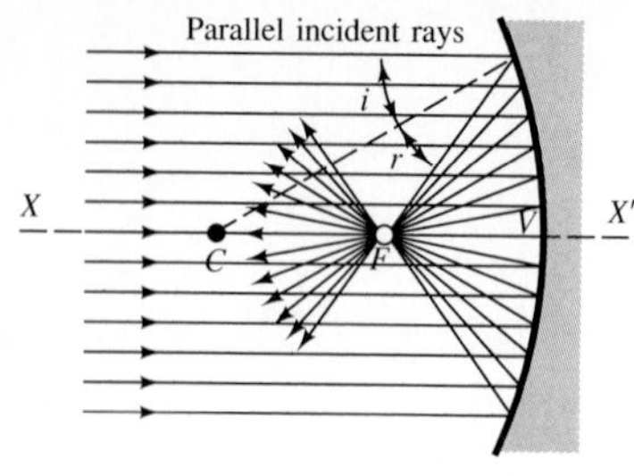

Fig. 23.7 Reflection of parallel light rays from a concave mirror.

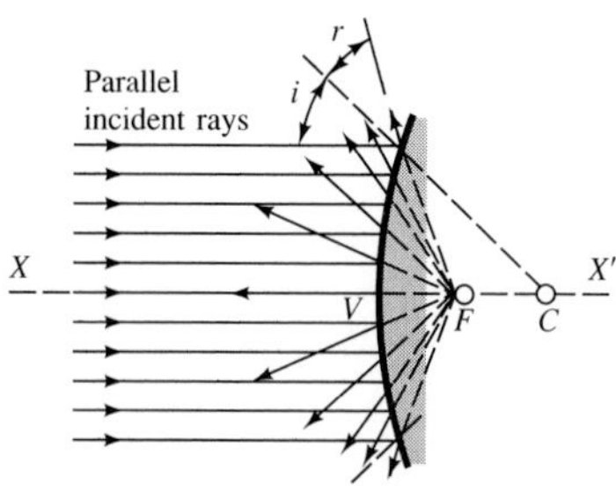

Fig. 23.8 Reflection of parallel light rays from a convex mirror.

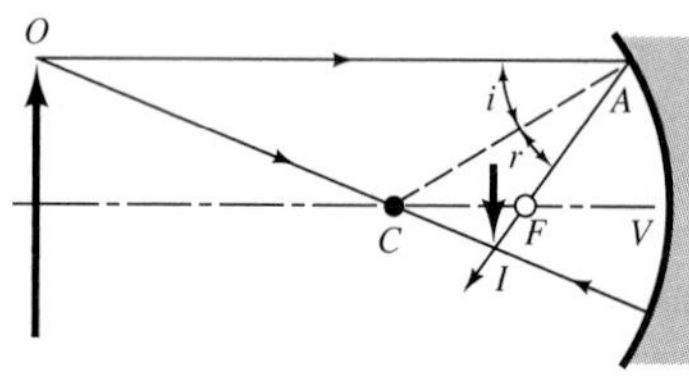

Fig. 23.9 Location of image formed by a concave mirror.

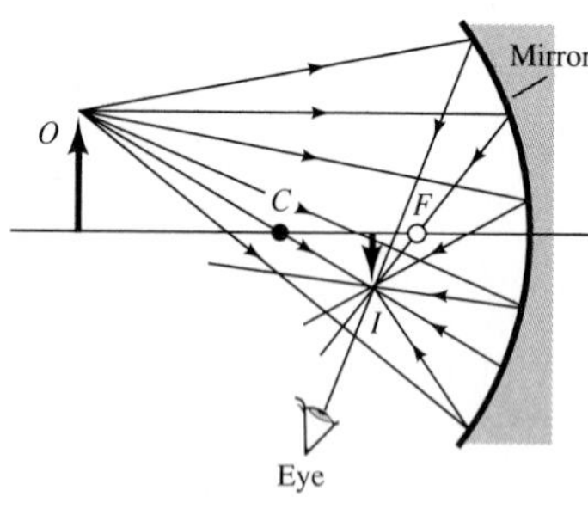

Fig. 23.10 All rays from O that strike the mirror contribute to the image.

sphere. The *vertex* of the mirror is the central point V of the mirror, while a line XX' passing through the center of curvature and the vertex is called the *principal axis* of the mirror. The *aperture* of the mirror is the diameter MM' of the small circle which is the base of the spherical segment.

Most spherical mirrors are comparatively flat; i.e., the size of the aperture is small in comparison with the radius of curvature. With a concave mirror of small aperture, rays of light parallel to the principal axis will strike the mirror and be so reflected as to pass through, or very close to, a single point F, called the *principal focus* of the mirror (Fig. 23.7). The distance FV from the principal focus to the mirror is called the *focal length* f.

The ability of a concave mirror to gather parallel rays from the sun and focus them at a point is used in one type of solar furnace. Temperatures at the focus as high as 5000°C have been produced in this way.*

Parallel rays which strike a *convex* mirror will diverge after reflection as though they originated at a common focal point F behind the mirror (Fig.23.8).

It can be shown geometrically that *for apertures which are small compared to the radius of curvature*, the principal focus F lies approximately halfway between the vertex and the center of curvature of the mirror. Stated mathematically

$$f = R/2 \qquad (23.2)$$

23.6 ■ Images Formed by Spherical Mirrors—Real and Virtual Images

Let us consider a concave mirror (Fig. 23.9). Any ray OA parallel to the axis is reflected back through the focal point F, while the ray OC passing through the center of curvature C will be perpendicular to the mirror and will be reflected back upon itself. After reflection these rays coming from O will intersect at I, forming a *real image* of O at point I. A real image is formed by converging rays which *actually pass through and radiate from the image location*. A real image can therefore be projected on a screen located at the apparent image position.

In addition to the two specific rays we have been using to locate the image, all other rays from point O which strike the mirror contribute to the image (see Fig. 23.10). The larger the aperture of the mirror, the brighter the image will be, although a large aperture causes the image to be less sharply focused.

A similar procedure is followed in locating the image in a convex mirror (Fig. 23.11). From the figure it is apparent that the rays, after reflection, will appear as if they came from behind the mirror at I. For all positions of O the image will be smaller than the object. An image that is formed by reflected rays that only appear to converge is called a *virtual image*. Virtual images will not show up on a screen.

Whether an image will be real or virtual depends not only on whether the mirror is concave or convex but also on the position of the object with respect to the mirror. Six cases may occur, as represented in Fig. 23.12*a* to 23.12*f*.

CONCAVE-MIRROR IMAGES

1. When the object lies at a distance from the vertex greater than the radius of the mirror, the image is located between the center and the focus. It is real, inverted, and smaller than the object (Fig. 23.12*a*). Reflecting telescopes use such an arrangement.
2. When the object is at the exact center of curvature, the image is also at the center. It is real, inverted, and the same size as the object (Fig. 23.12*b*).
3. When the object is located between the center and the focus, the image lies beyond the center. It is real, inverted, and larger than the object (Fig. 23.12*c*). Floodlights have the light source between the center and the focus of a concave reflector in order to spread the rays.

* Although the solar energy received at the earth's surface is immense (approx. 320 Btu/h/ft^2 or 1000 watt /m^2 in bright sunlight), to effectively and economically harness it requires complex equipment and large surface areas.

4. When the object is at the focus, the reflected rays are parallel. Hence there is no image formed by the mirror (Fig. 23.12*d*). Searchlights, car headlights, and flashlights have the light source near the focus of a concave mirror to use this property. The common slide projector of the amateur photographer has a lamp placed at the focal point of a concave mirror in order that the light energy will not be dissipated by spreading out in all directions. Lenses in front of the lamp direct the rays through the slide. A similar arrangement is used in a photographic enlarger.

5. When the object is between the focus and the vertex, the reflected rays do not actually meet but appear to meet behind the mirror. Thus the image is virtual, erect, and enlarged (Fig. 23.12*e*). Shaving mirrors and dental mirrors use this arrangement to get virtual magnified images.

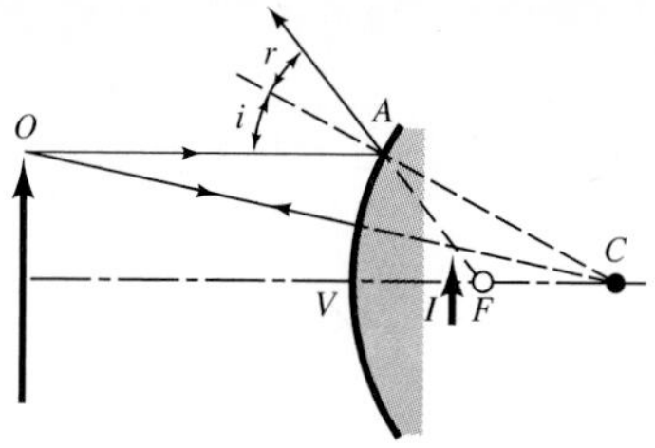

Fig. 23.11 Virtual image formed by a convex mirror.

CONVEX-MIRROR IMAGES

6. No matter where the object is located with respect to a convex mirror, the image is always virtual, erect, and smaller than the object (Fig. 23.12*f*). Large convex

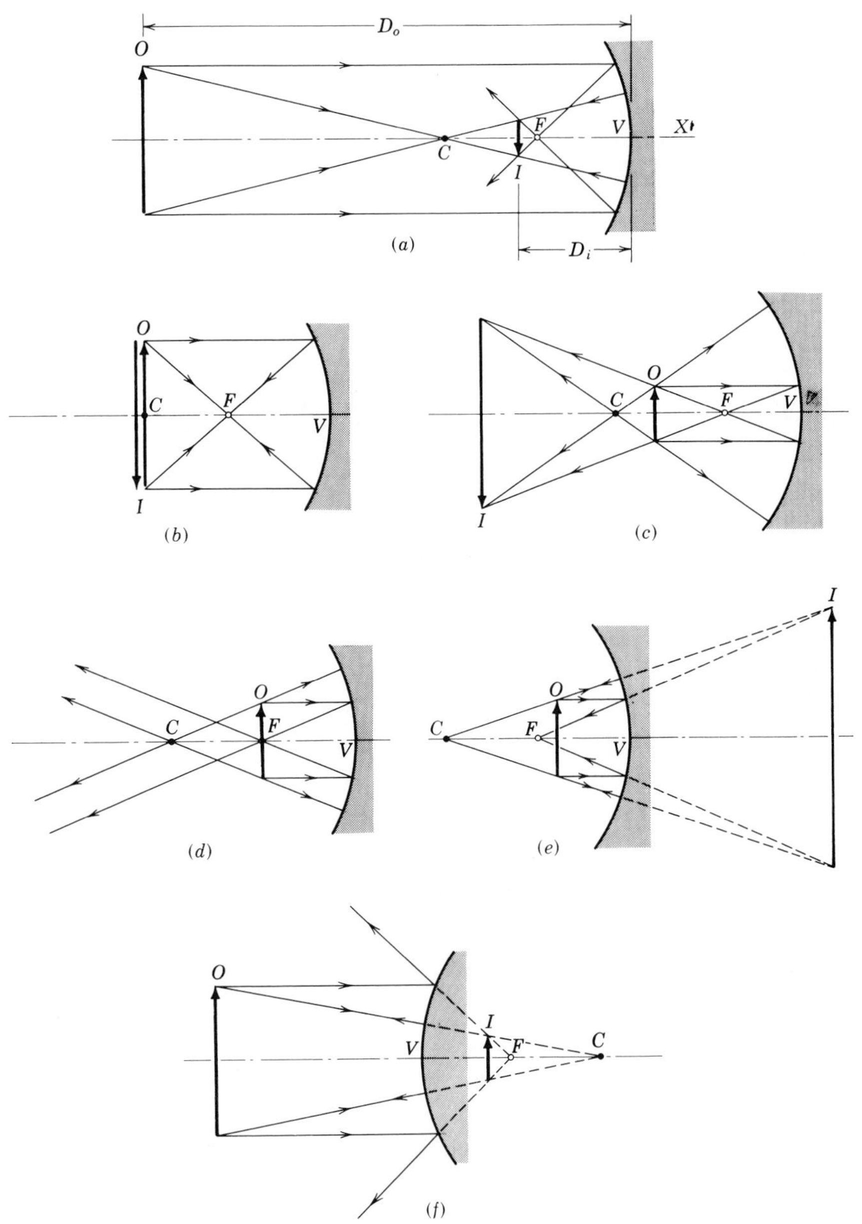

Fig. 23.12 Images formed by concave and convex mirrors. (*a*) Image is real, inverted, smaller than object. (*b*) Image is real, inverted, same size as object. (*c*) Image is real, inverted, larger than object. (*d*) No image, reflected rays are parallel. (*e*) Image is virtual, erect, larger than object. (*f*) Image is virtual, erect, smaller than object.

mirrors are sometimes used in stores to protect against shoplifters because they make it possible to see the entire room in miniature. Rear-view mirrors for autos are often slightly convex to afford a wider field of view.

23.7 ■ The Mirror Equation

The relation between the distance of object D_o, the distance of image D_i, and the focal length f, for any spherical mirror, is given by the formula

$$\frac{1}{D_o} + \frac{1}{D_i} = \frac{1}{f} = \frac{2}{R} \tag{23.3}$$

The convention of signs for spherical mirror calculations.

If any two of these quantities are known, the third can be computed. In order to make the equation applicable to both concave and convex mirrors, *distances back of the mirror must be considered negative*. Thus D_i, f and R for convex mirrors will always have negative values. A negative value for D_i with a concave mirror will therefore also signify that the image is behind the mirror and virtual. Distances in front of both concave and convex mirrors are positive.

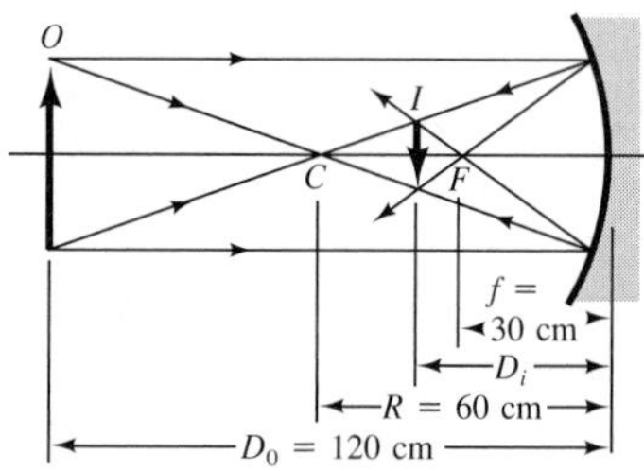

Fig. 23.13 Ray diagram for Illustrative Problem 23.3.

Illustrative Problem 23.3 An object is placed 120 cm in front of a concave mirror whose focal length is 30 cm. Find the location of the image (see Fig. 23.13).

Solution Since the mirror is concave, f is positive. The image will be real, since D_o is greater than f; hence D_o is taken positive and D_i should solve to be positive. Substitute known values in Eq. (23.3) and obtain

$$\frac{1}{120} + \frac{1}{D_i} = \frac{1}{30}$$

Multiply each term by $120D_i$ to clear fractions.

$$D_i + 120 = 4D_i$$

Transpose, collect terms, and solve.

$$3D_i = 120$$
$$D_i = 40 \text{ cm} \qquad \textit{answer}$$

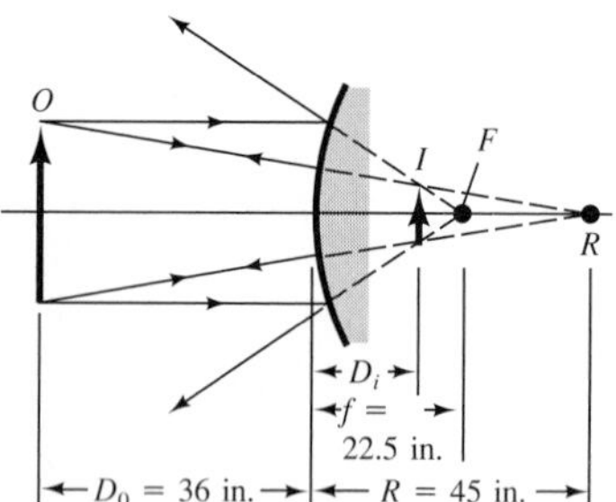

Fig. 23.14 Ray diagram for Illustrative Problem 23.4.

Illustrative Problem 23.4 An object is placed 36 in. in front of a convex spherical mirror of 45-in. radius. Locate the image (see Fig. 23.14).

Solution In this problem take f negative, since the mirror is convex and the image is virtual and solve for D_i (which should be negative). The focal length is equal to one-half the radius of curvature, or $\frac{45}{2}$ in. Substituting known values in Eq. (23.3), we get

$$\frac{1}{D_o} + \frac{1}{D_i} = \frac{1}{f}$$

$$\frac{1}{36} + \frac{1}{D_i} = \frac{-2}{45}$$

Multiplying each term by the least common denominator $180D_i$, transposing, and solving for D_i, we get

$$5D_i + 180 = -8D_i$$
$$13D_i = -180$$
$$D_i = -13.8 \text{ in. (behind the mirror)} \qquad \textit{answer}$$

23.8 ■ Size of the Image

By using any of the diagrams of Fig. 23.12*a* through 23.12*f* it can be readily shown that the size of the image I is to the size of the object O as the image distance D_i is to the object distance D_o, neglecting signs. The *magnification* is expressed as the ratio of the size of the image to the size of the object. Thus for any spherical mirror this relationship can be expressed algebraically as

$$\text{Magnification} = \frac{I}{O} = \frac{D_i}{D_o} \qquad (23.4)$$

The magnification equation for spherical mirrors.

Illustrative Problem 23.5 An object 8 cm tall is placed 40 cm from a concave mirror of focal length 60 cm. Locate and describe the image, and find its size (see Fig. 23.15).

Solution Substitute known values in Eq. (23.3), and solve for the image distance.

$$\frac{1}{D_o} + \frac{1}{D_i} = \frac{1}{f}$$

$$\frac{1}{40} + \frac{1}{D_i} = \frac{1}{60}$$

$$3D_i + 120 = 2D_i$$

$$D_i = -120 \text{ cm} \qquad \textit{answer}$$

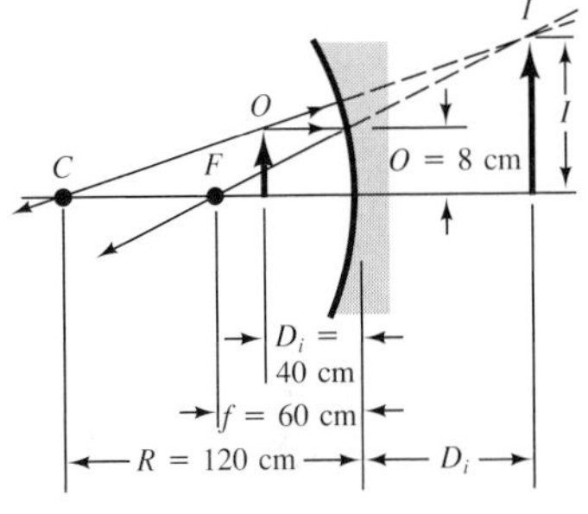

Fig. 23.15 Ray diagram for Illustrative Problem 23.5.

The minus sign indicates a virtual image behind the mirror. Substituting the value for D_i and other known values in Eq. (23.4), we can determine the magnification.

$$\frac{I}{O} = \frac{D_i}{D_o}$$

$$\frac{I}{8} = \frac{120}{40}$$

$$I = 24 \text{ cm} \qquad \textit{answer}$$

Hence the image is virtual, situated 120 cm behind the mirror, and 24 cm tall.

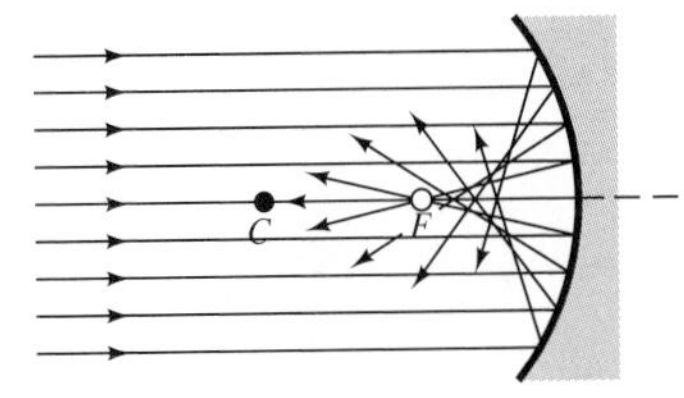

Fig. 23.16 Spherical aberration produced by a spherical mirror with a large aperture.

23.9 ■ Spherical Aberration with Mirrors

It should be reemphasized that the foregoing discussion on spherical mirrors and their images applies *only to mirrors whose apertures are small compared to their radii of curvature* and for objects on or near the principal axis of the mirror. As the aperture becomes relatively larger, the images become blurred and imperfect. The effect of a large aperture on parallel rays can be illustrated by Fig. 23.16. Here it will be noticed that only those rays close to the principal axis converge in the immediate vicinity of the focal point. The farther the incident ray is from the axis the more distant from the focal point (and the closer to the vertex) does its reflection pass. This effect is called *spherical aberration*. If we use a parabolic mirror, however, the parallel rays from a distant object will all be focused at a common focal point. Conversely, light radiating from a point source at the focal point F of a parabolic mirror is reflected as parallel rays (Fig. 23.17). Parabolic mirrors are used for both purposes—for forming sharp images of distant objects and for reflecting intense beams of parallel rays. Auto headlights and military and display searchlights are equipped with parabolic mirrors. Reflecting-type astronomical telescopes (see Fig. 24.18) have concave parabolic mirrors. Since it is difficult and expensive to grind parabolic mirrors, spherical mirrors with small apertures are sometimes used on commercially produced, inexpensive equipment.

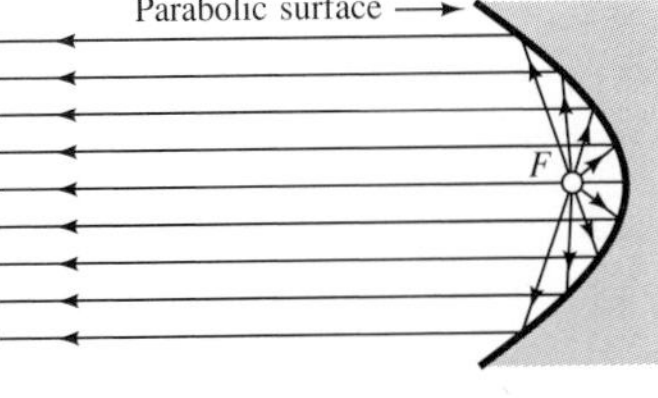

Fig. 23.17 Reflection of light rays emanating from a point source of light located at the focal point of a parabolic mirror.

Table 23.1 Summary of Image Properties for *Concave* Spherical Mirrors

Object location	*Image location*	*Magnification**	*Real or virtual*	*Erect or inverted*	*Example of application*	*See ray diagram in Fig. 23.12:*
	Image Properties					
Region I	Region II	<1	Real	Inverted	Reflecting telescope	(*a*)
At center of curvature *C*	At center of curvature *C*	=1	Real	Inverted	Magic tricks; stage illusions	(*b*)
Region II	Region I	>1	Real	Inverted	Floodlights	(*c*)
At focal point F	Reflected rays are parallel; no image				Searchlights, car headlights, flashlights	(*d*)
Region III	Region IV	>1	Virtual	Erect	Magnifying shaving mirrors and dental mirrors	(*e*)

* When the magnification is <1, the image is smaller than the object; when it is =1, the image and the object are the same size; when it is >1, the image is larger than the object.

23.10 ■ Summary of Properties of Images Formed by Spherical Mirrors

Concave Mirrors The image-forming possibilities for concave mirrors are much greater than for convex mirrors. Concave mirrors can produce images that are real or virtual, magnified or made smaller, and erect or inverted. These various types of images do not, however, appear in all possible combinations. For example, a concave mirror cannot produce a magnified, erect, real image of an object. Table 23.1 summarizes the image-forming properties for concave spherical mirrors. In order to more simply and concisely present information in that table, some of the entries refer to one or more of four regions along the principal axis of a concave mirror. The regions, identified by Roman numerals I through IV, are defined in Fig. 23.18.

Convex Mirrors The image that you can see of your face in a shiny brass doorknob is an example of the type of image that can be formed by a convex spherical mirror. That image is an erect, virtual image that is smaller than the object. No table is needed to summarize the image-forming possibilities for convex mirrors because, regardless of the object distance, any convex spherical mirror forms only one type of image—an erect, virtual image that is smaller than the object.

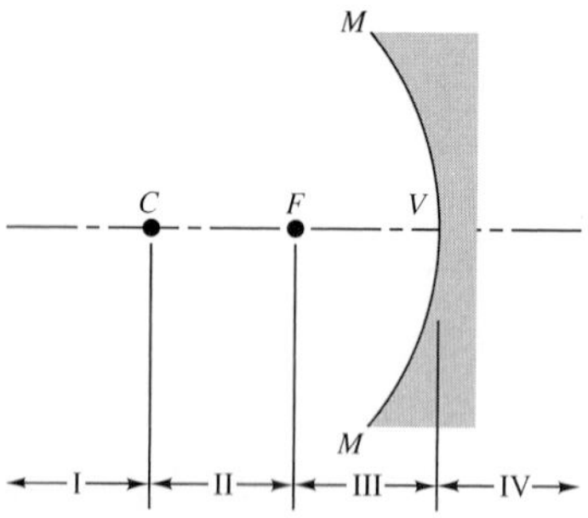

Fig. 23.18 Four regions along the principal axis of a concave mirror as referred to in Table 23.1.

REFRACTION OF LIGHT

One of the most interesting and useful properties of light is that its rays bend as it passes obliquely from one transparent substance into another. Were it not for this bending, we could not have lenses, magnifiers, telescopes, binoculars, microscopes, eyeglasses, cameras, and the many other instruments which extend the usefulness and power of our eyes.

The bending of the path of light at the boundary between substances of different density is called refraction.

Refraction is a matter of common observance. For example, a stick that is partially submerged in clear water will often appear sharply bent where it enters the water; the water in a swimming pool will appear less deep than it actually is. A coin placed on the bottom of an empty cup where it is just out of sight behind the rim will appear to be lifted into view and closer to the observer when the cup is filled with water (Fig. 23.19).

23.11 ■ Refraction of a Plane Wave at a Plane Surface

Let us consider one of the simplest cases of refraction, that of a plane wave of light passing obliquely from air into water (Fig. 23.20). The beam of light in passing from air into water from an incident direction IO is bent in the direction OR in such a way as to make the angle of refraction r' less than the angle of incidence i.

Refraction of light is caused by a change in the speed of light as it leaves one medium and enters another. The speed of light in water is about three-fourths that in air. Let us consider the wavefront AC of a bundle of parallel rays represented by AO, BD, and CE. The rays will advance in the air with a constant speed, making OE parallel to AC.

When ray IO enters the water, its speed is reduced. This ray will travel a distance OT in the water, while the ray BD will move in air a greater distance DS ($OT \approx \frac{3}{4}DS$). In like manner, while these two rays travel equal distances TK and SL in water, ray CE will travel the greater distance PM in air. The new wavefront will be represented by the straight line KM. The effect will be a bending of the beam of rays toward the lines that are perpendicular (called normals) to the water surface at the points where the rays enter the water.

Thus, when a ray of light passes from one material into another in which the speed is less, it is bent *toward the normal*. In like manner, it can be shown that when the speed of light is greater in the second medium, the ray is bent *away from the normal*. For example, if ray RO in Fig. 23.20 were an incident ray, the direction of refraction would be OI. That is, the angles i and r' in the figure can be interchanged, if the direction of the ray is reversed.

23.12 ■ Refraction Caused by Earth's Atmosphere

Light from the sun can be refracted by the earth's atmosphere. Whenever you watch a sunrise you will see the sun several minutes (time) before it actually rises above the horizon (see Fig. 23.21) because the earth's atmosphere bends the rays of the sun. Light from the stars seen at night moves through varying masses of air, some hotter, some cooler, some more dense and some less dense. The light is bent from side to side as it moves from one mass to another. These masses of air are in motion and give

Star twinkling is caused by changing atmospheric refraction of the starlight.

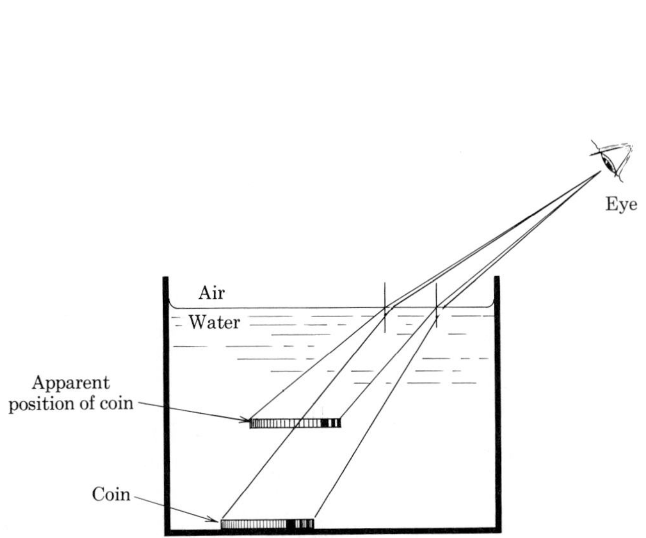

Fig. 23.19 The coin is visible because light is bent in passing from water to air.

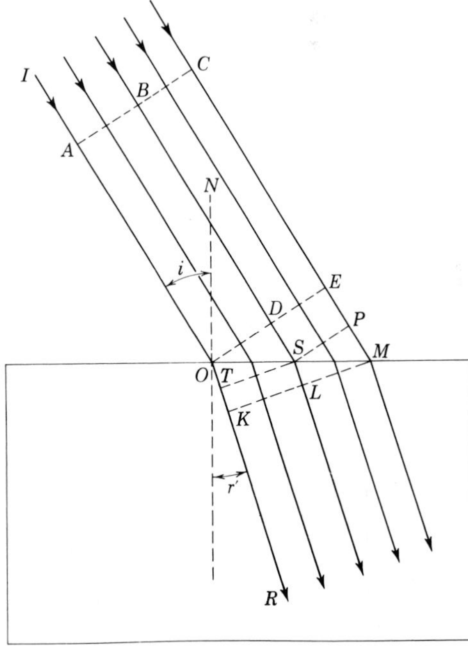

Fig. 23.20 Refraction of a plane wave passing from air into water.

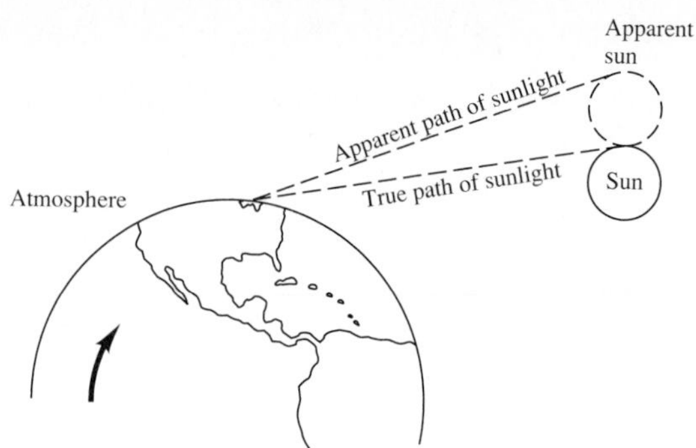

Fig. 23.21 Refraction of the sun's rays at sunrise.

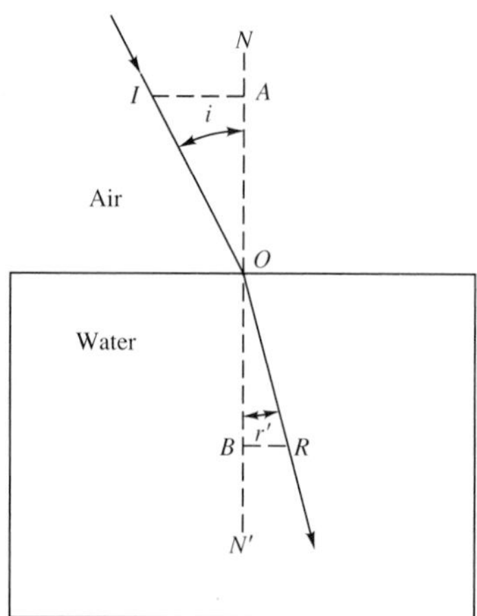

Fig. 23.22 Measurements for computing the index of refraction.

the starlight its twinkling effect. Astronomers get their best photographs on cold, still nights or by sending instruments aloft in high-altitude balloons or spacecraft. Space stations outside the region of atmospheric disturbances show great promise for making detailed and precise astronomical observations.

Objects viewed through the air above a pavement on a hot summer day have a wavy appearance because the convection air currents of varying densities cause uneven refraction of light. Often a still hot layer of air above a paved surface or the floor of the desert refracts the sky light to the observer at ground level. The surface thus seems to be covered with a layer of water. These optical illusions are called *mirages*.

23.13 ■ Index of Refraction

If in Fig. 23.22, we mark off equal distances IO and OR on the incident and refracted rays and draw perpendiculars IA and RB to the normal NN', we can show that no matter what the angle of incidence, the segment IA is always a definite number of times segment RB. If the light travels from air to water, IA will equal $\frac{4}{3}RB$ or $IA/RB = \frac{4}{3}$. Similarly, for light traveling from any material to another, we shall get $IA/RB = \mu$, which is a constant for each pair of materials. Since $IO = RO$, we can divide numerator and denominator by equals and get

$$\frac{IA/IO}{RB/RO} = \frac{\sin i}{\sin r'} = \mu \tag{23.5}$$

This simple relationship, now known as *Snell's law*, was discovered by the Dutch mathematician Willebrord Snell in 1621.

Snell's law of refraction—$\sin i/\sin r$ = constant. The astronomer Ptolemy (probably born in Egypt, about 75 A.D.) believed that the ratio i/r was equal to a constant.

Snell's Law of Refraction

The ratio of the sine of the angle of incidence to the sine of the angle of refraction is a constant, for any two given media.

By application of the laws of geometry, it can be shown that IA and RB (Fig. 23.22) are proportional, respectively, to the speed of light in the air v_1 and the speed of light in water v_2. Hence Eq. (23.5) can be written

$$\frac{\sin i}{\sin r'} = \frac{v_1}{v_2} \tag{23.6}$$

Equation (23.6) can be written in the form

$$\left(\frac{1}{v_1}\right) \sin i = \left(\frac{1}{v_2}\right) \sin r'$$

Multiplying the above equation by the speed of light in a vacuum c gives

$$\left(\frac{c}{v_1}\right) \sin i = \left(\frac{c}{v_2}\right) \sin r'$$

The ratio c/v_1 is called the *index of refraction* μ_1 of air, and the ratio c/v_2 is called the index of refraction μ_2 of water. The index of refraction of a substance is independent of the angle of incidence, but it does depend on the characteristics of the substance and on *the wavelength of the light being transmitted.* Snell's law of refraction can be written in terms of the refractive indexes μ_1, and μ_2:

$$\mu_1 \sin i = \mu_2 \sin r' \qquad (23.7)$$

Although this discussion involves a ray of light traveling from air into water, Eq. (23.7) is valid for any pair of transparent substances.

The refractive index of a substance is defined as the speed of light in a vacuum divided by the speed of light in the substance. It has no units.

$$\mu = c/v \qquad (23.8)$$

The index of refraction of a vacuum ($\mu = c/c$) must be equal to exactly 1. The refractive index of air (1.000293) is only slightly greater. Since refractive indexes are computed to only three or four significant figures for most practical calculations, the value 1.000 is usually used for the refractive index of air.

The refractive index of air is the same as that of a vacuum to four significant digits (1.000).

In passing through two transparent media, the greater the difference between the index of refraction of the second medium and that of the first, the greater the amount of refraction (bending). Thus, light passing from air into glass will be refracted more than if it passes from air into water.

The index of refraction of a substance can be measured by passing a narrow beam of light into it as suggested by Fig. 23.22, measuring the angles of incidence and refraction, and applying Eq. (23.7). A yellow light source of wavelength 589.3 nm is used as a standard in determining indices of refraction.

Illustrative Problem 23.6 A narrow beam of light strikes a plane glass surface at an angle of incidence of 52°. The refracted ray makes an angle of 31° with the normal to the surface (Fig. 23.23). What is the index of refraction of the glass?

Solution Substitute the known angles (Fig. 23.23) for i and r' in Eq. (23.7) and solve the trigonometric equation:

$$\mu_1 \sin i = \mu_2 \sin r'$$

$$\mu_2 = \frac{\mu_1 \sin i}{\sin r'}$$

$$= \frac{(1.000) \sin 52°}{\sin 31°} = \frac{0.788}{0.515}$$

$$= 1.53 \qquad \textit{answer}$$

Table 23.2 gives values of the index of refraction μ for a number of common substances.

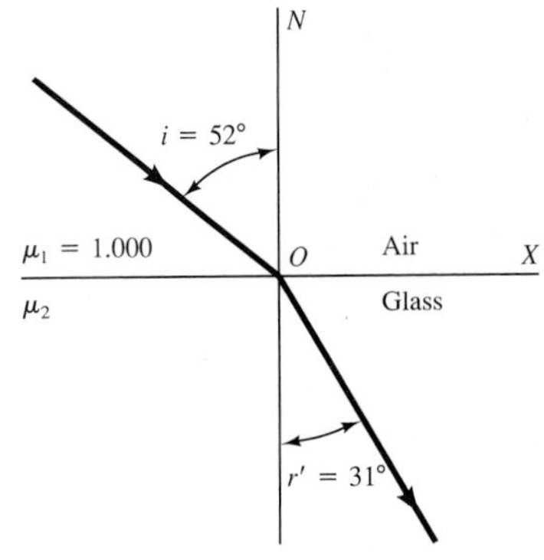

Fig. 23.23 Ray diagram for Illustrative Problem 23.6.

Illustrative Problem 23.7 Light, in air, is incident at an angle of 40° on the surface of a light flint-glass plate (Fig. 23.24). Through what angle is the light deviated from its incident direction by refraction?

Solution From Table 23.2 the index of refraction of light flint glass is 1.575. Solve for r' in Eq. (23.7). Substituting known quantities, we then have

Table 23.2 Index of refraction μ, for various substances, for light of wavelength 589.3 nm (the yellow line of sodium)

Substance	*Index*
Calcite	1.486–1.658
Diamond	2.417
Glass, ordinary crown	1.517
Light flint	1.575
Dense flint	1.656
Ice at −8°C	1.31
Quartz, fused	1.458
Carbon disulfide at 20°C	1.625
Carbon tetrachloride at 20°C	1.461
Ethyl alcohol at 20°C	1.360
Water at 0°C	1.334
At 20°C	1.333
At 40°C	1.331
At 60°C	1.327
Air	1.000293
Carbon dioxide	1.000450
Ethyl ether	1.001521

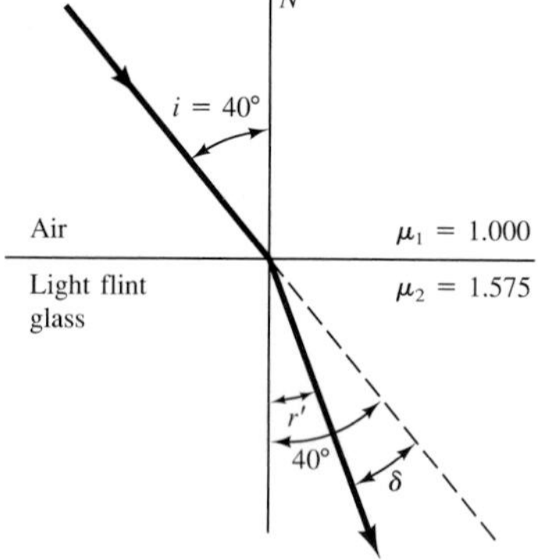

Fig. 23.24 Ray diagram for Illustrative Problem 23.7.

$$\sin r' = \frac{\mu_1 \sin i}{\mu_2}$$
$$= \frac{(1.000)\sin 40°}{1.575}$$
$$= 0.408$$

The angle whose sine is 0.408 is 24° (determined using the $\sin^{-1}$ key or the INV SIN keys on a pocket calculator), so

$$r' = 24°$$

The deviation angle δ is equal to the difference between the angles i and r' (see Fig. 23.24).

$$\delta = 40° - 24°$$
$$= 16° \qquad \textit{answer}$$

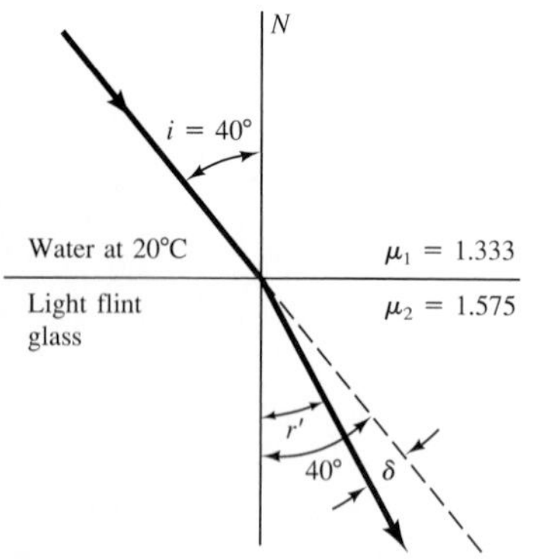

Fig. 23.25 Ray diagram for Illustrative Problem 23.8.

Illustrative Problem 23.8 Light, in 20°C water, is incident at an angle of 40° on the surface of a light flint-glass plate (see Fig. 23.25). Through what angle is the light deviated from its incident direction by refraction?

Solution This problem is similar to Illustrative Problem 23.7, but now the incident ray travels in water rather than in air. From Table 23.2 the index of refraction of water at 20°C is 1.333, and the index of refraction of light flint glass is 1.575. First solve for r' in Eq. (23.7), and then substitute known values.

$$\sin r' = \frac{\mu_1 \sin i}{\mu_2}$$
$$= \frac{(1.333)\sin 40°}{1.575}$$
$$= 0.544$$
$$r' = 33°$$

The deviation angle δ is equal to the difference between the angles i and r' (see Fig. 23.25).

$$\delta = 40° - 33°$$
$$= 7° \qquad \textit{answer}$$

Note that a reduction in the difference in refractive indexes (compared to that in Illustrative Problem 23.7) has resulted in a reduction in the angle of deviation.

23.14 ■ Reflection and Refraction Combined—Total Reflection

Discussions of reflection and refraction have been considered separately in previous sections. But reflection and refraction often occur simultaneously. For example, whenever a light beam in air strikes a transparent surface, some of the light is reflected and some is refracted (see Fig. 23.26*a*). The energy in the incident beam is divided into the reflected and refracted beams in a way that depends on the angle of incidence. As Fig. 23.26*b* illustrates, when the angle of incidence is small the reflected beam is dim and the refracted beam is bright. For large angles of incidence (see Fig. 23.26*c*), the reverse is true—the reflected beam is bright and the refracted beam is dim.

When the incident ray is on the side of the boundary with the lower refractive index, reflected and refracted rays exist for all possible angles of incidence. But when the incident ray is on the high-refractive-index side of the boundary, it is possible for all of the incident light to be reflected, as explained below.

Consider light in a medium of high refractive index and directed toward one of lower index, e.g., light traveling from water into air (Fig. 23.27). Light traveling from a point O along OA is mostly refracted into the air along AR' in such a direction that (sin i)/(sin r') = $1/\mu$ (since the light rays are passing from a medium of lower light speed to one of higher light speed), but a small part is reflected along AR. As the angle of incidence at the water-air surface increases, the angle of refraction in the less dense medium increases until we finally come to ray OC, which is refracted along CT' and just grazes the surface ($r' = 90°$). The angle i_c for which this occurs is called the *critical angle*. Any ray, such as OD, for which the angle of incidence is greater than i_c, does not emerge in air at all but is *totally reflected* along some line DW—just as if it had fallen on a highly polished metal surface at D. The critical angle must not be exceeded if the ray is to pass out of the denser medium. If we substitute 90° for r' in the formula (sin i)/(sin r') = $1/\mu$, we get the value for the critical angle:

$$\sin i_c = \frac{1}{\mu} \qquad (23.9)$$

Illustrative Problem 23.9 What is the critical angle for crown glass and air?

Solution From Table 23.2 the index of refraction for crown glass is 1.517. Substituting this value in Eq. (23.9) and solving for i_c, we get

$$\sin i_c = 1/\mu$$
$$= 1/1.517 = 0.659$$
$$i_c = 41°14' \qquad \textit{answer}$$

In the above discussion, total reflection occurred at an interface of two media with air on the low-refractive-index side of the boundary. In some applications (see Illustrative Problem 23.11 in Sec. 23.15 below) total reflection occurs at the boundary between two solid transparent substances both of which have a refractive index greater than 1.00. Equation (23.7) can be used to write a more general equation for the critical angle i_c. If we substitute 90° for r' and i_c for i in Eq. (23.7), the result is

$$\mu_1 \sin i_c = \mu_2 \sin 90°$$

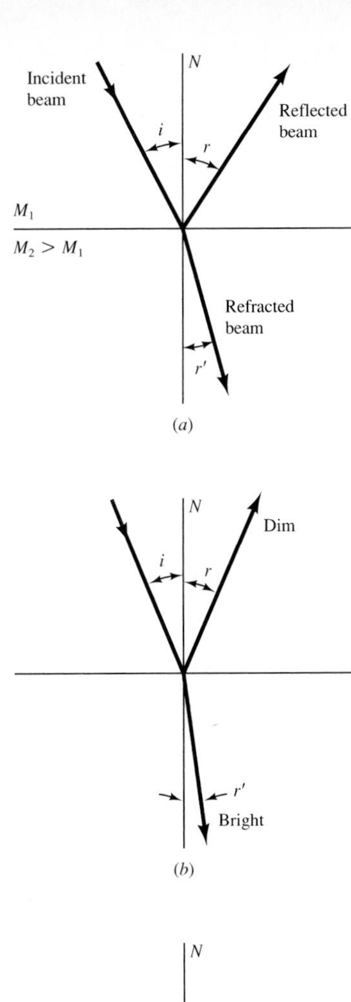

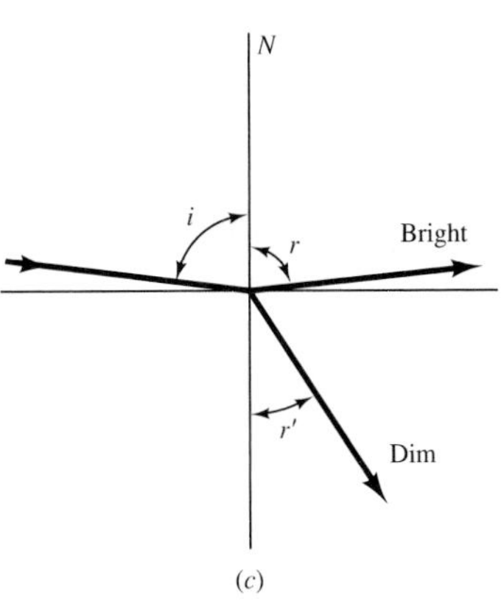

Fig. 23.26 (*a*) When a light beam strikes a transparent surface, it is divided into a reflected beam and a refracted beam. (*b*) The reflected beam is dimmest and the refracted beam is brightest when the angle of incidence is small. (*c*) The reflected beam is brightest and the refracted beam is dimmest when the angle of incidence is very large.

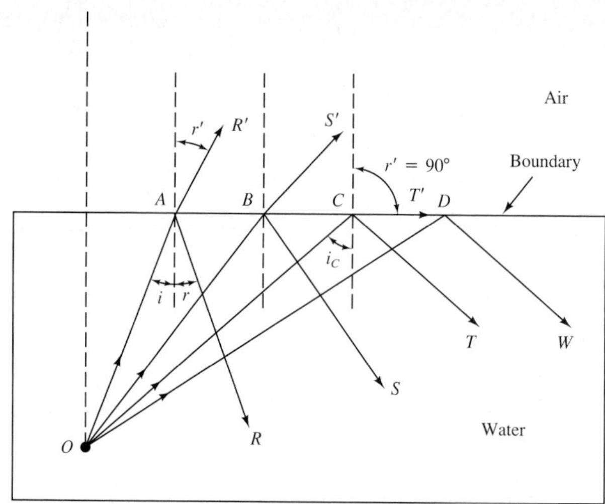

Fig. 23.27 When the angle of incidence is equal to the critical angle i_c, the angle of refraction is 90°. For angles of incidence greater than the critical angle, the incident ray is totally reflected.

and

$$\sin i_c = \frac{\mu_2}{\mu_1} \tag{23.10}$$

Note that Eq. (23.10) is identical to Eq. (23.9) for the case where $\mu_2 = 1$ and $\mu_1 = \mu$.

Total reflection is employed in optical instruments such as telescopes, microscopes, periscopes, prism binoculars, and spectroscopes. A *right-angle prism* with polished sides is used in optical instruments where a nearly perfect reflector is necessary. Figure 23.28 shows how a crown-glass prism serves as a *reflector*. When ray *AB* strikes the side of *RT* of such a prism at right angles, it suffers no refraction but passes on through the glass to *B* on the side *ST*, where its angle of incidence is 45°. Since the critical angle for crown glass is less than 45°, the ray does not emerge from the glass but is totally reflected in the direction *BE*. It is not refracted at *E* since it strikes the surface *RS* at right angles. Prisms are used in this manner in periscopes and optical range finders.

Figure 23.29 illustrates two other ways a right-angle prism is used. In Fig. 23.29*a*, we see how a prism can be used to invert the image. Projection lanterns frequently

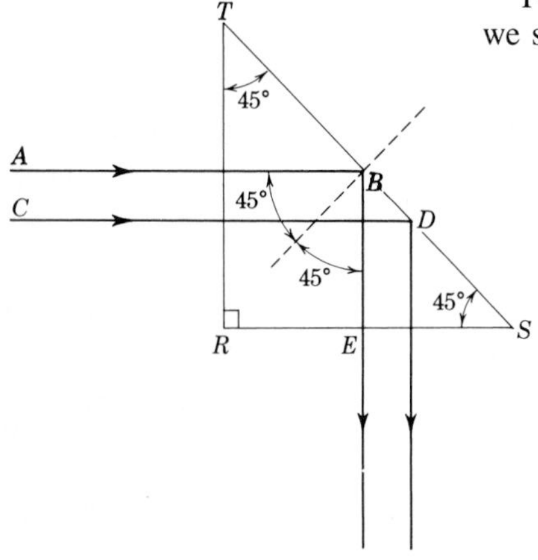

Fig. 23.28 Total reflection of light by a right-angle prism.

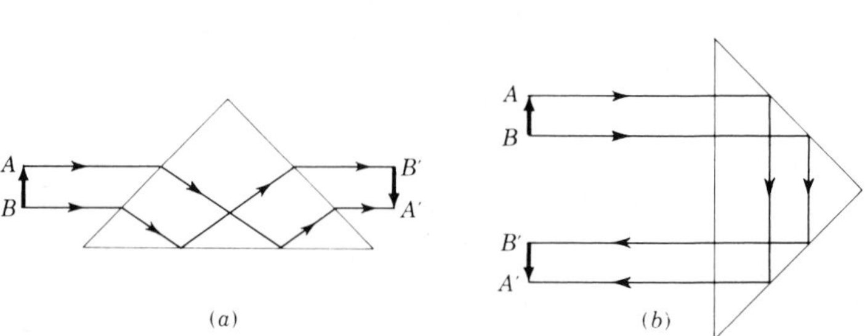

Fig. 23.29 Reflections by right-angle prisms.

Fig. 23.30 Prism binocular. The double total reflections give an optical path length between lenses considerably greater than the overall length of the instrument. (Bausch & Lomb Co.)

use prisms in this manner. When the prism is rotated to the position of Fig. 23.29*b*, it will reverse the beam and invert the image. Two such prisms are used for each eye in *prism binoculars*—one to reverse the image *up* and *down,* one to reverse it *left* to *right* (Fig. 23.30).

Total reflection is also used in enhancing the brilliance of cut diamonds. By taking advantage of the fact that the critical angle of diamond is only 24°, the skilled artisan will cut the facets of a diamond in such a way that most of the light entering it will be totally reflected many times before emerging through the outer surfaces. The effect will be to give the diamond added sparkle.

23.15 ■ Fiber Optics

Light can be "piped" from one point to another by using total internal reflection of light in transparent cylindrical fibers of plastic or glass (Fig. 23.31). On entering one end of an optical fiber, light undergoes repeated total internal reflections at the cylindrical boundary of the fiber until it emerges at the other end. As Illustrative Problem 23.10 shows, rays incident on the end face of a fiber at large angles of incidence experience total internal reflection after entering the fiber.

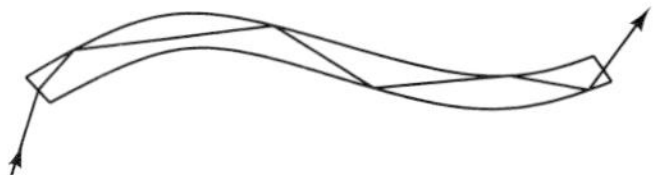

Fig. 23.31 A light ray experiences total internal reflection as it travels through a single optical fiber.

Illustrative Problem 23.10 The critical angle for total internal reflection for glass with a refractive index of 1.50 is 41.8°. Figure 23.32 shows a ray of light entering the end face of an optical fiber with an angle of incidence of 85°. If the refractive index of the glass fiber is 1.50, will the angle of incidence θ for the first internal reflection be greater or less than the critical angle (41.8°)?

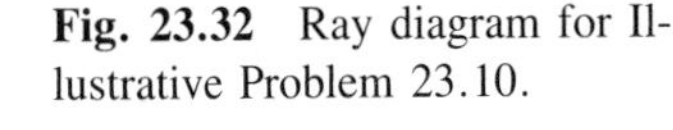

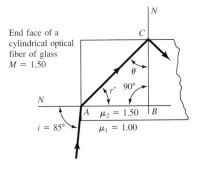

Fig. 23.32 Ray diagram for Illustrative Problem 23.10.

Solution First use Eq. (23.7) to calculate the angle of refraction r'.

$$\mu_1 \sin i = \mu_2 \sin r'$$

$$\sin r' = \frac{\mu_1 \sin i}{\mu_2} = \frac{(1.00) \sin 85°}{1.50} = 0.664$$

$$r' = 41.6°$$

The sum of the angles in the triangle *ABC* must equal 180°.

$$r' + \theta + 90° = 180°$$

Solving for θ gives

$$\begin{aligned}\theta &= 90° - r' \\ &= 90° - 41.6° \\ &= 48.4° \qquad \textit{answer}\end{aligned}$$

which is *greater* than the critical angle. Therefore, total internal reflection will occur at point C in Fig. 23.32.

Similar calculations can be used to show that θ would be greater than the critical angle for all angles of incidence between 0° and 90° if the fiber is straight. However, bending the cable could cause a reduction in the size of the angle θ, possibly causing θ to be smaller than the critical angle which would result in a loss of light out the side of the fiber.

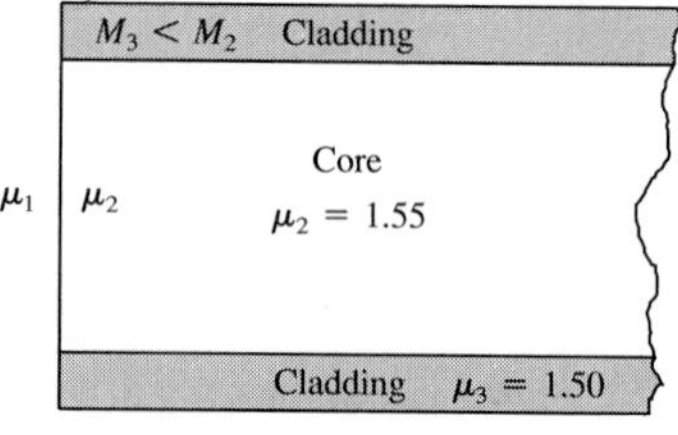

Fig. 23.33 An optical fiber with a core fiber surrounded by a cladding material of lower refractive index is called a *step-index fiber*.

If the surface of a simple optical fiber makes contact with something other than air, the value for the critical angle for total internal reflection is reduced, and light may escape at the point of contact. This makes the success with which light can be transmitted to the far end of the fiber sensitive to the methods used for holding or supporting the fiber. This problem can be solved by enclosing the fiber in a cylindrical shell of *cladding* material that has a lower refractive index than the central core fiber (see Fig. 23.33). This type of optical fiber is called a *step-index fiber* because of the abrupt change in refractive index at the boundary between the central core and the outer cladding material. The presence of the cladding material reduces the range of angles of incidence which will be totally reflected after entering the central fiber, but that is a desirable effect for communications applications of optical fibers because it reduces the range of times required for rays taking different paths to travel through a fiber.

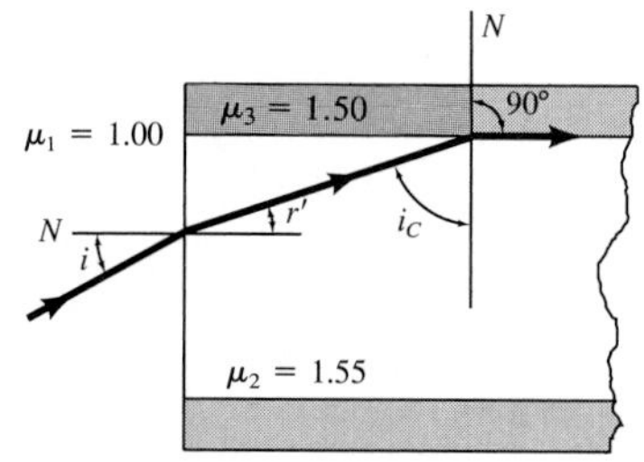

Fig. 23.34 Ray diagram for Illustrative Problem 23.11.

Illustrative Problem 23.11 Figure 23.34 shows an optical fiber with a central core having a refractive index of 1.55 that is surrounded by cladding material with refractive index 1.50. What is the critical angle i_c for a light ray incident at the boundary between the central core of the optical fiber and the outer cladding?

Solution The critical angle can be calculated using Eq. (23.10)

$$\begin{aligned}\sin i_c &= \frac{\mu_3}{\mu_2} = \frac{1.50}{1.55} \\ &= 0.968 \\ i_c &= \sin^{-1}(0.968) \\ &= 75.5° \qquad \textit{answer}\end{aligned}$$

This relatively large required angle for total reflection means that only those rays that make an angle of 14.6° or less with the centerline of the central fiber core will be totally reflected at the cladding boundaries and transmitted to the far end of the fiber.

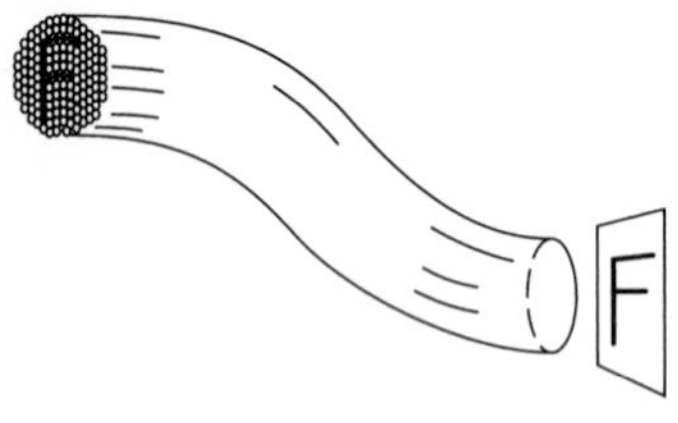

Fig. 23.35 A bundle of a large number of very small diameter optical fibers can be used to transmit images.

Practical applications in the rapidly developing field of fiber optics can be divided into four categories. The first is the use of bundles of fine, flexible optical fibers for the transmission of light to provide illumination in hard-to-reach locations. The second is the transmission of images using bundles of fine, flexible optical fibers (Fig. 23.35). These two applications are combined in an instrument called a *fiberscope* in which some of the fibers in a bundle are used to illuminate the object to be viewed, and other fibers in the bundle bring an image of the object back to the observer (see Fig. 23.36). The fiberscope is used for viewing internal body cavities for medical diagnoses and for monitoring internal surgical procedures that are carried out using laser light.

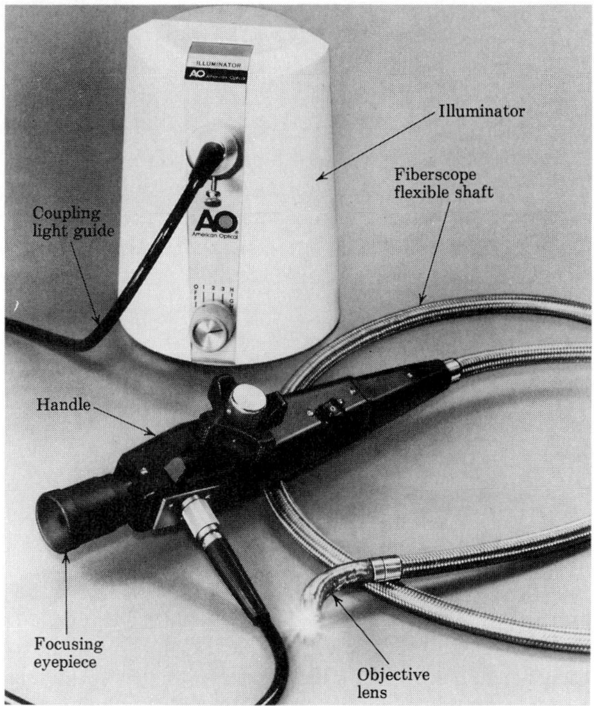

Fig. 23.36 One model of a fiberscope for medical diagnostic uses. The fiberscope is used to produce images of internal body cavities such as the esophagus, stomach, bronchi, bladder, and the large bowel. Fiberscopes are also used for industrial inspection of internal or remote areas hidden from direct view. (Courtesy American Optical Co.)

The third category of applications is communications. Information previously transmitted over long distances using electric currents in copper wires (e.g., telephone conversations) can now be transmitted with laser light in optical fibers. Rapid progress was made in this area during the 1970s and 1980s.

Among the advantages of optical fibers over electrical conductors for communications applications are the following. Optical fibers weigh less, cost less, can transmit much more information for a given diameter cable, and are not subject to electrical interference. Finally, because it is much more difficult to intercept a signal from a fiber-optic communications system than from an electrical conductor system, the fiber-optic system is the choice where a secure communications system is required.

A fourth category is guidance. ''Smart bombs'' and TOW missiles use trailing optical fibers. The video cameras in their noses ''see'' the target. This information goes back to the fire-control computer in the launch vehicle (tank, aircraft, ship) and target acquisition corrections are made during the missile's flight.

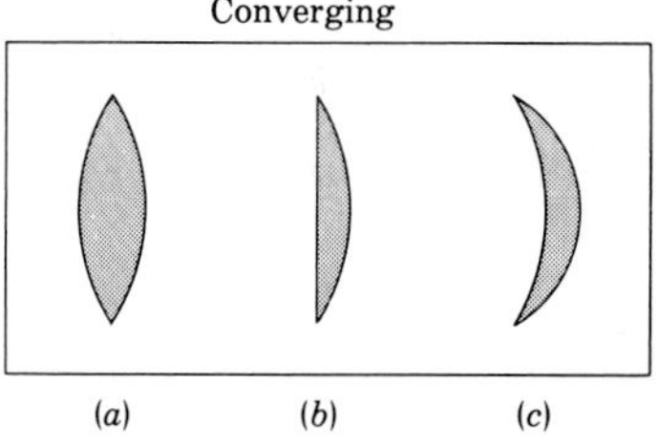

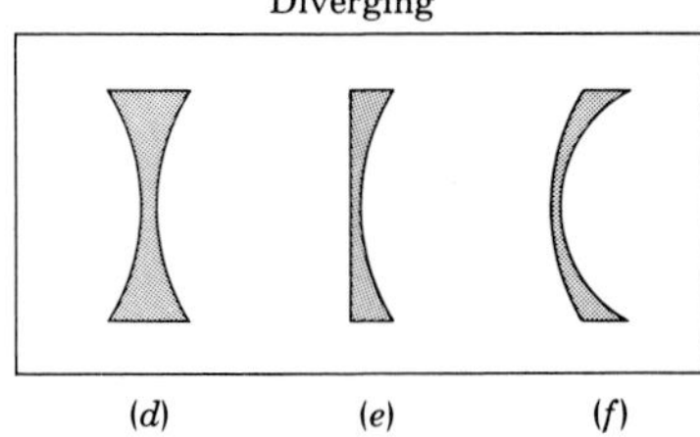

Fig. 23.37 Types of converging and diverging lenses: (*a*) double convex, (*b*) plano-convex, (*c*) concavo-convex, (*d*) double concave, (*e*) plano-concave, (*f*) convexo-concave.

LENSES

A lens is any transparent medium, usually glass, with smooth spherical surfaces. We see because images are formed by lenses of the human eye. Defective eye lenses can often be compensated for by the proper use of other lenses made of glass or plastic. By the use of combinations of lenses, scientists delve into the realm of astronomical research on the one hand and into the secrets of extremely small microsopic life on the other.

23.16 ■ Types of Lenses

Lenses are divided into two classes: *converging lenses*, which are thicker at the center than at the edge, and *diverging lenses*, which are thinner at the center than at the edge. Typical lens forms are shown in Fig. 23.37. Special names associated with each of the six types are the following: (*a*) double-convex, (*b*) plano-convex, (*c*) concavo-convex, (*d*) double-concave, (*e*) plano-concave, and (*f*) convexo-concave.

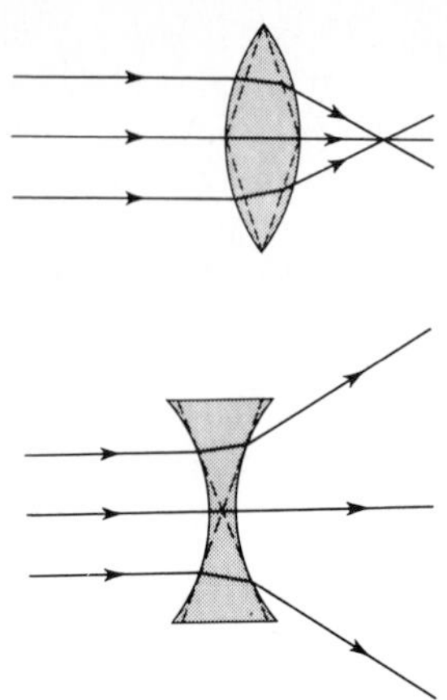

Fig. 23.38 Lenses act like prisms.

By referring to Fig. 23.38 it can be seen that a double-convex lens is in cross section essentially equivalent to two triangular prisms placed base to base. Light rays entering the prisms are bent toward the bases, causing them to converge. The double-concave lens can be represented in cross section as essentially equivalent to two similar prisms placed apex to apex. Light rays falling on these prisms will in like manner be bent toward the bases, causing them to diverge.

The *principal axis* of a lens is the straight line passing through the centers of curvatures of the surfaces of the lens. (It must be remembered that a plane is a sphere with an infinite radius.) The point at which rays parallel to the principal axis converge after passing through the convex lens is called the *principal focus* of the lens (see Fig. 23.39).

Parallel rays striking a lens from either side of the lens will converge or appear to converge at a given point after passing through the lens. Hence every lens has two such *focal points* F and F' equidistant from the lens. The distance from the center of the lens to either focal point is called the *focal length* f of the lens. The *aperture* of a lens is the diameter of the uncovered part of the lens.

23.17 ■ Formation of Images by Lenses

If rays from a light source *converge* after passing through a lens, an image of the source can be formed and viewed on a screen. Such an image is called a *real* image. If the rays *diverge* after passing through a lens, a real image cannot be formed on a screen, but the image can be seen by looking *through* the lens. The image viewed by looking through such a lens is called a *virtual* image.

23.18 ■ Determination of Image Location and Size by Means of Rays

Procedure for locating the lens-formed image of an object using two rays.

It is possible by the application of a few simple geometric constructions to determine the position and size of images formed by lenses. The method is similar to that used for spherical mirrors. If we trace any two rays from a point on the object to their intersection after refraction, we shall have the position of the corresponding point on the image. The two rays usually selected are (1) a ray from the object parallel to the principal axis and (2) a ray from the object through the center of the lens. The first ray will, after refraction, pass through the focal point, while the second will pass directly through the lens without deviation.

Figure 23.40 illustrates four cases of image formation by *converging lenses*.

1. In Fig. 23.40*a*, the object AB, if a considerable distance from the lens, will appear as a small, real, inverted image $A'B'$ at a relatively short distance from the lens. The eye lens and the camera lens form this type of image.

2. When the object is placed at a distance equal to twice the focal distance ($D_o = 2f$), the image of the object is real, inverted, and the same size as the object and is the same distance from the lens (Fig. 23.40*b*).

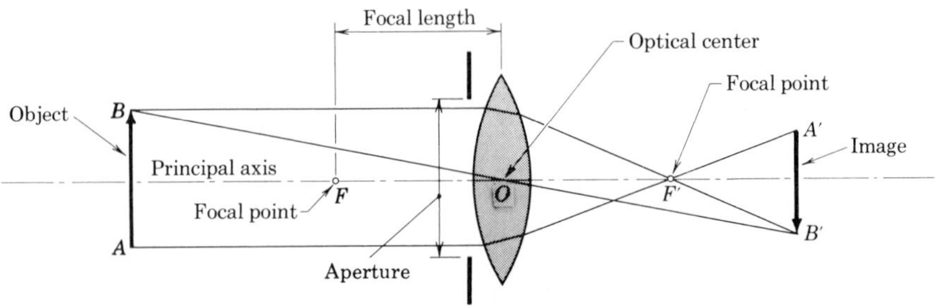

Fig. 23.39 Diagram illustrating terminology used with lenses.

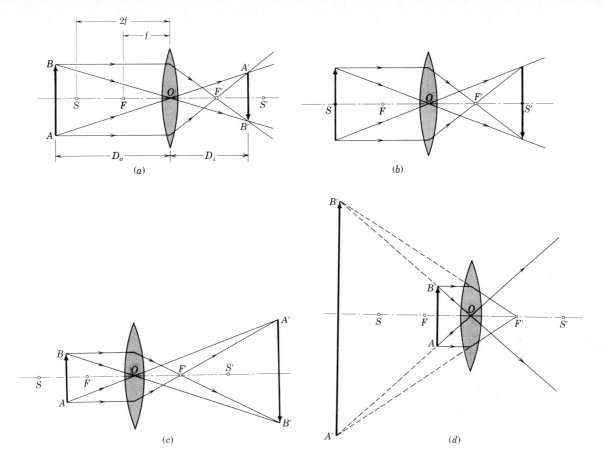

Fig. 23.40 Images formed by double convex (converging) lenses: (*a*) Image is real, inverted, smaller than object. (*b*) Image is real, inverted, same size as object. (*c*) Image is real, inverted, larger than object. (*d*) Image is virtual, erect, larger than object.

3. If the object is placed between $2f$ and f, the image will appear beyond $2f$ on the other side, as an enlarged, real, and inverted image (Fig. 23.40*c*). This effect is used by film and slide projectors, and photographic enlargers. It should be evident from the diagram why the operator of the projector must put the slides upside down and inverted in the machine.

4. If the object is placed between the lens and its principal focus (Fig. 23.40*d*), the rays will not intersect on the opposite side of the lens but will appear to intersect on the same side of the lens. The eye can see their *apparent* intersection as a virtual, erect image larger than the object. Ordinary magnifying glasses and the eyepieces on microscopes and telescopes are used in this manner.

The ray diagram of Fig. 23.41 indicates that no matter where the object location with reference to a *concave* (*diverging*) *lens*, the image will appear as a virtual, erect, small image, nearer to the lens than the object. An important use of concave lenses is in eyeglasses for the correction of nearsighted vision. They are also used as eyepieces in telescopes, field glasses, and binoculars.

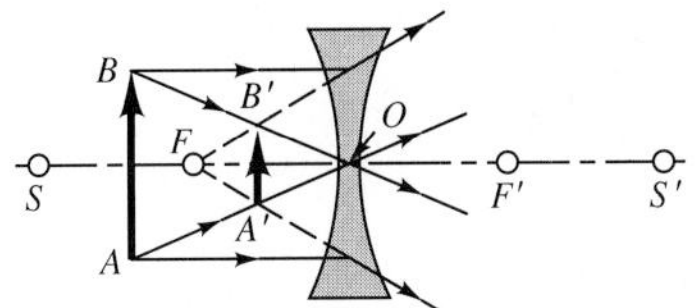

Fig. 23.41 Image formed by double concave (diverging) lens is virtual, erect, and smaller than object.

23.19 ■ The Lens Equation

It can be shown that the relation between the distance of the object from a lens D_o, the distance of its image from the lens D_i, and the focal length f is given by the same equation as the one previously given for spherical mirrors.

Algebraic sign convention for lenses.

$$\frac{1}{D_o} + \frac{1}{D_i} = \frac{1}{f} \tag{23.3}$$

In order to make Eq. (23.3) applicable to all types of lenses, it is necessary to adopt the following *convention of algebraic signs:*

1. Take f positive for converging lenses and negative for diverging lenses.
2. Take D_o positive when the object is real and negative when the object is virtual. (In optical instruments the image formed by one lens often serves as the virtual object for another lens.)
3. Take D_i positive when the image is real and negative when the image is virtual.

Illustrative Problem 23.12 A double-concave lens has a focal length of 60 cm. Where does it form an image of an object placed 75 cm from it?

Solution Since the object is real and the lens is diverging, take D_o as positive and f as negative in accordance with the above convention of signs. Equation (23.3) then becomes

$$\frac{1}{75} + \frac{1}{D_i} = \frac{1}{-60}$$

Multiplying through by the lowest common denominator, transposing, and solving for D_i yields

$$\begin{aligned} 4D_i + 300 &= -5D_i \\ 9D_i &= -300 \\ D_i &= -33\tfrac{1}{3}\text{ cm} \qquad \textit{answer} \end{aligned}$$

The minus sign indicates that the image is virtual, and it is therefore on the same side of the lens as the object.

Illustrative Problem 23.13 An object placed 100 cm from a lens will form an image that can be focused on a screen placed 22.3 cm on the other side of the lens. What kind of lens is it? Is the image real or virtual? Find the focal length of the lens.

Solution Since the image is formed on the opposite side of the lens, the lens must be a converging one and the image is real (see Fig. 23.40*a*). Hence take positive values for D_o and D_i. We should get a positive answer for f. Substitute known values for D_o and D_i in Eq. (23.3) and solve for f:

$$\begin{aligned} 1/D_o + 1/D_i &= 1/f \\ 1/100 + 1/22.3 &= 1/f \\ 0.01 + 0.0448 &= 1/f \\ f &= 1/0.0548 \\ &= 18.2\text{ cm} \qquad \textit{answer} \end{aligned}$$

23.20 ■ Size of Image

By comparing similar triangles AOB and $A'OB'$ in Figs. 23.40 and 23.41 made by the rays AA' and BB' passing through the center of the lens O, it is seen that

$$\frac{I}{O} = \frac{\text{Size of image}}{\text{Size of object}} = \frac{\text{image distance}}{\text{object distance}} = \frac{D_i}{D_o} \tag{23.11}$$

The absolute value of this ratio, as in the case of curved mirrors, is called *magnification*.

Illustrative Problem 23.14 Determine the size of the image of an object 12.5 cm tall with the arrangement of Illustrative Problem 23.13 above.

Table 23.3 Summary of Image Properties for Converging Lenses

		Image Properties				
Object location	*Image location*	*Magnification**	*Real or virtual*	*Erect or inverted*	*Example of application*	*See ray diagrams in Fig. 23.40:*
Region I	Region V	<1	Real	Inverted	Eyepiece for reflecting telescope or objective lens for refracting telescope	(*a*)
At $2F$	$2F'$	=1	Real	Inverted		(*b*)
Region II	Region VI	>1	Real	Inverted	Slide projectors, photo enlargers	(*c*)
At F	Refracted rays are parallel; no image				Small light source at F to produce narrow beam of light	
Region III	Region I, II, or III	>1	Virtual	Erect	Magnifying glasses, eyepieces for telescopes and microscopes	(*d*)

* When the magnification is <1, the image is smaller than the object; when it is =1, the images and the object are the same size; when it is >1, the image is larger than the object.

Solution

$$\begin{aligned} \text{Magnification} &= D_i/D_o \qquad \text{from Eq. (23.10)} \\ &= 22.3/100 = 0.223 \\ \text{Size of image} &= 0.223 \times 12.5 = 2.79 \text{ cm high} \qquad \textit{answer} \end{aligned}$$

23.21 ■ Summary of Properties of Images Formed by Lenses

The image-forming possibilities for converging lenses are much greater than for diverging lenses which, like convex mirrors, can produce only virtual, erect, images that are smaller than the object. Converging lenses can produce images that are real or virtual, magnified or made smaller, and erect or inverted. These various types of images do not, however, appear in all possible combinations. For example, a converging lens cannot produce a magnified, erect, real image of an object. Table 23.3 summarizes the image-forming properties for converging lenses. In order to present more simply and concisely the information in that table, some of the entries refer to one or more of six regions along the principal axis of a converging lens. The regions, identified by Roman numerals I through VI, are defined in Fig. 23.42.

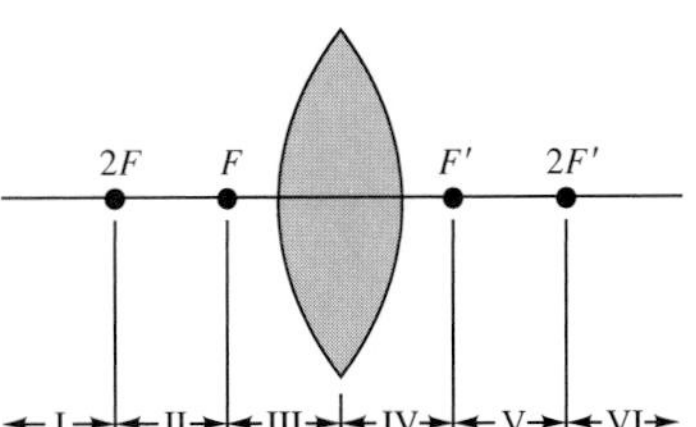

Fig. 23.42 Six regions along the principal axis of a converging lens as referred to in Table 23.3.

23.22 ■ The Lens Maker's Formula

Criteria for lens design—the lens maker's formula.

In order to manufacture lenses which will refract light by exactly the right amounts to meet specifications of optical equipment, the lens maker must deal with the following variables:

The refractive index of the glass to be used
The radii of curvature to be used for the lens surfaces
The desired focal length for the lens

These variables are all related in the *lens maker's formula*

$$\frac{1}{f} = (\mu - 1)\left(\frac{1}{R_1} + \frac{1}{R_2}\right) \tag{23.12}$$

where R_1, R_2 = radii of curvature of the two lens surfaces
f = focal length of lens
μ = index of refraction of glass to be used

A fourth factor, the *aperture*, determines the light-gathering power of the lens and consequently the brightness of the image. It should be pointed out, however, that increasing the aperture in order to increase brightness of the image will also increase the spherical aberration and therefore reduce the sharpness of focus of the image. (See Sec. 23.24 and Fig. 23.43.)

Illustrative Problem 23.15 A manufacturer produces a quantity of double-convex lenses of 30-in. focal length from optical glass whose refractive index is 1.52. Both faces are to have the same radius of curvature. Calculate the radius to be used.

Solution Since the effect of both surfaces of a double-convex lens is to *converge* the rays, the algebraic sign of both R_1 and R_2 will be positive. Since, for this problem, $R_1 = R_2$, Eq. (23.12) becomes

$$\frac{1}{f} = (\mu - 1)\frac{2}{R}$$

Solving for R, we have

$$R = 2f(\mu - 1)$$

Substituting values, we get

$$R = 2 \times 30 \text{ in. } (1.52 - 1)$$
$$= R_1 = R_2 = 31.2 \text{ in.} \qquad \textit{answer}$$

23.23 ■ Power of a Lens

Opticians in dealing with optical instruments employ a term called the power of a lens or of a combination of lenses. The power P of a lens, measured in units called *diopters,* is the reciprocal of its focal length f measured in meters.

$$P \text{ (diopters)} = \frac{1}{f\text{(meters)}} \tag{23.13}$$

Thus, the shorter the focal length, the greater the power of a lens. Converging lenses are said to have positive (+) powers and diverging lenses negative (−) powers.

The chief advantage of expressing lens power in diopters is that *for thin lenses in contact,* (if the thickness of the lens is small compared to its diameter) *the power in diopters of a combination of lenses is equal to the sum of the powers of the separate lenses.*

Thus

$$1/f_c = 1/f_1 + 1/f_2 \tag{23.14}$$

where f_1 and f_2 are the focal lengths of the two thin lenses in contact and f_c is the focal length of the combination.

Illustrative Problem 23.16 A convex lens of focal length 0.25 m is combined with a concave lens of −1.25 m focal length. Find the lens power of the combination. What would be the focal length of an equivalent single lens?

Solution The converging lens will have a power of 1/0.25 or +4 diopters, while the diverging lens will have a power of 1/(−1.25) or −0.80 diopters. Hence the power of the combination is

$$P = +4 - 0.80 = +3.20 \text{ diopters} \qquad \textit{answer}$$

To solve for the focal length of a single equivalent lens, we solve for f_c in Eq. (23.13):

$$f_c = 1/P = 1/3.20$$
$$= 0.31 \text{ m} = 31 \text{ cm} \qquad \textit{answer}$$

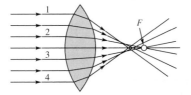

23.24 ■ Spherical Aberration with Lenses

In the geometric construction of images formed by spherical lenses (Fig. 23.40), it was assumed that all rays parallel to the principal axis converge at a common point. This is not exactly true. Actually, if the surfaces of the lens in Fig. 23.43 are spherical, rays such as 1 and 4 which strike the lens near the edges are focused nearer the lens than rays such as 2 and 3 which strike the lens near the focal point. This imperfect focusing of spherical lenses is called *spherical aberration*. The effect of this aberration is an indistinct and distorted image. This defect can be partly overcome by special grinding of the outer thin portion of the lens (called *aspherizing*), or by using a diaphragm with a circular aperture in front of the lens which will eliminate the outer rays. This reduction of aperture is called *stopping down*. Decreasing the effective aperture will result in a sharper image but at the same time will diminish the brightness of the image. Generally, in practice, combinations of lenses are used instead of single lenses. These combinations are so designed that any imperfection in one lens is compensated for or nullified by the optical properties of another lens.

Fig. 23.43 Spherical aberration, a defect of lenses. Not all rays are focused at one point. Light rays that strike the outer portion of the lens are refracted more than the rays that fall on the central portion.

The images formed by simple lenses suffer from a defect called spherical aberration which causes imperfect focusing of images.

QUESTIONS AND EXERCISES

1. What will the image of the letter F look like when reflected from a plane mirror?

2. Describe two applications for plane mirrors that do not involve image formation.

3. In which of the following is the optical lever used? (*a*) Telescope. (*b*) Microscope. (*c*) To magnify linear displacements. (*d*) To magnify angular displacements.

4. Describe an application for an optical lever not mentioned in this chapter.

5. If you take a picture of your image in a plane mirror while standing 8 ft away from the mirror, for what distance should your camera be focused?

6. Under what conditions will the image formed by a mirror be smaller than the object? Is the image real or virtual?

7. Why is there is no problem like spherical aberration when a parabolic mirror is used?

8. Plane mirrors, concave spherical mirrors, and convex spherical mirrors can all produce virtual images. Can they also all produce real images?

9. Are the images formed by convex rear-view mirrors for automobiles real or virtual images?

10. Is it possible to form an inverted, magnified, virtual image using either a concave mirror or a convex mirror?

11. At sunset, does the atmospheric refraction of light from the sun cause the sun to appear higher or lower in the sky than it would appear if no refraction occurred? Is your answer the same or different for sunrise?

12. Why would you expect the atmospheric refraction of light from the sun to have a greater influence on the sun's apparent position at sunrise or sunset than at noon?

13. The index of refraction of ice is 1.31 and that of glass is 1.50. Is the speed of light greater in ice or in glass?

14. If the refractive index of a gas is proportional to the density of the gas, which has a greater density, air or carbon dioxide? (See Table 23.2.)

15. Describe the path of a ray of light which passes (obliquely) through a rectangular plate of glass. Illustrate with a diagram.

16. A person swimming under the surface of water looks diagonally up at the diving board. Does the diving board appear higher or lower than it actually is?

17. Under what conditions can total reflection of a light beam occur at the boundary between two transparent regions?

18. Does all the light that enters one end of an optical fiber exit the other end if total reflection occurs at all points where the light strikes the cylindrical surface on the trip from one end of the fiber to the other end? (The answer is not *yes*. Can you explain why?)

19. How would you expect the diameters of the individual fibers in a bundle of many optical fibers to be related to the quality of the image transmitted from one end of the bundle to the other end?

20. What is the reason for the cladding material that surrounds an optical fiber core?

21. A fiberscope uses a bundle of very small diameter optical fibers to transfer images from inaccessible locations in medicine and engineering. How would a broken fiber within the bundle affect the produced image? (HINT: Think about the opportunity for reflection produced by an air-gap at the break.)

22. Are the image-forming capabilities of a diverging lens similar to those of a concave or a convex mirror?

23. Why can a virtual image be seen with the eye but cannot be projected on a screen?

24. Is it possible to get an image formed by a lens that is virtual and smaller than the object? If so, how?

25. An application for bringing parallel rays of visible light together can be accomplished with a converging lens. Consider an application requiring the bringing together of parallel rays of infrared light. Since glass is not transparent in the infrared, the converging of the rays must be accomplished with a mirror. Should a concave or a convex mirror be used?

26. Is it possible to form an inverted, magnified, virtual image using a converging or a diverging lens?

27. What causes spherical aberration in an image formed by a lens? How can the aberration be decreased?

28. How can an adjustable aperture be used to reduce spherical aberration and improve the quality of the images formed by the lens? Would the same method work for improving the quality of the images formed by spherical mirrors?

29. Which of the following will increase the sharpness of a photograph? (*a*) Use a longer exposure time. (*b*) Use a smaller diaphragm opening. (*c*) Use flood lights. (*d*) None of these.

30. Which of the following will be true when two parallel rays of light pass through a combination of thin lenses with respective focal lengths of 30 cm and -120 cm? (*a*) The rays diverge. (*b*) The rays will converge. (*c*) The focal length of the combination is -90 cm. (*d*) The power of the combination is $(\frac{1}{30} - \frac{1}{120})$ diopters.

PROBLEMS

NOTE: *Geometric optics problems will yield more accurate answers if graph paper is used.*

Group One

1. Figure 23.44 illustrates a pair of plane mirrors with an angular separation of 50°. A ray of light strikes mirror M_1 with an angle of incidence of 30°. The reflected ray from mirror M_1 strikes mirror M_2 with an angle of incidence θ. What is the value of θ?

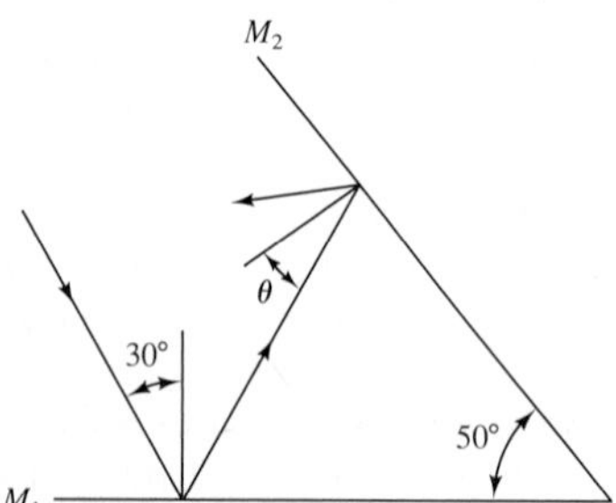

Fig. 23.44 Diagram for Problem 1.

2. An optical lever arrangement, like that illustrated in Fig. 23.4, is set up with a mirror-to-scale distance of 1.2 m. If the mirror rotates through an angle of 2°, through what distance (in centimeters) will a reflected laser beam spot move along the scale?

3. An object is located 10 cm in front of a convex mirror with a radius of curvature of 50 cm. (*a*) Use a ray diagram to find the approximate location of the image. (*b*) Is the image real or virtual?

4. An object 1 cm tall is placed 75 cm in front of a concave mirror with a focal length of 25 cm. (*a*) Where is the image located? (*b*) What is the image size?

5. What is the approximate speed of light in water at 20°C? (HINT: Use Table 23.2 and Eq. 23.8.)

6. Light strikes a plane of ordinary crown glass at an angle of incidence of 30°. What is (*a*) the angle of reflection? (*b*) the angle of refraction?

7. A narrow beam of light strikes a smooth liquid surface with an angle of incidence of 55°. The refracted ray makes an angle of 37° with the normal to the surface. What is the index of refraction of the liquid?

8. What is the critical angle for rays passing from fused quartz to air?

9. Calculate the critical angle for rays passing from a piece of dense flint glass to air.

10. A lens which has a focal length of −25 cm will have what power, expressed in diopters?

11. Find the focal length of a combination of two thin lenses placed in contact if their focal lengths are 30 cm and −15 cm.

Group Two

12. An optical lever arrangement, like that illustrated in Fig. 23.4, is to be set up to meet the following condition: When the mirror is rotated through an angle of 0.5°, the laser beam spot on the scale should move 4 cm along the scale. What must be the distance from the mirror to the scale?

13. A woman 1.8 m tall faces a vertical plane mirror. What is the minimum height of the mirror in which she can see her full-length image?

14. An incandescent bulb placed 20 in. in front of a concave mirror has its image formed 50 in. in front of the mirror. Determine the radius of curvature of the mirror.

15. A concave spherical mirror whose radius of curvature is 75 cm is used as a shaving mirror. If the mirror is held 30 cm from the face, how far behind the mirror does the image appear to be? What is the magnification?

16. An object 4 in. high is placed 20 in. in front of a concave spherical mirror. The image is sharp on a screen 100 in. away from the mirror. Find the radius of curvature of the mirror and the size of the image.

17. A convex mirror has a focal length of 30 cm. An object 10 cm high is placed 20 cm in front of the mirror. Calculate the position and the height of the image formed.

18. A dentist uses a concave mirror whose radius of curvature is 6 cm. What will be the apparent magnification of a tooth when the mirror is held 2 cm from the tooth?

19. Two concave mirrors use the sun to heat objects placed at their foci. The mirrors have equal focal lengths, but the first has a 2-m aperture while the second has a 3-m aperture. Which will produce more intense heat from the sun? How much more (Btu/h)?

20. An object is located 45 cm in front of a concave spherical mirror with a radius of curvature of 60 cm. Use both a ray diagram and an algebraic calculation to determine the location of the image.

21. A beam of laser light strikes a plane glass surface with an angle of incidence θ. The angle of refraction is 27°. If the refractive index of the glass is 1.575, what is the value of the angle of incidence θ?

22. A beam of light strikes the surface of a liquid with an angle of incidence of 47.0°. If the angle of refraction is 30.0°, and if the temperature of the liquid is 20°C, is the liquid carbon tetrachloride or ethyl alcohol? (See Table 23.2.)

23. A beam of laser light is incident at an angle of 72° on the surface of a glass plate that has a refractive index of 1.520. Through what angle is the light deviated from its incident direction by refraction?

24. A ray of light strikes the surface of carbon disulfide at an angle of 50°. By what angle is the ray deviated in passing into the liquid?

25. A light ray strikes a glass plate with an angle of incidence of 75°. What will be the angle θ between the reflected and refracted rays if the index of refraction of the glass is 1.50? (See Fig. 23.45.)

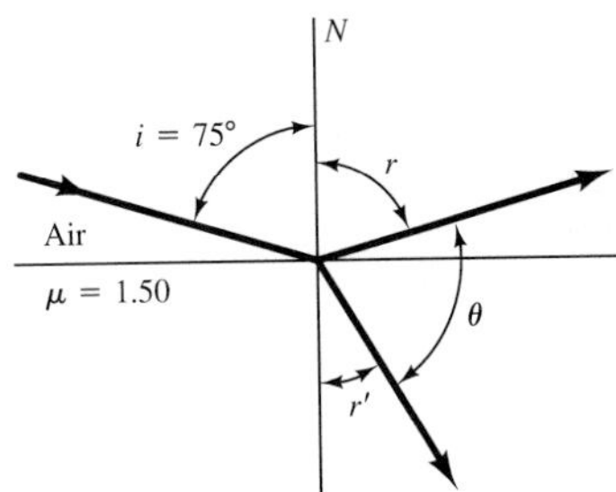

Fig. 23.45 Diagram for Problem 25.

26. The critical angle for total internal reflection in a glass plate surrounded by air is 40.0°. What is the index of refraction of the glass plate?

27. A certain optical fiber has a core with a refractive index of 1.58 and a cladding with a refractive index of 1.52. (*a*) What is the critical angle for total internal reflection at the core-cladding interface? (*b*) What is the largest angle that a ray of light can make with the central axis of the cylindrical core and still be totally reflected at the core-cladding interface?

28. A concave mirror has a focal length of 20 cm. Determine the image distance when the object is (*a*) 40 cm from the mirror and (*b*) 10 cm from the mirror.

29. How large is the image in each case of Problem 28 if the object in each case is 2.5 cm high?

30. A convex mirror has a focal length of 20 cm. Determine the image distance when the objects is (*a*) 40 cm from the mirror and (*b*) 10 cm from the mirror.

31. How large is the image in each case of Problem 30 if the object in each case is 3 cm high?

32. A double-convex lens is used in a projector for colored slides. If the slides are placed in the carrier 20 cm behind the center of the lens and the image is sharp on a screen 9 m away on the other side of the lens, what focal-length lens must be used? If the slide is 5 by 5 cm, what will be the size of the image on the screen?

33. A plano-convex lens is to be made from glass whose refractive index is 1.57. A focal length of 75 cm is desired. What must be the radius of curvature of the convex face of the lens?

Group Three

34. An optical lever arrangement, like that illustrated in Fig. 23.4, is set up with a mirror-to-scale distance of 3 m. If the minimum measurable distance for movement of the laser beam spot on the scale is 0.5 mm, what is the minimum measurable rotation angle of the mirror?

35. A ray of monochromatic light strikes the first face of a prism at an angle of 30° with the normal to that face. The vertex angle of the prism is 50°, and the refractive index of the prism material is 1.55. Find the total angle through which the ray is bent in passing through the prism and emerging into the air again.

36. An object is 4 in. high. What type of mirror would have to be used to obtain a real image 1 in. high 100 in. from the object? What is the focal length of the mirror which will do this? [HINT: Let $D_i = x$ and $D_o = 100 + x$ in Eq. (23.4); solve for D_i and D_o and then use Eq. (23.3).]

37. A narrow beam of light in a liquid is incident on a glass plate in the liquid with an angle of incidence of 50°. The incident beam is deviated through an angle of 7.1° by refraction as it enters the plate. If the index of refraction of the glass plate is 1.500, what is the index of refraction of the liquid?

38. A farsighted person sees distant objects clearly (rays assumed parallel) with eyeglasses whose lenses have a power of 1.35 diopters. What power lenses should be prescribed to enable him to read without strain when the printed page is 40 cm away?

39. In aerial mapping a camera uses a lens with a 50-in. focal length. How high must the airplane fly in order to photograph a strip of terrain 1 mi long so that its image will fit exactly on the filmstrip, which is 10 in. long?

40. A real image of an object is to be produced 250 cm away from the object. This can be accomplished with a converging lens of focal length 60 cm if the correct object distance is chosen. What should the object distance be?

CHAPTER 24 DISPERSION, OPTICAL INSTRUMENTS, AND POLARIZED LIGHT

Nearly a thousand years ago the Arabian physicist Alhazen (965–1038) constructed pinhole cameras and parabolic mirrors and correctly explained how lenses form images. During the thirteenth century, the use of convex lenses to correct the vision of farsighted people was first suggested, and in the fifteenth century concave lenses were first used to correct the vision of nearsighted people. No other important developments in optics beyond these uses of simple lenses to correct vision occurred until Kepler (1571–1630) and others read about the work of Alhazen that was published in Latin in the sixteenth century. During the early years of the seventeenth century, the age of modern optical instruments began with the inventions of the refracting and reflecting telescopes and the compound microscope.

24.1 ■ Dispersion by a Prism

Triangular glass prisms can be used as *dispersing prisms,* to separate a beam of light into its constituent wavelengths (colors) because the refractive index for glass decreases with increasing wavelength of light (see Table 24.1 and the spectrum chart on the front endpapers).

Thus, when a thin band or beam of light travels through a prism, the shorter (violet) wavelengths are refracted the most, while the longer (red) wavelengths are refracted the least (Fig. 24.1). The separation of light into its component colors is called *dispersion*. The angular spread ϕ of all the colors is called the *angle of dispersion*. The band of colors produced by dispersion is called a *spectrum*. A spectroscope (Fig. 24.5) is used to produce and observe the spectrum. Figure 24.6 shows the spectrum of the sun, for example. Spectra are used to study the structure of atoms and molecules, for chemical analyses, and for the study of the composition of the stars in our universe.

The angle δ (delta) between the entering ray and the emerging ray of a given color is termed the *angle of deviation* for that color. Deviations for a given color will

Table 24.1 Refractive indices of various wavelengths of light for some transparent media

	Color and Wavelength					
Substance	*Red 670 nm*	*Orange 620 nm*	*Yellow 570 nm*	*Green 520 nm*	*Blue 470 nm*	*Violet 410 nm*
Crown glass	1.515	1.522	1.523	1.526	1.532	1.538
Diamond	2.410	2.415	2.417	2.426	2.444	2.458
Flint glass	1.624	1.626	1.627	1.632	1.640	1.651
Ice	1.306	1.308	1.309	1.311	1.314	1.317

Fig. 24.1 Dispersion of white light by a prism. The hues in the spectrum blend imperceptibly into one another, but they can be grouped into six principal colors.

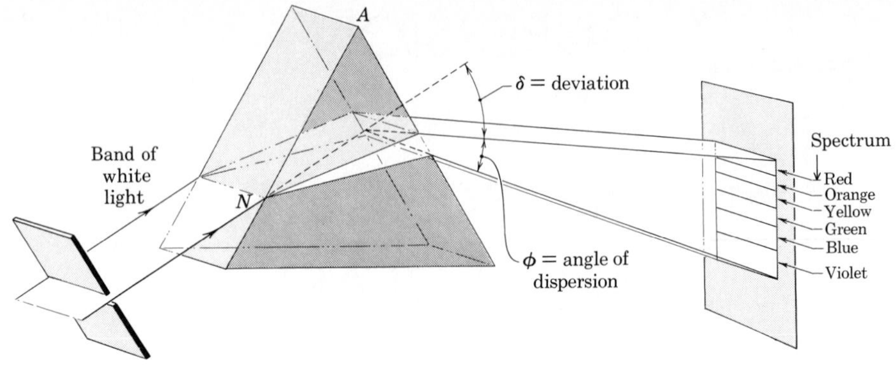

Relationship between the refractive index μ and the angle of minimum deviation δ for a prism with apex angle A.

depend upon the index of refraction of the prism, the angle of incidence of the ray, and the size of the apex angle A of the prism. The angle of deviation for a given color will have its smallest value when the angle of incidence i_1 at the first surface separating air and glass, and the angle of refraction i_2 at the second surface separating glass and air are equal (Fig. 24.2). When the deviation is a minimum, the index of refraction μ is equal to

$$\mu = \frac{\sin[(A + \delta)/2]}{\sin A/2} \tag{24.1}$$

Equation (24.1) also holds reasonably well when the deviation is *near* the minimum.

The refractive indices for various wavelengths of light for some common substances are given in Table 24.1. Note the relatively high indices of refraction for diamond and the low indices for ice. Media that produce large deviations also produce large dispersions but the two properties are not proportional to each other.

Illustrative Problem 24.1 In a glass prism with angle $A = 60°$, the angle of minimum deviation for blue light is found to be 40°. What is the index of refraction of the prism for blue light?

Solution Substitute known values in Eq. (24.1):

$$\begin{aligned}\mu &= \frac{\sin[(A + \delta)/2]}{\sin(A/2)}\\ &= \frac{\sin[(60° + 40°)/2]}{\sin(60°/2)}\\ &= \frac{0.766}{0.500}\\ &= 1.53 \qquad \textit{answer}\end{aligned}$$

Fig. 24.2 Angle of minimum deviation occurs when $i_1 = i_2$.

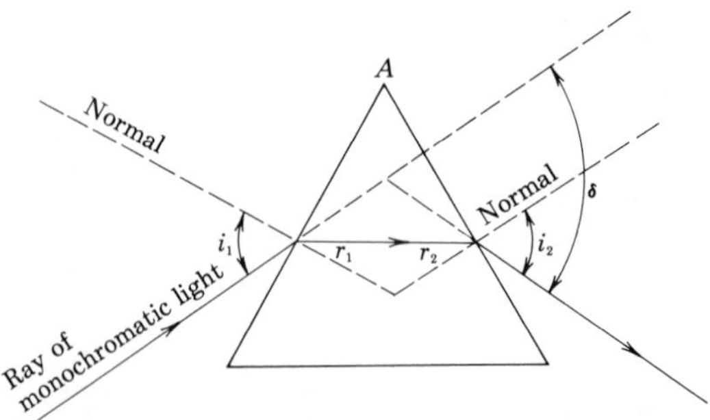

Illustrative Problem 24.2 Find the angle of dispersion from the red rays of the spectrum to the blue rays produced by a flint-glass prism if the refracting angle is 60°. (Assume minimum deviation.)

Solution From Table 24.1, we find $\mu_r = 1.624$, $\mu_b = 1.640$. Solve Eq. (24.1) for $\sin \frac{1}{2}(A + \delta_r)$, to get

$$\begin{aligned} \sin \tfrac{1}{2}(A + \delta_r) &= \mu \sin A/2 \\ \sin \tfrac{1}{2}(60 + \delta_r) &= 1.624 \sin \tfrac{1}{2}(60°) \\ &= 0.812 \end{aligned}$$

Using the $\sin^{-1}$ or Inv Sin key on an electronic calculator to find the angle whose sine is 0.812, we get

$$\tfrac{1}{2}(60° + \delta_r) = 54.3°$$

Then

$$\delta_r = 48.6°$$

In like manner,

$$\begin{aligned} \sin \tfrac{1}{2}(60° + \delta_b) &= 1.640 \sin 30° \\ &= 0.820 \\ \tfrac{1}{2}(60° + \delta_b) &= 55.1° \\ \delta_b &= 50.2° \\ \phi = \delta_b - \delta_r &= 1.6° \text{ (see Fig. 24.1)} \qquad \textit{answer} \end{aligned}$$

24.2 ■ Chromatic Aberration

When white light passes through a single converging lens, it is dispersed in much the same way that it is when passing through a prism. Since violet light is refracted most, it comes to a focus nearer the lens than red light does (Fig. 24.3). This is one reason why the image formed by a single lens with white light is not sharply defined. (A second reason is the *spherical aberration* associated with a single lens, which was discussed in Sec. 23.24.) For example, the images formed by a cheap pair of binoculars often are surrounded by a faint halo of color, usually red or blue. The focusing of light of different colors at different locations is called *chromatic aberration.*

The word "*chromatic*" is derived from the Greek word *khroma*, which means "color."

The presence of chromatic aberration in the early (refracting) telescopes led Newton to invent the reflecting telescope (see Fig. 24.18) in which all wavelengths are reflected to the same focal point, since the law of reflection does not depend on wavelength.

To correct a lens for chromatic aberration a converging lens of crown glass is combined with a diverging lens of flint glass, as illustrated in Fig. 24.4.

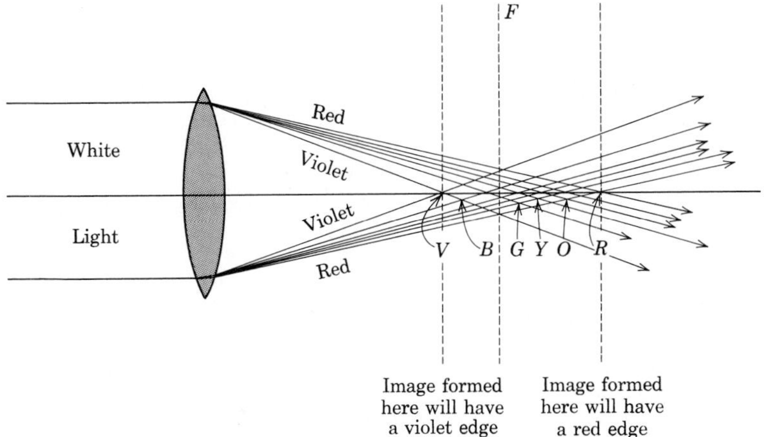

Fig. 24.3 Diagram illustrating chromatic aberration (exaggerated).

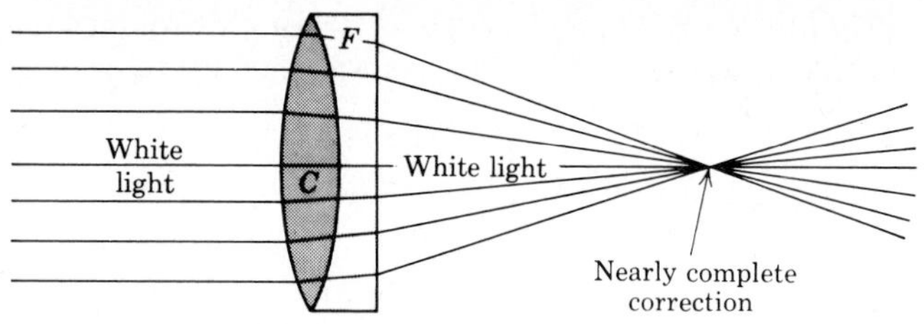

Fig. 24.4 Achromatic lens combination of a double convex crown-glass lens and a plano-concave flint-glass lens.

The effects of chromatic aberration can be reduced by using two lenses of different refractive index in contact, in place of a single lens.

With this arrangement the dispersion of white light into its spectrum caused by one of the lenses is partly canceled by the other. To accomplish this, the refractive indexes and focal lengths for the two lenses must be chosen so that the focal length for the combined lenses (see Eq. 23.14) is the same for red and blue light. Lenses made in this way are called *achromatic lenses,* but because the dispersion and deviation of the lens for each color are not exactly proportional, the correction is not perfect. The best achromatic lenses are made of more than two components (see, for example, the camera lens in Fig. 24.7).

24.3 ■ The Spectroscope

Sir Isaac Newton's prism experiments in 1665–66 made him famous.

The principle of dispersion is used in the prism spectroscope (Fig. 24.5), which is used to produce and examine a spectrum. It consists essentially of four main parts: a *collimator,* which has a small adjustable slit S at one end, placed at the principal focus of a convex lens L_1 at the other end; a *prism*; a *circular scale* with a vernier; and a *telescope*, whose objective lens L_2 forms real images of the slit S. Since the slit is at the principal focus of the collimator, the rays are made parallel before they reach the prism. The prism refracts and disperses the various colors of the light source. The spectrum, as it emerges from the prism, is magnified by the lenses of the telescope until each wavelength of the source of light produces an image of the slit at the focal point of the objective lens. In this manner all the constituent colors which come from the source of light can be separated and analyzed. If the telescope is replaced by a camera so that a photograph of the spectrum can be made, the instrument is called a *spectrograph*.

A spectroscope shows that different sources of light produce spectra that vary greatly. Spectra can be classified into five groups:

Five different types of spectra.

1. Continuous emission spectra
2. Bright-line spectra
3. Continuous absorption spectra
4. Line absorption spectra
5. Band spectra

24.4 ■ Continuous Emission Spectra

When a spectrum is produced by light from a hot glowing solid, a glowing liquid, or glowing gas under high pressure, a continuous band of color from red to violet is observed. The spectrum of an incandescent lamp and that of a carbon arc are of this type and are known as *continuous spectra*. The intensity of the colors depends upon the temperature and upon the hot body itself.

A study of the spectrum of a star will reveal its temperature even though it may be thousands of light-years away.

24.5 ■ Bright-Line Spectra

When vaporized under moderate or low pressure, each of the chemical elements produces a unique spectrum. When the narrow slit of a spectrograph is illuminated by atoms of glowing vapors, a number of bright lines appear on the photographic plate in place of a continuous spectrum. Bright-line spectra of various elements can be seen in

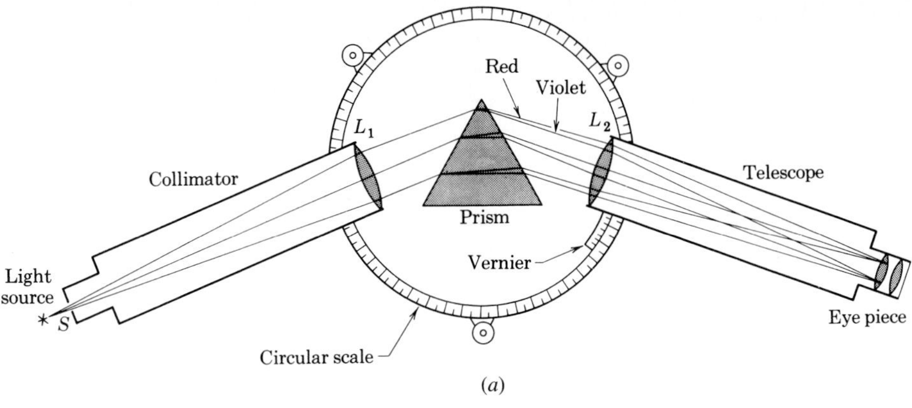

(*a*)

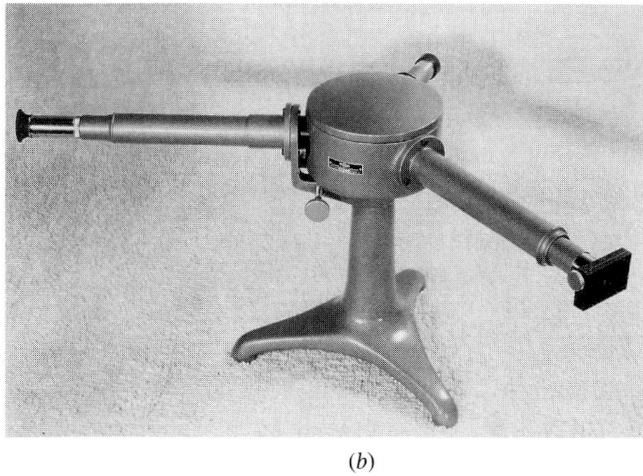

(*b*)

Fig. 24.5 A prism spectroscope. (*a*) Diagram of optical components. (*b*) A typical physics laboratory model. (Gaertner Scientific Corp.)

Fig. 24.6 which appears on the endpapers at the front of this book. Since each element has a unique bright-line spectrum, the spectrum becomes the "fingerprint" of the element. One way of identifying the elements present in a bit of material is to vaporize some of it and study the spectrum of the light emitted.

Bright-line spectroscopic analyses are used in criminology, archeology, medicine, chemistry, and engineering.

The name *line spectra* is derived from the fact that a narrow slit in the collimator is used and its image is a line. If a small circular opening had been used in the collimator, a circular image would appear in place of each line.

24.6 ■ Continuous Absorption Spectra

When white light is passed through various transparent solids or liquids and then examined in a spectroscope, some of the colors present in the spectrum of white light are often found to be missing. A common example occurs when white light is allowed to pass through a colored glass and then to the prism. The missing colors may cover a wide band of wavelengths. For example, a piece of red glass absorbs all visible light except the red, and a magenta-colored glass absorbs the central part of the visible spectrum.

24.7 ■ Line Absorption Spectra

When white light is passed through a gas or vapor, an absorption spectrum is obtained that consists of individual black lines interspersed throughout the continuous spec-

Helium was discovered in the atmosphere of the sun before it was discovered on earth.

trum because a gas absorbs the very same wavelengths of light that it emits in its own spectrum when it is incandescent. Hence, the dark absorption lines tell us which gases are doing the absorbing. When the spectrum of the sun is examined in the spectroscope, it is found that there are black lines distributed throughout the spectrum because the sun is surrounded by an atmosphere of cooler gases that absorb some of the wavelengths coming from the main body of the sun. By analyzing the sun's spectrum in 1868, scientists were able to identify helium in the sun's atmosphere 27 years before helium was first discovered on earth.

24.8 ■ Band Spectra

All the line spectra discussed thus far arise from single free atoms of a gas. Molecules of two or more atoms also give rise to spectrum lines. A *band spectrum* is produced by glowing molecules of a gas. The bands turn out to be a series of many bright lines very close together. As for the atomic case, a molecular absorption spectrum is also possible. A chemical compound can be identified by studying its band spectrum.

OPTICAL INSTRUMENTS

The number of optical instruments based on the principles of light studied thus far is much too large to discuss in an introductory physics text. We shall, however, describe the more important features of cameras, spectacles, projectors, microscopes, and telescopes.

24.9 ■ The Camera

The simplest camera employs a single lens, as illustrated in Fig. 24.7. It is essentially a lightproof box with the converging lens at one end and a light-sensitive film at the opposite end. When the shutter is opened for a fraction of a second, a real, inverted image is formed on the film by the lens. The amount of light reaching the film depends upon the time the shutter is open and on the effective area of the lens aperture which is controlled by an adjustable diaphragm. The effective area of the lens aperture is given as a fraction of the focal length. An *f*/8 lens setting means that the diameter of the aperture is one-eighth of the focal length of the lens. Thus, the *f*-number of a lens is equal to its focal length F divided by its effective diameter D:

Definition of *f*-number, as used with photographic lenses.

$$f\text{-number} = F/D \tag{24.2}$$

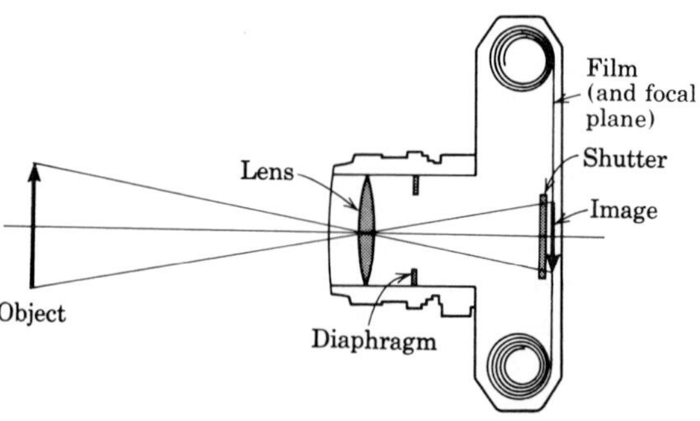

Fig. 24.7 Essential features of a simple camera.

The amount of light reaching the film is proportional to the square of the diameter of the lens opening and is inversely proportional to the square of the focal length of the lens. This is equivalent to saying that the amount of light that reaches the film is inversely proportional to the square of the *f*-number for the lens aperture setting used. Thus an *f*/8 lens will let four times as much light reach the film as an *f*/16 lens and only one-fourth as much light as an *f*/4 lens. We say that an *f*/4 lens is "4 times as fast" as an *f*/8 lens.

Illustrative Problem 24.3 A photographer first has set his camera lens with an *f*/8 stop and a correct film-exposure time of $\frac{1}{60}$ s. What exposure time should he use to get the same amount of light exposure with a second lens aperture setting of *f*/5.6?

Solution The amount of light striking the film is proportional to the area of the aperture and to the exposure time. Since the aperture area $A = \pi D^2/4$ and since $D = F/f\text{-number}$,

$$A_1 = \pi D_1^2/4 \text{ with } D_1 = F/8$$

and

$$A_2 = \pi D_2^2/4 \text{ with } D_2 = F/5.6$$

The ratio of areas

$$\frac{A_2}{A_1} = \frac{D_2^2}{D_1^2} = \frac{F^2/(5.6)^2}{F^2/(8)^2} = \frac{8^2}{5.6^2} \approx 2$$

Or twice the amount of light is available through *f*/5.6 as was available through *f*/8.

Let t_1 be the original $\frac{1}{60}$-s exposure time, and t_2 the new required exposure time. Then since, for equal exposures, $t_1A_1 = t_2A_2$

$$\frac{t_2}{t_1} = \frac{A_1}{A_2} = \frac{1}{2}$$

and

$$t_2 = \frac{t_1}{2} = \frac{\frac{1}{60}\text{ s}}{2} = \frac{1}{120}\text{ s} \qquad \textit{answer}$$

Cheaper cameras use only a single converging lens, which means that the camera will be subject to the spherical aberration discussed in Chap. 23 and the chromatic aberration discussed in this chapter (Sec. 24.2). More expensive cameras correct much of the spherical and chromatic aberrations by using a three- to five-element compound lens. Figure 24.8 illustrates the essential features of a single-lens reflex (SLR) camera that uses such a lens.

The images formed by simple lenses in inexpensive cameras suffer from spherical and chromatic aberration.

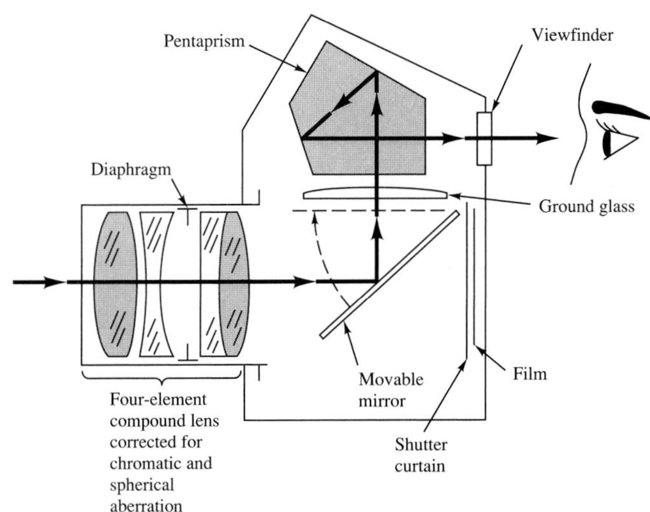

Fig. 24.8 Diagram of a modern single-lens reflex (SLR) camera. The same lens that produces an image on the film also can send an image, through the pentaprism, to the viewfinder. The position of the movable mirror determines whether the image goes to the film or to the viewfinder.

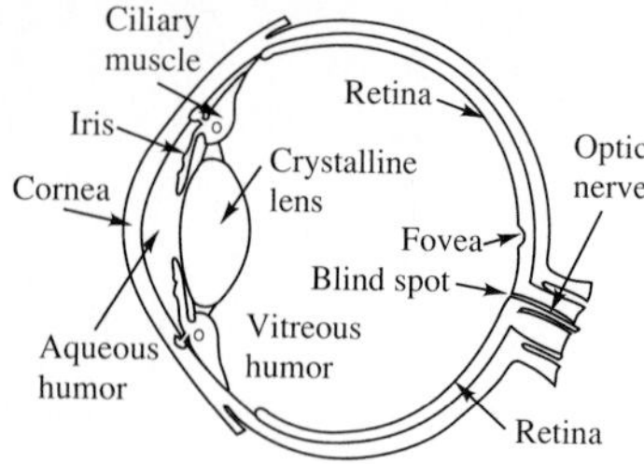

Fig. 24.9 Cross section of the human eye.

24.10 ■ The Eye

The human eye is in some ways like a camera. Optically it consists of a light-tight enclosure having an elaborate lens system at one end and a light-sensitive "film" of nerve fibers at the other. The eyeball (Fig. 24.9) is a nearly spherical opaque chamber of about 2.5 cm diameter filled with a transparent semifluid substance, called the *vitreous humor*. Light enters through a rather firm transparent tissue called the *cornea* into a clear fluid known as the *aqueous humor*. Behind the cornea is a circular diaphragm, the *iris,* with a central hole called the *pupil*. The iris has muscles which can shrink or dilate the diameter of the pupil as the light intensity increases or decreases. The iris contains the pigment noticeable as the color of the eye. Immediately behind the iris is a biconvex *crystalline lens,* composed of microscopic glassy fibers which can change shape by sliding over each other under the control of the *ciliary muscle*. In the interior of the back wall of the eye is a light-sensitive surface, the *retina*. Attached to the retina is the *optic nerve,* which carries information from the retina to the brain.

The cornea is more effective than the eye's lens in bending light that enters the eye.

The refracting media of the eye are the cornea, the aqueous humor, the lens, and the vitreous humor. The effect of all the refractions is to form real images of external objects on the retina. The principal bending of light occurs as the light enters the cornea because of the cornea's relatively small radius of curvature, and because of the relatively large difference between the refractive index of air (1.000) and the refractive index of the cornea (1.376). The lens of the eye has a refractive index of 1.406 at its center, and 1.375 at its outer edge. Both the aqueous humor and the vitreous humor have a refractive index of 1.336. Because the maximum difference between the refractive index of the lens and the aqueous or vitreous humor is 0.070, the lens causes less bending of light than the cornea does. The cornea, which has a converging power of about 43 diopters (see Sec. 23.23), provides roughly two-thirds of the refraction required to focus the image of a distant object on the retina. When a normal eye is viewing intermediate and distant objects, it is relaxed, and the lens has its thinnest shape, with a converging power of about 20 diopters. To bring the image of a near object in focus on the retina, the ciliary muscles contract and thicken the lens so that the radii of curvature of the lens both decrease. Thus, the focal length of the lens is decreased and the image is focused on the retina. The ability of the eye to adjust its focal length is called *accommodation*.

The relaxed lens of the eye has the same converging power as a converging glass lens with a focal length of 5 cm; in other words, 20 diopters.

Accommodation is the ability of the eye to change its focal length.

The ability of the eye to bring the image of an object to a sharp focus upon the retina (accommodation) decreases as a person gets older. Many 60-year-old persons cannot read a book held at a convenient distance of about 40 cm (16 in.). Thus, the extra power required must be provided by convex lenses in front of the eye (see Sec. 24.11 below).

The normal eye is most relaxed when it is focused for parallel rays, i.e., for distant objects, but an object to be studied in detail must often be brought closer to the eye. The reason, as we discussed in Chap. 23, is that the closer the object is to the eye, the larger the image is formed on the retina. For normal eyes, vision is most distinct when they are focused on objects about 25 cm (10 in.) away. Prolonged observations at distances of 25 cm or less result in considerable fatigue and eyestrain.

It is interesting to note that the image on the retina is upside-down (see Fig. 24.10). It is the brain that interprets the image as being erect.

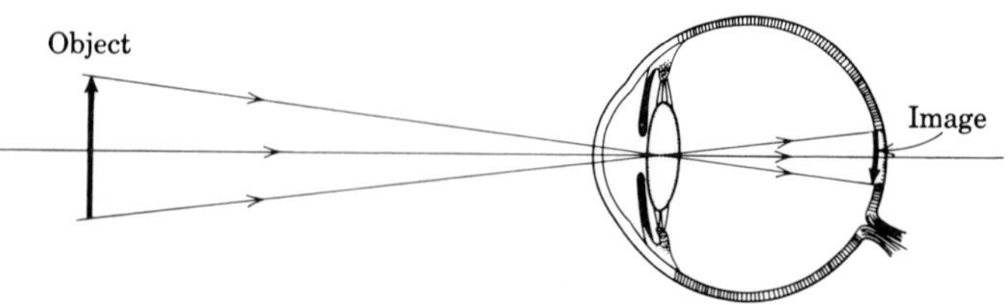

Fig. 24.10 Retinal images are real and inverted.

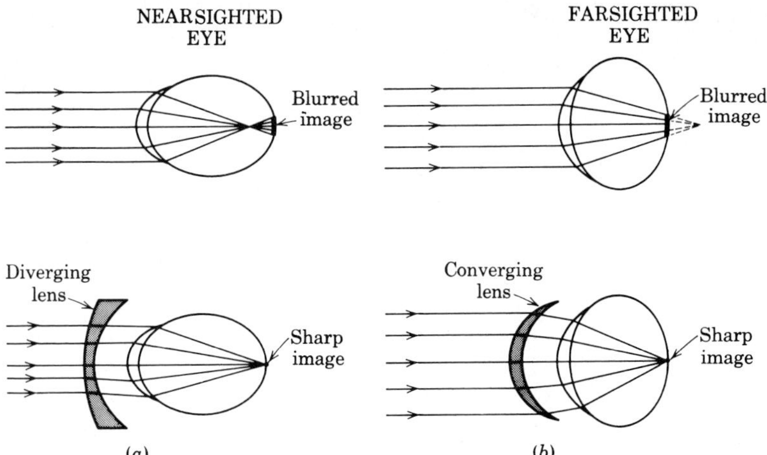

Fig. 24.11 Eyeglasses correct defects of vision. (*a*) Myopia (nearsightedness) and (*b*) hyperopia (farsightedness).

24.11 ■ Defects of Vision

The defective eye may be nearsighted, farsighted, or astigmatic. If the image formed by a distant object falls in front of the retina, as shown in Fig. 24.11*a*, the eye is *nearsighted*, or *myopic*. This may occur because the eyeball is too long, because the lens has too short a focal length, or for some other reason. In such cases, the ciliary muscle lacks the power to reduce the curvature of the eye lens sufficiently to focus the rays from a distant object on the retina. Only objects close to the eye can be focused sharply on the retina; all others appear blurred.

Nearsighted, or myopic, eyes can be corrected by using a suitable diverging lens (see Fig. 24.11*a*).

The eyes of a *farsighted*, or *hyperopic*, person form images behind the retina, as shown in Fig. 24.11*b*. This defect arises because the eyeball is too short or the lens has too long a focal length. As an object is moved farther away from the eye, the image moves nearer the retina. Hence, the farsighted person can see distant objects clearly, while images of nearer objects appear blurred.

Farsighted vision can be corrected by using a suitable converging lens in front of the eye. As an object is brought still closer to the hyperopic eye, the same eye will require the use of a converging lens of still greater power. Bifocal spectacles have lenses with different focal lengths in the upper and lower halves of the glasses. Trifocal spectacles are quite common today.

Another common defect of the eye is *astigmatism*, which occurs when at least one of the refracting surfaces (cornea or lens) of the eye is not spherical. Since such an eye has varying focal lengths in different planes, the image formed may be distinct in one direction and blurred in another (see Fig. 24.12). Astigmatism can be corrected by glasses that have greater curvature in the plane in which the cornea or lens has less curvature.

Millions of people wear contact lenses made of various soft and pliable plastics. These lenses are worn to correct such conditions as nearsightedness, farsightedness and astigmatism. Athletes wear them in sports in which eyeglasses are unsafe.

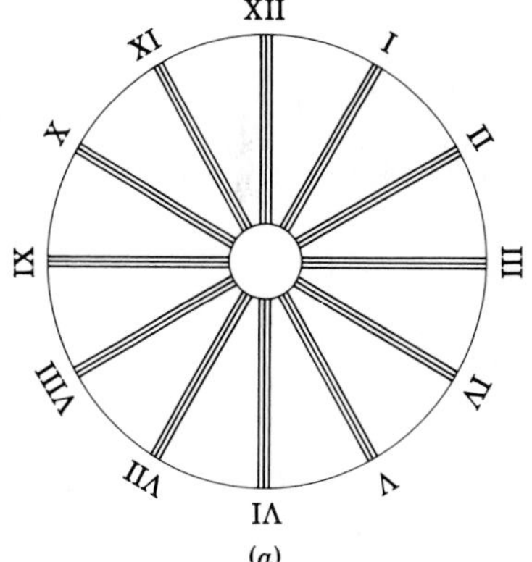

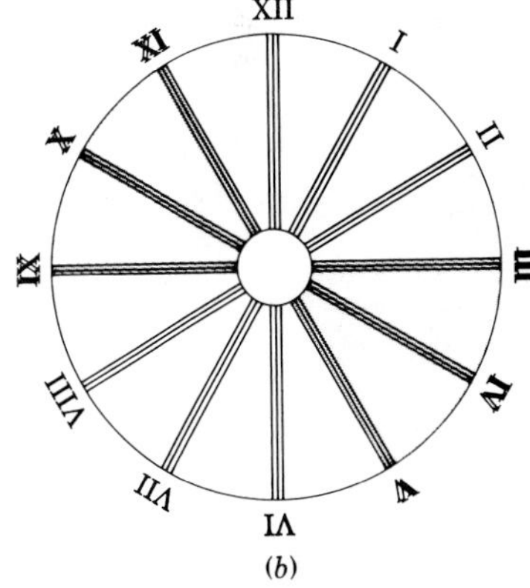

Fig. 24.12 Astigmatism. (*a*) Each set of parallels appears equally sharp to the normal eye. (*b*) To the astigmatic eye, the sets of radial lines differ in sharpness.

24.12 ■ The Projector

The projector for slides and films consists optically of the *illuminating system* and the *projection lens* (Fig. 24.13). The illuminating system consists of a light source and a condensing pair of lenses which collects light from the source and concentrates it upon the film or slide. The slide is placed slightly beyond the focal length of the projection lens, so that an enlarged inverted image of the slide or film is formed on the screen.

Fig. 24.13 Optical system of the type of projector used for 35-mm color slides.

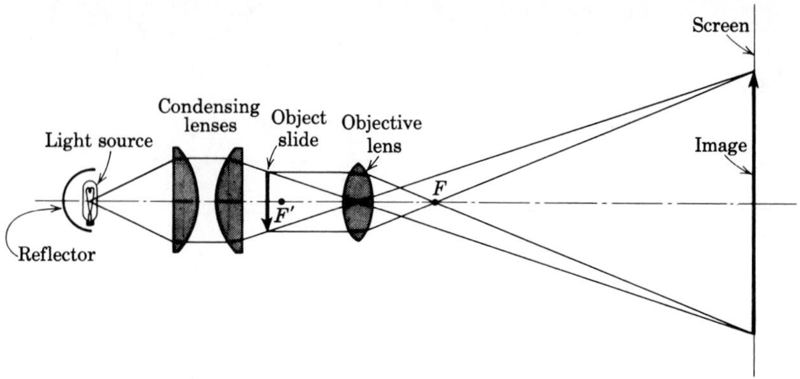

Since all the light rays forming the image must come from the object (the slide or film), the object must be strongly illuminated if a bright image is to be projected. This is accomplished by having a powerful source of light with a reflector behind it and condensing lenses in front of the object to direct the rays through the slide.

In order to get an erect image on the screen, the slide must be placed in the projector upside-down, with left and right interchanged.

Slide projectors and motion picture projectors are mechanically quite different, but optically very similar.

Motion-picture projection is essentially the same as slide projection, but mechanical means must be provided to change from one picture, or *frame,* to another many times a second. A rate of 18 frames per second is sufficient to produce a satisfactory illusion of motion. "Slow motion" effects are obtained by running the film *in the camera* at faster rates when the movie is being filmed.

24.13 ■ The Simple Magnifier

If you looked at the print on this page from across a room, you probably would not be able to read it but as the book is brought nearer and nearer the eye, the image formed on the retina gets larger and larger, until you can read the print. For the normal eye the print is most easily read at a distance of about 25 cm (10 in.) from the eye. If it were possible for the normal eye to accommodate to much nearer distances by making the lens of the eye thicker and thicker, you could bring an object very close to the eye and have the object enormously magnified. However, unless you are very nearsighted, you will see a blurred image if the object is brought closer than about 10 cm.

A magnifying glass produces magnified virtual images of objects nearer to it than its focal length.

If a magnifying glass of short focal length is now placed before the eye, as shown in Fig. 24.14, the object can be brought much closer to the eye and can be observed in detail. If the object to be examined is placed a little nearer than one focal length to the lens, then an enlarged, erect, virtual image of the object can be seen. The magnifying glass is also called a simple microscope.

The linear magnification M of a simple microscope is the ratio of the image size I to the object size O:

$$M = I/O \tag{24.3}$$

Fig. 24.14 A simple magnifier.

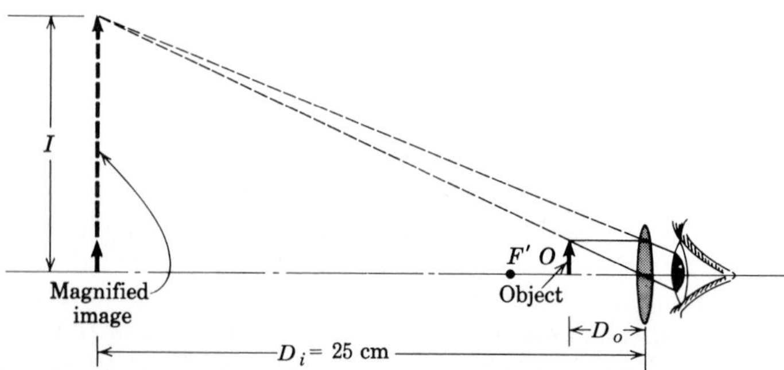

Using the properties of similar triangles, we can prove that $I/O = D_i/D_o$, where D_o is the object distance and D_i is the image distance. Then

$$M = D_i/D_o \tag{24.4}$$

If we use Eq. (23.3) to solve for D_i/D_o, we get

$$\frac{1}{D_o} + \frac{1}{D_i} = \frac{1}{f}$$

Multiplying each term by D_i yields

$$\frac{D_i}{D_o} + 1 = \frac{D_i}{f}$$

or

$$\frac{D_i}{D_o} = \frac{D_i}{f} - 1 \tag{24.5}$$

Using $D_i = -25$ cm (or -10 in.), the distance of most distinct vision for the normal eye, Eq. (24.5), becomes

$$\frac{D_i}{D_o} = \frac{-25\text{ cm}}{f\text{ cm}} - 1 \tag{24.6}$$

or

$$\frac{D_i}{D_o} = \frac{-10\text{ in.}}{f\text{ in.}} - 1 \tag{24.7}$$

Illustrative Problem 24.4 A converging lens has a focal length of 6.0 cm. If it is to be used as a simple magnifier, how far from the object should the lens be placed to produce a virtual image 25 cm from the eye? What is the magnification?

Solution Solve Eq. (24.5) for D_o:

$$\frac{D_i}{D_o} = \frac{D_i}{f} - 1$$

$$D_o = \frac{D_i}{D_i/f - 1}$$

Substitute $D_i = -25$ cm and $f = 6.0$ cm in the above equation:

$$D_o = \frac{-25\text{ cm}}{-25\text{ cm}/6\text{ cm} - 1} = 4.8\text{ cm} \qquad \textit{answer}$$

$$M = \frac{D_i}{D_o} = \frac{-25\text{ cm}}{4.8\text{ cm}} = -5.2 \qquad \textit{answer}$$

The minus sign merely indicates that the image is virtual.

24.14 ■ The Compound Microscope

When a simple magnifier is used to its best advantage, the object is placed just inside the focus of the lens so that an image is formed at about 25 cm from the lens. The magnification M produced is the ratio of the image distance, 25 cm, to the object distance from the lens, which is approximately equal to f, the focal length of the lens. Hence, the magnifying power of a simple microscope is equal to $25/f$ (if f is given in centimeters) or $10/f$ (if f is given in inches). Thus, a converging lens with a focal length of 5 cm can magnify an object $\frac{25}{5}$ or 5 times, and a converging lens with a focal length of 2.5 in. will have a magnifying power of 4.

To achieve a magnification greater than that obtainable with a single converging lens a compound microscope is used. It consists of a group of lenses acting as a single

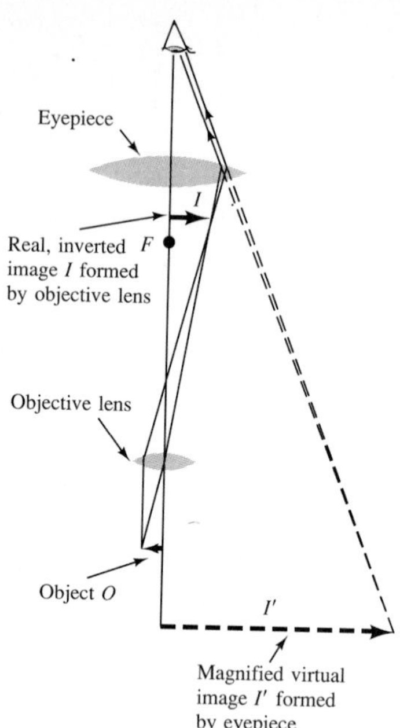

Fig. 24.15 Optical system (not drawn to scale) of the compound microscope (much simplified).

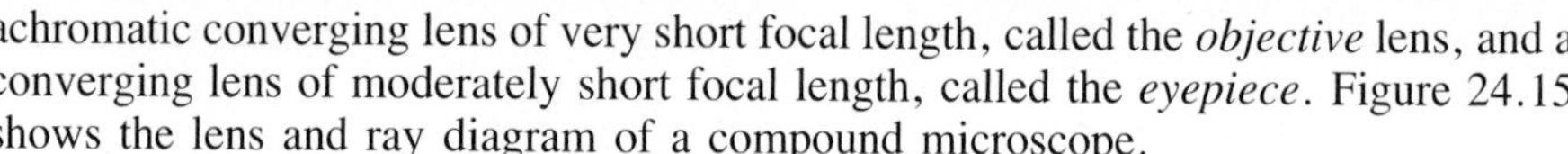

achromatic converging lens of very short focal length, called the *objective* lens, and a converging lens of moderately short focal length, called the *eyepiece*. Figure 24.15 shows the lens and ray diagram of a compound microscope.

The object is placed just beyond the principal focus of the objective lens; this position produces a real, somewhat magnified image I of the object. The image is then magnified again by the eyepiece, which acts as a simple microscope. The eyepiece produces an enlarged virtual image I' of the real image I.

The magnification M of a compound microscope is equal to the product of the magnifying power M_o of its objective and the magnifying power M_e of its eyepiece.

$$M = M_o M_e \tag{24.8}$$

Illustrative Problem 24.5 A compound microscope has an objective lens of 7.50 mm focal length and an eyepiece of 30.0 mm focal length. What is the magnifying power of the microscope if the object is in sharp focus when it is 8.00 mm from the objective?

Solution Consider the objective lens alone and use Eq. (23.3):

$$\frac{1}{D_o} + \frac{1}{D_i} = \frac{1}{f}$$

$$\frac{1}{8.00 \text{ mm}} + \frac{1}{D_i} = \frac{1}{7.50 \text{ mm}}$$

$$D_i = 120 \text{ mm}$$

and

$$M_o = \frac{D_i}{D_o} = \frac{120 \text{ mm}}{8 \text{ mm}} = 15$$

Next consider the eyepiece alone. Let $D_i = -250$ mm, the distance for most distinct vision:

$$\frac{1}{D'_o} + \frac{1}{-250 \text{ mm}} = \frac{1}{30.0 \text{ mm}}$$

$$D'_o = 26.8 \text{ mm}$$

and

$$M_e = \frac{D_i}{D'_o} = \frac{-250 \text{ mm}}{26.8 \text{ mm}} = -9.33$$

The total magnification is equal to

$$M = M_o M_e = (15)(-9.33)$$
$$= -140 \qquad \textit{answer}$$

Again, the minus sign merely indicates a virtual image.

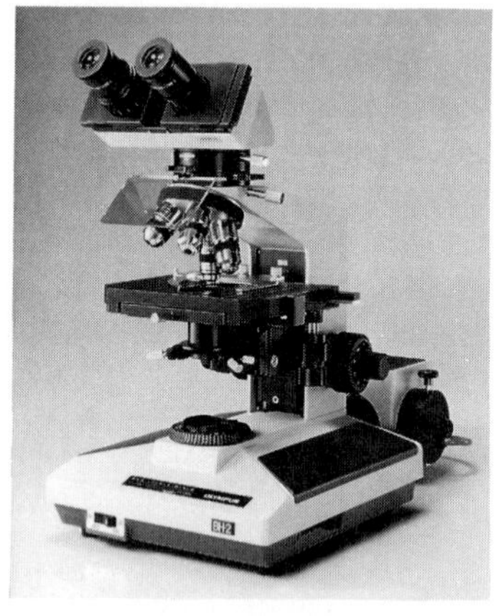

Fig. 24.16 Modern high-quality microscope with a turret carrying four different-focal-length objective lenses. (Olympus Corporation, Lake Success, New York)

Most high-quality microscopes have a set of interchangeable eyepieces and/or a turret carrying three or more objective lenses, each of a different magnifying power (see Fig. 24.16). Turning the turret gives a variety of magnifying powers. Microscopes with magnifying powers of a little more than 2000 are quite common today. This is near the upper limit of magnification for *optical* microscopes.

For greater magnification, an instrument called the *electron microscope*, which employs electromagnetic lenses, can be used. Magnifications as high as 100,000 have been achieved with these instruments.(See Chap. 33.)

24.15 ■ Refracting Telescopes

The refracting telescope like the compound microscope uses an objective lens system and an eyepiece. The objective lens in this instrument, however, is a large converging lens of long focal length. A diagram of a small telescope is shown in Fig. 24.17.

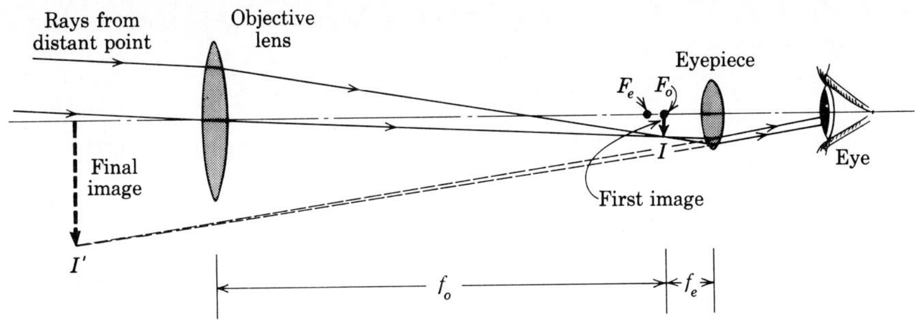

Fig. 24.17 The optical system of a refracting telescope.

Parallel light rays from a distant object enter the objective, and a real, inverted image of the object is formed at I.

The eyepiece of the telescope magnifies the real image formed by the objective. The eyepiece, which has a short focal length, is moved until the image I is just inside its focal plane, a plane through F_e. The eyepiece, used as a simple magnifier, leaves the final image I' inverted. For normal observations, the focal planes through F_o and F_e are the same, and light emerges from the eyepiece in "parallel" rays. Hence, the image I' will appear to be far off, "at infinity."

Magnifying power of a refracting telescope.

The magnifying power of a refracting telescope is defined as the ratio of the angle subtended at the eye by the final image I' to the angle subtended at the eye by the object when it is viewed without the telescope. By applying the geometry of similar triangles to Fig. 24.17 for the case where the focal points F_o and F_e are at the same location, it can be shown, (assuming $\tan \theta = \theta$ for small angles) that the magnifying power is equal to the ratio of the focal lengths of the two lenses. Thus,

$$\text{Magnification } M = \frac{f_o}{f_e} \tag{24.9}$$

where

f_o = focal length of the objective
f_e = focal length of the eyepiece

Illustrative Problem 24.6 The objective lens of a small telescope has a focal length of 120 cm, and the focal length of the eyepiece is 2 cm. Find the magnifying power of the telescope for distant objects.

Solution By Eq. (24.9),

$$M = f_o/f_e$$

Hence

$$M = 120\text{ cm}/2\text{ cm} = 60$$

The magnification is 60. *answer*

Equation (24.9) suggests that a magnification as large as desired can be obtained simply by using an eyepiece with a sufficiently short focal length. For example, if a refracting telescope with an objective lens of focal length 1200mm is used with a 1.2-mm focal length eyepiece, the magnification predicted by Eq. (24.9) would be 1000. Such a combination of lenses would, however, produce unacceptably blurry images unless the objective lens had a diameter of at least 20 inches, because of the influence of *diffraction,** which causes light passing through the objective lens near its edge to bend. The maximum usable (practicable) power P_{max} for a telescope under

* Any bending of light by means other than reflection or refraction is called *diffraction*. (See Question 27 at the end of this chapter.)

Rule-of-thumb for maximum usable telescope power.

ideal atmospheric conditions is given by a rule-of-thumb that can be stated mathematically as follows:

$$P_{max} = 50 \times (\text{diameter of objective lens in inches}) \qquad (24.10)$$

This equation also holds for reflecting telescopes, to be discussed in the next section.

24.16 ■ Reflecting Telescopes

Sir Isaac Newton invented the reflecting telescope in 1668.

Most of the very large astronomical telescopes in the world are *reflecting telescopes,* such instruments consist essentially of a tube, at the bottom of which is a concave parabolic mirror of long focal length (Fig. 24.18).

Parallel rays of light from a star enter the tube and are brought to a focus at F, where images can be viewed through an eyepiece or photographed. Often a small plane mirror m reflects the convergent rays to a focus at F, out of the path of the incoming light. The small mirror prevents only a small percent of the light from the star from reaching the objective mirror. Reflecting telescopes have several advantages over refracting telescopes.

Advantages of reflecting telescopes.

1. Large light-gathering elements are needed in both instruments. A large glass lens is very heavy and tends to distort under its own weight since it must be supported only at its thin outer edges to prevent any blocking of the incoming light. In contrast, the structural supports to hold a mirror in place can be distributed over the entire back surface of the mirror.

2. Both surfaces of a lens must be ground to the desired shape, and the glass between the ground surfaces must be of high optical quality. Only one surface of the mirror needs to be ground, and only the glass near the surface must be free of bubbles and other optical defects.

3. The problems of chromatic and spherical aberration must be dealt with for the objective lens in a refracting telescope. There is no chromatic aberration with a mirror, and the problem of spherical aberration is solved by making the mirror surface parabolic.

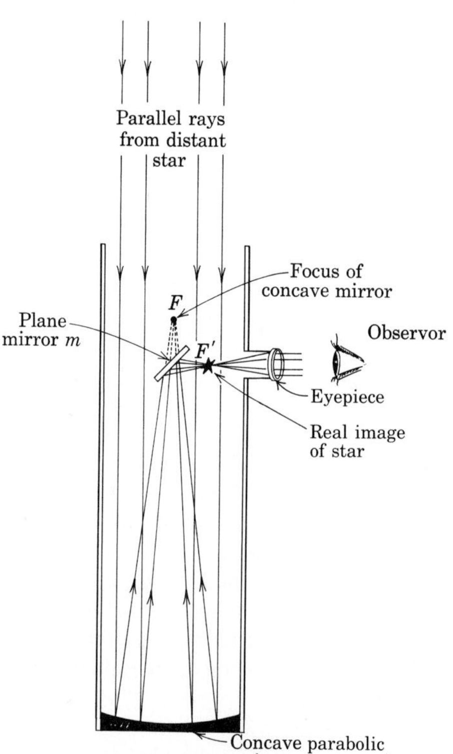

Fig. 24.18 Diagram of a reflecting telescope. This particular design for a reflecting telescope is usually called a newtonian reflector, because it was invented by Isaac Newton.

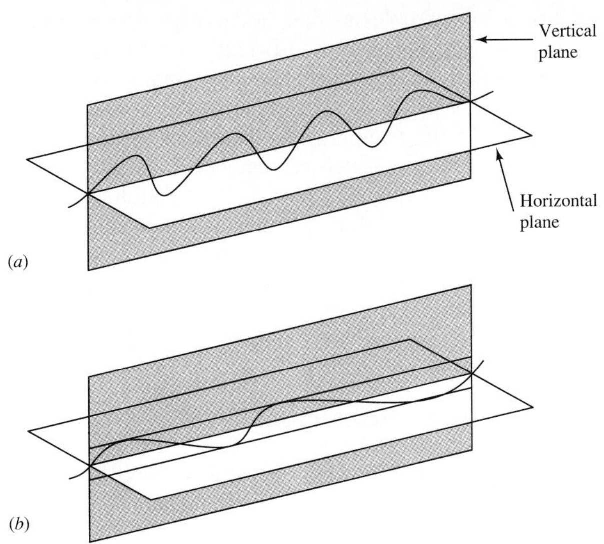

Fig. 24.19 (*a*) Vertically plane-polarized transverse wave on a rope. (*b*) Horizontally plane-polarized wave on rope.

4. In the refracting telescope the large, heavy lens is near the top of the instrument, while in the reflecting telescope the large, heavy mirror is at the bottom of the instrument. This results in greater stability for the reflecting telescope. Achieving stability is particularly important when the telescope is used to make long time-exposure photographs.

POLARIZATION OF LIGHT

Transverse waves can be *plane-polarized.*

All types of waves, whether mechanical or electromagnetic, can be classified as longitudinal waves, transverse waves, or combinations of longitudinal and transverse waves. For example, sound waves are longitudinal waves and waves on vibrating strings and ropes are transverse waves. In a transverse mechanical wave, the moving particles of the vibrating system oscillate in a direction perpendicular to the direction of motion of the wave. If the vibration continues in a single plane as time passes, the wave is said to be *plane-polarized*. Since, in longitudinal waves the vibration is parallel to the direction of the wave motion, longitudinal waves cannot be polarized. Only transverse waves can be polarized. Figure 24.19 illustrates two examples of plane-polarized waves on a rope. The wave on the rope in Fig. 24.19*a* is vertically polarized, and the wave in the rope in Fig. 24.19*b* is horizontally polarized. There are an infinite number of other possible planes of polarization for the rope.

Light, which is a transverse (electromagnetic) wave, can also be polarized. The possibility of polarizing light has resulted in a variety of practical applications in many industries. In order to understand what is meant by polarization of light, let us graphically represent an ordinary beam of light as composed of many transverse waves whose vibrations are along straight lines perpendicular to the direction of propagation of the light beam. A beam of light consists of millions of such waves, each with its own plane of vibration. These vibrations occur in all planes around the axis of propagation (Fig. 24.20).

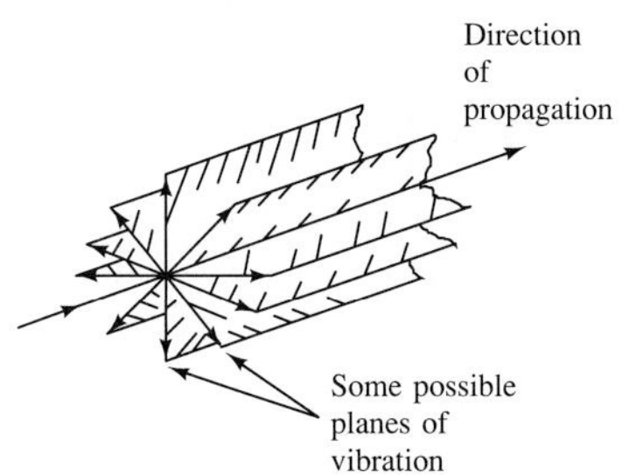

Fig. 24.20 Schematic diagram showing a beam of light with waves vibrating in all planes.

24.17 ■ Production of Polarized Light

By suitable interaction with matter, it is possible to eliminate components of the waves in all but one given plane or in parallel planes of vibration. Such a beam is said to be *plane-polarized*. There are several ways in which light can be polarized. When light is reflected from a transparent surface at a given angle, called the *polarizing angle,* the reflected light is polarized. Crystals such as quartz, Iceland spar, and tourmaline, when cut and arranged properly, are effective in polarizing light. The scattering of light by small particles of dust and smoke produces partially polarized light. However, each of these methods is difficult to control, is too expensive, absorbs too much light, or is limited to too small a light beam. Most of these limitations were removed with the introduction of the commercial polarizing film called *Polaroid* (not to be confused with the photographic process and camera of the same name). This film comes in thin sheets in which are embedded millions of ultramicroscopic needlelike crystals (sulfate of iodoquinine), which have all been made to align in one direction during the process of manufacture. When ordinary light passes through such a film, nearly all the components of the light vibrating in one direction are transmitted by the sheet, while nearly all the other components of the light are absorbed. Such sheets will transmit about 40 percent of the energy of the incident beam as polarized light vibrating in parallel planes; the remainder of the light energy is absorbed.

24.18 ■ Detection of Polarized Light

To the unaided human eye, polarized light will appear no different from unpolarized light. However, the presence of polarized light can be easily detected by the use of a second sheet of Polaroid. Let us set up a simple demonstration (Fig. 24.21) involving a source of light S and two polarizers A and B. Unpolarized light from S will be polarized by the *polarizer* A. If the polarizing axis of the second polarizer B, called the *analyzer*, is parallel to that of the polarizer A (Fig. 24.21*a*), the light beam polarized by the first sheet will pass through the second sheet without appreciable loss in intensity. If the analyzer is rotated through 90° about an axis parallel to the beam (Fig. 24.21*b*), the crystals of the analyzer will absorb the polarized light from A, and almost no light will be transmitted through it. If the anaylzer is rotated another 90°, the polarized light from A will again be transmitted through B.

24.19 ■ Applications of Polarized Light

A considerable amount of the glare of daylight driving comes from sunlight that is partially horizontally polarized as it is reflected from roads and other surfaces. By using Polaroid sunglasses with the lenses aligned to transmit vertically polarized light, a great deal of this glare can be eliminated by the absorption of some of the horizontally polarized reflected rays.

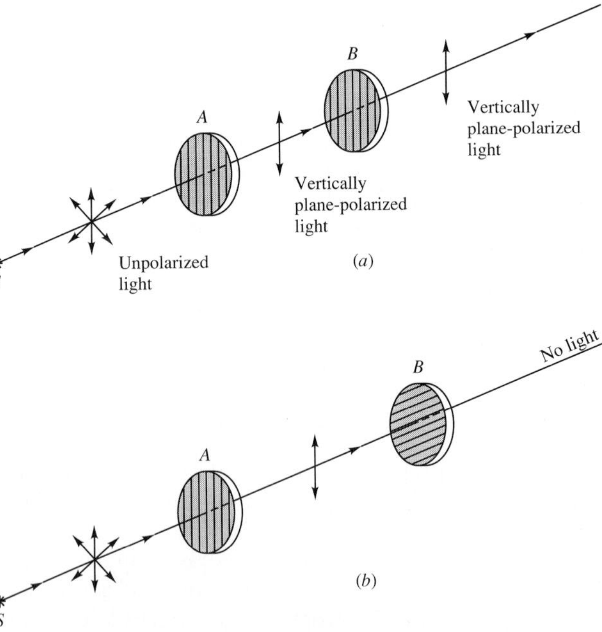

Fig. 24.21 Production and detection of polarized light.

Fig. 24.22 Strain pattern in a plastic model of meshing gears detected by polarized light. (Polaroid Corp.)

Industry has used polarized light in analyzing strains set up in transparent models of complex engineering structures (Fig. 24.22). Certain transparent materials (such as glass or celluloid) possess polarizing qualities when they are subjected to mechanical strains. With proper polarizing equipment, these regions of strain in the models can be detected. In this manner it can be determined whether excessive strains might occur at critical points in the real structure.

Photographic applications of polarizing filters.

Reflected light from nonmetallic surfaces tends to be partially polarized. Photographers sometimes use polarizing filters to reduce the intensity of unwanted reflected light from windows (see Fig. 24.23). In black-and-white photography, a yellow filter can be used to darken the blue sky to get effective pictures of clouds, but such a filter cannot be used in color photography. Since the light from the sky is partially polarized, a polarizing filter can be used in color photography to darken the blue sky.

(*a*)

(*b*)

Fig. 24.23 (*a*) Picture taken without a filter. (*b*) Picture taken with a polarizing filter.

24.20 ■ Rotation of Plane of Polarization

Liquid Crystal Displays The ability to polarize light and to rotate the plane of polarization of polarized light are used in the familiar liquid crystal displays (LCDs) used in electronic calculators, digital watches, portable computers, digital electronic multimeters, gas pump displays, automotive instrument panels, cash register displays, and many other applications. Because they use essentially no electric power, LCDs are excellent choices for displaying information in battery-powered equipment.

The six-layer sandwich of materials that makes up an LCD is illustrated in Fig. 24.24*a* with the layers separated from one another to allow displaying of the state of polarization of the light between the layers. The arrows that indicate whether the light is horizontally or vertically polarized at locations B, C, D, E, and F are correctly drawn for light travelling to the right or to the left. Note that for light traveling in either direction, the plane of polarization of the light is rotated by 90° as it passes through the liquid crystal.

A *liquid crystal* can rotate the plane of polarization of plane-polarized light.

The liquid crystal, which is made up of long parallel strings of molecules, is able to rotate the plane of polarization of polarized light because of the way the orientation

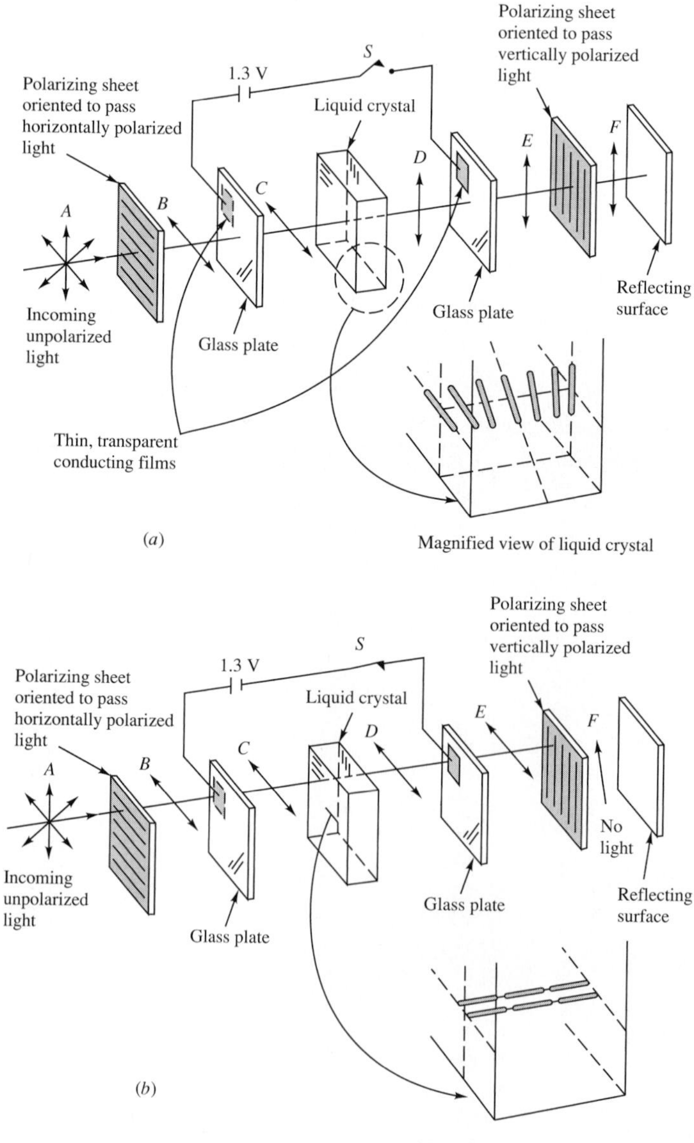

Fig. 24.24 Exploded-view diagram of the sandwich of materials that make up a liquid crystal display (LCD). In (*a*) some of the incoming light goes all the way to the reflecting surface, and after reflection travels back through locations *F* to *B* and leaves the crystal. In (*b*) the incoming light gets no farther than location *E*.

of those strings of molecules gradually changes between the left and right faces of the crystal. As the magnified view of the liquid crystal in Fig. 23.24*a* suggests, the molecular strings at the left face are horizontally oriented, and therefore parallel to the plane of polarization of the incoming light that strikes that face. At the right face of the liquid crystal, the molecular strings are vertically oriented. In between the left and right faces, the molecular strings gradually change from being horizontally oriented to being vertically oriented. The interaction of the horizontally polarized light with these strings of molecules causes the light to gradually change from being horizontally polarized as it enters the crystal, to being vertically polarized as it leaves the right side of the crystal. Vertically polarized light traveling in the opposite direction becomes horizontally polarized as it leaves the left side of the liquid crystal. Although the liquid crystal is drawn relatively thick in Fig. 24.24 to show some of these details, in an actual LCD the liquid crystal may be only 0.01 mm thick.

Now consider the problem of displaying symbols on the LCD. The shaded rectangular regions on each of the two glass plates in Fig. 24.24*a* represent areas where there is a thin, electrically conducting, transparent film. These films are on the faces of the glass plates nearest the liquid crystal. If a voltage is applied between those two regions by closing switch *S* (as shown in Fig. 24.24*b*), the strings of molecules in the liquid crystal between those two regions will become reoriented perpendicular to the faces of the glass plates because of the electric forces that act on the molecules when the voltage is applied. The liquid crystal will therefore lose its ability to change the plane of polarization of the light in the region between the transparent conducting films. Now consider light traveling through the LCD sandwich, in the region where the transparent conducting films are located (see Fig. 24.24*b*).

Unpolarized light at *A* arrives at location *C*, as before, horizontally polarized. But now after passing through the crystal to location *D*, the light is still horizontally polarized. It remains horizontally polarized after it passes through the second glass plate, but it cannot then pass through the second polarizing sheet which is oriented to transmit vertically polarized light. No light strikes the reflecting surface, and therefore no light makes the trip to the left to leave the LCD sandwich. The display therefore appears dark in the region where the electrically conducting transparent films are located. If switch *S* is opened, the strings of molecules reorient themselves (as illustrated in the magnified view in Fig. 24.24*a*), and the rectangular region becomes bright again. Symbols, such as numbers and letters can be displayed if the singular rectangular region is replaced by an array of small, isolated, square regions, each of which can be connected with a switch to a voltage supply. Figure 24.25 shows an example of such an LCD display.

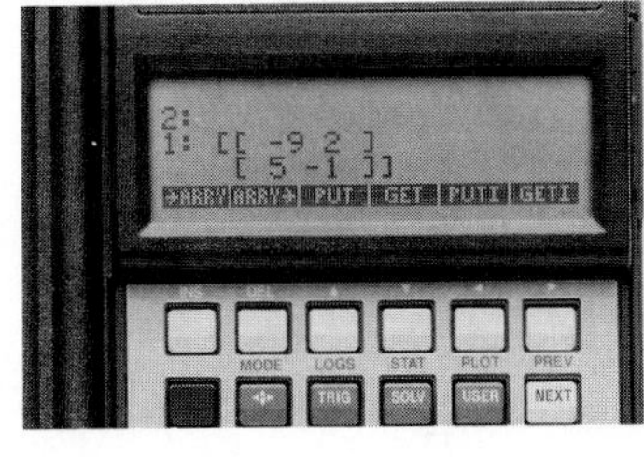

Fig. 24.25 Liquid crystal display (LCD) for a scientific professional calculator. (Courtesy Hewlett-Packard Company)

Optically Active Substances—the Polarimeter Certain materials, such as sugar and cellophane, have the ability to rotate the plane of polarization while transmitting the polarized light. They are called *optically active substances*. This can be demonstrated by arranging an apparatus as shown in Fig. 24.26, in which light from a source *S* is passed through a polarizer P_1, then through a tube *T* with glass ends

Dissolved sugars can also rotate the plane of polarization of plane-polarized light.

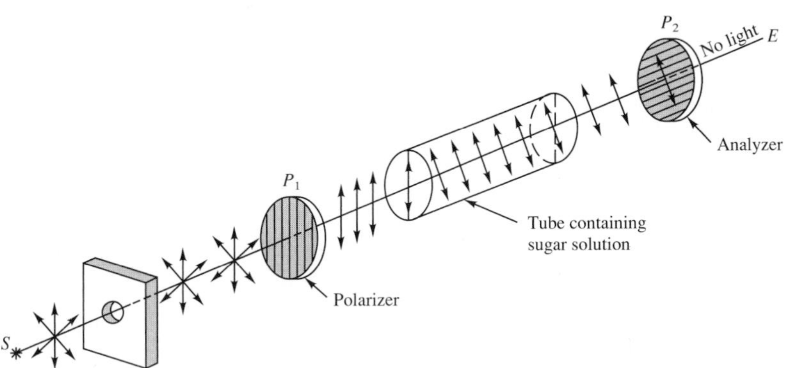

Fig. 24.26 Illustration of the principle of the saccharimeter.

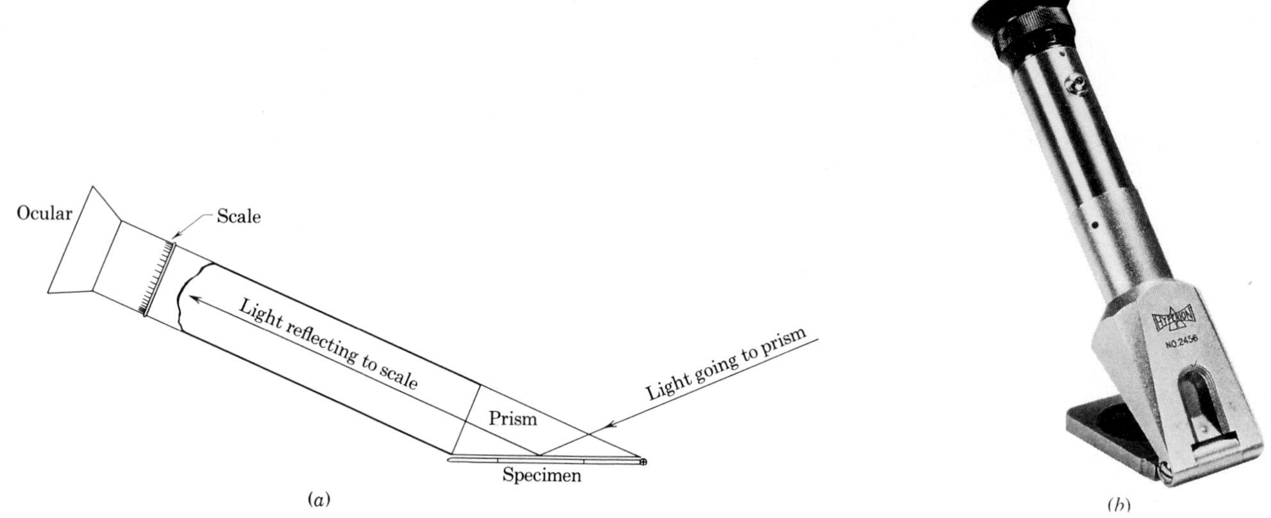

Fig. 24.27 (*a*) Cutaway diagram of juice refractometer. (*b*) Refractometer. (Kaufmann & Jost.)

containing a sugar solution, and finally through an analyzer P_2 to the eye E. With the tube T removed, we turn the analyzer about an axis joining S and E until no light passes through P_2. If the tube with the sugar solution is then placed between the two Polaroids, some light will once more be transmitted. This is true because the plane of vibration of the polarized light from P_1 has been rotated around the direction SE. The light can once more be extinguished by rotating the analyzer. The amount of rotation required for a given light source is proportional to the length of the tube T and proportional to the concentration of the solution. By adopting a standard length of the tube T, one can very readily determine the concentration of an optically active substance. This method is used quite extensively in determining the amount of sugar present in syrups. The instrument is called a *saccharimeter*. Grape growers use such a device to determine in the field the sugar content of their grapes (Fig. 24.27), and thus decide on the proper time to harvest them.

QUESTIONS AND EXERCISES

1. Will an increase in the temperature of a hot metal shift the brightness of its spectrum from the blue to the red region or from the red to the blue region? Give reasons for your answer.

2. Give two ways in which the speed of light in a vacuum differs from that in a transparent solid.

3. Which experiences a greater deviation when passing through a triangular glass prism, red light or blue light?

4. Is the spectrum of light emitted by a hot glowing solid, a continuous spectrum or a bright-line spectrum?

5. Which is hotter, a blue star or a yellow star?

6. Distinguish between spherical aberration and chromatic aberration. Describe methods for minimizing each.

7. A simple converging lens that exhibits chromatic aberration produces a series of images of different colors at different locations along the principal axis. Discuss how the magnification of the different color images depends on the color (or location) of the image. Which is larger, the red or the blue image of an object?

8. For which of the following is an achromatic lens designed? (*a*) To produce deviation without dispersion. (*b*) To produce dispersion without deviation. (*c*) To eliminate spherical aberration. (*d*) To produce a continuous spectrum.

9. Why is chromatic aberration not a problem with spherical or parabolic mirrors?

10. Is it possible to photograph a virtual image of an object? Explain your answer.

11. If you constructed a camera using a simple converging lens, what part of a photograph taken with the camera would you expect to be the sharpest—the center or the edges of the photograph?

12. An *f*/2.8 lens and an *f*/1.7 lens have the same focal length. Which lens has the greater diameter?

13. Which of the following will decrease blurring of an image in a camera? (*a*) Decrease *f*-stop of the lens. (*b*) Decrease the aperture diameter. (*c*) Increase the time of exposure. (*d*) Decrease the time of exposure.

14. What influence does the diameter of a camera lens have on the size of the real image projected on the film if all other dimensions remain the same?

15. Does the cornea or the lens of the eye cause greater bending of light entering the eye? Can you give an argument to support your answer?

16. Is a person nearsighted or farsighted if her eyeglasses use diverging lenses?

17. What is the name of the vision defect caused by a nonspherical cornea or lens?

18. The images on the retina of the eye are inverted. Why then do things not appear upside down to the observer?

19. Which of the following methods of inserting a slide in a common projector will produce an erect image? (*a*) Upside down with right and left not interchanged. (*b*) Upside down with right and left interchanged. (*c*) Right side up with right and left interchanged. (*d*) Right side up with right and left not interchanged.

20. Under what conditions can a simple magnifier produce a real image of an object?

21. Is the magnified image of an object that you can see by looking through a magnifying glass a real image or a virtual image?

22. Describe a simple experimental procedure for quickly obtaining an approximate value for the focal length of a magnifying glass.

23. Given the magnifying powers of the objective lens and the eyepiece, how is the magnification of a compound microscope determined?

24. The parabolic mirror in the world's largest reflecting telescope has a diameter of 236 in. Why is the diameter of the objective lens in the world's largest refracting telescope only 40 in.?

25. Although all the world's large astronomical telescopes are reflecting telescopes, very small telescopes are usually refracting telescopes. For example, although many manufacturers produce refracting telescopes with 2-in.-diameter objective lenses, no one manufactures a reflecting telescope with a 2-in.-diameter mirror. Can you explain why?

26. An amateur astronomer with a reflecting telescope that uses a very high quality 6-in.-diameter parabolic mirror claims to have used the telescope at a magnifying power of 420 to see a sharp, detailed image of the rings of Saturn. Do you believe this claim?

27. Look up the meaning of diffraction and then describe what happens when light is diffracted.

28. Plane-polarized light can be vertically polarized or horizontally polarized. Are there any other possibilities?

29. Describe two photographic applications of polarized light.

30. What do liquid crystal displays and saccharimeters have in common?

PROBLEMS

Group One

1. An equilateral prism with an apex angle A of 60° causes a beam of red laser light to be deviated by 48.68° when the angle of incidence is adjusted to produce minimum deviation. What is the index of refraction of the prism glass?

2. A 58° prism produces an angle of minimum deviation of 48° for a given color. Calculate the refractive index for that color.

3. The lens of a camera has an effective diameter of 32 mm and a focal length of 180 mm. What is the *f*-number?

4. The lens aperture for a portrait camera is 2.5 cm when the *f*-stop is 2.8. What is the focal length of the lens?

5. A certain 35-mm camera has a lens with an *f*-number of *f*/1.4. If the focal length of the lens is 58 mm, what must be the effective diameter of the lens at the *f*/1.4 setting?

6. The lens for a 35-mm slide projector has a focal length of 100 mm and an *f*-number of *f*/3.5. What is the diameter of the lens at that setting?

7. A converging reading glass has a focal length of 8.0 cm. What is the magnification if the lens is to produce a sharp image 25 cm from the eye?

8. The objective lens of a telescope has a focal length of 180 cm. The eyepiece has a focal length of 6.0 cm. What is the magnification of the telescope?

9. Under ideal atmospheric conditions, what is the approximate value for the maximum usable magnifying power for a refracting telescope whose objective lens has a diameter of 5 in.?

10. The objective lens of a refracting telescope has a focal length of 45 in. What should the focal length of the eyepiece be to produce a telescope with a magnification of 25?

11. The objective lens of a telescope has a focal length of 150 cm and the eyepiece has a focal length of 2.5 cm. Find the magnification of the telescope.

12. The focal length of the 200-in. reflecting telescope on Mount Palomar is 17 m. What focal length eyepiece (in millimeters) must be used with that telescope to give a magnifying power of 600?

Group Two

13. A beam of laser light of wavelength 632.8 nm experiences a minimum deviation of 38.93° as it passes through a prism with an apex angle A of 60°. Is the prism made of crown glass or flint glass?

14. A 62° prism has refractive indices of 1.572 for orange light and 1.597 for violet light. Find the angle of dispersion from the violet to the orange refracted rays.

15. A camera lens has a focal length of 40 mm. Calculate the image distances for the following object distances: (*a*) infinity, (*b*) 5 m, (*c*) 1 m, (*d*) 0.4 m.

16. A 35-mm camera, with its standard lens, can focus on objects at a distance from the lens between 2 ft and infinity. It is possible to photograph objects at distances closer than 2 ft from the camera by placing a simple converging lens in front of the camera's standard lens and photographing the more distant virtual image of the nearby object. What focal length converging lens is needed to produce a virtual image with an image distance of 2 ft if the object distance is 1 ft? (See Fig. 24.28.)

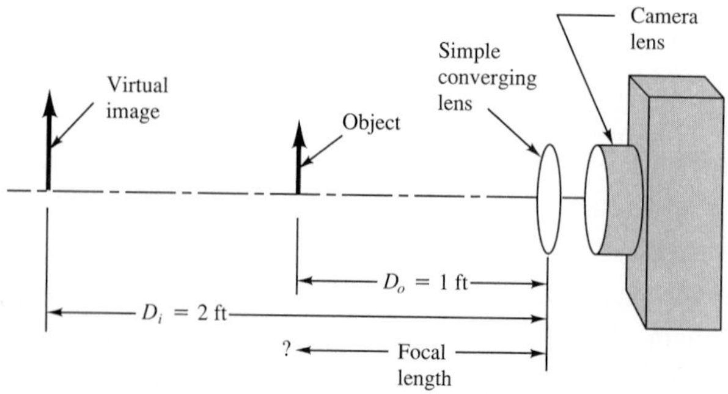

Fig. 24.28 Diagram for Problem 16.

17. An object is placed 4.0 cm from a concave lens of focal length −16.0 cm. Find (*a*) the image distance and (*b*) the magnification.

18. Where should the object be with respect to a +6.0-cm-focal length magnifier in order to produce a magnification of 8?

19. The total power of the cornea-lens combination in typical human eye is about 60 diopters. Consider a single glass converging lens in air with equal radii of curvature ($R_1 = R_2 = R$). If such a lens had a refractive index of 1.500, what value for R would make its power 60 diopters?

20. A nearsighted person needs spectacle lenses that will make an object 25 cm away appear to be 60 cm away. What focal length lenses are needed?

21. A nearsighted person's greatest range of clear vision is 80 cm. (*a*) What power spectacles must he wear to see distant objects clearly? (*b*) What would be the greatest distance he could see clearly if he wore spectacles with a power of −0.60 diopters?

22. A farsighted person cannot form sharp images of objects closer than 60.0 cm. (*a*) What power spectacles will enable him to see clearly objects 20.0 cm distant? (*b*) If he is fitted with lenses of 5.5-diopter power, what is the nearest distance which allows him to observe objects clearly?

23. The lens of a slide projector has a focal length of 15 cm. It is set 6 m from a screen. If the height of the slide is 22 mm, what is the height of the image on the screen?

24. A slide projector uses 22- by 32-mm slides and projects the slides upon a screen at a distance of 12 m from the projector. What must the focal length of the lens of the projector be to produce images with dimensions 1.54 by 2.24 m?

25. A double-convex lens, having a 7.5-cm focal length, is used as a simple microscope. If the lens is held close to the eye, where should the object be located in order to form an image 25 cm from the eye? What will be the magnification?

26. A microscope has an objective lens of focal length 1.25 cm and an eyepiece of focal length 3.60 cm. A specimen is placed 1.50 cm from the objective. What is the magnification if the virtual image formed by the eyepiece is 25 cm from the eye?

27. A pocket microscope has an objective lens with a focal length of 5.0 mm, and an eyepiece with a focal length of 20 mm. When the microscope is adjusted so that the real image formed by the objective lens is located at the focal point of the eyepiece, the distance between the two lenses is 8.0 cm. What is the distance of the object from the objective lens?

28. A telescope used for monitoring hazardous chemical reactions from a distance has an objective lens with a focal length of 30 cm and an eyepiece of power 50 diopters. What is the magnifying power of the telescope?

29. The area of the circular aperture of a certain lens is 0.125 cm² when the diaphragm is set for *f*/16. What will be the aperture area when the lens is set for *f*/4?

30. The area of the circular aperture of a certain lens is 1.250 in.² when the diaphragm is set for *f*/2. What will be the aperture area when the lens is set for *f*/22?

31. When a certain 35-mm camera lens is set with an *f*/4 stop, the correct exposure time is $\frac{1}{125}$ s. What exposure time should be used to get the same amount of light exposure with the lens diaphragm set for *f*/5.6?

32. A photographer has set her camera lens with an *f*/16 stop and a correct film-exposure time of $\frac{1}{30}$ s. What exposure time should she use to get the same amount of light exposure with the lens diaphragm set for *f*/2.8?

33. An amateur astronomer uses a reflecting telescope that has a 6-in.-diameter parabolic mirror of focal length 1200 mm. What is the approximate focal length (in millimeters) of the shortest-focal-length eyepiece capable of producing sharp images under ideal atmospheric conditions?

Group Three

34. A beam of red laser light passes through a triangular prism at minimum deviation. The apex angle *A* for the prism is 60°, and the refractive index of the prism glass is 1.625. What is the angle of minimum deviation δ for the laser beam?

35. A crown-glass lens with radii of +10 cm and +8 cm is combined with a flint-glass lens with one surface radius of −8 cm. What must be the radius of the second surface of the flint-glass lens to produce a lens that is achromatic for violet and red light?

36. A nearsighted person can see objects clearly only if they are not more than 75 cm distant. (*a*) What is the power of the weakest lens that will permit him to see distant ($D_o = \infty$) objects distinctly? (*b*) If he wears a pair of −1.25-diopter glasses, what will be his greatest distance of clear vision?

37. The vertical dimension of the image on a 35-mm color slide is 23 mm. The slide is to be projected so that the vertical dimension of the image equals the 92 cm height of the projection screen. If the distance from the projector lens to the screen is 8 m, and if the focal length of the projector lens is 100 mm, (*a*) what object distance (in millimeters) is required for the slide? (*b*) How far from the focal point of the lens must the slide be located?

38. The 200-in. Hale telescope on Mount Palomar has a 72-in.-diameter circular cross-section observer's cage located on its optical axis at the focal point of the mirror. What percent of the incoming light is blocked by the observer's cage?

39. The diameter of the circular aperture of a certain lens is 15 mm when the diaphragm is set for *f*/2.8. What will be the diameter of the aperture when the diaphragm is set for *f*/16?

40. A lens diaphragm setting of *f*/4 and a shutter speed of $\frac{1}{125}$ s give the correct film exposure for a certain camera. If the exposure time is increased to $\frac{1}{30}$ s, what new *f*-stop should be used to give the correct film exposure? Assume the available *f*-stops are *f*/1.4, *f*/2, *f*/2.8, *f*/4, *f*/8, *f*/11, and *f*/16.

PART 7

ELECTRICITY AND MAGNETISM

Turbine-driven alternator at an electric power generating plant. (Power Authority of the State of New York)

CHAPTER 25 ELECTROSTATICS

Most of the common applications of electric energy involve electric charges in motion. However, we will begin our discussion of electricity by studying *electrostatics*, the science of stationary electric charges. Historically, electric charges at rest were discovered long before the discovery of electric charges in motion. The principles that govern electric charges at rest underlie those for current electricity.

Today we find an increasing number of devices that are based upon the principles of electrostatics. Among them are high-energy atom-smashing accelerators, electrostatic precipitators for removing dust and smoke particles from the air, electronic copiers and printers, and devices for reducing static charges on rolls of printing paper.

25.1 ■ Electric Charges

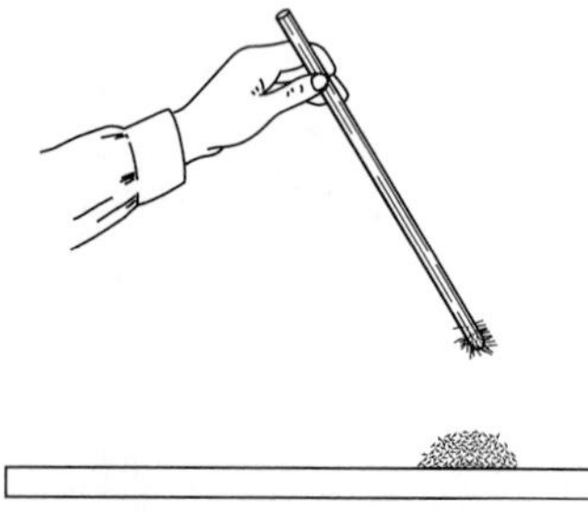

Fig. 25.1 Small particles of paper and straw will be attracted to a rod that has been electrified.

The discovery of electric charges is credited to Thales of ancient Greece (about 600 B.C.). He observed that amber, when rubbed with a cloth, would pick up small fibrous materials such as straw or feathers. We can easily perform like experiments to illustrate electrification. First rub a hard-rubber rod briskly with cat's fur. Then place the rod near small pieces of paper, dry grass, or aluminum foil. The particles will "jump" toward the rod and cling to it (Fig. 25.1). The experiment can be repeated with a glass rod rubbed with silk. The same effects will be noted.

Electrical phenomena, such as lightning and the northern lights (aurora borealis), have been known for thousands of years. At the beginning of the seventeenth century William Gilbert (1540–1603) of England announced that many substances could be electrified by friction or contact. Gilbert called these substances "electrics," which comes from the Greek word *elektron*, meaning "amber." Our modern terms *electron* and *electricity* are derived from this Greek word.

Static electric charges frequently develop when leather belts travel over iron pulleys. Severe electric shocks can occur when paper in a printing press comes off the rollers. Such charges are also developed when hair is combed on a dry day with a hard-rubber comb.

25.2 ■ Positive and Negative Charges

When objects are electrified by rubbing, sometimes there is attraction and sometimes repulsion of electric nearby objects.

There are two kinds of electric charge. To illustrate this, rub a hard-rubber rod with cat's fur. Suspend it with a silk thread. Next bring the tip of a similarly charged rubber rod near one end of the fixed rod. The two charged rods repel each other as shown in Fig. 25.2*a*. Repeat the experiment, using glass rods rubbed with silk. These rods, too, will repel each other (Fig. 25.2*b*). When a charged glass rod is brought near the tip of a charged rubber rod, however, the rods are attracted to each other

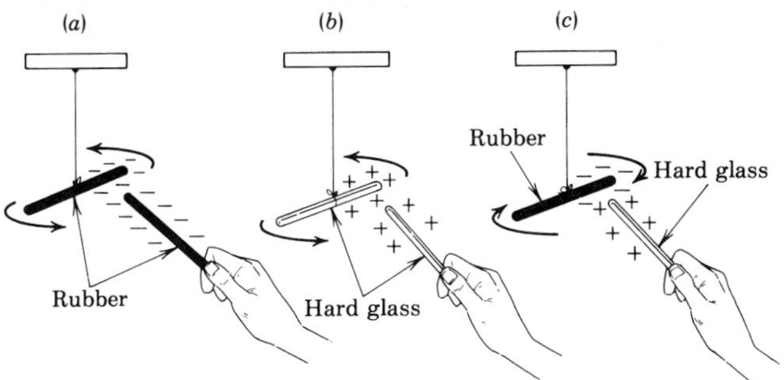

Fig. 25.2 Like charges repel at (*a*) and (*b*). Unlike charges attract at (*c*).

(Fig. 25.2*c*). Evidently, the kind of charge on the rubber rod must be different from that on the glass rod.

Scientists have conducted similar experiments with many kinds of materials but they have never found more than the two kinds of charge. Benjamin Franklin (1706–1790) arbitrarily introduced the terms *negative* and *positive* to designate these two kinds of electric charge. We apply the term *negative* to substances that behave like the electrified rubber rod and the term *positive* to those that behave like the electrified glass rod. Thus, we can deduce the first law of interaction between electrified bodies:

Benjamin Franklin was an American scientist, printer, and statesman. Among his many contributions to the science of his day are his fluid theory of electricity, lightning experiments, the use of bifocal lenses, telegraphy experiments, electrical chimes, and the harmonica.

Like electric charges repel each other, and unlike charges attract each other.

25.3 ■ Electrical Structure of Matter

In Chap. 10 we learned that matter is made up of *molecules*. Molecules, in turn, are built of more elemental particles, called *atoms*. Atoms, in general, have quite complex structures. We will use a simplified, idealized structure to explain the electrical nature of matter. The following discussion assumes a "planetary model" of the atom.

Basically, each atom has a small, dense, positively charged mass at its center, called the *nucleus*. This nucleus is surrounded by larger and much lighter negatively charged particles called *electrons*. The nucleus consists of one or more *protons*. Each proton has a single unit of positive charge. The nucleus (except that of hydrogen) also contains one or more *neutrons*. A neutron has a mass slightly (0.14 percent) greater than that of a proton. It has no charge. The electrons can be idealized as revolving about the positive nucleus, moving in elliptical orbits much as planets revolve around the sun. The proton and neutron have masses about 1840 times that of an electron. An electron has a rest mass of 9.11×10^{-31} kg.

More is known about the nature of invisible atoms and electrons and how they behave than of the nature and behavior of human beings.

Each electron carries a single negative charge. The proton carries a single positive charge of the same magnitude. The atom itself is mostly empty space. Almost all the mass of the atom resides in the small, but extremely dense, central nucleus. The weight density of a proton is of the order of 8 billion tons per cubic inch! In the planetary model, the electrons revolve about the nucleus at tremendous speeds. The centripetal forces needed to keep the electrons in their orbits are provided by the electrical attraction forces between the negative electrons and the positive nucleus (Fig. 25.3).

There are many particles, in addition to the electron, proton, and neutron, that make up matter. They will be introduced and discussed in Chap. 34.

Normally each atom exists in a neutral or uncharged state. It has an equal number of electrons and protons. The outermost electrons are less strongly bound to the atom than the inner ones are, and that is why it is the outer electrons that take part in chemical reactions between atoms. They are responsible for accumulation of charges on bodies. Atoms of different substances have varying degrees of affinity or attraction for outer electrons. (See Chap. 34 for further discussions on atomic structure.) For example, when hard rubber is rubbed with fur, the atoms of rubber have a greater attraction for electrons than the atoms of the fur do; hence they attract some of the

Fig. 25.3 Two-dimensional planetary model of a copper atom.

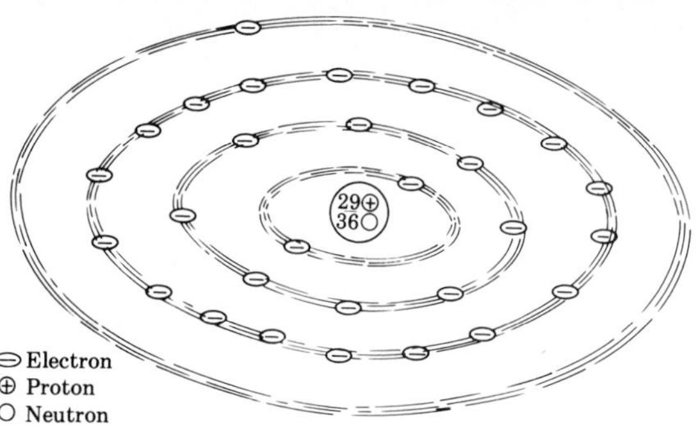

nearby electrons from the atoms of cat's fur. Since the rubber atoms now have a greater number of (negative) electrons than (positive) protons, the rubber rod will have acquired a net negative charge. The atoms of the fur, having lost electrons, will have acquired a resultant positive charge.

> ***A negatively charged body has an excess of electrons; a positively charged body has a deficiency of electrons.***

Electric charge is conserved. No electrons are created; they just move from one substance to another.

When a glass rod is rubbed with silk cloth, the glass will have a positive charge and the silk an equal negative charge. This happens because the silk fibers have a greater affinity for electrons than the glass does. When two bodies are rubbed together to produce a given charge on one of the bodies, *an equal but opposite charge* will be left on the other. This concept can be expressed as the *law of conservation of charge*, which states:

> ***The net amount of charge produced in any electrostatic process is zero.***

In solid materials, only some of the outer electrons may have freedom to move. Other electrons are firmly fixed in their orbits (unless they are bombarded by highly energized charged particles).

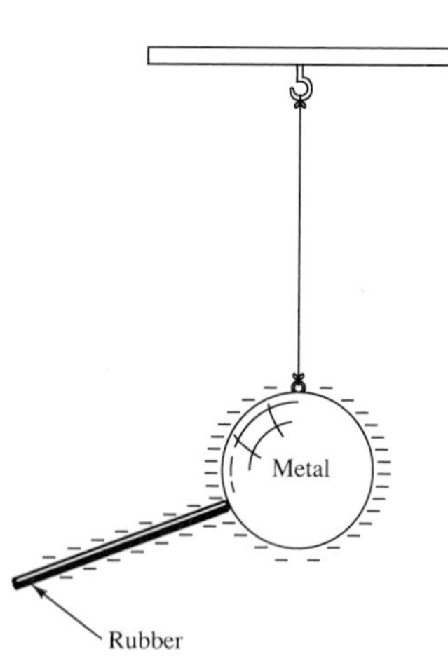

Fig. 25.4 The metal sphere is charged by contact with the charged rubber rod.

25.4 ■ Conductors and Insulators

Some substances can be charged by simply bringing a charged body in contact with them. For example, if a neutral metal sphere is suspended by a silk string and touched with a negatively charged rubber rod (Fig. 25.4), it can be shown that the sphere will acquire a negative charge. Some of the surplus electrons from the rubber will have left the rod and attached themselves to the sphere. If the uncharged sphere had been touched with a positively charged glass rod, the sphere would have become positively charged because some of its electrons would have been attracted away by the positively charged glass rod. This manner of charging a body is called *conduction*. Some substances will conduct electricity quite readily, while others will not. The metal in Fig. 25.4 will conduct the electricity quite readily, while the silk will not.

Those substances in which electrical charge flows quite freely are called *conductors*. Materials in which charges are not conducted freely are termed *insulators*. Insulators of porcelain or glass are used to prevent leakage of electricity from wires and electric equipment. There is no sharp line dividing conductors and insulators. Table 25.1 gives a few common substances arranged according to their insulating ability.

Table 25.1 Insulators and conductors

Insulators	*Poor Conductors*	*Conductors*
Hard rubber	Dry wood	Metals
Dry air	Paper	The earth
Paraffin	Oil	Moist materials
Porcelain	Distilled water	Water solutions of salts
Sulfur		The human body
Sealing wax		Graphite
Glass		
Dry silk		
Bakelite and similar plastics		

Substances which are the most easily charged by friction are all classified as insulators. When an electric charge is developed on an insulator, it remains localized and does not flow freely throughout its length and leak away.

We will discuss a third type of conductor in Chap. 34 when we consider transistors. Such materials are called *semiconductors*. Silicon and germanium are examples of semiconductors. Although they have few electrons available for moving electric charges, the conductivity can be greatly increased by the addition of certain impurities or (at room temperatures) by increasing the temperature.

(*a*)

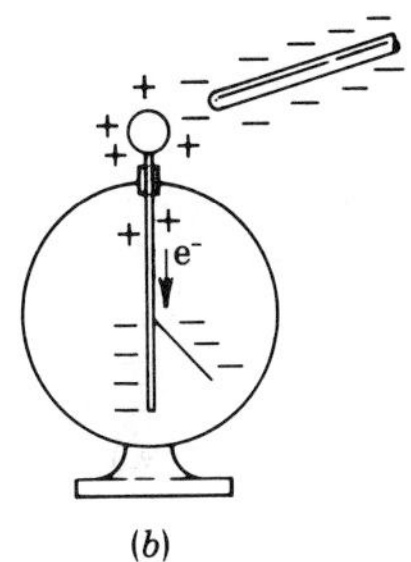

(*b*)

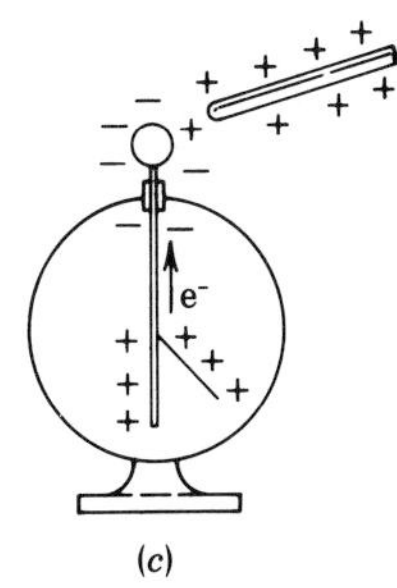

(*c*)

Fig. 25.5 The goldleaf electroscope is a sensitive instrument used to detect and identify small electric charges.

25.5 ■ The Electroscope

An instrument which can be used to detect the presence of an electric charge is called an *electroscope*. A common and very sensitive type is the *gold-leaf* electroscope. The essential features of the instrument are shown in Fig. 25.5*a*. It consists of a metal (copper or brass) rod that has a brass knob or plate attached at one end and one or two strips of aluminum or gold foil at the other. This rod passes vertically through the center of an insulating ring of sulfur or plastic at the top of the instrument. It is housed in a metal case with glass windows front and back.

When a charged rod is brought near the knob of the electroscope, the leaves will repel each other and diverge. When the charged rod is removed, the leaves will come together again. It will be noticed that *either a positive or a negative charge will cause the leaves to diverge*. When a negative charge is placed near, but not touching, the knob of the electroscope, some of the electrons (represented by e^- in the sketches) in the knob are repelled into the leaves. This makes both leaves predominantly negative (Fig. 25.5*b*) so that they repel each other. If a positive charge is placed near the knob, it will attract some of the electrons from the leaves to the knob. The result will be a predominant positive charge on the leaves, which again will cause them to repel each other (Fig. 25.5*c*). The greater the charge, the greater the divergence of the leaves; also, the closer the charging rod, the greater the divergence.

The electroscope (or any insulated conductor) can be charged by conduction. Touch the knob of an uncharged electroscope with a negatively charged rod (Fig. 25.6*a*). Electrons from the rod will enter the knob of the electroscope. In addition, the excess electrons left on parts of the rod that are not in contact with the knob will repel electrons down to the leaves. The leaves will diverge.

Remove the rod and with it the force of repulsion. The leaves will collapse slightly, but not completely. A residual charge of a somewhat lower density will be left on the electroscope.

The electroscope can be charged *positively* by using a positively charged glass rod in a similar manner (Fig. 25.6*b*).

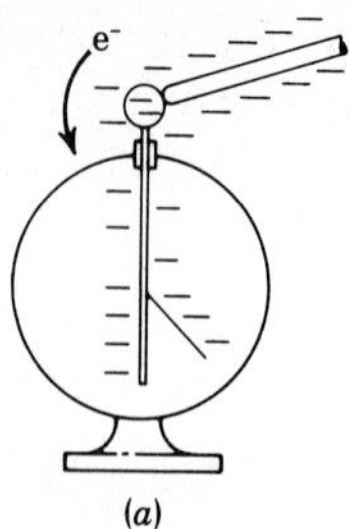

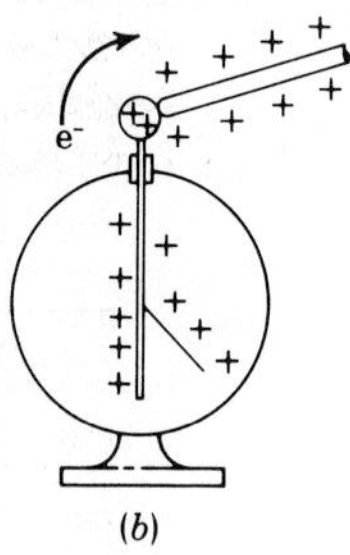

Fig. 25.6 Charging an electroscope by conduction.

> **Charge Acquired by Conduction**
>
> ***An insulated conductor charged by conduction acquires a residual charge of the same sign as that of the body touching it.***

25.6 ■ Electrostatic Induction

We have considered the electrification of insulated bodies by friction and by conduction. There is still another way of placing a charge on a body. In Figure 25.7 there are three brass spheres, each insulated by a glass support. Assume that sphere *A* has a positive charge, while spheres *B* and *C* are neutral or uncharged. If sphere *B* is brought near *A* (diagram *a*), the free electrons in *B* will be attracted to the surface nearest *A*. This will leave that side of *B* charged negatively. The other will have an equal positive charge on it. Now bring sphere *C* in contact with *B* (diagram *b*). Some of the electrons of *C* will be attracted by the positive side of sphere *B* and will pass on to sphere *B*. When the two spheres are separated (diagram *c*), *B* will have a net negative charge, and sphere *C* will have an equal net positive charge. Now remove sphere *A*. The charges on *B* and *C* (negative and positive, respectively), since the spheres are conductors, will distribute themselves uniformly as shown in diagram *d*. Spheres *B* and *C* are said to have been charged by *induction*. The presence of sphere *A* in close proximity has *induced* a negative charge on *B* and a positive charge on *C*. Sphere *A* has lost none of its charge in the process.

25.7 ■ Electrical Ground and Charging by Induction

In Fig. 25.7 charges were induced on two spheres, but a charge can also be induced on a single sphere. This can be done by using the earth as a reservoir for charge.

Moist earth is a good conductor of electric charges.

The earth is a large massive object and hence contains an extremely large number of atoms with free electrons. Electrons can flow from or to the earth without appreciably changing the earth's relatively uncharged state. When a charged object is connected to the earth by a conductor, it is said to be grounded. *Grounding* is represented schematically by the symbol: ⏚

Charged objects when grounded to the earth will cause electrons to flow between the earth and the object in a direction which tends to neutralize the charge.

When a negatively charged rod is brought near an uncharged electroscope, electrons in the knob will be repelled to the leaves of the electroscope. The excess electrons will cause the leaves to diverge (Fig. 25.8*a*), and a deficiency of electrons will be left on the knob of the electroscope. Upon grounding the electroscope, free electrons will escape to the ground, and the leaves will converge (Fig. 25.8*b*). When the ground is removed, the positive charge on the knob will be held by the mutual attraction with the negatively charged rod (Fig. 25.8*c*), and the leaves will remain converged. Removal of the rod will then permit an even distribution of the charges in the electron-deficient electroscope (Fig. 25.8*d*), and the leaves will once more diverge. The residual charge on the electroscope will be positive. Note that in the process, the negatively charged rod loses none of its charge.

The electroscope may also be given a net negative charge by a similar process using a positively charged object in place of the negatively charged rod.

> **Charge Acquired by Induction**
>
> ***The induced charge on a single object has a charge opposite in sign to that of the charging body.***

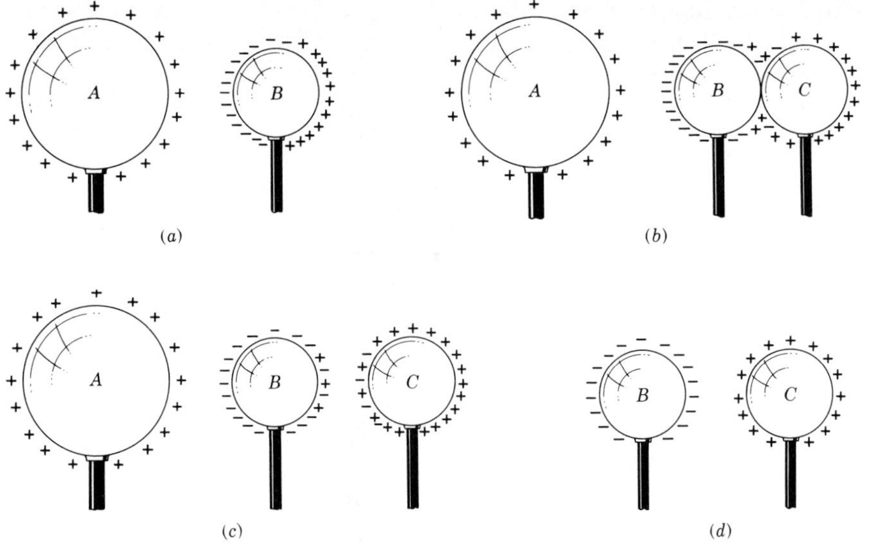

Fig. 25.7 (*a*) The positively charged sphere A when brought near a second isolated conducting sphere B induces charges of the same sign in the far end of the sphere. (*b*), (*c*), and (*d*) Spheres B and C are charged by induction.

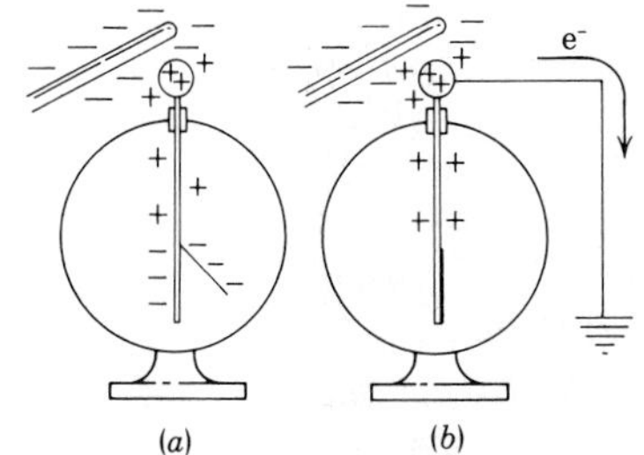

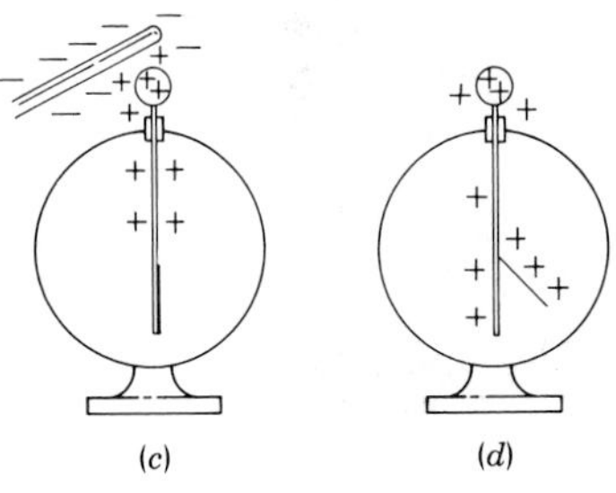

Fig. 25.8 Steps in placing a charge on an electroscope by induction.

25.8 ■ Distribution of Charge on Conductors

The distribution of charges on conductors of various shapes can be determined by means of a small metal disk or ball mounted on an insulating handle (Fig. 25.9) and known as a *proof plane*. Consider metallic surfaces of different shapes, each with a negative charge on them (Fig. 25.10). If we touch the proof plane to the surface of the sphere (*a*) in several places and test the intensity of the charge by touching the proof plane to an electroscope, the charge on the sphere will be found to be uniformly distributed. When the same test is applied to surfaces like those shown in diagrams (*b*) and (*c*), it will be found that the charge density is greatest in the regions of greatest curvature. If the curvature is sharp enough, the surface density of the charge may be so great that the charge will actually leak off the sharp surface into the air. Such a leakage, called a *corona discharge*, is often observed in nature and in experiments conducted in electrical laboratories. It is used in electrostatic precipitators and copiers. The air in the vicinity of corona discharges is said to be *ionized*.

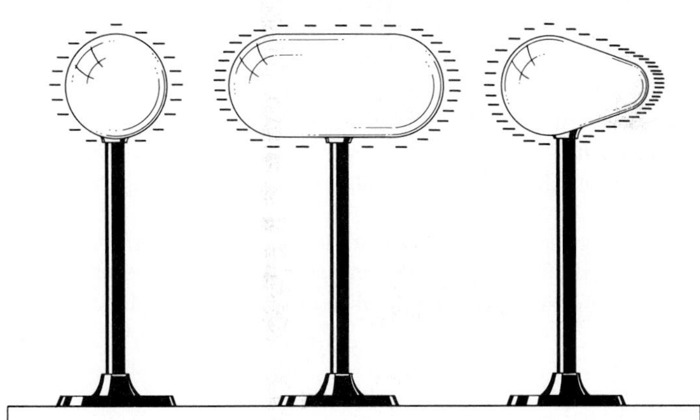

Fig. 25.10 Distribution of electric charge on various metallic surfaces. The charge density is greatest at the point of greatest curvature.

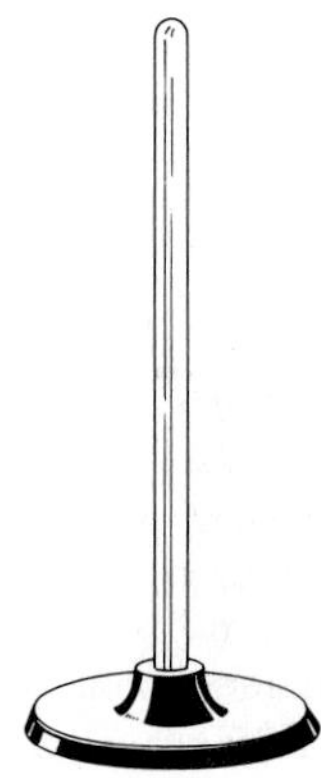

Fig. 25.9 A proof plane can be used to remove a small charge from another charged body.

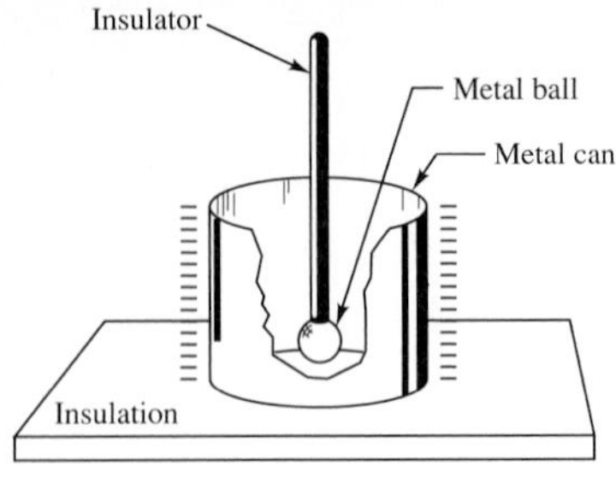

Fig. 25.11 Electric charges exist only on the outside of the metal can.

An atom or molecule that carries a net charge is called an *ion*. In a corona discharge a molecule of gas becomes a positive ion when it loses one or more of its electrons; if it gains one or more electrons in addition to its normal complement, it becomes a negative ion.

On testing the intensity of charge at various places on a *hollow* metal can which has had a charge placed on it (Fig. 25.11), it will be found that *the charge exists entirely on the outer surface*. An electroscope placed in a metal cage which has a large charge on it will show that no charge exists on the inside. This suggests that a safe place in case of an electrical storm might well be inside an automobile. Even though lightning might possibly hit the car, the charge is not likely to reach the occupants inside. (It is very unlikely that an auto or an airplane would be struck in the first place, for neither is grounded, and lightning takes the easiest path to the ground.) Protecting a region from electric fields in this manner is called *electrostatic shielding*.

25.9 ■ Coulomb's Law

Charles Coulomb was a French physicist. He founded the mathematical theory of the interaction of electric charges.

Charles Augustin de Coulomb (1736–1806) in 1785 was the first to measure electrical attractions and repulsions quantitatively. As a result of his experimentation, he deduced the law that governs these forces. It is called

Coulomb's Law

The force between two stationary point charges is directly proportional to the product of their charges and inversely proportional to the square of the distance between them.

Coulomb's law can be expressed algebraically as

$$F = k\frac{Q_1 Q_2}{r^2} \tag{25.1}$$

where F = magnitude of the force experienced by each of the two charges
Q_1, Q_2 = magnitudes of charges
r = distance between charges
k = constant of proportionality

The value of k depends upon the nature of the medium and the units used for the several quantities. Equation (25.1) assumes that the dimensions of the charged objects are small (point charges, actually) compared with the distance r between the bodies.

The restriction to stationary charges is made, for the present, in order to exclude magnetic forces that arise from moving charges, which we will study in Chap. 26.

The coulomb is a quantity of electric charge equal to 6.242×10^{18} electrons.

Coulomb's law resembles the inverse square law of gravitation (see Eq. 8.19). In gravity, however, the forces are always attractive. In the law of gravitation we are dealing with only one kind of *mass*, but in Coulomb's law there are two kinds of *charges* to be considered.

Electrostatic forces between charged bodies are much greater than gravitational forces between the bodies.

The unit of charge used in the SI system is called the *coulomb* (C). The formal definition of the coulomb will be given in Sec. 28.22, after we have studied electric currents. At present, we can give its value in terms of the electron charge:

$$1\ \text{C} = 6.242 \times 10^{18}\ \text{electrons}$$

Therefore,

$$1\ \text{electron charge} = 1.602 \times 10^{-19}\ \text{C}$$

Similarly, the charge on a proton is 1.602×10^{-19} C. The sign of Q is + for a proton, − for an electron.

Table 25.2 presents some data on the charge and mass of atomic particles.

Table 25.2 Charge and Mass of the Electron, Proton, and Neutron

Particle	*Symbol*	*Charge, C*	*Mass, kg*
Electron	e^-	$-e = 1.602 \times 10^{-19}$	9.110×10^{-31}
Proton	p	$+e = 1.602 \times 10^{-19}$	1.673×10^{-27}
Neutron	n	0 0	1.675×10^{-27}

If Q_1 and Q_2 are measured in coulombs, and if the distance r is measured in meters and the force in newtons, the value of k for free space (vacuum) in Eq. (25.1) has been found by experiment to be

$$k_0 = 8.988 \times 10^9 \text{ N} \cdot \text{m}^2/\text{C}^2$$

For most calculations involving charges in vacuum or in air, k may be rounded off to 9.0×10^9. The value of k for air is only 0.06 percent less than that for a vacuum. Note that $8.988 \times 10^9 \text{m}^2/\text{s}^2 = 10^{-7}\ c^2$, where c is the speed of light in free space in meters per second.

One coulomb is that point charge which repels a like charge 1 m away in a vacuum with a force of approximately 9.0 × 10⁹ N.

The force expressed by Coulomb's law is a *vector* quantity. Equation (25.1) gives only the *magnitude* of the force. The *direction* of the force is along the line joining Q_1 and Q_2. The plus and minus signs affixed to the charges have no mathematical meaning in Eq. (25.1). A positive value for F indicates a repulsive force between like charges; a negative value of F indicates an attractive force between unlike charges.

Illustrative Problem 25.1 Two insulated small objects have charges of 1.0 C and −2.0 C and are 50 cm apart. What will be the electrostatic force acting on each object?

Solution Given:

$$Q_1 = 1.0 \text{ C} \qquad r = 50 \text{ cm} = 0.50 \text{ m}$$
$$Q_2 = -2.0 \text{ C} \qquad k = 9.0 \times 10^9 \text{ N} \cdot \text{m}^2/\text{C}^2$$

To find F, use Eq. (25.1)

$$F = k\frac{Q_1Q_2}{r^2}$$
$$= 9.0 \times 10^9 \text{ N} \cdot \text{m}^2/\text{C}^2 \times \frac{1.0 \text{ C} \times (-2.0 \text{ C})}{(0.50 \text{ m})^2}$$
$$= -7.2 \times 10^{10} \text{ N} \qquad \textit{answer}$$

The negative sign indicates an attractive force between the charges.

The answer to Illustrative Problem 25.1 is more than 16 billion pounds of force! Obviously, the coulomb is a very large unit of charge, much too large for most electrostatic situations. The highest charges that can be produced on bodies seldom contain more than a very small fraction of a coulomb. Frequently it is more convenient to work with the unit called the microcoulomb (μC):

$$1\ \mu\text{C} = 10^{-6}\ \text{C}$$

When two or more forces act on a charge, the resultant force is the vector sum of the separate forces.

Illustrative Problem 25.2 Two very small spheres, each weighing 10^{-4} kg, are supported from the same point by silk strings (assumed weightless) of length 1.2 m. Each carries a positive charge Q. The mutual repulsion of the charges forces the balls apart so that each string makes an angle of 8° with the vertical (see Fig. 25.12*a*). What is the charge Q?

Solution

$$\begin{aligned} x &= 1.2\ \text{m} \times \sin 8° \\ &= 1.2\ \text{m}\ (0.1392) \\ &= 0.167\ \text{m} \end{aligned}$$

The forces acting on each sphere are the weight mg, the electric force F_e, the force of gravitational attraction between the spheres F_g, and the tension force T in the string (see Fig. 25.12*b*).

For equilibrium, $\Sigma F_x = 0$ and $\Sigma F_y = 0$

$$\begin{aligned} \Sigma F_x &= F_g - F_e + T \sin\theta = 0 \\ \Sigma F_y &= T \cos\theta - mg = 0 \end{aligned}$$

From which we get

$$T \sin\theta = F_e - F_g \qquad (1)$$
$$T \cos\theta = mg \qquad (2)$$

Dividing Eq. (1) by Eq. (2) gives

$$\tan\theta = \frac{F_e - F_g}{mg}$$

or

$$F_e = F_g + mg \tan\theta$$

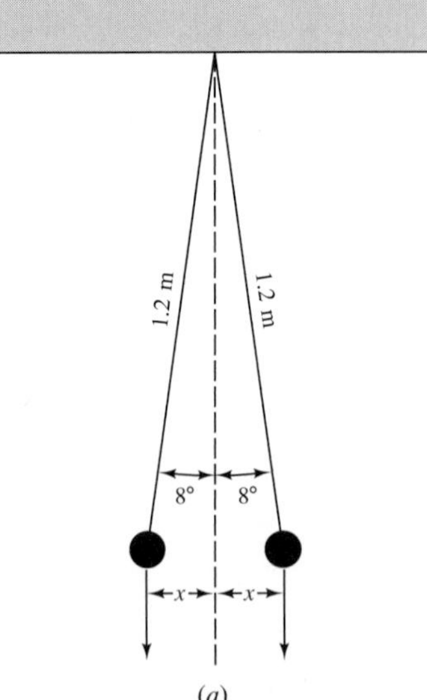

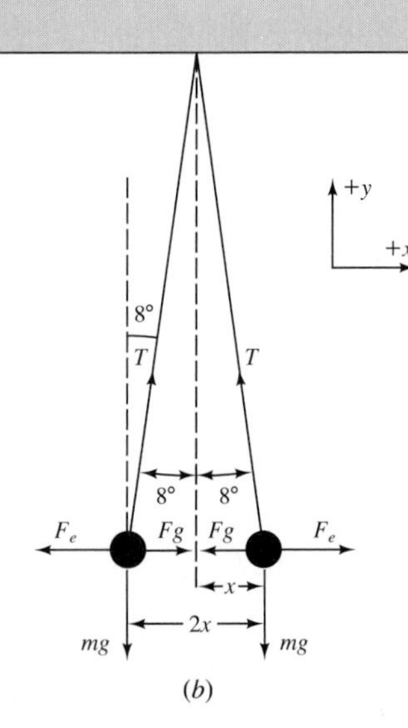

Fig. 25.12 Illustrative Problem 25.2.

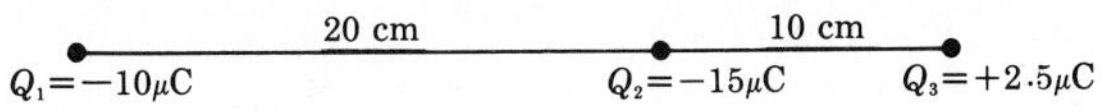

20 cm 10 cm

$Q_1 = -10\mu C$ $Q_2 = -15\mu C$ $Q_3 = +2.5\mu C$

Fig. 25.13 Illustrative Problem 25.3.

We know that

$$F_e = \frac{kQ^2}{4x^2} \quad \text{and} \quad F_g = \frac{Gm^2}{4x^2}$$

Then

$$\frac{kQ^2}{4x^2} = \frac{Gm^2}{4x^2} + mg \tan \theta$$

Multiply each term by $4x^2/k$ to get

$$Q^2 = \frac{Gm^2 + 4x^2\ mg \tan \theta}{k} \tag{3}$$

Substitute known values in Eq. (3)

$$Q^2 = \frac{6.67 \times 10^{-11}\ \text{N} \cdot \text{m/kg}^2 \times (10^{-4}\ \text{kg})^2 + 4(0.167\ \text{m})^2 (10^{-4}\ \text{kg})(9.81\ \text{m/s}^2)(0.141)}{8.988 \times 10^9\ \text{N} \cdot \text{m}^2/\text{C}^2}$$

$$= \frac{6.67 \times 10^{-19} + 1.543 \times 10^{-5}}{8.988 \times 10^9}\ \text{C}^2$$

$$\text{Q} = \sqrt{17.167 \times 10^{-16}\ \text{C}^2}$$
$$= 4.14 \times 10^{-8}\ \text{C} \qquad \textit{answer}$$

Note that F_g is too small to be significant.

Illustrative Problem 25.3 A charge of $-10\ \mu$C is located 30.0 cm from a charge of $+2.5\ \mu$C. A charge of $-15\ \mu$C is placed on the line joining the two charges. It is 20 cm from the charge of $-10\ \mu$C (Fig. 25.13). What is the force on the $-15\ \mu$C-charge?

Solution The force F_1 on Q_2 due to Q_1 will be

$$F_1 = \frac{kQ_1Q_2}{r^2} = 9.0 \times 10^9\ \frac{\text{N} \cdot \text{m}^2}{\text{C}^2} \times \frac{(-10 \times 10^{-6}\ \text{C}) \times (-15 \times 10^{-6}\ \text{C})}{(0.20\ \text{m})^2}$$

$= 33.75$ N (Q_2 is repelled to the right)

The force F_2 on Q_2 due to Q_3 will be

$$F_2 = 9.0 \times 10^9 \frac{\text{N} \cdot \text{m}^2}{\text{C}^2} \times \frac{(-15 \times 10^{-6}\ \text{C}) \times (2.5 \times 10^{-6}\ \text{C})}{(0.10\ \text{m})^2}$$

$= -37.5$ N (Q_2 is attracted to the right)

$$F = F_1 + F_2$$
$$= 33.75\ \text{N} + 37.5\ \text{N}$$
$$= 71\ \text{N directed to the right} \qquad \textit{answer}$$

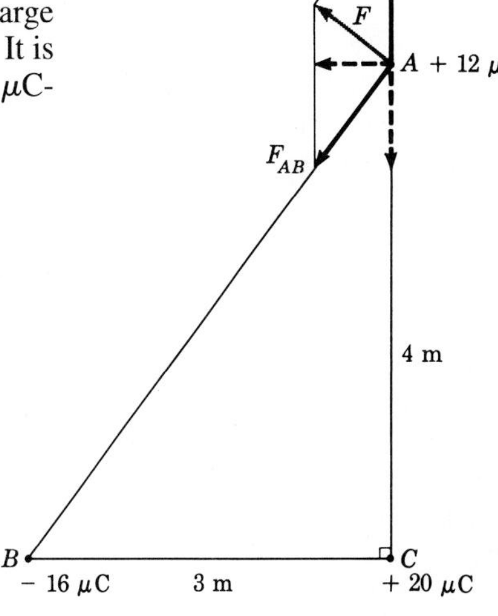

Fig. 25.14 Illustrative Problem 25.4.

Illustrative Problem 25.4 Charges A, B, and C of $+12$, -16, and $+20\ \mu$C, respectively, are arranged as shown in Fig. 25.14. Find the magnitude of the force on charge A. Angle C is a right angle.

Solution Given

$$BC = 3 \text{ m} \qquad AC = 4 \text{ m}$$
$$Q_A = +12\ \mu\text{C} \qquad Q_B = -16\ \mu\text{C}$$
$$Q_C = +20\ \mu\text{C} \qquad \angle C = \text{right angle}$$

Use the Pythagorean theorem to find

$$AB = \sqrt{(3 \text{ m})^2 + (4 \text{ m})^2} = 5 \text{ m}$$

$$F_{AB} = 9.0 \times 10^9 \text{ N} \cdot \text{m}^2/\text{C}^2 \times \frac{(12 \times 10^{-6} \text{ C})(-16 \times 10^{-6} \text{ C})}{(5 \text{ m})^2}$$

$$= 0.0691 \text{ N}$$

$$F_{AC} = 9.0 \times 10^9 \text{ N} \cdot \text{m}^2/\text{C}^2 \times \frac{(12 \times 10^{-6} \text{ C})(20 \times 10^{-6} \text{ C})}{(4 \text{ m})^2}$$

$$= 0.135 \text{ N}$$
$$(F_{AB})_x = \tfrac{3}{5}F_{AB} = -0.0415 \text{ N}$$
$$(F_{AB})_y = \tfrac{4}{5}F_{AB} = -0.0553 \text{ N}$$
$$F_x = (F_{AB})_x = -0.0415 \text{ N}$$
$$F_y = F_{AC} + (F_{AB})_y = 0.135 \text{ N} - 0.0553 \text{ N}$$
$$= 0.0797 \text{ N}$$
$$F = \sqrt{(F_x)^2 + (F_y)^2}$$
$$= \sqrt{(0.0415 \text{ N})^2 + (0.0797 \text{ N})^2}$$
$$= 0.090 \text{ N}$$ *answer*

Permittivity may be considered as a measure of the medium surrounding two electric charges which determines the force between the charges.

Scientists have found it convenient to introduce another constant ϵ_0 (epsilon) where $\epsilon_0 = 8.854 \times 10^{-12}$ C^2/N $\cdot$ m^2 and is called the *permittivity of free space* (a vacuum). The constant k_0 and the constant ϵ_0 are related by the equation

$$k_0 = \frac{1}{4\pi\epsilon_0} \tag{25.2}$$

While replacing k_0 by $1/4\pi\epsilon_0$ seems only to complicate Eq. (25.1), it will lead to simpler expressions for capacitances to be studied later in this chapter. An alternative expression for Coulomb's law is:

$$F = \frac{1}{4\pi\epsilon_0} \times \frac{Q_1 Q_2}{r^2} \tag{25.3}$$

for free space.

25.10 ■ The Electric Field

Michael Faraday was an English physicist and chemist. He is best known for discovering the ion and for his work on magnetic induction (which will be covered in Chap. 29).

The notion of forces acting across empty space has puzzled scientists for many years. The effect of gravitational forces acting on bodies in space was studied in Chap. 4. In the study of electrostatics and magnetism we are again faced with examples of forces "acting at a distance".

Let Fig. 25.15*a* represent two positively charged bodies *A* and *B*. We know there

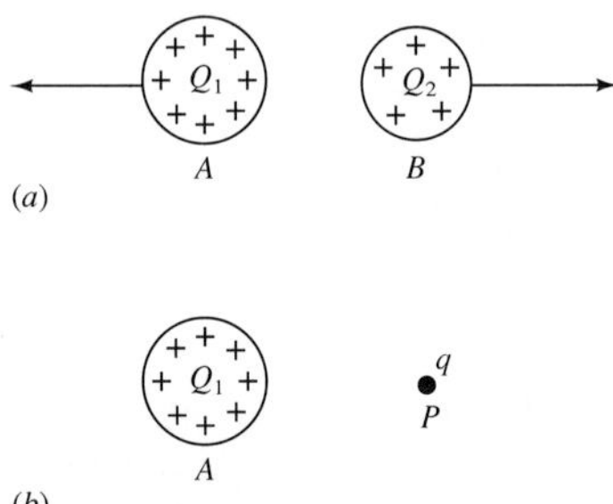

Fig. 25.15 An electric field exists around a charged body.

will be an electrical force of repulsion between A and B. These forces exist even though the charged objects do not touch. In order to explain why the charged bodies behave this way the idea of an *electric field* was developed by the British scientist Michael Faraday (1791–1867). He conceived of an electric field extending outward from every charge and permeating all of space. Thus, each charge modifies the space around it from what it would be if there were no charge present.

When a charged body is placed in an existing electric field, the force it experiences is ascribed to the field itself, rather than to any direct action of the field-producing charges. The electric field produced by body A at the location of B interacts directly with the charge Q_2 on B and, in like manner, the electric field produced by body B interacts directly with charge Q_1 on A. It must be emphasized, however, that an electric field is not a kind of matter, but just a convenient concept which aids in explaining electrostatic phenomena.

Since a force would be experienced by body B at all points around body A, we know that the whole space around A is an electric field.

The electric field surrounding a charged body can be investigated by placing a small positive *test charge* q in the field (Fig. 25.15*b*). The test charge must be so small that its field does not significantly alter the field being investigated.

An electric field is said to exist at a point if a small positive charge placed at that point experiences an electric force on it.

A unit positive charge is an imaginary charge used to aid in defining field intensity.

25.11 ■ Electric-Field Intensity

The electric field **E** has magnitude and direction. Thus it is a vector quantity. The *electric-field intensity* (or *strength*) E at any point is defined as the force per unit positive charge acting on a charged body at that point:

$$\mathbf{E} = \frac{\mathbf{F}}{Q} \tag{25.4}$$

The SI unit of field intensity is the *newton per coulomb* (N/C).

Direction of Electric Field

The direction of the electric field at a point is defined as the direction of the force on a positive charge placed at that point.

E is a vector quantity that has the direction of the force experienced by a positive charge.

The *magnitude* E of the intensity of the electric field at a distance r from an isolated point charge Q can be found by using Eqs. (25.3) and (25.4). If we let the test charge be q, Eq. (25.3) becomes

$$F = \frac{1}{4\pi\epsilon_0} \times \frac{qQ}{r^2}$$

Substituting this value for F in Eq. (25.4), gives

$$E = \frac{(1/4\pi\epsilon_0)\,(qQ/r^2)}{q} = \frac{1}{4\pi\epsilon_0} \times \frac{Q}{r^2} \tag{25.5}$$

Electric fields set up by several charges superpose to form a single net field. The vector specifying the field intensity E at any point in such a field is the vector sum of the fields due to each charge taken separately.

Illustrative Problem 25.5 Find the magnitude of the electric field intensity at *A* in Fig. 25.14 due to the charges at *B* and *C*.

Solution In Problem 25.4 we found the magnitude of the force acting on the charge of $+12\ \mu\text{C}$ at *A* to be 0.090 N. Then by Eq. (25.4),

$$E = \frac{F}{Q} = \frac{0.090\ \text{N}}{+12\ \mu\text{C}} = 7500\ \text{N/C} \qquad \textit{answer}$$

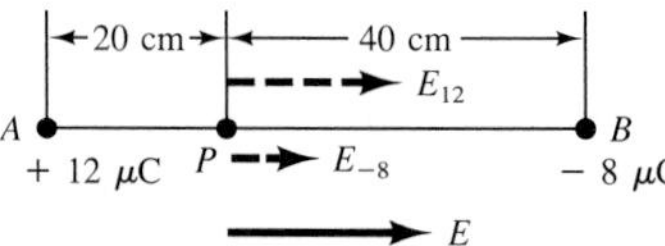

Fig. 25.16 Illustrative Problem 25.6.

Illustrative Problem 25.6 In Fig. 25.16 the charge of body *A* is $+12\ \mu\text{C}$ and the charge of body *B* is $-8\ \mu\text{C}$. Find the electric field intensity at point *P*. Assume free space.

Solution The field intensity at *P* due to the $+12\ \mu\text{C}$ charge has a magnitude of

$$E_{12} = \frac{1}{4\pi\epsilon_0} \times \frac{Q_{12}}{r^2}$$

$$= 9.0 \times 10^9\ \text{N}\cdot\text{m}^2/\text{C}^2 \times \frac{12 \times 10^{-6}\ \text{C}}{(0.20\ \text{m})^2}$$

$$= 27 \times 10^5\ \text{N/C}$$

Similarly,

$$E_{-8} = 9.0 \times 10^9\ \text{N}\cdot\text{m}^2/\text{C}^2 \times \frac{8 \times 10^{-6}\ \text{C}}{(0.40\ \text{m})^2}$$

$$= 4.5 \times 10^5\ \text{N/C}$$

Consider a positive unit charge at *P*. The fields due to both charges will be to the right. Therefore the magnitude of the total field intensity *E* at *P* will be $27 \times 10^5 + 4.5 \times 10^5 = 31.5 \times 10^5$ N/C. It is directed toward the right. *answer*

Illustrative Problem 25.7 Use Fig. 25.16 of the previous problem and the same data, except change the charge at *B* to $+8\ \mu\text{C}$. Where in the neighborhood of the two positive charges is the electric field zero?

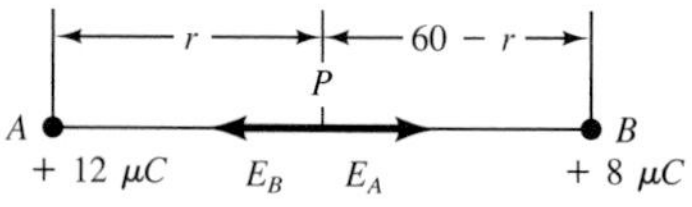

Fig. 25.17 Illustrative Problem 25.7.

Solution Figure 25.17 represents the fields due to the two positive charges. Let *P* be the point on the line joining *A* and *B* where the resultant of the two fields is zero. The direction of the field E_A of *A* is opposite to the field E_B of *B*. By convention, E_A will be to the right in the figure and E_B will be to the left. Point *P* will be located where $E_A = E_B$.

$$k\frac{Q_A}{r^2} = k\frac{Q_B}{(60 - r)^2}$$

Divide both sides of the equation by *k* and solve for *r*

$$\frac{(60\ \text{cm} - r)^2}{r^2} = \frac{Q_B}{Q_A}$$

$$\frac{60\ \text{cm} - r}{r} = \sqrt{\frac{Q_B}{Q_A}} = \sqrt{\frac{8\ \mu\text{C}}{12\ \mu\text{C}}} = 0.816$$

$$0.816\, r = 60\ \text{cm} - r$$

$$r = 33\ \text{cm from } A \text{ along line } AB \qquad \textit{answer}$$

If the charges on *A* and *B* were negative rather than positive, the point where $E = 0$ would be the same. But, if one charge were negative and the other positive, the point where $E = 0$ would be on the extension of the line joining the two bodies and beyond the smaller charge.

25.12 ■ Electric Lines of Force

The electric field near one or more charged bodies can be represented by drawing "lines of force" as an aid in visualizing the field. The relationship between the (imaginary) lines of force and the electric field intensity vector is this: (1) the direction of the line of force at any point is that in which a positive charge would move if placed at that point, and (2) the lines of force are drawn so that the number of lines per unit cross-sectional area is proportional to the magnitude of **E**.

Figure 25.18*a* shows the lines of force near a negatively charged sphere, and Fig. 25.18*b* shows the lines of force near a positively charged sphere. The lines of force lie along radii of the spheres. By convention, the lines of force leave the surface of a positively charged body and terminate at the surface of a negatively charged body. A line of force is normal to the surface of the charged body where it joins that surface. The lines of force do not enter the charged spheres and no two field lines can cross. We must remember that the two-dimensional drawings show only the field lines that lie in the plane of the page. The lines of Fig. 25.18*a* are actually directed radially outward from the charge in *all* directions in space, while the lines of Fig. 25.18*b* are directed radially inward in all directions to the charge. For a single charge the lines are considered as extending to infinity. You are reminded again, however, that the electric lines of force are not material objects but merely a convenient way to provide a quantitative description of an electric field. Note how the magnitude E of the electric field intensity is not constant but decreases with increasing distance from the charge. This is shown by the lines being spaced farther apart at greater distances. From symmetry, we know that the magnitude E is the same for all points that are a given distance from the center of the charge.

Figure 25.19 shows the lines of force around two equal like charges and two equal unlike charges. The electric field for two point charges of equal magnitude but opposite signs (Fig. 25.19*a*) will show the number of lines that begin at the positive charge

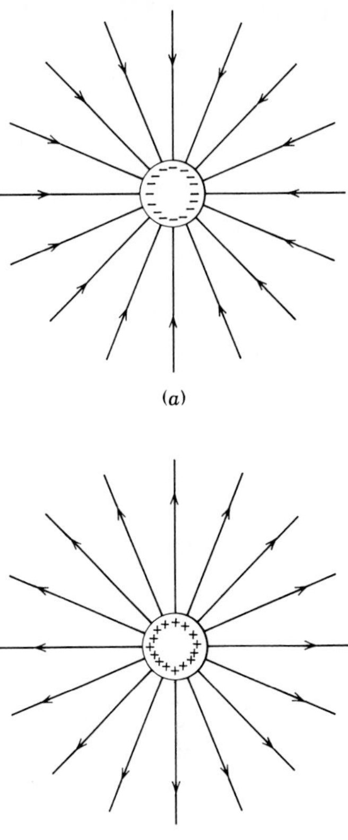

Fig. 25.18 Lines of force around a single charged sphere.

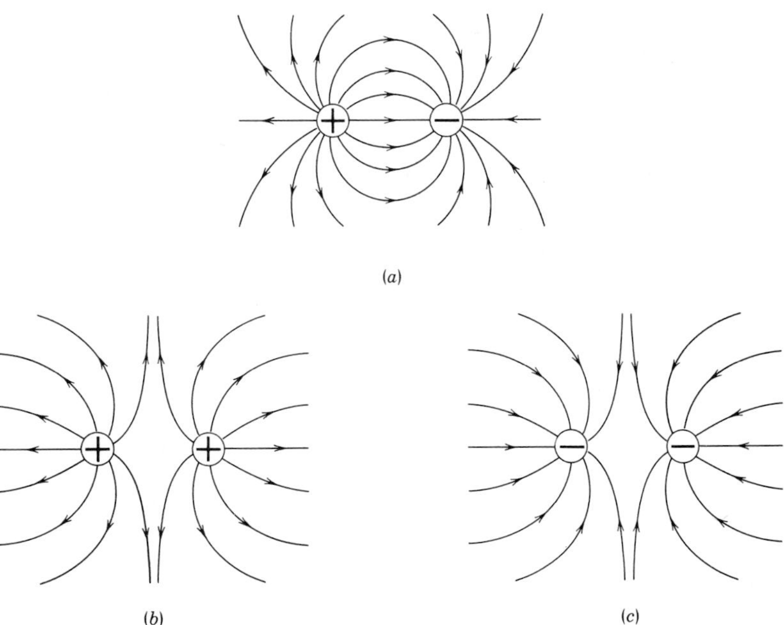

Fig. 25.19 Electric field lines between (*a*) two unlike charges, (*b*) two like positive charges, (*c*) two like negative charges.

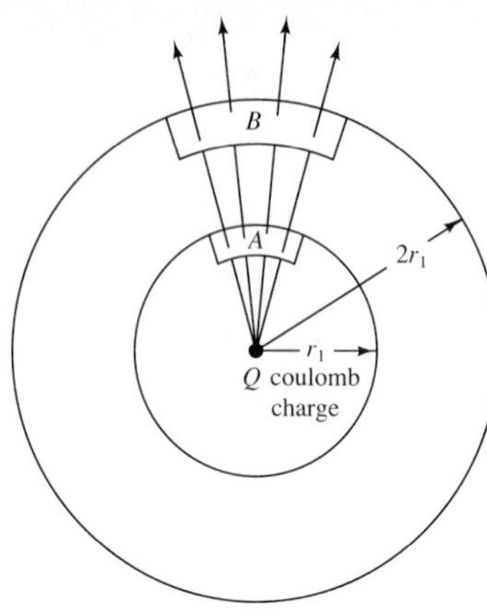

Fig. 25.20 The electric field intensity is proportional to the number of lines per unit area.

equal to the number that terminate at the negative charge. Thus, one can readily visualize the attractive nature of the force between the two charges.

Again, if the point charges have the same signs and equal magnitudes (Fig. 25.19*b*,*c*), the lines emerge (or enter) from each charge and are nearly radial at points close to the charges. The lines are considered as extending to infinity. Here, too, one can visualize the repulsive nature of the electric forces between like charges.

25.13 ■ Lines of Force and Field Intensity

Consider the field (lines of force) directed radially outward from a positive point charge Q (Fig. 25.20). Let N equal the number of lines drawn to represent the field. Next imagine a spherical surface surrounding the point charge at a distance r from the charge. Whatever number of lines of force we choose to represent the electric field around a point charge, the same number must pass through *any* spherical surface around the charge.

We will define electric *flux density E* as the number of lines per unit area crossing a surface at right angles to the direction of the field. Let A be the surface area of the sphere. Then

$$N = EA \tag{25.6}$$

the magnitude of the electric field intensity near the point charge Q is

$$E = \frac{Q}{4\pi\epsilon_0 r^2}$$

The net flux passing out of any closed surface is equal to the net charge enclosed by the surface divided by ϵ_0.

Recall the formula for the surface area of a sphere, $A = 4\pi r^2$, and substitute values in Eq. (25.6) to get

$$N = \frac{Q}{4\pi\epsilon_0 r^2} \times 4\pi r^2$$

from which,

$$N = \frac{Q}{\epsilon_0} \tag{25.7}$$

The number N is independent of the radius of the sphere. Thus, the same number of lines must cross every imaginary sphere concentric with the charge.

We derived Eq. (25.7) for a point charge and a spherical surface. However, it holds for any collection of charges Q surrounded by a closed surface of any shape. A generalized statement of Eq. (25.7) is known as

When a point charge is located outside a closed surface of an arbitrary shape, the number of electric field lines entering the surface equals the number leaving the surface. The net flux through the closed surface is zero, if there is no charge inside.

> ***Gauss' Law***
>
> ***The number of electric lines of force that pass through a closed surface surrounding a charge Q is Q/ϵ_0.***

All static charge on a conductor lies on its surface.

Illustrative Problem 25.8 The electric field intensity in the region between a pair of oppositely charged plane parallel conducting plates, each 72 cm^2 in area, is 12 newtons/coulomb. What is the charge on each plate?

Solution Imagine a closed surface which will enclose either of the two plates. Gauss' law states that the number of lines passing from either charged plate through the Gaussian surface toward (or from) the second plate is equal to Q/ϵ_0. $N = EA$ (Eq. 25.6) and $N = Q/\epsilon_0$ (Eq. 25.7). The effective area A is 144 in.^2 (the sum of the top *and* bottom surface areas of a plate of negligible thickness). Solve for Q and substitute known values.

$$Q = \epsilon_0 EA$$

$$= 8.854 \times 10^{-12} \frac{C^2}{N \cdot m^2} \times \frac{12\ N}{C} \times 144\ in.^2 \times \left(\frac{1\ m}{100\ cm}\right)^2$$

$$= 1.5 \times 10^{-12} C$$ *answer*

25.14 ■ Potential Energy in an Electric Field

In Chap. 5 we studied the work done moving a mass against forces in a gravitational field. This work was expressed as an increase in gravitational potential energy (GPE). Often problems in mechanics are simplified by considering the potential energies of masses. This will also be found true for forces in electric fields.

Work must be done in moving a mass against a gravitational field. In like manner, work must be done in moving an electric charge in an electric field. This work can be expressed in terms of energy units.

Consider a positive charge $+q$ resting on line l_1 in a uniform field E, as shown in Fig. 25.21*a*. The electric force on the charge is qE

$$F = qE \tag{25.4'}$$

An external force F equal to qE is required to move the charge against the field E. The work to move the charge to line l_2, a distance d, against the field, is qEd. This work can be expressed as an increase in the potential energy of the charge with respect to line l_1. Thus the potential energy of the charge at line l_2 is

$$PE = qEd \tag{25.8}$$

When the charge q is then released ($F = 0$), the electric field will perform work on the charge. The field can do so because the work is stored in the field. As the charge is accelerated back toward line l_1, the potential energy decreases and the charged particle gains an equivalent kinetic energy. If the mass of the charged particle is m, the kinetic energy of the charged mass will be

$$KE = \tfrac{1}{2}mv^2 = qEd \tag{25.9}$$

when it returns to its initial position (Fig. 25.21*b*).

The electric potential energy when $+q$ is at l_2 is independent of the path taken to reach l_2. This result is similar to that of a mass being moved in a gravitational force field.

Gravitational forces are always forces of attraction. A mass moved against the gravitational field always increases the potential energy. Moving an electric charge against an electric field does not always result in a gain of potential energy because there are two kinds of charges. Just as there is no change in potential energy of a mass moved perpendicular to a gravitational field, so there is no change in the potential energy of a charge that is moved perpendicular to an electric field.

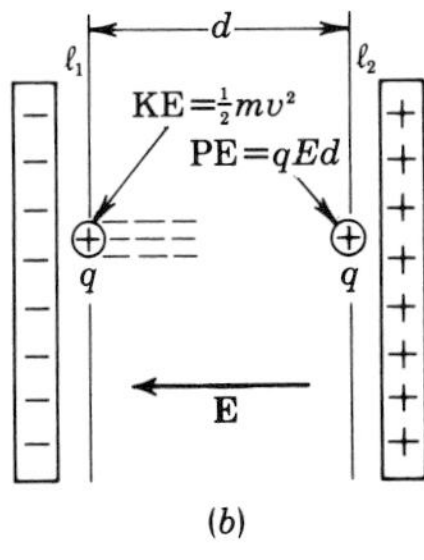

Fig. 25.21 The gain in potential energy = qEd when a positive charge $+q$ is moved a distance d against a field E. When the restraining force is released, the charge will gain kinetic energy $=\frac{1}{2}mv^2 = qEd$.

Illustrative Problem 25.9 What is the increase in the potential energy of a positive charge of 12 μC if it is moved 83 cm against an electric field intensity of 2740 N/C?

Solution $12\ \mu C = 12 \times 10^{-6}$ C

Use Eq. (25.9)

$$\begin{aligned} \Delta PE &= QEd \\ &= 12 \times 10^{-6}\ C \times 2740\ N/C \times 0.83\ m \\ &= 0.027\ N \cdot m \\ &= 0.027\ J \end{aligned}$$ *answer*

25.15 ■ Potential Difference

If work is required to move a charge from one point to another in an electric field or if work is done as a charge moves from one point to another in the field, these two points are said to differ in electric potential.

In dealing with gravitational potential we are often concerned with differences in potential energies. For falling bodies, losses in potential energies were equated to gains in kinetic energies. In like manner, in this discussion we are concerned with differences in electrical potential energies.

The *electric potential difference* V between two points is defined as the *work* W *done per unit positive charge* q^* *to move the charge from A to B.*

$$V = V_A - V_B = W/q \tag{25.10}$$

The work can be positive, negative, or zero. In these cases the electric potential at B will be greater, less, or the same (respectively) as the electric potential at A.

The unit for electrical potential difference in the SI system is the *volt.*

Definition of the Volt

The potential difference between two points is one volt (V) if it requires one joule of external work to move one coulomb of charge from one point to the other.

1 volt = 1 joule/1 coulomb.

$$1 \text{ volt (V)} = 1 \text{ joule/coulomb (1 J/C)}$$

A "twelve-volt" battery is one that has a potential difference of 12 V between its terminals. The two electric wires connected to a household outlet plug have an effective difference of potential of 110 to 120 V. On the average, 110 to 120 J of work must be expended for each coulomb of electricity which is transferred through an appliance connected between the wires.

Small potential differences are often expressed in *millivolts* or *microvolts.*

$$1 \text{ millivolt (1 mV)} = 10^{-3} \text{ V}$$
$$1 \text{ microvolt (1 } \mu\text{V)} = 10^{-6} \text{ V}$$

Large potential differences are expressed in *kilovolts* or *megavolts.*

$$1 \text{ kilovolt (1 kV)} = 10^{3} \text{ V}$$
$$1 \text{ megavolt (1 MV)} = 10^{6} \text{ V}$$

Illustrative Problem 25.10 What is the energy in joules needed to accelerate an electron (or proton) through a potential difference of one volt?

Solution

$$1 \text{ V} = 1 \text{ J/C}$$

The charge of an electron $= 1.6 \times 10^{-19}$ C

$$W = qV \qquad \text{(Eq. 25.11)}$$

$$= 1.6 \times 10^{-19} CV \times \frac{1 \text{ J/C}}{1 \text{ V}}$$

$$= 1.6 \times 10^{-19} \text{J} \qquad \textit{answer}$$

This amount of energy is called the *electron-volt* (eV). Electron-volt units are commonly used in atomic and nuclear physics. Thus, 1 eV $= 1.602 \times 10^{-19}$ J.

* In this text, hereafter, we will let q represent a *positive* charge, thus making it clear what the algebraic sign of the potential difference may be in any case.

Illustrative Problem 25.11 An alpha particle moves in a particle accelerator through the field where the potential difference is 2.5 MV. The mass of the particle is 6.64×10^{-27} kg. An alpha particle has a positive charge of 3.204×10^{-19} C. (*a*) What is the energy of the particle when it reaches its final speed? (*b*) What is that speed?

Solution
(***a***) Substitute known values in Eq. (25.11).

$$
\begin{aligned}
W &= qV \\
&= (3.204 \times 10^{-19}\ \text{C})(2.5 \times 10^{6}\ \text{V})\left(\frac{1\ \text{J/C}}{1\ \text{V}}\right) \\
&= 8.01 \times 10^{-13}\ \text{J}
\end{aligned}
$$

The kinetic energy of the particle is equal to the work done on it, namely, 8.01×10^{-13} J. *answer*

(***b***) $\text{KE} = \frac{1}{2}mv^2 = W$

or

$$v = \sqrt{\frac{2W}{m}} = \sqrt{\frac{2(8.01 \times 10^{-13}\ \text{J})}{6.64 \times 10^{-27}\ \text{kg}}}$$

Remembering that 1 J = 1 N · m and 1 N = 1 kg · m/s², we have

$$
\begin{aligned}
v &= \sqrt{\frac{16.02 \times 10^{-13}\ \text{kg} \cdot \text{m/s}^2 \times \text{m}}{6.64 \times 10^{-27}\ \text{kg}}} \\
&= \sqrt{2.413 \times 10^{14}\ \text{m}^2/\text{s}^2} \\
&= 1.55 \times 10^{7}\ \text{m/s} \qquad \textit{answer}
\end{aligned}
$$

25.16 ■ Potential Difference in a Uniform Electric Field

Consider two large plates *A* and *B* separated by a distance *d* (Fig. 25.22). Let plate *A* bear a charge $+Q$ and plate *B* a charge $-Q$. The electric field E between the two plates is constant in both magnitude and direction. Such a field is said to be *uniform*. A charge $+q$ placed in the region between the plates will experience a force F = qE (Eq. 25.4). The work W done by the field E in moving the $+q$ charge a distance d from *A* to *B* is $Fd = qEd$.

The potential difference V between plates *A* and *B* is W/q (Eq. 25.10). Thus

$$V = V_A - V_B = \frac{W}{q} = \frac{qEd}{q}$$

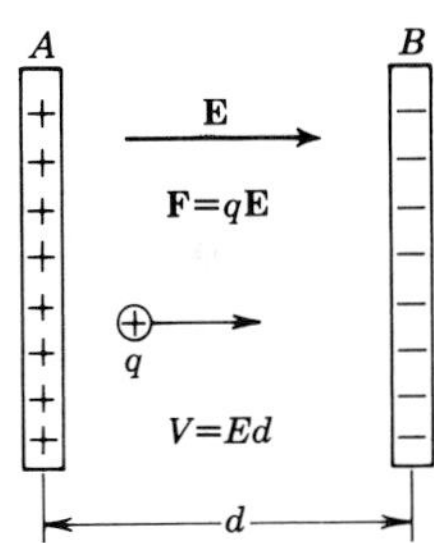

Fig. 25.22 The potential difference between two oppositely charged plates is equal to *Ed*.

If we divide both numerator and denominator by q, we get

$$V = Ed \tag{25.12}$$

> ***The potential difference between two oppositely charged plates is equal to the product of the electric field intensity and the distance between the plates.***

Potential difference is a scalar quantity.

Potential difference, like work, is a scalar quantity. The defining equation for potential difference (Eq. 25.10) assumes q to be a *positive* charge. Thus, a positive potential difference means that the electric potential energy of the charge is greater at *A* than at *B*. A negative potential difference means that the electric potential energy at *A* is less than that at *B*. If V is positive, the positive charge at *A* will tend to return to *B*. If V is negative, the positive charge will tend to move farther away from *B*.

The effects of a charge distribution can be described either in terms of an electric field or in terms of electric potential.

Equation (25.12) can be solved for E.

$$E = V/d \tag{25.13}$$

The SI metric unit for E in Eq. (25.13) is the *volt per meter*. It is left as an exercise for the student to prove that the volt per meter is equivalent to the *newton per coulomb*.

The units for field intensity can be written as volts/meter or newtons/coulomb.

Illustrative Problem 25.12 Two large plates are 1.5 cm apart. The potential difference between the plates is 20,000 V. Determine the field intensity E between the plates.

Solution Use Eq. (25.13) and substitute given values:

$$E = \frac{V}{d} = \frac{2 \times 10^4 \text{ V}}{1.5 \times 10^{-2} \text{ m}}$$

$$= 1.3 \times 10^6 \text{ V/m} \qquad \textit{answer}$$

25.17 ■ Potential Difference in a Nonuniform Field

We have been considering the potential differences in a *uniform* electric field. We will next consider a case where the field is not uniform.

Consider the field around an isolated charge $+Q$ (Fig. 25.23). The field is directed radially away from the charge. The field intensity varies inversely as the square of the distance from the center of the charge. At points A and B, the field intensities are

$$E_A = kQ/r_A^2 \qquad E_B = kQ/r_B^2$$

where r_A and r_B are respective distances from the charge to A and B.

The forces on a test charge q at A and B are

$$F_A = qE_A = kqQ/r_A^2 \qquad F_B = qE_B = kqQ/r_B^2$$

It can be shown that the average of forces F_A and F_B can be approximated by

$$F = kqQ/r_A r_B$$

Then, the work done by an external agent in moving the test charge the distance $r_A - r_B$ is

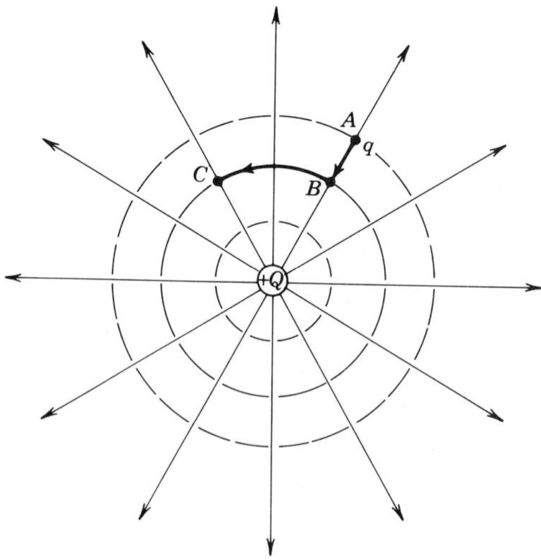

Fig. 25.23 Motion of test charge q in a field due to charge $+Q$. Work is done in moving the test charge from A to B. No work is done in moving the test charge from B to C.

$$W_{A\to B} = \frac{kqQ}{r_A r_B}(r_A - r_B)$$

$$= kqQ\left(\frac{1}{r_B} - \frac{1}{r_A}\right) \tag{25.14}$$

Using the defining equation (25.10) for potential difference and Eq. (25.14), we get

Electrical potentials due to a single point charge.

$$V = V_A - V_B = \frac{W_{A\to B}}{q}$$

$$= kQ\left(\frac{1}{r_B} - \frac{1}{r_A}\right) \tag{25.15}$$

Potential differences are independent of the path followed in moving the test charge. Thus the potential difference between two points at distances r_A and r_B from a point charge depends only on the magnitude of the charge and the distances of the points from the charge.

25.18 ■ Electric Potential

The *potential at a point* is defined as the work done per unit charge in moving a positive charge from some zero potential to the point. This work results in an increase in the electric potential energy. Let us choose infinity as the reference point for zero potential.

Referring to the past example (Fig. 25.23), we can choose A at a very large (actually an infinite) distance from Q. We will arbitrarily assign zero electric potential V_A at this infinite distance. This then permits us to define the electric potential at a point. Putting $V_A = 0$, $r_B = r$, and $r_A = \infty$ leads to the equation

$$V = V_B - V_A = kQ\left(\frac{1}{r} - \frac{1}{\infty}\right)$$

$$= \frac{kQ}{r} \tag{25.16}$$

or

$$V = \frac{1}{4\pi\epsilon_0}\frac{Q}{r} \tag{25.17}$$

In the SI system V is in volts, Q in coulombs, and r in meters. The value of k is $9.0 \times 10^9\ \text{N}\cdot\text{m}^2/\text{C}^2$. The value of ϵ_0 is $8.9 \times 10^{-12}\ \text{C}^2/\text{N}\cdot\text{m}^2$ (see Sec. 25.9). The units are consistent because

$$1\ \text{V} = 1\ \frac{\text{J}}{\text{C}} = 1\ \frac{\text{N}\cdot\text{m}}{\text{C}}$$

The student must keep in mind that choosing infinity as the reference point for zero potential is purely arbitrary. Any other reference position could well have been chosen for zero potential. In subsequent chapters, however, *potential differences* will generally be our fundamental concern.

Illustrative Problem 25.14 Point charges of $Q_1 = +0.015\ \mu\text{C}$ and $Q_2 = -0.015\ \mu\text{C}$ are placed 10 cm apart as in Fig. 25.24. Compute the potentials at A, B, and C.

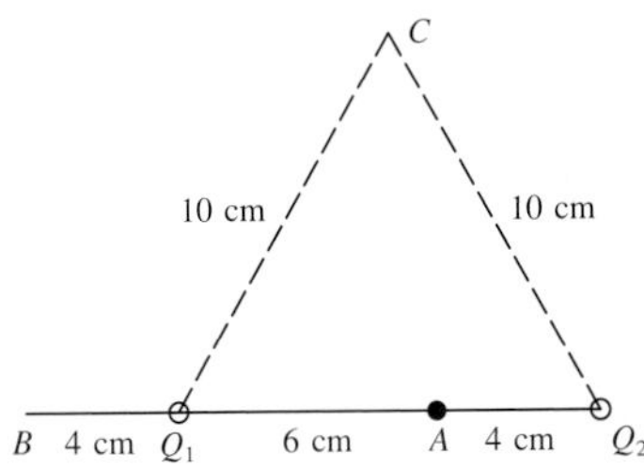

Fig. 25.24 Illustrative Problem 25.14.

Solution We must evaluate the algebraic sum of the voltages at each point due to two charges.

(***a***) At point A the potential due to the positive charge is

$$V = kQ/r \tag{25.16}$$

$$= 9.0 \times 10^9 \text{ N} \cdot \text{m}^2/\text{C}^2 \times \frac{+0.015 \times 10^{-6}\text{C}}{0.06 \text{ m}}$$
$$= 2250 \text{ N} \cdot \text{m/C}$$
$$= 2250 \text{ V}$$

At point A the potential due to the negative charge is

$$V = 9.0 \times 10^9 \frac{\text{N} \cdot \text{m}^2}{\text{C}^2} \times \frac{-0.015 \times 10^{-6}\text{C}}{0.04 \text{ m}}$$
$$= -3375 \text{ V}$$

Hence the potential at $A = 2250 \text{ V} - 3375 \text{ V} = -1100 \text{ V}$. *answer*

(***b***) At B the potential due to the positive charge is $+3375$ V and that due to the negative charge is

$$4/14 \times -3375 \text{ V} = -960 \text{ V}$$

Hence the potential at $B = +3375 \text{ V} - 960 \text{ V} = 2400 \text{ V}$. *answer*

(***c***) The potential at C is $1350 \text{ V} - 1350 \text{ V} = 0$ *answer*

Illustrative Problem 25.14 The diameter of the nucleus of the copper atom (see Fig. 25.3) is roughly 6×10^{-13} cm. What is the electric potential at the surface of the copper nucleus?

Solution The surface of the nucleus is assumed spherical. The charge on the nucleus for external purposes behaves as if it were a point charge at the center. Recall that the positive charge on a proton is 1.6×10^{-19} C (see Sec. 25.9). There are 29 protons in the nucleus. Substitute known values in Eq. (25.16).

$$V = k\frac{Q}{r}$$
$$= \frac{(9.0 \times 10^9 \text{ N} \cdot \text{m}^2/\text{C}^2)(29)(1.6 \times 10^{-19} \text{ C})}{3 \times 10^{-15} \text{ m}}$$
$$= 1.4 \times 10^7 \frac{\text{N} \cdot \text{m}}{\text{C}} \left(\frac{1 \text{ V/m}}{1 \text{ N/C}}\right) = 1.4 \times 10^7 \text{ V}$$
$$= 14 \text{ MV}$$

answer

25.19 ■ Equipotential Surfaces

Equipotential surfaces are perpendicular to the electric field lines of force.

If all the points which have the same potential in an electric field near a charged object are joined, an *equipotential line* or *surface* within the field is formed. No work is done when a test charge is moved along an equipotential surface in an electric field.

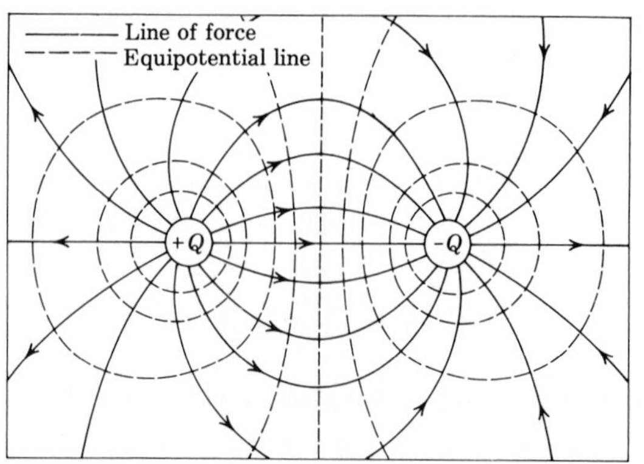

Fig. 25.25 Lines of force and equipotential lines in the field formed by two oppositely charged bodies.

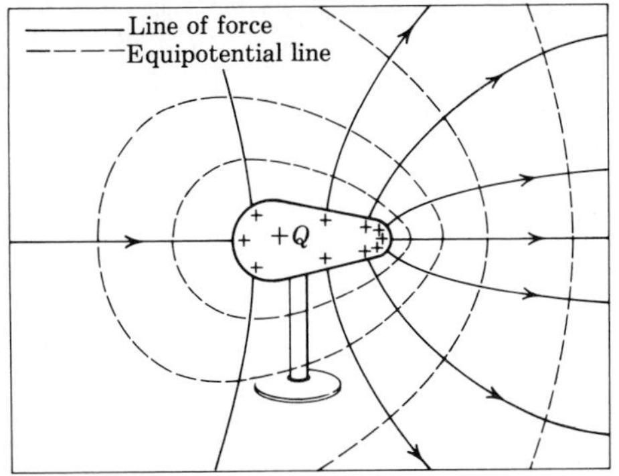

Fig. 25.26 Lines of force and equipotential lines in a nonuniform field. Note that the equipotential lines are perpendicular to the electric field lines.

Thus, in Fig. 25.23, no work would be done in moving the charge from B to C.

Equipotential lines (dashed lines) and lines of force (solid lines) are illustrated in Fig. 25.25 and Fig. 25.26. The actual field, of course, is three dimensional. Note that the electric lines of force are normal to equipotential surfaces.

The *electric potential gradient* is the rate of change of potential with distance along a line of force. If ΔV is a potential change that corresponds to a change Δr along a line of force, the potential gradient E_r is equal to $-\Delta V/\Delta r$.

Potential gradients, in the mks system, are expressed in volts per meter. Since the electric intensity in any direction is numerically equal to the potential gradient in that direction, it follows that the units of electric field intensity and potential gradient are equivalent. It is common practice, however, to use the volt/meter rather than newton/coulomb to express field intensities. The electric intensity E is the negative potential gradient. Note that the greater the electric intensity, the smaller the distance between the equipotential lines. The equipotential lines are therefore crowded closer in a strong field and more widely separated in a weak field.

25.20 ■ Electric Potential Referred to the Earth

In Chap. 5 sea level was used as an arbitrary choice for the zero of gravitational potential energy. Heights above sea level were used in determining potential energies. In other problems we used the difference of level between two points, neither of which was at sea level.

In Sec. 25.18 zero potential is assumed to be at infinity. It is frequently convenient to take the electric potential of the earth as zero. The earth can be considered as an (almost) infinite reservoir of positive and negative charges.

If a charged body connected to the earth by a metallic conductor of electricity allows electrons to flow to that body from the ground, the body is at a *positive potential* (see Fig. 25.27*a*). Conversely, if the connection to the ground allows electrons to flow from the body into the ground, the body is at a *negative potential* (see Fig. 25.27*b*). If the negative terminal of a 12-V battery is connected to the ground, it will be at 0-V potential, and the potential at the positive terminal will be +12 V. In practice, the negative terminal is connected to the frame of an automobile and the

In determining potential difference the potential at *any* point can be assigned *any value* we please.

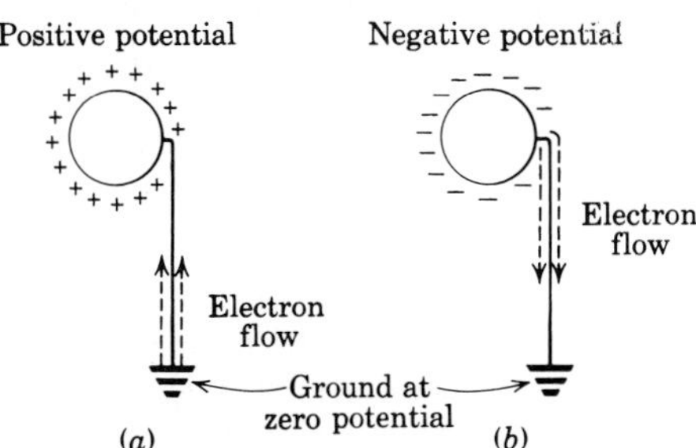

Fig. 25.27 Direction of flow of electrons when a positively charged body and a negatively charged body are grounded.

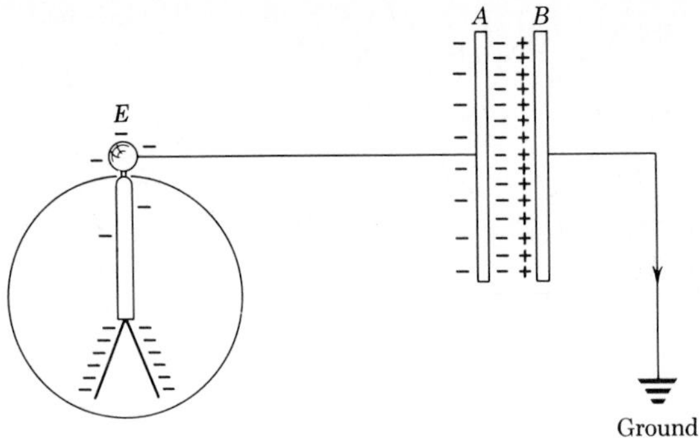

Fig. 25.28 The principle of a capacitor.

frame is regarded as being at zero potential. In radios and television receivers the chassis is usually regarded as being at zero potential. Electric tools and appliances are grounded for safety, so there is no potential difference between them and ground.

25.21 ■ The Capacitor

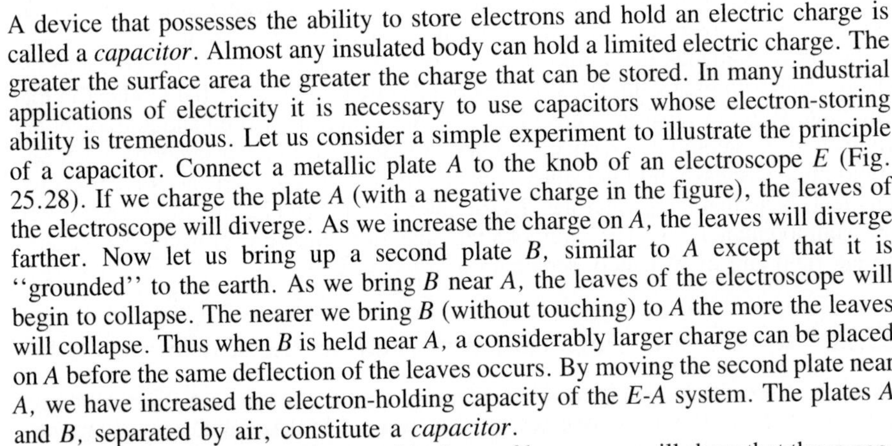

A device that possesses the ability to store electrons and hold an electric charge is called a *capacitor*. Almost any insulated body can hold a limited electric charge. The greater the surface area the greater the charge that can be stored. In many industrial applications of electricity it is necessary to use capacitors whose electron-storing ability is tremendous. Let us consider a simple experiment to illustrate the principle of a capacitor. Connect a metallic plate A to the knob of an electroscope E (Fig. 25.28). If we charge the plate A (with a negative charge in the figure), the leaves of the electroscope will diverge. As we increase the charge on A, the leaves will diverge farther. Now let us bring up a second plate B, similar to A except that it is "grounded" to the earth. As we bring B near A, the leaves of the electroscope will begin to collapse. The nearer we bring B (without touching) to A the more the leaves will collapse. Thus when B is held near A, a considerably larger charge can be placed on A before the same deflection of the leaves occurs. By moving the second plate near A, we have increased the electron-holding capacity of the E-A system. The plates A and B, separated by air, constitute a *capacitor*.

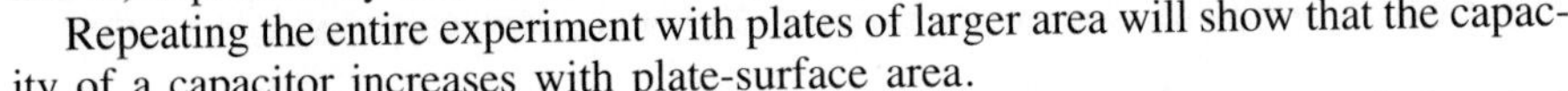

Repeating the entire experiment with plates of larger area will show that the capacity of a capacitor increases with plate-surface area.

Capacitors, as we shall find in later chapters, are used in a variety of electrical circuits. For example, they are used as energy-storing devices in electronic flashing lights, as filters in power supplies, to tune in frequencies received by radio receivers, to eliminate sparking in high voltage systems, and to improve "power factor" on alternating current power lines.

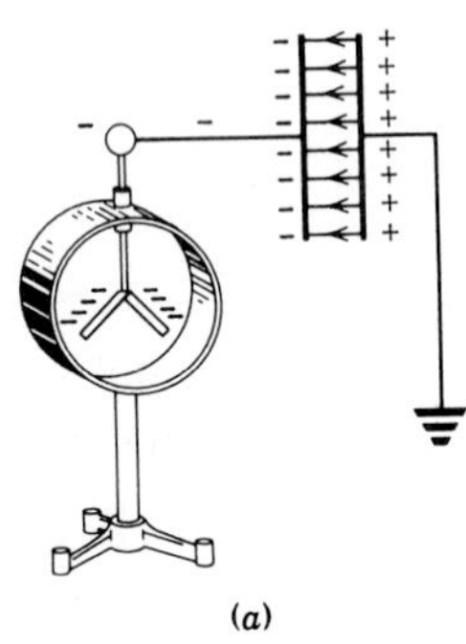

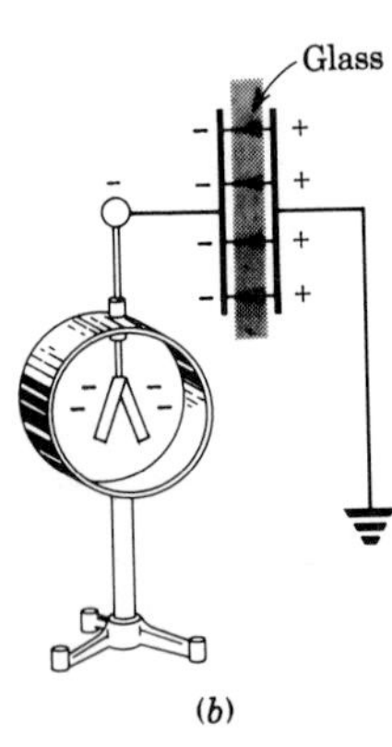

Fig. 25.29 Insertion of a glass slab between the plates of a charged capacitor lowers the potential difference between the plates. This will allow a new, larger charge to exist on the metal plates for the same potential difference across the plates. The effect of the dielectric is to increase the capacitance of the capacitor.

25.22 ■ Capacitance

The capacitance of a capacitor is a measure of its ability to store up electric charge. We have discussed two ways to increase capacitance: (1) the area of the plates can be increased; (2) the plates can be brought closer together. A third way to increase the capacitance would be to insert a thin slab of nonconducting materials (other than air) between the plates. These materials are called *dielectrics*. Common dielectrics are air, glass, mica, oil, and waxed paper.

The effectiveness of the dielectric in increasing the ability of a capacitor to hold a charge can be illustrated in the laboratory. Figure 25.29*a* shows a parallel-plate capacitor with an electroscope connected across the plates. If a charge is placed on the capacitor, a difference of potential between the plates will be indicated by the diverging leaves of the electroscope. If a slab of glass or plastic is inserted between the plates (Fig. 25.29*b*), the leaves will partially collapse, indicating a smaller potential difference between the plates. When the slab is removed, the original potential difference will be indicated by the electroscope.

If the glass plate is kept between the plates, a greater charge must be put on the plates to produce the potential difference exhibited with air as the dielectric medium. The *capacitance* of a capacitor is defined as the *ratio of the charge on either plate to the potential difference between the two plates*.

$$C = Q/V \tag{25.18}$$

where Q = charge on either plate, coulombs
V = potential difference between the conducting plates, volts
C = capacitance of capacitor, coulombs/volt

The Unit of Capacitance

The coulomb per volt unit has been given a special name, the farad (F). Thus a capacitor whose capacitance is one farad will hold a charge of one coulomb if a difference of one volt is applied between the plates.

1 farad = 1 coulomb/1 volt.

In practice, the farad proves to be extremely large. For example, a capacitor of 1-F capacity, with two parallel plates 1 mm apart in air would require each plate to have an area of over 40 sq. mi!

Practical capacitors are usually calibrated in microfarads and picofarads:

$$1 \text{ microfarad}(\mu\text{F}) = 10^{-6} \text{ F}$$
$$1 \text{ picofarad(pF)} = 10^{-12} \text{ F}$$

Illustrative Problem 25.15 What is the capacitance of a two-plate capacitor that holds a charge of 7.5×10^{-4} μC when the potential difference between the plates is 400 V?

Solution Using Eq. (25.18), we have

$$\begin{aligned} C &= Q/V \\ &= \frac{7.5 \times 10^{-10} \text{ C}}{400 \text{ V}} \\ &= 1.875 \times 10^{-12} \text{ F} \\ &= 1.875 \text{ pF} \qquad \textit{answer} \end{aligned}$$

25.23 ■ Dielectrics

In Fig. 25.3 the atom is pictured as consisting of a small, but extremely dense, positive nucleus surrounded by electrons whirling in orbits around the nucleus. When a conductor is placed in an electric field, some of the free electrons move from atom to neighboring atom in the conductor.

Materials which make good dielectrics are generally those whose atoms have few outer orbit electrons free to migrate. Atoms of dielectrics, as atoms of all substances, are electrically neutral. Normally the centers of their orbital electrons also coincide with the center of the nucleus. However, in an electric field, the electrons and protons are somewhat displaced from their normal positions. An electric field will cause the electrons to shift slightly against the field. The protons will shift slightly in the direction of the field.

Figure 25.30 illustrates schematically what happens when a dielectric slab is placed between the charged plates of a capacitor. The atom of the dielectric becomes an *induced electric dipole* with one end positively charged and the other negatively charged.

Fig. 25.30 The surfaces of the dielectric slab in an electric field produces an electric field $\mathbf{E}_d$ which opposes the external field $\mathbf{E}_0$ due to the charged plates.

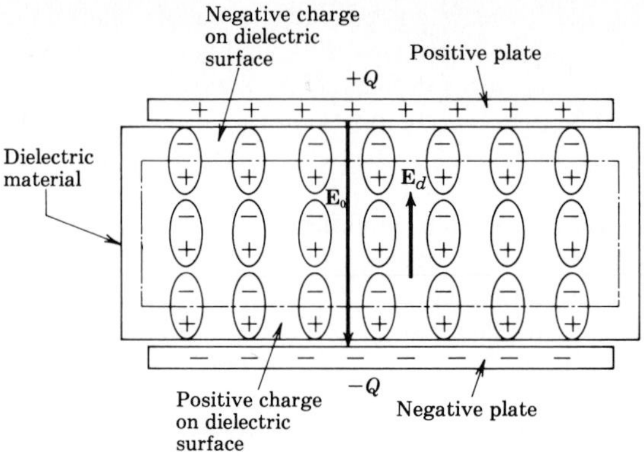

In the electric field the tiny dipoles tend to line up with their positive ends pointing toward the negative plate. When the dielectric is removed from the field, the atoms quickly cease to be aligned as dipoles. This is due to the constant state of thermal agitation or internal energy of the molecules.

The charges on the surfaces of dielectric material in a capacitor result from dipole moments of the dielectric molecules and not from the transfer of electrons as in the case of metallic conductors.

The degree of alignment of the dipoles depends upon the intensity of the electric field, the temperature, and the nature of the dielectric. The greater the field intensity, the greater the alignment. As the temperature is decreased, the alignment will increase. In Fig. 25.30, note that all the positive and negative charges in the interior of the dielectric (within the dashed lines) neutralize each other. There are left, however, layers of positive and negative charges on opposite surfaces of the dielectric. These give the dielectric a positive charge on one surface and a negative bound charge on the other.

The electric field established in a dielectric slab by charges opposes the external field of the capacitor.

These surface charges produce an electric field $\mathbf{E}_d$ in the dielectric slab which opposes the external field $\mathbf{E}_0$ of the capacitor. The effect is to weaken the original field. The net electric field $\mathbf{E}$ is the vector sum of the two fields $\mathbf{E}_0$ and $\mathbf{E}_d$. Thus, $\mathbf{E}$ will be a weaker field in the direction of $\mathbf{E}_0$.

$$\mathbf{E} = \mathbf{E}_0 + \mathbf{E}_d \tag{25.19}$$

The net effect of the dielectric in the charged capacitor is to lower the potential gradient of the electric field (see Eq. 25.13). Since $C = Q/V$ for the capacitor, we know that lowering V for a constant Q results in an increase in C. Or, in other words, if a dielectric is placed between the plates of a capacitor, Q will be larger for the same V.

The presence of a dielectric increases the capacitance of a capacitor.

25.24 ■ Dielectric Constant

The ratio of the capacitance of a capacitor with the dielectric between the plates to that with plates separated by a vacuum is the *dielectric constant* K of the material.

$$K = \frac{C}{C_0} \tag{25.20}$$

The dielectric allows the plates of a capacitor to be placed closer together, thereby increasing the capacitance without fear of the plates touching each other.

The relation between the three factors which determine the capacitance of a parallel-plate capacitor can be expressed by the formula

$$C = \epsilon \frac{A}{d} \tag{25.21}$$

where A = area of either parallel plate, square meters
d = distance between plates, meters
ϵ = constant of separating medium
C = capacitance, farads

The constant ϵ is called the *permittivity of the separating medium*. The value of ϵ is found by

$$\epsilon = \epsilon_0 K \tag{25.22}$$

where $\epsilon_0 = 8.85 \times 10^{-12}\ C^2/N \cdot m^2$ (see Sec. 25.9) and K is the *dielectric constant*. Recall that ϵ_0 is the *permittivity of free space*.

Values of the dielectric constant for a few substances are given in Table 25.3. Note that K is dimensionless since it equals the quotient of like quantities.

Table 25.3 Dielectric constants K of some common dielectric materials

Material	K
Air (1 atm)	1.00059
Ammonia, liquid at −34°C	22
Glass	5–10
Mica	3–6
Paraffin wax at 20°C	2.1–2.5
Polyethylene at 20°C	2.25–2.30
Porcelain	6.0–8.0
Rubber	2.5–3.0
Sulfer at 20°C	4.0
Transformer oil at 20°C	2.24
Vacuum	1.00000
Water at 20°C	80
Strontium titanite	223

Illustrative Problem 25.16 Two rectangular sheets of copper foil 16 by 20 cm are separated by a thin layer of paraffin wax 0.2 mm thick. Calculate the capacitance if the dielectric constant for the wax is 2.4.

Solution Substitute $K = 2.4$, $d = 2 \times 10^{-4}$ m, and $A = 0.16$ m by 0.20 m in Eqs. (25.21) and (25.22):

$$C = K\epsilon_0 \frac{A}{d} = 2.4 \times 8.85 \times 10^{-12}\ C^2/N \cdot m^2 \times \frac{0.032\ m^2}{2 \times 10^{-4}\ m}$$

$$= 3.4 \times 10^{-9}\ C^2/N \cdot m \times \frac{1\ N \cdot m/V}{C} \times \frac{1\ F}{1\ C/V}$$

$$= 3.4 \times 10^{-9}\ F = 3400\ pF \qquad \textit{answer}$$

The dielectric strength equals the maximum field intensity that can exist in the dielectric without electrical breakdown.

There is a limit to how much charge can be placed on a given capacitor. When sufficiently large charges are placed on a capacitor, the electrons break loose from the parent atoms of the dielectric. The electric field stress applied will make the dielectric temporarily conductive. Breakdown occurs. The breakdown consists in the passage of a spark from plate to plate through the dielectric. In dielectrics such as air, oil, and those used in electrolytic capacitors the breakdown is only temporary. The capacitive properties are restored after the spark disappears. But, in glass, mica, and other solid dielectrics, the dielectrics are punctured or shattered when the breakdown occurs. After breakdown the dielectrics have only the strength of an equal thickness of air. The *dielectric strength* of a dielectric is the field intensity, measured in megavolts per meter (MV/m), required to breakdown the dielectric. Table 25.4 lists approximate dielectric strengths for some common dielectrics.

Table 25.4 Approximate dielectric strengths

Material	*Dielectric Strength (MV/m to breakdown)*
Air	3.0
Bakelite	10–30
Glass	20–60
Mica	50–220
Paraffin, solid	25–45
Paraffined paper	30–50
Porcelain	5.5
Rubber	15–50
Transformer oil	5–15
Strontium titanite	8

Illustrative Problem 25.17 The plates of a parallel-plate capacitor are separated by paraffined paper 0.015 mm thick. What voltage applied to the capacitor will cause the capacitor to break down? Assume the dielectric strength of the paper to be 35 MV/m.

Solution Let the answer be x volts.

$$35 \text{ MV/m} = 35 \times 10^6 \frac{\text{V}}{\text{m}} \times \frac{1 \text{ m}}{1000 \text{ mm}}$$

$$= 35 \times 10^3 \frac{\text{V}}{\text{mm}}$$

Then

$$x/0.015 \text{ mm} = 35 \times 10^3 \text{ V/mm}$$

and

$$x = 35 \times 10^3 \text{ V/mm} \times 0.015 \text{ mm}$$
$$= 525 \text{ V} \qquad \textit{answer}$$

25.25 ■ Growth and Decay in RC Circuits

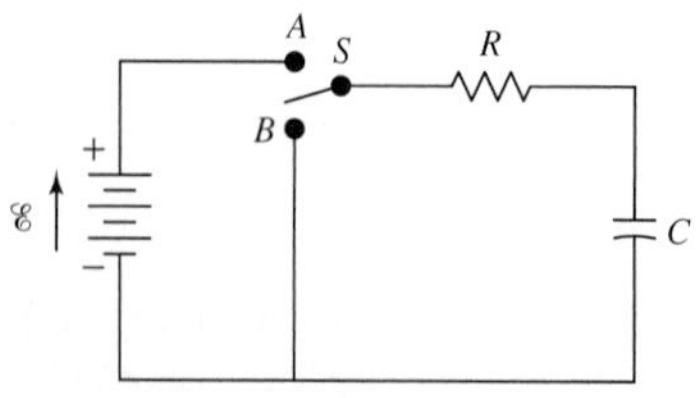

Fig. 25.31 An *RC* circuit.

Capacitors and resistors are frequently found in the same circuit. Consider the simple circuit illustrated in Fig. 25.31. When the switch is closed at A the circuit will include a source of potential difference $\mathscr{E}$, a resistor R, a capacitor C, and a switch S. When the switch is closed at B, the circuit will include the capacitor, the resistor, and the switch.

Let us start with the switch S open, and with the capacitor having no charge on it ($Q = 0$). When the switch is thrown to position A, current will be established in the circuit and the capacitor will gain a charge. At time $t = 0$, only the resistor holds current back. As a result, the charge flows rapidly into the capacitor at first. As the charge Q in the capacitor grows, the opposing potential difference Q/C between the plates of the capacitor increases. The result will be a decrease in current i (and in the rate of growth of the charge in the capacitor). This process will continue until eventually the voltage across the capacitor equals but opposes the emf of the battery and no more current exists in the circuit.

The charge q on the capacitor expressed as a function of time t can be found by using the equation (given here, not derived)

$$q = C\xi(1 - e^{-t/RC}) \qquad (25.23)$$

In charging and discharging a capacitor, the charge on the capacitor varies exponentially with elapsed time.

where q = the charge on the capacitor (in coulombs)
t = the time elapsed after the switch is closed (in seconds)
C = the capacitance of the capacitor (in farads)
R = the resistance in ohms
ξ = the emf of the battery (in volts)

Since $C\xi$ equals the final charge Q on the capacitor (from Eq. 25.18), we can write the equation

$$q = Q(1 - e^{-t/RC}) \qquad (25.24)$$

The constant e in Eqs. (25.23) and (25.24) is the *Naperian base of logarithms*. It can be found to any desired degree of accuracy by using the defining equation

$$e = 1 + \frac{1}{1!} + \frac{1}{2!} + \frac{1}{3!} + \frac{1}{4!} + \cdots$$

$$= 1 + \frac{1}{1} + \frac{1}{2 \cdot 1} + \frac{1}{3 \cdot 2 \cdot 1} + \frac{1}{4 \cdot 3 \cdot 2 \cdot 1} + \cdots$$

The value of e is 2.71828 accurate to six significant figures. The student will usually need a table of values for e^x, a table of natural logarithms, or a hand-held calculator to solve RC circuit growth or decay problems.

A cursory study of Eq. (25.24) will reveal that when $t = 0$, $q = Q(1 - e^{\circ}) = Q(1 - 1) = 0$, which are the initial conditions in our example. Also, when t is very large ($t \rightarrow \infty$), $e^{-t/RC}$ will get extremely small (approach zero) and q will approach Q as a limiting value. In between $t = 0$ and $t \rightarrow \infty$ the value of q will grow exponentially as is shown in the graph of Fig. 25.32.

The quantity RC in Eqs. (25.23) and (25.24) is called the *time constant* (τ) of the circuit. (τ is the Greek letter *tau*.) When $t = RC$, Eq. (25.24) becomes $q = Q(1 - e^{-1}) \approx 0.63Q$.

The time constant (τ = RC) represents the time required for the charge on a capacitor to reach 63 percent of its final value.

The time constant RC is an indication of how rapidly a capacitor charges (or discharges).

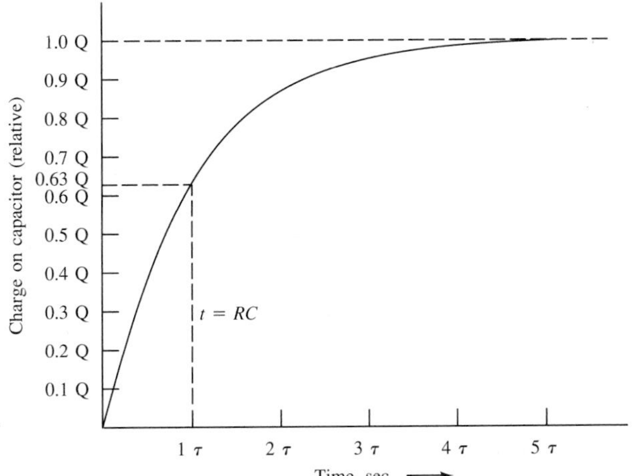

Fig. 25.32 Graph of the charge on a capacitor as a function of the charging time in an RC circuit.

Fig. 25.33 Graph of the decay of charge in a circuit containing capacitance and resistance when the circuit is closed.

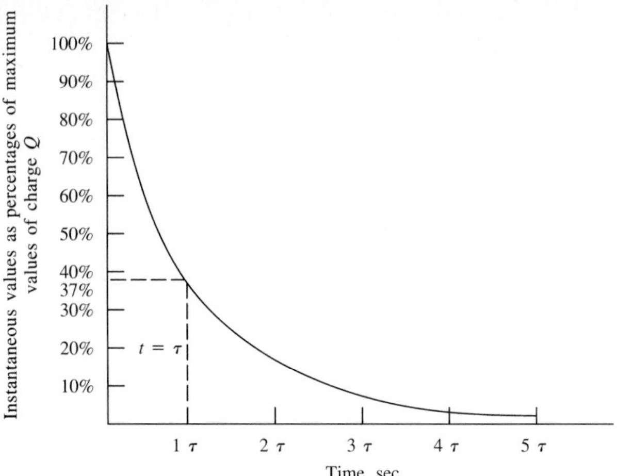

Now let us return to the circuit of Fig. 25.31. By closing the switch to B the charged capacitor C is connected across the resistor R and is allowed to discharge. As soon as the switch is closed, charge will flow from the positive plate of the capacitor through the resistor R to the negative plate of the capacitor. The charge on the capacitor decreases exponentially with time t (see Fig. 25.33). The charge q left in the capacitor at any time t seconds can be expressed by the equation

$$q = Qe^{-t/RC}. \tag{25.25}$$

The constant RC gives the time in seconds for the capacitor charge q to go to Qe^{-1} or (approximately) 37 percent of the original charge. In n time constants the charge remaining on the capacitor will be $0.37^n Q$.

Illustrative Problem 25.18 The capacitance in the circuit in Fig. 25.31 is 0.75 μF, the total resistance R is 1200 Ω, and $\mathscr{E}=12$ V. Determine (*a*) The time constant, (*b*) the maximum charge the capacitor acquires, and (*c*) the charge on the capacitor 1.8 ms after the switch is closed.

Solution

(***a***) The time constant

$$\begin{aligned} &= RC \\ &= 1200\ \Omega \times 0.75 \times 10^{-6}\ \text{F} \\ &= 9.0 \times 10^{-4}\ \text{s} \qquad \textit{answer} \end{aligned}$$

(***b***) The maximum charge

$$\begin{aligned} Q &= CV \\ &= 0.75 \times 10^{-6}\ \text{F} \times 12\ \text{V} \\ &= 9.0 \times 10^{-6}\ \text{C} \\ &= 9.0\ \mu\text{C} \qquad \textit{answer} \end{aligned}$$

(***c***)

$$\begin{aligned} q &= Q(1 - e^{-t/RC}) \\ &= 9.0 \times 10^{-6}\ \text{C}\ (1 - \text{e}^{-1.8\times10^{-3}\ \text{s}/9.0\times10^{-4}\ \text{s}}) \\ &= 9.0 \times 10^{-6}\ \text{C}\ (1 - \text{e}^{-2}) \end{aligned}$$

Recalling that $e = 2.71828$, and rounding off,

$$\begin{aligned} q &= 9.0 \times 10^{-6}\ \text{C} \left(1 - \frac{1}{(2.7)^2}\right) \\ &= 7.8 \times 10^{-6}\ \text{C} \qquad \textit{answer} \end{aligned}$$

The same graph (Fig. 25.33) can be used to illustrate the relationship expressed by Eqs. (25.24) and (25.25).

Illustrative Problem 25.19 In the circuit of Fig. 25.31*b* (switch closed at *B*), if the charge on the capacitor is 9.0 μC, what will the charge be reduced to after four time constants?

Solution The charge will have decreased to $(0.37)^4Q$

$$(0.37)^4(9.0\ \mu\text{C}) = 0.170\ \mu\text{C} = 170\ \text{nC} \qquad \textit{answer}$$

25.26 ■ Combination of Capacitors

Capacitors are connected in *parallel* when one plate of each capacitor is connected to a common conductor while the remaining plates are connected to a second conductor (see Fig. 25.34*a*). (The symbol ∋⊢ or ⊣⊢ is used to indicate a fixed capacitor in the diagram.) Essentially this arrangement is equivalent to a single capacitor with two larger plates each equal in area to the sum of one set of three smaller areas and with the same spacing and dielectric. If the capacitors are charged, they must have the same potential difference across their plates. The charge on each capacitor will be

$$Q_1 = C_1V \qquad Q_2 = C_2V \qquad Q_3 = C_3V$$

The total charge Q_T of the system must be the sum of the separate charges on the three capacitors.

$$Q_T = Q_1 + Q_2 + Q_3$$

Then

$$Q_T = C_1V + C_2V + C_3V$$

But the total charge is equivalent to the product of the total capacitance C_T and the common potential difference V:

$$Q_T = C_TV$$

Substituting, we obtain

$$C_TV = C_1V + C_2V + C_3V$$

and dividing through by V yields

$$C_T = C_1 + C_2 + C_3 \tag{25.26}$$

Capacitors in Parallel

When capacitors are connected in parallel their total capacitance is the sum of the individual capacitances.

The equivalent capacitance of capacitors in parallel is larger than that of any of the individual capacitances.

When capacitors are connected in *series,* as shown in Fig. 25.34*b*, a positive charge on one plate, for example, of C_3 induces a negative charge on the adjacent plate. The electrons will be attracted away from the plate of C_2 that is connected to C_3. A positive charge of the same magnitude will be left on the other plate of C_2. This plate will in turn draw electrons from a plate of C_1, leaving it positive. All three capacitors will have equal charges. Thus,

$$Q_T = Q_1 = Q_2 = Q_3$$

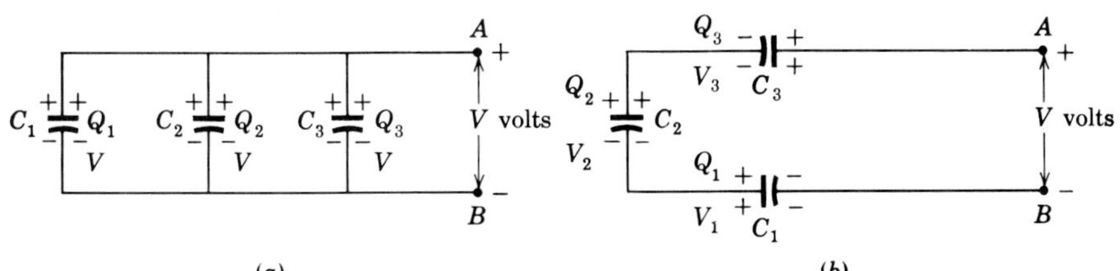

Fig. 25.34 Capacitors connected in (*a*) parallel and (*b*) series.

The total difference in potential V_T across the series combination must equal the sum of the separate potential differences across each capacitor:

$$V_T = V_1 + V_2 + V_3$$

But

$$V_T = \frac{Q_T}{C_T} \quad V_1 = \frac{Q_1}{C_1} \quad V_2 = \frac{Q_2}{C_2} \quad V_3 = \frac{Q_3}{C_3}$$

Substituting, we have

$$\frac{Q_T}{C_T} = \frac{Q_T}{C_1} + \frac{Q_T}{C_2} + \frac{Q_T}{C_3}$$

and dividing by Q_T, we obtain

$$\frac{1}{C_T} = \frac{1}{C_1} + \frac{1}{C_2} + \frac{1}{C_3} \tag{25.27}$$

The equivalent capacitance of capacitors in series is always less than that of any one of the individual capacitances.

Capacitors in Series

When capacitors are connected in series the reciprocal of their total capacitance is equal to the sum of the reciprocals of their individual capacitances.

Illustrative Problem 25.21 Three capacitors having capacitances of 2.0, 3.0, and 5.0 μF are connected in parallel to a 12-V source. (*a*) Find the charge on each capacitor. (*b*) Find the total charge of the combination. (*c*) If the capacitors are discharged and connected in series to the 12-V source, find the total charge of the combination.

Solution

(***a***)

$$V_T = V_1 = V_2 = V_3 = 12 \text{ V}$$

Then

$$Q_1 = C_1V_1 = 2.0\ \mu\text{F} \times 12\ \text{V} = 24\ \mu\text{C}$$
$$Q_2 = C_2V_2 = 3.0\ \mu\text{F} \times 12\ \text{V} = 36\ \mu\text{C}$$
$$Q_3 = C_3V_3 = 5.0\ \mu\text{F} \times 12\ \text{V} = 60\ \mu\text{C} \qquad \textit{answer}$$

(***b***)

$$Q_T = Q_1 + Q_2 + Q_3 = 120\ \mu\text{C} \qquad \textit{answer}$$

(***c***)

$$\frac{1}{C_T} = \frac{1}{C_1} + \frac{1}{C_2} + \frac{1}{C_3}$$

$$= \frac{1}{2.0\ \mu\text{F}} + \frac{1}{3.0\ \mu\text{F}} + \frac{1}{5.0\ \mu\text{F}}$$

$$C_T = \tfrac{30}{31}\ \mu\text{F}$$

$$V_T = \frac{Q_T}{C_T} \quad \text{or} \quad Q_T = V_TC_T$$

$$Q_T = 12\ \text{V} \times \tfrac{30}{31}\ \mu\text{F} \times \frac{1\ \text{F}}{10^6\ \mu\text{F}}$$

$$= 1.2 \times 10^{-6}\ \text{C}$$
$$= 1.2\ \mu\text{C} \qquad \textit{answer}$$

25.27 ■ Commercial Capacitors

Fig. 25.35 Common multiplate capacitor.

Most capacitors are modifications of the parallel-plate capacitor. Increased capacitance in a capacitor can be achieved by using two sets of plates connected in parallel (idealized in Fig. 25.35). Such a capacitor is called a *multiple-plate capacitor*.

Commercial capacitors vary considerably in shape and size (Fig. 25.36 and Fig. 25.37). They are usually classified in terms of the dielectric used.

The *variable-air capacitor* consists of two sets of parallel aluminum plates that are separated by air. One set is fixed while the other is mounted on a shaft between the fixed plates (Fig. 25.37). Only the overlapping areas contribute to the capacitance, which can be varied by turning the shaft. The capacitance will be maximum when the movable plates are completely meshed within the fixed plates. This type of capacitor is used in tuning radio receiving sets and in many other types of electronic equipment where tuning one circuit to the frequency of another is necessary.

Paper capacitors are constructed by rolling two long strips of metal foil with a long thin strip of a paper dielectric. The paper is first treated with wax or oil. The rolled strips are then sealed in a metal container. Paper capacitors are used in radios and the ignition systems of automobiles.

The *plastic-film capacitor* is similar to the paper capacitor. The dielectric is one of the newer plastics on the market, such as teflon, mylar, or polystyrene.

The sheets of metal foil in a *mica capacitor* are separated by sheets of mica dielectric. The entire capacitor is sealed in a metal or Bakelite case.

In the *electrolytic capacitor* a dielectric layer is formed by chemical action on the metal plates of the capacitor. The slight space between the layers is filled with an electrolyte in liquid or paste form. The electrolyte constitutes one of the plates. Electrolytic capacitors usually have relatively large capacitances.

We will study the effect of capacitors in various circuits in Chap. 30.

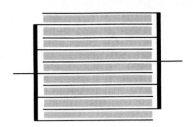

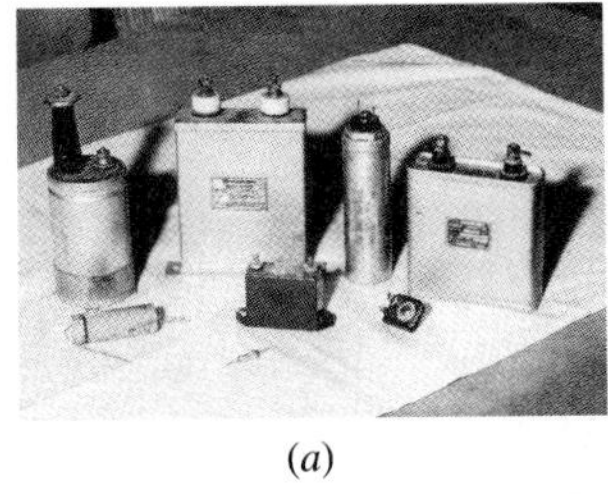

(*a*)

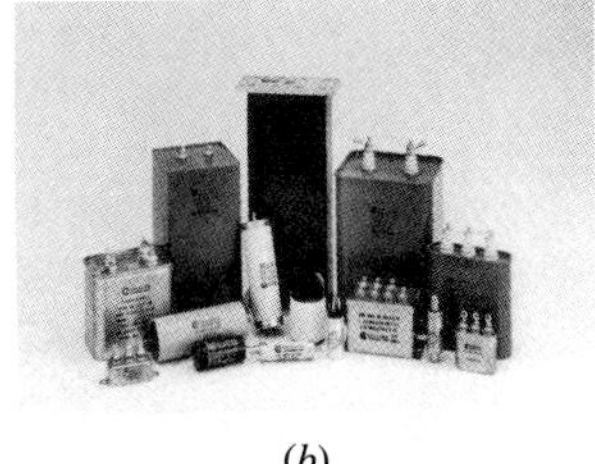

(*b*)

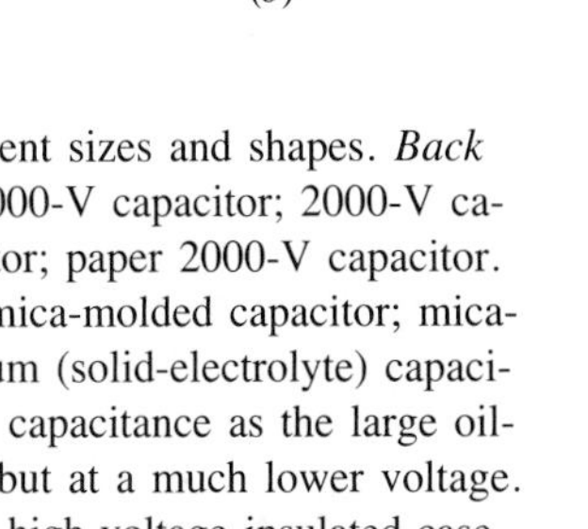

Fig. 25.36 Capacitors of different sizes and shapes. *Back row* (left to right): oil-filled 10,000-V capacitor; 2000-V capacitor; liquid electrolyte capacitor; paper 2000-V capacitor. *Second row:* liquid electrolyte, mica-molded capacitor; mica-variable capacitor. *Front:* tantalum (solid-electrolyte) capacitor. This capacitor has the same capacitance as the large oil-filled capacitor in the back row but at a much lower voltage. (*b*) High-voltage power supplies high-voltage insulated-case capacitors, multiple-cased capacitors wall-feed-through capacitors, high-voltage eramic capacitors. (Chicago Condenser Products)

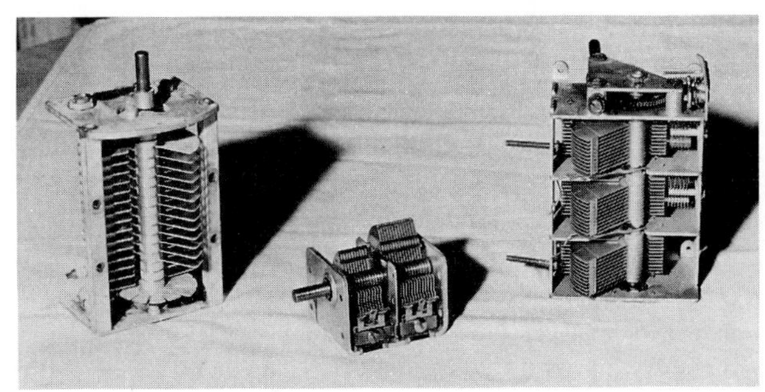

Fig. 25.37 Variable capacitors. On the left is a single-section capacitor with air as the dielectric. In the middle are two capacitors of different sizes mounted on the same shaft. In front of the two capacitors can be seen two small mica-variable capacitors. At the right is a gear-driven capacitor (really three separate capacitors mounted on the same shaft). To the right of the two capacitors are two small variable air capacitors, called *trimmers,* used to adjust for minor variations in the larger capacitors and to bring the two up to the third capacitance.

QUESTIONS AND EXERCISES

1. When a small neutral pith ball suspended by a silk thread is brought near a negatively charged sphere, the ball is first attracted to the sphere until contact is made; then the ball is repelled. Explain in terms of the electron theory the reason for the action.

2. An electroscope is charged by induction from a positively charged body. (*a*) What charge is left on the electroscope? (*b*) Explain how the induced charge is produced.

3. How would you determine the sign of the charge obtained on sulfur when it is rubbed with leather?

4. Do some research on lightning rods. Describe how they are installed and why they may reduce the chances of a building being struck by lightning in a storm.

5. A pith ball is suspended by a silk thread. When a charged body is near it, which would offer the more conclusive proof that a charge is on the ball—a force of attraction or one of repulsion? Why?

6. Where would be the safest place for you to be if you were caught in an electrical storm in some wooded hills? Why?

7. Discuss the relative masses of the electron, proton, and the neutron? What are their relative charges?

8. Which of the following is not a measure of electric field intensity: newton/coulomb, coulomb/meter, volt/meter, farad/coulomb?

9. When a charged rod is brought near bits of paper, the paper will first cling to the rod. Soon thereafter the paper will drop from the rod. Why?

10. When a negatively charged rod is brought near a positively charged electroscope, the leaves collapse. As the rod is brought still closer (but not touching), the leaves again diverge. Give reasons for this behavior.

11. What is 1 coulomb/volt called?

12. What is the equation which relates capacitance, voltage, and charge? Give the proper units for each variable in the equation.

13. Distinguish between the electric field intensity of a charged capacitor and the dielectric field strength of the capacitor.

14. What factors determine the capacitance of a capacitor?

15. List three functions of the dielectric in a capacitor.

16. To which of the following is the farad unit not equivalent? (*a*) coulomb^2/joule (*b*) coulomb/volt (*c*) coulomb/volt^2 (*d*) joule/volt^2.

17. When the separation between the plates of a capacitor is halved, what happens to its capacitance?

18. What is meant when we speak of a positive potential of +24 V at point *P*?

19. How many lines of flux come into or emanate from a charge *Q* of one coulomb?

20. What is meant when it is stated that there is a potential difference of 12 V between points *A* and *B* with *B* being at the higher potential?

PROBLEMS

Group One

1. Find the force between charges of +120 μC and −25 μC located 50 cm apart in air.

2. Two charges attract each other with a force of 18×10^{-6} N when they are 40 cm apart. Find the force between them when their separation is 60 cm.

3. Find the force on an electron placed in an electric field whose intensity is 5.25×10^4 N/C.

4. A 50-μF capacitor is connected to a 12-V battery. What is the charge on each plate of the capacitor?

5. A capacitor has a capacitance of 5000 μF and is given a charge of 100 C. What is the potential difference between the plates?

6. A 0.4-μF capacitor consists of two parallel plates, each with an area of 144 in.^{2} and an air gap of 0.0020 in. between them. If the gap is increased to 0.0025 in., what will be the resulting capacitance?

7. What is the potential difference between the plates of a capacitor that has a capacitance of 5000 μF when given a charge of 250 C?

8. What is the capacitance of a capacitor if it acquires a charge of 5.4×10^{-4}C when connected across a 12-V battery?

9. A 75-μF capacitor has a potential difference of 12 V across it. What is the charge of the capacitor?

10. What voltage must be impressed across two parallel plates 2 mm apart in order to obtain an electric field of 2×10^5 N/C between the plates?

11. A plastic film capacitor is made from two strips of aluminum foil each 5 cm wide and 10 m long, separated by a 0.25-mm-thick film dielectric of Mylar. If the dielectric constant for Mylar is 3.0, what is the capacitance of the capacitor?

12. A parallel-plate air capacitor has a capacitance of 72 pF. When a sheet of Bakelite is inserted between the plates the capacitance is 432 pF. What is the dielectric constant of bakelite?

Group Two

13. What is the electric potential 15 cm from a point charge of 2.5 μC?

14. An electrostatic generator accelerates protons through a potential difference of 9.00×10^6 V. What is the energy of a proton when it reaches its maximum velocity?

15. The potential difference between the two terminals of a battery is 12 V. How much work is required to transfer 15 C of electricity from one terminal to the other? Where does the energy come from to do this work?

16. Four equal positive charges $+Q_1$ are placed at the four corners of a square which has sides of length s. What is the net force on a negative charge Q_2 placed at the center of the square?

17. A parallel-plate capacitor is made of two metal sheets 10×10 cm, 1.6 mm apart. The plates are connected to the terminals of a 32-V battery. What is the electric field intensity between the plates in newtons/coulomb?

18. Two charges of +60 and −20 μC are 40 cm apart. A third charge of +30 μC is located halfway between the two charges. (*a*) What is the magnitude and direction of the force on the 30-μC charge? (*b*) What is the field intensity at the midway point?

19. Two charges of 40 and 30 μC are 50 cm apart. Find the electric field strength at a point 20 cm from the smaller charge on a line connecting the two charges.

20. The field intensity between two deflecting plates of an electrostatic-deflection television tube is 50,000 N/C. (*a*) What will be the force on an electron passing between the two plates? (*b*) What acceleration does the electron experience? Assume a mass of 9.11×10^{-31} kg for the electron.

21. What is the potential difference between two points if 450 J of work is required to move 75 C from one point to the other?

22. An electric potential of 150 V exists at all points 12 cm from an isolated positive charge. What is the magnitude of the charge?

23. What potential difference must be applied across a 160-μF capacitor to give it a potential energy of 6.4 J?

24. The plates of a parallel capacitor are 5 mm apart and each plate has an area of 2.5 m^2. A potential difference of 12,000 V is applied across the capacitor. If the plates are separated by a vacuum, compute (*a*) the capacitance, and (*b*) the charge on each plate.

25. A 2-μF and a 4-μF capacitor are connected in series and the series is connected to a 24-V battery. (*a*) What is the equivalent capacitance of the combination? (*b*) What is the charge on each capacitor? and (*c*) calculate the voltage across each capacitor.

26. A drop of oil with a mass of 2×10^{-11} g is suspended (neither rising or falling) when placed in a uniform electric field **E** of 4×10^5 N/C directed vertically downward. What is the charge on the drop of oil?

27. A single-gang variable capacitor (see Fig. 25.37) consists of 11 rotating plates and 12 fixed plates. The gap between adjacent plates is 1 mm. The dielectric is air. What must be the area of the field between adjacent plates if the maximum capacitance is to be 250 pF?

28. A capacitor-resistance circuit has a time constant of 2×10^{-6}. What fraction of the total final charge is on the capacitor 0.20×10^{-6} s after starting charging? (HINT: You will need a calculator or a table for values for e^{-x}.)

29. A 20-μF capacitor is connected to a 6.0-V storage battery. The resistance of the circuit is 8 Ω. (*a*) What is the final charge on the capacitor? (*b*) How long does it take to reach 63 percent of its final charge? The capacitor is then disconnected from the battery. (*c*) If the dielectric between the capacitor plates has a resistance of 10^{10} Ω, how long will it take for the charge in the capacitor to drop to 37 percent of its initial value?

30. Find the capacity of the combination of capacitors shown in Fig. 25.38.

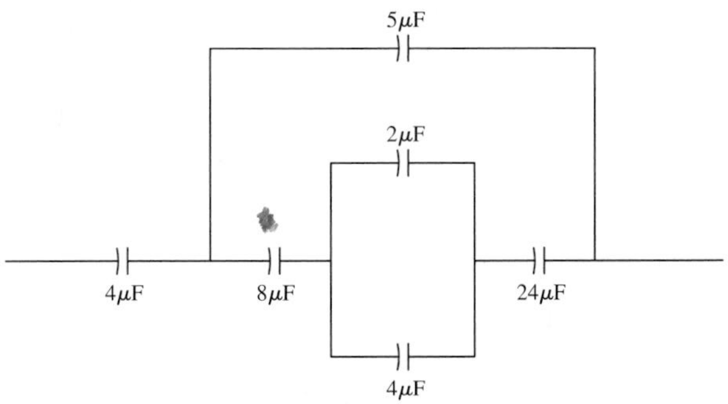

Fig. 25.38 Problem 30.

31. A charge of 5.0×10^{-9} C is placed in a uniform field of intensity 75,000 N/C. (*a*) How much work is done if the charge is moved 0.40 m in the direction of the field? (*b*) How much work is done if the charge is moved 0.40 m perpendicular to the field?

32. Three capacitors 4, 8, and 12 μF, respectively, are connected in series. Find the capacitance of the combination.

33. Three capacitors 3, 6, and 12 μF, respectively, are connected in series to a 360-V battery. Find (*a*) the capacitance of the series combination, (*b*) the charge on each capacitor, and (*c*) the voltage across each capacitor.

34. Given three capacitors connected as shown in Fig. 25.39. Find the total capacitance if $C_1 = 2.0\ \mu$F, $C_2 = 3.0\ \mu$F, and $C_3 = 5.0\ \mu$F.

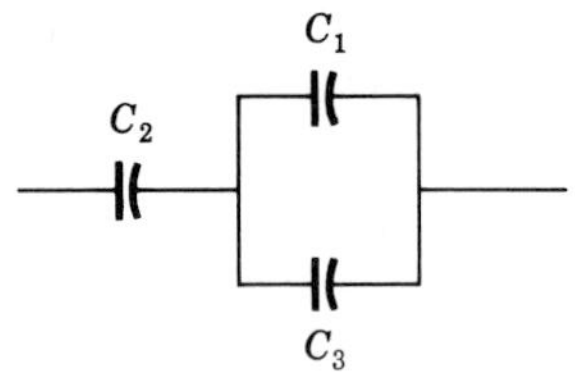

Fig. 25.39 Problem 34.

35. An electron is accelerated between two plates that are 1.5 cm apart. One of the plates is grounded. The other has a potential of +240 V with respect to the earth. What is (*a*) the field intensity between the plates and (*b*) the force acting on the electron? (*c*) If the electron starts at rest at one of the plates, what is the speed with which it strikes the other plate? (*d*) Calculate the gravitational force at the earth's surface on the electron. Compare the answers to (*b*) and (*d*).

Group Three

36. It is desired to combine 1.0-μF capacitors to make a combination with a capacity of 0.75 μF. How should they be connected to use as few 1.0-μF capacitors as possible?

37. Point charges of +0.24 μC and −0.18 μC are 60 cm apart in a vacuum. Point *A* is midway between the charges. Point *B* is 20 cm from the 0.24 μC charge and 40 cm from the −0.18 μC charge. What is the potential difference between *A* and *B*?

38. Four equal 30-μC positive charges are placed at the four corners of a square. The length of each side of the square is 300 cm. Find the force on one of the charges.

39. Electrons are accelerated in a uniform electric field whose field intensity is 1.8×10^{-4} N/C for a distance of 2.4 cm. If the electrons were initially at rest, find (*a*) their final speed, and (*b*) the time taken to travel the 2.4 cm. Assume an electron mass of 9.109×10^{-31} kg. [HINT: You will need to use Eqs. (25.4), (5.1), (4.8), and (4.5) in that order.]

40. A beam of electrons is produced by allowing electrons to fall through a potential difference of 2500 V. What is the final speed of the electrons if they start from rest?

41. A point charge of *q* μC is to be placed on a line between two charges of 3 μC and 4 μC which are separated by a distance of 20 cm. Where should the charge be placed so that the force on the point charge is zero? (You will need to know how to solve a quadratic equation here.)

42. How long before a capacitor will lose 99 percent of its charge if the time constant for the *RC* circuit is 0.25 μs? What is the current in the circuit at that time, as a fraction of its initial current? (A calculator or a table for e^{-x} will be required in order to solve this problem.)

43. In Fig. 25.40, if point *B* is grounded and point *A* has a potential of 1500 *V*, find the charge on each capacitor.

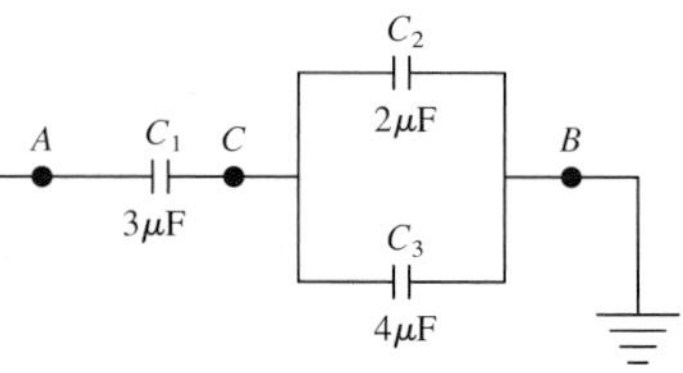

Fig. 25.40 Problem 43.

44. An atom of hydrogen has a proton nucleus and an orbital electron. Assume the orbit of the electron is circular and that the electron and the proton are 5.6×10^{-11} m apart. Calculate (*a*) the mutual force of attraction between the particles, (*b*) the electron's orbital speed, and (*c*) the gravitational attraction between the proton and the electron. [HINT: Use Eq. (8.13) to solve (*b*).]

45. Using the data given in Fig. 25.41, find the capacitance C_x if the voltmeter reads 10 *V*.

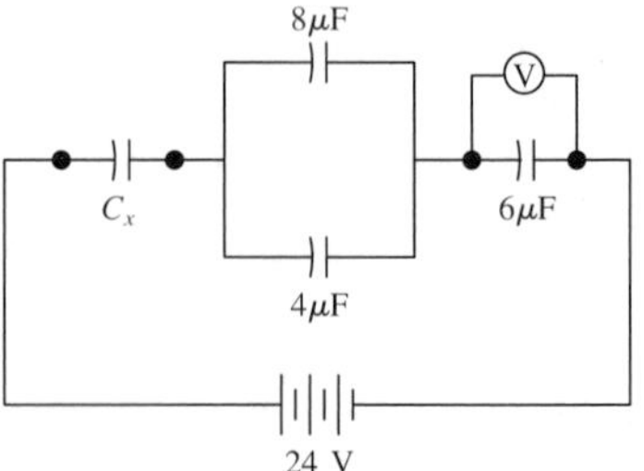

Fig. 25.41 Problem 45.

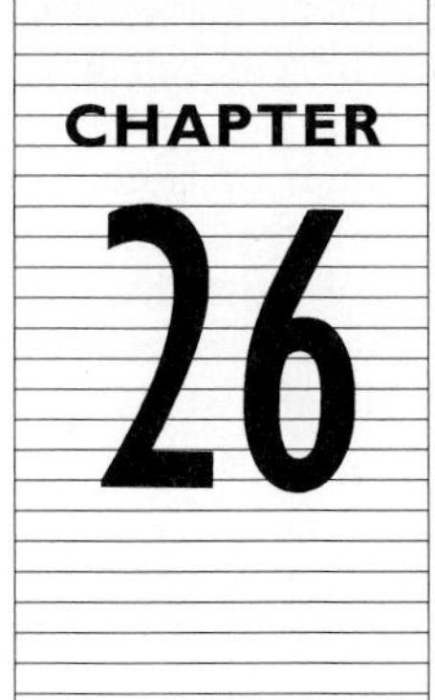

BASIC ELECTRIC CIRCUITS

A flow of electric charges constitutes *current*. An electric current can occur when there is a source of electric energy, such as a dry cell, a photo cell, a battery, or a generator to continually supply the energy. The electric charges flow through a *closed electric circuit*. The circuit exists for the purposes of transferring energy and doing work.

The movement of charged particles resulting from the pressure of a potential gradient is called an electric current.

In this chapter, we discuss elements that make up an electric circuit and develop laws governing such circuits. In developing circuit theory in this and the next three chapters, we are primarily concerned with *direct currents*. A direct current (dc) is a unidirectional current, and it is assumed to have a constant magnitude. *Alternating current* (ac), in which the direction of the current keeps reversing, will be discussed in Chap. 31.

26.1 ■ Electric Current

A metallic conductor, such as a copper wire, consists of a tremendously large number of atoms, each having electrons from outer orbits that are free to move through the material. In the absence of an electric field the rate at which electrons pass a point in the wire from left to right is the same as the rate at which they pass the point from right to left. The *net* rate is zero.

If the ends of the wire are connected to the terminals of a battery or generator, an electric field or potential gradient will be set up within the wire. This potential gradient will cause a continuous motion of electrons through the metallic conductor. For example, if the ends of a copper wire 1-m long are connected to the terminals of a 12-V battery, a potential gradient or electric field intensity of 12 V/m or 12 N/C will be established and maintained along the length of the wire. Thus, if the wire is 4 m long, the strength of the field at every point within the wire will be 3 V/m. The field **E** will act on the electrons to give them a motion in the direction of $-\mathbf{E}$.

If the terminals of the battery or generator are reversed, the motion of the electrons will be reversed. We therefore speak of the sources of electric energy as having fixed polarity. One terminal is called *positive* and the other *negative*.

A person might think it logical to describe the direction of the current as the direction of electron flow. However, when the convention of positive and negative charges was first conceived two centuries ago, it was assumed that positive charges flowed in a conductor. Also, we now know that although in metals the charge carriers are electrons, in electrolytes and in gaseous conductors the carriers may be either positive or negative.

As far as the *external* effects are concerned, it makes no difference if we consider hypothetical positive charges moving in one direction or real negative charges (electrons) moving in the opposite direction. The two ideas are equivalent. In this and

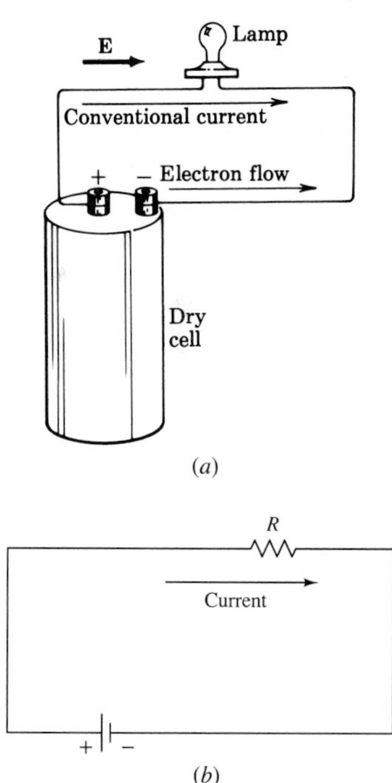

Fig. 26.1 (*a*) In a metallic conductor, the conventional current is in the direction of the electric field (from positive to negative). Electron flow is in the direction opposite that of the conventional current. (*b*) Schematic diagram of the circuit in part (*a*).

subsequent chapters, we will adopt the *conventional direction* of current, that is, from positive (+) to negative (−). We will even speak of the current in a metal conductor as though it consisted of motion of positive charges (see Fig. 26.1).

Current carriers may be electrons, positive charges, or ions.

In later chapters we will discuss the motion of atoms (or groups of atoms) in solutions or gases that have lost or gained negative charges. These charged atoms are called *ions*. Those atoms that gain electrons are negative ions and those that lose electrons are positive ions. Positive ions do truly move in the direction of the conventional current in ionic conduction in solutions and gases. We will also study how positive "holes" in semiconductors move in the direction of the conventional current.

Adding an electron to a neutral atom gives it a negative charge; removing an electron from a neutral atom gives it a positive charge.

The Institute of Electrical and Electronics Engineers (IEEE) uses conventional current direction in its standards and in its publications.

The conventional direction of current is the same as the direction in which a positive charge will move.

For most purposes, positive charge flowing in one direction is exactly equivalent to negative charge flowing in the opposite direction.

Thus our conventional current direction will agree with the concepts developed in the chapter on electrostatics. You will remember that electric field, potential energy, and potential difference were defined in terms of positive charges. We will learn later, also, that the conventional direction will result in loss in electric energy "down a potential gradient" when it flows through the external circuit from + to −.

In the cases where we will need to consider actual electron movement (as in transistors), we will speak of *electron flow* and not current.

In this and subsequent chapters schematic drawings (see Fig. 26.1*b*) will be used to represent various electrical circuits. A single cell will be represented by the symbol |ı with the longer line segment of the symbol representing the positive and the shorter segment the negative terminal. Notice carefully that the *conventional* current is from "plus to minus" *in the external circuit only*. Within the cell, the direction of the *flow of charge* is from "minus to plus." A battery made up of multiple cells will be represented by the symbol |ı|ı|ı.

26.2 ■ The Magnitude of Electric Current—the Ampere

If a coulomb of charge passes a given point in a conductor in one second, there is one ampere of current at that point.

André Marie Ampère (1775–1836), a French mathematician and physicist, was a professor of mathematics at École Polytechnique in Paris. He is considered the founder of the science of electrodynamics.

An electric current is a flow of charge.* The electron is an extremely small electric charge. Consequently, a larger unit of electric charge or quantity of electricity is used for practical measurements. This unit, the *coulomb,* equals approximately 6.3×10^{18} (6.3 billion billion) electrons. The magnitude of an electric current in a circuit is the *time rate of flow of electric charge*. Current is represented by the letter I (or i). The SI-metric unit of current is the *coulomb per second* and is called the *ampere* (A), named after the French physicist André Ampère. Thus the formula relating amperes I, coulombs Q, and seconds t is:

$$I = \frac{Q}{t} \tag{26.1}$$

We speak of a 75-W, 120-V incandescent lamp as using about 0.62 A of current. This means that about 0.62 C of electricity passes through the lamp each second. Small currents are often expressed in milliamperes (mA) or microamperes (μA) where

$$1\ \text{mA} = 10^{-3}\ \text{A}$$
$$1\ \mu\text{A} = 10^{-6}\ \text{A}$$

* Since current is a flow of charge, the common expression "flow of current" should be avoided, since it would literally mean "flow of flow of charge."

Illustrative Problem 26.1 What quantity of electricity passes through the filament of an incandescent lamp in 3 h if it draws 0.41 A?

Solution

$$\begin{aligned} Q &= I \times t \\ &= 0.41 \text{ A} \times 3 \text{ h} \times 60 \text{ min/h} \times 60 \text{ s/min} \\ &= 4400 \text{ C} \qquad \textit{answer} \end{aligned}$$

Illustrative Problem 26.2 The current in a fuse is 95 mA. How long does it take 22 coulombs to pass through the fuse?

Solution From Eq. 26.1, we get

$$\begin{aligned} t &= \frac{Q}{I} \\ &= \frac{22 \text{ C}}{95 \times 10^{-3} \text{ A}} \\ &= 2.3 \times 10^{2} \text{ s} \qquad \textit{answer} \end{aligned}$$

In actual practice the term *coulomb* is seldom used, for we are less interested in the quantity of charge supplied than in the *electric energy* available. The ampere has another (*official*) definition. That definition will be given in Chap. 28.

26.3 ■ MKSA Units in Electricity

The *fundamental* units in the mks system are the meter, the kilogram, and the second. Up to now, we have been able to define all other physical quantities in terms of these three fundamental units. Now, for electrical purposes, we will enlarge the number of fundamental units to include the ampere (A) unit. The system then is known as the MKSA system. The MKSA system was adopted in 1965 by the Institute of Electrical and Electronics Engineers.

The fundamental units in the MKSA system are the meter, the kilogram, the second, and the ampere.

Each electrical unit can be defined in terms of these four MKSA units. This system was adopted at the General Conference on Weights and Measures in 1960. Today most of the scientific and engineering organizations throughout the world have adopted the SI system of units which, in addition to the above four, includes the degree kelvin, *K* and the candela, cd. The table in Appendix I lists symbols of SI quantities and units.

26.4 ■ Hydraulic Analogy of a Circuit

When a wire is connected to the terminals of a continuous source of potential difference, electrons will flow through the wire from the negative to the positive terminal. The electrons flowing through the wire are repeatedly deflected or stopped as they strike neighboring atoms. The kinetic energy lost by the electrons in these collisions is gained by the atoms. The flow of electrons through the conductor will generate heat in the wire and raise its temperature. Thus, the electron motion is not a constant accelerated motion from one end of the wire to the other. Rather there is a slow *drift flow* (about 0.2 mm/s average speed) of electrons. However, the *changes* in flow rate are propagated with a speed approaching the speed of light. The propagation is similar to that which occurs when a water valve connected to a full hose of water is opened or closed. Almost as soon as the water flow is altered at the valve, an equal change occurs at the other end of the hose. Yet each molecule of water is likely to move very little in the interval of time it takes for the change to occur at the open end.

We can illustrate the flow of electricity by comparing it to water flow. The water pipe corresponds to the electric conductor. The amount of water per second (gallons per second) flowing past a given point in the pipe corresponds to the electric current (coulombs per second). When one ampere flows in a wire it means that 6.3×10^{18} electrons per second enter one end of the wire and an equal number of electrons per second leave the wire at the other end. Chances are that *very few* are the same electrons.

26.5 ■ Difference of Potential—the Volt

Both a conducting path and a potential difference are needed for a current to occur.

EMF is the no-load potential difference of a source.

Let us carry our hydraulic analogy further. Water will not move in pipes unless the pressure behind it is greater than the pressure in front of it. Similarly, electrons move along a wire only if there is a difference of electric pressure (called *potential difference*) along the conductor. This difference is produced by a battery or a generator, which acts as a pressure pump to produce a potential difference, just as a water pump (Fig. 26.2) builds up water pressure. Points along the pipe R are at different pressures; hence, when the valve V is open, water will flow. The pump P maintains the pressure by lifting water to reservoir A. The unit difference of mechanical pressure is in newtons per square meter (pascals) or in pounds per square inch. In a similar manner (see Fig. 26.3) a generator G builds up electric potential difference so that electric charges will flow in the circuit when the switch K is closed. The unit of electric pressure is the *volt*. The common dry cell develops a voltage of 1.5 V. Most household circuits are supplied at 120 or 220 V. Industrial circuits are typically supplied at 220 or 440 V.

A source of electric current, such as a battery or a generator, produces a potential difference across its terminals (battery) or windings (generator).

***The potential difference across a source* is known as *electromotive force* (emf).* It is the zero-current potential difference supplied by the source.**

Electromotive force is also measured in volts, the same as potential difference. We speak of the emf (pronounced "ee-em-eff") of a circuit as the total voltage applied to the circuit, or as the total potential drop across the entire circuit. The term *potential drop* may be used to describe the voltage drop across any portion of a circuit. Thus we see that though emf and potential drop are measured in the same units (volts), they are not at all synonymous terms. Hereafter, *emf* or $\mathscr{E}$ will be used to designate the potential difference that exists across a battery, generator, or other *source* of electrical energy that is not connected to an external circuit. On the other hand, V will

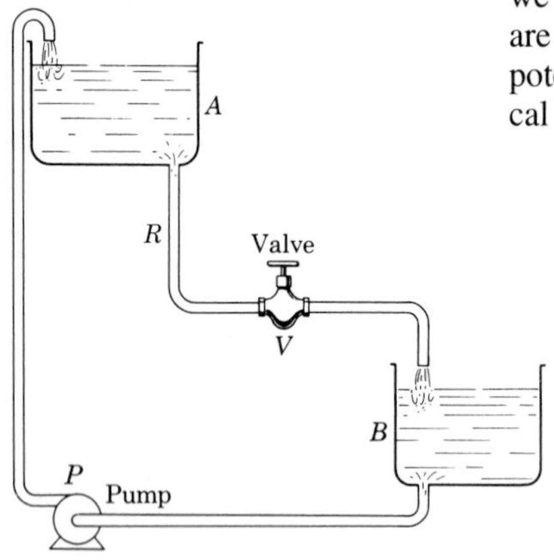

Fig. 26.2 Hydraulic analogy of an electric circuit.

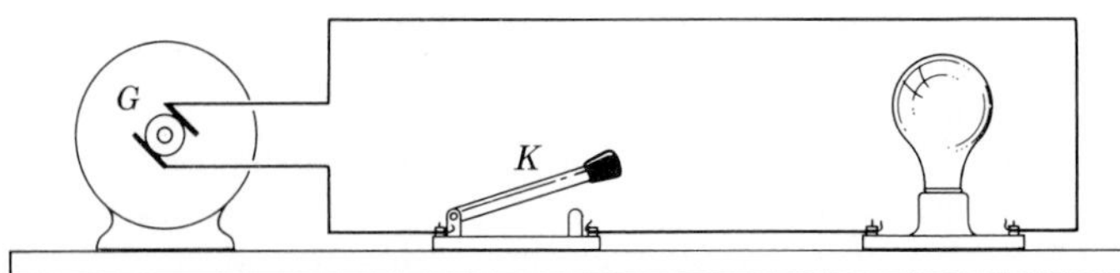

Fig. 26.3 Simple electric circuit.

* While "electromotive force" conveys the idea of a force which causes electrons to move, it should be noted that voltage is not a true force. The unit does not have the same mksa dimension as the newton, the SI unit of force.

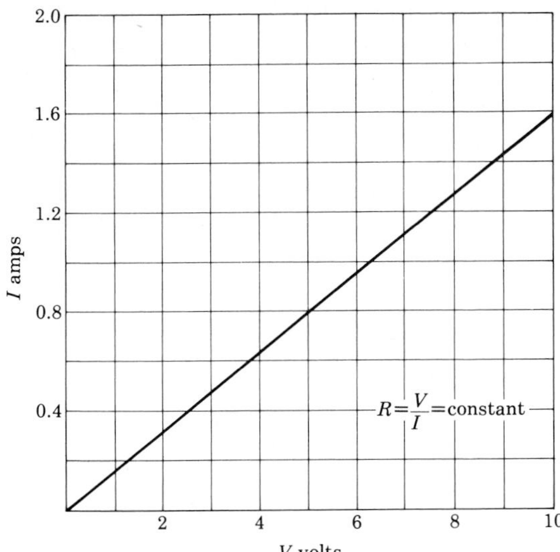

Fig. 26.4 When a potential difference exists between the ends of a metal conductor at constant temperature, the potential difference–current graph is a straight line. V/I is a constant (resistance R).

designate the potential difference between any two points of a closed electric circuit. An electromotive force $\mathscr{E}$ causes differences of potential V to exist between points in a closed circuit. The concept of emf will be covered in greater detail in Chap. 27.

26.6 ■ The Unit of Resistance Is the Ohm

If there were no resistance to a current in a conductor, no potential difference would be needed along its length to sustain the current. But just as the stream of water flowing in the pipe of Fig. 26.2 is retarded by frictional resistances, so various materials offer resistances to electric currents. The degree to which a substance offers resistance to an electric current divides substances into two broad classes: *conductors* and *nonconductors* (or *insulators*). All conductors, except superconductors, offer some resistance to the flow of electricity. *Superconductors* are a group of metals, alloys, and compounds that, when cooled to temperatures below a certain low critical temperature, offer essentially zero resistance to the passage of an electric current. (See Sec 26.12).

The ohm is the unit of electrical resistance.

George Simon Ohm (1787–1854), a German scientist, studied the relationship between potential differences applied to the ends of a metal wire and a current applied through it. He applied increasing potential differences to the ends of a metal wire whose temperature was kept constant. He then measured the current in the wire and discovered that the current increased in direct proportion to the difference of potential. The potential difference-current graph for a given wire at a fixed temperature is a straight line (see Fig. 26.4).

The constant ratio of potential difference to current is called the *resistance* of the conductor. The relationship between potential difference and current in a metal conductor is known as *Ohm's law*.

Ohm's Law

$$\frac{\textbf{Potential difference}}{\textbf{Current}} = \textbf{Resistance}$$

$$\frac{V}{I} = R \qquad \textbf{(26.2)}$$

When V is expressed in volts and I is expressed in amperes, R is expressed in the unit called *ohm* (Ω). Thus, the ohm has the dimension volt/ampere.

Unit of Electrical Resistance

The ohm is the resistance of a conductor such that a potential difference of one volt will maintain a current of one ampere in the conductor.

Large resistances are usually expressed in megohms (MΩ). Small resistances are expressed in micro-ohms ($\mu\Omega$).

$$1 \text{ megohm} = 10^6 \ \Omega$$
$$1 \text{ micro-ohm} = 10^{-6} \ \Omega$$

Illustrative Problem 26.3 A 1000-W toaster when plugged into a 120-V outlet draws 12 A. What is the operating resistance of the toaster?

Solution

$$V = 120 \text{ V}, \ I = 12 \text{ A}$$
$$R = \text{V/I}$$
$$= \frac{120 \text{ V}}{12 \text{ A}} = 10 \ \Omega \qquad \textit{answer}$$

The student should verify the following identities

$$\Omega = \frac{V}{A} = \frac{J/C}{C/s}$$

$$\Omega = \frac{J\cdot s}{C^2} = \frac{N\cdot m\cdot s}{C^2}$$

$$\Omega = \frac{kg\cdot m^2\cdot s}{C^2\cdot s^2} = \frac{kg\cdot m^2}{C^2\cdot s}$$

26.7 ■ Ohm's Law

Ohm's law relates current, potential difference, and resistance.

Ohm's law (Eq. 26.2) can be rewritten in the forms

$$I = \frac{V}{R} \qquad (26.2')$$

and

$$V = IR \qquad (26.2'')$$

Ohm's law has limitations.

Equation (26.2′) expresses the fact that the current passing through a metal conductor is directly proportional to the potential difference and inversely proportional to the resistance of the conductor.

Equation (26.2″) is used to find the voltage V required to cause a current I in a conductor whose resistance is R.

Two limitations to Ohm's law should be noted: (1) the law is true only if external conditions, e.g., the temperature of the conductor, are maintained constant; (2) the law is strictly valid only for ordinary metallic conductors. The current is not proportional to the potential difference, except over limited ranges, when electricity is conducted in liquids, gases, or solid-state devices.*

Illustrative Problem 26.4 How much current will an electric heater draw from a 120-V line if the resistance of the heater (when hot) is 26.7 Ω?

Solution Substitute known values in Eq. (26.2′) and solve:

$$I = \frac{V}{R}$$

$$= \frac{120\ \text{V}}{26.7\ \Omega}$$

$$= 4.5\ \text{A} \qquad \textit{answer}$$

Illustrative Problem 26.5 A 2.6-MΩ resistor has a current of 8μA through it. What is the potential difference across the resistor?

Solution

$$V = IR \qquad \text{(Eq. 26.2″)}$$

$$= 8\mu\text{A} \times \frac{10^{-6}\ \text{A}}{1\ \mu\text{A}} \times 2.6\ \text{M}\Omega \times \frac{10^{6}\ \Omega}{1\ \text{M}\Omega}$$

$$= 20.8\ \text{V} \qquad \textit{answer}$$

It is essential to distinguish carefully between the application of Ohm's law to a part of a circuit and to a complete circuit. Equation 26.2 expresses the relationship between the resistance R, the current I, and the potential difference V for a *part of a closed circuit*. When a *complete circuit* is to be considered, one must take into account all the potential differences (which are due to the emfs $\mathscr{E}$ of the sources of energy) and the total of all the resistances (R_t) in the circuit. In equation form, Ohm's law, then, becomes

Every source of electric energy has a certain amount of internal resistance which must be recognized in using Ohm's law.

$$\frac{\text{Net emf}}{\text{Current}} = \text{total resistance}$$

$$\frac{\mathscr{E}}{I} = R_t \qquad (26.3)$$

26.8 ■ Wire Measurements

The electrical resistance of a metallic conductor depends on four factors: (1) the kind of material, (2) the length, (3) the cross-sectional area, and (4) the temperature of the wire. As a general rule, pure metals offer smaller resistances, while alloys offer considerably greater resistances to current flow. A conductor 2 ft long offers twice as much resistance as a similar conductor 1 ft long. Also, the larger the cross-sectional area of a conductor of a given material and of a fixed length, the lower its resistance.

* Many conductors do not obey Ohm's law. The thermistor, for example, is a nonohmic device. Modern electronics in many respects depends in a fundamental way on the fact that many conductors, such as vacuum tubes, transistors, and thermistors, do not obey Ohm's law. In Chap. 31 we will find that Ohm's law will need to be modified for some alternating-current circuits.

Standardized conductor wire sizes are based on the cross-sectional area of the wires measured in units called *circular mils*

This is true because there are more electrons free to move in the larger conductor. The resistance of most metal conductors increases with an increase in temperature. Since most common wires in the United States still have diameters which are fractions of an inch, it is more convenient to express these diameters in terms of a unit called the *mil*. A mil is equal to 0.001 in. Thus a wire whose diameter is 0.072 in. is said to have a diameter of 72 mils.

Since most electric wires are circular in cross section, it is simpler to specify the area in *circular mils* than in square units. A *circular mil* (cmil) is defined as *the area of a circle whose diameter is* 1 *mil*. Using circular units eliminates the use of π. Area in circular mils equals the square of the diameter in mils. The area of a circle, $A = \pi d^2/4$, is proportional to the diameter squared. Consequently the area of a circle whose diameter is d mils is d^2 cmils.

Illustrative Problem 26.6 What is the area (a) in circular mils and (b) in square inches of a wire 0.032 in. in diameter?

Solution

(*a*)
$$\text{Diameter} = 0.032 \text{ in.} \left(\frac{1 \text{ mil}}{0.001 \text{ in.}}\right)$$
$$= 32 \text{ mils}$$

Then
$$\text{Area} = (32 \text{ mil})^2 \left(\frac{1 \text{ cmil}}{1 \text{ mil}^2}\right)$$
$$= 1024 \text{ cmils} \qquad \textit{answer}$$

(*b*)
$$\text{Area} = \frac{\pi d^2}{4} \text{ in conventional square units}$$
$$= \frac{3.14(0.032 \text{ in.})^2}{4}$$
$$= 0.000804 \text{ in.}^2 \qquad \textit{answer}$$

26.9 ■ Resistivity

The resistivity, measured in ohm-meter units, is numerically equal to the resistance of a material with a length of 1 m and a cross-sectional area of 1 m^2.

Resistivity depends only on the material used and on its temperature.

The practical (English system) unit of length for wires is the foot. The resistances of most conductors are generally standardized in terms of the resistance of a wire having an area of 1 cmil and a length of 1 ft. Such a unit wire is called the *circular mil/foot*. The *resistivity* of a substance is equal numerically to the resistance of 1 cmil/ft of the substance. The unit of resistivity is the *ohm-circular mil per foot*. In the metric system the resistivity of a material is numerically equal to the resistance of a piece of the material 1 m long and 1 m^2 in cross-sectional area. The unit of resistivity in the SI-metric system is called the *ohm-meter* (ohm-meter^2/meter = ohm-meter). The symbol ρ is used to denote resistivity units. To obtain ρ in ohm-cmils per foot, multiply ρ in ohm-meters by 6.02×10^8; to obtain ρ in ohm-meters, multiply ρ in ohm-cmils per foot by 0.166×10^8. Table 26.1 lists resistivities of some common materials.

Table 26.1 shows that the pure metals have the lower resistivities. Copper and aluminum are generally used for electric-transmission purposes. Silver is an even better conductor of electricity than copper, but it is not used extensively because of its high cost. Gold, however, is used frequently especially for electric contacts. Because of its low resistivity it doesn't oxidize (burn) readily and is quite ductile. Tungsten is used almost exclusively for filaments of incandescent lamps. Alloys have relatively higher resistances than the pure metals. They are used in heating units because they have high resistances and because they do not oxidize (burn) readily at high temperatures. The alloys listed are quite hard and durable.

Table 26.1 Resistivities of common conductors (and some insulators) at 20°C (68°F)

	Resistivity ρ	
	ohm-meter ($\Omega \cdot m$)	*ohm-circular mils/foot*
Elements:		
Aluminum, commercial	2.8×10^{-8}	16.9
Carbon	3500×10^{-8}	21,000
Copper, commercial	1.7×10^{-8}	10.3
Gold	2.4×10^{-8}	14
Iron	$12\text{-}14 \times 10^{-8}$	72-84
Lead	22×10^{-8}	132
Mercury	96×10^{-8}	575
Silver	1.6×10^{-8}	9.7
Sulfur	10^{5}	
Tungsten	5.5×10^{-8}	33
Zinc	6.1×10^{-8}	36.7
Alloys:		
Brass	$6.3\text{-}8.6 \times 10^{-8}$	38-52
Chromel	$104\text{-}109 \times 10^{-8}$	625-655
Constantan (Cu 60%, Ni 40%)	49×10^{-8}	295
German silver	33×10^{-8}	199
Manganin (Cu 84%, Mn 12%, Ni 4%)	43×10^{-8}	260
Nichrome (Ni 60%, Cr 12%, Mn 2%, Fe 26%)	110×10^{-8}	660
Paraffin	3×10^{24}	

26.10 ■ Calculation of the Electric Resistance of a Conductor

The resistance of a conductor of uniform cross section is directly proportional to the resistivity and the length and inversely proportional to the cross-sectional area of the conductor. In the English system

$$R = \rho \frac{L}{d^2} \tag{26.4}$$

The resistance of a conductor depends on the material, its length, and its cross-sectional area.

where R = resistance, ohms
ρ = resistivity, ohm-cmils/foot
L = length, feet
d^2 = cross-sectional area of conductor, cmils

In the SI system

$$R = \rho \frac{L}{A} \tag{26.5}$$

$1\ \Omega \cdot \text{cm} = 10^{-2}\ \Omega \cdot \text{m}$.

If R is expressed in ohms, A in square centimeters, and L in centimeters, then ρ is in ohm · cm²/cm, or ohm-centimeters. This unit is usually more convenient than the ohm-meter. The value of ρ in ohm-meters is multiplied by 10^2 to get ρ in ohm · cm.

Illustrative Problem 26.7 Find the resistance of 2000 ft of commercial copper wire having a diameter of 0.102 in.

Solution The cross-sectional area of a wire whose diameter is 0.102 in. is

$$d^2 = (102 \text{ mils})^2 = 10{,}404 \text{ cmils}$$

From Table 26.1, ρ (for commercial copper) = 10.3 Ω · cmils/ft. Substituting in Eq. (26.4) gives

$$R = 10.3\ \Omega \cdot \text{cmils/ft} \times \frac{2000\ \text{ft}}{10{,}404\ \text{cmils}}$$

$$= 1.98\ \Omega \qquad \textit{answer}$$

Illustrative Problem 26.8 How much nichrome wire of diameter 0.081 cm will have to be used in order to build a resistor of 5 Ω (20°C)?

Solution Solving for L in Eq. (26.5), we have

$$L = R\frac{A}{\rho}$$

From Table 26.1,

$$\rho = 110 \times 10^{-8}\ \Omega \cdot \text{m}$$

Substituting yields

$$L = \frac{5\ \Omega \times (\pi/4)(8.1 \times 10^{-4}\ \text{m})^2}{110 \times 10^{-8}\ \Omega \cdot \text{m}}$$

$$= 2.34\ \text{m} \qquad \textit{answer}$$

Illustrative Problem 26.9 Find the resistivity of a wire made of an alloy of copper, zinc, and nickel whose mean diameter, measured by a micrometer caliper, is 0.072 in. The wire is 100 ft long and has a total resistance of 3.88 Ω.

Solution Solve Eq. (26.4) for ρ to get

$$\rho = R\frac{d^2}{L}$$

$$d^2 = (72\ \text{mils})^2 = 5184\ \text{cmils}$$

Then

$$\rho = 3.88\ \Omega \times \frac{5184\ \text{cmils}}{100\ \text{ft}}$$

$$= 201\ \Omega \cdot \text{cmils/ft} \qquad \textit{answer}$$

26.11 ■ Temperature and Resistivity

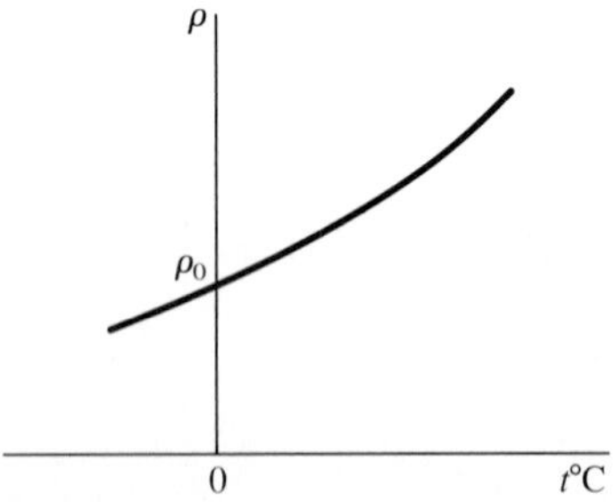

Fig. 26.5 Variation of resistivity with temperature for a typical metal conductor.

The resistivity of most pure metals increases with a rise in temperature. In nonmetallic conductors the resistivity decreases as the temperature rises. Most alloys not only have relatively high resistivities, but their resistivities are much less affected by temperature changes.

The relationship between resistivity and temperature for a typical metallic conductor is shown in Fig. 26.5. For temperatures which are not too great, the graph is approximately a straight line. Thus we say that resistivity varies directly with the temperature. We can express this relationship by the equation

$$\rho = \rho_0(1 + \alpha t) \tag{26.6}$$

where ρ_0 is the resistivity at 0°C and ρ is the resistivity at t°C. The quantity α (Greek letter *alpha*) is called the *temperature coefficient of resistivity*. The temperature coefficient of resistivity is the fractional increase in resistivity per Celsius degree increase in temperature. Its units are deg^{-1} or "reciprocal degrees." Table 26.2 lists temperature coefficients of resistivity for several common materials.

Table 26.2 Temperature coefficients of resistivity

Material	α (°C^{-1})
Metals:	
Aluminum	0.0039
Carbon	−0.0005
Copper, annealed	0.0039
Gold	0.0034
Iron	0.0050
Lead	0.0043
Mercury	0.0089
Platinum	0.0030
Tin	0.0042
Tungsten	0.0045
Zinc	0.0020
Alloys:	
Brass	0.0020
Constantin	0.00003
Nichrome	0.0004

In general, metals increase in resistivity with an increase in temperature, while nonmetals decrease in resistivity with an increase in temperature.

Since the resistance of a given conductor is proportional to its resistivity, and its length and cross sectional area change little with temperature changes, its resistance at any temperature t may be written

$$R_t = R_0(1 + \alpha t) \tag{26.7}$$

where R_0 is the resistance of the conductor at 0°C and R_t is the resistance at t°C.

Illustrative Problem 26.10 The resistance of a coil of copper wire at 20°C is 2.4 Ω. What is the resistance of the coil at 50°C?

Solution Solve Eq. (26.7) for R_0 and substitute values for R_t, α, and t.

$$\begin{aligned} R_0 &= \frac{R_t}{1 + \alpha t} \\ &= \frac{2.4\ \Omega}{1 + (0.0039/°\text{C})(20°\text{C})} \\ &= 2.23\ \Omega \text{ (the resistance of the coil at 0°C)} \end{aligned}$$

Then at 50°C,

$$\begin{aligned} R_{50} &= R_0(1 + \alpha t) \\ &= 2.23\ \Omega(1 + 0.0039°\text{C}^{-1} \times 50°\text{C}) \\ &= 2.66\ \Omega \qquad \textit{answer} \end{aligned}$$

Illustrative Problem 26.11 The resistance of a tungsten filament at 20°C is 2.5 Ω. What is the temperature of the filament if its resistance is 17.5 Ω? Assume α is constant over the temperature range.

Solution From Table 26.2, we find $\alpha = 0.0045°C^{-1}$.

$$R_0 = \frac{R_t}{1 + \alpha t} = \frac{2.5\ \Omega}{1 + 0.0045°C^{-1} \times 20°C}$$

$$= 2.29\ \Omega \text{ at } 0°C$$

Solve Eq. (26.7) for t to get

$$t = \frac{R_t - R_0}{\alpha R_0}$$

$$= \frac{17.5\Omega - 2.29\Omega}{(0.0045°C^{-1})(2.29\Omega)}$$

$$= 1500°C \qquad \textit{answer}$$

Scientists have known since 1911 that some metals cooled to near absolute zero will become superconductors.

26.12 ■ Superconductivity

Since the resistivity of a conductor decreases with a decrease in temperature, it would be reasonable to expect that if the temperature were decreased to absolute zero (0 K), a conductor would offer no resistance to an electric current. Ultralow temperature research with conductors began with the liquefaction of helium, accomplished by the Dutch physicist Heike Kamerlingh Onnes (1853–1926).

The boiling point of liquid helium is about 4 K (−269°C). Onnes found that the resistivity of conductors in general truly approximated a linear variation with decreasing temperatures. However, he found that some materials had a specific temperature below which the resistivity dropped suddenly to zero. Materials that possess this property are called *superconductors*. The transition temperature at which superconductivity occurs is called the *critical temperature* for that material. Onnes found (about 1911) that the critical temperature for mercury is 4.2 K (see Fig. 26.6). The critical temperatures for zinc, aluminum, tin, lead, and niobium are 0.88, 1.19, 3.72, 7.18, and 9.46 K, respectively. Electric current in a ring-shaped *superconducting* material has been observed to flow for years (in the absence of a potential difference) with no appreciable decrease in the current.

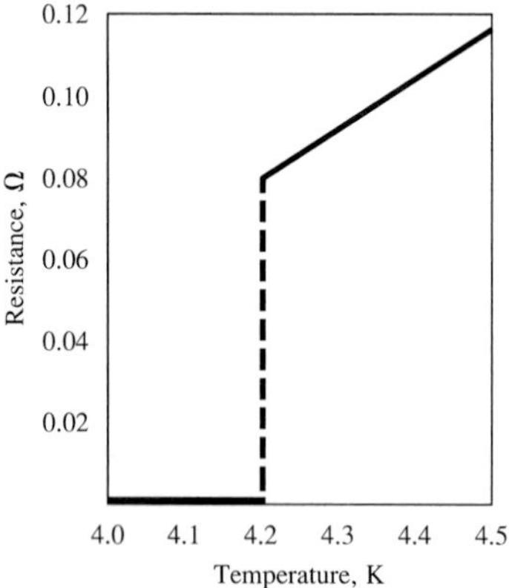

Fig. 26.6 Resistance versus temperature for mercury. Above 4.2 K the graph follows the normal curve given by Eq. (26.7). The resistance drops to zero at the critical temperature of 4.2 K.

Although known to physicists for three-quarters of a century, superconductors have found very few commercial applications, because of the extremely low temperatures required. Liquid helium is very expensive and temperamental to use. Since Onne's discovery, scientists have been experimenting with various mixtures, seeking to find a material which would exhibit superconductivity at higher temperatures.

In 1973 scientists managed to push up the temperature at which a known material became superconducting to 23 K (−418 °F). In 1986, K. Alex Mueller, a U.S. research physicist, announced that he had produced a compound of barium, copper, oxygen, and the rare earth element lanthanum which pushed the critical temperature up to 30 K. In 1987 Paul Ching-Wu Chu and a team of scientists from the University of Houston developed a compound which pushed the superconducting temperature up to 94 K. Chu's team crossed a crucial barrier. A superconductor could now be cooled by the more readily available and cheaper liquid nitrogen. Liquid nitrogen is 10 times cheaper and 20 times more effective as a coolant (within its temperature range) than liquid helium.

Superconductivity research may someday result in more efficient and cheaper electric power generation and distribution.

The production of superconducting materials with higher critical temperatures has caused scientists around the world to experiment feverishly with many new mixtures in an effort to raise the threshold temperatures to levels that can open new areas of commercial applications.

Perhaps the most important use of superconductivity will be in the operation of electromagnets to produce stronger magnetic fields. (Electromagnetism will be discussed in Chap. 28.) Electromagnets employing superconducting wires are in use today that are 1000 times as strong as conventional types of electromagnets of the same size. Medicine already uses superconducting magnets cooled by helium in "magnetic resonance imaging" used to "peer" deep in the body. If liquid nitrogen is used, the operating costs of such machines will be slashed and the size of the equipment can be reduced. The exposure times for patients would also be measurably shortened.

Superconducting electromagnets for particle accelerators can be used in basic research to study subatomic particles. Magnets could cushion (levitate) high-speed passenger trains off the roadbed so there will be essentially no friction. The levitation would arise from the repulsive force between the magnet (on the train) and the eddy currents produced in the track below. (Eddy currents will be discussed in Chap. 31.)

Electric-generating stations could make power more efficiently if their generator magnets could be shrunk in half by being wound with superconducting wire. Electricity could be generated at night and stored in gigantic underground coils of superconducting wire, ready to be tapped later the next day when the demand for electrical power is greater. Superconductors could be used for purposes of carrying an electric current almost indefinitely and for more efficient operation of electrical devices. Military applications of superconductivity could also be numerous.

Many extremely difficult problems remain to be solved before any of the new superconductors can be made practical. For example, the new superconductor materials are ceramics, prepared by grinding a blend of metallic elements and heating them at a high temperature. In its current form the rocklike material becomes very brittle on cooling. Scientists and engineers will have to discover how to fashion these materials into wire, tapes, and very thin films, a difficult task with brittle ceramics.

26.13 ■ Wire Gauge

Copper wire is commonly manufactured in accordance with standard specifications. For convenience, wire sizes are indicated by gauge numbers. In the United States the commonly used gauge is the *Brown & Sharpe* (B & S), often called the *American Wire Gauge* (AWG). The largest commercial-size round wire has an AWG number of 0000. The next smaller size is No. 000, then No. 00, then No. 0. The sizes then vary in whole-number values ranging from No. 1 to 40, which is the smallest size. Standard copper wire sizes and their gauge numbers plus other valuable information will be found in Appendix VI. This table will simplify wire computations. It reveals that diameters of the wires grow smaller as their gauge numbers increase and that the ratio of the areas of any two consecutive standard sizes is about 1.26. It also shows that every third gauge number halves the area of cross section and thus doubles the resistance.

In the United States, wire is manufactured in standard sizes indicated by their American Wire Gauge (AWG) numbers (see Appendix VI).

In checking the table, you will note that the resistance of a copper wire 1000 ft long and 0.1 in. in diameter is about 1 Ω. This will be convenient to remember in estimating resistances of other copper wires. For example, you should be able to determine readily that the resistance of a copper wire 4000 ft long and 0.2 in. in diameter will also be 1 Ω.

26.14 ■ Measurement of Current and Voltage

The two instruments most commonly used to measure current and voltage in a given circuit are the *ammeter* and the *voltmeter* respectively (Fig. 26.7). The details of construction of these two meters will be covered in Chap. 28. They are both direct-reading instruments; i.e., by reading the deflection of a pointer on each instrument or a digital readout when the current is flowing, one can read directly the amperage and the voltage. The ammeter has an extremely low resistance. Thus, when an ammeter is placed *in a circuit,* the total resistance of the circuit is not appreciably changed. On the other hand, the voltmeter has a very high internal resistance. When it is connected *across a circuit,* the total current flow in the circuit is not appreciably affected.

Scientists generally use schematic diagrams to represent the essential parts of a circuit. Standard symbols are used to represent specific apparatus. Many of the standard wiring diagram symbols are shown in Appendix VII. The student will become

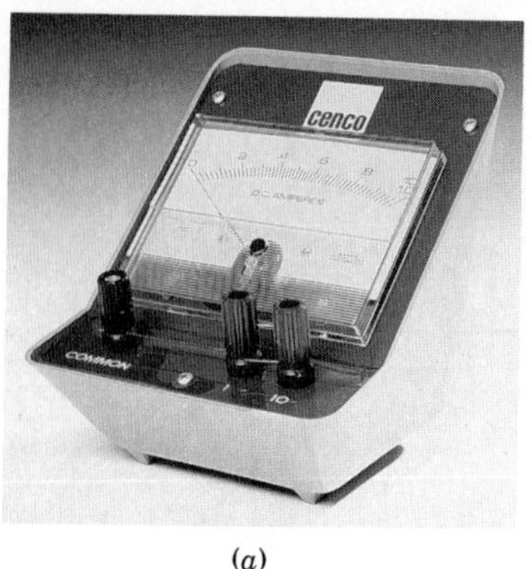

(a)

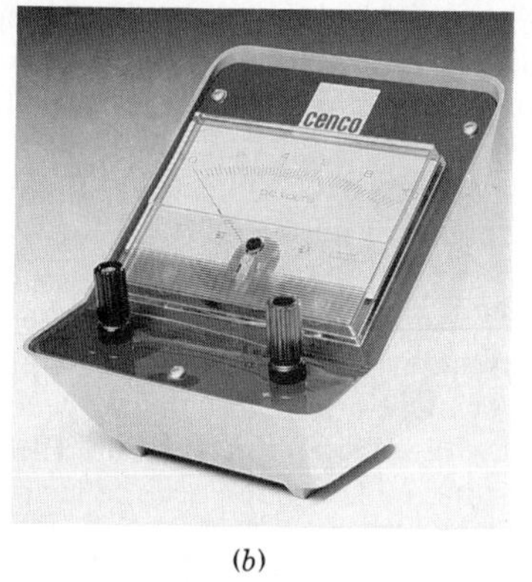

(b)

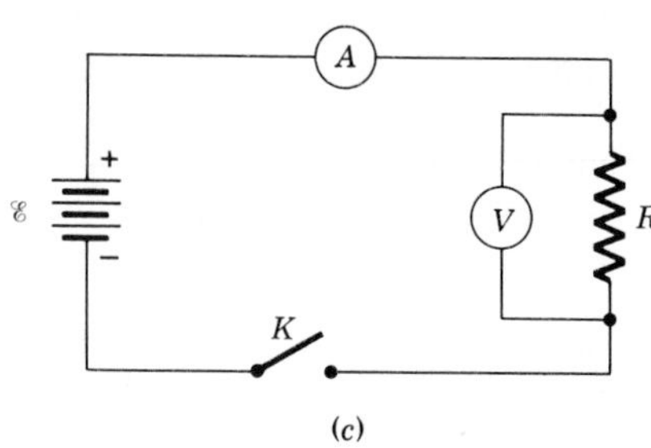

(c)

Fig. 26.7 (*a*) Dual range ammeter, and (*b*) single range voltmeter, commonly used in college physics laboratories. (*c*) Circuit used for measurement of amperes and volts. (Courtesy of Central Scientific Company)

acquainted with the more common ones as we proceed. Note the symbols for the ammeter and the voltmeter in Fig. 26.7*c*. The figure illustrates how the ammeter and the voltmeter are connected in a single circuit. Note that the ammeter is inserted directly *in the line* and the voltmeter is hooked *across* the portion of the circuit whose voltage drop is to be measured.

The ammeter should be connected in series, while the voltmeter should be connected in parallel with the resistance *R*.

In Fig. 26.7*c* the source of the emf is a battery *E*. Note its symbol. The ammeter *A* is connected *in* the circuit, and the voltmeter *V across* the load *R*. *K* is a single-pole switch. The ammeter is said to be connected in *series,* while the voltmeter is connected in *parallel* with resistor *R*. Thus if we read *A* and *V*, we will get the current *through* and the potential difference *across R*.

It should be pointed out that in a schematic diagram, the conductors connecting elements of a circuit are idealized to have no resistance. Such conductors are usually represented by straight lines. Thus, for example in Fig. 26.7*c*, all parts of the conductor connecting the battery and the ammeter to one side of the load *R*, are assumed to be at the same potential. There is a potential drop in *R*, in the conventional direction of current. All parts of the conductor from the other end of the load *R*, through the switch and back to the battery, are at the same potential also. The potential in this second part of the circuit is lower than that in the first by the magnitude of the potential drop measured by the voltmeter, *V*.

26.15 ■ Resistors

Fig. 26.8 The slide-wire rheostat provides a means of limiting the current in a circuit. (Sargent-Welch Scientific Co.)

In Tables 26.1 and 26.2 it can be noted that the resistivity of nichrome is approximately 60 times the resistivity of copper, and its temperature coefficient is about 10 percent that of copper. These properties make nichrome a good metal from which to make resistors. Spools wound with nichrome (or chromel) wire are used in an electric circuit to provide either *fixed* or *variable* amounts of resistance. Fixed nichrome resistors are used in electric heaters, electric ranges, and toasters. Such commercial resistors are manufactured in sizes ranging from a fraction of 1 Ω up to more than 100,000 Ω. The *slide wire rheostat* (Fig. 26.8) consists of a spiral coil of nichrome or chromel wire wound on a porcelain or enamel tube. Connections are made to one end and to a sliding contact. In Fig. 26.8, the current enters the coil at the terminal on the left, passes through the coil, and leaves it at the moving contact. From there it travels along a massive copper bar (assumed to have zero resistance) and leaves the rheostat at the right terminal. By moving the slider, the resistance can be varied within the range of the resistance of the entire length of the coil. In electronics, variable resistors are smaller. Some are made by winding the wire around a toroid (doughnut-shaped) form. The moving contact in this case moves along a circular path.

Dial-type variable resistance boxes are commonly used today in electrical laboratories. One such type is shown in Fig. 26.9. Step values of resistance are obtained by turning the dials. Each dial controls four or more series-connected resistors. The resistance of each set in the box increases consecutively by a factor of 10, hence it is called a decade resistance box.

Some decade resistors have as many as 10 series-connected resistors and can provide an accuracy of measurement to within 0.1 percent. It is important, in using a variable resistance box, that the current capacity of the highest-valued resistance not be exceeded. For that reason, the maximum permissible current values are usually shown on each dial.

Each switch has a multiple leaf phosphor bronze spring which provides a positive, wiping action against heavy brass plates to which are attached the various resistance coils (or high-grade carbon resistors). Each dial has ten positions, with each dial representing a multiple decade value of resistance. For example, in the four dial resistance box with a range of 0.1–999.9 Ω, the dials are marked from left to right: 100, 10, 1, 0.1. The resistance is indicated by reading, from left to right across the box, numbers appearing next to the dials on the plastic box which houses the mechanism. Thus, if the position numbers of the four dials (left to right) are 7, 0, 5, and 3, a resistance of 705.3 Ω would be indicated.

Small commercial resistors have wide applications in electronics; one widely used type is the *carbon resistor*. A mixture of carbon (or graphite) granules and a binding material can be varied to yield a wide range of high resistances. The hardened mixture is enclosed in a ceramic housing with a wire connector attached to each end of the resistor. Since the carbon resistor is quite small in size, it is easier to use a color code to indicate the resistance value than to print the resistance on the housing. The carbon resistor has four color-code bands. The EIA (Electronics Industries Association) has adopted the color code described in Table 26.3 and Fig. 26.10. Bands *A* and

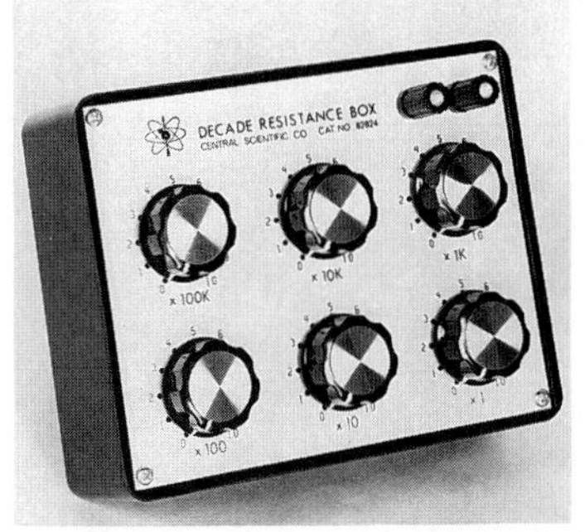

Fig. 26.9 Decade resistance box. Step values of resistance can be made by turning the dials to total resistance from 0 to 1,111,110 ohms in 1-ohm increments. (Courtesy Central Scientific Company)

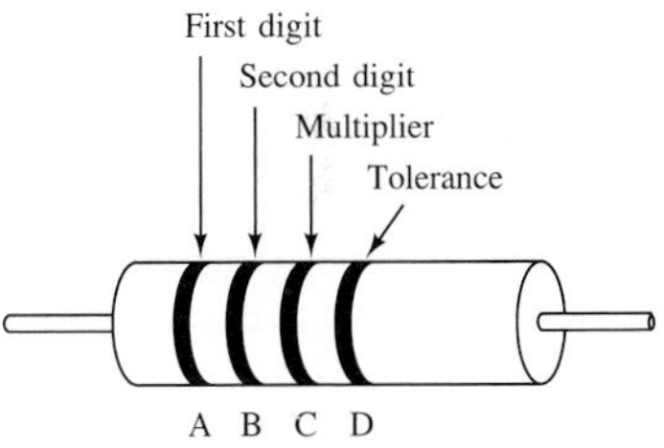

Fig. 26.10 EIA color-coded bands of a carbon resistor. As an example, if the colors of the bands *A*, *B*, *C*, *D* are yellow, violet, orange, and silver, the resistance value is 47×10^3 Ω and the tolerance is 10 percent.

Table 26.3 EIA resistor color code

Color	*Band A (1st significant figure)*	*Band B (2nd significant figure)*	*Band C (multiplier)*	*Band D (tolerance, %)*
Black	0	0	10^0	
Brown	1	1	10^1	
Red	2	2	10^2	
Orange	3	3	10^3	
Yellow	4	4	10^4	
Green	5	5	10^5	
Blue	6	6	10^6	
Purple	7	7	10^7	
Gray	8	8	10^8	
White	9	9	10^9	
Silver				10
Gold				5

Illustrated examples

Color-coded resistors.

Color bands	*Resistance, Ω*
orange-green-red-silver	$35 \times 10^2 \pm 10\%$
brown-black-orange-gold	$10 \times 10^3 \pm 5\%$
purple-yellow-black-gold	$74 \pm 5\%$
purple-yellow-black	$74 \pm 20\%$

B indicate the first two significant figures of the resistance value. The first two significant numbers are multiplied by the multiplier indicated by band C. Band D indicates the accuracy of the indicated resistance. A missing fourth (D) band indicates that the resistance value is correct within ± 20 percent of the coded value.

Composition resistors are generally less expensive than metal ones. They are commonly used in low-power situations. Wire-wound or metal-film resistors are usually more accurate than carbon resistors.

SERIES AND PARALLEL CIRCUITS

26.16 ■ Series Circuits

Most electrical circuits consist not merely of a single source of emf and a single external resistor. They usually comprise a number of emfs, resistors, or other elements such as capacitors, interconnected in various ways. To begin with, we will discuss a few of the simpler ways resistors may be interconnected in a circuit.

Consider the circuit illustrated in Fig. 26.11. In it there is one source of emf, and the resistors provide a single path between the end points of the emf source. In such a circuit the resistors are said to be connected in *series*. Other circuit elements such as capacitors, cells, motors, etc., are also said to be connected in series with one another if they are similarly connected *so as to provide a single path for the current*.

In a series circuit the current is the same through all the resistors in the circuit.

A series combination of elements in a circuit is one in which the current is the same in every part of the circuit.

An ammeter can be connected anywhere in a series circuit because the current is the same at every point in the circuit. Thus, in Fig. 26.11,

$$I = I_1 = I_2 = I_3 \tag{26.8}$$

where I, I_1, I_2, I_3 represent the current from the source, and the current in the resistors R_1, R_2, R_3, respectively.

The total *resistance R* of the circuit is equal to the *sum of the resistances of all the parts*. If we neglect the resistance of the wires connecting the loads, we get

$$R_T = R_1 + R_2 + R_3 \tag{26.9}$$

The emf, measured by the voltmeter V, must be equal to the sum of the potential drops measured by V_1, V_2, and V_3. If we express the voltages as V, V_1, V_2, V_3, respectively, we get

$$V = V_1 + V_2 + V_3 \tag{26.10}$$

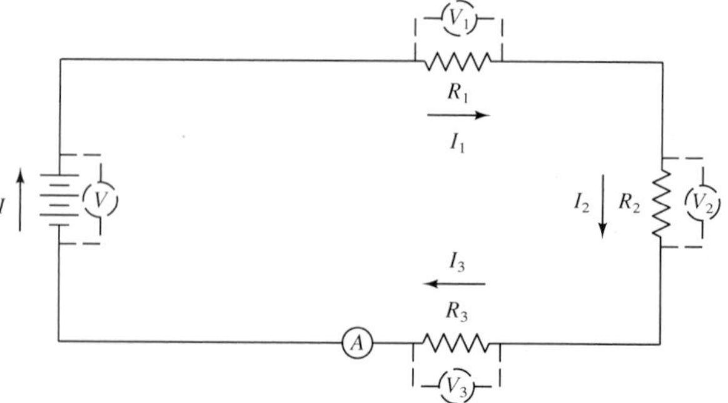

Fig. 26.11 A circuit with resistors in series.

Equation (26.10) can also be expressed in another manner. Since from Ohm's law, $V_1 = IR_1$, $V_2 = IR_2$, and $V_3 = IR_3$, we can substitute these values in Eq. (26.10) and get

$$V = IR_1 + IR_2 + IR_3$$

or

$$V = I(R_1 + R_2 + R_3) \qquad (26.10')$$

For a series circuit (1) the current is the same in every part of the circuit, (2) the resistance of the combination of resistors is equal to the sum of the resistances of the individual resistors, and (3) the total voltage across the combination is equal to the sum of the voltage drops across the separate resistors.

The voltage drop across any individual resistor in the circuit is directly proportional to the resistance of that portion of the circuit. Such a circuit as that in Fig. 26.11 is referred to as a *voltage divider network.*

Illustrative Problem 26.12 A series circuit consisting of three resistors having resistances of 40, 50, and 20 Ω, respectively, is connected across a 120-V line (Fig. 26.12). Find (*a*) the current in the circuit and (*b*) the voltage drop across each resistor.

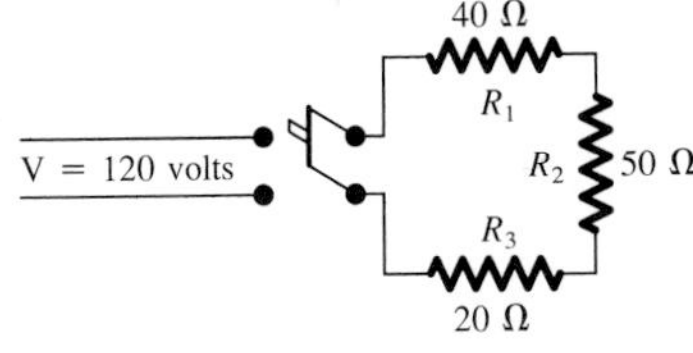

Fig. 26.12 Illustrative Problem 26.12.

Solution

(*a*) The total resistance R of the circuit is equal to the sum of the individual resistances.

$$\begin{aligned} R &= R_1 + R_2 + R_3 \\ &= 40\ \Omega + 50\ \Omega + 20\ \Omega \\ &= 110\ \Omega \end{aligned}$$

The current is equal to the total voltage divided by the total resistance.

$$\begin{aligned} I &= \frac{V}{R} \\ &= \frac{120\ \text{V}}{110\ \Omega} \\ &= 1.09\ \text{A} \qquad \textit{answer} \end{aligned}$$

(*b*) The amount of current in each resistor is then equal to 1.09 A. Then

$$\begin{aligned} V_1 &= IR_1 = 1.09\ \text{A} \times 40\ \Omega = 43.6\ \text{V} \\ V_2 &= IR_2 = 1.09\ \text{A} \times 50\ \Omega = 54.5\ \text{V} \\ V_3 &= IR_3 = 1.09\ \text{A} \times 20\ \Omega = 21.8\ \text{V} \qquad \textit{answer} \end{aligned}$$

The answers can be checked by showing that

$$V = V_1 + V_2 + V_3$$

26.17 ■ Parallel Circuits

When two or more resistors are joined in *parallel* (Fig. 26.13*a*), the effect is similar to that which occurs when two or more water pipes are joined in parallel (Fig. 26.13*b*). The pressure difference (water) or the potential difference (electricity) between A and B will be the same for each branch of the circuit. Hence, in the electric circuit,

$$\mathscr{E} = V_1 = V_2 = V_3 \qquad (26.11)$$

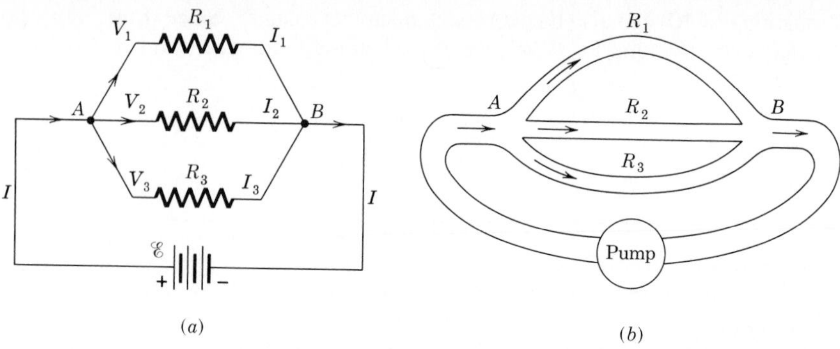

Fig. 26.13 Hydraulic analogy of resistors connected in parallel.

In a parallel circuit there is more than one path for the current and each branch has the same potential difference across it.

The total amount of current that will flow from A to B is equal to the sum of the separate currents passing through each channel. For a parallel circuit, then,

$$I = I_1 + I_2 + I_3 \tag{26.12}$$

If we denote the total resistance of the combination of parallel resistors of Fig. 26.13*a* as R_T, we know from Eq. (26.3) that $R_T = \mathscr{E}/I$; that is, the total resistance of any system is equal to the quotient of the potential difference divided by the current for that system. Thus, if we invert Eq. (26.3),

$$\frac{1}{R_T} = \frac{I}{\mathscr{E}}$$

Replacing I by its equivalent $I_1 + I_2 + I_3$, gives

$$\frac{1}{R_T} = \frac{I_1 + I_2 + I_3}{\mathscr{E}}$$

$$= \frac{I_1}{\mathscr{E}} + \frac{I_2}{\mathscr{E}} + \frac{I_3}{\mathscr{E}}$$

$$= \frac{I_1}{V_1} + \frac{I_2}{V_2} + \frac{I_3}{V_3} \qquad \text{from Eq. (26.11)}$$

or

$$\frac{1}{R_T} = \frac{1}{R_1} + \frac{1}{R_2} + \frac{1}{R_3} \tag{26.13}$$

The reciprocal of the total resistance of a combination of resistors in parallel is equal to the sum of the reciprocals of the separate resistors.

Illustrative Problem 26.13 A 100-W incandescent lamp, when hot, has a resistance of 144 Ω, while a hot 60-W lamp has a resistance of 240 Ω. Compute (*a*) the current in each lamp, (*b*) the total current in the circuit, and (*c*) the total circuit resistance when two 100-W lamps and four 60-W lamps, all connected in parallel, operate from a 120-V line.

Solution

(*a*) The potential difference across each lamp is 120 V. Hence, we can readily compute the current through each lamp by using Ohm's law.

For each 100-W lamp: $I_1 = \frac{120}{144} = 0.83$ A
For each 60-W lamp: $I_2 = \frac{120}{240} = 0.50$ A *ans.*

(*b*) The total current I will equal the sum of the individual currents.

$$I = 2I_1 + 4I_2$$
$$= 2(0.83 \text{ A}) + 4(0.50 \text{ A})$$
$$= 3.66 \text{ A} \qquad \textit{answer}$$

(*c*) To determine the total resistance use Ohm's law as follows:

$$R_T = \frac{V}{I_T}$$

$$= \frac{120 \text{ V}}{3.66 \text{ A}}$$

$$= 32.8\ \Omega \qquad \textit{answer}$$

26.18 ■ Conductance and Conductivity

Conductivity is the reciprocal of resistivity.

Since the reciprocal of resistance ($1/R$) occurs so frequently in parallel circuits, it is frequently convenient to designate the notion by a new unit called the conductance of the resistor. The *conductance of a resistor is the reciprocal of the resistance*. It is a measure of how readily the resistor will permit a current to be established. A circuit with a high resistance will have a low conductance, and vice versa. The symbol for conductance is G.

$$G = \frac{1}{R} \tag{26.14}$$

Thus, Eq. (26.13) can be replaced by

$$G_T = G_1 + G_2 + G_3 \tag{26.15}$$

The SI unit for conductance is called the *siemen*, S.* Since the ohm unit is equivalent to 1 V/A, one siemen is equivalent to 1 A/V.

In Sec. 26.10 we found that the resistance R in ohms of a conductor with resistivity measured in ohm-circular-mils per foot is determined by the formula $R = \rho L/A$. If we define the *electrical conductivity* σ as equivalent to $1/\rho$, the formula for conductance will become

$$G(\text{siemens}) = \frac{\sigma(\text{ft/ohm} \cdot \text{cmils})\text{A(cmils)}}{\text{L(ft)}} \tag{26.16}$$

Illustrative Problem 26.14 Find the conductance of 250 ft of number 14 copper wire at 20°.

Solution From the table in Appendix VI, we note that the resistivity ρ of No. 14 copper wire at 20° is 10.3 ohm · cmils/ft (from Table 16.1).

$$\sigma = \frac{1}{\rho}$$

$$= \frac{1 \text{ ft}}{10.3 \text{ ohm} \cdot \text{cmils}}$$

Substitute in Eq. (26.16). The cross-sectional area of No. 14 wire is 4107 cmils

$$G = \frac{\sigma A}{L}$$

$$= \frac{1 \text{ ft}}{10.3 \text{ ohm} \cdot \text{cmils}} \times \frac{4107 \text{ cmils}}{250 \text{ ft}}$$

$$= 1.6 \text{ siemens} \qquad \textit{answer}$$

* The siemen is equivalent to the older unit mho. The symbol for the mho unit is ℧ (an inverted Ω).

26.19 ■ Series-Parallel Circuits

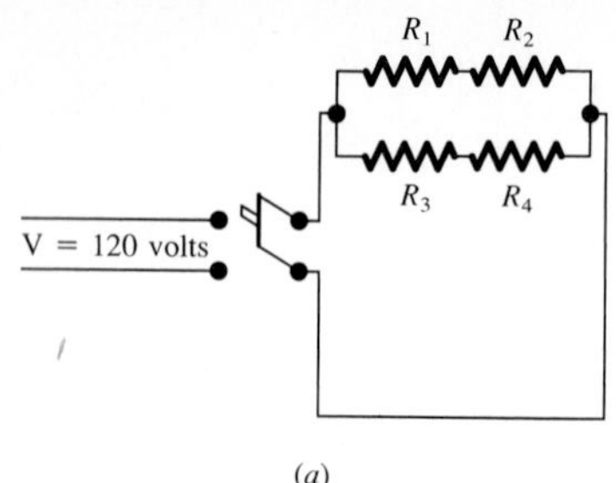

(a)

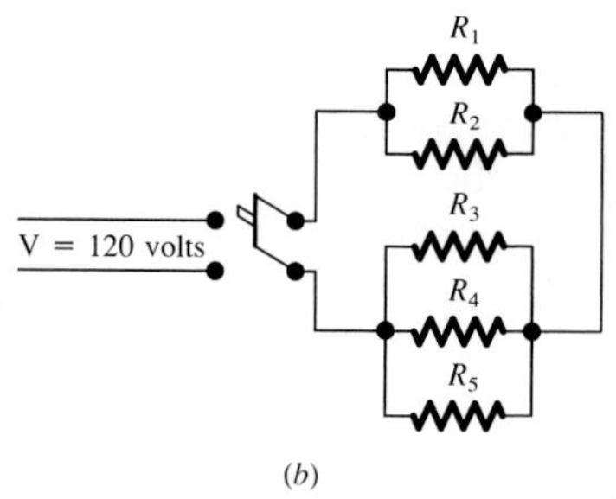

(b)

Fig. 26.14 Diagrams of typical series-parallel circuits.

Thus far we have dealt only with separate series and parallel circuits. Practical electric circuits, however, very often consist of combinations of the two basic types. There are a great number of combinations possible. Figure 26.14 illustrates two of the more elementary types. In type (*a*) R_1 and R_2 are in series, R_3 and R_4 are in series, and the two groups are connected in parallel. In type (*b*) R_1 and R_2 are in parallel, R_3, R_4, and R_5 are in parallel, and these two groups are connected in series. There is no definite procedure that can be followed in solving combination series-parallel circuits. The method of approach depends upon the arrangement. However, whenever possible, it is a good practice to replace each parallel branch by an equivalent series branch.

Illustrative Problem 26.15 Compute the total resistance of the circuit of Fig. 26.14*a* if $R_1 = 10\ \Omega$, $R_2 = 30\ \Omega$, $R_3 = 20\ \Omega$, and $R_4 = 40\ \Omega$. What is the total current in the circuit?

Solution The resistors R_1 and R_2 can be replaced by a single resistor R_x, where

$$\begin{aligned} R_x &= R_1 + R_2 \\ &= 10\ \Omega + 30\ \Omega \\ &= 40\ \Omega \end{aligned}$$

In like manner, R_3 and R_4 can be replaced by R_y, where

$$\begin{aligned} R_y &= R_3 + R_4 \\ &= 20\ \Omega + 40\ \Omega \\ &= 60\ \Omega \end{aligned}$$

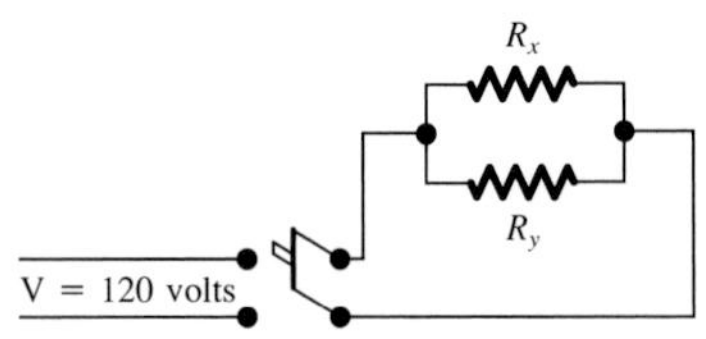

Fig. 26.15 Illustrative Problem 26.15.

The total resistance of the system (Fig. 26.15) can then be found by applying Eq. (26.13):

$$\frac{1}{R_T} = \frac{1}{R_x} + \frac{1}{R_y}$$

$$= \frac{1}{40\ \Omega} + \frac{1}{60\ \Omega} = \frac{5}{120\ \Omega}$$

$$R_T = \frac{120}{5} = 24\ \Omega \qquad \textit{answer}$$

The current can then be found by using Ohm's law.

$$\begin{aligned} I &= V/R_T \\ &= 120\ \text{V}/24\ \Omega \\ &= 5\ \text{A} \qquad \textit{answer} \end{aligned}$$

Illustrative Problem 26.16 If, in Fig. 26.14*b*, $R_1 = 5\ \Omega$, $R_2 = 40\ \Omega$, $R_3 = 30\ \Omega$, $R_4 = 20\ \Omega$, and $R_5 = 60\ \Omega$, compute (*a*) the total resistance R_T and (*b*) the total current I in the circuit.

Solution Replace the parallel resistors R_1 and R_2 by one resistor R_x (Fig. 26.16) solved as follows:

Fig. 26.16 Illustrative Problem 26.16.

(*a*)

$$\frac{1}{R_x} = \frac{1}{5\ \Omega} + \frac{1}{40\ \Omega} = \frac{9}{40\ \Omega}$$

$$R_x = \tfrac{40}{9} = 4.4\ \Omega$$

In like manner replace R_3, R_4, and R_5, by one resistor R_y:

$$\frac{1}{R_y} = \frac{1}{30\ \Omega} + \frac{1}{20\ \Omega} + \frac{1}{60\ \Omega} = \frac{6}{60\ \Omega}$$

$$R_y = 10\ \Omega$$

Hence

$$R_T = R_x + R_y$$
$$= 4.4\ \Omega + 10\ \Omega = 14.4\ \Omega \qquad \textit{answer}$$

(*b*) Applying Ohm's law for I, yields

$$I = V/R_T$$
$$= 120\ \text{V}/14.4\ \Omega$$
$$= 8.3\ \text{A} \qquad \textit{answer}$$

26.20 ■ Kirchhoff's Laws

Many circuits contain more than one source of emf and several devices which dissipate electric energy by converting it into heat. These more intricate networks are not easily solved by the method used in the preceding pages. In more complicated networks it is necessary to employ two laws given by Gustav Robert Kirchhoff (1824–1887), an eminent German physicist. Kirchhoff's laws refer to conditions that exist in a network of conductors where currents are either steady-state or alternating.

In addition to his work on electric current analysis, Kirchhoff developed spectrum analyses, discovered cesium and rubidium, and explained the Fraunhofer lines, in the solar spectrum.

First, let us define two terms. A *junction* is a point of a circuit at which three or more conductors are joined. A *loop* is a closed conducting path in an electric circuit.

Kirchhoff's first law is essentially a restatement of the conservation of electric charge. It states that

Kirchhoff's laws enable a person to analyze the more complex circuits into simple branch circuits.

At any junction in an electric circuit, the sum of the currents directed toward the junction equals the sum of the currents directed away from the junction.

Currents toward a junction are considered positive and those away from the junction negative. With this convention applied, the first law can be rephrased as follows:

The algebraic sum of the currents at a junction equals zero.

$$\Sigma I = 0 \qquad (26.17)$$

Figure 26.17 will be used to illustrate Kirchhoff's first law. Remember that conventional current will travel from higher to lower potential in the external circuit (from + to −) and from lower to higher potential in the emf source (from − to +). The equivalent resistance R_T for the three-branch system is found by using the equation

$$\frac{1}{R_T} = \frac{1}{2\Omega} + \frac{1}{3\Omega} + \frac{1}{4\Omega}$$

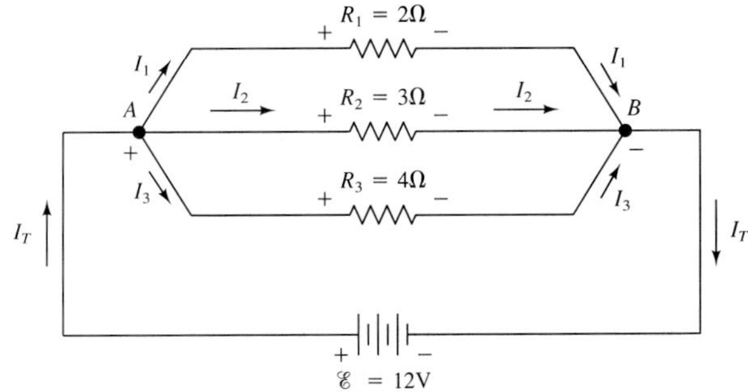

Fig. 26.17 In this figure, the current I_T divides into three branches at junction A. $I_T = I_1 + I_2 + I_3$. Similarly, at junction B, $I_1 + I_2 + I_3 = I_T$.

from which $R_T = \frac{12}{13}\ \Omega$. The total current I_T is determined from Eq. 26.2′.

$$I_T = \frac{\mathscr{E}}{R_T} = \frac{12\ \text{V}}{12/13\ \Omega} = 13\ \text{A}$$

Again, using Eq. 26.2′, we find the current in each of the three branches.

$$I_1 = \frac{\mathscr{E}}{R_1} = \frac{12\ \text{V}}{2\ \Omega} = 6\ \text{A}$$

$$I_2 = \frac{\mathscr{E}}{R_2} = 4\ \text{A and } I_3 = \frac{\mathscr{E}}{R_3} = 3\ \text{A}$$

NOTE: $I_T = I_1 + I_2 + I_3$

Kirchhoff's second law is essentially a restatement of the conservation of energy. It states that

In any closed loop of a circuit, the algebraic sum of the emfs equals the algebraic sum of the potential drops in the same loop.

$$\Sigma\mathscr{E} = \Sigma\ IR \qquad \textbf{(26.18)}$$

A rise in potential will be considered positive. A drop in potential will be negative. Remember that a current I in a resistance R corresponds to a potential difference IR. An alternative equation for the second law is

$$\Sigma\mathscr{E} - \Sigma\ IR = 0 \qquad (26.18')$$

The second law simply states that the total energy supplied by the sources of emf is equal to the total energy dissipated as the charges move around the loop.

The second law is illustrated in Fig. 26.18. Here we must remember that in the external circuit, going in the direction of the current a potential drop of magnitude IR occurs in the resistor. At the emf source a potential rise of magnitude $\mathscr{E}$ occurs, in going from the negative to the positive terminal. Use Eq. (26.9) to find the total resistance R_T of the three resistors in series.

$$R_T = R_1 + R_2 + R_3 = 9\ \Omega$$

Then

$$I = \frac{\mathscr{E}}{R_T} = \frac{12\ V}{9\ \Omega} = \frac{4}{3} A$$

The potential difference (drop) from A to B is

$$IR_1 = \tfrac{4}{3}\ \text{A} \times 2\Omega = -\tfrac{8}{3}\ \text{V}$$

From B to C the potential difference is

$$IR_2 = -4\ \text{V}$$

From C to D the potential difference is

$$IR_3 = -\tfrac{16}{3}\ \text{V}$$

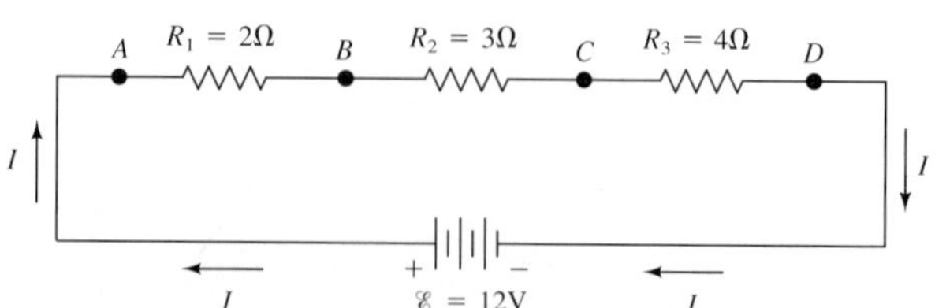

Fig. 26.18 Illustration of Kirchhoff's second law $\mathscr{E} + IR_1 + IR_2 + IR_3 = 0$.

Here, we note that

$$\mathscr{E} + IR_1 + IR_2 + IR_3 = 0$$

In applying Kirchhoff's laws in circuit analysis it is convenient to carry out the following steps:

1. Arbitrarily assign a direction and a symbol to the current in each independent loop of the network. It is not essential that the selected direction be correct. If an incorrect direction is chosen, the current will simply show up as a negative number.
2. Place appropriate positive and negative signs at the terminals of each source of emf and at every resistor for each loop. It must be remembered that the current external to the emf source is from the plus terminal to the negative terminal.
3. Use the two laws to develop as many independent equations as there are unknowns in the circuit. Be sure that no two of these equations have exactly the same unknowns in them.
4. Solve the simultaneous equations for the desired unknowns.

Illustrative Problem 26.17 Find the currents I_1, I_2, and I_3 in each of the branches of Fig. 26.19.

Solution Arbitrary directions have been assigned to the current in each branch. If the directions are incorrect, our answers will tell us so. Since there are three unknowns, we will need three independent equations in the three unknowns.

First, apply Kirchhoff's first law at junction A. We will omit units in the equations since it is clear that I will be in amps, V in volts, and R in ohms.

$$I_2 = I_1 + I_3 \tag{1}$$

Equation (1) also holds at junction D.

Now apply Kirchhoff's second law for the two closed loops. The voltage drop from A to $C = -I_1(40\Omega)$. From C to D there is no change in potential; but from D to F the potential changes $-I_2$ ($50\Omega + 10\Omega$). From F to A the potential increases by 60 V.

Thus, we have

$$V_{AC} + V_{DF} + \mathscr{E}_1 = 0 \qquad \text{(leaving off units)}$$
$$-40\ I_1 - 60\ I_2 + 60 = 0$$

or

$$2I_1 + 3I_2 = 3 \tag{2}$$

We will take the loop $ABCDHGA$ for a second application of the second law. Then $V_{ABCD} = -I_1$ (40Ω). We are assuming negligible resistance for the connecting conductors.

In assuming our direction for I_3, the potential at H will be higher than at D. Thus, the voltage drop from D to H is $I_3(20\ \Omega)$. We have assumed current *through* the

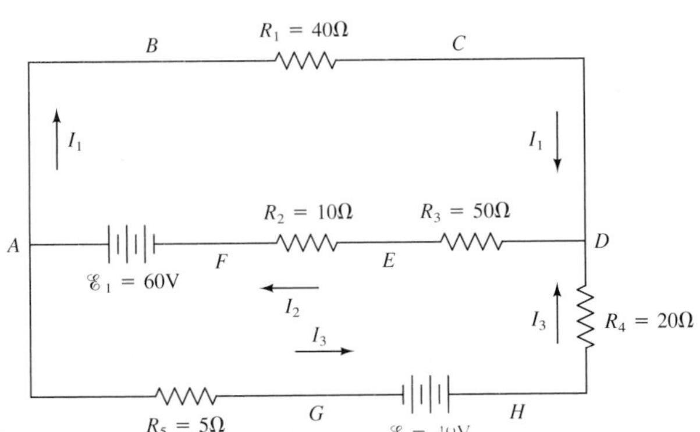

Fig. 26.19 Illustrative Problem 26.17.

battery GH to be from positive to negative. Then, there will be a decrease in potential of 40 V and $V_{HG} = -40V$. Also, $V_{GA} = +I_3\,(5\Omega)$.

Our equation for the loop becomes

$$-40I_1 + (20 + 5)I_3 - 40 = 0 \qquad (3)$$

We now have three equations in three unknowns. Substitute I_2 from (1) into (2).

$$-40I_1 - 60(I_1 + I_3) + 60 = 0 \qquad (4)$$

$$-100\,I_1 - 60\,I_3 + 60 = 0 \qquad (5)$$

$$I_1 = \frac{60 - 60\,I_3}{100} = \frac{3 - 3I_3}{5} \qquad (6)$$

Substitute (6) in (3)

$$-40\left(\frac{3 - 3I_3}{5}\right) + 25\,I_3 = 40$$

$$-24 + 24\,I_3 + 25\,I_3 = 40$$

$$I_3 = \tfrac{64}{49}\text{ A} = 1.31\text{ A} \qquad \textit{answer}$$

Substitute $I_3 = \frac{64}{49}$ in Eq. (3)

$$-40\,I_1 + 25(\tfrac{64}{49}) - 40 = 0$$

$$I_1 = -\tfrac{9}{49}\text{ amp} = -0.18\text{ A} \qquad \textit{answer}$$

NOTE: *The negative sign means our arbitrarily selected direction for I_1 was wrong.*

Use Eq. (2) and the value for I_1 to get I_2.

$$2\left(-\frac{9}{49}\right) + 3\,I_2 = 3$$

$$I_2 = \tfrac{55}{49}\,A = 1.12\,A \qquad \textit{answer}$$

The same answers would be obtained should we use the second law with loop *AFEDHGA* and either of the other loops. It will be left as an exercise for the student to try other combinations of the three loops.

THE MEASUREMENT OF RESISTANCE

26.21 ■ Ammeter-Voltmeter Method of Measuring Resistance

When a high degree of accuracy is not required, an easy and simple method of measuring resistance is with an ammeter and a voltmeter. Figure 26.20 illustrates the circuit employed. A direct current is passed through the unknown resistance R. An ammeter A is connected in series with R, while a voltmeter V is connected across (parallel with) the resistance. Simultaneous readings are taken on the meters, and R is found by applying Ohm's law. This method is well adapted for general use in the electrical repair shop, where the electrician is interested in measuring field-coil resistances of generators and motors, resistances of armature windings, and resistances of various electric appliances while in operation.

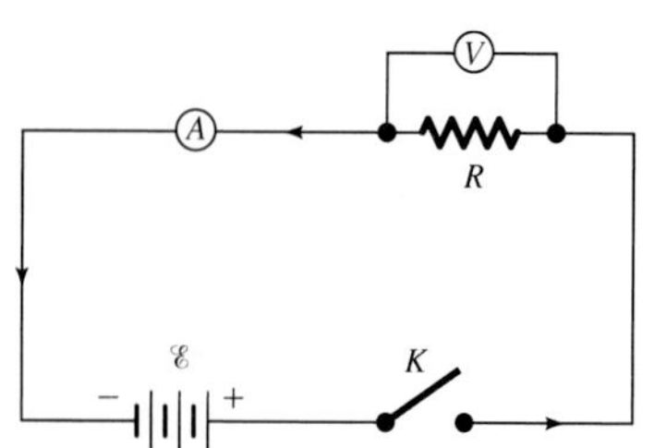

Fig. 26.20 The circuit for the ammeter-voltmeter method of measuring resistance.

Several precautions must be taken in using the ammeter-voltmeter method. Too much current should not be used, because if the resistance to be measured is heated, the computed results will be too high. (Of course, if it is the heated resistance which is desired, as in the case of lamps and heating appliances, the full rated current should be used.)

It will be noticed from Fig. 26.20 that A measures not only the current which passes through the unknown resistance R but also the current passing through V. However, it will be recalled that the voltmeter is an instrument of high internal resistance; hence relatively little current will pass through it, unless R is also of very high resistance. If R is a high resistance, the ammeter-voltmeter method does not provide very accurate results.

When a very low resistance is to be measured, the potential drop across R will be small. Hence a sensitive voltmeter (*millivoltmeter*) is used for V. The millivoltmeter measures potential differences of thousandths of one volt. On the other hand, when the resistance is high, the amount of current passing through R may be small. In this case, a *milliammeter* can be used to good advantage. Generally, when the resistance is high and the current is small, the voltmeter is connected across both the unknown resistance and the ammeter, as shown in Fig. 26.21. In this case the ammeter will measure only the current going through R. Since the current is small, the potential drop across the ammeter (which has a very low resistance) will not be appreciable.

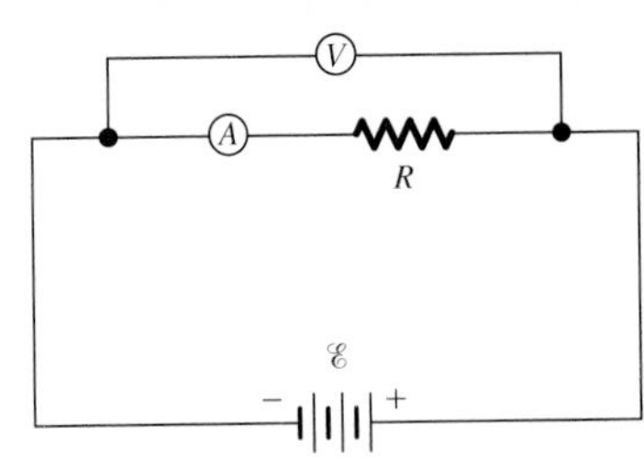

Fig. 26.21 When the resistance in a circuit is high and the current is small, the voltmeter is connected across both the unknown resistance and the ammeter.

26.22 ■ The Wheatstone Bridge

The Wheatstone bridge is used to measure an unknown resistance R_x in terms of known resistances R_1, R_2, R_3.

A more precise instrument for measuring resistances is the *Wheatstone bridge*. It consists essentially of a network of three known adjustable resistances R_1, R_2, R_3, and the unknown resistance R_x, joined by heavy connecting wires of very low resistance, as shown in Fig. 26.22 *a*. The resistors are arranged in two parallel circuits. A sensitive current-measuring device, called a *galvanometer*, is bridged across the two circuits at points C and D. The construction and operation of the galvanometer will be discussed in Chap. 28.

In general, the current at A divides unequally into the two channels. As a rule, there will also be a potential difference between C and D as indicated by a deflection on the galvanometer G. However, by adjusting the values of the known resistances, it is possible to get the same potential at C and at D. When this occurs, no deflection will be shown by the galvanometer. This adjustment is known as *balancing the bridge*.

When the bridge is balanced, the potential difference between A and D will equal the potential difference between A and C; also the potential difference between D and B will equal the potential difference between C and B. Let us call I_1 the current in the ADB circuit and I_2 the current in the ACB circuit when the bridge is balanced. Equating these potential differences, we write

$$I_1R_1 = I_2R_2$$

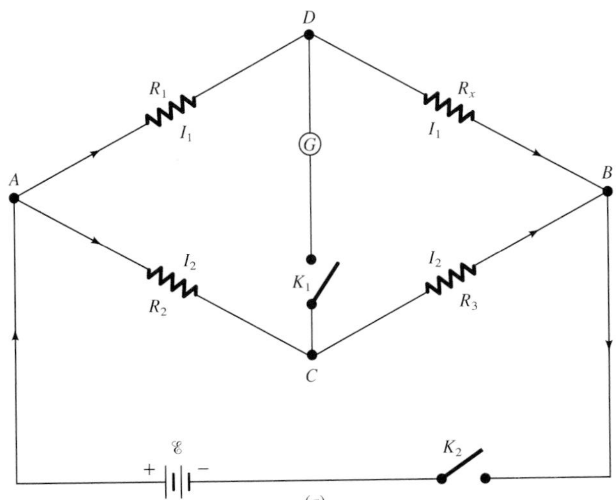

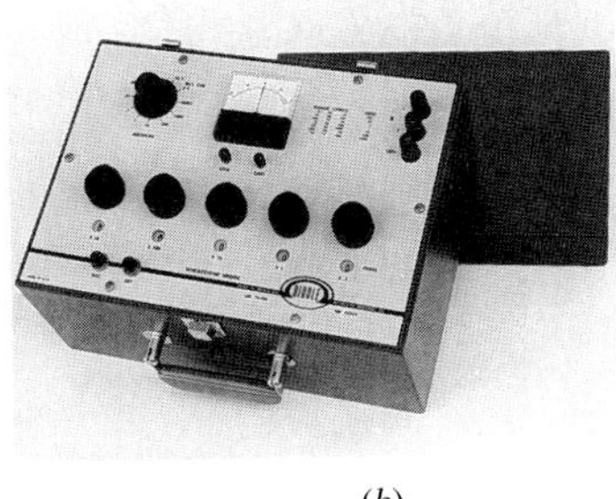

Fig. 26.22 (*a*) Circuit of a Wheatstone bridge. (*b*) Portable laboratory Wheatstone bridge, featuring a multiplier dial which selects the optimum operating voltage for each range. (Courtesy of Biddle Instruments)

and $$I_1 R_x = I_2 R_3$$

Dividing the first equation by the second results in

$$\frac{R_1}{R_x} = \frac{R_2}{R_3}$$

or $$R_x = R_1 \frac{R_3}{R_2} \tag{26.19}$$

This is the basic equation for the Wheatstone bridge.

Illustrative Problem 26.18 The Wheatstone bridge diagramed in Fig. 26.22 *a* is used to determine an unknown resistance R_x. It is balanced when $R_3 = 1000\ \Omega$, $R_2 = 10\ \Omega$, and $R_1 = 156\ \Omega$. Compute the value of R_x.

Solution

$$R_x = R_1 \frac{R_3}{R_2}$$

$$= 156\ \Omega \times \frac{1000\ \Omega}{10\ \Omega}$$

$$= 15{,}600\ \Omega \qquad \textit{answer}$$

It will be observed from Eq. (26.19) that it is not necessary to know the values of R_2 and R_3 in order to solve for the unknown R_x. As long as the value of R_1 and the ratio of R_3 to R_2 are known, we can compute the value of R_x. Commercial bridges (Fig. 26.22*b*) are so constructed that the ratio of R_3 to R_2 is selected by means of a rotary switch. These ratios generally include multiples of 10 from 0.0001 to 1000. The values for R_1 can be adjusted by a system of four or more rotary switches, one of which varies the resistance in steps of 1 Ω from 1 to 10, the second varies in steps of 10 Ω from 10 to 100, the third varies in steps of 100 Ω from 100 to 1000, etc. Thus the value of R_1 can be read directly by summing up the resistances (in series) of these four or more rotary switches. R_x can then be readily computed by multiplying R_1 by the value indicated by the ratio switch.

26.23 ■ The Potentiometer

Internal resistance of a cell will cause a potential drop in the cell.

The emf of a cell is not determined exactly by the reading of a voltmeter connected across the terminals of the cell, because every voltmeter draws *some current* from the cell even though it does have a high resistance. The potentiometer is an instrument that permits the measurement of an unknown emf of a cell without drawing any current from that cell, and hence there can be no voltage drop across the internal resistance.

The elements of a simple potentiometer circuit are shown in Fig. 26.23. The currents and emfs are marked in the figure. A working battery *W* with a constant emf $\mathscr{E}_0$ produces a current I_0 which is divided into two branches. One branch (the upper one in Fig. 26.23) provides a uniform potential drop between the ends of a variable resistor *AH*. The other branch (the lower one) includes a standard cell *S* (or the cell *X* under test), a galvanometer *G* and part *AE* (or *AF*) of the resistor *AH*. The circuit is completed by a slider which can touch different points, such as *E* and *F*, on the variable resistor *AH*. The emf $\mathscr{E}_0$ should be considerably larger than either $\mathscr{E}_s$ or $\mathscr{E}_x$.

The cells *S* and *X* both have their positive terminals connected to *A*. Let us assume that the double-throw switch is connected to the unknown cell *X* in the circuit. There will be a point *E* where the potential drop along *AE* will be equal to the emf of the cell which is in the circuit. This will be evident when the galvanometer *G* shows no deflection ($I_x = 0$) when the slider makes contact with the resistor at *E*.

Kirchhoff's first law states that at junction A the current I_0 going into the junction equals the sum of the currents (I_x and I_y) leaving the junction. Thus, $I_0 = I_x + I_y$. Solving for I_y, yields $I_y = I_0 - I_x$.

Kirchhoff's second law states that, for the lower loop

$$-\mathscr{E}_x + (I_0 - I_x)R_x = 0$$

When the galvanometer reads 0 the equation becomes

$$\mathscr{E}_x = I_0 R_x$$

Next, the cell of the unknown emf $\mathscr{E}_x$ is replaced by a standard cell whose emf $\mathscr{E}_s$ has previously been certified by a test laboratory. The slider, again, is moved until a spot F is found at which the galvanometer indicates no current. Then, we have

$$\mathscr{E}_s = I_0 R_s$$

Combining the last two equations will determine that

$$\mathscr{E}_x = \frac{R}{R_s}\mathscr{E}_s \qquad (26.20)$$

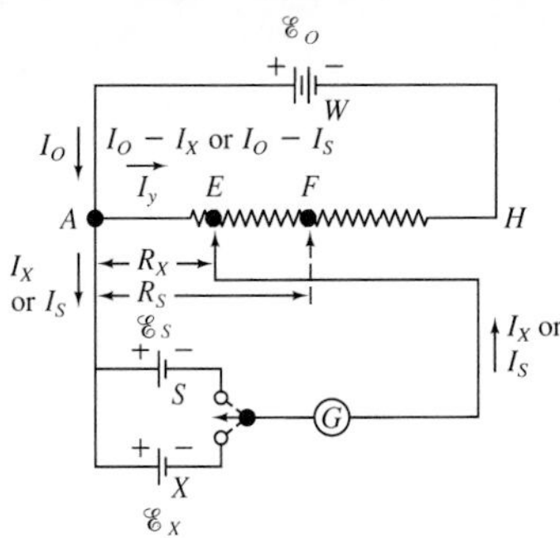

Fig. 26.23 Circuit diagram for a potentiometer. The circuit is used to measure an unknown emf ($\mathscr{E}_x$) by comparing it with a known emf ($\mathscr{E}_s$) provided by a standard cell.

Equation (26.20) enables a person to determine an unknown emf from a knowledge of the emf of a standard cell and the ratio of the resistance of two resistors.

A common potentiometer in the physics laboratory has a wire of resistivity ρ as the resistor AH. Again, the resistances R_x and R_s can be obtained by sliding the contact along the wire to vary the length of the resistive element. From Eq. (26.5),

$$R_x = \rho\frac{L_x}{A} \qquad \text{and} \qquad R_s = \rho\frac{L_s}{A}.$$

Substituting these values for R_s and R_x in Eq. (26.20) will give

$$\mathscr{E}_x = \frac{L_x}{L_s}\mathscr{E}_s \qquad (26.21)$$

There is no basic difference between a potentiometer and a rheostat (except in how the sliding contact operates).

The unknown emf can then be determined by measuring two wire lengths and from the magnitude of the emf of a standard cell.

Any quantity that can be made to produce or control a potential difference may be measured by a potentiometer. Temperature, stress, acidity, frequency, and numerous other quantities are both measured and controlled by the use of a potentiometer.

26.24 ■ The Ohmmeter

The *ohmmeter* is a convenient, portable, direct-reading instrument for measuring resistances in ohms. A diagram of the essential parts of a simple single-range ohmmeter is shown in Fig. 26.24*a*. It consists of a series circuit which includes a dry cell C, a fixed known resistance R, and a specially calibrated milliammeter G. To this circuit is added the unknown resistance R_x across the terminals A and B. A variable resistor R_v is connected in parallel with the coil of the milliammeter. To operate the instrument, a conductor of extremely low resistance, called a *shunt*, is first connected to terminals A and B. The milliammeter will then reach its maximum deflection (in a clockwise direction). Next the resistance R_v is varied until the pointer of the meter reaches zero on the scale. This will indicate no external resistance between terminals A and B. The shunt is then replaced by the unknown resistance R_x. This added resistance in the series circuit will diminish the current flowing through the sensitive milliammeter. Since the current varies inversely as the resistance of the circuit, the meter can be calibrated to read directly in ohms. It should be observed that because of this inverse relationship, the scale readings increase as the scale is read to the left and also that the scale divisions are not uniform. Unlike the voltmeter or ammeter, the ohmmeter requires a battery (1.5 V and up). It is used in circuits with no other power applied. The instrument must never be used in a circuit with an external applied voltage. Any external voltage may damage the meter and /or the ohmmeter circuitry. Figure 26.24*b* shows one of the commercial types of multirange ohmmeters.

Since the calibrations crowd up at the left-hand side of an ohmmeter scale, the degree of accuracy diminishes as the deflections move to the left side of the meter.

The ohmmeter generally is not a very precise instrument, but it is convenient and often precise enough for ordinary measurement of resistances. Digital ohmmeters are commonly available now.

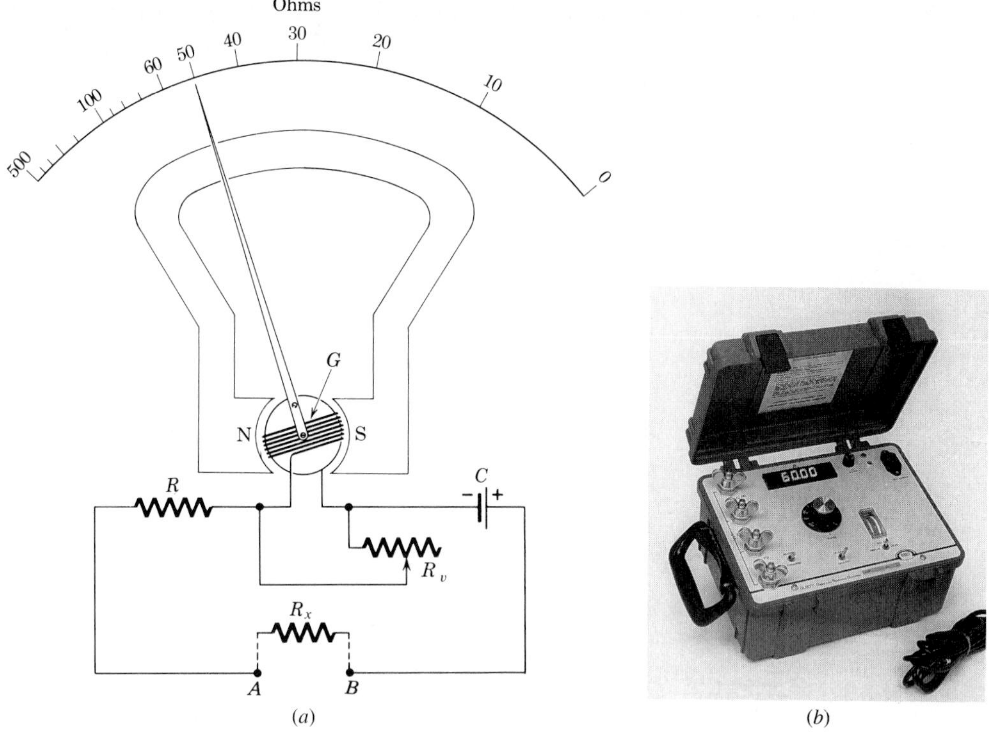

Fig. 26.24 (*a*) Circuit of a single-range ohmmeter. (*b*) Commercial digital resistance ohmmeter featuring a rechargeable NiCd battery and charger in one compact unit. The instrument will give readings across the entire 1 $\mu\Omega$ to 60 Ω range. It is used for low-resistance measurements such as switch/circuit breaker contacts, soldered joints, cable splices, motor and transformer windings, welds, and fuses. (Courtesy of Biddle Instruments)

QUESTIONS AND EXERCISES

1. A Wheatstone bridge is commonly used to determine what electrical quantity?

2. What happens to the resistance of a copper wire if its temperature is raised?

3. If the diameter of a wire is *d* mils what is its area expressed in circular mils?

4. How is an ammeter connected in a circuit? a voltmeter? Show both in a circuit diagram.

5. What is meant by the term *resistivity*?

6. Name the factors that influence the resistance of a piece of wire and explain how each factor affects the resistance.

7. An ammeter and a voltmeter are connected in a dc circuit as shown in Fig. 26.25. Each instrument has a positive reading. Show the direction of the current in each circuit. Which (*A* or *B*) is the supply and which is the load?

8. Explain why such conductors as *chromel* and *nichrome* are desirable as heating elements.

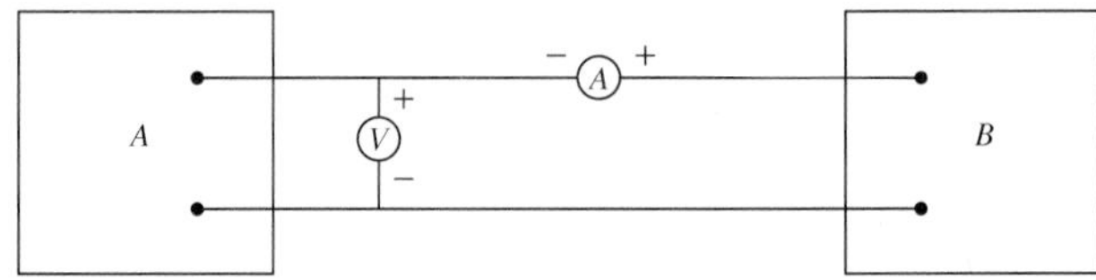

Fig. 26.25 Question 7.

9. Give as many reasons as you can for connecting house lights in parallel instead of series.

10. In what ways is the current in a conductor similar to the flow of water through a pipe? In what ways is it different?

11. Explain how the voltmeter-ammeter method is used for measuring resistance. What are some reasons why the method is not a particularly accurate one to measure resistance?

12. When a Wheatstone bridge is balanced, will interchanging the galvanometer and the battery affect the balance? Give reasons for your answer.

13. What is the name given to a variable resistor used for current control?

14. What are the advantages of a carbon resistor?

15. What are the advantages of Christmas lights that are in parallel against those that are connected in series?

16. Is it possible to use 6-V lamps hooked up to a 24-V line without burning them out? Explain.

17. What is the structural difference between an ammeter and a voltmeter?

18. When, if ever, does the potential difference across the terminals of a battery equal its emf?

19. Is it possible for the (conventional) direction of the current through a battery to go from positive to the negative terminal? Give reason for your answer.

20. What is the advantage of a potentiometer over a voltmeter in measuring emfs? What are the advantages of a voltmeter over the potentiometer?

PROBLEMS

Group One

1. What is the diameter in mils of a wire $\frac{3}{4}$ in. in diameter?

2. What is the area in circular mils of a wire whose diameter is $\frac{1}{4}$ in.?

3. How many coulombs per hour pass a point in a circuit that is carrying 5 A of current?

4. How much does 1 mi of AWG No. 16 copper wire weigh? Consult Appendix VI.

5. What would be the likely AWG number of a copper wire if 10 ft of the wire had a resistance of 1.67 Ω? Consult Appendix VI.

6. The resistance coil of an electric ironer is 12 ft long and has a resistance of 14.4 Ω. The coil is made of nichrome wire. What is the area of the wire in circular mils?

7. Manganin wire of diameter 0.051 cm is used to make a resistor of 15 Ω. What length of wire will be needed?

8. A wire having a cross-sectional area of 808 cmils has a resistance of 13.24 Ω. What is the resistance of a wire of the same length and the same material if its cross-sectional area is 101 cmils?

9. What is the resistance of 4500 ft of annealed copper wire at 20°C if the wire has a diameter of 0.064 in.?

10. What size wire on the AWG scale would have an area twice that of a No. 24 wire? What size wire would have an area four times that of a No. 24 wire?

11. How much current does a 50-Ω electric iron draw when operating from a 115-V line?

12. What is the electrical resistance of 12 m of commercial copper wire at 20°C if it has a diameter of 1.5 mm?

13. What is the total resistance of the circuit in Fig. 26.26?

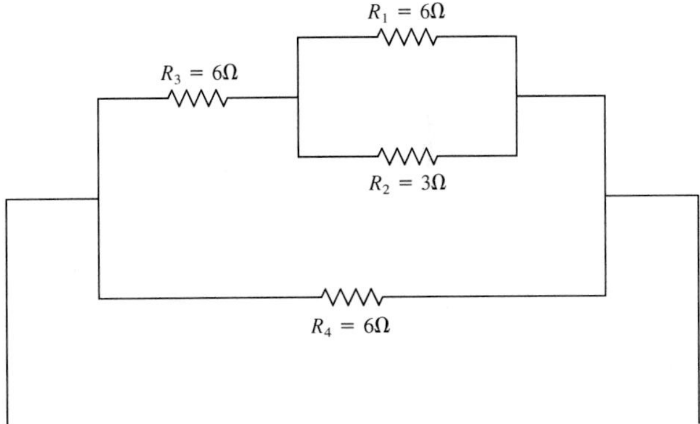

Fig. 26.26 Problem 13.

14. The Wheatstone bridge illustrated in Fig. 26.22*a* is balanced when $R_1 = 1.25\ \Omega$, $R_2 = 2.50\ \Omega$, and $R_3 = 770\ \Omega$. Calculate the value of R_x.

15. What should be the resistance of a conductor if 110 V is required to force 2.5 A of current through it?

16. Determine the current through the 6-Ω resistor in Fig. 26.27.

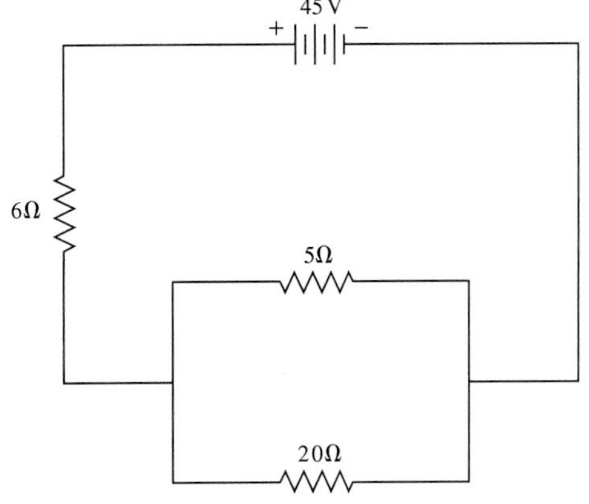

Fig. 26.27 Problem 16.

Group Two

17. What is the charge in coulombs passing through a 6-Ω resistance in 24 s if the potential difference across the resistor is 40 V?

18. The resistance of a 100-W tungsten filament lamp is 12 Ω at 20°C. When it is operating as a light the resistance is 152 Ω. Calculate the temperature of the hot filament. Assume that the increase in resistance varies directly as the temperature. (Actually, this is not exactly true for such a wide range of temperatures, however.)

19. A length of wire has a resistance of 12.80 Ω at 30°C and 11.36 Ω at 0°C. Compute (*a*) the temperature coefficient and (*b*) its resistance if heated to 350°C.

20. The resistance of a platinum wire is 8.00 Ω at 25°C. Determine its resistance at 100°C.

21. A storage battery for farm lighting produces an emf of 62 V across its terminals on open circuit. The battery has 0.032 Ω internal resistance and is connected through a pair of copper wires which have a specific resistance of 1.835×10^{-3} Ω/ft, to a load 200 ft distant. The load draws 14 A. (*a*) What is the potential at the load 200 ft from the battery? (*b*) What is the potential difference across the battery terminals when supplying this load?

22. Two cells are connected to three resistance as shown in Fig. 26.28. (*a*) Determine the current through each battery. (*b*) Determine the current through R_3.

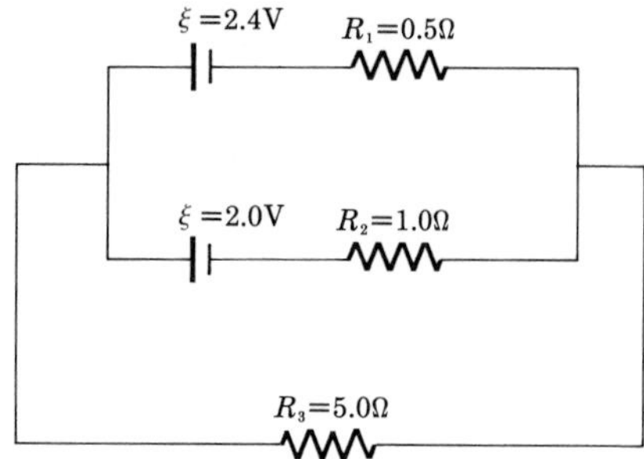

Fig. 26.28 Problem 22.

23. Six batteries, each with an emf of ℰ volts and internal resistance *r* ohms, are connected in a series-parallel combination as shown in Fig. 26.29. This arrangement is connected in series with a resistance *R* ohms. Develop an equation for *I* as a function of ℰ, *r*, and *R*.

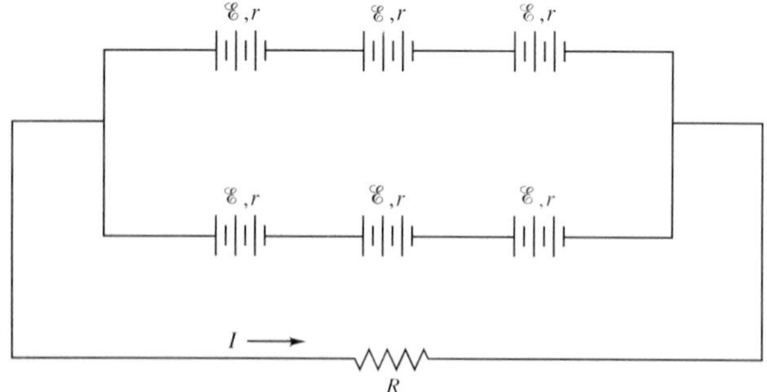

Fig. 26.29 Problem 23.

24. What emf must be impressed on the circuit of Fig. 26.30 in order to have a current of 12 A in a 5-Ω resistor?

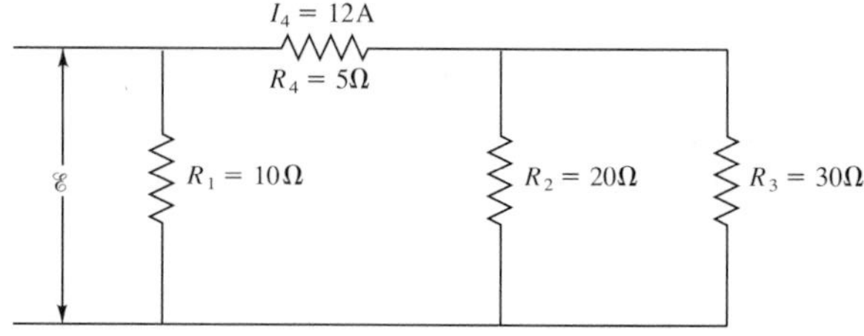

Fig. 26.30 Problem 24.

25. Find the resistance of the circuit shown in Fig. 26.31.

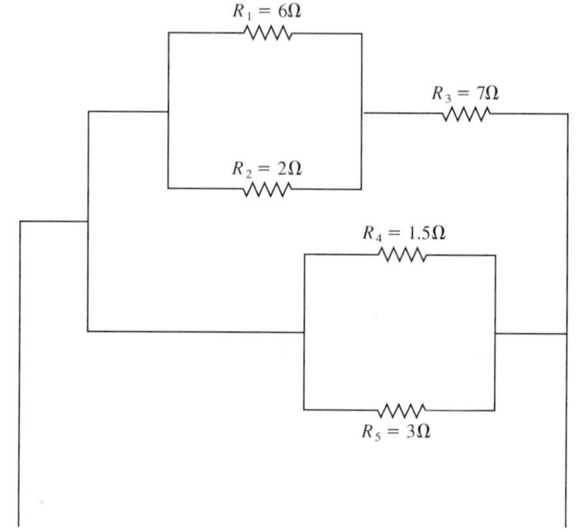

Fig. 26.31 Problem 25.

26. Determine the current *I* in the circuit shown in Fig. 26.32.

27. A wire 500 ft long with a diameter of 102 mils has a resistance of 0.5 Ω. What would be the resistance of 500 ft of wire 51 mils in diameter and made of the same material?

28. A farmer has a generator that generates electricity at 120 V. He intends to run two wires from the generator to an appliance 300 ft from the generator. If the appliance is to be operated at 15.0 A and 110 V, what is the diameter of the smallest copper wire (in inches) that he can use? (Assume no drop in terminal potential difference at the generator when it is delivering current to the appliance.)

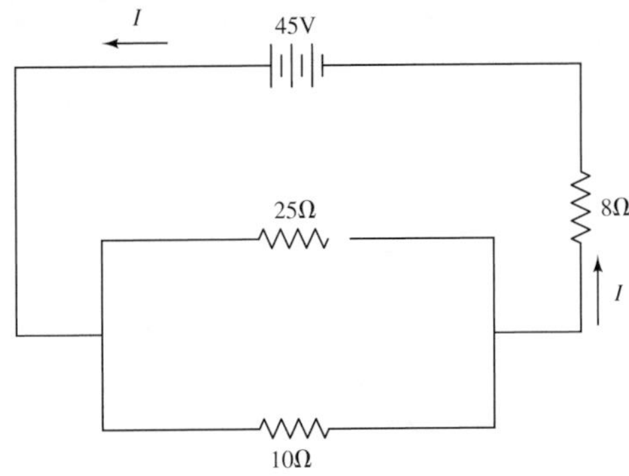

Fig. 26.32 Problem 26.

29. Four resistors having resistances of 1, 2, 4, and 5 Ω are connected in series with a battery of 24 V. (*a*) What is the total resistance of the circuit? (*b*) What is the current flow in the circuit?

30. Three resistors of 5, 10, and 20 Ω, respectively, are connected in parallel across a 12-V battery. (*a*) Find the combined resistance. (*b*) Find the current flowing through each resistor. (*c*) Find the total current flow.

31. Three coils R_1, R_2, and R_3 having resistances of 4, 8, and 16 Ω, respectively, are connected in parallel. These are connected in series with two other coils R_4 and R_5 of 12 and 20 Ω, respectively, as shown in Fig. 26.33. Find the total resistance of the combination.

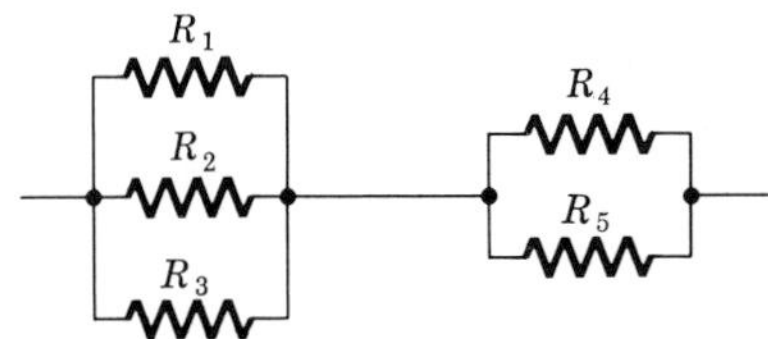

Fig. 26.33 Problem 31.

32. If 2 A is flowing through R_2 of Problem 31, find the amount of current passing through each of the other four coils.

33. What voltage must be impressed on the circuit shown in Fig. 26.34 in order to have a current of 6 A in the 4-Ω resistor?

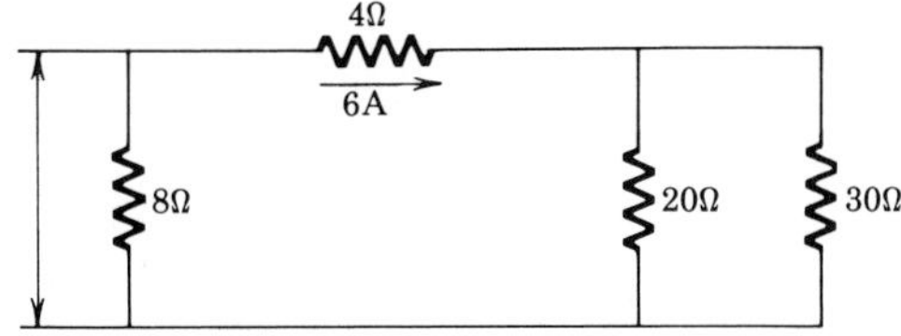

Fig. 26.34 Problem 33.

34. What is the magnitude of the total resistance R_T for the arrangement shown in Fig. 26.35?

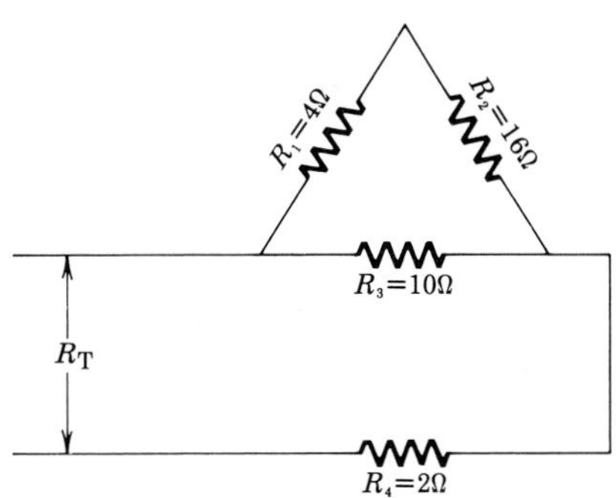

Fig. 26.35 Problem 34.

35. The wiring diagram for the circuit in a direct-reading voltmeter is shown in Fig. 26.36. When 0.612 V is placed across *AB*, the voltmeter registers a full-scale deflection calibrated to read 15 V. What will the voltmeter read when 20 V is applied across *OP*? The internal resistance of the meter is 50 Ω, R_1 = 600 Ω, and R_2 = 1800 Ω.

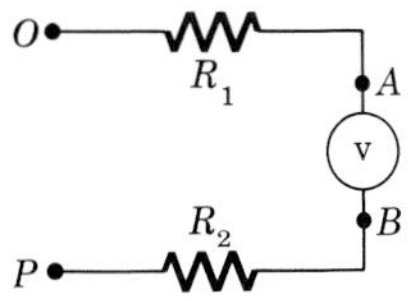

Fig. 26.36 Problem 35.

Group Three

36. What is the resistance across *AB* of Fig. 26.37 if each edge of the cube has a resistance of 10 Ω?

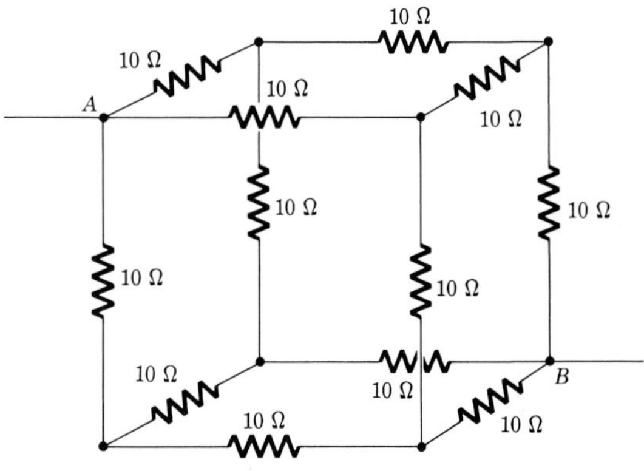

Fig. 26.37 Problem 36.

37. Two dissimilar batteries are connected to three resistors as shown in Fig. 26.38. Determine the current through the three resistors.

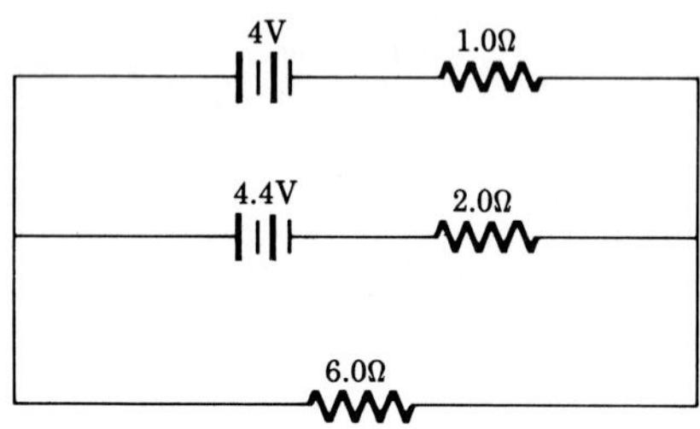

Fig. 26.38 Problem 37.

38. Use the data in Fig 26.39 to determine the currents in each of the resistors.

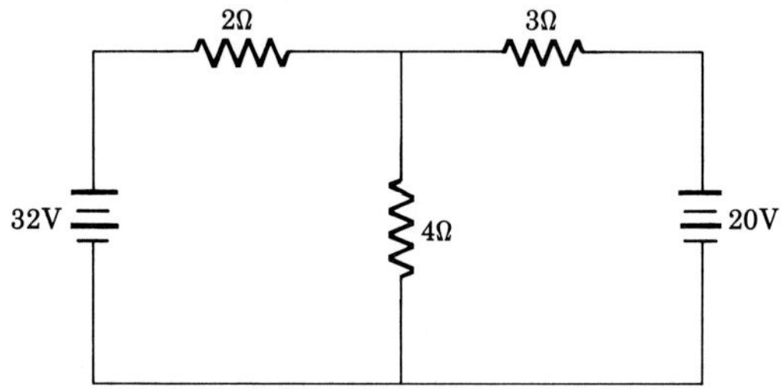

Fig. 26.39 Problem 38.

39. Solve for the current in the 6-, 4-, 5-Ω resistors of Fig. 26.40.

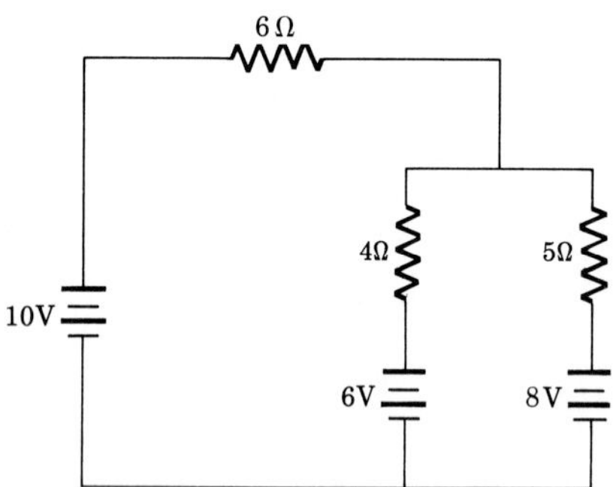

Fig. 26.40 Problem 39.

40. Solve for the current I through the resistor R shown in the circuit of Fig. 26.41.

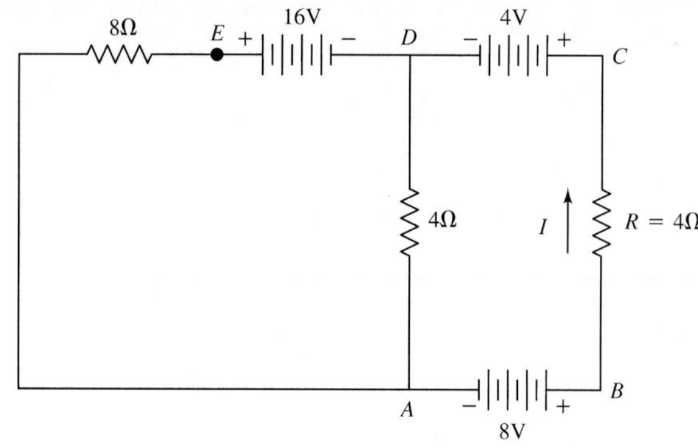

Fig. 26.41 Problem 40.

41. When two identical batteries are connected in series across a 5-Ω resistor the current is 0.2 A. When the same two batteries are connected in parallel with the same resistor, the current across the resistor is 0.16 A. (*a*) What is the internal resistance of each battery? (*b*) What is the emf of each battery?

42. In Fig. 26.42, the resistances of the resistors R_1, R_2, R_3, and R_4 are equal. Which resistor has the greatest current? Which has the least? Show how you arrived at your answers.

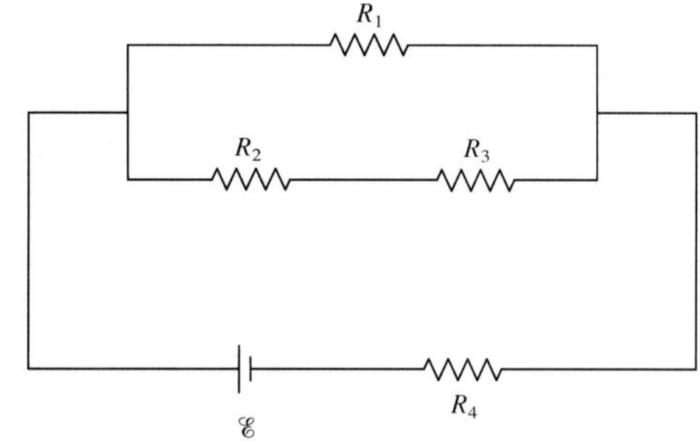

Fig. 26.42 Problem 42.

43. What is the current input for the circuit of Fig. 26.43?

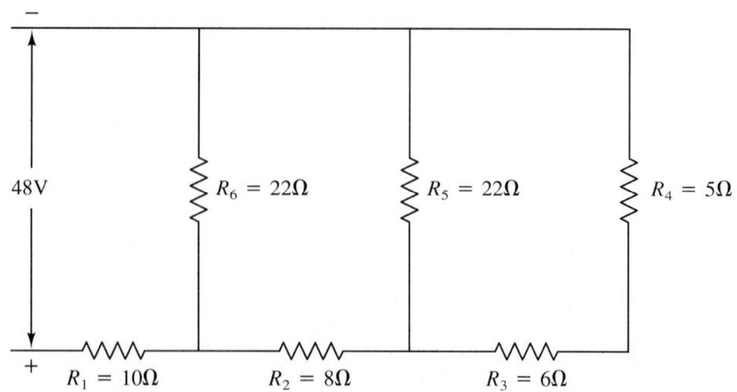

Fig. 26.43 Problem 43.

44. Find the total current in the circuit of Fig. 26.44.

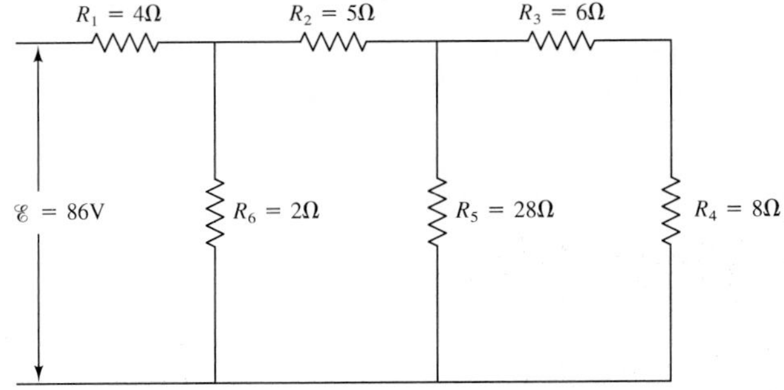

Fig. 26.44 Problem 44.

45. A tetrahedron *ABCD* is constructed from Gauge 32 annealed copper wire (see Fig. 26.45). Each side is 5 ft long. Find the resistance from *A* to *B*. Use the table of Appendix VI.

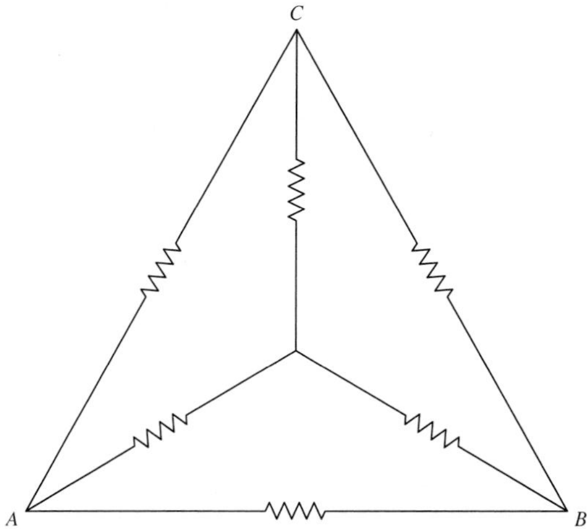

Fig. 26.45 Problem 45.

46. Given the delta and wye circuits shown in Fig. 26.46, determine the resistance of R_1, R_2, and R_3 so that $R_{AB} = R_{DE}$, $R_{BC} = R_{EF}$, and $R_{AC} = R_{FD}$.

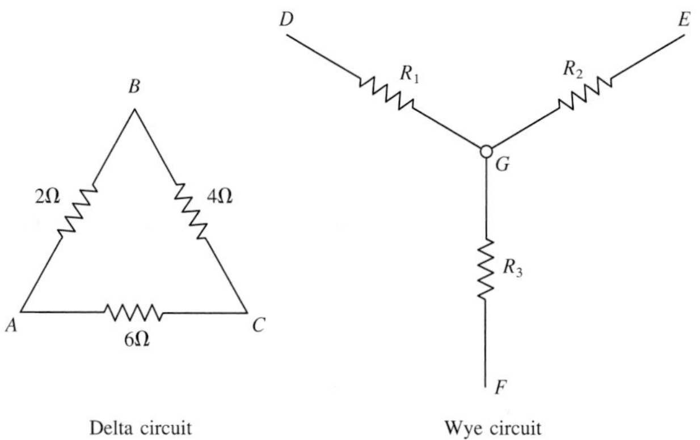

Fig. 26.46 Problem 46.

47. Battery *A* (Fig. 26.47) has an emf of 9 V and an internal resistance of 2 Ω. Battery *B* has an emf of 6.5 V and an internal resistance of 1 Ω. The batteries are connected in parallel (the positive terminals connected together and the negative terminals connected together). A 3-Ω resistor is connected between the positive and negative terminals. What will be the current through each battery?

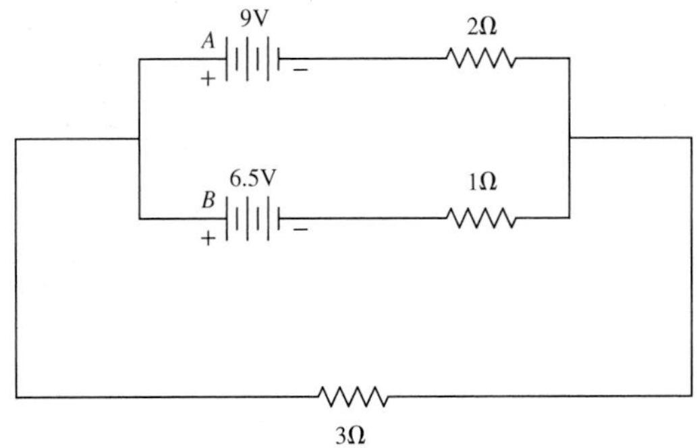

Fig. 26.47 Problem 47.

48. Determine the current *I* through the resistor *R* in the circuit of Fig. 26.48.

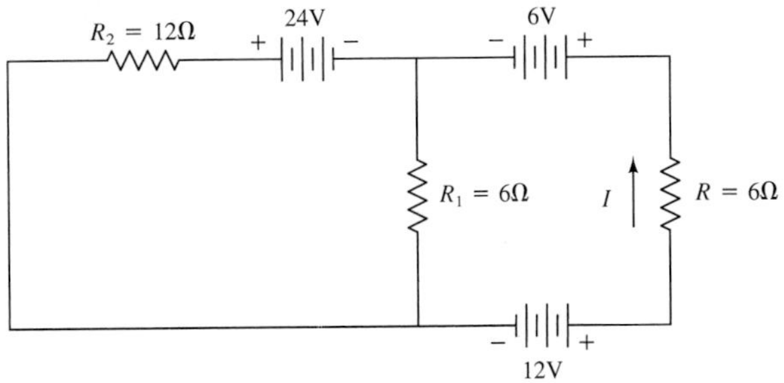

Fig. 26.48 Problem 48.

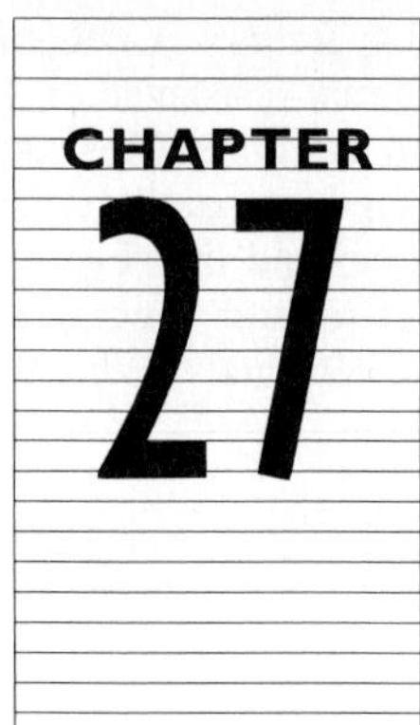

SOURCES AND EFFECTS OF ELECTRIC CURRENT

In this chapter we will discuss some of the more common devices for producing a steady difference of potential. It will be noted that each device provides two terminals between which a potential difference exists. And in each such device, when a charge flows out from one terminal, an equal charge will flow into the other terminal. In each device, energy in some other form must first be changed into electric energy. In batteries, chemical reactions take place, converting chemical energy into electric energy. In the generators at power plants, mechanical energy is converted into electric energy. Kinetic energy of wind is converted into electric energy in the wind generator. Heat energy is converted to electric energy in an instrument called the thermocouple. Solar energy is converted into electric energy in the photoelectric cell and the solar (photovoltaic) cell.

27.1 ■ Electromotive Force and Potential Difference

Electromotive force (emf) is the work done per unit charge that passes through the generating source of electric energy.

A device that produces electric energy is called a *source of electromotive force*. It should be emphasized that such a device does not manufacture electrical charge. It simply moves the charge through a circuit. The use of the term "electromotive force" is unfortunate; "electromotive force" is not a force at all. It is the work per unit charge done as the charge is moved through the generating source. Hereafter we shall avoid the words "electromotive force" and use instead the abbreviation *emf* (see Sec. 26.5).

As the charge moves through a source of emf, work is done *by the source* in raising the electrical potential energy of the charge. This emf causes differences of potential energy to exist between points in the external circuit. As the charge moves through the external circuit, work is done *by* the electric field on the charge. The electric potential energy supplied by the emf source is dissipated as heat energy in a pure resistor. In an electric motor, the energy is dissipated as useful work plus heat energy. In some circuits, the emf results in chemical energy. In each of the above instances the energy provided by the source of the emf exactly equals the energy expended in the external circuit. A source of emf has an emf of 1 volt if 1 J of energy is gained for each coulomb of charge passing through it. We shall use the script letter $\mathscr{E}$ for the value of the emf. (Do not confuse it with E for the electric field.) Thus, *the value of the emf* ($\mathscr{E}$), *in volts* is defined by

$$\mathscr{E} = \frac{W}{Q} \qquad (27.1)$$

where W is the amount of energy, in joules, gained by a charge Q, in coulombs, passing through the source.

In a source of emf, not only is some other form of energy transformed into electric energy, but the transformation is *reversible*. For example, a storage battery is charged by forcing electric current in a direction opposite to that in which it travels when the battery is discharging. By the same token, a generator can be converted into a motor. On the other hand, a pure resistor is not a source of emf. A potential difference across the resistor will establish a current in the circuit which will heat up the resistor. But building a fire under the resistor does not produce a current.

EMF voltage is associated with *generated energy*. Potential difference is associated with *dissipated energy*. Thus, if between two points in a circuit there exists the capability of obtaining electrical energy from some other form of energy, an emf exists between those two points. However, if between two points in a circuit electrical energy is *converted* to some other form of energy, a *potential difference* exists between those two points.

27.2 ■ Resistance in a Source of EMF

The potential difference across a battery, generator, or other *source* of electric energy *when it is not connected to any external circuit, is equal to its emf*. When the source of electric energy is a part of a closed circuit, the circuit carries a current, and the potential difference across the terminals of the source is always less than the emf because of the *internal resistance* of the source. The circuit diagram of Fig. 27.1 will illustrate the situation.

Assume a 6.0-V battery connected across a resistance R of 3Ω. When the switch K is open, the circuit will carry no current. A voltmeter connected across the terminals of the battery will register an emf of 6.0 V. When the switch is closed, about 2 A of current will be established through R, while the voltmeter will read a potential difference V across the battery terminals of about 5.7 V.

The drop in voltage from 6.0 V in the open circuit to 5.7 V in the closed circuit is due to the internal resistance r of the battery. Hence the actual terminal voltage V across a source of emf $\mathscr{E}$ and its internal resistance r is

$$V = \mathscr{E} - Ir \tag{27.2}$$

Terminal voltage = emf − potential drop within the source. On an open circuit, no current exists, and $V = \mathscr{E}$; on a closed circuit the current that is established through the battery lowers the value of V by an amount equal to Ir.

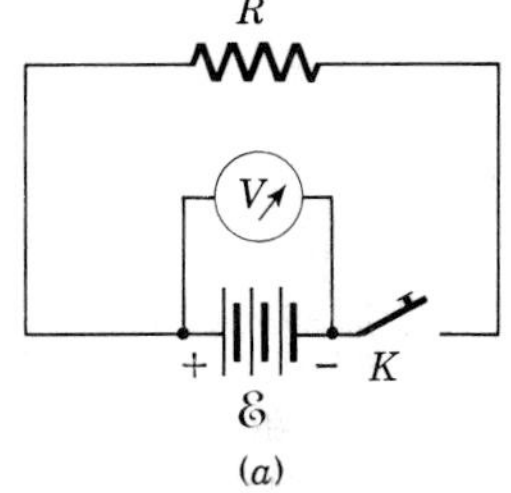

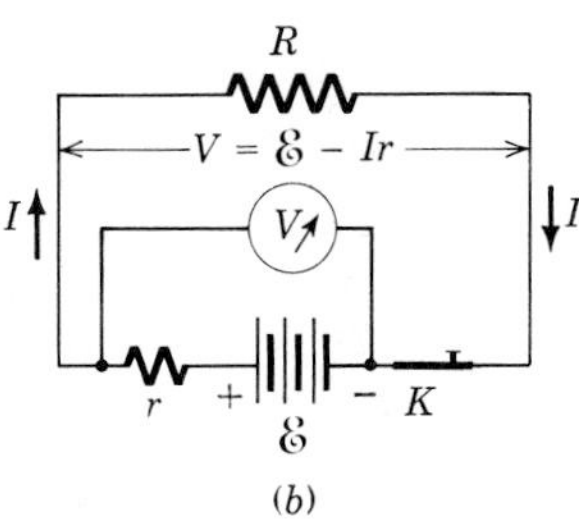

Fig. 27.1 Potential drop within a battery. The terminal voltage in (*b*) is always less than in (*a*) because of the potential drop *Ir* within the battery when a current is established.

Illustrative Problem 27.1 The voltmeter shown in Fig. 27.1*a* reads 6.0 V (no current flowing). When the voltmeter is removed and an ammeter is placed in series with a 4.5-Ω resistor and the battery, the ammeter reads 1.3 A when the key K is closed. What is the internal resistance of the battery?

Solution It is known that $\mathscr{E}$=6.0 V. The potential difference V across the resistor (also at the terminals of the battery) to cause a current of 1.3 A in the closed circuit is

$$\begin{aligned} V &= IR \\ &= 1.3 \text{ A} \times 4.5\Omega \\ &= 5.85 \text{ V} \end{aligned}$$

Solve Eq. (27.2) for r and substitute known values.

$$\begin{aligned} V &= \mathscr{E} - Ir \\ r &= \frac{\mathscr{E} - V}{I} \\ &= \frac{6.0 \text{ V} - 5.85 \text{ V}}{1.3 \text{ A}} \\ &= 0.12\ \Omega \qquad \textit{answer} \end{aligned}$$

Illustrative Problem 27.2 A 12-V battery has an internal resistance of 0.15 Ω. It is connected in a circuit as shown in Fig. 27.2. Find the terminal voltage and the current in each resistor.

Solution Replace R_2 and R_3 by R, where

$$\frac{1}{R} = \frac{1}{R_2} + \frac{1}{R_3} \tag{26.13}$$

Substitute

$$\frac{1}{R} = \frac{1}{3.0\ \Omega} + \frac{1}{4.0\ \Omega}$$

$$R = \frac{12}{7}\ \Omega = 1.71\ \Omega$$

Use Ohm's law to solve for I in the complete circuit.

$$I = \frac{\mathscr{E}}{R_T} = \frac{12\ \text{V}}{0.15\ \Omega + 1.71\ \Omega + 2.0\ \Omega}$$

$$= 3.11\ \text{A}$$

The current through R_1 and r is 3.11 A. *answer*

Use Eq. (27.2) to find the terminal voltage.

$$V = \mathscr{E} - Ir$$
$$= 12\ \text{V} - 3.11\ \text{A} \times 0.15\ \Omega$$
$$= 11.53\ \text{V}$$ *answer*

The potential difference across R_2 and R_3 equals

$$V = IR$$
$$= 3.11\ \text{A} \times 1.71\ \Omega$$
$$= 5.32\ \text{V}$$

Therefore,

$$I_2 = \frac{V_2}{R_2}$$
$$= \frac{5.32\ \text{V}}{3.0\ \Omega}$$
$$= 1.77\ \text{A}$$ *answer*

$$I_3 = \frac{5.32\ \text{V}}{4.0\ \Omega}$$
$$= 1.33\ \text{A}$$ *answer*

NOTE: $I_2 + I_3 = I_1 = I$

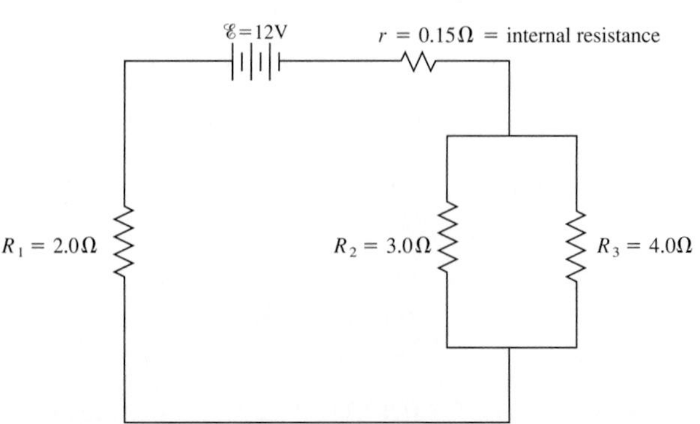

Fig. 27.2 Illustrative Problem 27.2.

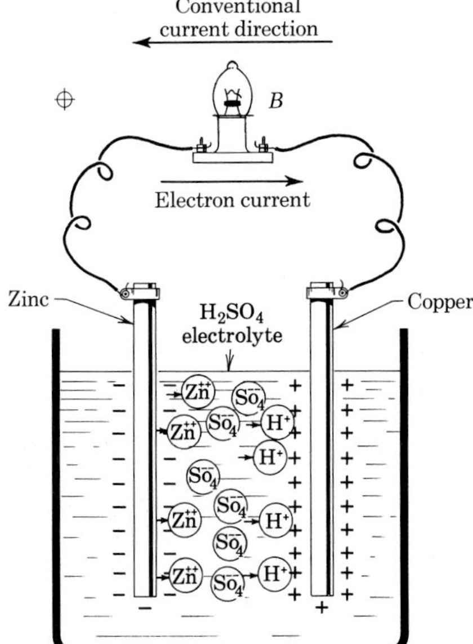

Fig. 27.3 Simple voltaic cell.

27.3 ■ The Voltaic Cell

In 1798 Alexander Volta discovered that if two dissimilar metal strips are placed in an acid solution, an electromotive force appears between the two metals. If a conductor is connected between the two strips, a current will exist in it. As a consequence of his findings, he invented the first chemical generator of electricity, now called a *voltaic cell* in his honor. His first simple cell can readily be duplicated in the laboratory.

Count Alessandro Volta (1745–1827) was an Italian physicist. He was a professor of physics at Como and Pavia. He was honored by Napoleon, who made him a count and a senator. Volta was noted for his experiments and inventions in electricity.

Although the voltaic cell is not available in commercial form, it illustrates the principles of how chemical reactions produce emfs in various types of practical cells. Place a strip of zinc and a strip of copper in a glass vessel containing a solution of dilute sulfuric acid. If the zinc and copper strips, called *electrodes,* are kept apart in the solution and a voltmeter is connected to the electrodes, a difference of potential between the electrodes will be indicated. The emf will light a small flashlight bulb *B* (Fig. 27.3).

In the voltaic cell chemical energy becomes electrical energy.

Considerable chemical action is observed between the acid and the zinc electrode as the zinc is gradually "eaten away" by the acid. During the reaction zinc atoms leave the strip and enter the solution as zinc *ions*. (Ions are atoms or groups of atoms that have a net negative or positive charge, due to an excess or deficiency of electrons.) Each zinc atom that leaves the strip leaves behind two electrons. Thus, the zinc atom when in solution becomes a positive zinc ion (Zn^{2+}). Because the electrons are left behind, the zinc strip becomes negatively charged. The zinc strip is called the *cathode*. The chemical reaction can be written

The negative electrode of a cell is called the *cathode*.

$$Zn^{\circ} \longrightarrow Zn^{2+} + 2e^{-}$$

Each molecule of sulfuric acid (chemical formula, H_2SO_4) separates into two positive hydrogen ions ($2H^+$) and one sulfate ion ($SO_4{}^{2-}$). The ionic equation is written

$$H_2SO_4 \longrightarrow 2H^+ + SO_4{}^{2-}$$

When substances in solution form ions, they are said to be *ionized*. Such a solution is called an *electrolyte*.

An electrolyte is a substance whose water solution conducts an electric current.

The positively charged zinc ions will repel the positive hydrogen ions toward the copper strip. As each positive ion reaches the copper strip, it takes an electron from a copper atom. The neutralized hydrogen will then form hydrogen molecules and bubble off into the air.

$$2H^+ + 2e^- \longrightarrow H_2^{\circ} \uparrow$$

The positive electrode of a cell is called the *anode*.

The copper strip, having lost electrons, becomes positively charged. The copper strip becomes the (positive) *anode*. Thus, a potential difference is created between the zinc and copper strips.

Many substances, when dissolved in water or another solvent, break up into charged ions.

It should be noted that the anode is that terminal which loses electrons to the electrolyte during discharge. The cathode terminal gains electrons from Zn atoms becoming ions. To complete the chemical reaction in the cell, the zinc (Zn^{2+}) ions unite with the sulfate (SO_4^{2-}) ions to form zinc sulfate ($ZnSO_4$) in water solution.

$$Zn^{2+} + SO_4^{2-} \longrightarrow ZnSO_4$$

A potential difference of approximately 1 V is established in the copper-zinc voltaic cell. It is done by raising the potential of the copper electrode 0.5 V and lowering the potential of the zinc electrode 0.5 V.

Thus, the overall chemical reaction (without referring to the exchanges of electrons) can be written

$$Zn + H_2SO_4 \longrightarrow ZnSO_4 + H_2$$

The battery will be ''dead'' when all of the zinc will have been converted to zinc sulfate.

The difference in potential that exists between the terminals of a chemical cell depends on what the electrodes are made of and on their ability to react chemically with the electrolyte or to give up electrons.

Almost any two conductors could have been used for the strips (electrodes), but the strips must not be identical. A carbon rod could have been used in place of the copper strip, since its function would be merely to supply electrons within the cell. Also, many other solutions (such as hydrochloric acid or sal ammoniac) could be used as the electrolyte. All that is necessary is that the solution should attack one of the electrodes.

The dry cells used in flashlights and the storage batteries used in automobiles are common examples of commercial sources of continuous electric energy that result from chemical action.

27.4 ■ Local Action

Local action is caused by impurities.

In the simple voltaic cell just described the commercial zinc used contains many impurities, such as small particles of carbon, iron, and lead. Even though no current is being drawn from the cell, the zinc may rapidly be eaten away. The small particles of impurities, when in contact with the zinc in the presence of an acid solution, will form small local cells with closed circuits. The result is a wasting away of the zinc even though no current is flowing in the external circuit. This process is called *local*

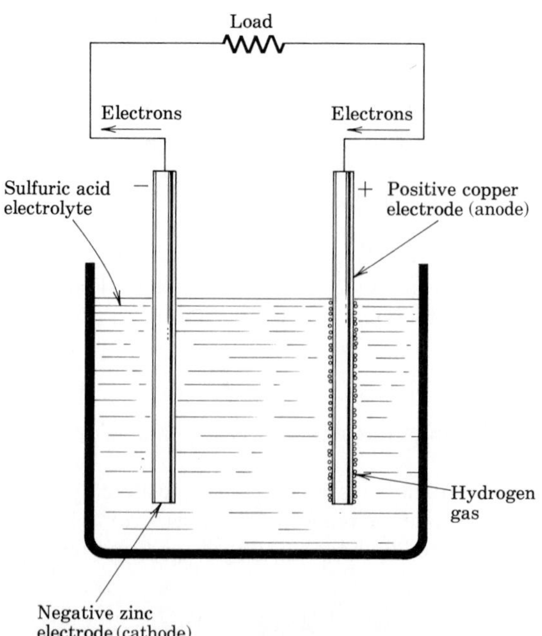

Fig. 27.4 Polarization in a simple voltaic cell. The formation of a gas at the anode of the cell increases the internal resistance of the cell.

action. Local action will continue as long as the impurity is in contact with the zinc and the acid. The effect of local action in a cell can be prevented by *amalgamating* the zinc, i.e., by polishing the zinc and then rubbing it with mercury until a thin coating of mercury surrounds the zinc. The mercury will dissolve some of the zinc but none of the impurities. In this way the impurities are prevented from making contact with the acid. The amalgamated zinc will then react with the acid only when the cell is in operation.

27.5 ■ Polarization

When a heavy current is drawn from a simple voltaic cell, the current will quickly fall off in intensity because hydrogen bubbles, which result from the chemical action that takes place, collect on the positive electrode (Fig. 27.4). These bubbles not only reduce the effective surface area of the copper but alter the material of the electrode, thus reducing the effective emf of the cell. The effect of such action is called *polarization*. Polarization increases the *internal resistance* of a cell. In commercial cells where polarization may occur, another substance, called a *depolarizer,* is introduced to combine with the hydrogen and remove it from the electrode. The oxygen of the depolarizer combines with the hydrogen to form water.

Polarization insulates the anode because hydrogen is a poor conductor of electricity.

27.6 ■ The Dry Cell

A dozen or more types of voltaic cells have been devised for commercial purposes, but the most widely used is the common *dry cell*. Actually the cell is not dry.

Carbon-Zinc Cell The negative electrode of the carbon-zinc dry cell (see Fig. 27.5) consists of a zinc cylinder which forms the walls of the cell. The zinc is amalgamated with traces of mercury. The cylinder contains the positive electrode (a carbon rod), the electrolyte (ammonium chloride), and the depolarizer (manganese

The most common primary cell is the carbon-zinc cell. It has a relatively low cost and is suited primarily for intermittent use.

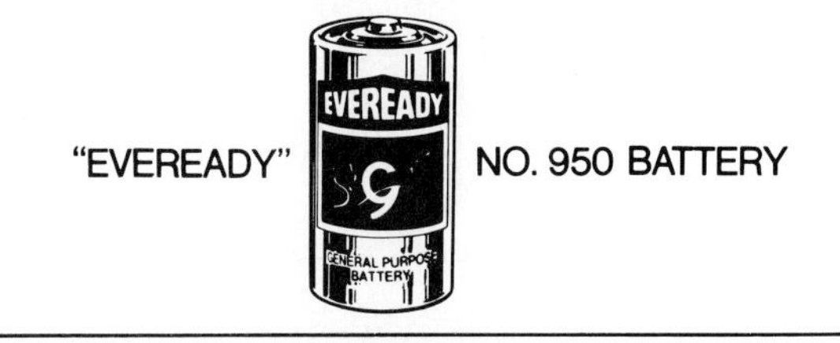

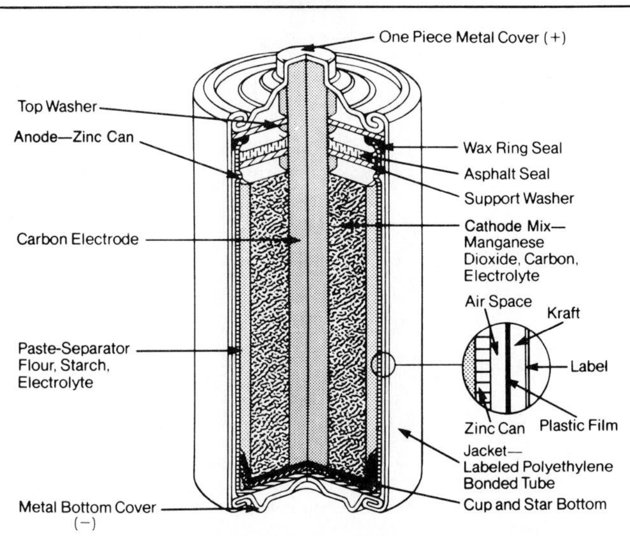

Fig. 27.5 Cutaway of a cylindrical general-purpose carbon-zinc cell. (Union Carbide Corp.)

dioxide). The manganese dioxide (MnO_2) is a chemical compound rich in oxygen. Whenever hydrogen tries to collect around the carbon rod, the oxygen reacts chemically with the hydrogen to form water. The chemical reaction is shown in the following equation.

$$MnO_2 + 2H_2 \longrightarrow Mn + 2H_2O$$

Mercury is coated on the interior zinc surface during manufacture to reduce local action due to impurities within the zinc.

The cell also contains small amounts of zinc chloride ($ZnCl_2$) to aid in depolarization. The electrolyte and the depolarizers are combined in the form of a paste. Powdered coke and graphite are added to the paste to reduce the internal resistance of the cell. The carbon electrode is placed in the center of the zinc cylinder, the space around it is filled with the paste, and the cell is then sealed with an asphalt sealant to make it watertight and airtight. Thus, the cell can be easily transported and can be used in any position.

Manganese dioxide acts as a depolarizing agent in a dry cell by reacting with hydrogen to form water.

More than 2 billion dry cells are sold annually and are used in devices such as flashlights, toys, watches, radios, calculators, and in a variety of electrochemical systems. Sizes vary from the small penlight cell up through AA, C, and D, each with an emf of approximately 1.5 V. The size of the cell determines the amount of current each cell can deliver and the amount of energy stored.

Although individual AA, C, and D cells are not batteries, common usage allows the term battery to be applied to them.

A carbon-zinc cell should be used only *intermittently* or with very low current drain, or it will polarize faster than the manganese dioxide and zinc chloride can depolarize it. The emf of the cell gradually declines even when not in use.

Alkaline Cell About one-half of the more than 2 billion dry cells sold annually are of the alkaline type (see Fig. 27.6). The alkaline cell is designed to provide an economic power source for devices that require a relatively high power input and/or frequent use. The cathode of the alkaline-manganese cell is a high-purity, high-density manganese dioxide mixed with a conductive carbon matrix. The anode is a mixture of amalgamated zinc powder and an electrolyte. The electrolyte is a highly conductive, low-freezing-point paste of potassium hydroxide (KOH) and water.

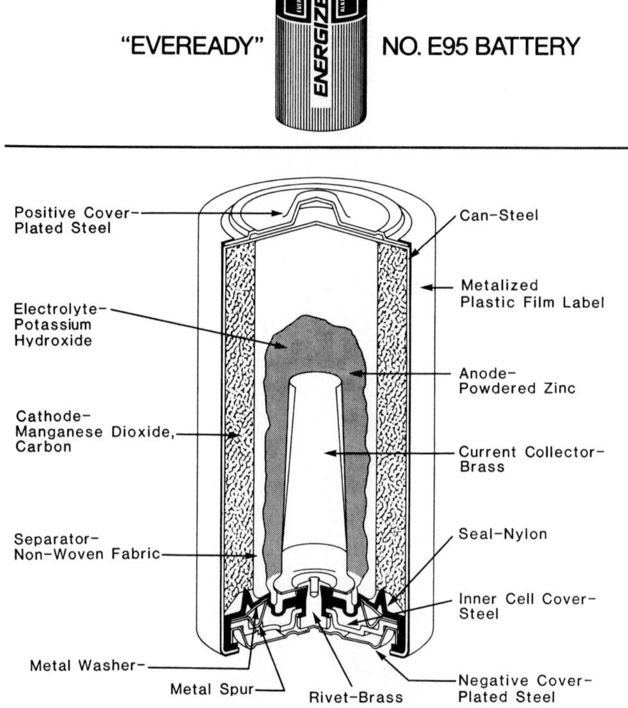

Fig. 27.6 Cutaway of a cylindrical alkaline-manganese cell. (Union Carbide Corp.)

An alkaline cell will usually last four times as long as a similar carbon-zinc cell. Whereas the shelf-life of an alkaline cell (the time it will retain its energy during storage) is 30 to 36 months, that of the carbon-zinc cell is only 6 to 12 months.

27.7 ■ Secondary or Storage Cells

The cells we have discussed thus far are termed *primary cells*. When the cell is "dead," it cannot be regenerated, because the active materials have been consumed. Primary cells are nonreversible cells. *Secondary cells* are reversible cells that may be recharged by reversing the direction of the electric current. The *storage cell* is one in which electric energy is *stored* in the form of chemical energy. In this cell two electrodes of dissimilar materials are immersed in an electrolyte. After they have been supplying current to an external circuit, they can be restored to their original condition by sending an electric current from an external source through the cell in a direction opposite to that of current flow from the cell. Only those electrodes and electrolytes can be used in a storage cell in which the chemical action can be readily reversed by changing the direction of the current. When the cell is delivering a current to an external circuit, it is said to be *discharging*, and while energy is being restored, it is *charging*. A *storage battery* consists of two or more such cells connected in series-parallel combinations.

Two or more cells connected together form a battery.

The most commonly used storage cell is the lead storage cell (Fig. 27.7), the type used in automobiles and for many other industrial purposes. The active material of the *positive* electrode of a lead storage cell is a chemical compound called lead oxide (PbO_2), while the active material of the *negative* electrode consists of finely divided, spongy, metallic lead (Pb). The electrolyte is dilute sulfuric acid (H_2SO_4). The positive and negative electrodes are held apart by separators of wood, rubber, or glass. The emf of such a cell is approximately 2.1 V. When the cell is discharging, the active materials of both plates change to lead sulfate ($PbSO_4$) while the electrolyte is converted to water (H_2O). As the cell continues to discharge, both plates eventually are covered with the same material ($PbSO_4$), and in this condition no emf is developed. The cell is said to be run down, dead, or discharged. Also the acid becomes more and more dilute as the battery discharges (see Sec. 12.12).

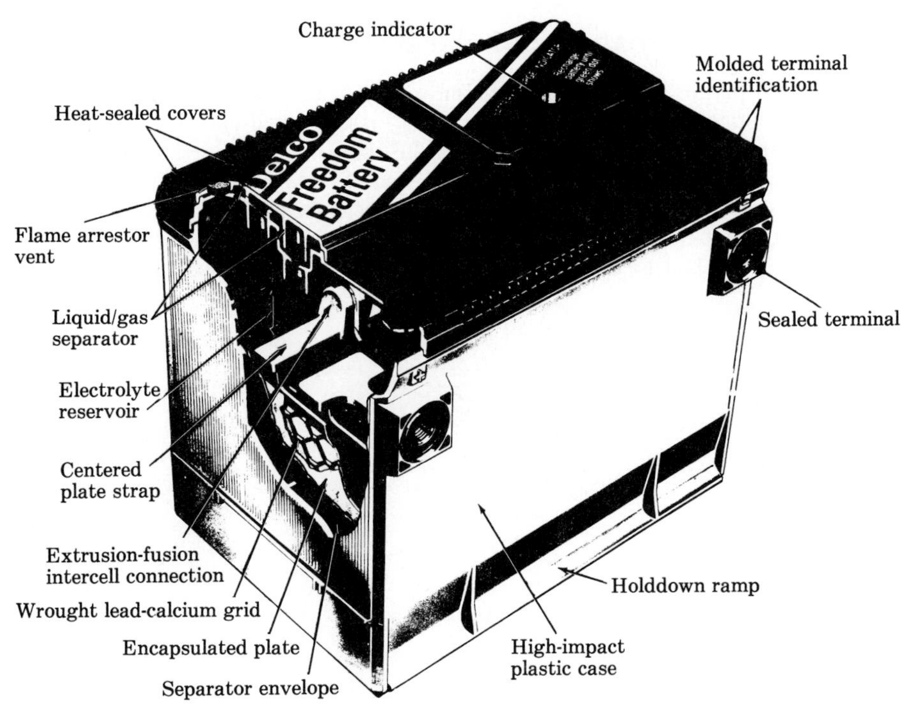

Fig. 27.7 Cutaway showing internal construction of a lead storage battery. A built-in charge indicator replaces the hydrometer test used on many other batteries. (Delco-Remy)

When a battery is being recharged, work is done by a source of emf to reverse the direction of the electric current within the battery.

To charge a lead storage battery, a current is passed through it in a direction opposite to that of the normal flow of the battery. This reverses the chemical action until the electrolyte and the electrodes are restored to their original condition. The chemical reaction that takes place in the storage battery can be expressed by the reversible reaction:

$$PbO_2 + Pb + 2H_2SO_4 \underset{\text{Charging}}{\overset{\text{Discharging}}{\rightleftharpoons}} 2PbSO_4 + 2H_2O + \text{Electrical Energy} \quad (27.3)$$

It should be evident that the battery charger should have an emf greater than that of the battery. A charger will require an emf of approximately 14.1 V to recharge a 12-V battery at currents up to 30 A. There is a danger in "hot charging" a battery at excessively high currents. High charging amperage can cause the electrolyte to "boil," lowering the liquid level as well as disintegrating the electrodes.

27.8 ■ Capacity Rating of a Battery

The capacity rating of a battery is usually given in ampere-hour units based upon an 8-h discharge period.

The capacity of a battery is usually rated in *ampere-hours* (Ah). It is a measure of the energy contained in a battery when fully charged. Stated another way, the capacity of a battery expressed in ampere-hours is a measure of the total charge or energy that the battery can deliver under normal operating conditions. Thus, a battery having a rating of 120 ampere-hours (A · h) will deliver one ampere of current for 120 hours, 2 A for 60 h, 3 A for 40 h, etc. However, if the battery is discharged slowly it may show a capacity greater than its rated 120-A · h capacity. On the other hand, if the battery is discharged quite rapidly, its capacity is reduced. It is common practice to give the rating of a battery in ampere-hour units based on an 8-h discharge period. Thus, a 120-A · h battery should provide a current of 15 A for a period of 8 hours.

At low temperatures the mobility of the ions in a lead storage battery decreases considerably. This appears as an increase in the internal resistance of the battery, thereby causing the terminal voltage to drop abnormally when current is drawn. For example, the emf of a battery at 0°C is less than 50 percent of its value at 80°C.

27.9 ■ Electromagnetic Generators

For industrial purposes, the most important source of electric energy is the electromagnetic generator, in which mechanical energy is converted into electrical energy. The essential parts of a simple generator are illustrated in Fig. 27.8. The simple form diagramed consists of a set of field magnets (N and S) and a rotating loop of wire *A*. If the loop is rotated in the magnetic field, an electric current will be established in the loop. This current can be taken off on brushes *B*, which touch a pair of collecting rings *R*, one attached to each end of the loop. A complete discussion of electric

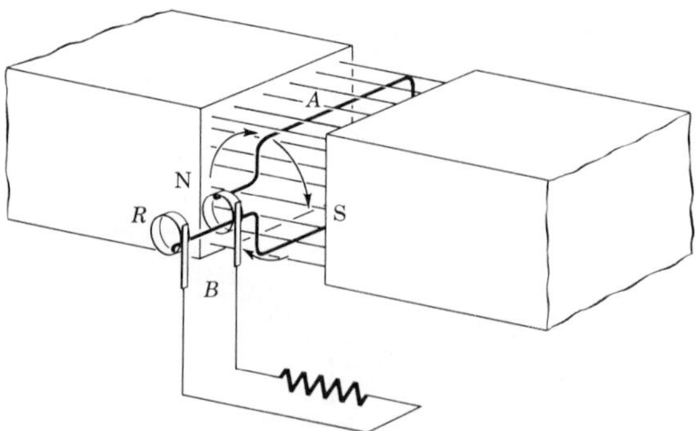

Fig. 27.8 A simple generator—schematic representation.

generators will be given in later chapters, after the principles of electromagnetism have been explained.

27.10 ■ Thermoelectricity

A means of converting heat energy into electric energy was discussed in Sec. 14.9 and illustrated in Fig. 14.8, which shows a circuit of two different wires, copper and iron. They have been fused or brazed together at both ends to form a closed loop. If one of these junctions is kept at a temperature different from that of the other, a small electric current will flow around the circuit, passing from the iron to the copper at one junction and from the copper to the iron at the other. The current will continue to flow as long as the temperature at one junction is higher than that at the other, and the intensity of the current is a measure of the difference in the temperatures. Such a device is called a *thermocouple*. Thermocouples are used for the measurement and control of temperature. A number of thermocouples may be joined in series, to form a *thermopile*. Thermopiles can produce voltages sufficient to operate radio receivers, relays, and other low-power devices.

Temperature differentials are used to generate emf in the thermocouple.

27.11 ■ Photoelectricity

Certain metals, such as selenium, cesium, and germanium have the ability to convert light energy into electric energy. When light falls on the surface of these metals, electrons are emitted from the surface. Such a phenomenon is called the *photoelectric effect* (see Chap. 34). An arrangement which permits the electric current emitted by the light-sensitive metal to be included in a closed circuit is called a *photoelectric cell*. The photoelectric current is generally extremely feeble but can be amplified by means of other electronic devices. The photoelectric cell has numerous applications in industry because (1) the photoelectric effect is almost instantaneous and (2) the intensity of the current is proportional to the intensity of the light beam. These properties have led to its application in sound motion pictures, in television, footcandle meters, exposure meters, automatic sorting machines, automatic burglar-alarm systems, door openers, drinking fountains, radio transmission of pictures, and many industrial devices for counting and control.

27.12 ■ The Solar Cell

The *solar cell* is a special type of photoelectric cell which converts light energy directly to electric energy. A typical solar cell (photovoltaic cell, or PVC) consists of a silicon crystal "sandwich" (Fig. 27.9). One layer of the sandwich consists of a crystal of pure silicon, a semiconductor that has been "doped" with a slight amount of phosphorus on the inside layer. Since phosphorus is an electron donor, the wafer contains an excess of free electrons and is called "negative-type" or *N*-type silicon.

In the solar cell, photovoltaic solar energy is converted directly to electric energy.

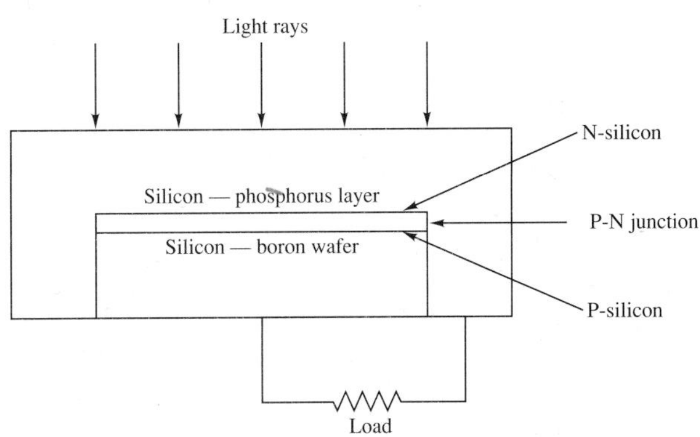

Fig. 27.9 Simplified diagram of a solar cell (see also Fig. 33.15).

The other layer consists of a very thin layer of silicon to which has been added a slight amount of boron, an electron acceptor. Boron-doped silicon is called "positive" or *P*-type silicon. When the cell is exposed to light, the silicon absorbs some of the light energy. This energy frees a quantity of electric charge carriers (both positive and negative). Some of these are collected and separated at the *P-N* junction, giving the silicon-phosphorus layer a negative charge and the silicon-boron wafer a positive charge. Thus an electric potential difference (emf) is produced between the wafer and the layer, which "sweeps" the photoelectrons across the interface between the two types of silicon, and on into the external circuit.

The solar cell is light in weight, maintenance-free, long-lasting and pollutant-free. However, solar cells are quite expensive. The area of a single solar cell cannot be very large, since it is difficult to produce large silicon crystals. A common silicon solar cell of current manufacture has a diameter of 4 in., is 0.015 in. thick, and weighs about 6 g. One such cell can produce an output of about 1 W (2 A at a potential of about 0.5 V). Solar cells can be connected in series to achieve higher voltages or in parallel for higher currents. In a typical installation, the solar panels or arrays will be used to charge up storage batteries. Thus, power is available even when the solar cells are not exposed to the sun.

Solar cells are used to power electronic equipment on space satellites. They develop power to transmit radio messages from millions of miles in space. Solar-electric home electric power systems are now available, up to 5-kW capacity. Commercial electric power from huge photovoltaic installations have been "on line" since 1986. (See also Chapter 33.)

27.13 ■ Fuel Cells

Considerable effort has been expended in the research laboratories of industry, universities, and government to develop a type of electrochemical cell known as the *fuel cell*. In the fuel cell, the energy of a conventional fuel is converted directly into electric energy. In the fuel cell active materials are supplied, and the reaction products are removed continuously. Thus the fuel cell can operate as long as the required materials are supplied and the chemical by-products removed.

Fuel cells generate an emf as long as the reactant fuels are fed to them.

Many fuels have been experimented with in the research on fuel cells. The reactants consist of an oxidizing substance (one that furnishes oxygen molecules) and a reducing material (one that removes oxygen molecules). Perhaps the simplest fuel cell to describe, construct, and to understand is one that uses hydrogen gas as the fuel and oxygen as the oxidizing agent—the *hydrogen-oxygen cell*. Fuel cells of this type which develop 2 kW of power have been used very effectively in spacecraft. A

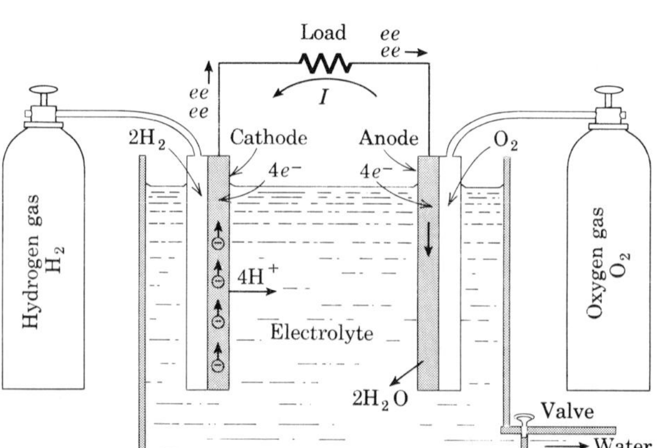

Fig. 27.10 Hydrogen-oxygen fuel cell.

schematic diagram of the hydrogen-oxygen fuel cell is shown in Fig. 27.10. The cell consists of three basic parts: fuel and oxidizer, electrodes, and an electrolyte. The electrodes are made of inert conducting materials and are porous to permit the gases to come gradually in contact with the electrolyte. The electrolyte of the cell is a solution of potassium hydroxide (KOH). The solution breaks up into K^+ and OH^- ions. (An *ion* is an atom or group of atoms that carries an electric charge.) At the negative electrode, hydrogen molecules combine with the hydroxide ions to form water. In the process electrons are released to the external current. The ionic reaction is

$$2H_2 + 4OH^- \longrightarrow 4H_2O + 4e^- \qquad (27.4)$$

The electrons move through the external circuit to the positive electrode, where oxygen molecules combine with water molecules and the incoming electrons to produce new hydroxide ions:

$$O_2 + 2H_2O + 4e^- \longrightarrow 4OH^- \qquad (27.5)$$

The hydroxide ions thus produced react with more hydrogen molecules at the anode. Thus for the entire cell the reaction is

$$2H_2 + O_2 \longrightarrow 2H_2O + \text{the transfer of } 4e^- \text{ from one electrode to the other.} \qquad (27.6)$$

The electrons can do useful work in passing from the cathode to the anode in the external circuit: 1 lb of hydrogen and 8 lb of oxygen in a fuel cell will produce 9 lb of water and enough electric power to light a 100 W bulb for nearly two weeks. The water produced must be removed from the electrolyte to maintain the proper electrolytic concentration.

A single fuel cell will generate from 0.5 V to 1 V. By a proper series-parallel arrangements of cells, 2 kW of electric power can be developed for spacecraft utilization.

An attractive feature of fuel cells is their high efficiency. They are thermodynamically not heat engines and therefore are not subject to the (Carnot) cycle limitations of heat engines. Efficiencies close to 100 percent are theoretically possible with fuel cells. The cell has no moving parts. The only energy loss in a fuel cell is that lost in heat by the electric current within the cell. But those losses are minimal compared with energy losses in conventional generating plants. Another advantage of fuel cells is that the waste products of the conversion, if any, are not pollutants. In the hydrogen-oxygen cell, the by-product is water. Fuel cells have found applications in space vehicles, where efficiency and reliability are critically important. Neither the electrodes nor the electrolyte are affected in the fuel-cell process. Also, the water can be used in manned spacecraft for drinking and cooling purposes.

COMBINATIONS OF CELLS

27.14 ■ Cells in Series

Often a single cell does not provide enough voltage or enough current for a specific need. It then becomes necessary to group two or more cells in combinations (to form a battery) to gain the desired results. Let us consider three dry cells connected in series (Fig. 27.11*a*). It will be noted that the negative electrode of the first is connected to the positive of the second; the negative of the second is connected to the positive of the third; and the negative of the third is connected through the external resistance R to the carbon electrode of the first cell. If the emf of each cell is 1.5 V, the emf of the three cells will be $3 \times 1.5 = 4.5$ V.

For *cells arranged in series* the following statements apply:

To provide for a higher emf, cells are connected in series. At the same time, an increase in the overall internal resistance will result.

1. The emf of the combination is equal to the sum of the emfs of the cells.
2. The current in each cell and in the external circuit is the same.
3. The total internal resistance of the combination is equal to the sum of the internal resistances of the individual cells.

The ampere-hour capacity of a series grouping of cells is the same as that of each cell.

Figure 27.11*b* is the wiring diagram, which illustrates the series arrangement of Fig. 27.11*a*. If we refer again to a hydraulic analogy (Fig. 27.11*c*), we would think

Fig. 27.11 (*a*) Three cells connected in series. (*b*) Wiring diagram of the series circuit. (*c*) Hydraulic analogy of three cells in series.

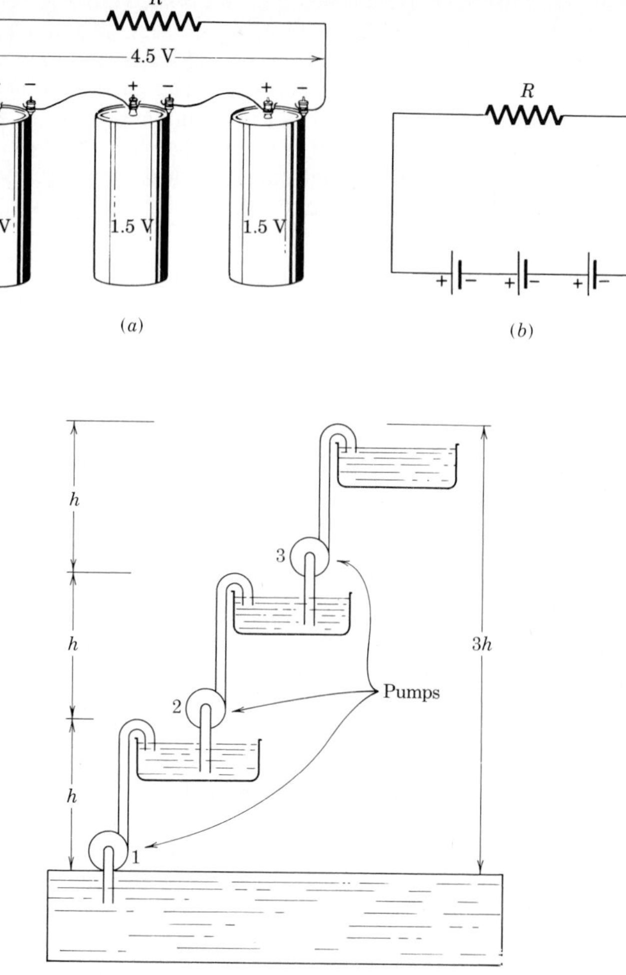

of the first cell as pumping electricity up to a certain potential (level), and the second cell as pumping it to a still higher potential, and the third to a potential three times as high as the potential of the first cell, just as the water pumps shown in the figure raise water to successively higher levels.

An example of cells connected in series is found in the use of six lead-acid cells (2 V each) in series to make the 12-V battery of a car.

Illustrative Problem 27.3

Three D cells each with an emf of 1.5 V and an internal resistance of 0.25 Ω are connected in series to a flashlight bulb whose resistance is 4.5 Ω. Find (*a*) the current in the circuit and (*b*) the terminal voltage of the three-cell combination.

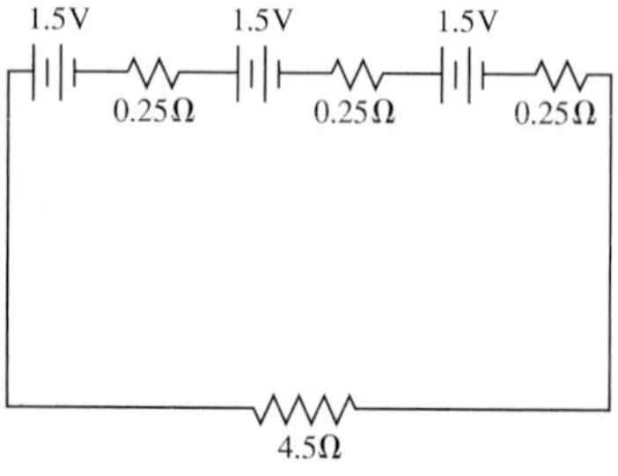

Fig. 27.12 Illustrative Problem 27.3.

Solution See Fig. 27.12.

(*a*)

$$I = \frac{\mathcal{E}}{R}$$

$$= \frac{1.5\ \text{V} + 1.5\ \text{V} + 1.5\ \text{V}}{0.25\ \Omega + 0.25\ \Omega + 0.25\ \Omega + 4.5}$$

$$= 0.86\ \text{A}$$ *answer*

(*b*) The terminal voltage V of each cell is

$$\begin{aligned} V &= \mathscr{E} - Ir \\ &= 1.5 \text{ V} - 0.86 \text{ A} \times 0.25\ \Omega \\ &= 1.29 \text{ V} \end{aligned}$$

For the three cells in series the terminal voltage is

$$\begin{aligned} 3 \text{ V} &= 3(1.29 \text{ V}) \\ &= 3.9 \text{ V} \qquad \textit{answer} \end{aligned}$$

Illustrative Problem 27.4

A battery with an emf of 24 V and an internal resistance of 1.2 Ω is connected in series with a 6-Ω resistor, an opposing battery with an emf of 12 V and an internal resistance of 0.6 Ω, and a group of resistors of 18 Ω and 12 Ω as shown in Fig. 27.13. Find (*a*) the current in each resistor and (*b*) the potential difference of each battery.

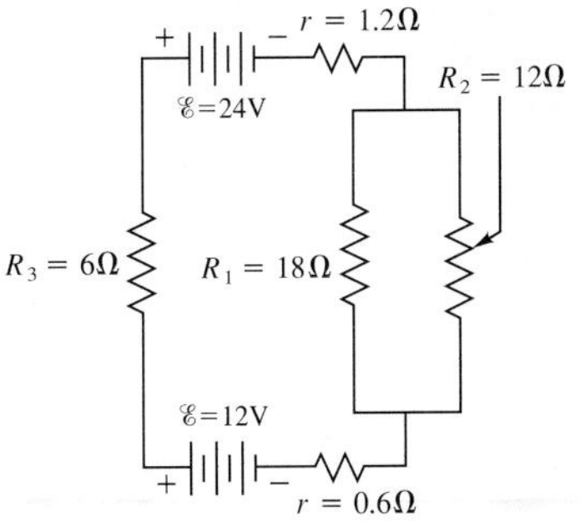

Fig. 27.13 Illustrative Problem 27.4.

Solution

(*a*) Replace the parallel combination of resistors by a single equivalent resistor R_{1-2}. Then

$$\begin{aligned} \frac{1}{R_{1-2}} &= \frac{1}{R_1} + \frac{1}{R_2} \\ &= \frac{1}{18\ \Omega} + \frac{1}{12\ \Omega} \\ R_{1-2} &= 7.2\ \Omega \end{aligned}$$

Solve for the current I in the circuit using Ohm's law

$$\begin{aligned} I &= \frac{24 \text{ V} - 12 \text{ V}}{6\ \Omega + 7.2\ \Omega + 1.2\ \Omega + 0.6\ \Omega} \\ &= 0.8 \text{ A} \end{aligned}$$

This is the current in each battery and in the 6-Ω resistor. The potential drop across the parallel resistors is equal to

$$IR_{1-2} = 0.8 \text{ A} \times 7.2\ \Omega = 5.8 \text{ V}$$

The currents in R_1 and R_2 are

$$I_1 = \frac{5.8 \text{ V}}{18\ \Omega} = 0.32 \text{ A} \qquad \textit{answer}$$

$$I_2 = \frac{5.8 \text{ V}}{12\ \Omega} = 0.48 \text{ A} \qquad \textit{answer}$$

(NOTE: $I_1 + I_2 = I$)

(*b*) The terminal potential differences for the batteries are

$$\begin{aligned} V_1 &= \mathscr{E}_1 - Ir_1 \\ &= 24 \text{ V} - 0.8 \text{ A} \times 1.2\ \Omega \\ &= 23.0 \text{ V} \qquad \textit{answer} \\ V_2 &= 12 \text{ V} - 0.8 \text{ A} \times 0.6\ \Omega \\ &= 11.5 \text{ V} \qquad \textit{answer} \end{aligned}$$

27.15 ■ Cells in Parallel

When *cells are connected in parallel* (Fig. 27.14*a*), the positive electrodes are joined together and so are all the negative electrodes. With this arrangement (identical cells in parallel), the following statements apply:

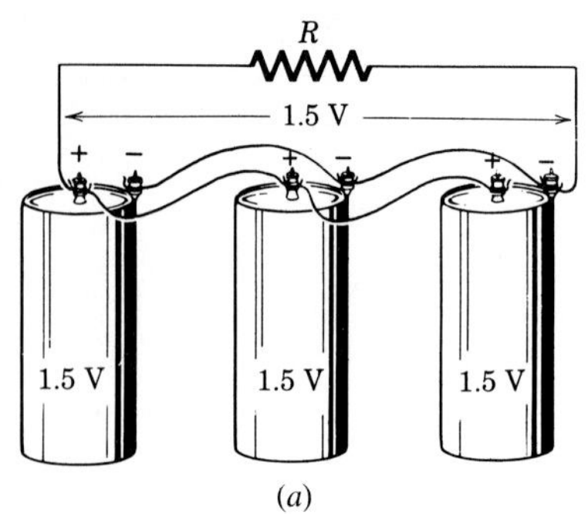

(a)

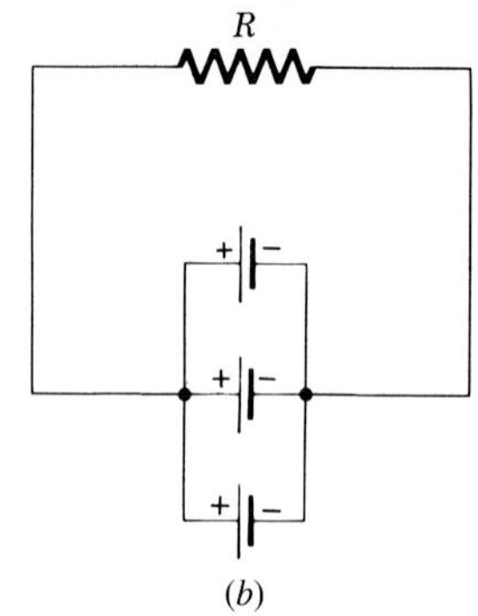

(b)

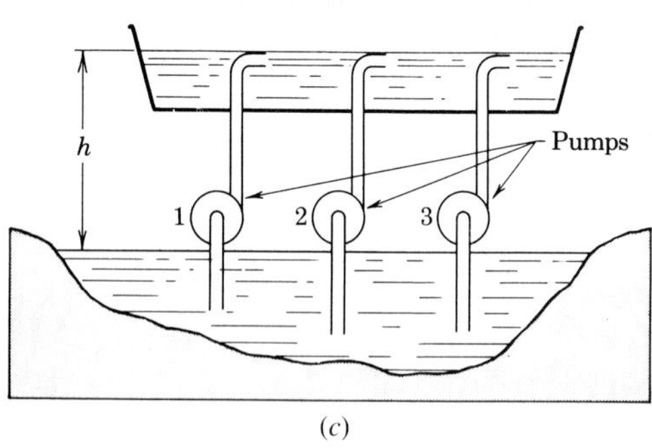

(c)

Fig. 27.14 (*a*) Three cells connected in parallel. (*b*) Wiring diagram of the parallel circuit. (*c*) Hydraulic analogy of the three cells in parallel.

1. The emf of the combination is the same as that for each cell.
2. The total internal resistance is equal to the resistance of one cell divided by the number of cells. (Assuming that the cells all have equal internal resistances.)
3. The current in the external circuit is the sum of the currents in each cell.

Current capacity is increased with a parallel arrangement of cells. The overall internal resistance will be reduced.

Again, a hydraulic analogy (Fig. 27.14*c*) may help in understanding these relationships. Think of each cell as pumping the same amount of electricity to the same potential (level). The potential difference created by the battery of three cells is just the same as that contributed by one cell. The pumps of Fig. 27.14*c*, since they also act in parallel, discharge the water into a common reservoir, at the same *level* (potential). Thus, the total amount of water pumped equals the sum of the amounts contributed by each pump.

The ampere-hour capacity of a parallel set of cells is the sum of the capacities of the individual cells.

When the emf of a battery is sufficient, but the capacity is too small for the desired application, the capacity can be increased by connecting two or more batteries with the same emf in parallel. This practice is often used in starting a car when its battery is too weak to turn the engine of the car. By using jumper cables, another battery connected in parallel with the weak battery will increase the available current.

"Jumping" a dead battery on your car can be done properly only if each battery has the same emf.

It is important in ''jumping'' a battery with a second battery that both batteries have the same emf. A 6-V battery, for example, should not be jumped by a 12-V battery. Not only, in this instance, will the current circulate between the two batteries, but more importantly, the resistances of the conducting cable being very small will allow very large currents to occur in the batteries and cables. Such a practice could result in the cables getting very hot and one or both of the batteries exploding. An extra danger would be the spilling of battery acid on anyone near the batteries.

Warning: When jumping batteries, the positive of one battery is connected to the positive of the other; and the two negative terminals are connected by a cable. If the positive terminal of each battery is connected to the negative terminal of the other, the batteries will be in series and an explosion is a definite possibility. The reaction in the batteries will produce hydrogen gas, which when heated can ignite explosively.

Fig. 27.15 Illustrative Problem 27.5.

Illustrative Problem 27.5

Three identical dry cells each with an emf of 1.5 V and an internal resistance of 0.12 Ω are connected in parallel to a circuit as shown in Fig. 27.15. If $R_1 = 10\ \Omega$, $R_2 = 20\ \Omega$, and $R_3 = 6\ \Omega$, (*a*) what is the current in each cell, and (*b*) what is the current in each resistor?

Solution

(*a*) Replace the parallel resistors R_1 and R_2 by R_{1-2} where

$$\frac{1}{R_{1-2}} = \frac{1}{R_1} + \frac{1}{R_2}$$

$$= \frac{1}{10\ \Omega} + \frac{1}{20\ \Omega}$$

Then,
$$R_{1-2} = \frac{20}{3}\Omega$$

The combined resistance R_{3r} of the three parallel internal resistances is found by

$$\frac{1}{R_{3r}} = \frac{1}{r} + \frac{1}{r} + \frac{1}{r}$$

$$= \frac{1}{0.12\ \Omega} + \frac{1}{0.12\ \Omega} + \frac{1}{0.12\ \Omega}$$

$$R_{3r} = \frac{0.12\ \Omega}{3} = 0.04\ \Omega$$

The total current
$$I = \frac{1.5\ \text{V}}{0.04\ \Omega + \frac{20}{3}\ \Omega + 6\ \Omega}$$

$$= 0.118\ \text{A}$$

Hence, the current in each cell is

$$\frac{0.118\ \text{A}}{3} = 0.039\ \text{A}$$ *answer*

(***b***) The potential drop across R_{1-2} equals

$$IR_{1-2} = 0.118\ \text{A} \times \frac{20}{3}\ \Omega$$

$$= 0.79\ \text{V}$$

Therefore,

$$I_1 = \frac{0.79\ \text{V}}{10\ \Omega} = 0.079\ \text{A}$$ *answer*

$$I_2 = \frac{0.79\ \text{V}}{20\ \Omega} = 0.039\ \text{A}$$ *answer*

$$I_3 = I = 0.118\ \text{A}$$ *answer*

EFFECTS OF ELECTRIC CURRENTS

Electric energy becomes useful only after it is converted to some other form of energy.

Scientists conceive of current electricity as a motion of large numbers of charges; yet the exact nature of electric current is still not completely understood. However, the effects of electric current flow are in most cases quite clearly understood. It is because of these effects that electricity has become such a universally used source of energy. The electric current in itself has no value; it is useful only after the electric energy has been converted into some other form of energy, e.g., heat energy, chemical energy, light energy, mechanical energy. It is only because we have been able to understand these effects and develop basic laws for them that we now have the numerous applications of electricity. In our study of electricity, we shall not be so much interested in the electric current itself as in its effects. The individual effects are not many or difficult to understand. Even complicated electric devices use combinations of several simple effects.

27.16 ■ The Heating Effect of an Electric Current

The heat developed by an electric current may be desirable in many instances; in others the heat may be undesirable.

Conductors carrying an electric current are heated by the current. In some cases the heating is desirable, while in many other cases, such as in electric motors, generators, household circuits, and transformers, it is highly undesirable. The amount of heating in an electric conductor depends on (1) the intensity of current flowing, (2) the time the current flows, (3) the kind and size of the conductor, and (4) the nature of the surrounding media. It is possible to regulate the controlling factors so that optimum heating effect can be obtained. Some devices in which the heating effects of an electric current are desirable are toasters, waffle irons, electric stoves, percolators, irons, heating pads, and electric furnaces. The tungsten filament of an incandescent lamp operates at a temperature of about 2700°C. At this temperature, tungsten would evaporate rapidly and would gradually burn (oxidize) if the lamps were not filled with an inert gas. Here we see electric energy converted into both heat and light energy. The heating elements of percolators, flatirons, waffle irons, and radiant heaters are embedded in a refractory (heat-resistant) material which serves to keep the element in place and to retard oxidation. Special conductors such as nickel-chromium alloys must be used if the heating element is exposed to air, as in toasters. The electric arc used in welding develops temperatures of around 4000°C.

Heat energy developed in motors, generators, and lighting circuits not only may represent an expensive loss of energy but also may be dangerous should the temperature developed be too high. Protection from the danger of too high temperature is provided by *fuses* or *circuit breakers* which will interrupt the current when too "heavy" a load is carried by the circuit. The fuse is a strip of a low-resistance, low-melting-point alloy. The alloy melts when the current increases above a certain value, thus "breaking" the circuit. Fuses may be *plug type* (Fig. 27.16*a*) or *cartridge type* (Fig. 27.16*b*). After the cause of the overload is removed, it is a simple task to replace the burned-out fuse with a good one. Fuses are rated by the number of amperes of current permitted in a circuit before the alloy in the fuse will melt to break the circuit. For example, a 20-A fuse will permit 20 A of current in the circuit. Should the current exceed 20 A, the fuse will melt. Electromagnetic and thermal devices (called *circuit breakers*) that open the circuit when the current exceeds an established value and that can be reset when the overload is removed have almost completely replaced low-melting-point fuses (see Sec. 28.19).

Circuit breakers are used for protection against damage caused by overloads or short circuits.

27.17 ■ The Chemical Effects of Electric Current

The chemical effects of an electric current depend on the movement of ions through a liquid toward charged electrodes.

There is a close connection between electricity and chemical action. To illustrate the electrochemical action of a current, we use a device called an *electrolytic cell* (Fig. 27.17*a*). It consists essentially of two metallic strips *A* and *B*, called *electrodes,* immersed in a solution called the *electrolyte,* which is a solution of an acid, an alkali, or a salt. When a current of electricity from an external source is passed through such

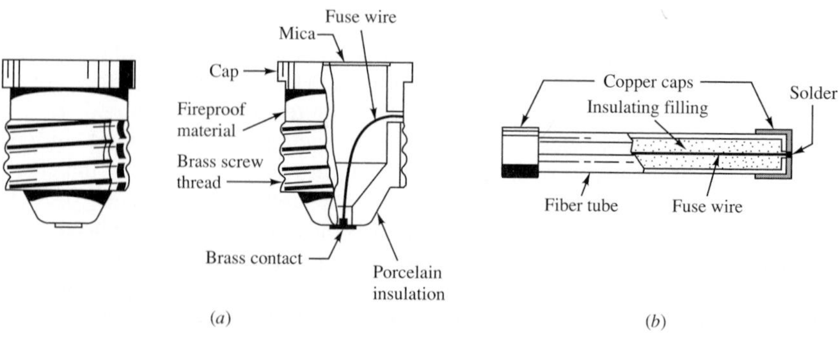

Fig. 27.16 Fuses. (*a*) Plug type. (*b*) Cartridge type.

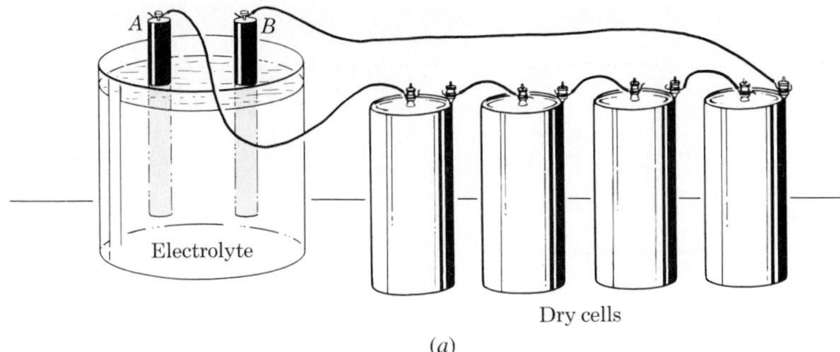

Fig. 27.17 (*a*) Electrolytic cell. (*b*) Electrolytic refining of aluminum. Molten aluminum is tapped from a series of smelters. Metal from this tap will be delivered in molten form to a nearby foundry. (Reynolds Metals Co.)

a cell, a chemical change takes place. Sometimes the solution is decomposed into its constituent parts and can be collected as gases at the electrodes. Oxygen, hydrogen, and chlorine are gases that are commercially obtained in this manner. In some cases, metal is removed from one electrode or from the solution and deposited on the other electrode. The refining of many metals, such as copper, gold, aluminum, magnesium, sodium, and potassium, is achieved by putting into solution (or melting to a liquid) the compound containing the desired element and then depositing it at the negative electrode (see Fig. 27.17*b*). Many important modern processes such as copper plating, silver plating, nickel plating, and chromium plating depend on electrochemical action for their success.

27.18 ■ The Magnetic Effect of an Electric Current

For many years scientists knew of no relationship between electricity and magnetism. In the year 1820, however, Hans Christian Oersted (1777–1851), a Danish physicist, accidentally discovered that a flow of electricity produces magnetic effects. After one of his lectures at the University of Copenhagen he quite by accident placed a current-carrying wire *parallel to* and directly above a suspended magnetic compass needle. To his surprise, he saw the compass turn and assume a position perpendicular to the conductor carrying the current.

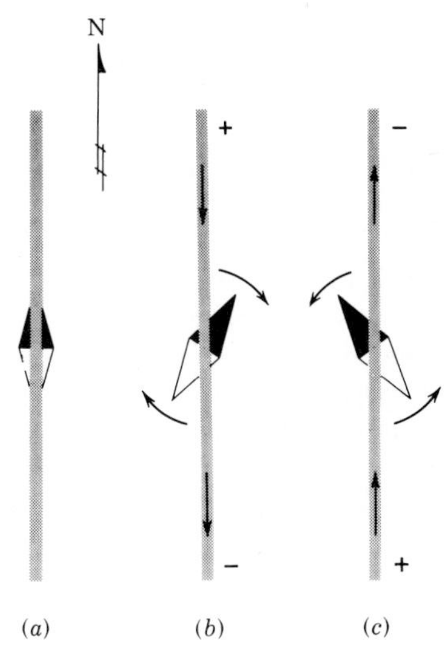

Fig. 27.18 Oersted's discovery: The effect of a current-carrying conductor on a suspended compass needle.

If we place a conductor *parallel to* and above a compass (Fig. 27.18), the compass needle will swing in a direction dependent on the direction of the current until it is perpendicular to the conductor. If we change the direction of the current, the compass

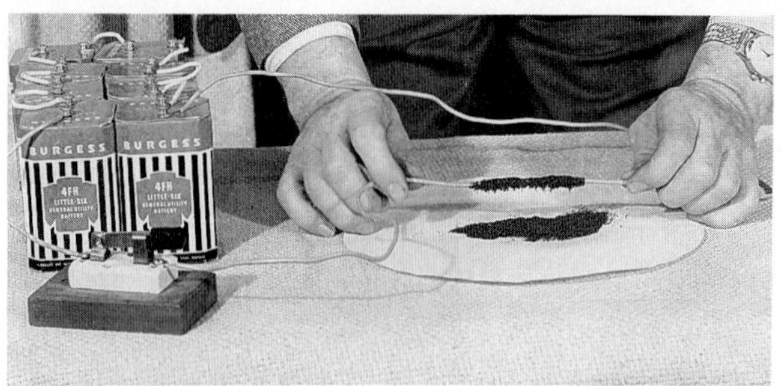

Fig. 27.19 Iron filings cling to a current-carrying conductor because of the magnetic field induced about the conductor.

will turn in the opposite direction. We could get the same effect by placing the conductor under the compass instead of changing the direction of the current. Hence it is evident that there is a magnetic field around a current-carrying conductor.

If a current-carrying conductor is dipped into a pile of iron filings (Fig. 27.19), the filings cluster around and cling to the wire. When the current is cut off, the filings drop off. We shall learn in Chap. 28 how this *electromagnetic* effect can be greatly strengthened by using coils of wire instead of straight wires and by employing cores of iron. We shall also study the principles and laws which relate to the phenomenon of *electromagnetism*. These principles apply in the effective use of electromagnetic forces in such common devices and machines as electric bells, telephones, telegraphs, relays, solenoids, loudspeakers, motors, meters, generators, and atomic particle accelerators.

27.19 ■ Electric Power

In Chap. 6 power was defined as the time rate of doing work

$$\text{Power} = \frac{\text{work}}{\text{time}} \tag{6.16}$$

The SI-metric system unit of power is the joule per second, called the *watt*. The time rate at which electrical energy is delivered or consumed is called *electric power*.

In earlier chapters the following relations were developed.

$$W\ (\text{joules}) = V\ (\text{volts}) \times \text{Q}\ (\text{coulombs}) \tag{25.11}$$

$$\text{Q}\ (\text{coulombs}) = I\ (\text{amperes}) \times t\ (\text{seconds}) \tag{26.1}$$

The watt of electric power is the rate at which electric energy is supplied when there is a current of one ampere across a potential difference of one volt.

Substituting Eq. (26.1) in Eq. (25.11) gives

$$W = VIt \tag{27.7}$$

Therefore
$$P = \frac{W}{t} = VI$$

$$\text{Electric power (watts)} = V\ (\text{volts}) \times I\ (\text{amperes}) \tag{27.8}$$

Since the watt is a quite small unit of power, a larger unit, the *kilowatt*, is often used. The kilowatt (kW) is equal to 1000 W.

$$\text{kW} = \frac{VI}{1000} \tag{27.9}$$

Most electrical appliances have labels which state the voltage at which the appliance is to operate and the power it consumes. Thus, a $\frac{1}{2}$-hp electric motor may be rated for 114 V at 370 W.

Illustrative Problem 27.6 What is the power consumption of a car radio if, when it is operating from a 12 V battery, it draws 1.5 A of current?

Solution Substituting in Eq. (27.8), we get

$$\begin{aligned} P &= VI \\ &= 12 \text{ V} \times 1.5 \text{ A} \\ &= 18 \text{ W} \qquad \textit{answer} \end{aligned}$$

Since $V = IR$, we can substitute for V in Eq. (27.8) and get an equation for power in terms of current and resistance. Thus

$$\begin{aligned} P &= VI \\ &= (IR)I \\ P &= I^2R \end{aligned} \qquad (27.10)$$

Substituting V/R for I in Eq. (27.8), we get an equation for power in terms of voltage and resistance.

$$\begin{aligned} P &= VI \\ &= V(V/R) \end{aligned}$$

or

$$P = V^2/R \qquad (27.11)$$

Because *electric* energy is frequently used to do *mechanical* work, it is important to relate the unit of electric power to the English-engineering unit of mechanical power. Experience (and calculation) has shown that 1 hp *is equivalent to* 746 W or that 1 kW *is equivalent to* 1.34 hp.

$$\begin{aligned} 1 \text{ hp} &= 746 \text{ W} \\ 1 \text{ kW} &= 1.34 \text{ hp} \end{aligned}$$

Illustrative Problem 27.7 How much current will a 25-hp (output) electric motor draw from a 220-V source of electricity if 80 percent of the electric energy used by the motor does useful mechanical work?

Solution The power input of the motor equals VI. The power output equals

$$(25 \text{ hp})\left(\frac{746 \text{ W}}{1 \text{ hp}}\right) = 18{,}650 \text{ W}$$

Hence,

$$\text{Eff.} = \frac{18{,}650 \text{ W}}{(220 \text{ V}) \times I} = 0.80$$

$$I = 106 \text{ A} \qquad \textit{answer}$$

Illustrative Problem 27.8 A dc electric motor operating from a 440 V line is used to hoist automobiles from a dock to the deck of a ship. The motor and winch which operate the hoist have a combined efficiency of 60 percent. (*a*) What is the power, in kilowatts, necessary to raise a car with a mass of 1600 kg a distance of 12 m in 90 s? (*b*) What current will the motor draw from the line which is operating the hoist?

Solution The work done to raise the car 12 m is equal to

$$\begin{aligned} mgh &= 1600 \text{ kg} \times 9.81 \text{ m/s}^2 \times 12 \text{ m} \\ &= 188{,}000 \text{ J} \end{aligned}$$

Since the efficiency is only 60 percent, the motor will expend 188,000/0.60 or 313,000 J.

(***a***) The power of the motor will be

$$\text{Power} = \frac{\text{work}}{\text{time}}$$

$$= \frac{313{,}000 \text{ J}}{90 \text{ s}}$$

$$= 3480 \text{ J/s}$$
$$= 3480 \text{ W}$$
$$= 3.48 \text{ kW} \qquad \textit{answer}$$

(***b***) Substitute known values in Eq. (27.9) and solve for I.

$$\text{Power in kW} = \frac{VI}{1000}$$

$$3.48 = \frac{440\, I}{1000}$$

$$I = \frac{3.48 \text{ kW} \times 1000 \text{ VA/kW}}{440 \text{ V}}$$

$$= 7.91 \text{ A} \qquad \textit{answer}$$

27.20 ■ Electric Power and EMF

In Sec. 27.1 it was explained that the terminal voltage V across a battery source of emf $\mathscr{E}$ and internal resistance r (which is also the voltage drop across the external circuit), is

$$V = \mathscr{E} - Ir \tag{27.2}$$

Thus the power delivered by the battery is

$$P = VI$$

or

$$P = \mathscr{E}I - I^2r \tag{27.12}$$

It will be noted that the first term on the right is the rate at which chemical energy is converted into electric energy and the second term is the rate of heating.

In charging a storage battery the terminal voltage of the battery during the charging process is greater than the emf by the amount of the internal voltage drop within the battery. Thus, if the battery charger produces a terminal voltage V in sending a charging current I through the internal resistance r, we have

$$V = \mathscr{E} + Ir$$

The power P to charge is found by

$$P = VI$$
$$P = \mathscr{E}I + I^2r \tag{27.13}$$

The first term on the right side of the equation is the rate at which electric energy is being transformed and stored as chemical energy. The second term is the rate at which electric energy is being converted into unusable thermal energy in the battery electrolyte and plates.

Table 27.1 summarizes the relationships between various electrical quantities in a circuit.

Illustrative Problem 27.9 A battery charger sets up a potential difference of 13.7 V at the terminals of a storage battery with an emf of 12.6 V and internal resistance of 0.221 Ω. (*a*) What is the charging current? (*b*) What is the rate at which electric energy is being converted into chemical energy? (*c*) At what rate is heat produced?

Table 27.1 Relationships between Electrical Quantities

Unknown Quantity	Known Quantities					
	P and I	P and V	P and R	V and I	I and R	V and R
P, watts				VI	I^2R	$\frac{V^2}{R}$
V, volts	$\frac{P}{I}$		$\sqrt{PR}$		IR	
I, amps		$\frac{P}{V}$	$\sqrt{\frac{P}{R}}$			$\frac{V}{R}$
R, ohms	$\frac{P}{I^2}$	$\frac{V^2}{P}$		$\frac{V}{I}$		

Solution

(***a***) On charging,

$$V = \mathscr{E} + Ir$$
$$13.7\ \text{V} = 12.6\ \text{V} + I\text{A} \times 0.221\ \Omega$$
$$I = 4.98\ \text{A} \qquad \textit{answer}$$

(***b***)
$$P = \mathscr{E}I + I^2r$$
$$\mathscr{E}I = 12.6\ \text{V} \times 4.98\ \text{A} = 62.7\ \text{W} \qquad \textit{answer}$$

(***c***)
$$I^2r = (4.98\ \text{A})^2 \times 0.221\ \Omega = 5.48\ \text{W} \qquad \textit{answer}$$

27.21 ■ Electric Energy

Energy was defined in Chap. 6 as the *capacity* (under stipulated conditions) *for doing work*. Since *power* = *work/time*, *energy* (*work*) = *power* × *time*. Electric-energy units are therefore the *watt-hour* and the *kilowatt-hour* (kWh). One watt-hour of energy is expended when 1 W of power is used for 1 h. Electric bills are paid on the basis of the kilowatt-hours used.

Electric services are bought and sold, not in terms of power, but in terms of energy or work. Electric energy is expressed in watt-seconds (joules), watt-hours, or kilowatt-hours.

In the laboratory we frequently use the smaller unit, the watt-second, equivalent to the *joule*.

$$\text{Joules} = \text{volts} \times \text{amperes} \times \text{seconds}$$
$$\text{J} = VIt \tag{27.14}$$

The electric energy in a circuit of resistance *R* ohms, carrying a current of *I* amperes can be converted to mechanical energy by the relation $J = I^2Rt$. This conversion is known as *Joule's law*.

Illustrative Problem 27.10 How much will it cost to operate a $\frac{1}{2}$-hp swimming pool filter motor continuously for 30 days if the rate for electric energy is 8.65 cents per kilowatt-hour? (Assume 75 percent efficiency.)

Solution

$$\frac{P_{\text{out}}}{P_{\text{in}}} = \text{efficiency}$$

$$\frac{\frac{1}{2}\ \text{hp} \times 746\ \text{W/1 hp}}{P_{\text{in}}} = 0.75$$

$$P_{\text{in}} = 497\ \text{W} = 0.497\ \text{kW}$$

The energy consumed by the motor in 30 days will be

$$\text{kWh} = 0.497 \text{ kW} \times 30 \text{ days} \times \frac{24 \text{ h}}{1 \text{ day}}$$

$$= 357.84 \text{ kWh}$$

The total cost will be

$$357.84 \text{ kWh} \times \$0.0865/\text{kWh} = \$30.95 \qquad \textit{answer}$$

27.22 ■ Relationship between Electric Energy and Heat Energy

Electric energy of one joule (1 W·s) is equivalent to 0.24 calorie or 3.41 Btu of heat. Thus, Joule's law becomes $H(\text{cal}) = 0.24\ I^2Rt$ or $H(\text{Btu}) = 3.41\ I^2Rt$.

We have seen that one equation relating electric energy to mechanical energy is $J = VIt = I^2Rt$, where J is measured in wattseconds (joules), V in volts, I in amperes, R in ohms, and t in seconds. Many thorough and careful experiments have been conducted to relate electric energy to heat energy. The following relationship is the accepted outcome of these experiments: The *heat in calories* is equal to 0.24 times the joules (wattseconds) of electrical energy expended. Let

H = heat, cal $\quad V$ = voltage, V
I = current, A $\quad t$ = time, s

then

$$H = 0.24VIt \tag{27.15}$$

The value of V in Eq. (27.14) can be replaced by its equivalent IR to get the equation

$$H = 0.24I^2Rt \tag{27.16}$$

or I can be replaced by V/R to yield

$$H = 0.24 \frac{V^2}{R} t \tag{27.17}$$

Eqs. (27.15), (27.16), and (27.17) can be used to measure the *heat energy* developed in heating devices and also the *heat losses* in transmission lines, dynamos, and other electrical instruments whose function it is to do work not related to heating. Such energy losses (in nonheating equipment) are often referred to as "I^2R losses." If H, V, I, and t are in Btu, volt, ampere, and hour units, the following equations can be derived:

$$H = 3.41\ VIt \tag{27.15'}$$
$$H = 3.41\ I^2Rt \tag{27.16'}$$
$$H = 3.41 \frac{V^2}{R} t \tag{27.17'}$$

In the English (engineering) system, heat and electrical energy are related by $H_{\text{Btu}} = 2.93 \times 10^{-4}$ kWh.

Illustrative Problem 27.11 The heating element in a percolator has a resistance of 23.5Ω. It is hooked in to a 115-V house circuit. How long will it take to heat 1.25 kg of water at 22.4°C to the boiling point, 100.0°C? Assume no loss of heat to the surroundings.

Solution Use Eq. (15.2): $H = mc(t_2 - t_1)$, where H is in calories, I in amperes, t is in degrees Celsius, and c is the specific heat of water. Equation (27.17) is needed also

$$H = 0.24 \frac{V^2}{R} t$$

where H is in calories, I is in amperes, R is in ohms, and t is in seconds.

Equate the expressions for H in the two equations

$$\mathrm{mc(t_2 - t_1) = 0.24 \frac{V^2}{R} t}$$

Solve for t and substitute known values.

$$\mathrm{t} = \frac{mc(t_2 - t_1)}{0.24\ V^2/R}$$

$$= \frac{1.25 \text{ kg} \times \dfrac{1 \text{ kcal}}{\text{kg}\cdot\text{C}^\circ} \times (100.0^\circ\text{C} - 22.4^\circ\text{C}) \times \dfrac{1000 \text{ cal}}{1 \text{ kcal}}}{0.24 \times (115 \text{ V})^2/23.5\Omega}$$

$$= 718 \text{ s}$$
$$= 11 \text{ min } 58 \text{ s.}$$ *answer*

Illustrative Problem 27.12 A 5-hp 220-V dc motor having an efficiency of 90 percent is installed 210 ft from a power line with a two-wire service using AWG No. 12 copper wire. Calculate (*a*) the current used by the motor, (*b*) the resistance of the wires connecting the motor to the power line, (*c*) the I^2R loss.

Solution

(***a***) Changing 5 hp to watts, gives

$$\text{Power output} = 5 \text{ hp} \times 746 \text{ W/hp} = 3730 \text{ W}$$

Since the motor is only 90 percent efficient, the power input will be

$$\text{Input} = \frac{3730 \text{ W}}{0.90} = 4144 \text{ W}$$

Using Eq. (27.8), solve for the current used by the motor,

$$P = VI$$

$$I = \frac{P}{V}$$

$$= \frac{4144 \text{ VA}}{220 \text{ volts}}$$

$$= 18.84 \text{ A}$$ *answer*

(***b***) From Appendix VI, the resistance of No. 12 wire is 1.588 Ω/1000 ft. Then the resistance of the two lead-in wires will be

$$R = \frac{2(210)\text{ft} \times 1.588\ \Omega}{1000 \text{ ft}}$$

$$= 0.667\ \Omega$$ *answer*

(***c***) The power loss will be

$$I^2R = (18.84 \text{ A})^2 \times 0.667\ \Omega = 237 \text{ W}$$ *answer*

Note that 18.84 A × 0.667 Ω > 12 V; so the power line must carry more than 232 V to maintain 220 V at the motor.

QUESTIONS AND EXERCISES

1. How do primary and secondary cells differ?

2. Explain the difference between "potential difference" and "emf."

3. Which needs replacement more often in a voltaic cell, the negative or the positive electrode? Why?

4. Give the advantages of (*a*) the dry cell, (*b*) the storage cell, as sources of electrical energy.

5. How does the depolarizer of a cell function?

6. How does the specific gravity of the electrolyte in a lead-acid battery indicate the state of charge of the battery?

7. Why is it a bad practice to replace a 15-A fuse that frequently blows out in a circuit by a 25-A fuse?

8. Explain why the wires leading to the electric bulb in a reading lamp do not get as hot as the filament, even though they carry the same amount of current. Give a justification based on mathematical formulas.

9. What would happen if you tried to charge a rundown battery with current from another battery of the same make and size?

10. The common 60-W incandescent lamp will draw about $\frac{1}{2}$ A when operating from the house lighting circuit (120 V). One new dry cell will deliver on short circuit a current of 30 A or more. Will this new cell light the lamp? Give (numerical) reasons for your answer.

11. Which of the following is not equal to the watt? (*a*) Volt-ampere; (*b*) ohms^2/volt; (*c*) amperes^2-ohm; (*d*) joule/second.

12. How are storage battery capacities rated?

13. How is it possible to determine whether a solution is an electrolyte or a nonelectrolyte?

14. Why should cells of different emfs *not* be connected in parallel?

15. Name five different sources of a continuous current.

16. What are the advantages of solar cells as sources of electric energy? What are the disadvantages?

17. What are the advantages of connecting cells in series?

18. What are the advantages of connecting cells in parallel?

19. A 60-, 75-, and 100-W lamp are connected in *parallel* in a house circuit. Which (*a*) carries the most current, (*b*) has the greatest resistance, (*c*) becomes the hottest?

20. A 60-, 75-, and 100-W lamp are connected in series to a source of potential difference. (*a*) Which lamp has the greatest potential difference across it? (*b*) Which lamp becomes the hottest?

PROBLEMS

Group One

1. Convert 30 hp to kilowatts.

2. A 32-V electric power plant has an output of 4.5 kW. What current (amperes) can be drawn from the generator?

3. Determine the resistance of a 1650-W, 120-V electric heater.

4. Compute (*a*) the work and (*b*) the average power required to transfer 112,000 C of charge in 3 h through a potential difference of 64 V.

5. A 12-V storage battery has a charge of 2.5×10^5 C. How much energy, in joules, does this represent?

6. What must be the emf of a battery which holds a charge of 5.0×10^5 C and has 3.0×10^6 J of stored energy?

7. A common commercial storage battery has a capacity rating of 140 Ah. What continuous steady current (amperes) should the battery deliver over a period of 8 h?

8. A battery has a capacity rating of 150 Ah. If the battery is "dead," how long will it take to fully charge the battery if it is charged at 6 C/s?

9. How much work (in joules) is done against the electric field in a 12-V battery as a proton is moved from the negative to the positive terminal? (See Sec 25.15.)

10. How long will it take a 500-W immersion heater to change the temperature of 400 L of water from 20 to 30°C? Neglect heat losses to air or holding tank.

Group Two

11. What is the power loss, in watts, in a circuit with a resistance of 400Ω and a current of 8 A? How much heat, in Btus, is generated in a 24-h period?

12. An electric iron has a resistance of 25Ω. It uses 5.8 A of current. Calculate the heat produced in 45 s (*a*) in joules, (*b*) in calories, (*c*) in Btu.

13. How much heat, in Btu, is produced by a 1250-W electric iron operating for 15 min?

14. An electric toaster operates on a 110-V potential difference and draws 7.5 A. (*a*) How many calories does the toaster produce in 1 min 20 s? (*b*) How many Btus?

15. A 24-V battery with an internal resistance of 1.5Ω is charged by connecting it in series with an opposing 60-V source of emf and a 6.0-Ω resistor. Find (*a*) the charging current and (*b*) the terminal potential difference of the battery when it is being charged.

16. A dc motor operating under full load draws 18 A at 240 V. What is the horsepower input to the motor?

17. What size fuse, rated in amperes, should be used in a house circuit that has a maximum load of 2100 W from a 120-V line?

18. If the cost of electric power is 8.5¢/kWh, how long can you operate a 250-W incandescent lamp for 1 ¢?

19. A soldering iron has a resistance (hot) of 350 Ω. What power does it draw from a 118-V line?

20. A 120-V potential difference maintains 2.4 A in a resistor. Find (*a*) the resistance of the resistor, and (*b*) the potential difference required to maintain 2.6 A of current in the resistor.

21. A dc motor rated at 5 hp, operating at full load, was found to draw 23 A at 240 V. What is its efficiency?

22. Given 20 dry cells, each with emf of 1.5 V and internal resistance of 0.2 Ω. Find the amount of current that will flow through an external load of 0.03 Ω when the cells are connected (*a*) in series; (*b*) in parallel.

23. Given 20 dry cells, each with emf of 1.5 V and internal resistance of 0.2 Ω. Find the current that will flow through an external resistance of 30 Ω when the cells are connected (*a*) in series; (*b*) in parallel.

24. The motor to a hoist operates at 15 A under a 240-V source of current. The hoist lifts a one-ton load at the rate of 30 ft/min. Determine (*a*) the power input to the motor, in kW; (*b*) the power output, in hp; (*c*) and the overall efficiency of the system.

25. An electric hot-water heater has a resistance of 45 Ω and operates at 220 V. How long will it take the heater to heat 100 gallons of water from 60 to 140°F?

26. Two groups of four cells in series are connected in parallel. The combination is connected in series with a resistance of 2.5 Ω (see Fig. 27.20). Each cell has an emf of 1.5 V and an internal resistance of 0.075 Ω. What is the current in the 2.5 Ω resistance?

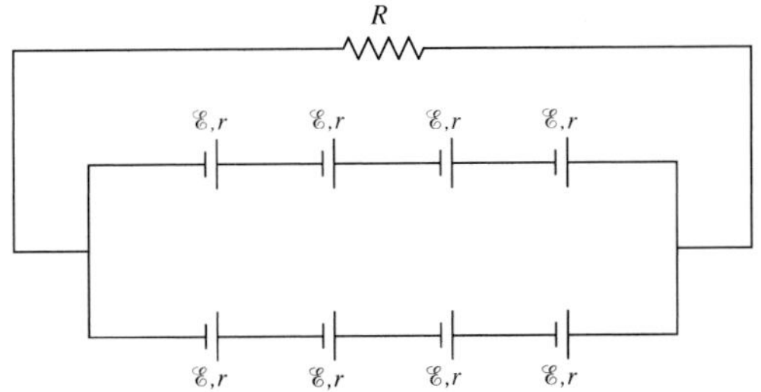

Fig. 27.20 Problem 26.

27. Determine the current through an external resistance R if it is connected to n cells in series, if each cell has an emf $\mathscr{E}$ and an internal resistance r.

28. Determine the current through an external resistance R if it is connected to n cells in parallel, if each cell has an emf $\mathscr{E}$ and an internal resistance r.

29. A storage battery has an open-circuit voltage of 13.2 V. When 100 A of current is delivered, the battery voltage drops to 10.6 V. Calculate the internal resistance of the battery.

30. How much will it cost per 40-h week to run a motor having an average load of 50 hp and an average efficiency of 90 percent if the power rate is 9.25 ¢/kWh?

31. A 6-V storage battery with 0.6-Ω internal resistance is to be charged from a 15-V circuit. What resistance must be placed in series with the battery if the charging rate is set at 5.0 A? What is the total energy supplied to the battery in 8 h?

32. A $\frac{1}{2}$-hp (output) furnace fan motor operates for an average of 7 h per day for a 30-day month. The motor is 92 percent efficient. If electric energy costs 7.8 ¢/kWh, what is the monthly bill?

33. A bank of 80 incandescent lamps (connected in parallel), each having a resistance of 240 Ω when hot, is connected to a 120-V power line. Find the cost of operating these lamps for an 8-h day at 8.52 ¢/kWh.

Group Three

34. A 12-V battery with an internal resistance of 0.67 Ω is connected to the circuit shown in Fig. 27.21. (*a*) What is the current drawn from the battery? (*b*) What is the terminal voltage when the current is steady state?

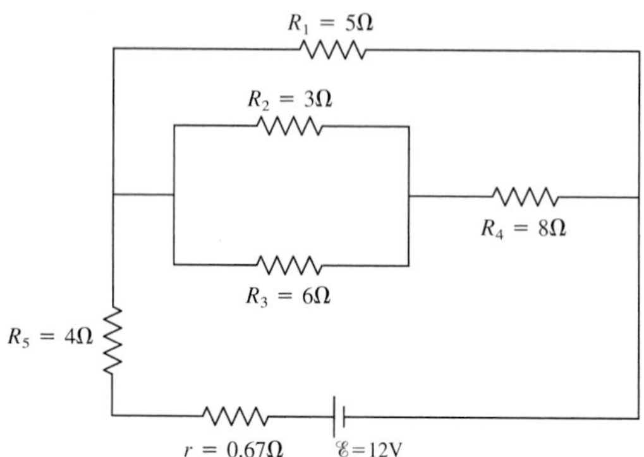

Fig. 27.21 Problem 34.

35. A storage battery for farm lighting produces 96 V across its terminals in open circuit. The battery has 0.034 Ω internal resistance and is connected through a pair of copper wires, which have a resistance of 1.835 Ω/1000 ft, to a load 250 ft distant. The load draws 21 A under these conditions. (*a*) What is the voltage at the load, 250 ft from the battery? (*b*) What will be the voltage across the battery terminals when delivering this load?

36. A battery charger applies a potential difference of 13.5 V at the terminals of a 12.0-V battery. The internal resistance of the battery is 0.95 Ω. (*a*) What is the charging current? (*b*) What is the rate at which electric energy is being converted into useful chemical energy? (*c*) At what rate is nonusable heat produced? (*d*) How long will it take to fully charge a dead battery if it is rated at 70 A · h?

37. A chicken brooder is to be built that will produce 220 cal of heat each second. Nichrome wire with a mean diameter of 0.180 in. is to be used as the heating element. The resistivity of nichrome is 660 Ω cmils/ft. What length of nichrome wire is needed if the brooder is to operate off a 120-V line?

38. In the circuit of Fig. 27.22 determine (*a*) the supply voltage, (*b*) the power supplied, and (*c*) the power dissipated in each resistor.

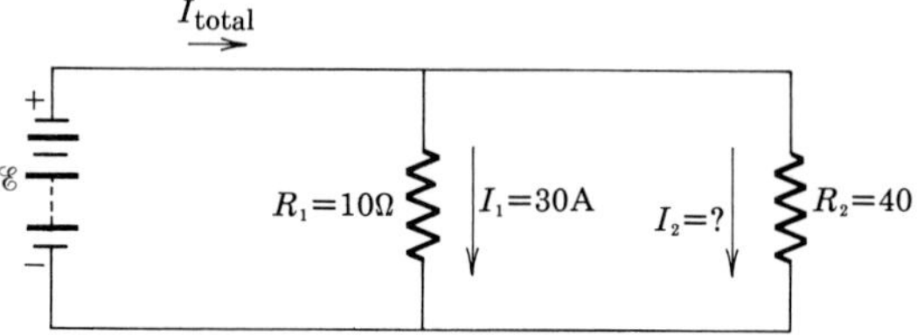

Fig. 27.22 Problem 38.

39. Two identical batteries are connected in series across a 5.0-Ω resistor, and the current is found to be 0.18 A. When the same batteries are connected in parallel with the same resistor, the current is 0.14 A. (*a*) What is the internal resistance of each battery? (*b*) What is the open circuit voltage of each battery?

40. Battery *A* has a no-load terminal voltage of 6.0 V and an internal resistance of 1.5 Ω. Battery *B* has a no-load terminal voltage of 8.5 V with an internal resistance of 2.1 Ω. The batteries are connected in parallel. A 3.5-Ω resistor is then connected to the terminals of the battery combination. What current will flow through each battery?

MAGNETISM AND ELECTROMAGNETISM

Magnetism and electricity have a great deal in common, both arising from the interaction of electric charges. All magnetic phenomena can ultimately be traced to moving electric charges. Both magnets and electric charges have force fields and they both display forces of attraction or repulsion under proper conditions.

Some phenomena of magnetism were known 2500 years ago.

The harnessing of magnetic forces has revolutionized human life by opening up the age of electricity and electronics.

28.1 ■ Artificial Magnets

Lodestone ore is a *natural magnet*. *Artificial magnets* can be made in a variety of ways. When an iron or steel bar is stroked with a natural (or artificial) magnet, the iron or steel bar itself becomes a magnet. Artificial magnets can also be produced by placing a bar of iron or steel in a coil of wire in which a heavy direct current of electricity is flowing. A magnet produced in this manner is generally a much stronger magnet than the one produced by stroking. Soft iron is rather easily magnetized, but such magnets readily lose their magnetism. They are called *temporary magnets*. If the magnet is made from hardened steel, it will retain its magnetic properties much longer. Such magnets are termed *permanent magnets*. One of the strongest and most permanent of artificial magnets is made of an alloy called *alnico* containing aluminum, nickel, and cobalt. As pure metals by themselves, however, aluminum, nickel, and cobalt are only very weakly attracted by a magnet. Most substances such as copper, gold, silver, lead, wood, glass, etc., are not noticeably attracted by a magnet. A strong magnet attracts iron filings at a considerable distance, even if nonmagnetic substances such as copper, wood, glass, etc., are placed between the magnet and the filings.

Only a few materials, other than iron, have the capability of becoming strong magnets.

Some substances are even repelled by magnets, although the repelling forces are very slight.

The following terms are used to describe magnetic materials.

1. *Paramagnetic*. Materials that are weakly attracted by a magnet. Examples of paramagnetic materials are: aluminum, rare earth metals, sodium, platinum, uranium.
2. *Ferromagnetic*. Materials that are strongly attracted by a magnet. Examples are: iron, cobalt, nickel, and alloys (such as permalloy and alnico) of ferromagnetic metals.
3. *Diamagnetic*. Materials that are weakly repelled by both poles of a magnet. Examples: bismuth, carbon, gold, zinc, salt.

Very strong and versatile artificial magnets can be made from ferromagnetic materials.

A magnet in the form shown in Fig. 28.1*a* is called a *horseshoe magnet*, the one in (*b*) is a *disk magnet*, and the type in (*c*) is a *bar magnet*. The horseshoe magnet is the most common form, since it gives a stronger field of magnetic force.

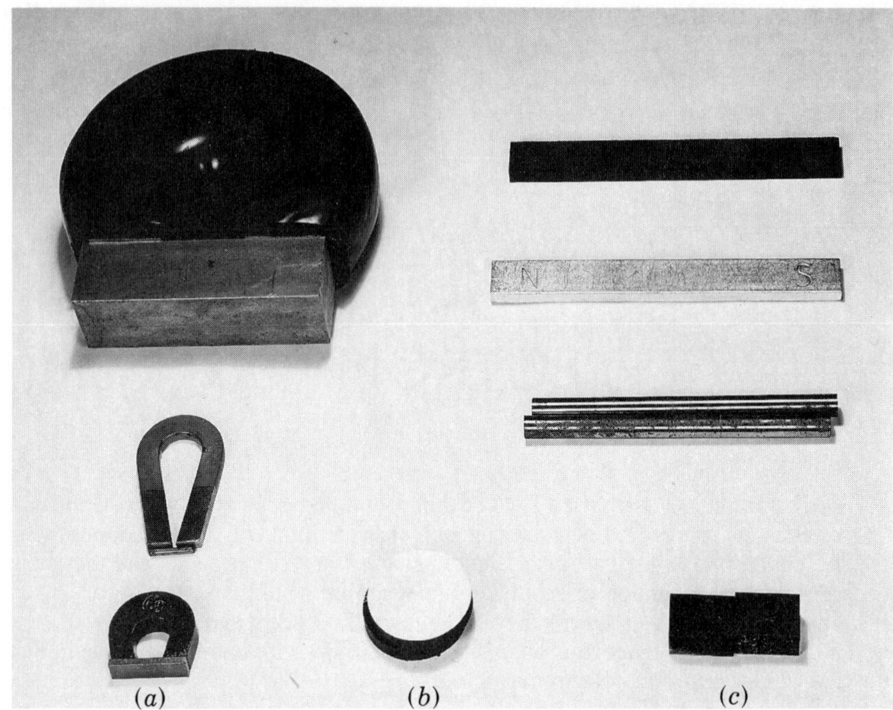

Fig. 28.1 Forms of permanent magnets. (*a*) Horseshoe magnets. (*b*) Disk magnet. (*c*) Bar magnets. (Central Scientific Co.)

Rare Earth Supermagnets Dependence on foreign countries for a supply of cobalt and a skyrocketing price increase for the metal prompted scientists in United States and Japan in 1973 to embark on an extensive search for other magnetic alloys. Eleven years later, working independently, scientists from the two countries were successful in developing a stronger magnetic alloy with the right combination of the same elements: iron, boron, and the rare earth element neodymium. Neodymium is sufficiently abundant and accessible to assure a stable price.

Neodymium-iron-boron magnets are lighter, more than ten times stronger, and retain their magnetism longer than ferrite magnets, up to now the most widely used type of magnet.

The main limitation of the "supermagnets" is their performance at high temperatures. Magnets of neodymium-iron-boron lose their magnetism at 585 K, which prohibits their use in certain high-temperature applications. Even with high temperature limitations, supermagnets made of neodymium-iron-boron abound. Supermagnets are currently being incorporated into a variety of dc motors and brushless servomotors. Automobile manufacturing factories now use six supermagnets, which weigh a total of 5 ounces, to replace hefty coils or permanent magnets that line the casings of ordinary starter motors, thus reducing the size and weight of the starter motors by half.

The supermagnets are ideal for applications where weight is critical, particularly in high-centrifugal-force situations. Conventional magnetic resonance imaging devices, which perform medical diagnostic scanning, utilize superconducting magnets (see Sec. 26.12). The conventional devices are extremely large and, because of the large stray magnetic field, require heavy steel shielding. Neodymium-iron-boron magnets make possible portable magnetic resonance imaging devices that have a small stray magnetic field. Many other potential applications of supermagnets include thin motors in medical pumps and valves, magnetic levitation, holding magnets for artificial teeth, and numerous military weapons.

28.2 ■ Magnetic Poles

If a magnet is lifted out of a pile of iron filings or small nails, the filings or nails cling to the magnet near the ends but scarcely at all near the middle (Fig. 28.2). These regions near the ends are called *magnetic poles* (called *dipoles*). Magnetic poles always exist in pairs; it is impossible to have a magnet with only one pole.

If a magnet is broken, as in Fig. 28.3, the fragments prove to be dipoles and not isolated poles. If magnets—whether natural or artificial, bar- or horseshoe-shaped—are suspended (on the earth) so as to turn freely about a vertical axis, they always come to rest in approximately a north-south position (Fig. 28.4). This fact was used by early mariners as they fashioned crude compasses by floating on water or otherwise suspending a lodestone magnet so that it might swing freely about a vertical axis. The pole of the suspended magnet that points toward the geographic north is called the *north pole*, while the other pole, which points toward the south, is called the *south pole*.

The fact that a magnet placed on a piece of cork floating on water will take a north-south position without moving longitudinally in either a north or a south direction shows that *the two poles of a magnet are of equal strength*.

Fig. 28.2 Iron filings cling near the poles of a magnet.

28.3 ■ Laws of Attraction and Repulsion of Magnetic Poles

If the north pole of a magnet is brought near the north pole of a second magnet which is free to rotate (Fig. 28.5), the poles will repel each other. But if a south pole is brought near the north pole of the second magnet, the two will be attracted toward each other. A general law of magnetic attraction states:

Fig. 28.3 When magnet (*a*) is broken in three pieces, three magnetic dipoles (*b*) result.

> ***Like magnetic poles repel each other, while unlike poles attract each other. These attractive and repulsive forces between magnetic poles vary directly as the product of the pole strengths and inversely as the square of the distance between them.***

This law was developed as a result of carefully conducted experiments by the French physicist Coulomb and is known as *Coulomb's law*. Pole strengths in the SI-metric system are expressed in *ampere-meter* (A · m) units.

> ***A magnetic pole has a pole strength of one ampere-meter if it repels a like pole of equal strength one meter distant with a force of 10^{-7} newtons.***

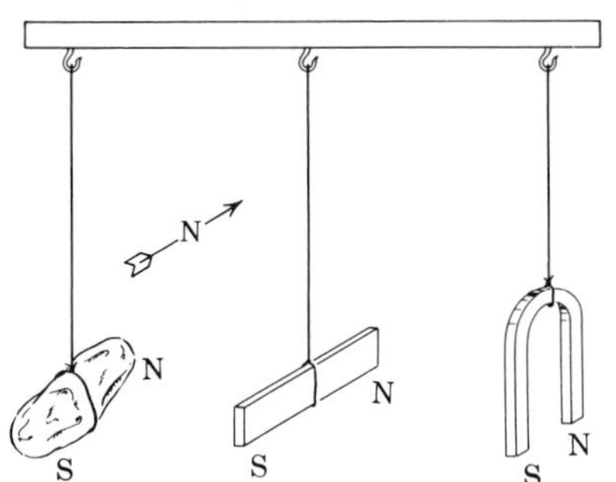

Fig. 28.4 Suspended magnets seek a north-south position (on earth) when free to rotate.

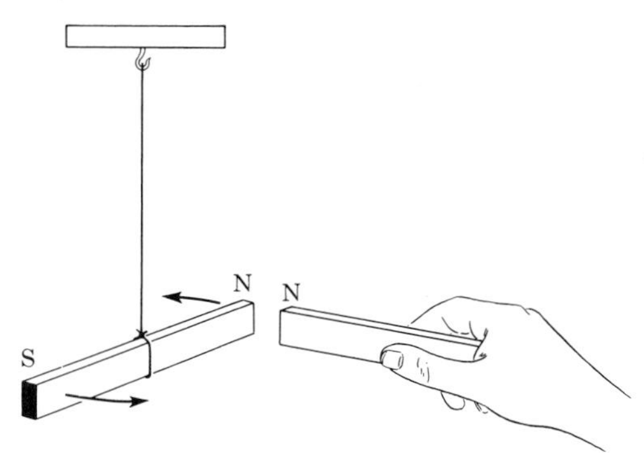

Fig. 28.5 Like magnetic poles repel each other.

Stated as a formula, Coulomb's law becomes

$$F = k\,\frac{m_1 \times m_2}{d^2} \tag{28.1}$$

where F = force in newtons
m_1, m_2 = pole strengths in ampere-meters
d = distance between the poles in meters
k = constant of the medium between poles in newtons/(ampere)^2
$k = 10^{-7}$ newtons/(ampere)^2 for a vacuum or air medium.

No magnet has only one pole.

Equation (28.1) for magnetic forces is similar to Eq. (25.1), which was developed for electrostatic forces. The difference lies in the fact that Eq. (28.1) was developed on the assumption that magnetic poles are points. Since magnetic poles really are not points, the equation can be used only for approximate calculations. This will not destroy the validity of the discussions that follow, however. Nonetheless, it should be emphasized that a ''unit magnetic pole'' does not actually exist. Since magnetic poles always exist in pairs, Eq. (28.1) can properly be applied only to long, thin magnets with well separated poles.

Illustrative Problem 28.1 A magnetic pole of 30 A · m strength exerts a force of 90×10^{-5} N upon a second pole placed 4 cm away. What is the strength of the second pole if they are located in air?

Solution Solve Eq. (28.1) for m_2.

$$m_2 = \frac{Fd^2}{km_1}$$

Substitute given values to get

$$m_2 = \frac{90 \times 10^{-5}\ \text{N} \times (0.04\ \text{m})^2}{10^{-7}\ \text{N/A}^2 \times 30\ \text{A} \cdot \text{m}}$$

$$= 0.48\ \text{A} \cdot \text{m} \qquad \textit{answer}$$

28.4 ■ Magnetic Lines of Force

Michael Faraday was an English chemist and physicist. He discovered ions and did considerable research on electromagnetic induction which led to his invention of the electric generator.

The space around a magnet is the site of a magnetic field, just as the space around a charged object is the site of an electric field. A *magnetic field* is said to exist in the region around a magnetic pole where the influence of the pole can be detected. If we can conceive of an isolated unit north pole placed in a magnetic field, the following law applies:

The direction of the magnetic field will be the same as the direction of the force acting on the isolated north pole.

Michael Faraday (1791–1867) was perhaps the first to study the configuration of a magnetic field. Although the magnetic field cannot be seen, it can be demonstrated and mapped in several ways. Place a sheet of glass or cardboard on top of a bar magnet whose field is to be studied. Then sprinkle the sheet with iron filings. Upon tapping the sheet, you will see that the filings align themselves with a magnetic field and form ''lines'' or ''strings'' of definite lengths and shapes (Fig. 28.6*a*). Repeating the experiment using the field between two like poles, you will obtain the configuration illustrated in Fig. 28.6*b*. The field between two unlike poles is shown in Fig. 28.6*c*. Faraday drew what he called *lines of force* around a magnet to correspond to the lines formed by the iron filings.

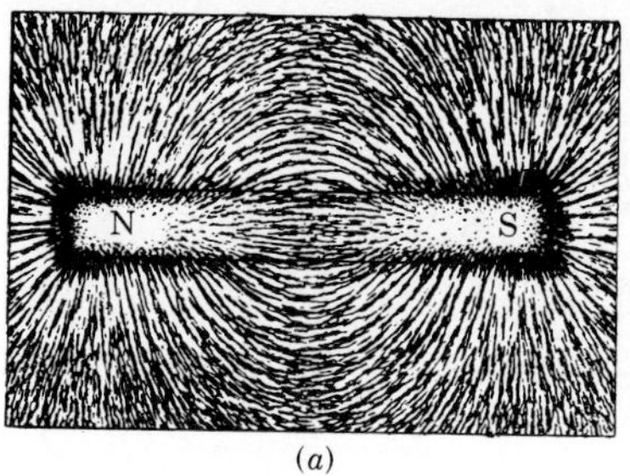

(a)

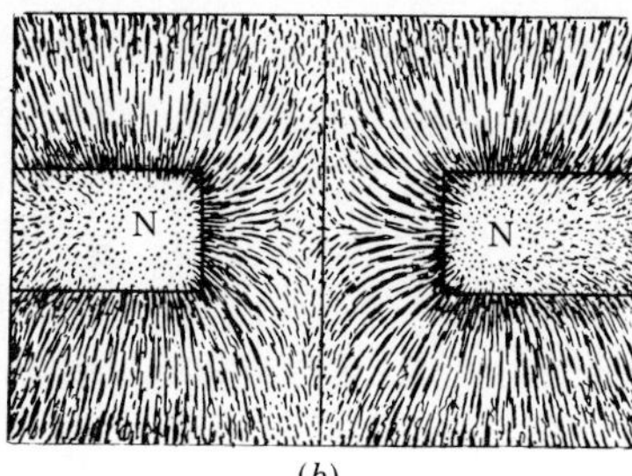

(b)

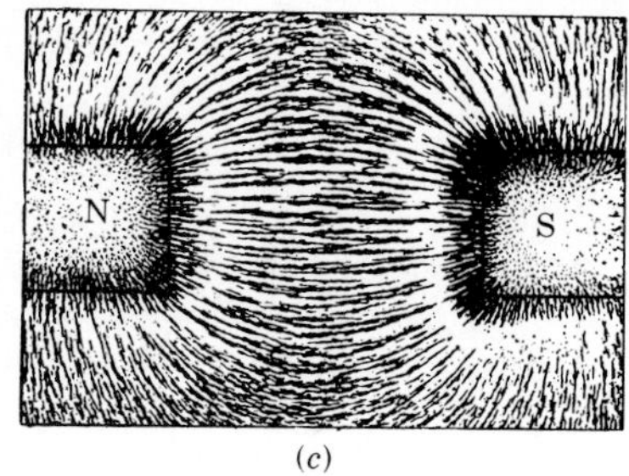

(c)

Fig. 28.6 Magnetic lines of force as shown by iron filings (*a*) about a bar magnet, (*b*) between like poles, and (*c*) between unlike poles.

Similar lines of force can be obtained by placing a small compass needle in the magnetic field and moving it always in the direction pointed to by its north pole while tracing its path.

A magnetic line of force (called a line of flux) can be more technically defined as a line which indicates at every point along its length the direction in which a north pole would be urged.

The lines should have small arrowheads on them to indicate the direction of the field (see Fig. 28.7).

Note that we arbitrarily assume that the lines of force point from the north pole to the south pole. It should also be noted that the pattern of lines of force represented in Fig. 28.7 represents the magnetic field only in the horizontal plane. To obtain a true picture, the lines of force should be drawn in three dimensions—showing the magnetic field over and under the magnet as well as on either side.

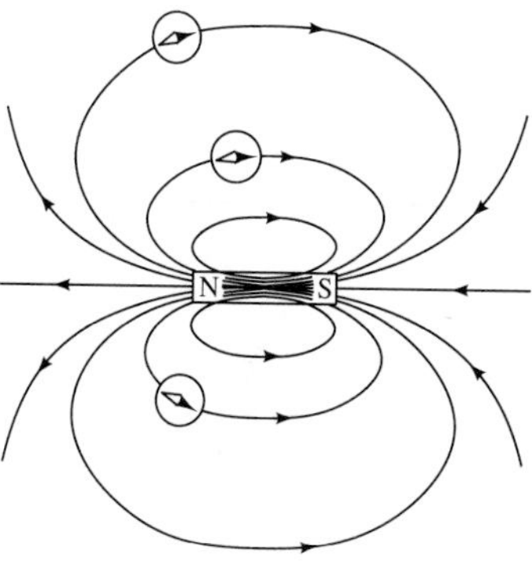

Fig. 28.7 Magnetic lines of force of a bar magnet.

A tangent to a magnetic line of force (line of flux) indicates the direction of the magnetic field.

28.5 ■ Theories of Magnetism

Although it was stated that the real cause of magnetism and magnetic fields is not clearly known, there are *theories* about the cause of magnetism, two of which are outlined here.

Ewing's Theory of Magnetism According to this theory of Sir Alfred Ewing (1855–1935), a piece of iron consists of millions of tiny elementary magnets. These submicroscopic magnets may be molecular in size, or they may possibly be made up of groups of molecules aligned to form tiny iron *domains*. In unmagnetized iron these elementary or molecular magnets are assumed to be pointing in all conceivable directions, with a random orientation which provides no external magnetic effect (Fig. 28.8*a*). When this unmagnetized iron is placed in a magnetic field, the tiny molecular magnets are forced to align themselves with the field (Fig. 28.8*b*), the degree of alignment depending on the intensity of the field in which the iron is placed. Soft iron, this theory assumes, is more easily magnetized because its tiny molecular magnets can be more readily turned than those of hard steel. Conversely, the molecular magnets of soft iron revert more readily to their random orientations when the magnetizing field is removed, and soft iron is therefore not as suitable as hard steel for making permanent magnets.

Sir Alfred (James) Ewing was a Scottish physicist. He was considered an authority in the fields of thermodynamics and magnetism. He was the inventor of several instruments for testing magnetic fields. During the First World War (1914–1918) he became famous for his ability for deciphering coded enemy radiograms.

The molecular theory of magnetism can also explain how a magnet will attract an unmagnetized piece of iron (see Fig. 28.2). When a piece of iron (or any ferromagnetic substance) is placed in the magnetic field of a magnet, it becomes magnetized

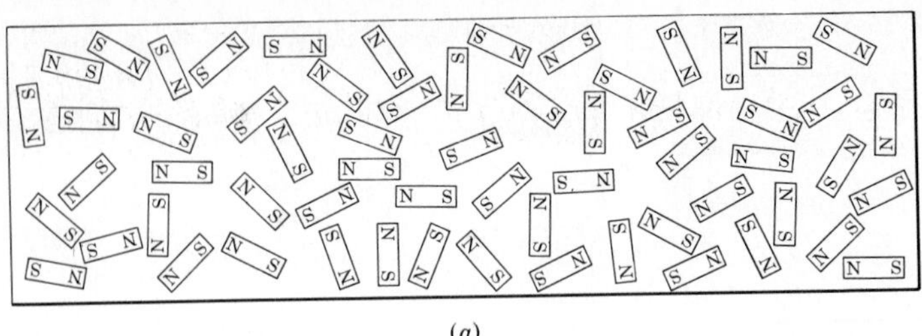

(*a*)

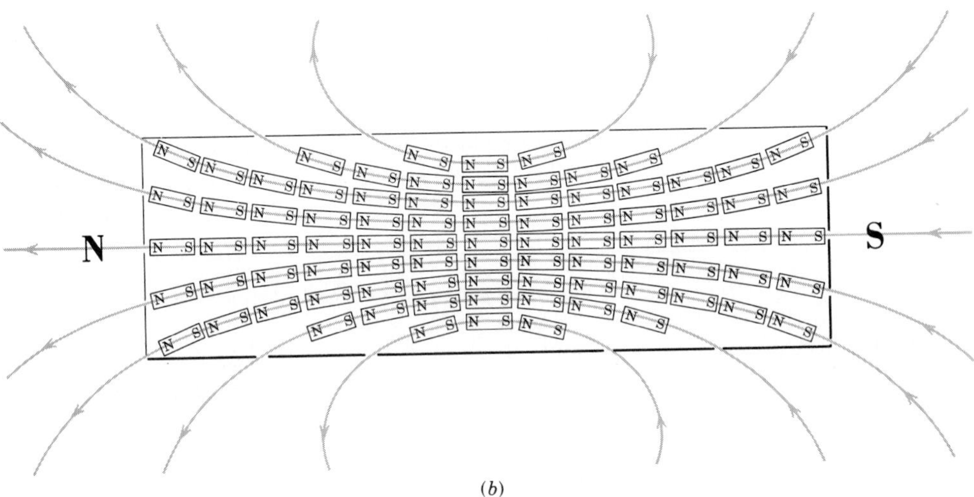

(*b*)

Fig. 28.8 Schematic diagram of Ewing's theory of magnetism. Elementary magnets or magnetic domains in (*a*) unmagnetized iron and (*b*) magnetized iron.

by induction, with poles at the ends. The pole nearest the permanent magnet will be an unlike pole. The attraction between the unlike poles will draw the piece of iron to the magnet.

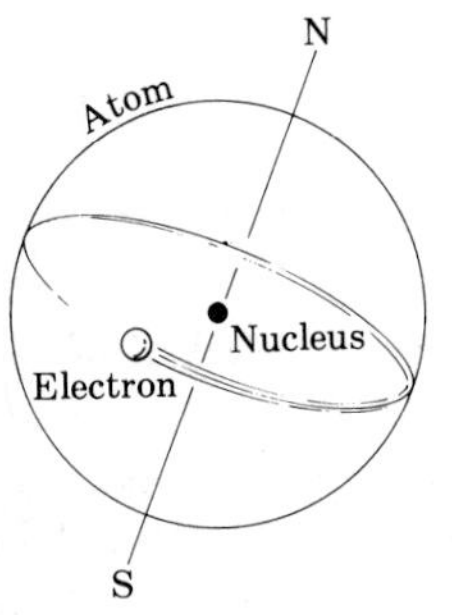

Fig. 28.9 Electrons revolving about the nucleus of an atom are thought to be one source of atomic magnetism.

The Modern Theory of Magnetism The present view attributes the phenomenon of magnetism to motion of electrons within the atoms of the molecules. Electron movements within the atom contribute to magnetic behavior in two ways. First, the electrons rotate in concentric shells around the nucleus (Fig. 28.9). We will see in Sec. 28.10 that a current in the shape of a loop will show magnetic polarity. Electric currents are made up of electrons in motion. Hence, the electron movement around a nucleus imparts a magnetic property to the atoms.

Second, and the major source of atomic magnetic properties, the electron is conceived as resembling a charged sphere spinning around an axis. The spinning electron is equivalent to an extremely small current loop.* Hence every electron, due to its spin, is considered to have a magnetic field equivalent to that of a tiny bar magnet (Fig. 28.10). It is believed that in a magnet each atom has many more electrons *spinning* in one direction than in another. In a submicroscopic region called a *domain*, many of these atoms with electron spin in one direction create a magnetic field which effectively supplements the field engendered by the *revolving* electrons. Each *domain* thus becomes a tiny permanent magnet. When the domains are in random orientation, the substance as a whole is not a magnet, but the presence of an external

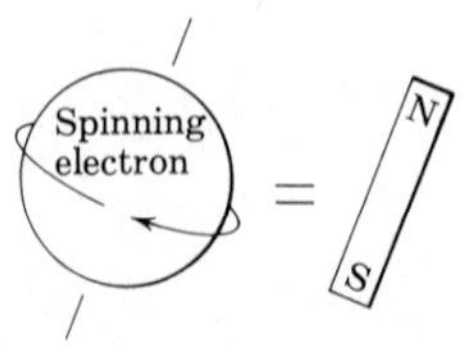

Fig. 28.10 The electron spinning on its axis is now considered to be the major source of atomic magnetism.

* Scientists believe that the phenomenon of magnetism is associated with the spin of the electrons within the third shell of the atoms of magnetic materials.

field will reorient the domains or make the favorably oriented domains grow in size, and produce a magnet in the manner described above. When, in the presence of a very strong field, all the domains have been aligned, the condition of *magnetic saturation* occurs and any further increase in strength of the external field will not increase the magnetization of the iron.

The domain theory is a modern interpretation of magnetic behavior.

The modern theory of magnetism is similar to that proposed by Ewing, except that now we talk of magnetic domains instead of tiny magnetic molecules.

28.6 ■ The Earth's Magnetic Field

The fact that a suspended magnet will seek a north-south direction if free to rotate is evidence that the earth itself is surrounded by a huge magnetic field. Thus the earth acts as a large magnet, with magnetic poles near the geographic poles.

Scientists today are convinced that the core of the earth is much too hot to be a permanent magnet. Walter M. Elsasser, professor of theoretical physics at the University of California, proposed the theory that the earth's magnetic field results from currents of charged particles in the earth's fluid core. Because the north pole of a compass points northward in the northern hemisphere, the magnetic field must be in the direction shown in Fig. 28.11. Since magnetic flux lines come out of the north pole and enter at the south pole, the earth's magnet has its south pole at the north geographical end of the earth and its north pole at the south end.

The earth acts like a large magnet.

Magnetic north, as indicated by a compass, in general differs from the true geographic north. This is due to the fact that the north magnetic pole of the earth does not coincide exactly with its geographic north pole. The angle between the magnetic north and the geographic north at a given point is called the *declination* at that point. In the United States the declination will vary roughly from 20°E in the northwestern section of the country to 20°W in the extreme northeastern section. Navigators and surveyors must continually make corrections for these declinations (see chart of Fig. 28.12), if magnetic compasses are being used. The word *variation* is used by navigators to denote declination.

28.7 ■ Magnetic Flux

The path which would be taken by an "independent north pole" in a magnetic field is called a *line of flux* (see Sec. 28.4). The total number of lines of flux used to represent the direction and the magnitude of a magnetic field is called *magnetic flux*. The SI

The collective lines of force in a magnetic field is called the magnetic flux.

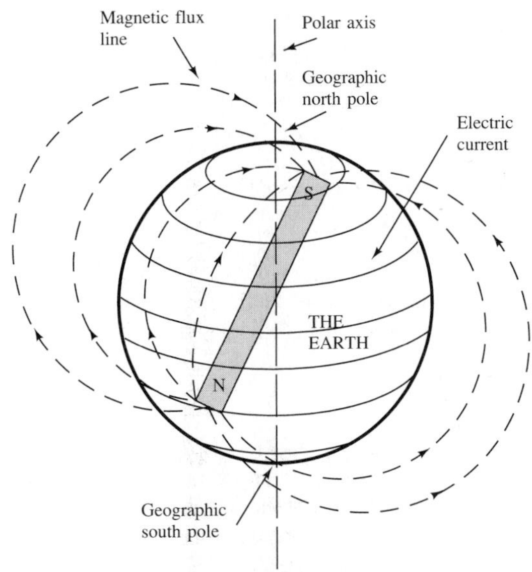

Fig. 28.11 Earth's magnetism is thought to be due to movement of charged particles in its fluid core.

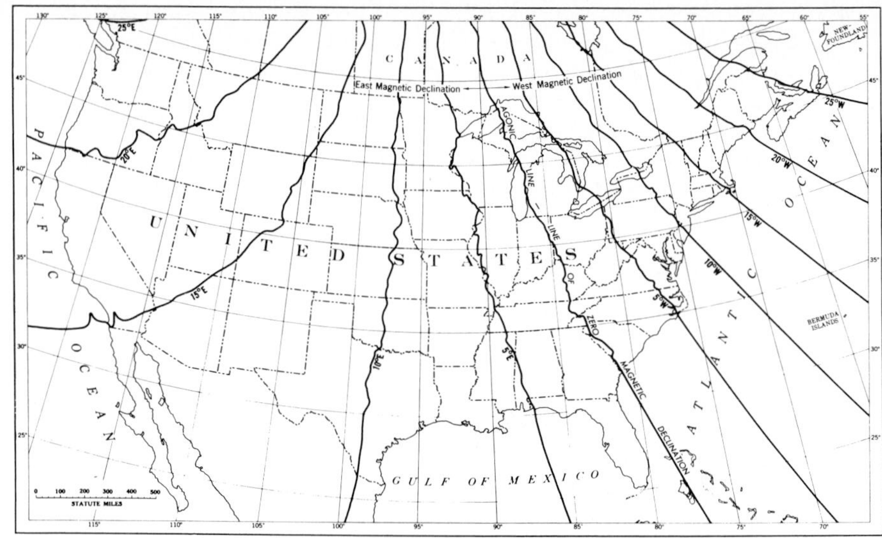

Fig. 28.12 Declination chart covering areas of continental United States. (U.S. Coast and Geodetic Survey)

unit of magnetic flux is the *weber* (Wb) and the symbol for magnetic flux is the Greek letter phi (ϕ). The question that naturally arises at this point should be, "How many lines of flux equal a weber?" Since the lines of flux are imaginary lines, the number is not important. They simply provide a quantitative means of discussing magnetic fields. The weber is defined as equal to one line of flux. In Sec. 28.21 we will give the scientific dimensions of the weber.

28.8 ■ Magnetic Flux Density

The number of flux lines per unit area that permeates a magnetic field is the magnetic flux density, *B*.

The total number of magnetic flux lines (ϕ) per unit area (A) perpendicular to the lines is a measure of the strength of the magnetic field at that point. The strength of a magnetic field is called *magnetic flux density* (**B**). It is also called *magnetic induction*. The flux density **B** is a vector quantity because it indicates the direction and the magnitude of the magnetic field at that point. As an equation

$$\mathbf{B} = \frac{\phi}{A} \tag{28.2}$$

The direction of **B** at a point in a magnetic field is the direction of the magnetic line of force at that point.

The unit of magnetic flux density is the weber per square meter (also called the *tesla* T). An older unit which remains in use today is the *gauss* (G) where 1 tesla = 10^4 gauss.

$$1\text{T} = 1 \text{ Wb/m}^2 = 10^4 \text{ G} \tag{28.3}$$

The earth's magnetic flux density is about $\frac{1}{2}$ G. Lodestone magnets possess a magnetic density of several hundred gauss. Magnetic flux densities of the order of 10 Wb/m^2 (10^5 G) have been developed in scientific laboratories. A more precise definition for magnetic flux density will be given in Sec. 28.13.

Illustrative Problem 28.2 A magnetic field with a flux density of magnitude $B = 0.275\ T$ passes through an area of 25 cm × 30 cm at an angle of 42° (see Fig. 28.13). What is the flux ϕ (in webers) through the area?

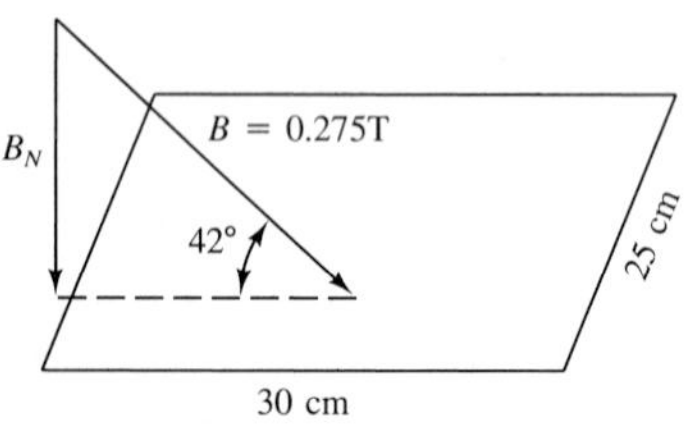

Fig. 28.13 Illustrative Problem 28.2.

Solution

$$B = 0.275 \text{ T}; \ \phi = 42°$$

The component (B_N) of **B** perpendicular to the plane of the area is found by the equation

$$\frac{B_N}{B} = \sin 42°$$

Then

$$B_N = 0.275 \text{ T } (0.669)$$
$$= 0.184 \text{ T}$$

If we solve Eq. (28.2) for ϕ and substitute known values,

$$\phi = B_N \times \text{A}$$
$$= 0.184 \text{ T } (0.25 \text{ m} \times 0.30 \text{ m})$$
$$= 0.0138 \text{ T} \cdot \text{m}^2 \qquad \textit{answer}$$

ELECTROMAGNETISM

The classic discovery by Oersted relating electricity and magnetism (Sec. 27.18) opened a vast new era of scientific advance. The fact that an electric current sets up a magnetic field at right angles to the conductor through which it is flowing has led to some of the most fruitful achievements in the entire history of industrial and technical development. The electromagnetic effect of a current is employed in electric bells, meters, telephones, solenoids, circuit breakers, radios and television, phonographs, electric motors, relays, electromagnetic control apparatus, radar, and nuclear devices. The relationship between electricity and magnetism is the basis for the development of most of our modern electrically energized machines.

An electric current in a wire sets up a magnetic field which encircles the wire.

28.9 ■ Magnetic Field around a Straight Conductor

If magnetism is attributed to motion of charge (the modern theory), a magnetic field should be expected to surround an electric current. Oersted discovered that a compass needle placed above and parallel to a current-bearing conductor will turn until the orientation of the magnetic N-S poles is perpendicular to the direction of the current. If the compass is then placed below the conductor (or the direction of the current is reversed), the compass will turn 180° to reverse the N-S poles' orientation (see Fig. 28.14).

The direction of the magnetic field around a straight current can readily be determined by noting the orientations of the compass in Oersted's experiment or by mounting a piece of cardboard horizontally and passing a heavy conductor through it (Fig. 28.15). When a strong current (say 30 A) is passed through the wire a moderate magnetic field will be established. Iron filings sprinkled on the cardboard will arrange themselves around the conductor in concentric rings, showing the pattern of the field. If small compasses are placed at various positions on the cardboard, the needles align

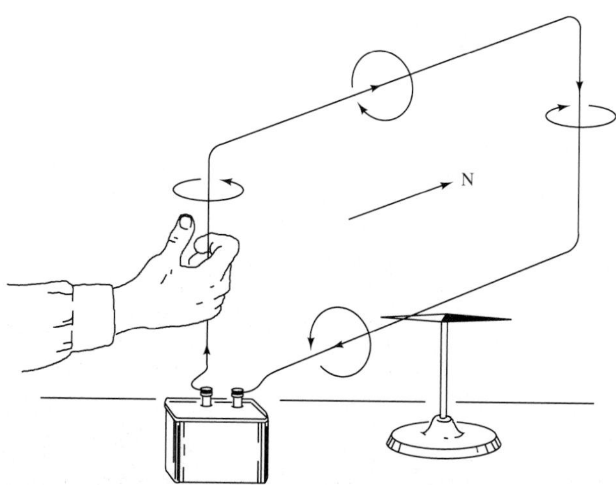

Fig. 28.14 Right-hand rule for determining the direction of the magnetic field around a current.

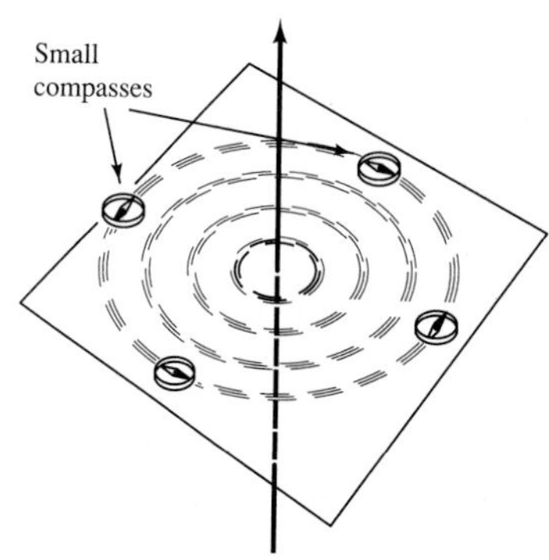

Fig. 28.15 A magnetic field exists around a wire carrying current. Deflections of a compass needle will show the presence and direction of the magnetic field.

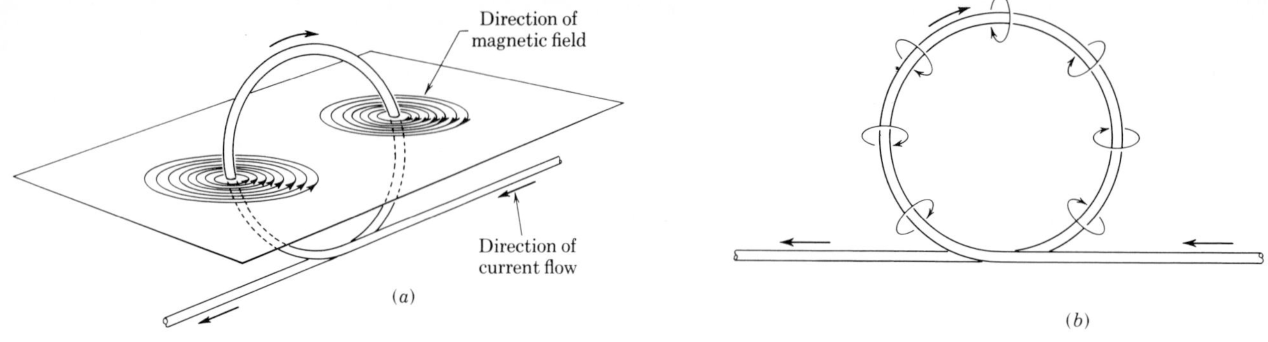

Fig. 28.16 Magnetic field about a circular current-carrying conductor.

themselves tangent to the rings of iron filings. When the current is flowing upward (as shown), the north poles of the compass needles will point in a counterclockwise direction. If the current is then reversed, the north poles will point in a clockwise direction around the wire. If the direction of a current is known, a convenient way to remember the direction of the circular magnetic field is the so-called *right-hand rule* (Fig. 28.14):

If a current-carrying wire is grasped with the right hand and the thumb is extended so it points in the conventional direction of the current flow (positive to negative), then the fingers will indicate the direction of the magnetic field.

28.10 ■ Magnetic Field of a Solenoid

If a current-carrying wire is bent into a loop, the faces of the loop will show magnetic polarity. Two views of such a single loop of wire with some of the lines of force around the wire are illustrated in Fig. 28.16. It can be seen that all the magnetic lines enter the loop at one face and leave at the other. In effect, the loop acts as a disk magnet. The polarity of the loop can be determined by applying the convention that magnetic lines leave a magnet from the north pole and enter at its south pole. Hence the left-hand faces of the illustrations would have north magnetic polarity and the right-hand faces south polarity. If such a loop of wire were free to turn, it would seek a north-south position when current flows through the wire.

The polarity becomes more pronounced and the magnetic field stronger if we wind a number of turns of wire side by side in a tight spiral. Such an arrangement is called a *solenoid* or *helix*. When current passes through a solenoid, the magnetic field developed resembles that of a bar magnet (Fig. 28.17). If the coils of wire are close together, most of the lines of flux around the loops merge and are in the same direction throughout the center of the solenoid (parallel to the axis). At the ends the lines of flux flare out, leaving the solenoid at one end and returning at the other. The solenoid has north polarity at the end where the magnetic lines of force leave and south polarity at the end where they return. The polarity of a solenoid can be determined by using a compass. That pole of the solenoid which repels the north pole of the compass is a north pole, and that pole which repels the south pole is a south pole. The polarity of a solenoid can be predicted by applying another *right-hand rule:*

The magnetic flux inside a current-carrying solenoid, except for points near the ends, is very nearly uniform.

The field lines of a solenoid resemble those of a bar magnet.

Grasp the solenoid with the right hand so that the fingers wrap around it in the conventional direction of the current; the outstretched thumb will then point to the north pole of the solenoid.

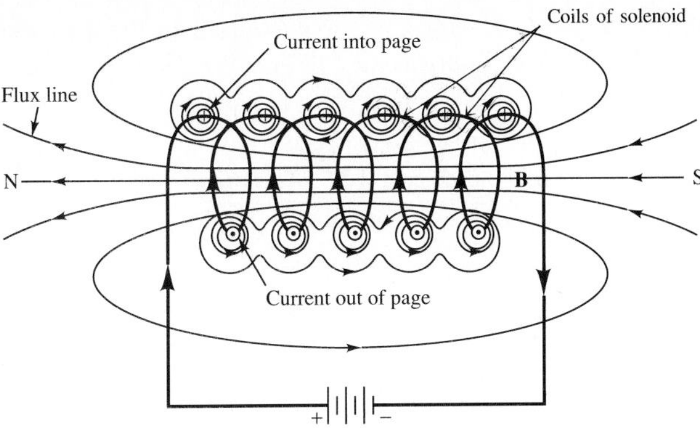

Fig. 28.17 Magnetic field about a current-carrying solenoid. The magnetic field near the axis is uniform. The magnetic field is similar to that produced by a bar magnet.

The converse of this rule can be applied to determine the direction of the current in the coil if the polarity is known.

28.11 ■ Magnetic Induction

The calculation of magnetic flux density **B** (induction) for circuits of various configurations is often quite difficult and involves using calculus methods. The following formulas can however be developed for some common and practical circuits. Each formula involves μ (the Greek letter mu), called the *permeability* of the particular medium in which the magnetic induction is occurring. The constant μ_0 is the permeability for a vacuum (also, approximately for air). Its value, in SI units, is

$$\mu_0 = 4\pi \times 10^{-7} \text{ Wb/A} \cdot \text{m}$$
$$= 4\pi \times 10^{-7} \text{ T} \cdot \text{m/A}$$

The permeability of a material (or vacuum) is a measure of the ease with which magnetic flux can be established in the material.

Induction Around a Single Straight Wire The *magnitude* of the magnetic flux density (magnetic induction B) at a distance r from a straight wire carrying a current I (Fig. 28.18a) is given by the formula

The magnetic field strength near an electric current in a conductor is inversely proportional to the distance from the current.

$$B = \frac{\mu_0 I}{2\pi r} \tag{28.4}$$

where B is in tesla (webers per square meter), I is in amperes, and r is in meters.

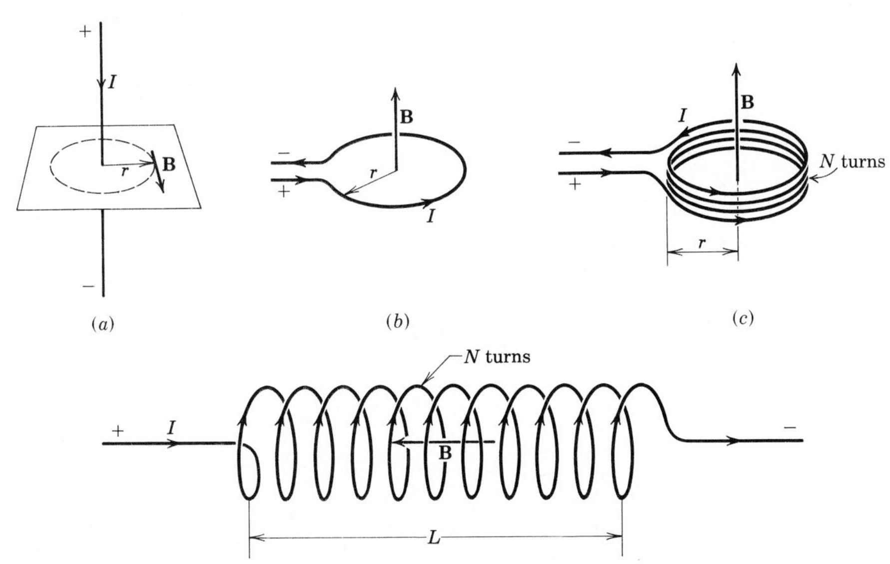

Fig. 28.18 Magnetic field intensity (a) near a single current-carrying wire; (b) at the center of a current loop; (c) at the center of a loop of N turns; (d) along the axis of a long solenoid. The magnetic field is indicated by the **B** vectors.

Illustrative Problem 28.3 Determine the magnitude of the magnetic flux density 2.5 cm from a wire carrying a current of 1.8 A.

Solution Use Fig. 28.18*a* and Eq. (28.4).

$$B = \frac{\mu_0 I}{2\pi r} = \frac{(4\pi \times 10^{-7}\ \mathrm{T \cdot m/A})(1.8\ \mathrm{A})}{2\pi \times 0.025\ \mathrm{m}}$$

$$= 1.4 \times 10^{-5}\ \mathrm{T} \qquad \textit{answer}$$

Illustrative Problem 28.4 Two parallel wires *a* and *b* a distance 25 cm apart carry equal currents of 2.0 A in opposite directions. Find the magnetic induction **B** at the point *P* between the wires at a distance 10 cm from one wire and 15 cm from the other. (See Fig. 28.19.)

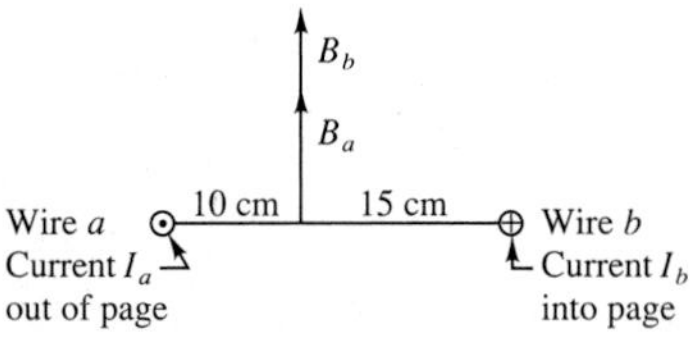

B_b

B_a

Wire *a* 10 cm 15 cm Wire *b*
Current I_a out of page
Current I_b into page

Fig. 28.19 Illustrative Problem 28.4.

Solution Studying Fig. 28.19 and using the right-hand rule for determining the direction of the magnetic field will help establish that the lines of force around conductor *a* (current out of the page) are counterclockwise while around conductor *b* (current into the page) the lines are clockwise. Therefore $\mathbf{B}_a$ due to current I_a and $\mathbf{B}_b$ due to current I_b both point in the same direction at *P*. Therefore, the magnetic induction is cumulative. That is,

$$\mathbf{B} = \mathbf{B}_a + \mathbf{B}_b$$

$$I_a = I_b = 2.0\ \mathrm{A}$$

$$B = \frac{\mu_0 I_a}{2\pi(0.10)} + \frac{\mu_0 I_b}{2\pi(0.15)}$$

$$= \frac{4\pi \times 10^{-7}\ \mathrm{Wb/A \cdot m} \times 2.0\ \mathrm{A}}{2\pi(0.10\mathrm{m})} + \frac{4\pi \times 10^{-7}\ \mathrm{Wb/A \cdot m} \times 2.0\ \mathrm{A}}{2\pi(0.15\mathrm{m})}$$

$$= 6.67 \times 10^{-6}\ \mathrm{Wb/m^2} \qquad \textit{answer}$$

The direction of **B** is up at *P*.

The magnetic flux at the center of a current-carrying loop is greatest at the center of the loop.

Induction Inside a Single Loop of Wire In Fig. 28.18*b*, if a circular loop of wire carries current *I* in amperes, the magnitude of the flux density *B* at the center of the loop of radius *r* meters is given by

$$B = \frac{\mu I}{2r} \tag{28.5}$$

The direction of **B** can be found by the right-hand rule for the solenoid. The magnetic induction is perpendicular to the plane of the loop. Note that the induction is not uniform at all points in the interior of the loop.

Induction Inside a Multiple Loop If a multiple loop of *N* turns is made, the magnitude of the flux density at the center is

$$B = \frac{\mu N I}{2r} \tag{28.6}$$

Induction at the Center of a Long Solenoid Figure 28.18*d* is a schematic diagram of a long solenoid, of length *L* meters and total number of turns *N*. The magnitude of the flux density *B* along the axis within the solenoid is

$$B = \frac{\mu N I}{L} \tag{28.7}$$

Equation 28.7 assumes that the radius of the loops is small compared with the length of the solenoid.

One type of solenoid consists of a coil of wire which has been wound around a doughnut-shaped form. Such a closed solenoid is called a *toroid* (Fig. 28.20). Here L in Eq. (28.7) is the mean circumference of the toroid ($2\pi r$). The magnetic field is nearly uniform within the ring or core material, and there is almost no field outside the core. Such toroids are frequently used for laboratory measurements where a source of a uniform field is needed or as magnetic-core memory devices in computers where the field in one core will not affect its neighbors.

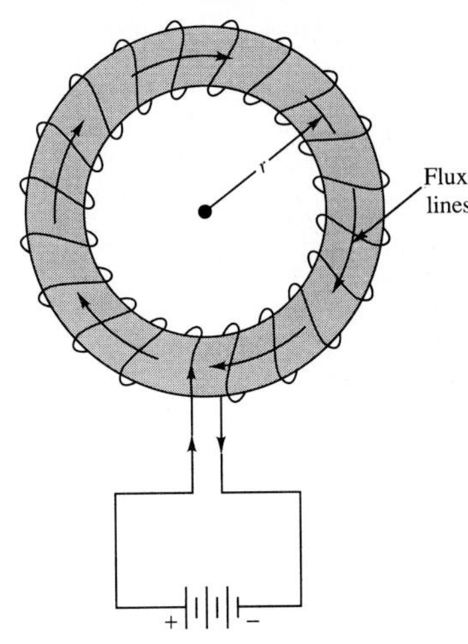

Fig. 28.20 Solenoid wound in the form of a toroid. All the magnetic flux is contained within the core of the solenoid.

Illustrative Problem 28.5 A solenoid 25 cm long has 840 turns of wire. What is the magnitude of the flux density of the magnetic field through the center of the solenoid when the current is 2.6 A?

Solution Substituting the known values in Eq. (28.7) and solving, we get

$$B = \frac{\mu NI}{L}$$

$$= \frac{(4\pi \times 10^{-7}\ \text{Wb/A}\cdot\text{m}) \times 840 \times 2.6\ \text{A}}{0.25\ \text{m}}$$

$$= 1.1 \times 10^{-2}\ \text{Wb/m}^2 \qquad \textit{answer}$$

Illustrative Problem 28.6 A horizontal power line carries a current of 80 A in a south-to-north direction. (*a*) What is the magnitude of the magnetic induction, due to the current, at a point 1 m below the wire? (*b*) What is the direction of the field at that point?

Solution

(***a***)

$$B = \frac{\mu I}{2\pi r}$$

$$= \frac{(4\pi \times 10^{-7}\ \text{Wb/A}\cdot\text{m}) \times 80\ \text{A}}{2\pi\ (1\ \text{m})}$$

$$= 1.6 \times 10^{-5}\ \text{Wb/m}^2 \qquad \textit{answer}$$

(***b***) By the right-hand rule, the magnetic field is found to be directed westward.

Illustrative Problem 28.7 The core of a toroid has inner and outer diameters of 24 cm and 28 cm, respectively. It has 4000 turns. When the current in the wire is 5 A the magnetic induction inside the core is 2.5 T. Determine the permeability of the core.

Solution The mean diameter is (24 cm + 28 cm)/2 = 26 cm. Solve Eq. (28.7) for μ and substitute known values.

$$\mu = \frac{BL}{NI} = \frac{(2.5\ \text{T})(\pi \times 0.26\ \text{m})}{(4000\ \text{turns})(5\ \text{A})}$$

$$= 1.02 \times 10^{-4}\ \frac{\text{T}\cdot\text{m}}{\text{A}} \qquad \textit{answer}$$

or

$$\mu = 1.02 \times 10^{-4}\ \frac{\text{T}\cdot\text{m}}{\text{A}}\left(\frac{\text{Wb/m}^2}{\text{T}}\right)$$

$$= 1.02 \times 10^{-4}\ \text{Wb/A}\cdot\text{m} \qquad \textit{answer}$$

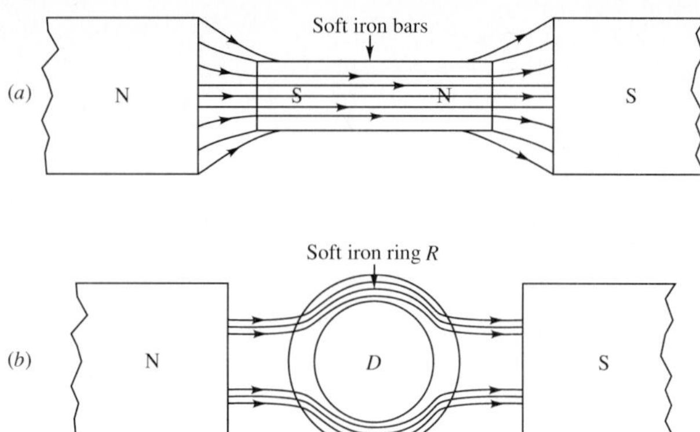

Fig. 28.21 Ferromagnetic materials become magnetized by induction. Ring R shields the area D against the magnetic field.

28.12 ■ Magnetic Shielding

If a bar of soft iron (Fig. 28.21*a*) is placed in a magnetic field, the lines of force are drawn away from the immediate region and concentrated so as to pass through the iron. The iron, in effect, draws the lines to it. The iron thus becomes magnetized by induction. Hence the term *magnetic induction* is used for **B**. On the other hand, a bar of nonmagnetic material will have little effect on the flux distribution in a given region.

If a soft iron ring (Fig. 28.21*b*) is placed in the magnetic field, the region D will be almost free of magnetic lines of force. The region D is said to be *shielded*. The property of magnetic shielding is used in many pieces of apparatus where magnetic fields are undesirable such as in color TV picture tubes. However, magnetic shielding is never as complete as electrostatic shielding (Sec. 25.8).

28.13 ■ Magnetic Field Intensity or Magnetic Field Strength

A magnetic field accompanies any system of moving charges. Magnetic flux density depends not only on the magnitude of the current and the geometric configuration of the conductor (e.g., loop, solenoid, toroid), but also on the material of the field (e.g., air, soft iron). For instance, the flux density produced by a current in a solenoid with a soft iron core will be hundreds of times greater than that produced in a solenoid with an air core.

A typical arrangement for studying magnetic properties of a ferromagnetic material when placed in a toroid-shaped electromagnet with N turns is shown in Fig. 28.22. The configuration is called a *Rowland ring*. A secondary coil is wrapped around part of the ring and is connected to a galvanometer which when properly calibrated, can indicate the rate of change of magnetic flux. Increasing the current in the ring from 0 to A amperes will result in a changing magnetic flux in the ring. The changing magnetic flux will induce an emf in the secondary coil which is proportional

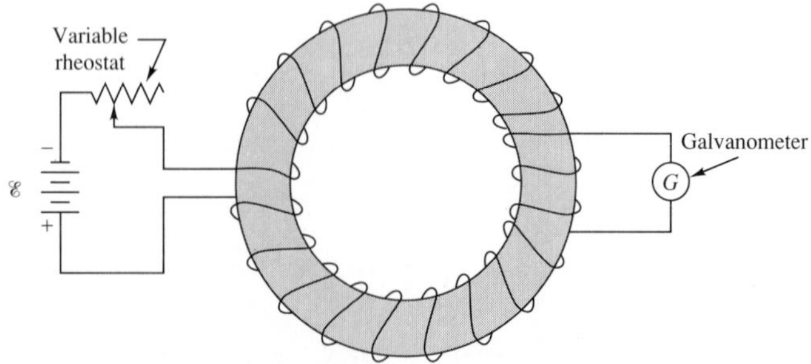

Fig. 28.22 Toroidal winding arrangement used to measure magnetic properties of a substance.

to the rate of change of magnetic flux. By properly calibrating the galvanometer, it is possible to obtain a value for the magnitude of the magnetic flux density B corresponding to any value of current in the ring. First the measurements of B are made with the coil empty (core of air); then the same coil is filled with a magnetic substance and B is measured. From a series of measurements the magnetic properties of the substance can be obtained.

No poles exist in a toroid. This does not present a problem, because poles are not essential to magnetic fields.

It was noted in Sec. 28.11 that the magnetic flux density B_0 inside an empty toroidal coil carrying a current I has a magnitude of $\mu_0 NI/L$, where N is the total number of turns in the coil, L is its length, and μ_0 is the permeability of free space.

When there is a magnetic core inside the coil, and a steady current is in the winding, the field of flux is increased. The additional flux in the magnetic field is due to the alignment of tiny magnetic dipoles in the magnetic material. The additional flux density B_m is proportional to the same quantities that produced B_0. That is,

Because of the iron core in the toroid, the actual value B of the toroid will exceed B_0 by a large factor in many cases, because elementary dipoles in the core line up with the field $\mathbf{B}_0$, thereby setting up their own field of intensity $\mathbf{B}_m$.

$$B_m = \chi \frac{NI}{L} \tag{28.8}$$

The proportionality constant χ (Greek letter chi) in the above is called *magnetic susceptibility*. It is not a constant, but varies, not only with the specimen, but also with the magnetic intensity.

Thus,

$$B = B_0 + B_m$$

$$= \mu_0 \frac{NI}{L} + \chi \frac{NI}{L} \tag{28.9}$$

The quantity NI/L is called the *magnetic field strength* or *magnetic field intensity* and is designated by $\mathbf{H}$.

The magnetic field intensity $\mathbf{H}$ *equals the number of ampere-turns per meter in the magnetizing windings.*

$$H = \frac{NI}{L} \tag{28.10}$$

Magnetic field strength $\mathbf{H}$ has SI units of ampere-turns per meter (A · turns/m). The permeability of the magnetic material is given by

$$\mu = \mu_0 + \chi \tag{28.11}$$

Illustrative Problem 28.8 A coil of 500 turns is wrapped around an iron core which has been bent into the shape of a toroid. The toroid has an outside diameter of 8 cm and an inside diameter of 6 cm. When a current of 2.5 A is established in the windings, a flux of 1.8×10^{-4} Wb is set up inside the core. Calculate (*a*) the magnetizing field intensity, (*b*) the flux density in the coil. Assume the coil has a circular cross section.

Solution

(***a***)

$$H = \frac{NI}{L}$$

$$= \frac{500 \text{ turns} \times 2.5 \text{ A}}{\pi\left(\frac{8+6}{2}\text{ cm}\right) \times \frac{1 \text{ m}}{100 \text{ cm}}}$$

$$= 5700 \text{ A} \cdot \text{turns/m} \qquad \textit{answer}$$

(***b***) $\phi = 1.8 \times 10^{-4}$Wb. The diameter of the core's cross section is 8 cm − 6 cm = 2 cm. Use Eq. (28.2) to find B.

$$B = \frac{\phi}{A} = \frac{1.8 \times 10^{-4}\text{Wb}}{\pi/4\ (0.02 \text{ m})^2} = 0.57 \text{ Wb/m}^2$$

$$= 0.57 \text{ T} \qquad \textit{answer}$$

The magnetic field intensity **H** might be thought of as the field intensity produced by the current, and the magnetic flux density **B** as the total field intensity including the contributions made by the material of the field.

28.14 ■ Magnetic Permeability

The permeability of a given material is a measure of the ease with which magnetic flux may be established in the material.

Iron and certain alloys of iron have the property of being able to multiply the strength of a magnetic field. In a highly permeable medium, the flux density may be thousands of times larger than the field intensity. Permeability may be thought of as a measure of how effective a ferromagnetic material is in multiplying a magnetic field intensity.

The equation relating magnetic flux density **B** and magnetic field intensity **H** can be obtained from Eqs. (28.9), (28.10), (28.11).

$$\mathbf{B} = (\mu_0 + \chi)\frac{NI}{L}$$

$$\mathbf{B} = \mu\mathbf{H} \tag{28.12}$$

Solving for μ in Eq. (28.12).

$$\mu = \frac{\mathbf{B}}{\mathbf{H}} = \frac{\text{magnetic flux density}}{\text{magnetic field intensity}} \tag{28.12'}$$

The permeability is not constant for any given ferromagnetic material, but varies markedly with field intensity. Using proper units for **B** and **H**, we can determine that the SI unit for μ is Wb/A · m or T · m/A.

Table 28.1 summarizes the electromagnetic units used in this chapter.

28.15 ■ Magnetization Curves—Hysteresis

Magnetization curves are used to show how the magnitude of the flux density B varies with field intensity H for an initially unmagnetized sample. Figure 28.23 shows a magnetization curve for a test sample and also shows how the permeability μ varies with the field intensity.

It will be noted from the magnetization curve that as the field intensity H (horizontal axis) (vertical axis) is increased, the flux density B increases slowly at first (section AB of the curve). This may be explained by assuming that there is considerable difficulty in bringing magnetic domains into alignment at first. Then, as the magnetized field is further increased, the magnetization curve rises steeply, showing that the magnetic domains are now rapidly and easily being realigned in the magnetic field. Section BC of the curve illustrates this phase of the process. Finally, as *magnetic saturation* is approached, further increase of the magnetizing field can realign fewer and fewer magnetic domains and the curve flattens out until finally (point D), no further increase in flux density can be obtained.

If a piece of iron is used as the core of a toroid and the current applied is slowly increased, a magnetization curve like that of $ABCD$ of Fig. 28.23 will be obtained. If

Table 28.1 Electromagnetic concepts and their units

Electromagnetic concepts	*SI units*
Total magnetic flux, ϕ	Wb or N · m/A
Magnetic field intensity, **H** or magnetic field strength	N/Wb or A/m
Magnetic flux density, **B** or magnetic induction	Wb/m^2 or T or N/A · m
Permeability, μ	Wb/A · m or T · m/A

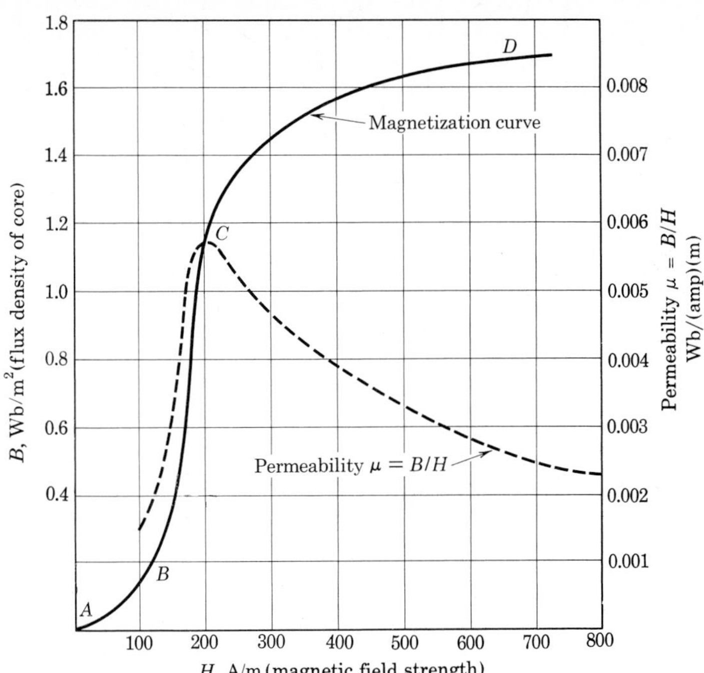

Fig. 28.23 Magnetism curve of a test sample. The variation of permeability with magnetic field strength is also shown.

the current to the toroid is then gradually decreased, it will be found that the field intensity and flux density decrease but the demagnetization curve does not follow back along the same path as the magnetization curve.

In Fig. 28.24 let the curve *abcd* represent the same magnetization curve as that of Fig. 28.23. As the current to the solenoid is decreased to zero, the curve of decreasing magnetization takes some such path as *de*. When current to the solenoid, and therefore the magnetic field intensity (H), has been reduced to zero, the flux density may still be as much as 1.3 Wb/m^2 (point *e*). A considerable amount of *residual flux density* (*residual inductance*) is thus indicated. This residual flux density corresponds to *permanent magnetization*.

A reversal of the current to the solenoid will reverse the direction of the magnetizing field, which will gradually bring the flux density in the test core back to zero at *f*. For the sample of Fig. 28.24, a reversed field of 120 A/m was required to reduce B to zero. If the current is further increased in the negative direction, the *fg* section of the curve will be obtained, the point *g* illustrating magnetic saturation with magnetic domains aligned in the opposite direction. Decreasing the *negative* current will result in the *gh* section of the curve, and finally, if the current is switched to its original (positive) direction and increased, the *hid* section of the curve is obtained.

It is seen that flux density B lags behind field intensity H. This lag is called *hysteresis* (from the Greek word meaning a *lag* or *coming behind*). The closed loop *dfgid* is called a *hysteresis loop*. As iron is magnetized, work must be done on the domains in order to align them, and this work shows up as heat. In alternating current apparatus (see Chap. 31), the magnetizing and demagnetizing process takes place many times a second, and the *hysteresis loss* (heat) may be considerable. For example, it may be shown by use of the calculus that the hysteresis loss (or heat loss) for one cycle in alternating current is proportional to the area enclosed within the hysteresis loop of a diagram like that just analyzed.

The hysteresis curve indicates how the magnetizing (and demagnetizing) effect lags behind the magnetizing (and demagnetizing) force.

The area of the hysteresis loop represents energy dissipated as heat.

The magnitude of the area of the hysteresis loop is important in the design of some electrical machinery. This is particularly true for alternating current machinery where the direction of the current is continually changing at the rate of 120 times per second. Not only is the loss of energy in the form of heat energy expensive, but the accumulation of too much heat energy in the apparatus can be dangerous as well as damaging to equipment.

Fig. 28.24 A typical hysteresis loop, showing the lagging of the magnetism of ferromagnetic material behind the magnetizing force.

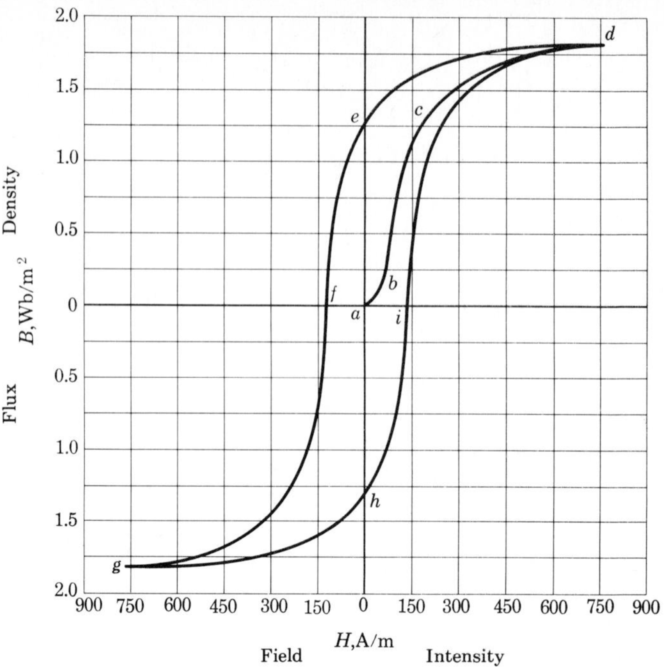

Materials whose hysteresis loops are relatively long and skinny are best suited for transformer cores, electric motors or generator pole pieces. Materials whose hysteresis loops are relatively short and fat are best suited for permanent magnets.

Since soft iron shows lower hysteresis losses than steel, it is used for the cores of rotors and stators in electric machinery. Soft iron also has high permeability, which is another factor in its favor for use where magnetic effects are desired.

28.16 ■ Lifting Magnets

The core of a lifting magnet is made of soft iron (high permeability).

Lifting magnets are constructed in various shapes. One of the more common types is illustrated in diagram form by Fig. 28.25. It consists of an energizing coil of copper wire placed around a soft-iron core. The soft-iron core is the center part of a larger iron shell which furnishes the permeable material to carry the magnetic lines of force. These powerful magnets are used in transferring scrap iron and steel from stockpiles to railroad cars or from railroad cars to river barges. Such magnets are also used in manufacturing plants, where heavy iron or steel products have to be hoisted in orderly fashion, e.g., kegs of nails, iron sheets, pipes, and steel billets (Fig. 28.26). Electromagnets have been made which will exert a force of 200 lb for every square inch of pole face. The electromagnet is effective so long as sufficient current flows through its coils. Thus it allows an operator to turn the magnetic field on or off, merely by the flick of a switch. Large electromagnets generally are equipped with portable dc generators that furnish the required current. Surgeons use a very small electromagnet for removing small splinters of steel from the eye or other parts of the body.

Electromagnets also are effective with alternating current. When alternating current is passed through the coils of an electromagnet, the polarity of the magnet will

Fig. 28.25 Cross-sectional view of a lifting magnet.

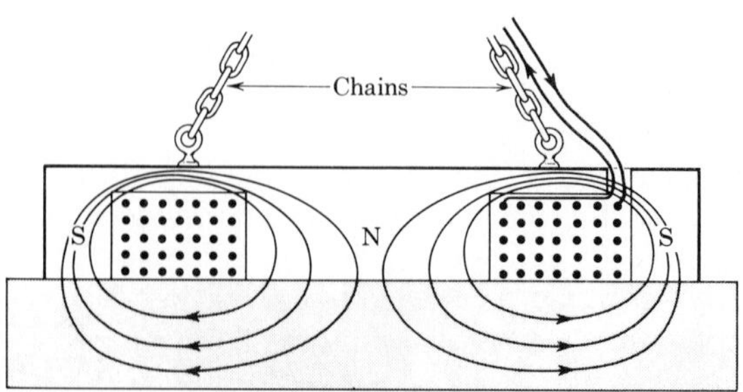

Fig. 28.26 Magnetic crane moving steel slab at steel plant. (Bethlehem Steel Corp.)

change periodically in step with the alternations of the current (60 cycles per second in the United States). However, it is usually not important whether it is a north or south pole that is doing the attracting. Since the alternations occur so rapidly, the interruptions of the magnetic field are, for most purposes, negligible.

There are some problems inherent in the use of ac electromagnets. The changing magnetic field will cause an electric current to be induced in the core. These currents are called *eddy currents*.

Eddy currents in the electromagnet are undesirable for two reasons: the eddy currents represent power losses and the core tends to get quite hot. Eddy currents and the means of eliminating them will be discussed in Chap. 31.

28.17 ■ Solenoids

The magnetic flux density along the axis of a solenoid was discussed in Sec. 28.11. If the solenoid is provided with a movable soft-iron core, called the *plunger,* and a current (either dc or ac) flows through the turns of the coil, the resultant magnetic field will tend to pull the plunger into the center of the solenoid. The field acts as though it sucked the plunger into the solenoid. Hence the coil and the plunger are termed a *sucking coil*. The sucking coil is also called a *solenoid,* even though only the coil itself is the solenoid.

The sucking coil can be used either as a *puller-type* solenoid (Fig. 28.27*a*) or a *pusher-type* solenoid (Fig. 28.27*b*).

Thus the solenoid can be used to provide either a mechanical push or pull to operate various devices. When the current does not flow in the solenoid, a spring forces the plunger back to its original position. Pusher-type solenoids properly motivated by an electric current find frequent use in removing objects from a moving belt. Small solenoids are also used in operating such devices as door chimes, gas valves, water valves, and pipe organ stops. The schematic diagram of such a device is left as an exercise for the student.

28.18 ■ Magnetic Separators

Electromagnets are used in removing iron from many industrial processes when its presence is undesirable. For example, the coal industry uses separator magnets,

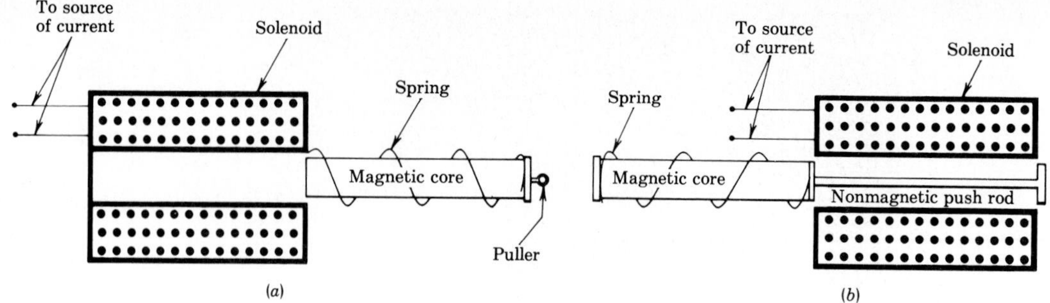

Fig. 28.27 (*a*) Puller-type solenoid; (*b*) Pusher-type solenoid.

which are quite similar to the lifting magnets discussed in Sec. 28.16, suspended over conveyor belts to remove "tramp" iron in the coal, which, if it were not removed, might break the rolls used to crush or pulverize the coal. Often *magnetic pulleys* replace the end pulleys on a belt-conveyor system. When material which contains iron reaches the magnetic pulley, all the tramp iron is attracted by the pulley and held against the belt on the bottom side of the magnetic pulley. When the tramp iron is carried past the magnetic field set up by the pulley, it falls harmlessly into a separate compartment or onto another conveyor belt.

Magnetic separators are used a great deal today in reclaiming iron and magnetic metal from foundry refuse or slag. They are also used to separate magnetic and nonmagnetic borings and turnings.

28.19 ■ Electric Relays

The *electric relay* is an electromagnetic device by means of which contacts in one electric circuit are operated by a change in current in the same circuit or in a different circuit. The essential parts of the relay are illustrated in Fig. 28.28. When the electromagnet M is energized by passing a current through it, the soft-iron armature A is attracted to it. The armature, which is a part of a second circuit, is provided with a tongue which can make contact with the fixed points C and D. As the armature is attracted to the electromagnet core, the tongue breaks contact with C and makes contact at D, thus *closing* the secondary circuit. When the current does not flow through the electromagnet, the attraction for A disappears and a spring S pulls the armature until the contact at D is broken.

Applications of electromagnetic relays are numerous. They are used extensively for local or remote control of circuits of all types. Electric motors at inaccessible locations, such as the motor of an elevator, can be operated by a system of relays.

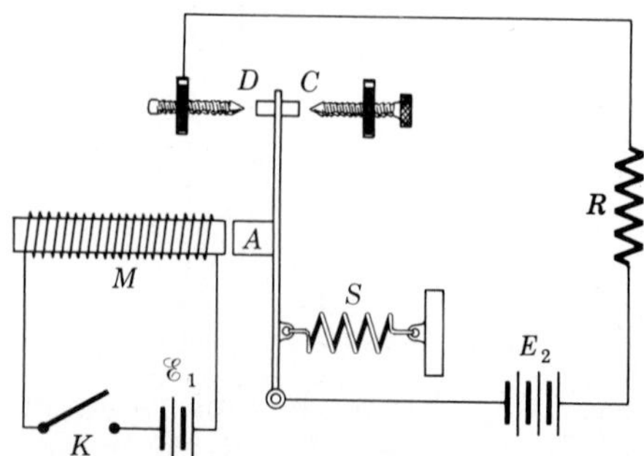

Fig. 28.28 Electric relay or magnetic switch.

Thermostats and other control devices can actuate solenoids which close or open magnetic switches to start or stop electric motors. Small relay currents can be used to control large currents through relay controls. *Electric protective relays* are used to protect apparatus and entire electric systems from dangerous conditions. When these conditions arise, the relays disconnect the apparatus from the system or shut off the main-line current. Some of the protective relays include *circuit relays, voltage relays, power relays, frequency relays,* and *temperature relays*. They can be set to operate as either the *over* or *under* type, i.e., if the factor being guarded is too great, the circuit can be disconnected, or if it is too small, it can be closed.

28.20 ■ Uses of Electromagnetism

We have discussed only a few uses of electromagnets and electromagnetic principles. There are hundreds that could be listed. The following partial list should serve to illustrate the great variety of uses to which electromagnetism has been applied.

1. Generators
2. Motors
3. Telephone receivers
4. Magnetic brakes
5. Magnetic switches
6. Circuit breakers
7. Magnetic clutches
8. Magnetic couplings
9. Ammeters
10. Voltmeters
11. Loudspeakers for radio, TV, and audio equipment
12. Electric chimes
13. Brakes for hoists, cranes, and elevators
14. Magnetic chucks
15. Electric clocks
16. Cyclotrons and other ''atom smashers''
17. Holding coils for motor starters
18. Relays for automatic control of equipment
19. Track switches
20. Solenoid valves
21. Magnetic containment for thermonuclear fusion research
22. Body ''scanners'' in medical diagnosis—*magnetic resonance imaging*

ELECTRICAL MEASURING INSTRUMENTS

There are many and diverse types of instruments designed to measure electrical quantities. Practically all of them, either directly or indirectly, involve the measurement or detection of electric current. These instruments measure electric quantities by using one of the following effects of an electric current: (1) chemical effect, (2) heating effect, or (3) magnetic effect. The magnetic effect is the most universally used in these instruments. The following pages present a discussion of a few common electrical measuring instruments which employ the magnetic effect of an electric current.

28.21 ■ Force of a Magnetic Field on a Current-Carrying Conductor

In Sec. 28.10, we studied the magnetic field around a wire carrying a current. Let us place such a wire in the field established between the opposite poles of a magnet. In Fig. 28.29*a* the current, represented by a dot, is directed up out of the paper toward the reader. The lines of force of its field are in a counter-clockwise direction (apply the right-hand thumb rule). In Fig. 28.29*b,* lines of force between the poles of a

Fig. 28.29 Lines of force (*a*) around a current-carrying conductor; (*b*) between magnetic poles; (*c*) when current-carrying conductor is in a magnetic field.

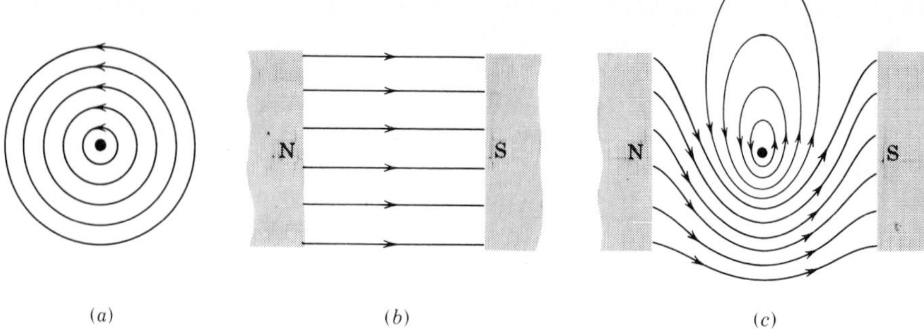

A current-carrying conductor is acted upon by a magnetic force when it is placed in a magnetic field. The effect is caused by the interaction between the magnetic field around the conductor and the magnetic field between the magnetic poles.

magnet are shown pointing from the north to the south pole. When the current-carrying wire is placed in the magnetic field (*c*), it can be seen that the lines of force (produced by the poles and by the current in the wire) below the conductor all are directed from left to right, while those above the conductor point in opposite directions. The two fields add vectorially, resulting in a strong magnetic field below the conductor and a weak field above. The effect of this distribution of lines on the conductor can best be visualized if we conceive of the magnetic lines of force as having elastic properties. As the lines contract to become as short as possible, the conductor will be pushed upward. An easy way to remember the direction of motion of the conductor is to apply what is commonly called the *right-hand* or *motor rule* (also called Ampere's rule).

A magnetic force interacts with a charged particle only when the particle is in motion.

Place the thumb and the first and second fingers of the right hand at right angles to each other (as in Fig. 28.30). Then hold the hand so that the first finger points in the direction of the conventional current, the second finger in the direction of the magnetic field; then the thumb will point in the direction in which the conductor will tend to move.

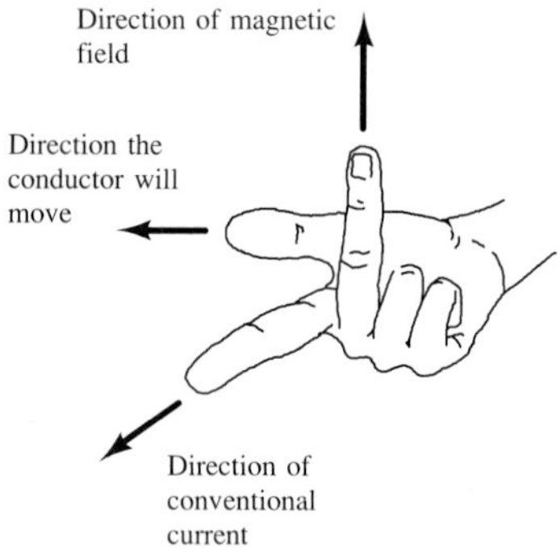

Fig. 28.30 Right-hand rule for finding direction of force on a conductor.

When the wire and magnetic field are at right angles to each other, the force exerted upon the wire by the field is directly proportional to each of three factors: (1) *the magnetic induction* **B** of the field, (2) *the current I* flowing in the wire, and (3) *the length L* of the part of the wire that is in the field. Thus, the magnitude of the force is given by

$$F = ILB \tag{28.13}$$

where F is expressed in newtons if I is in amperes, L in meters, and B in newtons per ampere-meter.

Flux density (magnetic induction) **B** is a vector quantity. In Secs. 28.7 and 28.8 the direction of **B** at any point in a magnetic field was given as the direction of the magnetic lines of force at that point. Equation (28.13) can now be used to define B in the SI system. Solving the equation for the magnitude of B, we find that

$$B = \frac{F}{IL} \tag{28.14}$$

Magnetic induction B is the force per unit length per ampere exerted by a magnetic field on a current-carrying conductor when the conductor is oriented so that it is perpendicular to the lines of force of the magnetic field.

Using the same basic units given for Eq. (28.13), the units for B become newtons per ampere-meter (N/A · m or T). In summary,

$$\begin{aligned} 1 \text{ N/A} \cdot \text{m} &= 1 \text{ magnetic line per square meter} \\ &= 1 \text{ Wb/m}^2 \\ &= 1 \text{ T} \end{aligned}$$

By equating 1 Wb/m^2 = 1 N/A · m, it can be shown that dimensionally *1 Wb is equivalent to 1 N · m/A.*

In discussing the development of Eq. (28.13), we assumed that the wire must be perpendicular to the direction of the magnetic field. If the wire makes an angle θ with the field, the magnitude of the force is given by

$$F = ILB \sin \theta \tag{28.15}$$

Whereas the electric force is always in the direction of the electric field, the magnetic force is perpendicular to the magnetic field.

Illustrative Problem 28.9 A wire carrying a current of 40 A is at right angles to a uniform magnetic field in which the magnetic induction is 0.50 N/A · m. If the length of the wire is 15 cm, what is the force on the wire?

Solution

$$I = 4.0 \text{ A} \qquad L = 0.15 \text{ m} \qquad B = 0.50 \text{ N/A} \cdot \text{m}$$

$$\begin{aligned} F &= ILB \\ &= 4.0 \text{ A} \times 0.15 \text{ m} \times 0.50 \text{ N/A} \cdot \text{m} \\ &= 0.30 \text{ N} \end{aligned}$$ *answer*

Illustrative Problem 28.10 A wire 40 cm long carries 25 A at an angle of 60° with a uniform magnetic field of flux density 12×10^{-4} N/A · m (see Fig. 28.31). What is the magnitude and direction of the force on the wire?

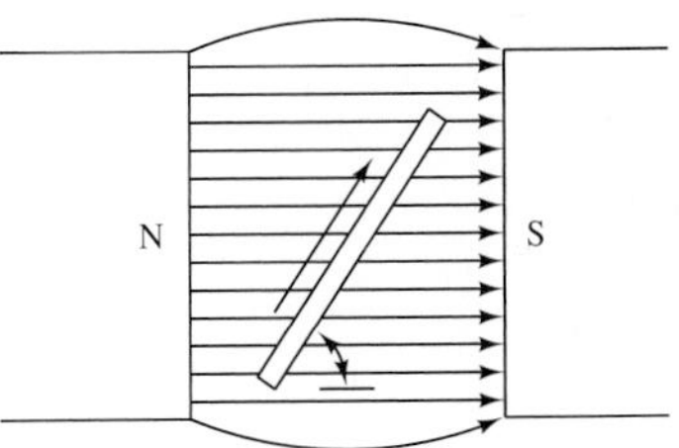

Fig. 28.31 Illustrative Problem 28.10.

Solution

$$I = 25 \text{ A} \qquad L = 0.40 \text{ m}$$
$$B = 12 \times 10^{-4} \text{ N/A} \cdot \text{m} \qquad \theta = 60°$$

$$\begin{aligned} F &= ILB \sin \theta \\ &= 25 \text{ A} \times 0.40 \text{ m} \times (12 \times 10^{-4} \text{ N/A} \cdot \text{m}) \times 0.866 \\ &= 0.010 \text{ N} \end{aligned}$$ *answer*

Using the motor rule, we find that the wire will be pushed away from the reader (into the page).

28.22 ■ Force between Two Parallel Conductors

When two long parallel current-carrying conductors are adjacent to each other, each gives rise to a magnetic field and the interaction of the fields gives rise to magnetic forces which act on both wires. The two circular fields combine as shown in Fig. 28.32. Note that two long straight parallel wires carrying current in the same direction attract each other and the same two wires carrying current in opposite directions repel each other.

We can use Eq. (28.4) to determine the field caused by I_1:

$$B = \mu \frac{I_1}{2\pi d}$$

From Eq. (28.13), the force on current I_2 by B will be

$$F = I_2 LB = I_2 L \mu \frac{I_1}{2\pi d}$$

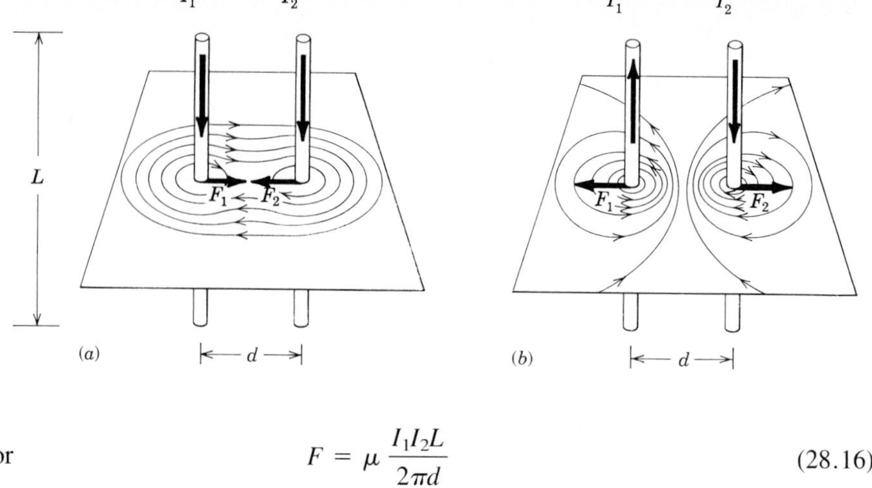

Fig. 28.32 (*a*) Parallel wires carrying current in the same direction attract each other. (*b*) Parallel wires carrying current in opposite directions repel each other.

or

$$F = \mu \frac{I_1 I_2 L}{2\pi d} \tag{28.16}$$

Definition of the Ampere (A) and the Coulomb (C) The force of attraction between long current-carrying parallel conductors, derived from Eq. (28.16), was officially adopted in 1947 by the U.S. National Bureau of Standards in defining the ampere unit.

The ampere is defined as the current in each of two long parallel conductors 1 m apart in free space which causes them to exert a force on each other of exactly 2 × 10⁻⁷ N for each meter of length.

Having defined the ampere in this way we can now define the *coulomb*.

The coulomb is the charge transferred through any cross section of a conductor in 1 second by a current of exactly 1 ampere.

This may seem a circuitous way for arriving at the definition for the ampere and the coulomb. The reason behind this method is to have definitions that involve quantities that *can actually be measured* with relative ease. For example, the definition for the unit of charge (coulomb) given in Sec. 25.9, involving the number of electrons in the unit itself defies measurement in the laboratory. The ampere was at that point defined (Sec. 26.2) in terms of the quantity of charge per unit time.

With this precise definition for the ampere, it is possible to define the coulomb as *exactly* one ampere-second (1 C = 1 A · s).

On the other hand, it is quite simple to vary the amount of current in a wire by using a variable resistor. Thus, the force between two current-carrying conductors is much easier to measure with precision* than to measure the number of electrons which equal a coulomb. Since the coulomb is defined in terms of the ampere it, too, can be measured with precision.

Illustrative Problem 28.11 Two long straight wires are 6 cm apart. What force per meter length will the wires exert upon each other if one wire carries 2 A of current while the other carries 3 A?

* Actually, the National Bureau of Standards in Washington, D.C., measures the current using circular coils rather than straight lengths of wire because it is easier and more accurate. This does not affect our definition.

Solution

$$\mu = 4\pi \times 10^{-7}\ \text{Wb/A} \cdot \text{m}$$

$$I_1 = 2\ \text{A} \qquad I_2 = 3\ \text{A}$$

$$d = 0.06\ \text{m} \qquad F = \mu \frac{I_1 I_2 L}{2\pi d}$$

Solve for F/L.

$$\frac{F}{L} = \mu \frac{I_1 I_2}{2\pi d}$$

$$= \frac{4\pi \times 10^{-7} \dfrac{\text{Wb}}{\text{A} \cdot \text{m}} \left(\dfrac{\text{N} \cdot \text{m/A}}{1\ \text{Wb}}\right) (2\text{A})(3\text{A})}{2\pi(0.06\ \text{m})}$$

$$= 2 \times 10^{-5}\ \text{N/m}$$ *answer*

28.23 ■ Torque on a Current-Carrying Coil in a Uniform Magnetic Field

The force exerted on a current-carrying conductor when placed in a magnetic field was studied in Sec. 28.21. The results of that analysis can be used to show that a torque is exerted on a current-carrying coil placed in a magnetic field.

The motion of the conducting loop results from the interaction between the current field and the magnetic field. It takes no energy from the magnetic field.

Consider a rectangular coil with length l and width w carrying a current I in the presence of a magnetic field. When the plane of the coil is parallel to the magnetic field **B** (see Fig. 28.33*a,b*), there will be no forces on the sides of width w because they are parallel to **B**. But there are forces F_1 and F_2 acting on the sides of length l. These forces are equal in magnitude and in opposite directions (vertically upward and vertically downward). It was shown in Sec. 28.21 that $F_1 = F_2 = ILB$ (Eq. 28.13). Since F_1 and F_2 do not act along the same line, a torque τ (Greek letter tau) is produced. Hence, the magnitude of this maximum torque, τ_{max}, is

Many electric meters and motors make use of the torque on a current-carrying conductor placed in a magnetic field.

$$\tau_{max} = F_1 \times \frac{w}{2} + F_2 \times \frac{w}{2}$$

$$= (IlB) \times \frac{w}{2} + (IlB) \times \frac{w}{2}$$

$$= IlwB$$

Since the area A of the coil is lw, the torque can be expressed as

$$\tau_{max} = IAB \qquad (28.17)$$

Equation (28.17) holds only when the plane of the coil is parallel to the field **B**. To find the torque when the plane of the coil makes an angle θ with the field, consider the front view of the coil given in Fig. 28.33*c* and *d*. The lever arm for F_1 and F_2 becomes $w/2 \cos\theta$ and the torque for any value of θ is

The torque exerted on a current-carrying coil in a magnetic field is used as a current-measuring device in the galvanometer. The torque is directly proportional to the current in the coil.

$$\tau = IlB\left(\frac{w}{2} \cos\theta\right) + IlB\left(\frac{w}{2} \cos\theta\right)$$

$$= IlwB \cos\theta$$

or

$$\tau = IAB \cos\theta \qquad (28.18)$$

If the coil has N turns the torque is increased N-fold. The general equation is

$$\tau = NIAB \cos\theta \qquad (28.19)$$

where torque is expressed in meter-newtons (m · N) where B is in newtons per ampere-meter or Wb/m^2(T), I is in amperes (A), and A is in m^2.

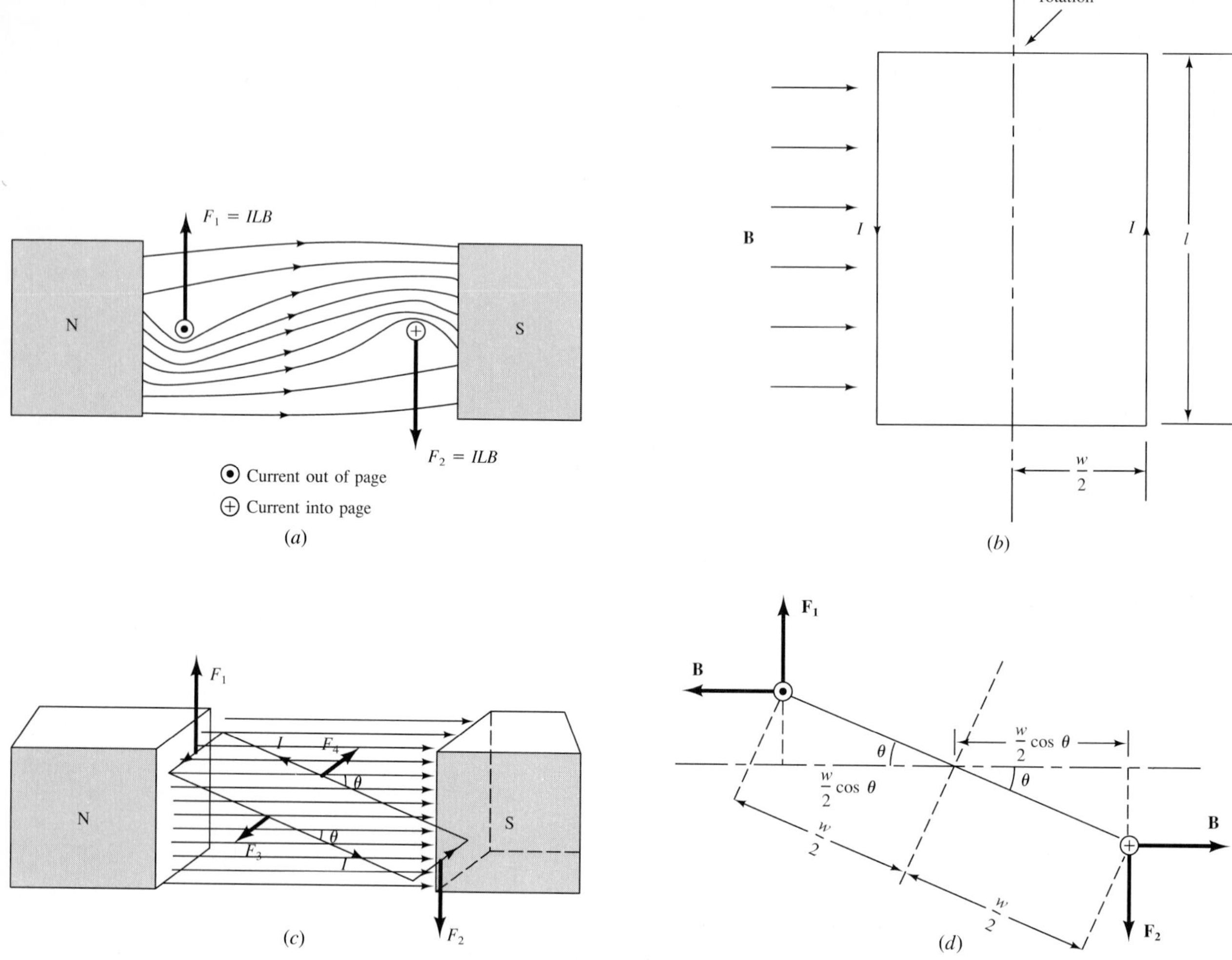

Fig. 28.33 Torque on a current-carrying loop in a uniform field. (*a*) Side view; (*b*) top view; (*c*) forces on the loop, (*d*) plane of loop makes an angle θ with the permanent magnetic field.

Although the equations for torque have been derived for a rectangular coil, it can be shown that they apply for a coil of any shape.

Illustrative Problem 28.12 A coil which has 25 turns with a width of 15 cm and a length of 30 cm is mounted in a uniform field of flux of 1.4×10^{-3} T. If the current in the coil is 15 A, find (*a*) the torque acting on the coil when the coil makes an angle of 30° with the direction of the field; (*b*) the maximum torque on the coil; and (*c*) the minimum torque on the coil?

Solution

(*a*)

$$\begin{aligned}\tau &= NIAB\cos\theta \\ &= 25 \times 15\text{ A} \times (0.15\text{m} \times 0.30\text{m}) \times 1.4 \times 10^{-3}\text{T} \times 0.866 \\ &= 2.5 \times 10^{-2}\text{ A}\cdot\text{m}^2\text{ T} \times \frac{1\text{ N/A}\cdot\text{m}}{1\text{ T}} \\ &= 2.0 \times 10^{-2}\text{ m}\cdot\text{N}\end{aligned}$$

answer

(*b*) The torque will be maximum when $\theta = 0°$. Then

$$\tau_{max} = 25 \times 15 \times 0.15 \times 0.30 \times 1.4 \times 10^{-3} \times 1$$
$$= 2.4 \times 10^{-2} \text{m} \cdot \text{N} \qquad \textit{answer}$$

(*c*) when $\theta = 90°$ ($\cos\theta = 0$), the torque will be 0 *answer*

ELECTRICAL MEASURING INSTRUMENTS

28.24 ■ Galvanometers

The basic current-measuring or current-detection instrument is the *galvanometer*. There are many kinds of galvanometers, of which we shall consider only two. The essential parts of the *d' Arsonval galvanometer* are shown in Fig. 28.34. It consists of a flat coil of wire *C* suspended by means of a light metallic ribbon between the poles of a permanent U-shaped magnet. Current is conducted to and from this coil by means of the light ribbon by which it is suspended from the top, and a helix of similar material below the coil. When a current flows through the coil, it produces a magnetic field with opposite poles on either side of the coils. In the figure (*top view*) the direction of the current in the coil is indicated by dots when it is toward the reader and by +'s when it is away from the reader. Thus it is seen that a north polarity is produced above the coil and a south polarity below. The interaction between this field and that of the permanent magnet will establish a clockwise torque which is proportional to the amount of current passing through the coil. The coil is provided with a

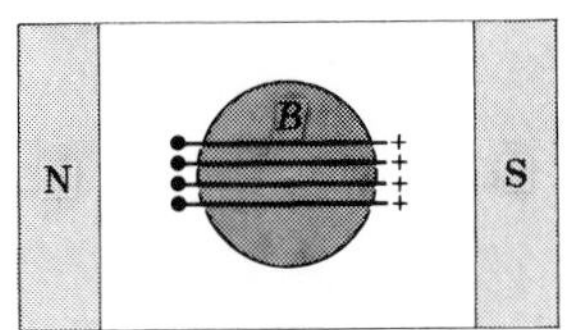

Top view

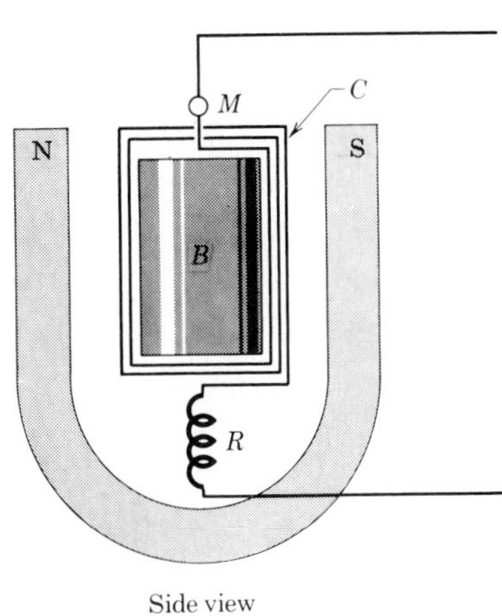

Side view
(*a*)

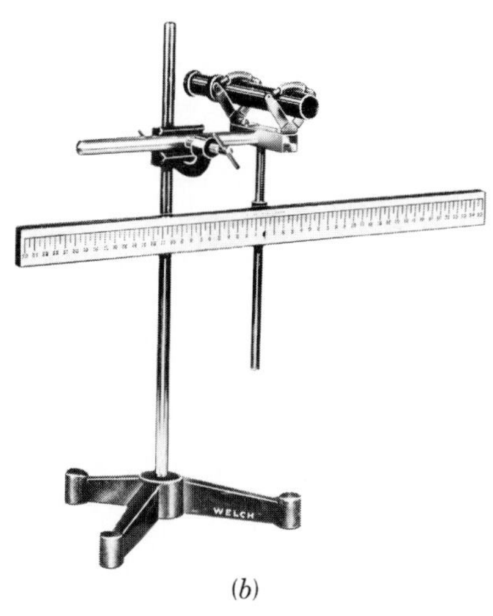

(*b*)

(*c*)

Fig. 28.34 (*a*) Moving system of d'Arsonval galvanometer; (*b*) d'Arsonval galvanometer telescope scale; (*c*) portable multipurpose instrument which can function as a zero center galvanometer, microammeter, millivoltmeter, voltmeter, ohmmeter, and polarity indicator. (Courtesy Central Scientific)

soft-iron core B to form a uniform magnetic field. The torque is opposed by the restraining force of the suspension ribbon R. The movable system carries a tiny mirror M which reflects the numbers from a distant scale (see Fig. 28.34*b*). The farther the scale is from the galvanometer, the greater the sensitivity of the instrument. A more rugged and portable (but less sensitive) instrument is obtained by replacing the optical (mirror) system of deflection by a pointer, which moves over a graduated scale, and by replacing the metallic-ribbon suspension by two spiral springs. Such an instrument is shown in Fig. 28.34*c*. The springs, besides balancing the magnetic deflection torque exerted on the coil, provide the external leads to the instrument. The coil is mounted in jeweled bearings.

28.25 ■ The Ammeter

The ammeter is a modified galvanometer.

It was emphasized in Sec. 26.13 that the ammeter is connected in *series* with the line in which the current is to be measured. The ammeter is simply a *low-resistance galvanometer* whose scale is calibrated to read current in amperes. Since the moving coil and the springs used in the galvanometer are designed to carry only very small currents (about 0.05 A), the instrument, if it is to serve as an ammeter, must be equipped with a *shunt of* such low resistance that only a small part of the total current flows through the moving coil. The greater part of the current in the main circuit bypasses the coil and travels through the shunt (see Fig. 28.35). Many ammeters are fitted with several shunts of different resistances in order to provide an instrument which will measure currents over several different ranges. These instruments require the use of several sets of figures on the scale (see Fig. 26.7*a*). Judicious selection of the proper range by the technician is required in using the instrument.

Several precautions must be observed in connecting an ammeter in a circuit. The movable coil is quite light in weight and is mounted on jeweled bearings. This increases its sensitivity but makes it more susceptible to damage from jarring. Since it is a *low*-resistance instrument, care must be taken never to connect it in a circuit

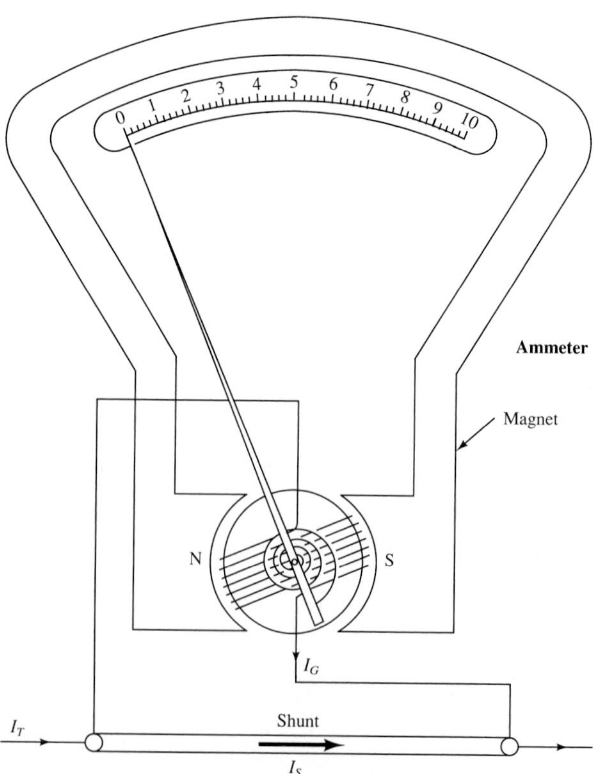

Fig. 28.35 Circuit of a single-scale ammeter. Most of the current passes through the shunt of an ammeter. Only a small part passes through the moving parts of the instrument.

which will cause too much current to flow through the moving coil. The instrument can easily *burn out* if subjected to too much current. Also, since most ammeters permit only clockwise deflection of the needle, it is imperative that the instrument terminal marked (+) be connected to the positive side of the current source, when used in dc circuits.

Illustrative Problem 28.13 The moving coil of an ammeter has a resistance of 5 Ω and the shunt a resistance of 0.05 Ω. Full-scale deflection of the meter results when 10 A flows through it. What is the actual current in the moving coil?

Solution Let

$$\begin{aligned} I_T &= \text{total current through ammeter} \\ \left.\begin{aligned} I_c &= \text{current in coil} \\ I_s &= \text{current in shunt} \end{aligned}\right\} &\text{ in parallel} \end{aligned}$$

Then

$$I_T = I_c + I_s = 10 \text{ A}$$

But, from Sec. 26.16

$$V_s = V_c$$

From Ohm's law

$$I_s R_s = I_c R_c$$

$$\frac{I_s}{I_c} = \frac{R_c}{R_s} = \frac{5\ \Omega}{0.05\ \Omega}$$

or

$$I_s = \frac{5}{0.05} I_c = 100 I_c$$

Substituting yields

$$I_c + 100 I_c = 10 \text{ A}$$

$$I_c = 0.099 \text{ A} \qquad \textit{answer}$$

28.26 ■ The Voltmeter

The galvanometer can also be used satisfactorily to measure voltages. For a given resistance, according to Ohm's law, the voltage is proportional to the current. The voltmeter is connected in *parallel*. Hence to measure voltage, we need only use a galvanometer of such high resistance that the instrument itself will draw a negligible amount of the current, not enough to affect appreciably the voltage drop between the points in the circuit to which it is connected (see Fig. 28.36). Multiple-range voltmeters are produced by providing the instrument with several coils of different resistances in series with the moving coil. The proper range is obtained by connecting across the correct terminals. With the voltmeter, as with the ammeter, it is important to avoid selecting a (voltage) range on the instrument below that which is to be measured. Also the positive (+) terminal of the voltmeter must be connected to the positive side of the line when used with dc circuits. The voltmeter must be connected *across* the load, not in series in the circuit.

28.27 ■ Alternating-Current Instruments

Although we do not discuss *alternating currents* in detail until Chap. 31, it should be made clear at this point that the instruments discussed thus far are used only in dc circuits (circuits in which the charges flow always in only one direction). For ac circuits the measuring instruments must be of different design.

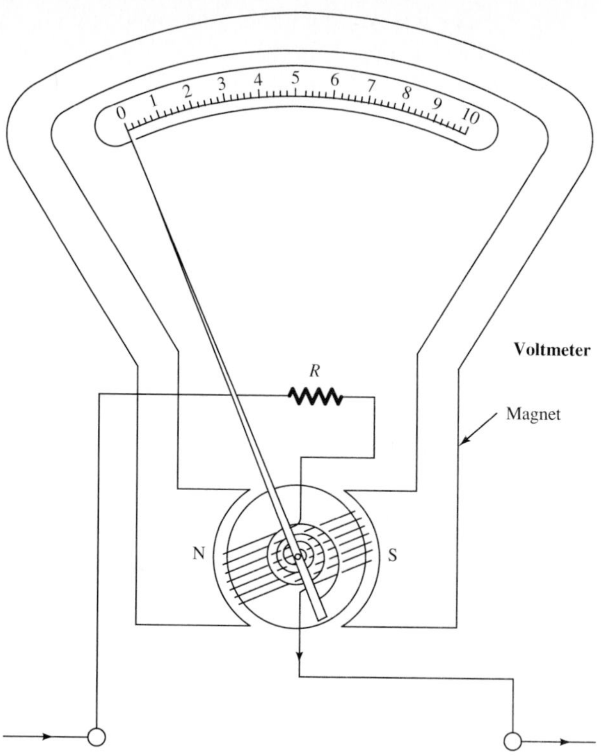

Fig. 28.36 Circuit of an ordinary voltmeter.

Alternating-current instruments for measurement of currents and voltages sometimes make use of the repulsion between two pieces of iron of like magnetic polarity. A simple demonstration will explain the principle involved. Support two pieces of soft iron side by side within a coil of wire (Fig. 28.37*a*). When current is allowed to pass through the coil, as shown in Fig. 28.37*b*, the two pieces of iron will acquire magnetic properties with north polarity at the tops and south polarity at the bottoms of the pieces. As a result the pieces will repel each other. The same result would be obtained if the direction of the current were reversed, even though the polarity, too, would be reversed. So, even though alternating current reverses its direction 120 times every second (60-cycle current), repulsion always takes place.

Thus if we use two soft-iron vanes (Fig. 28.38), a stationary one *A*, fixed at one side of the coil, and the other *B*, next to it and attached to a shaft which is mounted to turn at the center of the coil, a mutual repulsion between the two will result in a

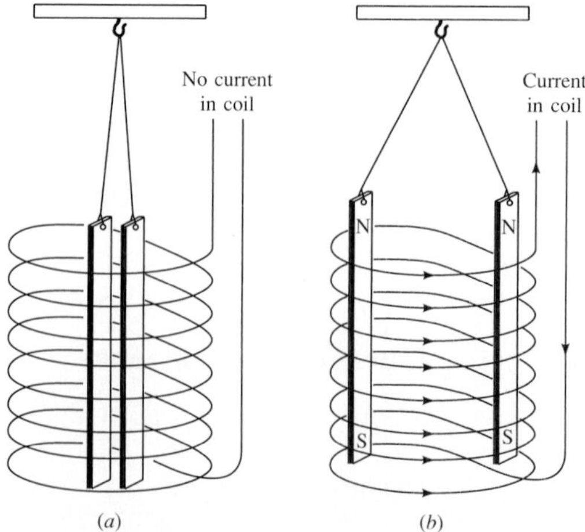

Fig. 28.37 (*a*) Two soft iron bars hung side by side within a coil when there is no current in the coil. (*b*) With current (in any direction) in the coil, the two pieces of iron will repel each other.

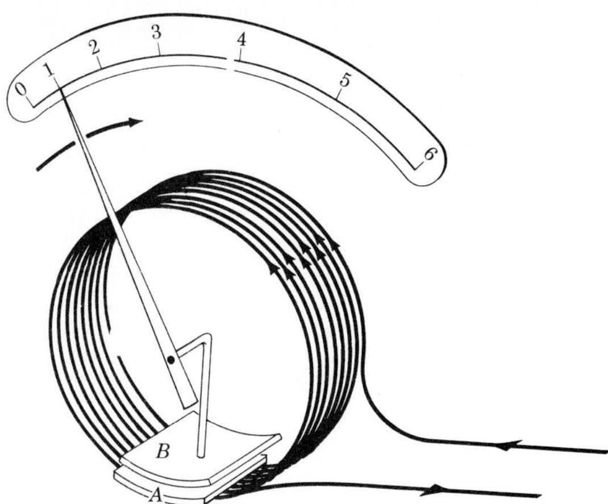

Fig. 28.38 Principle of moving-iron vane meter. When there is current in the coil, the two vanes *A* and *B* will repel each other.

rotation of the center shaft. When current flows in the coil, the two vanes will repel each other, causing the moving element to turn in a clockwise direction. The force of repulsion is proportional to the strength of the current and is restrained by a spiral spring (not shown) at the top of the moving element. By proper use of shunts in parallel or resistances in series, the instrument can be calibrated to read directly in amperes or in volts.

QUESTIONS AND EXERCISES

1. Distinguish between the magnetic field on the earth near a stationary electrostatic charge and the field near a charge that is moving.

2. One end of a surveyor's compass needle is made slightly heavier than the other. Why? Which should be made the heavier in the Northern Hemisphere? [HINT: Look in your library references for material on *magnetic dip*.]

3. List similarities and differences between magnetism and electrostatics.

4. Distinguish between a diamagnetic and paramagnetic substance.

5. The current in a conductor is directed westward. What is the direction of the lines of force of the magnetic field (*a*) above the conductor, (*b*) below the conductor?

6. Which of the following will serve best to shield a watch from a magnetic field: gold, lead, iron? Give reasons for your answer.

7. A method for magnetizing a bar of soft iron by tapping when it is placed parallel to the earth's magnetic field. In your locality at what angle from true north, clockwise or counterclockwise, should the rod be held while hitting it with the hammer? [HINT: Consult Fig. 28.12.]

8. Use the (modern) theory of magnetism to give an explanation of how a magnet will attract an unmagnetized piece of iron.

9. You are given a bar magnet. List at least two methods for determining which is the north pole of the magnet.

10. How can a galvanometer be made into an ammeter of a stipulated full-scale reading? into a voltmeter of a stipulated full-scale reading?

11. What precautions should be observed in using (*a*) an ammeter and (*b*) a voltmeter?

12. A stream of electrons is moving in a westerly direction. The stream passes through a uniform magnetic field directed downward. In what direction will the electrons be deflected?

13. Why is it impossible to magnetize an iron bar beyond a certain maximum flux density?

14. Show by a diagram how an ammeter can be wired to provide for two ranges of current.

15. Make a diagram to show how a voltmeter can be provided with two ranges.

16. What is the significance of the area of the hysteresis loop for a given ferromagnetic material?

17. If you were given three iron bars and were told that only two of them were magnets, how could you determine without using any additional objects which one is not a magnet?

18. Give a rule, using the fingers of the left hand, relating the direction of electron flow in a conductor, the direction of the magnetic field, and the direction of movement of the conductor when placed in the magnetic field.

19. A small compass rests on a table. A current-carrying wire is placed slightly above the compass and parallel to the needle. If the (conventional) current is directed toward the north, toward which direction will the south pole of the compass swing?

20. A beam of positively charged particles is moving horizontally toward an observer. As it approaches the observer, it passes through a magnetic field which is directed downward. In which direction is the beam deflected, if at all?

PROBLEMS

Group One

1. Two unlike magnetic poles 10 and 20 A · m units strong are placed 5 cm apart. Find the force of attraction between the two poles. (Assume $k = 10^{-7}$ N/A^2.)

2. What is the flux through an area of 575 cm^2 if a magnetic flux density of 0.25 Wb/m^2 is directed normal to the area?

3. The coil of an electromagnet has 2400 turns. If the resistance of the coil is 13.6 Ω, how many ampere-turns will it develop when it is connected to a 36-V dc source? (An ampere-turn is the product of the amperes and the number of turns in the coil.)

4. What is the flux density of the magnetic field at a distance of 10 cm from a straight wire carrying a current of 6 A?

5. Calculate the magnetic induction at a distance of 15 cm from a long straight wire carrying a current of 25 A.

6. A magnetic flux density of 12×10^{-3} Wb/m^2 is desired at the center of a solenoid 1.5 m long carrying a current of 2.5 A. How many turns should the solenoid have?

7. A field intensity of 300 A-turns/m produces a flux density in the iron core of a solenoid of 0.21 Wb/m^2. What is the permeability of the iron core?

8. What is the magnetic induction at a point 75 cm from a straight wire carrying a current of 5 A?

9. What is the current in each of two long parallel conductors 1 m apart if the current results in a mutually repulsive force of 8×10^{-7} N/m?

10. The current I in a conductor has a direction parallel to the X-axis in Fig. 28.39. Use the right-hand rule to determine the direction of **B** at points P, Q, R, S and T.

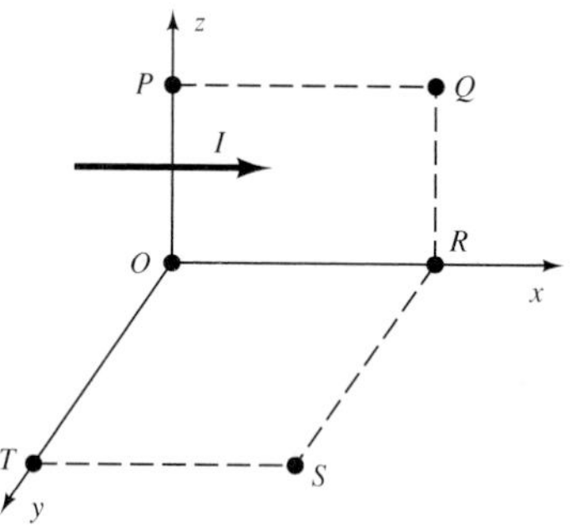

Fig. 28.39 Problem 10.

Group Two

11. A multiple loop coil of 50 turns has a diameter of 30 cm. What current must be in the coil to produce a flux density of 4×10^{-4}Wb/m^2 at its center?

12. The cord to an electric appliance is 2.0 m long and the two parallel wires are 3 mm apart. What is the force acting on the two wires when carrying a 5.0 A current?

13. A toroid has a cross-sectional area of 8 cm^2 and a magnetic flux density of 10^{-3} Wb/m^2. What is (*a*) the magnetic flux density and (*b*) the flux in the toroid if an iron core which increases the permeability 600-fold replaces the air core?

14. A circular coil of 16 turns and diameter 8 cm carries a current of 4 A. What is the torque on the coil when it is placed in a uniform magnetic field of flux density 12×10^{-4} T if its plane makes an angle of 45° with the magnetic flux?

15. The coil in Fig. 28.33*b* has a length L of 18 cm and a width w of 15 cm. It has 60 turns and carries a current of 3 A. Calculate the torque on the coil (*a*) if it is suspended with its plane parallel to the direction of a uniform magnetic field of flux density 0.38 Wb/m^2, and (*b*) if it is suspended so that the plane of the coil makes an angle of 60° with the direction of the field.

16. A 10 cm long conducting wire is located between two pole faces, each of which has a cross-sectional area of 10 cm by 10 cm. The flux between the two poles is 1.5×10^{-2} Wb. (*a*) What is the force in newtons acting on the wire when it is carrying 4.5 A and is perpendicular to the magnetic field? (*b*) What is the force acting on the wire when it is carrying 4.5 A and makes an angle of 45° with the field?

17. A solenoid is 20 in. long. It is wound with 5000 turns of copper wire. What current must flow through the windings to produce a magnetic field strength of 40 A/cm?

18. A current of 3 A is flowing through a long solenoid with 12 turns/cm. If the cross-sectional area of the solenoid is 4 cm^2, what is the magnetic flux density inside the solenoid? What is the total flux inside the solenoid?

19. A U-core electromagnet has two similar exciting coils. Each coil has 300 turns and a resistance of 8 Ω. What ampere-turns (the product of the amperage and the number of turns in the coil) are produced by the electromagnet if it is connected to a 32-V dc source and the coils are connected (*a*) in series; (*b*) in parallel?

20. What is the flux density of the magnetic field at the center of a circular coil of 75 turns having a radius of 8 cm if the current in the coil is 0.5 A?

21. A wire 50 cm long carries a current of 25 A at an angle of 30° with a magnetic field whose flux density is 7.5×10^{-4} Wb/m^2. What is the magnitude of the force on the wire?

22. Find the magnetic flux density inside a solenoid 38 cm long with 300 turns when it is carrying a current of 2 A. What would be the flux density in a core of soft iron ($\mu = 1.6 \times 10^{-3}$ Wb/A · m) inserted in the solenoid?

23. The cables connecting an auto battery with the starter motor carry 250 A. What would be the force between the cables if they are 50 cm long and 1 cm apart?

24. How many turns should a solenoid 1.5 m long have for a current of 2 A to produce a magnetic flux density of 8×10^{-3} Wb/m^2 at its center?

25. What force per meter do two parallel conductors 15 cm apart exert upon each other if one carries 2.4 A and the other carries 3.6 A of current?

26. The resistance of a milliammeter (range is 0-1 mA) is 18 Ω. What resistance must be added to convert it into a voltmeter with a 0 to 1.0 V range?

27. The resistance of a milliammeter is 20 Ω. What resistance must be shunted across the instrument to convert it to an ammeter, range 0 to 1.0 A?

28. A solenoid 30 cm long with 1000 turns carries a current of 2.5 A. (*a*) What is the magnitude of the flux density in tesla at its center? (*b*) What would be the magnetic flux density if a soft iron core were inserted in the solenoid if the iron has a permeability of 2.3×10^{-4} Wb/A · m? (*c*) How many times as large is the permeability of the iron as that for a vacuum? This ratio is termed *relative permeability*.

29. A solenoid 45 cm long, with a cross-sectional diameter of 3.6 cm is wound with 350 turns and carries a current of 1.5 A. The permeability of the core is 7.5×10^{-4} Wb/A · m. Compute (*a*) the magnetic field intensity (magnetic field strength) H; (*b*) the magnetic flux density (magnetic induction) B; and (*c*) the flux in the core.

30. Two long parallel wires 10 cm apart, carry 3 A and 4 A in opposite directions. Find the points where **B** will be zero. Consider only the magnetic flux produced by the two wires.

Group Three

31. A bar magnet has poles each of strength 75 A · m that are separated by a distance of 10 cm. Find the resultant force acting on a magnetic pole of 3 A · m strength placed on the axis of the magnet 5 cm from the north pole and 15 cm from the south pole.

32. Two identical magnets are 8 cm long and have pole strengths of 75 A · m. One magnet is placed on a table and the other is placed so that its north and south poles are directly above the north and south poles of the first magnet. The second magnet then "floats" 6 cm above the first magnet. What is the weight of each magnet (in newtons)?

33. The moving coil of a galvanometer has a resistance of 50 Ω and requires 0.0080 A to produce a full-scale deflection. What resistance must be added in series with the coil to convert the galvanometer into a voltmeter capable of reading 10 V at full-scale deflection? Find the shunt resistance to be added in series to give the meter a 15-V range.

34. The moving coil of a galvanometer requires 0.0045 A to produce a full-scale deflection. The coil has a resistance of 50 Ω. What must be the resistance shunted across the moving coil to convert the galvanometer into an ammeter reading 1 A at full-scale deflection? Find the shunt resistance to give the meter a 5-A range.

35. Two long wires carry currents of $I_1 = 5$ A and $I_2 = 3$ A at right angles to each other (say along the *X*-axis and *Y*-axis as shown in Fig. 28.40). What is the magnitude and direction of the magnetic induction at point *P* 10 cm from 0 (along the Z-axis)?

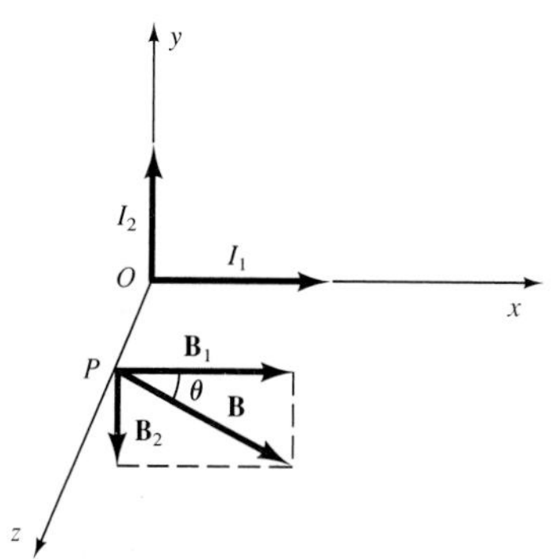

Fig. 28.40 Problem 35.

ELECTROMAGNETIC INDUCTION

Michael Faraday in England and Joseph Henry (1797–1878), an American scientist, working independently, discovered in 1831 that an emf is set up in a conductor that moves through a magnetic field or in a conductor that is cut by a moving magnetic field.

The findings of Oersted, Faraday, and Henry form the basis for the development of the entire electrical industry. The limitations inherent in the development of electric power by chemical means were overcome. With the discoveries of Faraday and Henry it became possible to convert mechanical energy from such sources as falling water, steam, nuclear reactions, petroleum, and fossil fuels to produce mechanical energy which in turn can be converted into electric energy. Electric energy today is transmitted with little loss over large distances to create light and heat, to drive motors, and to accomplish many functions in the home as well as in business and industry.

29.1 ■ Electromagnetic Induction

Joseph Henry pioneered in the field of electromagnetic induction. He discovered the oscillatory nature of capacitor discharges. He also developed methods for weather forecasting.

The experiments which led to Faraday's and Henry's discoveries can readily be repeated by the student in the laboratory. Consider the equipment arrangement depicted in Fig. 29.1. The ends of a wire loop are connected to a sensitive galvanometer. When a strong horseshoe magnet is moved down quickly over the wire (Fig. 29.1*a*) there will be a momentary deflection of the galvanometer, indicating the flow of an electric current in the conductor. The current is said to be an *induced current* and the process by which the current is produced is called *electromagnetic induction*. The galvanometer needle will show no deflection when there is no relative motion between the magnet and the conductor. Thus, it can be established that *the induced current is related not merely to the presence of the magnet, but to the motion of the magnet*.

If the magnet is next moved up from the wire (Fig. 29.1*b*), there will be another momentary deflection of the galvanometer. However, this time the needle will be deflected in the opposite direction, indicating a current whose direction is reversed with respect to that of the first current.

If the experiment is repeated, but with the polarity reversed (Fig. 29.1*c,d*), the directions of the current will also be reversed.

Thus, if the magnet is moved up and down past the wire, the induced current will alternate in direction with each change in direction of motion. Such a current is called an *alternating current* (ac).

The effects just described are also obtained if the magnet is held stationary and the wire is moved up and down so that it cuts across the flux of the magnet. *Relative motion* between magnetic field and conductor is necessary to initiate and sustain an induced current.

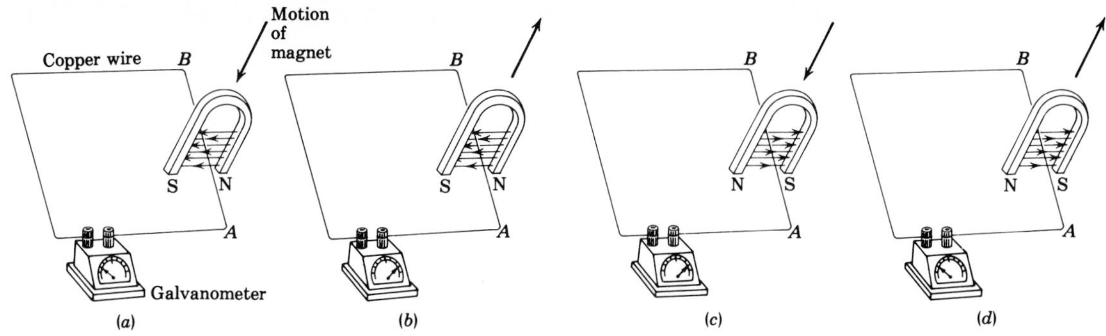

Fig. 29.1 Electromagnetic induction. A current is induced in the conduction loop whenever magnetic lines of force cut across it.

29.2 ■ Induced EMF

Relative motion between a conductor and a magnetic flux linked by the conductor will result in an induced emf in the conductor.

To make a current flow in a circuit there must be a source of energy or an emf. In Chap. 27, we studied how emfs are produced by chemical means in voltaic cells, storage batteries, and electrochemical cells. Attention was also given to thermoelectric and photoelectric sources of emf. With Faraday's and Henry's discoveries yet another, and perhaps the most important, means for producing emfs became available. By introducing a relative motion between a conductor and a magnetic field, which results in a change of magnetic flux linked by the conductor, an *induced emf* is produced in the conductor. The greater the change of magnetic flux or the faster the *relative motion of the wire and the magnet,* the greater the induced emf. In Sec. 29.5 we will discuss why emfs are induced in the circuit.

29.3 ■ Direction of Induced Current—Fleming's Rule

In Fig. 29.2, let a single straight wire *AB* be moved as shown in a magnetic field. A complete *electric circuit* is provided by connecting the ends of the moving wire to a sensitive galvanometer. If the wire is moved upward through the field, the current (in accordance with the positive-to-negative convention) will be from *A* to *B* and around the circuit in the direction shown by the arrows. If the wire is thrust downward through the field, the galvanometer will show a reversal of the current direction. Or, if the poles of the magnet are reversed, the current direction will be reversed.

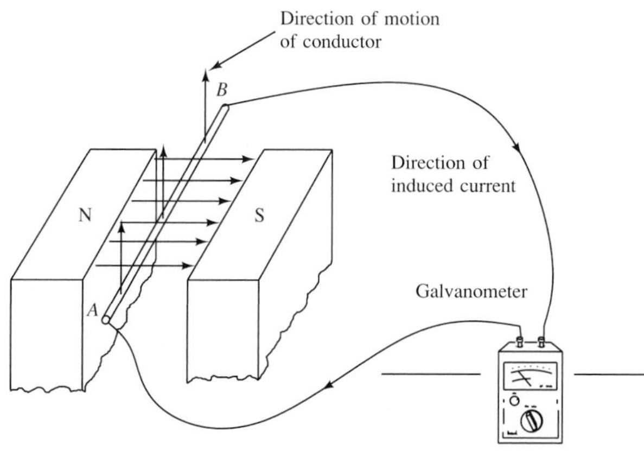

Fig. 29.2 Current is induced in a conductor whenever it cuts across magnetic lines of force.

If the moving wire segment AB is not part of a closed circuit, a negative charge will pile up at one of its ends. An equal positive charge will appear at the other end. The difference of potential $\mathscr{E}$ between the two ends will be similar to that between the terminals of a battery on open circuit.

The direction of the induced emf, then, depends on (1) the direction of motion of the conductor and (2) the direction of the magnetic field. *Fleming's right-hand rule* relates these three directions, and is used to predict the direction of magnetically induced electric currents.

Sir John Ambrose Fleming (1849–1945) was an English physicist and electrical engineer. He was active in the development of the telephone and in electric lighting.

Fleming's Rule

Position the thumb, forefinger, and middle finger of the right hand so that they are all at right angles to each other (Fig. 29.3). Let the thumb point in the direction of motion of the conductor, and the forefinger in the direction of the magnetic field (N to S). Then the middle finger will indicate the direction of the current induced in the conductor.

29.4 ■ Strength of Induced EMF

Induced emf equals the change in magnetic flux per second cutting across the conductor.

Faraday found experimentally that the magnitude of the emf $\mathscr{E}$ induced by a conductor moving at right angles to the lines of force of a magnetic field (or by the lines of force moving at right angles to the conductor) is proportional to the number of lines of force cut per second. In other words, $\mathscr{E}$ depends on the rate of change of magnetic flux. The *average* emf $\overline{\mathscr{E}}$ is given by

$$\overline{\mathscr{E}} = -\frac{\Delta\phi}{\Delta t} \tag{29.1}$$

Faraday's law gives the induced emf no matter whether the change in ϕ is produced by moving a magnet, moving a conductor, moving a coil, changing the strength of a magnetic field, changing the shape of a conducting loop, or in other ways.

Faraday's law of induction

The electromotive force induced in a conductor is equal to the rate of change of magnetic flux through the conductor.

In the SI system $\overline{\mathscr{E}}$ is measured in volts when $\Delta\phi$ is expressed in webers, and Δt in seconds. Thus, a flux change of 1 Wb/s will induce an emf of 1 V in the conductor. Equation (29.1) expresses what is known as *Faraday's law*. The reason for the negative sign will be explained in the next section.

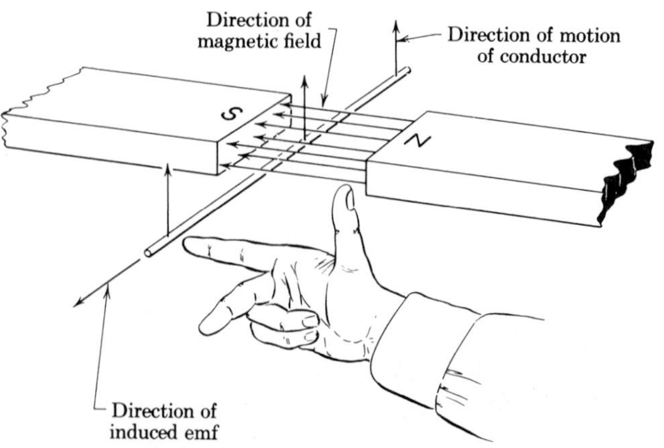

Fig. 29.3 Fleming's right-hand rule for determining the direction of induced emf.

Consider a U-shaped conductor *EGHF* with a second straight conductor *CD* moving with velocity v across a uniform magnetic field B, as shown in Fig. 29.4. When the straight conductor is in position *CD*, the total flux ϕ_1 linking the closed circuit *GHDC* is BA_1, where A_1 is the cross-sectional area of the magnetic field enclosed by *GHDC*. Assume that in time Δt the conductor reaches position *EF* and the total flux ϕ_2 linking circuit *EGHF* has increased to BA_2 enclosed by *EGHF*. The change in flux $\Delta\phi$ is equal to $\phi_2 - \phi_1$ or $B(A_2 - A_1)$. But $A_2 - A_1 = Lw$. Hence, for the single loop, the *average* emf $\overline{\mathscr{E}}$ will be

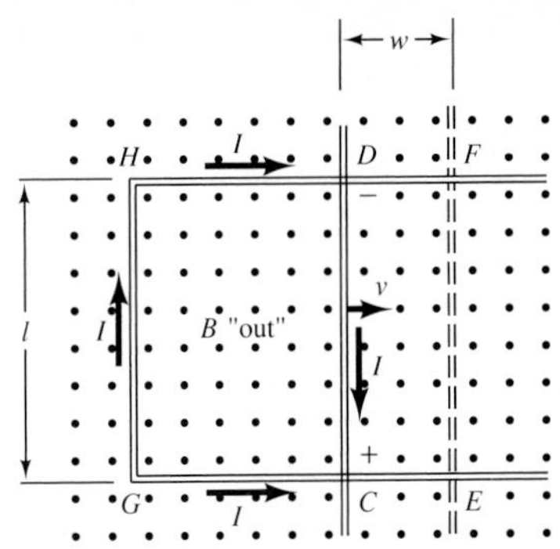

Fig. 29.4 EMF is induced in conductor *CD* moving across a magnetic field **B** (directed out of page).

$$\overline{\mathscr{E}} = -\frac{\Delta\phi}{\Delta t} = -\frac{BLw}{\Delta t} = -BL\frac{w}{\Delta t}$$

$$\overline{\mathscr{E}} = -BLv \qquad (29.2)$$

where B is in webers per square meter, L is in meters, and v is in meters per second. If a length L of the conductor is perpendicular to the velocity v and to the magnetic field, and the velocity is inclined at an angle θ (Greek theta) with respect to the magnetic field B, the induced emf $\overline{\mathscr{E}}$ is found by the equation

$$\overline{\mathscr{E}} = -BLv \sin\theta \qquad (29.2')$$

From Eq. (29.2′), it should be clear that an emf can be induced by (1) changing the magnetic field or (2) by changing the orientation of the loop with respect to the field. Methods for changing the orientation of loops in a generator to produce an emf at the terminals of the loop will be discussed in Chap. 30.

Equations (29.1) and (29.2) give the average emf induced by the relative motion of the conductor and the magnetic flux. Frequently, scientists are more concerned with the instantaneous value for the induced emf, which can be obtained by taking the rate of change of the magnetic flux over a very small time interval. As a formula, the instantaneous emf e is determined by

$$e = -\lim_{\Delta t \to 0} \frac{\Delta\phi}{\Delta t} \qquad (29.3)$$

Equation (29.3) is read "the instantaneous emf is equal to the negative of the limiting value of $\Delta\phi/\Delta t$ as Δt approaches zero." Hereafter the lower-case letters e and i will be used to represent the values of *instantaneous emf* and *instantaneous current*.

The induced emf depends upon the number of lines cut per second. Thus, to increase the emf induced to an outside circuit, a coil of many loops of wire can be used, the magnetic field can be made stronger, or the coil can be turned faster. An equal change of flux occurs in each loop of the coil, inducing the same emf in each loop of the coil. Since the loops are in series with each other, the total emf between the ends of the coil is the sum of the emfs for each loop. Thus, if there are N turns in the circuit, Eqs. (29.1) and (29.2) become

$$\overline{\mathscr{E}} = -NBLv \qquad (29.4)$$

and

$$\overline{\mathscr{E}} = -N\frac{\Delta\phi}{\Delta t} \qquad (29.5)$$

Illustrative Problem 29.1 A wire 60 cm long moves across a uniform magnetic field in a direction that makes an angle of 30° with respect to the field. The magnetic induction (magnetic flux density, B) of the field is 2.4×10^{-2} Wb/m². The velocity of the conductor is 5 m/s. (*a*) What is the average emf induced in the wire? (*b*) If the wire is connected to a circuit whose resistance is 0.3 Ω, what current flows through it?

Solution

$L = 0.60$ m $\qquad B = 2.4 \times 10^{-2}$ Wb/m²
$v = 5$ m/s $\qquad \theta = 30°$

No emf occurs from conductor motion parallel to a magnetic field.

(*a*)

$$\begin{aligned}\overline{\mathscr{E}} &= BLv \sin \theta \\ &= 2.4 \times 10^{-2} \text{ Wb/m}^2 \times 0.60 \text{ m} \times 5 \text{ m/s} \times \tfrac{1}{2} \\ &= 3.6 \times 10^{-2} \text{ Wb/s} \\ &= 36 \text{ mV}\end{aligned}$$

(*b*)

$$V = 0.036 \text{ V} \qquad R = 0.3\ \Omega$$

$$I = \frac{V}{R} = \frac{0.036 \text{ V}}{0.3\ \Omega}$$

$$= 0.12 \text{ A}$$ *answer*

Illustrative Problem 29.2 A small test coil has 250 loops, each with a diameter of 5.0 cm. It is placed in a uniform 0.38-T magnetic field with its plane perpendicular to the field. The coil is quickly pulled out of the field in 0.12 s. (*a*) What is the average emf developed? (*b*) How much energy is expended in removing the coil if the resistance of the coil is 100 Ω?

Solution

(***a***) The area of the coil is

$$A = \frac{\pi}{4}(0.050 \text{ m})^2 = 1.96 \times 10^{-3}\text{m}^2$$

The initial flux in the coil is

$$\begin{aligned}\phi &= BA = 0.38 \text{ T } (1.96 \times 10^{-3}\text{m}^2) \\ &= 7.45 \times 10^{-4} \text{ Wb}\end{aligned}$$

Using Eq. (29.4),

$$\overline{\mathscr{E}} = -N\frac{\Delta\phi}{\Delta t}$$

$$= -250\left(\frac{0 - 7.45 \times 10^{-4} \text{ Wb}}{0.12 \text{ s}}\right)$$

$$= 1.6 \text{ V}$$ *answer*

(***b***) The current through the coil is

$$I = \frac{\overline{\mathscr{E}}}{R} = \frac{1.6 \text{ V}}{100\ \Omega} = 1.6 \times 10^{-2} \text{ A}$$

The energy dissipated in the coil is

$$\begin{aligned}I^2Rt &= (1.6 \times 10^{-2} \text{ A})^2(100\ \Omega)(0.12 \text{ s}) \\ &= 3.1 \times 10^{-3} \text{ J}\end{aligned}$$

Thus, from the principle of the conservation of energy, the work needed to move the coil out of the field is 3.1×10^{-3} J. *answer*

29.5 ■ Lenz's Law and the Direction of Induced EMF

The German physicist Heinrich Lenz (1804–1865) conducted extensive experiments on induced electromotive forces which finally led to certain generalizations now ascribed to him.

Lenz's Law

Electromagnetically induced currents always have such a direction that the magnetic field set up by the induced currents tends to oppose the change in flux which produces them.

This law can be illustrated by using a simple diagram (Fig. 29.5). Figure 29.5*a* shows a wire located in a magnetic field. When the wire is pushed down through the field, the current induced in the conductor (see Sec. 28.21) will be directed into the page (represented by the tail of an arrow). The magnetic field set up by this current is shown in Fig. 29.5*b*. We see the two fields superimposed and interacting in Fig. 29.5*c*; the effect is the production of a stronger field below the wire than above it. We can conceive of the lines of force acting as taut rubber bands tending to straighten out with a resultant push upward on the wire. Work must be done against these forces to push the conductor downward through the magnetic field.

An induced current always produces a magnetic field which opposes the field causing the motion.

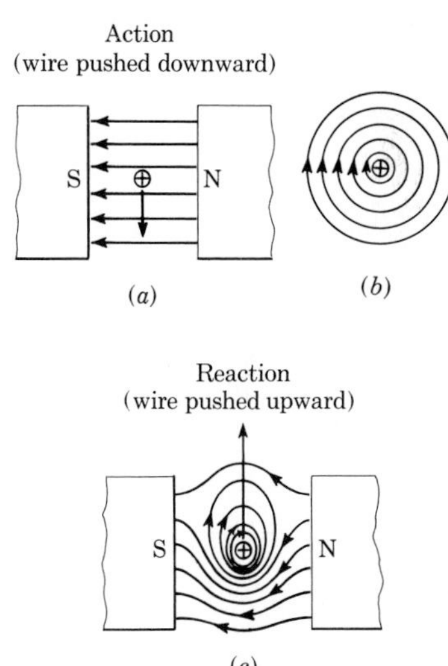

Fig. 29.5 When the wire in (*a*) is moved across a magnetic field an emf is induced in the conductor. (*b*) If the wire is made part of a closed circuit, a current will be induced in the wire. The magnetic field of the induced current will oppose the motion of the conductor, as shown in (*c*).

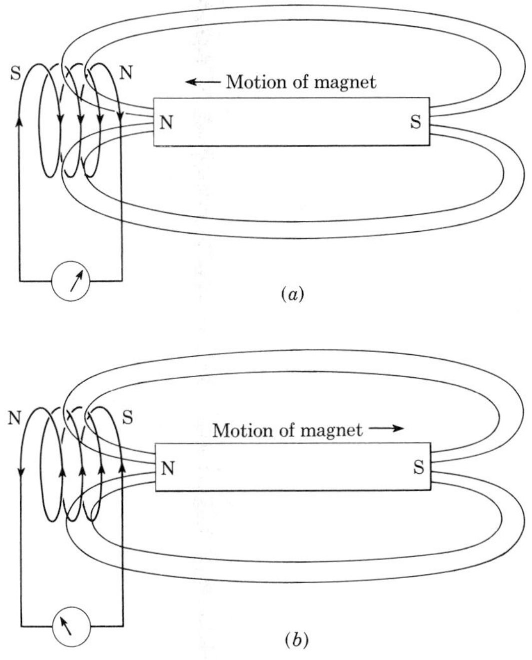

If a permanent bar magnet is thrust into a coil of wire, an emf is induced in the coil. If the coil is connected across a closed circuit, current will flow. According to Lenz's law, the direction of the current in the coil will be such as to establish a polarity which will oppose the direction of motion of the bar magnet. Thus in Fig. 29.6, if the N-pole of a magnet is moved toward the coil, the induced current in the coil will produce north polarity at the face nearest the magnet, resulting in a force of repulsion between magnet and coil. If the magnet is moved away from the coil, the polarity of the magnetic field of the coil will be reversed, and attraction between magnet and coil results. In either case, external work must be done to cause a current to flow in the coil. The negative signs of Eqs. (29.2) and (29.5) express the fact that the induced emf is of such direction as to oppose the change which produced it.

If the sign of $\overline{\mathscr{E}}$ and $\Delta\phi/\Delta t$ in Eq. 29.5 were the same, the induced electric current would be in such a direction that its own magnetic flux would be added to the external

Fig. 29.6 Induced emf in coil (*a*) when magnet moves toward coil; (*b*) when magnet moves away from coil.

flux **B** that produced it. If this were true, the additional changing magnetic field would increase the existing rate and more current would result. This would increase the value of **B**, producing still more current, and so on. The result would be an ever increasing current. We know that this is contrary to the law of *conservation of energy*.

Lenz's law is an example of the principle of the conservation of energy.

Energy must be expended to produce electric current by induction. The mechanical energy expended in moving the magnet against the forces of repulsion and attraction set up by the interacting magnetic fields of the permanent magnet and of the induced current is converted to the electric energy of the current. No force is required to move the magnet near the coil if the circuit is open, for no magnetic field is established to oppose the motion when no current exists. Lenz's law can be looked upon, then, as stating the law of conservation of energy for electromagnetic induction.

The work done in moving a conductor in a magnetic field is equal to the electrical energy required to cause the induced current.

In many electric generators (Chap. 30) emf is induced by moving conductors through a magnetic field at high speeds. In all electric generators, the direction of the induced emf is such that the resulting current through the conductors sets up a magnetic field which opposes the movement of the conductors across the original magnetic field. This opposition is overcome by the mechanical energy supplied by the hydraulic or steam turbine or other driving engine.

29.6 ■ Mutual Induction

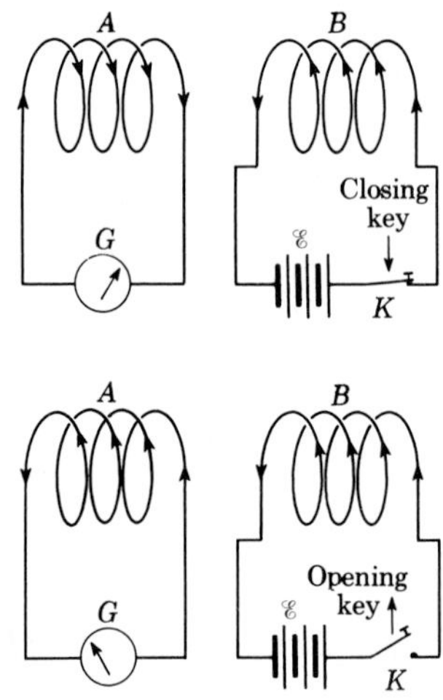

Fig. 29.7 A change in the amount of current in coil *B* will induce an emf in coil *A*.

We have seen how an emf can be induced either by moving magnetic lines of force past a conductor or by moving the conductor so it cuts lines of force. Electromotive forces can also be induced in a conductor, without moving either the magnet or the conductor, by *varying the strength or direction* of a magnetic field in which a conductor is placed. For example, we might replace the bar magnet of Fig. 29.6 by a second coil as shown in Fig. 29.7. When the key *K* is closed, a current builds in coil *B*, called the *primary coil*. The increasing current will produce an increasing magnetic flux density around the coil. When no current is in the primary (*B*) coil, there will be no lines of force around it, but as the current builds up, flux lines originate in the coil and continue to grow outward by expanding into larger and larger loops. The extent and intensity of this magnetic flux depend upon the current and the number of turns (*ampere-turns*) of the coil. While the current is building up, the expanding magnetic lines cut across the nearby loops of the coil *A*, called the *secondary coil*, thereby inducing an emf in the secondary which is indicated by the deflection of the galvanometer *G*.

An emf is induced in any circuit in which the magnetic flux is changing.

After the current in the primary has reached its maximum value, the flux in it will be steady; hence no emf will then be induced in the secondary. However, when the key is opened, the magnetic field of the primary will collapse until once more no lines are present in the coil. As these lines shrink inward, they cut the coils of the secondary again but in an opposite direction, inducing a reverse emf in the secondary. This will be evident by a deflection of the galvanometer opposite to that which occurred when the key was closed. The development of an induced emf in one circuit by the change of current in another is called *mutual induction*.

The instantaneous emf induced in the secondary coil is directly proportional to the rate of change of current in the primary coil, or, as an equation,

$$e_s = -M\frac{\Delta i_p}{\Delta t} \tag{29.6}$$

where M is a constant called the *mutual induction* of the system, and e and i represent *instantaneous values*.

The mutual induction M is defined as the ratio of the emf induced in the secondary to the time rate of change of the current in the primary.

$$M = \frac{-e_s}{\Delta i_p/\Delta t} \tag{29.7}$$

Here, again, the negative sign is used to show that the induced emf is in such a direction as to oppose the change of current in the primary.

Mutual inductance is expressed in henrys (H) when e_s is in volts and $\Delta i_p/\Delta t$ is in amperes per second. Thus a pair of adjacent circuits is said to have a mutual inductance of *one henry* if a time rate of change of current of one ampere per second in the primary induces an emf of one volt in the secondary. Where the henry is too large a unit, the millihenry (mH) or the microhenry (μH) are used: 1 mH is equal to 10^{-3} H; 1 μH is equal to 10^{-6} H.

1 henry = 1 volt per ampere per second, or 1 volt-second per ampere.

$$1 \text{ H} = \frac{1 \text{ V}}{1 \text{ A/s}} = \frac{1 \text{ V} \cdot \text{s}}{\text{A}}$$

Illustrative Problem 29.3 A magnetic flux density of 0.3 T flows through a coil having 400 turns and an area of 20 cm². If in $\frac{1}{100}$ s the flux density is reduced to 0.2 T, what is the average emf induced in the coil?

Solution

$$N = 400 \qquad \Delta B = 0.1 \text{ T}$$
$$A = 20 \text{ cm}^2 \qquad \Delta t = \tfrac{1}{100} \text{ s}$$

$$\bar{\mathscr{E}} = N\frac{\Delta\phi}{\Delta t} = N\frac{\Delta B \times A}{\Delta t}$$
$$= 400 \times 0.1 \text{ T} \times 20 \text{ cm}^2 \times \frac{1 \text{ m}^2}{10{,}000 \text{ cm}^2} \times \frac{1}{\frac{1}{100} \text{ s}}$$
$$= 8 \text{ Wb/s} = 8 \text{ V} \qquad \textit{answer}$$

Illustrative Problem 29.4 Determine the mutual inductance of a pair of adjacent coils if an average emf of 115 V is induced in the secondary by a current in the primary that changes at the rate of 25.0 A/s.

Solution

$$e_s = 115 \text{ V} \qquad \frac{\Delta i_p}{\Delta t} = 25.0 \text{ A/s}$$

Substitute the known values in Eq. (29.7):

$$M = \frac{e_s}{\Delta i_p/\Delta t} = \frac{115 \text{ V}}{25.0 \text{ A/s}} = 4.6 \text{ H} \qquad \textit{answer}$$

Illustrative Problem 29.5 The mutual inductance of a pair of adjacent coils is 2.5 H. The current in the primary coil changes from 0 to 30 A in 0.045 s. (*a*) What is the average emf induced in the secondary coil? (*b*) What is the change of flux in the secondary coil if it has 750 turns?

Solution

$$M = 2.5 \text{ H} \qquad \frac{\Delta i_p}{\Delta t} = \frac{30 \text{ A}}{0.045 \text{ s}} \qquad N_s = 750$$

(*a*)
Use Eq. (29.6) to find $\bar{\mathscr{E}}_s$.

$$\bar{\mathscr{E}}_s = M\frac{\Delta i_p}{\Delta t} = 2.5 \text{ H}\left(\frac{30 \text{ A}}{0.045 \text{ s}}\right) = 1670 \text{ V} \qquad \textit{answer}$$

(**b**) Now substitute this value for $\overline{\mathscr{E}}_s$ in Eq. (29.5)

$$\overline{\mathscr{E}}_s = N_s\frac{\Delta\phi}{\Delta t} = 750\left(\frac{\Delta\phi}{0.045\ \text{s}}\right) = 1670\ \text{V}$$

Then

$$\Delta\phi = \frac{1670\ \text{V}(0.045\ \text{s})}{750} = 0.100\ \text{Wb} \qquad \textit{answer}$$

An alternate expression for mutual inductance can be derived from Eq. (29.5) and Eq. (29.6).

$$e_s = -N_s\left(\frac{\Delta\phi}{\Delta t}\right)$$

$$e_s = -M\left(\frac{\Delta i_p}{\Delta t}\right)$$

The operation of a transformer (detailed discussion in Chap. 31) depends on mutual inductance.

Equating the expressions for e_s and canceling Δt's

$$N_s\ \Delta\phi = M\ \Delta i_p$$

or

$$M = \frac{N_s\ \Delta\phi_{sp}}{\Delta i_p} \tag{29.8}$$

and

$$M = \frac{N_p\ \Delta\phi_{ps}}{\Delta i_s} \tag{29.9}$$

where the subscripts p and s refer to primary and secondary. The mutual inductance will be large when the circuits are such that a large part of the flux in either circuit links with the other circuit, as for example when both coils are wound on the same core. Mutual inductances from Eqs. (29.8) and (29.9) are usually expressed in weber-turns per ampere.

Illustrative Problem 29.6 A mutual inductance coil consists of a long solenoid 25 cm long with a 5 cm^2 cross-sectional area and 600 turns, and a smaller coil 10 cm long, with the same cross-sectional area and 50 turns, all wound on a cardboard cylinder. Compute the mutual inductance of the two circuits.

Solution Either solenoid could be considered the primary. Let us call the longer solenoid the primary. A current i_p in it sets up a field at its center (Eq. 28.7)

$$B = \mu_0\frac{N_p i_p}{L}$$

From Eq. (28.2) and Eq. (28.7), we get the flux through the central section of the primary.

$$\phi_p = BA = \mu_0\frac{AN_p i_p}{L}$$

All this flux links with the secondary coil, $\Delta\phi_{ps} = \Delta\phi_{sp}$. Therefore, from Eq. (29.8),

$$M = \frac{N_s\ \Delta\phi_{sp}}{\Delta i_p} = \frac{N_s\left(\dfrac{\mu_0 AN_p\ \Delta i_p}{L}\right)}{\Delta i_p}$$

$$= \mu_0\frac{AN_pN_s}{L}$$

Substituting known values,

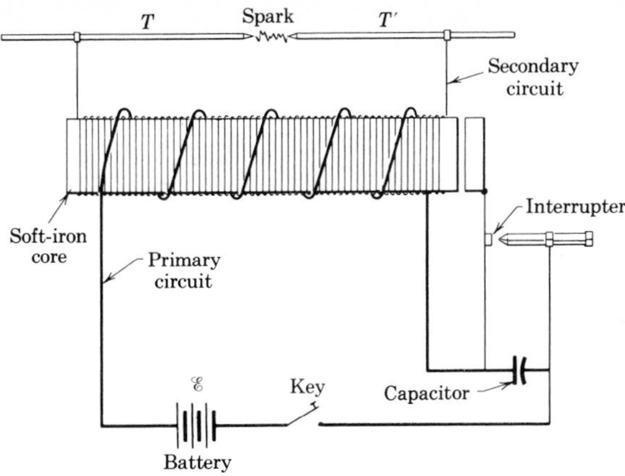

Fig. 29.8 Wiring diagram of an induction coil. A high voltage is induced in the secondary winding every time the primary circuit is completed or broken.

$$M = 4\pi \times 10^{-7}\frac{\text{Wb}}{\text{A}\cdot\text{m}} \times \frac{5\ \text{cm}^2\left(\frac{1\ \text{m}}{100\ \text{cm}}\right)^2 \times 600 \times 50}{1\ \text{m}}$$

$$= 18.8 \times 10^{-6}\ \text{Wb/A}$$
$$= 188 \times 10^{-6}\text{H}$$
$$= 188\ \mu\text{H} \qquad \textit{answer}$$

29.7 ■ The Induction Coil

Frequently a high voltage is desired when only a source of low voltage is available. One device for obtaining high voltages is the *induction coil* (see Fig. 29.8). It consists of a primary coil of a relatively few turns of heavy insulated wire and a secondary coil of many thousands of turns of insulated fine copper wire, both coils being wound around a cylindrical core of a bundle of soft-iron wires. The primary is connected through an automatic interrupter, similar to that of an electric bell or buzzer. When the switch to the *primary* circuit is closed and opened, current alternately builds up and diminishes in the coil, with a frequency controlled by the interrupter. This establishes a changing flux through and about the core. This rapidly changing magnetic flux cuts the many turns of the secondary, inducing a high emf across the terminals T and T'. Even in a small coil this emf is high enough to cause a spark of an inch or more to jump across the terminals of the secondary.

When the current in the primary is interrupted, the current tends to continue to flow just as if it possessed inertia. The result is a *jumping* of the current across the interrupter gap even after it has opened slightly. This, in time will pit and burn the contact points of the interrupter. A capacitor is usually connected across the interrupter gap to diminish the arcing at "make" and "break." Then, as the contact points are separated, the electric energy between them will charge up the capacitor, rather than produce a spark across the interrupter.

It should be made clear that the induction coil provides a high *voltage* at the sacrifice of *current*. Actually, the power (watts) which can be taken from the secondary is always a little less than that supplied to the primary, owing to losses within the coils.

29.8 ■ Induction Coils in Automobile Ignition Systems

An important practical application of the induction coil is the production of sparks in automobile-engine cylinders for ignition of the gasoline vapors. A potential differ-

Fig. 29.9 Conventional ignition system of a six-cylinder gasoline engine. (General Motors Corp.)

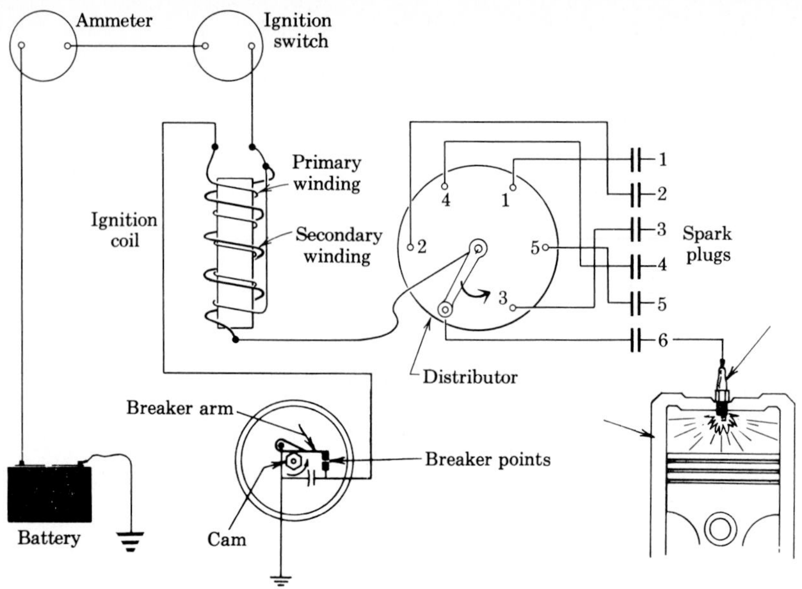

ence of several thousand volts is required between the terminals of the *spark plug* to ignite the vapors; the higher the cylinder pressure, the greater the voltage required. Figure 29.9 illustrates one type of ignition system still used for some six-cylinder engines. The primary (low-voltage) circuit includes the storage battery, the primary of the induction coil, the circuit-breaker mechanism (points), and the capacitor. It is completed through the frame of the automobile (ground). The secondary (high-voltage) circuit includes the secondary of the induction coil, the distributor, and the spark plugs of the engine. A cam in the circuit-breaker mechanism causes the making and breaking of the circuit in the primary of the induction coil, which in turn causes high induced emfs in the secondary circuit. The breaker is timed to break the circuit at just the instant a spark is needed in a cylinder. The distributor makes connection between the secondary of the induction coil and the proper cylinder. It is important that the ignition system be adjusted to give a spark at just the proper instant. Each automobile ignition system is provided with some means for advancing or retarding the spark to afford the most effective engine performance. Various automatic-control mechanisms are used to give the desired spark advance for varying engine speeds.

Electronic Ignition Electronic ignition systems (EIS) are now being used on most passenger cars and many trucks. The function of the electronic system is the same as that of the older system illustrated in Fig. 29.9—to produce a high-voltage spark and to distribute it at the proper time to the proper spark plug to fire the plug. But most of the newer ignition systems eliminate the conventional breaker points and the capacitor (condenser). These components are replaced by a magnetic pick-up coil and pulse-distributor assembly inside the distributor. The magnetic-pulse distributor eliminates the problem of pitting and burning of the contact points in the conventional interrupter.

In the EIS (electronic ignition) system, a transmitter is used as a solid-state switch in place of the breaker points. The signal from the electronic module is suitably amplified by information such as engine speed and atmospheric pressure.

The newer ignition system is a magnetic-pulse triggered, transistor-controlled, inductive discharge system. The magnetic pick-up assembly located inside the distributor contains a permanent magnet, a pole piece with internal teeth, and a pick-up coil. There are as many teeth in the pole piece as there are cylinders in the engine. When the teeth of the timer core, rotating inside the pole piece, line up with the teeth of the pole piece, an induced voltage in the pick-up coils signals an electronic device, called a module, to trigger the coil primary circuit. As the primary current decreases, a high voltage is induced in the ignition secondary coil which is directed through the rotor and secondary leads to fire the spark plugs (see Fig. 29.10). There is a capacitor

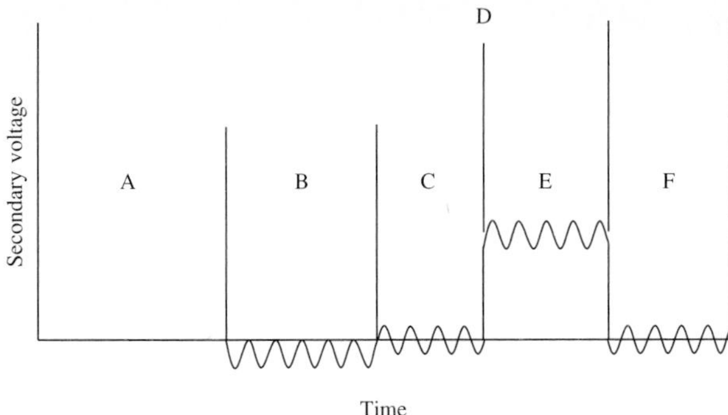

Fig. 29.10 A typical secondary voltage pattern for a good EIS. (*A*) No current flows in the secondary coil. (*B*) As the core teeth approach alignment with the pole piece teeth the primary current increases, and a small voltage is induced in the secondary. (*C*) When the teeth are aligned, the primary current reaches and sustains its maximum value. (*D*) As the timer core teeth just start to separate, the spark plug fires as the primary current decreases. (*E*) After the plug initially fires, the voltage induced in the secondary is lower and is sustained throughout the period to maintain the spark. (*F*) After the spark stops, the remaining energy in the ignition coil is dissipated in the form of a damped oscillation.

in the distributor, but it is not a part of the ignition system. It is used to suppress static for car radio reception.

All of the ignition components for EIS are housed within the distributor.

29.9 ■ Self-Induction

We have observed how a changing electric current in one circuit will induce emfs in neighboring circuits. Moreover, when the current varies in a single coil, the changing flux within the coil induces an emf in the coil itself. Such a coil is called a self-inductor, or, more commonly, simply an *inductor*. The symbol for an inductor with an air core is and with an iron core the symbol is . The direction of the emf, according to Lenz's law, is such that it opposes the changing current that causes it. Such a process is called *self-induction* and the emf produced is called *counter emf*.

Consider a coil with a soft-iron core and many turns of wire that is connected to a storage battery. When the switch in the circuit is closed the current does not instantly reach its full value as determined by Ohm's law ($I = V/R$). Nor does it instantly decrease to zero when the switch is opened. The explanation of this phenomenon can be illustrated by another diagram (Fig. 29.11). The latter figure illustrates the growth of magnetic flux in a coil as the current increases from left to right in the turns. As the current moves from left to right in the coil, a flux of circular lines of force radiates from each turn of the coil. These lines, as they expand outward, cut the turns of wire which produce them as well as all the other turns in the coil. There is thus induced in the coil a voltage in such a direction that it opposes the growth of the current. When the switch connecting the coil to the source of emf is opened, the current decreases, but as the current's own magnetic field dies away, the lines of force again cut the turns of wire—this time in a reverse direction. An emf is then induced which tends to *maintain* the current in the coil. When a coil has this property of opposing any change of current in a circuit, it is said to possess *self-inductance* or simply *inductance*. The self-inductance of an electric circuit might be thought of as an *electromagnetic inertia* because it behaves very much like the inertia of material bodies.

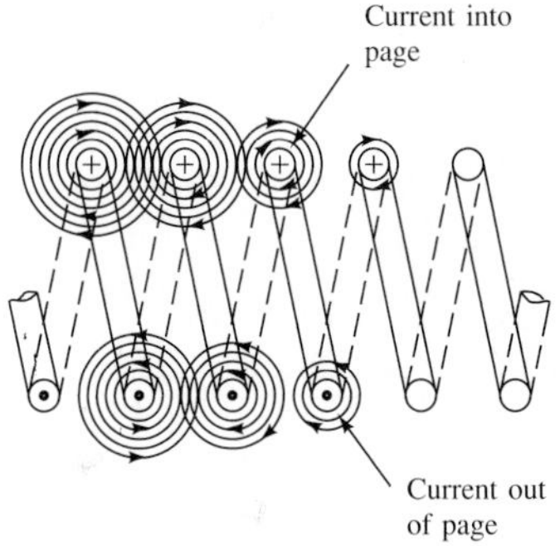

Fig. 29.11 Cross-sectional view of the growth of the flux lines in a solenoid. The current starts from the left and quickly moves right.

Self-induction opposes not only the increase of current in a coil but also the decrease of current in the coil.

The instantaneous induced counter emf (e_L) is directly proportional to the time rate of change of current $\Delta i/\Delta t$ in the coil.

$$e_L = -L\frac{\Delta i}{\Delta t} \tag{29.10}$$

or

$$L = -\frac{e_L}{\Delta i/\Delta L} \tag{29.11}$$

where L is called the *self-inductance* of the circuit and e and i represent *instantaneous values*. The unit of self-inductance is the henry (H). Self-inductance is defined (from Eq. 29.11) as the ratio of the emf of self-induction to the rate of change of the current in the coil. Thus, a circuit has a self-inductance of one henry if a time rate of change of current equal to one ampere per second induces a counter emf of one volt.

Illustrative Problem 29.7 The current in a 35 mH coil drops from 0.25 A to 0 A in 0.15 seconds. What is the average self-induced emf in the coil?

Solution

$$\mathscr{E} = -L\frac{\Delta i}{\Delta t}$$

$$= -35 \text{ mH} \times \frac{\text{H}}{1000 \text{ mH}} \times \frac{0.25 \text{ A} - 0 \text{ A}}{0.15 \text{ s}} \times \frac{\text{IV}}{1\,\frac{\text{H} \cdot \text{A}}{\text{s}}}$$

$$= -5.8 \times 10^{-2}\text{V}$$ *answer*

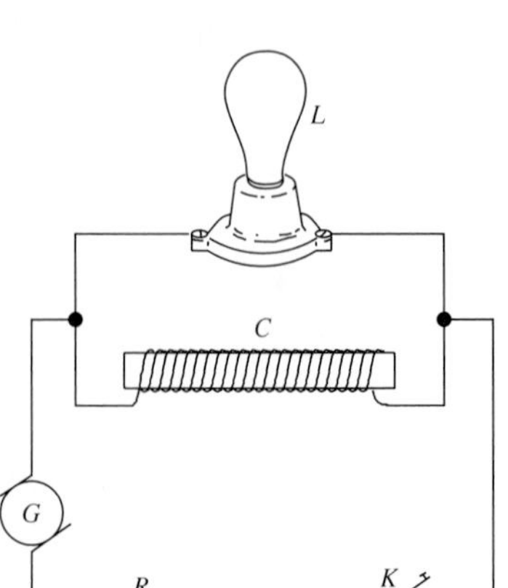

Fig. 29.12 Demonstration of self-inductance in a coil.

Illustrative Problem 29.8 It takes 75 μs for a current of 35 μA in a circuit to fall to zero. An average back emf of 20 V is induced in the circuit. Calculate the self-inductance of the circuit.

Solution

$$L = \frac{\overline{\mathscr{E}}_L}{\Delta i/\Delta t}$$

$$= \frac{-20 \text{ V}}{-35 \times 10^{-6} \text{ A}/75 \times 10^{-6}\text{s}}$$

$$= 43 \text{ H}$$ *answer*

The effect of self-induction can be demonstrated by connecting a coil C which has an iron core, with a lamp L in parallel, and this combination in series with a dc source G, a suitable resistance R, and a switch K, as shown in Fig. 29.12. When the switch is closed, the lamp will flash for a moment and then become quite dim. This is because the back emf built up in the coil temporarily retards the passage of current through the coil and hence most of the current flashed through the lamp. Then, since the actual ohmic resistance of the coil is considerably less than that of the lamp, as the self-induced emf diminishes, most of the current flows through the coil. When the switch is opened, the lamp once more momentarily lights up brightly. This is due to the large self-inductance in the coil and the rapidly decaying magnetic flux, causing the current to flow back through the lamp.

In some instruments it is necessary to have coils of wire that have little or no self-inductance. In order to eliminate self-inductance, some means must be devised to eliminate the magnetic flux around the wires. This is generally done by looping the wire back on itself so that two conductors side by side will have current going in opposite directions, thereby having the magnetic fields around them nullify each other (see Fig. 29.13*a*). The closer the wires A and B in the figure are together, the smaller the self-induction. Coils wound with such looped wires (Fig. 29.13*b*), even though they have soft-iron cores, have little self-induction.

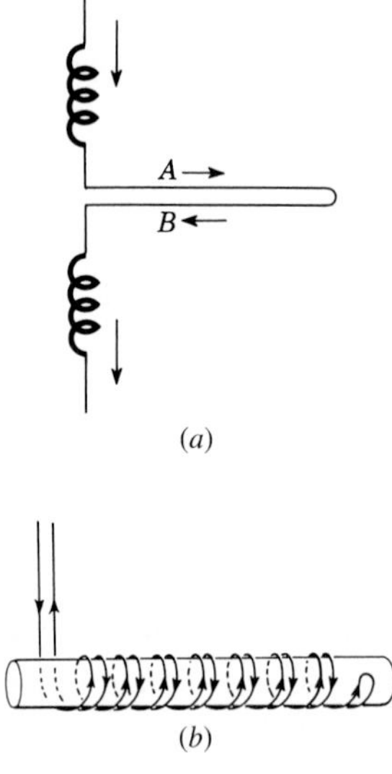

Fig. 29.13 Methods of eliminating self-inductance in coils.

Noninductive resistance coils are used in commercial ammeters and voltmeters, in most other electrical measuring instruments, and in some electronic communications circuits.

29.10 ■ Current Growth in an Inductive Circuit

In a purely resistive circuit (Fig. 29.14*a*) the current rises to its final Ohm's-law value virtually instantaneously (Fig. 29.14*b*) when the switch K is closed. Also, the current falls to zero almost instantaneously when the switch is opened. For the circuit of Fig. 29.15*a*, the final current will be $\mathscr{E}/R = 12 \text{ V}/120\ \Omega = 0.10$ A.

Now consider the circuit of Fig. 29.15. The circuit contains a 12 V source of emf, a 2.4-H coil of negligible resistance in series with a 120-Ω noninductive resistor. (Or we could consider the circuit having the 12-V source of emf connected to a 120 Ω coil with a *self-inductance* L of 2.4 H.)

When the switch is closed the increasing current in the coil becomes the cause of an emf whose direction, by Lenz's law, is opposite to that of the current. As a consequence of this back emf, at the closing of the switch, the current will not rise to its full value instantaneously. The rate of growth of the current will depend on the inductance and the resistance of the circuit. You will note the similarity between the increase of current in an inductor and the increase of charge on a capacitor (see Sec. 25.25). After a short time, however, the current level in the RL circuit reaches its final value.

The following formula for the current in an inductive circuit is given without derivation, since calculus methods are necessary to derive it.

$$i = I(1 - e^{-(R/L)t}) \qquad (29.12)$$

I = the maximum current in amperes,
t = elapsed time in seconds,
R = the resistance in ohms,
L = the inductance of the circuit in henries.
e = the Naperian base of logarithms.

For a given circuit, L/R is called its *time constant*. The time constant $\tau\ (= L/R)$ of an inductive circuit is the time, in seconds, for the current (or coil voltage) to reach 63 percent of its final value. It will be left as an exercise for the student to show that the dimension of τ is time.

Since I can be replaced by $\mathscr{E}/R$, Eq. (29.12) can be written

$$i = \frac{\mathscr{E}}{R}(1 - e^{-(R/L)t}) \qquad (29.12')$$

(increasing current)

Figure 29.16 shows graphically current growth in a circuit containing a source of emf, a resistance, and an inductance.

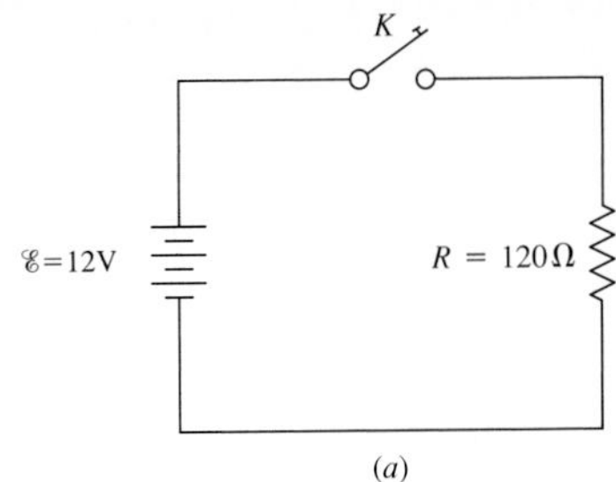

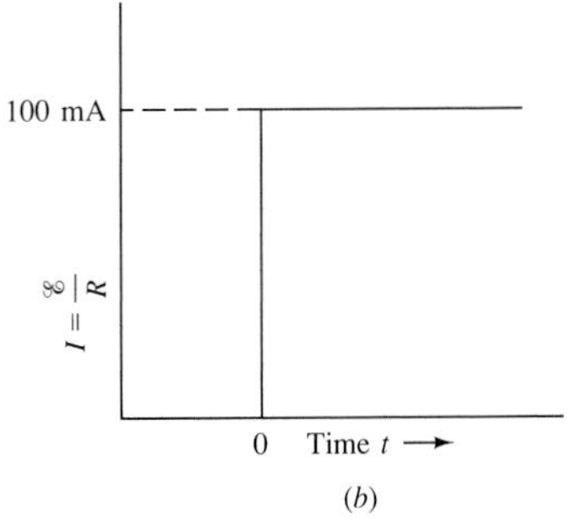

Fig. 29.14 (*a*) Pure resistance circuit. (*b*) Current reaches maximum value almost instantaneously.

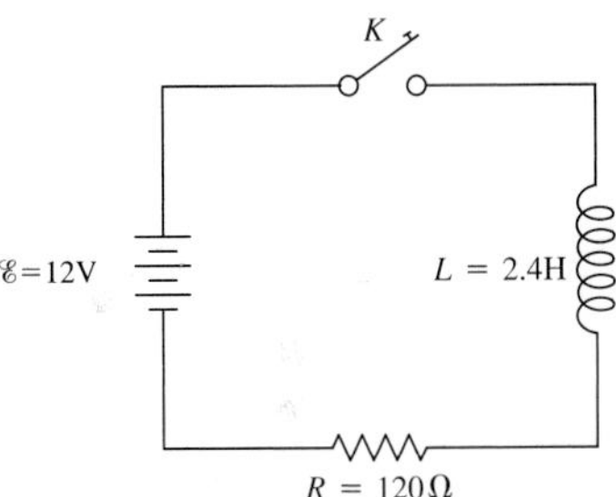

Fig. 29.15 Schematic of an RL circuit.

Illustrative Problem 29.9 Using the circuit shown in Fig. 29.15, determine (*a*) the time constant of the circuit, (*b*) the final value of the current in the coil, and (*c*) the instantaneous current 0.020 s after the switch is closed.

Solution

(*a*)
$$\tau = \frac{L}{R} = \frac{2.4\ \text{H}}{120\ \Omega} = 0.020\ \text{s} = 20\ \text{ms} \qquad \textit{answer}$$

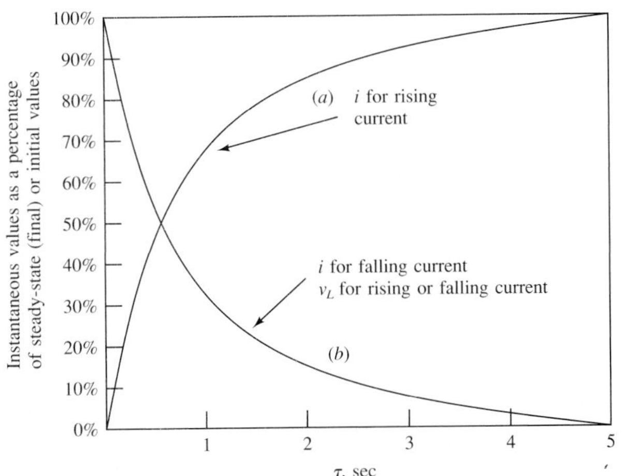

Fig. 29.16 Time constant curves for rise and decay in an inductive reactance-resistance circuit.

The final value of the current in an inductive circuit is determined by the resistance and the applied voltage, as given by Ohm's law.

(*b*)
$$I_{\text{final}} = \frac{\mathscr{E}}{R} = \frac{12\text{ V}}{120} = 0.10\text{ A} = 100\text{ mA} \qquad \textit{answer}$$

(*c*)
$$i = \frac{\mathscr{E}}{R}(1 - e^{-(R/L)t})$$
$$= \frac{12\text{ V}}{120\ \Omega}(1 - e^{-(120\ \Omega/2.4\text{ H})\times 0.020\text{ s}})$$
$$= 0.10\ (1 - 2.7^{-1})$$
$$= 0.063\text{ A} = 63\text{ mA} \qquad \textit{answer}$$

An inductive circuit, for all practical purposes, reaches a maximum (steady-state) value after 5τ ($=5L/R$) seconds has elapsed.

A rapid fall of current in a coil causes a high induced voltage.

The changes in current (and voltage) illustrated in Fig. 29.16 are called *transient* responses.

Once the current is established in the circuit of Fig. 29.15, the inductor will oppose any change in the magnitude of the current. When the switch is opened, the current will drop, creating a collapsing magnetic field in and near the inductor. The sudden drop in the field will create a counter emf across the inductor. The voltage produced by the suddenly collapsing magnetic field may be many times greater than the original source voltage. The result may be an arcing across the contacts of the switch. To avoid such arcing, a capacitor can be connected in parallel with the switch.

The decay of the current in an LR circuit (see Fig. 29.16) can be expressed by the equation

$$i = Ie^{-(R/L)t} \qquad (29.13)$$

(decreasing)

An alternate form of the equation is

$$i = \frac{\mathscr{E}}{R}(e^{-(R/L)t}) \qquad (29.13')$$

Illustrative Problem 29.10 An inductor with a self inductance of 60 mH and a resistance of 3.5 Ω is connected in series with a 8.5-Ω resistor and a 60-V source of emf. Calculate (*a*) the initial current in the circuit, and (*b*) the instantaneous current in the circuit 15 ms after the source of emf is shut off.

Solution

(*a*)
$$I_{\text{max}} = \frac{\mathscr{E}}{R_{\text{total}}} = \frac{60\text{ V}}{3.5\ \Omega + 8.5\ \Omega} = 5\text{ A} \qquad \textit{answer}$$

(*b*)
$$i = \frac{\mathscr{E}}{R_T} e^{-(R_T/L)t}$$
$$= \frac{60\text{ V}}{12\ \Omega}(e^{-(12.0\ \Omega/0.060\text{ H})\times 0.015\text{ s}})$$
$$= 5\ (e^{-3})\text{ A}$$
$$= 0.25\text{ A}$$

Hence, the instantaneous current after 15 ms equals 0.25 A. *answer*

29.11 ■ Inductors in Series and Parallel

Many circuits contain two or more inductors in series or parallel. The total inductances of such circuits can be calculated in the same manner as was done for combined resistances in Chap. 26. When inductors are connected in series (Fig. 29.17) the total inductance L_T is equal to the sum of the individual inductances, providing there is no mutual inductance (no magnetic coupling) between them. Thus,

L_1 L_2 L_3

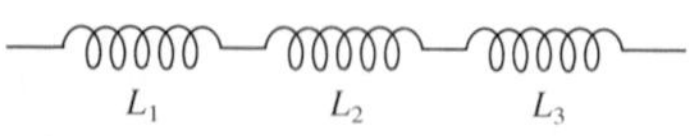

Fig. 29.17 Inductors in series.

$$L_T = L_1 + L_2 + L_3 + \cdots \qquad (29.14)$$

(Series of inductors with no magnetic coupling)

Two inductors will have no coupling when the coils are far apart.

When two inductors are in series, but are so arranged that the magnetic flux of each links with the other, the total inductance is

$$L_T = L_1 + L_2 \pm 2M \qquad (29.15)$$

(Two inductors with magnetic coupling)

where the positive sign is used for series-aiding mutual inductance and the negative sign is used for series-opposing mutual inductance.*

Illustrative Problem 29.11 An 8-H and a 16-H inductor are connected in series. What is the (*a*) maximum inductance and (*b*) the minimum inductance of an arrangement where the combination has a mutual inductance of 5 H?

(*a*) $L_{T_{\text{max}}} = L_1 + L_2 + 2M = 8\text{ H} + 16\text{ H} + 2(5\text{ H}) = 34\text{ H}$ *answer*

(*b*) $L_{T_{\text{min}}} = L_1 + L_2 - 2M = 8\text{ H} + 16\text{ H} - 2(5\text{ H}) = 14\text{ H}$ *answer*

Inductors connected in parallel (Fig. 29.18) so that no mutual inductance exists between them will have a total inductance determined by the equation

$$\frac{1}{L_T} = \frac{1}{L_1} + \frac{1}{L_2} + \frac{1}{L_3} + \cdots \qquad (29.16)$$

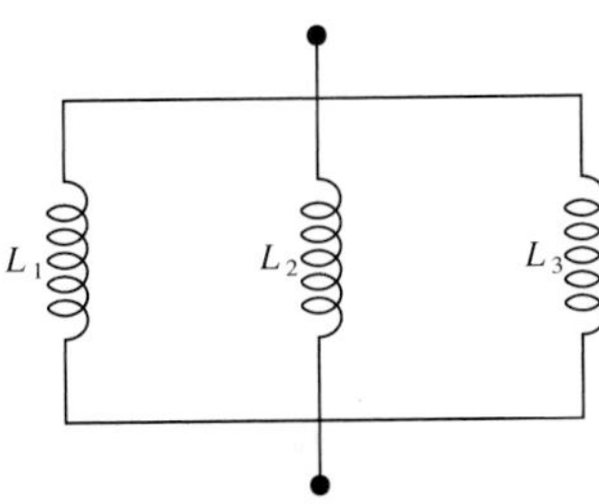

Fig. 29.18 Inductors in parallel.

Determining the total inductance of a circuit with coils in parallel and so arranged that there is mutual inductance between the coils may become quite complex, well beyond the scope of a basic physics course. Therefore, this text will consider only combinations of nonlinking inductances (no mutual inductance).

Illustrative Problem 29.12 Determine the total inductance of three noninteracting inductors with respective inductances of 2.0, 4.0, and 8.0 H when they are (*a*) in series, (*b*) in parallel, and (*c*) arranged so that the first two are in parallel and the combination is connected in series with the third inductor (see Fig. 29.19).

Solution

(*a*) $L_{T_{\text{series}}} = 2.0\text{ H} + 4.0\text{ H} + 8.0\text{ H} = 14.0\text{ H}$ *answer*

(*b*) $$\frac{1}{L_{T_{\text{parallel}}}} = \frac{1}{2.0\text{ H}} + \frac{1}{4.0\text{ H}} + \frac{1}{8.0\text{ H}}$$

$$L_{T_{\text{parallel}}} = \tfrac{8}{7}\text{ H} = 1.1\text{ H}$$ *answer*

(*c*) $$\frac{1}{L_{\text{parallel}}} = \frac{1}{L_1} + \frac{1}{L_2}$$

$$= \frac{1}{2.0\text{ H}} + \frac{1}{4.0\text{ H}}$$

$$L_{1-2} = \tfrac{4}{3}\text{ H}$$

Then $$L_T = L_{1-2} + L_3 = \frac{4\text{ H}}{3} + 8.0\text{ H}$$

$$= 9.3\text{ H}$$ *answer*

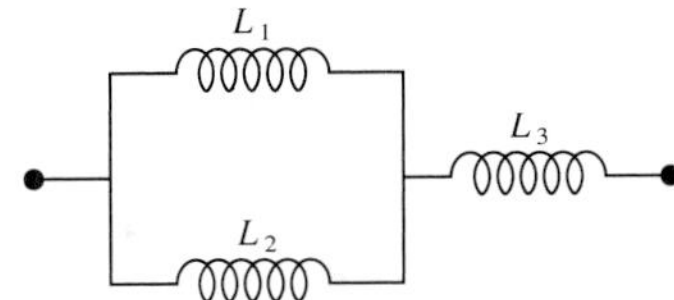

Fig. 29.19 Illustrative Problem 29.12.

QUESTIONS AND EXERCISES

1. Explain why the current in a secondary coil is zero when a steady (dc) current flows in a nearby primary coil.

2. What is the purpose of the capacitor in the older type ignition system of a gasoline engine?

3. List several ways in which an emf can be induced in a coil of wire.

4. Why is the self-inductance of an iron-core coil greater than a similar air-core coil?

* The mutual inductance of the two coils must be both aiding or both opposing. The polarity for M cannot be + for one coil and − for the other.

5. What are the factors upon which an induced emf depend?

6. Why is the secondary of the induction coil wound with many more turns than the primary?

7. Why must the wire in the primary of an induction coil be heavier than that of the secondary?

8. Discuss the difference in the flow of a charge from a battery and that from the secondary of an induction coil.

9. If the north pole of a bar magnet is dropped through a closed-loop coil whose plane is horizontal, what is the direction of the induced current as the north pole approaches the loop?

10. What physical quantities are measured in (*a*) webers; (*b*) webers per second; (*c*) webers per square meter; (*d*) webers per ampere-meter; (*e*) joules per coulomb; (*f*) newtons per ampere-meter?

11. Which of the following is true for Fig. 29.20? A current will be induced in B (*a*) only when the key is being closed, (*b*) only when the key is being opened, (*c*) whenever the key is either being closed or opened, (*d*) only when the key is kept closed, (*e*) none of these.

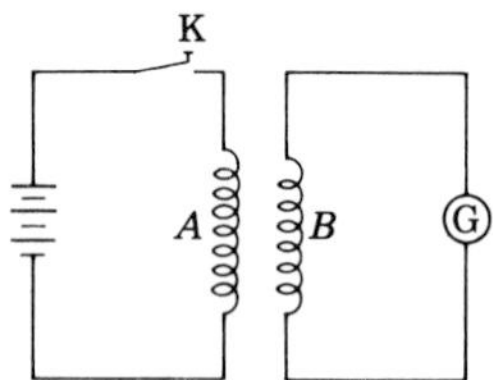

Fig. 29.20 Questions 11 and 12.

12. Which of the following is not true for Fig. 29.20? Whenever a current is induced in the right-hand circuit (*a*) Lenz's law applies to the circuit, (*b*) the current in coil B will have the same value as that in coil A, (*c*) the voltage of coil B can be more than that of coil A, (*d*) the voltage of coil B can be less than that of coil A, (*e*) the direction of the current in B will alternate on closing and opening the key.

13. The north pole of a bar magnet is moved away from a copper ring as shown in Fig. 29.21. In which direction does the conventional current point in the part of the ring nearest the reader?

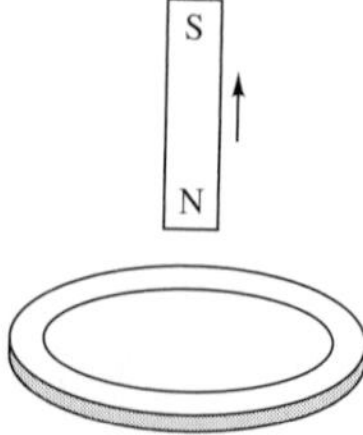

Fig. 29.21 Question 13.

14. A current-carrying solenoid is moved toward a conducting loop as shown in Fig. 29.22. Is the conventional current clockwise or counterclockwise as viewed by an observer at P?

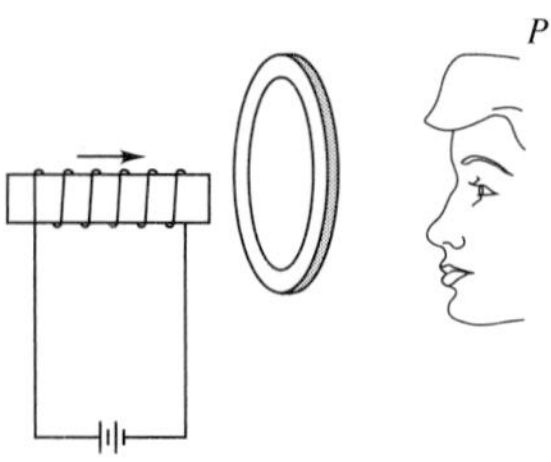

Fig. 29.22 Question 14.

15. What is the direction, in Fig. 29.23, of the conventional current through resistor R (*a*) immediately after the switch K is closed, (*b*) after the switch has been closed for some time, (*c*) immediately after the switch K is open, and (*d*) when the switch is closed, which end of the coil acts as a north pole?

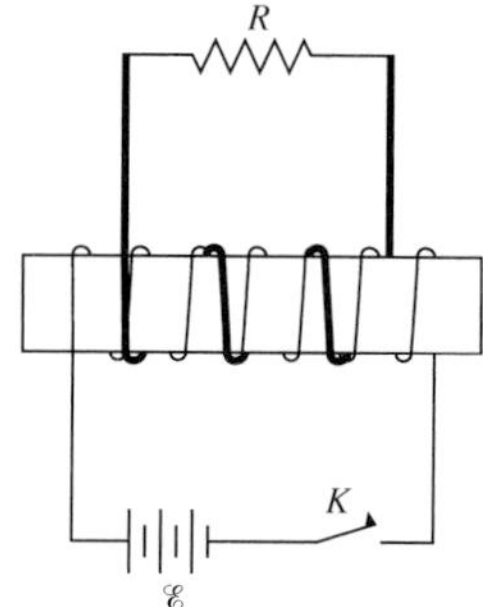

Fig. 29.23 Question 15.

16. What will be the direction, in Fig. 29.24, of the induced current in the right-hand circuit if the variable resistance R is gradually increased?

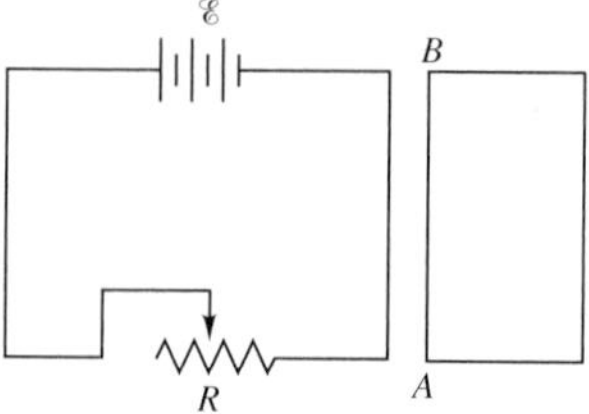

Fig. 29.24 Question 16.

17. In what direction will the induced current be in the resistors R_B and R_C when the switch K is (*a*) suddenly closed and (*b*) suddenly opened.

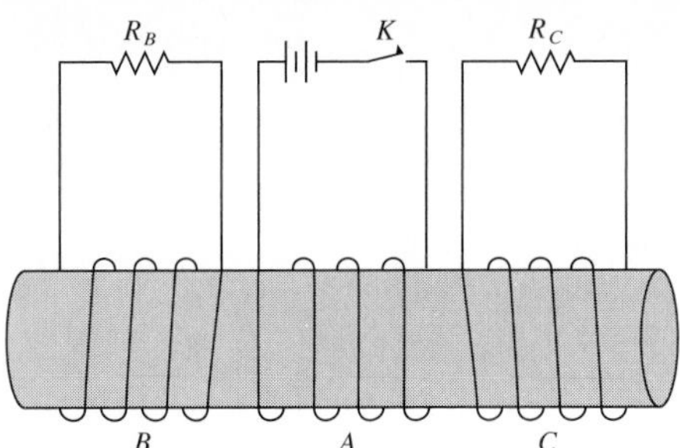

Fig. 29.25 Question 17.

PROBLEMS

Group One

1. When a 3.0-H solenoid is wound around a 5.0-H solenoid and the two are connected in a series-adding combination, the total inductance is 11.4 H. What is the mutual inductance of the combination?

2. A 3.0-H, a 6.0-H, and a 9.0-H inductor are connected in parallel. What is the equivalent inductance of the combination?

3. An 8-H and a 16-H coil have a mutual inductance of 6 H. Determine (*a*) the minimum inductance available and (*b*) the maximum inductance available from series connecting the coils.

4. A 6-H and a 12-H inductor are placed at right angles to each other. Determine the total inductance when they are connected (*a*) in series and (*b*) in parallel.

5. The current in a 220-mH coil increases from 1.0 A to 4.0 A in 0.10 s. Determine the self-induced emf.

6. A 400-H coil with a resistance of 300 Ω is connected across a 500-mV dc supply. Determine (*a*) the final current, (*b*) the time constant, (*c*) the time required for the current to reach 63 percent of its final value, and (*d*) the approximate time for the current to reach its final value.

7. What is the time constant for a circuit which has a resistance of 2 kΩ and an inductance of 2 H?

8. A straight wire 25 cm long moves perpendicular to and through a field of magnetic flux with a velocity of 75 cm/s. What emf is induced in the wire if the flux density is 0.065 Wb/m^2?

9. A flux of 1.5×10^{-3} Wb emanates from the north pole of a bar magnet. What emf will be induced at the ends of a coil of 400 turns when it takes 0.75 s to thrust the magnetic pole through the coil? Assume each turn of the coil is cut by the flux in that time.

10. A single conductor passes through a flux of 8.2 Wb. How long did it take the conductor to cut the magnetic flux to produce an emf of 18 V?

Group Two

11. A coil has 450 turns of wire and an area of 25 cm^2. What is the average emf induced in the coil if the magnetic flux density B increases from 0.10 to 0.65 Wb/m^2 in $\frac{1}{60}$ s?

12. Find the emf induced in a coil of 840 turns and 12 cm diameter if the magnetic flux density B inside the coil is changed from 0.15 to 0.55 Wb/m^2 in 1 s.

13. With what velocity must a straight wire 50 cm long cut across a magnetic field of flux density 0.5 Wb/m^2 in order to induce across the ends of the wire a potential difference of 0.2 V?

14. A coil having 150 turns and an area of 7.5 cm^2 is inserted between the poles of a magnet having a uniform flux density of 2.5 Wb/m^2 in 0.025 s. (*a*) What voltage is induced across the coil? (*b*) If the coil has 45 Ω resistance, what current will flow in it?

15. The secondary of an induction coil has 30,000 turns. If the total flux in the coil changes from 4.5×10^{-4} to 9.0×10^{-5} Wb in 1.8×10^{-4} s, what is the magnitude of the induced emf?

16. Find the potential difference between the tips of the 20-m wing of a jet plane that is traveling 1600 km/h east if the vertical component of the earth's magnetic field has a flux density of 7.5×10^{-5} Wb/m^2.

17. An induction coil has 2.5×10^4 turns. The number of flux lines ϕ through it changes from 5.0×10^{-3} to 0 in $\frac{1}{100}$ s. What is the average induced emf?

18. Show that the time constant τ $(=L/R)$ of an RL circuit has a dimension of time.

19. What is the original current in a 15-mH coil if, when the current falls to 0 A in 0.15 s, an average self-induced emf of −0.45 V is produced in the coil?

20. The time constant of an RL circuit is 1 ms. The circuit's steady-state (final) current is 15 mA. What is the current in the circuit (*a*) after 1 ms, and (*b*) after 2 ms?

21. Develop an explicit equation for the total inductance L_T for two inductors connected in parallel if the inductors have inductances of L_1 and L_2 and there is no mutual inductance.

22. The current in a coil increases from 0 to its final value of 0.45 A in 0.075 s when a 24-V battery is connected across it. Determine (*a*) the resistance of the coil, (*b*) the time constant, and (*c*) the inductance of the coil.

23. A coil has 800 turns. A flux of 2×10^{-4} Wb is established in the coil when a current of 4 A is in the coil. Determine (*a*) the average counter emf in the coil if the current drops to zero in 0.06 s and (*b*) the inductance L of the coil.

24. A coil has 15 turns and a diameter of 12 cm. It is placed in a field with its plane perpendicular to a magnetic field of 0.6 T which is produced by an adjacent electromagnet. When the current in the electromagnet is cut off, the collapsing magnetic field induces an average emf of 10 V in the coil. How long does it take the field to disappear? [HINT: Use Eqs. (28.2) and (29.5).]

25. The current in a coil of wire decreased from 6.2 to 0.5 A in 0.015 s. What is the self-inductance of the coil if the average induced emf was 1200 V?

26. The self-inductance of a coil of wire is 3.5 H. When a switch is opened, the current through the coils decreases from 8.2 to 0 amperes in 0.15 s. What is the average induced emf which opposes the decrease in the current?

27. Two separate coils are wound on the same iron core. When the current in one coil drops from 6.4 A to 2.6 A in 0.015 s, the potential difference between the terminals of the second coil is 64 V. What is the value of the mutual inductance?

28. Two circuits have a mutual inductance M of 85 mH. What is the average emf induced in the secondary by a change from 0 to 25 A in 46 μs in the primary?

29. Two coils placed side by side have a mutual inductance of 0.35 H. The secondary coil has 55 turns. The current in the primary increases from 1.2 A to 6.2 A in 15 μs. Calculate (*a*) the average induced emf in the secondary, and (*b*) the total change in the magnetic flux $\Delta\phi$ in the secondary during the time interval of 15 μs.

30. A small test coil has 200 turns, each with a cross-sectional area of 5 cm^2. It is placed in a magnetic field of unknown flux density with its plane perpendicular to the lines of induction. It is then quickly jerked out of the field in 0.08 s. The average emf developed during the process is 0.4 V. Find the flux density B of the magnetic field.

Group Three

31. Determine a formula for the self-inductance L of a solenoid that has N turns of conducting wire, a cross-sectional area A, and a length l.

32. Coil A with 300 turns is placed near coil B with 900 turns. When there is a current of 2 A in coil A it produces a flux of 3.5×10^{-4} Wb in A and a flux of 2.0×10^{-4} Wb in B. If the current in A drops to zero in 0.25 s, determine (*a*) the self-inductance in A, (*b*) the mutual inductance between A and B (see Eq. 29.8), and (*c*) the average emf induced in B.

33. A solenoid has 1500 evenly spaced turns of conducting wire wound on an iron core 75 cm long and 2.5 cm in diameter. A secondary coil of 12,000 turns is wound around the central portion of the solenoid. The iron has a permeability of 2.0×10^{-5} Wb/A · m. If the current in the iron-core solenoid is changing at the rate of 20 A per 0.10 s, what average emf is induced in the secondary coil? Assume that the flux through the secondary coil equals that through the primary solenoid.

34. A resistor and an inductor are connected in series to a 12-V dc source of current as shown in Fig. 29.26. How many seconds after the switch K is closed will the current reach 80 percent of its maximum value? (NOTE: You will need a table for values of e^{-x} or a hand-held calculator to solve this problem.)

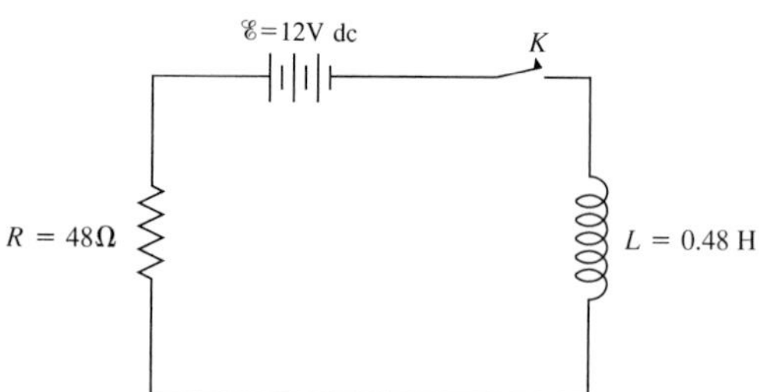

Fig. 29.26 Problem 34.

35. A 12-H inductor is connected in series with a 480-Ω resistor. A switch is closed to apply 120-V dc to the circuit. Determine (*a*) the current after 60 ms, (*b*) how long it will take for the current to reach 0.20 A, and (*c*) the voltage across the inductor after 75 ms. (NOTE: A table for values of e^{-x} or a calculator will be needed to solve this problem.)

36. A coil has a resistance of 600 Ω and an inductance of 12 H. A 24-V dc supply is connected across the coil by a switch. Determine (*a*) the time constant of the coil, (*b*) the final value of the current in the coil, (*c*) the current 50 ms after the switch is closed, and (*d*) how long it will take for the current to reach 20 mA. (NOTE: A table for values of e^{-x} or a calculator will be needed to solve this problem.)

CHAPTER 30 GENERATORS AND MOTORS

Motion of a conductor across a magnetic field produces a potential difference between the ends of the conductor. When the ends of the conductor are connected to a closed circuit, an electric current is produced. If a way could be provided for a continuous motion of the conductor across the field, then a continuous emf would be generated. It is much easier to rotate a coil in a fixed magnetic field than to move a conductor back and forth across the magnetic field. If we except automobile alternators, this coil rotation is done in more than 99 percent of the electric generators today. The mechanical energy to turn the coil can be provided by a *prime mover* such as a water turbine, steam turbine, wind turbine, or an internal-combustion engine.

The generator that uses a bicycle wheel for the prime mover and that provides electric energy for the bicycle lights is a familiar example of a small electric generator.

A machine that converts mechanical energy into electric energy is called a *generator*. It has already been pointed out that a generator does not actually *generate* electricity. It converts some other form of energy into electric energy. The study of electric generators is a highly technical field, and in this chapter, we deal only with some of the more basic principles of operation, design, and use of dc and ac generators and motors.

In an electric *motor*, electric energy is converted into mechanical energy. An electric motor is just a generator in reverse. The word *dynamo* is applied to machines that convert either electric energy into mechanical energy or mechanical energy into electric energy. Thus, the dynamo is a reversible machine capable of operation either as a generator or as a motor.

30.1 ■ The Simple AC Generator

The principles involved in the electric generator can probably best be explained in terms of a generator made up of one loop of wire turning in a uniform magnetic field established between two field magnets, as illustrated in Fig. 30.1. The current generated *in the coil* is always an *alternating* one. However, the current delivered *from the coil* to an external circuit can be either alternating or direct, depending on how it is picked up from the coil. If the current that is established in the coil is picked up by means of *brushes* and *slip rings* attached to the ends of the coil (Fig. 30.1), an alternating current will also flow through the external circuit R. The current will reverse direction twice in each complete revolution. The value of the emf, and hence that of the current, will not be constant as the coil turns with constant velocity but will vary with time in a manner which results in a plot which is a sine curve (see Fig. 30.2).

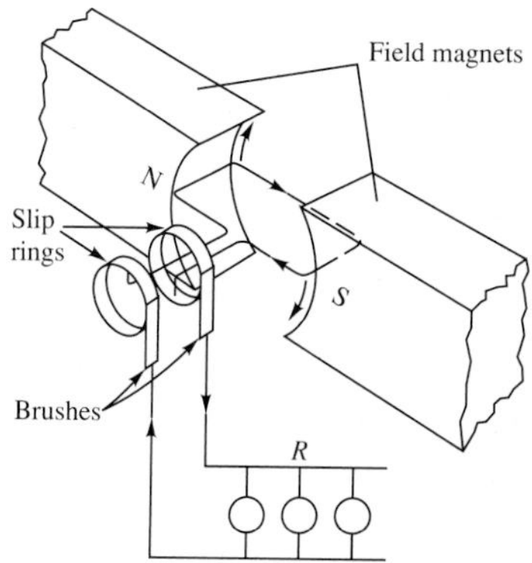

Fig. 30.1 Generator with slip rings—a schematic diagram.

30.2 ■ Nature of Alternating Current

An *alternating current* periodically reverses its direction. We saw in Sec. 29.5 how an alternating current can be produced by alternately thrusting a bar magnet into a coil of wire and then withdrawing it. The same effect can be obtained by rotating a coil of

Fig. 30.2 Graph of alternating emf produced by a simple generator like that of Fig 30.1.

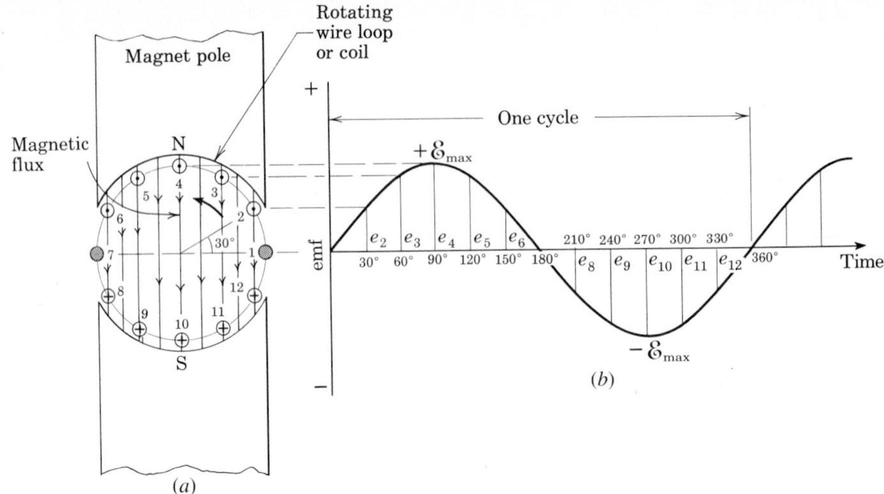

The coil in which the emf is induced is called an *armature.*

The induced current in a coil reverses itself every half-turn of the coil.

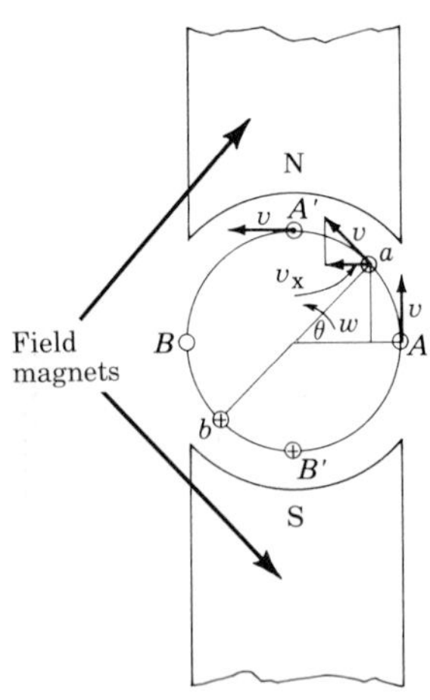

Fig. 30.3 The emf and current in the coil are proportional to that component of the linear velocity v which is perpendicular to the magnetic flux.

wire in a uniform magnetic field. In order to investigate how the emf varies as a coil turns with uniform angular velocity in a magnetic field, let us consider the motion of the coil from moment to moment. Figure 30.2*a* shows 12 positions of the rotating coil, 30° apart. The instantaneous emfs in the coil corresponding to these 12 positions are shown plotted in Fig. 30.2*b*. The horizontal axis is marked off in units of degrees of rotation (which also may be considered as units of time, since we assume uniform angular velocity) and the vertical axis shows the emfs produced.

When the rotating coil is in position 1, the coil motion is parallel to the lines of force; hence no lines are cut and no emf is induced in the coil. When the coil has reached position 2, lines of force are being cut and an emf is produced, giving rise to a current in a direction which is *out* from the page (indicated by a dot). The vertical line e_2 in part (*b*) represents the instantaneous emf in the coil when it is at position 2. In like manner, e_3, e_4, . . . represent the instantaneous emfs as the coil occupies positions 3, 4, . . . From Fig. 30.2*a* it is seen that as the coil changes from position 1 to position 4, the number of lines of force cut in a given period of time increases from zero at position 1 to a maximum at position 4. Then as the coil moves on from 4, the number of lines cut in the same period of time diminishes from that maximum to zero when the coil reaches position 7 again.

As soon as the coil passes position 7, it cuts the lines of force in an opposite direction. An emf will now be induced in the coil which results in a current directed toward the page (indicated by a cross). This induced emf will continue to rise in a negative direction (assuming the original direction as positive) until a maximum value at position 10 is reached. As the coil continues to rotate from position 10, the emf gradually decreases until it once more is zero at position 1. Thus we see that the emf of an alternating current is continually changing as the coil rotates.

To investigate exactly how the emf varies from moment to moment, consider the diagram of Fig. 30.3. Let A and B represent the cross sections of the two segments of the conductor which makes up the loop turning with angular velocity ω in a magnetic field. With the loop in the AB position no emf is produced in the loop, since the conductor segments are at that instant moving parallel to the magnetic lines of force.

When the coil is in the $A'B'$ position one-quarter turn (90°) later, lines of force are being cut at the greatest rate and the emf produced will be a maximum. Let this maximum emf be designated e_{max}. At some intermediate coil position such as ab the coil will have turned through an angle $\theta = \omega t$. The emf produced in the coil depends on the rate at which magnetic force lines are being cut, and this rate, in turn, is proportional to that component of the linear velocity v which is perpendicular to the flux.

The curve of Fig. 30.2*b* is called a *sine curve*, because the same curve would result in plotting the sines of angles from 0 to 360°. Although our graph has been used to illustrate emfs, a curve of the same general shape would illustrate the changing instantaneous values of *current* established in the coil when it is connected to a closed circuit.

The emf produced in a rotating coil varies sinusoidally with time.

When a single loop of wire is rotated with constant angular velocity ω (radians/second) in a uniform magnetic field $\mathbf{B}$ (Wb/m^2), the instantaneous emf e, in volts, induced in the loop is determined by the equation

$$e = BA\omega \sin \theta \tag{30.1}$$

or

$$e = BA\omega \sin \omega t \tag{30.2}$$

where A (m^2) is the area of the loop and both ω (rad) and t (s) are measured from the instant the plane of the coil is perpendicular to the magnetic flux.

Note that when $\theta = 90°$, $\sin \theta = 1$ and e has its maximum value. That is,

$$e_{max} = BA\,\omega \tag{30.3}$$

Thus, Eqs. (30.1) and (30.2) can also be written

$$e = e_{max} \sin \theta \tag{30.4}$$
$$e = e_{max} \sin \omega t \tag{30.5}$$

The current produced (in a pure resistance circuit) will follow the same pattern of variation as the emf. The instantaneous current i can be found by

$$i = \frac{e_{max}}{R} \sin \theta = i_{max} \sin \theta \tag{30.6}$$

or

$$i = i_{max} \sin \omega t \tag{30.7}$$

If we use ω (in rad/s) $= 2\pi f$ (cycles/s—Hertz), Eqs. (30.5) and (30.7) become

$\omega = 2\pi f$, where f is frequency in Hertz.

$$e = e_{max} \sin 2\pi f t \tag{30.5$'$}$$

$$i = i_{max} \sin 2\pi f t \tag{30.7$'$}$$

Illustrative Problem 30.1 A rectangular coil 12 cm long and 8 cm wide turns 50 times a second in a uniform magnetic field of 2000 G. What is the maximum emf induced in the coil?

Solution

$$\omega = \frac{50 \text{ rev}}{\text{s}} \times \frac{2\pi \text{ rad}}{1 \text{ rev}}$$
$$= 100\pi \text{ s}^{-1}$$

(Remember the radian is dimensionless.) Substitute in Eq. (30.3), to get

$$e_{max} = BA\omega$$
$$= 2000 \text{ G} \times \frac{1 \text{ Wb/m}^2}{10{,}000 \text{ G}} \times (0.12\text{m} \times 0.08\text{m}) \times 100\pi\text{s}^{-1}$$
$$= 0.603 \text{ Wb/s}$$
$$= 0.60 \text{ V} \qquad \textit{answer}$$

Illustrative Problem 30.2 The armature of an ac generator has 20 turns of wire, each with an area $A = 0.12$ m^2 and a total resistance of 24Ω. The loop rotates in a magnetic field $B = 0.25$ T at a constant frequency of 60 Hz (1 Hz = 1 cycle/s). Find the maximum induced current.

Solution

$$\omega = 2\pi f = 2\pi(60 \text{ Hz})$$
$$= 2\pi(60 \text{ s}^{-1})$$
$$= 120\pi \text{s}^{-1}$$

Then

$$e_{max} = NBA\omega$$
$$= 20(0.25 \text{ T})(0.12 \text{ m}^2)(120\pi \text{ s}^{-1})$$
$$= 226 \text{ V}$$

Alternating current in the armature can be converted to direct current by a split-ring commutator.

(See Illustrative Problem 30.1 for proper conversion units)

$$i_{max} = \frac{e_{max}}{R} = \frac{226 \text{ V}}{24}$$
$$= 9.4 \text{ A} \qquad \textit{answer}$$

DC GENERATORS

30.3 ■ The DC Generator

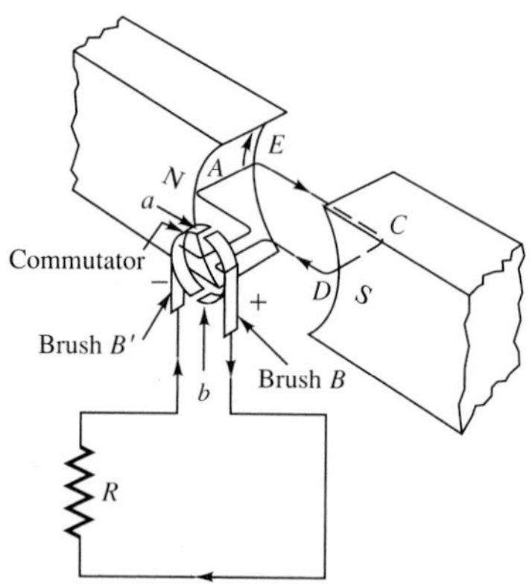

Fig. 30.4 Generator with commutator. The commutator rectifies the alternating current in the coil to a direct current in the external circuit.

Even though the current that flows in the rotating coil of a generator is always alternating, it is quite possible to obtain a unidirectional current through an external circuit. A current is said to be *rectified* if it is changed from alternating to direct. Figure 30.4 shows how an ac generator can be modified by means of a split ring, called a *commutator*, and two brushes, so that it will deliver a pulsating direct current to an external circuit.

In the dc generator the terminals of the coil are connected to the two halves of a *split-ring commutator*. The segments of the ring are insulated from each other. The brushes are placed on opposite sides of the ring so that when the emf changes direction (at zero emf), the brushes pass over the division between the ring segments.

When the conductor AE in Fig. 30.4 is moving upward, the induced emf will cause a current to flow in the direction from A to E. In like manner, as conductor CD moves downward, the induced emf will cause a current to flow in the direction from C to D. This will make the semiring a of the split-ring commutator negative and the semiring b positive. During the half-revolution when AE moves downward and CD moves upward, the signs of the charges on semirings a and b will be reversed. But the brushes are so adjusted that at the instant the signs of the charges on the semirings change, the segments will make contacts with the other brushes. Hence brush B will always have a positive charge and brush B' will always be charged negatively, with a resultant unidirectional flow through the external load R. Thus, although the emf still fluctuates, it always produces a current in the same direction in the *external* circuit.

The direction of current in the circuit *external* to the generator is from a higher or positive potential to a lower or negative potential, while in the *internal circuit* of the generator the current flows from a lower potential to a higher potential. Using the hydraulic analogy, work is done *on* the coil (in this case the rotating part) by a prime mover, and this will force current to flow ''uphill'' just as water is pumped from a lower to a higher level. It must be remembered that the real direction of flow of electricity in a circuit is the direction of flow of electrons, but in this discussion we

Fig. 30.5 Curve showing the unidirectional pulsating nature of the emf at the brushes of a single loop with commutator.

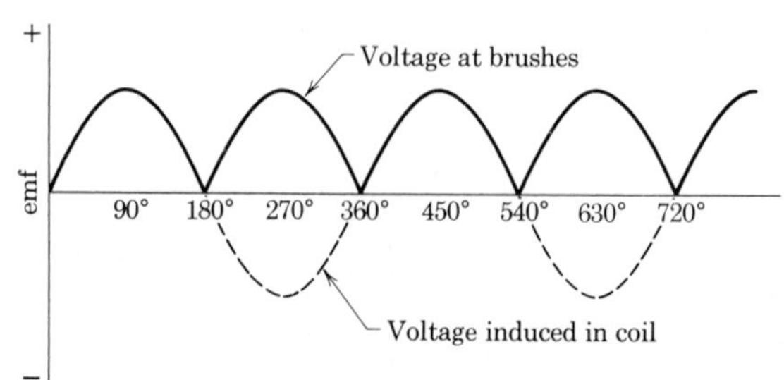

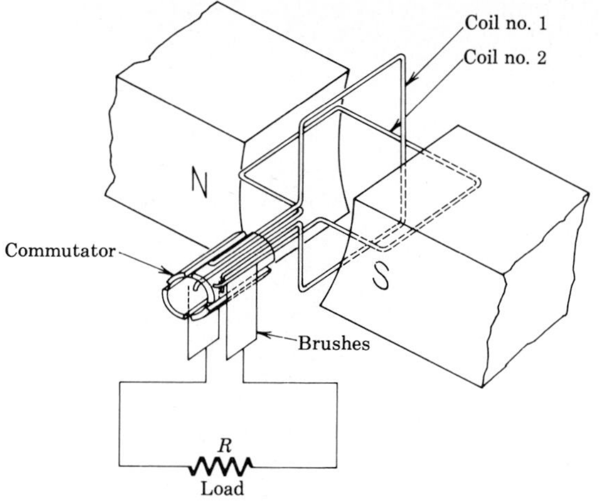

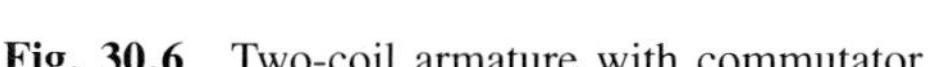

Fig. 30.6 Two-coil armature with commutator.

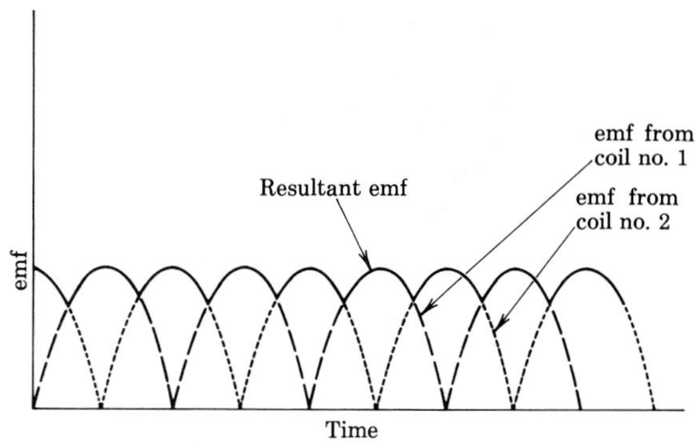

Fig. 30.7 EMF fluctuations from a two-coil armature with coils at right angles to each other. The coils are insulated from each other.

are assuming the conventional *direction of current is* opposite to the direction of electron flow.

Figure 30.5 illustrates graphically the pulsating nature of the rectified emf from a single-loop dc generator. The curve is similar to that in Fig. 30.2*b*, except that the second half of the cycle is inverted.

A pulsating direct current such as the one just described is not suitable for commercial users of direct current. A *steady* unidirectional current is necessary, and this is obtained in the commercial dc generator by using many armature coils distributed evenly around a soft-iron laminated drum. Each coil has a pair of commutator bars (segments) that are insulated from the other bars and from the shaft.

Figure 30.6 illustrates how two coils at right angles can be used to produce a pulsating direct current. Note that the commutator is divided into four segments or bars. Each coil terminates in two opposite commutator bars. The emf fluctuations in the two coils are shown with the resultant emf in Fig. 30.7. We see that when either coil has a maximum emf, the other has an emf of zero.

The arrangement shown in Fig. 30.6 is not a practical one. Note that at any given time only the armature coil that connects the bars in contact with the brushes produces current. The induced voltages in the other coil do not produce a current because it is not connected to the outside circuit. Thus only half the induced voltages are used to produce a current.

However, if the emfs coming from the two coils are joined in series, the instantaneous value of the combined emfs of the two coils will equal the sum of the individual emfs of each coil at that instant. This resultant is shown in Fig. 30.8. Thus the resultant emf is seen to fluctuate considerably less than the emfs of the individual coils. The resultant emf never falls to zero in this arrangement.

Most generators today use *drum-type armatures*, that is armatures in the form of a cylinder or drum. These are wound with a large number of coils of wire so that the current supplied to the external circuit will be almost free from fluctuation. The core of the armature is made of sheet-steel laminations that are keyed to a *spider*, which in

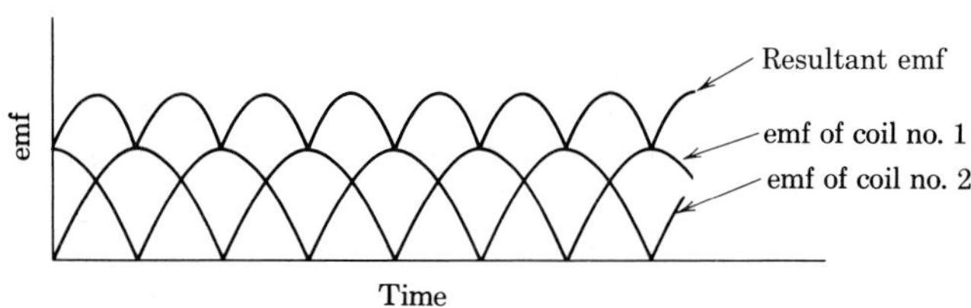

Fig. 30.8 EMF fluctuations from a two-coil armature with coils at right angles to each other and joined in series.

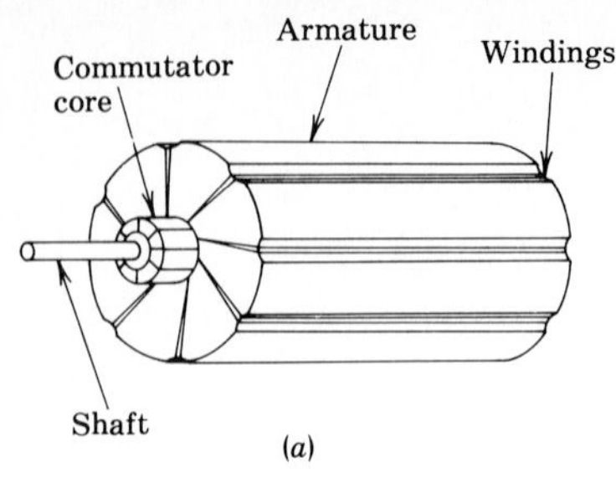

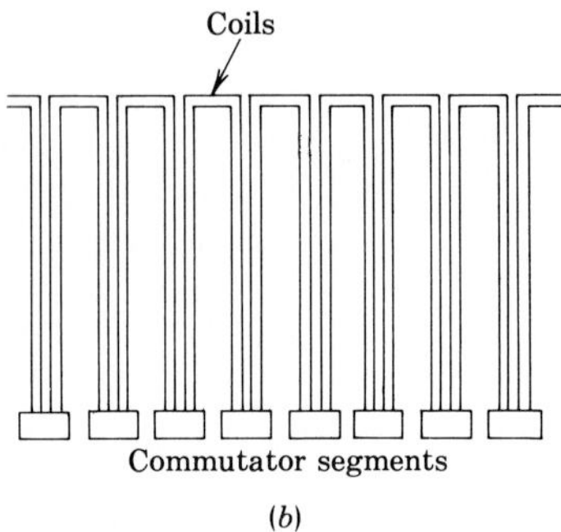

Fig. 30.9 Drum-type armature with eight armature coils and eight commutator bars.

Fig. 30.10 Partly wound armature of an 830-kW, 800-V dc generator. The formed coils can be seen joined to the commutator segments. (Fairbanks, Morse Co.)

turn is keyed to the shaft. The armature winding consists of coils of wire or bars connected in series and lying in slots on the surface of the core. Each end of a coil is soldered to a commutator segment. Figure 30.9*a* shows a drum-type armature with eight coils and segments. Figure 30.9*b* illustrates the commutator segments and armature coils laid on a flat surface to show how they are connected. Note that each slot contains conductors from two adjacent coils. These coils are inserted in the slots and kept in place by retaining wedges. A partly wound armature of a dc generator is shown in Fig. 30.10.

The use of formed coils provides for superior ventilation and permits speedier repair of the generator in case one of the coils needs to be removed. On small generators, the coils are generally not form-wound but are wound by hand or by machine directly into the slots.

30.4 ■ Field Excitation

The magnetic field of a dc generator is usually produced by using a portion of the induced current to energize the field magnets. Such generators are said to be *self-exciting*. If a permanent magnet is used for the field, the generator is commonly called a *magneto* (Fig. 30.11*a*).

There are three common ways of self-exciting the field magnets. When the field magnets are connected in series with the armature loops so that all the generator current passes through the field coil windings (Fig. 30.11*b*), the generator is called a *series-wound* generator. Since the entire current furnished by the generator flows through the field coils, the coils must be wound with heavy wire.

When no load is connected to the generator (that is, when it operates an *open circuit*), there is no current in the armature and the field windings. When a load is connected, the small emf induced because of the residual magnetism of the pole pieces will cause a current in the armature winding, the load, and the field winding. As the current flows, the magnetic flux in the field increases, causing an increase in the terminal emf. This, in turn, will increase the output current.

As the terminal voltage and output current build up, there is an accompanying increase in the back emf developing in the armature and field windings. This tends to reduce the terminal voltage. Soon a point will be reached where the two processes balance and the terminal voltage ceases to rise.

Since the terminal voltage varies so much with the load, series-wound generators are seldom used. They are used, however, in arc-welding because once the

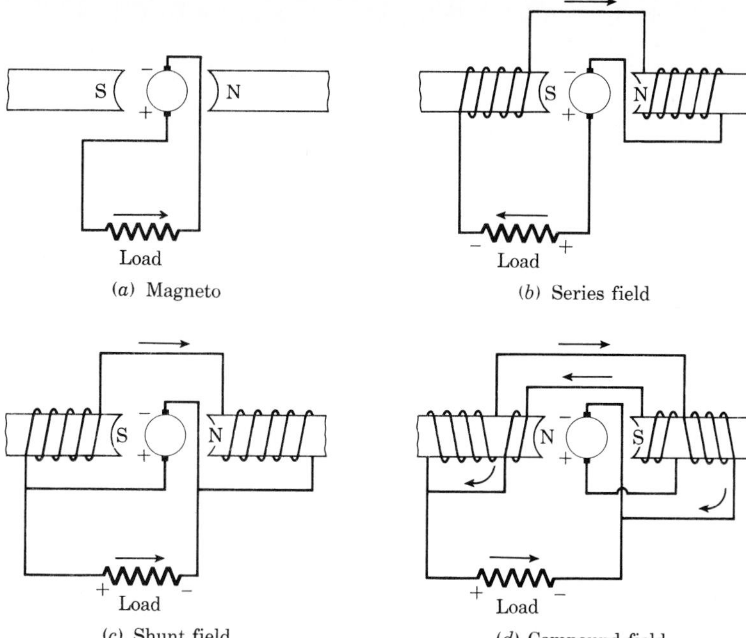

Fig. 30.11 Methods of connecting field and armature in dc generators.

generator reaches its peak terminal voltage, the output current remains fairly constant.

In the *shunt-wound* dc generator, the field magnets are connected in parallel with the armature so that only a portion of the generated current is used to excite the field (Fig. 30.11*c*). Usually less than 5 percent of the armature current is fed to the field coils. To keep the current low, the coils have a high resistance. In order to gain an adequate magnetic flux in the field coils, many turns of fine wire are used.

The principles involved in the shunt-wound generator are similar to those applying to the series-wound type. As the armature is rotated, lines of residual magnetic flux are cut to produce an emf at the brushes. A current in the armature windings and field windings, causes an increase in the magnetic flux of the field windings which, in turn, increases the current. The cumulative process continues until equilibrium is reached for a given load.

Most dc generators employ part of the induced current to energize their field magnets (coils) and are said to be *self-exciting*.

For light loads the terminal voltage remains quite constant for variations in load currents. However, if the load is too heavy, the terminal voltage drops sharply.

The series-wound generator and the shunt-wound generator have generally been replaced by the compound-wound generator. In the *compound-wound* generator both series and shunt field windings are employed. The series field windings consist of a few turns of heavy wire over a shunt winding of many turns of fine wire. The two windings are connected in series with the armature winding (Fig. 30.11*d*). In such a generator the potential difference across an external circuit may be kept fairly constant, since an increase in the load will cause an increase in the current in the series windings and a decrease in the shunt windings. By using the proper number of turns of each type of winding, a constant flux density can be maintained under varying loads, so that the generator voltage will be constant regardless of the load in the external circuit.

Illustrative Problem 30.3 A shunt-wound dc generator has a field resistance of 165Ω in parallel with an armature resistance of 1.5Ω. If the generator delivers an output power of 4.4 kW at a potential difference of 110 V, calculate (*a*) the total current in the external load, (*b*) the total resistance of the generator, (*c*) the field current, (*d*) the induced current in the armature, (*e*) the total induced emf, and (*f*) the efficiency of the generator.

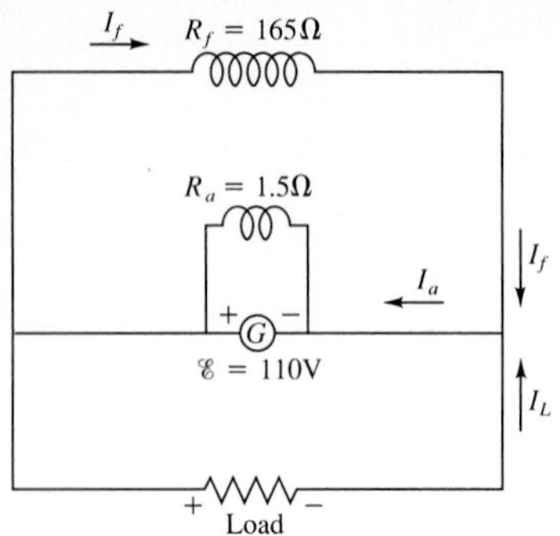

Fig. 30.12 Illustrative Problem 30.3.

Solution A schematic diagram of the circuits involved will aid in the solution (see Fig. 30.12).

(*a*)

$$110 \text{ V} \times I_L = 4400 \text{ W}$$

$$I_L = \frac{4400 \text{ W}}{110 \text{ V}}$$

$$= 40 \text{ A} \qquad \textit{answer}$$

(*b*)

$$\frac{1}{R_T} = \frac{1}{R_a} + \frac{1}{R_f}$$

$$= \frac{1}{1.5\Omega} + \frac{1}{165\Omega}$$

$$R_T = \frac{165\Omega}{111.5} = 1.48\Omega \qquad \textit{answer}$$

(*c*)

$$I_f = \frac{110 \text{ V}}{165\Omega} = \frac{2 \text{ A}}{3} = 0.67 \text{ A} \qquad \textit{answer}$$

(*d*) By Kirchhoff's first law:

$$I_a = I_f + I_L$$
$$= 0.67 \text{ A} + 40 \text{ A}$$
$$= 40.7 \text{ A} \qquad \textit{answer}$$

(*e*)

$$E_T = E_f + E_a$$
$$= I_f R_f + I_a R_a$$

$$\frac{2\text{A}}{3} \times 165\Omega + 40.7 \text{ A} \times 1.5\Omega$$

$$= 171 \text{ V} \qquad \textit{answer}$$

(*f*)

$$\text{Eff.} = \frac{P_{\text{out}}}{P_{\text{in}}} \times 100$$

$$= \frac{4400 \text{ W}}{171 \text{ V} \times 40.7 \text{ A}} \times 100$$

$$= 63 \text{ percent} \qquad \textit{answer}$$

The emf (and current resulting from it) can be made greater in three ways: (1) increase the magnetic flux between the poles, (2) increase the number of loops in the armature, and (c) increase the speed of relative motion between the magnetic field and the rotating conductors.

Fig. 30.13 Schematic diagram of a four-pole generator.

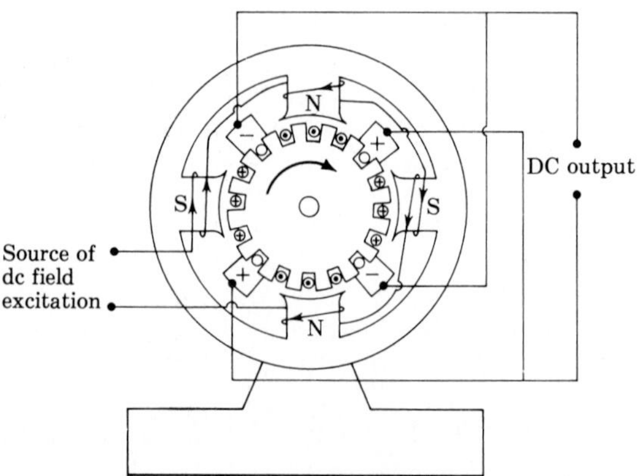

30.5 ■ Multipolar DC Generators

Generators such as the ones discussed so far, having one north pole and one south pole, are called *bipolar generators*. Commerical generators, especially large ones, commonly have four, six, eight, or more poles (an even number). The greater the number of poles, the slower the rotation necessary to produce a given emf. A four-pole machine rotates half as fast as a two-people machine to develop the same voltage. The poles of a four-pole generator (Fig. 30.13) are magnetized alternately north and south; alternate brushes are positive and connected to the same generator terminal. Each of the four flux paths leaves the north pole, passes through part of the armature, enters the south pole of an adjacent field pole, and completes the path through the pole cores and the yoke.

Brushes in the dc generator make contact with each coil when the coil is nearly parallel with the magnetic field and the torque on the armature is nearly constant.

30.6 ■ EMF Field of DC Generators

In Sec. 29.4 we discussed the *average* or *mean* voltage generated in a single wire moving at right angles to a magnetic field. The emf of any dc generator can be computed by finding the average emf induced in each conductor in the armature and multiplying this value by the number of conductors wound in series around the armature. The average emf of the generator can be calculated from the formula.

$$\mathscr{E}=\frac{\phi PnZ}{60b} \qquad (30.8)$$

where ϕ = flux density (webers) per pole
P = number of poles
n = angular speed of armature, rev/min
Z = total number of conductors on armature
b = number of paths in parallel through armature

For any given generator, the quantities P, Z, and b are fixed. The average emf for a given generator then depends only on the flux per pole ϕ and the speed n. If we substitute K for all the fixed values or *constants* of a given generator, Eq. (30.8) can be simplified to

$$\mathscr{E}=K\phi n \qquad (30.9)$$

Thus it can be seen that generated emf can be raised or lowered by increasing or decreasing the speed of rotation of the armature and/or increasing or decreasing the flux per unit pole.

Illustrative Problem 30.4 A four-pole generator rotates at 1500 rev/min. It has 240 armature conductors arranged in four parallel paths. A flux of 3.6×10^{-2} weber extends from each magnet pole. Compute the average emf of the generator.

Solution

$\phi = 3.6 \times 10^{-2}$ Wb
$P = 4 \qquad n = 1500$ rev/min
$Z = 240 \qquad b = 4$

$$\mathscr{E} = \frac{\phi PnZ}{60b}$$

$$= \frac{3.6 \times 10^{-2}\ \text{Wb} \times 4 \times 1500\ \text{rev/min} \times 240}{60\ \text{s/min} \times 4}$$

$= 216$ Wb/s
$= 216$ V *answer*

30.7 ■ Efficiency of Generator

The power output of a dc generator is rated in W or kW units; the power output of an ac generator is rated in kVA units.

The efficiency of a generator is defined as the ratio of the power output to the power input. The power output of a dc generator normally is rated in watts or kilowatts. However, in ac circuits the inductance and capacitance of the external loads affect the power. Both the capacitance and the inductance of a load are variable factors that cannot be determined in advance. These factors will be studied in Chap. 31. The power output of alternators usually is rated in voltamperes (VA) or kilovoltamperes (kVA). The efficiencies of commercial generators usually range between 80 and 95 percent.

$$\text{Percent efficiency} = \frac{\text{power output}}{\text{power input}} \times 100 \tag{30.10}$$

For the dc generator, the power output is the product of the terminal emf and the current to the external circuit. The input to the generator will always be greater than the output because of power losses such as the I^2R heat loss in the armature and field coils, hysteresis losses, eddy current losses, and frictional losses.

Illustrative Problem 30.5 A dc generator whose output is 1500 kW is turned by a 2200-hp engine and uses 50 kW to excite the field. What is its efficiency?

Solution

$$\text{Power input} = 2200 \text{ hp} \times \frac{0.746 \text{ kW}}{1 \text{ hp}} + 50 \text{ kW}$$

$$= 1690 \text{ kW}$$

$$\text{Percent efficiency} = \frac{\text{Power output}}{\text{Power input}} \times 100$$

$$\frac{1500 \text{ kW}}{1690 \text{ kW}} \times 100$$

$$= 88.8 \text{ percent} \qquad \textit{answer}$$

DC MOTORS

30.8 ■ Generators and Motors Compared

A dc generator can be used as a dc motor. The construction of each, however, is modified so as to improve its operation as a generator or a motor, whichever is desired.

We have seen that a generator converts mechanical energy to electric energy. An electric motor does just the reverse; i.e., it converts electric energy into mechanical energy. In fact, any dc generator will run as an electric motor, and conversely a dc motor will operate as a generator.

In most dc motors, the armature is the rotating inner part (the rotor); and the magnetic field is produced by electromagnets attached to the heavy outer part of the motor (the stator).

Structurally, both the dc generator and the dc motor consist essentially of field magnets, armature, commutator, and brushes. Both are constructed in bipolar or multipolar form. One significant structural difference between motors and generators lies in the housing of the two machines. Since generators ordinarily operate in more sheltered locations, their construction is of the *open* type. Motors often operate in locations that are exposed to moisture, dust, foreign particles, and corrosive gases. The housings of motors are designed to protect the machine against these elements. The motors are generally either semi-enclosed or totally enclosed by the housing. Ventilating openings are provided in the semi-enclosed type of motor but are so placed as to protect the motor against falling foreign particles or liquids. Totally enclosed motors (Fig. 30.14) are completely sealed in the housing. The heat developed in these motors must be entirely dissipated by radiation from the housing or by circulating cooling fluids.

The load of a generator consists of devices which convert electric energy into other forms of energy, while the load of a motor consists of countertorques that tend to

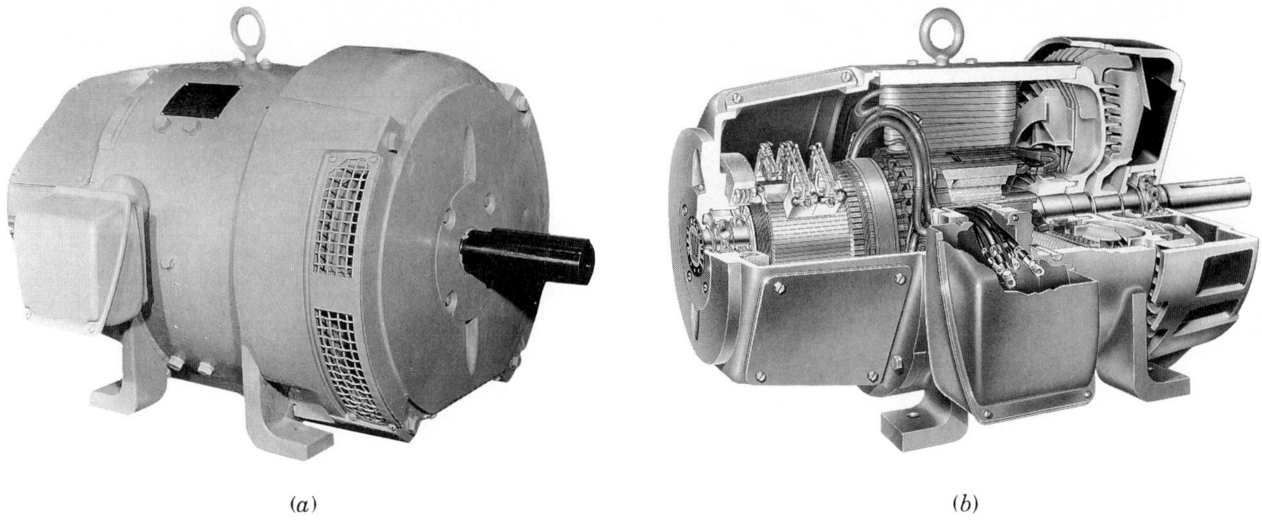

Fig. 30.14 (*a*) Totally enclosed fan-cooled dc motor. (*b*) Cutaway view showing the high-capacity system of controlled ventilation in which blowers and heat exchanger are mounted at the shaft end.

oppose the rotation of the moving part of the motor. When the load of a generator changes, there is a tendency for the emf of the generator to change; in the case of the motor, a change in load tends to result in a change in speed. The emf of a generator can be changed by adjusting the strength of the magnetic field or by changing the speed of the prime mover; the speed of a dc motor can be altered by changing the voltage impressed across the armature or by changing the field strength of the motor.

Generators are always started with no electric load in the external circuit; motors may be started either with or without an external mechanical load.

30.9 ■ Counter EMF of a Motor

The armature of a dc motor produces its own magnetic field, which reacts with the flux of the field coils to cause a rotary motion. However, the conductors on the rotating armature cut the lines of force of their own field just as if the machine were a generator driven by some prime mover. Consequently, there is induced a counter emf in the conductors of the armature which is opposite in direction to the flow of current in the conductors of the armature (Lenz's law). Thus every machine that is operating as a motor is at the same time a generator.

Lenz's law is a fundamental principle in the operation of electric motors and generators.

The effect of counter emf on the amount of current used by a motor can be demonstrated by a simple experiment. Connect an ammeter and an incandescent lamp in series with a small dc motor (Fig. 30.15). If we forcibly hold the armature station-

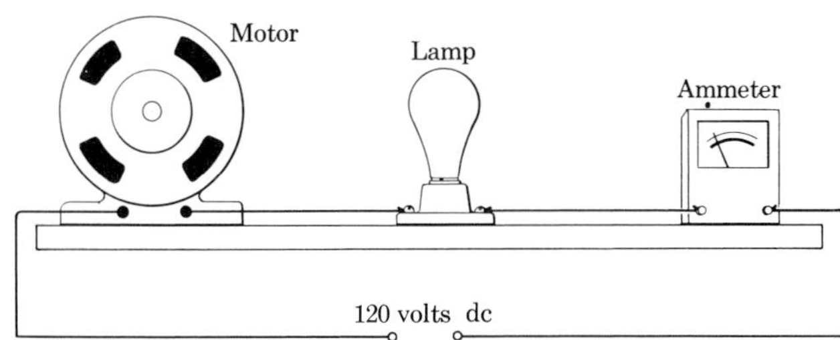

Fig. 30.15 An experiment to show the effect of counter emf. More current is required to start the motor than to keep it running.

ary while the line current is turned on, the lamp will glow with its normal brilliance; but when the armature is allowed to turn freely, the lamp grows dim and the ammeter reading drops considerably. Thus it is evident that the motor passes more current when the rotation is stopped or retarded than when it is allowed to run freely. Since the voltage of the line source has not changed in the experiment, the current must be diminished by the development of a counter emf which opposes that of the driving emf. The faster the motor turns, the greater the counter emf, and therefore the smaller the difference between the impressed emf and the counter emf. It should be clear that the counter emf can never be equal to, but must always be less than, the impressed voltage on the armature terminals. The difference in the two will be equal to the potential drop in the motor armature. The current flowing through the armature, then, depends upon and is controlled by the counter emf. Using Ohm's law, we find the dc armature current to be

Counter emf reduces the current that would otherwise be drawn by the armature of a dc motor.

$$I_a = \frac{V_a - \mathscr{E}_c}{R_a} \tag{30.11}$$

where I_a = armature current
V_a = impressed voltage on armature
$\mathscr{E}_c$ = counter emf generated in armature
R_a = armature-circuit resistance

The counter emf of a dc motor operating at any speed will be equal to the emf developed by the machine if it is operated as a generator at the same speed, provided the field strength of the machine remains the same in both cases. Hence the counter emf of a motor at any speed can be determined by operating the motor as a generator at that speed and measuring the emf generated.

Illustrative Problem 30.6 A small dc motor operates at 120 V and has an armature resistance of 2.8 Ω. When the motor is running at 3600 rev/min the armature current is 3.5 A. (*a*) What counter emf is developed in the armature? (*b*) What current would flow through the armature if it were held fast and not permitted to turn?

Solution

(*a*)

$$I_a = \frac{V_a - \mathscr{E}_c}{R_a}$$

$$3.5 = \frac{120 - \mathscr{E}_c}{2.8}$$

$$120 - \mathscr{E}_c = 2.8(3.5)$$

$$\mathscr{E}_c = 120\ \text{V} - (2.8\ \Omega)(3.5\ \text{A})$$

$$= 110\ \text{V} \qquad \textit{answer}$$

(*b*) If the armature is not permitted to rotate, the value of $\mathscr{E}$ will be 0.

$$I_a = \frac{V_a - \mathscr{E}_c}{R_a}$$

$$= \frac{120\ \text{V} - 0\ \text{V}}{2.8\ \Omega}$$

$$= 42.9\ \text{A} \qquad \textit{answer}$$

Such a large current would burn out the windings of a small motor.

The counter emf regulates the amount of current drawn by a motor; hence a motor is, to some extent, a self-regulating device. No counter emf exists in the motor until it begins to turn. When the motor first starts to turn, the current passing through the armature is very large. In order to avoid burning out the motor with these high starting currents, adjustable starting resistors to take the place of counter emfs are frequently inserted in series with the motor. As the speed increases, the resistance may be cut out gradually because of the increasing counter emf. When the motor attains its normal speed, all the adjustable resistance can be cut out. When the motor turns at constant speed, the impressed voltage will be balanced by the counter emf and by the drop in potential due to resistance. If the load upon the motor is increased, the torque being developed no longer is sufficient to overcome the larger load. Hence the speed of the motor falls. This decreases the counter emf, permitting a greater amount of current to flow through the armature, which in turn produces the larger torque required.

If a dc motor is stalled under a load, the counter emf will drop to zero. The resultant net voltage will cause an increased circuit current which could cause the motor to overheat and even burn out.

AC GENERATORS

30.10 ■ The AC Generator or Alternator

In dc generators the armature windings are always placed on the rotating part, called the *rotor*, while the field poles are placed on the stationary part, called the *stator*. This is necessary in order to provide a means of converting the ac voltage generated in the armature windings (the rotating coils) to dc voltage at the terminals of the windings by the use of a commutator. Almost any dc generator can be converted to an ac generator, commonly called *alternator*, by replacing the commutator by a pair of slip rings properly connected to the armature windings.

Large ac generating plants are usually of two kinds. In hydroelectric generating plants, slowly rotating water turbines furnish the mechanical energy to turn the alternator rotor. The alternators for hydroelectric generators use four or more poles to produce the 60-Hz current. In steam-generating plants, heat from burning oil, gas, or coal, or from nuclear reactors is used to produce steam. The steam drives a rapidly rotating steam turbine, which, in turn, rotates the rotor. The alternator in these generators uses only two or four poles.

In most dc generators, the field is stationary and the armature rotates, while in ac generators the armature usually is stationary and the field rotates.

The larger generators generate emfs of thousands of volts and may produce thousands of amperes of current in the armature. Consequently the armature coil must be very heavy and well insulated. If the armature is the rotor, the contact between the slip rings and brushes tends to produce serious losses of energy at these high voltages and currents. For this reason it is better to have the armature remain stationary and to rotate the field coils within it.

You will remember that it is the relative motion between the armature windings and field magnets that is essential to any generator; it does not matter which of the two revolves. Alternators can be classified into two types: (1) *field stator*, in which the field windings are stationary and the armature revolves in the field; and (2) *field rotor*, in which the armature windings are stationary and the field poles revolve. Slip rings are used in either type to conduct current to and from the rotor.

All but the small low-voltage alternators are built with a *stationary armature* and a *revolving field*. The construction of this type of alternator is more economical and simplifies the problems of insulation of the alternator. Perhaps the chief advantage of this type of construction lies in the fact that relatively high voltages can be drawn from the *stationary* armature to the switchboard without the use of sliding-contact slip rings and brushes. Also the problem of centrifugal forces acting on the armature windings is eliminated by making them stationary. Every alternator must have its field coils energized with direct current from some kind of separate dc generator, called an *exciter* (see also Chap. 32).

30.11 ■ Revolving-Field Alternator

The field structure of a revolving-field alternator may be of two distinct types—the *salient-pole type* and the *cylindrical type*. In the salient-pole type the field poles are laminated pole pieces which, with their field coils, are mounted on the outer rim of a

Armature cores are laminated to reduce losses from eddy currents.

steel *spider*, which in turn is keyed to the shaft. This type of construction is illustrated in Fig. 30.16. The field windings are generally form-wound, and the completed coils are then fitted around the pole pieces and attached rigidly in place. The coils are connected in series in such a way as to give alternate polarity to the poles. The ends of these field windings are then connected to the slip rings. The salient-pole-type alternator generally is a slow-speed generator driven by diesel engines or water turbines and may have 30 or more poles in the rotor.

The *stationary* armature of the alternator consists of carefully insulated form-wound coils placed in iron core slots supported by a circular cast-iron or fabricated steel frame. The core of the armature is formed of sheet-steel laminations placed together and held rigidly in place by suitable wedges. A partly wound stationary armature with some of the coils placed in their core slots is shown in Fig. 30.17.

High-speed alternators have cylindrical rotors whose diameters are considerably shorter than their axial lengths. The diameters must be relatively small to withstand the large centrifugal stresses that are set up at high speed. The prime mover of this type of alternator is usually a steam turbine. The rotor must be an extremely well-constructed and rigid structure, and is generally forged with the shaft in a single piece to the proper dimensions. The rotor is then slotted longitudinally by a milling machine to a depth sufficient to house the field windings. The windings for the rotor are form-wound and fitted in the slots and held in place by steel wedges placed at the outer ends of the slots. The ends of these windings are further held in place by a steel retaining ring attached to the rotor. These ends are then attached to slip rings from which the direct current for excitation is drawn through brushes riding on the surface of the rings. The rotor of a turbo-alternator of this type is shown in Fig. 30.18.

30.12 ■ Frequency of Alternators

The frequency of an ac generator depends upon the number of field poles and the speed of the rotor. A complete cycle is generated in a given armature coil when a pair

Fig. 30.16 Rotor of a salient-pole alternator. The rotor, containing 18 poles, is driven by a diesel engine at 333 rev/min and delivers three-phase 60-Hz current rated at 6300 and 3640 V. (General Electric Co.)

Fig. 30.17 Coils being placed in slots of stationary armature for a steam-driven generator. (General Electric Co.)

of poles (one north and one south) is moved past the coil. Thus a complete cycle is developed in a bipolar alternator when the rotor makes one complete revolution. In a four-pole alternator two cycles will be developed per revolution.

Let

N = rotor speed, rev/min
p = number of poles
f = frequency, Hz

Then the frequency can be found by the equation

$$f = pN/120 \tag{30.12}$$

Fig. 30.18 The rotor of a steam-turbine generator is being machine-slotted on a huge slot miller. (General Electric Co.)

Illustrative Problem 30.7 Find the frequency of an alternator with 30 poles that is revolving at 240 rev/min.

Solution This alternator will have 30 poles, so $p = 30$.

$$f = \frac{pN}{120}$$

$$= \frac{30 \text{ poles} \times 240 \text{ rev/min}}{120}$$

$$= 60 \text{ Hz} \qquad \textit{answer}$$

Illustrative Problem 30.8 How many revolutions per minute must an eight-pole alternator make to generate a 60-Hz current?

Solution The alternator has eight poles. So $p = 8$. Then

$$N = \frac{120\,f}{p} = \frac{120 \times 60}{8} = 900 \text{ rev/min} \qquad \textit{answer}$$

As previously noted, the most common frequency in the United States is 60 Hz. In some localities 50-Hz currents are in use, notably in most European countries.

In an automobile alternator, the frequencies vary a great deal because of the variation of engine speeds. This poses no problem, since the output is rectified (using diodes). The voltage, however, must be controlled to avoid large output voltages. This is done by using a voltage regulator which uses electric relays or semiconductor devices.

Illustrative Problem 30.9 A 60-cycle generator develops a maximum voltage of 155 V when the rotor is turning 600 rev/min. (*a*) How many poles does it have? (*b*) What is the instantaneous voltage when $t = 5$ ms? (Assume $e = 0$ when $t = 0$)

Solution

(***a***)

$$f = \frac{pN}{120}$$

Then

$$p = \frac{120\,f}{N} = \frac{120(60)}{600} = 12 \text{ poles} \qquad \textit{answer}$$

(***b***) After 0.005 s

$$\theta = \frac{600 \text{ rev}}{1 \text{ min}} \times \frac{1 \text{ min}}{60 \text{ s}} \times 0.005 \text{ s} \times \frac{360°}{1 \text{ rev}} = 18°$$

Then

$$\begin{aligned} e &= e_{max} \sin 18° \\ &= 155 \text{ V} \times 0.309 \\ &= 48 \text{ V} \qquad \textit{answer} \end{aligned}$$

30.13 ■ Polyphase Alternators

Alternators may be grouped not only according to their design and construction but also as to the type of connections of their armature windings. When all the armature windings are connected in series so that their individual emfs will add together, the alternator is a *single-phase* machine. This is distinguished from the *polyphase* alternator, which generates two or more alternating emfs at the same time.

The *two-phase alternator* has two separate and independent armature windings which are mounted 90° apart. The alternating emfs developed in the two windings are said to be 90° apart in phase. Figure 30.19*a* illustrates the emf curves of a two-phase alternator. It can be seen from the figure that at no time during a cycle will the emf be zero if a two-phase alternator is used.

Engineers have found the operating characteristics of three-phase machinery to be far superior to those of either the single-phase or two-phase equivalent. The three-phase alternator has three separate sets of windings symmetrically placed on the

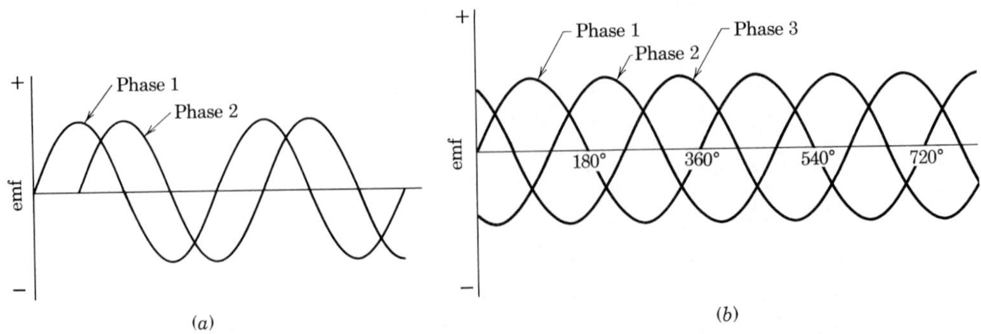

Fig. 30.19 EMF curves of a (*a*) two-phase alternator, and (*b*) three-phase alternator.

armature so that three alternating emfs of the same frequency, differing by 120 electrical degrees, are obtained. The emf curves for such an alternator are shown in Fig. 30.19*b*. The three sets of coils that are mounted 120° apart may each be brought out from the alternator to form three separate single-phase circuits. However, these coils are generally interconnected in such a way as to bring out only three, or sometimes four, terminals from the alternator. (See Sec. 32.10.)

AC MOTORS

30.14 ■ AC Motors vs. DC Motors

AC motors are used much more widely in industry than dc motors. There are several reasons for this. Alternating current can be transmitted over long distances with little loss of power; hence most utility companies offer only ac service to the consumer. The commutators of dc motors are subject to sources of trouble and must be frequently checked and serviced. Sparks passing between the commutator and brushes of dc motors not only present a source of power loss but may also be a serious hazard in areas filled with combustible gases or dust particles. Most ac motors do not require brushes and a commutator.

Although the ac motor is generally less expensive to build, is more rugged, and can stand more abuse, dc motors are used in a large number of applications where large torques at slow speed are required. Speed in the dc motor can be regulated with greater accuracy than in the ac motor.

30.15 ■ Universal (Series) Motors

The series motor is essentially a dc motor (with commutator segments) in which the field coils and the armature are connected in series. Such a motor will operate effectively on either an alternating or a direct current because the current in the field and that in the armature reverse at the same instant. Since both the field polarity and the direction of the current through the armature change together, the armature and the field poles repel and attract each other in such a way as to give a continuous rotary motion to the armature. Since the windings of the stator and the rotor are connected in series, the speed of the motor increases with decrease in load and the starting torque is high. This type of motor is called a *universal motor*. It is used when it is desirable to use a motor that operates satisfactorily on either ac or dc circuits. It has application where high speeds are required. Small universal motors with load speeds of 10,000 rev/min are not uncommon. It also is used when a motor is required which automatically adjusts itself to the magnitude of the load—high speed for light loads and low speeds for heavy loads.

Universal motors find wide application in vacuum cleaners, sewing machines, electric shavers, electric fans, hair dryers, kitchen food mixers, portable drills, power saws, pipe threaders, small grinders, and certain woodworking lathes.

30.16 ■ Induction Motors

The simplest and most common type of ac motor is the *induction motor*, invented in 1888 by Nikola Tesla (1856–1943). Like any other motor, it consists of a stationary part and a rotating part. However, the principle upon which the induction motor operates is quite different from those on which other motors are based. In all the motors discussed thus far the current has been "fed" into the *rotating* part of the machine through brushes and a commutator. That is, the electric energy has been conducted directly to that part of the machine where it is converted into mechanical energy. In the induction motor the *stator* is connected to the ac supply. The rotor is not connected to any supply but has current *induced* in it by inductive action from the stator. The induction motor will operate satisfactorily only when it is connected to an ac source of the frequency for which it was designed.

Nikola Tesla was an Austrian-born scientist and inventor, who became an American citizen in 1884. He discovered the principle of the rotating electric field and the induction motor. He invented radio control of ships and many other electrical devices.

Fig. 30.20 Typical small-horsepower induction motor disassembled, showing main components. (Siemens-Allis, Inc.)

Fig. 30.21 Squirrel-cage rotor assembly. The rotor fans are cast as an integral part of the squirrel-cage winding and are used to provide air turbulence and better heat transfer. (Robbins & Myers, Inc.)

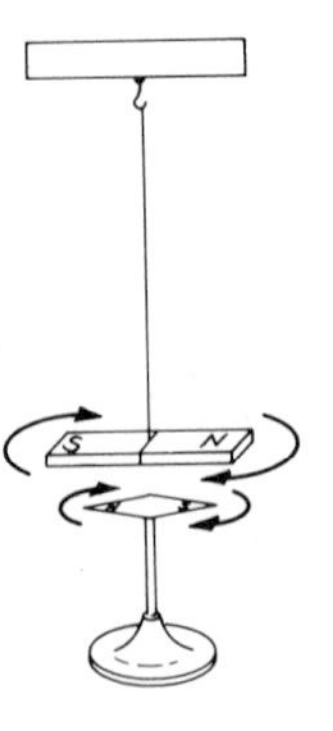

Fig. 30.22 Illustration of a rotating magnetic field.

The stator core is built of a stack of circular sheet-steel laminations with slots on the inner surface and parallel to the axis (Fig. 30.20). This core is attached to a cast-iron or cast-steel yoke for support. Stator windings are placed in these slots according to the number of phases and the required speed of the motor. The stator windings are very similar to those of the alternator from which it will draw its current.

The rotor of an induction motor is built up of a laminated-steel core attached to a cast spider that is then pressed onto a shaft. The rotor windings are placed in the rotor slots with the conductors parallel or approximately parallel to the shaft. These conductors are not insulated from the core since the current will flow more readily through the copper conductors than through the iron. The ends of the rotor conductors are joined (short-circuited) together by a cast ring. Thus the rotor of an induction motor is essentially a number of parallel copper bars cast into slots and joined together by a ring. Because the appearance of such a rotor is similar to a revolving squirrel cage, it is commonly called a *squirrel-cage rotor*. Figure 30.21 shows a typical squirrel-cage rotor with the copper windings cast into the steel rotor core.

The operating principle of the induction motor is based on a *rotating magnetic field*. This concept can be illustrated by suspending, with a piece of string, a horizontal bar magnet over a compass (Fig. 30.22). The compass needle will align itself parallel to the bar magnet. If the magnet is now rotated, its rotating magnetic field will cause the needle of the compass to follow. In the induction motor, the magnetic field revolves even though the windings which produce it do not move. The production of a rotating magnetic field in an induction motor can be illustrated by considering a two-phase alternator from which we may have two alternating currents of the same frequency but differing in phase by 90 electrical degrees. Curves of the two currents which differ in phase by 90° are shown in Fig. 30.23. Let us connect these currents to coils of a four-pole circular ring, as shown in Fig. 30.24

When the current of phase 1 has maximum value, the current of phase 2 will be zero. As a result, poles B and B_1 will be fully magnetized and poles A and A_1 will be unmagnetized. The flux will then be directed from north to south, as shown by the arrow in Fig. 30.24*a*. One-eighth of a cycle later (45 electrical degrees) the current in phase 1 will have diminished by as much as the current in phase 2 will have increased, and the magnetic flux will take the direction indicated in Fig. 30.24*b*. Another eighth of a cycle later the current in phase 1 will be zero and the current in

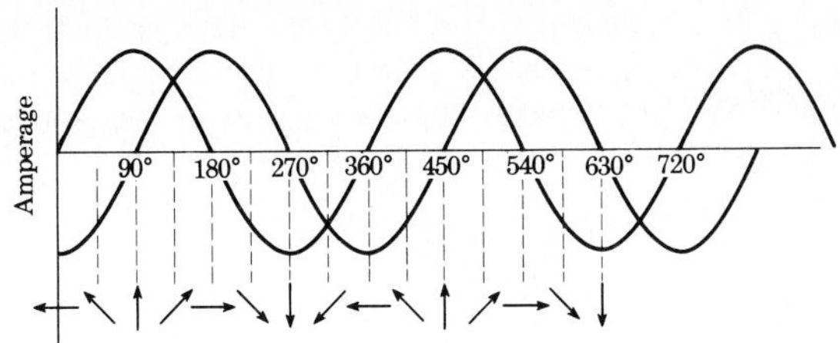

Fig. 30.23 Curves of two alternating currents 90° out of phase.

phase 2 will be at a maximum. At this stage, poles A and A_1 will be fully magnetized, poles B and B_1 will be unmagnetized, and the flux will have the direction indicated by the arrow in Fig. 30.24*c*. As the two-phase current continues to flow through the coils, the magnetic flux will continue to rotate, as shown by the arrows drawn in Fig. 30.23.

When the induction-motor rotor is placed in this rotating field, there will be emfs induced in the rotor conductors. Since the rotor windings form a series of closed paths, large currents will flow in the conductors. This current will produce a magnetic field about the rotor which will react with the magnetic field of the stator and cause rotation. It should be clear that the rotor can never spin as fast as the magnetic field. If it did, no lines of force would be cut by the rotor conductors, no current would be induced, and hence no torque would be exerted by the conductor.

Commercial induction motors have multipolar stators. These poles are not projecting poles, as shown in Fig. 30.24, but are embedded in slots in steel-laminated plates. Figure 30.25 shows a typical commercial stator of an induction motor. Most large ac motors in use today are induction-type motors. They have fair starting torques and fairly constant speeds with varying loads.

The most widely used type of ac motor is the induction motor.

30.17 ■ Synchronous Speed and Slip

The speed at which the *magnetic field* revolves is called the synchronous speed of the motor. It corresponds to the value of N in Eq. (30.12) and is determined by the frequency of the source of alternating current and by the number of poles in the stator. Thus, solving Eq. (30.12) for N yields

$$\text{Synchronous speed } N_s = \frac{120 \times \text{frequency } f}{\text{number of poles}} \qquad (30.13)$$

or

$$N_s = 120\, f/p$$

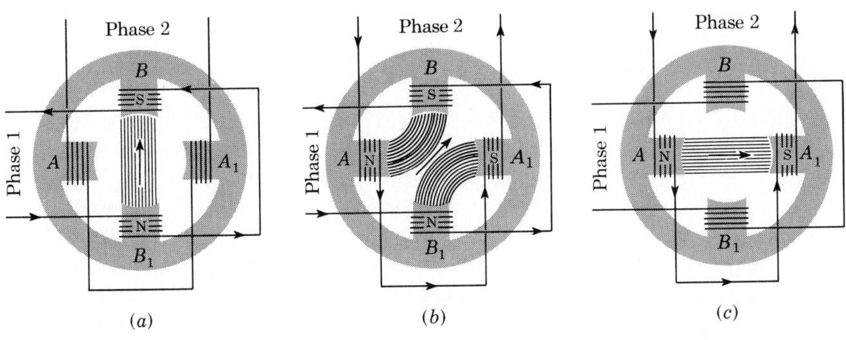

Fig. 30.24 Fields produced in the stator of an induction motor by a two-phase alternator.

Fig. 30.25 Wound stator of a polyphase induction motor. (General Electric Co.)

where N_s is expressed in revolutions per minute. We have pointed out that an induction motor will not function unless the speed of the rotor is less than that of the revolving field. The difference between the rotor speed and the synchronous speed is known as the *slip* of the motor. It can be expressed in revolutions per minute but is usually expressed as a percentage of the synchronous speed.

$$\text{Percent slip } S = \frac{\text{synchronous speed} - \text{rotor speed}}{\text{synchronous speed}} \times 100$$

Or, written as an equation,

$$S = \frac{N_s - N_r}{N_s} \times 100 \tag{30.14}$$

The equation for the speed of the induction motor then becomes

$$N_r = \frac{120\, f(1 - S)}{p} \tag{30.15}$$

where N_r is the rotor speed expressed in revolutions per minute.

Illustrative Problem 30.10 Calculate the speed of (*a*) a two-pole, (*b*) a four-pole, and (*c*) a six-pole induction motor connected to a 60-Hz source of current, assuming a 2.8 percent slippage in each case.

Solution
(***a***) Two-pole motor

$$N_r = \frac{120 \times 60}{2}(1 - 0.028)$$

$$= 3500 \text{ rev/min} \qquad \textit{answer}$$

(***b***) Four-pole motor

$$N_r = \frac{120 \times 60}{4}(1 - 0.028)$$

$$= 1750 \text{ rev/min} \qquad \textit{answer}$$

(***c***) Six-pole motor

$$N_r = \frac{120 \times 60}{6}(1 - 0.028)$$

$$= 1170 \text{ rev/min} \qquad \textit{answer}$$

The most common speeds for large commercial motors are 1750 and 1150 rev/min.

30.18 ■ Synchronous Motors

In the induction motor the current necessary to produce torque is induced in the rotor only when the rotor rotates at a speed slower than the rotating magnetic field. The speed of most electric motors changes as the load changes; the greater the load the less the speed, and vice versa. But if we made the rotor an electromagnet by a means other than the induced current, there would be no need for the rotor to slip behind the rotating field. One way this could be done would be to use a dc source to magnetize the rotor. Then, if by some external means the rotor is brought up to the speed of the rotating field, its poles would lock in step with the rotating field. The rotor would then be *synchronized* with the rotating field. Such a motor is called a *synchronous motor*. It is the only truly constant-speed ac motor. Let us consider two alternators connected in parallel and supplying current with the same frequency and voltage and in phase. If the source of mechanical energy (the prime mover) of one of the generators is removed, that generator will continue to turn, but as a motor, and will draw its electric energy from the second alternator. The speed of the motor will be exactly the same as when operating as a generator but will slip back a few degrees from the position it held when operating as a generator. Such a synchronous motor is essentially a generator operating as a motor.

If the synchronous motor has the same number of poles as the alternator from which it is drawing its electric energy, it will rotate with exactly the same speed as the alternator. If the motor has twice as many poles as the generator, it will turn exactly half as fast as the generator; if it has half as many poles, it will turn twice as fast.

Fig. 30.26 A 4,150-hp synchronous motor (right foreground) powers a grinding mill in a copper mining project in Arizona. (Siemens-Allis, Inc.)

Modern synchronous motors do not need external starting motors. Their rotors are constructed as a combination squirrel cage induction motor and a wound-type motor with slip rings. The motor is started as an induction motor. The induction motor brings the rotor speed to about 95 percent of the speed of rotation of the stator field. The direct current is then applied to the wound portion of the rotor through the slip rings. This sets up north and south magnetic poles in the rotor which lock in step with the poles of the rotating field. Synchronization results, with the rotor revolving at the same speed as the rotating field.

Once synchronization is established the squirrel-cage portion of the rotor will not cut any lines of force of the rotating field. Hence, in effect, it is removed from the operation of the motor.

The speed of rotation of a synchronous motor depends upon (1) the frequency of the alternating current supplied to it and (2) the number of poles on the stator. As a formula $N = 120f/p$ (see Eq. 30.12). The number of poles in commercial synchronous motors generally ranges from four to 100 poles.

Synchronous motors are used where constant speed is essential. Some of the common applications of this type of motor are driving extremely large air and refrigerant compressors, large blowers, fans, pulverizers, and dc generators which need a large source of direct current. These motors are usually larger than 100 hp (Fig. 30.26). Small, nonexcited single-phase synchronous motors are used to drive electric clocks, phonograph turntables, and other devices where constant speed is essential.

30.19 ■ Measurement of AC Energy Consumption

Power companies charge the consumer according to the amount of electric energy used. This electric energy is measured by means of a watt-hour meter. *Energy* is the *product of power and time*, and consequently an instrument must be used which takes into account both the power and the time. Electric power is measured in terms of watts, which is the product of amperes and volts. The common commercial *ac watt-hour meter* (Fig. 30.27*a*) is essentially a small induction motor in which the speed of rotation is a measure of the power being consumed at that instant. The total number of revolutions of the meter rotor in a given time interval is then a measure of the total energy used during that interval. Figure 30.27*b* illustrates the essential parts of the meter: the *electromagnetic field*, a *rotor*, a *damping system*, and a *register*. The field is set up by a voltage coil and two current coils mounted on an iron core. The voltage coil, wound with many turns of fine wire in order to have a very high inductance, is connected directly across the line. The current in this coil is proportional to the line voltage V and lags behind the voltage by nearly 90°. In order to make the voltage-coil flux lag exactly 90° behind the line voltage, a small "lagging coil" is used. The current coils consist of only a few turns of heavy wire and are connected in series with the line. The flux produced by the current coils is proportional to, and is in phase with, the line current I.

(*a*)

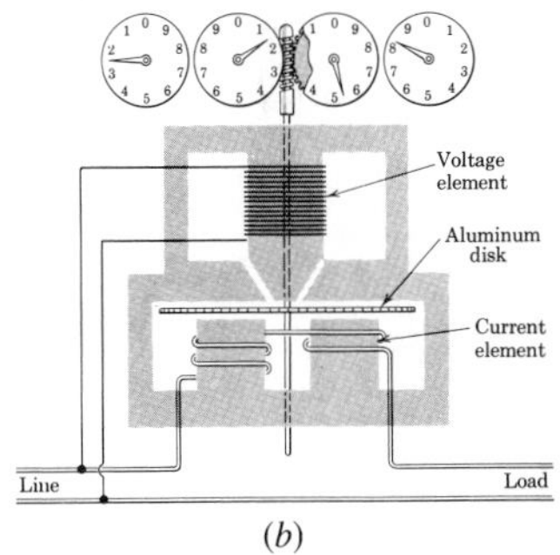

(*b*)

Fig. 30.27 (*a*) Single-phase watt-hour meter. (*b*) Essential parts of the meter. (General Electric Co.)

The rotating element (rotor) of the meter is an aluminum disk mounted on a vertical shaft. The shaft rotates on jeweled bearings to reduce friction to a minimum. Since the fluxes produced by the voltage and current coils are out of phase, *eddy currents* will be induced in the disk. These eddy currents will set up their own field, which will react with that of the stator to produce a motor action. The speed of the rotating element is proportional to the product of V and I. The motor would run too fast with a given load or, because of its inertia, would continue to run for some time when the load is removed, unless some damping or braking device were provided. This is done by placing permanent magnets above and below the aluminum disk so that their fluxes are cut by the disk. This arrangement develops eddy currents in the disk which are proportional to the speed and provides a braking torque proportional to that speed.

QUESTIONS AND EXERCISES

1. What is a dynamo?

2. Name the essential parts of a dc generator. What are the essential parts of an alternator?

3. Upon what factors does the emf of a generator depend? Which factors are constant and which are variable?

4. Why does a generator require more power to turn it when it is delivering current to a load than when there is no load in the external circuit?

5. Why does a motor take more current as it starts from rest than when it is turning at its rated speed?

6. List the advantages and disadvantages of dc series and dc shunt motors. List instances where each type could be used.

7. Sketch the windings for a dc compound motor. What are its advantages?

8. Give three practical applications of electromagnetic induction.

9. Under what condition is the induced voltage in a conductor constant?

10. What determines the speed of an induction motor?

11. What is a universal motor? How is it constructed?

12. Name three *prime movers* used to drive commercial alternators.

13. What is meant by the *generator action* of a *motor*?

14. What is meant by *synchronous speed* and *slip*?

15. What conditions must be fulfilled before a synchronous motor will function properly?

16. What is the effect on the induced emf of a generator if (*a*) the flux per pole is doubled, (*b*) the speed of the armature is doubled?

17. Which of the following is true? The magnitude of the generated voltage (induced in the armature) in a shunt-wound dc generator is (*a*) directly proportional to the armature speed, (*b*) directly proportional to the power generated, (*c*) inversely proportional to the field current, (*d*) inversely proportional to the field flux, (*e*) none of these.

PROBLEMS

Group One

1. A 25-hp motors draws 70 A at 440 V. What is its efficiency?

2. What is the frequency of a sine wave that has a period of 1 ms?

3. What emf is induced in a 24-turn coil subjected to a flux change of a 3 Wb in 2 min?

4. A four-pole dc generator with a fixed field excitation develops an emf of 100 V when operating with an armature speed of 1800 rev/min. At what speed must the armature rotate to develop 120 V?

5. A 16-pole alternator is driven at 240 rev/min. What is the frequency of the current?

6. What will be the synchronous speed of a four-pole synchronous motor operating from a 50-Hz power supply?

7. At what speed must a 16-pole 60-Hz generator be driven to develop its rated frequency?

8. A four-pole synchronous motor operates from a 50-Hz power supply. What will be its synchronous speed?

9. What is the frequency of an alternator that has 24 poles if it is rotating at 250 rev/min?

10. How fast must the rotor of a two-pole turbo-alternator rotate to produce current with 60-Hz frequency?

11. How many poles does a 60-Hz generator have if it rotates at 300 rev/min?

12. What should be the speed (revolutions per minute) of an ac generator with four pairs of poles to develop 60 Hz?

13. What is the speed of a synchronous motor operating from a 60-Hz dc supply if it has 24 poles?

Group Two

14. Each of the 60 coils of a dc generator is 48 cm long and 24 cm wide. The generator runs at 1200 rev/min and uses a flux of 0.6 T. What is the peak voltage generated?

15. What is the average emf generated by a two-pole generator that has 200 parallel paths in the armature and the conductors cut a flux of 8×10^{-3} Wb per pole when the armature is turning 1800 rev/min?

16. A dc shunt-wound motor is connected to a 120-V line. The armature generates a back emf of 115 V when it is carrying a current of 15 A. What is the resistance of the armature?

17. A shunt-wound generator has an armature resistance of 0.10Ω and develops a total emf of 120 V when driven at its rated speed. What is the terminal voltage of the generator if the armature current is 55 A?

18. A 400-cycle alternator generates a maximum voltage of 340 V at 4000 rev/min. (*a*) How many poles does it have? (*b*) If $e = 0$ when $t = 0$, find the generated emf when $t = 375 \times 10^{-6}$ s.

19. The armature of an ac generator turns at 3600 rev/min, has a length of 60 cm, a diameter of 30 cm, and has 25 turns. If the field strength is 0.5 T, (*a*) What is the peak voltage? (*b*) the frequency of the generated voltage if the generator has one pair of poles?

20. A four-pole generator has a drum armature with a cylindrical core 30 cm long, 15 cm in diameter, and 100 armature conductors arranged in four parallel paths. The flux density in the air gap is 0.6 T. The armature is driven at 12 rev/min. What is the average emf generated?

21. The coil of an ac generator consists of 24 turns of wire each with an area $A = 0.10\ m^2$ and a total resistance of 36Ω. The coil rotates in a magnetic field of $B = 0.6$ T at a constant frequency of 60 Hz. Determine (*a*) the maximum induced emf, (*b*) the maximum induced current.

22. A six-pole generator has 300 armature conductors arranged in four parallel paths. Compute the average emf generated if the armature turns at the rate of 1750 rev/min in a field flux of 4.2×10^{-2} Wb.

23. The armature of a motor has a resistance of 1.5Ω and operates on a 120-V line. In starting the motor, a rheostat is connected in series with the armature. What resistance should the rheostat add to the circuit to limit the starting current to not more than 6 A?

24. Compute the slip of an eight-pole induction motor operating from a 60-Hz ac supply if the motor turns at 875 rev/min.

25. What is the synchronous speed of a six-pole 60-Hz induction motor? If the motor has a slip of 5 percent, what is the speed of the motor in revolutions per minute?

Group Three

26. A two-pole shunt generator has an armature of 450 conductors around the armature core arranged in two parallel paths. Compute the back emf it develops when run as a motor at 1500 rev/min if there is a flux of 12×10^{-3} Wb per pole.

27. Show how to connect a double-throw double-pole switch to a dc shunt-wound motor with a separately excited field coil so that the rotation of the motor can be reversed by the switch. (Draw a circuit diagram.)

28. A shunt-wound dc generator has an emf of 110 V when it delivers 1.98 kW to the line. The resistance of the field coils is 220Ω and that of the armature is 0.450Ω. What is (*a*) the emf of the generator, and (*b*) the efficiency of the generator?

29. A shunt-wound generator has an armature resistance of 0.08Ω and a shunt-field resistance of 120Ω. The generator delivers 50 kW at 240 V to an external load. What power is being developed in the armature?

30. A four-pole shunt-wound motor has 480 conductors arranged in four parallel paths on the surface of the armature core. Each conductor cuts a flux of 0.030 Wb per pole. (*a*) What is the back emf developed by the armature when rotating at 1000 rev/min? (*b*) What is the impressed voltage if the armature resistance is 0.14Ω and the armature current is 50 A?

CHAPTER 31 ALTERNATING CURRENT

Resistance is the factor that controls current, voltage, power, and energy in dc circuits. Circuits containing resistors, capacitors, and inductors have been studied, but only in circuits connected to a dc source of emf.

Alternating current behaves quite differently from direct current in circuits containing capacitors and/or inductors. In this chapter we will study how series combinations of resistors, inductors, and capacitors affect the characteristics of alternating current.

Since alternating-current theory requires for its understanding more advanced mathematics than can be developed in this text, only the elemental aspects of ac circuits will be given. All the basic laws developed for steady dc conditions apply to ac circuits, but the methods of applying them can be different. Some of the study will involve stating mathematical equations without formal proofs.

31.1 ■ The AC Cycle

Each revolution of the coil in Fig. 30.1 results in one complete cycle of emf and current. The number of cycles generated per second is called the *frequency* (f) of the alternating emf (or current). The *period* is the time to complete one cycle. One-half cycle is called an *alternation*. Often the word *cycle*, with the words *per second* being omitted, but understood, is used to express frequency. Thus, we speak of a 60-cycle current. The 60-cycle current will have a period of $\frac{1}{60}$ second (0.0167 s). A 60-cycle current completes 120 alternations per second.

In the mksa system, the unit of frequency is the hertz (Hz).

Today the word *hertz* (Hz) is more commonly used in place of cycles or cycles per second. The most prevalent ac frequency generated for commercial power in the United States is 60 Hz. In Europe, Asia, and Africa 50 Hz is a common frequency. Lower frequencies, such as 25 Hz and $16\frac{2}{3}$ Hz, are used for electric railways in many countries. In aircraft and guided missiles 400-Hz generators are widely used. Radiobroadcast frequencies are of the order of a million hertz; and some superhigh frequency transmitters operate at several billion hertz.

Instantaneous emf's (or currents) in terms of maximum emf's (or currents) and frequencies were established as follows, in Chapter 30:

$$e = e_{\text{max}} \sin 2\pi f t \qquad (30.5')$$

and

$$i = i_{\text{max}} \sin 2\pi f t \qquad (30.7')$$

Illustrative Problem 31.1 An alternator with 24 poles has a speed of 3600 rev/min and develops a maximum voltage of 325 V. What is (a) the frequency of the alternating current, (*b*) the period of the alternating current, and (*c*) the equation for the instantaneous emf e at any time t?

Solution

(***a***) Using Eq. (30.12),

$$f = \frac{pN}{120}$$

$$= \frac{24 \times 3600}{120}$$

$$= 720 \text{ cps (Hz)} \qquad \textit{answer}$$

(***b***)

$$\text{Period } T = \frac{1}{f}$$

$$= \frac{1}{720 \text{ s}^{-1}}$$

$$= 1.4 \text{ ms} \qquad \textit{answer}$$

(***c***) Use Eq. (30.5′) to find e

$$e = 325 \sin 2\pi(720)t$$

$$= 325 \sin 1440\pi t \qquad \textit{answer}$$

31.2 ■ Effective Current and Effective Voltage

A dc voltmeter or ammeter will read zero if it is used to measure alternating currents. A glance at the curve of Fig. 30.2 will reveal that the mathematical average of the values for emf (and, hence, for current also) for one complete cycle is zero. As much of the curve lies above emf = 0 as lies below it. We defined one ampere of current in terms of the force of mutual attraction or repulsion between two parallel conductors carrying direct current. This definition cannot be used for alternating current because the forces alternate 120 times a second. We know that alternating currents do possess energy. Hence, the ampere for alternating current must be defined in terms of some property which is independent of the direction of the current.

The heat produced by an electric current is independent of the direction of the current. Thus

An alternating current of one ampere is defined as that current which produces the same average heating effect as a dc ampere under the same conditions.

The heating effect of all types of current is given by the equation: Power loss = i^2R.

The ampere thus defined is the unit of *effective value* of the alternating current.

The heat (power) dissipated at any instant in a metallic resistor which has a resistance R and which carries an instantaneous current i is (from Eqs. 27.10 and 30.7′)

$$P = i^2R = i_{\max}^2 R \sin^2 (2\pi ft)$$

Squaring i will have the effect of eliminating the negative sign for i. Figure 31.1 shows how i, which is equal to $i_{\max} \sin 2\pi ft$, and i^2 vary with angle θ (where $\theta = 2\pi ft$). The dashed line represents the value of $\frac{1}{2}(i \max)^2$. It should be evident from the graph for i^2 that as much of the curve lies above as below the dashed line. Thus, the average (mean) power for a complete cycle of alternating current is

$$P = \frac{(I_{\max})^2}{2} R \qquad (31.1)$$

The effective value I_{eff} of the alternating current is equivalent to the direct current that produces the same power dissipation in R.

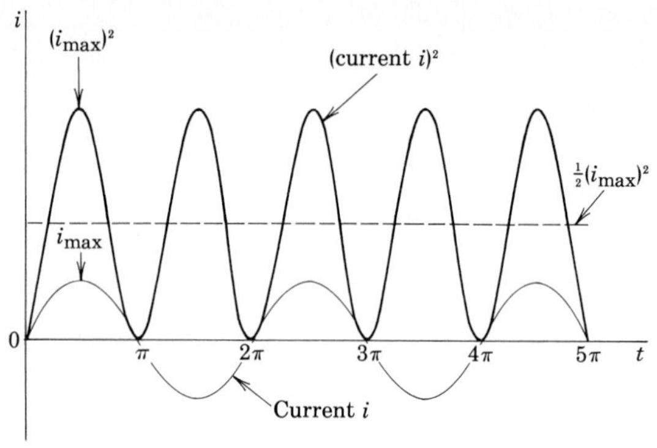

Fig. 31.1 For alternating current $i_{eff} = \sqrt{\overline{i^2}} = \sqrt{\dfrac{(i_{max})^2}{2}} = 0.707 i_{max}$.

$$P = I_{eff}^2 R = \frac{(I_{max})^2}{2} R$$

or

$$I_{eff}^2 = \frac{(I_{max})^2}{2}$$

and

$$I_{eff} = \frac{I_{max}}{\sqrt{2}} = 0.707 I_{max} \tag{31.2}$$

The effective value of an ac current (or voltage) is the square root of the average (mean) of the instantaneous values squared of the current (or voltage).

Thus, a direct current with a value of 0.707 of the maximum value of an alternating current has the same power output or heating effect as the alternating current.

The electric power for direct current can also be expressed in terms of V and R as $P = V^2/R$. Thus we can follow the same reasoning as above by drawing graphs of V and V_{max} as a function of t to determine that the average power

$$P = \frac{(V_{max})^2/2}{R} \tag{31.3}$$

Effective currents and effective voltages are equal to 70.7 percent ($1/\sqrt{2}$) of their maximum values.

from which it can be shown that

$$V_{eff} = \frac{V_{max}}{\sqrt{2}} = 0.707 V_{max} \tag{31.4}$$

Equations (31.2) and (31.4) can also be written

$$I_{max} = \sqrt{2} I_{eff} = 1.414 I_{eff} \tag{31.5}$$

and

$$V_{max} = \sqrt{2} V_{eff} = 1.414 V_{eff} \tag{31.6}$$

It should be noted that when a house is wired to use 120 V for its appliances, the peak value of the household supply from the power lines is ($\sqrt{2} \times 120V$) or about 170 V. Thus, the wiring and insulation must be able to withstand this peak voltage.

The effective value of the current is a measure of its ability to do work or to produce heat.

In measuring alternating currents we are interested in measuring effective values. Hence, unless otherwise noted, we will assume the terms *ampere* and *volt* in ac circuits to mean effective values. We will also drop the subscript *eff*. I and V for ac circuits will refer to effective current and voltage.

Most ac meters are so made that they read directly the values of $\sqrt{(i^2)_{av}}$ and $\sqrt{(e^2)_{av}}$ which we will label I and $\mathcal{E}$ (or V).

NOTE: $\sqrt{(i^2)_{av}}$ is the square *root* of the *mean* (average) of the *square* of i. It is called the *root mean square* (abbreviated "*rms*") of i.

Ilustrative Problem 31.2 A house circuit is connected to a two-wire 60-Hz power line. The potential difference between the wires, as measured by the ac voltmeter, is found to be 110 V. (*a*) What is the maximum voltage between the wires? (*b*) Write the equation for the line voltage difference as a function of *t*.

Unless marked to the contrary, all ac meters are calibrated in effective (rms) values.

(*a*) Remember that the meter reads effective voltages.

$$\begin{aligned} V_{max} &= \sqrt{2}\ V \\ &= \sqrt{2}\ (110\ \text{V}) \\ &= 156\ \text{V} \qquad \textit{answer} \end{aligned}$$

(*b*)

$$\begin{aligned} V &= V_{max} \sin 2\pi ft \\ &= 156 \sin 120\pi t \qquad \textit{answer} \end{aligned}$$

Illustrative Problem 31.3 When a purely resistive radiant heater is connected to a standard 120-V, 60-Hz line, it dissipates power at the rate of 1320 W. Calculate (*a*) the maximum voltage, (*b*) the effective current, and (*c*) the maximum current.

Solution

(*a*) $V_{max} = 1.414V = 1.414(120\ V) = 170\ V$ *answer*

(*b*) Since average power $\bar{P} = IV$, we know that

$$\begin{aligned} I = \frac{\bar{P}}{V} &= \frac{1320\ \text{W}}{120\ \text{V}}\left(\frac{1\ \text{VA}}{1\ \text{W}}\right) \\ &= 11.0\ \text{A} \qquad \textit{answer} \end{aligned}$$

(*c*) $I_{max} = 1.414I = 1.414(11.0\ \text{A}) = 15.6\ \text{A}$ *answer*

PROPERTIES OF AC CIRCUITS

31.3 ■ Pure-Resistance Circuit

A pure-resistance circuit (Fig. 31.2*a*) is one in which there is no inductance or capacitance. In such a circuit the current and voltage curves would both be sine curves, so related that when the voltage is zero, the current is zero; when the voltage is a positive maximum, the current is a positive maximum; and when the voltage is a negative maximum, the current is a negative maximum. When this condition exists, the current is said to be *in phase* with the voltage. Figure 31.2*b* illustrates this condition graphically.*

Pure-resistance circuits can be treated as if they were dc circuits. Thus the following laws apply, if we use *effective values* for the ac units:

$$I(\text{amperes}) = \frac{V(\text{volts})}{R(\text{ohms})} \tag{31.7}$$

and

$$\begin{aligned} P(\text{watts}) &= V(\text{volts}) \times I(\text{amperes}) && (31.8) \\ &= I^2(\text{amperes}^2) \times R(\text{ohms}) && (31.9) \end{aligned}$$

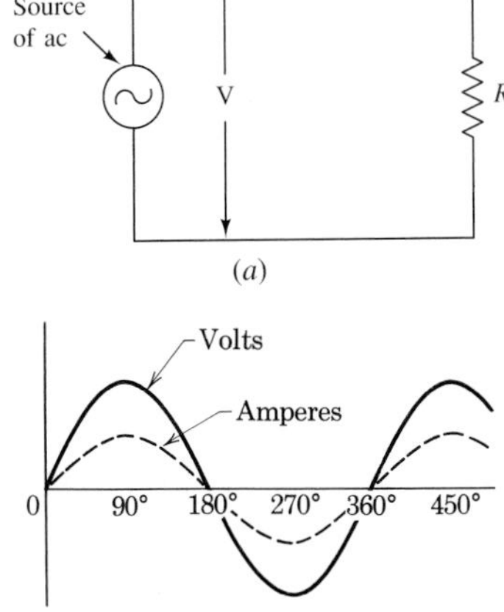

Fig. 31.2 (*a*) A circuit with ac generator and pure resistance in series. (*b*) The current in the resistor and the voltage across it are in phase.

Devices whose sole function it is to provide heat generally use pure-resistance circuits. Examples of such devices are electric irons, electric blankets, water heaters, radiant heaters, and toasters.

The ac power consumed by a resistance load is equal to the product of the effective voltage and the effective current. In other words, ac power is expressed in volt-amperes (VA) or kilovolt-amperes (KVA).

* The relative amplitudes of the curves relating *V* and *I* in this and subsequent figures in this chapter have no significance. They are plotted with different amplitudes primarily for the sake of clarity in the presentations.

The term "power" in an ac circuit usually refers to the true average power taken over a complete cycle.

Illustrative Problem 31.4 Two resistors of 25Ω and 30Ω are connected in series with a 60-Hz source of 220 V emf (Fig. 31.3). What is the power dissipated by the two resistors?

Solution

$$R_T = R_1 + R_2 = 55\Omega$$

$$I = \frac{V}{R_T} = \frac{220\text{ V}}{55\Omega} = 4\text{ A}$$

Then

$$V_1 = IR_1 = 4\text{ A }(25\Omega) = 100\text{ V}$$
$$V_2 = IR_2 = 4\text{ A }(30\Omega) = 120\text{ V}$$
$$P_1 = I^2R_1 = (4\text{ A})^2(25\Omega) = 400\text{ W} \qquad \textit{answer}$$
$$P_2 = I^2R_2 = (4\text{ A})^2(30\Omega) = 480\text{ W} \qquad \textit{answer}$$
$$P_T = P_1 + P_2 = 400\text{ W} + 480\text{ W} = 880\text{ W}$$

Check: $P_T = I^2R_T = (4\text{ A})^2(55\text{ A}) = 880\text{ W}$

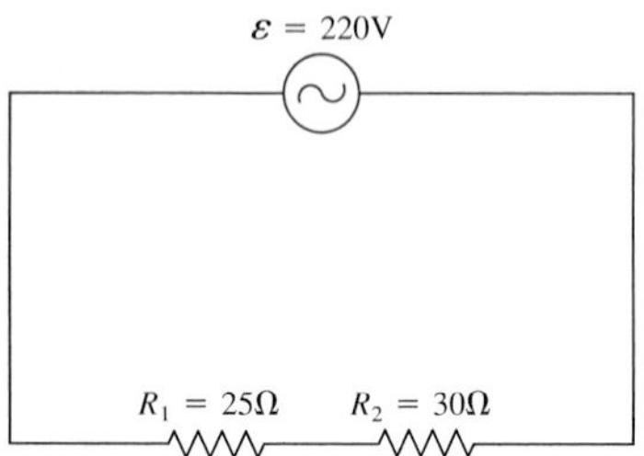

Fig. 31.3 Illustrative Problem 31.4.

31.4 ■ Circuits Containing Inductance

Rarely does a circuit contain only pure resistance. Most circuits also contain inductance, and frequently they contain capacitance. In such circuits the maximum value for the applied voltage and the maximum value for the current will not occur at the same time. This time difference is usually expressed in terms of a phase angle difference ϕ expressed in degrees. Consider that an ac voltage source is connected across an inductor L as shown in Fig. 31.4. The sinusoidal applied voltage will cause a sinusoidal current in the inductor. Note the symbol for an inductor. But the inductor will induce a counter emf which will oppose the change in the current. That property of the circuit which opposes the change in value of the current is termed *self-induction* or *inductance* (see Sec. 29.9). The electrical property of inductance is analogous to the mechanical property of inertia. It will be recalled (Chap. 29) that inductance is measured in henrys or millihenrys. A circuit has an inductance of one henry if, when the current changes at the rate of one ampere per second, there is induced in it a counter emf of one volt. The inductance L is found by the formula

$$L = \frac{-e}{\Delta i/\Delta t} \tag{29.11}$$

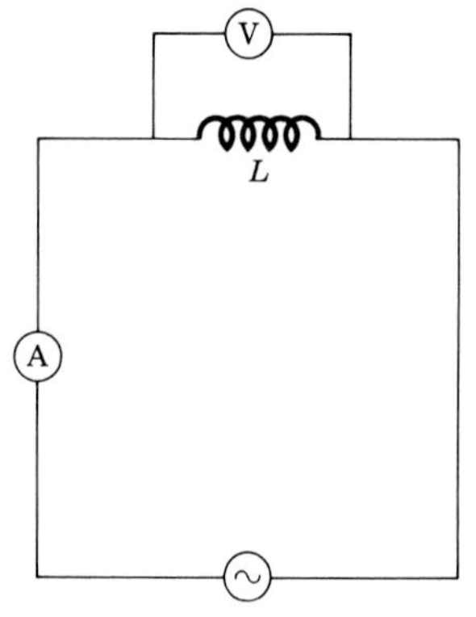

Fig. 31.4 Arrangement for measuring voltage across and current through an inductor.

An inductive circuit usually has in it a coil. The windings of motors, generators, and transformers are examples of circuits that have inductance. The inductance of a circuit is determined by (1) the number of turns in the coil, (2) the cross-sectional area, (3) the magnetic permeability of the core, (4) the transverse length of the coil, and (5) the shape of the coil and its core. The property of inductance can be illustrated by connecting a lamp in series with a "choke" coil first to a 120-V dc line and then to a 120-V ac line, as shown in Fig. 31.5. When the lamp and coil are connected to the dc line, the lamp will burn brightly. If we then insert a soft-iron core in the coil, there

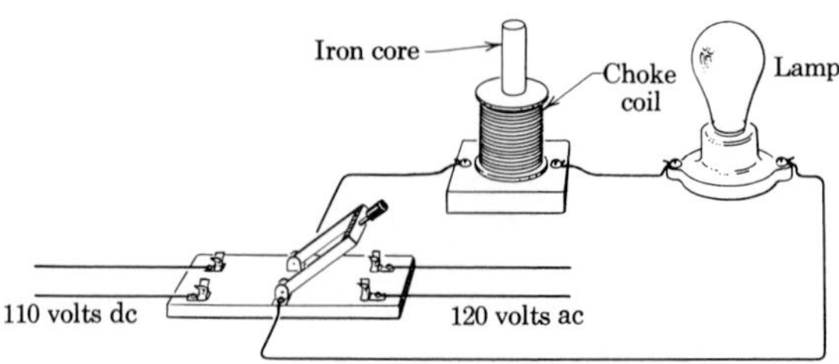

Fig. 31.5 A choke coil and lamp connected in series to a dc or ac circuit can be used to show the effect of inductance in the two circuits.

will be no change in intensity of light coming from the lamp. If we next remove the iron core and connect the lamp and coil to the ac line, the lamp lights only dimly. Inserting the iron core in the coil will dim the lamp a great deal more.

31.5 ■ Pure-Inductance Circuit

If we could have a circuit which includes only a coil of many turns of wire of negligible resistance wrapped around a good quality of magnetic steel, we would have what is known as a *pure-inductance circuit*. We know that such a circuit is theoretically impossible, because every wire (except superconductors) has some resistance; but a zero-resistance circuit can be approximated in practice. In such a circuit the counter emf, or self-inductance, will offer continual opposition to any change in current. Thus if an *alternating current* is sent through such a circuit, since the current is continually changing, there will be constant opposition to the flow of current. This opposition, called *inductive reactance*, is an *impedance* to current and is measured in ohms. It is generally represented by X_L. It can be proved both experimentally and mathematically that

The term "pure inductance" is used when the resistance of the inductor is negligible.

Every inductor has some resistance. We may treat the resistance and the pure inductance of the inductor as being connected in series.

$$X_L = 2\pi f L \qquad (31.10)$$

where X_L = inductive reactance, ohms
f = frequency, hertz
L = inductance, henrys

As Eq. (31.10) shows, X_L depends on the frequency, and the impeding effect of an inductor is greatest at high frequencies. For low frequencies the inductive reactance is low and dc conditions are approximated.

Illustrative Problem 31.5 The "choke" coil (illustrated in Fig. 31.5) has an inductance of 400 mH and a negligible resistance. If it is connected across a 120-V, 60-Hz line, find (*a*) the inductive reactance of the coil and (*b*) the current which flows through the coil.

Solution

(*a*)

$$L = 400 \text{ mH} = 0.4 \text{ H}$$
$$X_L = 2\pi f L$$
$$= 2(3.14) \times 60 \text{ Hz} \times 0.4 \text{ H}$$
$$= 151\ \Omega \qquad \textit{answer}$$

(*b*)

$$I = \frac{V}{X_L}$$
$$= \frac{120 \text{ V}}{151\ \Omega}$$
$$= 0.79 \text{ A} \qquad \textit{answer}$$

A graph relating the current and voltage in a pure-inductance circuit is shown in Fig. 31.6*b*. When the current is increasing, the induced emf tends to oppose the increase, and when the current is decreasing, the induced emf opposes the decrease and tends to keep the current flowing. Hence the rise in the current takes place later than does the rise in the source voltage. The fall of the current will take place later than does the fall of the source voltage. Thus the current lags behind the voltage throughout the entire cycle. From the graph it will be seen that the current lags 90° behind the voltage. The current and the voltage are said to be 90° *out of phase*. Thus a *pure-inductance circuit* will not only offer opposition (inductive reactance in ohms) to the current but the inductance will also cause the current to lag behind the voltage by exactly 90°. The self-induced emf will be opposite to the applied voltage at all times.

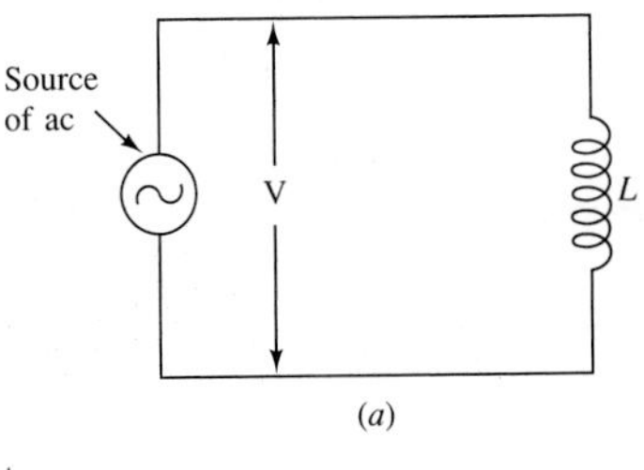

(*a*)

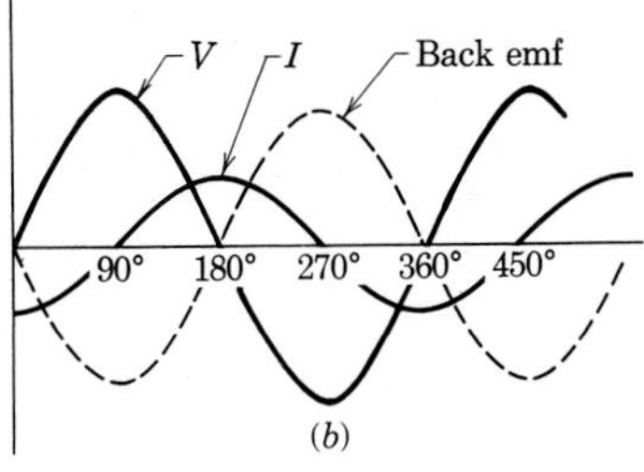

(*b*)

Fig. 31.6 (*a*) Pure inductor connected to ac source. (*b*) The voltage leads the current by one-quarter cycle or 90°.

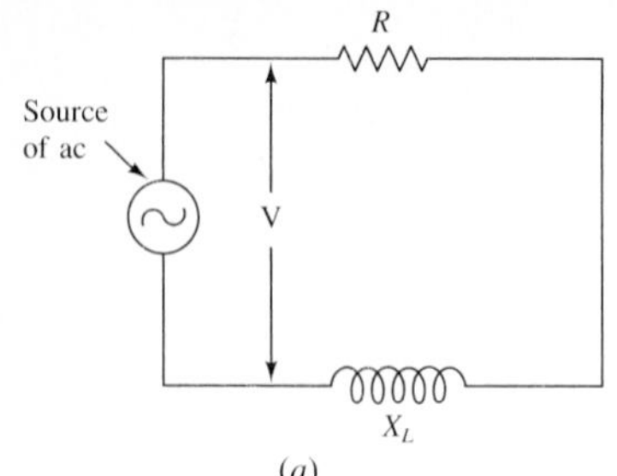

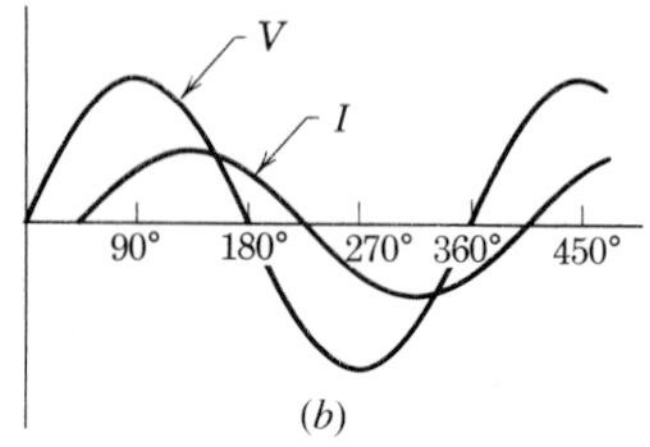

Fig. 31.7 (a) AC circuit containing a resistor and an inductor in series. (b) Current and voltage curves when $R = X_L$.

Current lags behind voltage in an ac inductive circuit.*

At 0° the current is momentarily neither increasing nor decreasing, and so the induced voltage is zero. At 90° the current is increasing at its greatest rate and so the back emf is at its most negative to oppose this increase. The voltage drop V across the coil has the opposite sign of the back emf at its maximum value. At 180° the current is again momentarily neither increasing nor decreasing, and so the induced voltage is zero. At 270° the current is decreasing at its greatest rate and so the back emf has its most positive value to oppose this decrease; V then is most negative. At 360° the cycle starts over.

31.6 ■ Resistance and Inductance in Series—Impedance

Usually we have to consider circuits in which both resistance and inductive reactance are present as illustrated in (Fig. 31.7*a*). In such cases, the current will lag behind the voltage by an angle greater than 0° but less than 90°. If, for example, the ohmic resistance equals the inductive reactance, the current will lag behind the voltage by 45° (see Fig. 31.7*b*). It should be made clear that although the ohmic resistance and the reactance are shown as separate entities in Fig. 31.7*a* they are actually properties of the same inductor.

The total resistance to current in an *RL* series circuit is called *impedance.*

The combined effect of a resistance and a reactance is known as *apparent resistance* or *impedance*. The letter Z is generally used to indicate impedance. Thus, in an ac circuit, we can write Ohm's law as

$$I(\text{amperes}) = \frac{V(\text{volts})}{Z(\text{ohms})} \tag{31.11}$$

Impedance is the vector sum of the ohmic resistance R and the inductive reactance X_L.

Impedance can be represented *vectorially* as the hypotenuse of the right triangle whose two sides are the ohmic resistance and the reactance (see Fig. 31.8). Thus

$$Z = \sqrt{R^2 + X_L^2} \tag{31.12}$$

Eq. (31.13) is often referred to as Ohm's law for an ac circuit.

And from Eq. (31.10) and Eq. (31.11) Ohm's law can be written

$$I = \frac{V}{Z} = \frac{V}{\sqrt{R^2 + (2\pi f L)^2}} \tag{31.13}$$

The angle ø between R and Z is called the *phase angle* and is equal to the lag of the current behind the voltage, in degrees. The phase angle can be calculated (see Fig. 31.8) by using the equations.

$$\tan ø = \frac{X_L}{R} \tag{31.14}$$

$$\tan ø = \frac{V_L}{V_R} \tag{31.15}$$

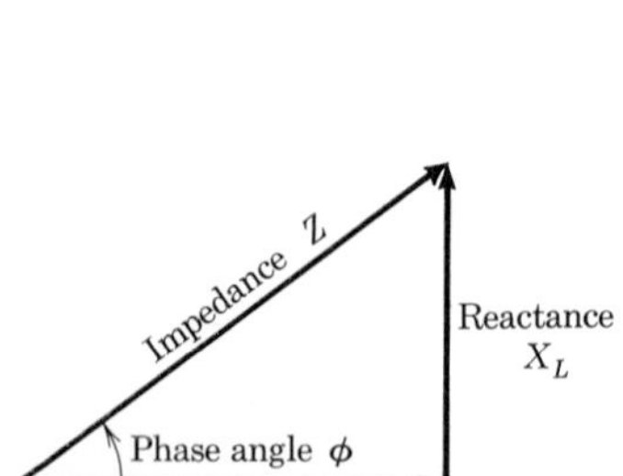

Fig. 31.8 Vector triangle relating resistance, reactance, and impedance in an ac circuit.

Illustrative Problem 31.6 A coil has a resistance of 2.4 Ω and an inductance of 5.8 mH. If it is connected to a 120-V, 60-Hz source, find (*a*) the reactance, (*b*) the impedance, and (*c*) the current in the coil. (*d*) What will be the current in the coil when it is connected to a 120-V, 6400-Hz source?

* Some people use the "code letters" *ELI* to remember that "ℰ (voltage) in an inductive circuit (*L*) leads *I* (current)."

Solution

(*a*)

$$\begin{aligned} X_L &= 2\pi f L \\ &= 2(3.14) \times 60 \text{ cycles/s} \times 0.0058 \text{ H} \\ &= 2.19\ \Omega \qquad \textit{answer} \end{aligned}$$

(*b*)

$$\begin{aligned} Z &= \sqrt{R^2 + X_L^2} \\ &= \sqrt{(2.4\ \Omega)^2 + (2.19\ \Omega)^2} \\ &= 3.25\ \Omega \qquad \textit{answer} \end{aligned}$$

(*c*)

$$\begin{aligned} I &= \frac{V}{Z} \\ &= \frac{120 \text{ V}}{3.25\ \Omega} \\ &= 36.9 \text{ A} \qquad \textit{answer} \end{aligned}$$

(*d*)

$$I = \frac{120 \text{ V}}{\sqrt{(2.4\ \Omega)^2 + (2\pi \times 6400 \times 0.0058\ \Omega)^2}}$$

$$= 0.51 \text{ A} \qquad \textit{answer}$$

Note that the effect of the resistance is negligible compared to the effect of the inductance when the frequency is as great as 6400 Hz. This example illustrates how an inductor can be used to keep currents low at high frequencies.

Illustrative Problem 31.7 When a coil whose resistance is 15 Ω is connected to a 120-V, 60-Hz source, it draws 5 A. Find (*a*) the impedance, (*b*) the inductance of the coil, and (*c*) the phase angle by which current lags behind voltage.

Solution

(*a*) We first solve for the impedance of the circuit by using Eq. (31.13)

$$\begin{aligned} Z &= \frac{V}{I} \\ &= \frac{120 \text{ V}}{5 \text{ A}} \\ &= 24\ \Omega \qquad \textit{answer} \end{aligned}$$

(*b*) Knowing values for R and Z, we can solve for X_L, using Eq. (31.12):

$$\begin{aligned} Z^2 &= R^2 + X_L^2 \\ X_L &= \sqrt{Z^2 - R^2} \\ &= \sqrt{(24\ \Omega)^2 - (15\ \Omega)^2} \\ &= 19\ \Omega \end{aligned}$$

But since $X_L = 2\pi f L$, or $L = X_L/2\pi f$,

$$L = \frac{19\ \Omega}{6.28 \times 60 \text{ cycles/s}} = 0.0504 \text{ H}$$

$$= 50.4 \text{ mH} \qquad \textit{answer}$$

Such problems can also be solved by vector methods, with vectors drawn to scale on graph paper

(*c*) From Fig. 31.8, we can determine ϕ from the relationship

$$\sin \phi = \frac{X_L}{Z}$$

Thus,

$$\sin \phi = \frac{19\Omega}{24\Omega}$$
$$= 0.7917$$

Using a calculator, we find

$$\phi \approx 52° \qquad \textit{answer}$$

31.7 ■ Capacitive Reactance

A capacitor in a circuit does not permit a dc current but does allow an ac current.

The effect of a capacitor in a dc circuit is to stop the flow of current entirely. In an ac circuit the effect requires some study. If we connect an incandescent lamp L in series with a capacitor C (about 10 μF capacity), then connect the two across a 110-V direct-current line (Fig. 31.9*a*), the lamp will not light because the circuit is *open* at the capacitor plates. When the lamp and capacitor are connected to an alternating-current line, however, the lamp will light despite the fact that the circuit is open between the plates of the capacitor. If the capacity of the capacitor is varied, it will be noted that as the capacity is made smaller the lamp will grow dimmer.

The action of the capacitor can be illustrated by again referring to a hydraulic analogy (Fig. 31.9*b*). The alternating current is analogous to the back-and-forth surging of the water through the line as the reciprocating pump P moves the water first in one direction and then in the other. The box C corresponds to the capacitor; the diaphragm D stretched across the box would have the same effect as the dielectric of the capacitor. The elastic strength of the diaphragm and the size of the box determine the "capacity" of the capacitor. The water wheel L on which the surging current does work corresponds to the incandescent lamp in the previous example.

Although it may appear that there is a current *through* a capacitor, the charge is being transferred back and forth from one plate to another.

Applying these principles learned from the water circuit to an electric circuit, it should be clear that no electric charges actually flow *through* the capacitor, but electricity does flow *into* and *out of* the capacitor. As the current moves in one direction in the circuit, one set of plates is charged positively while the other is charged negatively. When the current changes direction, the charges on the plates reverse. Hence there is an oscillation of electric charge flow in the circuit which depends on the frequency of the ac source. A 60-Hz source will cause 120 changes per second in the direction of current into and out of the capacitor.

The capacitive reactance to an ac current is inversely proportional to the current frequency.

Every capacitor placed in a circuit will offer some opposition to the motion of an alternating current in the circuit. The effect may be regarded as similar to that offered by resistance and by inductive reactance. The opposition to electric charge flow offered by a capacitor is termed *capacitive reactance* to distinguish it from inductive reactance. It, too, is measured in ohms. The symbol for capacitive reactance is X_c. Its value is given by the formula

$$X_c = \frac{1}{2\pi f C} \tag{31.16}$$

where X_c = capacitive reactance, ohms
f = frequency, cycles/s (Hz)
C = capacitance, farads

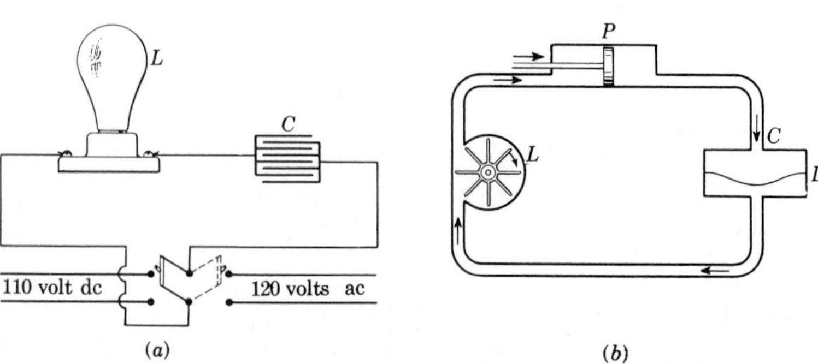

Fig. 31.9 (*a*) Effect of a capacitor in dc and ac circuits. (*b*) Hydraulic analogy of a capacitor in an ac circuit.

Illustrative Problem 31.8 What is the capacitive reactance of a capacitor of 50 μF capacitance when an alternating current of 60 Hz is impressed on it?

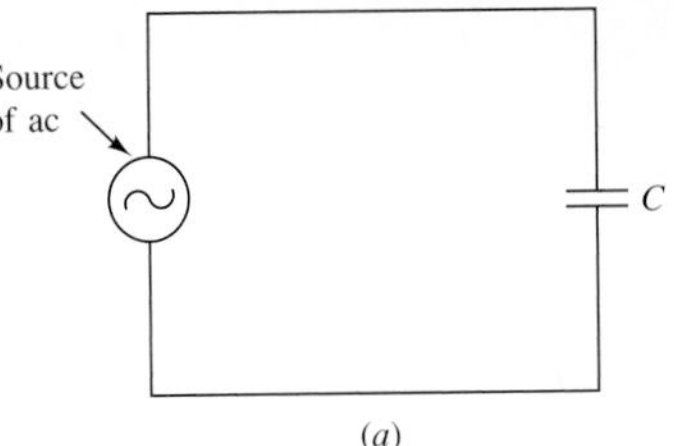

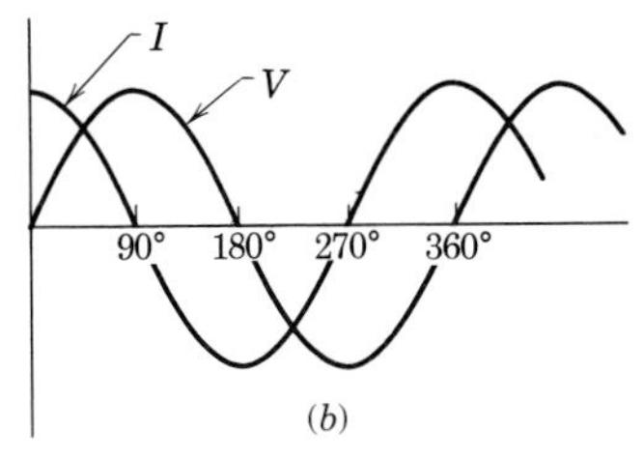

Fig. 31.10 (*a*) Simple ac source connected to a capacitor. (*b*) The current leads the voltage by one-quarter cycle or 90°.

Solution

$$1 \text{ F} = 1{,}000{,}000\ \mu\text{F}$$

Hence

$$50\ \mu\text{F} = 50 \times 10^{-6}\text{ F}$$

$$X_C = \frac{1}{2\pi f C}$$

$$= \frac{1}{2(3.14) \times 60 \text{ cycles/s} \times (50 \times 10^{-6}\text{ F})}$$

$$= 53\ \Omega$$ *answer*

31.8 ■ Comparison of Inductive Reactance and Capacitive Reactance

The *lagging* of current behind voltage in a *pure-inductive* circuit was illustrated in Fig. 31.6*b*. It can be shown that in a *pure-capacitive* circuit the *current will lead the voltage* by 90°* (see Fig. 31.10*b*).

Current leads voltage in a capacitive circuit

Thus the effect of capacitance is just the reverse of that of inductance. When a resistor and a capacitor are in series in an ac circuit, the current will lead the voltage by amounts varying from 0 to 90°. The combined effect Z (*impedance*) of resistance R and capacitive reactance X_c is found by the equation

$$Z = \sqrt{R^2 + X_C^2} \qquad (31.17)$$

Fig. 31.11 Vector diagram of the relation between impedance, resistance, and capacitance in an ac circuit.

This impedance can be represented vectorially as the hypotenuse of a right triangle whose sides are R and X_c (Fig. 31.11). The angle ø between R and Z is the *phase angle* and is equal to the "lead" of the current over the voltage. The phase angle can be calculated from the equations

$$\tan ø = \frac{X_C}{R} \qquad (31.18)$$

$$\tan ø = \frac{V_C}{V_R} \qquad (31.19)$$

The current I in such resistance-capacitive-reactance circuits can be expressed by the equation

$$I = \frac{V}{Z} = \frac{V}{\sqrt{R^2 + X_C^2}} \qquad (31.20)$$

At low frequencies a capacitor has a high impedance, and an inductor has a low impedance. At high frequencies a capacitor offers little impedance to current, but the impedance of an inductor is large.

Illustrative Problem 31.9 What will be the current in a 60-Hz ac circuit in which an emf of 120 V is impressed upon an 80-μF capacitor in series with a 24-Ω resistor? Replace the 60-Hz ac source with a 6400-Hz source and determine I.

* The code letters *ICE* are used by some individuals to recall that "current leads voltage in a capacitive circuit." (It should be noted that the voltage in this case should be represented by V rather than $\mathscr{E}$.)

Solution X_c is found by using Eq. (31.16), as follows:

$$X_c = \frac{1}{2\pi fC}$$
$$= \frac{1}{2(3.14) \times 60 \text{ Hz} \times (80 \times 10^{-6} \text{ F})}$$
$$= 33.2\ \Omega$$

Then

$$I = \frac{V}{\sqrt{R^2 + X_C^2}}$$
$$= \frac{120 \text{ V}}{\sqrt{(24\ \Omega)^2 + (33.2\ \Omega)^2}}$$
$$= 2.93 \text{ A}$$ *answer*

In a dc circuit a voltmeter reads V_L when placed across an inductor and V_C when placed across a capacitor. But in an ac circuit a voltmeter placed across an inductor and a capacitor in series gives the magnitude of $|V_L - V_C|$.

For a 6400-Hz alternating source,

$$I = \frac{120 \text{ V}}{\sqrt{(24\ \Omega)^2 + \left(\dfrac{1}{2\pi \times 6400 \text{ Hz} \times 80 \times 10^{-6} \text{ F}}\right)^2}}$$
$$= 5.00 \text{ A}$$ *answer*

Note how the *current increases with increasing frequency*. In this latter case the capacitive reactance is almost negligible.

When both inductance and capacitance are present in a (zero-resistance) series circuit, one tends to neutralize the other since their effects are opposite. The net reactance in such a circuit is the difference between them. If we let X equal the net reactance, X_L the inductive reactance, and X_C the capacitive reactance, then

$$X = X_L - X_C \tag{31.21}$$

or

$$X = X_C - X_L \tag{31.22}$$

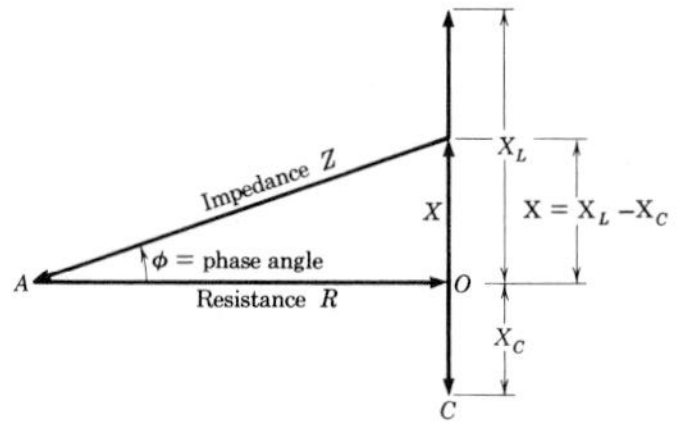

Fig. 31.12 Vector diagram relating impedance, resistance, inductance, and capacitance in an ac circuit.

31.9 ■ Resistance, Inductance, and Capacitance in Series—*RCL* Circuits

It is impossible to have a pure-inductance circuit, a pure-capacitance circuit, or simply a combination of the two, since any circuit must have some resistance (except for superconductors). Hence we must consider all three in practical computations. Often, however, one of the effects may be negligible. We can represent the relationship between the three geometrically by the vector diagram of Fig. 31.12. From the diagram we can deduce the relation

$$Z = \sqrt{R^2 + X^2}$$
$$= \sqrt{R^2 + (X_L - X_C)^2} \tag{31.23}$$

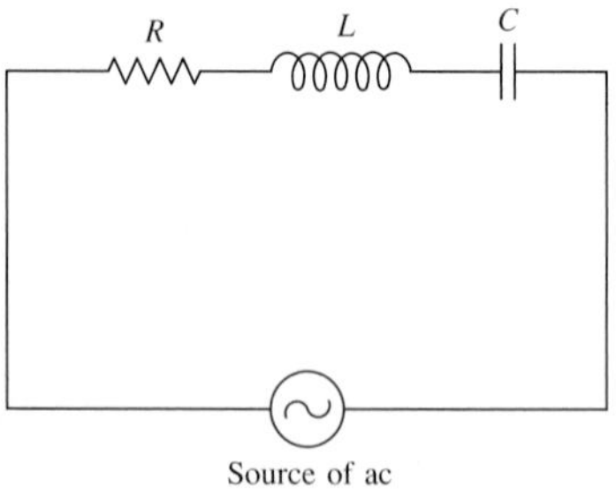

Fig. 31.13 A series circuit consisting of a resistor, an inductor, and a capacitor connected to a source of ac.

Illustrative Problem 31.10 What is the impedance of a circuit which contains an inductance of 0.7 H in series with a capacitance of 40 μF if the resistance of the circuit is 50 Ω and the frequency of the current is 60 Hz? (See Fig. 31.13.)

Solution

$$X_L = 2\pi fL$$
$$= 2(3.14) \times 60 \text{ cycles/s} \times 0.7 \text{ H}$$
$$= 264\ \Omega$$

$$X_C = \frac{1}{2(3.14) \times 60 \text{ cycles/s} \times (40 \times 10^{-6} \text{ F})}$$
$$= 66.3\ \Omega$$

$$Z = \sqrt{(50\ \Omega)^2 + (264 - 66.3\ \Omega)^2}$$
$$= 204\ \Omega$$ *answer*

We may now substitute from Eqs. (31.10) and (31.16) in Eq. (31.23) and write Ohm's law for ac circuits containing resistance, inductance, and capacitance in series:

$$I = \frac{V}{Z} = \frac{V}{\sqrt{R^2 + (2\pi fL - 1/2\pi fC)^2}} \tag{31.24}$$

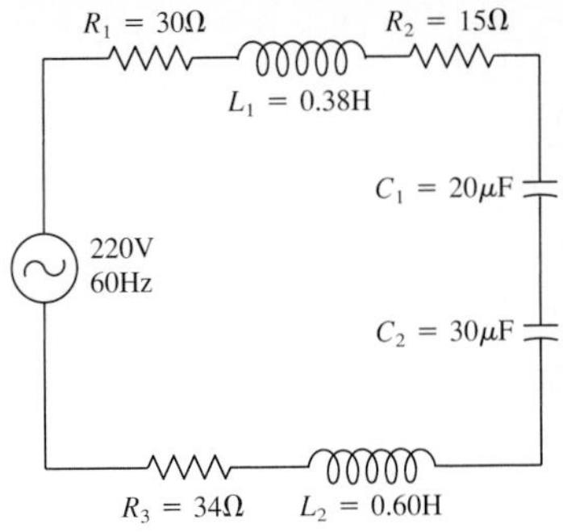

$R_1 = 30\Omega$ $R_2 = 15\Omega$ $L_1 = 0.38\text{H}$ $C_1 = 20\mu\text{F}$ 220V 60Hz $C_2 = 30\mu\text{F}$ $R_3 = 34\Omega$ $L_2 = 0.60\text{H}$

Fig. 31.14 Illustrative Problem 31.11.

Illustrative Problem 31.11 Given the circuit illustrated in Fig. 31.14, find (*a*) the total resistance R and the reactance X of the circuit, (*b*) the impedance Z, (*c*) the current I in the circuit, (*d*) the voltage drop V_{C_1} across C_1, (*e*) the voltage drop V_L across L_1, and (*f*) the phase angle ϕ.

Solution

(*a*)

$$R = R_1 + R_2 + R_3 = 30\ \Omega + 15\ \Omega + 34\ \Omega = 79\ \Omega \qquad \textit{answer}$$

$$L_T = L_1 + L_2 = 0.38\ \text{H} + 0.60\ \text{H} = 0.98\ \text{H}$$

$$\frac{1}{C_T} = \frac{1}{C_1} + \frac{1}{C_2}$$

$$= \frac{1}{20\ \mu\text{F}} + \frac{1}{30\ \mu\text{F}} \qquad C_T = 12\ \mu\text{F}$$

$$X_{L_T} = 2\pi fL_T = 2\pi(60\ \text{Hz})(0.98\ \text{H}) = 369\ \Omega$$

$$X_{C_T} = \frac{1}{2\pi fC_T} = \frac{1}{2\pi(60\ \text{H})(12\ \mu\text{F})} = 221\ \Omega$$

$$X = X_{LT} - X_{CT} = 369\ \Omega - 221\ \Omega = 148\ \Omega \qquad \textit{answer}$$

(*b*) The impedance Z can be found by using Eq. (31.23)

$$Z = \sqrt{R^2 + X^2} = \sqrt{(79\ \Omega)^2 + (148\ \Omega)^2} = 168\ \Omega \qquad \textit{answer}$$

(*c*)

$$I = \frac{220\ \text{V}}{168\ \Omega} = 1.31\ \text{A} \qquad \textit{answer}$$

(*d*)

$$V_{C_1} = IX_{C_1} = I \times \frac{1}{2\pi fC_1} = 1.31 \times \frac{1}{2\pi \times 60\ \text{H} \times 20\ \mu\text{F}}$$

$$= 174\ \text{V} \qquad \textit{answer}$$

(*e*)

$$V_{L_1} = IX_{L_1} = I(2\pi fL) = 1.31\ \text{A} \times 2\pi(60\ \text{Hz})(0.38\ \text{H})$$
$$= 188\ \text{V} \qquad \textit{answer}$$

(*f*) The vector diagram relating ϕ, X, and R is given in Fig. 31.15.

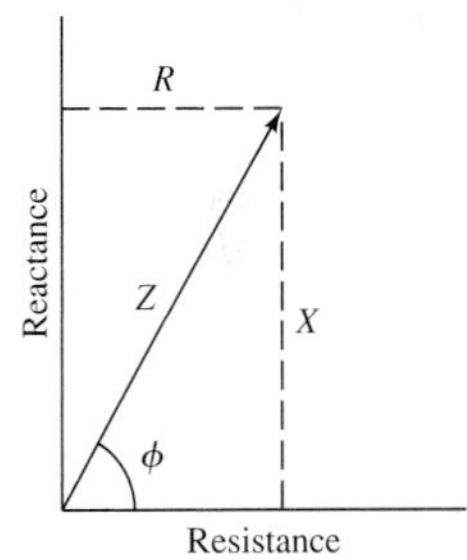

Fig. 31.15 Vector diagram relating phase angle ϕ, reactance X, and resistance R.

$$\tan\phi = \frac{X}{R} = \frac{148\ \Omega}{79\ \Omega} = 1.873$$

From a hand-held calculator, $\phi \approx 62°$ *answer*

NOTE: *Since the product* $2\pi f$ is common to all reaction equations, considerable time can be saved by memorizing its value for 60-Hz circuits ($2\pi f \approx 377$).

31.10 ■ Series Resonance

The impedance of an ac circuit is minimum when $X_L = X_C$.

It can be seen from Eq. (31.10) and Eq. (31.16) that the inductive reactance will increase and the capacitive reactance will decrease as the frequency is increased and vice versa. Then, for any set of values for inductance and capacitance, there must be a frequency at which the inductive reactance will equal the capacitive reactance. This is called the *resonant frequency* of the circuit. At the resonant frequency the inductive and capacitive reactances will nullify each other and Z will equal R. The current in such a circuit is in phase with the applied voltage ($\phi = 0$) and will have its maximum value (equal to V/R). The effect of resonance can be clearly demonstrated by using a

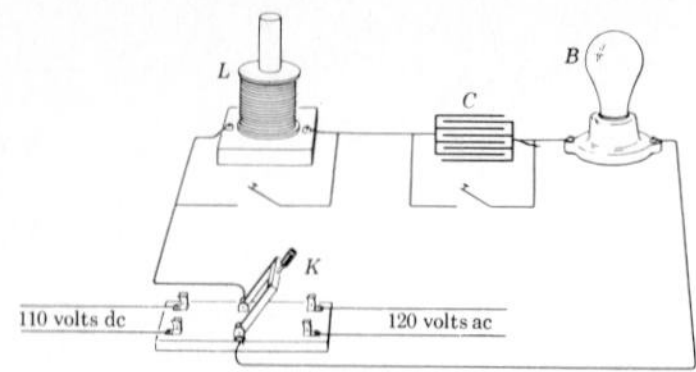

Fig. 31.16 Series resonance demonstration. The lamp will glow brightly when $X_L = X_C$.

circuit like that illustrated in Fig. 31.16. Let L be a choke coil of many turns, C a fixed capacitor of about 20 μF capacity, B an incandescent lamp, and K a double-pole double-throw switch connected to both a 120-V dc source and a 120-V ac source of current. The circuit is provided with switches to bypass either the choke coil or the capacitor. If the capacitor is left out of the circuit, the lamp will glow brightly when the switch is thrown to the dc line but only faintly when connected to the ac line. (Why?) If the choke coil is now bypassed but the capacitor included, the lamp will burn brightly when the switch is thrown to the ac line and not at all when connected with the dc line. (Why?)

It is seen that the coil partially blocks the passage of alternating current, while the capacitor stops the direct current completely. If we now include both the coil and capacitor in series in the circuit and pass alternating current through the circuit, the lamp will glow. We can adjust the inductive reactance of the coil by inserting a moveable soft-iron core in the coil (or we could vary the capacitance) until the lamp gives the normal illumination that it would give if only the resistance of the circuit were considered. We would then have obtained the condition of *series resonance,* when $X_L = X_c$.

Resonating circuits are used to tune radios and TV sets.

Resonant circuits have great usefulness in the field of electronics. A particular frequency is *tuned in* by establishing a resonant circuit for that frequency. This is often done by the use of a variable capacitor (see Fig. 25.39) to change the capacitive reactance of the circuit until resonance is established. We can find the frequency which will give resonance in a series ac circuit by recalling that resonance occurs when $X_L = X_C$. Set

At resonant frequency the series ac circuit behaves as one containing only pure resistance. At frequencies above the resonant frequency, the circuit is inductive (current lags behind voltage). At frequencies below the resonant frequency, the current is capacitive (current leads voltage).

$$X_L = X_C$$

Then

$$2\pi fL = \frac{1}{2\pi fC}$$

and

$$\underset{\text{(resonance)}}{f} = \frac{1}{2\pi\sqrt{LC}} \tag{31.25}$$

Thus, by altering the value of C (or L), the circuit can be made resonant for any fixed frequency.

Illustrative Problem 31.12 Find the resonant frequency of a circuit in which a 0.03-mH inductance is in series with a 0.005-μF capacitor.

Solution From Eq. (31.25) for series resonance

$$\begin{aligned} f &= \frac{1}{2\pi\sqrt{LC}} \\ &= \frac{1}{6.28\sqrt{3 \times 10^{-5}\ \text{H} \times 5 \times 10^{-9}\ \text{F}}} \\ &= \frac{1}{6.28 \times 3.87 \times 10^{-7}} \\ &= 4.11 \times 10^5 \text{ or } 411 \text{ kilohertz (kHz)} \qquad \textit{answer} \end{aligned}$$

Illustrative Problem 31.13 What capacitance must be added in series to get a resonant circuit with an inductance of 58μH if the frequency of the current is 1200 kHz?

Solution

$$f(\text{resonance}) = \frac{1}{2\pi\sqrt{LC}}$$

Solve for C and substitute known values for f and L. Squaring Eq. (31.25) gives

$$f^2 = \frac{1}{4\pi^2\,(LC)}$$

Then

$$C = \frac{1}{4\pi^2 f^2 L}$$

$$= \frac{1}{4\pi^2(1200 \times 10^3\ \text{Hz})^2 \times 58 \times 10^{-6}}$$

$$= 303 \times 10^{-12}\ \text{F}$$

$$= 303\ \text{PF (picofarads)} \qquad \textit{answer}$$

POWER IN AC CIRCUITS

31.11 ■ Power in a Pure Resistance Circuit

In Sec. 31.3, we discussed how the current and voltage in a purely resistive ac circuit are in phase with each other ($\phi = 0$). The power formulas for the circuit are the same as those for a dc resistive circuit. Thus,

Power is dissipated only in the resistive components of a circuit.

$$P = V_R I_R \qquad (31.26)$$

$$P = (I_R)^2 R \qquad (31.27)$$

$$P = \frac{V_R}{R} \qquad (31.28)$$

True power is the power which converts electrical energy into another form, such as heat, light, torque, etc.

where P = the *real* or *true power* in watts
V_R = the voltage, in volts, across R, the resistance
I_R = the current, in amperes, in R

The symbol P (with no subscript) and the unit watt (W) are used exclusively for *real* or *true power* as distinguished from reactive units to be discussed in Sec. 31.12 and Sec. 31.13. True power is what is measured by a properly connected wattmeter.

31.12 ■ Power in a Pure Inductance Circuit

A pure inductance does not dissipate power in the same sense as does a resistor (as heat). The inductor does not become warm because no power is dissipated in the coil. The average true power P is zero in a pure inductance. As the magnetic field builds up in the inductor, energy is stored in the magnetic field. This energy is returned to the circuit when the magnetic field collapses. Therefore, the inductor does not dissipate true power. The inductor is said to dissipate *reactive power*. Reactive power is a measure of the energy that an inductor cyclically absorbs and gives back to the circuit. Reactive power is measured in *volt-amperes reactive* (*vars*) and is given by the following equations

There is no (true) power loss in a pure inductance circuit.

$$P_q = V_L I_L \qquad (31.29)$$

$$P_q = (I_L)^2 X_L \qquad (31.30)$$

$$P_q = \frac{(V_L)^2}{X_L} \qquad (31.31)$$

The power *returned* by the inductor during one-quarter cycle is the same as that delivered during the previous quarter-cycle. Hence, the subscript q in P_q represents quadrature power.

where P_q = the reactive power in vars
V_L = the voltage across the inductor in volts
I_L = the current in the inductor in amperes
X_L = the inductive reactance in ohms

Illustrative Problem 31.14 What is the reactive power of a circuit that has a 4.5-H inductor and draws 1.5 A from a 60-Hz supply?

Solution

$$X_L = 2\pi fL = 2\pi(60 \text{ Hz}) \times 4.5 \text{ H} = 1700\ \Omega$$

$$P_q = (I_L)^2 X_L = (1.5 \text{ A})^2(1700\ \Omega) = 3.8 \text{ kvars} \qquad \textit{answer}$$

31.13 ■ Power in a Pure Capacitance Circuit

A pure capacitance circuit also shows no (true) power loss.

When a circuit contains both inductance and capacitance, there is a transfer of reactive power back and forth between the inductor and the capacitor.

In a pure capacitive circuit, the current leads the voltage by 90°. Here, again, the average true power is zero. The energy delivered to the capacitor is stored in the electric field. Ninety degrees later all this energy is returned to the source as the capacitor discharges. Consequently, no power is dissipated in a pure capacitive circuit and no heat is developed. Here, as in a pure inductive circuit, note the similarity between reactive and true power.

$$P_q = V_C I_C \tag{31.32}$$

$$P_q = (I_C)^2 X_C \tag{31.33}$$

$$P_q = \frac{(V_C)^2}{X_C} \tag{31.34}$$

where P_q = the reactive power in voltamperes reactive (vars)
V_C = the voltage across the capacitor in volts
I_C = the current in the capacitor in amperes
X_C = the capacitive reactance in ohms

Illustrative Problem 31.15 What is the frequency of the current which will produce a reactive power of 120 vars in a 9.34-μF capacitor due to a current of 0.92 A?

Solution

$$P_q = (I_C)^2 X_C$$

or

$$X_C = \frac{P_q}{(I_C)^2} = \frac{120 \text{ vars}}{(0.92 \text{ A})^2} = 142\ \Omega$$

$$X_C = \frac{1}{2\pi fC}$$

Then

$$f = \frac{1}{2\pi C X_C} = \frac{1}{2\pi \times 9.34 \times 10^{-6} \text{ F} \times 142\ \Omega} = 120 \text{ Hz} \qquad \textit{answer}$$

31.14 ■ Power in RLC Circuit. Power Triangle

In a circuit which has inductance and/or capacitance the apparent (reactive) power is expressed in voltamperes.

Power in a dc circuit is equal to the product of volts V and amperes I. In a pure resistance ac circuit, i.e., one in which the current and voltage are in phase, the average power is equal to the product of the (effective) voltage and (effective) current. If one connects a voltmeter and an ammeter in a circuit that has inductance and/or capacitance, the product of the two readings is called *apparent power* and the unit is called the *voltampere* (*VA*), since it is neither true power in *watts* nor reactive power in *vars*.

Apparent power can be calculated by using a voltmeter to measure the effective voltage and an ammeter to measure the effective amperage. The product of the effective values gives the apparent power.

$$P_s = V_T I_T \tag{31.35}$$

where P_s = the apparent power in voltamperes
V_T = the total applied voltage in volts
I_T = the total current in amperes.

Since $I_T = V_T/Z$, where Z is the circuit impedance in ohms,

$$P_s = \frac{(V_T)^2}{Z} \tag{31.36}$$

and

$$P_s = (I_T)^2 Z \tag{31.37}$$

We have identified three different types of power in an ac circuit: true power in watts, reactive power in vars, and apparent power in voltamperes. These three types can be related to the sides of a right (Pythagorean) triangle (see Fig. 31.17). The relation can then be expressed by the equation

$$P_s = \sqrt{P^2 + (P_{qC} - P_{qL})^2} \tag{31.38}$$

where P_s = the apparent power in voltamperes, VA
P = the true power in watts, W
P_{qC} = the capacitive reactive power in vars
P_{qL} = the inductive reactive power in vars
ϕ = the phase angle.

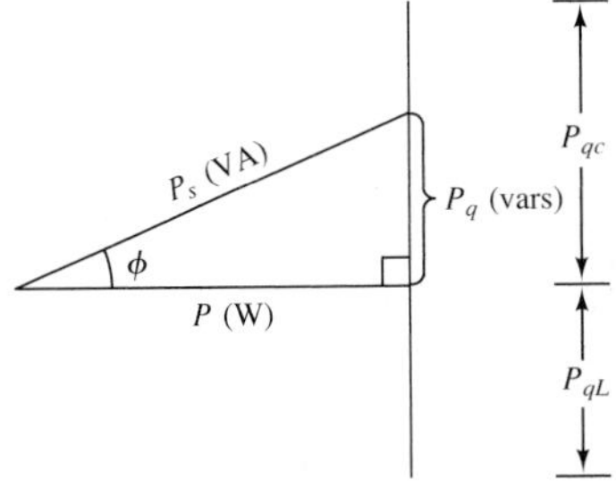

Fig. 31.17 Power triangle for an *RLC* circuit.

Fig. 31.17 arbitrarily assumes that P_{qC} is greater than P_{qL}. The following illustrative problem should assist in summarizing the power relations.

Illustrative Problem 31.16 A 0.30-H coil and a 15-μF capacitor are connected in series across a 240-V, 60-Hz power line (see Fig. 31.18). The coil has a resistance of 85 Ω. Determine (*a*) the true power, (*b*) the net reactive power, (*c*) the apparent power, and (*d*) the total current in the circuit.

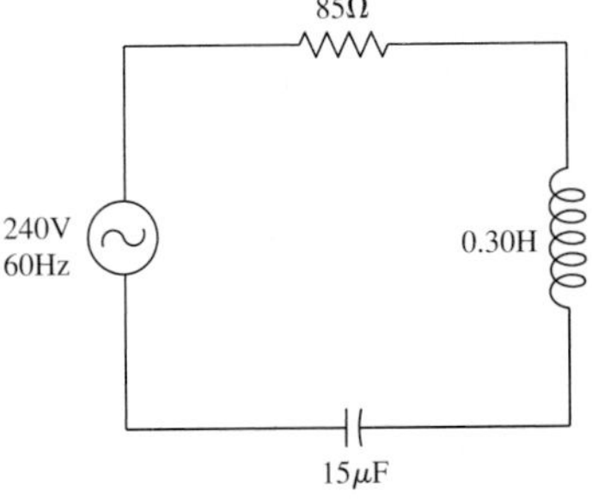

Fig. 31.18 Series *RLC* circuit.

Solution
The inductive reactance

(*a*)

$$\begin{aligned} X_L &= 2\pi f L \\ &= 2\pi(60 \text{ Hz})(0.30 \text{ H}) \\ &= 113 \ \Omega \end{aligned}$$

The capacitive reactance

$$\begin{aligned} X_C &= \frac{1}{2\pi f C} = \frac{1}{2\pi \times 60 \text{ Hz} \times 15 \times 10^{-6} \text{ F}} \\ &= 177 \ \Omega \end{aligned}$$

The impedance

$$\begin{aligned} Z &= \sqrt{R^2 + (X_C - X_L)^2} \\ &= \sqrt{(85 \ \Omega)^2 + (177 \ \Omega - 113 \ \Omega)^2} = 106 \ \Omega \end{aligned}$$

The current through the coil

Equation (31.38) can be used for all ac circuit combinations, series or parallel, *RL*, *RC*, or *RLC* circuit.

$$I_C = I_R = I_L = \frac{V}{Z} = \frac{240 \text{ V}}{106 \ \Omega} = 2.26 \text{ A}$$

Then, the true power

$$\begin{aligned} P &= (I_R)^2 R = (2.26 \text{ A})^2 (85 \ \Omega) \\ &= 434 \text{ W} \end{aligned}$$ *answer*

(*b*) The capacitive reactive power

$$P_{qC} = (I_c)^2 X_c = (2.26\ \text{A})^2 \times (177\ \Omega) = 904 \text{ vars (leading)}$$

The inductive reactive power

$$\begin{aligned} P_{qL} &= (I_L)^2 X_L = (2.26\ \text{A})^2 \times 113\ \Omega \\ &= 577 \text{ vars (lagging)} \end{aligned}$$

Then, the net reactive power equals

$$\begin{aligned} P_{qC} - P_{qL} &= 904 \text{ vars (leading)} - 577 \text{ vars (lagging)} \\ &= 327 \text{ vars capacitive} \qquad \textit{answer} \end{aligned}$$

(*c*) The apparent power

$$\begin{aligned} P_s &= \sqrt{P^2 + (P_{qC} - P_{qL})^2} \\ &= \sqrt{(250\ \text{W})^2 + (327\ \text{vars})^2} \\ &= 412\ \text{VA} \qquad \textit{answer} \end{aligned}$$

(*d*) In a *series* circuit, $I_T = I_C = I_R = I_L$

$$I_T = 2.26\ \text{A} \qquad \textit{answer}$$

31.15 ■ Power Factor

The power factor is the ratio of the power (in watts) *dissipated* by the circuit to the product of the voltage and amperage (in voltamperes) *supplied* to the circuit.

By convention, the inductive reactive power vector is drawn vertically downward (current lags voltage). The capacitive reactive power vector is drawn vertically upward (current leads the voltage).

The power factor (cos ϕ) is the factor by which the apparent power (VA) must be multiplied to obtain the true power (W).

The ratio of the true power delivered to an ac circuit divided by the apparent power that the source must supply is called the *power factor* (pf) of the load. A study of Fig. 31.17 will affirm that

$$\text{pf} = \frac{P}{P_s} = \cos\phi \qquad (31.39)$$

where P = the true power dissipated in the load in watts
P_s = the apparent power drawn by the load in voltamperes
$\cos\phi$ = the power factor (dimensionless)
Thus,

The power consumed in a portion of an ac circuit is given by

Power (W) = $V_T I_T$ (VA) × pf (31.40)

It can also be shown that the power factor is equal to the ratio of the resistance to the impedance. This ratio is equivalent to the cosine of the phase angle ϕ (see Fig. 31.12). Thus, we have the following equations for ac power

$$P\ (\text{W}) = V_T I_T \frac{R}{Z} \qquad (31.41)$$

$$P\ (\text{W}) = V_T I_T \frac{R}{\sqrt{R^2 + (R_L - R_C)^2}} \qquad (31.42)$$

and

$$\text{pf} = \cos\phi = \frac{R}{Z} \qquad (31.43)$$

We see that the power factor depends upon how much the voltage and current are out of phase. In a pure-resistance circuit the power factor is unity and power equals *VI*. In a pure-inductance or in a pure-capacitance circuit the current and voltage are 90° out of phase. In such case the power factor is zero, *resulting in a zero value for the true power*. In circuits which contain both resistance and reactance, the power factor will have a value which lies somewhere between 1 and 0, depending on the relative values of the resistance and the reactance of the circuit. Power factors may be expressed as decimals or as percentages.

In power distribution systems it is desirable to keep the power factor as close to 1.0 as possible, so that the line current will be minimum for a given true power dissipation.

In industrial systems, it is generally desirable to obtain true power at minimum costs. This objective is achieved by designing the system to make the power factor as close to unity as possible (make $P = V_T I_T$ or $R = Z$).

Illustrative Problem 31.17 A wattmeter connected to an *RLC* circuit indicates 1.5 kW. A voltmeter across the supply indicates 240 V when an ammeter connected in series with the circuit reads 7.5 A. What is the power factor of the circuit?

Solution

$$\text{Apparent power is } P_s = V_T I_T = 240\text{ V} \times 7.5\text{ A} = 1800\text{ VA}$$

The true power $P = 1500$ W.

$$\text{Then, to calculate the power factor, pf} = \frac{P}{P_s} = \frac{1500\text{ W}}{1800\text{ VA}}$$

$$= 0.83 \text{ or } 83 \text{ percent} \qquad \textit{answer}$$

Illustrative Problem 31.18 A 120-V, 60-Hz voltage is impressed upon a series circuit consisting of an 18-Ω resistor, a 0.36-H inductor, and a 24-μF capacitor. (*a*) What is the impedance of the circuit? (*b*) What is the power factor of the circuit? (*c*) What will be the current in the circuit? (*d*) How much (true) power is consumed by the circuit?

Solution

(*a*)

$$X_L = 2\pi f L = 377 \times 0.36\text{ H} = 135.7\ \Omega$$

$$X_C = \frac{1}{2\pi f C} = \frac{1}{377 \times 24 \times 10^{-6}\text{ F}} = 110.5\ \Omega$$

$$Z = \sqrt{R^2 + (X_L - X_C)^2} = \sqrt{(18\ \Omega)^2 + (135.7\ \Omega - 110.5\ \Omega)^2} = 31.0\ \Omega \qquad \textit{answer}$$

(*b*)

$$\text{Pf} = \frac{R}{Z} = \frac{18\ \Omega}{31.0\ \Omega} = 0.581 \qquad \textit{answer}$$

(*c*)

$$I = \frac{V}{Z} = \frac{120\text{ V}}{31.0\ \Omega} = 3.87\text{ A} \qquad \textit{answer}$$

(*d*)

$$P = VI \times \text{power factor} = 120\text{ V} \times 3.87\text{ A} \times 0.581 = 270\text{ W} \qquad \textit{answer}$$

TRANSFORMERS

We have noted that alternating currents are used more widely than direct currents because of the ease with which voltage can be "stepped up" or down. Voltage changes for alternating currents are accomplished by means of a device called a *transformer*. A transformer uses a direct application of mutual inductance (see Sec. 29.6). Its construction and principles of operation will now be explained.

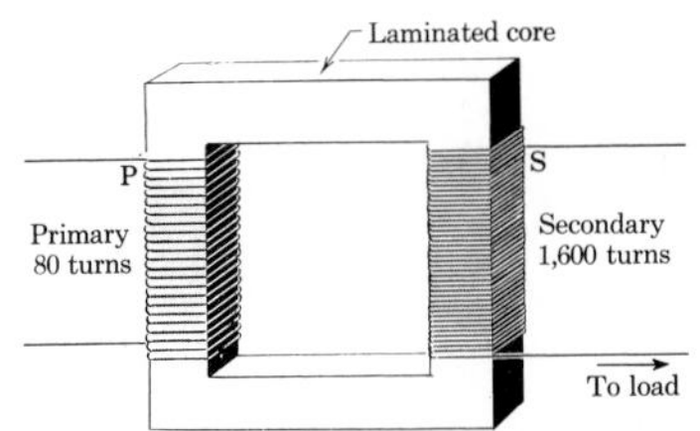

Fig. 31.19 Step-up transformer schematic.

31.16 ■ Action of the Transformer

A simple transformer usually consists of two coils of conducting wire wound on a closed iron core (Fig. 31.19). Electric energy is supplied to one of the coils, called the *primary coil,* and is delivered to the load from the other coil, called the *secondary coil*. Either coil may be used as the primary. The symbol for an iron core transformer is [transformer symbol].

If the primary coil P is connected to a source of alternating potential difference, the current in the coil will cause an alternating magnetic flux ϕ in the iron core which will surge first in one direction and then in the other. This alternating magnetic flux will also cut the secondary coil S, inducing an alternating emf in it which usually differs in magnitude from that applied to the primary coil, depending on the relative number of turns in the two coils.

If the number of turns in the primary coil is N_p, then the self-induced emf in the primary is

$$V_p = (\text{emf})_p = -N_p\frac{\Delta\phi}{\Delta t}$$

Transformers cannot operate with a steady direct current.

If we assume that there is *no flux leakage* in the iron core, then the same flux passes through the secondary coil as through the primary coil. If we let N_s equal the number of turns in the secondary coil, according to Faraday's law, the emf in the secondary is

$$V_s = (\text{emf})_s = -N_s\frac{\Delta\phi}{\Delta t}$$

Since $\Delta\phi/\Delta t$ is the same for both the secondary and primary coils, we can divide the second equation by the first one. Then

$$\frac{V_s}{V_p} = \frac{(\text{emf})_s}{(\text{emf})_p} = \frac{N_s}{N_p} \tag{31.44}$$

or

$$\frac{\text{output voltage}}{\text{input voltage}} = \frac{\text{total secondary turns}}{\text{total primary turns}}$$

Equation (31.44) is strictly true only when no current is flowing in the secondary coil. When current flows in the secondary, counter emfs are set up in the primary (and the secondary) which will make V_p somewhat less than the voltage impressed on the primary from the outside source. The voltage at the terminals of the secondary will therefore be slightly less than V_s, the value predicted from Eq. (31.44).

Thus we see how an input voltage at the primary can be converted to a higher or lower output voltage at the secondary. In a *step-up transformer*, N_s/N_p is greater than 1.00 In a *step-down transformer*, N_s/N_p is less than 1.00.

Illustrative Problem 31.19 Voltage from a power transmission line is reduced from 2400 V to 220 V by a transformer. The primary winding of the transformer has 4800 turns. (*a*) How many turns are there in the secondary? (*b*) If the power output of the transformer is 12 kW, what is the current in each of the two coils? Assume 100% efficiency and a power factor of 1.

Solution Use Eq. (31.44) to solve for N_s.

(***a***)

$$\frac{V_s}{V_p} = \frac{N_s}{N_p}$$

or

$$N_s = N_p\left(\frac{V_s}{V_p}\right) = 4800\left(\frac{220\text{ V}}{2400\text{ V}}\right)$$

$$= 440 \text{ turns} \qquad \textit{answer}$$

(***b***) The power input = the power output because we assume 100% efficiency. Then

$$P = V_pI_p = V_sI_s = 12{,}000\text{ W}$$

$$I_p = \frac{12{,}000\text{ W}}{2400\text{ V}} = 5.0\text{ A} \qquad \textit{answer}$$

$$I_s = \frac{12{,}000 \text{ W}}{220 \text{ V}} = 55 \text{ A} \qquad \textit{answer}$$

If we assume 100 percent efficiency, and a power factor of 1, we can equate the power input to power output; or $V_pI_p = V_sI_s$. Thus, we can write

$$\frac{V_s}{V_p} = \frac{I_p}{I_s} \tag{31.45}$$

In other words, a stepped-up voltage means a consequent reduction in current, and vice versa. This can be verified by checking the answers in Illustrated Problem 31.19.

31.17 ■ Uses of Transformers

One may wonder what advantages are to be gained by stepping up or stepping down voltages. Let us consider a few examples. When an electric current flows through a conductor, a certain amount of heat is developed, depending on the current and the resistance of the circuit. We have determined its value in Chap. 26 as I^2R (watts). For electric power transmission purposes it is desirable to keep the heat developed at a minimum, because whatever heat is developed in the transmission lines will be just that much energy loss. The rate of electric energy produced by a generator is fixed and equal to the product of emf generated times the current furnished. Let us consider a generator which furnishes 10 A of current at 550 V. The power output of the generator is $P = VI = 5500$ W. Let us assume further that this energy is to be transmitted for a considerable distance over wires that have a resistance of 20 Ω. The rate of energy loss in transmission will equal

$$I^2R = 10^2 \times 20 = 2000 \text{ W}$$

The transformer does not, nor can it, step up power.

which is 36 percent of the original power. By stepping up the voltage to 5500 V before transmission, the current will then be only 1 A and the I^2R loss will equal $1^2 \times 20$, or 20 W, which is only 0.36 percent of the original power. Thus by stepping up the voltage 10 times, the efficiency of transmission is increased from 64 to 99.64 percent.

By stepping up the voltage, it is possible to furnish electric power over long distances with little loss in energy. A step-down transformer (or series of transformers) at the other end of the line can be used to deliver the electric power at any desired voltage.

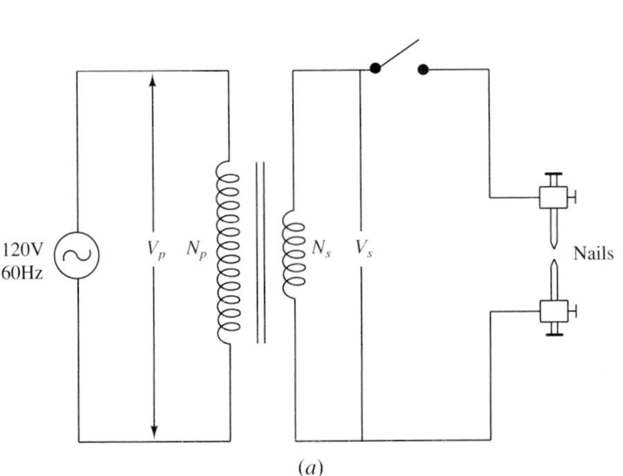

(a)

(b)

Fig. 31.20 (a) Circuit for a step-down transformer. (b) Small power transformer for radio use made with laminated closed core, with steel covers protecting the windings. Marked binding posts on the cover provide connections to the input and output currents. (Central Scientific Co.)

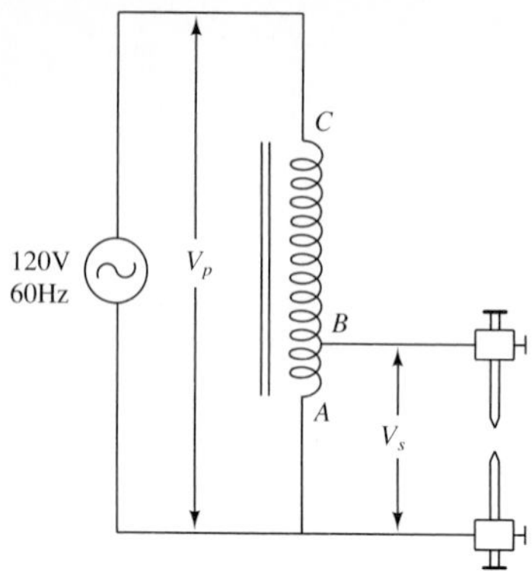

Fig. 31.21 Schematic diagram of a stepdown autotransformer.

When large currents are desired, as in electric welding, *step-down transformers* are used. The action of such a transformer in electric welding can be illustrated by winding the primary of a transformer with many turns of light wire and the secondary with only a turn or two of very large copper wire. If the primary is connected to a 120-V ac source and a pair of nails are attached to the ends of the secondary coil (Fig. 31.20*a*), the current flowing through the secondary when the nail tips are brought together will be sufficient to melt the nail at the tips and weld them together.

Small step-down transformers are also used when low voltages are needed. These transformers are used for operating electric bells (see Fig. 31.20*b*), radios, thermostatic controls, toy electric trains, and the like. Picture tubes in TV sets generally require potential differences of 15,000 to 30,000 V. A step-up transformer is used for that purpose.

In some small transformers the same coil is used for the primary and secondary. Such a transformer is called an *autotransformer*. A schematic diagram of a step-down autotransformer is shown in Fig. 31.21. The entire coil *AC* is the primary and the part between *A* and *B* is the secondary. Equation (31.44) can be applied to autotransformers as well as to other types of transformers.

31.18 ■ Operation of a Transformer Under Load

When no current is drawn from the secondary of a transformer (secondary circuit "open"), the impedance set up by the counter emf in the primary circuit is such that *practically no current will flow in the primary*. However, when the secondary circuit is closed and current is drawn from the secondary, a magnetic field is set up around it which, according to Lenz's law, tends to neutralize some of the magnetic field around the primary winding. This reduces the counter emf force established by the primary current. As a result, more current will flow through the primary coil. The primary current will build up until its counter emf will once more exactly balance the impressed emf due to the secondary current. If more current is then drawn from the secondary, the counter emf of the primary is further reduced and more current is drawn from the source through the primary circuit. Conversely, if the load in the secondary is decreased, less current will be drawn from the primary circuit. Thus the transformer will adjust itself to changes in load in the secondary circuit. A transformer primary, therefore, does not draw appreciable power from the line unless the secondary has an appreciable load. However, if the load in the secondary becomes too great, the resultant heavy current in the primary could burn out the primary windings.

31.19 ■ Eddy Currents

It is desirable that the core of an induction coil, the core of a transformer, and the armatures of ac motors and generators be made of material of high magnetic permeability, in order to increase their magnetic flux densities when current flows in the coils surrounding them. However, these highly permeable substances are also good conductors of electricity. Hence when there is any change in flux density in them, induced currents will be sent through portions of their mass. Such induced currents, set up in cores and armatures, are referred to as *eddy currents*. Eddy currents produce heat which may be harmful to the electrical apparatus as well as a source of energy loss.

To eliminate this wasted energy, cores and armatures of electric equipment are *laminated;* that is, they are made up of many very thin sheets of iron, with insulation between them, in a direction perpendicular to the eddy currents.

Eddy currents are desirable in induction heating and in shielding from electromagnetic radiation.

Ferromagnetic materials, such as iron oxide (Fe_3O_4) are good insulators. As insulators, virtually no eddy currents are induced in them. They are used for coil cores, as well as memory and microwave devices.

For the sake of clarity, in Figs. 31.19 and 31.20 we have shown the primary and secondary windings on separate legs of the core. Commercial transformers are not constructed in this manner because with this arrangement a considerable amount of the magnetic flux produced by the primary current does not cut the secondary winding and the transformer is said to have a large leakage flux. In order to keep this leakage flux to a minimum, the two coils are wound around a common core (or cores). The low-voltage coil is placed next to the core, with the high-voltage coil placed around the low-voltage coil.

Most transformers have relatively high efficiencies. Even with the energy losses due to flux leakage, resistance of the windings, hysteresis of the iron core, and eddy currents, transformers usually have efficiencies in excess of 97 percent.

Large power transformers usually have efficiencies between 98 and 99.5 percent. However, the heat generated within the transformer is still appreciable and can be a problem unless removed by a cooling system.

QUESTIONS AND EXERCISES

1. List as many applications as you can (*a*) where direct current is preferable to alternating current and (*b*) where alternating current is preferable to direct current.

2. A single-phase circuit contains only resistance, no reactance. What effect would doubling the resistance have on the power factor?

3. What is the power factor of a single-phase ac circuit containing only (*a*) resistance, (*b*) inductance, (*c*) capacitance?

4. A circuit contains a capacitor. Describe the difference between the currents in the circuit when (*a*) dc voltages and (*b*) ac voltages are impressed across the capacitor.

5. What is meant by the effective value of an alternating current?

6. What is meant by (*a*) inductance, (*b*) capacitance, (*c*) reactance, and (*d*) impedance? (*e*) Give the unit in which each of the above is measured.

7. What is meant by resonant frequency? How can it be varied in an *RCL* circuit?

8. Develop the equation which relates reactive power P_q to apparent power P_s and the phase angle ϕ.

9. A common demonstration for the physics laboratory is illustrated in Fig. 31.22. An aluminum ring is slipped over an extra long iron core set in a solenoid. At the instant an alternating emf is applied to the coil, the ring will be flipped violently into the air (see Fig. 31.22). Explain.

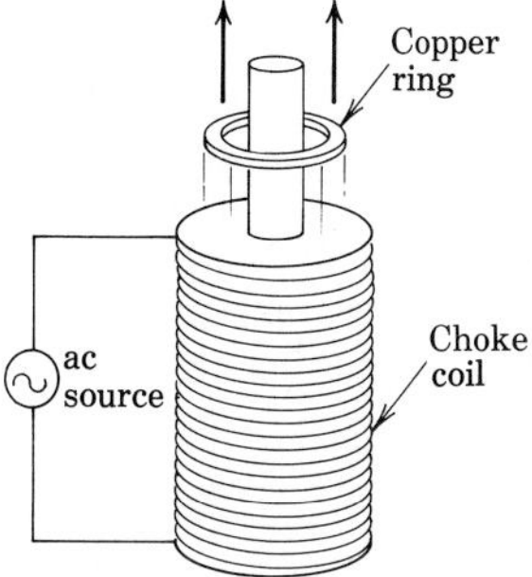

Fig. 31.22 Question 9.

10. A transformer used for operating an electric buzzer is designed for 120-V, 60-Hz current. Explain why the transformer is likely to burn out if it is connected to a 120-V dc source of current.

11. Check definitions for the henry (unit of inductance for *L*) and the farad (unit of capacitance for *C*). Show how $X_L = 2\pi fL$ *and* $X_C = 1/2\pi fC$ will give ohms of resistance, even though the *L* appears in the numerator while *C* appears in the denominator.

12. Under what conditions will the total power of an ac circuit equal the product *VI*?

13. Explain why a small step-down transformer connected to a door bell will take only very little, if any, electric energy from the supply line when the bell is not ringing, even though the primary is connected directly across the line.

14. Plot a rough graph of X_L as a function of frequency f for an inductance *L* of (*a*) 1.5 H and (*b*) 3.0 H. (*c*) What is the significance of the fact that $X_L = 0$ when $f = 0$ regardless of the value for *L*?

15. Plot a rough graph of X_C as a function of frequency f for a capacitance *C* of (*a*) 1.5 *F* and (*b*) 3.0 *F*. (*c*) What is the significance of the fact that when $f = 0$, X_C is extremely high regardless of the value for *C*?

16. When a sine-wave alternating current is applied to an *RC* circuit, what is true of the current value when the instantaneous voltage is zero?

17. If a sine-wave alternating current is applied to an *RL* circuit, what is true of the current value when the instantaneous voltage is zero?

18. A 10-kVA single-phase transformer operating on 60-Hz reduces 2200 V to 440 V at no load. Which of the following is true? (*a*) From these figures the efficiency of the transformer can be determined. (*b*) The current in the primary at full load will be 44 A. (*c*) The no-load current in the primary is 4.5 A. (*d*) The ratio of the turns on the primary to the turns on the secondary is 5:1. (*e*) None of these.

19. To increase the resonant frequency of a series circuit containing resistance, capacitance, and inductance, which of the following must be done? (*a*) Increase the resistance, (*b*) increase the capacitance, (*c*) increase the voltage, (*d*) decrease the inductance, (*e*) decrease the resistance.

20. Which of the following will make the statement true? A series ac circuit consisting of a resistor, capacitor, and inductor is in resonance when (*a*) the in-phase current equals the out-of-phase current, (*b*) no current flows through the circuit, (*c*) maximum current is established through the circuit, (*d*) minimum current is established through the circuit, (*e*) the voltage across the resistor is equal to the combined impedance across the capacitor and the inductor.

PROBLEMS

Group One

1. If the effective voltage of an ac circuit, as shown by an ac voltmeter, is 115 V, what is the maximum voltage?

2. What is the capacitive reactance of a 1.5-μF capacitor in a 60-Hz circuit?

3. What is the inductive reactance of a coil with an inductance of 2.5 H when it carries a 60-Hz alternating current?

4. A door chimes transformer is connected to a 120-V house circuit. The primary coil has 720 turns. The secondary coil has 180 turns. What is the voltage delivered to the door chimes? Assume 100 percent efficiency.

5. An ac generator with emf of 550 V supplies electric energy to an 11,000-V line. If the primary of the step-up transformer has 80 turns, how many turns are there in the secondary? Assume 100 percent efficiency.

6. A spot welder operates on a current of 300 A. The step-down transformer has a secondary of a single loop. If the transformer draws 1.5 A from the power line, how many turns must there be in the primary coil? Assume 100 percent efficiency.

7. What is the capacitive reactance of a 4.0-μF capacitor when connected to an alternating current of 60 Hz?

8. What is the reactance of a capacitor of 12 μF capacitance when it is connected to an alternating emf which has a frequency of 1800 Hz?

9. A wattmeter connected in a purely resistive ac circuit reads 55 W. An ammeter connected in series in the whole circuit reads 1.5 A. What is the resistance of the circuit?

10. The current in a 750-pF capacitor is 1.5 mA at 1.2×10^6 Hz. What is the magnitude of the voltage across the capacitor?

11. What is the resonant frequency of a series circuit consisting of a 0.075-μF capacitor and a 15-mH inductor?

12. What is the resonant frequency of a series circuit with negligible resistance containing a 60-mH inductor and a 900-pF capacitor?

Group Two

13. A series circuit carries 3.2 A through a 450-Ω resistor and a 600-Ω inductive reactance. What is the apparent power drawn by the circuit?

14. What is the reactive power drawn by a 20-μF capacitor due to a 0.45-A, 120-Hz current?

15. A 120-V, 60-Hz ac circuit consists of a capacitive reactance of 45Ω and a resistance of 12Ω connected in series. (*a*) What current will exist in the circuit? (*b*) What size inductor, in millihenrys must be connected in series so that the current in the circuit will be a maximum?

16. A 5-Ω resistor, a 6-μF capacitor, and an inductor with unknown inductance are connected in series across a 120-V, 60-Hz ac source. (*a*) What must be the inductance of the inductor to cause a maximum current to be in the circuit? (*b*) What is the maximum current? (*c*) What is the voltage across the capacitor?

17. What is the total impedance of a 60-Hz series circuit consisting of a 300-Ω resistor, a 4-H inductor, and a 30-μF capacitor?

18. What is the impedance of a 60-Hz ac circuit containing, in series, a 3-H inductor, a 5-H inductor, a 1500-Ω resistor, and a 3000-Ω resistor?

19. A single-phase, 60-Hz motor draws 5.4 A at 115 V and has an inductive power factor of 0.88 at this load. How much power in watts does the motor use?

20. An impedance coil has a negligible resistance and an inductance of 0.0036 H. What is the frequency of the emf which will maintain a current of 0.15 A through the coil at 5.5 V?

21. A 50-kW ac generator delivers power at 12.5 kV to a step-up transformer which has a primary coil with 60 turns and a secondary coil with 900 turns. Find (*a*) the emf induced in the secondary and (*b*) the current in the secondary. Assume 100 percent efficiency.

22. A resistance of 15 Ω, a coil with a reactance of 25 Ω, and a capacitor with a reactance of 45 Ω are connected in series to an alternating 60-Hz voltage. What is the voltage required to maintain a current of 3.5 A through the circuit?

23. What inductance must be placed in series with a 32-μF capacitor to produce a circuit with a resonant frequency of 4000 Hz?

24. What is the resonant frequency of an ac series antenna circuit which has a 5.7-Ω resistance, an inductance of 25 μH, and a capacitance of 2.8×10^{-10} F?

25. A capacitor, a 20-Ω resistor, and a 0.20 H coil are connected in series to a 60-Hz applied voltage of 120 V. What capacitance will produce resonance in the circuit?

26. A series circuit consists of a 120-V, 60-Hz source of emf, a resistor of 15 Ω, an inductor of 0.25 H, and a capacitor of 35 μF. (*a*) What is the impedance of the circuit? (*b*) What is the current in the circuit?

27. The transmitter of a television station transmits at a frequency of 78 megahertz (MHz). What inductance is needed with a capacitance of 16 *p*F to tune in the station? (Remember that 1 *p*F = 10^{-12} F.)

28. Radio waves from the transmitter of a radio station are broadcast with a frequency of 1200 kHz. What capacitance is needed with an inductance of 4×10^{-8} H to form a circuit resonant to this frequency?

29. A coil having a resistance of 2 Ω and an inductance of 0.15 H is connected to a 30-V, 60-Hz source. Calculate the current in the coil.

30. A coil with a resistance of 15 Ω and an inductance of 0.25 H is connected to a 220-V, 60-Hz line. Compute (*a*) the impedance of the coil, (*b*) the current through it, (*c*) the phase angle, and (*d*) the power factor.

31. A 45-Ω resistor and a capacitor are connected in series. When the combination is connected across a 220-V ac line the reactance of the capacitor is 60Ω. Determine (*a*) the current in the circuit, (*b*) the power factor, and (*c*) the phase angle between the current and the supply voltage.

32. A 0.15-H coil with a resistance of 12Ω is connected across a 120-V, 25-Hz line. Determine (*a*) the current in the coil, (*b*) the power factor, (*c*) the phase angle between the current and the supply voltage, and (*d*) the power absorbed in the coil.

33. A series circuit of a noninductive resistance of 125Ω, a 0.15-H coil (of negligible resistance), and a 25-μF capacitor are connected across a 120-V, 60-Hz power line. Determine (*a*) the current in the circuit, (*b*) the power factor, and (*c*) the phase angle between the current and the supply voltage.

34. A 45-mH coil (with negligible resistance) and a 500-Ω resistor are connected in series across a 120-V, 60-Hz power supply. Determine (*a*) the impedance of the circuit, (*b*) the current in the circuit, and (*c*) the phase angle.

35. A 0.15-μF capacitor is connected to a 15-V, 800-Hz sinusoidal power supply. (*a*) How much resistance must be added in series with the capacitor to limit the current to 7.5 mA? (*b*) What will be the phase angle between the applied voltage and the current?

36. For an ac power supply rated at 120-V, 60-Hz, what are (*a*) the maximum voltage value, and (*b*) the instantaneous value of the voltage 15 ms after $t = 0$?

37. From the data given in Fig. 31.23, determine the following. (*a*) The capacitive reactance, (*b*) the impedance, (*c*) the current, (*d*) the voltage across R, (*e*) the voltage across C, (*f*) the reactive power, (*g*) the true power, and (*h*) the phase angle.

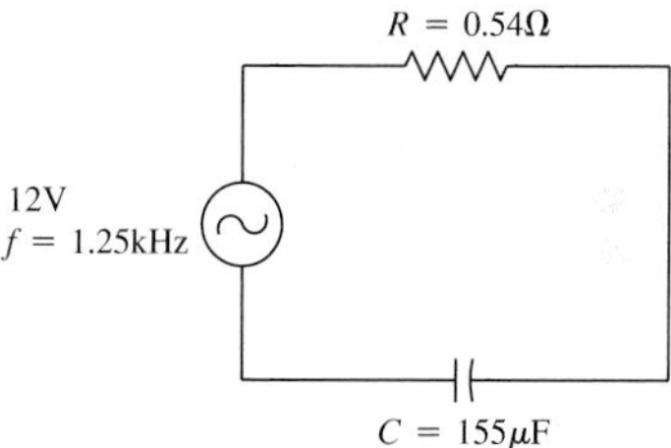

Fig. 31.23 Problem 37.

Group Three

38. A wattmeter reads 475 W when a coil draws 4.5 A from a 110-V, 120-Hz power line. What is the inductance of the coil?

39. An electric load takes 875 W and 9.85 A from a 120-V, 60-Hz power source. It is known that the current lags the voltage in the circuit. Determine (*a*) the power factor of the load, (*b*) the impedance of the load, (*c*) the resistance of the load, (*d*) the reactance of the load, (*e*) and the phase angle.

40. In the circuit illustrated by Fig. 31.24, determine (*a*) the impedance of the circuit, (*b*) the magnitude of the line current, (*c*) the magnitude of the voltage across *AB*, and (*d*) the power factor of the circuit.

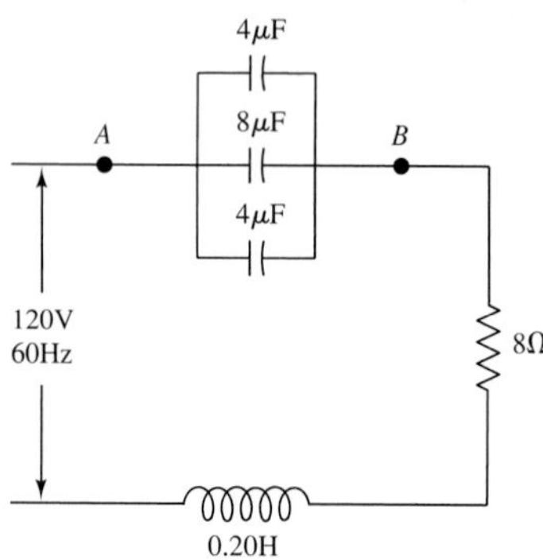

Fig. 31.24 Problem 40.

41. A series ac circuit consisting of a 24-Ω resistor, a 0.30-H inductor and a 36-μF capacitor has a 120-V, 60-Hz voltage impressed upon it. (*a*) What are the inductive and capacitive reactances in the circuit? (*b*) What is the impedance of the circuit? (*c*) Show by a vector diagram the relation between the resistance, the inductive and capacitive reactances, and the impedance. What is the value of the phase angle? (*d*) What is the power factor of the circuit? (*e*) How much current will flow in the circuit? (*f*) How much power is consumed in the circuit?

42. A 60-Hz current *I* is established in the series circuit shown in Fig. 31.25. The voltages across each element are as follows: $\mathbf{V}_1 = 120$ V leading 25°, $\mathbf{V}_2 = 240$ V lagging 75°, $\mathbf{V}_3 = 150$ V leading 50°. What is the total voltage **V** and the power factor of the circuit. (NOTE: It will be necessary to use a vector diagram to solve this problem since **V** is the vector sum of the vectors $\mathbf{V}_1$, $\mathbf{V}_2$, and $\mathbf{V}_3$.)

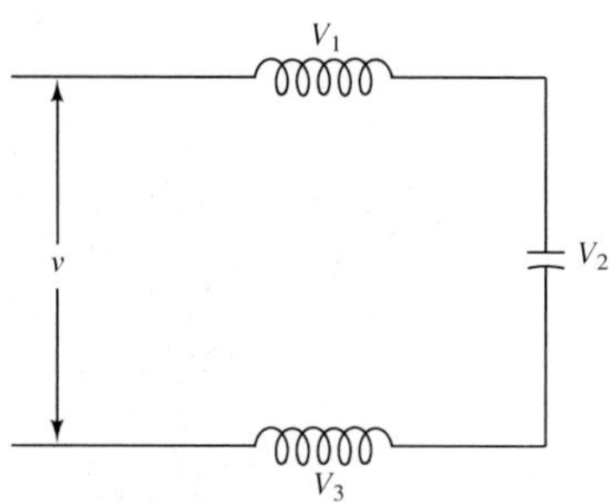

Fig. 31.25 Problem 42.

43. A 0.50-H coil and a 36-μF capacitor are connected in parallel across a 240-V, 60-Hz power line. The coil has a resistance of 150Ω. Calculate (*a*) the true power, (*b*) the net reactive power, (*c*) the apparent power, (*d*) the total line current, and (*e*) the power factor. Draw the circuit diagram and the power triangle.

PRODUCTION AND DISTRIBUTION OF ELECTRIC ENERGY

Electricity is a *commodity*. It is produced, transported, and sold at wholesale and retail to both jobbers and ultimate consumers. However, except in the case of batteries and fuel cells, electricity cannot be stored awaiting sale, nor can it be produced during an off-season for consumption later when load demands are higher. When consumers close a switch, they want electricity immediately, not later in the day or the following week. Electricity must ordinarily be used as it is produced, and there must be enough production to satisfy instantaneous demand.

The purpose of this chapter is to describe the commercial production of alternating-current electric energy and the methods by which it is transmitted and distributed to consumers.

Every complete electrical circuit contains (1) a *generator* (or source) of electric energy, (2) a *load* where the electrical energy is put to use, and (3) a *network* which conveys the energy from the generator to the load. Most circuits also have *control* devices such as switches, transformers, relays and circuit breakers to control the flow of electrical energy to the load.

PRODUCTION OF ELECTRIC POWER

Generating electricity usually involves the conversion of mechanical energy to electric energy through the medium of electromagnetism.

The *production* of ac power is rated in kilovoltamperes (kVA), and the capacity of generators, transformers, transmission lines, switches, and associated equipment is also so rated. *At the point of use* of power the power rating is in watts, kilowatts or megawatts. Alternating-current voltmeters and ammeters read *effective values* of voltage and current, and wattmeters read *true power*.

Most of the electric power currently produced comes from steam plants, internal-combustion engine plants, hydroelectric plants, and nuclear plants. Steam plants would include those burning coal, natural gas or oil to produce steam. In an ongoing effort to develop new resources of energy, smaller amounts of electric power are being supplied by wind turbines and by solar and geothermal power plants.

One of the latest trends in the electric power production field is the development of *cogeneration,* which involves the production of two useful forms of energy such as electricity and processed steam from a single energy input. All the above types of power production plants will be briefly discussed here, and nuclear power generation will be discussed later in Chap. 34.

32.1 ■ Steam Plants

At one time reciprocating steam engines were widely used for generating electricity, but they have been almost entirely replaced by steam-turbine installations in the United States. The present discussion will be limited to steam-turbine plants. Steam turbines (see Chap. 17) are now in use for generating electricity in capacities from 500 to more than 700,000 hp.

High steam pressures are the rule with turbine plants, and a high degree of superheat is also used. Pressures of 2000 lb/in^2 and temperatures of 1050°F are not at all

Fig. 32.1 Control room of a steam-electric generating plant. Combustion and turbine-generator operation are monitored and controlled by switches, relays, gauges, and other equipment shown. Computers are also used to control certain operations. The control center shown governs a large plant which produces 3,580,000 lb of steam per hour at 1000°F and 2400 lb/in.^{2} (gauge) pressure. Electric power produced is 530 MW. (Detroit Edison Company)

uncommon. All such turbines discharge into condensers which are designed to create a high degree of vacuum, usually 1 to 2 mm Hg absolute.

The turbines themselves were briefly described in Chap. 17. Small sizes (up to 50,000-kVA capacity) normally operate at 3600 rev/min, while the larger units usually run at 1800 rev/min, though with modern improvements in metallurgy some 100,000-kVA units now operate at 3600 rev/min.

There is, of course, a maze of complex equipment for the control and automatic operation of a steam-generating plant. We cannot describe in an elementary book of this kind the full details of the interconnected series of operations required at a large plant. The photo of Fig. 32.1 and the diagram of Fig. 32.2 will, however, give some idea of the scope of operations in such a plant.

32.2 ■ Internal-Combustion-Engine Plants

The use of internal-combustion engines for generating electric power is generally limited to two types of plants: (1) very small plants for intermittent operation, as for mountain resorts, or for portable industrial plants and arc-welding machines; and (2) larger plants where the peculiar local circumstances, e.g., abundance of natural gas or fuel oil combined with a scarcity of water suitable for boilers, indicate the relative economy of diesel or natural-gas engines over a steam-turbine installation.

Small generators may be V-belt driven, but larger installations are either direct drive (750 to 1800 rev/min) or gear-driven to obtain the proper generator speed. Several engines may be connected to a single generator through a torque-converter drive. For larger installations 16-cylinder diesel engines are often used as the prime mover. Hot-gas turbines are also used (see Sec. 17.18). Electric utilities usually use these as "peak plants" to be connected to the system during peak load conditions.

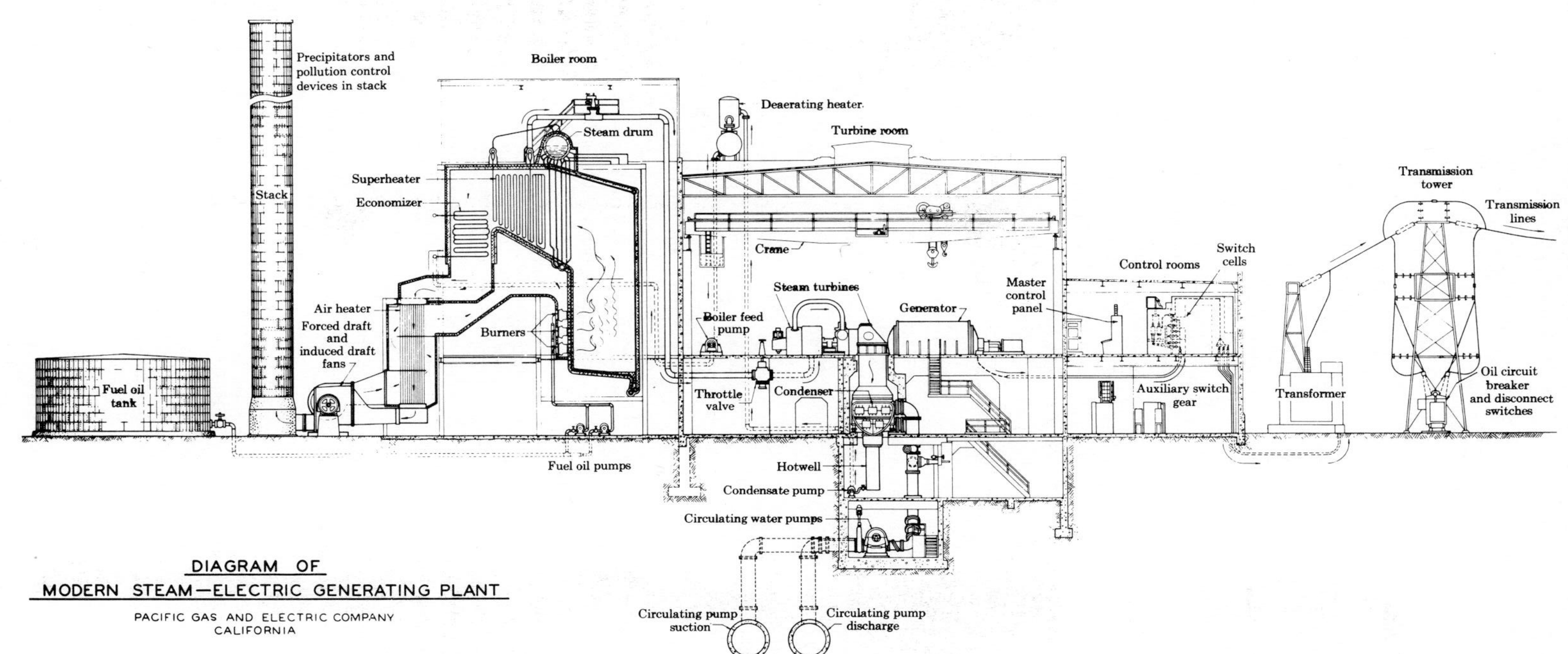

Fig. 32.2 Diagram of the physical layout of a steam-powered electric generating plant, using fuel oil as the energy source. (Pacific Gas and Electric Co.)

32.3 ■ Hydroelectric Plants

In a hydroelectric system, the *potential energy* (energy of position) of water is changed to *kinetic energy* (energy of motion) and the kinetic energy causes the shaft of the generator to turn, which produces electrical energy.

In hydroelectric generators, the alternator is of the slow-speed type, employing four or more poles to produce the 60-Hz current.

Where a steady flow of water is available reasonably close to the consumer load, and if reservoir development costs are not excessive, hydroelectric plants are an economical method of generating electricity. Many more hydropower sites will have to be developed in the future, as fossil fuels get scarcer.

Classification of Hydroelectric Plants Hydroelectric plants are of three types:

1. *Low-head plants,* heads up to about 30 ft of water
2. *Medium-head plants,* 30 to 300 ft
3. *High-head plants,* 300 to 2000 ft or more.

Classification of Water Turbines Water turbines are of two general types: *impulse turbines* and *reaction turbines.*

Impulse turbines develop power from the impact of a stream of high-velocity water against the blades or buckets of the turbine. For greatest efficiency, the water emerging from the *outlet* of the turbine should have a velocity as near zero as possible, since this means that all the kinetic energy of the water has been absorbed by the turbine. Impulse turbines are best adapted to high pressure heads with relatively small quantities of water. Their efficiencies may run as high as 80 percent. They find frequent application on the mountain streams of the West, where extremely high heads are available and water flow, especially in the late summer and fall, may be quite limited.

Reaction turbines are commonly employed with medium- and low-head plants where there is an ample water flow the year around. They are also used for high-head installations where there is a large volume of water flow. The rivers of the Northwest and of the Mississippi Valley have many reaction-turbine power-plant installations. Reaction turbines develop their power in accordance with the *third law of motion—action and reaction.* The turbine blades are curved in such a way that as the water leaves the turbine, its momentum gives a reactive kick to the turbine rotor. Large reaction turbines may have efficiencies ranging from 85 to 92 percent. The turbine and generator are usually mounted vertically on the same shaft, with the generator above the turbine. The weight of the rotating parts and the force of the downward thrust of the water are carried by thrust bearings in the generator. Figure 32.3 shows, in diagram form, a hydroelectric development in a mountain setting. Figure 32.4 illustrates the arrangement of turbo-generators in a powerhouse.

32.4 ■ Power from Water

If the water pressure head and the quantity of water flow are known, it is easy to compute the theoretical power available at a hydroelectric plant. Let h be the head in feet and Q the rate of water flow in cubic feet per second. Since the weight density of water $D = 62.4\ \text{lb/ft}^3$, the horsepower available may be obtained by starting with Eq. (6.17), $P = Fv$, from which we obtain

$$P_{\text{hp}} = \frac{F(\text{lb}) \times v(\text{ft/s})}{550 \dfrac{\text{ft} \cdot \text{lb}}{\text{s} \cdot \text{hp}}}$$

But *force F* equals *pressure* times *area,* and pressure equals hD. Making these substitutions, we have

$$P_{\text{hp}} = \frac{hDAv}{550}$$

The quantity Av (area × velocity) equals the water flow Q in cubic feet per second. Therefore, in terms of the units above,

$$P_{\text{hp}} = \frac{62.4\ \text{lb/ft}^3 \times Q\ \text{ft}^3\text{/s} \times h\ \text{ft}}{(550\ \text{ft} \cdot \text{lb/s})/\text{hp}} \tag{32.1}$$

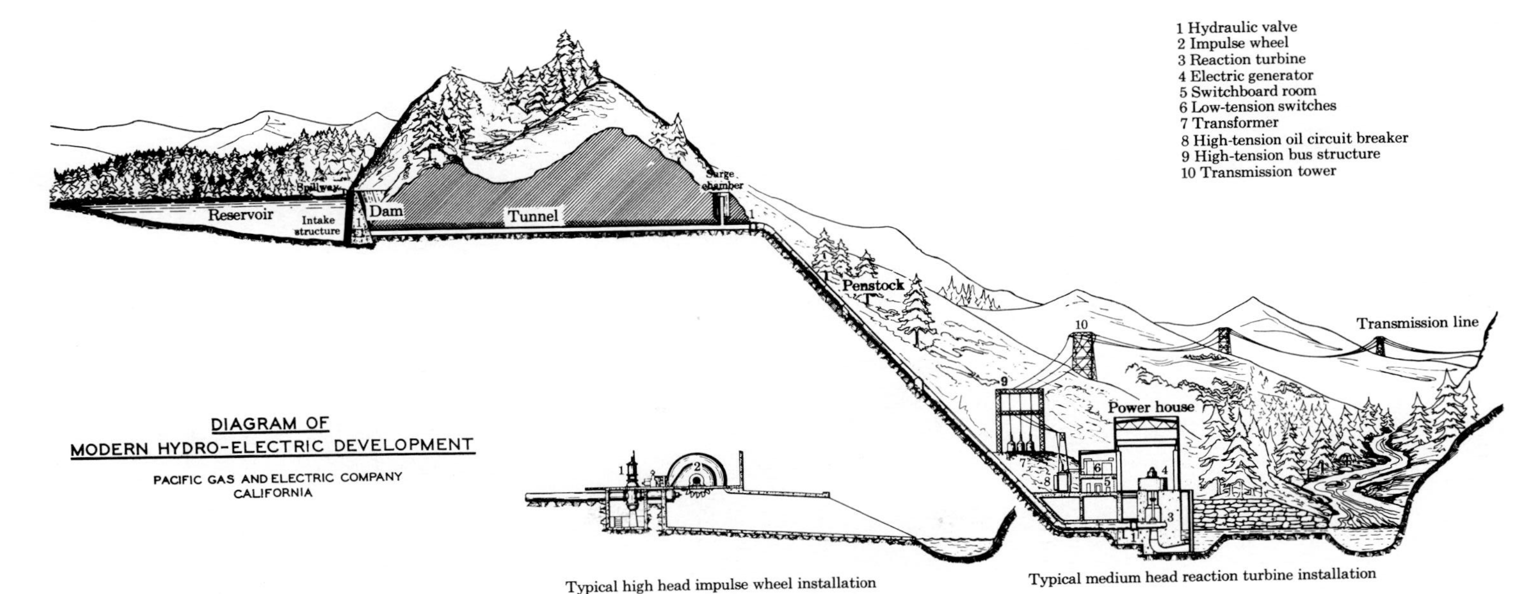

Fig. 32.3 Diagram of a hydroelectric development in a mountain setting. The insert at the bottom left shows the modification of design that would be required for a high-head impulse turbine installation. (Pacific Gas and Electric Co.)

Fig. 32.4 Buried deep within a solid mountain of granite, the three turbine-generators of the Helms Pumped Storage Project, shown in the picture, can produce enough electricity to supply a city the size of San Francisco with all its electrical needs.

During periods of high electrical demand the Helms Project produces more than one million kilowatts of electricity. At night, surplus power is used to reverse the turbine-generator units and pump water back up through the plant's tunnel to an elevated storage reservoir for use during the following peak electric period. The high-speed turbines operate under a pressure head of 2444 ft of water. (Pacific Gas and Electric Co.)

Fig. 32.5 A large wind turbine. The blades are 300 ft from tip to tip that rest on top of towers that are 200 ft tall. Each turbine generates 2.5 MW of electricity in winds of 14 to 45 mi/h while the blades rotate at 17.5 rev/min. Each turbine has a mechanism to turn the blades into the wind and to control the blade rotation in high winds.

Since 1 hp is equivalent to 0.746 kW, the electric power theoretically available at any site will be

$$P_{\text{kW}} = \frac{46.55Qh}{550} \tag{32.2}$$

To these theoretical values the overall water turbine-generator efficiency must be applied to obtain the actual power output. The value of this overall efficiency is usually in the range of 65 to 80 percent.

32.5 ■ Pumped Storage Power Plant

Pumped storage is a special form of hydroelectric power that is utilized by several utilities as an efficient means of providing adequate generation to meet peak loads (see Fig. 32.4). Through basic hydro principles, power generation is achieved as the water is released *downhill* through the generators.

Pumped storage installations differ from conventional hydro plants in that they have the ability to reverse their turbines and pump water back uphill to the upper reservoir to be used again when needed (see Fig. 32.4). It obviously takes energy to do this. However, it is feasible, considering the nature of the demand for electricity on some systems. In the off-peak hours, available capacity from base load plants is used to supply power to pump the water back uphill for use the next day to help meet the peak load. Like any other hydro operation, pumped storage can respond quickly—from a few seconds to no more than a few minutes—to any sudden change in demand for energy.

32.6 ■ Wind Turbine Power

As fuel oil supplies decrease and/or become more expensive, electric power utilities have been searching for other sources of electric energy. One of these involves the harnessing of wind power. Wind-powered generators have been installed in areas where there are continuous prevailing winds (see Fig. 32.5 and Fig. 32.6). In selected mountain passes, for example, "wind farmers" have erected more than 10,000 wind turbines which convert wind power into electric energy. The electric energy is sold to utility companies which distribute the electricity to the consumer. Wind generators are currently producing more than 675 million kWh of electric energy, enough power to meet the annual electricity needs of more than 110,000 typical homes.

Wind-powered generators utilize a free nonpolluting, renewable source of energy.

Several types of wind turbines are used to harness wind power (see Fig. 32.5 and Fig. 32.6). Under typical operating conditions wind turbines have capacity factors that vary between 4 and 11 percent. A wind turbine would have a 100 percent capacity if the wind were blowing at rates which cause the generator to turn at its top-rated speed all day, every day. With improved turbine designs, power companies anticipate the development of turbines with a 30-percent capacity factor.

Fig. 32.6 An egg-beater-shaped vertical-axis wind turbine. The blades on this 55-ft-diameter turbine are shaped like the cross-section of an airplane wing. In a 30 mi/hr wind the 24-in. blades produce 60 kW of electricity. An advantage of this turbine is that the generating equipment is at ground level and no control mechanism is needed to turn it into the wind. (U.S. Department of Energy)

32.7 ■ Solar-Electric Power

Solar power shows considerable promise for commercial development. Two approaches (designs) are used: (1) Huge arrays of solar cells (PVCs) covering acres and acres of area in desert regions where solar radiation is maximum; and (2) mammoth arrays of heliostats (sun reflectors) that concentrate the solar radiation falling on many acres of area and focus it on a receiver-boiler, where steam is generated to energize conventional steam-turbine-driven generators. Solar electric power is described in greater detail in Sec. 16.11.

32.8 ■ Geothermal Power

In certain locations of the world there exist geological conditions in which molten rock called magma is close enough to the surface so that it heats the layers of rock above it and if underground water is present, the heat rises to the surface as hot springs, geysers, and fumaroles. In these areas wells are drilled into the earth to tap the superheated geothermal steam and conduct it to the surface to be used to drive steam turbines, which in turn drive generators to produce electricity (see Fig. 32.7).

Two-thousand year-old Roman documents tell of a steam field at what is now Lardenello, Italy. Starting in the nineteenth century, this natural steam was harnessed for industrial heating, mechanical power, and later for the generation of electricity. The United States, Italy, New Zealand, Russia, Japan, Iceland, The Philippines, and Mexico are using geothermal energy as a source to generate electricity, with an additional 50 nations considering the development of their geothermal resources.

32.9 ■ Cogeneration

Cogeneration is an emerging technology which is being utilized by the manufacturing industry for on-site electricity and process steam production in a combined cycle. It involves the production of two useful forms of energy, such as electricity and process steam from a single energy input. This technology is now in full swing with major activity in petroleum-producing regions where steam can be used for enhanced oil recovery.

DESIGN AND OPERATION OF ALTERNATORS

The principles and underlying theory of the operation of ac generators have been described in Chap. 30. In this section some fundamentals of design and construction of large alternators will be discussed, such as speed of rotation, number of poles, types of armature and field windings, and problems of synchronizing two or more alternators.

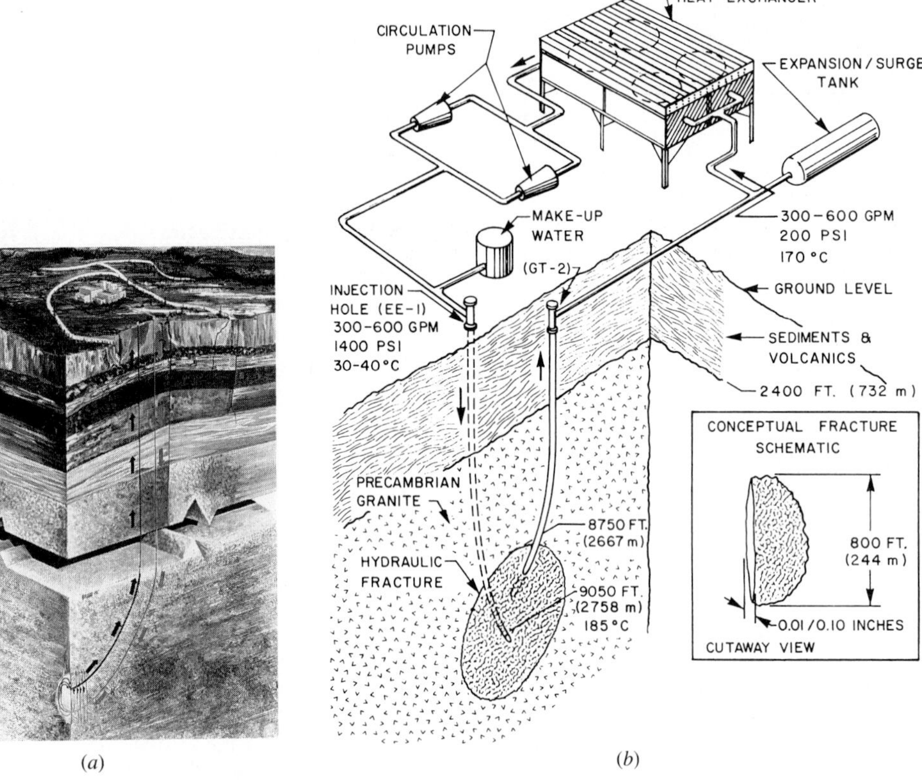

Fig. 32.7 (*a*) Schematic of the hot dry rock thermal power concept. (*b*) A closed-loop system for pumping water in and out of the ground after being heated by the hot rocks to produce steam. Scientists used hydraulic fracturing to create a heat extraction loop in the hot granite under the Jemez plateau in New Mexico. The two wells were 8500 ft deep and the surrounding rock was at a temperature of 185°C (365°F). Water is pumped into one well, absorbs heat as it circulates throughout the fracture, and is brought to the surface through the second well under sufficient pressure to prevent boiling. This water is cooled and injected into the well to bring up more heat. Part of the heat from the water is used to boil Freon 114, whose vapor drives a small turbine which rotates an alternator producing 60 kVA of electrical power. (U.S. Department of Energy)

32.10 ■ AC Generators

Less copper is needed for three-phase transmission than for an equivalent-capacity single-phase power transmission.

Three-phase motors are simpler in construction and have higher efficiencies. Large-sized motors operate more smoothly if they are three-phase motors.

Generators for the production of commercial electric power are, for the most part, *three-phase alternators,* because (1) alternating current is desired in most commercial applications, and (2) three-phase power is more economical to produce than two-phase or single-phase power. It is also more efficient to transport large amounts of power by utilizing three-phase current. If two-phase or single-phase power is desired by the consumer, it is easily obtained by merely tapping the appropriate wires of the three-phase distribution system.

Three-phase generators are usually of the *delta* or *wye* type. We will consider only the wye system of generating three-phase current. The wye-connected generator (Fig. 32.8) is a four-wire system. The generator has three identical sets of armature coils, each separated by 120 electrical degrees in a "Y" configuration. The generator uses a rotating magnetic field (the *rotor*), which induces voltages in the three sets of

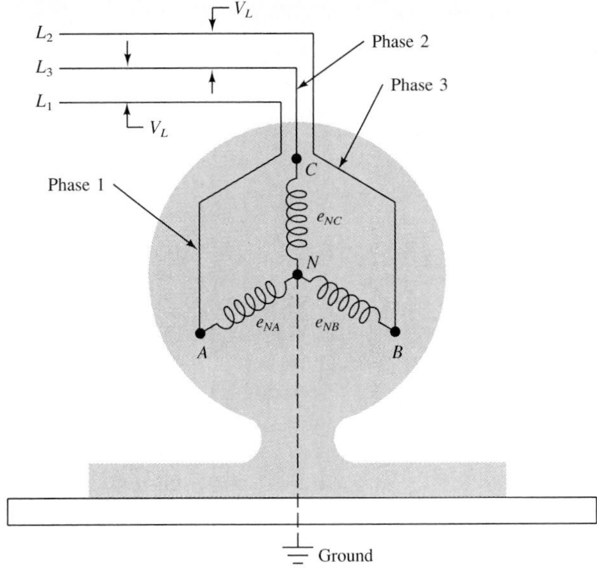

Fig. 32.8 The wiring diagram of a wye-connected four-wire, three-phase alternator. The armature windings of such alternators are on the stator. The ground becomes the fourth (neutral) wire in the system.

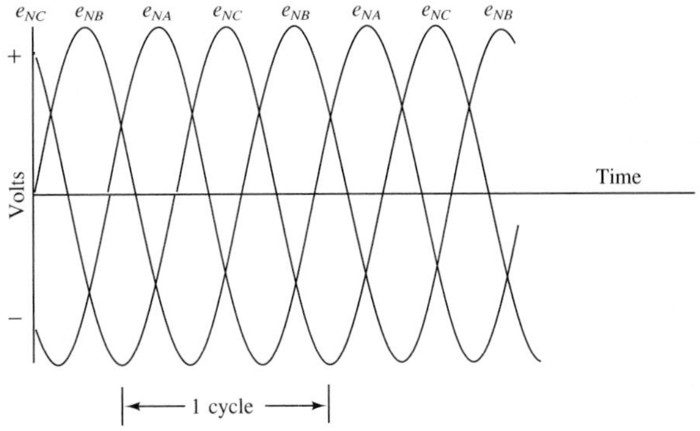

Fig. 32.9 Waveforms of the instantaneous emfs of a three-phase generator.

stationary windings (the *stator*). The field windings in the rotor are excited by direct current from two stationary brushes riding on two slip rings. The waveforms of the instantaneous voltages from each of the coils are shown in Fig. 32.9. The waves for each of the three coil voltages have the same amplitude and frequency, but there is a time difference between the three wave forms.

The three coils are joined together at a common point called the *neutral, N*. The neutral is ordinarily grounded as shown in Fig. 32.8. Leads *A, B,* and *C* are brought out to line terminals called phase 1, phase 2, and phase 3.

The wye arrangement allows the generator to act as a single-phase generator by using one of the three ''hot'' wires and the ground. Single-phase current is used for most household appliances and for small single-phase motors. Should a three-phase current be required (as for a three-phase motor), the load is connected to the three hot wires (see Fig. 32.10). Three-phase motors run more smoothly than single-phase motors. They also are smaller and more efficient for the same power capacity.

The potential difference between the neutral and any line (*A, B,* or *C* in Fig. 32.8) represents the *phase voltage,* V_ϕ, of that particular line. The potential difference

Since the phase voltages are 120° apart, the vector sum of the voltages at any time is zero. If a vertical line is drawn at any point of time in Fig. 32.9, the sum of the instantaneous voltages is zero. Thus, we can conclude that *the sum of the instantaneous voltages in a balanced three-phase system is zero.*

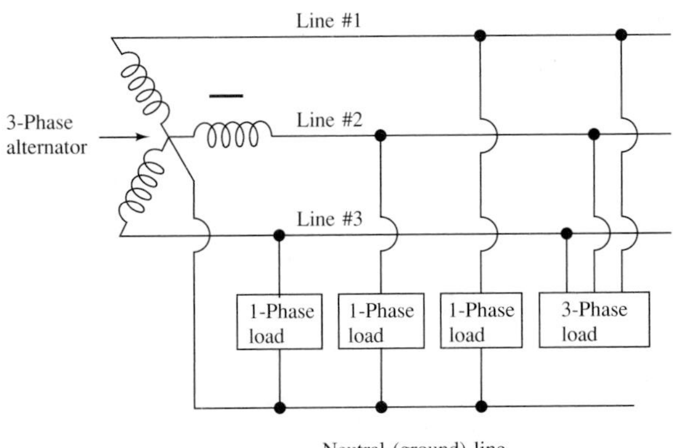

Fig. 32.10 Single-phase and three-phase loads connected to lines from a three-phase, four-line wye configuration.

between two hot lines is called the *line-to-line* or *line voltage*, V_L. It can be shown that

In a three-phase system the line-to-line voltage equals $\sqrt{3} \times$ line-to-neutral voltage.

$$V_L = \sqrt{3}V_\phi$$

$$\text{Line voltage} = \sqrt{3} \times \text{phase voltage} \tag{32.3}$$

Equation (32.3) is one of the fundamental relations in a three-phase system. Thus, if the line-to-neutral voltage is 120 V, the three lines (three phase) will operate at 208 V ($120 \times \sqrt{3} = 208$) (see Fig. 32.10).

32.11 ■ Residential Wiring System

The local transformer for a residential wiring system is a single-phase system. It produces a 120/240-V, three-wire system (one wire of which is grounded and is referred to as the "neutral"). It is *not* a three-phase system.

A typical wiring system for household installations is shown in Fig. 32.11. Three lines emerge from a transformer. Two lines will remain ungrounded. The third line (the neutral) is grounded just before the lines enter the house and at the transformer location. A two-pole switch (called *main switch*) is connected in series between each hot wire and the ground. Then all three conductors are attached to the watt-hour meter. From the watt-hour meter (which measures the electric energy consumed in the house) a 120-V circuit is taken off between each ungrounded wire and the ground (neutral) conductor. A 240-V circuit, such as is used in many electric ranges and central air conditioners, can be obtained by forming a circuit between the two ungrounded conductors. Smaller circuit breakers are placed in each 120-V circuit. Frequently additional fuses (or circuit breakers) are included in the circuits of the various household appliances on line.

Electrical appliances are connected in parallel across a hot line and the neutral (or another line)—hence all are at the same voltage.

32.12 ■ Standard Generating Voltages and Frequencies

When too much current is in a given line, a fuse will "blow" or a circuit breaker will open the circuit, cutting off all appliances on that line.

The voltage at which a plant is designed to operate depends on the nature of the connected load and the length of the distribution system. A small domestic plant designed for local lighting only might generate at 115 to 120 V. For industrial plants a local generating system might operate at 440 to 480 V and distribute on a three- or four-wire system. Motors, lights, and heating equipment are all well served by such a system. Small city generating plants where the distribution lines are only a few miles long may generate at 2400 V. In large public utility plants, the standard generating voltage is 13,200 V, and if the distribution system is short, the same voltage may be used for transmission. If, as is usually the case, longer transmission lines are required to reach the consumer load, the voltage is stepped up before sending the power out on the line (see Fig. 32.17).

House conductors are normally color-coded for easy identification. The neutral conductor is always white. The ungrounded conductors are red and/or black.

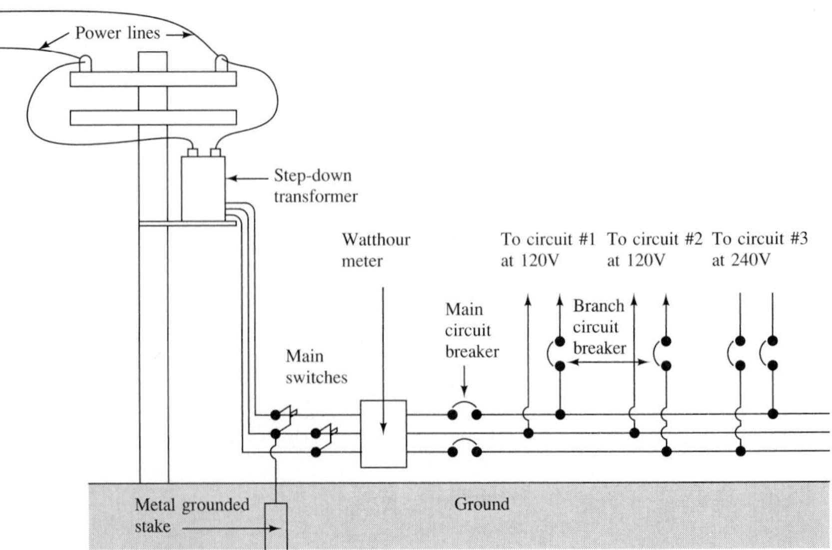

Fig. 32.11 Wiring diagram from power lines to circuits in a house.

Frequency It will be recalled that the frequency of an ac generator is a function of the number of poles on the armature and the speed of rotation of the rotor. In general, if f is the frequency, p the total number of poles, and N the number of revolutions per minute, then the frequency in cycles/s (Hz) is

$$f = \frac{pN}{120} \tag{30.12}$$

Illustrative Problem 32.1

What must the speed of the rotor of an ac generator be to produce a 440-V, 3-phase, 60-Hz current?

Solution Solve Eq. (30.12) for N.

$$f = \frac{pN}{120}$$

$$N = \frac{120\,f}{p}$$

Substitute the known values of f and p in the equation.

$$N = \frac{120 \times 60}{6}$$

$$= 1200 \text{ rev/min} \qquad \textit{answer}$$

Frequency is one of the rigid standards of operation of a power plant for two reasons: (1) electric clocks of thousands of consumers must keep accurate time; and (2) when generators are connected in parallel to the same distribution system, their outputs must be *in phase and remain at the same frequency*. Automatic controls govern the revolutions per minute of all three types of prime movers to give constant generator speed.

Engine-driven alternators may turn at 300 to 500 rev/min. Hydroplant alternators vary from 50 to 750 rev/min, while steam-turbine-driven units range from 1500 to 3600 rev/min, with four-pole 1800 rev/min machines being most common for 60-cycle operation. We have already seen that 60 cycles (Hz) is the standard frequency in the United States. In Europe 50-cycle (Hz) frequency is standard in many countries.

Commercial power stations maintain the frequency of 60 Hz within ± 0.02 Hz. Since the speed of rotation for a given alternator determines the frequency (and amplitude) of the generated voltage, the speed must be closely regulated. The generated voltage is controlled by varying the dc current in the field windings of the rotor. When the output of the system is not sufficient to meet the peak power demand, utilities normally react in one of two ways.

Fig. 32.12 Rotor for a large hydroalternator being lowered by a giant crane into position in the stator. This hydroelectric generator is rated at 600,000 kW. Three identical units have been installed at Grand Coulee Dam on the Columbia River in Washington. (Westinghouse Electric Corp.)

1. Customers who are given special rates if they voluntarily curtail and/or interrupt service; they are notified to either curtail or interrupt their loads immediately.

2. If the peak load exceeds the available capacity, frequency will start to drop, and if it drops for any length of time, automatic load sheddings of customer loads via under-frequency relays at substations occur. This will continue until the frequency recovers, at which time their loads will be reconnected to the system either manually or automatically. This is commonly referred to as a "blow-out" condition.

Details of alternator construction were given in Chap. 30. Figures 30.16, 30.17, and 30.18 show some of these construction features for engine- and steam-turbine-driven alternators. Figure 32.12 shows the rotor for one of the world's largest hydroelectric generators.

32.13 ■ Connecting Alternators to the Line

Many different methods are used to connect ac generators to the line. On a small city system where no voltage step-up is required the alternators are connected directly to the line through the necessary control and protection equipment. Larger systems, where voltage step-up is required for long-distance transmission and where several alternators may be required to serve the load, must have a very flexible connection system. One or several alternators may be operating, according to load demand, and provision for *synchronizing* them, i.e., keeping them in phase with each other and paralleling them onto the line must be made. Power may be sent to several substations, some near, some far away. This situation may require feeding a portion of the power being generated to local consumers at 13,200 V (the generator voltage); another portion to transformers for step-up to, say, 115,000 V for transmission to a consumer area 30 to 50 mi away; and possibly another portion to transformers for step-up to 500,000 V or more for a long-distance transmission line (see Fig. 32.17).

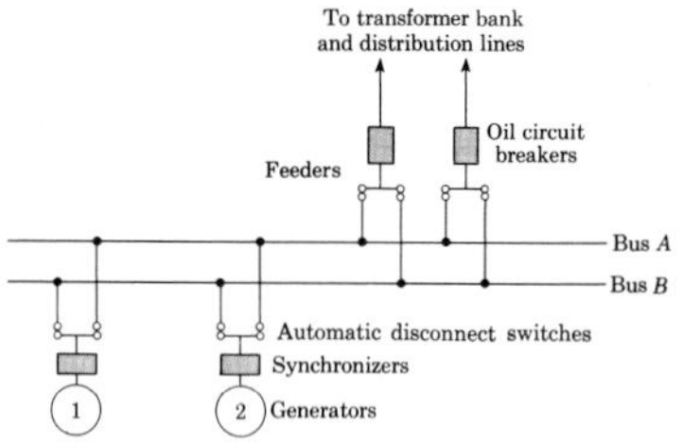

In the larger ac power-generating systems the alternators are connected in parallel to heavy copper bars called *busses*. Provision for synchronizing each alternator with those already in operation is made at the master switchboard. The main busses run the entire length of the switchboard behind the panel, but they are separated from the switchboard, to lessen the danger of injury to personnel. Each generator has a main switch to connect it to the bus. This switch opens and closes in oil or vacuum, to prevent arcing.

Wires or cables leading out from the busses to the powerhouse distributing station are called *feeders*. Each feeder has its associated meters, switchboard, and overload circuit breaker. The feeders supply the step-up transformers for the transmission lines. *Lightning arresters* are installed just beyond the transformers. Figure 32.13 is a simplified schematic diagram of a small system for two alternators. Figure 32.14 shows equipment installed at a large generating station.

Fig. 32.13 One possible system for connecting a small power plant to its distribution lines.

TRANSMISSION OF ELECTRIC POWER

Most of the electric power produced in the United States today is 60-Hz, three-phase, alternating current. Our discussion of transmission and distribution systems will therefore concentrate on this form of electric power.

Commercial generators are usually built to develop ac (instead of dc) because of the ability of a transformer to step up the voltage and distribute the power economically.

The reason why alternating current is preferred is that its voltage may be stepped up or down, i.e., *transformed,* with very little loss at any stage in the distribution system.

Since the *apparent* electric power ($P = VI$) is the product of the current and voltage, the same amount of power may be transmitted by using a high voltage and a low current or a low voltage and a high current. But, since the ohmic (nonreactive) power loss is the product of the line resistance and the square of the current ($P = I^2R$), the advantage of transmitting the electric power at lower current and higher voltage is easily seen. As a result, electric power is generally transmitted at the highest practical voltages.

High-voltage generation and transmission, however, pose some problems. High voltages require suitable and relatively expensive insulation. The higher the generated voltage, the thicker and heavier must be the insulation of the wires of the generator. Then, too, with high-voltage transmission lines, large insulators are required to insulate the individual phase conductor (nongrounded) lines from the metal towers which support the lines.

With less than ideal conditions (balanced voltages and balanced loads), the sum of the three line currents is not zero. Even though not zero, it is quite small, and the neutral wire can actually be much smaller than the other three.

Three-phase power is preferred over two-phase or single-phase power because it is far more economical of copper, both in the alternator construction and in the transmission and distribution system. A three-phase three-wire system may require only 75 percent as much copper to transmit a given amount of power as a single-phase or two-phase system; and a three-phase, four-wire system only one-third as much copper.

(*a*)

(*b*)

(*c*)

Fig. 32.14 Generating station equipment. (*a*) Busses, to lead current away from generators. (Detroit Edison Company) (*b*) Bank of transformers. (Pacific Gas and Electric Company) (*c*) Oil circuit breakers. (Southern California Edison Company)

32.14 ■ Transmission Lines and Equipment

Not all the power sent out on the transmission system reaches its destination. The principal losses are classified as follows:

1. *Line losses,* due to actual ohmic resistance and reactance of the lines themselves. Line losses are a major factor in transmission losses and may amount to 10 percent or more of the power put on the line at the generating station.

2. *Leakage losses*—to the air, to ground, to trees and buildings, and from wire to wire. This factor is small at low voltages, but at high voltages (200,000 V and up) it becomes a serious factor. Weather conditions appreciably affect this loss.

3. *Transformer losses*. With well-designed transformers this loss should not exceed 2 to 4 percent.

The first two of these losses will be discussed in detail in this section. Transformer losses take the form of heat produced in the windings from eddy currents and hysteresis losses.

Line Losses Since the power loss varies as the square of the current, it is imperative that the current be kept at a low value. This is, of course, equivalent to saying that the voltage should be at a high level. A much simplified numerical example will serve to illustrate the principle.

Suppose a manufacturing plant is to be supplied electric power from a hydroelectric plant in the mountains nearby. Let the average demand be 500 kW and the voltage desired at the plant 440 V. The line is 20 mi of No. 2 copper, whose resistance is about 0.85 ohm/mi (see Fig. 32.15). For simplicity a two-wire line is shown. A power factor of unity is assumed, and other simplifications have been made to allow for easy computation.

In diagram (*a*) the 500-kW load is shown drawing power from the line through a 4400-440-volt step-down transformer. The line current is therefore

$$I = \frac{P}{V} = \frac{500{,}000 \text{ W}}{4400 \text{ V}} = 114 \text{ A}$$

The voltage drop over the 20 mi of line is then, by Ohm's law

$$V = IR = 114 \text{ A} \times 20 \text{ mi} \times 0.85 \text{ ohm/mi} = 1940 \text{ V}$$

The step-up transformer at the generating station must supply the power at 6340 V (4400 + 1940 = 6340).

The line power loss is

$$\begin{aligned} P &= I^2R = (114)^2 \times 20 \times 0.85 \\ &= 220{,}000 \text{ W} = 220 \text{ kW} \end{aligned}$$

The generator must therefore produce 720 kW to supply a 500-kW load. This represents a very inefficient operation (30.6 percent line loss), one which can be materially improved by using a higher voltage for transmission, as will be shown.

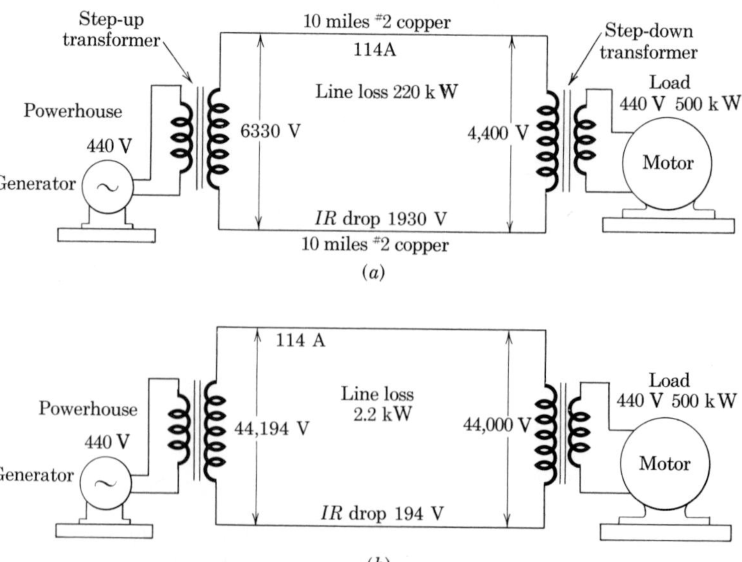

Fig. 32.15 High-voltage transmission line losses. The diagram is schematic only and assumes unity power factor.

In diagram (*b*) the same generator, line, and load are shown. In this case, however, the load draws power from the line through a 44,000-440-V step-down transformer. The line current in this case will be

$$I = \frac{500{,}000}{44{,}000} = 11.4 \text{ A}$$

The voltage drop over the 20 mi line is

$$V = 11.4 \times 20 \times 0.85 = 194 \text{ V}$$

The generating-station transformer must supply the power at 44,194 V to compensate for the line-voltage drop. The line power loss in this case is

$$\begin{aligned} P &= (11.4)^2 \times 20 \times 0.85 \\ &= 2200 \text{ W} = 2.2 \text{ kW} \qquad \text{only 0.4 percent} \end{aligned}$$

This example illustrates the economic necessity for high-voltage transmission.

Leakage Losses The higher the transmission voltage, the greater the tendency for electrons to leak off the wires. Alternating current tends to flow in the outer layers of the wire (this is called *skin effect*), and this crowding of the surface of the wire, together with the very large potential difference with respect to ground, ionizes the surrounding air so that a conducting path to ground, trees, nearby buildings, or other wires is encouraged. This type of leakage is accentuated during stormy weather because moist air is a better conductor than dry air and also because the air may already be ionized due to atmospheric electric disturbances.

Insulators several feet long support high-voltage wires, holding them away from the towers and away from the other wires of the system, to minimize leakage losses within the system itself.

With extremely high voltages (300,000 and up), atmospheric conditions sometimes set up an ionization of the surrounding air which permits an extremely large line leakage known as *corona loss*. The discharge from the wire, in such cases, can actually be seen as a bluish-purple haze surrounding the wire. It can also be heard as a persistent 60-cycle tone, accompanied by spitting and crackling noises. It is this loss which sets a practical upper limit on transmission voltages. This practical upper limit is generally considered to be in the region of 500,000 V, although a few lines have been designed for and are operating at one million volts.

DC Transmission Lines With the availability of high-capacity electronic rectifiers and oscillators, and rotary converters, it is feasible to first step up the ac voltage to 500,000 V or more with transformers, and then rectify it to direct current. It is transmitted as direct current at the high voltage (one power company now has a 750,000 V dc line), and then converted back to alternating current at substations where local distribution begins. Direct-current (dc) transmission materially reduces line losses due to corona problems, and also those resulting from inductive and capacitive reactance. Appreciable operating economies have been reported from dc transmission systems, especially for large loads over distances of from 40 to 400 mi. DC transmission lines use two conductors—one positive and one negative—to transmit power; and in the event of an emergency, some power can be transmitted with one conductor and the earth. Two conductors as opposed to three reduce the initial cost of the line. Although still not very common, dc transmission will probably increase in the future, especially if ''superconducting'' wires can be developed and utilized.

32.15 ■ Transformers for Power Transmission

The basic theory of transformer construction and operation has been explained in a previous chapter.

For small stations transmitting at medium and low voltages to several load areas, the best practice is to provide a transformer bank for each generator at the power

In the ideal transformer, no power is consumed in the primary circuit. Electrical energy is changed to magnetic energy and back again.

station. For large plants, especially hydroelectric plants at remote locations, where all the power generated is to be sent for a long distance at high voltage, the entire bank of transformers is connected to the main busses through automatic-overload circuit breakers (see Fig. 32.14*c*).

It is not possible for transformers to step up power. They only step up the voltage.

Transformer Windings Power-station transformers are usually three-phase and may be of either the core type or the shell type. In the core type the copper windings encircle the iron core, and in the shell type the iron circuit is placed around the copper windings. Provision must be made for removing the heat generated in the transformer. Heat removal can be accomplished by circulating oil and then cooling the oil; by "finning" the outer shell of the transformer; or by providing fans to produce an air stream over the transformer case.

Fig. 32.16 Thyrite (patented) lightning arrestors mounted on a large three-phase transformer bank. (General Electric Co.)

32.16 ■ Lightning Arresters

A high-voltage transmission line is very likely to transmit to the generating station, or to substations, extremely high-voltage high-frequency electric discharges if the line is struck by lightning. Even if the line is not actually struck, passing clouds may induce large charges in the wires, and when the lightning bolt discharges the cloud, high-frequency oscillations are set up in the transmission line and a heavy surge travels to the stations along the line. These surges may start arcing of generators and similar equipment if the surges are not channeled off to ground. Disastrous "blackouts" (power outages) initiated by lightning-induced surges, with consequent arcing and equipment breakdown, are problems power companies have to contend with. The devices which provide lightning protection are called *lightning arresters* (see Fig. 32.16).

32.17 ■ Overall Arrangement of Power-Station Equipment

One possible arrangement of a coal-fired power station, showing the disposition of the units and equipment discussed in the preceding paragraphs, is shown in Fig. 32.17. A three-phase three-wire system is shown, and a medium-length (20 to 60 mi) transmission line is assumed. The diagram also shows a typical distribution system for industrial, commercial, and residential consumers.

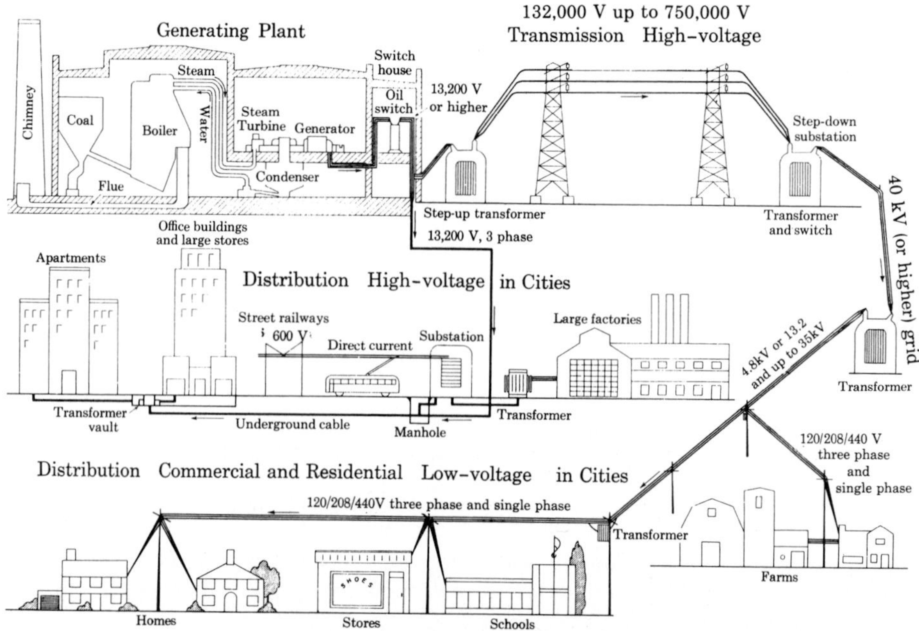

Fig. 32.17 Schematic arrangement of a power production and distribution system. Coal is the energy source of the system. (Bureau of Labor Statistics, U.S. Department of Labor)

Interconnected Transmission Systems In order to make full utilization of available generation, many utilities are linked together with high-voltage transmission lines. In the western United States, for example, the Western Systems Coordinating Council is made up of many utilities. Through interconnected transmission lines, these utilities are in constant communication with each other to determine supply of and demand for electric power. The result is that surplus power can be sold and bought with the overall objective of utilizing the most economical power available. Another significant benefit is the increased reliability that the system provides in the event of a major transmission line or equipment failure.

This concept has prompted the construction and utilization of extra-high-voltage transmission lines for these interconnections so that large amounts of power can be transmitted with a minimum of line losses. These lines are terminated at various key locations such as generating station switchyards and transmission switching stations. Figure 32.18 shows an extra-high-voltage switchyard with associated circuit breakers.

Fig. 32.18 Extra-high-voltage transmission switching station showing buss work, disconnects, and circuit breakers. (Pacific Gas and Electric Co.)

SUBSTATIONS

The function of a substation in an electric-energy distribution system is to modify certain properties of the transmission current to conform to the needs of the consumer load which the substation serves. There are substations designed to accomplish a variety of purposes, and we shall discuss only a few of the more common types. A single station may, of course, accomplish two or more of the functions discussed.

32.18 ■ Substations for Reducing Voltage

This type of substation is located near the city or load area to be served and steps down the voltage from the high transmission voltage to an intermediate voltage (say 40,000, or 13,200 or 4800 V) for local distribution. Such substations are sometimes known as step-down substations, or *transformer substations,* since transformers are the key equipment involved.

Feeders lead out from the substation, and they branch into *subfeeders, mains, submains,* and finally into services, which are the final connections to the premises of the consumer.

In general, the lines from a generating plant to a step-down substation are called the *transmission system,* and the network from the substation to the ultimate consumers is called the *distribution system* (see Fig. 32.19).

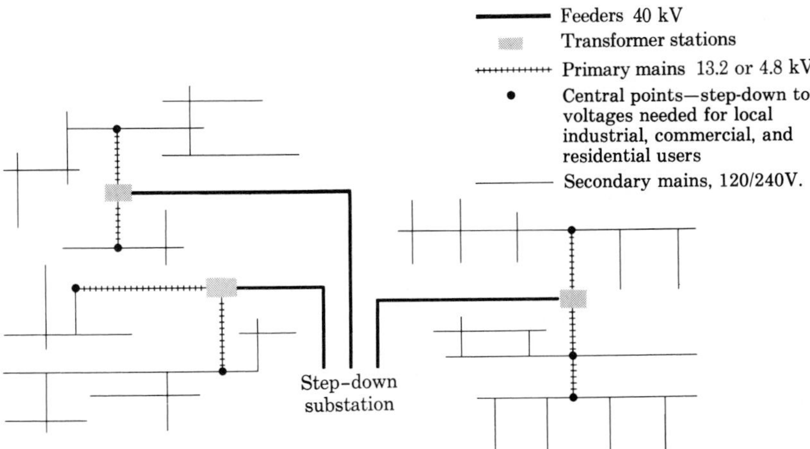

Fig. 32.19 A possible arrangement of a feeder-and-main multiple distribution system.

32.19 ■ Frequency-Changer Substations

In cases where a different frequency is required by an industrial consumer, a substation may be built primarily for frequency changing. The principal components of such a substation are synchronous machines in pairs, one of which runs as a motor on the applied frequency and the other as a generator delivering the desired frequency. If the frequency-changer substation operates off a high-voltage transmission line, the high voltage must first be stepped down to a value suitable for synchronous machine operation—usually 13,200 V or less.

It is common practice to tie in the facilities of one power company with those of another so that electric power can be shared between two localities as load demands require. If the two systems are of different frequencies, frequency-changer substations are required to tie in the two systems. The United States and Canada have extensive facilities for tying in electric power networks.

32.20 ■ Substations for Supplying Direct Current

Many industrial operations require large quantities of dc power. Substations for rectifying alternating to direct current may use either *rotary converters* or huge *electronic rectifiers*. The rotary converter is merely a synchronous ac motor, direct-connected to a dc generator. Electronic equipment is also used for rectification. Street-railway systems, trolley-bus systems, metal refining companies, metal rolling mills, and paper and textile mills are heavy users of dc power.

32.21 ■ Distribution Substations

Substations whose principal function is to act as the distributing center for electric power in a given area are called *distribution substations*. They may incorporate one or all of the above functions. In addition, they have a variety of complex associated equipment for switching, phasing, protection, voltage regulation, power-factor control, and emergency service in case of casualty.

DISTRIBUTION SYSTEMS

As previously defined, the *distribution system* is the network which has to do with local distribution of electric power. Distribution systems are of many types, but we shall discuss here, in an elementary way, a simple one which has the following specifications:

Current: alternating
Connection system: multiple, feeder, and main
Phases: three-phase, and single-phase
Voltages: 40 kV, 13.2 kV, 120/240 volts
Frequency: 60 Hz (cycles/s)

32.22 ■ Feeder-and-Main Multiple Distribution System

In order to minimize power losses in a three-phase system, power companies attempt to balance the load lines (make the power factor = 1). Load lines are balanced if all the line currents are equal and have the same phase angle with respect to the line voltage.

In this type of distribution system, four-wire three-phase *feeders* from the central substation supply other smaller transformer stations at relatively high voltages, perhaps 40,000 V. From these stations, *primary mains* carry current to other central points at 13,200 V. Local distribution to commercial and residential consumers at 240 and 120 V is effected from these central points. Industrial consumers are supplied with 440-V three-phase service from individual transformers off 13.2-kV lines. Ordinarily 440 V is not used as a distribution voltage. Figure 32.19 shows in diagram form how such a multiple distribution system might be organized. For 13.2- and 4.8-kV feeders and mains the trend is to underground installation. In many states the law now requires that all residential, commercial, and light industrial distribution networks be underground in specified urban areas.

Load Factors In planning the distribution network a great deal of attention must be given to balancing out the load served from a substation. The greatest diversification possible (in type of consumer load) is desired, in order to keep the power factor as high as possible. If the load is largely of an *inductive* nature (induction heating, induction motors, etc.), *the current will lag the voltage,* while loads with elements of *capacitance* predominant will cause the *current to lead the voltage*. In either case, the power factor (cos ϕ, where ϕ is the angle of lead or lag, the so-called *phase angle*) will be reduced, and the distribution of the power becomes an uneconomical operation (see Sec. 31.14). Residential-type loads tend to be largely resistance loads, and these circuits tend to have a comparatively high power factor. Industrial-type loads, made up as they usually are of a predominant motor load, are highly inductive in nature, and it requires continuous attention and planning by power-company engineers to keep power factors in industrial areas as high as possible. In general, a residential network may have a power factor as high as 90 to 95 percent; industrial loads, one of 60 to 75 percent; and a combined residential and industrial network, one of 75 to 90 percent. The increasing use of electric motors in homes (air conditioners, workshops, etc.) is tending to cause lower power factors on these lines also. Banks of capacitors may be installed to correct this condition (see Fig. 32.20). Capacitor banks are ordinarily installed with time switches which cut them out of the line after about 11 p.m. Otherwise, as the inductive load falls, current may lead voltage, with a resultant low power factor from too much capacitive reactance.

Fig. 32.20 Bank of capacitors on a rural electrification line to balance the induction load of electric motors. The capacitors bring current and voltage more nearly in phase and this will result in a better power factor. (Westinghouse Electric Corp.)

Illustrative Problem 32.2 An industrial-plant service line has an impressed power of 100 kVA on a 60-cycle line at 440 V. An installed motor offers a total ohmic resistance of 350 ohms and an inductance of 1 H. Find (*a*) the inductive reactance X_L of the circuit, (*b*) the impedance Z, (*c*) the current to the motor, (*d*) the power factor, (*e*) the power (in kilowatts) being used by the motor, and (*f*) the size capacitor (microfarads) which would have to be installed to increase the power factor to 95 percent.

It can be shown that the greatest transfer of electric energy from source to load occurs when the impedance of the load is equal to that of the source.

Solution

(*a*) $$X_L = 2\pi fL = 6.28 \times 60 \times 1 = 377\ \Omega \qquad \textit{answer}$$

(*b*) $$Z = \sqrt{R^2 + X_L^2} = \sqrt{(350)^2 + (377)^2} = 514\ \Omega \qquad \textit{answer}$$

(*c*) $$I = \frac{V}{Z} = \frac{440\ \text{V}}{514\ \Omega} = 0.856\ \text{A} \qquad \textit{answer}$$

(*d*) $$\text{Power factor} = \frac{\text{kW}}{\text{kVA}} = \cos\phi = \frac{R}{Z}$$
$$= \frac{350\ \Omega}{514\ \Omega} = 0.681 \text{ or } 68 \text{ percent} \qquad \textit{answer}$$

(*e*) $$\text{Power used} = 100\ \text{kVA} \times 0.68$$
$$= 68\ \text{kW} \qquad \textit{answer}$$

(*f*) For the power factor to be 95 percent, the phase angle can be found from cos $\phi = 0.95$. From a calculator,

$$\phi = 18°$$

Then $$\frac{X}{R} = \tan 18°, \text{ where } X \text{ is the net reactance}$$

or $$X = 350 \tan 18°$$
$$= 113.7\ \Omega$$

Now $$X = X_L - X_C = 113.7\ \Omega$$

or $X_C = X_L - 113.7\ \Omega$
$= 377\ \Omega - 113.7\ \Omega$
$= 263.3\ \Omega$, the *capacitive reactance* required in the circuit

But

$$X_C = \frac{1}{2\pi f C} \quad \text{and} \quad C = \frac{1}{2\pi f X_C}$$

Substituting gives

$$C = \frac{1}{2\pi \times 60 \times 263.3}$$

$$= 10.1 \times 10^{-6} \text{ farad (F)} = 10.1\ \mu\text{F} \qquad \textit{answer}$$

QUESTIONS AND EXERCISES

1. Distinguish clearly between the terms *kilowatt* and *kilovoltampere*. Under what conditions would the kilowatts used by a customer be equal to the kilovoltamperes available on the line?

2. Why are high voltages used for long-distance transmission of electric power?

3. What phase difference exists between the currents in a three-phase ac generator?

4. When electrical energy is converted to other forms, not all the energy serves a useful purpose. What undesirable forms of energy may be present in (1) an electric motor, (2) an electric light bulb?

5. Name four forms of energy which are readily converted into electric energy.

6. What term is ordinarily used to describe an electric circuit that contains a switch with its contacts not touching?

7. If transformers cannot operate on pure dc, how is it possible for an automotive induction coil (transformer) to induce such a high voltage in the secondary from a 12-V dc primary source?

8. Could a step-up transformer be used as a step-down transformer? Give a reason for your answer.

9. Recalling that 1 W equals 1 J/s, show that the kilowatt-hour is a unit of energy.

10. Discuss the factors which often make steam plants a more economical installation than hydroelectric plants, even though the "fuel" for the latter is "free," once the installation is completed.

11. Contrast the basic principle of operation of the hydro-plant impulse turbine with that of the reaction turbine. What laws of motion are involved? What conditions at the power-plant site would influence the decision as to which of the two types to use?

12. Summarize the reasons for winding most commercial alternators as three-phase wye-connected machines.

13. Enumerate the kinds of leakage losses suffered by electric-power transmission lines. Discuss the causes of each loss, and some of the measures taken to minimize the loss.

14. Name and describe the principal function of four types of substations. See if you can get permission from the power company to visit a nearby substation and get an explanation of the operations there.

15. Why is it of critical importance to balance inductive loads with capacitors in electrical distribution and service systems?

16. Explain why cos ϕ (where ϕ is the *phase angle*) is equal to the ratio kW/kVA.

17. Why is it necessary to synchronize two or more alternators before connecting them to the busses and feeders? Use a sketch in your explanation.

18. What is the basic electrical principle on which transformers operate? Why is alternating current necessary?

19. Draw a wiring diagram for a circuit arrangement that can be applied in controlling a ceiling light from opposite sides of the room.

For questions 20 to 23, indicate which expression makes the statement true.

20. A consumer of electric energy pays for (*a*) ampere-hours, (*b*) voltamperes, (*c*) voltamperes reactive, (*d*) kilowatts, (*e*) kilowatt-hours.

21. Alternating current is not suitable for (*a*) heating electric toasters, (*b*) transmitting electric power, (*c*) charging storage batteries, (*d*) running electric motors, (*e*) operating a lifting magnet.

22. We can calculate the number of kilowatt-hours by means of the formula: kilowatt-hours equals (*a*) (volts × hours)/(amperes × 1000), (*b*) volts × amperes × hours × 1000, (*c*) (volts × amperes × hours)/1000, (*d*) (volts × hours × 1000)/amperes, (*e*) (volts × amperes)/(hours × 1000).

23. The following would probably cause the circuit breaker in a house circuit to open: (*a*) a great decrease in the resistance of the circuit, (*b*) a great increase in the circuit resistance, (*c*) connecting all the household lights in series, (*d*) a great decrease in voltage applied to the circuit, (*e*) none of these.

24. An electric motor has a nameplate rating as follows: 240 V, 25 A, 5 hp, 3 phase, 60 Hz, 1750 rev/min. Which of the following is true? (*a*) the motor uses 240 V, 25 A at no load, (*b*) the power input of the motor is 5 hp, (*c*) the no load speed is 1750 rev/min, (*d*) the motor will deliver 5 hp at full load, (*e*) none of the above.

PROBLEMS

Group One

1. What is the frequency of the ac voltage produced by a 13,200-V, three-phase, 12-pole alternator rotating at 1800 rev/min?

2. How many cycles per second are generated by a 24-pole alternator when driven at 250 rev/min?

3. Determine the period of the waveform when the frequency of the waveform is (*a*) 450 kHz, (*b*) 60 Hz, and (*c*) 250 MHz.

4. Determine the frequency of the waveform when the period of the waveform is : (*a*) 450 μs, (*b*) 0.75 s, and (*c*) 250 ms.

5. At what speed, in revolutions per minute, should the rotor of a water turbine turn if it is to produce a 60-Hz voltage in a generator having 30 poles?

6. An electric toaster uses 5 A when connected to a 120-V line. If the cost of electric power is 8.5¢/kilowatt-hour, what is the cost of using the toaster for 3.5 min?

7. A transformer which is 100 percent efficient has 300 turns in the primary coil and 75 turns in the secondary coil. (*a*) What is the voltage across the secondary if the voltage across the primary is 12 V? (*b*) If the current in the primary is 12 A, what is the current in the secondary?

8. A step-down transformer is used on a 4800-V distribution line to deliver 80 A at 120 V. About what current is drawn from the distribution line?

9. A step-down transformer operates on a 4400-V line and supplies a load with 150 A. The ratio of the primary to the secondary windings is 25:1. Determine (*a*) the secondary voltage, (*b*) the primary current, and (*c*) the power output. Assume 100 percent efficiency.

10. A power company delivers 580 kVA to the line serving an industrial plant during a period when the power *used* by the plant is metered at 500 kW. What is the power factor and the phase angle on the factory's circuit?

11. An alternator has 20 field poles and is used with a reaction-type water turbine. At what speed, in rev/min, must it turn to generate 60 Hz?

12. A steam turbine-generator turns at 1500 rev/min and has four poles. What is the frequency of the alternating current generated?

Group Two

13. What is the speed, in revolutions per minute, of a 440-V, three-phase, 60-Hz, four-pole induction motor with a 5 percent slippage?

14. A separately excited dc generator has a power output rated at 1200 kW. It is turned by a 1800-hp diesel engine and uses 60 kW to excite the field. What is its efficiency?

15. A small stream runs through your mountain ranch with an average flow of 15 s-ft (ft^3/s). It has a total drop in elevation of more than 100 ft as it runs through your property. What pressure head (feet of water) must you provide for a small turbogenerator (efficiency 70 percent) if it is to generate 35 kW?

16. A hydroelectric-plant site has a steady flow of 1200 ft^3/s under a head of 1600 ft. Calculate (*a*) the theoretical horsepower available and (*b*) the theoretical kilowatt output of the plant.

17. A shunt-wound dc generator has a field resistance of 100Ω and an armature resistance of 0.05Ω. It delivers 24 kW at 120 V to an external circuit (see Fig. 32.21). What is (*a*) the current in the external circuit? (*b*) The field current? (*c*) The armature current? (*d*) The total induced emf in the armature (terminal voltage + voltage drop in armature)? (*e*) The power developed in the armature?

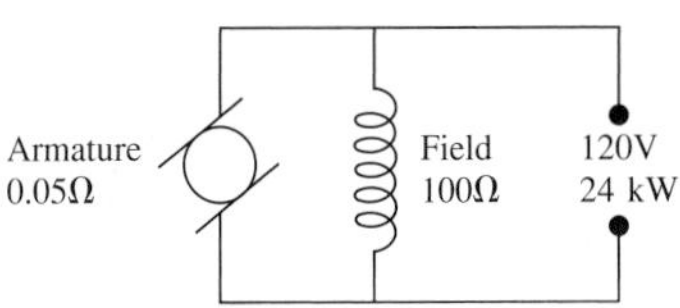

Fig. 32.21 Problem 17.

18. The emf induced in the armature (terminal voltage + voltage drop in armature) of a shunt generator is 451 V. The armature resistance is 0.15Ω. (*a*) What is the terminal voltage when the armature current is 340 A? (*b*) If the field resistance is 100Ω, what is the field current, the current in the external circuit, and the power delivered to the external circuit?

19. A step-down autotransformer (see Fig. 31.21), having an overall efficiency of 80 percent, provides 80 V output across a 10-Ω load resistor when the input is connected to a 120-V, 60-Hz source. What is the current, in amperes, through the common winding?

20. Two separately excited 240-V generators are to be connected in parallel to share a 750-A load. The first generator has a no-load terminal voltage of 250 V and a total armature resistance of 0.08Ω. The second generator has a no-load terminal voltage of 245 V and an armature resistance of 0.12Ω. If we neglect contact resistances and brush voltage drops, what must the current be in the armature of the first generator? (Recall that in load sharing, the terminal voltage of the generators must be equal.)

21. A manufacturing plant is supplied electric power from a 4400-V hydroelectric generator 20 mi distant (see Fig. 32.15). The plant needs 660 kW of energy delivered at 440 V; No 2 copper wire, with a resistivity of approximately 0.85Ω/mi is used for the transmission lines. (For simplicity, assume a two-wire system and a power factor of unity.) A 10:1 step-down transformer is used at the plant. (*a*) Determine the line current. (*b*) What is the voltage drop between the generator and the plant? (*c*) What is the power loss in the line? (*d*) How much electric energy must the generator develop to provide the required power at the plant? (*e*) What is the power efficiency of this arrangement?

22. For the data given in Problem 21, determine the answers to the questions listed if the voltage at the generator were stepped up to 44,000 V and sent over the lines and then stepped down using a 100:1 transformer at the plant.

Group Three

23. An industrial plant has a three-phase, 440-V, 60-Hz power supply. Part of the plant draws 80 kW at a power factor of 0.68 lagging. The other part draws 50 kW at a power factor of unity. What is the line current drawn by the plant?

24. A steam generator produces one million pounds of saturated steam per hour (pressure 500 lb/in.^{2} abs) from boiler feedwater whose temperature is 80°F. (Steam turbines are actually operated on *superheated* steam. This example uses saturated steam to avoid undue complexity of the problem.) (*a*) What is the input in Btu/hour to this boiler? (*b*) What horsepower turbine would it operate at 35 percent efficiency? (*c*) What is the fuel-consumption in gallons per hour if burner-boiler efficiency is 60 percent? Take fuel-oil density as 6.25 lb/gal. (HINT: See Table 15.6 for properties of saturated steam.)

25. A transmission line is solid aluminum, diameter 0.875 in., and spans a distance of 22.5 mi. Assume a two-wire line and unity power factor for simplicity. A 10,000-kVA generator is generating at 13,800 V, which is the voltage impressed on the line. (*a*) What is the *IR* drop along the line? (*b*) What is the line loss in kilowatts? (*c*) If the voltage is stepped up to 85,000 V before transmission, what is the line loss?

26. A steam plant produces ac power at 13,200 V and has been transmitting its energy to nearby towns at that voltage. A new industrial plant and town site have located 20 mi away, and the average demand for power at the new site will be 40,000 kW. The new town wants to purchase electric energy delivered to its distribution substation. Assume (for simplicity) a two-wire line whose total length is 40 mi and whose ohmic resistance is 0.15 Ω/mi. Assume unity power factor. (*a*) Compare the ohmic line losses under two possibilities: (1) transmitting at a voltage which provides 100,000 V at the town's substation; or (2) transmitting at a voltage which provides 320,000 V at the town's substation. (*b*) If electric energy costs $0.038 per kVA to produce, and the wholesale price delivered to the town's substation is $0.052 per kVA, how many dollars per day would the power company gain from transmitting at the higher voltage?

27. The power factor on a 60-cycle, 440-V three-phase industrial service line is unsatisfactory, and it is desired to build it up to 90 percent. The plant load is found to have a total inductance of 2.5 H and an ohmic resistance of 450 Ω. What should be the capacitance installed on this line?

28. After several years of operation, the power factor on a 230-V residential distribution circuit has fallen below acceptable levels. When the load impressed on the line is 350 kVA (60-cycle), the power *actually being consumed is* 200 kW. An estimate of the total ohmic resistance of the line and its load is 500 Ω. Find (*a*) the power factor, (*b*) the impedance, (*c*) the inductive reactance, and (*d*) the amount of capacitance in microfarads which must be installed to bring the power factor up to 85 percent.

PART 8 MODERN PHYSICS

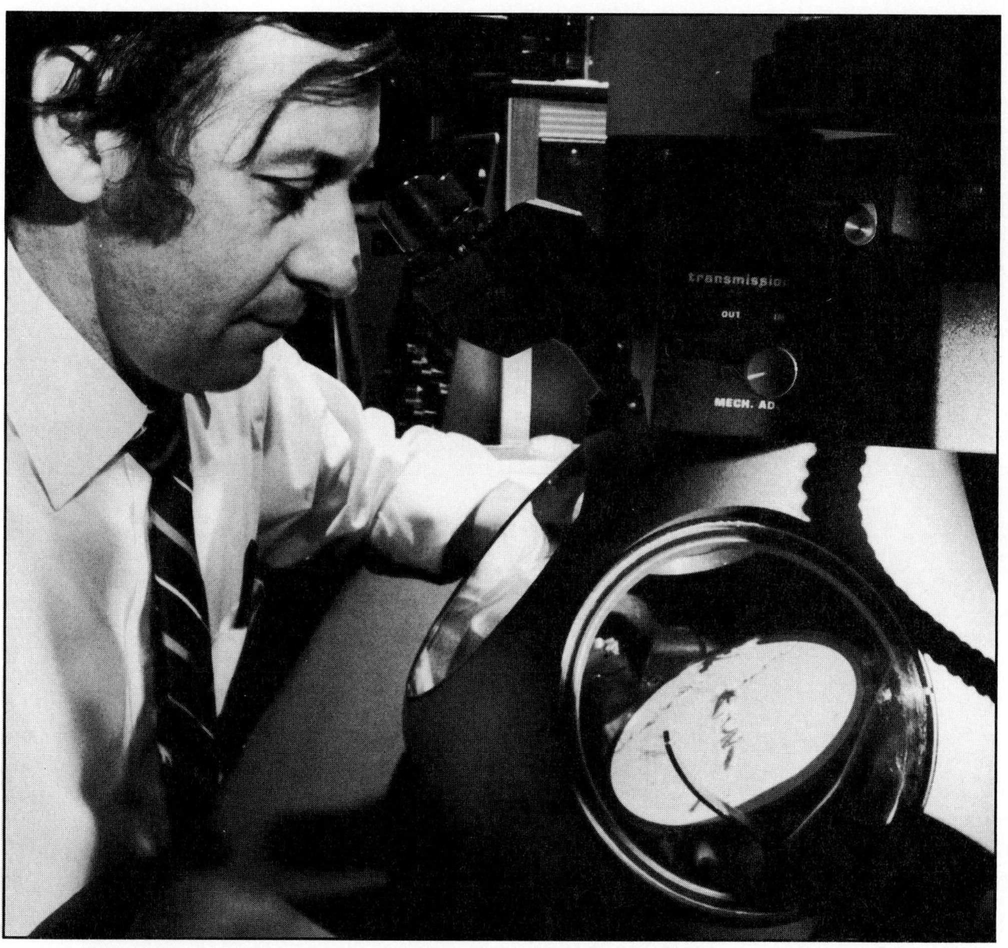

Transmission electron microscope. (United Technologies—Pratt and Whitney)

ELECTRONICS

In this chapter we take a brief look at electronics, which is partly based in electrical theories of classical physics of the nineteenth century, but which mainly depends on the theories and discoveries of *modern physics* developed after 1900. Much of the progress in the development of semiconductor devices and integrated circuits has required an understanding of how electrons behave in solids, which is an extension of the theories and experimental findings of modern physics as they relate to atomic structure (see Chap. 34).

Electronics, defined.

> ***Electronics is the science of freeing electrons from atoms of matter and controlling their flow in conductors and semiconductors and between the metallic electrodes in vacuum tubes or gas-filled tubes.***

The elementary introduction to the field of electronics provided in this chapter will include discussions of electrical conduction in solids, individual semiconductor devices and integrated circuits, and applications of electronics in industry, communications, and medicine.

SOLID-STATE ELECTRONICS

Before considering some specific electronic devices, it will be useful to consider some of the electrical properties of solids.

33.1 ■ Conductivity

The electrical conductivity of a material is proportional to the number of free electrons per unit volume in the material.

All materials at temperatures above absolute zero contain some "free electrons" that do not "belong" to any atom. In Eq. (26.5), $R = \rho L/A$, the value for the electrical resistivity ρ (or its reciprocal the electrical conductivity σ) for a material is mainly a measure of the number of free electrons per unit volume that are available in the material. As we shall see in the next section, it is the *availability of free electrons* that determines whether a material is a conductor, a semiconductor, or an insulator.

33.2 ■ Conductors, Semiconductors, and Insulators

The best conductors of heat are also good conductors of electricity.

The range of electrical conductivities for solids is very great compared to the range of the numerical values of most other properties of solids. Figure 33.1 shows values of electrical conductivity σ and electrical resistivity ρ for some selected solids. As indicated thereon, the best conductors (the metals) are called *conductors*, the poorest

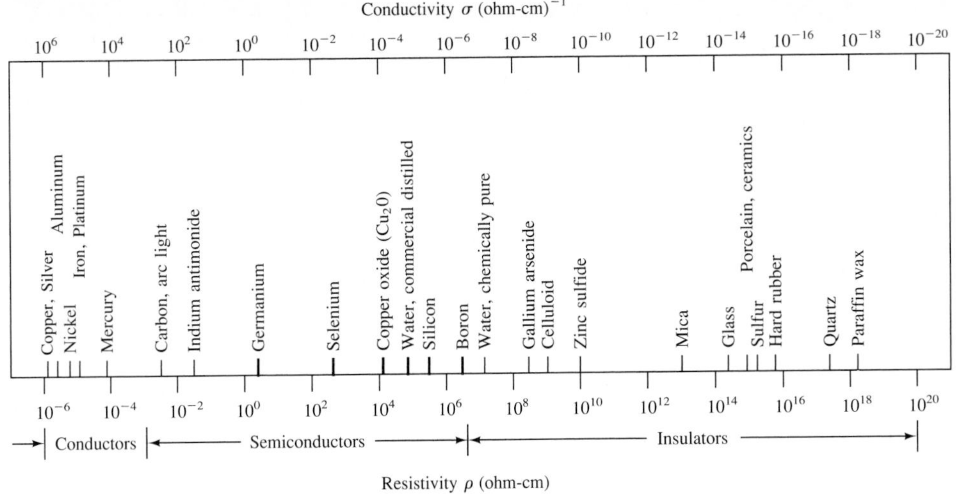

Fig. 33.1 Comparative electrical resistivities and conductivities of conductors, semiconductors, and insulators.

conductors are called *insulators,* and those of intermediate ability to conduct electricity are called *semiconductors.*

The ease with which the outer electrons of the atoms in a solid can be made *free electrons* determines whether the material will be a conductor, semiconductor, or insulator. Since one source of energy to free outer electrons from the atoms in a solid is thermal energy, the number of free electrons in a material is likely to depend on temperature. However, conductors contain free electrons at all temperatures down to absolute zero. For example, in copper there is one free electron for each atom, and in one cubic centimeter of copper there are approximately 8.4×10^{22} free electrons. At very low temperatures some conductors become perfect conductors (superconductors) with zero resistance (see discussion in Secs. 18.8 and 26.12).

For semiconductors and insulators, energy must be provided to free the outer electrons from the atoms. The main difference between insulators and semiconductors is that more energy is required to remove an outer electron from an atom in an insulator than in a semiconductor. For example, 5 electron-volts (eV) of energy is required to remove an outer electron from a carbon atom in diamond (a good insulator), while only 1.1 eV of energy is required to remove an outer electron from an atom in silicon (a semiconductor). An *electron-volt* of energy is acquired by a single electronic charge as it is accelerated through a potential difference of one volt.

The energy required to remove an outer electron from an atom is greater for an insulator than for a semiconductor.

$$1 \text{ eV} = 1.60 \times 10^{-19} \text{ C} \times 1 \text{ V} = 1.60 \times 10^{-19} \text{ J}$$

33.3 ■ Pure Semiconductors

Each atom in silicon has four outer *valence electrons,* located farther from the nucleus than the other ten electrons, and each atom has four "nearest-neighbor" atoms. The atoms in silicon are bonded to one another by an electron-sharing mechanism called *covalent bonding* in which a valence electron sometimes orbits the nucleus of its parent atom, and sometimes orbits the nucleus of one of the four "nearest-neighbor" atoms. A two-dimensional diagram showing an arrangement of silicon atoms with these features is shown in Fig. 33.2. At a temperature of absolute zero, all the valence electrons are involved in the covalent bonding process, and there are no free electrons in the material to make up an electric current. Because of this, silicon (and every other semiconductor) is a perfect insulator at absolute zero. At temperatures above absolute zero, however, some of the available thermal energy may be used to free valence electrons from their parent atoms. The number of free electrons per unit

Covalent bonding is the electron-sharing, chemical-bonding mechanism that holds the atoms together in the semiconductor materials silicon and gallium arsenide.

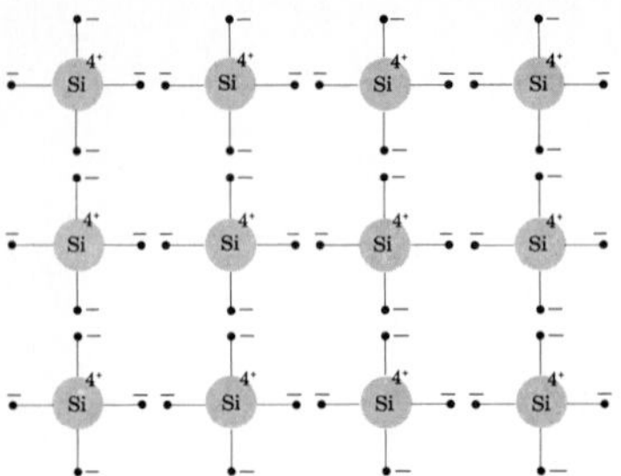

Fig. 33.2 Two-dimensional diagrammatic representation of the arrangement of atoms in a crystal of pure silicon. Each shaded circle represents the nucleus and ten inner electrons of a silicon atom. The − signs represent the four outer (valence) electrons of the atoms. Actually the bonds exist in three dimensions.

volume in a sample of material depends both on the temperature of the material and on the quantity of energy needed to remove a valence electron from its parent atom. The energy required to remove a valence electron from an atom is different for different semiconductors, but for all semiconductors the number of free electrons per unit volume increases as the temperature of the material increases.

When a valence electron becomes a free electron, it can respond to an electric force and become one of the moving charges needed to establish a current in the material. But there is another important contribution to the electric current that is not so obvious. When a valence electron is freed from a silicon atom, only three valence electrons remain, and the silicon atom that lost the valence electron now *has a net positive charge*. A valence electron from a nearby silicon atom can move to the site of the missing valence electron (called a *hole*) without requiring the 1.1 eV usually needed to remove a valence electron from a silicon atom. But in the process, the *site* of the hole moves in the opposite direction. The sequence of events is shown in Fig. 33.3. The moving holes in a semiconductor behave like free *positive charges*.

Thermistors Since the number of free electrons (and free holes) per unit volume in a semiconductor changes with temperature, the electrical conductivity and resistivity of the semiconductor also change with temperature. Consider a cylindrical piece of semiconductor material with a length L and a cross-sectional area A. If the resistivity ρ of the material changes with temperature, the resistance R between the ends of the cylinder will also change with temperature since $R = \rho L/A$. If wire leads are attached to the ends of the cylinder (see Fig. 33.4a), a device is obtained that is very similar in appearance to the carbon resistor used in electronic circuits, but in this case, the resistance between the two wire leads depends on temperature. Such a temperature-dependent resistor is called a *thermistor*. Typical thermistors do not have the cylindrical geometry described here, but usually consist of two wire leads connected to a small bead of semiconductor material as shown in Fig. 33.4*b*. Thermistor manufacturers usually provide tables and graphs to show how the thermistor resistances change with temperature (see Table 33.1 and Fig. 33.5 for some typical data).

Illustrative Problem 33.1 A dishwashing machine in a restaurant uses a thermistor in its control system to keep the wash water temperature between 65 and 75°C. What is the associated range of thermistor resistances (approximately)? (See Fig. 33.5 for the graphical relationship between resistance and temperature for the thermistor.)

Some examples of applications of thermistors.

Solution From Fig. 33.5, the thermistor resistance is approximately 1800 Ω at 75°C and 2500 Ω at 65°C. So, for thermistor resistances between 2500 and 1800 Ω, the water temperature should be between 65 and 75°C (approximately). *answer*

In a pure semiconductor, since a "hole" appears each time a valence electron becomes a free electron, the process is called *electron-hole pair* production.

The electrical conductivity of a pure semiconductor depends on the availability of free electrons and free holes.

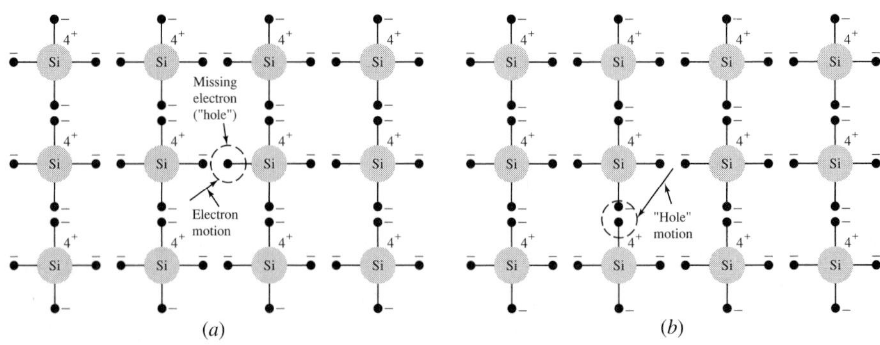

Fig. 33.3 (*a*) Electron motion and (*b*) associated "hole" motion in silicon.

Table 33.1 Temperature-resistance data for a typical thermistor

Temperature, °C	*Resistance, Ω*
0	33.60K
10	20.82K
20	13.33K
30	8794
40	5955
50	4132
60	2930
70	2120
80	1562
90	1171
100	891.0
110	687.8
120	538.0
130	426.0
140	341.2

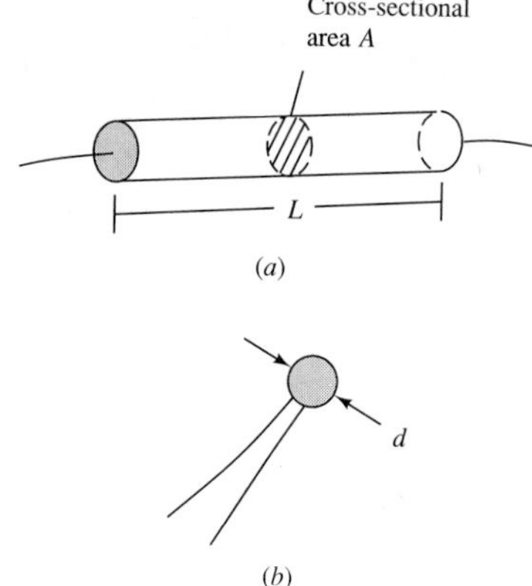

Fig. 33.4 (*a*) The resistance of a thermistor with cylindrical geometry depends on the length L and cross-sectional area A of the cylinder, and on the resistivity of the material (which changes with temperature). (*b*) Most commercial thermistors have a "bead" geometry. Thermistors are available with bead diameters as small as 0.020 in.

Thermistors are used in some medical thermometers, as detectors to sense overheating in electronic equipment, and as temperature-compensating components in electronic circuits. A thermistor placed in the throat of a medical patient in intensive care can be used to monitor the patient's breathing rate because the temperature of the inhaled air is lower than that of the exhaled air. Thermistors attached to the spindle axes of computer numerically controlled (CNC) milling machines are used to provide temperature information to the computer so that thermal expansion can be taken into account during the process of machining a part.

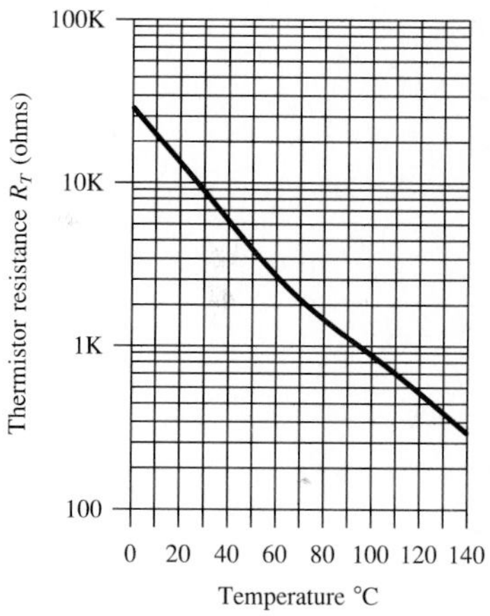

Fig. 33.5 Graphical display of thermistor resistance R_T versus temperature in °C for the thermistor whose resistance-temperature data is tabulated in Table 33.1. (vertical scale is logarithmic.)

33.4 ■ Doped Semiconductors

Pure silicon is a poor conductor at room temperature because so few of its valence electrons receive enough thermal energy (1.1 eV) to become free electrons. At room temperature each cubic centimeter of pure silicon contains approximately 1.5×10^{10} or 15 billion free electrons. Although this may seem to be a large number, it is only 7×10^{-12} percent of the total number of valence electrons in a cubic centimeter of silicon which contains 5.0×10^{22} atoms. Moreover, if 1.5×10^{10} electrons were to move past a point in a material each second, the current would be only 2.4×10^{-9} A.

The electrical conductivity of a semiconductor can be significantly increased by the addition of appropriate impurities. Two different ways in which this can be accomplished are discussed below.

N-Type Silicon If pure silicon is contaminated ("doped") with a mere trace of some element whose atoms have five valence electrons, such as arsenic (As), the fifth electron in each arsenic impurity atom will become a free electron if the temperature of the silicon is above about 100 K (100 K = −279° F). The arsenic impurity is called a *donor* impurity because it donates an electron to the crystal. These extra electrons are available within the silicon crystal structure for carrying an electric current (see Fig. 33.6*a*). Since there are 5.0×10^{22} silicon atoms per cubic centimeter, adding one arsenic impurity atom per million silicon atoms will result in the addition of 5×10^{16} free electrons per cubic centimeter. This is more than three million times as great as the number of free electrons per cubic centimeter in pure silicon at room temperature. Since electrical conductivity is proportional to the number of free charges per unit volume, this means that the conductivity of silicon is increased by a factor of approximately three million by the addition of one arsenic atom for every million silicon atoms.

The number of free electrons per unit volume in a semiconductor can be increased by adding *donor* impurities that have one more valence electron than the atoms of the semiconductor material.

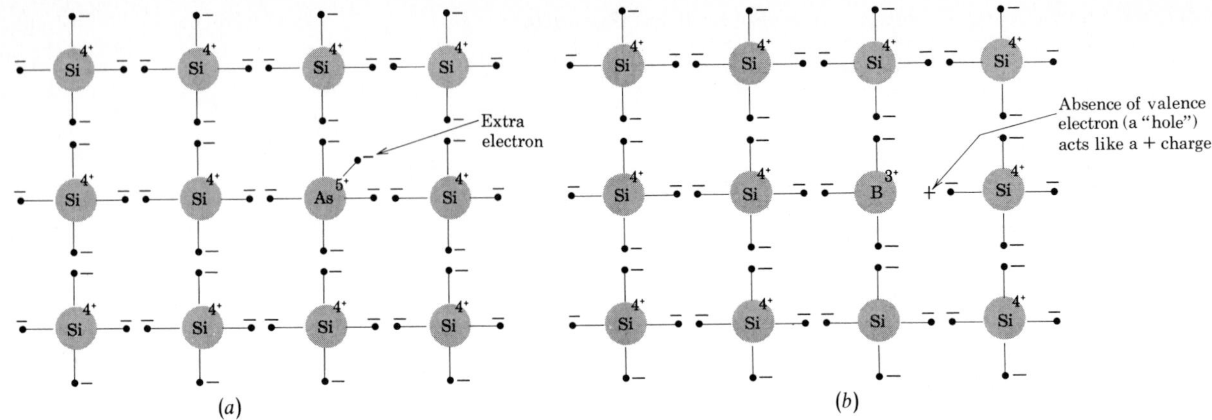

Fig. 33.6 Two-dimensional diagrams of two types of silicon. (*a*) *N*-type silicon showing one of the arsenic (As) impurity atoms. (*b*) *P*-type silicon showing one of the boron (B) impurity atoms.

Adding donor impurities to silicon to make it an *N*-type semiconductor causes an increase in the conductivity and a decrease in the temperature sensitivity of the conductivity of the silicon.

But there is another change in the electrical conductivity of the silicon due to the doping process discussed above. Since the free electrons provided by the arsenic impurities are present over a wide range of temperatures, the temperature dependence of the conductivity (which was important for the thermistor) is essentially eliminated. However, if the silicon were heated enough to free significant numbers of valence electrons from silicon atoms, the temperature dependence of the conductivity would reappear.

Since there is an excess of electrons (negative charge) in the arsenic-doped silicon, is it called *N*-type silicon. The arsenic impurities that have donated their fifth, outer electron to the crystal are positive ions. *N*-type silicon has free electrons and fixed positive impurity ions distributed throughout its volume.

***P*-Type Silicon** If pure silicon is doped with a trace of some element whose atoms have three valence electrons, such as boron (B) (see Fig. 33.6*b*), valence electrons from silicon atoms will move to the sites of the boron impurities to become the fourth valence electron for those impurities. This results in a ''hole'' in the crystal structure at each site where a silicon atom gave up an electron, and each boron atom that receives a fourth electron becomes a negative ion. Although 1.1 eV of energy is required to free a valence electron from a silicon atom, only about 0.05 eV of energy is needed to allow a valence electron to move from a silicon atom to the site of a boron impurity. That small quantity of energy is likely to be available to valence electrons in silicon atoms at temperatures as low as 100 K. Once ionized by receiving a fourth outer electron, the boron impurities, present in typical concentrations of one boron atom per million silicon atoms, are likely to stay ionized. Because of this, *P*-type silicon has fixed negative impurity atoms distributed throughout its volume. The boron impurities are called *acceptor* impurities because they accept a fourth electron from a silicon atom.

The number of free holes per unit volume in a semiconductor can be increased by adding *acceptor* impurities that have one fewer valence electron than the atoms of the semiconductor material.

As for the case of pure semiconductors discussed in Sec. 33.3, the holes behave like free positive charges that can move from place to place in the crystal under the influence of an externally impressed emf. Holes flow in the opposite direction from the electron flow, since the motion of an electron in one direction to fill a hole produces motion of the hole (or positive charge) in the opposite direction. Holes behave as if they were positive charges, since their effective motion is toward a negative electrode.

Since there is an excess of (positive) holes in the boron-doped silicon, it is called *P*-type silicon. The influence of the boron impurities on the conductivity and on the temperature sensitivity of the conductivity of the silicon is similar to that for the case of arsenic-doped *N*-type silicon.

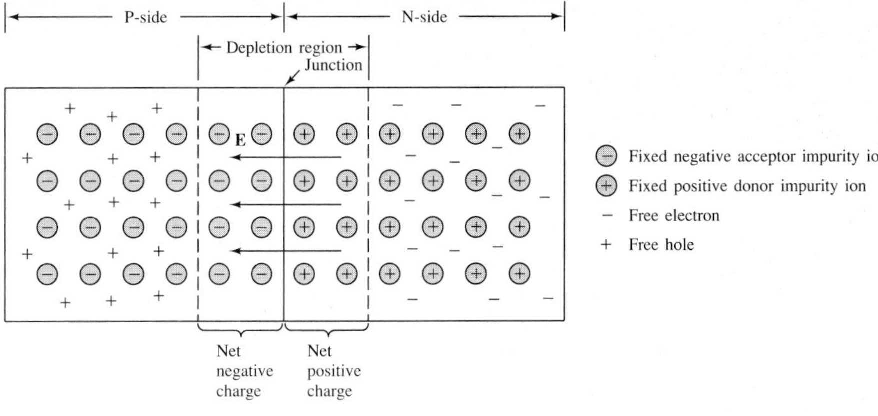

Fig. 33.7 Diagram showing the depletion region between the *P* and *N* regions of a *P-N* junction diode.

33.5 ■ Semiconductor Diodes

The positive and negative fixed impurity ions on the *N* and *P* sides of the *P-N* junction are responsible for the existence of an electric field in the depletion region.

If a block of *P*-type silicon and a block of *N*-type silicon are formed or compressed together, holes from the *P* silicon combine with the free electrons of the *N* silicon at the interface between the *N*- and *P*-type regions and effectively deplete the junction region of free charge carriers. As Fig. 33.7 shows, on the *P* side of the junction in the depletion region there is a net negative charge due to the presence of the fixed negative acceptor impurity ions. Similarly, on the *N* side of the junction in the depletion region there is a net positive charge due to the presence of the fixed positive donor impurity ions. These fixed impurity ions are responsible for the existence of an electric field in the depletion region that points from the *N* side to the *P* side. This electric field causes electrons in the depletion region to experience a force directed toward the *N* side of the junction, and therefore aids electrons, thermally generated in or near the depletion region, in their motion from the *P* side to the *N* side of the junction. The same field, however, prevents all but the most energetic electrons from moving from the *N* side to the *P* side of the junction. The resulting small current of electrons from the *N* side to the *P* side is called a *diffusion* current because the randomly moving free electrons on the *N* side tend to move or *diffuse* to the *P* side where there are fewer electrons.

A similar situation exists for the current of thermally generated holes from the *N* side to the *P* side and the diffusion current of free holes from the *P* side to the *N* side of the junction.

The free holes on the *P* side of the junction, and the free electrons on the *N* side of the junction are called *majority* carriers. The free holes on the *N* side of the junction and the free electrons on the *P* side of the junction are called *minority* carriers. For an isolated *P-N* junction, the strength of the electric field in the depletion region automatically adjusts itself so that the net current of minority and majority electrons and holes across the junction is equal to zero (see Fig. 33.8*a*). However, if an external emf is applied to the *P* and *N* regions, the net current across the junction will not be zero.

Biasing If an external emf is applied positive to the *P* material and negative to the *N* material (Fig. 33.8*b*), the strength of the electric field in the depletion region will be reduced and majority carriers will diffuse across the junction in greater quantities than occurs for the isolated junction. There will be no important change in the currents of thermally generated minority carriers across the junction. Applying an emf in the manner described is called *forward biasing* the junction.

In Fig. 33.8*c reverse biasing* is shown. The strength of the electric field in the depletion region is increased, and the diffusion of majority carriers across the junction is essentially eliminated. Under reverse-bias conditions the net current across the junction is due almost entirely to thermally generated minority carriers. These are the currents of thermally generated electrons from the *P* side to the *N* side and of ther-

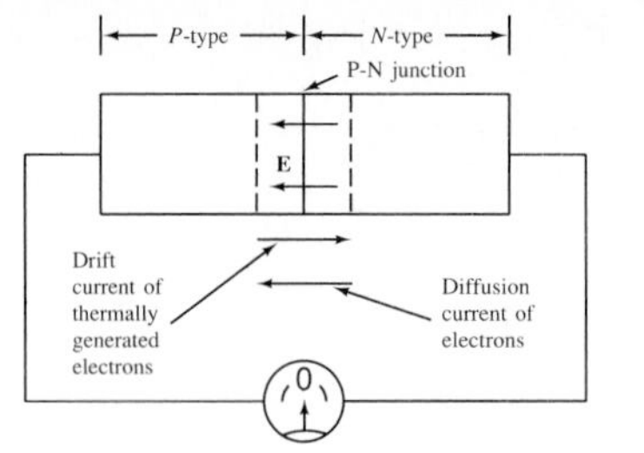

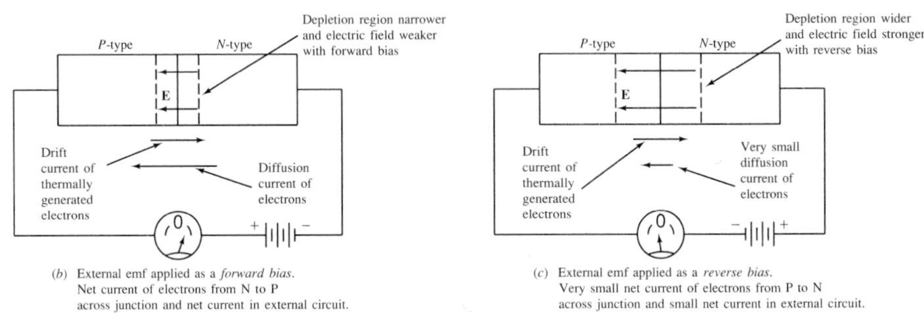

Fig. 33.8 Diagram explaining semiconductor diode operation. (*a*) zero bias, (*b*) forward bias, (*c*) reverse bias.

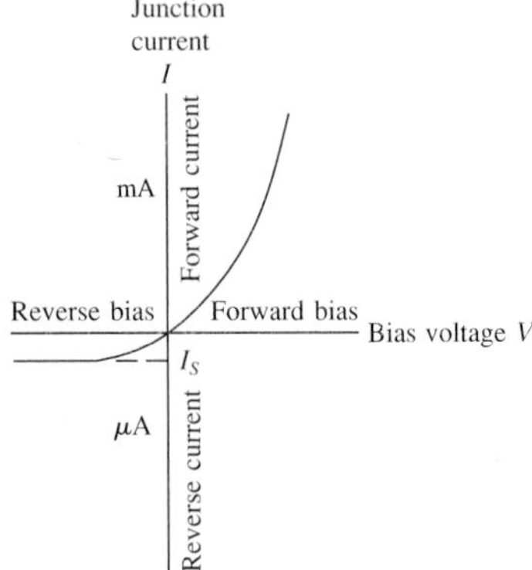

Fig. 33.9 Typical plot of junction current versus bias voltage (characteristic curve) for a *P-N* junction diode.

mally generated holes from the *N* side to the *P* side. This reverse current is a temperature-dependent current of very small magnitude.

A semiconductor diode is, then, a one-way valve for current flow. Figure 33.9 shows how the junction current depends on the magnitude and polarity of the externally applied emf. When a semiconductor diode is connected to an external *alternating* emf, current will flow for the half of the cycle during which the junction is *forward-biased* and will hardly flow at all during the half cycle of *reverse bias*. Semiconductor diodes are therefore widely used as rectifiers in direct-current power supplies (see Fig. 33.10). Below are additional examples of semiconductor devices that have one or two *P-N* junctions.

Zener Diodes For any *P-N* junction diode, there is some value of reverse bias for which the reverse current becomes abruptly very large. The magnitude of the reverse bias for which this large current appears is called the *reverse breakdown voltage*.

Diodes that are designed, by appropriate doping, to have the reverse breakdown occur at a specific, desired (usually low) voltage, are called *zener diodes*. Figure 33.11 shows the *I-V* characteristic curve for a zener diode. Zener diodes, which are operated in the reverse-bias mode, are available with breakdown voltages ranging from about 2 V to hundreds of volts.

Zener diodes are used in simple voltage-regulation circuits in direct-current power supplies. An example of such a circuit is shown in Fig. 33.12.

A varactor diode is a voltage-controlled capacitor.

Varactor Diodes A *P-N* junction diode has some properties similar to a parallel-plate capacitor with a dielectric material between the plates. From Eqs. (25.21) and (25.22), $C = k\epsilon_0 A/d$. The *P* and *N* regions of a diode behave like the conducting plates of a capacitor, and the depletion region behaves like the dielectric between the plates. In a charged parallel-plate capacitor there is an electric field in the dielectric material that is perpendicular to the conducting plates. Similarly, in a *P-N* junction diode there is an electric field in the depletion region that is perpendicular to the *P* and *N* boundaries of the depletion region. A diode designed to take advantage of its junction capacitance in circuit applications is called a *varactor diode*.

One way to change the capacitance of a parallel-plate capacitor is by changing the distance *d* between the conducting plates. The equivalent control for a varactor diode can be accomplished by changing the magnitude of the reverse bias for the *P-N* junction, since changing the reverse bias causes the width of the depletion region to change. That is, a reverse-biased *P-N* junction can be used as a voltage-controlled capacitor. The relationship between the capacitance C_j of the junction and the magnitude V_R of the reverse bias for one type of varactor diode is given approximately by an equation of the form

$$C_j = \frac{B}{\sqrt{V_R}} \tag{33.1}$$

where B is a constant that depends on the properties of the semiconductor material, and C_j is the capacitance, in picofarads.

Varactor diodes, which are used in tuning circuits for radio and television receivers, have the advantages of small size and high reliability.

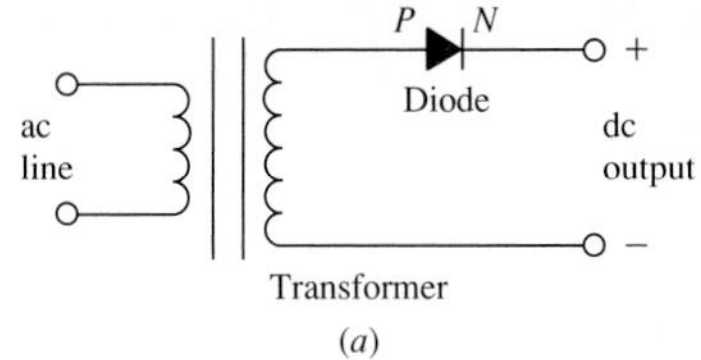

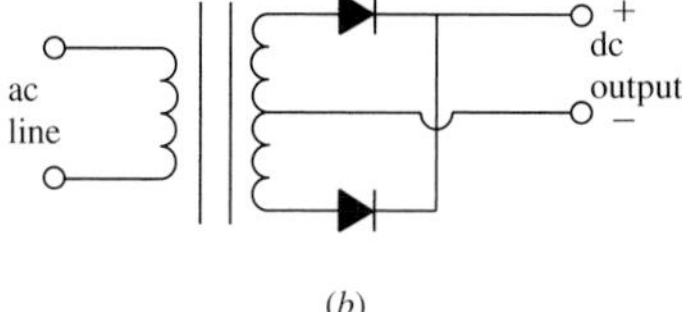

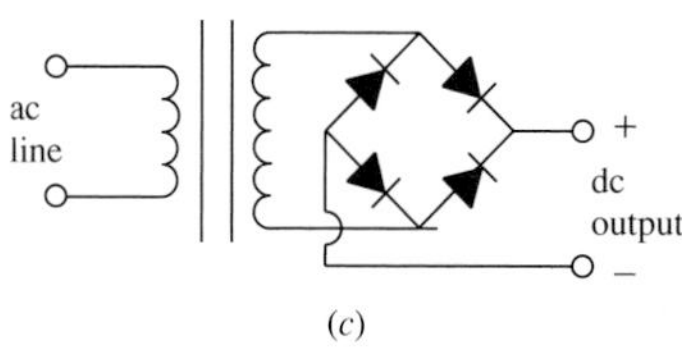

Fig. 33.10 Schematic diagrams showing how semiconductor diodes can be used as rectifiers in half-wave (*a*), full-wave (*b*), and full-wave bridge rectifiers (*c*).

Illustrative Problem 33.2 The junction capacitance, in picofarads (pF), of a 1N3495 diode can be calculated using the equation $C_j = 317/\sqrt{V_R}$, where V_R is the magnitude of the applied reverse bias (in volts). (*a*) What is the junction capacitance when V_R is equal to 25 V? (*b*) In a certain circuit application, a capacitance of 54 pF is required. For what value of V_R will the junction capacitance of a 1N3495 diode be equal to 54 pF?

Solution

(***a***) Substitute 25 V for V_R in the equation

$$C_j = \frac{317}{\sqrt{V_R}}$$

to obtain a value for the junction capacitance C_j in pF.

$$C_j = \frac{317}{\sqrt{25}}$$

$$C_j = 63 \text{ pF} \qquad \textit{answer}$$

(***b***) Solve the equation

$$C_j = \frac{317}{\sqrt{V_R}}$$

for V_R and substitute 54 pF for C_j.

$$\sqrt{V_R} = \frac{317}{C_j}$$

$$V_R = \left(\frac{317}{54}\right)^2$$

$$= 34 \ V \qquad \textit{answer}$$

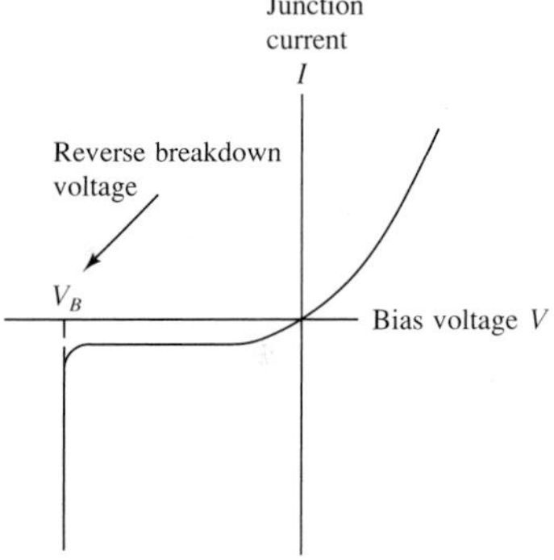

Fig. 33.11 Typical plot of junction current versus bias voltage for a zener diode.

Light-Emitting Diodes A light-emitting diode (LED) is a *P-N* junction device capable of emitting photons when forward-biased (see Fig. 33.13). When a free electron combines with a hole (i.e., becomes a valence electron in an atom missing one of its electrons), it loses a certain amount of energy. The amount of energy lost by the free electron is equal to the energy that would have to be given to a valence electron to make it a free electron. For example, when an electron combines with a hole in gallium arsenide, it loses 1.42 eV of energy, which appears in the form of an *infrared* photon of that energy. Gallium arsenide phosphide LEDs can, however, be designed to emit light in the *visible* spectrum. Integrated circuit devices containing an infrared LED and a light-sensitive detector (called *optoelectronic isolators*) are available for applications where electrical isolation between input and output circuits is required. LEDs are used as indicator lamps for electronic and medical instruments because of their high efficiency, long operating lifetimes, mechanical sturdiness, and

Light-emitting diodes (LEDs) can be designed to emit either infrared or visible light.

ability to operate from voltages typically available in solid-state electronic equipment. Arrays of LEDs can be used to display numbers and letters as shown in Fig. 33.14.

33.6 ■ Solar Cells or Photovoltaic Cells (PVCs)

A solar cell converts the energy of photons of sunlight to electric energy.

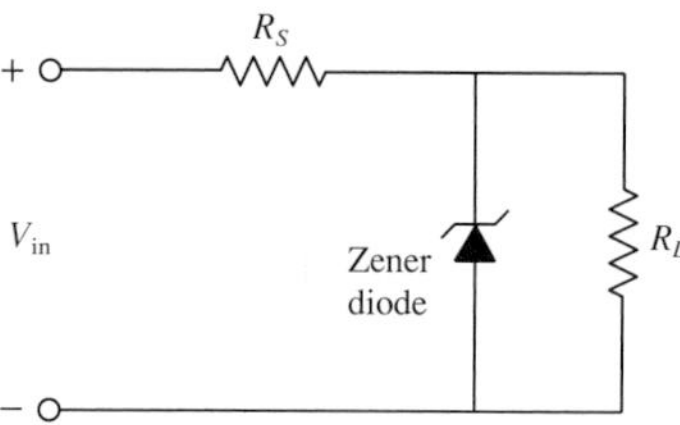

Fig. 33.12 Schematic diagram of a simple voltage-regulator circuit using a zener diode. If the unregulated input voltage V_{in} is greater than the breakdown voltage V_B for the zener diode, the potential difference across the load resistor R_L will be equal to the magnitude of the reverse breakdown voltage V_B even if the values of V_{in} or R_L fluctuate. The resistor R_s limits the current through the zener diode to a safe value.

Solar cells are large-area *P-N* junction devices capable of direct conversion of sunlight to electricity (see also Sec. 16.11). Figure 33.15 shows a schematic diagram for a solar cell with a magnified view of a segment of the depletion region. The sunlight is incident on the surface of the *N* region of the device which is so thin ($\cong 0.5\ \mu$m) that the photons of sunlight have a good chance of penetrating to the *P* side of the depletion region. For solar cells, as for other *P-N* junction devices, there is an electric field in the depletion region that is established by the positive and negative fixed impurity ions in the depletion region (see Fig. 33.7). No sunlight or photon activity is involved in establishing this electric field. When electron-hole pairs are produced by photons of sunlight in the depletion region, the minority carriers are swept across the junction by the electric field in the depletion region as shown in Fig. 33.15. This *photocurrent* of minority carriers across the junction is independent of junction bias. If the solar cell is short-circuited, the current will be equal to the photocurrent. A 14 percent efficient, 4-in diameter circular single-crystal silicon solar cell in bright (directly overhead) sunlight, will deliver a dc current of about 2.0 A at a potential difference of about 0.5 V. The short-circuit current for the same solar cell in bright, direct sunlight is over 2.5 A.

Solar cells have been used since the 1960s to power electronic equipment on earth satellites. Other uses include providing power for pocket calculators and electronic watches, and, on a larger scale, home power systems, and prototype commercial power systems capable of providing as much as 10 MW of power (see Sec. 16.11). Solar cells are ideal for providing power at remote locations to operate navigational aids at sea, signal lights for railroads, irrigation systems for farms, and microwave repeater stations.

Illustrative Problem 33.3 What electric power would be provided in bright, direct sunlight (intensity $I = 1000\ \text{W/m}^2$) by 100, 12-percent-efficient solar cells each with an area of 12.5 in.^{2}?

Solution The power *P*, the intensity *I* of the incident sunlight, the total area *A* of the solar cells, and the efficiency *e* of the solar cells are related by the equation $P = eIA$, since intensity is equal to power per unit area. For this calculation, the 12 percent efficiency must be expressed as a decimal fraction, so $e = 0.12$. Also, since the intensity of the sunlight is given in W/m^2, the total area of the solar cells must be expressed in m^2.

$$\begin{aligned} A &= (\text{total number of cells}) \times (\text{area per cell in m}^2) \\ &= (100)\,(12.5\ \text{in.}^2)\left(2.54^2\ \frac{\text{cm}^2}{\text{in.}^2}\right)\left(\frac{1\ \text{m}^2}{10^4\ \text{cm}^2}\right) \\ &= 0.806\ \text{m}^2 \end{aligned}$$

Now substitute the numerical values for *e*, *I*, and *A* into the equation $P = eIA$ to determine the electric power provided by the 100 solar cells in bright, direct sunlight.

$$\begin{aligned} P &= (0.12)\left(1000\ \frac{\text{W}}{\text{m}^2}\right)(0.806\ \text{m}^2) \\ &= 97\ \text{W} \qquad \textit{answer} \end{aligned}$$

33.7 ■ Transistors

The word *transistor* comes from the words tran(sfer) (re)sistor.

The transistor, developed at Bell Telephone Laboratories in 1947, has since 1960 assumed a position of commanding importance in the electronics industry.

> ***A transistor is a semiconductor device consisting of a combination of P-type and N-type materials, that is capable of switching, amplifying, and detecting electronic signals.***

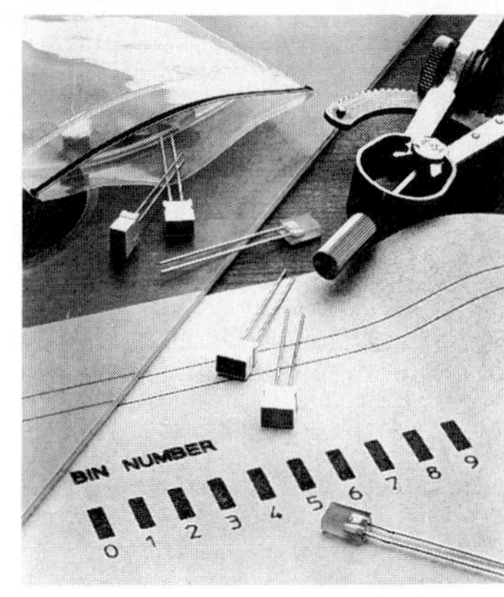

Fig. 33.13 Examples of miniature light-emitting diodes (LEDs) of the type used for indicator lamps in the panels of electronic equipment. (Hewlett-Packard Company)

Transistors (often in integrated circuit form) are used in radio and television transmitting and receiving equipment, in all types of computers, in medical diagnostic and imaging equipment, and in essentially all other applications of modern electronics.

Brief discussions follow of two types of transistors, the *bipolar junction transistor* and the *junction field-effect transistor*.

Bipolar Junction Transistors The bipolar junction transistor (BJT), sometimes simply called a bipolar transistor, is a three-terminal device that has two *P-N* junctions. It is called a *bipolar* device because both electrons and holes participate in the conduction processes. The field-effect transistor, to be discussed next, is an example of a *unipolar* device.

Bipolar transistors are composed of a thin section of one type (*N* or *P*) of silicon between two sections of the opposite type giving either an *N-P-N* or a *P-N-P* structure. The middle section of this sandwich, called the ''base,'' acts to control the flow of electrons (or holes) across it. One outer section acts as an emitter, and the other as a collector of free charges.

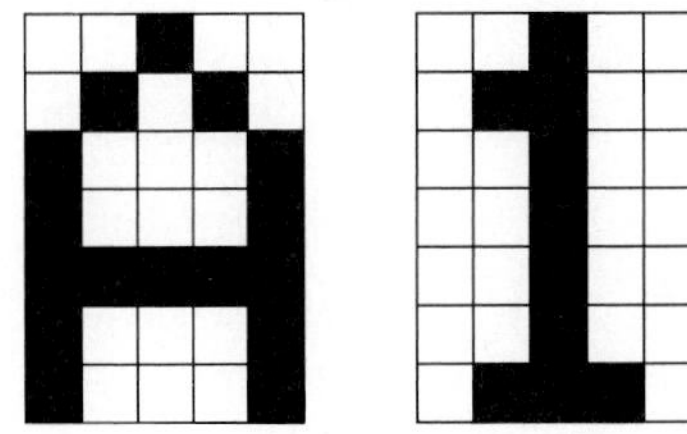

Fig. 33.14 An example of how a 5 by 7 array of light-emitting diodes can be used to display letters and numbers.

Figure 33.16*a* is a diagrammatic representation of a typical *N-P-N* junction transistor in a circuit called a common-base amplifier. Figures 33.16*b* and 33.16*c* show the schematic symbols for *N-P-N* and *P-N-P* bipolar transistors.

In a common-base amplifier, as Fig. 33.16*a* shows, the emitter-base junction is forward-biased and the base-collector junction is reverse-biased. Because the emitter-base junction is forward-biased, the strength of the electric field in the depletion region of that junction is reduced and the majority electrons diffuse from the emitter into the base region relatively easily. But, because the base region is made very thin, more than 95 percent of those electrons soon find themselves in the presence of the depletion region of the reverse-biased base-collector junction. The electric field in the base-collector depletion region pulls those electrons into the collector region. The result of this sequence of events is that there is a relatively large current across the reverse-biased base-collector junction. That current continues through the load resistor R_L in the collector circuit.

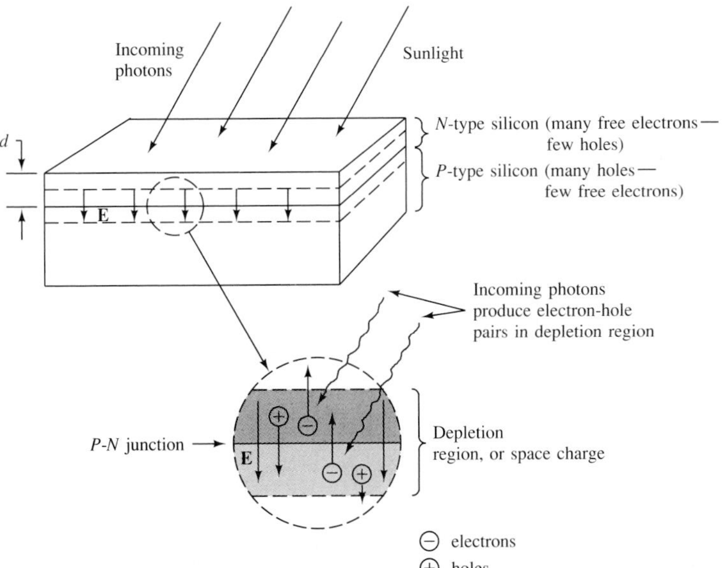

Fig. 33.15 Diagram of a *P-N* junction solar cell. Photons of light that pass through the very thin *N* region provide energy to free valence electrons from atoms in the depletion region at the boundary between the *P* and *N* regions of the device.

The conversion of light energy into an electric voltage is called the *photovoltaic effect*, and solar cells are commonly called *photovoltaic cells* (PVCs).

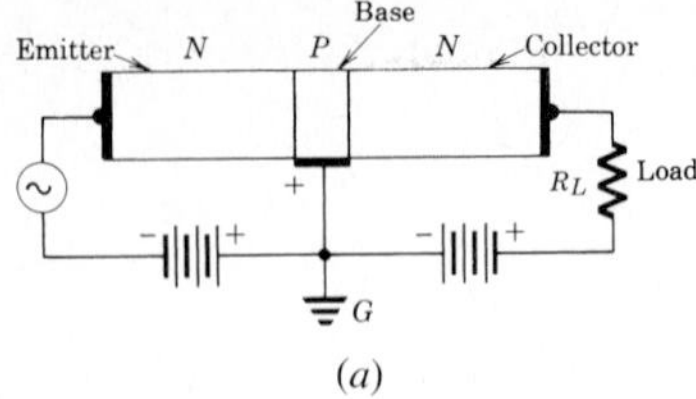

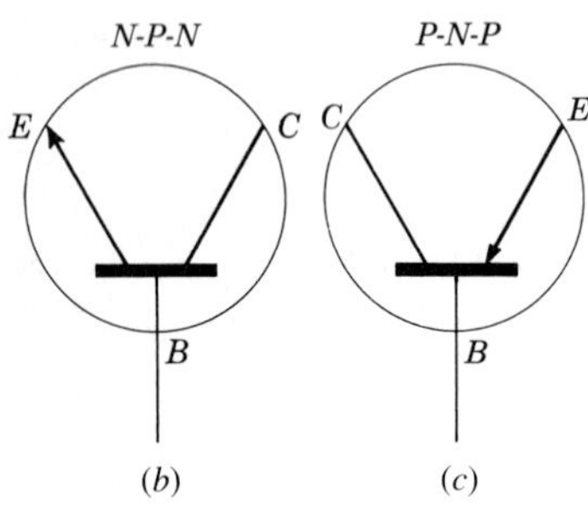

Fig. 33.16 Bipolar junction transistors. (*a*) Diagrammatic representation of an *N-P-N*-type bipolar transistor, showing forward bias on emitter-to-base and reverse bias on collector-to-base. (*b*) Symbol used to indicate an *N-P-N* transistor in circuit diagrams. (*c*) Symbol for *P-N-P* transistor. (*d*) Examples of surface mount transistors designed for use with communications equipment, instrumentation, and in the consumer electronics industry. Sharpened pencil point for size comparison. (Hewlett-Packard Company)

There is also a diffusion current of holes from the base to the emitter, but that is a smaller current because the base region is physically smaller, and because it is lightly doped compared to the emitter.

The bipolar transistor can be used as an amplifier because a small change in the emitter-base voltage can produce a large change in collector current. Figure 33.17 shows a simplified diagram of an audio amplifier. The input alternating voltage from a microphone causes variations in the emitter-base voltage, which produces large variations in the current through the loudspeaker load resistance R_L.

In addition to the common-base bipolar transistor amplifier, there are two other amplifier designs. These are the common-collector amplifier (sometimes called an emitter-follower) shown in Figure 33.18*a*, and the common emitter amplifier shown in Fig. 33.18*b*. The common-base amplifier can produce a voltage gain, but not a current gain. The common-collector can produce a current gain, but not a voltage gain. The common-emitter amplifier, which can produce both current gain and voltage gain, is the most widely used design for a bipolar transistor amplifier.

Although *N-P-N* bipolar transistors have been shown in the applications presented here, *P-N-P* transistors can also be used if the applied voltages are reversed.

Junction Field-Effect Transistors The geometry for a junction field-effect transistor (JFET) is shown in Fig. 33.19. Like the BJT, the JFET is a three-terminal device. Two of the terminals, called *source* and *drain,* are located at the ends of an *N*-type channel that has two heavily doped *P*-type regions at the outer boundaries of its sides. The third terminal, called a *gate,* is connected to the two *P*-type regions. When a potential difference V_{DD} is connected between the source and drain terminals as shown in Fig. 33.19, the free electrons in the *N* channel flow from the source to the drain, which explains the reason for the choice of names for those two terminals. This current, which continues in the external circuit, is called the drain current. Because the *P* regions are heavily doped compared to the *N* region, the depletion regions associated with the *P-N* junctions extend mainly into the *N* region, and the boundaries of these depletion regions in the *N* region define the boundaries of the *N* channel available to electrons moving from the source to the drain. The *P-N* junctions are normally reverse-biased by applying a voltage V_{GG} between the gate and source terminals as shown in Fig. 33.19. Variations of the gate-to-source voltage V_{GS} can be used to control the width of the depletion regions, and, therefore, the effective area available to electrons moving from the source to the drain. In different words, the resistance of the *N* channel to a current of electrons in the channel (the drain current) can be controlled by a voltage applied to the gate.

If a signal source is connected in series with the gate supply V_{GG}, the reverse bias, the channel resistance, and the drain current will all fluctuate in step with the signal voltage. Under these conditions, the voltage across the load resistor R_L in the drain circuit will be an amplified version of the input signal. Figure 33.20 shows a schematic diagram for a JFET amplifier with a schematic symbol used to represent the JFET.

P-channel JFETs with *N*-type gates are also used. Their operation is the same as that of the *N*-channel devices, except that polarities of power sources must be reversed.

One of the advantages of field-effect transistors over bipolar transistors is the high input resistance of typically 100 MΩ or greater. They are also easier to fabricate, and they are particularly suited for use in integrated circuits because they occupy less space than bipolar transistors.

33.8 ■ Integrated Circuits

The first integrated circuit (IC) chips contained only one transistor, but by the late 1980s it was possible to fabricate chips with an area of less than 1 cm^2 that contained several million components (diodes, transistors, resistors, capacitors, etc.).

Integrated circuits (Fig. 33.21) are used in most kinds of electronic equipment because the large number of components per chip allows complex electronic functions to be performed by a single IC chip, and because of the following additional desirable properties.

- Low power requirements. For example, the operating power for a typical hand-held computer is only 0.01 W.
- Low unit cost. Hundreds of circuits can be produced simultaneously on a single silicon wafer in a process called *batch fabrication,* resulting in a low cost per individual circuit.
- High reliability. Low operating power, the elimination of the need for soldered wire connections between components, and the ability to include extra backup circuitry, are some of the reasons for the high reliability of integrated circuits.

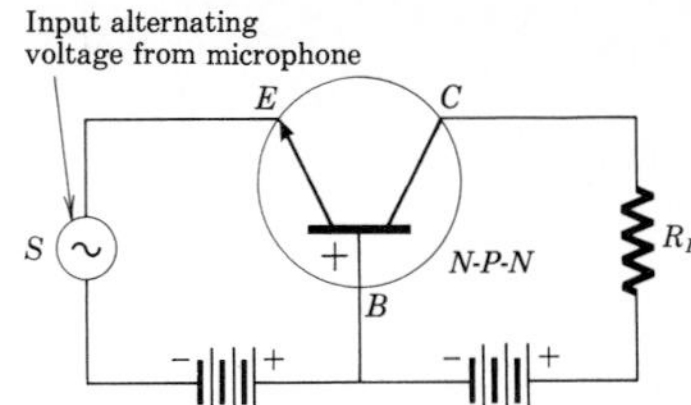

Fig. 33.17 A simplified transistor amplifier circuit. S is any source of weak ac pulses, as from a microphone. R_L is a resistance load, such as a loudspeaker. Note that the emitter-base junction is forward-biased, and the base-collector junction is reverse-biased.

INDUSTRIAL ELECTRONICS

Industry uses electronics in nearly every step of research, design, manufacturing, testing, packaging, marketing, and shipping. Only a few of these applications can be described here. Any standard text on industrial electronics would give ample coverage of this topic.

33.9 ■ The Electronic Computer

Modern computers are of two general types, the *digital computer* and the *analog computer*. The former is the type universally used in business and government and in the solution of problems where data can be reduced to some form of binary arithmetic.

Analog computers translate information about physical variables being investigated into electrical signals. They are less versatile than digital computers, and have become less important in recent years, partially because programs that simulate analog computers are available for digital computers.

The newest digital computers are smaller, faster, and more complex and sophisticated than earlier generations of digital computers (Fig. 33.22). In the mid-1960s a computer that sold for $280,000, occupied an 18 by 18 ft room, required air conditioning, and could perform 33,000 additions per second. By the late-1980s, desktop personal computers that sold for $1000 could perform 700,000 additions per second. Much of the computing work previously performed by large mainframe computers is today accomplished by desktop personal computers. Personal computers can also be used as computer terminals to operate a large mainframe computer for those cases where the memory capacity or speed of a large mainframe computer is required. Applications of digital computers include the many mathematical calculations associated with scientific and engineering projects, word processing, and data-base searches such as those performed at many libraries to identify titles and sources of periodical literature on specific topics.

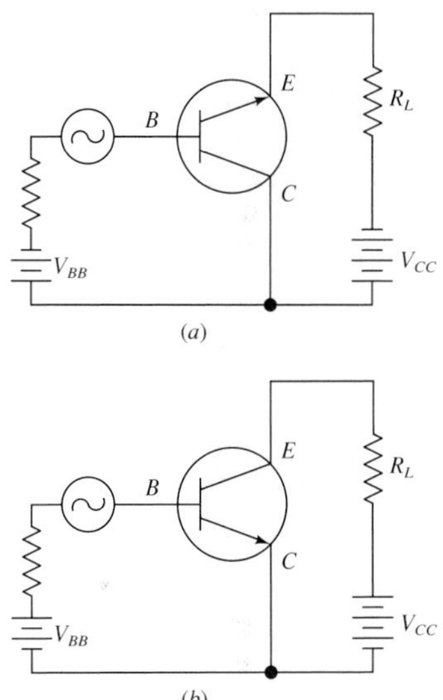

Fig. 33.18 Simplified schematic diagrams for (*a*) a common-collector amplifier and (*b*) a common-emitter amplifier. In the common-collector amplifier the collector terminal of the transistor is *common* to both the input and output circuits. The same idea holds for the common-base and common-emitter amplifiers.

33.10 ■ Automated Industrial Processes

The introduction of the electronic computer to control automated manufacturing processes during the decade 1960 to 1970, has been called the "second industrial revolution." During the 1970s, partially because of the advent of low-cost computer capabilities, rapid progress was made in *computer-aided design* (CAD) and *computer-aided manufacturing* (CAM). In computer-aided design, the ability of computers to display information graphically and pictorially, and to process large quantities of numerical data rapidly, is used to aid in the design of parts and equipment to be manufactured. In computer-aided manufacturing, computers are used to control equipment such as *computer numerical control (CNC) machine tools* and *robots* to fabricate parts and to assemble, test, and package products. Manufacturing systems that use both computer-aided design and computer-aided manufacturing are called CAD/CAM systems.

***Computer-aided design* (CAD) and *computer-aided manufacturing* (CAM), defined.**

Fig. 33.19 Simplified schematic diagram for an amplifier using a junction field-effect transistor (JFET) showing the internal structure of the JFET. The depletion regions (shown shaded) are wider near the drain end of the device because the reverse bias is greater near the drain end.

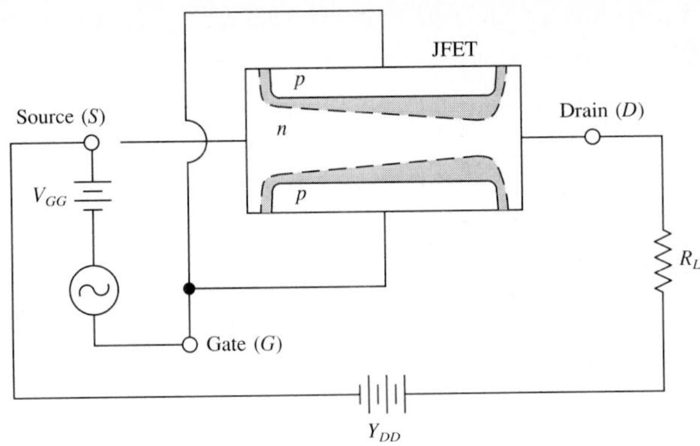

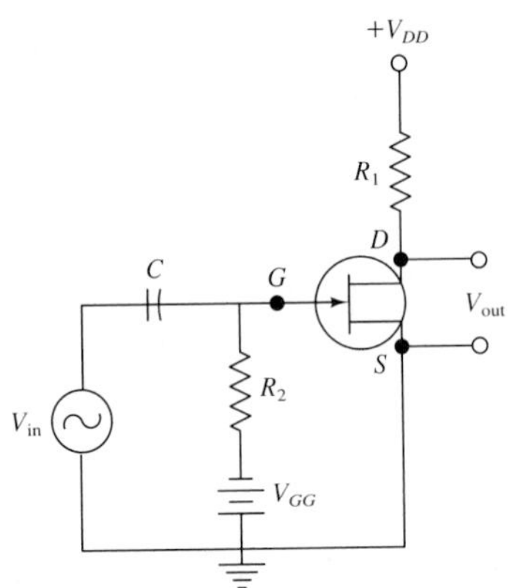

Fig. 33.20 Diagram for a JFET amplifier showing the schematic symbol for a JFET.

Numerical Control Machine Tools The first numerical control (NC) machine tools were developed in the 1950s to fabricate contoured aircraft parts. Since then numerically controlled lathes, machining centers (milling machines), drilling machines, and precision grinders have been developed, and the technology has spread throughout industry (see Fig. 33.23). CNC machine tools have higher accuracy and better repeatability than standard machine tools, and they can process products that have complex shapes. Typical CNC machine tools can produce parts with an accuracy of 0.0005 in. by using lasers to check machining error, and by using thermistors to monitor temperatures to provide feedback to the computer so that adjustments can be made in the machining process to compensate for thermal expansion.

33.11 ■ Robotics and Machine Vision

Another type of computer-controlled automated industrial process involves the use of industrial robots. The Robotics Industries Association has formulated the following definition of a robot:

Definition of a robot.

A robot is a reprogrammable, multifunctional manipulator designed to move material, parts, tools, or specialized devices through variable programmed motions for the performance of a variety of tasks.

Note that a robot is defined as a *reprogrammable, multifunctional manipulator*. Not every manipulating mechanism is a robot. Moreover, according to the definition, a robot can perform a *variety of tasks*. A mechanism designed to accomplish a single industrial task, no matter how intricately performed and cleverly accomplished, is not a robot. Nonrobotic automation is sometimes called *fixed automation* or *hard automation*. Note also that robots are not humanoids, and a robot need not ''look like a robot'' to meet the demands of the above definition (see Fig. 33.24). However, some robots have parts that correspond to the human arm, hand, and brain. The robot arm, called a *manipulator,* has flexible joints similar to the human shoulder, elbow, and wrist. The robot hand, called an *end effector* or *gripper,* uses a mechanical, vacuum, or magnetic device for parts handling. The robot ''brain'' is a digital computer.

The manufacturing of robots involves many areas of technology.

The proper design and operation of industrial robots can be accomplished only by taking advantage of a wide variety of areas of technology, including mechanics, hydraulics, pneumatics, electronics, optics, and computers. Some robots perform their tasks by assuming a sequence of programmed positions with an accuracy of better than 0.0004 in. This approach is successful if the part or device to be manipulated always has the same orientation when it arrives at the robot's location. But sometimes it is impossible or uneconomical to have the workpiece repeatedly positioned for the robot. In many such cases the robot's task can still be accomplished if a *machine vision* system (sometimes called *robot vision*) is made part of the robot. The image formed by an optical system can be computer-processed to provide instructions to the robot to allow it to carry out its assigned task.

Some examples of applications of industrial robots include:

- Welding and soldering
- Water-jet cutting of plastics, textiles, and paper
- Handling of semiconductor wafers in manufacturing steps involving hazardous chemicals
- Optical inspection of parts to meet quality control objectives
- Spray painting
- Handling radioactive materials and heated parts
- Performing nighttime security tasks for factories, warehouses, and museums
- Inserting and removing parts from the chuck of a numerically controlled machine
- Production line assembly of many different types of products including electronic watches, printed circuit boards, typewriter keyboards, tape recorder mechanisms, floppy disk drives, automobile alternators, and automobile bodies

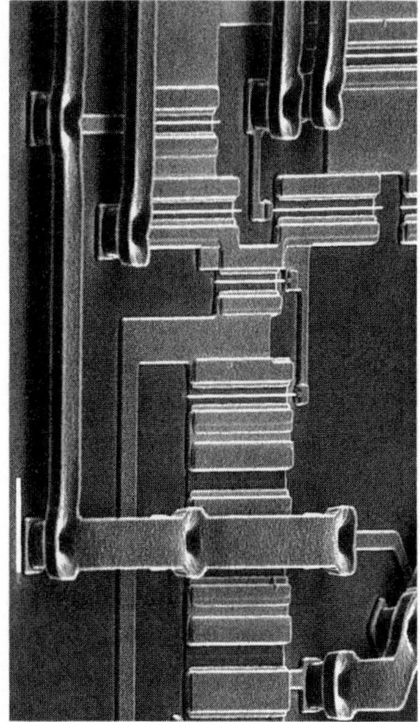

Fig. 33.21 Microphotograph of a portion of a gallium arsenide integrated circuit. The photo shows the circuit magnified 1400 times. The actual length of the white line at the left of the photo is 0.01 mm. (Hughes Aircraft Company)

Description of a completely automated production line.

Figure 33.25 shows a portion of an automated production line that is used to manufacture *contactors* which are electrically operated, high-current-capacity switches used for starting electric motors. The production line uses both fixed automation and robots to assemble 125 possible variations of the contactor on a common steel-frame base. At the beginning of the production line, a bar code (like that on supermarket products) on the steel frame base is used to identify the contactor variation to be assembled. The 26 machines on the production line are then given instructions to select the proper components and to perform the required tasks to assemble that variation of contactor. The 26 machines are synchronized to select and assemble the needed parts (some of which are extremely small) to turn out a completely manufactured, tested, labeled (using a laser printer), and packaged product.

Fig. 33.22 A desktop computer used for computer-aided engineering (CAE) applications. CAE systems are used to test proposed designs of equipment and systems without actually building them first. (Hewlett-Packard Company)

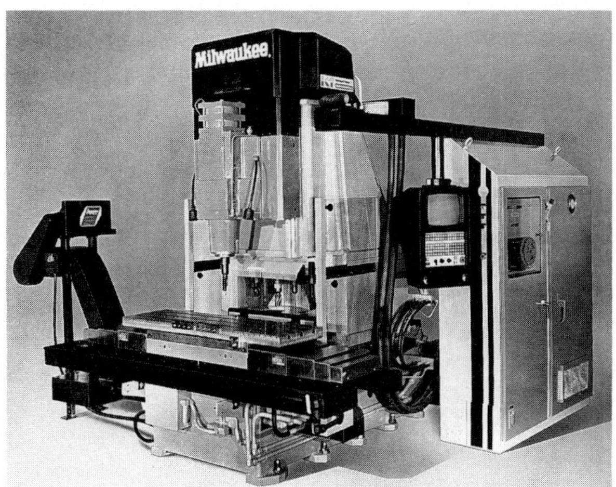

Fig. 33.23 An example of a computer numerical control (CNC) machine tool called a *machining center*. (Kearney & Trecker Corporation)

Fig. 33.24 Arc welding workcell using robot in production environment. (CIMCORP, Inc., St. Paul, MN)

ELECTRONICS IN COMMUNICATIONS

Electronic communications equipment has been revolutionized during the past 35 years thanks to the transistor, solid-state physics, and integrated circuits. It is the purpose of this section to explain some of the fundamental principles of electronic communications and to discuss briefly radio, television, radar, and satellite communications systems.

RADIO

Light and radio waves both travel 1000 mi in just over 0.005 s.

In 1856 the brilliant theoretical physicist James Clerk Maxwell proposed the theory, on purely mathematical grounds, that light and many other types of such waves, differing only in wavelength and frequency, were *electromagnetic* in nature and that all would travel with the speed of light.

33.12 ■ Electromagnetic Waves

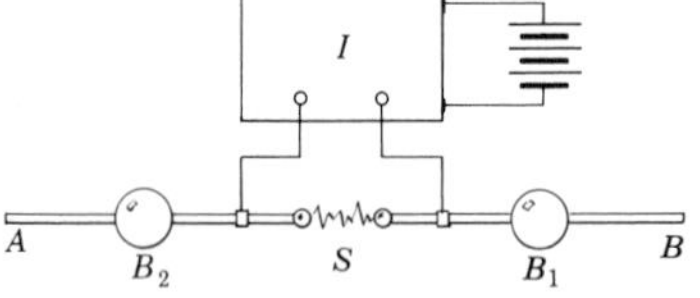

Fig. 33.26 Hertz's electromagnetic wave apparatus.

It was some 30 years later (1888) that Heinrich Hertz (1857–1894) devised an apparatus that was to prove Maxwell's electromagnetic wave theory. He worked with a spark-gap apparatus like that diagramed in Fig. 33.26. An induction coil I supplies high voltage to the spark-gap or transmitting circuit AB_2SB_1B. Now, the natural (resonant) frequency of oscillation of such a circuit is given by

$$f = \frac{1}{2\pi\sqrt{LC}} \qquad (31.25)$$

where f = frequency, cycles/s (Hz)
L = inductance, henrys
C = capacitance, farads

By moving the large metal balls B_1 and B_2, and by adjusting the length of the air gap at S, the capacitance and inductance of the circuit can be varied. It is therefore possible to vary, within limits, the frequency of oscillation of the sparking circuit.

Fig. 33.25 A portion of a totally automated assembly line used to manufacture *contactors,* which are electrically operated switches used for starting large electric motors. (Allen-Bradley, a Rockwell International Company)

Hertz found that if he brought a loop L with another spark gap D within a few meters of the spark gap S, sparks could also be observed jumping the gap D when the two circuits were adjusted to have the same natural *resonant* frequency. Hertz measured the wavelength of these radiations quite accurately, and found, in every case, that multiplying the wavelength by the frequency gave a velocity equal to the speed of light. Hertz's experiments, then, definitely supported Maxwell's original proposals. These new waves of energy were called *electromagnetic waves.*

NOTE: The velocity of any wave motion is given by the expression $v = f\lambda$ where f is the frequency of the motion and λ is the wavelength. By international agreement the term *hertz* (Hz) is now used as the unit of frequency; 1 Hz equals 1 cycle/s. The terms *kilohertz* (kHz), *megahertz* (MHz), and *gigahertz* (GHz) are used also, for 1000, 1 million, and 1 billion cycles/s, respectively.

Illustrative Problem 33.4 What is the wavelength in meters of the radio waves from a transmitter broadcasting on a frequency of 970 kHz?

The wavelength range of the commercial FM broadcast band is 2.8 to 3.4 m.

Solution Use the equation

$$c = f\lambda$$

Solve for λ:

$$\lambda = \frac{c}{f}$$

Substitute given values:

$$\lambda = \frac{3.00 \times 10^8 \text{ m/s}}{970 \times 10^3 \text{ cycles/s}}$$

$$= 309 \text{m} \qquad \textit{answer}$$

Fig. 33.27 Circuit diagram for crystal-controlled JFET Colpitts oscillator for producing radio-frequency (RF) waves of constant amplitude.

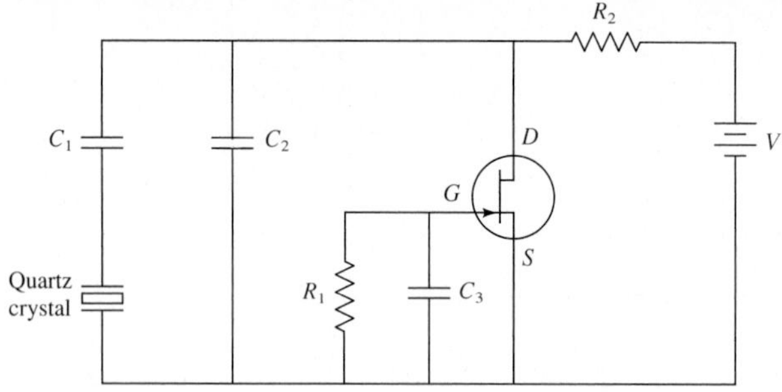

Reginald A. Fessenden, who worked as chief chemist for Thomas Edison in the 1880s, obtained 500 patents for his inventions.

33.13 ■ Early Radio Messages

In 1895 Guglielmo Marconi (1874–1937), a young Italian inventor, became interested in Hertz's work and its possibilities for wireless communication. He found that he could greatly increase the distance of reception of electromagnetic radiations by increasing transmitter power, by using an antenna high in the air, and by improved detection methods. By 1900 he had sent and received signals at distances of 300 km, and wireless telegraphy was born. Early transmitters used spark coils to actuate the oscillatory circuit, but with the development of the vacuum tube and its application to oscillatory circuits, the science of radio began its rapid rise. At first mere signals, as in telegraphy, could be sent, but in 1906 Reginald A. Fessenden (1866–1932) of the University of Pittsburgh was successful in *modulating* a radio-frequency carrier wave with audio-frequency currents from a microphone, and the first transmission of human speech was carried out. We shall explain here only the basic principles of radiotelephony, not the design and operation of actual radio transmitters and receivers.

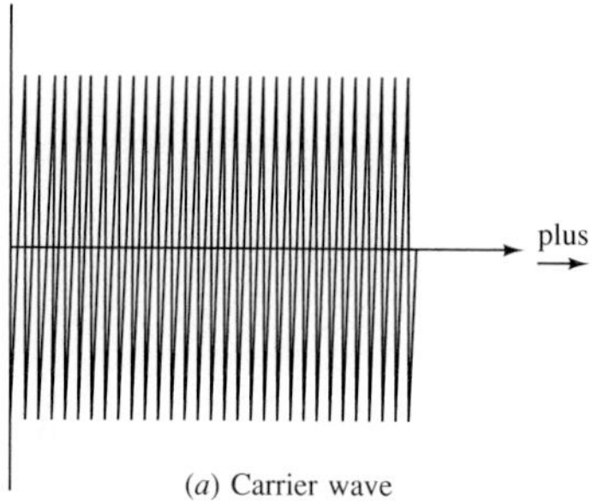

(*a*) Carrier wave

33.14 ■ Amplitude Modulation

Voice transmissions by electromagnetic waves require a carrier wave of high frequency and constant amplitude. A JFET can be employed to produce these oscillations. A circuit like that shown in Fig. 33.27 may be used. The frequency of the oscillator is determined by the vibrations of the quartz crystal connected in series with the capacitor between the gate (*G*) and drain (*D*) leads. Similar crystals are used in electronic watches.

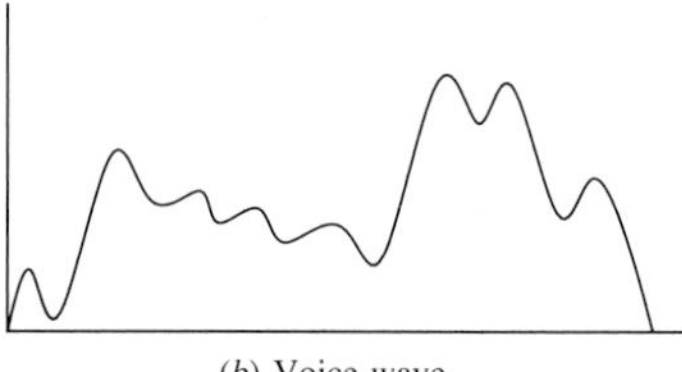

(*b*) Voice wave

In a radio transmitter, audio-frequency signals, ranging from 20 to about 10,000 Hz are superimposed on the radio-frequency oscillations, and these modulated voltages are amplified, transferred to the antenna, and broadcast as electromagnetic waves with the speed of light. Figure 33.28 illustrates graphically this process of *amplitude modulation*.

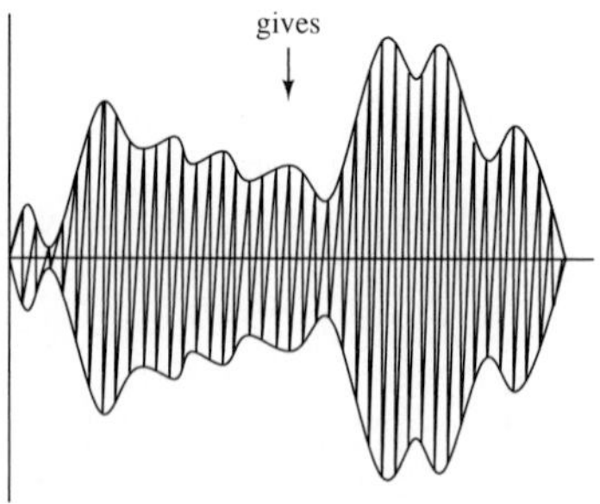

(*c*) Amplitude-modulated wave

Fig. 33.28 Graphic representation of amplitude modulation of a carrier wave for radio broadcasting. Carrier wave (*a*) plus voice wave (*b*) combine to produce amplitude-modulated wave (*c*).

33.15 ■ Frequency Modulation

Audio-frequency signals can be used to modulate the carrier *frequency* rather than the *amplitude* of the carrier wave. With this system of *frequency modulation,* the amplitude of the carrier wave remains constant and depends on the power output of the transmitter. Because FM receivers are sensitive to frequency changes but not to amplitude changes, they successfully discriminate against most forms of noise such as the static from electrical storms, power lines, electric machinery, and the like. Also, frequency modulation is more adaptable than amplitude modulation to the higher audio frequencies (8000 and up), and therefore frequency modulation is considerably better than amplitude modulation for the high-fidelity broadcasting of programs of fine music. Figure 33.29 shows graphically how frequency modulation functions.

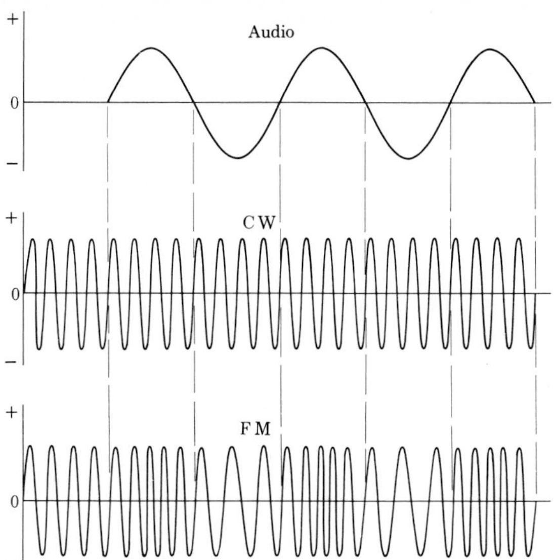

Fig. 33.29 Graphic representation of frequency modulation. An audio sine waveform is shown modulating the frequency of the carrier wave.

33.16 ■ Radio Broadcasting

The complete frequency range of radio waves (10 kHz to 300 GHz) offers a wide selection of frequencies for communication purposes. Since radio communication affects so vitally the life and security of a nation, most governments exercise considerable control over radio transmissions. All transmitters in the United States must be licensed and must operate on the frequency band assigned by the Federal Communications Commission (FCC). The usual classification of frequency bands is listed in Table 33.2, along with the typical type of assignment for each.

Table 33.2 Designation and assignment of radio frequencies

Frequency f, kHz except as noted	Wavelength λ, m except as noted	Designation	Partial list of frequency assignments
10–30	30,000–10,000	Very low frequency (VLF)	Radio navigation, maritime mobile, submarine communication
30–300	10,000–1000	Low frequency (LF)	Maritime mobile, maritime mobile satellite, intersatellite
300–3000	1000–100	Medium frequency (MF)	Standard radio broadcast band (AM), amateur, mobile
3–30 MHz	100–10	High frequency (HF)	Amateur, amateur satellite, citizens' band, maritime mobile
30–300 MHz	10–1	Very high frequency (VHF)	VHF television, FM radio, radio astronomy, amateur
300–3000	1 m–10 cm	Ultrahigh frequency (UHF)	UHF television, amateur, synchronous satellites, radar, meteorological aids, cellular telephones
3000–30,000 MHz (3–30 GHz)	10 cm–1 cm	Superhigh frequency (SHF)	Synchronous satellites, radio astronomy, earth exploration satellites, amateur radio astronomy, aeronautical radio-navigation satellites
30–300 GHz	1 cm–1 mm	Extremely high frequency (EHF)	

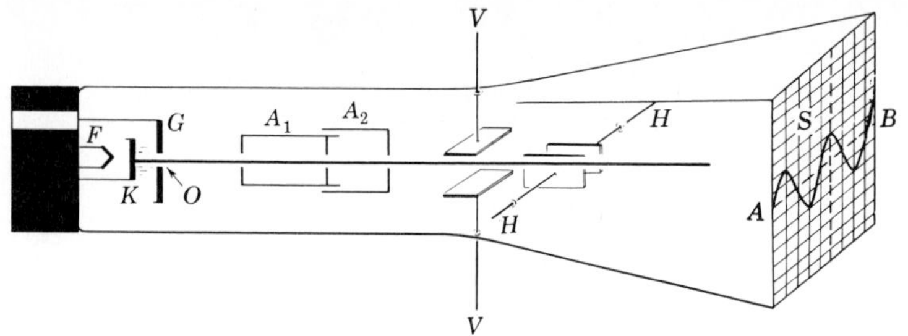

Fig. 33.30 Simplified diagram of a cathode-ray tube. The $FKGA_1A_2$ system is called an *electron gun*. V and H are pairs of electrostatic deflection plates. When voltages are connected cross these pairs of plates, they can cause the beam of electrons to be deflected in the vertical and horizontal planes.

THE CATHODE-RAY TUBE (CRT)

In the following section we discuss the cathode-ray tube, which converts electricity into radiation to display information on a screen for a wide variety of applications.

33.17 ■ The Cathode-Ray Tube and the Oscilloscope

The cathode-ray tube (CRT), diagramed in Fig. 33.30, is an electronic tube of almost unlimited potential. Its use in television receivers, personal computers, and video games is well known. Other important applications include the display of information on the CRT screens of radar and sonar receivers, oscilloscopes, aircraft instrument panels, electrocardiographs (Fig. 33.31), several different types of medical imaging systems, and automotive test instruments (Fig. 33.32).

The "cathode ray" in a cathode-ray tube is a beam of electrons.

An idea of how an electron beam is produced and controlled in a CRT can be obtained by referring to Fig. 33.30. Electrons emitted from the heated cathode K emerge only through a small hole in the grid G. The slender beam of electrons is accelerated by an *electrostatic plate system* (A_1 and A_2). Electric potentials applied to these anodes cause the electrons to experience a great acceleration and narrow the beam down to a slender "pencil" of electrons heading straight down the center of the tube. The portion of the CRT so far discussed is called the *electron gun*.

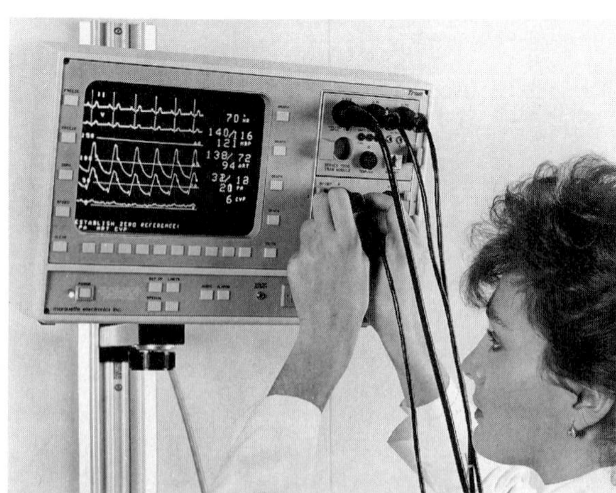

Fig. 33.31 A bedside surgical monitor for displaying information from a multilead electrocardiograph (ECG). (Marquette Electronics, Inc.)

Fig. 33.32 The power and versatility of the CRT in displaying information is used in this automotive test instrument for auto engine tune-up and analysis. (Bear Automotive Service Equipment Company)

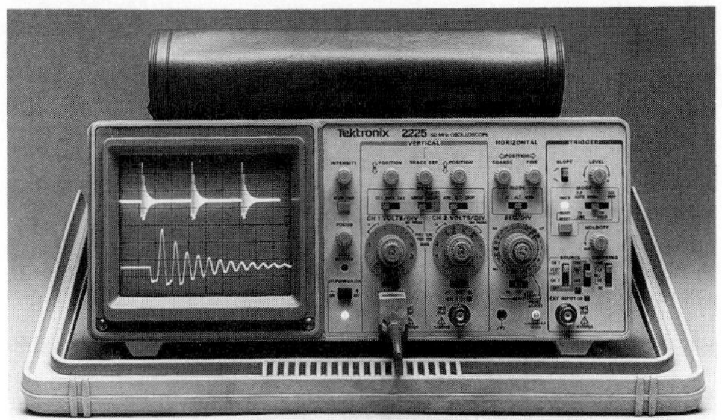

Fig. 33.33 A dual-channel oscilloscope typical of the type used by engineers and technicians in industry. (Tektronix, Inc.)

The beam next passes first through a pair of vertical deflection plates *V* and then through a pair of horizontal deflection plates *H*. Note that the vertical deflection plates are horizontally oriented, and the horizontal deflection plates are vertically oriented. The names given to the deflection plates depend not on their orientation, but on the influence they have on the electron beam when a voltage is applied to the plates.

For example, as used in a cathode-ray oscilloscope (Fig. 33.33), a high "sweep frequency" ac potential is applied to the horizontal deflection plates, which causes the electron beam to sweep horizontally across the fluorescent screen *S* (Fig. 33.30) with constant speed so rapidly that the visual result is a continuous straight line. The electron beam sweeps from *A* to *B* and then jumps very rapidly back to *A* to repeat the sweep. Now, if a variable potential of some sort is to be studied, it is applied (usually after amplification) across the vertical deflection plates, and the net result of the two motions is a plot on the screen of voltage versus time for the variable input potential.

Oscilloscope applications include measurement of voltage, frequency, distortion, and time intervals between electrical events.

For the cathode-ray tubes used in television sets, computer monitors, and other instruments that require a pictorial display, such as medical imaging instruments, the *electrostatic* deflection system described above is not used. Instead the internal deflection plates are replaced by external coils. Electric currents in the coils cause magnetic fields inside the tube, and the beam direction is controlled by magnetic forces acting on the electrons.

33.18 ■ The Cathode-Ray Tube and the Television Receiver

The television *receiver* has as its principal element a kind of cathode-ray tube, called the *kinescope* or *picture tube*. It uses a magnetic deflection system and has one electron beam for each of the three primary colors—red, blue, and green. The electron beams sweep and scan in exact synchronism with video information from the television camera. Electronic information for this synchronization is transmitted with the carrier wave from the transmitter, along with the *video* and *audio information*.

The incoming signals have to be amplified, the audio portion routed to the audio circuits, the sweep and scan information to the magnetic deflection coils, and the video information to the electron gun, all in proper sequence and timing, with time intervals which may be as short as a few hundredths of a microsecond.

The surface of the picture tube has a phosphor coating of red, blue, and green dots or lines which emit light as the beams from the electron gun hits them. The successive sweeps in the scan and the number of complete scans per second reproduce a "picture" of the original scene. In the United States 525 horizontal scans of the electron beams are uniformly distributed across the face of the picture tube 30 times per second. Both the persistence of human vision and the persistence of the fluorescing process on the picture-tube surface help make for satisfactory viewing.

New, "high-tech" television systems, now under development (1989) will have different specifications than those listed above.

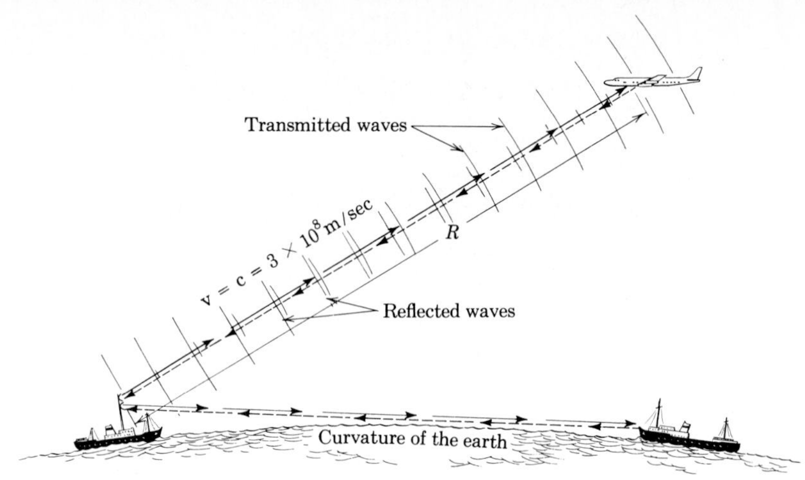

Fig. 33.34 Sketch showing the basic principles of radar. Radar waves are electromagnetic waves which travel in straight lines like light and are therefore not very effective beyond the visual horizon. They obey the inverse-square law as concerns intensity of reflected energy. If R is the range to a target, the reflected signal strength back at the ship is inversely proportional to R^4.

ELECTRONICS IN SPECIALIZED OPERATIONS

33.19 ■ Radar

In *radar* receivers a cathode-ray tube is modified to give a visual indication of surrounding land masses, other ships on the surface of the sea, and aircraft in the surrounding skies.

A complete radar system (the word *radar* stands for *ra*dio *d*etection *a*nd *r*anging) has a *transmitter*, a *receiver*, and various *indicating devices*. The high-frequency radar waves (600 MHz and up) that are sent out from the transmitter in a narrow cone-shaped beam penetrate fog, haze, smoke, and clouds and are therefore very useful for navigational purposes. They travel outward with the velocity of light (c = 3 × 10^8 m/s) and are reflected when they strike a distant object (Fig. 33.34). The very small part of the reflected wave energy which returns and strikes the receiving antenna is amplified in the radar receiver, and these strengthened signals are fed to the desired indicating devices.

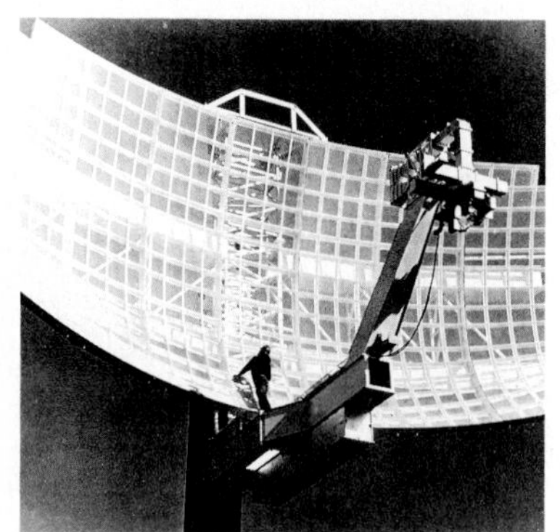

Fig. 33.35 Giant antenna for air route surveillance radar. The detection range of the system served by this antenna is 234 nautical miles (433 km), with a coverage altitude of 100,000 ft (30.5 km). (Westinghouse Electric Corp.)

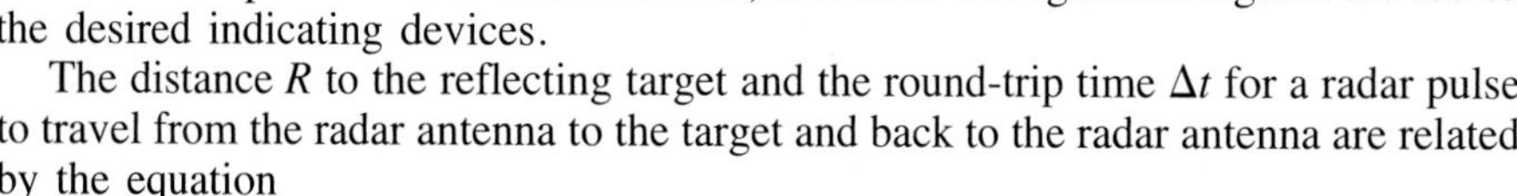

The distance R to the reflecting target and the round-trip time Δt for a radar pulse to travel from the radar antenna to the target and back to the radar antenna are related by the equation

$$R = \tfrac{1}{2}c\ \Delta t \qquad (33.2)$$

where c is the speed of light ($c = 3.00 \times 10^8$m/s $= 1.86 \times 10^5$ mi/s).

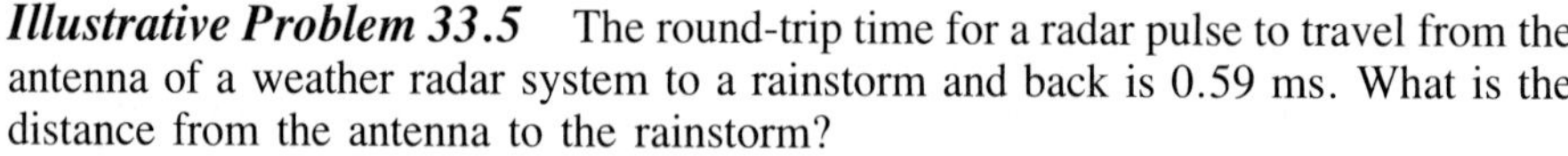

Illustrative Problem 33.5 The round-trip time for a radar pulse to travel from the antenna of a weather radar system to a rainstorm and back is 0.59 ms. What is the distance from the antenna to the rainstorm?

Solution Substitute 0.59 ms for Δt in Eq. (33.2) and solve for R.

$$\begin{aligned} R &= \tfrac{1}{2}c\ \Delta t \\ &= (0.5)\ (1.86 \times 10^5 \text{ mi/s})\ (0.59 \times 10^{-3} \text{ s}) \\ &= 54.9 \text{ mi.} \end{aligned}$$

answer

Radar uses the Doppler effect to determine the speed of a moving target.

Range Rate In the case of an aircraft, ship, space vehicle, or speeding car, which may be coming toward or going away from the transmitter station, its *relative velocity* will cause an apparent frequency difference between the outgoing waves and the returning echo waves. This is a manifestation of the *Doppler effect* (see Sec. 20.11).

By accurately measuring this frequency change, and applying the proper Doppler formula, the velocity of approach or escape can be determined. This velocity of approach or escape, known as the *range rate*, is a necessary factor for navigation and for the use of radar in the monitoring and control of air traffic (Fig. 33.35).

33.20 ■ Communications by Satellite

Mention was made of communications satellites in Chap. 8, in connection with an explanation of the dynamics of orbital motion. Since Telstar I was first launched in 1962 and operated for nearly a year, a series of communications satellites has been giving more or less continuous intercontinental communications service.

The term *synchronous* means that a satellite's motion is synchronized with the earth's rotation so that the satellite will "hover" over the same location on the earth's equator, the altitude required being about 22,250 mi. Modern communications satellites (see Fig. 8.19) are often powered by solar cells which convert the sun's radiant energy into electricity.

Design of modern satellites provides for the reception and retransmission of thousands of telephone conversations simultaneously. They are also designed for television transmissions, and live television programs between the United States and other countries by communications satellites are now a daily occurrence. Satellites are also used for navigation control, military surveillance, and weather monitoring; and for gathering information on crops, minerals, forests, ocean currents, and other earth data.

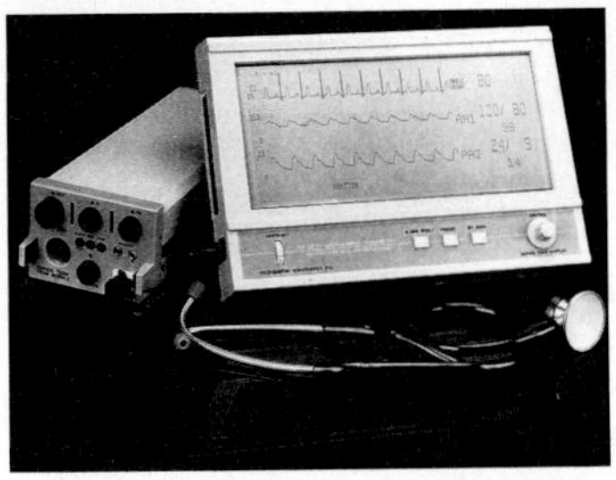

Fig. 33.36 A liquid crystal display monitor used to provide bedside or in-transit monitoring of the vital signs of critically ill patients. (Marquette Electronics, Inc.)

33.21 ■ Production and Uses of X-Rays

A discovery made in 1895 has had tremendous applications to medical science and industrial technology in the 20th century. Wilhelm Röntgen (1845–1923), while experimenting with high-voltage discharges through gas-filled tubes, noticed a bright fluorescence in some crystals which happened to be close by (see also Sec. 34.9). In follow-up experiments Röntgen soon discovered that radiation of some sort seemed to be emanating from the tube and that these rays had extraordinary penetrating power, for they would pass readily through paper, cloth, the experimenter's hand, sheets of wood, and other like materials. Since the rays were of unknown origin, Röntgen called them x-rays. Within a few weeks he discovered that x-ray penetration was inversely proportional to the density of the material and, using photographic plates, took the first "x-ray picture" of a human hand. In less than a year, x-rays were being used throughout Europe as an aid to surgeons in setting broken bones.

In addition to producing x-ray images on photographic plates, computer-processed data from x-ray detectors can be used to display an x-ray image on the screen of a cathode-ray tube. The technology for using computer techniques to produce x-ray images is discussed in the next section.

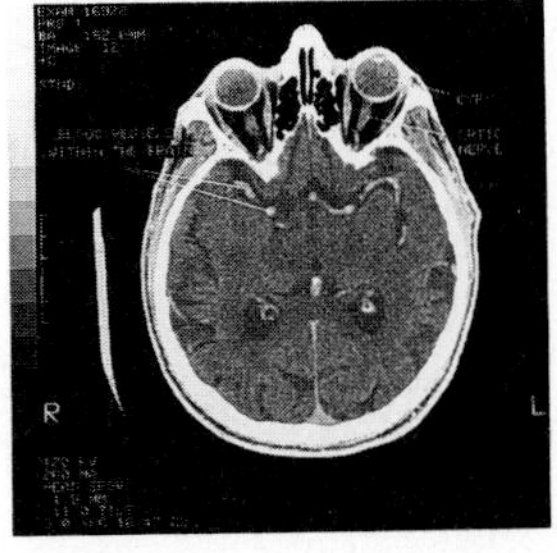

Fig. 33.37 A computed tomography (CT) image of a human head showing structures such as the optic nerve, blood vessels within the brain, and gray and white matter differences in the brain. Computed tomography also permits the visualization of minute variations in tissue density within the body, which is often critical in detecting the early stages of abscesses, tumors, and cancers. (GE Medical Systems)

33.22 ■ Electronics in Medicine

For many years there was just one important imaging system for diagnostic purposes in medicine. That system involved radiating a patient with x-rays (see Chap. 34) and recording on photographic film the x-rays not absorbed by the patient. The resulting image on the film, called a radiograph, or more commonly an x-ray "picture," was then used by physicians and surgeons to make a diagnosis and then decide how to treat the patient.

In the early 1970s rapid progress was made in the design of imaging systems that eliminated the photographic film, and instead used a digital computer to process the signals from detectors of radiation to obtain an image of the body, or of a portion of the body for diagnosis. Some, but not all, of the imaging systems involved an interaction of x-rays with the body. Table 33.3 provides a summary of information about five medical imaging systems, all of which use a digital computer to produce an image on a cathode-ray tube for analysis by health care professionals.

Some of the imaging systems designed for medical uses are now also being used for scientific and industrial purposes.

Other examples of electronics in medicine include the use of a liquid crystal display to monitor the vital signs of patients under intensive care (see Fig. 33.36), and the use of a cathode-ray tube to display information from an electrocardiograph machine (see Fig. 33.31). Kidney machines, heart-lung machines, artificial hearts, respirators, computed tomography equipment (CAT) (Fig. 33.37), magnetic reso-

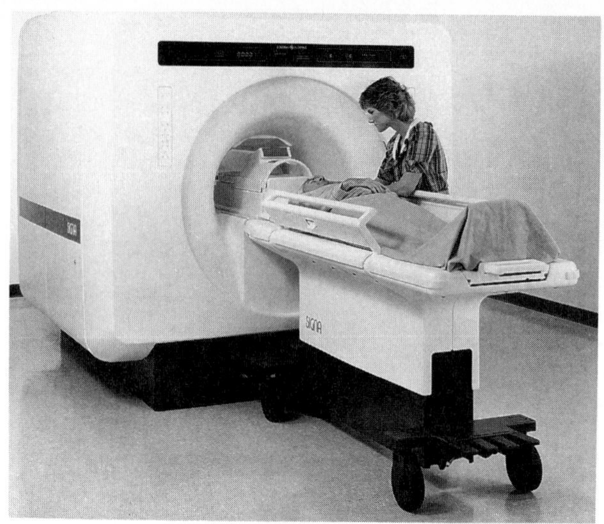

Fig. 33.38 A magnetic resonance imaging (MRI) system that uses a superconducting electromagnet to produce a magnetic field 25,000 times as strong as the earth's magnetic field. When a patient is centered in the magnetic field, the hydrogen atoms in the body align themselves with the field. A series of radio-frequency waves is then introduced while the patient is inside the magnet. These cause hydrogen atoms to give off faint signals which are then detected and computer-processed. The system is capable of acquiring images so detailed that they approach the fine-line anatomical drawings found in medical textbooks. (GE Medical Systems)

nance imaging systems (Fig. 33.38), ultrasound imaging systems (Fig. 33.39), and other operating room equipment depend on electronic circuits and components for their operation.

The advent of the transistor and microcomputers on chips have made possible an electric pulse generator known as a *pacemaker* which is used to increase to normal the rate at which a heart beats. The pacemaker stimulates the heart muscle with small electric shocks, typically provided by 5-V, 10-mA pulses of 2-ms duration. The combination of a computer with sophisticated equipment for chemical analysis has made available to hospitals technological aids like the *multichannel analyzer* for making a number of different but simultaneous laboratory analyses of a patient's blood sample.

Table 33.3 Six computerized body-scanning systems for producing images for diagnostic purposes

Imaging system	*Type of radiation or substance into body*	*Features and capabilities*
Computed tomography (CT) (See Fig. 33.37)	X-rays	Can locate tumors with great precision
Magnetic resonance imaging (MRI) (See Fig. 33.38)	Radio waves and a very strong magnetic field	Able to produce images of soft tissue
Digital subtraction angiography (DSA)	X-rays and injection into blood of substance opaque to x-rays	Provides an image of the blood flow in blood vessels to locate problems involving narrowed or blocked blood vessels
Single-photon-emission computed tomography (SPECT)	Injection of trace amounts of radioisotopes	Provides images and information about blood flow
Positron emission tomography (PET)	Injection of trace amounts of radioisotopes	Produces images showing distribution of blood in tissues; images more accurate than in SPECT system
Sonography (SONO) (or Echosound) (See Fig. 33.39)	Ultrasound	Recommended for use on pregnant women instead of x-ray methods for obtaining an image of a fetus. Also useful for examination of the breasts, heart, liver, and gall bladder

QUESTIONS AND EXERCISES

1. List all the devices and/or appliances in your home that have any kind of electronic component. Do not include devices that merely use electricity in ordinary wired circuits. Many of these have electronic components, however, so mention the components.

2. Does the resistance of a thermistor increase or decrease when its temperature decreases?

3. Suggest an application for a thermistor not mentioned in this chapter.

4. Would a doped, *N*-type semiconductor be a good choice of material for a thermistor? Why or why not?

5. Which is a better conductor of electricity, pure silicon or *N*-type silicon? Explain your answer.

6. Discuss two ways in which adding *N*-type impurities to a pure semiconductor changes the electrical conductivity of the material.

7. Would you expect the current through a reverse-biased *P-N* junction diode to be sensitive to temperature? Why or why not?

8. If the magnitude of the reverse bias of a varactor diode is decreased, does the junction capacitance of the device increase or decrease?

9. In what ways are bipolar junction transistors (BJTs) and junction field-effect transistors (JFETs) the same? In what ways are they different?

10. Which type transistor, the BJT or the JFET, can be thought of as a voltage-controlled variable resistor?

11. Is the current across the forward-biased emitter-base junction in an *N-P-N* bipolar transistor mainly a current of electrons or a current of holes? Answer the same question for a *P-N-P* transistor.

12. Explain what is meant by the terms: *semiconductor, valence electrons, holes, emitter, collector, forward bias, reverse bias.*

13. Explain the operation of a cathode-ray oscilloscope as it might be used in analyzing a musical sound, such as a note from a trumpet.

14. How is modulation of a radio-frequency carrier wave accomplished in a radio transmitter? Explain with diagrams.

15. Suggest several specific applications for an industrial robot.

16. What advantages do the images produced by a computed tomography (CT) system have over the older technique of producing x-ray images using photographic film?

PROBLEMS

Group One

1. Microwave telephony uses waves as short as 1 cm. What would the frequency (megahertz) of such waves be?

2. What is the range of wavelengths for the commercial AM broadcast band if the frequency range is 550 to 1600 kHz?

3. A broadcasting station transmits on a frequency of 1410 kHz. What is the wavelength (meters) of its radio waves?

4. A television station broadcasts on a carrier frequency of 500 MHz. What is the wavelength (meters) of these waves?

5. One of the citizens' band (CB) channels is on a frequency of 27 MHz. How long (meters) are these waves?

6. A radar receiver shows an elapsed time of 10^{-3} s for the pulse to travel to an airborne target and return. How far away (miles) is the target?

7. A radio signal is sent from "Houston Control" to a space vehicle in a low-altitude orbit around Mars. How long does it take for the signal to reach the vehicle if Mars is 85 million miles away at the time?

8. A transistor oscillator circuit has a frequency of 150 MHz. What is the period (time for one complete oscillation) for this oscillator?

9. A thermistor whose resistance varies with temperature as shown in Table 33.1 and Fig. 33.5 is used as a medical thermometer. What would the approximate thermistor resistance be when its temperature is (*a*) 22°C (typical room temperature), and (*b*) 37°C (normal body temperature)?

10. A thermistor is used to monitor the spindle-axis temperature on a computer numerical control (CNC) machining center. If the thermistor resistance is 500Ω, what is the approximate spindle-axis temperature? (Refer to Table 33.1 and Fig. 33.5 for the thermistor data.)

11. The junction capacitance C_j of a certain varactor diode depends on the magnitude of the reverse bias V_R in a way predicted by the equation

$$C_j = \frac{116}{\sqrt{V_R}}$$

where C_j is in picofarads (pF) and V_R is in volts. What is the junction capacitance when the magnitude of the reverse bias is 3.0 V?

12. What electric power would be provided in bright, direct sunlight (intensity $I = 1000\ \text{W/m}^2$) by six 10-percent-efficient solar cells each with an area of 12.5 in.^{2}?

13. The round-trip time for a radar pulse to travel from the radar antenna at an airport to a jet aircraft is 0.054 ms. What is the distance from the antenna to the aircraft?

14. The frequency of a transistor oscillator is determined by a series resonant circuit consisting of 0.0025-μF capacitor and a 0.06-mH inductor. What is the natural frequency of the oscillator?

15. An oscillator for an induction heating machine has a resonance circuit consisting of an inductance L of 0.3 mH and a capacitance C of 0.04 μF. What is the frequency of oscillation?

Group Two

16. The junction capacitance, in picofarads (pF), of a certain varactor diode used in tuning circuits in television sets is given by the equation $C_j = 105/\sqrt{V_R}$ where V_R is the magnitude of the reverse bias in volts. For what value of V_R will the junction capacitance C_j be equal to 25 pF?

17. The junction capacitance of a certain reverse-biased varactor diode is given by the equation

$$C_j = \frac{B}{\sqrt{V_R}}$$

where C_j is in pF, V_R is in volts, and B is a constant. The junction capacitance is 20 pF when the magnitude of the reverse bias is 4.0 V. (*a*) What is the constant B equal to? (*b*) What is the junction capacitance in pF when the magnitude of the reverse bias is 8.0 V?

18. How many 10-percent-efficient solar cells are required to provide an electric power of 50 W in bright sunlight (intensity $I = 1000\ \text{W/m}^2$), if the area of each individual cell is 12.5 in.^{2}?

19. One of the frequencies assigned for use in radio astronomy is 1665 MHz. What is the wavelength of electromagnetic waves of that frequency?

20. The radar antenna for a weather station uses 10-cm-wavelength radar pulses to keep track of the progress of intense storms. (*a*) What is the frequency of the emitted radiation? (*b*) How much time is required for a radar pulse to travel from the antenna to a storm 110 mi away and back to the antenna again?

21. A microwave telephone signal is transmitted from Los Angeles to a synchronous communications satellite 22,250 mi above the equator. The signal is then instantly relayed from the satellite to a receiving station in London. Is the time required for the telephone signal to travel from Los Angeles to the satellite to London greater or less than 0.20 s?

22. A certain microcomputer can perform 1 million mathematical operations per second. A technician with a pocket electronic calculator can perform one mathematical operation per second. How many 8-h days would the technician have to work to perform the same number of mathematical operations as the computer could perform in one minute?

Group Three

23. The number of ohms the resistance of a thermistor changes when its temperature changes by 1 Celsius degree is a measure of the temperature sensitivity of the thermistor. On the average, for temperatures between 10 and 20°C, what is the resistance change in ohms per degree change in temperature for the thermistor whose data is tabulated in Table 33.1? For the same thermistor, what is the average resistance change in ohms per degree change in temperature for temperatures between 90 and 100°C? Is the thermistor resistance more sensitive to changes in temperature at high or at low temperatures?

24. The resonant frequency f of an LC circuit depends on the inductance L and the capacitance C in accordance with the equation $f = 1/(2\pi\sqrt{LC})$. Assume the desired resonant frequency is 1.60 MHz and that the inductance is provided by a 0.330-mH coil. (*a*) What is the required value for the capacitor C? (*b*) If the capacitance is provided by a varactor diode for which $C_j = 105/\sqrt{V_R}$, what magnitude of reverse bias is required to provide the needed junction capacitance?

25. What is the lowest intensity of sunlight (in watts per square meter) for which two hundred 11-percent-efficient solar cells will provide 100 W of electric power if the area of each cell is 12.5 in.^{2}?

26. Use unit conversion techniques to show that for electromagnetic waves (such as those used for radar, microwaves, and television) the wavelength in meters is equal to 300 divided by the frequency in megahertz. Show also that the wavelength in feet is equal to 984 divided by the frequency in megahertz.

27. How many times must an input signal be amplified (voltage amplification) to have a gain of 40 dB?

CHAPTER 34 ATOMIC AND NUCLEAR PHYSICS

During the last years of the nineteenth century, and during the first third of the twentieth century, there occurred a series of developments that provided convincing evidence for the existence of atoms. These developments, and other advances that occurred later, constitute a segment of physics that is called *modern physics*, even though some of the developments occurred nearly 100 years ago.

We now know that the atoms that make up the elements are not indestructible spheres but are themselves made of yet smaller particles. This chapter deals with the properties of particles that make up atoms, with the properties of photons of electromagnetic radiation, with the structure of atoms, and with examples of applications of the principles of modern physics.

34.1 ■ Discovery of the Electron and Proton

In the late 1870s William Crookes, while experimenting with high-voltage electric discharges through evacuated tubes, discovered a visible discharge from the negative electrode, or *cathode,* of the tube. The discovery of this visible stream of energy, which was given the name *cathode rays* by the German physicist Eugen Goldstein, gave rise to speculation that atoms were not indivisible after all. Crookes was able to deflect the radiation with a magnet (see Fig. 34.1), and in 1897 J. J. Thompson was able to show cathode-ray deflection in an electric field. These experiments proved that the cathode rays were a stream of charged particles. Although Thompson chose to call the cathode-ray particles *corpuscles*, the Dutch physicist, Hendrik A. Lorentz chose to call them *electrons*—the name that is still used today.

Sir William Crookes (1832-1919) was an English physicist who developed methods for producing a high vacuum in glass tubes. This discovery played an important role in making possible the later mass production of Edison's incandescent light bulb.

The electron has a charge of 1.602×10^{-19} coulombs (C) and a mass of 9.110×10^{-31} kg, about 1/1836 the mass of the hydrogen atom. The electron, whose mass and charge were determined by 1909, is still believed to be one of the fundamental particles of the universe—a particle not made up of yet smaller particles.

The English physicist Sir Joseph John Thompson (1856-1940) is usually credited with the discovery of the electron. He was awarded the Nobel prize in physics in 1906.

In later experiments with cathode rays, Eugen Goldstein observed in 1886 a stream of invisible particles that seemed to be flowing *counter to the cathode-ray stream*

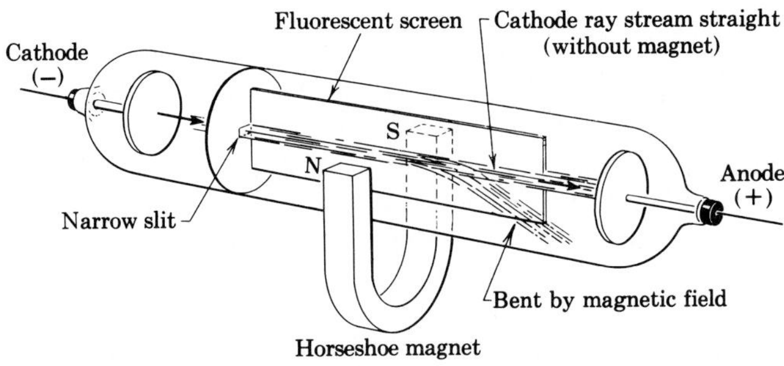

Fig. 34.1 Diagram showing the bending of cathode rays in a magnetic field. If the polarity of the magnet is reversed, the stream of rays will bend up instead of down.

The mass of an electron is approximately 10^{-30}kg and the mass of the sun is approximately 10^{30}kg.

through the tube. The name *proton* was given to these particles. The proton has a mass of about 1836 times that of the electron, or approximately 1.673×10^{-27} kg. Every atom of every element contains one or more electrons that orbit about a central positively charged nucleus containing one or more protons (see Sec. 34.8).

34.2 ■ The Quantum Idea

In the 1890s scientists searched for a fundamental, *theoretical* answer to the question: How does the intensity of light emitted from the surface of an object heated to incandescence depend on the wavelength of the emitted light? The *experimental* answer to this question was already available.

Blackbody radiation—the perfect radiator.

The ideal version of such an incandescent object is called a *blackbody*. A blackbody is a perfect radiator, and a perfect absorber at all wavelengths. According to classical theory, the radiation emitted from a blackbody is due to oscillations of the heated atoms in the body. (See also Sec. 16.9.)

The investigators of blackbody radiation in the period from 1885 to 1890 assumed that there was no lower limit to the size of the possible increase or decrease of the energy of an oscillating atom in a blackbody radiator. Their efforts, however, resulted in predictions that did not agree with the observed variation of intensity of radiation with wavelength for an object heated to incandescence.

The German physicist Max Planck (1858-1947) is responsible for introducing the quantum theory in 1900. That contribution is usually viewed as marking the dividing line between *classical physics* and *modern physics.* Planck won the Nobel prize in physics in 1918 for his work on the quantum theory.

The German physicist Max Planck was able to account for this discrepancy by assuming that the changes in energy of the oscillating atoms occurred in steps of a certain minimum size. He called each of these steps a *quantum* (plural *quanta*).

Planck derived a simple equation for the minimum possible energy change or *quantum*. If E is the quantum of energy, and f is the frequency of oscillation, then

$$E = hf \qquad \text{(Planck's quantum condition)} \qquad (34.1)$$

The proportionality constant h, known as *Planck's constant,* is equal to 6.626×10^{-34} J · s.

Quantum theory indicates that an oscillator of frequency f can only absorb or emit an amount of energy that is a whole-number multiples of hf, that is, $1hf$, $2hf$, $3hf$, . . . Planck's quantum equation can therefore be written in the form

$$E = nhf \qquad (34.2)$$

where n is restricted to integral values, 1, 2, 3, . . . n is called a *quantum number.*

Illustrative Problem 34.1 An atomic oscillator has a frequency of 3.28×10^{14} Hz. What are its three lowest possible levels of energy, assuming Planck's quantum condition?

Solution Use Planck's quantum equation (34.2), where

$$\begin{aligned} n &= 1, 2, \text{ and } 3 \\ E_1 &= 1 \times 6.626 \times 10^{-34}\ \text{J}\cdot\text{s} \\ &\quad \times 3.28 \times 10^{14}\ \text{Hz} \\ &= 2.17 \times 10^{-19}\ \text{J} \qquad \textit{answer} \\ E_2 &= 2E_1 = 4.34 \times 10^{-19}\ \text{J} \qquad \textit{answer} \\ E_3 &= 3E_1 = 6.51 \times 10^{-19}\ \text{J} \qquad \textit{answer} \end{aligned}$$

34.3 ■ The Photoelectric Effect

The spark-gap circuit used by the German physicist Heinrich Hertz (1857-1894) in 1888 was the first radio transmitter. Hertz used a simple loop of wire with an air gap for a receiver.

In 1888 Heinrich Hertz experimented with an oscillating electric circuit that caused a spark to periodically jump across a small air gap between two metal balls when the potential difference between the balls became great enough. He noticed that the spark was more easily produced when the negatively charged metal ball was illuminated with ultraviolet light. This was the first observation of what is now called the *photoelectric effect,* which refers to the ability of light striking the surface of a substance to cause the ejection of electrons from that surface. The ejected electrons are called *photoelectrons*.

The wave theory of light (see Chap. 22) suggests that light waves might transmit energy to electrons in the same manner that a water wave transmits motion to a floating body. With water, the more energy there is in the water wave, the more violently a floating body will be agitated. Reasoning along this line, the energy imparted by a beam of light to a single electron should be related to the intensity of the light. But this was not found to be the case. The following rather unexpected observations were confirmed by many investigators of photoelectricity at about the turn of the century:

1. For each metallic substance, there is a certain minimum *frequency* of the light required, before *any electron emission occurs, no matter how great the intensity of illumination*. This frequency is referred to as the *threshold frequency* of that metal.
2. When the threshold frequency is reached, emission of electrons begins immediately, *no matter how weak the intensity of illumination*.
3. For radiation of a given frequency (above the threshold), the rate of emission of electrons from the metal surface is directly proportional to the intensity of the illumination.
4. The light intensity has no effect on the maximum kinetic energy of the emitted electrons. The kinetic energy is related instead to the *frequency* of the light.

Light of extremely low intensity, but above the threshold frequency, can cause the ejection of photoelectrons from a metal surface in less than 10^{-9}s after the light strikes the surface.

These findings were for the most part incompatible with the wave theory of light. According to the wave theory, the energy in a light beam would be proportional to the intensity of illumination, not the frequency, and ejected electrons should therefore have kinetic energies related to light intensity, not frequency. Also, at low light intensities the wave theory would predict either no electron ejection at all, or a significant time delay before each electron would absorb enough energy to be ejected. A new explanation was needed to account for these experimental findings.

Light as a Stream of Photons Albert Einstein (1879–1955) in 1905 proposed a theoretical explanation for the photoelectric effect. He revived the old particle (or "corpuscular") theory of light and suggested that:

Albert Einstein (1879-1955), who ranks with Archimedes, Galileo, and Newton in the history of science, was awarded the Nobel prize in physics in 1921 for his services to theoretical physics, and especially for his explanation of the photoelectric effect.

1. Light travels as a stream of individual packets of energy, called *photons*.
2. The amount of energy carried by a *single photon* is

$$E_{\text{photon}} = hf \tag{34.3}$$

a form of Planck's quantum equation, where

$$f = \text{the frequency of the light}$$
$$h = \text{Planck's constant } (6.626 \times 10^{-34}\ \text{J}\cdot\text{s})$$

3. Each electron emitted acquires its energy from a single photon, never from two or more. In other words, the photon-to-electron interaction is a one-to-one proposition.
4. As a photon gives its total energy hf to one electron, part of the energy is required to do the work of removing the electron from the metal, and the remainder will show up as kinetic energy of the emitted photoelectron. Stated another way, maximum electron KE = photon energy minus work to remove the electron,

$$\tfrac{1}{2}mv_{\text{max}}^2 = hf - \phi$$

Therefore,

$$E = hf = \phi + \tfrac{1}{2}mv_{\text{max}}^2 \tag{34.4}$$

(*the photoelectric equation*)

where $E = hf$ is the energy of a single photon, $\frac{1}{2}mv_{\text{max}}^2$ = the KE_{max} of the photoelectron produced, ϕ is the *work function*—the minimum amount of work required to remove an electron from the metal, m is the mass of the photoelectron, and v is its velocity.

Work function defined

Table 34.1 Photoelectric Work Functions

Metal	*Symbol*	*Work function ϕ, eV*
Calcium	Ca	2.76
Cesium	Cs	1.81
Copper	Cu	4.6
Germanium	Ge	4.56
Gold	Au	5.22
Platinum	Pt	5.27
Potassium	K	2.22
Silicon	Si	4.1
Silver	Ag	4.73
Tungsten	W	4.56
Zinc	Zn	3.74

In terms of the wavelength of (monochromatic) light,

$$E = h\left(\frac{c}{\lambda}\right) = \phi + \tfrac{1}{2}mv_{max}^2 \tag{34.5}$$

The work function ϕ is a property of the material from which the electrons are ejected.

The photoelectric effect is, then, an energy-conversion process. The light photon with energy $E = hf$ ceases to exist after it transfers its energy to an electron.

If the photon's total energy is less than the work function for that metal ($hf < \phi$), no electron emission will occur. If $hf > \phi$, the fastest photoelectrons will move away from the illuminated surface with a kinetic energy equal to $hf - \phi$. The *threshold frequency* is thus found when

$$hf = \phi \tag{34.6}$$

High-intensity light means that more photons are striking the metal surface, causing more electrons to be emitted. The kinetic energy of the photoelectrons is not increased by greater light intensity as would be expected from the wave theory.

The work function of most metals is between 1 and 5 electron-volts (eV) (see Table 34.1).

The electron-volt is a unit of energy equal to that acquired or lost by an electron when it is accelerated through a potential difference of 1 V. It is a very useful energy unit in atomic and nuclear physics.

$$\begin{aligned} 1\ \text{eV} &= 1.602 \times 10^{-19}\ \text{C} \times 1\ \text{V} \\ &= 1.602 \times 10^{-19}\ \text{J} \end{aligned} \tag{34.7}$$

The photoelectric effect constitutes one of the major pieces of evidence that light must be considered as having particle-like characteristics. A photon, like a particle, carries a concentrated amount of energy. *Photon* means a "particle of light."

Illustrative Problem 34.2 The frequency of light in the middle of the visible spectrum (wavelength = 550 nm) is approximately 5.45×10^{14} Hz. What is the energy of a photon of light of that frequency?

Solution The photon energy can be calculated using Eq. (34.3):

$$E_{photon} = hf$$

where h = Planck's constant (6.626×10^{-34} J · s). For this problem, $f = 5.45 \times 10^{14}$ Hz, and

$$E = (6.626 \times 10^{-34}\ \text{J} \cdot \text{s})(5.45 \times 10^{14}\ \text{Hz})$$

Since Hz = s^{-1},

$$(\text{J} \cdot \text{s})(\text{Hz}) = (\text{J} \cdot \text{s})(\text{s}^{-1}) = \text{J}$$

and

$$E = 3.61 \times 10^{-19}\ \text{J} \qquad \textit{answer}$$

This photon energy can also be expressed in *electron-volts* (eV)

$$\frac{3.61 \times 10^{-19}\ \text{J}}{1.602 \times 10^{-19}\ \text{J/eV}}$$
$$= 2.25\ \text{eV} \qquad \textit{answer}$$

Notice that the *electron-volt* (eV) is a more convenient unit than the *joule* (J) for the energy of a visible light photon. (Measuring the energy of a photon in joules is a little like measuring the thickness of a human hair in miles.)

The ratio of a joule to an electron volt (eV) is approximately the same as the ratio of the kinetic energy of a 77,000-lb truck traveling 60 mi/h to the kinetic energy of a mosquito moving with a speed of 3 in./min.

Illustrative Problem 34.3 Light of wavelength 600 nm (1 nm = 10^{-9} m) shines on a metal whose work function is 1.3 eV. What is the velocity of the emitted photons of maximum kinetic energy? (NOTE: Some electrons deeper in the metal may also escape, but the work necessary to separate them will be greater than the work function of 1.3 eV and their KE will be less than maximum. We are concerned here only with the electrons of *maximum* KE.)

Solution Solve Eq. (34.4) for KE_{max}:

$$KE_{max} = \tfrac{1}{2}mv^2 = h\frac{c}{\lambda} - \phi$$

$$= 6.626 \times 10^{-34}\ \text{J} \cdot \text{s}\ \frac{3 \times 10^8\ \text{m/s}}{6 \times 10^{-7}\ \text{m}}$$

$$- 1.3\ \text{eV} \times 1.602 \times 10^{-19}\ \text{J/eV}$$

$$= 1.23 \times 10^{-19}\ \text{J}$$

Rearranging algebraically yields

$$v^2 = \frac{2 \times 1.23 \times 10^{-19}\ \text{J}}{\text{m}}$$

But (see tables inside back cover), m = rest mass of the electron = 9.11×10^{-31} kg. Substituting,

$$v^2 = \frac{2.46 \times 10^{-19}\ \text{kg} \cdot \text{m}^2/\text{s}^2}{9.11 \times 10^{-31}\ \text{kg}}$$

from which,

$$v = 5.2 \times 10^5\ \text{m/s} \qquad \textit{answer}$$

Illustrative Problem 34.4 Zinc has a work function of 3.74 eV. What threshold wavelength is needed to produce electrons?

Solution Use Eq. (34.5) and recall that $f = c/\lambda$. Solving for the threshold wavelength, we get

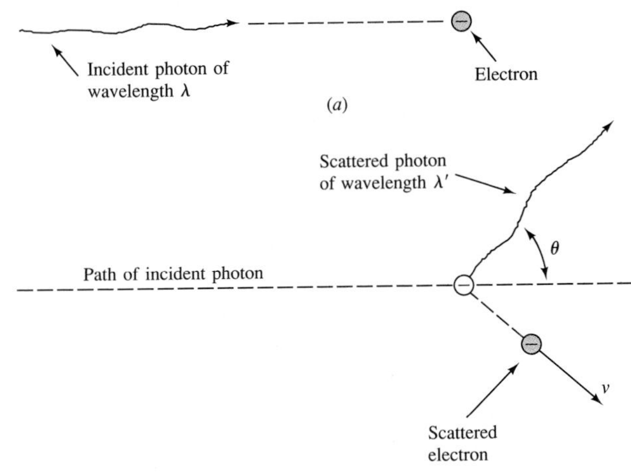

Fig. 34.2 (*a*) A photon of wavelength λ heads toward an electron at rest. (*b*) After the collision, the electron moves away with speed v and a lower energy photon of longer wavelength λ' moves away from the site of the collision. This type of scattering of a photon by an electron is called the Compton effect.

$$\lambda = \frac{hc}{\phi}$$

$$= \frac{6.626 \times 10^{-34}\ \text{J} \cdot \text{s} \times 3 \times 10^8\ \text{m/s}}{3.74\ \text{eV} \times 1.602 \times 10^{-19}\ \text{J/eV}}$$

$$= 3.32 \times 10^{-7}\ \text{m}$$

$$= 332\ \text{nm}$$

answer

This frequency is in the ultraviolet region. (See Chap. 22.) Photons of *visible* light do not have enough energy to cause photoelectrons to be emitted from zinc.

The American physicist Arthur H. Compton (1892-1962) received the Nobel prize in 1927 for his discovery of the Compton effect.

34.4 ■ The Compton Effect

When a photon of electromagnetic radiation interacts with an electron, there are two possibilities: The photon could give all of its energy to an electron (photoelectric effect), or it could give some fraction of its energy to an electron in an event called a *Compton scattering event*. Compton scattering is named after the American physicist Arthur Compton who studied scattering of x-ray photons by electrons in 1923.

Figure 34.2 shows the sequence of events for a Compton scattering event. Before the interaction, a photon of wavelength λ heads toward an essentially stationary electron (Fig. 34.2*a*). After the interaction (see Fig. 34.2*b*), the electron moves away from its original location with kinetic energy acquired from the "collision," and a "scattered" photon with wavelength λ' moves away in a direction that makes an angle Θ with the direction of travel of the incident photon. The wavelength λ' of the scattered photon is slightly greater than the wavelength λ of the incident photon. This is consistent with the principle of conservation of energy and the fact that $E = hc/\lambda$ for a photon. The increase in wavelength experienced by the radiation when the photons are scattered by electrons is called the *Compton effect*.

The Compton effect is one of the mechanisms by which high energy gamma ray photons from a radioactive source are absorbed by a lead barrier.

The observed wavelength increase does not agree with classical theory, which would predict that when an electron is "struck" by an electromagnetic wave, the electron would oscillate with the same frequency as the wave and would radiate electromagnetic waves of the same frequency and wavelength.

By assuming a photon model for the radiation, and by assuming conservation of energy and conservation of momentum, Compton was able to write the following relationship between λ, λ', Θ, and the mass m of the electron:

$$\lambda' - \lambda = \frac{h}{mc}(1 - \cos\Theta) \tag{34.8}$$

(Compton effect equation)

Illustrative Problem 34.5 An x-ray photon of wavelength 0.139 nm is Compton-scattered by an electron. After the interaction between the photon and the electron, the photon travels in a direction that makes an angle of 60.0° with the direction that the photon was traveling before the interaction (see Fig. 34.3). What is the wavelength of the scattered photon in nanometers?

Solution Solve Eq. (34.8) for the wavelength λ' of the scattered photon and then substitute numerical values for h, m, c, Θ, and the wavelength λ of the incident photon.

$$\lambda' = \frac{h}{mc}(1 - \cos\Theta) + \lambda$$

$$= \frac{6.63 \times 10^{-34}\ \text{J} \cdot \text{s}}{(9.11 \times 10^{-31}\ \text{kg})(3.00 \times 10^8\ \text{m/s})}(1 - \cos 60.0^\circ) + 0.139 \times 10^{-9}\ \text{m}$$

$$= 2.43 \times 10^{-12}\ \text{m}\ (1 - 0.50) + 0.139 \times 10^{-9}\ \text{m}$$

$$= 1.40 \times 10^{-10}\ \text{m} = 0.140\ \text{nm}$$

answer

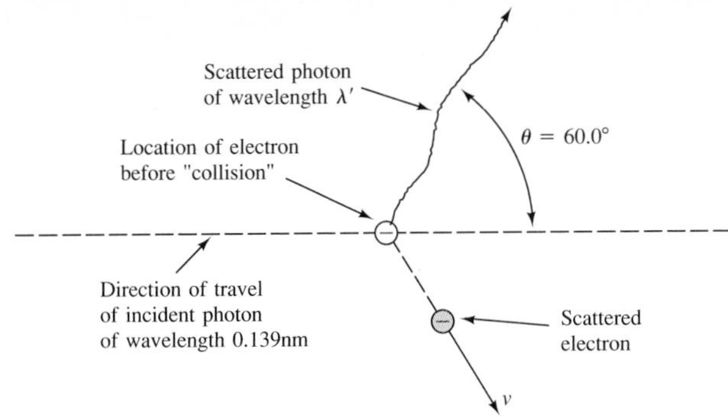

Fig. 34.3 Diagram for Illustrative Problem 34.5.

Note that the scattered photon has a wavelength that is only 0.001 nm greater than the wavelength of the incident photon. This is less than a one percent increase in wavelength. If the incident photon had a longer wavelength, the percent increase in wavelength of the scattered photon would have been even smaller.

34.5 ■ Wave-Particle Duality

Both the photoelectric effect and the Compton effect provide convincing evidence that light is made up of *particles* (called photons), each of which has an energy hf or hc/λ. Other experiments involving interference and diffraction, on the contrary, seem to provide equally convincing evidence that light is *not made up of particles,* but that it is an *electromagnetic wave* that has many of the properties of sound waves and water waves.

Light and other forms of electromagnetic radiation have both particle properties and wave properties.

Since the wave and particle characteristics of light do not seem to occur simultaneously or in the same process, two generalizations are possible: (1) Concerning *propagation* of light (and electromagnetic radiation in general), the wave theory seems to be obeyed. (2) When light (or other electromagnetic radiation) strikes or *interacts* with atoms, molecules, electrons, neutrons, or other subatomic units, it acts like particles or photons.

34.6 ■ Matter Waves—The de Broglie Hypothesis

Albert Einstein is responsible for two equations for the energy of a photon. The first, which comes from his explanation of the photoelectric effect, relates the energy of a photon to its wavelength λ by the equation $E = hc/\lambda$. The second, which comes from the special theory of relativity, relates the energy of a photon to its momentum p by the equation $E = pc$. These two equations can be used to find a relationship between the momentum and wavelength of the photon.

Since $E = hc/\lambda$ and $E = pc$, it must be true that $pc = hc/\lambda$. Solving this last equation for p gives

$$p = \frac{h}{\lambda} \tag{34.9}$$

In 1923 Louis de Broglie suggested that Eq. (34.9) should hold not only for photons but for any particle of matter. He predicted that a particle such as an electron or proton should have a wavelength that is related to its momentum by the equation

$$\lambda = \frac{h}{p} = \frac{h}{mv} \tag{34.10}$$

(the de Broglie wave equation)

The French physicist Prince Louis Victor Pierre Raymond de Broglie (1892-1987) was awarded the 1927 Nobel prize in physics for his discovery of the wave nature of electrons.

The symbol λ in the above equations refers to what is called the *de Broglie wavelength* of the matter waves associated with the particle. These matter waves are not electromagnetic waves, but rather are waves whose properties are related to the probability of finding the particle at a specific location.

Only for particles that have an extremely small mass are the wavelengths of the matter waves long enough to be measurable. As the following problem shows, a particle that is just large enough to be visible will have a de Broglie wavelength that is smaller than the diameter of the nucleus of an atom.

Illustrative Problem 34.6 The small specks of dust that can be seen floating in the air in a bright beam of sunlight are among the smallest objects visible to the unaided eye. Compute the de Broglie wavelength of a speck of dust having a mass of 10^{-14}kg, and a speed of 1 cm/s.

Solution The de Broglie wavelength can be calculated using Eq. (34.10):

$$\lambda = \frac{h}{p}$$

which can be written $\lambda = h/mv$, since $p = mv$. For the speck of dust, $m = 10^{-14}$kg and $v = 10^{-2}$m/s. Planck's constant h is equal to 6.63×10^{-34}J · s, which can be written 6.63×10^{-34}kg · m²/s. Substitution of these values into the above equation gives

$$\lambda = \frac{6.63 \times 10^{-34}\text{kg} \cdot \text{m}^2/\text{s}}{(10^{-14}\text{kg})\,(10^{-2}\text{m/s})}$$

$$= 6.63 \times 10^{-18}\text{m} \qquad \textit{answer}$$

The de Broglie wavelength of the dust particle is about 100 times smaller than the diameter of the nucleus of a typical atom. It is difficult to imagine an *experiment* that could confirm the calculated value for the de Broglie wavelength of the dust particle.

Illustrative Problem 34.7 An electron accelerated through a potential difference of 6.4 V has a speed of 1.5×10^6m/s. What is the de Broglie wavelength for the electron?

Solution As in Illustrative Problem 34.6, the de Broglie wave equation

$$\lambda = \frac{h}{mv}$$

can be used to determine the de Broglie wavelength. For the electron, $m = 9.11 \times 10^{-31}$kg and $v = 1.50 \times 10^6$m/s. Substituting these values and the value of Planck's constant into the above equation gives

$$\lambda = \frac{6.63 \times 10^{-34}\text{kg} \cdot \text{m}^2/\text{s}}{(9.11 \times 10^{-31}\text{kg})\,(1.50 \times 10^6\text{m/s})}$$

$$= 4.85 \times 10^{-10}\text{m}$$

$$= 0.485 \text{ nm} \qquad \textit{answer}$$

This de Broglie wavelength is equal to the typical distance between atoms in solids. An experimental technique called low-energy electron diffraction takes advantage of the wave properties of electrons to study the surfaces of solid materials on an atomic scale.

The study of surfaces using electron diffraction techniques is important in modern microelectronics because a significant fraction of the atoms that make up semiconductor devices are at or near the surface of the semiconductor material because the devices are so small. The electron microscope, to be discussed in Sec. 34.16, also depends on electron diffraction.

34.7 ■ The Rutherford Scattering Experiment

Although Democritus had suggested 2400 years ago that matter is made of small particles called atoms, it was not until 1906 that the British physicist Ernest Rutherford devised an experiment to determine, in some detail, information about the structure of atoms. This experiment, now known as the Rutherford scattering experiment, involved shooting a beam of positively charged alpha particles at a very thin gold foil, and then studying the directions traveled by the alpha particles after they passed through the foil. (An alpha particle, which is identical to the nucleus of a helium atom, is made up of two neutrons and two protons. Alpha particles are emitted from the nuclei of some radioactive atoms.) The main features of the experimental arrangement are shown in Fig. 34.4.

Ernest Rutherford (1871-1931) was a British physicist who won the 1908 Nobel prize for chemistry. Although he made important contributions to our understanding of the structure of atoms, he did not believe that the energy of the atomic nucleus could ever be controlled.

Rutherford was able to accurately predict the observed distribution of directions of travel for the deflected alpha particles by assuming that all the positive charge of the atom is concentrated in a very small volume compared to the volume occupied by the electrons (see Fig. 34.5). The very small volume that contains all the positive charge of the atom is called the *nucleus* of the atom.

We now know that the nucleus contains positively charged particles called *protons*. The nuclear model for the atom established by the Rutherford scattering experiment is still accepted today.

34.8 ■ The Bohr-Rutherford Atom Model

On the basis of experimental evidence available from the Rutherford scattering experiment, Lord Rutherford proposed that an atom looks like a miniature solar system with a heavy center, which he called the *nucleus*, surrounded by electrons revolving in various orbits, much as the planets of our solar system revolve around the sun. Niels Bohr, a Danish physicist and a student of Rutherford's, was also instrumental in the development of this *planetary* model of the atom (see Fig. 34.5). Since ordinary hydrogen is the simplest atom—one (negative) electron revolving around a nucleus of one (positive) proton, according to the model, we will use it for a brief discussion of energy levels in the atom.

The first hint of quantized energy states in atoms came from observations of spectral lines. Since hydrogen is the simplest atom, it has a simple, orderly spectral

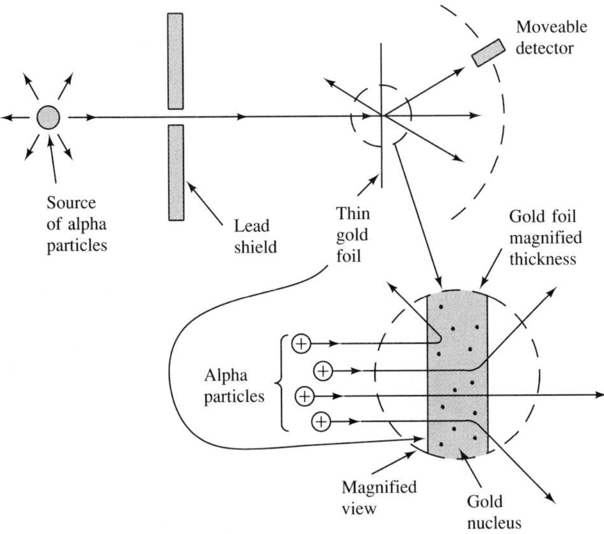

Fig. 34.4 Apparatus arrangement for the Rutherford scattering experiment showing how a collimated beam of positively charged alpha particles is arranged to strike a thin sheet of gold foil.

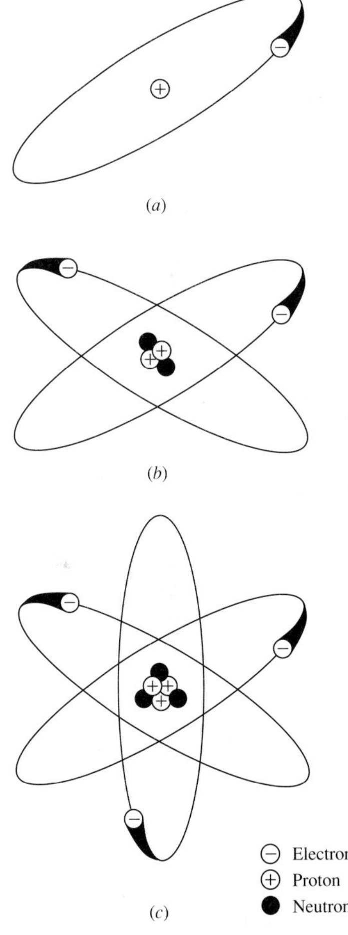

Fig. 34.5 Schematic (two-dimensional) diagrams of the Rutherford-Bohr planetary models for the three lightest atoms, (*a*) hydrogen, (*b*) helium, and (*c*) lithium. The nuclei of the atoms are not drawn to scale here. If the nuclei and the planetary orbits were drawn to the same scale, the nuclei would have to be drawn so small that they would not be visible.

pattern as Fig. 34.6*a* shows. J. J. Balmer (1825–1898) derived the following equation in 1884 for calculating the wavelength of the 4 visible bright lines in the spectrum of hydrogen.

$$\frac{1}{\lambda} = R\left(\frac{1}{2^2} - \frac{1}{n^2}\right) \quad \text{where } n = 3,\ 4,\ 5\cdots \tag{34.11}$$

(*Balmer series equation*)

R is called *Rydberg's constant* and is equal to $1.097 \times 10^7 \text{m}^{-1}$, and λ is the wavelength in meters of the bright lines in the *Balmer series* of spectral lines in the hydrogen spectrum (often called the *visible series* even though λ is shorter than the wavelength of visible light for n greater than 6).

Illustrative Problem 34.8 Find the wavelength of the third line in the visible (Balmer) series for hydrogen.

Solution In the Balmer series equation (34.11), substitute $n = 5$, the third permitted value, obtaining

$$\frac{1}{\lambda} = (1.097 \times 10^7 \text{ m}^{-1})\left(\frac{1}{2^2} - \frac{1}{5^2}\right)$$

$$= (1.097 \times 10^7 \text{ m}^{-1})(0.210)$$
$$= 2.30 \times 10^6 \text{ m}^{-1}$$

and

$$\lambda = \frac{1 \text{ m}}{2.30 \times 10^6} = 4.34 \times 10^{-7} \text{ m}$$

$$= 434 \text{ nm} \qquad \textit{answer}$$

Other series, with no lines in the visible spectrum, were soon discovered for hydrogen, by Lyman, Paschen, and others. The Lyman series lies in the ultraviolet region and the Paschen series in the infrared (see Fig. 34.6*b*). Each has an equation, similar to the Balmer equation, and involving the Rydberg constant, for predicting the wavelength of the series lines. By making certain assumptions that take into account both the particle and wave properties of the electron, it is possible to calculate theoretical values for the possible total energies of the orbiting electron and the wavelengths of the spectral lines for hydrogen. These assumptions are:

1. The electron moves with constant speed in a circular orbit under the influence of the electrical force of attraction to the proton, which is the nucleus for the atom (Fig. 34.7). The electron experiences no other forces, and the electron does not

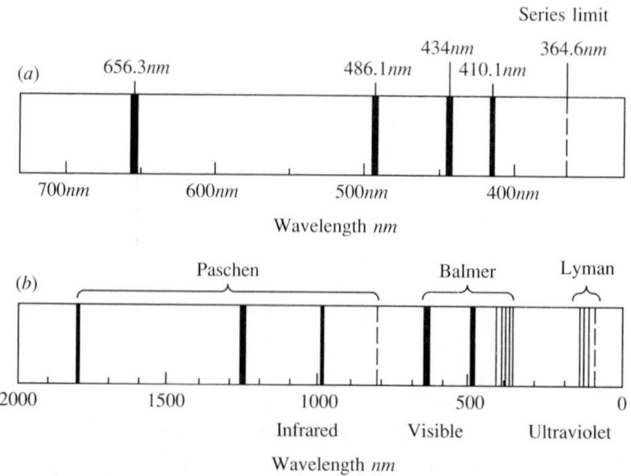

Fig. 34.6 Diagrams of three series of spectral lines of hydrogen. (*a*) The four lines of the Balmer series that lie in the visible spectrum. (*b*) The Lyman, Balmer, and Paschen series (stronger lines of each only) on the same scale of wavelengths, showing their relative positions.

radiate energy in the form of electromagnetic waves even though it is experiencing a centripetal acceleration while it moves in the orbit.

2. *The electron can exist only in orbits for which the circumference is an integral number of de Broglie wavelengths of the electron.* That is, if the circumference of the orbit is equal to $2\pi r$, and if the de Broglie wavelength λ of the electron is equal to h/mv, then the allowed orbits are those for which

$$2\pi r = n\left(\frac{h}{mv}\right) \qquad (34.12)$$

$$\text{Orbit circumference} = n \times \text{de Broglie wavelength} \left(\lambda = \frac{h}{mv}\right)$$

where
r = the radius of the orbit
n = 1, 2, 3, . . .
h = Planck's constant
m = the electron mass
v = the speed of the orbiting electron

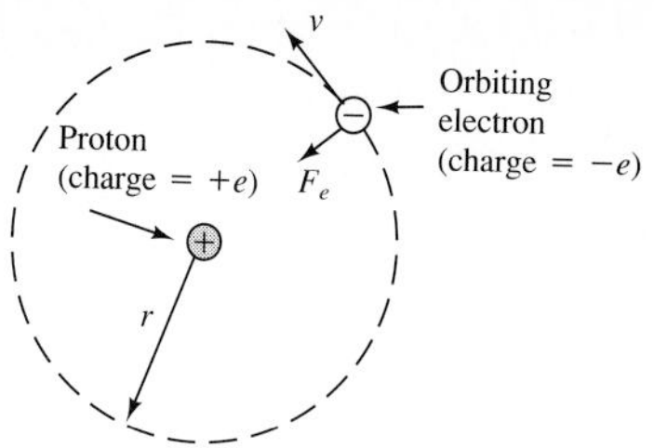

Fig. 34.7 In the Rutherford-Bohr model for the hydrogen atom, the electron is assumed to move with constant speed v in a circular orbit of radius r. The electron experiences a centripetal (Coulomb) force due to its electrical attraction to the proton.

3. The total energy of the orbiting electron is the sum of its kinetic energy and its electrical potential energy. Although the electron is restricted to certain orbits as specified above, the electron can move from one allowed orbit to another.

If an electron moves from an orbit where its total energy is E_A to a smaller orbit where it has a lower total energy E_B, a photon of light will be emitted by the atom. The energy of the photon will be equal to the energy lost by the electron and since the photon energy is equal to hf or hc/λ, it follows that

A photon is emitted by a hydrogen atom when the electron spontaneously moves from a larger to a smaller orbit.

$$E_A - E_B = hf = hc/\lambda \qquad (34.13)$$

If E_B is greater than E_A, the electron will move to the orbit where its total energy is greater only if it gains the required energy from some source outside the atom. Two possible ways for the electron to gain the required energy are by absorbing energy from a photon or by experiencing a collision with another electron.

If the electrical potential energy of the electron is taken to be zero when it is infinitely far from the proton, then the allowed total energy E_n (in eV) of an orbiting electron in a hydrogen atom can be calculated using the following equation.

$$E_n = -13.6 \text{ eV}/n^2 \qquad n = 1, 2, 3, \ldots, \qquad (34.14)$$

In the above equation, $n = 1$ for an electron in the smallest allowed orbit, $n = 2$ for the second smallest orbit, etc. The total energy of the orbiting electron is negative because the potential energy of the electron was chosen to be zero when the electron is infinitely far from the nucleus, and because, in any given orbit, the (negative) potential energy of the electron is twice as great as its kinetic energy.

Equation (34.14) predicts that the energies allowed to an electron in a hydrogen atom are limited to values with steplike or incremental (discrete) differences from one value to the next. We say the electron energy is *quantized*. A "quantum" of energy is "a small increment" of energy.

A number of quantities you have already studied are quantized variables. Some examples are: (1) the electric charge on a particle or object, which can have only those values that are multiples of 1.602×10^{-19} C; (2) the atomic number (Z) of an element, which can only be a positive integer; and (3) the harmonic frequencies of a vibrating string or air column, which are whole multiples of the first harmonic. Figure 34.8 shows a "quantized man," that is, what it would be like if gravitational potential energy were always quantized, the way the energy of atomic particles is quantized.

Stair steps are not foreign to our experience, but suppose that steps of a specified (discrete) height were the only way to change gravitational potential energy level. No ramps, no rolling hills, and no slides would be possible.

The allowed energies for an electron can be displayed in an energy-level diagram as shown in Fig. 34.9*a*. Figure 34.9*b* shows the paths of the four smallest electron orbits. When an electron is in the smallest ($n = 1$) allowed orbit, its total energy is

Fig. 34.8 A "quantized" person. True if gravitational potential energy (GPE) could exist only in discrete increments, or "quanta," of 1 mgh. Steps 1, 2, 3, 4 . . . could be regarded as energy levels or energy states.

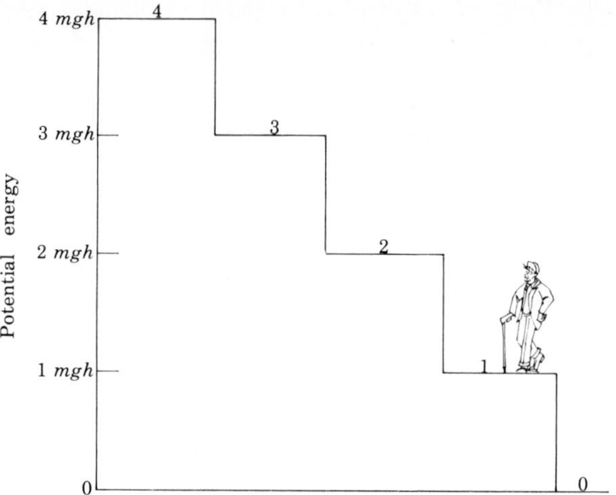

-13.6 eV. When the electron is in the second smallest ($n = 2$) allowed orbit, its energy is -3.40 eV, etc. An electron initially in the $n = 2$ energy state with energy $E_2 = -3.40$ eV will lose 10.2 eV of energy if it moves to the $n = 1$ energy state—usually called the "ground state." That is, the energy change, $E_f - E_i$, experienced by the electron is given by

$$\begin{aligned} E_1 - E_2 &= -13.6 \text{ eV} - (-3.40 \text{ eV}) \\ &= -10.2 \text{ eV}. \end{aligned}$$

Illustrative Problem 34.9 What is the wavelength of the photon emitted when an electron moves from the $n = 5$ to the $n = 2$ orbit in a hydrogen atom?

Solution When the electron is in the $n = 5$ orbit, its energy E_5 (from Eq. 34.14) is equal to $-13.6 \text{ eV}/5^2 = -0.54$ eV. Similarly, the energy of the electron in the $n = 2$

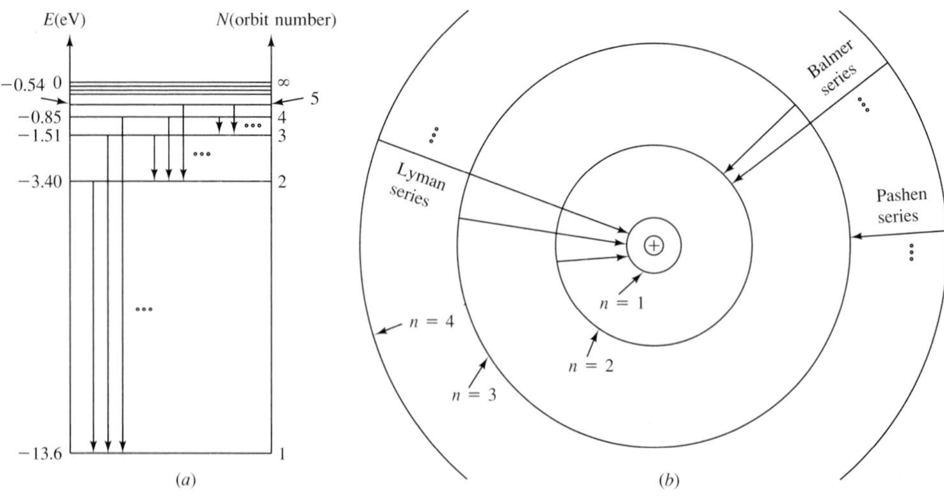

Fig. 34.9 (*a*) Energy level diagram for the electron in a hydrogen atom. The set of horizontal lines represents the different possible values of total energy for the orbiting electron. (*b*) The four smallest allowed orbital paths for an electron in a hydrogen atom. The nucleus (proton) is not drawn to scale here. Its diameter is actually about 10^{-5} times the diameter of the $n = 1$ orbit for the electron.

orbit, $E_2 = -13.6 \text{ eV}/2^2 = -3.40$ eV. The magnitude of the energy lost by the electron is equal to its initial energy E_5 minus its final energy E_2. That is, $E_5 - E_2 = -0.54 \text{ eV} - (-3.40 \text{ eV}) = 2.86$ eV.

When the electron moves to the smaller orbit and loses energy, a photon is emitted whose energy is equal to the energy lost by the electron. For a photon, $E = hc/\lambda$ and therefore, $\lambda = hc/E$. A photon whose energy is 2.86 eV will have a wavelength

$$\lambda = \frac{(6.63 \times 10^{-34}\text{J} \cdot \text{s})(3.00 \times 10^{8}\text{m/s})}{(2.86 \text{ eV})(1.60 \times 10^{-19}\text{J/eV})}$$
$$= 4.35 \times 10^{-7} \text{ m}$$
$$= 435 \text{ nm} \qquad \textit{answer}$$

Note that this wavelength is only 0.2 percent larger than the wavelength of one of the spectral lines in the Balmer series for hydrogen shown in Fig. 34.6. (See also color plate on end papers, front of book.)

34.9 ■ The Production of X-Rays

In the photoelectric effect, photons strike a metal surface and electrons are ejected from that surface. Production of x-rays, which have wavelengths in the range 0.001 nm to 10 nm, is accomplished by the inverse of this process. Rapidly moving electrons strike a metal surface and photons are ejected from the surface. X-rays were first produced this way by Wilhelm Roentgen in 1895.

Figure 34.10 shows the details of an x-ray tube. The filament in the x-ray tube is heated to incandescence, and the electrons that are boiled off the surface of the hot filament are drawn to the metal target which is part of the anode for the tube. The speed with which the electrons strike the target depends on the potential difference between the anode and the filament, which for medical diagnostic purposes may be from 20,000 to 100,000 V, but for industrial purposes or medical therapy may be 1 MV or higher. Metals used for the target materials in x-ray tubes include copper, molybdenum, chromium, and tungsten.

The operating voltage for an x-ray tube may be greater than one million volts.

Braking Radiation (Bremsstrahlung) X-rays can be produced in two ways. The first process is called *bremsstrahlung* (from the German, meaning *braking radiation*). It is depicted in Fig. 34.11. As an electron at *A* moves at high speed in the immediate vicinity of a heavy nucleus (i.e. impacting on a solid metal target), it has rapid negative acceleration, and loses energy. Wave theory predicts that radiation must be emitted for energy to be conserved. In quantum terms, the energy emitted is in the form of photons, each having a definite energy. The negative acceleration

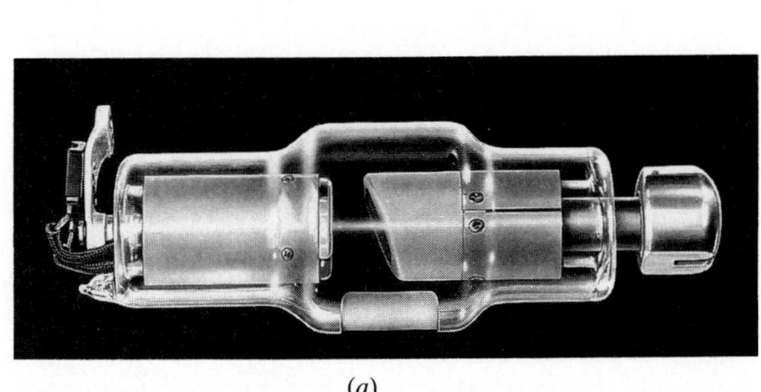

(*a*)

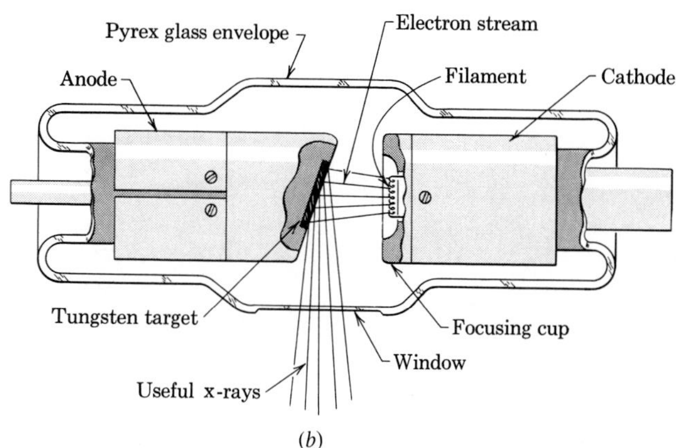

(*b*)

Fig. 34.10 (*a*) The Coolidge x-ray tube, as used in diagnostic medicine. (*b*) Diagram of Coolidge x-ray tube with parts labeled. (X-ray Department, General Electric Co.)

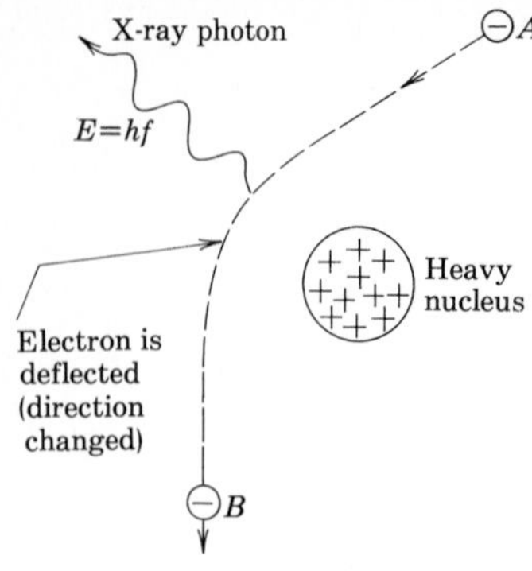

Fig. 34.11 Schematic diagram illustrating braking radiation (*bremsstrahlung*). A high-velocity electron at A is slowed and redirected as it passes near an atomic nucleus. The change in speed *and direction* represents a large negative acceleration. The electron's loss in energy shows up as an x-ray photon whose energy is $E = hf$.

(braking) of the electron can occur either a result of its actually being stopped by collision with nuclei of the target material, or merely by having its direction changed (velocity is a vector quantity) as shown in Fig. 34.11.

An electron accelerated through a potential difference V as it accelerates from the filament to the target in an x-ray tube will have a kinetic energy equal to Ve just as it strikes the target.

$$\text{KE} = \frac{1}{2}mv^2 = Ve \tag{34.15}$$

The minimum wavelength photon is produced when the electron is completely stopped by the target. In that case the kinetic energy of the electron before it strikes the target is equal to the energy of the minimum wavelength photon produced. Since the photon energy is equal to hc/λ,

$$Ve = \frac{hc}{\lambda} \qquad \textit{from which} \qquad \lambda = \frac{hc}{Ve}$$

Substituting known physical constants the *minimum* x-ray wavelength is given by

$$\lambda_{\min} = \frac{(6.626 \times 10^{-34}\ \text{J} \cdot \text{s})\,(3 \times 10^8\ \text{m/s})}{(1.602 \times 10^{-19}\ \text{C}) \times \text{V}}$$

Since coulombs = joules/volts, or C = J/V, and 1 nm = 10^{-9} m, it follows that

$$\lambda_{\min}(\text{nm}) = \frac{1240}{V(\text{volts})} \tag{34.16}$$

Equation (34.16) says that an applied voltage V of 1240 V produces x-ray photons with a minimum wavelength of 1.0 nm. To obtain the minimum wavelength in nanometers of the x-rays from an x-ray tube, divide 1240 by the voltage applied across the tube. Typical accelerating voltages range from 10 kV up to 2 MV or more.

Illustrative Problem 34.10 An x-ray machine uses an accelerating voltage V of 25,000 V. Find (*a*) the maximum electron kinetic energy produced, and (*b*) the minimum x-ray wavelength.

Solution

(***a***) Using Eq. (34.15) gives

$$\frac{1}{2}mv^2 = Ve = (1.602 \times 10^{-19}\ \text{C})\,(25 \times 10^3\ \text{V})$$
$$= 4 \times 10^{-15}\ \text{J} \qquad \textit{answer}$$

(***b***) Use Eq. (34.16) to solve for wavelength:

$$\lambda_{\min}(\text{nm}) = \frac{1.24 \times 10^3}{2.5 \times 10^4\ \text{V}}$$
$$= 0.050\ \text{nm} \qquad \textit{answer}$$

Characteristic X-rays The second mechanism for creating x-rays involves electron transitions between energy levels, as an electron "jumps" from one orbit to another orbit of an atom. In heavy atoms (i.e. with heavy nuclei), *energy shells* have been identified and labeled K, L, M, N, etc. When a high-speed electron collides with a tightly bound electron in the innermost (K) shell, it may transfer enough energy to knock that electron out of the atom. A vacancy is thus created, and an electron from a higher-energy shell may move to fill the vacancy, thereby creating a new vacancy (Fig. 34.12). Other electrons change shell locations until the vacancy moves to the outside of the atom. A number of photons are emitted, but only orbital changes by the innermost electrons involve enough energy to produce x-ray photons. These x-rays are called *characteristic x-rays*, because they are characteristic of the energy differences between shells involved in the electron jumps in a particular atom.

Applications of X-rays Two main areas of practical applications of x-rays are in the study of the properties of materials and in medical diagnostic studies.

Of all the different possible *wavelengths* of electromagnetic waves, only x-rays have wavelengths that are about equal to the diameter of typical atoms. It is the possibility of generating x-rays of this ideal range of wavelengths that makes x-ray analysis of materials a technique used by essentially all large industrial and university research groups.

The analysis of the interaction of x-rays with a solid material can be used to determine how the atoms are arranged in the material (the crystal lattice structure) and to determine the distance between adjacent atoms (the lattice constant). In addition, measurement of the energies of x-rays that can be induced to be emitted from a material can be used to identify the elements that make up the material.

Of all the different possible *energy-possessing* electromagnetic photons, only x-ray photons pass quite easily through low density human tissue, but not so easily through high density tissue or bones, which makes it possible to produce x-ray images that show detailed features of the interior of the body. Although the first x-ray image, which showed the bones in a human hand, was produced in 1896, scientists today continue to search for new, better, and safer ways to produce images of the interior of the human body using x-rays.

Medical diagnosis using x-rays took a significant step forward very recently when a diagnostic system called *computed tomography* was developed (see Fig. 34.13 and Sec. 33.22). A computed tomography system can be used to pinpoint the precise location of tumors and other abnormalities, and thereby possibly eliminate the need for exploratory surgery or other more dangerous diagnostic procedures.

A *computed tomography* (CT) system is sometimes called a *CAT scanner,* where CAT stands for *computed axial tomography.* Tomography is the technique of producing an x-ray image of a plane section of a solid object.

NUCLEAR PHYSICS

Up to this point, no greater knowledge about the nucleus of an atom was required than that provided by the Rutherford scattering experiment (Sec. 34.6), which showed that all the positive charge of an atom was contained in a very small volume (the *nucleus*) at the center of the atom.

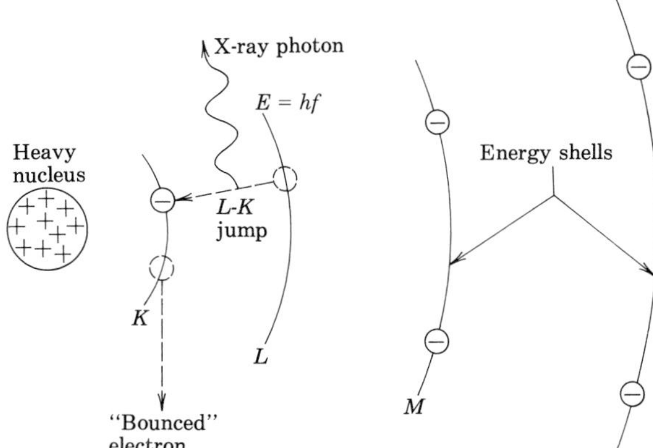

Fig. 34.12 Schematic diagram of the production of characteristic x-rays. Only those electron "jumps" that occur in the innermost shells or orbits involve a great enough electron energy loss to produce x-ray photons.

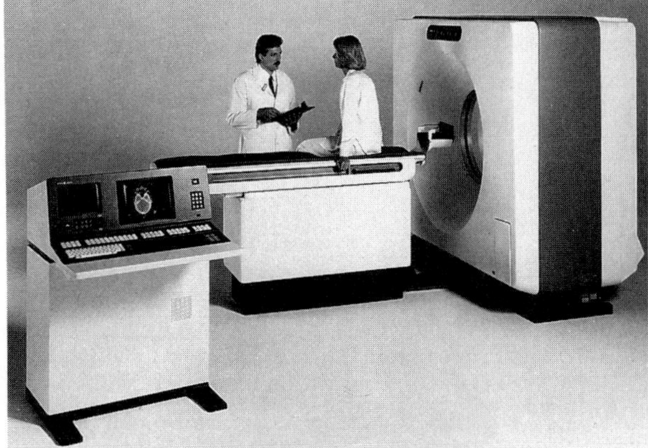

Fig. 34.13 A computed tomography (CT) system, which combines an x-ray scanning system with a computer, is capable of pinpointing the precise location of tumors and other abnormalities in virtually any portion of the human anatomy. (Courtesy of GE Medical systems)

This simple view of the nucleus cannot, however, be used to explain why different atoms of the same element sometimes have different masses, why some elements are radioactive, or how it is possible to obtain energy from the nuclei of atoms.

Now, we look inward to the nucleus of the atom where most of the mass of the atom is located, and where most of the energy is. Two nuclear particles have already been mentioned—the *proton*, a particle with unit positive charge and a mass about 1836 times that of the electron; and the *neutron*, a particle with no charge, only slightly more massive than the proton.

34.10 ■ Nuclear Structure and Charge

The nucleus is very small compared to the diameter of the atom's outer electronic shell. The atom is mostly empty space. A typical atom has a diameter in the range of 10^{-10}m (0.1 nm), with the diameter of its nucleus being of the order of 10^{-14}m (10^{-5}nm), only one ten-thousandth as large.

Here are some definitions pertaining to the structure and charge of atomic nuclei. The notation introduced in these definitions will be used in later sections.

Definitions of *nucleon, atomic number, neutron number,* and *mass number.*

1. Both neutrons and protons are called *nucleons*.

2. The *atomic number* Z of an atom is given by the number of protons in the nucleus. Since each proton has a unit charge of ^{+}e, the charge on the nucleus is ^{+}Ze. Also, for a *neutral* atom, Z indicates the total number of electrons in all the orbits of that atom. Their total charge is ^{-}Ze.

3. The *neutron number* N, of an atom is given by the number of neutrons in the nucleus.

4. The *mass number* A of an atom is equal to the number of *nucleons*, that is to the sum of the numbers of protons and neutrons. The mass number A is the nearest whole number to the actual *atomic mass*, expressed in atomic mass units (see Sec. 34.12 and Table 34.2). As an equation,

$$A = Z + N \tag{34.17}$$

Illustrative Problem 34.11 Which atom has a nucleus with the larger number of neutrons—an iron atom with atomic number $Z = 26$ and mass number $A = 59$, or a nickel atom with atomic number $Z = 28$ and mass number $A = 58$?

Solution First solve Eq. (34.17) for the number of neutrons N in the nucleus.

$$N = A - Z$$

Now substitute values for A and Z for each of the elements. First, for iron, $A = 59$ and $Z = 26$, so

$$\begin{aligned} N &= 59 - 26 \\ &= 33 \end{aligned}$$

Now, for nickel, $A = 58$ and $Z = 28$, so

$$\begin{aligned} N &= 58 - 28 \\ &= 30 \end{aligned}$$

Therefore the nucleus of the iron atom contains 33 neutrons and the nucleus of the nickel atom contains 30 neutrons. The nucleus of the iron atom contains the larger number of neutrons. *answer*

Nuclear Forces or Interactions Scientists were very early concerned with what holds the atomic nucleus together. Protons, for example, all being positively charged, repel each other. Moreover, they would not be attracted to neutrons *electrically* (the *Coulomb force*), since neutrons have no charge. For the same reason, neutrons would not attract each other. *Gravitational* attractive forces could not possibly hold the nucleus together, since they are extremely weak compared to the electrical forces of repulsion tending to force the protons apart. In order to account for stable nuclei, it has been necessary to resort once again to an hypothesis: The nucleons in the nucleus are attracted to one another by a *nuclear force*, the exact nature of which is still unknown.

This attractive nuclear force is usually called the *strong nuclear force*, or more simply, the *strong force*. Sometimes it is called the *meson force*. [There is also another nuclear force called the *weak nuclear force*. Although this *weak force*, as it is called, is one of the four fundamental forces in nature (the gravitational force, the Coulomb force, the strong nuclear force, and the weak nuclear force), it will not be important for our discussions in this chapter.] The strong force is a very short-range force. That is, two particles must be very close together in order to be influenced by the strong force. For example, two adjacent protons in the nucleus experience an attractive, strong nuclear force whose strength is great enough to overcome the very strong repulsive Coulomb (electrical) force that each of them experiences. If the protons are sufficiently close, the attractive nuclear force can be 100 times as strong as the Coulomb repulsive force. However, the Coulomb force is a much longer-range force than the strong nuclear force. Because of this, it is possible for the distance between two protons in the same nucleus to be great enough so that the repulsive Coulomb force is stronger than the attractive strong nuclear force between those two protons. Each proton in a nucleus affects only its nearest neighbors with the strong nuclear force, but every proton in the nucleus experiences a Coulomb repulsive force due to every other proton in the nucleus.

The *strong* and *weak* nuclear forces.

Since the neutron is not a charged particle, no Coulomb forces act between a proton and a neutron, or between two neutrons. However, the attractive strong nuclear force has the same character between *any pair of nucleons*—between two protons, two neutrons, or a proton and a neutron. However, a pair of neutrons or a pair of protons does not exist in isolation. Except for hydrogen, every atomic nucleus contains both protons and neutrons.

Here is a summary of the four fundamental forces or interactions believed to be responsible for all the forces observed in nature.

Force or interaction	*Relative strength (no units)*	*Particles affected*
Strong nuclear force	1	Nucleons (neutrons and protons)
Coulomb force	10^{-2}	All electrically charged particles
Weak nuclear force	10^{-13}	Protons, neutrons, electrons, neutrinos, and others
Gravitational force	10^{-43}	All particles that have mass

34.11 ■ Isotopes

Most elements have been found to exist in nature with atoms of more than one atomic mass. The *atomic number* Z of all atoms of the same element is the same, but the *listed atomic mass* of most elements is merely the average of the masses of the several types of atoms possessing the same atomic number. Since the atomic number specifies the number of protons in the nucleus (and also the number of electrons in the orbits), it must be the number of *neutrons* in the nucleus which varies, in order to allow for atoms of the same element to have different atomic masses. Such atoms, called *isotopes*, have the same *chemical* properties as the ''ordinary'' atom, but they differ in mass and in nuclear structure. ''Ordinary'' uranium, for example, has a mass number of 238 and is designated ^{238}U. One isotope of uranium is ^{235}U.

Not all atoms of the same element have the same mass.

Definition of *isotopes*.

In order to distinguish between isotopes of the same element, it is conventional to write the symbol of the element, as C (carbon), with a subscript in front which is its *atomic number* Z and a superscript in front, which is its *mass number* A. The notation ${}^{12}_{6}C$, for example, designates "natural" carbon whose nucleus contains six protons ($Z = 6$) and six neutrons for a mass number of $12(6 + 6 = 12)$. The notation ${}^{14}_{6}C$ indicates an isotope of carbon with eight neutrons in the nucleus for an atomic mass of 14. In like manner, ${}^{238}_{92}U$ is "natural" uranium, and ${}^{235}_{92}U$ is an isotope of uranium that contains three fewer neutrons.

34.12 ■ Atomic Mass Unit

The reference standard for atomic mass units is the "natural" carbon atom ${}^{12}_{6}C$. One-twelfth of the mass of the natural carbon atom is defined as 1.0000 *atomic mass unit* (u). The *actual mass* of 1u is

$$1\text{u} = 1.6605 \times 10^{-27}\ \text{kg}.$$

Table 34.2 gives some atomic data on a few of the lighter elements.

The three major components of the atom, the *electron*, the *proton*, and the *neutron*, have the following rest masses and charges:

	Rest mass		Electric charge
Electron	$m_e = 0.000549$u	$=9.109 \times 10^{-31}$ kg	$^{-}e = 1.602 \times 10^{19}$ C
Proton	$m_p = 1.007277$u	$=1.6725 \times 10^{-27}$ kg	$^{+}e = 1.602 \times 10^{19}$ C
Neutron	$m_n = 1.008665$u	$=1.6748 \times 10^{-27}$ kg	Zero

Table 34.2 Atomic data on isotopes of some light elements

Isotope	Atomic Mass, u	Mass Number A	Atomic Number Z	Neutron Number N
${}^{1}_{1}H$	1.007825	1	1	0
${}^{2}_{1}H$	2.014102	2	1	1
${}^{3}_{2}H$	3.016030	3	2	1
${}^{4}_{2}H$	4.002604	4	2	2
${}^{6}_{3}Li$	6.015126	6	3	3
${}^{10}_{5}B$	10.012939	10	5	5
${}^{12}_{6}C$	12.00000 (defined as the standard reference)	12	6	6
${}^{14}_{7}N$	14.003074	14	7	7
${}^{16}_{8}O$	15.994915	16	8	8

NOTE: ${}^{2}_{1}H$ is called deuterium.

34.13 ■ Atomic Disintegration—Natural Radioactivity

Early in the development of nuclear physics, it was discovered that some extremely heavy atoms, e.g., *radium* and *uranium*, have relatively *unstable* nuclei and that certain of these atoms undergo natural disintegration. The disintegration results in lighter nuclei and energy liberated in the form of radiations. These radiations were studied, their energies were determined, and they were named alpha (α), beta (β), and gamma (γ), for the first three letters of the Greek alphabet.

Alpha, beta, and gamma radiation.

Alpha particles are made up of two protons and two neutrons and are identical to helium nuclei. They are emitted from radioactive nuclei at speeds of about 10^7 m/s with typical energies of several million electron-volts (MeV). Alpha particles have relatively low penetrating power. They can travel only about 1 cm in air for each MeV of energy and they are easily stopped by a very thin metal foil. However, their double charge gives them high *ionizing power*. A neutral atom becomes a positive ion if it loses one of its electrons. One way an atom can lose an electron is if that electron experiences a sufficiently strong electric (Coulomb) force due to some particle external to the atom. Alpha particles are more effective at causing the required electric force than particles with a single unit charge because the electric force between the external particle and the electron is proportional to the product of the charges of the interacting particles [see Coulomb's law, Eq. (25.1)].

Beta particles are electrons with speeds very nearly equal to that of light. Being of such small mass, their KE is ordinarily less than that of alpha particles. They can penetrate about 1 mm of lead. Since it is known that there are no electrons in the nucleus, the question as to the source of beta radiation is a puzzling one. At present it is believed that a neutron initiates beta radiation by changing to a proton. The proton stays in the nucleus, and an electron (beta particle) and a new particle called a *neutrino* are simultaneously ejected.

Gamma rays are highly penetrating electromagnetic waves usually of wavelength shorter than 10^{-10}m (0.1 nm). They can penetrate several centimeters of lead, and are very dangerous to humans in uncontrolled situations. Like x-rays, however, they are used for both diagnosis and therapeutic treatment in medicine. They are also used in industry.

Gamma rays can penetrate matter more deeply than beta particles, and beta particles can penetrate more deeply than alpha particles.

Medical and Industrial Uses of Radioisotopes If trace amounts of radioisotopes are injected into the bloodstream, detectors (see Sec. 34.14) can be used to measure the intensity of radioactive particles emitted from different parts of the body to obtain information about blood flow. Two computerized medical imaging systems use this approach to produce images that provide information about blood flow in the body and about the distribution of blood in tissues. One of these systems is called *single-photon-emission computed tomography* (SPECT), and the other is *positron-emission tomography* (PET).

In industry the flow of chemicals, gas, or oil through pipes is checked by introducing a radioactive isotope in the fluid and checking the flow with a Geiger counter. Gamma radiation has been used in petroleum exploration to locate oil-bearing strata deep under the surface of the earth. Cobalt 60 is extensively used for making industrial radiographs.

Radioisotopes are used in many automated devices for the control of industrial processes, such as thickness control, tests of compaction, and density control.

Half-life A characteristic of radioactive elements and isotopes is that each has a particular period of time during which half of any given initial amount will undergo radioactive decay. The *half-life* of a radioactive element or isotope is the time required for the number of radioactive nuclei in the sample to decrease to one-half the original number. Half-lives can be determined by measuring how the number of radioactive particles emitted per second decreases with time (the *rate of decay*) with a suitable radiation counter, such as a *Geiger counter* (see Sec. 34.14). The half-life of $^{238}_{92}$U (natural uranium) is about 4.5 billion years. That of $^{224}_{88}$Ra (an isotope of radium), 3.64 days; of $^{60}_{27}$Co (cobalt 60), commonly used in industry as a gamma-radiation source), 5.25 years; and $^{14}_{6}$C (carbon 14), about 5700 years. Half-lives of isotopes are often used for dating events of the past. Carbon 14 dating is very commonly used with objects or artifacts containing carbon (all matter that was once plant or animal), to establish the time of historic or prehistoric events.

Although the number of particles emitted by a radioactive isotope always decreases as time passes, for some isotopes the emission process continues for billions of years.

Illustrative Problem 34.12 One of the isotopes of silver ($^{109}_{47}$Ag) becomes the radioactive isotope $^{110}_{47}$Ag when a sample of silver is bombarded with neutrons. The half-life of the activated isotope is 24 s. If there are 4.0×10^6 radioactive nuclei in the sample initially, how many will there be at the end of (*a*) 48 s, (*b*) 72 s?

Solution During each 24-s period (the half-life) the number of radioactive nuclei present decreases to one-half the number present at the beginning of the 24-s period. This idea can be used to construct the table below

Elapsed time, s	*Number of half-lives elapsed*	*Number of radioactive nuclei present*
0	0	4.0×10^6
24	1	2.0×10^6
48	2	1.0×10^6
72	3	5.0×10^5

(***a***) From the above table, at the end of 48 s, 1.0×10^6 radioactive nuclei remain in the sample. *answer*

(***b***) At the end of 72 s, 5.0×10^5 radioactive nuclei remain in the sample. *answer*

Illustrative Problem 34.13 A certain radioactive sample has a half-life of 2.00 years. Initially there are 2.0×10^8 radioactive nuclei present in the sample. After approximately how many years will the number of radioactive nuclei in the sample be reduced by a factor of 20, to 10^7?

Solution This problem can also be solved by constructing a table like that in Illustrative Problem 34.12:

Elapsed time, years	*Number of half-lives elapsed*	*Number of radioactive nuclei present*
0	0	2.0×10^8
2	1	1.0×10^8
4	2	5.0×10^7
6	3	2.5×10^7
8	4	1.25×10^7
10	5	6.25×10^6

From the above table, sometime between the passage of 8 and 10 years, the number of radioactive nuclei in the sample would be reduced to 10^7. *answer*

34.14 ■ Radiation Detectors

Although the particles and photons emitted by radioactive sources typically have far more energy than photons of visible light, human senses usually cannot detect the presence of such radiation. Medical or dental x-rays, for example, produce no noticeable sensations. Therefore, some sort of radiation detector must be used to indicate the presence of potentially harmful radioactive particles. Radiation detectors are also essential components in some medical imaging instruments. A brief discussion follows of three different types of detectors capable of registering the presence of radioactive particles.

The Geiger Counter One of the oldest and simplest radiation detectors is the Geiger counter, diagramed schematically in Fig. 34.14*a*. The Geiger tube is made up of an outer metal cylindrical electrode and a central thin wire. The ends of the

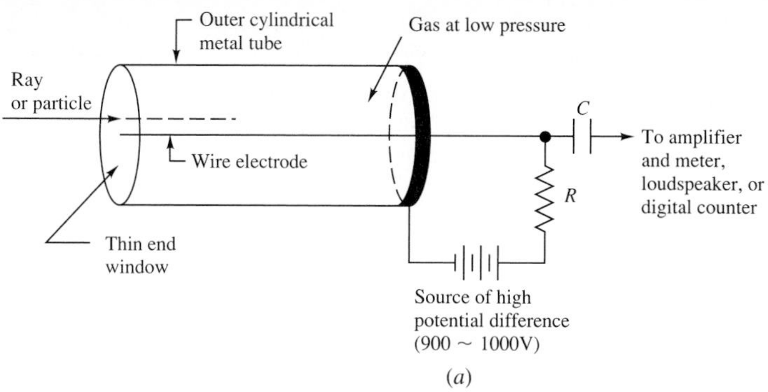

(b)

Fig. 34.14 (*a*) Diagram of the operation of a Geiger counter. (*b*) A Geiger counter being used to indicate plant absorption of a fertilizer that was "tagged" with small amounts of a "tracer" radioactive substance.

cylindrical assembly are closed and the interior is filled with gas (for example, argon plus a little hydrogen) at low pressure. One end is closed with a very thin window that allows radioactive particles to pass through it.

The Geiger tube was invented by the German physicist Hans Geiger (1882-1945), who was Rutherford's assistant for the Rutherford scattering experiment.

Approximately 1000 V is connected between the central wire and the outer cylindrical electrode with a series resistor. When a radioactive particle or ray enters the tube, it ionizes one or more of the gas atoms. The positive ions and free electrons, in the presence of the electric field between the electrodes, accelerate rapidly and produce additional positive ions and free electrons. The result is a brief pulse of current, with the electrons moving toward the central wire electrode, and the positive ions moving toward the outer electrode. This brief pulse of current continues in the external circuit and produces a voltage pulse across the series resistor. This voltage pulse can be amplified and used to increase by one the numerical value of the digital display on an electronic counter to keep track of the number of entering particles.

The Geiger counter, which is best at counting beta particles (electrons), has some disadvantages. If the intensity of the incoming radiation is too great, not all the particles will be counted. For example, a particle will not be counted if it enters the Geiger tube before the current pulse produced by the previous particle has ended. Another disadvantage is that the Geiger counter is not capable of providing information about the energy of the incoming particles. Fig. 34.14*b* shows a Geiger counter in use.

A Geiger counter is a good detector of beta particles, but it can also detect gamma rays.

The Scintillation Counter A scintillation detector is made up of a photomultiplier tube with a scintillation crystal cemented to its light-sensitive face. The word *scintillate* means to give off sparks or to flash. As discussed below, radioactive particles that enter the scintillation crystal cause the production of flashes (photons) of light as they lose energy while moving through the crystal. A photomultiplier tube is a multielement vacuum tube with a photosensitive (light-sensitive) face called a *photocathode* that emits electrons by the photoelectric effect (see Sec. 34.3) when incident photons strike it. As Fig. 34.15 shows, the scintillation detector is noticeably more complex than the simple two-element Geiger tube. A scintillation detector is particularly useful for detecting gamma rays.

Scintillation counters are often used in crime laboratories to identify sources of materials (by energy analysis of gamma rays) from samples made radioactive by bombardment with neutrons.

When a gamma ray enters the scintillation crystal, it loses its energy in a rapid sequence of photoelectric-effect and Compton-scattering events. The lost energy appears in the form of photons that strike the photocathode causing the ejection of photoelectrons that are then attracted to the nearest element in the photomultiplier tube, which is called a *dynode*. When an electron strikes the surface of the first dynode, two or more electrons are ejected, and these electrons are attracted to a second dynode which is kept at a higher voltage than the first dynode. Each electron that strikes the second dynode again causes the ejection of two or more electrons.

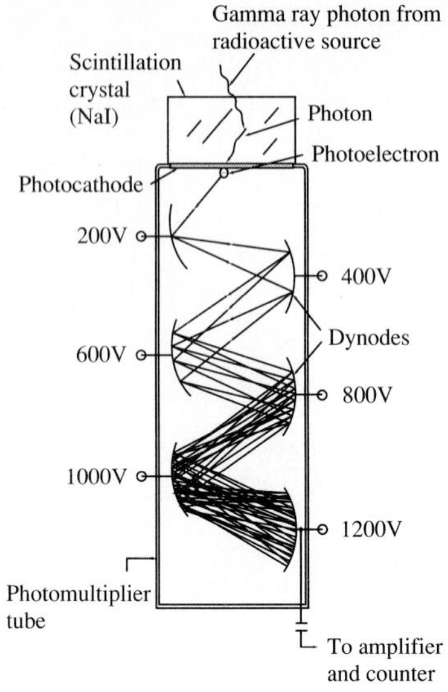

Fig. 34.15 Diagram of a scintillation detector, consisting of a scintillation crystal and a photomultiplier tube.

This electron multiplication process continues for a total of 10 or more dynodes. A single electron striking the first dynode may cause a million electrons to strike the tenth dynode. The output signal from the photomultiplier tube is proportional to the energy of the gamma ray that entered the scintillation crystal. Usually a fairly complex signal-processing system called a pulse-height analyzer is connected to the output of the photomultiplier tube and a computer is used to display information about the abundance and energies of particles in the incident radiation.

The Semiconductor Detector A reverse-biased *P-N* junction device (see Sec. 33.5) can be used as a detector of radioactive particles (see Fig. 34.16). When a radioactive particle produces an electron-hole pair in the depletion region, the electric field in the depletion region causes those free charges to experience forces that result in a current across the junction. That current, which continues in the external circuit, produces a voltage pulse across the series resistor R shown in Fig. 34.16. That voltage pulse can be amplified and used to cause an electronic counter to register the presence of the radioactive particle. Like scintillation detectors, semiconductor detectors can provide information about the energy of the incident radioactive particles.

34.15 ■ Units of Radiation From Radioactive Sources

The *quantity of radioactive material present* in a source is determined by the rate at which atoms are disintegrating. The more atoms that disintegrate per unit time, the greater the *activity* of the source. The amount of radioactive material that undergoes 3.7×10^{10} disintegrations (events) per second (eps) is called the *curie* (Ci), in honor of Marie and Pierre Curie, who first identified the element, *radium*.

$$\begin{aligned} 1 \text{ curie} &= 3.7 \times 10^{10} \text{ eps} \\ &= 1000 \text{ millicuries (mCi)} \\ &= 1{,}000{,}000 \text{ microcuries } (\mu\text{Ci}) \end{aligned}$$

The French physicist Antoine Henri Becquerel (1852-1908) discovered the radioactivity of uranium in 1896.

The SI unit for activity is named after the French physicist Henri Becquerel:

$$\begin{aligned} 1 \text{ becquerel} &= 1\text{Bq} = 1 \text{ event/s (eps)} \\ 1 \text{ curie} &= 3.7 \times 10^{10}\text{Bq} = 37\text{GBq (gigabecquerel)} \end{aligned}$$

One kilogram of ordinary potassium has an activity of approximately 1 mCi because it contains 0.01 percent of a radioactive isotope of ^{40}K.

The activities of sources used for most applications are usually in the megabecquerel ($1\text{MBq} = 10^6\text{Bq}$) or the gigabecquerel ($1\text{GBq} = 10^9\text{Bq}$) range. The curie is an extremely large unit of radioactivity, approximately equivalent to the radioactivity of 1 g of pure radium. Only in industrial radiography and in cancer therapy are sources of multicurie strength used. Most sources used as tracers in industry and medicine, or for laboratory work in colleges are in the millicurie or microcurie range.

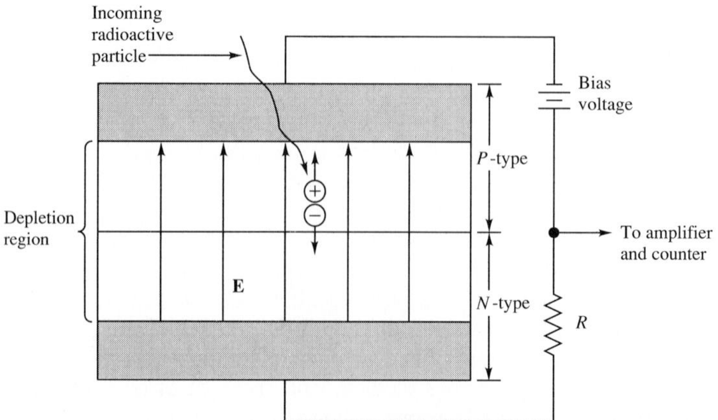

Fig. 34.16 A *P-N* junction semiconductor radiation detector. There are some similarities between the operation of this detector and the operation of a photovoltaic solar cell (PVC) (see Sec. 33.6).

Units of Absorbed Radiation—Damaging Effects Only the radiation that is *absorbed* by living cells damages them. Radiations which *pass through* cells do no damage. Absorbed radiation results in *ionization,* and this ionization in turn promotes cell damage and bone marrow damage and initiates changes in body chemistry which all too frequently are irreversible and do not respond to treatment.

The *rad* and the *gray* defined.

The *absorbed dose* of radiation can be measured in terms of the quantity of energy absorbed. For example, a unit called a *rad* (*r*adiation *a*bsorbed *d*ose) is used to represent the quantity of radiation that has the same biological effect as the absorption of 0.01 J of x-rays or gamma rays by 1 kg of body tissue. The SI unit of absorbed dose, called the *gray* (Gy), is equal to 100 rad.

The safe limit for persons exposed to low-level radiation of x-rays or gamma rays over the entire body is thought to be about 0.05 rad/week. The body can repair and keep ahead of cell damage which occurs from doses below this rate. Acute radiation exposure for short intervals of time is another and far more serious matter. Exposure over the whole body of a 20- to 50-rad dose will cause blood-cell damage; 100 to 200 rad will cause severe illness from which most persons might recover; 400 rad is fatal to 50 percent of persons exposed; and 500 rad is fatal to nearly all persons. The effects of radiation are ordinarily cumulative, so small doses are also dangerous if they are repeated daily or weekly.

The *rem* defined.

Since the various kinds of radiation (alpha and beta particles and gamma rays, protons and neutrons) are absorbed differently, and cause different amounts of biological damage, the *dose equivalent* is measured with a unit called the *rem* (*r*oentgen *e*quivalent *m*an), which takes into account the *relative biological effectiveness* (RBE) of the absorbed radiation. The *rem* and the *rad* are related as follows:

$$\text{rem} = \text{RBE} \times \text{rad} \qquad (34.18)$$

The RBE is rated as 1 (unity), for x-rays and gamma rays of 1 MeV. The *rad* and the *rem* are equivalent for these conditions. Other rays and high-energy particles do more biological damage and have RBEs approximately as follows: fast neutrons (and protons), 10; 1-MeV alpha particle, 20; beta particle, 1.0 to 1.7.

In the U.S., the average annual radiation dose due to unavoidable (natural) radiation present in the atmosphere and earth's crust, is 100 millirem (mrem). The radiation dose associated with a typical chest x-ray is 20 mrem. The suggested maximum permissible dose equivalent for occupational exposure is 5 rems in any one year (P).* The SI unit for dose equivalent, called the *sievert* (Sv), is equal to 100 rem.

Table 34.3 Summary of radiation and dose units

Quantity	*Customary unit*	*SI unit*	*Conversion*
Activity	curie (Ci)	becquerel (Bq)	$1\text{Ci} = 3.7 \times 10^{10}$decays/s $= 3.7 \times 10^{10}$Bq
Absorbed dose	rad	gray (Gy)	$1 \text{ rad} = 10^{-2}$J/kg $= 10^{-2}$Gy
Dose equivalent	rem	sievert (Sv)	1 Sv = 100 rem

Table 34.4 Summary of relative biological effectiveness (RBE) values for different types of radiation

Radiation type	*RBE*
Gamma rays, 1 MeV	1.0
Gamma rays > 4 MeV	0.7
X-rays	1.0
Beta particles > 30 keV	1.0
Beta particles < 30 keV	1.7
Fast neutrons (and protons)	10
Alpha particles	20

Illustrative Problem 34.14 A worker in a plant that processes radioactive materials is exposed to 0.015 rad of alpha radiation. How many rems is this? How many sieverts?

Solution From Eq. (34.18), rem = RBE × rad. The RBE for alpha-particle radiation is 20. So, for an absorbed dose of 0.015 rad,

* Preliminary estimates of the absorbed dose in the vicinity of the recent Chernobyl atomic power plant disaster in Russia were that as many as 20,000 people probably absorbed radiation in amounts up to 45 rem—the recommended maximum permissible dose for a *period of nine years*—over a period of only a few days.

$$\begin{aligned} \text{rem} &= 20 \times 0.015 \text{ rad} \\ &= 0.30 \text{ rem} \\ &= 300 \text{ millirem} \qquad \textit{answer} \\ &= .30 \text{ rem} \frac{1 \text{ Sv}}{100 \text{ rem}} \\ &= 3.0 \times 10^{-3} \text{ Sv} \qquad \textit{answer} \end{aligned}$$

SOME APPLICATIONS OF MODERN PHYSICS

This chapter closes with a discussion of several applications of some principles of modern physics.

34.16 ■ The Electron Microscope

Table 34.5 Sizes of selected micro-objects

Chromosome, inside a cell	10^{-6} m
Small bacterium	2×10^{-7}
Virus	10^{-8}
Atom diameter	1 to 2×10^{-10}
Bohr radius, hydrogen atom	5.292×10^{-11}
de Broglie wavelength of a 1-MeV electron	1.23×10^{-12}
Diameter, atomic nucleus	10^{-14}

Optical microscopes are useful for viewing and magnifying only those objects whose dimensions are greater than about 10^{-6} m (1000 nm). Below this object size, the wavelength of visible light begins to become comparable to the size of the object being viewed. (The visible spectrum ranges from about 400 to about 700 nm.) The problem of *diffraction* then enters in, and the image becomes so fuzzy as to be useless. In optical terms we say that the *resolution* is unsatisfactory.

Many objects that we need to "see" and magnify for the purposes of modern technology, are much too small to be viewed and magnified by light. Some examples, beginning at about the limit of satisfactory resolution of an optical microscope and getting smaller, are listed in Table 34.5.

Since light is useless for viewing objects smaller than its own wavelength, another source, of shorter wavelength, has to be used. The electron itself can be this source.

The de Broglie hypothesis predicts that the electron has an associated wavelength which depends on its velocity—the higher the velocity, the shorter the wavelength.

$$\lambda = \frac{h}{mv} \tag{34.10}$$

Using a high voltage to accelerate electrons, one can obtain an electron beam which will give good resolution (sharp definition) of objects of molecular and atomic size. The electron microscope makes use of the *wave nature* of the electron, another example of the *wave-particle duality* concept discussed earlier. An example of how the accelerating voltage and associated wavelength are related follows.

Illustrative Problem 34.15 An electron beam is given an accelerating voltage of 10,000 V. What is the associated wavelength of the electrons?

Solution Use Eq. (34.15), relating KE to the accelerating voltage.

$$Ve = \tfrac{1}{2}mv^2 \qquad \text{and} \qquad v = \sqrt{\frac{2Ve}{m}}$$

Substituting,

$$\begin{aligned} v &= \sqrt{\frac{2 \times 1.602 \times 10^{-19} \text{ C} \times 10^4 \text{ V}}{9.11 \times 10^{-31} \text{ kg}}} \\ &= 5.93 \times 10^7 \text{ m/s} \end{aligned}$$

Now use the de Broglie wave equation (34.10):

$$\begin{aligned} \lambda = \frac{h}{mv} &= \frac{6.626 \times 10^{-34} \text{ J} \cdot \text{s}}{9.11 \times 10^{-31} \text{ kg} \times 5.93 \times 10^7 \text{ m/s}} \\ &= 0.123 \times 10^{-10} \text{ m} = 0.0123 \text{ nm} \qquad \textit{answer} \end{aligned}$$

Electron beams can be focused by electrostatic and/or magnetic "lenses" (*electron optics*) in a manner analogous to the use of lenses to refract (bend) light rays and bring them to a focus. The electron microscope brings its "image" to a focus on a fluorescent screen, where it can be viewed or photographed. The electrostatic or magnetic "lenses" are charged plates or coils whose field lines interact with the charged electrons, producing the bending needed to bring them to a focus. The electron stream itself is supplied by an electron gun, like that used in a cathode-ray tube.

Magnifications of greater than 10^6 have been achieved with electron microscopes.

In a *transmission* electron microscope, the "object" must be in the form of a thin section, so the electron beam will pass through it. Tissues are often embedded in plastic with the slice being less than 20 nm in thickness.

In a *scanning* electron microscope or SEM, a sharply focused electron beam scans back and forth across the sample (object) to be examined, much like a television-tube "sweep" or scan. This type of instrument was first developed in the late 1960s and is used to study surface details of objects such as blood cells. The great detail and the depth of field possible with the SEM gives an almost three-dimensional quality to the image.

The electron microscope has contributed important new knowledge in such fields as atomic structure of metals, molecular biology, studies of the human cell, and research on viruses, cancer, and leukemia over the past three decades. Precautions must be taken by the operators of these instruments. High-speed electrons often produce x-rays, and operating personnel should be adequately protected.

34.17 ■ The Laser

A laser is a device that emits a concentrated, narrow beam of light. The word *laser* is an acronym made up of the first letters of the words, *l*ight *a*mplification by *s*timulated *e*mission of *r*adiation. Lasers come in several varieties—ruby lasers, helium-neon lasers, etc. They can generally be classified as either a *solid, liquid,* or a *gas* laser, depending on the nature of the material used for the production of light. An introductory discussion of lasers was presented in Chap. 22. A more detailed and quantum-related explanation of laser operation follows.

The first successful laser was developed in 1960 by the American physicist Theodore H. Maiman. The intensity (power per unit area) of a 1 milliwatt laser beam is approximately the same as the intensity of bright sunlight. *Never look down a laser beam!* Serious retinal damage could result.

Solid lasers are usually *pulsed* lasers. They are operated to emit bursts of energy with a pulse time of (typically) one nanosecond (10^{-9} s) and may have power ratings of up to 100 trillion (10^{14}) watts. Gas lasers, on the other hand, emit a *continuous* beam of light energy. Helium-neon gas lasers, having a power rating in the milliwatt range, are commonly used for classroom demonstrations and experiments. Carbon dioxide industrial gas lasers have power ratings as great as 10 kW.

Properties of Laser Light A laser produces a *monochromatic*, or single frequency (wavelength), source of light. Ordinary light sources, on the other hand, emit photons of many different wavelengths. Laser light is *coherent* light. That is, all of the light waves are in a fixed phase relationship everywhere in the beam, and the phase remains steady over time. By way of contrast, an incandescent lamp produces random-phase *incoherent* light. Beams of ordinary light diverge, causing the intensity of the light beam to decrease with distance from the source. Coherent laser light results in a collimated narrow beam of great intensity, capable of traveling to the moon and beyond without undue divergence.

Because a laser beam is highly concentrated, safety precautions must be taken. *Never look down a laser beam!* A low-power, 1-mW laser can cause a retinal burn in a few seconds.

Quantum Explanation of Stimulated Emission Ordinary atoms have their electrons in stable orbits in "ground" energy states. An atom that has been "excited" for any reason has electrons in higher energy states. It may be about ready to emit light (photons). Electrons in higher energy states are not very stable and usually return to the $n = 1$ ground state, the most stable state, in a very short time, of the

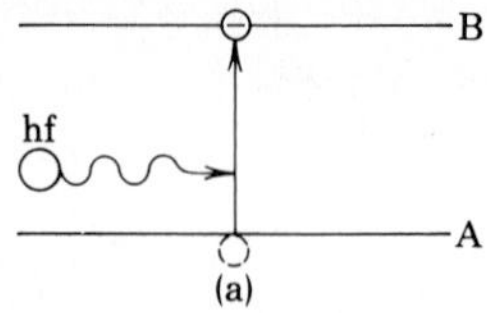

Absorption of a photon of energy $E=hf$ elevates the electron to a higher energy state.

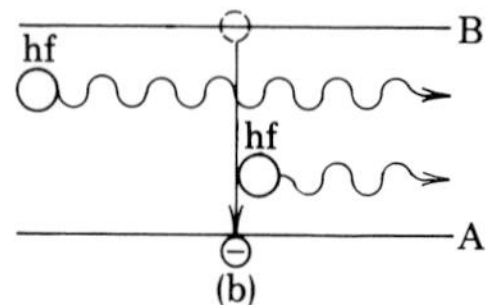

Stimulated emission. Photon of energy $E = hf$ causes electron in higher energy state to emit a photon of the same frequency and phase, and "jump" to a lower energy state.

Fig. 34.17 Schematic diagram of electron ''stimulation'' by a passing photon. (*a*) Photon is absorbed and electron moves to a higher energy state. (*b*) Photon stimulates electron in higher energy state to emit photon and ''jump'' to ground state.

order of 10^{-8} s. No prodding from an outside source is needed; they spontaneously jump back to the ground state, losing energy and giving off a photon in the process. Since many excited states are possible, many wavelengths result from electron jumps. The jumps do not ordinarily occur at the same time, so *spontaneous* emission results in multifrequency, noncoherent light emitted in all directions.

In *stimulated* emission, electrons change energy states due to ''stimulation'' from photons passing nearby. This can occur in two ways, as illustrated in Fig. 34.17. Consider a photon of energy *hf* (equal to the difference in energy between states *A* and *B*), passing close by the electron in Fig. 34.17*a*. If the electron is in the lower energy level *A*, *absorption* of the photon's energy results, and the electron moves to the higher energy state, *B*. If the electron is in the higher energy state *B*, the photon can stimulate it to emit a photon (Fig. 34.17*b*) and return to the lower state, *A*. This stimulation process has the unique characteristic of creating an additional photon, identical with the stimulating photon, and ''in step'' with it. The process is roughly analogous to a bystander ''falling in,'' in step with a marching band as it passes by. When the two photons in Fig. 34.17*b* encounter two more electrons in energy state *B*, four photons will become available. This chain reaction process quickly produces great amplification of light and a very intense beam.

Population Inversion and Metastable States Two conditions are necessary for the stimulated emission reaction to continue. First, there must be more electrons in level *B* than in level *A*, otherwise the photons will stimulate more absorptions than emissions. It happens, however, that normally there are more electrons in the lower energy state *A* than in state *B*, so a *population inversion* is essential. The second condition is that level *B* be semistable, or *metastable*. Electrons in a metastable condition do not immediately emit radiation spontaneously, but remain stable long enough (10^{-7} s or longer) for the stimulated emission process to build up the number of monochromatic (i.e. same wavelength) coherent (i.e. same phase) photons.

Population inversion in a ruby laser is accomplished by ''optical pumping'' with a bright flash of 550-nm yellow-green light. Chromium atoms in the ruby absorb the 550-nm photons and go to an ''excited'' state. From there they quickly move to a metastable state, in which condition, photons of 694.3-nm (red) will stimulate them to emit photons and return to ground state. Stimulated emission of photons is possible in the ruby laser because the chromium in the ruby has a well-defined metastable state.

In a helium-neon gas laser, helium is first excited by an electrical discharge to a metastable energy state 20.61 eV above its ground state. Neon happens to have a metastable energy state of 20.66 eV, very nearly equal to that of helium. The helium atoms transfer energy to neon atoms by collision, moving neon atoms to metastable state *B* (Fig. 34.18). Neon emits 632.8nm photons as stimulated emission, causing electron jumps from *B* to *A*. The population inversion is maintained between levels *B* and *A*, because state *A* contains few electrons to begin with (it is not the ground state), and spontaneous emission between *A* and ground occurs rapidly.

Fig. 34.18 Diagram illustrating metastable states and population inversion in a helium-neon gas laser.

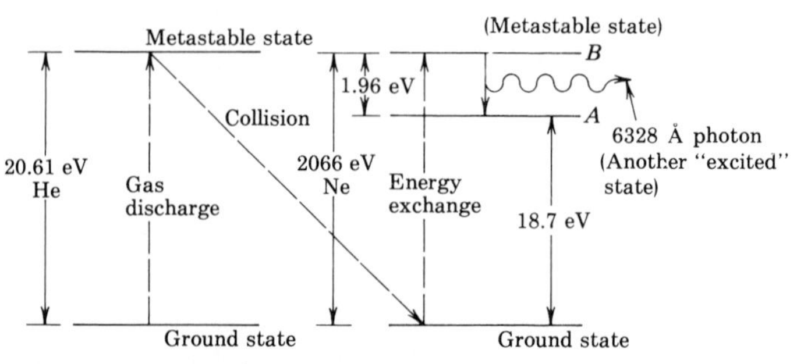

Helium atom Neon atom

Despite the high intensity of a laser beam, the efficiency of some lasers is rather low. The percent of energy input to a laser that finds its way into the coherent beam ranges from less than 0.01 percent to greater than 50 percent. Consequently, in order to accomplish some of the tasks for which lasers are suited (welding, cutting, initiating nuclear fusion, etc.) large amounts of energy must be supplied to the laser apparatus.

One example of the medical/surgical use of lasers is illustrated in Fig. 34.19. They are also commonly used in "welding" damaged or torn retinal tissue in eye surgery. Fig. 34.20 illustrates an important industrial application.

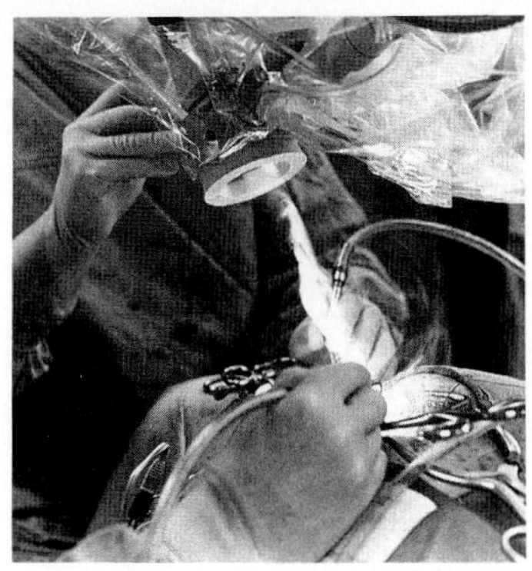

Fig. 34.19 High-power carbon-dioxide laser being used to vaporize a spinal cord tumor. (Printed with permission of St. Mary's Hospital, Milwaukee.)

34.18 ■ Low-Temperature Physics

At absolute zero, according to the original kinetic-molecular hypothesis, molecular motion ceases. But according to quantum theory, some energy is left in the particles making up the atoms of the molecule. Evidently the amount remaining is less than that discrete amount which can be radiated, the quantum itself. Physicists and technicians in low-temperature research are discovering many strange and important facts about molecules, atoms, and atomic particles at or near this zero-energy boundary.

As discussed in section 26.12, one of the important areas of application of low-temperature physics involves superconducting materials—materials that have zero electrical resistance when cooled to a sufficiently low temperature.

The combined application of the principles of quantum physics and superconductivity technology are responsible for the development of a device called a SQUID (*s*uperconducting *qu*antum *i*nterference *d*evice). SQUIDs are capable of measuring magnetic fields as weak as 10^{-15} tesla (T). To get some feeling for how weak such a magnetic field is, consider the following example. The strength of the magnetic field associated with a long straight wire carrying a current of 1 mA is 10^{-15} T at a location more than 124 mi away from the wire. One of the medical applications for SQUIDs in the field of neurobiology is the measurement of the extremely weak magnetic fields associated with the brain.

In 1987 a period of rapid progress began in the development of materials that become superconductors at significantly higher temperatures than the very low temperatures (<25 K) required earlier. (See Sec. 26.12.)

Quantum Tunneling From probability theory and quantum mechanics theory, a bound particle needing energy E_o to escape from a system through or over a boundary has a small but finite probability of escaping even though its energy is less than E_o. This effect is called *tunneling* (it "tunnels" through the boundary). Quantum tunneling explains alpha particle emission in radioactive decay and it explains the properties of a semiconductor device known as a tunnel diode that is used in low-power microwave applications.

Quantum tunneling also explains the operation of the *scanning tunneling microscope* (STM) which is capable of distinguishing features on the surface of a sample of material separated by 0.1 nm (one atomic diameter). The operation of the STM involves measurement of a quantum tunneling current of electrons between the surface being studied and an extremely sharp metallic needle point located a very short distance above the surface. By moving the needle around the surface, an image of the surface can be generated on a cathode-ray tube that shows how the atoms are arranged on the surface.

The first scanning tunneling microscope (STM) was constructed in 1981 by Gerd Binnig and Heinrich Rohrer in Zurich Switzerland.

QUESTIONS AND EXERCISES

1. Express Planck's hypothesis in your own words and then show how the SI-metric units of Planck's constant are obtained.

2. Use a simple diagram and explain the *photoelectric effect*. What was Einstein's hypothesis with respect to the photoelectric effect? What is the meaning of the *work function* ϕ?

3. Why is it not possible to compute the wavelength of a photon of white light?

4. Which has more energy, a photon of blue light or a photon of red light?

5. There are some properties of particles that are associated with both electrons and photons, and some that are not. Discuss.

6. Explain the *Compton effect* and explain how it supports the photon theory for light.

7. Describe one way in which the photoelectric effect and the Compton effect are the same, and one way in which they are different.

8. Summarize the half-dozen or so essential findings of both classical and modern physics that make it necessary, at least for the present, to accept the concept of wave-particle duality in explaining electromagnetic radiation in general and light in particular.

9. What information about the structure of atoms was obtained from analysis of data from the Rutherford scattering experiment?

10. A proton and an electron have identical de Broglie wavelengths. Which particle has the greater speed?

11. If, in observing the spectrum of an incandescent gaseous element, you detected *bright lines*, how would you explain this in terms of the electron behavior in the energy levels or orbits of the atom?

12. The visible spectrum extends from 400 to 700 nm. How many different possible wavelengths in this spectral region are associated with the photons that can be emitted by a hydrogen atom when its electron moves from a larger to a smaller orbit?

13. How can it be for a hydrogen atom that the total energy of an electron increases when it moves to a larger orbit, even though the kinetic energy of the electron decreases when the electron moves to a larger orbit?

14. What determines the wavelength of the shortest waves produced in an x-ray tube? How are the braking-radiation x-rays different from characteristic x-rays?

15. An electron microscope is capable of magnifying objects smaller than the wavelength of visible light. Would it be possible to make a proton microscope? Would it have any advantages or disadvantages?

16. The power of a 100-W light bulb is 10^5 times as great as that of a 1-mW laser. But the intensity of the laser beam is approximately the same as the intensity of bright sunlight, while the intensity of the light 1 ft away from a 100-W light bulb is less than one-tenth the intensity of bright sunlight. Can you explain why?

17. What are some of the industrial and medical applications of lasers? Can you suggest one or more new applications? Military uses?

18. Are electrons, protons, or neutrons the most abundant particles in the sheet of paper on which these words are printed?

19. Define the following: alpha particle, beta particle, gamma ray.

20. Which is most likely to expose photographic film that is kept in a cardboard box: alpha particles, beta particles, or gamma rays? Which of these three types of radiation is least likely to expose the film?

21. Why is it unlikely that Newtonian gravitational forces hold nuclear particles together? Why is it even more unlikely that electrical attraction (the *Coulomb force*) holds them together?

22. What are isotopes? How can isotopes of the same element be separated?

23. The two naturally occurring isotopes of copper are ${}^{63}_{29}Cu$ and ${}^{65}_{29}Cu$. In what way are the nuclei of these two isotopes different?

24. A certain radioactive isotope has a half-life of 6 months. Will it be completely decayed at the end of one year?

25. A certain radioactive isotope has a half-life of one day. After how many days will seven-eighths of the original sample have decayed?

26. Two radioactive sources emit the same number of radioactive particles per unit time. Are they likely to be equally dangerous health hazards?

27. Based on some library research (read about the work of Prof. Willard Libby) explain how carbon-14 dating could be used to establish the date of a prehistoric event.

28. Describe two or more ways in which the strong nuclear force and the electrostatic Coulomb force are different.

PROBLEMS

Group One

1. An atomic oscillator has a frequency of 2.5×10^{15} Hz. What is the lowest energy it can have according to Planck's quantum hypothesis? If it emits radiation of the same frequency, what wavelength would this radiation have?

2. Find the energy (in joules) of a photon whose frequency is 5.7×10^{14} Hz.

3. What is the energy (in electron-volts) of a photon of ultraviolet light of wavelength 300 nm?

4. Light of wavelength 500 nm shines on a metal whose work function is 1.0 eV. Calculate the maximum kinetic energy of ejected photoelectrons.

5. Copper has a work function of 4.5 eV. What threshold wavelength is needed to cause the ejection of photoelectrons from the surface of copper?

6. The threshold frequency f_o for ejection of photoelectrons from the surface of silicon is 9.91×10^{14}Hz. If ultraviolet light of frequency 1.53×10^{15}Hz is incident on the surface of silicon, what is the maximum kinetic energy of the ejected photoelectrons in electron-volts?

7. An x-ray photon of wavelength 0.1540 nm is Compton-scattered by an electron. The scattered photon travels in a direction perpendicular to the direction of the incident photon. What is the wavelength of the scattered photon?

8. Calculate the de Broglie wavelength of an electron and for a proton, each traveling at 2.5×10^6 m/s.

9. A helium nucleus (alpha particle) is accelerated to a speed of 2.4×10^7 m/s. Calculate its de Broglie wavelength. The mass of an alpha particle is approximately 6.64×10^{-27} kg.

10. What is the momentum of a high-speed proton whose de Broglie wavelength is 0.065 nm?

11. Only four of the bright lines in the Balmer series for hydrogen lie in the visible spectrum. Use the Balmer-series formula to calculate the longest wavelength visible line (use $n = 3$), and the shortest wavelength visible line (use $n = 6$).

12. What is the wavelength of the photon emitted when an electron moves from the $n = 2$ to the $n = 1$ orbit in a hydrogen atom? Is the photon an infrared, a visible, or an ultraviolet photon?

13. What is the minimum wavelength of x-rays produced in a 1.5 MV x-ray tube?

14. The atomic number Z for copper is 29. The mass number A for one of the isotopes of copper is 65. What is the neutron number N for that isotope?

15. The mass number A for an isotope of a certain metallic element is 56, and the neutron number N is 30. (*a*) What is the atomic number Z for that element? (*b*) What is the name of the element?

16. Which atom has a nucleus with a larger number of neutrons—a platinum atom with atomic number $Z = 78$ and mass number $A = 198$, or a mercury atom with atomic number $Z = 80$ and mass number $A = 196$?

17. An element has atomic number 82. It has 125 neutrons in the nucleus. What is its atomic mass to the nearest whole number?

18. A certain radioactive isotope has a half-life of 5.0 min. If there are 3.2×10^7 radioactive nuclei in the sample initially, how many will there be at the end of (*a*) 5 min., (*b*) 20 min., (*c*) 30 min.?

19. A certain radioactive sample has a half-life of 10 days. Initially there are 1.6×10^7 radioactive nuclei present in the sample. After how many days will there be 1.0×10^6 radioactive nuclei in the sample?

20. A worker is exposed to 14 rads of x-ray radiation. How many rems is this? Is this a dangerous exposure?

Group Two

21. The second lowest possible energy for a certain atomic oscillator is 2.0×10^{-19}J. What is the frequency of the atomic oscillator?

22. What is the wavelength of a photon whose energy is 2.00 eV?

23. Light of wavelength 420 nm shines on a metal with a work function of 3.8 eV. No electrons are emitted. Show mathematically why this is the case.

24. The threshold wavelength for emission of photoelectrons in tungsten is 273 nm. (*a*) What is tungsten's work function? (*b*) What is the maximum kinetic energy (in electron-volts) of the electrons ejected from tungsten by monochromatic light of wavelength 160 nm?

25. When the surface of a material is illuminated with laser light of wavelength 632.8 nm, the maximum kinetic energy of the ejected photoelectrons is 0.15 eV. What is the material being illuminated? (HINT: See Table 34.1.)

26. The ratio h/mc in Eq. 34.8 is called the Compton wavelength λ_c of the electron. (*a*) What is the Compton wavelength for the electron in nanometers? (*b*) Show that the greatest possible difference between the wavelength λ' of the scattered photon and the wavelength λ of the incident photon is equal to $2\lambda_c$.

27. A certain photon is Compton-scattered by an electron. The scattering angle is 130°. If the wavelength of the scattered photon is 2.9×10^{-11}m, what must have been the wavelength of the incident photon?

28. The de Broglie wavelength of a certain electron is 10^{-10}m. What must be the speed of that electron?

29. A particle moving with a speed of 200m/s has a de Broglie wavelength of approximately 2 nm. Is the particle an electron or a proton?

30. What wavelength photon is emitted when an electron moves from the $n = 6$ to the $n = 3$ orbit in a hydrogen atom?

31. The accelerating voltage V for an x-ray tube is 50,000 volts. (*a*) What is the maximum kinetic energy of the accelerated electrons in the tube just before they strike the target? (*b*) What is the minimum x-ray wavelength produced by the tube?

32. The maximum kinetic energy of the electrons in an x-ray tube is 6.41×10^{-15}J. (*a*) What is the accelerating voltage for the tube? (*b*) What is the minimum x-ray wavelength produced by the tube?

33. Electrons in an electron microscope have a de Broglie wavelength of 0.02 nm. What is the accelerating voltage for the electrons?

34. Iodine 131, which is used to study thyroid metabolism, has a half-life of 8.0 days. If you start with 1 g of this isotope, how much will be left 32 days later?

35. Cobalt 60, which is used to provide radiation to kill cancer cells, has a half-life of about 5.3 years. Approximately how much of any initial amount would exist after 16 years?

36. A worker is exposed to 40 rads of 1-MeV alpha radiation. (*a*) How many rems is this? (*b*) How many sieverts?

37. A person whose mass is 65 kg absorbs 1 J of 1 MeV gamma rays. How many rads of gamma radiation are absorbed? How many grays?

38. A worker is exposed to 5 millirads of neutron radiation. What is the dose equivalent (*a*) in rems? (*b*) in sieverts?

Group Three

39. The three lowest levels of energy for an atomic oscillator are $E_1 = 1.47 \times 10^{-19}$J, $E_2 = 2.94 \times 10^{-19}$J, and $E_3 = 4.41 \times 10^{-19}$J. What is the frequency of the atomic oscillator?

40. The earth's surface receives solar radiation at the rate of about 1.1 kW/m^2 on a sunny day if the sun is directly overhead. If the average wavelength of this radiation is 670 nm, compute the number of photons passing through a 1 cm^2 area each second.

41. What is the value for the scattering angle θ when the maximum possible energy is transferred to an electron in a Compton scattering event?

42. An incident x-ray photon of wavelength 1.60×10^{-11}m is Compton-scattered by an electron. The wavelength of the scattered photon is 1.76×10^{-11}m. What is the magnitude of the scattering angle θ?

43. An x-ray photon of wavelength 2.0×10^{-11}m is Compton-scattered through 180° by an electron. (*a*) What is the wavelength of the scattered x-ray photon? (*b*) What fraction of the incident photon energy was transferred to the electron?

44. X-rays of wavelength 10^{-11}m are Compton-scattered by electrons in a carbon sample. What is the range (minimum to maximum) of the wavelengths of the scattered x-ray photons?

45. An electron that has been accelerated from rest through a potential difference V has a de Broglie wavelength given by $\lambda = h/\sqrt{2meV}$ where h is Planck's constant, m is the mass of the electron, and e is the magnitude of the charge of the electron. Since h, m, and e are constants, this equation can be written in the simpler form $\lambda = B/\sqrt{V}$ where B is a constant. (*a*) What is the numerical value for the constant B in SI units? (*b*) What is the de Broglie wavelength of an electron accelerated through a potential difference of 10 V? (*c*) Through what potential difference must an electron be accelerated if its de Broglie wavelength is to be 1 nm?

46. What is the longest wavelength photon that can be absorbed by an electron in the ground ($n = 1$) state of a hydrogen atom?

47. In 1 cm^3 of a certain material containing 5×10^{22} atoms, 1 percent of the nuclei are initially radioactive. The half-life for the material is 30 min. (*a*) What is the initial number of radioactive nuclei in the material? (*b*) How many radioactive nuclei will be in the sample at the end of 1, 5, and 10 half-lives?

48. A sample of a certain radioactive isotope emits 4000 beta particles per second at a certain point in time. Two minutes later the same sample emits only 125 beta particles per second. What is the half-life (in seconds) of the isotope?

Appendixes

APPENDIX I Fundamental and Derived Units in the SI-Metric System

Quantity	*Symbol*	*Unit*	*Unit Symbol*	*Dimensions*
Six Fundamental Units				
Length	L or ℓ	meter	m	same as unit
Mass	M or m	kilogram	kg	
Time	T or t	second	s	
Electric current	I	ampere	A	
Temperature	t or T	kelvins	K	
Luminous intensity	I	candela	cd	
Angular Measure				
Plane angle	Greek	radian	rad	dimensionless
Solid angle	letters	steradian	sr	
Some Derived Units				
Acceleration	a	meter/second/second	a	m/s^2
Angular acceleration	α	radian/second/second	α	rad/s^2
Angular velocity	ω	radian/second	ω	rad/s
Area	A	square meter	A	m^2
Capacitance	C	farad	F	A · s/V
Charge	Q	coulomb	C	A · s
Density	ρ	kilogram/cu. meter	ρ	kg/m^3
Electric field intensity	E	newton/coulomb		N/C or V/m
Energy, heat	Q	kilocalorie	kcal	kg · C° or kg · K
Energy, mech.	PE or KE	joule	J	N · m or kg · m^2/s^2
Entropy	ΔS	joule/kelvin	ΔS	J/K
Force	F	newton	N	kg · m/s^2
Frequency	f	hertz	Hz	s^{-1}
Illumination	E	lux	lx	lm/m^2
Impedance (elect.)	Z	ohm	Ω	V/A or kg · $m^2/(A^2 \cdot s^3)$
Inductance	L	henry	H	V · s/A
Luminous intensity	I	candela	cd	cd
Luminous flux	F	lumen	lm	cd-sr
Magnetic field strength (intensity)	H	newtons/weber		N/wb or A/m
Magnetic flux (total)	Ø	weber	Wb	V · s
Magnetic flux density	B	tesla	T	Wb/m^2 or N/(A · m)
Moment of inertia	I			kg · m^2
Momentum	p			kg · m/s
Momentum (angular)	L			kg · m^2/s
Permeability (magn.)	μ	Wb/A · m or T · m/A or H/m		kg · m/C^2
Potential difference	V or $\mathscr{E}$	volt	V	watts/amp (joules/coul)
Power	P	watt	W	J/s
Pressure	P	pascal	Pa	N/m^2
Resistance, elect.	R	ohm	Ω	V/A
Resistivity (elec.)	ρ	ohm-meter or ohm · cmils/ft	Ω · m	
Stress	S	newton/sq. meter		N/m^2
Torque	τ	N · m		kg · m^2/s^2
Velocity, speed	v	meter/second	v	m/s
Volume	V	cubic meter	V	m^3
Work	W	joule	J	N · m

Note that it is necessary in a few instances to use the same letter to symbolize two or more units.

APPENDIX II Numerical Prefixes and Their Symbols

Factors	*Prefixes*	*Symbols*
10^{12}	tera	T
10^{9}	giga	G
10^{6}	mega	M
10^{3}	kilo	k
10^{2}	hecto	h
10	deka	da
10^{-1}	deci	d
10^{-2}	centi	c
10^{-3}	milli	m
10^{-6}	micro	μ
10^{-9}	nano	n
10^{-12}	pico	p
10^{-15}	femto	f
10^{-18}	atto	a

APPENDIX III Letters of the Greek Alphabet

Capital	*Small*		*English Equivalent*
A	α	Alpha	a
B	β	Beta	b
Γ	γ	Gamma	g
Δ	δ	Delta	d
E	ϵ	Epsilon	e
Z	ζ	Zeta	z
H	η	Eta	ē
Θ	θ	Theta	th
I	ι	Iota	i
K	κ	Kappa	k
Λ	λ	Lambda	l
M	μ	Mu	m
N	ν	Nu	n
Ξ	ξ	Xi	x
O	o	Omicron	o
Π	π	Pi	p
P	ρ	Rho	r
Σ	σ	Sigma	s
T	τ	Tau	t
Υ	υ	Upsilon	u
Φ	ϕ	Phi	ph
X	χ	Chi	ch
Ψ	ψ	Psi	ps
Ω	ω	Omega	ō

APPENDIX IV Table of Natural Trigonometric Functions

Degrees	Sine	Cosine	Tangent	Degrees	Sine	Cosine	Tangent
0	0.000	1.000	0.000	46	0.719	0.695	1.03
1	0.017	1.000	0.017	47	0.731	0.682	1.07
2	0.035	0.999	0.035	48	0.743	0.669	1.11
3	0.052	0.999	0.052	49	0.755	0.656	1.15
4	0.070	0.998	0.070	50	0.766	0.643	1.19
5	0.087	0.996	0.087	51	0.777	0.629	1.23
6	0.105	0.995	0.105	52	0.788	0.616	1.28
7	0.122	0.993	0.123	53	0.799	0.602	1.33
8	0.139	0.990	0.141	54	0.809	0.588	1.38
9	0.156	0.988	0.158	55	0.819	0.574	1.43
10	0.174	0.985	0.176	56	0.829	0.559	1.48
11	0.191	0.982	0.194	57	0.839	0.545	1.54
12	0.208	0.978	0.213	58	0.848	0.530	1.60
13	0.225	0.974	0.231	59	0.857	0.515	1.66
14	0.242	0.970	0.249	60	0.866	0.500	1.73
15	0.259	0.966	0.268	61	0.875	0.485	1.80
16	0.276	0.961	0.287	62	0.883	0.470	1.88
17	0.292	0.956	0.306	63	0.891	0.454	1.96
18	0.309	0.951	0.325	64	0.899	0.438	2.05
19	0.326	0.946	0.344	65	0.906	0.423	2.14
20	0.342	0.940	0.364	66	0.914	0.407	2.25
21	0.358	0.934	0.384	67	0.920	0.391	2.36
22	0.375	0.927	0.404	68	0.927	0.375	2.48
23	0.391	0.920	0.424	69	0.934	0.358	2.61
24	0.407	0.914	0.445	70	0.940	0.342	2.75
25	0.423	0.906	0.466	71	0.946	0.326	2.90
26	0.438	0.899	0.488	72	0.951	0.309	3.08
27	0.454	0.891	0.510	73	0.956	0.292	3.27
28	0.470	0.883	0.532	74	0.961	0.276	3.49
29	0.485	0.875	0.554	75	0.966	0.259	3.73
30	0.500	0.866	0.577	76	0.970	0.242	4.01
31	0.515	0.857	0.601	77	0.974	0.225	4.33
32	0.530	0.848	0.625	78	0.978	0.208	4.70
33	0.545	0.839	0.649	79	0.982	0.191	5.14
34	0.559	0.829	0.674	80	0.985	0.174	5.67
35	0.574	0.819	0.700	81	0.988	0.156	6.31
36	0.588	0.809	0.726	82	0.990	0.139	7.11
37	0.602	0.799	0.754	83	0.993	0.122	8.14
38	0.616	0.788	0.781	84	0.995	0.105	9.51
39	0.629	0.777	0.810	85	0.996	0.087	11.4
40	0.643	0.766	0.839	86	0.998	0.070	14.3
41	0.656	0.755	0.869	87	0.999	0.052	19.1
42	0.669	0.743	0.900	88	0.999	0.035	28.6
43	0.682	0.731	0.932	89	1.000	0.017	57.3
44	0.695	0.719	0.966	90	1.000	0.000	∞
45	0.707	0.707	1.000				

APPENDIX V Table of Common Logarithms of Numbers

No.	0	1	2	3	4	5	6	7	8	9
10	0000	0043	0086	0128	0170	0212	0253	0294	0334	0374
11	0414	0453	0492	0531	0569	0607	0645	0682	0719	0755
12	0792	0828	0864	0899	0934	0969	1004	1038	1072	1106
13	1139	1173	1206	1239	1271	1303	1335	1367	1399	1430
14	1461	1492	1523	1553	1584	1614	1644	1673	1703	1732
15	1761	1790	1818	1847	1875	1903	1931	1959	1987	2014
16	2041	2068	2095	2122	2148	2175	2201	2227	2253	2279
17	2304	2330	2355	2380	2405	2430	2455	2480	2504	2529
18	2553	2577	2601	2625	2648	2672	2695	2718	2742	2765
19	2788	2810	2833	2856	2878	2900	2923	2945	2967	2989
20	3010	3032	3054	3075	3096	3118	3139	3160	3181	3201
21	3222	3243	3263	3284	3304	3324	3345	3365	3385	3404
22	3424	3444	3464	3483	3502	3522	3541	3560	3579	3598
23	3617	3636	3655	3674	3692	3711	3729	3747	3766	3784
24	3802	3820	3838	3856	3874	3892	3909	3927	3945	3962
25	3979	3997	4014	4031	4048	4065	4082	4099	4116	4133
26	4150	4166	4183	4200	4216	4232	4249	4265	4281	4298
27	4314	4330	4346	4362	4378	4393	4409	4425	4440	4456
28	4472	4487	4502	4518	4533	4548	4564	4579	4594	4609
29	4624	4639	4654	4669	4683	4698	4713	4728	4742	4757
30	4771	4786	4800	4814	4829	4843	4857	4871	4886	4900
31	4914	4928	4942	4955	4969	4983	4997	5011	5024	5038
32	5051	5065	5079	5092	5105	5119	5132	5145	5159	5172
33	5185	5198	5211	5224	5237	5250	5263	5276	5289	5302
34	5315	5328	5340	5353	5366	5378	5391	5403	5416	5248
35	5441	5453	5465	5478	5490	5502	5514	5527	5539	5551
36	5563	5575	5587	5599	5611	5623	5635	5647	5658	5670
37	5682	5694	5705	5717	5729	5740	5752	5763	5775	5786
38	5798	5809	5821	5832	5843	5855	5866	5877	5888	5899
39	5911	5922	5933	5944	5955	5966	5977	5988	5999	6010
40	6021	6031	6042	6053	6064	6075	6085	6096	6107	6117
41	6128	6138	6149	6160	6170	6180	6191	6201	6212	6222
42	6232	6243	6253	6263	6274	6284	6294	6304	6314	6325
43	6335	6345	6355	6365	6375	6385	6395	6405	6415	6425
44	6435	6444	6454	6464	6474	6484	6493	6503	6513	6522
45	6532	6542	6551	6561	6571	6580	6590	6599	6609	6618
46	6628	6637	6646	6656	6665	6675	6684	6693	6702	6712
47	6721	6730	6739	6749	6758	6767	6776	6785	6794	6803
48	6812	6821	6830	6839	6848	6857	6866	6875	6884	6893
49	6902	6911	6920	6928	6937	6946	6955	6964	6972	6981
50	6990	6998	7007	7016	7024	7033	7042	7050	7059	7067
51	7076	7084	7093	7101	7110	7118	7126	7135	7143	7152
52	7160	7168	7177	7185	7193	7202	7210	7218	7226	7235
53	7243	7251	7259	7267	7275	7284	7292	7300	7308	7316
54	7324	7332	7340	7348	7356	7364	7372	7380	7388	7396
No.	0	1	2	3	4	5	6	7	8	9

APPENDIX V Table of Common Logarithms of Numbers (*continued*)

No.	0	1	2	3	4	5	6	7	8	9
55	7404	7412	7419	7427	7435	7443	7451	7459	7466	7474
56	7482	7490	7497	7505	7513	7520	7528	7536	7543	7551
57	7559	7566	7574	7582	7589	7597	7604	7612	7619	7627
58	7643	7642	7649	7657	7664	7672	7697	7686	7694	7701
59	7709	7716	7723	7731	7738	7745	7752	7760	7767	7774
60	7782	7789	7796	7803	7810	7818	7825	7832	7839	7846
61	7853	7860	7868	7875	7882	7889	7896	7903	7910	7917
62	7924	7931	7938	7945	7952	7959	7966	7973	7980	7987
63	7993	8000	8008	8014	8021	8028	8035	8041	8048	8055
64	8062	8069	8075	8082	8089	8096	8102	8109	8116	8122
65	8129	8136	8142	8149	8156	8162	8169	8176	8182	8189
66	8195	8202	8209	8215	8222	8228	8235	8241	8248	8254
67	8261	8267	8274	8280	8287	8293	8299	8306	8312	8319
68	8325	8331	8338	8344	8351	9357	8363	8370	8376	8382
69	8388	8395	8401	8407	8414	8420	8426	8432	8439	8445
70	8451	8457	8463	8470	8476	8482	8488	8494	8500	8506
71	8513	8519	8525	8531	8537	8543	8549	8555	8561	8567
72	8573	8579	8585	8591	8597	8603	8609	8615	8621	8627
73	8633	8639	8645	8651	8657	8663	8669	8675	8681	8686
74	8692	8698	8704	8710	8716	8722	8727	8733	8739	8745
75	8751	8756	8762	8768	8774	8779	8785	8791	8797	8802
76	8808	8814	8820	8825	8831	8837	8842	8848	8854	8859
77	8865	8871	8876	8882	8887	8893	8899	8904	8910	8915
78	8921	8927	8932	8938	8943	8949	8954	8960	8965	8971
79	8976	8982	8987	8993	8998	9004	9009	9015	9020	9025
80	9031	9036	9042	9047	9053	9058	9063	9069	9074	9079
81	9085	9090	9096	9101	9106	9112	9117	9122	9128	9133
82	9138	9143	9149	9154	9159	9165	9170	9175	9180	9186
83	9191	9196	9201	9206	9212	9217	9222	9227	9232	9238
84	9243	9248	9253	9258	9263	9269	9274	9279	9284	9289
85	9294	9299	9304	9309	9315	9320	9325	9330	9335	9340
86	9345	9350	9355	9360	9365	9370	9375	9380	9385	9390
87	9395	9400	9405	9410	9415	9420	9425	9430	9435	9440
88	9445	9450	9455	9460	9465	9469	9474	9479	9484	9489
89	9494	9499	9504	9509	9513	9518	9523	9528	9533	9538
90	9542	9547	9552	9557	9562	9566	9571	9576	9581	9586
91	9590	9595	9600	9605	9609	9614	9619	9624	9628	9633
92	9638	9643	9647	9652	9657	9661	9666	9671	9675	9680
93	9685	9689	9694	9699	9703	9708	9713	9717	9722	9727
94	9731	9736	9741	9745	9750	9754	9759	9763	9768	9773
95	9777	9782	9786	9791	9795	9800	9805	9809	9814	9818
96	9823	9827	9832	9836	9841	9845	9850	9854	9859	9863
97	9868	9872	9877	9881	9886	9890	9894	9899	9903	9908
98	9912	9917	9921	9926	9930	9934	9939	9943	9948	9952
99	9956	9961	9965	9969	9974	9978	9983	9987	9991	9996
No.	0	1	2	3	4	5	6	7	8	9

APPENDIX VI Electrical Tables Wire Data for Standard Annealed Copper at 20°C

AWG Gauge	Diameter d, Mils	Area d^2, Cir Mils	Ohms/ 1000 ft	Pounds/ 1000 ft
0000	460.0	211,600	0.04901	640.5
000	409.6	167,800	0.06180	508.0
00	364.8	133,100	0.07793	402.8
0	324.9	105,500	0.09827	319.5
1	289.3	83,690	0.1239	253.3
2	257.6	66,370	0.1563	200.9
3	229.4	52,640	0.1970	159.3
4	204.3	41,740	0.2485	126.4
5	181.9	33,100	0.3133	100.2
6	162.0	26,250	0.3951	79.46
7	144.3	20,820	0.4982	63.02
8	128.5	16,510	0.6282	49.98
9	114.4	13,090	0.7921	39.63
10	101.9	10,380	0.9989	31.43
11	90.74	8,234	1.260	24.93
12	80.81	6,530	1.588	19.77
13	71.96	5,178	2.003	15.68
14	64.08	4,107	2.525	12.43
15	57.07	3,257	3.184	9.858
16	50.82	2,583	4.016	7.818
17	45.26	2,048	5.064	6.200
18	40.30	1,624	6.385	4.917
19	35.89	1,288	8.051	3.899
20	31.96	1,022	10.15	3.092
21	28.45	810.1	12.80	2.452
22	25.35	642.4	16.14	1.945
23	22.57	509.5	20.36	1.542
24	20.10	404.0	25.67	1.223
25	17.90	320.4	32.37	0.9699
26	15.94	254.1	40.81	0.7692
27	14.20	201.5	51.47	0.6100
28	12.64	159.8	64.90	0.4837
29	11.26	126.7	81.83	0.3836
30	10.03	100.5	103.2	0.3042
31	8.928	79.70	130.1	0.2413
32	7.950	62.31	164.1	0.1913
33	7.080	50.13	206.9	0.1517
34	6.305	39.75	260.9	0.1203
35	5.615	31.52	329.0	0.0954
36	5.000	25.00	414.8	0.0757
37	4.453	19.83	523.1	0.0600
38	3.965	15.72	659.6	0.0476
39	3.531	12.47	831.8	0.0377
40	3.145	9.888	1,049	0.0299

APPENDIX VII Symbols Used in Electricity and Electronics

Anode, or plate

Antenna

Battery

Capacitors

Capacitor, fixed

Capacitor, variable

Cell, single

Circuit breaker

Fuse

Generators

A-C generator

D-C generator

Grid

Ground connection

Inductors

Inductor, air core

Inductor, iron core

Inductor, variable

Lamp

Meters

Ammeter

Galvanometer

Voltmeter

Resistors

Resistor, fixed

Resistor, variable (Rheostat)

Switches

Single-pole, single-throw

Single-pole, double-throw

Double-pole, single-throw

Double pole, double-throw

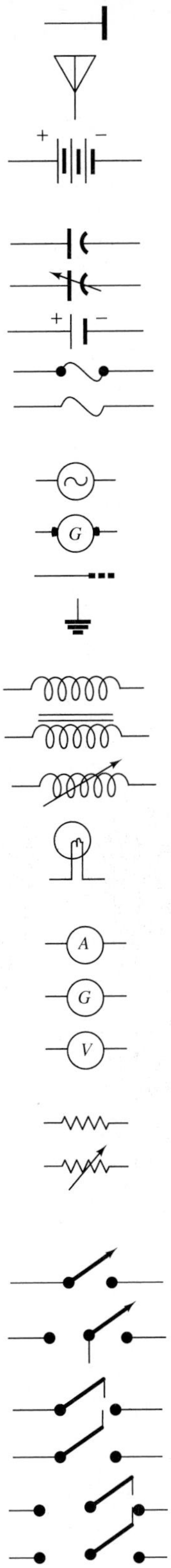

Transformers

Transformer, air-core

Transformer, iron-core

Thermistor

Solid-state devices

P-N junction diode

Zener diode

Bipolar junction transistor (NPN)

Bipolar junction transistor (PNP)

N-channel junction field-effect transistor

P-channel junction field-effect transistor

Vacuum tubes

Diode

Triode

Three phase A-C configurations

Delta configuration

Wye configuration

Four-wire configuration

Wires, connected

Wires, crossing, not connected

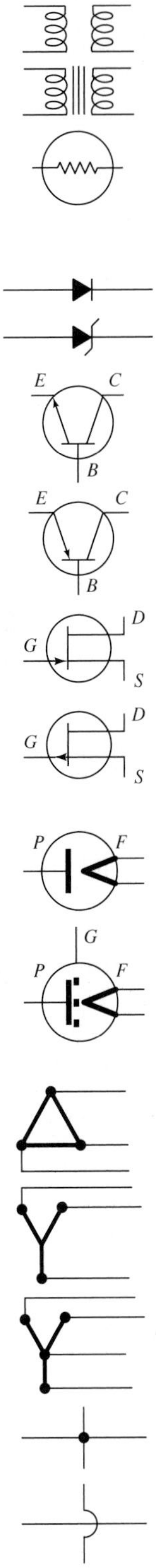

APPENDIX VIII Table of the Elements—Alphabetical Listing

Element	Symbol	Atomic number	Average atomic mass*	Element	Symbol	Atomic number	Average atomic mass*
Actinium	Ac	89	(227)	Mercury	Hg	80	200.59
Aluminum	Al	13	26.9815	Molybdenum	Mo	42	95.94
Americum	Am	95	(243)	Neodymium	Nd	60	144.24
Antimony	Sb	51	121.75	Neon	Ne	10	20.183
Argon	Ar	18	39.948	Neptunium	Np	93	(237)
Arsenic	As	33	74.9216	Nickel	Ni	28	58.71
Astitine	At	85	(210)	Niobium	Nb	41	92.906
Barium	Ba	56	137.34	Nitrogen	N	7	14.0067
Berkelium	Bk	97	(247)	Nobelium	No	102	(254)
Beryllium	Be	4	9.0122	Osmium	Os	76	190.2
Bismuth	Bi	83	208.980	Oxygen	O	8	15.9994
Boron	B	5	10.811	Palladium	Pd	46	106.4
Bromine	Br	35	79.904	Phosphorus	P	14	30.9738
Cadmium	Cd	48	112.40	Platinum	Pt	78	195.09
Calcium	Ca	20	40.08	Plutonium	Pu	94	(244)
Californium	Cf	98	(251)	Polonium	Po	84	(209)
Carbon	C	6	12.01115	Potassium	K	19	39.102
Cerium	Ce	58	140.12	Praseodymium	Pr	59	140.907
Cesium	Cs	55	132.905	Promethium	Pm	61	(145)
Chlorine	Cl	17	35.453	Protactinium	Pa	91	(231)
Chromium	Cr	24	51.996	Radium	Ra	88	(226.0)
Cobalt	Co	27	58.9332	Radon	Rn	86	222
Copper	Cu	29	63.546	Rhenium	Re	75	186.2
Curium	Cm	96	(247)	Rhodium	Rh	45	102.905
Dysprosium	Dy	66	162.50	Rubidium	Rb	37	85.47
Einsteinium	Es	99	(254)	Ruthenium	Ru	44	101.07
Erbium	Er	68	167.26	Samarium	Sm	62	150.35
Europium	Eu	63	151.96	Scandium	Sc	21	44.956
Fermium	Fm	100	(257)	Selenium	Se	34	78.96
Fluorine	F	9	18.9984	Silicon	Si	14	28.086
Francium	Fr	87	(223)	Silver	Ag	47	107.868
Gadolinium	Gd	64	157.25	Sodium	Na	11	22.9898
Gallium	Ga	31	69.72	Strontium	Sr	38	87.62
Germanium	Ge	32	72.59	Sulfur	S	16	32.064
Gold	Au	79	196.967	Tantalum	Ta	73	180.948
Hafnium	Hf	72	178.49	Technetium	Te	43	(97)
Helium	He	2	4.0026	Tellerium	Te	52	127.60
Holmium	Ho	67	164.930	Terbium	Tb	65	158.924
Hydrogen	H	1	1.00797	Thallium	Tl	81	204.37
Indium	In	49	114.82	Thorium	Th	90	232.0381
Iodine	I	53	126.9044	Thulium	Tm	69	168.934
Iridium	Ir	77	192.2	Tin	Sn	50	118.69
Iron	Fe	26	55.847	Titanium	Ti	22	47.90
Krypton	Kr	36	83.80	Tungsten	W	74	183.85
Lanthanum	La	57	138.91	Uranium	U	92	238.03
Lawrencium	Lw	103	(257)	Vanadium	V	23	50.942
Lead	Pb	82	207.19	Xenon	Xe	54	131.30
Lithium	Li	3	6.939	Ytterbium	Yb	70	173.04
Lutetium	Lu	71	174.97	Yttrium	Y	39	88.905
Magnesium	Mg	12	24.312	Zinc	Zn	30	65.37
Manganese	Mn	25	54.9380	Zirconium	Zr	40	91.22
Mendelevium	Md	101	(256)				

*The atomic masses are based upon ${}^{12}_{6}C = 12$ **u**, where **u** is the unit of atomic mass. An atomic mass in parenthesis is the mass number of the most stable (longest lived) isotope of the element.

APPENDIX IX Answers to Odd-Numbered Problems

Chapter 1

1. *a*) 0.0006
b) 0.0035
c) 0.000005
d) 0.075

3. *a*) 2.225×10^4
b) 2.585×10^7
c) 6.29×10^2
d) 6.29×10^{-3}
e) 7.85×10^{-5}
f) 8.4×10^{-2}

5. 2180 ft/s

7. 12,950 ft^2

9. 7.85 ft

11. *a*) $15\ x^2$
b) $-x^3y^2z^2$
c) x^2y^6
d) $8\ a^6b^3$
e) $-3 \times y^2$
f) x^{10}

13. *a*) 31.006
b) 8.988×10^{16}
c) 17.9813
d) 0.20973

15. 3,180 in.

17. 235.5 ft

19. 63.44 ft

21. *a*) Hollow shell—597 in.^{3}
b) Solid sphere—524 in.^{3}

23. 9.44 ft

25. 358 in.^{2}

27. *a*) 27.2°
b) 31.13 ft

29. 115.5 ft

31. 9.9 (or 10 loads)

33. 482 ft

35. 2510 ft^2

37. Graph this problem

39. 776 revs

Chapter 2

1. 88.6 km/h

3. 68.13 L

5. No. Range exceeds allowed tolerances

7. *a*) 201 km
b) 0.01 mm $<$ 0.001 in.

9. 23.4 km/L

11. 323 mg (wt)

13. *a*) 0.001 in.
b) 0.0001 in.
c) 0.000001 in.
d) 0.000001 m
e) 0.625 in.
f) $3.1429 \approx \pi$

15. 2.49 in.^{3}

17. 37 gr = 2398 mg (wt)
39 gr = 2527 mg (wt)

19. 1 in. = 19.05 mm
A worn 19-mm wrench might be used, but a 20-mm size is more probable.

21. 232 in.^{3}

23. $5.7 \times 10^{-1} \mu$ m
= 570 nm
= 22.44 μin.

25. 418 rev

27. 2.993×10^4 cm/μs

29. 0.403 s

31. 14,930 mi

33. *a*) 16.7 gal
b) 10.8 in.

35. 316 in.^{3}

37. 1.283 s

39. 0.0001 in.

Chapter 3

1. R = 19.8 km at 340°T

3. R = 1825 mi from starting point

5. x = 57.3 m
y = 40.2 m

7. F_x = 321 lb
F_y = 383 lb

9. *a*) $F = N$ = 86.6 lb at 30° with vertical
b) 94 lb at 20° with vertical

11. 3400 lb

13. R = 84.5 lb
Θ = 20.2°

15. *a*) T_D = 1200 lb
b) $T_B = T_A$ = 1000 lb

17. AB = 5.4 tons
BC = 4.4 tons
CD = 5.9 tons

19. F_{Right} = 92.9 lb
F_{Left} = 37.1 lb

21. F = 1200 lb
x = 4.92 ft

23. 29.9 m/s at 41° angle with path of auto

25. 28.4 knots at 93° with ship's heading

27. Compass heading—065°
Time of flight—4.6 h

29. R_B = 26,900 lb
R_A = 15,100 lb

31. 3.5 ft from right end of board

Chapter 4

1. 1.57 h

3. 60 mi/h = 88 ft/s.
Therefore, second car is faster.

5. 0.136 s

7. *a*) 403 ft
b) 161 ft/s

9. 240 s (4 min "flat")

11. 120 ft/s^2

13. 112 ft

15. Elapsed time = 3.983 s

17. 242 m/s

19. 794 ft/s

21. -3.02×10^6 ft/s^2
(negative acceleration)

23. 18,200 yd

25. *a*) 75 ft/s
b) 21.8 ft

27. 64.8 ft/s

29. Plot analysis—non-numerical

31. -1520 ft/s^2 (deceleration)

33. 18.5 mi/s

35. Graphical solution

Chapter 5

1. 370 lb force

3. 14,400 N

5. 57.5 slugs (sl)

7. 15,700 N

9. *a*) Up. 49.6 N
b) Down. 10.4 N

11. 67.9×10^3 N

13. 61.6 m

15. *a*) 134.7 lb (up)
b) 113.4 lb (down)
c) 125 lb (the woman's weight)

17. *a*) -7.14 ft/s^2 (deceleration)
b) -621 lb (retarding force)
c) 175 ft

19. *a*) 1.56×10^6 ft/s^2
b) 1.04×10^3 lb

21. *a*) -10.5×10^6 m/s^2 (deceleration)
b) -12.6×10^4 N (retarding force)
c) 7.6×10^{-5} s

23. 3.83 s

25. *a*) 3.33 slugs
b) 60 ft/s
c) -66.7 lb (retarding force)

27. 12.4 kg

29. P = 11.04 N; T = 3.75 N

31. 23,720 lb

33. *a*) 1.05 m/s^2
b) 27.2 N

35. *a*) 2.15×10^5 m/s^2
b) 2.19×10^4 "g's"

Chapter 6

1. 2400 ft · lb

3. 240 ft · lb

5. 1.275×10^6 N · m (joules, J)

7. $p = 44{,}440$ kg · m/s
9. 110 lb/min
11. 1.962×10^6 J
13. −4.76 m/s
15. 88.3 kW
17. 3.17 hp
19. 1.88 m/s
21. 88.9 kW
23. 23.4 hp
25. 12 kg · m/s
27. 10,000 lb
29. 4.75×10^3 kW
31. 1225 km/h = 760 mi/h
33. *a*) 32.4 kJ
    b) 6.69×10^2 m/s or 2410 km/h
35. *a*) 24.6 slugs/s (loss of mass)
    b) $a_{20} = 75.9$ ft/s^2
    = 2.36 "g's"
37. 12.5 m
39. 827 m/s

Chapter 7

1. 8.7 ft from fulcrum
3. Right hand—104 N
   left hand—300 N
5. 3750 N
7. 858 N
9. 280 rev/min
11. 1730 lb
13. *a*) 112.5 lb
    b) 435 lb
15. 16,600 N
17. 7.88 ft/s
19. *a*) 255 kg
    b) 78 percent
21. 59.3 ft/s
23. 160 rev/min
25. 120 ft/s
27. 8.57 cm
29. 33.5 m
31. 0.148
33. *a*) 38.8 ft/s
    b) 38.78 ft/s
35. 4.9 m/s^2
37. 49 in.
39. 81.8 rpm

Chapter 8

1. 23 rad/s
3. 14,620 rad
5. 183 rad/s
7. 3.14 m/s
9. 66.6 N
11. 0.218 rad/s^2
13. 29,700 rpm
15. 5.54 in.
17. 298 km/h
19. *a*) 86.4 min
    b) Same time. Mass makes no difference in orbital period.
21. *a*) 6030 km/h
    b) 1.84 h
23. 19.85×10^{19} N
25. 797 km/h
27. 156,900 km/h
29. 5310 mi/h
31. *a*) $a_{tan} = 62.5$ m/s^2
    b) $a_{tot} = 258$ m/s^2
33. Any speed below 16,100 mi/h.
35. 30,840 N
37. 2640 km
39. Requires a mathematical proof, not a numerical answer.
41. 5.27×10^6 kW · h

Chapter 9

1. 251 rad/s
3. 25 N · m (torque, not joules)
5. 22.9 hp
7. 2270 rev/min
9. 119 hp
11. *a*) 1.46×10^{-3} kg · m^2
    b) 0.009 J
13. *a*) 10 rad/s
    b) 122.6 kW
15. 22.25 kg · m^2
17. 163 N · m (torque, not joules)
19. 50.9 lb · ft
21. 5.47 lb · ft
23. 21,060 N · m (torque)
25. 1.53×10^3 rad/s
27. *a*) 1310×10^{-6} J
    b) 6.67×10^{-4} N · m (torque)
    c) 4.69×10^{-2} N
29. 3.75 rev/s
31. *a*) $KE_{trans} = 158$ J
    b) $KE_{rot} = 63.4$ J
    c) $KE_{tot} = 221.4$ J
33. 7.54 N · m (torque)
35. *a*) 37.2 rad/s = 355 rpm
    b) 0.917 kW
37. *a*) 1.96 m/s^2
    b) 3.92 m

Chapter 10

1. 5
3. *a*) 2
   b) 3
   c) 12
5. Approx. 2:1
7. 342.148
9. 13.5 oz
11. 1.12 percent

Chapter 11

1. 27 cm
3. 563 g
5. *a*) 1700 lb
   b) 15,000 lb
7. *a*) 8.34 lb
   b) 5.61 lb
9. 2.75 in.
11. Approx. 3.2×10^{18} kg/m^3
13. 18,000 lb/in^2
15. 6870 lb
17. −0.39
19. 2.2×10^4 N
21. 2.2 in.
23. 3.91×10^2 cm^2
25. 2.34×10^4 N
27. 0.00264 in.
29. $Ø = 8.58 \times 10^{-7}$ rad
    $x = 2.18 \times 10^{-5}$ cm
    or 2.18×10^{-3} mm
31. 0.0148 in.
33. 1.72 cm
35. 0.84 rad, or approx. 48 deg.
37. 11.1 cm
39. 0.60 rad, or approx. 34 deg.
41. 16,450 lb/in^2

Chapter 12

1. 1:1 The forces are equal.
3. 1800 kg/m^3
5. 52 lb/in^2
7. 53.3 lb
9. $y = 3$ cm
11. No buoyant upward force
13. 4.9×10^{-3} N
15. 26.7 m
17. 51.2 lb/ft^3
19. *a*) $V = 0.019$ ft^3
    b) $D = 439$ lb/ft^3
21. 18,700 lb
23. 0.82
25. 6.3×10^{-3} N/m
27. 986 N/cm^2
29. 87 hp
31. 0.867
33. 356 lb/ft^3
35. $F = \pi r\ h^2 D$
37. 2.80 kg
39. $V_B = 3.42$ m/s
    $V_C = 0.855$ m/s
    $h_A = 0.118$ m
    $h_B = 0.149$ m
41. *a*) 139 ft/s
    b) 3.03 ft^3/s
    c) 101 hp
43. 31.9 gm gold

45. $Q = 12.0$ ft³/s

47. $x = 3.7$ cm

Chapter 13

1. *a*) 5000 atoms *H*
   b) 2500 atoms *S*
   c) 10,000 atoms *O*

3. 25 ft

5. 14.4 lb/in.^{2}

7. 5.68×10^{-25} kg

9. Not numerical

11. 1.33 ft^3

13. 2.25 g oxygen left uncombined

15. 2.656×10^{-23} g

17. 1.72×10^5 N/m^2 (abs)

19. 1.34×10^{23} molecules

21. 5.17×10^{24} molecules

23. 9.797×10^4 Pa

25. *a*) 111 cm Hg (abs)
    b) 21.5 lb/in.^{2} (abs)
    c) 1.48×10^5 Pa (abs)

27. 0.36 lb

29. 15.3 ft^3

31. 28.6 tons

33. 0.0054 percent

35. $V_2 = 1.62$ ft^3

37. $P = 68$ ft of water

39. *a*) 36.4 lb/in.^{2}
    b) $V_2 = \frac{147}{364} V_1$
    c) $D = 0.199$ lb/ft^3

41. 1900 kg/m^3

43. 1.34 10^4 N

45. 1.51 m

47. 2570 Pa

49. Simplest formula for the compound is C_2H_6O (ethyl alcohol).

Chapter 14

1. *a*) −38°F
   b) 234.1 K

3. 20°C

5. 185°F

7. +10.4°F

9. 350 K

11. 5279.16 ft (About 0.84 ft short)

13. 3.92 ft

15. 0.146 mm

17. 6.84 ft too long

19. 4.06 gal

21. 13.36 g/cm^3

23. 1.497 cm

25. 131 lb

27. 0.125 ft

29. 542°F

31. 990 ft

33. 336°C

35. 35 psig; 2.41×10^5 Pa

37. 25.6 psig

39. 14.25×10^5 Pa

41. A mathematical proof is required.

Chapter 15

1. 5.99×10^6 Btu

3. 5.16×10^4 Btu

5. 1500 Btu

7. 69.3×10^3 kcal

9. 66°F

11. 9270 ft

13. 35.4 min

15. 7790 kcal/kg

17. 2.12×10^4 ft^3/day

19. 319,800 kcal

21. 195 s, or 3.25 min

23. 125 min (2 h, 5 min)

25. 374.7×10^6 Btu

27. 19.86 mi

29. 28 lb

31. 277 gal/h

33. 66.8°C—all water

35. 13°C—all water

37. 1031 Btu/lb

39. $1370.00

Chapter 16

1. 5×10^{-5} m

3. 8.13 kW

5. Pine wall—12.44 in.
   Brick wall—74 in.
   Concr. wall—89 in.

7. 20,200 Btu/h

9. 38,500 Btu/h

11. *a*) 10,800 Btu/h
    b) 3180 W

13. 0.085 Btu/h · ft^2 · F°

15. 19 lb/day

17. 2.53 kcal/min

19. 2350 W

21. $R_T = 11.1$

23. 5.1 lb/h or 0.61 gal/h

25. 4.435×10^{23} kW

27. 8 Btu/h · ft^2

29. 0.833 lb/h

31. 3.31×10^{-5} m^2

33. 2.62×10^{-5} m^2 or 26.2 mm^2

35. 7.1 lb sat. steam/h

37. 0.429 cm/h

Chapter 17

1. 53 percent

3. 27 percent

5. 2.24×10^7 ft · lb

7. 51 percent

9. 22.1 percent

11. 30.6°F

13. *a*) 324 Btu
    b) 2.52×10^5 ft

15. 235°C

17. 15.91 Btu/min

19. 167 kJ

21. *a*) None
    b) Internal energy increases

23. 215 kJ

25. 1.5 kJ

27. 35 percent

29. *a*) 30.8 percent
    b) 927°R

31. *a*) 56.5 percent
    b) 4.125 kcal/cycle
    c) 1.79 kcal/cycle

33. *a*) 450 J on gas
    b) 350 J on gas
    c) 1800 J on piston
    d) No work done
    e) Work output/cycle, 1000 J

35. 181 hp

37. 819 m/s

39. 21 percent

41. 17,000 hp

Chapter 18

1. 8350 Btu/h

3. 7.96×10^6 Btu/h

5. c.o.p equals S.17

7. 0.15 tons of refrig.

9. 130.5 tons of refrig.

11. 890 ft/min

13. 44.2 gr/lb

15. 13,900 Btu/h

17. 20.9 tons

19. 35.8 percent

21. 36,000 Btu/h

23. *a*) 8.42 kW
    b) 9.04 Btu/h · W

25. *a*) 5.62
    b) 2.83
    c) 9.64 Btu/h · W

27. *a*) 10.1 (water source)
    b) 6.05 (air source)

29. *a*) 11.4 kW
    b) 3.14

31. *a*) 97,800 Btu/dollar of energy cost
    b) 129,500 Btu/dollar of energy cost
    Furnace more energy efficient

Chapter 19

1. $T_2 = \sqrt{2}\, T_1$ Doubling the pendulum length multiplies the period by a factor of $\sqrt{2}$.

3. 2.5×10^{-3} s

5. 7 cm

7. 1.8 m

9. 3 ft/s

11. *a*) 208 m
b) 2.88 m

13. 4.9 s

15. *a*) 18 cm
b) 39.6 m/s

17. 25 cm/s

19. *a*) 3π in./s or 9.42 in./s
b) 29.6 in./s^2

21. *a*) 5.7 cm
b) 350 cm/s^2

23. *a*) 0.25/s
b) 4.7 cm/s

25. *a*) 0.5 s
b) 32 cm

27. *a*) $T = 0.77$ s
$f = 1.30$ Hz
b) $v_{max} = 0.755$ m/s
c) $a_{max} = 470$ m/s^2
d) $v = 49.6$ cm/s
e) $t = 0.13$ s

29. 7.1 kg

31. 4.2 s

33. 0.14 cm/s

35. KE/PE = 3:1

37. 1.95 s^{-1}

39. $g = 8.95$ m/s^2

41. 2.8×10^9 N/m^2

43. 0.44 s

45. 176.9 s/day

47. *a*) $v = 245$ m/s
$\lambda = 16.33$ m
b) 21 Hz

Chapter 20

1. 32.5 cm

3. 6 million people

5. 1st, $f_1 = 330$ Hz, $f_2 = 660$ Hz, $f_3 = 990$ Hz, $f_4 = 1320$ Hz, $f_5 = 1650$ Hz

7. 69 dB

9. *a*) 969 ft/s
b) 1160 ft/s

11. 3.0 dB

13. Proofs required; not numerical.

15. 910 Hz

17. 4030 Hz

19. *a*) 396 Hz
First overtone,
$f_2 = 792$ Hz
Second overtone,
$f_3 = 1188$ Hz
b) 198 Hz
$f_3 = 594$ Hz
$f_5 = 990$ Hz

21. *a*) $L = 65.1$ cm
b) $f_1 = 269$ Hz

23. *a*) 3220 ft
b) 16,100 ft/s

25. 10^{-7} W/m^2

27. 30 dB

29. *a*) 0.0026 W/m^2
b) 94 dB

31. 55 dB

33. $L_{open}/L_{closed} = 6/5$

Chapter 21

1. 520 m

3. 17.27 m

5. 2460 sabins

7. *a*) 0.753 m
b) 3.23 m

9. *a*) Hardwood floor
80 sabins
b) Carpet—800 sabins

11. 3000 m

13. 455 persons

15. 2130 mi/h

17. 1.2 mm

19. 790 ft^2

21. *a*) $t = 1.62$ s
b) $t = 1.27$ s

23. *a*) 0.79 s
b) 1.22 s
c) For $t = 1.5$, replace 2835 ft^2 of acoustic plaster with smooth plaster.

25. 2450 ft/s

27. 2.1 mi

Chapter 22

1. 930 mi

3. 6.00×10^5 m or 600 km

5. 4.74×10^{14} s^{-1}

7. 84 cd

9. 0.005 W

11. 14 lm/W

13. 16 lm/W for 75-W bulb
18.6 lm/W for 150-W bulb

15. 190 lm

17. *a*) 70 cm
b) 233 cm

19. 2.33×10^5 mi

21. *a*) 932 nm
b) Infrared
c) Rel. visib. = zero
d) Not visible. Human eye not sensitive to infrared.

23. 24.5 s

25. Four 60-W bulbs are needed

27. 1.8 m

29. *a*) 1.6 ft-c = 17 lux
b) 0.40 ft-c = 4.3 lux

31. 216 cd

33. 55 lux

35. 230 lux

37. 41 ft

39. *a*) 25 ft-c
b) 20 ft-c

Chapter 23

1. 20 deg.

3. *a*) not a numerical answer
b) virtual image

5. 2.25×10^8 m/s

7. 1.361

9. 37.15 deg

11. −30 cm

13. Min. ht. mirror = 0.9 m

15. *a*) −150 cm; virtual, erect, and enlarged
b) $M = 5$

17. *a*) $D_i = -12$ cm
b) $I = 6$ cm high

19. 9/4

21. 45.6 deg

23. 33.3 deg

25. 65 deg

27. *a*) $i_c = 74.2$ deg.
b) 15.8 deg.

29. *a*) $I = 2.5$ cm
b) $I = 5.0$ cm

31. *a*) $I = 1$ cm
b) $I = 0.5$ cm

33. 42.8 cm

35. 33.4 deg

37. 1.333

39. 5.0 mi

Chapter 24

1. 1.625

3. 5.6

5. 41 mm

7. $M = 4.1$

9. 250

11. 60

13. $\mu = 1.520$
Prism is crown glass

15. *a*) 4.0 cm
b) 4.03 cm
c) 4.17 cm
d) 4.44 cm

17. *a*) −3.2 cm
b) 0.80

19. 1.7 cm

21. *a*) −1.25 diopters
b) $D_O = 1.54$ m

23. 85.8 cm

25. $D_O = 5.8$ cm
$M = 4.3$

27. $D_O = 5.5$ mm

29. 2 cm^2

31. 0.016 s (≅ 1/60 s)

33. 4 mm

35. $R_2 = -15.0$ cm (concave surf.)

37. *a*) $D_O = 102.5$ mm
b) 2.5 mm

39. 2.6 mm

Chapter 25

1. 1.1×10^2 N (attraction)

3. 8.40×10^{-15}

5. 20,000 V

7. 500,000 V

9. 9.0×10^{-4} C

11. 0.531 μF

13. 1.5×10^5 V

15. 180 J

17. 20,000 N/C

19. $E = 2.75 \times 10^6$ N/C toward the 40 μC charge

21. 6.0 V

23. 200 V

25. *a*) 1.33 μF
b) 32 μC
c) $V_1 = 16$ V
$V_2 = 8$ V

27. Area (one pair) = 13 cm^2

29. *a*) $Q = 1.2 \times 10^{-4}$ C
b) $RC = 1.6 \times 10^{-4}$ s
c) $RC = 2.0 \times 10^5$ s
= 56 h

31. *a*) 1.5×10^{-4} J
b) zero

33. *a*) $C_T = 1.71$ μF
b) 6.17×10^{-4} C
c) $V_3 = 206$ V; $V_6 = 103$ V
$V_{12} = 51$ V

35. *a*) 1.60×10^4 N/C toward the 250-V plate
b) $F = 2.57 \times 10^{-15}$ N
c) $v = 9.21 \times 10^6$ m/s
d) Grav. force = 8.93×10^{-30} N (The elec. force is approx. 3×10^{14} times as large as the gravitational force).

37. 38,000 V

39. *a*) $v = 1230$ m/s
b) $t = 3.89 \times 10^{-5}$ s

41. 9.3 cm from the 3 μC charge

43. $Q = Q_1 = 3.0 \times 10^{-3}$ C
$Q_2 = 1.0 \times 10^{-3}$ C
$Q_3 = 2.0 \times 10^{-3}$ C

45. 6.7 V

Chapter 26

1. 750 mils

3. 18,000 C/h

5. 167 Ω The wire is probably 32-gauge.

7. 7.13 m

9. 11.4 Ω

11. 2.3 A

13. 3.16 Ω

15. 275 Ω

17. 80 C

19. *a*) $\propto \cong 0.0042$ Ω/C°
b) 28.16 Ω

21. *a*) 51.3 V
b) 61.6 V

23. $I = \dfrac{3\,\mathscr{E}}{\dfrac{3}{2}r + R}$

25. $R_T = 0.89$ Ω

27. $R_2 = 2.0$ Ω

29. *a*) 12 Ω
b) 2 A

31. $R = 9.79$ Ω

33. 96 V

35. Reading No. 1 = 10.0 V

37. $I_1 = 0.28$ A
$I_2 = 0.34$ A
$I_3 = 0.62$ A

39. $I_1 = 14/37$ A
$I_2 = 16/37$ A
$I_3 = 2/37$ A

41. $r = 2.5$ Ω
$\mathscr{E} = 1.0$ V

43. $I = 0.5$ A

45. $R_T = 2.4$ Ω

47. $I_1 = 1.50$ A
$I_2 = 0.50$ A

Chapter 27

1. 22.4 kW

3. 8.73 Ω

5. 3.0×10^6 J

7. 17.5 A

9. 1.9×10^{-18} J

11. $P = 25.6$ kW
$H = 2.10 \times 10^6$ Btu

13. $H = 1070$ Btu

15. *a*) 4.8 A
b) 16.8 V

17. $I = 17.5$ A
Use a 20-A fuse

19. 40 W

21. 67.6 percent

23. *a*) 0.88 A
b) 0.050 A

25. 18.2 h

27. $I = \dfrac{n\,\mathscr{E}}{nr + R}$

29. $r = 0.026$ Ω

31. $R = 1.2$ Ω
Elec. energy = 0.36 kW · h

33. $3.27

35. *a*) 76 V
b) 95.3 V

37. 770 ft

39. *a*) 2.3 Ω
b) 0.86 V

Chapter 28

1. 8×10^{-3} N

3. 6360 A-turns

5. 3.3×10^{-5} Wb/m^2 or, 3.3×10^{-5} T

7. 7×10^{-4} Wb/A · m

9. $I = 2$ A

11. $I = 2.0$ A

13. *a*) 0.6 T
b) $\phi = 4.8 \times 10^{-4}$ Wb

15. *a*) 1.8 m · N
b) 0.9 m · N

17. 0.406 A

19. *a*) 1200 A-turns
b) 4800 A-turns

21. 4.7×10^{-3} N

23. 0.625 N

25. 1.15×10^{-5} N/m

27. 0.02 Ω

29. *a*) 1200 A-turns/m
b) 0.90 T
c) 9.2×10^{-4} Wb

31. $F = 8.0 \times 10^{-3}$ N
directed away from both poles along the axis of the magnet

33. *a*) 1200 Ω
b) 1825 Ω

35. 1.17×10^{-7} T
in a direction 30° below the *x*-axis

Chapter 29

1. 1.7 H

3. *a*) 30 H
b) 18 H

5. 6.6 V

7. 1 m s

9. 0.80 V

11. 37 V

13. 0.80 s

15. 60,000 V

17. 12,500 V

19. 4.5 A

21. $L_T = \dfrac{L_1 \times L_2}{L_1 + L_2}$

23. *a*) 2.7 V
b) L = 40 mH

25. 3.2 H

27. M = 0.25 H

29. *a*) 1.2×10^5 V
b) 3.3×10^{-2} Wb

31. $L = \dfrac{\mu N^2 A}{l}$

33. 47 V

35. *a*) 0.23 A
b) 40 m s
c) 6 V

Chapter 30

1. 60.6 percent

3. 0.6 V

5. 32 Hz

7. 450 rpm

9. 50 Hz

11. 24 poles

13. 300 rpm

15. 48 V

17. 114.5 V

19. *a*) 850 V
b) 60 Hz

21. *a*) 540 V
b) 15 A

23. 18.5 Ω

25. *a*) 1200 rpm
b) 1140 rpm

27. Diagram—not numerical

29. $P = 54$ kW

Chapter 31

1. 163 V

3. 940 Ω

5. 1600

7. 660 Ω

9. 24 Ω

11. 4750 Hz

13. 7.68 kVA

15. *a*) 2.4 A
b) 119 mH

17. 1450 Ω

19. 550 W

21. *a*) 187.5 kV
b) 0.267 A

23. 49 μH

25. 35 μF

27. 2.6×10^{-7} H

29. 0.53 A

31. *a*) 2.9 A
b) PF = 0.60
c) $\phi = -53°$
(voltage lagging current)

33. *a*) 0.90 A
b) 0.93
c) −21° (voltage lagging)

35. −42° (voltage lagging)

37. *a*) 1.22 Ω
b) 1.33 Ω
c) 9.02 A
d) 4.87 V
e) 11.0 V
f) 99.3 vars
g) 43.9 W
h) −66°
(voltage lagging)

39. *a*) 0.74
b) 12.2 Ω
c) 9.03 Ω
d) 8.20 Ω
e) 42°

41. *a*) $X_L = 113$ Ω; $X_c = 73.7$Ω
b) 46.0 Ω
c) $\phi = 59°$
d) 0.52
e) 2.61 A
f) 163 W

43. *a*) 150 W
b) 593 vars (capacitive)
c) 610 VA
d) 2.54 A
e) PF = 0.25

Chapter 32

1. 180 Hz (cycles/s)

3. *a*) 2.2 μs
b) 16.7 m s
c) 4 s

5. 240 rpm

7. *a*) 3 V
b) 48 A

9. *a*) 176 V
b) 6.0 A
c) 26.4 kW

11. 360 rpm

13. 1710 rpm

15. 39.4 ft

17. *a*) 200 A
b) 1.2 A
c) 201.2 A
d) 130 V
e) 26.2 kW

19. 1.3 A

21. *a*) 150 A
b) 5100 V
c) 765 kW
d) 46 percent

23. 188 A

25. *a*) 3800 V
b) 2750 kW
c) 72.5 kW

27. 3.66 μF

Chapter 33

1. 30,000 MHz

3. 213 m

5. 11 m

7. 7.62 min

9. *a*) 13 KΩ
b) 7 KΩ

11. 67 pF

13. 5.0 mi

15. 46 KHz

17. *a*) 40 pF · $V^{1/2}$
b) 14 pF

19. 18 cm

21. 0.24 s

23. 749 Ω/°C between 10 and 20°C
28 Ω/°C between 90 and 100°C

25. 565 W/m^2

27. 10^4

Chapter 34

1. 1.657×10^{-18} J
120 nm

3. 4.14 eV

5. 276 nm

7. 0.1564 nm

9. 4.158×10^{-6} nm

11. For $n = 3$, $\lambda = 656$ nm
For $n = 6$, $\lambda = 410$ nm

13. 8.27×10^{-4} nm

15. *a*) Z = 26
b) Iron

17. A = 207 u

19. 40 days

21. 1.51×10^{14} Hz

23. Photon energy = 2.95 eV. Since the photon energy is smaller than the work function, no electron emission occurs.

25. ø=1.81eV. Identify *cesium* from Table 34.1

27. 2.5×10^{-11} m

29. $m = 1.66 \times 10^{-27}$ kg (proton)

31. *a*) 5.0×10^4 eV or 8.0×10^{-15} J
b) 0.0248 nm

33. 3.7×10^3 V

35. Approximately ⅛ of the original amount

37. 1.54 rad; 0.0154 Gy

39. 2.22×10^{14} Hz

41. $\Theta = 180°$

43. *a*) 2.48×10^{-11} m
b) 0.19

45. *a*) $B = 1.23 \times 10^{-9}$ kg$^{1/2}$ m^2 s^{-1} c$^{-1/2}$
b) 0.388 nm
c) 1.50 V

47. *a*) 5×10^{20}
b) 2.5×10^{20}
1.56×10^{19}
4.88×10^{17}

Index

Definitions, Conversion Factors, Equivalents, and Physical Constants

(continued from front endpapers)

Mass

1 kilogram = 10^3 gm
= 10^6 mg
= 2.205 lb (mass equivalent)
= 0.06852 slug

1 slug = 14.59 kg = 32.17 lb (mass equivalent)
1 international atomic mass unit (amu) = 1.661×10^{-27} kg

Time

1 year = 365.3 days
1 day = 24 h = 86,400 s

1 min = 60 s
1 h = 60 min

Frequency: 1 hertz (Hz) = 1 cycle/s = s^{-1}

Force

1 newton (N) = 0.1020 kg force
= 102.0 g force
= 0.2248 lb force

1 pound force (wt) = 16 oz force
= 7000 grains (gr)
= 4.448 newtons
= 453.6 g force
= 0.4536 kg force

1 ton force = 2000 lb force (wt)
1 gram force = 0.009807 newton (N)
1 kilogram force (wt) = 2.205 lb force
= 9.807 newtons
1000 kg force = 1 metric ton
1 metric ton = 1.103 English tons

Density of "Standard" Substances

Mass density of mercury, ρ = 13,600 kg/m^3
= 13.6 g/cm^3

Mass density of water (4° C), ρ = 1000 kg/m^3
= 1 g/cm^3
= 1 kg/liter

Weight density of water (4° C), D = 62.4 lb/ft^3
= 8.34 lb/gal

Mass density of dry air (STP), ρ = 1.293 kg/m^3
= 1.293×10^{-3} g/cm^3

Weight density of dry air (STP), D = 0.08074 lb/ft^3

Pressure

1 newton/m^2 = 1 pascal (Pa)
= 1.450×10^{-4} lb/in.^{2}
= 9.869×10^{-6} atm
= 7.501×10^{-4} cm Hg
= 10^{-5} bar
= 0.007501 torr

1 lb/in.^{2} = 144 lb/ft^2
= 6895×10^3 newtons/m^2
= 5.171 cm Hg
= 27.68 in. water

1 atmosphere (atm) = 76 cm Hg = 760 mm Hg
= 1.013×10^5 N/m^2 (Pa)
= 1.013 bar = 1013 millibars
= 14.70 lb/in.^{2}
= 29.95 in. Hg
= 33.90 ft water

1 bar = 10^5 Pa
= 14.50 lb/in.^{2}

1 millibar = 10^2 Pa

1 torr = 1 mm Hg = $^1/_{760}$ atm = 133.3 Pa

Mechanical Energy

1 joule (J) = 1 newton-meter (N•m)
= 1 watt-second
= 0.7376 ft•lb
= 2.778×10^{-7} kW•h
= 6.241×10^{18} electron volts (eV)

1 foot-pound = 1.356 joules (J)
= 3.766×10^{-7} kW•h

1 kilowatt-hour (kW•h) = 3.600×10^6 J
= 2.655×10^6 ft•lb

1 horsepower-hour (hp•h) = 1.980×10^6 ft•lb
= 0.7457 kW•h

1 electron-volt (eV) = 1.602×10^{-19} J

Heat Energy

1 kilocalorie (kcal) = 1000 cal
= 3.968 Btu
= 2.613×10^{22} eV
1 British thermal unit (Btu) = 252 cal
= 0.252 kcal

1 Quad = 10^{15} Btu
= 1.8×10^{8} Bbls of oil
= 10^{12} ft^3 natural gas
1 therm = 10^5 Btu
1 ton of refrigeration = 12,000 Btu/h
1 Btu/lb = 0.555 kcal/kg

Power

1 watt = 1 joule/s
1 kilowatt (kW) = 1000 watts
= 737.6 ft•lb/s
= 1.341 hp
1 megawatt (MW) = 10^6 watts

1 horsepower (hp) = 550 ft•lb/s
= 33,000 ft•lb/min
= 2545 Btu/h
= 745.7 W
= 0.7457 kW

Electricity and Magnetism

Electric Charge, Q
1 coulomb (C) = 6.241×10^{18} electron charges (e)
1 electron charge (e) = 1.602×10^{-19} C
1 ampere-hour (A•h) = 3600 C
1 coulomb = 2.778×10^{-4} A•h
1 faraday = 96,487 C

Magnetic Flux, Ø
1 weber (wb) = 1 tesla•m^2
= 1 V•s
= 1 N•m/A
= 10^8 maxwells
= 10^4 gauss•m^2

Magnetic Flux Density or Magnetic Induction, B
1 tesla (T) = 1 weber/m^2
= 1 N/A•m
= 10^4 gauss
1 gauss = 10^{-4} weber/m^2

Magnetic Field Intensity, H
1 amp-turn/m = 1 N/wb
= 1.257×10^{-2} gauss
1 gauss = 79.55 amp-turn/m

Electrical Relationships
1 ampere = 1 C/s
1 coulomb = 1 A•s
= 1 F•V
1 ohm = 1 V/A
1 farad (F) = 1 C/V
1 microfarad (μF) = 10^{-6} F
1 picofarad (pF) = 10^{-12} F
1 volt = 1 J/C
1 henry = 1V•s/A
= 1 ohm•s
= 10^6 microhenrys (μH)
= 10^3 millihenrys (mH)
1 joule = 1V•C
= 1 wb•A
1 eV = 1.602×10^{-19} J
1 MeV = 1.602×10^{-13} J

Conversion Factors for Heat Energy and Mechanical Energy

1 Btu = 778.0 ft•lb (Joule's constant J, English System)
= 1055 joules
= 2.930×10^{-4} kW•h
1 therm = 10^5 Btu
= 29.3 kW•h
1 kcal = 4.186×10^3 joules (Joule's constant J, mks system)
= 3.086×10^3 ft•lb
= 1.163×10^{-3} kW•h
= 3.968 Btu
1 calorie (cal) = 4.186 joules
1 ft•lb = 1.285×10^{-3} Btu
= 3.239×10^{-4} kcal
1 joule = 2.389×10^{-4} kcal
= 9.480×10^{-4} Btu
= 2.778×10^{-7} kW•h

1 kilowatt (kW) = 0.2389 kcal/s
= 3.413×10^3 Btu/h
1 horsepower (hp) = 0.1781 kcal/s
= 2.545×10^3 Btu/h
1 Btu/h = 0.2160 ft•lb/s
= 3.928×10^{-4} hp
= 6.998×10^{-5} kcal/s
= 2.930×10^{-4} kW
1 kcal/s = 4.186 kW
= 1.429×10^4 Btu/h
1 kilowatt hour (kW•h) = 1.341 hp•h
= 3.600×10^6 J
= 3413 Btu
= 860.6 kcal